Manufacturing Processes
for Engineering Materials

Manufacturing Processes for Engineering Materials

FIFTH EDITION

Serope Kalpakjian

Illinois Institute of Technology, Chicago, Illinois

Steven R. Schmid

University of Notre Dame, Notre Dame, Indiana

Prentice Hall

Pearson Education, Inc.
Upper Saddle River, New Jersey 07458

Library of Congress Cataloging-in-Publication Data

Kalpakjian, Serope, 1928–
    Manufacturing processes for engineering materials / Serope Kalpakjian, Steven R. Schmid.—5th ed.
        p. cm.
    ISBN 0-13-227271-7
    1. Manufacturing processes.   I. Schmid, Steven R.   II. Title.
TS183.K34 2007
670.42—dc22

                                                                    2007023982

Vice President and Editorial Director, ECS: *Marcia J. Horton*
Senior Editor: *Holly Stark*
Editorial Assistant: *Jennifer Lonschein*
Director of Team-Based Project Management: *Vince O'Brien*
Senior Managing Editor: *Scott Disanno*
Production Editor: *Winifred Sanchez*
Art Editor: *Greg Dulles*
Cover Designer: *Kenny Beck*
Manufacturing Buyer: *Lisa McDowell*
Senior Marketing Manager: *Holly Stark*

© 2008 by Pearson Education, Inc.
Pearson Education, Inc.
Upper Saddle River, NJ 07458

The author and publisher of this book have used their best efforts in preparing this book. These efforts include the development,
research, and testing of the theories and programs to determine their effectiveness. The author and publisher make no warranty
of any kind, expressed or implied, with regard to these programs or the documentation contained in this book. The author
and publisher shall not be liable in any event for incidental or consequential damages in connection with, or arising out of,
the furnishing, performance, or use of these programs.

Printed in the United States of America.

10 9 8 7 6 5 4 3 2 1

ISBN 0-13-227271-7

Pearson Education Ltd., *London*
Pearson Education Australia Pty. Ltd., *Sydney*
Pearson Education Singapore, Pte. Ltd.
Pearson Education North Asia Ltd., *Hong Kong*
Pearson Education Canada, Inc., *Toronto*
Pearson Educación de Mexico, S.A. de C.V.
Pearson Education—Japan, *Tokyo*
Pearson Education Malaysia, Pte. Ltd.
Pearson Education, Inc., *Upper Saddle River, New Jersey*

To

M. Eugene Merchant (1913–2006)

and

Milton C. Shaw (1915–2006)

in grateful memory

CONTENTS

6   Bulk Deformation Processes   265

7   Sheet-Metal Forming Processes   346

8   Material-Removal Processes: Cutting   416

9   Material-Removal Processes: Abrasive, Chemical, Electrical, and High-Energy Beams   523

10 Properties and Processing of Polymers and Reinforced Plastics; Rapid Prototyping and Rapid Tooling   584

13 Fabrication of Microelectronic, Micromechanical, and Microelectromechanical Devices; Nanomanufacturing   803

14 Automation of Manufacturing Processes and Operations   869

15 Computer-Integrated Manufacturing Systems   914

PREFACE

With rapid advances in all aspects of manufacturing, the authors continue to present an up-to-date, comprehensive, and balanced coverage of the science, engineering, and technology of manufacturing processes and operations. As in previous editions, this text maintains the same number of chapters while continuing to emphasize the complexity and interdisciplinary nature of manufacturing activities, including the interactions between materials, design, and manufacturing processes, and the numerous factors involved in their selection.

Every attempt has been made to motivate and challenge students to understand and develop an appreciation of the importance of manufacturing in modern global economies in all nations. The large number of questions and problems at the end of each chapter is designed to allow students to explore viable solutions to a wide variety of challenges, thus giving them an opportunity to assess the capabilities as well as limitations of all manufacturing processes and operations. These challenges include economic considerations and highly competitive aspects in a global marketplace. The numerous examples and case studies in the book also help give students a perspective on the real-world applications of the topics described.

What's new in this edition

- To better introduce the subject matter, each chapter now begins with an outline describing briefly its objectives and contents.
- The text has been completely updated, with numerous new and relevant materials and illustrations on all aspects of manufacturing.
- Among new or expanded topics in this edition are:

  Communications networks
  Design considerations in manufacturing
  Fabrication of micromechanical and microelectromechanical devices
  Holonic manufacturing systems
  Incremental forming
  Life-cycle engineering and sustainable manufacturing
  Mechanics of polymer processing
  Micromachining
  Nanomanufacturing and nanomaterials
  Rapid prototyping and rapid tooling
  Taguchi methods
  Web sites of various organizations for information relevant to manufacturing

- The fifth edition has been thoroughly edited for improved readability and clarity.
- Numerous cross-references have been added throughout the text as an aid to students and to offer a broader perspective on the complex interrelationships of the topics described.
- Many figures have been added and others improved for better graphic impact.
- The number of questions and problems is now 1445, an 18 percent increase from the previous edition. The answers to many numerical problems are now given at the end of the book.
- The bibliographies at the end of each chapter have been thoroughly updated.

Instructor's Supplements

To aid in teaching the subject matter in this textbook, visual aides have been prepared in Powerpoint, Keynote and pdf format, and are available for download at (www.prenhall.com/kalpakjian). In addition, an extensive solutions manual has been prepared and is also available for download by registered instructors. The companion website has additional supplementary material and links to information sources.

Audience

This text has been written for undergraduate as well as graduate students and practitioners in mechanical, metallurgical and materials, and industrial engineering. We hope that by studying the broad topic of manufacturing science and engineering in this text, the students will come to appreciate the importance and significance of manufacturing to the economic health of a country and the well-being of its people.

We would be grateful to receive any comments from instructors and students alike regarding suggestions for improvements, as well as any errors, deficiencies, or incorrect information that may have escaped our attention during the preparation of this text.

Acknowledgments

We are happy to present the following comprehensive list of all those individuals, in academic institutions as well as in industrial and research organizations, who, in one way or another, have made various contributions to this and the previous editions of the book.

B.J. Aaronson
R. Abella
D. Adams
K. Anderson
S. Arellano
R.A. Arlt
D.D. Arola
V. Aronov
A. Bagchi
E.D. Baker
J. Barak
J. Ben-Ari
G.F. Benedict
S. Bhattacharyya
J T. Black
W. Blanchard
C. Blathras
G. Boothroyd
D. Bourell
B. Bozak
N.N. Breyer
C.A. Brown
R.G. Bruce
J. Cesarone
T.-C. Chang
R.L. Cheaney
A. Cheda
S. Chelikani

S. Chen
S.-W. Choi
A. Cinar
R.O. Colantonio
P. Cotnoir
P.J. Courtney
P. Demers
D. Descoteaux
M.F. De Vries
R.C. Dix
M. Dollar
D.A. Dornfeld
H.I. Douglas
M. Dugger
D.R. Durham
D. Duvall
S.A. Dynan
J. El Gomayel
M.G. Elliott
E.C. Feldy
J. Field
G.W. Fischer
D.A. Fowley
R.L. French
B.R. Fruchter
D. Furrer
R. Giese
E. Goode

K.L. Graham
P. Grigg
M. Grujicic
P.J. Guichelaar
B. Harriger
D. Harry
M. Hawkins
R.J. Hocken
E.M. Honig, Jr.
S. Imam
R. Jaeger
C. Johnson
K. Jones
D. Kalisz
J. Kamman
S.G. Kapoor
R. Kassing
R.L. Kegg
W.J. Kennedy
B.D. King
J.E. Kopf
R.J. Koronkowski
J. Kotowski
S. Krishnamachari
K.M. Kulkarni
T. Lach
L. Langseth
M. Laurent

M. Levine
B.S. Levy
X.Z. Li
Z. Liang
B.W. Lilly
D.A. Lucca
M. Madou
S. Mantell
L. Mapa
A. Marsan
R. J. Mattice
C. Maziar
T. McClelland
W. McClurg
L. McGuire
K.E. McKee
K.P. Meade
M.H. Miller
R. Miller
T.S. Milo
J. Moller
D.J. Morrison
S. Mostovoy
C. Nair
P.G. Nash
J. Nazemetz
E.M. Odom
U. Pal
N. Pacelli

S. Paolucci
S.J. Parelukar
J. Penaluna
C. Petronis
S. Petronis
M. Philpott
M. Pradheeradhi
J.M. Prince
D.W. Radford
W.J. Riffe
R.J. Rogalla
Y. Rong
A.A. Runyan
P. Saha
G.S. Saletta
M. Salimian
M. Savic
W.J. Schoech
S.A. Schwartz
S. Shepel
R. Shivpuri
M.T. Siniawski
J.E. Smallwood
J.P. Sobczak
L. Soisson
P. Stewart
J. Stocker
L. Strom
A.B. Strong

K. Subramanian
T. Sweeney
W.G. Switalski
T. Taglialavore
M. Tarabishy
K.S. Taraman
R. Taylor
B.S. Thakkar
A. Trager
A. Tseng
C. Tszang
M. Tuttle
S. Vaze
J. Vigneau
G.A. Volk
G. Wallace
J.E. Wang
K.J. Weinmann
R. Wertheim
K. West
J. Widmoyer
K.R. Williams
G. Williamson
B. Wiltjer
J. Wingfield
P.K. Wright
N. Zabaras

We are also thankful to the following reviewers:

Z.J. Pei, Kansas State University
John Lewandowski, Case Western Reserve University
Yong Huang, Clemson University
T. Kesavadas, University at Buffalo
Nicholas X. Fang, University of Illinois
Philip J. Guichelaar, Western Michigan University
Zhongming Liang, Indiana University-Purdue University
Klaus J. Weinmann, University of California at Berkeley

We would like to thank Kent M. Kalpakjian (Micron Technology, Inc.) as the author of the sections on fabrication of microelectronic devices, and Robert Kerr (also at Micron) for his review of this material. We would also like to acknowledge the dedication and continued help and cooperation of our editor Holly Stark, Senior Editor at Pearson Prentice Hall, and the editorial staff at Prentice-Hall, including Scott Disanno, Winifred Sanchez and Xiahong Zhu.

We are grateful to numerous organizations that supplied us with many illustrations and materials for the case studies. These contributions have specifically been acknowledged throughout the text.

SEROPE KALPAKJIAN
STEVEN R. SCHMID

ABOUT
THE AUTHORS ...

Serope Kalpakjian is a professor emeritus of mechanical and materials engineering at the Illinois Institute of Technology. He is the author of *Mechanical Processing of Materials* (Van Nostrand, 1967) and a co-author of *Lubricants and Lubrication in Metalworking Operations* (with E.S. Nachtman; Dekker, 1985). Both of the first editions of his textbooks *Manufacturing Processes for Engineering Materials* (1984) and *Manufacturing Engineering and Technology* (1989, now in its fifth edition) have received the M. Eugene Merchant Manufacturing Textbook Award. He has conducted research in various areas of manufacturing, is the author of numerous technical papers and articles in handbooks and encyclopedias, and has edited several conference proceedings. He also has been editor and co-editor of various technical journals and has served on the editorial board of *Encyclopedia Americana*.

Among other awards, Professor Kalpakjian has received the Forging Industry Educational and Research Foundation Best Paper Award (1966); the Excellence in Teaching Award from IIT (1970); the ASME Centennial Medallion (1980); the International Education Award from SME (1989); A Person of the Millennium Award from IIT (1999); and the Albert Easton White Outstanding Teacher Award from ASM International (2000). The SME Outstanding Young Manufacturing Engineer Award for 2002 was named after him. Professor Kalpakjian is a Life Fellow of ASME; Fellow of SME; Fellow and Life Member of ASM International; Fellow Emeritus of The International Academy for Production (CIRP); and is a founding member and past president of NAMRI/SME. He is a high-honor graduate of Robert College (Istanbul), Harvard University, and the Massachusetts Institute of Technology.

Steven R. Schmid is an associate professor in the Department of Aerospace and Mechanical Engineering at the University of Notre Dame, where he teaches and conducts research in the general areas of manufacturing, machine design, and tribology. He received his Bachelor's degree in mechanical engineering from Illinois Institute of Technology (with Honors) and Master's and Ph.D. degrees, both in mechanical engineering, from Northwestern University. He has received numerous awards, including the John T. Parsons Award from the Society of Manufacturing Engineers (2000), the Newkirk Award from the American Society Mechanical Engineers (2000), the Kaneb Center Teaching Award (2000 and 2003), and the Ruth and Joel Spira Award for Excellence in Teaching (2005).

Professor Schmid is the author of over ninety technical papers, has co-authored the texts *Fundamentals of Machine Elements* (McGraw-Hill), *Fundamentals of Fluid Film Lubrication* (Dekker), and *Manufacturing Engineering and Technology*, and has contributed two chapters to the *CRC Handbook of Modern Tribology*. He is an Associate Editor of the ASME *Journal of Manufacturing Science and Engineering*, and is a registered Professional Engineer and a Certified Manufacturing Engineer.

Introduction

The objectives of this chapter are to

- Define manufacturing and describe the technical and economic considerations involved in manufacturing successful products.
- Explain the relationships among product design and engineering and factors such as materials, process selection and the various costs involved.
- Describe the important trends in modern manufacturing and how they can be utilized in a highly competitive global marketplace to minimize production costs.

1.1 | What Is Manufacturing?

As you read this Introduction, take a few moments to inspect the different objects around you: pencil, paper clip, table, light bulb, door knob, and cell phone. You will soon realize that these objects have been transformed from various raw materials into individual parts and assembled into specific products. Some objects, such as nails, bolts, and paper clips, are made of one material; the vast majority of objects (such as toasters, bicycles, computers, washers and dryers, automobiles, and farm tractors) are, however, made of numerous parts from a wide variety of materials (Fig. 1.1). A ballpoint pen, for example, consists of about a dozen parts, a lawn mower about 300 parts, a grand piano 12,000 parts, a typical automobile 15,000 parts, a C-5A transport plane more than 4 million parts, and a Boeing 747-400 about 6 million parts. All are produced by a combination of various processes called manufacturing.

Manufacturing, in its broadest sense, is the process of converting raw materials into products; it encompasses the design and manufacturing of goods using various production methods and techniques. Manufacturing began about 5000 to 4000 B.C. with the production of various articles of wood, ceramic, stone, and metal (Table 1.1). The word *manufacturing* is derived from the Latin *manu factus,* meaning "made by hand"; the word *manufacture* first appeared in 1567, and the word *manufacturing,* in 1683. The word **production** is also used interchangeably with the word *manufacturing.*

Manufacturing may produce *discrete products,* meaning individual parts or pieces, such as nails, rivets, gears, steel balls, and beverage cans. On the other hand,

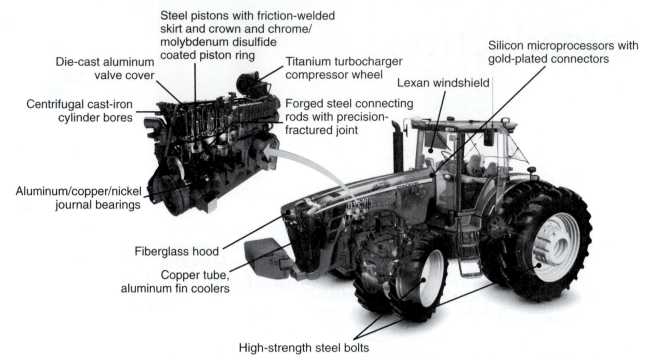

Steel pistons with friction-welded
skirt and crown and chrome/
molybdenum disulfide
coated piston ring

Die-cast aluminum
valve cover

Centrifugal cast-iron
cylinder bores

Aluminum/copper/nickel
journal bearings

Titanium turbocharger
compressor wheel

Forged steel connecting
rods with precision-
fractured joint

Silicon microprocessors with
gold-plated connectors

Lexan windshield

Fiberglass hood

Copper tube,
aluminum fin coolers

High-strength steel bolts

FIGURE 1.1   Model 8430 tractor, with detailed illustration of its diesel engine, showing the variety of materials and processes incorporated. *Source:* Courtesy of John Deere Company.

wire, sheet metal, tubing, and pipe are *continuous products* that may be cut into individual pieces and thus become discrete products.

Because a manufactured item has undergone a number of changes in which raw material has become a useful product, it has **added value,** defined as monetary worth. For example, clay has a certain value when mined. When the clay is used to make a ceramic dinner plate, cutting tool, or electrical insulator, value is added to the clay. Similarly, a wire coat hanger or a nail has added value over and above the cost of a piece of wire from which it is made.

Manufacturing is extremely important for national and global economies. Consider Fig. 1.2, which depicts the Gross Domestic Product (GDP) per capita in a country as a function of manufacturing activity within that country. The curves show the trends from 1982 until 2006. The following observations can be made:

1. At the start of the curves, that is, in 1982, the wealth of countries was closely tied to the level of manufacturing activity, as shown by the shaded area.

2. In 2006, the dependence of wealth on an active manufacturing sector is not as clear and may be attributed to a number of contributing factors, including:

   - Some nations have natural resources that provide their citizens with higher standards of living. This is most clearly seen for Kuwait and Mexico, where petroleum exports contribute significantly to wealth. Most nations without abundant natural resources must, instead, generate wealth in order to have robust economies.

   - Even if the level of manufacturing in a country remains constant or increases slightly, its contribution as a percent of the national economy will decrease if the economy grows. It should, therefore, be recognized that the absolute level of manufacturing activity in a country can increase, even though the relative contribution of manufacturing to the country's entire economy decreases.

TABLE 1.1

Historical Development of Materials and Manufacturing Processes

Period:
Egypt: ~3100 B.C. to ~300 B.C.
Greece: ~1100 B.C. to ~146 B.C.
Roman Empire: ~500 B.C. to 476 A.D.
Middle Ages: ~476 to 1492
Renaissance: 14th to 16th centuries.

Dates	Metals and casting	Various materials and composites	Forming and shaping	Joining	Tools, machining and manufacturing systems
Before 4000 B.C.	Gold, copper, meteoric iron	Earthenware, glazing, natural fibers	Hammering		Tools of stone, flint, wood, bone, ivory, composite tools
4000–3000 B.C.	Copper casting, stone and metal molds, lost-wax process, silver, lead, tin, bronze		Stamping, jewelry	Soldering (Cu-Au, Cu-Pb, Pb-Sn)	Corundum (alumina, emory)
3000–2000 B.C.	Bronze casting and drawing, gold leaf	Glass beads, potter's wheel, glass vessels	Wire by slitting sheet metal	Riveting, brazing	Hoe making, hammered axes, tools for ironmaking and carpentry
2000–1000 B.C.	Wrought iron, brass				
1000–1 B.C.	Cast iron, cast steel	Glass pressing and blowing	Stamping of coins	Forge welding of iron and steel, gluing	Improved chisels, saws, files, woodworking lathes
1–1000 A.D.	Zinc, steel	Venetian glass	Armor, coining, forging, steel swords		Etching of armor
1000–1500	Blast-furnace type metals, casting of bells, pewter	Crystal glass	Wire drawing, gold- and silversmith work		Sandpaper, windmill-driven saw
1500–1600	Cast-iron cannon, tinplate	Cast plate glass, flint glass	Water power for metalworking, rolling mill for coinage strips		Hand lathe for wood

(continued)

3

TABLE 1.1

Historical Development of Materials and Manufacturing Processes (*continued*)

Period	Dates	Metals and casting	Various materials and composites	Forming and shaping	Joining	Tools, machining and manufacturing systems
	1600–1700	Permanent-mold casting, brass from copper and metallic zinc	Porcelain	Rolling (lead, gold, silver), shape rolling (lead)		Boring, turning, screw-cutting lathe, drill press
	1700–1800	Malleable cast iron, crucible steel (iron bars and rods)		Extrusion (lead pipe), deep drawing, rolling		
	1800–1900	Centrifugal casting, Bessemer process, electrolytic aluminum, nickel steel, babbitt, galvanized steel, powder metallurgy, open-hearth steel	Window glass from slit cylinder, light bulb, vulcanization, rubber processing, polyester, styrene, celluloid, rubber extrusion, molding	Steam hammer, steel rolling, seamless tube, steel-rail rolling, continuous rolling, electroplating		Shaping, milling, copying lathe for gunstocks, turret lathe, universal milling machine, vitrified grinding wheel
	1900–1920		Automatic bottle making, bakelite, borosilicate glass	Tube rolling, hot extrusion	Oxyacetylene; arc, electrical-resistance, and thermit welding	Geared lathe, automatic screw machine, hobbing, high-speed steel tools, aluminum oxide and silicon carbide (synthetic)
	1920–1940	Die casting	Development of plastics, casting, molding, polyvinyl chloride, cellulose acetate, polyethylene, glass fibers	Tungsten wire from metal powder	Coated electrodes	Tungsten carbide, mass production, transfer machines
	1940–1950	Lost-wax process for engineering parts	Acrylics, synthetic rubber, epoxies, photosensitive glass	Extrusion (steel), swaging, powder metals for engineering parts	Submerged arc welding	Phosphate conversion coatings, total quality control

Industrial Revolution: ~1750 to 1850

WW I

WW II

Era	Dates	Casting	Forming and shaping	Joining	Tool and die materials, and machining	Various materials and composites
Space Age	1950–1960	Ceramic mold, nodular iron, semiconductors, continuous casting	Cold extrusion (steel), explosive forming, thermochemical processing	Gas metal arc, gas tungsten arc, and electroslag welding; explosion welding	Electrical and chemical machining, automatic control	Acrylonitrile-butadiene-styrene, silicones, fluorocarbons, polyurethane, float glass, tempered glass, glass ceramics
	1960–1970	Squeeze casting, single-crystal turbine blades	Hydroforming, hydrostatic extrusion, electroforming	Plasma-arc and electron-beam welding, adhesive bonding	Titanium carbide, synthetic diamond, numerical control, integrated circuit chip	Acetals, polycarbonate, cold forming of plastics, reinforced plastics, filament winding
	1970–1990	Compacted graphite, vacuum casting, organically bonded sand, automation of molding and pouring, rapid solidification, metal-matrix composites, semisolid metalworking, amorphous metals, shape-memory alloys (smart materials), computer simulation	Precision forging, isothermal forging, superplastic forming, dies made by computer-aided design and manufacturing, net-shape forging and forming, computer simulation	Laser beam, diffusion bonding (also combined with superplastic forming), surface-mount soldering	Cubic boron nitride, coated tools, diamond turning, ultraprecision machining, computer-integrated manufacturing, industrial robots, machining and turning centers, flexible-manufacturing systems, sensor technology, automated inspection, expert systems, artificial intelligence, computer simulation and optimization	Adhesives, composite materials, semiconductors, optical fibers, structural ceramics, ceramic-matrix composites, biodegradable plastics, electrically conducting polymers
Information Age	1990–2000s	Rheocasting, computer-aided design of molds and dies, rapid tooling	Rapid prototyping, rapid tooling, environmentally friendly metalworking fluids	Friction stir welding, lead-free solders, laser butt-welded (tailored) sheet-metal blanks, electrically conducting adhesives	Micro- and nanofabrication, LIGA (German acronym for a process involving lithography, electroplating, and molding), dry etching, linear motor drives, artificial neural networks, six sigma	Nanophase materials, metal foams, advanced coatings, high-temperature superconductors, machinable ceramics, diamondlike carbon

Source: J.A. Schey, C.S. Smith, R.F. Tylecote, T.K. Derry, T.I. Williams, S.R. Schmid, and S. Kalpakjian.

FIGURE 1.2 Importance of
manufacturing to national
economies. The trends
shown are from 1982 until
2006. *Source:* After J.A.
Schey with data from the
World Development Report,
World Bank, various years.

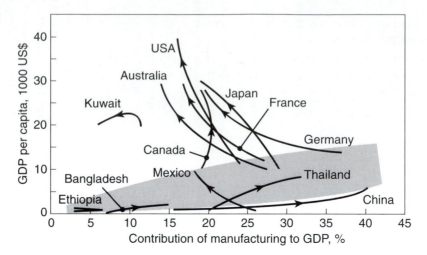

- The emergence of global economies is often perceived as detrimental in the media, but global trading leads to an increase in wealth in all participating nations, and this increase dominates the effects of manufacturing on economic health.

3. Nations with the largest GDP growth have their economic activity concentrated on high value-added products, such as automobiles, airplanes, medical devices, computers, electronics, and machinery. Other products, such as clothing, toys, and hand-held tools, are labor intensive and are concentrated in countries where labor rates are lower. Such labor-intensive manufacturing is associated with the traditional curve shown as a shaded area in Fig. 1.2.

It becomes clear when reviewing Fig. 1.2 that to achieve a standard of living at a level that generally is taken for granted in the West, a healthy and vibrant manufacturing sector is required, and high value-added manufacturing activity is essential.

Manufacturing is generally a complex activity involving people who have a broad range of disciplines and skills, together with a wide variety of machinery, equipment, and tools with various levels of automation and controls, including computers, robots, and material-handling equipment. Manufacturing activities must be responsive to several demands and trends:

1. A product must fully meet **design requirements** and **specifications** and **standards**.

2. It must be manufactured by the most **economical** and **environmentally friendly** methods.

3. **Quality** must be built into the product at each stage, from design to assembly, rather than relying on quality testing after the product is made.

4. In a highly competitive and global environment, production methods must be sufficiently **flexible** to respond to changing market demands, types of products, production rates and quantities, and on-time delivery to the customer.

5. New developments in **materials, production methods,** and **computer integration** of both technological and managerial activities in a manufacturing organization must constantly be evaluated with a view to their timely and economic implementation.

6. Manufacturing activities must be viewed as a large **system,** each part of which is interrelated. Such systems can be modeled in order to study the effect of various factors, such as changes in market demand, product design, materials, costs, and production methods, on product quality and cost.

7. A manufacturer must work with the customer to get timely feedback for **continuous product improvement**.

8. A manufacturing organization must constantly strive for higher **productivity**, defined as the optimum use of all its resources: materials, machines, energy, capital, labor, and technology. Output per employee per hour in all phases must be maximized.

1.2 | Product Design and Concurrent Engineering

Product design is a critical activity because it has been estimated that, generally, 70 to 80% of the cost of product development and manufacture is determined at the initial design stages. The design process for a particular product first requires a clear understanding of the functions and the performance expected of that product. The product may be new or it may be an improved model of an existing product. The market for the product and its anticipated uses must be defined clearly, with the assistance of sales personnel, market analysts, and others in the organization.

Traditionally, design and manufacturing activities have taken place sequentially rather than concurrently or simultaneously (Fig. 1.3a). Designers would spend considerable effort and time in analyzing components and preparing detailed part drawings. These drawings would then be forwarded or "sent over the wall" to other departments in the organization, where, for example, particular materials and vendors would be identified. The product specifications would then be sent to the manufacturing department, where the detailed drawings would be reviewed and processes selected for efficient production. While this approach at first appears to be logical and straightforward, this practice has been found to be extremely wasteful of resources.

In theory, a product can flow (Fig. 1.3a) from one department in an organization to another and then directly to the marketplace, but in practice there usually are difficulties encountered. For example, a manufacturing engineer may wish to make a design change to improve the castability of a part or he or she may decide that a different alloy is preferable. Such changes necessitate a repeat of the design analysis stage in order to ensure that the product will still function satisfactorily. These iterations, also shown in Fig. 1.3a, waste resources, but, more importantly, they waste time.

A more advanced product development approach is shown in Fig. 1.3b; while it still has a general product flow from market analysis to design to manufacturing, it contains deliberate iterations. The main difference from the older approach is that all disciplines are now involved in the earliest stages of product design; they progress concurrently, so that the iterations (which by nature occur) result in less wasted effort and lost time. A key to this approach is the well-recognized importance of *communication* among and within disciplines. While there must be communication between engineering, marketing, and service functions, so too must there be avenues of interaction between engineering subdisciplines; for example, design for manufacture, design for recyclability, and design for safety.

Concurrent engineering, also called **simultaneous engineering**, is a systematic approach integrating the design and manufacture of products with the view toward optimizing all elements involved in the life cycle of the product (see Section 1.4). The basic goals of concurrent engineering are to minimize product design and engineering changes, as well as the time and costs involved in taking the product from design concept to production and introduction of the product into the marketplace. An extension of concurrent engineering, called **direct engineering**, utilizes a database

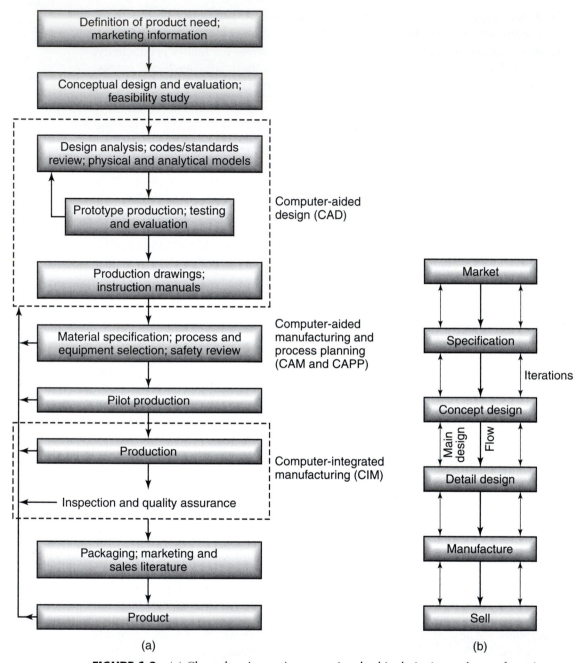

FIGURE 1.3   (a) Chart showing various steps involved in designing and manufacturing a product. Depending on the complexity of the product and the type of materials used, the time span between the original concept and the marketing of a product may range from a few months to many years. (b) Chart showing general product flow, from market analysis to selling the product, and depicting concurrent engineering. *Source:* After S. Pugh.

representing the engineering logic used in the design of each component of a product. For example, if a design modification is made on a part, direct engineering will determine the manufacturing consequences of that change.

Although the concept of concurrent engineering appears to be logical and efficient, its implementation can require considerable time and effort, especially when those using it either are not able to work as a team or fail to appreciate its real benefits. It is apparent that for concurrent engineering to succeed it must (1) have the full support of an organization's top management, (2) have multifunctional and interacting work teams, including support groups, and (3) utilize all available state-of-the-art technologies.

For both large and small companies, product design often involves preparing analytical and physical models of the product as an aid to analyzing factors such as forces, stresses, deflections, and optimal part shape. Although the necessity for such models depends on product complexity, constructing and studying analytical models is now highly simplified through the use of **computer-aided design, engineering,** and **manufacturing** techniques.

On the basis of these models, the product designer selects and specifies the final shape and the dimensions, dimensional tolerances and surface finish of the parts, and the materials to be used. The selection of materials generally is made with the input and cooperation of materials engineers, unless the design engineer is qualified to do so. An important design consideration is how a particular component is to be assembled into the final product. Take apart a ballpoint pen or a toaster, or lift the hood of a car and observe how hundreds of components are assembled in a generally confined space.

A powerful and effective tool, particularly for complex production systems, is **computer simulation** in evaluating the performance of the product and planning the manufacturing system to produce it. Computer simulation also helps in early detection of design flaws, identification of possible problems in a particular production system, and optimization of manufacturing lines for minimum product cost. Several computer simulation languages using animated graphics and with various capabilities are now widely available.

The next step in the production process is to make and test a **prototype,** that is, an original working model of the product. An important technique is **rapid prototyping** (Chapter 10), which relies on CAD/CAM and various manufacturing techniques (typically using polymers or metal powders) to rapidly produce prototypes in the form of a solid physical model of a part. Rapid prototyping can significantly reduce development times and hence costs. These techniques are now advanced to such an extent that they can be used for low-volume economical production of actual parts.

Virtual prototyping is a software form of prototyping and uses advanced graphics and virtual-reality environments to allow designers to examine a part. In a way, this technology is used by CAD packages to render a part so that designers can observe and evaluate the part as it is drawn. However, virtual prototyping systems should be recognized as highly demanding cases of rendering part details.

During the prototype stage, modifications of the original design, the materials selected, or production methods may be necessary. After this phase has been completed, appropriate process plans, manufacturing methods (Table 1.2), equipment, and tooling are selected with the cooperation of manufacturing engineers, process planners, and all other involved in production.

1.3 | Design for Manufacture, Assembly, Disassembly, and Service

It is apparent that design and manufacturing must be closely interrelated; they should never be viewed as separate disciplines or activities. Each part or component of a product must be designed so that it not only meets design requirements and

TABLE 1.2

Shapes and Some Common Methods of Production

Shape or feature	Production method[a]
Flat surfaces	Rolling, planing, broaching, milling, shaping, grinding
Parts with cavities	End milling, electrical-discharge machining, electrochemical machining, ultrasonic machining, blanking, casting, forging, extrusion, injection molding, metal injection molding
Parts with sharp features	Permanent-mold casting, machining, grinding, fabricating,[b] powder metallurgy, coining
Thin hollow shapes	Slush casting, electroforming, fabricating, filament winding, blow molding, sheet forming, spinning
Tubular shapes	Extrusion, drawing, filament winding, roll forming, spinning, centrifugal casting
Tubular parts	Rubber forming, tube hydroforming, explosive forming, spinning, blow molding, sand casting, filament winding
Curvature on thin sheets	Stretch forming, peen forming, fabricating, thermoforming
Openings in thin sheets	Blanking, chemical blanking, photochemical blanking, laser machining
Cross sections	Drawing, extrusion, shaving, turning, centerless grinding, swaging, roll forming
Square edges	Fine blanking, machining, shaving, belt grinding
Small holes	Laser or electron-beam machining, electrical-discharge machining, electrochemical machining, chemical blanking
Surface textures	Knurling, wire brushing, grinding, belt grinding, shot blasting, etching, laser texturing, injection molding, compression molding
Detailed surface features	Coining, investment casting, permanent-mold casting, machining, injection molding, compression molding
Threaded parts	Thread cutting, thread rolling, thread grinding, injection molding
Very large parts	Casting, forging, fabricating, assembly
Very small parts	Investment casting, etching, powder metallurgy, nanofabrication, LIGA, micromachining

Notes:
[a]Rapid prototyping operations can produce all of these features to some degree.
[b]*Fabricating* refers to assembly from separately manufactured components.

specifications but also can be manufactured economically and with relative ease. This approach improves productivity and allows a manufacturer to remain competitive. This broad concept, known as **design for manufacture** (DFM), is a comprehensive approach to production of goods. It integrates the product design process with materials, manufacturing methods, process planning, assembly, testing, and quality assurance.

Effectively implementing design for manufacture requires that designers acquire a fundamental understanding of the characteristics, capabilities, and limitations of materials, production methods, and related operations, machinery, and equipment. This knowledge includes characteristics such as variability in machine performance, dimensional accuracy and surface finish of the parts produced, processing time, and the effect of processing method on part quality.

Designers and product engineers must assess the impact of design modifications on manufacturing process selection, tools and dies, assembly, inspection, and especially product cost. Establishing quantitative relationships is essential in order to optimize the design for ease of manufacturing and assembly at *minimum cost*

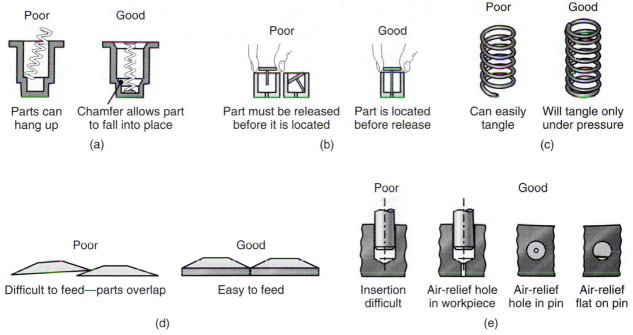

FIGURE 1.4   Redesign of parts to facilitate automated assembly. *Source:* Reprinted from G. Boothroyd and P. Dewhurst, *Product Design for Assembly*, 1989, by courtesy of Marcel Dekker, Inc.

(also called *producibility*). Computer-aided design, engineering, manufacturing, and process planning techniques, using powerful computer programs, are now indispensable to those conducting such analysis. They include **expert systems,** which are computer programs with optimization capabilities, thus expediting the traditional iterative process in design optimization.

After individual parts have been manufactured, they are assembled into a product. **Assembly** is an important phase of the overall manufacturing operation and requires considerations of the ease, speed, and cost of putting parts together (Fig. 1.4). Products must be designed so that **disassembly** is possible with relative ease and require little time, enabling the products to be taken apart for maintenance, servicing, or recycling of their components.

Because assembly operations can contribute significantly to product cost, **design for assembly** (DFA) and **design for disassembly** are important aspects of manufacturing. Typically, a product that is easy to assemble is also easy to disassemble. **Design for service** is another important aspect, ensuring that individual parts in a product are easy to reach and service. These activities are now combined into **design for manufacture and assembly** (DFMA), which recognizes the inherent and important interrelationships among design, manufacturing, and assembly.

Design principles for economic production may be summarized as follows:

- Designs should be as simple as possible to manufacture, assemble, disassemble, service, and recycle.
- Materials should be chosen for their appropriate design and manufacturing characteristics as well as for their service life.
- Dimensional accuracy and surface finish specified should be as broad as permissible.
- Because they can add significantly to cost, secondary and finishing operations should be avoided or minimized.

1.4 | Environmentally Conscious Design, Sustainable Manufacturing, and Product Life Cycle

In the United States alone, more than 24 billion kilograms of plastic products and 75 billion kilograms of paper products are discarded each year. Every three months, U.S. industries and consumers discard enough aluminum to rebuild the country's commercial air fleet. Globally, countless tons of automobiles, television sets, appliances, and computer equipment are discarded each year. Metalworking fluids such as lubricants and coolants, and fluids and solvents used in cleaning manufactured products can pollute the air and waters, unless recycled or disposed of properly.

Likewise, there are many byproducts from manufacturing plants: sand with additives used in metal-casting processes; water, oil, and other fluids from heat-treating facilities and plating operations; slag from foundries and welding operations; and a wide variety of metallic and nonmetallic scrap produced in operations such as sheet forming, casting, and molding. Consider also the effects of water and air pollution, acid rain, ozone depletion, hazardous wastes, landfill seepage, and global warming. Recycling efforts have gained increasing momentum over the years: Aluminum is now recycled at a rate between 21 and 59%, depending on the product, and plastics at around 5%.

The present and potential adverse effects of these activities, their damage to our environment and to the earth's ecosystem, and, ultimately, their effect on the quality of human life are now well recognized by the public as well as local and federal governments. In response, a wide range of laws and regulations have been and continue to be promulgated by local, state, and federal governments as well as international organizations. These regulations are generally stringent, and their implementation can have a major impact on the economic operation of manufacturing organizations. These efforts have been most successful when there is value added, such as in reducing energy requirements (and associated costs) or substituting materials that have both cost and environmental design benefits.

Much progress has also taken place regarding **design for recycling** (DFR) and **design for the environment** (DFE) or **green design** (*green* meaning "environmentally safe and friendly"), indicating universal awareness of the problems outlined above, and waste has become unacceptable. This comprehensive approach anticipates the possible negative environmental impact of materials, products, and processes so that they can be considered at the earliest stages of design and production.

Among other developments is **sustainable manufacturing,** which refers to the realization that natural resources are vital to economic activity and that energy and materials management are essential to ensure that resources are available for future generations. It is thus necessary to conduct a thorough analysis of the product, materials used, and the manufacturing processes and practices employed. The basic guidelines to be followed in this regard are

- Reducing waste of materials *at their source* by refinements in product design and the amount of materials used.
- Reducing the use of hazardous materials in products and processes.
- Ensuring proper handling and disposal of all waste.
- Making improvements in waste treatment and in recycling and reuse of materials.

The **cradle-to-cradle** philosophy encourages the use of environmentally friendly materials and designs. By considering the entire life cycle of a product, materials can be selected and employed that have minimal real waste. Environmentally friendly materials can be

- Part of a *biological cycle,* where (usually organic) materials are used in design, function properly for their intended life, and can then be safely disposed of. Such materials degrade naturally and, in the simplest version, lead to new soil that can sustain life.
- Part of an *industrial cycle,* as aluminum in beverage containers that serve an intended purpose and are then recycled, so that the same material is reused continuously.

Product Life Cycle (PLC). The *product life cycle* consists of the stages that a product goes through, from design, development, production, distribution, and use, to its ultimate disposal and recycling. A product typically goes through five stages:

1. Product development stage, involving much time and high costs.
2. Market introduction stage, in which the acceptance of the product in the marketplace is closely watched.
3. Growth stage, with increasing sales volume, lower manufacturing cost per unit, and hence higher profitability to the manufacturer.
4. Maturation stage, where sales volume begins to peak and competitive products begin to appear in the marketplace.
5. Decline stage, with decreasing sales volume and profitability.

Product Life Cycle Management (PLCM). While the life cycle consists of stages, from development to the ultimate disposal or recycling of the product, *product life cycle management* is generally defined as the strategies employed by the manufacturer as the product goes through its life cycle. Various strategies can be employed in PLCM, depending on the type of product, customer response, and market conditions.

1.5 | Selecting Materials

An increasingly wide variety of materials is now available, each having its own characteristics, composition, applications, costs, advantages, and limitations. The types of materials generally used in manufacturing today are

1. **Ferrous metals:** Carbon steels, alloy steels, stainless steels, and tool and die steels (Chapter 3).
2. **Nonferrous metals and alloys:** Aluminum, magnesium, copper, nickel, super-alloys, titanium, refractory metals (molybdenum, niobium, tungsten, and tantalum), beryllium, zirconium, low-melting alloys (lead, zinc, and tin), and precious metals (Chapter 3).
3. **Plastics:** Thermoplastics, thermosets, and elastomers (Chapter 10).
4. **Ceramics:** Glass ceramics, glasses, graphite, and diamond (Chapter 11).
5. **Composite materials:** Reinforced plastics, metal-matrix and ceramic-matrix composites, and honeycomb structures; these are also known as engineered materials (Chapters 10 and 11).
6. **Nanomaterials, shape-memory alloys, metal foams, amorphous alloys, super-conductors,** and **semiconductors** (Chapters 3 and 13).

Material Substitution. As new or improved materials continue to be developed, there are important trends in their selection and application. Aerospace structures,

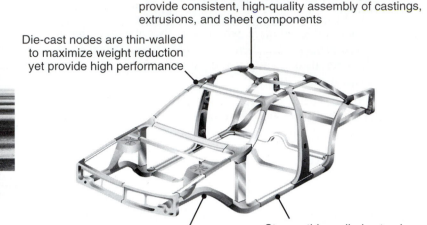

Robotically applied, advanced arc-welding processes provide consistent, high-quality assembly of castings, extrusions, and sheet components

Die-cast nodes are thin-walled to maximize weight reduction yet provide high performance

Advanced extrusion bending processes support complex shapes and tight radii

Strong, thin-walled extrusions exhibit high ductility, energy absorption, and toughness

(a)   (b)

FIGURE 1.5   (a) The Audi A8 automobile, an example of advanced materials construction. (b) The aluminum body structure, showing various components made by extrusion, sheet forming, and casting processes. *Source*: Courtesy of ALCOA, Inc.

sporting goods, and numerous high-tech products have especially been at the forefront of new material usage. Because of the vested interests of the producers of different types of natural and engineered materials, there are constantly shifting trends in the usage of these materials, driven principally by economics. For example, by demonstrating steel's technical and economic advantages, steel producers are countering the increased use of plastics in automobiles and aluminum in beverage cans. Likewise, aluminum producers are countering the use of various materials in automobiles (Fig. 1.5). Obviously, making materials-selection choices that facilitate recycling is a fundamental aspect of these considerations.

Some examples of the use or substitution of materials in common products are (a) steel vs. plastic paper clips, (b) plastic vs. sheet metal light-switch plates, (c) wood vs. metal handles for hammers, (d) glass vs. metal water pitchers, (e) plastic vs. leather car seats, (f) sheet metal vs. reinforced plastic chairs, (g) galvanized steel vs. copper nails, and (h) aluminum vs. cast-iron frying pans.

Material Properties.   When selecting materials for products, the first consideration generally involves **mechanical properties** (Chapter 2), typically strength, toughness, ductility, hardness, elasticity, fatigue, and creep. These properties can significantly be modified by various heat treatment methods, as described in Chapter 5. The *strength-to-weight* and *stiffness-to-weight ratios* of materials are also important considerations, particularly for aerospace and automotive applications. Aluminum, titanium, and reinforced plastics, for example, have higher strength-to-weight ratios than steels and cast irons. The mechanical properties specified for a product and its components should, of course, be appropriate for the conditions under which the product is expected to function.

Physical properties (Chapter 3), such as density, specific heat, thermal expansion and conductivity, melting point, and electrical and magnetic properties, also need to be considered. **Chemical properties** can also play a significant role in hostile as well as normal environments. Oxidation, corrosion, general degradation of properties, and flammability of materials are among the important factors to be considered, as is toxicity [note, for example, the development of lead-free solders (Chapter 13)]. Both physical and chemical properties are important in advanced machining processes (Chapter 9). Also, the **manufacturing properties** of materials determine whether they can be processed (cast, formed, shaped, machined, welded, or heat treated for property enhancement) with relative ease. Finally, the methods used to process materials to the desired shapes should not adversely affect the product's final properties, service life, and cost.

Cost and Availability. The economic aspects of material selection are as important as the technological considerations of properties and characteristics of materials. Cost and availability of raw and processed materials are a major concern in manufacturing. If raw or processed materials are not commercially available in the desired shapes, dimensions, tolerances, and quantities, substitutes or additional processing may be required; these steps can contribute significantly to product cost, as described in Chapter 16. For example, if we need a round bar of a certain diameter and it is not commercially available, then we have to purchase a larger rod and reduce its diameter, by such processes as machining, drawing through a die, or grinding.

Reliability of supply as well as demand affect material costs. Most countries import numerous raw materials that are essential for production. The United States, for example, imports the majority of such raw materials as natural rubber, diamond, cobalt, titanium, chromium, aluminum, and nickel. The geopolitical implications of such reliance on other countries are self-evident.

Various costs are involved in processing materials by different methods. Some methods require expensive machinery, others require extensive labor (known as *labor intensive*), and still others require personnel with special skills, high levels of formal education, or specialized training.

Service Life and Recycling. Time- and service-dependent phenomena such as wear, fatigue, creep, and dimensional stability are important considerations as they can significantly affect a product's performance and, if not controlled, can lead to failure of the product. The corrosion caused by compatibility of the different materials used in a product is also important; an example is galvanic action between mating parts made of dissimilar metals. Recycling or proper disposal of the individual components in a product at the end of its useful life is important as we become increasingly aware of conserving materials and energy so we can live in a clean and healthy environment. The proper treatment and disposal of toxic wastes is also a crucial consideration.

1.6 | Selecting Manufacturing Processes

As can be seen in Table 1.2, a wide range of manufacturing processes are used to produce a variety of parts, shapes, and sizes. Also note that there is usually more than one method of manufacturing a part from a given material (Fig. 1.6). As

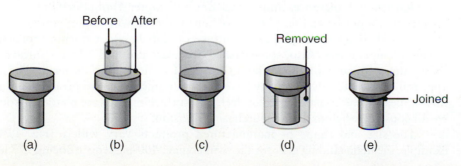

FIGURE 1.6 Various methods of making a simple part: (a) casting or powder metallurgy, (b) forging or upsetting, (c) extrusion, (d) machining, (e) joining two pieces.

expected, each of these processes has its own advantages, limitations, production rates, and cost. The broad categories of processing methods for materials can be listed as follows:

- **Casting:** Expendable molding and permanent molding (Chapter 5).
- **Forming and shaping:** Rolling, forging, extrusion, drawing, sheet forming, powder metallurgy, and molding (Chapters 6, 7, 10, and 11).
- **Machining:** Turning, boring, drilling, milling, planing, shaping, broaching, grinding, ultrasonic machining; chemical, electrical, and electrochemical machining; and high-energy beam machining (Chapters 8 and 9).
- **Joining:** Welding, brazing, soldering, diffusion bonding, adhesive bonding, and mechanical joining (Chapter 12).
- **Micromanufacturing and nanomanufacturing:** Surface micromachining, dry and wet etching, and electroforming (Chapter 13).
- **Finishing:** Honing, lapping, polishing, burnishing, deburring, surface treating, coating, and plating (Chapter 9).

Selection of a particular manufacturing process, or a series of processes, depends not only on the component or part shape to be produced, but also on many other factors. Brittle and hard materials, for example, cannot easily be shaped whereas they can be cast or machined by various methods. The manufacturing process usually alters the properties of materials; metals that are formed at room temperature, for example, become stronger, harder, and less ductile than they were before processing. Thus, characteristics such as *castability, formability, machinability,* and *weldability* of materials have to be studied. Some examples of manufacturing processes used or substituted for common products are (a) forging vs. casting of crankshafts, (b) sheet metal vs. cast hubcaps, (c) casting vs. stamping sheet metal for frying pans, (d) machined vs. powder-metallurgy gears, (e) thread rolling vs. machining of bolts, and (f) casting vs. welding of machine structures.

Manufacturing engineers are constantly challenged to find new solutions to production problems as well as finding means for significant cost reduction. For example, sheet-metal parts have typically been cut and formed by using traditional tools such as punches and dies. Although they are still widely used, these operations can be replaced by laser-cutting techniques. With advances in computer controls, the laser path can automatically be controlled, thus producing a wide variety of shapes accurately, repeatedly, and economically without the use of expensive tooling, which requires maintenance due to wear.

Part Size and Dimensional Accuracy. The size, thickness, and shape complexity of a part have a major bearing on the process selected. Complex parts, for example, may not be formed easily and economically, whereas they may be produced by casting, injection molding, powder metallurgy, or may be fabricated and assembled from individual pieces. Likewise, flat parts with thin cross sections may not be cast properly. Dimensional tolerances and surface finish (Chapter 4) obtained in hot-working operations cannot be as fine as those obtained in cold-working operations because dimensional changes, warping, and surface oxidation occur during processing at elevated temperatures. Also, some casting processes produce a better surface finish than others because of the different types of mold materials used. The visual appearance of materials after they have been manufactured into products influences their appeal to the consumer; color, feel, and surface texture are characteristics that we all consider when making a purchasing decision.

The size and shape of manufactured products vary widely (Fig. 1.7). For example, the main landing gear for the twin-engine, 400-passenger Boeing 777 jetliner

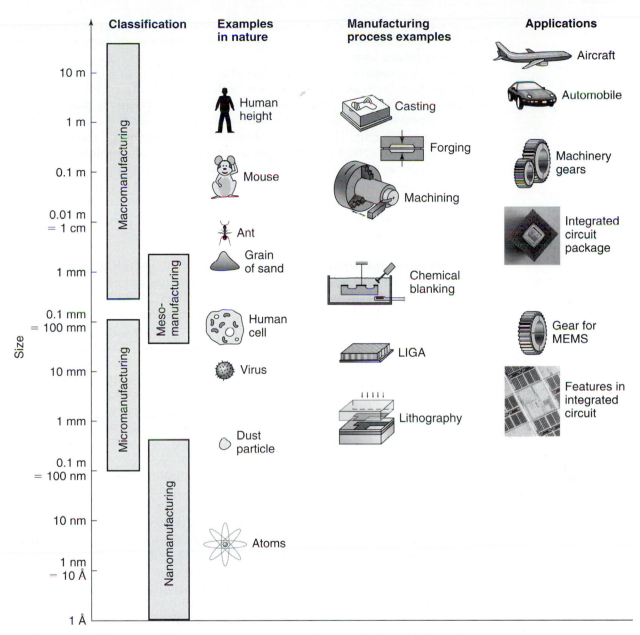

FIGURE 1.7   Illustration of the range of common sizes of parts and the capabilities of manufacturing processes in producing these parts.

is 4.3 m (14 ft) tall and has three axles and six wheels; the main structure is made by forging and followed by various machining processes (Chapters 6, 8, and 9). At the other extreme in part size is the manufacturing of microscopic parts and mechanisms. These components are produced through surface micromachining operations, typically using electron beam, laser beam, and wet and dry etching techniques on materials such as silicon.

Ultraprecision manufacturing techniques and related machinery are coming into common use. For machining mirrorlike surfaces, for example, the cutting tool is a very sharp diamond tip, and the equipment has very high stiffness and must be operated in a room where the temperature is controlled within less than one degree.

Highly sophisticated techniques such as molecular-beam epitaxy and scanning-tunneling engineering are being implemented to reach dimensional accuracies on the order of the atomic lattice (nanometer).

Microelectromechanical systems (MEMS; Chapter 13) are micromechanisms with an integrated circuit. MEMS are now widely used in sensors, ink jet printing mechanisms, and magnetic storage devices, and they can potentially power micro-robots to repair human cells, produce microknives for surgery and camera shutters for precise photography. The most recent trend is the development of **nanoelectromechanical systems** (NEMS), which operate at the same scale as biological molecules. Much effort has been directed at using nanoscale materials and developing entire new classes of materials. An example is the great interest in carbon nanotubes (see Section 13.18) that can reinforce high-performance composite materials, assist in developing nanometer-sized electronic devices, and can store hydrogen in next-generation fuel cells.

Manufacturing and Operational Costs. Considerations such as the design and cost of tooling, the lead time required to begin production, and the effect of workpiece materials on tool and die life are of major importance (Chapter 16). Tooling costs can be very significant, depending on tool size, design, and expected life; a set of steel dies for stamping sheet-metal fenders for automobiles, for example, may cost about $2 million or more. For components made from expensive materials (such as titanium landing gear for aircraft or tantalum-based capacitors), the lower the scrap rate, the lower the production cost will be. Also, because machining (Chapter 8) takes longer time and wastes material by producing chips, it may not be as economical as forming operations, all other factors being equal.

The quantity of parts required and the desired production rate (pieces per hour) help determine the processes to be used and the economics of production. Beverage cans or transistors, for example, are consumed in numbers and at rates much higher than propellers for ships or large gears for heavy machinery. Availability of machines and equipment, operating experience, and economic considerations within the manufacturing facility are also important cost factors. If certain parts cannot be produced within a manufacturing facility, they have to be made by outside firms (an example of *outsourcing*). Automobile manufacturers, for example, purchase numerous parts from outside vendors or have them made according to their specifications.

The operation of production machinery has significant environmental and safety implications. Depending on the type of operation and the machinery involved, some processes adversely affect the environment. For example, chemical vapor deposition and electroplating of coatings involve very hazardous chemicals such as chlorine gases and cyanide solutions, respectively. Metalworking operations usually require the use of lubrication, the disposal of which has potential environmental hazards (see Section 4.4.4). Unless properly controlled, such processes may cause air, water, and noise pollution. The safe use of machinery is another important consideration, requiring precautions to eliminate hazards in the workplace.

Net-Shape Manufacturing. Because not all manufacturing operations produce finished parts or products to desired specifications, additional finishing operations may be necessary. For example, a forged part may not have the desired dimensional accuracy or surface finish; thus additional operations such as machining or grinding may be necessary. Likewise, it may be difficult, impossible, or uneconomical to produce by using only one manufacturing process a part that, by design, has a number of holes in it, necessitating additional processes such as drilling. Also, the holes produced by a particular process may not have the proper roundness, dimensional accuracy, or surface finish, thus necessitating the need for additional operations, such as honing.

These additional operations can contribute significantly to the cost of a product. Consequently, *net-shape* or *near-net-shape manufacturing* has become an important concept in which the part is made as close to the final desired dimensions, tolerances, and specifications as possible. Typical examples of such manufacturing methods are near-net-shape forging and casting of parts, powder-metallurgy techniques, metal-injection molding of metal powders, and injection molding of plastics and ceramics (Chapters 5, 6, 10, and 11).

1.7 | Computer-Integrated Manufacturing

Few developments in history have had a more significant impact on manufacturing than have computers. *Computer-integrated manufacturing* (CIM) now has a very broad range of applications, including control and optimization of manufacturing processes, material handling, assembly, automated inspection and testing of products, inventory control, and numerous management activities. Computer-integrated manufacturing has the capability for (1) improved responsiveness to rapid changes in market demand and product modification; (2) better use of materials, machinery, and personnel, and reduction in inventory; (3) better control of production and management of the total manufacturing operation; and (4) manufacturing high-quality products at low cost.

The following is an outline of the major applications of computers in manufacturing (Chapters 14 and 15):

   a. **Computer numerical control** (CNC). This is a method of controlling the movements of machine components by direct insertion of coded instructions in the form of numerical data (Fig. 1.8). Numerical control was first implemented in the early 1950s and was a major advance in automation of all types of machines.

   b. **Adaptive control** (AC). In *adaptive control,* the parameters in a manufacturing process are adjusted automatically to optimize production rate and product quality and to minimize cost. For example, forces, temperatures, surface finish, and dimensions of the part can constantly be monitored. If these parameters shift outside the acceptable range, the AC system automatically adjusts the process variables until these parameters again fall within the acceptable range.

   c. **Industrial robots.** Introduced in the early 1960s, industrial robots have been replacing humans in operations that are repetitive, boring, and dangerous, thus reducing the possibility of human error, decreasing variability in product quality, and improving productivity. Robots with sensory-perception capabilities are being developed (intelligent robots), with movements that simulate those of humans (Chapter 14).

   d. **Automated handling.** Computers have made possible highly efficient handling of materials and products in various stages of completion (work in progress), such as when being moved from storage to machines or from machine to machine and when products are at the points of inspection, inventory, and shipment (Section 14.6).

   e. **Automated and robotic assembly systems** can replace costly assembly by human operators. Products must be designed or redesigned so that they can be assembled more easily by machine.

   f. **Computer-aided process planning** (CAPP). This method is capable of improving plant productivity by optimizing process plans, reducing planning costs,

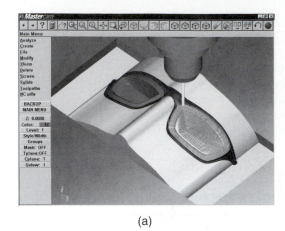

(a)

(b)

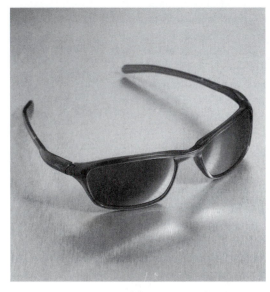

(c)

FIGURE 1.8   Machining a mold cavity for making sunglasses. (a) Computer model of the sunglasses as designed and viewed on the monitor. (b) Machining the die cavity using a computer numerical control milling machine. (c) Final product produced from the mold. *Source:* Courtesy Mastercam/CNC Software, Inc.

and improving the consistency of product quality and reliability. Functions such as cost estimating and monitoring of work standards (time required to perform a certain operation) can also be incorporated into the system (Section 15.6).

g. **Group technology** (GT). The main concept in *group technology* is that parts can be organized by classifying them into families and according to similarities in (a) design and (b) manufacturing processes employed to produce the part. Part designs and process plans can thus be standardized, and families of similar parts can be produced more efficiently and economically (Section 15.8).

h. **Just-in-time production** (JIT). The principal of JIT is that supplies are delivered just in time to be used for production, parts are produced just in time to be made into subassemblies and assemblies, and products are finished just in time to be delivered to the customer. In this method, inventory-carrying costs are low, any defects are detected right away, productivity is increased, and high-quality products are made at low cost (Section 15.12).

FIGURE 1.9   General view of a flexible manufacturing system, showing several machines (machining centers) and an automated guided vehicle (AGV) moving along the aisle. *Source:* Courtesy of Cincinnati Milacron, Inc.

i. **Cellular manufacturing.** *Cellular manufacturing* involves workstations, called manufacturing cells, typically containing several machines that are controlled by a central robot, each machine performing a different operation on the part (Section 15.9).

j. **Flexible manufacturing systems** (FMS). This methodology integrates manufacturing cells into a large unit, all interfaced with a central computer. Flexible manufacturing systems have the highest level of efficiency, sophistication, and productivity among manufacturing systems (Fig. 1.9). Although very costly, they are capable of efficiently producing parts in small runs and of quickly changing production sequences on different parts. This flexibility enables FMS to meet rapid changes in market demand for a wide variety of products.

k. **Expert systems.** These systems are basically complex computer programs. They have the capability to perform tasks and solve difficult real-life problems much as human experts would.

l. **Artificial intelligence** (AI). This important field involves the use of machines and computers to replace human intelligence. Computer-controlled systems are capable of learning from experience and can make decisions that optimize operations and minimize costs. **Artificial neural networks** (ANNs), which are designed to simulate the thought processes of the human brain, have the capability of modeling and simulating production facilities, monitoring and controlling manufacturing processes, diagnosing problems in machine performance, conducting financial planning, and managing a company's manufacturing strategy.

In view of the important advances outlined above, some experts have envisioned the *factory of the future.* Although highly controversial, and viewed as unrealistic by some, this is a system in which production will take place with little or no direct human intervention. The human role is expected to be confined to the supervision, maintenance, and upgrading of machines, computers, and software.

The implementation of some of the modern technologies outlined above requires significant technical and economic expertise, time, and capital investment. Some of the advanced technology can be applied improperly, or it can be implemented on too large or ambitious a scale involving major expenditures with questionable *return on investment* (ROI). Consequently, it is essential to perform a comprehensive analysis and assessment of the real and specific needs of a company and of the market for its products, as well as of whether there is good communication among the parties involved, including vendors.

1.8 | Lean Production and Agile Manufacturing

Lean production, also called *lean manufacturing,* basically involves (a) a major assessment of each activity of a company regarding the efficiency and effectiveness of its operations, (b) the efficiency of the machinery and equipment used in the operation while maintaining and improving quality, (c) the number of personnel involved in a particular operation, and (d) a thorough analysis in order to reduce the cost of each activity, including both productive and nonproductive labor. This concept, although not novel, may require a fundamental change in corporate culture as well as require cooperation and teamwork between management and the work force. Lean production does not necessarily mean cutting back resources but aims at *continuously improving the efficiency and profitability* of a company by removing all types of waste from its operations (zero-base waste) and dealing with problems as soon as they arise.

Agile manufacturing is a term coined to indicate the use of the principles of lean production on a broader scale. The principle behind agile manufacturing is ensuring *flexibility* (agility) in the manufacturing enterprise so that it can quickly respond to changes in product variety and demand and customer needs. Agility is to be achieved through machines and equipment with built-in flexibility (**reconfigurable machines**), using *modular* components that can be arranged and rearranged in different ways, advanced computer hardware and software, reduced changeover time, and implementing advanced communications systems. It has been predicted, for example, that the automotive industry will be able to configure and build a custom car in three days and that, eventually, the traditional assembly line will be replaced by a system in which a nearly custom-made car will be produced by connecting modules.

1.9 | Quality Assurance and Total Quality Management

Product quality has always been one of the most important concerns in manufacturing, as it directly influences the marketability of a product and customer satisfaction. Traditionally, quality assurance has been obtained by inspecting parts after they have been manufactured; parts are inspected to ensure that they conform to a detailed set of specifications and standards such as dimensional tolerances, surface finish, and mechanical and physical properties. Quality, however, cannot be inspected into a product after it is made; it must be built into a product, from the early design stage through all subsequent stages of manufacturing and assembly. Because products are typically manufactured using several different processes, each of which can have significant variations in its performance throughout the day, the control of processes is a critical factor in product quality. Thus, *we control processes and not products.*

Producing defective products can be very costly to the manufacturer, causing difficulties in assembly operations, necessitating repairs in the field, and resulting in customer dissatisfaction. **Product integrity** is a term that can be defined by the degree to which a product (a) is suitable for its intended purpose, (b) fills a real market need, (c) functions reliably during its life expectancy, and (d) can be maintained with relative ease.

Total quality management (TQM) and *quality assurance* are the responsibility of everyone involved in the design and manufacture of a product. Our global awareness of the technological and economic importance of product quality has been guided by pioneers such as Deming, Taguchi, and Juran (see Section 16.3). They pointed out the importance of management's commitment to product quality, pride of workmanship at all levels of production, and use of powerful techniques such as **statistical process control** (SPC) and *control charts* for on-line monitoring of part production and rapid identification of the sources of quality problems (Chapter 4). Ultimately, the major goal here is to prevent defects from occurring rather than to detect defects in products. As a consequence, computer chips, for example, are now produced in such a way that only a few chips out of a million may be defective.

Quality assurance now includes the implementation of **design of experiments,** a technique in which the factors involved in a production process and their interactions are studied simultaneously. Thus, for example, variables affecting dimensional accuracy or surface finish in a machining operation can readily be identified, allowing appropriate actions to be taken.

Global competitiveness has created the need for international conformity in the use of (and for consensus regarding the establishment of) quality control methods, resulting in the International Organization for Standardization ISO 9000 series on Quality Management and Quality Assurance Standards, as well as QS 9000. This standard is a *quality process certification* and *not* a product certification. A company's registration for this standard means that the company conforms to consistent practices as specified by its own quality system. These two standards have permanently influenced the manner in which companies conduct business in world trade, and they are now the world standard for quality.

Product Liability.   We are all familiar with the consequence of using a product that has malfunctioned, causing bodily injury or even death, and the financial loss to a person as well as the organization manufacturing that product. This important topic is referred to as **product liability.** Because of the related technical and legal aspects in which laws can vary from state to state and from country to country, this complex subject can have a major economic impact on all the parties involved.

Designing and manufacturing safe products are important and integral parts of a manufacturer's responsibilities. All those involved with product design, manufacture, and marketing must fully recognize the possible consequences of product failure, including failures occurring during possible misuse of the product. Numerous examples of products that could involve liability may be cited: (1) a grinding wheel that shatters during its use and injures a worker, (2) a supporting cable that snaps, allowing a platform with workers on it to drop, (3) a brake that becomes inoperative because one of its components has failed, (4) a machine with no guards or has inappropriate guards around its gears or moving belts, and (5) an electric or pneumatic tool without appropriate warnings on its proper use and the dangers involved.

Human-factors engineering and **ergonomics** (human–machine interactions) are also important aspects of the design and manufacture of safe products. Examples of products in which such considerations are important include (1) an uncomfortable or unstable workbench or chair, whose design leads to fatigue or permanent injury; and (2) a mechanism that is difficult to operate manually or a poorly designed keyboard, causing pain to the user's hands and arms as a result of repetitive use (which can lead to *repetitive stress injury* or *carpal tunnel syndrom*).

1.10 | Manufacturing Costs and Global Competitiveness

The cost of a product is often the overriding consideration in its marketability and general customer satisfaction. Manufacturing costs typically represent about 40% of a product's selling price. The total cost of manufacturing a product consists of costs of materials, tooling, and labor, as well as fixed and capital costs; several factors are involved in each cost category. Manufacturing costs can be minimized by analyzing the product design to determine whether part size and shape are optimal and the materials selected are the least costly, while still possessing the desired properties and characteristics. The possibility of substituting materials is also an important consideration in minimizing costs (Chapter 16).

The economics of manufacturing have always been a major consideration and have become even more so as **global competitiveness** for high-quality products (**world-class manufacturing**) and *low prices* have become a necessity in worldwide markets. Beginning with the 1960s, the following trends developed which have had a major impact on manufacturing:

- Global competition increased rapidly, and the markets became multinational and dynamic.
- Market conditions fluctuated widely.
- Customers demanded high-quality, low-cost products and on-time delivery.
- Product variety increased substantially, products became more complex, and product life cycles became shorter.

A further important trend has been the wide disparity in manufacturing labor costs (by an order of magnitude) among various countries. Table 1.3 shows the estimated relative hourly compensation for production workers in manufacturing, based on a scale of 100 for the United States. These estimates are approximate because of such

TABLE 1.3

Approximate Relative Hourly Compensation for Production Workers, for 2003. United States = 100. Compensation Costs Vary Depending on Benefits and Allowance			
Denmark	147	Ireland, Italy	85
Norway	144	Spain	67
Germany	136	Israel	53
Belgium, Switzerland	127	New Zealand, Korea	48
Finland, Netherlands	123	Singapore	33
Austria, Sweden	116	Portugal, Taiwan	27
United States	100	Czech Republic	20
France	96	Brazil, Mexico	11
United Kingdom	93	China, India	10
Australia, Canada, Japan	90		
European countries	111		
Asian countries	33		

Source: Courtesy of U.S. Department of Labor, November 2004.

factors as various benefits and housing allowances, which vary from country to country and are not calculated consistently.

Outsourcing is the practice of taking internal company activities and paying an outside firm to perform them. With multinational companies, outsourcing can involve shifting activities to other divisions in different countries. Outsourcing of manufacturing tasks became viable only when the communication and shipping infrastructure developed sufficiently, a trend that started in the early 1990s. For example, an Indian software firm could not effectively collaborate with European or American counterparts until fiber optic communication lines and high-speed Internet access became possible, even if the labor costs were lower.

It is not surprising that many of the products one purchases today are either made or assembled in countries such as China or Mexico, where labor costs, thus far, are the lowest, but are bound to increase as the living standards in these countries rise. Likewise, software development and information technology can be far more economical to implement in India than in Western countries. Keeping costs at a minimum is a constant challenge to manufacturing companies and an issue crucial to their very survival. The cost of a product is often the overriding consideration in its marketability and in general customer satisfaction.

To respond to these needs, while keeping costs low, these approaches require that manufacturers **benchmark** their operations. Benchmarking means understanding the competitive position of a company with respect to that of others and setting realistic goals for its future. It is thus a *reference* from which various measurements can be made and compared.

1.11 | General Trends in Manufacturing

With rapid advances in all aspects of materials, processes, and production control, there are several important trends in manufacturing, as briefly outlined below.

Materials. The trend is for better control of material compositions, purity, and defects (impurities, inclusions, flaws) in order to enhance their overall properties, manufacturing characteristics, reliability, and service life while keeping costs low. Developments are continuing on superconductors, semiconductors, nanomaterials and nanopowders, amorphous alloys, shape-memory alloys (*smart materials*), coatings, and various other engineered metallic and nonmetallic materials. Testing methods and equipment are being improved, including use of advanced computers and software, particularly for materials such as ceramics, carbides, and various composites.

Concerns over energy and material savings are leading to better recyclability and higher strength- and stiffness-to-weight ratios. Thermal treatment of materials is being conducted under better control of relevant variables for more predictable and reliable results, and surface treatment methods are being advanced rapidly. Included in these developments are advances in tool, die, and mold materials, with better resistance to a wide variety of process variables, thus improving the efficiency and economics of manufacturing processes. As a result of these developments, production of goods has become more efficient, with higher-quality products at low cost.

Processes, Equipment, and Systems. Continuing developments in computers, controls, industrial robots, automated inspection, handling and assembly, and sensor technology are having a major impact on the efficiency and reliability of all manufacturing processes and equipment. Advances in computer hardware and software, communications systems, adaptive control, expert systems, and artificial intelligence and neural networks have all helped enable the effective implementation of concepts such as group technology, cellular manufacturing, and flexible manufacturing systems, as well as modern practices in the efficient administration of manufacturing organizations.

Computer simulation and modeling are becoming widely used in design and manufacturing, resulting in the optimization of processes and production systems and better prediction of the effects of relevant variables on product integrity. As a result of such efforts, the speed and efficiency of product design and manufacturing are improving greatly, also affecting the overall economics of production and reducing product cost in an increasingly competitive marketplace.

SUMMARY

- Manufacturing is the process of converting raw materials into products using a variety of processes and methods. (Section 1.1)

- Product design is an integral part of manufacturing, as evidenced by trends in concurrent engineering, design for manufacture, design for assembly, disassembly, and service. (Sections 1.2 and 1.3)

- Designing and manufacturing safe and environmentally friendly products are an important and integral part of a manufacturer's responsibilities. (Section 1.4)

- A key task is to select appropriate materials and an optimal manufacturing method(s) among several possible alternatives, given product design goals, process capabilities, and cost considerations. (Sections 1.5 and 1.6)

- Computer-integrated manufacturing technologies utilize computers to automate a wide range of design, analysis, manufacturing, and quality-control tasks. (Section 1.7)

- Lean production and agile manufacturing are approaches that focus on the efficiency and flexibility of the entire organization in order to help manufacturers respond to global competitiveness and economic challenges. (Section 1.8)

- Ensuring product quality is now a concurrent engineering process rather than a last step in the manufacture of a product. Total quality management and statistical process control techniques have increased our ability to build quality into a product at every step of the design and manufacturing process. (Section 1.9)

- Manufacturing costs and global competitiveness are critical considerations for a manufacturing enterprise. (Section 1.10)

- There are several general trends that have important bearing on manufacturing processes, materials, and systems, including demands on quality, environmental sustainability, and continued proliferation of computer technology. (Section 1.11)

REFERENCES

The following are the Web sites for selected professional societies and technical and industrial organizations that can be used as a source for information on various aspects of manufacturing science and engineering.

Abrasive Engineering Society
www.abrasiveengineering.com

Aluminum Extruders Council
www.aec.org

The Aluminum Association
www.aluminum.org

The American Ceramic Society
www.ceramics.org

American Foundrymen's Society
www.afsinc.org

American Gear Manufacturers
Association
www.agma.org

American Institute of Mining,
Metallurgical and Petroleum Engineers
www.aimeny.org

American National Standards Institute
www.ansi.org

American Plastics Council
www.americanplasticscouncil.org

American Society of Mechanical
Engineers
www.asme.org

American Society for Metals (ASM)
International
www.asminternational.org

American Society for Nondestructive
Testing
www.asnt.org

American Society for Precision
Engineering
www.aspe.net

American Society for Quality
www.asq.org

American Society for Testing and
Materials International
www.astm.org

American Welding Society
www.aws.org

Association for Iron and Steel
Technology
www.aist.org

Association for Manufacturing
Technology
www.mfgtech.org

Computer Aided Manufacturing
International
www.intota.com

Copper Development Association
www.copper.org

Edison Welding Institute
www.ewi.org

Electronic Industries Association
www.eia.org

The Fastener Engineering and Research
Association
www.fera.org.uk

Federation of Materials Societies
www.materialsocieties.org

Forging Industry Association
www.forging.org

Forging Industry Educational and
Research Foundation
www.forgings.org

Grinding Wheel Institute
www.abrasiveengineering.com

Industrial Fasteners Institute
www.industrial-fasteners.org

Institute of Electrical and Electronics
Engineers
www.ieee.org

Institute of Industrial Engineers
www.iienet.org

The International Academy for
Production Engineering (CIRP)
www.cirp.net

International Copper Research
Association
www.copper.org

International Lead Zinc Research
Association
www.recycle.net

International Magnesium Association
www.intlmag.org

International Organization for
Standardization
www.iso.org

International Society for Measurement
and Control
www.isa.org

International Titanium Association
www.titanium.org

Investment Casting Institute
www.investmentcasting.org

Materials Research Society
www.mrs.org

Metal Powder Industries Federation
www.mpif.org

National Association of Corrosion
Engineers (NACE) International
www.nace.org

National Association of Manufacturers
www.nam.org

National Electrical Manufacturers
Association
www.nema.org

National Institute of Standards and
Technology
www.nist.gov

National Science Foundation
www.nsf.gov

North American Die Casting
Association
www.diecasting.org

Occupational Safety and Health
Administration
www.osha.gov

Robotic Industries Association
www.roboticsonline.com

Society for the Advancement of
Material and Process Engineers
www.sampe.org

Society of Automotive Engineers (SAE)
International
www.sae.org

Society of Manufacturing Engineers
www.sme.org

Society of Plastics Engineers
www.4spe.org

Society of Tribologists and Lubrication
Engineers
www.stle.org

Steel Founders' Society of America
www.sfsa.org

The Welding Institute
www.twi.co.uk

BIBLIOGRAPHY

Alukai, G., and Manos, A., *Lean Kaizen: A Simplified Approach to Product Improvements,* ASQ Quality Press, 2006.

Ashby, M., *Materials Selection in Mechanical Design,* 3rd ed., Butterworth-Heineman, 2005.

Boothroyd, G., Dewhurst, P., and Knight, W.A., *Product Design for Manufacture and Assembly,* 2nd ed., CRC Press, 2001.

Bralla, J.G., *Design for Manufacturability Handbook,* 2nd ed., McGraw-Hill, 1999.

Chang, T-C., Wysk, R.A., and Wang, H-P, *Computer-Aided Manufacturing,* 3rd ed., Prentice Hall, 2005.

Cheng, T.C., Podolsky, S., and Podolsky, S., *Just-in-Time Manufacturing—An Introduction,* Springer, 1996.

Clausing, D.P., *Total Quality Development,* American Society of Mechanical Engineers, 1994.

Craig, J.J., *Introduction to Robotics,* 3rd ed., Prentice Hall, 2003.

DeGarmo, E.P., Black, J T., and Kohser, R.A., *Materials and Processes in Manufacturing,* 9th ed., Industrial Press, 2004.

Deming, W.E., *Out of the Crisis,* MIT Press, 1982.

Friedman, T.L., *The World Is Flat,* Farrar, Straus and Giroux, 2006.

Groover, M.P., *Fundamentals of Modern Manufacturing,* 3rd ed., Wiley, 2007.

Gunasekaran, A., *Agile Manufacturing: The 21st Century Competitive Strategy,* Elsevier, 2001.

Hugos, M., *Essentials of Supply Chain Management,* 2nd ed., Wiley, 2006.

Hyer, N., and Wemmerlov, U., *Reorganizing the Factory: Competing Through Cellular Manufacturing,* Productivity Press, 2002.

Irani, S.A., *Handbook of Cellular Manufacturing Systems,* Wiley, 1999.

Juran, J.M., and Godfrey, A.B., *Juran's Quality Handbook,* McGraw-Hill, 1998.

Kalpakjian, S., and Schmid, S.R., *Manufacturing Engineering and Technology,* 5th ed., Prentice Hall, 2006.

Madou, M.J., *Fundamentals of Microfabrication,* 2nd ed., CRC Press, 2002.

Madu, C., *Handbook of Environmentally Conscious Manufacturing,* Springer, 2001.

McDonough, W., and Braungart, M., *Cradle to Cradle,* North Point Press, 2002.

Montgomery, D.C., *Introduction to Statistical Quality Control,* Wiley, 2004.

Ortiz, C.A., *Kaizen Assembly,* CRC Press, 2006.

Pugh, S., *Creating Innovative Products Using Total Design,* Addison-Wesley Longman, 1996.

Pyzdek, T., *The Six Sigma Handbook,* 2nd ed., McGraw-Hill, 2003.

Schey, J.A., *Introduction to Manufacturing Processes,* 3rd ed., McGraw-Hill, 2000.

Taguchi, G., Chowdhury, S., and Wu, Y., *Taguchi's Quality Engineering Handbook,* Wiley, 2004.

Fundamentals of the Mechanical Behavior of Materials

This chapter describes

- The types of tests commonly used to determine the mechanical properties of materials.
- Stress–strain curves, their features, significance, and dependence on parameters such as temperature and deformation rate.
- Characteristics and the roles of hardness, fatigue, creep, impact, and residual stresses in materials processing.
- Yield criteria and their applications in determining forces and energies required in processing metals.

2.1 | Introduction

The manufacturing methods and techniques by which materials can be shaped into useful products were outlined in Chapter 1. One of the oldest and most important groups of manufacturing processes is **plastic deformation**, namely, shaping materials by applying forces by various means. Also known as **deformation processing**, it includes *bulk deformation processes* (forging, rolling, extrusion, and rod and wire drawing) and *sheet-forming processes* (bending, drawing, spinning, and general pressworking). This chapter deals with the fundamental aspects of the mechanical behavior of materials during plastic deformation; the individual topics described are deformation modes, stresses, forces, work of deformation, effects of rate of deformation and temperature, hardness, residual stresses, and yield criteria.

In stretching a piece of metal to make an object such as an automobile fender or a length of wire, the material is subjected to *tension*. A solid cylindrical piece of metal is forged in the making of a turbine disk, subjecting the material to *compression*. Sheet metal undergoes shearing stresses when, for example, a hole is punched through its cross section. A piece of plastic tubing is expanded by internal pressure to make a beverage bottle, subjecting the material to tension in various directions.

In all these processes, the material is subjected to one or more of the three basic modes of deformation shown in Fig. 2.1, namely, tension, compression, and shear. The degree of deformation to which the material is subjected is defined as

FIGURE 2.1 Types of strain: (a) tensile, (b) compressive, (c) shear. All deformation processes in manufacturing involve strains of these types. Tensile strains are involved in stretching sheet metal to make car bodies, compressive strains in forging metals to make turbine disks, and shear strains in making holes by punching.

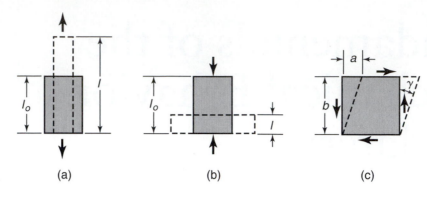

strain. For tension or compression, the **engineering strain,** or **nominal strain,** is defined as

$$e = \frac{l - l_o}{l_o}. \tag{2.1}$$

Note that in tension the strain is positive, and in compression it is negative. In the deformation shown in Fig. 2.1c, the **shear strain** is defined as

$$\gamma = \frac{a}{b}. \tag{2.2}$$

In order to change the shape of the elements, or bodies, shown in Fig. 2.1, *forces* must be applied to them, as shown by the arrows. The determination of these forces as a function of strain is an important aspect in the study of manufacturing processes. A knowledge of these forces is essential in order to design the proper equipment to be used, to select the tool and die materials for proper strength, and to determine whether a specific metalworking operation can be accomplished on certain equipment.

2.2 | Tension

Because of its relative simplicity, the **tension test** is the most common test for determining the *strength-deformation characteristics of materials*. It involves the preparation of a test specimen according to ASTM (American Society for Testing and Materials) standards, and testing it in tension on any of a variety of available testing equipment.

The test specimen has an original length l_o and an original cross-sectional area A_o (Fig. 2.2a). Although most specimens are solid and round, flat sheet or tubular specimens are also tested under tension. The original length is the distance between **gage marks** on the specimen and typically is 50 mm (2 in.). Longer lengths may be used for larger specimens, such as structural members, as well as shorter lengths for specific applications on small parts.

Typical results from a tension test are shown in Fig. 2.2. The **engineering stress,** or **nominal stress,** is defined as the ratio of the applied load to the original area of the specimen,

$$\sigma = \frac{P}{A_o}, \tag{2.3}$$

and the engineering strain is given by Eq. (2.1).

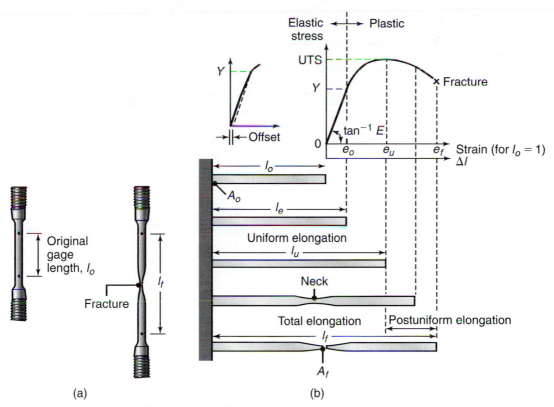

FIGURE 2.2   (a) Original and final shape of a standard tensile-test specimen. (b) Outline of a tensile-test sequence showing different stages in the elongation of the specimen.

When the load is first applied, the specimen elongates proportionately to the load up to the **proportional limit;** this is the range of **linear elastic behavior.** The material will continue to deform elastically, although not strictly linearly, up to the **yield point,** Y. If the load is removed before the yield point is reached, the specimen will return to its original length. The **modulus of elasticity,** or **Young's modulus,** E, is defined as

$$E = \frac{\sigma}{e}. \tag{2.4}$$

This linear relationship between stress and strain is known as **Hooke's law,** the more generalized forms of which are given in Section 2.11. The elongation of the specimen is accompanied by a contraction of its lateral dimensions. The *absolute* value of the ratio of the lateral strain to longitudinal strain is known as **Poisson's ratio,** ν. Typical values for E and ν for various materials are given in Table 2.1.

The area under the stress–strain curve up to the yield point, Y, of the material is known as the **modulus of resilience:**

$$\text{Modulus of resilience} = \frac{Ye_o}{2} = \frac{Y^2}{2E}. \tag{2.5}$$

This area has the units of **energy per unit volume** and indicates the **specific energy** that the material can store elastically. Typical values for modulus of resilience are, for example, 2.1×10^4 N-m/m³ (3 in.-lb/in³) for annealed copper, 1.9×10^5 (28) for annealed medium-carbon steel, and 2.7×10^6 (385) for spring steel.

TABLE 2.1

Typical Mechanical Properties of Various Materials at Room Temperature (See also Tables 10.1, 10.4, 10.8, 11.3, and 11.7)

	E (GPa)	Y (MPa)	UTS (MPa)	Elongation in 50 mm (%)	Poisson's ratio (ν)
METALS (WROUGHT)					
Aluminum and its alloys	69–79	35–550	90–600	45–5	0.31–0.34
Copper and its alloys	105–150	76–1100	140–1310	65–3	0.33–0.35
Lead and its alloys	14	14	20–55	50–9	0.43
Magnesium and its alloys	41–45	130–305	240–380	21–5	0.29–0.35
Molybdenum and its alloys	330–360	80–2070	90–2340	40–30	0.32
Nickel and its alloys	180–214	105–1200	345–1450	60–5	0.31
Steels	190–200	205–1725	415–1750	65–2	0.28–0.33
Stainless steels	190–200	240–480	480–760	60–20	0.28–0.30
Titanium and its alloys	80–130	344–1380	415–1450	25–7	0.31–0.34
Tungsten and its alloys	350–400	550–690	620–760	0	0.27
NONMETALLIC MATERIALS					
Ceramics	70–1000	—	140–2600	0	0.2
Diamond	820–1050	—	—	—	—
Glass and porcelain	70–80	—	140	0	0.24
Rubbers	0.01–0.1	—	—	—	0.5
Thermoplastics	1.4–3.4	—	7–80	1000–5	0.32–0.40
Thermoplastics, reinforced	2–50	—	20–120	10–1	—
Thermosets	3.5–17	—	35–170	0	0.34
Boron fibers	380	—	3500	0	—
Carbon fibers	275–415	—	2000–5300	1–2	—
Glass fibers (S, E)	73–85	—	3500–4600	5	—
Kevlar fibers (29, 49, 129)	70–113	—	3000–3400	3–4	—
Spectra fibers (900, 1000)	73–100	—	2400–2800	3	—

Note: In the upper table, the lowest values for E, Y, and UTS and the highest values for elongation are for the pure metals. Multiply GPa by 145,000 to obtain psi, and MPa by 145 to obtain psi. For example, 100 GPa = 14,500 ksi, and 100 MPa = 14,500 psi.

With increasing load, the specimen begins to yield; that is, it begins to undergo **plastic (permanent) deformation,** and the relationship between stress and strain is no longer linear. For most materials the *rate of change* in the slope of the stress–strain curve beyond the yield point is very small, thus the determination of Y can be difficult. The usual practice is to define the yield stress as the point on the curve that is *offset* by a strain of (usually) 0.2%, or 0.002 (Fig. 2.2b). Other offset strains may also be used and should be specified in reporting the yield stress of a material.

It is important to note that yielding does not necessarily mean failure. In the design of structures and load-bearing members, yielding is not acceptable since it leads to permanent deformation. However, yielding is necessary in metalworking processes, such as forging, rolling, and sheet-metal forming operations, where materials have to be subjected to permanent deformation to develop the desired part shape.

As the specimen continues to elongate under increasing load beyond Y, its cross-sectional area decreases *permanently and uniformly* throughout its gage length. If the specimen is unloaded from a stress level higher than Y, the curve follows a straight line downward and parallel to the original elastic slope, as shown in Fig. 2.3. As the load, hence the engineering stress, is further increased, the curve eventually reaches a maximum and then begins to decrease. The maximum stress is

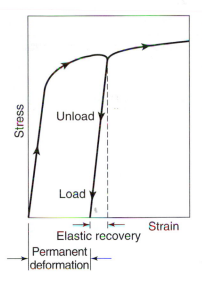

FIGURE 2.3 Schematic illustration of loading and unloading of a tensile-test specimen. Note that during unloading the curve follows a path parallel to the original elastic slope.

known as the **tensile strength** or **ultimate tensile strength** (UTS) of the material (Table 2.1). Ultimate tensile strength is thus a simple and practical measure of the overall strength of a material.

When the specimen is loaded beyond its UTS, it begins to *neck* (Fig. 2.2a) and the elongation between the gage marks is no longer uniform. That is, the change in the cross-sectional area of the specimen is no longer uniform but is concentrated locally in a "neck" formed in the specimen (called **necking**, or *necking down*). As the test progresses, the engineering stress drops further and the specimen finally fractures within the necked region. The final stress level (marked by an **x** in Fig. 2.2b) at fracture is known as **breaking** or **fracture stress.**

2.2.1 Ductility

The strain in the specimen at fracture is a measure of *ductility*, that is, how large a strain the material withstands before fracture. Note from Fig. 2.2b that until the UTS is reached, elongation is *uniform*. The strain up to the UTS is called **uniform strain.** The elongation at fracture is known as the **total elongation** and is measured between the original gage marks after the two pieces of the broken specimen are placed together.

Two quantities that are commonly used to define ductility in a tension test are *elongation* and *reduction of area*. **Elongation** is defined as

$$\text{Elongation} = \frac{l_f - l_o}{l_o} \times 100 \tag{2.6}$$

and is based on the total elongation (Table 2.1).

Necking is a *local* phenomenon. If we put a series of gage marks at different points on the specimen, pull and break it under tension, and then calculate the percent elongation for each pair of gage marks, we find that with decreasing gage length, the percent elongation increases (Fig. 2.4). It should be noted that the closest pair of gage marks have undergone the largest elongation because they are closest to the necked and fractured region. (Note that the curves will not approach zero elongation because the specimen has already undergone some finite permanent elongation before fracture.) Consequently, it is important to include the gage length in reporting elongation data. Other tensile properties are generally found to be independent of gage length.

FIGURE 2.4 Total elongation in a tensile test as a function of original gage length for various metals. Because necking is a local phenomenon, elongation decreases with gage length. Standard gage length is usually 2 in. (50 mm), although shorter ones can be used if larger specimens are not available.

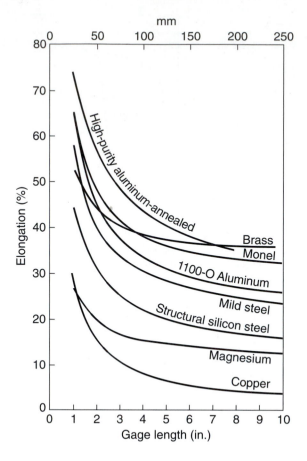

A second measure of ductility is **reduction of area,** defined as

$$\text{Reduction of area} = \frac{A_o - A_f}{A_o} \times 100. \qquad (2.7)$$

Note that a material that necks down to a point at fracture, such as a glass rod at elevated temperature, has a reduction of area of 100%.

Elongation and reduction of area are generally related to each other for many engineering metals and alloys. Elongation ranges approximately between 10% and 60%, and values between 20% and 90% are typical for reduction of area for most materials. *Thermoplastics* (Chapter 10) and *superplastic* materials (Section 2.2.7) exhibit much higher ductility. Brittle materials, by definition, have little or no ductility; typical examples are common glass at room temperature, chalk, and gray cast iron.

2.2.2 True stress and true strain

Because stress is defined as the ratio of force to area, *true stress* is likewise defined as

$$\sigma = \frac{P}{A}, \qquad (2.8)$$

where A is the actual (hence true) or instantaneous area supporting the load.

The complete tension test may be regarded as a series of incremental tension tests where, for each succeeding increment, the specimen is a little longer than at

TABLE 2.2

Comparison of Engineering and True Strains in Tension								
e 0.01	0.05	0.1	0.2	0.5	1	2	5	10
ϵ 0.01	0.049	0.095	0.18	0.4	0.69	1.1	1.8	2.4

the preceding stage. Thus, *true strain* (or *natural* or *logarithmic strain*), ϵ, can be defined as

$$\epsilon = \int_{l_o}^{l} \frac{dl}{l} = \ln\left(\frac{l}{l_o}\right).\tag{2.9}$$

Note that, for small values of engineering strain, we have $e = \epsilon$ since $\ln(1 + e) = \epsilon$. For larger strains, however, the values rapidly diverge, as can be seen in Table 2.2.

The *volume* of a metal specimen remains constant in the plastic region of the test (see *volume constancy*, Section 2.11.5). Thus the true strain within the uniform elongation range can be expressed as

$$\epsilon = \ln\left(\frac{l}{l_o}\right) = \ln\left(\frac{A_o}{A}\right) = \ln\left(\frac{D_o}{D}\right)^2 = 2\ln\left(\frac{D_o}{D}\right).\tag{2.10}$$

Once necking begins, the true strain at any point between the gage marks of the specimen can be calculated from the reduction in the cross-sectional area at that point. Thus, by definition, the largest strain is at the narrowest region of the neck.

We have seen that, at small strains, the engineering and true strains are very close and, therefore, either one can be used in calculations. However, for the large strains encountered in metalworking, the true strain should be used because it is the true measure of the strain, as can be illustrated by the following two examples.

1. Assume that a tension specimen is elongated to twice its original length. This deformation is equivalent to compressing a specimen to one-half its original height. Using the subscripts t and c for tension and compression, respectively, it can be seen that $\epsilon_t = 0.69$ and $\epsilon_c = -0.69$, whereas $e_t = 1$ and $e_c = -0.5$. Thus, true strain is a correct measure of strain.

2. Assume that a specimen 10 mm in height is compressed to a final thickness of zero; thus, $\epsilon_c = -\infty$, whereas $e_c = -1$. Note that we have deformed the specimen infinitely (because its final thickness is zero); this is exactly what the value of the true strain indicates.

From these two examples, it can be seen that true strains are consistent with the actual physical phenomenon whereas engineering strains are not.

2.2.3 True stress–true strain curves

The relationship between engineering and true values for stress and strain, respectively, can now be used to construct *true stress–true strain curves* from a curve such as that shown in Fig. 2.2b. A typical true stress–true strain curve is shown in Fig. 2.5a. For convenience, such a curve is typically approximated by the equation

$$\sigma = K\epsilon^n.\tag{2.11}$$

Note that Eq. (2.11) indicates neither the elastic region nor the yield point, Y, of the material, but these quantities are readily available from the engineering stress–strain curve. (Since the strains at the yield point are very small, the difference between true yield stress and engineering yield stress is negligible for metals. The reason is that, at yielding, the difference in the cross-sectional areas A_o and A is negligible.)

FIGURE 2.5 (a) True stress–true strain curve in tension. Note that, unlike in an engineering stress–strain curve, the slope is always positive and that the slope decreases with increasing strain. Although in the elastic range stress and strain are proportional, the total curve can be approximated by the power expression shown. On this curve, Y is the yield stress and Y_f is the flow stress. (b) True stress–true strain curve plotted on a log–log scale. (c) True stress–true strain curve in tension for 1100-O aluminum plotted on a log–log scale. Note the large difference in the slopes in the elastic and plastic ranges. *Source:* After R.M. Caddell and R. Sowerby.

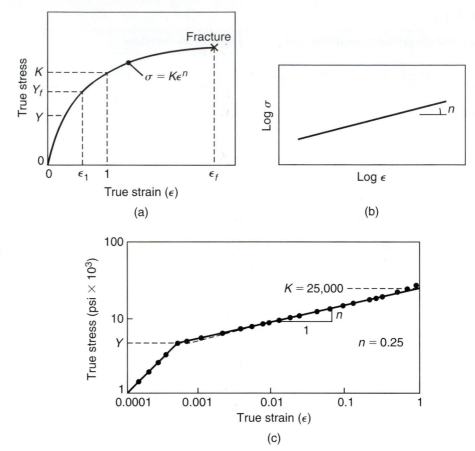

We can now rewrite Eq. (2.11) as

$$\log \sigma = \log K + n \log \epsilon.$$

If we plot the true stress–true strain curve on a log–log scale, we obtain Fig. 2.5b. The slope n is known as the **strain-hardening exponent,** and K is known as the **strength coefficient.** Note that K is the true stress at a true strain of unity. Values of K and n for a variety of engineering materials are given in Table 2.3. The true stress–true strain curves for several materials are given in Fig. 2.6. Some differences between Table 2.3 and these curves exist because of different sources of data and test conditions employed.

In Fig. 2.5a, Y_f is known as the **flow stress** and is defined as the true stress required to continue plastic deformation at a particular true strain, ϵ_1. For strain-hardening materials, the flow stress increases with increasing strain. Note also from Fig. 2.5c that the elastic strains are much smaller than plastic strains. Consequently, and although both effects exist, we will ignore elastic strains in our calculations for forming processes throughout the rest of this text, and the plastic strain will thus be the total strain that the material undergoes.

Toughness. The area under the true stress–true strain curve is known as **toughness** and can be expressed as

$$\text{Toughness} = \int_0^{\epsilon_f} \sigma \, d\epsilon, \tag{2.12}$$

where ϵ_f is the true strain at fracture. Note that toughness is the energy per unit volume (*specific energy*) that has been dissipated up to the point of fracture. It is

TABLE 2.3

Typical Values for *K* and *n* in Eq. (2.11) at Room Temperature		
Material	*K* (MPa)	*n*
Aluminum, 1100-O	180	0.20
2024-T4	690	0.16
5052-O	210	0.13
6061-O	205	0.20
6061-T6	410	0.05
7075-O	400	0.17
Brass, 7030, annealed	895	0.49
85–15, cold rolled	580	0.34
Bronze (phosphor), annealed	720	0.46
Cobalt-base alloy, heat treated	2070	0.50
Copper, annealed	315	0.54
Molybdenum, annealed	725	0.13
Steel, low carbon, annealed	530	0.26
1045 hot rolled	965	0.14
1112 annealed	760	0.19
1112 cold rolled	760	0.08
4135 annealed	1015	0.17
4135 cold rolled	1100	0.14
4340 annealed	640	0.15
17–4 P-H, annealed	1200	0.05
52100, annealed	1450	0.07
304 stainless, annealed	1275	0.45
410 stainless, annealed	960	0.10

Note: 100 MPa = 14,500 psi.

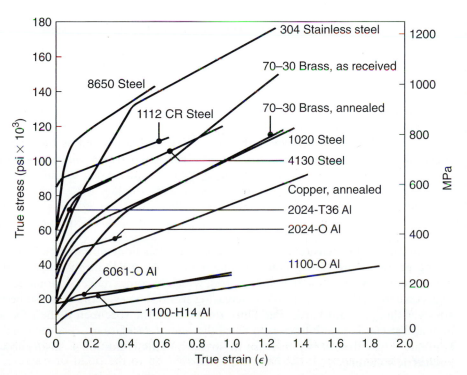

FIGURE 2.6 True stress–true strain curves in tension at room temperature for various metals. The point of intersection of each curve at the ordinate is the yield stress *Y*; thus, the elastic portions of the curves are not indicated. When the *K* and *n* values are determined from these curves, they may not exactly agree with those given in Table 2.3 because of the different sources from which they were collected. *Source:* S. Kalpakjian.

important to also note that this specific energy pertains only to the volume of material at the narrowest region of the neck; any volume of material away from the neck has undergone lower strain and hence has dissipated less energy than that in the fracture zone.

As defined here, toughness is different from the concept of *fracture toughness,* as treated in textbooks on fracture mechanics. Fracture mechanics (the study of the initiation and propagation of cracks in a solid medium) is beyond the scope of this text. Except in die design and die life, fracture toughness is of limited relevance to metalworking processes.

2.2.4 Instability in tension

We have observed that once ultimate tensile strength is reached, the specimen will begin to neck and thus deformation is no longer uniform. This phenomenon has important significance because nonuniform deformation will cause part thickness variation and localization in processing of materials, particularly in sheet-forming operations (Chapter 7) where the materials are subjected to tension. In this section it will be shown that the true strain at the onset of necking is numerically equal to the strain-hardening exponent, n.

Note in Fig. 2.2 that the slope of the load-elongation curve at UTS is zero (or $dP = 0$). It is here that *instability* begins; that is, the specimen begins to neck and cannot support the load because the cross-sectional area of the necked region is becoming smaller as the test progresses. Using the relationships

$$\epsilon = \ln\left(\frac{A_o}{A}\right), \quad A = A_o e^{-\epsilon}, \quad \text{and} \quad P = \sigma A = \sigma A_o e^{-\epsilon}$$

we can determine $dP/d\epsilon$ by noting that

$$\frac{dP}{d\epsilon} = \frac{d}{d\epsilon}(\sigma A_o e^{-\epsilon}) = A_o \left(\frac{d\sigma}{d\epsilon} e^{-\epsilon} - \sigma e^{-\epsilon}\right).$$

Because $dP = 0$ at the UTS where necking begins, we set this expression equal to zero. Thus,

$$\frac{d\sigma}{d\epsilon} = \sigma.$$

However, since

$$\sigma = K\epsilon^n,$$

thus,

$$nK\epsilon^{n-1} = K\epsilon^n$$

and therefore

$$\epsilon = n. \tag{2.13}$$

Instability in a tension test can be viewed as a phenomenon in which two competing processes are taking place simultaneously. As the load on the specimen is increased, its cross-sectional area decreases, which becomes more pronounced in the region where necking begins. With increasing strain, however, the material becomes stronger due to strain hardening. Since the load on the specimen is the product of area and strength, instability sets in when the *rate of decrease* in cross-sectional area is greater than the *rate of increase* in strength. This condition is also known as **geometric softening.**

EXAMPLE 2.1 Calculation of ultimate tensile strength

A material has a true stress–true strain curve given by

$$\sigma = 100{,}000\epsilon^{0.5} \text{ psi.}$$

Calculate the true ultimate tensile strength and the engineering UTS of this material.

Solution. Since the necking strain corresponds to the maximum load and the necking strain for this material is given as

$$\epsilon = n = 0.5,$$

we have as the true ultimate tensile strength,

$$\sigma = Kn^n$$

$$\text{UTS}_{\text{true}} = 100{,}000\,(0.5)^{0.5} = 70{,}710 \text{ psi.}$$

The cross-sectional area at the onset of necking is obtained from

$$\ln\left(\frac{A_o}{A_{\text{neck}}}\right) = n = 0.5.$$

Consequently,

$$A_{\text{neck}} = A_o e^{-0.5},$$

and the maximum load P is

$$P = \sigma A = \sigma A_o e^{-0.5},$$

where σ is the true ultimate tensile strength. Hence,

$$P = (70{,}710)(0.606)(A_o) = 42{,}850 A_o \text{ lb.}$$

Since $\text{UTS} = P/A_o$, we have

$$\text{UTS} = 42{,}850 \text{ psi.}$$

2.2.5 Types of stress–strain curves

Each material has a differently shaped stress–strain curve, its shape depending on its composition and many other factors to be treated in detail later in this chapter. In addition to the power law given in Eq. (2.11), some of the major types of curves are shown in Fig. 2.7, with their associated stress–strain equations, and have the following characteristics:

1. A **perfectly elastic material** displays linear behavior with slope E. The behavior of brittle materials, such as common glass, most ceramics, and some cast irons, may be represented by such a curve (Fig. 2.7a). There is a limit to the stress the material can sustain, after which it fractures. Permanent deformation, if any, is negligible.

2. A **rigid, perfectly plastic** material has, by definition, an infinite value of E. Once the stress reaches the yield stress, Y, it continues to undergo deformation at the same stress level. When the load is released, the material has undergone permanent deformation; there is no elastic recovery (Fig. 2.7b).

FIGURE 2.7 Schematic illustration of various types of idealized stress–strain curves: (a) perfectly elastic, (b) rigid, perfectly plastic, (c) elastic, perfectly plastic, (d) rigid, linearly strain hardening, (e) elastic, linearly strain hardening. The broken lines and arrows indicate unloading and reloading during the test.

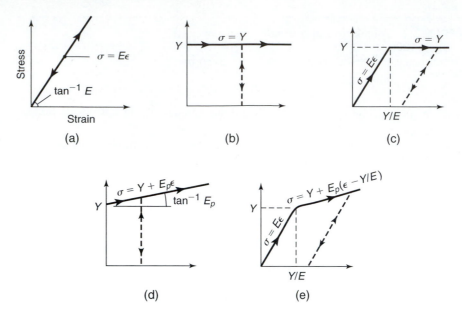

3. The behavior of an **elastic, perfectly plastic** material is a combination of the first two: It has a finite elastic modulus and its undergoes elastic recovery when the load is released (Fig. 2.7c).

4. A **rigid, linearly strain-hardening** material requires an increasing stress level to undergo further strain; thus, its **flow stress** (magnitude of the stress required to maintain plastic deformation at a given strain; see Fig. 2.5a) increases with increasing strain. It has no elastic recovery upon unloading (Fig. 2.7d).

5. An **elastic, linearly strain-hardening** curve (Fig. 2.7e) is an acceptable approximation of the behavior of most engineering materials, with the modification that the plastic portion of the curve has a decreasing slope with increasing strain (Fig. 2.5a).

Note that some of these curves can be expressed by Eq. (2.11) by changing the value of n (Fig. 2.8) or by other equations of a similar nature.

2.2.6 Effects of temperature

In this and subsequent sections, we will discuss various factors that have an influence on the shape of stress–strain curves. The first factor is temperature. Although somewhat difficult to generalize, increasing temperature usually lowers the modulus of elasticity, yield stress, and ultimate tensile strength and increases ductility and toughness (Fig. 2.9). Temperature also affects the strain-hardening exponent, n, of most metals, in that n decreases with increasing temperature. Depending on the type of material and

FIGURE 2.8 The effect of strain-hardening exponent n on the shape of true stress–true strain curves. When $n = 1$, the material is elastic, and when $n = 0$, it is rigid and perfectly plastic.

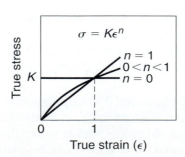

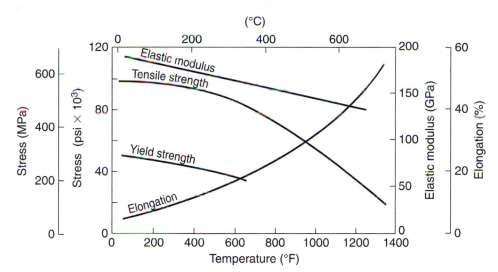

FIGURE 2.9 Effect of temperature on mechanical properties of a carbon steel. Most materials display similar temperature sensitivity for elastic modulus, yield strength, ultimate strength, and ductility.

its composition and level of impurities present, elevated temperatures can have other significant effects, as detailed in Chapter 3. The influence of temperature is best discussed in conjunction with strain rate for the reasons explained in the next section.

2.2.7 Effects of strain rate

Depending on the particular manufacturing operation and equipment, a piece of material may be formed at speeds from very low to very high. Whereas the *deformation speed* or *rate* is typically defined as the speed at which a tension test is being carried out (such as m/s), the *strain rate* (such a 10^2 s^{-1}, 10^4 s^{-1}, etc.) is also a function of the shape of the specimen, as described below. Typical deformation speeds and strain rates in various metalworking processes and the strain rates are shown in Table 2.4. In order to simulate the actual metalworking processes, the specimen in a tension test (as well as in compression and torsion) can be strained at different rates.

The engineering strain rate, $\dot{e}$, is defined as

$$\dot{e} = \frac{de}{dt} = \frac{d\left(\dfrac{l - l_o}{l_o}\right)}{dt} = \frac{1}{l_o}\frac{dl}{dt} = \frac{v}{l_o} \tag{2.14}$$

TABLE 2.4

Typical Ranges of Strain, Deformation Speed, and Strain Rates in Metalworking Processes			
Process	True strain	Deformation speed (m/s)	Strain rate (s^{-1})
Cold working			
Forging, rolling	0.1–0.5	0.1–100	1–10^3
Wire and tube drawing	0.05–0.5	0.1–100	1–10^4
Explosive forming	0.05–0.2	10–100	10–10^5
Hot working and warm working			
Forging, rolling	0.1–0.5	0.1–30	1–10^3
Extrusion	2–5	0.1–1	10^{-1}–10^2
Machining	1–10	0.1–100	10^3–10^6
Sheet-metal forming	0.1–0.5	0.05–2	1–10^2
Superplastic forming	0.2–3	10^{-4}–10^{-2}	10^{-4}–10^{-2}

and the true strain rate, $\dot{\epsilon}$, as

$$\dot{\epsilon} = \frac{d\epsilon}{dt} = \frac{d\left[\ln\left(\frac{l}{l_o}\right)\right]}{dt} = \frac{1}{l}\frac{dl}{dt} = \frac{v}{l}, \tag{2.15}$$

where v is the rate of deformation, for example, the speed of the jaws of the testing machine in which the specimen is clamped.

It can be seen from the preceding equations that although the deformation rate, v, and engineering strain rate, $\dot{e}$, are proportional, the true strain rate, $\dot{\epsilon}$, is not proportional to v. Thus, in a tension test and with v constant, the true strain rate decreases as the specimen becomes longer. Therefore, in order to maintain a constant $\dot{\epsilon}$, the speed must be increased accordingly. (Note that for small changes in length of the specimen during a test, this difference is not significant.)

The typical effects of temperature and strain rate on the strength of metals are shown in Fig. 2.10. It clearly indicates that increasing strain rate increases strength and that the sensitivity of strength to the strain rate increases with temperature, as shown in Fig. 2.11. Note, however, that this effect is relatively small at room temperature. From Fig. 2.10 it can be noted that the same strength can be obtained either at low temperature and low strain rate or at high temperature and high strain rate. These relationships are important in estimating the resistance of materials to deformation when processing them at various strain rates and temperatures.

The effect of strain rate on the strength of materials is generally expressed by

$$\sigma = C\dot{\epsilon}^m, \tag{2.16}$$

where C is the **strength coefficient**, similar to K in Eq. (2.11), and m is the **strain-rate sensitivity exponent** of the material. A general range of values for m is up to 0.05 for cold working and 0.05 to 0.4 for hot working of metals, and 0.3 to 0.85 for superplastic materials (see below). Some specific values for C and m are given in

FIGURE 2.10 The effect of strain rate on the ultimate tensile strength of aluminum. Note that as temperature increases, the slope increases. Thus, tensile strength becomes more and more sensitive to strain rate as temperature increases. *Source:* After J.H. Hollomon.

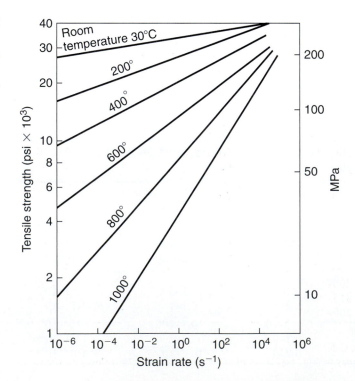

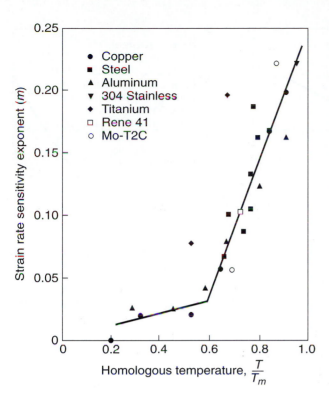

FIGURE 2.11 Dependence of the strain-rate sensitivity exponent m on the homologous temperature T/T_m for various materials. T is the testing temperature and T_m is the melting point of the metal, both on the absolute scale. The transition in the slopes of the curve occurs at about the recrystallization temperature of the metals. *Source:* After F.W. Boulger.

Table 2.5. It has been observed that the value of m decreases with metals of increasing strength.

The magnitude of m has a significant effect on necking in a tension test. Experimental observations have shown that with higher m values, the material stretches to a greater length before it fails, an indication that necking is delayed with increasing m.

TABLE 2.5

Approximate Range of Values for C and m in Eq. (2.16) for Various Annealed Metals at True Strains Ranging from 0.2 to 1.0

Material	Temperature, °C	C (psi × 10³)	C (MPa)	m
Aluminum	200–500	12–2	82–14	0.07–0.23
Aluminum alloys	200–500	45–5	310–35	0–0.20
Copper	300–900	35–3	240–20	0.06–0.17
Copper alloys (brasses)	200–800	60–2	415–14	0.02–0.3
Lead	100–300	1.6–0.3	11–2	0.1–0.2
Magnesium	200–400	20–2	140–14	0.07–0.43
Steel				
Low carbon	900–1200	24–7	165–48	0.08–0.22
Medium carbon	900–1200	23–7	160–48	0.07–0.24
Stainless	600–1200	60–5	415–35	0.02–0.4
Titanium	200–1000	135–2	930–14	0.04–0.3
Titanium alloys	200–1000	130–5	900–35	0.02–0.3
Ti-6Al-4V*	815–930	9.5–1.6	65–11	0.50–0.80
Zirconium	200–1000	120–4	830–27	0.04–0.4

*At a strain rate of 2×10^{-4} s⁻¹.

Note: As temperature increases, C decreases and m increases. As strain increases, C increases and m may increase or decrease, or it may become negative within certain ranges of temperature and strain.

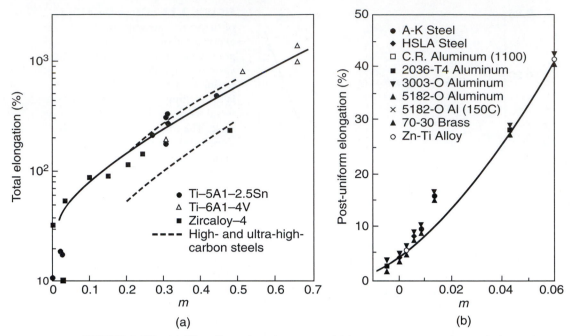

FIGURE 2.12   (a) The effect of strain-rate sensitivity exponent m on the total elongation for various metals. Note that elongation at high values of m approaches 1000%. *Source: After D. Lee and W.A. Backofen.* (b) The effect of strain-rate sensitivity exponent m on the post-uniform (after necking) elongation for various metals. *Source: After A.K. Ghosh.*

When necking is about to begin, the region's strength with respect to the rest of the specimen increases because of strain hardening. However, the strain rate in the neck region is also higher than in the rest of the specimen because the material is elongating faster there. Since the material in the necked region is becoming stronger as it is strained at a higher rate, this region exhibits a higher resistance to necking.

The increase in the resistance to necking thus depends on the magnitude of m. As the test progresses, necking becomes more *diffuse* and the specimen becomes longer before it fractures; hence, total elongation increases with increasing m value (Fig. 2.12). As expected, the elongation after necking (*postuniform elongation*) also increases with increasing m.

The effect of strain rate on strength also depends on the particular level of strain: It increases with strain. The strain rate also affects the strain-hardening exponent, n, because it decreases with increasing strain rate.

Because the formability of materials depends largely on their ductility, it is important to recognize the effect of temperature and strain rate on ductility. Generally, higher strain rates have an adverse effect on the ductility of materials. The increase in ductility due to the strain-rate sensitivity of materials has been exploited in **superplastic forming of metals**, as described in Section 7.5.5. The term *superplastic* refers to the capability of some materials to undergo large uniform elongation prior to failure. The elongation may be on the order of a few hundred percent to over 2000%. An example of such behavior is hot glass, thermoplastic polymers at elevated temperatures, very fine grain alloys of zinc-aluminum, and titanium alloys. Nickel alloys can display superplastic behavior when in a nanocrystalline form (see Section 3.11.9).

2.2.8 Effects of hydrostatic pressure

Although most tests are generally carried out at ambient pressure, experiments also have been performed under hydrostatic conditions, with pressures ranging up

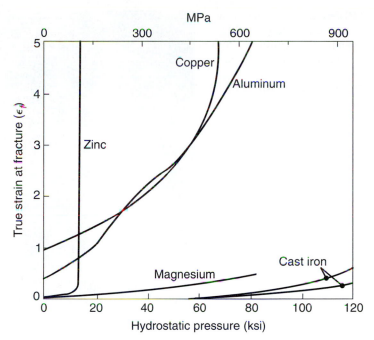

to 10^3 MPa (10^5 psi). Three important observations have been made concerning the effects of high hydrostatic pressure on the behavior of materials: (a) It substantially increases the strain at fracture (Fig. 2.13); (b) it has little or no effect on the shape of the true stress–true strain curve but only extends it; (c) it has no effect on the strain or the maximum load at which necking begins; and (d) the mechanical properties of metals are generally not altered after being subjected to hydrostatic pressure.

The increase in ductility due to hydrostatic pressure has also been observed in other tests, such as compression and torsion tests. The increase has been observed not only with ductile metals, but also with brittle metals and nonmetallic materials. Materials such as grey cast iron, marble, and various other apparently brittle materials acquire some ductility (or an increase in ductility) under pressure, and deform plastically when subjected to hydrostatic pressure. The level of pressure required to enhance ductility depends on the particular material.

2.2.9 Effects of radiation

In view of the nuclear applications of various metals and alloys, studies have been conducted on the effects of radiation on material properties. Typical changes in the mechanical properties of steels and other metals exposed to high-energy radiation are (a) higher yield stress, (b) higher tensile strength, (c) higher hardness, and (d) lower ductility and toughness. The magnitudes of these changes depend on the particular material and its condition, as well as the temperature and level of radiation to which it is subjected.

2.3 | Compression

Many operations in metalworking, such as forging, rolling, and extrusion, are performed with the workpieces under externally applied compressive forces. The **compression test,** in which the specimen is subjected to a compressive load as shown

FIGURE 2.14 Barreling in compressing a round solid cylindrical specimen (7075-O aluminum) between flat dies. Barreling is caused by friction at the die-specimen interfaces, which retards the free flow of the material. See also Figs. 6.1 and 6.2. *Source:* K.M. Kulkarni and S. Kalpakjian.

in Fig. 2.1b, can give useful information for these processes, such as the stresses required and the behavior of the material under compression. The deformation shown in Fig. 2.1b is ideal. This test is usually carried out by compressing (*upsetting*) a solid cylindrical specimen between two flat platens. The friction between the specimen and the dies is an important factor, in that it causes **barreling** (Fig. 2.14) because friction prevents the top and bottom surfaces from expanding freely.

This phenomenon makes it difficult to obtain relevant data and to properly construct a compressive stress–strain curve because (a) the cross-sectional area of the specimen changes along its height and (b) friction dissipates energy, and this energy is supplied through an increased compressive force. With effective lubrication or other means (see Section 4.4.3) it is, however, possible to minimize friction, and hence barreling, to obtain a reasonably constant cross-sectional area during this test.

The engineering strain rate, $\dot{e}$, in compression is given by

$$\dot{e} = -\frac{v}{h_o},\tag{2.17}$$

where v is the speed of the die and h_o is the original height of the specimen. The true strain rate, $\dot{\epsilon}$, is given by

$$\dot{\epsilon} = -\frac{v}{h},\tag{2.18}$$

where h is the instantaneous height of the specimen. Note that if v is constant, the true strain rate increases as the test progresses. In order to conduct this test at a constant true strain rate, a *cam plastometer* has been designed that, through a cam action, reduces the magnitude of v proportionately as the specimen height h decreases during the test.

The compression test can also be used to determine the ductility of a metal by observing the cracks that form on the barreled cylindrical surfaces of the specimen (see Fig. 3.21d). Hydrostatic pressure has a beneficial effect in delaying the formation of these cracks. With a sufficiently ductile material and effective lubrication, compression tests can be carried out uniformly to large strains. This behavior is unlike that in the tension test, where, even for very ductile materials, necking can set in after relatively little elongation of the specimen.

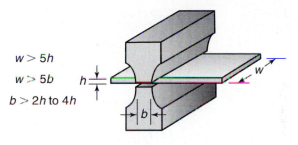

$w > 5h$

$w > 5b$

$b > 2h$ to $4h$

FIGURE 2.15 Schematic illustration of the plane-strain compression test. The dimensional relationships shown should be satisfied for this test to be useful and reproducible. This test gives the yield stress of the material in plane strain, Y'. *Source:* After A. Nadai and H. Ford.

2.3.1 Plane-strain compression test

The plane-strain compression test (Fig. 2.15) is designed to simulate bulk-deformation processes such as forging and rolling (as described in Chapter 6). In this test, the die and workpiece geometries are such that the width of the specimen does not undergo any significant change during compression; that is, the material under the dies is in the condition of *plane strain* (see Section 2.11.3). The yield stress of a material in plane strain, Y', is given by

$$Y' = \frac{2}{\sqrt{3}} Y = 1.15 Y, \tag{2.19}$$

according to the distortion-energy yield criterion (see Section 2.11.2).

As the geometric relationships in Fig. 2.15 indicate, the test parameters must be chosen properly to make the results meaningful. Furthermore, caution should be exercised in test procedures, such as preparing the die surfaces, aligning the dies, lubricating the surfaces, and accurately measuring the load.

When the results of tension and compression tests on the same material are compared, it is found that for *ductile* metals, the true stress–true strain curves for both tests coincide (Fig. 2.16). However, this is not true for brittle materials, particularly with respect to ductility (see also Section 3.8).

2.3.2 Bauschinger effect

In deformation processing of materials, a workpiece is sometimes first subjected to tension and then to compression, or vice versa; examples are bending and unbending sheet metal, roller leveling in sheet production (see Section 6.3.4), and reverse drawing in making cup-shaped parts (see Section 7.6.2). When a metal with a tensile yield stress, Y, is subjected to tension into the plastic range, and the load is then

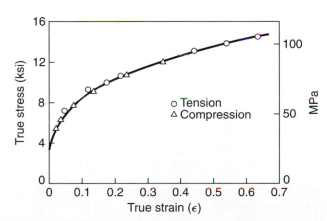

FIGURE 2.16 True stress–true strain curve in tension and compression for aluminum. For ductile metals, the curves for tension and compression are identical. *Source:* After A.H. Cottrell.

FIGURE 2.17 Schematic illustration of the Bauschinger effect. Arrows show loading and unloading paths. Note the decrease in the yield stress in compression after the specimen has been subjected to tension. The same result is obtained if compression is applied first, followed by tension, whereby the yield stress in tension decreases.

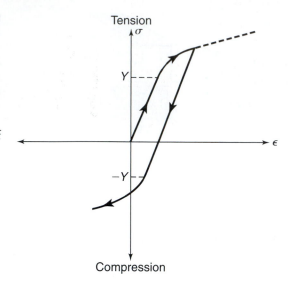

released and applied in compression, the yield stress in compression is found to be lower than that in tension (Fig. 2.17). This phenomenon, known as the *Bauschinger effect*, is exhibited to varying degrees by all metals and alloys. This effect is also observed when the loading path is reversed, that is, compression followed by tension. Because of the lowered yield stress in the reverse direction of load application, this phenomenon is also called **strain softening** or **work softening.** This behavior is also observed in torsion (Section 2.4).

2.3.3 The disk test

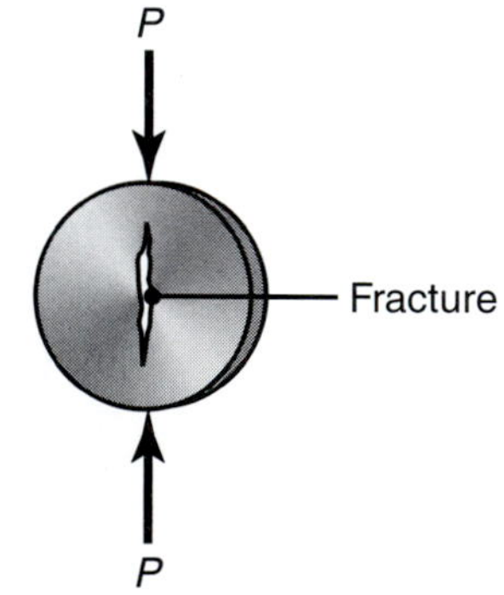

For brittle materials such as ceramics and glasses, a **disk test** has been developed in which the disk is subjected to diametral compression between two hardened flat platens (Fig. 2.18). When loaded as shown, tensile stresses develop perpendicular to the vertical centerline along the disk, fracture begins, and the disk splits vertically in half. (See also *rotary tube piercing* in Section 6.3.5.)

The tensile stress, σ, in the disk is uniform along the centerline and can be calculated from the formula

$$\sigma = \frac{2P}{\pi dt},$$ (2.20)

where P is the load at fracture, d is the diameter of the disk, and t is its thickness. In order to avoid premature failure of the disk at the top and bottom contact points, thin strips of soft metal are placed between the disk and the platens. These strips also protect the platens from being damaged during the test.

FIGURE 2.18 Disk test on a brittle material, showing the direction of loading and the fracture path. This test is useful for brittle materials, such as ceramics and carbides.

2.4 | Torsion

Another method of determining material properties is the **torsion test.** In order to obtain an approximately uniform stress and strain distribution along the cross section, this test is generally carried out on a tubular specimen with a reduced midsection (Fig. 2.19). The *shear stress, τ*, can be determined from the equation

$$\tau = \frac{T}{2\pi r^2 t},$$ (2.21)

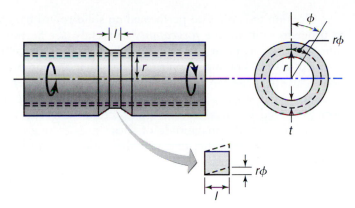

FIGURE 2.19 A typical torsion-test specimen. It is mounted between the two heads of a machine and is twisted. Note the shear deformation of an element in the reduced section.

where T is the torque applied, r is the mean radius, and t is the thickness of the reduced section in the middle of the tube. The shear strain, γ, is determined from the equation

$$\gamma = \frac{r\phi}{l}, \qquad (2.22)$$

where l in the length of the reduced section and ϕ is the angle of twist, in radians. With the shear stress and shear strain thus obtained from this test, we can construct the shear stress–shear strain curve of the material (see also Section 2.11.7).

In the elastic range, the ratio of the shear stress to shear strain is known as the **shear modulus** or the **modulus of rigidity, G**:

$$G = \frac{\tau}{\gamma}. \qquad (2.23)$$

The shear modulus and the modulus of elasticity are related by the formula

$$G = \frac{E}{2(1 + \nu)}, \qquad (2.24)$$

which is based on a comparison of **simple shear** and **pure shear** strains (Fig. 2.20). Note from this figure that the difference between the two strains is that simple shear is equivalent to pure shear plus a rotation of $\gamma/2$ degrees. (See also Section 2.11.7.)

In Example 2.2 we show that a thin-walled tube does not neck in torsion, unlike that in a tension-test specimen described in Section 2.2.4. Consequently, we do not have to be concerned with changes in the cross-sectional area of the specimen in torsion testing. The shear stress–shear strain curves obtained from torsion tests increase monotonically just as they do in true stress–true strain curves.

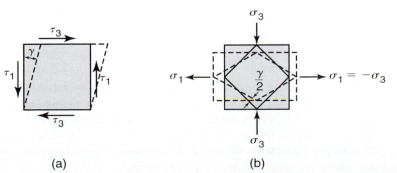

(a)

(b)

FIGURE 2.20 Comparison of (a) simple shear and (b) pure shear. Note that simple shear is equivalent to pure shear plus a rotation.

Torsion tests are also performed on solid round bars at elevated temperatures in order to estimate the *forgeability* of metals (see Section 6.2.6). The greater the number of twists prior to failure, the better the forgeability of the metal. Torsion tests can also be conducted on round bars that are compressed axially to indicate a beneficial effect similar to that of hydrostatic pressure (see Section 2.2.8). The effect of compressive stresses in increasing the maximum shear strain at fracture has also been observed in metal cutting (see Section 8.2). The normal compressive stress has no effect on the magnitude of shear stresses required to cause yielding or to continue the deformation, just as hydrostatic pressure has no effect on the general shape of the stress–strain curve.

EXAMPLE 2.2 Instability in torsion of a thin-walled tube

Show that necking cannot take place in the torsion of a thin-walled tube made of a material whose true stress–true strain curve is given by $\sigma = K\epsilon^n$.

Solution. According to Eq. (2.21), the expression for torque, T, is

$$T = 2\pi r^2 t \tau,$$

where for this case the *shear stress*, τ, can be related to the normal stress, σ, using Eq. (2.56) for the distortion-energy criterion as $\tau = \sigma/\sqrt{3}$ (see Section 2.11.7 for details). The criterion for instability in torsion would be

$$\frac{dT}{d\epsilon} = 0.$$

Because r and t are constant, we have

$$\frac{dT}{d\epsilon} = \left(\frac{2}{\sqrt{3}}\right)\pi r^2 \tau\, \frac{d\sigma}{d\epsilon}.$$

For a material represented by $\sigma = K\epsilon^n$,

$$\frac{d\sigma}{d\epsilon} = nK\epsilon^{n-1}.$$

Therefore,

$$\frac{dT}{d\epsilon} = \left(\frac{2}{\sqrt{3}}\right)\pi r^2 \tau nK\epsilon^{n-1}.$$

Since none of the quantities in this expression are zero, $dT/d\epsilon$ cannot be zero. Consequently, a tube in torsion does not undergo instability.

2.5 | Bending

Preparing specimens from brittle materials such as ceramics and carbides can be difficult because (a) shaping and machining them to proper dimensions can be challenging; (b) brittle materials are sensitive to surface defects, scratches, and imperfections; (c) clamping brittle test specimens for testing can be difficult; and (d) improper alignment of the test specimen may result in nonuniform stress distribution along the cross section of the specimen.

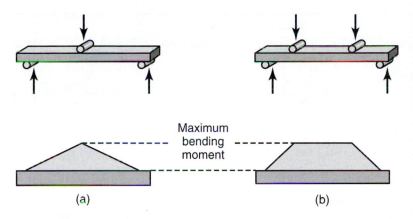

Maximum
bending
moment

(a)   (b)

FIGURE 2.21 Two bend-test methods for brittle materials: (a) three-point bending; (b) four-point bending. The shaded areas on the beams represent the bending-moment diagrams, described in texts on the mechanics of solids. Note the region of constant maximum bending moment in (b), whereas the maximum bending moment occurs only at the center of the specimen in (a).

A common test method for brittle materials is the **bend (flexure) test,** usually involving a specimen with rectangular cross section and supported at both ends (Fig. 2.21). The load is applied vertically, either at one or two points; hence these tests are referred to as **three-point** or **four-point** bending, respectively. The stresses developed in these specimens are tensile at their lower surfaces and compressive at their upper surfaces. They can be calculated using simple beam equations, described in texts on the mechanics of solids. The stress at fracture in bending is known as the **modulus of rupture,** or **transverse rupture strength,** and is obtained from the formula

$$\sigma = \frac{Mc}{I}, \tag{2.25}$$

where M is the bending moment, c is one-half of the specimen depth, and I is the moment of inertia of the cross section.

Note that there is a basic difference between the two loading conditions in Fig. 2.21. In three-point bending, the maximum stress is at the center of the beam, whereas in the four-point test, the maximum stress is constant between the two loading points. The stress magnitude is the same in both situations when all other parameters are maintained. There is, however, a higher probability for defects and imperfections to be present in the larger volume of material between the loading points in the four-point test than in the much smaller volume under the single load in the three-point test. This means that, as also verified by experiments, the four-point test is likely to result in a lower modulus of rupture than the three-point test. Similarly, the results of the four-point test also show less scatter than those in the three-point test.

2.6 | Hardness

One of the most common tests for assessing the mechanical properties of materials is the **hardness test.** *Hardness* of a material is generally defined as its resistance to permanent indentation; it can also be defined as its resistance to scratching or to wear (see Chapter 4). Several techniques have been developed to measure the hardness of materials using various indenter geometries and materials. Hardness is not a fundamental property because resistance to indentation depends on the shape of the indenter and the load applied. The most common standardized hardness tests are described next and summarized in Fig. 2.22. Hardness tests are commonly performed on dedicated equipment in a laboratory, but portable hardness testers are

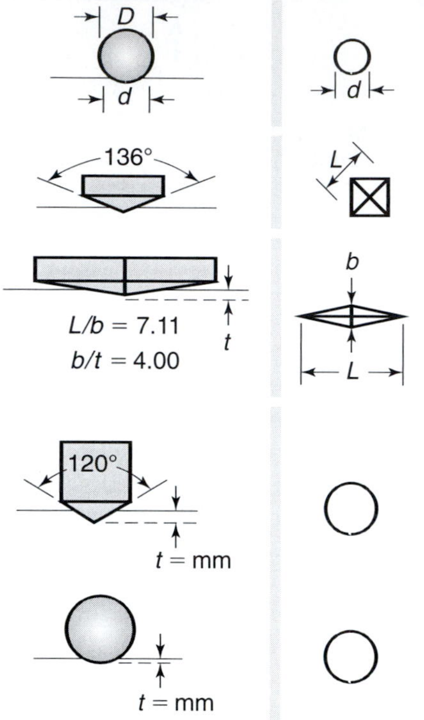

Test	Indenter	Shape of indentation		Load, P	Hardness number
		Side view	Top view		
Brinell	10-mm steel or tungsten carbide ball	$\rightarrow\mid D\mid\leftarrow$ $\rightarrow\mid d \mid\leftarrow$	$\rightarrow\mid d\mid\leftarrow$	500 kg 1500 kg 3000 kg	$HB = \dfrac{2P}{(\pi D)(D - \sqrt{D^2 - d^2})}$
Vickers	Diamond pyramid	$136°$	L	1–120 kg	$HV = \dfrac{1.854P}{L^2}$
Knoop	Diamond pyramid	$L/b = 7.11$ $b/t = 4.00$ t	b $\leftarrow L \rightarrow$	25 g–5 kg	$HK = \dfrac{14.2P}{L^2}$
Rockwell A C D	Diamond cone	$120°$ $t = mm$	◯	60 kg 150 kg 100 kg	HRA HRC HRD $\Big\} = 100 - 500t$
B F G	$\frac{1}{16}$- in. diameter steel ball	$t = mm$	◯	100 kg 60 kg 150 kg	HRB HRF HRG $\Big\} = 130 - 500t$
E	$\frac{1}{8}$- in. diameter steel ball			100 kg	HRE

FIGURE 2.22 General characteristics of hardness testing methods. The Knoop test is known as a microhardness test because of the light load and small impressions. *Source: After H.W. Hayden, W.G. Moffatt, and V. Wulff.*

also available. Portable hardness testers can generally run any hardness test described below and can be specially configured for particular geometric features, such as hole interiors, gear teeth, and so on, or tailored for specific classes of material.

2.6.1 Brinell test

In the **Brinell test**, a steel or tungsten carbide ball 10 mm in diameter is pressed against a surface with a load of 500, 1500, or 3000 kg. The *Brinell hardness number* (HB) is defined as the ratio of the load P to the curved area of indentation, or

$$HB = \frac{2P}{(\pi D)(D - \sqrt{D^2 - d^2})}, \tag{2.26}$$

where D is the diameter of the ball and d is the diameter of the impression, in millimeters.

Depending on the condition of the material tested, different types of impressions are obtained on the surface after a Brinell hardness test has been performed. As shown in Fig. 2.23, annealed materials, for example, generally have a rounded profile,

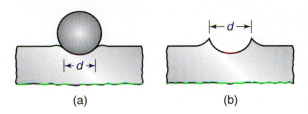

FIGURE 2.23 Indentation geometry for Brinell hardness testing: (a) annealed metal; (b) work-hardened metal. Note the difference in metal flow at the periphery of the impressions.

whereas cold-worked (strain-hardened) materials have a sharp profile. The correct method of measuring the indentation diameter d for both cases is shown in the figure.

Because the indenter has a finite elastic modulus, it also undergoes elastic deformation under the applied load, P; thus, hardness measurements may not be as accurate as expected. A common method of minimizing this effect is to use tungsten carbide balls, which, because of their high modulus of elasticity, deform less than steel balls. Tungsten carbide balls are generally recommended for Brinell hardness numbers higher than 500. In reporting the test results for these high hardnesses, the type of ball used should be cited. Since harder workpiece materials produce very small impressions, a 1500-kg or 3000-kg load is recommended in order to obtain impressions that are sufficiently large for accurate measurement.

Because the impressions made by the same indenter at different loads are not geometrically similar, the Brinell hardness number depends on the load used. Consequently, the load employed should also be cited with the test results. The Brinell test is generally suitable for materials of low to medium hardness.

Brinelling is a term used to describe permanent indentations on a surface between contacting bodies, for example, a ball bearing indenting a flat surface under fluctuating loads or vibrations. Such loads commonly occur during transportation or from dynamic loads associated with vibration.

2.6.2 Rockwell test

In the **Rockwell test**, the *depth* of penetration is measured. The indenter is pressed on the surface, first with a *minor* load and then a *major* load. The difference in the depth of penetration is a measure of the hardness. There are several Rockwell hardness scales that employ different loads, indenter materials, and indenter geometries. Some of the more common hardness scales and the indenters used are listed in Fig. 2.22. The Rockwell hardness number, which is read directly from a dial on the testing machine, is expressed as follows: If, for example, the hardness number is 55 using the C scale, then it is written as 55 HRC. *Rockwell superficial hardness* tests have also been developed using lighter loads and the same type of indenters.

2.6.3 Vickers test

The **Vickers test**, formerly known as the *diamond pyramid hardness test,* uses a pyramid-shaped diamond indenter (see Fig. 2.22) with loads ranging from 1 to 120 kg. The *Vickers hardness number* (HV) is given by the formula

$$\text{HV} = \frac{1.854P}{L^2}. \tag{2.27}$$

The impressions are typically less than 0.5 mm on the diagonal. The Vickers test gives essentially the same hardness number regardless of the load and is suitable for testing materials with a wide range of hardness, including very hard steels.

2.6.4 Knoop test

The **Knoop test** uses a diamond indenter in the shape of an elongated pyramid (see Fig. 2.22), using loads ranging generally from 25 g to 5 kg. The *Knoop hardness number* (HK) is given by the formula

$$\text{HK} = \frac{14.2P}{L^2}. \tag{2.28}$$

The size of the indentation is generally in the range of 0.01 to 0.10 mm; surface preparation is therefore very important. Since the hardness number obtained depends on the applied load, test results should always cite the load applied. The Knoop test is a *microhardness* test because of the light loads utilized; hence it is suitable for very small or thin specimens and for brittle materials such as gemstones, carbides, and glass. Because the impressions are very small, this test is also used to measure the hardness of individual grains in a metal.

2.6.5 Scleroscope

The **scleroscope** is an instrument in which a diamond-tipped indenter (called a *hammer*), which is enclosed in a glass tube, is dropped on the specimen from a certain height. The hardness is determined by the *rebound* of the indenter; the higher the rebound, the harder the specimen. Indentation on the workpiece surface is slight. Because the instrument is portable, it is useful for measuring the hardness of large objects.

2.6.6 Mohs test

The **Mohs test** is based on the capability of one material to scratch another. The Mohs hardness is expressed on a scale of 10, with 1 for talc and 10 for diamond (the hardest substance known); thus, a material with a higher Mohs hardness can scratch materials with a lower hardness. Soft metals have a Mohs hardness of 2 to 3, hardened steels about 6, and aluminum oxide 9. The Mohs scale is generally used by mineralogists and geologists; however, some of the materials also are of interest in manufacturing. Although the Mohs scale is qualitative, good correlation is obtained with Knoop hardness.

2.6.7 Durometer

The hardness of rubbers, plastics, and similar soft and elastic materials is generally measured with an instrument called a **durometer,** in which (a) an indenter is pressed against the surface, with a constant load that is applied rapidly; and (b) the depth of penetration is measured after one second. This is an empirical test and there are two different scales. Type A has a blunt indenter, a load of 1 kg, and is used for softer materials. Type D has a sharper indenter, a load of 5 kg, and is used for harder materials. The hardness numbers in these tests range from 0 to 100.

2.6.8 Relationship between hardness and strength

Since hardness is the resistance to permanent indentation, hardness testing is equivalent to performing a compression test on a small volume of a material's surface. We would then expect some correlation between hardness and yield stress, Y, in the form of

$$\text{Hardness} = cY, \tag{2.29}$$

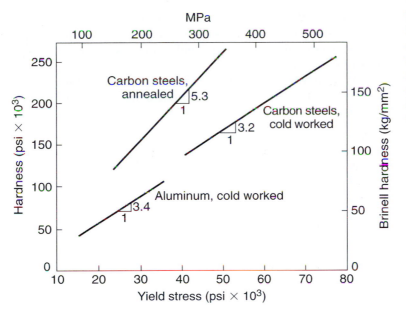

FIGURE 2.24 Relation between Brinell hardness and yield stress for aluminum and steels. For comparison, the Brinell hardness (which is always measured in kg/mm²) is converted to psi units on the left scale.

where c is a proportionality constant. The magnitude of c is not a constant for a material or a class of materials because the hardness of a given material depends upon its processing history, such as cold or hot working and surface treatments.

Theoretical studies, based on plane strain slip-line analysis with a smooth flat punch indenting the surface of a semi-infinite body (see Fig. 6.12), have shown that for a perfectly plastic material, the magnitude of c is about 3. This is in reasonable agreement with experimental data, as can be seen in Fig. 2.24. Note that cold-worked metals (which are close to being perfectly plastic in their behavior) show better agreement than annealed metals. The higher value of c for annealed materials is explained by the fact that, due to strain hardening, the average yield stress they exhibit during indentation is higher than their initial yield stress.

The reason that hardness, as a compression test, gives higher values than the uniaxial yield stress, Y, of the material can be seen in the following analysis. If we assume that the volume under the indenter is a column of material, it would exhibit a uniaxial compressive yield stress, Y. However, the volume being deformed under the indenter is, in reality, surrounded by a rigid mass (Fig. 2.25). The surrounding

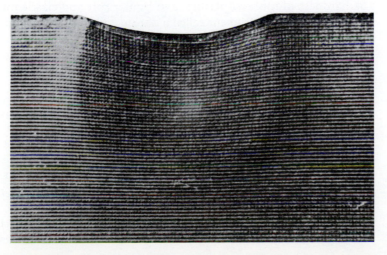

FIGURE 2.25 Bulk deformation in mild steel under a spherical indenter. Note that the depth of the deformed zone is about one order of magnitude larger than the depth of indentation. For a hardness test to be valid, the material should be allowed to fully develop this zone. This is why thinner specimens require smaller indentations. *Source:* Courtesy of M.C. Shaw and C.T. Yang.

mass prevents this volume of material from deforming freely. In fact, this volume is under *triaxial compression*. As shown in Section 2.11 on yield criteria, this material requires a normal compressive yield stress that is higher than the uniaxial yield stress of the material.

More practically, a relationship has also been observed between the ultimate tensile strength (UTS) and Brinell hardness number (HB) for steels:

$$UTS = 500(HB),\qquad(2.30)$$

where UTS is in psi and HB in kg/mm^2 as measured with a load of 3000 kg. In SI units, the relationship is given by

$$UTS = 3.5(HB),\qquad(2.31)$$

where UTS is in MPa.

Hot hardness tests can be carried out using conventional testers with certain modifications, such as enclosing the specimen and indenter in a small electric furnace. The hot hardness of materials is important in applications where the materials are subjected to elevated temperatures, such as in cutting tools in machining and dies for hot metalworking operations.

EXAMPLE 2.3 Calculation of modulus of resilience from hardness

A piece of steel is highly deformed at room temperature. Its hardness is found to be 300 HB. Estimate the modulus of resilience for this material.

Solution. Since the steel has been subjected to large strains at room temperature, it may be assumed that its stress–strain curve has flattened considerably, thus approaching the shape of a perfectly plastic curve. According to Eq. (2.30) and using a value of $c = 3$, we obtain

$$Y = \frac{300}{3} = 100 \text{ kg/mm}^2 = 142{,}250 \text{ psi.}$$

The modulus of resilience is defined by Eq. (2.5):

$$\text{Modulus of resilience} = \frac{Y^2}{2E}.$$

From Table 2.1, $E = 30 \times 10^6$ psi for steel. Hence,

$$\text{Modulus of resilience} = \frac{(142{,}250)^2}{2 \times 30 \times 10^6} = 337 \text{ in.-lb/in}^3.$$

2.7 | Fatigue

Gears, cams, shafts, springs, and tools and dies are typically subjected to rapidly fluctuating (cyclic or periodic) loads. These stresses may be caused by fluctuating mechanical loads (such as gear teeth, dies, and cutters), or by thermal stresses (such as a cool die coming into repeated contact with hot workpieces). Under these conditions, the part fails at a stress level below which failure would occur under static loading. This phenomenon is known as **fatigue failure** and is responsible for the majority of failures in mechanical components.

Fatigue tests repeatedly subject specimens to various states of stress, usually in a combination of tension and compression, or torsion. The test is carried out at

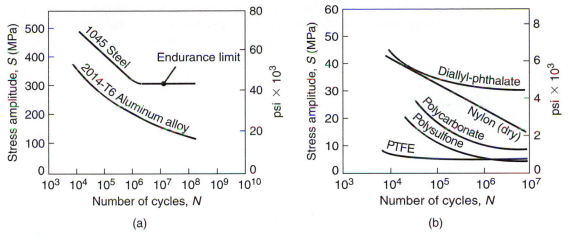

FIGURE 2.26 Typical *S-N* curves for two metals. Note that, unlike steel, aluminum does not have an endurance limit.

various stress amplitudes (*S*) and the number of cycles (*N*) to cause total failure of the specimen or part is recorded. Stress amplitude is the maximum stress, in tension and compression, to which the specimen is subjected.

A typical plot of the data obtained, known as *S-N curves,* is shown in Fig. 2.26. These curves are based on complete reversal of the stress, that is, maximum tension, maximum compression, maximum tension, and so on, such as that obtained by bending a piece of wire alternately in one direction, then the other. The fatigue test can also be performed on a rotating shaft with a constant downward load. The maximum stress to which the material can be subjected without fatigue failure, regardless of the number of cycles, is known as the **endurance limit** or **fatigue limit.**

The fatigue strength for metals has been found to be related to their ultimate tensile strength, UTS, as shown in Fig. 2.27. Note that for steels, the endurance limit is about one-half their tensile strength. Although most metals, especially steels, have a definite endurance limit, aluminum alloys do not have one and the *S-N* curve

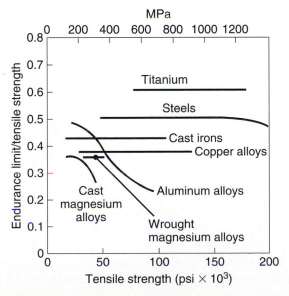

FIGURE 2.27 Ratio of fatigue strength to tensile strength for various metals, as a function of tensile strength.

continues its downward trend. For metals exhibiting such behavior (most face-centered cubic metals), the fatigue strength is specified at a specific number of cycles, such as 10^7. In this way, the useful service life of the component can be specified.

2.8 | Creep

Creep is the permanent elongation of a material under a static load maintained for a period of time. It is a phenomenon of metals and some nonmetallic materials, such as thermoplastics and rubbers, and it can occur at any temperature. Lead, for example, creeps under a constant tensile load at room temperature. Also, the thickness of window glass in old houses has been found to be greater at the bottom than at the top of windows, the glass having undergone creep by its own weight over many years. The mechanism of creep at elevated temperature in metals is generally attributed to *grain-boundary sliding* (see Section 3.4.2). For metals and their alloys, creep of any engineering significance occurs at elevated temperatures, beginning at about 200°C (400°F) for aluminum alloys, and up to about 1500°C (2800°F) for refractory alloys.

Creep is especially important in high-temperature applications, such as gas-turbine blades and similar components in jet engines and rocket motors. High-pressure steam lines and nuclear-fuel elements are also subject to creep. Creep deformation also can occur in tools and dies that are subjected to high stresses at elevated temperatures during metalworking operations such as hot forging and extrusion.

A **creep test** typically consists of subjecting a specimen to a constant tensile load (hence constant engineering stress) at a certain temperature, and measuring the change in length over a period of time. A typical creep curve usually consists of primary, secondary, and tertiary stages (Fig. 2.28). The specimen eventually fails by necking and fracture, as in the tension test, which is called **rupture** or **creep rupture.** As expected, the creep rate increases with temperature and the applied load.

Design against creep usually requires a knowledge of the secondary (linear) range and its slope, because the creep rate can be determined reliably when the curve has a constant slope. Generally, resistance to creep increases with the melting temperature of a material; this serves as a general guideline for design purposes. Thus, stainless steels, superalloys, and refractory metals and their alloys are commonly used in applications where creep resistance is required.

Stress relaxation is closely related to creep. In this phenomenon, the stresses resulting from external loading of a structural component decrease in magnitude

FIGURE 2.28 Schematic illustration of a typical creep curve. The linear segment of the curve (constant slope) is useful in designing components for a specific creep life.

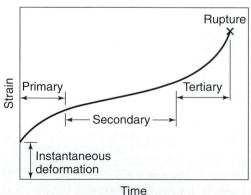

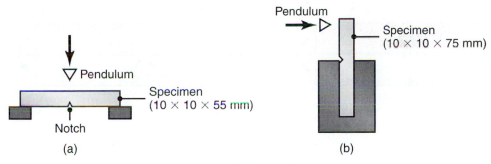

FIGURE 2.29   Impact test specimens: (a) Charpy; (b) Izod.

over a period of time, even though the dimensions of the component remain constant. Examples of stress relaxation are rivets, bolts, guy wires, and parts under tension, compression, or bending; it is particularly common and important in thermoplastics (see Section 10.3).

2.9 | Impact

In many manufacturing operations, as well as during their service life, various components are subjected to *impact* (or *dynamic*) *loading*. A typical **impact test** consists of placing a *notched specimen* in an impact tester and breaking it with a swinging pendulum (Fig. 2.29). In the **Charpy test** the specimen is supported at both ends, whereas in the **Izod test** it is supported at one end, like a cantilever beam. From the amount of swing of the pendulum, the *energy dissipated* in breaking the specimen is obtained; this energy is the **impact toughness** of the material.

Impact tests are particularly useful in determining the ductile-brittle *transition temperature* of materials (see Fig. 3.26). Generally, materials that have high impact resistance are those that also have high strength and high ductility. Sensitivity of materials to surface defects (**notch sensitivity**) is important, as it lowers their impact toughness.

2.10 | Residual Stresses

In this section we show that inhomogeneous deformation during processing leads to *residual stresses;* these are stresses that remain within a part after it has been deformed and all external forces have been removed. A typical example of inhomogeneous deformation is the bending of a beam (Fig. 2.30). The bending moment first produces a linear elastic stress distribution. As the moment is increased, the outer fibers begin to yield and, for a typical strain-hardening material, the stress distribution shown in Fig. 2.30b is eventually obtained. After the part is bent (permanently, since it has now undergone plastic deformation), the moment is removed when the part is unloaded. Unloading is equivalent to applying an equal and opposite moment to the beam.

As previously shown in Fig. 2.3, all recovery is elastic; therefore, the moments of the areas *oab* and *oac* about the neutral axis in Fig. 2.30c must be equal. (For purposes of this treatment, let's assume that the neutral axis does not shift.) The difference between the two stress distributions produces the residual stress pattern shown in the figure. Note that there are compressive residual stresses in layers *ad* and *oe,* and tensile residual stresses in layers *do* and *ef.* With no external forces

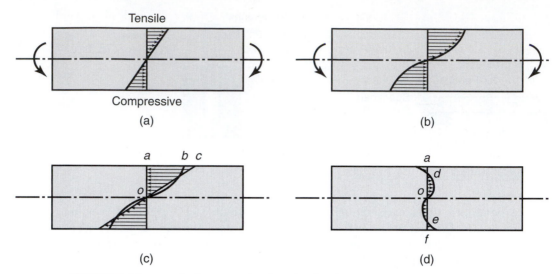

(a)

(b)

(c)

(d)

FIGURE 2.30 Residual stresses developed in bending a beam made of an elastic, strain-hardening material. Note that unloading is equivalent to applying an equal and opposite moment to the part, as shown in (b). Because of nonuniform deformation, most parts made by plastic deformation processes contain residual stresses. Note that the forces and moments due to residual stresses must be internally balanced.

present, the residual stresses in the beam must be in static equilibrium. Although this example involves stresses in one direction only, in most situations in deformation processing of materials the residual stresses are three dimensional.

The equilibrium of residual stresses will be disturbed when the shape of the beam is altered, such as by removing a layer of material by machining. The beam will then acquire a new radius of curvature in order to balance the internal forces. Another example of the effect of residual stresses is the drilling of round holes on surfaces of parts that have residual stresses. It may be found that, as a result of removing this material by drilling, the equilibrium of the residual stresses is disturbed and the hole now becomes elliptical. Such disturbances of residual stresses lead to *warping*, some common examples of which are given in Fig. 2.31.

The equilibrium of internal stresses may also be disturbed by *relaxation* of residual stresses over a period of time, which results in instability of the dimensions and shape of the component. These dimensional changes are an important consideration for precision machinery and measuring equipment.

Residual stresses also may be caused by *phase changes* in metals during or after processing due to density differences among different phases, such as between

FIGURE 2.31 Distortion of parts with residual stresses after cutting or slitting: (a) rolled sheet or plate; (b) drawn rod; (c) thin-walled tubing. Because of the presence of residual stresses on the surfaces of parts, a round drill may produce an oval-shaped hole because of relaxation of stresses when a portion is removed.

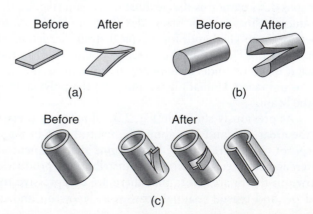

ferrite and martensite in steels. Phase changes cause microscopic volumetric changes and thus result in residual stresses. This phenomenon is important in warm and hot working of metals and in their heat treatment. Residual stresses also can be caused by *temperature gradients* within a body, as (a) during the cooling cycle of a casting (Chapter 5), (b) when applying brakes to a railroad wheel, or (c) in a grinding operation (Section 9.4.3).

2.10.1 Effects of residual stresses

Because they lower the fatigue life and fracture strength, tensile residual stresses on the surface of a part are generally undesirable. A surface with tensile residual stresses will sustain lower additional tensile stresses (due to external loading) than can a surface that is free from residual stresses. This is particularly true for relatively brittle materials, where fracture can occur with little or no plastic deformation. Tensile residual stresses in manufactured products can also lead to **stress cracking** or **stress-corrosion cracking** (Section 3.8.2) over a period of time.

Conversely, compressive residual stresses on a surface are generally desirable. In fact, in order to increase the fatigue life of components, compressive residual stresses are imparted on surfaces by common techniques such as **shot peening** and **surface rolling** (see Section 4.5.1).

2.10.2 Reduction of residual stresses

Residual stresses may be reduced or eliminated either by **stress-relief annealing** (see Section 5.11.4) or by further *plastic deformation*. Given sufficient time, residual stresses may also be diminished at room temperature by *relaxation*; the time required for relaxation can be greatly reduced by increasing the temperature of the component. However, relaxation of residual stresses by stress-relief annealing is generally accompanied by warpage of the part; hence a *machining allowance* is commonly provided to compensate for dimensional changes during stress relieving.

The mechanism of residual stress reduction or elimination by plastic deformation can be described as follows: Assume that a piece of metal has the residual stresses shown in Fig. 2.32a, namely, tensile on the outside and compressive on the inside; these stresses are in equilibrium. Also assume that the material is elastic and perfectly plastic, as shown in Fig. 2.32d. The levels of the residual stresses are shown

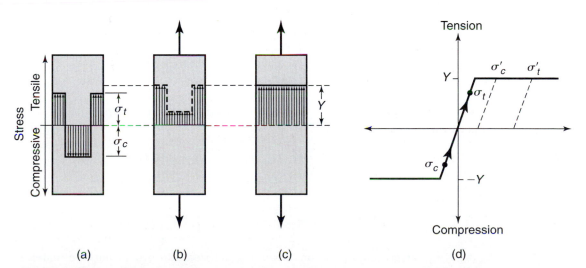

FIGURE 2.32 Elimination of residual stresses by stretching. Residual stresses can be also reduced or eliminated by thermal treatments, such as stress relieving or annealing.

on the stress–strain diagram; note that both are at a level below the yield stress, Y (since all residual stresses have to be in the elastic range).

If a uniformly distributed tension is now applied to this part, points σ_c and σ_t in the diagram move up on the stress–strain curve, as shown by the arrows. The maximum level that these stresses can reach is the tensile yield stress, Y. With sufficiently high loading, the stress distribution eventually becomes uniform throughout the part, as shown in Fig. 2.32c. If the load is then removed, the stresses recover elastically and the part is free of residual stresses. Note also that very little stretching is required to relieve these residual stresses. The reason is that the elastic portions of the stress–strain curves for metals are very steep; hence the elastic stresses can be raised to the yield stress with very little strain.

The technique for reducing or relieving residual stresses by plastic deformation, such as by stretching as described, requires sufficient straining to establish a uniformly distributed stress within the part. Consequently, a material such as the elastic, linearly strain-hardening type (Fig. 2.7e) can never reach this condition since the compressive stress, σ_c', will always lag behind σ_t'. If the slope of the stress–strain curve in the plastic region is small, the difference between σ_c' and σ_t' will be rather small and little residual stress will be left in the part after unloading.

EXAMPLE 2.4 Elimination of residual stresses by tension

Refer to Fig. 2.32a and assume that $\sigma_t = 140$ MPa and $\sigma_c = -140$ MPa. The material is aluminum and the length of the specimen is 0.25 m. Calculate the length to which this specimen should be stretched so that, when unloaded, it will be free from residual stresses. Assume that the yield stress of the material is 150 MPa.

Solution. Stretching should be to the extent that σ_c reaches the yield stress in tension, Y. Therefore, the total strain should be equal to the sum of the strain required to bring the compressive residual stress to zero and the strain required to bring it to the tensile yield stress. Hence,

$$\epsilon_{\text{total}} = \frac{\sigma_c}{E} + \frac{Y}{E}. \tag{2.32}$$

For aluminum, let $E = 70$ GPa, as obtained from Table 2.1. Thus,

$$\epsilon_{\text{total}} = \frac{140}{70 \times 10^3} + \frac{150}{70 \times 10^3} = 0.00414.$$

Hence, the stretched length should be

$$\ln\left(\frac{l_f}{0.25}\right) = 0.00414 \qquad \text{or} \qquad l_f = 0.2510 \text{ m}.$$

As the strains are very small, we may use engineering strains in these calculations. Hence,

$$\frac{l_f - 0.25}{0.25} = 0.00414 \qquad \text{or} \qquad l_f = 0.2510 \text{ m}.$$

2.11 Triaxial Stresses and Yield Criteria

In most manufacturing operations involving deformation processing, the material, unlike that in a simple tension or compression test specimen, is generally subjected to triaxial stresses. For example, (a) in the expansion of a thin-walled spherical shell under internal pressure, an element in the shell is subjected to equal biaxial tensile

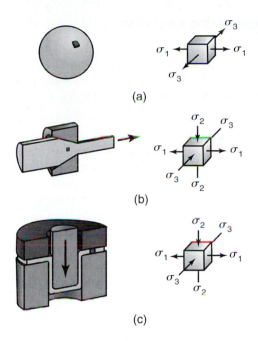

(a)

(b)

(c)

FIGURE 2.33 The state of stress in various metalworking operations. (a) Expansion of a thin-walled spherical shell under internal pressure. (b) Drawing of round rod or wire through a conical die to reduce its diameter (see Section 6.5). (c) Deep drawing of sheet metal with a punch and die to make a cup (see Section 7.6).

stresses (Fig. 2.33a); (b) in drawing a rod or wire through a conical die (Chapter 6), an element in the deformation zone is subjected to tension in its length direction and to compression on its conical surface (Fig. 2.33b); and (c) an element in the flange in deep drawing of sheet metal (Section 7.6) is subjected to a tensile radial stress and compressive stresses on its surface and in the circumferential direction (Fig. 2.33c). As shown in subsequent chapters, several other examples can be given in which the material is subjected to various normal and shear stresses during processing. The three-dimensional stresses shown in Fig. 2.33 cause distortion of the elements.

In the *elastic range*, the strains in the elements are represented by the *generalized Hooke's law* equations:

$$\epsilon_1 = \frac{1}{E}[\sigma_1 - \nu(\sigma_2 + \sigma_3)] \tag{2.33a}$$

$$\epsilon_2 = \frac{1}{E}[\sigma_2 - \nu(\sigma_1 + \sigma_3)] \tag{2.33b}$$

$$\epsilon_3 = \frac{1}{E}[\sigma_3 - \nu(\sigma_1 + \sigma_2)]. \tag{2.33c}$$

Thus, for simple tension where $\sigma_2 = \sigma_3 = 0$,

$$\epsilon_1 = \frac{\sigma_1}{E},$$

and

$$\epsilon_2 = \epsilon_3 = -\nu\frac{\sigma_1}{E}.$$

The negative sign indicates a contraction of the element in the 2 and 3 directions.

In a simple tension or compression test, when the applied stress reaches the uniaxial yield stress, Y, the material will deform *plastically*. However, if it is subjected to a more complex state of stress, relationships have been developed among these stresses that will predict yielding; these relationships are known as *yield criteria*. The most commonly used ones are the maximum-shear-stress criterion and the distortion-energy criterion.

2.11.1 Maximum-shear-stress criterion

The **maximum-shear-stress criterion**, also known as the **Tresca criterion**, states that yielding occurs when the maximum shear stress within an element is equal to or exceeds a critical value. As shown in Section 3.3, this critical value of the shear stress is a *material property* and is called **shear yield stress,** k. Hence, for yielding to occur,

$$\tau_{\text{max}} \geq k. \tag{2.34}$$

A convenient way of determining the stresses acting on an element is by using **Mohr's circles for stresses,** described in detail in textbooks on mechanics of solids. The **principal stresses** and their *directions* can be determined easily from Mohr's circle or from stress-transformation equations.

If the maximum shear stress is equal to or exceeds k, then yielding will occur. It can be seen that there are many combinations of stresses (known as *states of stress*) that can produce the same maximum shear stress. From the simple tension test, we note that

$$k = \frac{Y}{2}, \tag{2.35}$$

where Y is the uniaxial yield stress of the material. If, for some reason, we are unable to increase the stresses on the element in order to cause yielding, we can simply lower Y by raising the temperature of the material. This simple technique is the basis and one major reason for hot working of materials. We can now write the maximum-shear-stress criterion as

$$\sigma_{\text{max}} - \sigma_{\text{min}} = Y, \tag{2.36}$$

which indicates that the maximum and minimum normal stresses produce the largest Mohr's circle of stress, hence the *largest* shear stress. Consequently, *the intermediate stress has no effect on yielding.* It is important to note that the left-hand side of Eq. (2.36) represents the *applied stresses* and that the right-hand side is a *material property.* Also, we have assumed that (a) the material is *continuous, homogeneous,* and *isotropic* (it has the same properties in all directions); and (b) the yield stress in tension and in compression are equal (see *Bauschinger effect* in Section 2.3.2).

2.11.2 Distortion-energy criterion

The **distortion-energy criterion**, also called the **von Mises criterion**, states that yielding occurs when the relationship between the principal stresses and uniaxial yield stress, Y, of the material is

$$(\sigma_1 - \sigma_2)^2 + (\sigma_2 - \sigma_3)^2 + (\sigma_3 - \sigma_1)^2 = 2Y^2. \tag{2.37}$$

Note that, unlike the maximum-shear-stress criterion, the intermediate principal stress is included in this expression. Here again, the left-hand side of the equation represents the applied stresses and the right-hand side a material property.

EXAMPLE 2.5 Yielding of a thin-walled shell

A thin-walled spherical shell is under internal pressure, p. The shell is 20 in. in diameter and 0.1 in. thick. It is made of a perfectly plastic material with a yield stress of 20,000 psi. Calculate the pressure required to cause yielding of the shell according to both yield criteria.

Solution. For this shell under internal pressure, the membrane stresses are given by

$$\sigma_1 = \sigma_2 = \frac{pr}{2t}, \tag{2.38}$$

where $r = 10$ in. and $t = 0.1$ in. The stress in the thickness direction, σ_3, is negligible because of the high r/t ratio of the shell. Thus, according to the maximum-shear-stress criterion,

$$\sigma_{max} - \sigma_{min} = Y$$

or

$$\sigma_1 - 0 = Y$$

and

$$\sigma_2 - 0 = Y.$$

Hence $\sigma_1 = \sigma_2 = 20,000$ psi. The pressure required is then

$$p = \frac{2tY}{r} = \frac{(2)(0.1)(20,000)}{10} = 400 \text{ psi.}$$

According to the distortion-energy criterion,

$$(\sigma_1 - \sigma_2)^2 + (\sigma_2 - \sigma_3)^2 + (\sigma_3 - \sigma_1)^2 = 2Y^2$$

or

$$0 + \sigma_2^2 + \sigma_1^2 = 2Y^2.$$

Hence $\sigma_1 = \sigma_2 = Y$. Therefore, the answer is the same, or $p = 400$ psi.

EXAMPLE 2.6 Correction factor for true stress–true strain curves

Explain why a correction factor has to be applied in the construction of a true stress–true strain curve, based on tensile-test data.

Solution. The neck of a specimen is subjected to a triaxial state of stress, as shown in Fig. 2.34. The reason for this stress state is that each element in the region has a different cross-sectional area; the smaller the area, the greater is the tensile stress on the element. Hence element 1 will contract laterally more than element 2, and so on; however, element 1 is restrained from contracting freely by element 2, and element 2 is restrained by element 3, and so on. This restraint causes radial and circumferential tensile stresses in the necked region, thus resulting in an axial tensile stress distribution as shown in Fig. 2.34.

The *true* uniaxial stress in tension is σ, whereas the *calculated* value of true stress at fracture is the *average* stress; hence a correction has to be made. A mathematical analysis by P.W. Bridgman gives the ratio of true stress to average stress as

$$\frac{\sigma}{\sigma_{av}} = \frac{1}{\left(1 + \frac{2R}{a}\right)\left[\left(1 + \frac{a}{2R}\right)\right]}, \tag{2.39}$$

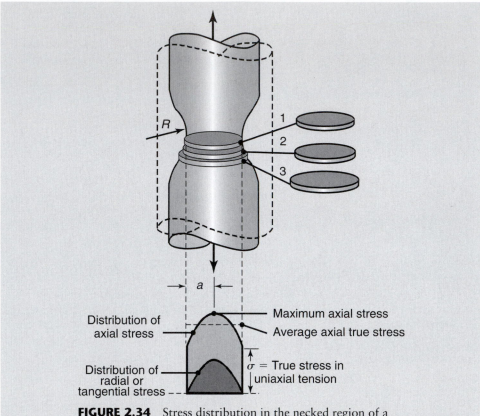

Distribution of axial stress

Maximum axial stress

Average axial true stress

σ = True stress in uniaxial tension

Distribution of radial or tangential stress

FIGURE 2.34   Stress distribution in the necked region of a tension-test specimen.

where R is the radius of curvature of the neck and a is the radius of the specimen at the neck. However, because R is difficult to measure during a test, an empirical relation has been established between a/R and the true strain at the neck.

2.11.3 Plane stress and plane strain

Plane stress and plane strain are important in the application of yield criteria. **Plane stress** is the state of stress in which one or two of the pairs of faces on an elemental cube are free from stress. An example is a thin-walled tube in torsion: There are no stresses normal to the inside or outside surface of the tube, thus plane stress. Other examples are given in Fig. 2.35.

Plane strain is the state of stress where one of the pairs of faces on an element undergoes zero strain (Figs. 2.35c and d). The **plane-strain compression test** described earlier and shown in Fig. 2.15 is an example of this. By proper selection of specimen dimensions, the width of the specimen is kept essentially constant during deformation. Note that an element does not have to be physically constrained on the pair of faces for plane-strain conditions to exist, an example being the torsion of a thin-walled tube in which it can be shown that the wall thickness remains constant (see Section 2.11.7).

A review of the two yield criteria described above indicates that the plane-stress condition (in which $\sigma_2 = 0$) can be represented by the diagram in Fig. 2.36. The maximum-shear-stress criterion gives an envelope of straight lines. In the first quadrant, where $\sigma_1 > 0$ and $\sigma_3 > 0$, and σ_2 is always zero (for plane stress),

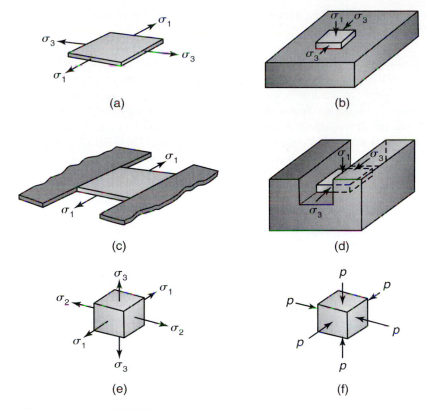

(a)

(b)

(c)

(d)

(e)

(f)

FIGURE 2.35 Examples of states of stress. (a) Plane stress in sheet stretching; there are no stresses acting on the surfaces of the sheet. (b) Plane stress in compression; there are no stresses acting on the sides of the specimen being compressed. (c) Plane strain in tension; the width of the sheet remains constant while being stretched. (d) Plane strain in compression (see also Fig. 2.15); the width of the specimen remains constant due to the restraint by the groove. (e) Triaxial tensile stresses acting on an element. (f) Hydrostatic compression of an element. Note also that an element on the cylindrical portion of a thin-walled tube in torsion is in the condition of both plane stress and plane strain (see also Section 2.11.7).

Eq. (2.36) reduces to $\sigma_{max} = Y$. The maximum value that either σ_1 or σ_3 can acquire is Y; hence the straight lines in the diagram.

In the third quadrant the same situation exists because σ_1 and σ_3 are both compressive. In the second and fourth quadrants, σ_2 (which is zero for the plane-stress condition) is the intermediate stress. Thus, for the second quadrant, Eq. (2.36) reduces to

$$\sigma_3 - \sigma_1 = Y, \tag{2.40}$$

and for the fourth quadrant it reduces to

$$\sigma_1 - \sigma_3 = Y. \tag{2.41}$$

Note that Eqs. (2.40) and (2.41) represent the 45° lines in Fig. 2.36.

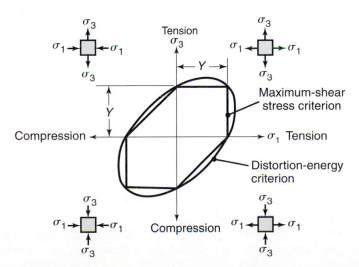

FIGURE 2.36 Plane-stress diagrams for maximum-shear-stress and distortion-energy criteria. Note that $\sigma_2 = 0$.

Plane stress. The distortion-energy criterion for *plane stress* reduces to

$$\sigma_1^2 + \sigma_3^2 - \sigma_1\sigma_3 = Y^2 \tag{2.42}$$

and is shown graphically in the figure. Whenever a point (with its coordinates representing the two principal stresses) falls on these boundaries, the element will yield.

The three-dimensional elastic stress–strain relationships were given by Eqs. (2.33a to c). When the stresses are sufficiently high to cause *plastic deformation,* the stress–strain relationships are obtained from **flow rules** (*Lévy-Mises equations*), described in detail in texts on plasticity. These relationships relate stress and strain increment as follows:

$$d\epsilon_1 = \frac{d\bar{\epsilon}}{\bar{\sigma}}\left[\sigma_1 - \frac{1}{2}(\sigma_2 + \sigma_3)\right]; \tag{2.43a}$$

$$d\epsilon_2 = \frac{d\bar{\epsilon}}{\bar{\sigma}}\left[\sigma_2 - \frac{1}{2}(\sigma_1 + \sigma_3)\right]; \tag{2.43b}$$

$$d\epsilon_3 = \frac{d\bar{\epsilon}}{\bar{\sigma}}\left[\sigma_3 - \frac{1}{2}(\sigma_1 + \sigma_2)\right]. \tag{2.43c}$$

Note that these expressions are similar to those given by the elastic stress–strain relationships of the generalized Hooke's law, Eqs. (2.33).

Plane strain. For the *plane-strain* condition shown in Figs. 2.35c and d, we have $\epsilon_2 = 0$. Therefore,

$$\sigma_2 = \frac{\sigma_1 + \sigma_3}{2}. \tag{2.44}$$

Note that σ_2 is now an intermediate stress.

For plane-strain compression shown in Figs. 2.15 and 2.35d, the distortion-energy criterion (which includes the intermediate stress) reduces to

$$\sigma_1 - \sigma_3 = \frac{2}{\sqrt{3}}Y \approx 1.15Y = Y'. \tag{2.45}$$

Note that, whereas for the maximum-shear-stress criterion $k = Y/2$, for the distortion-energy criterion for the plane–strain condition we have $k = Y/\sqrt{3}$.

2.11.4 Experimental verification of yield criteria

The yield criteria described have been tested experimentally, typically using a specimen in the shape of a thin-walled tube under internal pressure and/or torsion. Under such loading, it is possible to generate different states of plane stress. Experiments using a variety of *ductile* materials have shown that the distortion-energy criterion agrees better with the experimental data than does the maximum-shear-stress criterion. This is why the distortion-energy criterion is often used for the analysis of metalworking processes (Chapters 6 and 7). The simpler maximum-shear-stress criterion, however, can also be used; the difference between the two criteria is negligible for most practical applications.

2.11.5 Volume strain

By summing the three equations of the generalized Hooke's law [Eqs. (2.33)], we obtain

$$\epsilon_1 + \epsilon_2 + \epsilon_3 = \frac{1 - 2\nu}{E}(\sigma_1 + \sigma_2 + \sigma_3), \tag{2.46}$$

where the left-hand side of the equation can be shown to be the **volume strain** or **dilatation**, Δ. Thus,

$$\Delta = \frac{\text{Volume change}}{\text{Original volume}} = \frac{1 - 2\nu}{E}(\sigma_1 + \sigma_2 + \sigma_3). \tag{2.47}$$

Note that in the plastic range, where $\nu = 0.5$, the volume change is zero. Thus, in plastic working of metals,

$$\epsilon_1 + \epsilon_2 + \epsilon_3 = 0, \tag{2.48}$$

which is a convenient means of determining a third strain if two strains are known. The **bulk modulus** is defined as

$$\text{Bulk modulus} = \frac{\sigma_m}{\Delta} = \frac{E}{3(1 - 2\nu)}, \tag{2.49}$$

where σ_m is the *mean stress*, defined as

$$\sigma_m = \frac{1}{3}(\sigma_1 + \sigma_2 + \sigma_3). \tag{2.50}$$

In the *elastic* range, where $0 < \nu < 0.5$, it can be seen from Eq. (2.47) that the volume of a tension-test specimen increases and that of a compression-test specimen decreases during the test.

2.11.6 Effective stress and effective strain

A convenient means of expressing the state of stress on an element is the **effective stress** (or *equivalent* or *representative stress*), $\bar{\sigma}$, and **effective strain**, $\bar{\epsilon}$. For the maximum-shear-stress criterion, the effective stress is

$$\bar{\sigma} = \sigma_1 - \sigma_3, \tag{2.51}$$

and for the distortion-energy criterion, it is

$$\bar{\sigma} = \frac{1}{\sqrt{2}}\left[(\sigma_1 - \sigma_2)^2 + (\sigma_2 - \sigma_3)^2 + (\sigma_3 - \sigma_1)^2\right]^{1/2}. \tag{2.52}$$

The factor $1/\sqrt{2}$ is chosen so that, for simple tension, the effective stress is equal to the uniaxial yield stress, Y.

The strains are likewise related to the effective strain. For the maximum-shear-stress criterion, the effective strain is

$$\bar{\epsilon} = \frac{2}{3}(\epsilon_1 - \epsilon_3), \tag{2.53}$$

and for the distortion-energy criterion, it is

$$\bar{\epsilon} = \frac{\sqrt{2}}{3}\left[(\epsilon_1 - \epsilon_2)^2 + (\epsilon_2 - \epsilon_3)^2 + (\epsilon_3 - \epsilon_1)^2\right]^{1/2}. \tag{2.54}$$

Again, the factors $\frac{2}{3}$ and $\frac{\sqrt{2}}{3}$ are chosen so that for simple tension, the effective strain is equal to the uniaxial tensile strain. It is apparent that stress–strain curves may also be called effective stress-effective strain curves. (See also Example 2.7.)

Effective stresses and strains given by Eqs. (2.51) through (2.54) can readily be calculated by mathematics software packages. Further, advanced software for stress analysis and manufacturing process simulation, such as finite element analysis software, often presents results in terms of effective stress and strains. With modern computational tools, it is often not necessary to manually calculate effective or principal stresses and strains, but a physical understanding of these concepts is nonetheless essential.

2.11.7 Comparison of normal stress–normal strain and shear stress–shear strain

Stress–strain curves in tension and torsion for ductile materials are, as expected, comparable. Thus, it is possible to construct one curve from the other since the material is the same, the procedure for which is outlined below. The following observations are made with regard to tension and torsional states of stress:

- In the tension test, the uniaxial stress σ_1 is also the effective stress as well as the principal stress.
- In the torsion test, the principal stresses occur on planes whose normals are at $45°$ to the longitudinal axis, and the principal stresses σ_1 and σ_3 are equal in magnitude, but opposite in sign.
- The magnitude of the principal stress in torsion is the same as the maximum shear stress.

We now can write the following relationships:

$$\sigma_1 = -\sigma_3, \quad \sigma_2 = 0, \quad \text{and} \quad \sigma_1 = \tau_1.$$

Substituting these stresses into Eqs. (2.51) and (2.52) for effective stress, we obtain the following: For the maximum-shear-stress criterion,

$$\overline{\sigma} = \sigma_1 - \sigma_3 = \sigma_1 + \sigma_1 = 2\sigma_1 = 2\tau_1 \tag{2.55}$$

and for the distortion-energy criterion,

$$\overline{\sigma} = \frac{1}{\sqrt{2}}\left[(\sigma_1 - 0)^2 + (0 + \sigma_1)^2 + (-\sigma_1 - \sigma_1)^2\right]^{1/2} = \sqrt{3}\sigma_1 = \sqrt{3}\tau_1. \tag{2.56}$$

The following observations can be made with regard to strains:

- In the tension test, $\epsilon_2 = \epsilon_3 = -\epsilon_1/2$.
- In the torsion test, $\epsilon_1 = -\epsilon_3 = \gamma/2$.
- The strain in the thickness direction of the tube is zero, hence $\epsilon_2 = 0$.

The third observation above is correct because the thinning caused by the principal tensile stress is countered by the thickening effect under the principal compressive stress which is of the same magnitude, hence $\epsilon_2 = 0$. Since σ_2 is also zero, a thin-walled tube under torsion is under the conditions of both a plane stress and a shear strain.

Substituting these strains into Eqs. (2.53) and (2.54) for effective strain, the following relationships can be obtained: For the maximum-shear-stress criterion,

$$\overline{\epsilon} = \frac{2}{3}(\epsilon_1 - \epsilon_3) = \frac{2}{3}(\epsilon_1 + \epsilon_1) = \frac{4}{3}\epsilon_1 = \frac{2}{3}\gamma, \tag{2.57}$$

and for the distortion-energy criterion,

$$\overline{\epsilon} = \frac{\sqrt{2}}{3}\left[(\epsilon_1 - 0)^2 + (0 + \epsilon_1)^2 + (\epsilon_1 - \epsilon_1)^2\right]^{1/2} = \frac{2}{\sqrt{3}}\epsilon_1 = \frac{1}{\sqrt{3}}\gamma. \tag{2.58}$$

This set of equations provides a means by which tensile-test data can be converted to torsion-test data, and vice versa.

2.12 | Work of Deformation

Work is defined as the product of collinear force and distance; therefore, a quantity equivalent to work per unit volume is the product of stress and strain. Because the relationship between stress and strain in the plastic range depends on the particular

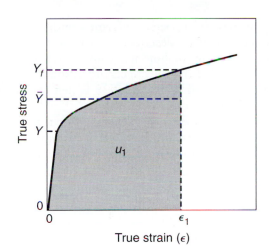

FIGURE 2.37 Schematic illustration of true stress–true strain curve showing yield stress, Y, average flow stress, $\overline{Y}$, specific energy, u_1, and flow stress, Y_f.

stress–strain curve of a material, this work is best calculated by referring to Fig. 2.37.

Note that the area under the true stress–true strain curve for any strain ϵ is the **energy per unit volume**, u (**specific energy**), of the material deformed, and is expressed as

$$u = \int_0^{\epsilon_1} \sigma \, d\epsilon. \tag{2.59}$$

As described in Section 2.2.3, true stress–true strain curves generally can be represented by the simple expression

$$\sigma = K\epsilon^n.$$

Hence Eq. (2.59) can be written as

$$u = K \int_0^{\epsilon_1} \epsilon^n \, d\epsilon,$$

or

$$u = \frac{K\epsilon_1^{n+1}}{n+1} = \overline{Y}\epsilon_1, \tag{2.60}$$

where $\overline{Y}$ is the **average flow stress** of the material.

This energy represents the work dissipated in uniaxial deformation. For triaxial states of stress, a general expression is given by

$$du = \sigma_1 \, d\epsilon_1 + \sigma_2 \, d\epsilon_2 + \sigma_3 \, d\epsilon_3.$$

For an application of this equation, see Example 2.7. For a more general condition, the effective stress and effective strain can be used. The energy per unit volume is then expressed by

$$u = \int_0^{\overline{\epsilon}} \overline{\sigma} \, d\overline{\epsilon}. \tag{2.61}$$

To obtain the work expended, we multiply u by the volume of the material deformed:

$$\text{Work} = (u)(\text{Volume}). \tag{2.62}$$

The energy represented by Eq. (2.62) is the *minimum energy* or the *ideal energy* required for uniform (homogeneous) deformation. The energy required for actual deformation involves two additional factors: (a) the energy required to overcome *friction* at the die–workpiece interfaces, and (b) the **redundant work** of deformation, which is described as follows.

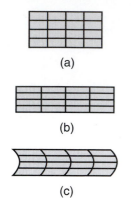

(a)

(b)

(c)

FIGURE 2.38 Deformation of grid patterns in a workpiece: (a) original pattern; (b) after ideal deformation; (c) after inhomogeneous deformation, requiring redundant work of deformation. Note that (c) is basically (b) with additional shearing, especially at the outer layers. Thus (c) requires greater work of deformation than (b). See also Figs. 6.3 and 6.49.

In Fig. 2.38a, a block of material is being deformed into shape by forging, extrusion, or drawing through a die, as described in Chapter 6. As shown in Fig. 2.38b, this deformation is uniform, or homogeneous. In reality, however, the material often deforms as in Fig. 2.38c from the effects of friction and die geometry. Note that the difference between (b) and (c) in Fig. 2.38 is that (c) has undergone additional shearing along horizontal planes.

Shearing requires expenditure of energy, because additional plastic work has to be done in subjecting the various layers to undergo shear strains. This is known as *redundant work*; the word *redundant* indicates the fact that this additional energy does not contribute to the shape change of the part. Note also that grid patterns (b) and (c) in Fig. 2.38 have the same overall shape and dimensions.

The *total specific energy* required can now be written as

$$u_{\text{total}} = u_{\text{ideal}} + u_{\text{friction}} + u_{\text{redundant}}. \tag{2.63}$$

The *efficiency*, η, of a process is then defined as

$$\eta = \frac{u_{\text{ideal}}}{u_{\text{total}}}. \tag{2.64}$$

Depending on the particular process, frictional conditions, die geometry, and other process parameters, the magnitude of η varies widely, with typical estimated values being 30% to 60% for extrusion and 75% to 95% for rolling.

EXAMPLE 2.7 Expansion of a thin-walled spherical shell

A thin-walled spherical shell made of a perfectly plastic material of yield stress, Y, original radius, r_o, and thickness, t_o, is being expanded by internal pressure. (a) Calculate the work done in expanding this shell to a radius of r_f. (b) If the diameter expands at a constant rate, what changes take place in the power consumed as the radius increases?

Solution. The membrane stresses are given by

$$\sigma_1 = \sigma_2 = Y$$

(from Example 2.6), where r and t are instantaneous dimensions. The true strains in the membrane are given by

$$\epsilon_1 = \epsilon_2 = \ln\left(\frac{2\pi r_f}{2\pi r_o}\right) = \ln\left(\frac{r_f}{r_o}\right).$$

Because an element in this shell is subjected to equal biaxial stretching, the specific energy is

$$u = \int_0^{\epsilon_1} \sigma_1 \, d\epsilon_1 + \int_0^{\epsilon_2} \sigma_2 \, d\epsilon_2 = 2\sigma_1 \epsilon_1 = 2Y \ln\left(\frac{r_f}{f_o}\right).$$

Since the volume of the shell material is $4\pi r_o^2 t_o$, the work done is

$$W = (u)(\text{Volume}) = 8\pi Y r_o^2 t_o \ln\left(\frac{r_f}{r_o}\right).$$

The specific energy can also be calculated from the effective stresses and strains. Thus, according to the distortion-energy criterion,

$$\bar{\sigma} = \frac{1}{\sqrt{2}}\left[(0)^2 + (\sigma_2)^2 + (-\sigma_1)^2\right]^{-1/2} = \sigma_1 = \sigma_2$$

and

$$\bar{\epsilon} = \frac{\sqrt{2}}{3}\left[(0)^2 + (\epsilon_2 + 2\epsilon_2)^2 + (-2\epsilon_2 - \epsilon_2)^2\right]^{1/2} = 2\epsilon_2 = 2\epsilon_1.$$

(The thickness strain $\epsilon_3 = -2\epsilon_2 = -2\epsilon_1$ because of volume constancy in plastic deformation, where $\epsilon_1 + \epsilon_2 + \epsilon_3 = 0$.) Hence

$$u = \int_0^\epsilon \bar{\sigma}\,\bar{\epsilon} = \int_0^{2\epsilon_1} \sigma_1\,d\epsilon_1 = 2\sigma_1\epsilon_1.$$

Thus, the answer is the same.

Power is defined as the rate of work; thus

$$\text{Power} = \frac{dW}{dt}.$$

Since all other factors in the expression are constant, the expression for work can be written as being proportional to strain

$$W \propto \ln\left(\frac{r}{r_o}\right) \propto (\ln r - \ln r_o).$$

Hence

$$\text{Power} \propto \frac{1}{r}\frac{dr}{dt}.$$

Because the shell is expanding at a constant rate, $dr/dt =$ constant. Hence the power is related to the instantaneous radius r by

$$\text{Power} \propto \frac{1}{r}.$$

2.12.1 Work, heat, and temperature rise

Almost all the mechanical work of deformation in plastic deformation is converted into heat. This conversion is not 100% because a small portion of this energy is stored within the deformed material as elastic energy, known as *stored energy* (see Section 3.6). Typically this energy represents 5% to 10% of the total energy input, although it may be as high as 30% in some alloys. In a simple frictionless process and assuming that all work of deformation is completely converted into heat, the *temperature rise*, ΔT, is given by

$$\Delta T = \frac{u_{\text{total}}}{\rho c}, \tag{2.65}$$

where u_{total} is the specific energy from Eq. (2.63), ρ is the density, and c is the specific heat of the material.

It will be noted that higher temperatures are associated with large areas under the stress–strain curve of the material and smaller values of its specific heat. The rise in temperature should be calculated using the stress–strain curve obtained at the appropriate strain rate level (see Section 2.2.7). It should also be noted that physical properties such as specific heat and thermal conductivity also depend on temperature and they should be taken into account in the calculations.

The theoretical temperature rise for a true strain of 1 (such as a 27-mm-high specimen compressed down to 10 mm) has been calculated to be as follows: aluminum, 75°C (165°F); copper, 140°C (285°F); low-carbon steel, 280°C (535°F); and titanium, 570°C (1060°F). Note also that the temperature rise given by Eq. (2.65) is for an ideal situation where there is no heat loss. In actual operations, however, heat is lost to the environment, to tools and dies, and to any lubricants or coolants used.

Thus, if deformation is carried out slowly, the actual temperature rise will be a small portion of the value calculated from the equation. Conversely, if the operation is performed very rapidly, these losses are relatively small. Under extreme conditions, an *adiabatic* state is approached, whereby the temperature rise is very high, leading to **incipient melting.**

EXAMPLE 2.8 Temperature rise in simple deformation

A cylindrical specimen 1 in. in diameter and 1 in. high is being compressed by dropping a weight of 100 lb on it from a certain height. The material has the following properties: $K = 15,000$ psi, $n = 0.5$, density $= 0.1$ lb/in^3, and specific heat $= 0.3$ Btu/lb-°F. Assuming no heat loss and no friction, calculate the final height of the specimen if the temperature rise is 100°F.

Solution. The expression for heat is given by

$$\text{Heat} = (c_p)(\rho)(\text{Volume})(\Delta T),$$

or

$$\text{Heat} = (0.3)(0.10)\left[\frac{(\pi)(1^2)}{4}\right](100)(778),$$

where 778 ft-lb = 1 BTU. Thus,

$$\text{Heat} = 1830 \text{ ft-lb} = 22,000 \text{ in.-lb.}$$

Also, ideally

$$\text{Heat} = \text{Work} = (u)(\text{Volume}) = \frac{\pi}{4}\frac{15,000\epsilon^{1.5}}{1.5}$$

and

$$\epsilon^{1.5} = \frac{(22,000)(1.5)(4)}{(\pi)(15,000)} = 2.8;$$

hence,

$$\epsilon = 1.99.$$

Using absolute values, we have

$$\ln\left(\frac{h_o}{h_f}\right) = \ln\left(\frac{1}{h_f}\right) = 1.99,$$

and therefore, $h_f = 0.137$ in.

SUMMARY

- Many manufacturing processes involve shaping materials by plastic deformation; consequently, mechanical properties such as strength, elasticity, ductility, hardness, toughness, and the energy required for plastic deformation are important factors. The behavior of materials, in turn, depends on the particular material and its condition, as well as other variables, particularly temperature, strain rate, and the state of stress. (Section 2.1)

- Mechanical properties measured by tension tests include modulus of elasticity (E), yield stress (Y), ultimate tensile strength (UTS), and Poisson's ratio (ν).

Ductility, as measured by elongation and reduction of area, can also be determined by tension tests. True stress–true strain curves are important in determining such mechanical properties as strength coefficient (K), strain-hardening exponent (n), strain-rate sensitivity exponent (m), and toughness. (Section 2.2)

- Compression tests closely simulate manufacturing processes such as forging, rolling, and extrusion. Properties measured by compression tests are subject to inaccuracy due to the presence of friction and barreling. (Section 2.3)

- Torsion tests are typically conducted on tubular specimens that are subjected to twisting. These tests model such manufacturing processes as shearing, cutting, and various machining processes. (Section 2.4)

- Bend or flexure tests are commonly used for brittle materials; the outer fiber stress at fracture in bending is known as the modulus of rupture or transverse rupture strength. The forces applied in bend tests model those incurred in such manufacturing processes as forming of sheet and plate, as well as testing of tool and die materials. (Section 2.5)

- A variety of hardness tests are available to test the resistance of a material to permanent indentation. Hardness is related to strength and wear resistance but is itself not a fundamental property of a material. (Section 2.6)

- Fatigue tests model manufacturing processes whereby components are subjected to rapidly fluctuating loads. The quantity measured is the endurance limit or fatigue limit; that is, the maximum stress to which a material can be subjected without fatigue failure, regardless of the number of cycles. (Section 2.7)

- Creep is the permanent elongation of a component under a static load maintained for a period of time. A creep test typically consists of subjecting a specimen to a constant tensile load at a certain temperature and measuring its change in length over a period of time. The specimen eventually fails by necking and rupture. (Section 2.8)

- Impact tests model some high-rate manufacturing operations as well as the service conditions in which materials are subjected to impact (or dynamic) loading, such as drop forging or the behavior of tool and die materials in interrupted cutting or high-rate deformation. Impact tests determine the energy required to fracture the specimen, known as impact toughness. Impact tests are also useful in determining the ductile-brittle transition temperature of materials. (Section 2.9)

- Residual stresses are those which remain in a part after it has been deformed and all external forces have been removed. The nature and level of residual stresses depend on the manner in which the part has been plastically deformed. Residual stresses may be reduced or eliminated by stress-relief annealing, by further plastic deformation, or by relaxation. (Section 2.10)

- In metalworking operations, the workpiece material is generally subjected to three-dimensional stresses through various tools and dies. Yield criteria establish relationships between the uniaxial yield stress of the material and the stresses applied; the two most widely used are the maximum-shear-stress (Tresca) and the distortion-energy (von Mises) criteria. (Section 2.11)

- Because it requires energy to deform materials, the work of deformation per unit volume of material (u) is an important parameter and is comprised of ideal, frictional, and redundant work components. In addition to supplying information on force and energy requirements, the work of deformation also indicates the amount of heat developed and, hence, the temperature rise in the workpiece during plastic deformation. (Section 2.12)

SUMMARY OF EQUATIONS

Engineering strain: $e = \dfrac{l - l_o}{l_o}$

Engineering strain rate: $\dot{e} = \dfrac{v}{l_o}$

Engineering stress: $\sigma = \dfrac{P}{A_o}$

True strain: $\epsilon = \ln\left(\dfrac{l}{l_o}\right)$

True strain rate: $\dot{\epsilon} = \dfrac{v}{l}$

True stress: $\sigma = \dfrac{P}{A}$

Modulus of elasticity: $E = \dfrac{\sigma}{e}$

Shear modulus: $G = \dfrac{E}{2(1 + \nu)}$

Modulus of resilience $= \dfrac{Y^2}{2E}$

Elongation $= \dfrac{l_f - l_o}{l_o} \times 100$

Reduction of area $= \dfrac{A_o - A_f}{A_o} \times 100$

Shear strain in torsion: $\gamma = \dfrac{r\phi}{l}$

Hooke's law: $\epsilon_1 = \dfrac{1}{E}[\sigma_1 - \nu(\sigma_2 + \sigma_3)]$, etc.

Effective strain (Tresca): $\bar{\epsilon} = \dfrac{2}{3}(\epsilon_1 - \epsilon_3)$

Effective strain (von Mises): $\bar{\epsilon} = \dfrac{\sqrt{2}}{3}\left[(\epsilon_1 - \epsilon_2)^2 + (\epsilon_2 - \epsilon_3)^2 + (\epsilon_3 - \epsilon_1)^2\right]^{1/2}$

Effective stress (Tresca): $\bar{\sigma} = \sigma_1 - \sigma_3$

Effective stress (von Mises): $\bar{\sigma} = \dfrac{1}{\sqrt{2}}\left[(\sigma_1 - \sigma_2)^2 + (\sigma_2 - \sigma_3)^2 + (\sigma_3 - \sigma_1)^2\right]^{1/2}$

True stress–true strain relationship (power law): $\sigma = K\epsilon^n$

True stress–true strain rate relationship: $\sigma = C\dot{\epsilon}^m$

Flow rules: $d\epsilon_1 = \dfrac{d\bar{\epsilon}}{\bar{\sigma}}\left[\sigma_1 - \dfrac{1}{2}(\sigma_2 + \sigma_3)\right]$, etc.

Maximum-shear-stress criterion (Tresca): $\sigma_{max} - \sigma_{min} = Y$

Distortion-energy criterion (von Mises): $(\sigma_1 - \sigma_2)^2 + (\sigma_2 - \sigma_3)^2 + (\sigma_3 - \sigma_1)^2 = 2Y^2$

Shear yield stress: $k = Y/2$ for Tresca and $k = Y/\sqrt{3}$ for von Mises (plane strain)

Volume strain (dilatation): $\Delta = \dfrac{1 - 2\nu}{E}(\sigma_1 + \sigma_2 + \sigma_3)$

Bulk modulus $= \dfrac{E}{3(1 - 2\nu)}$

BIBLIOGRAPHY

Ashby, M.F., *Materials Selection in Mechanical Design,* 3rd ed., Pergamon, 2005.

ASM Handbook, Vol. 8: *Mechanical Testing,* ASM International, 2000.

ASM Handbook, Vol. 10: *Materials Characterization,* ASM International, 1986.

ASM Handbook, Vol. 20: *Materials Selection and Design,* ASM International, 1997.

Beer, F.P., Johnston, E.R., and DeWolf, J.T., *Mechanics of Materials,* McGraw-Hill, 2005.

Boyer, H.E. (ed.), *Atlas of Creep and Stress-Rupture Curves,* ASM International, 1986.

————, *Atlas of Fatigue Curves,* ASM International, 1986.

————, *Atlas of Stress-Strain Curves,* ASM International, 1986.

Budinski, K.G., *Engineering Materials: Properties and Selection,* 8th ed., Prentice Hall, 2004.

Chandler, H., (ed.), *Hardness Testing,* 2nd ed., ASM International, 1999.

Cheremisinoff, N.P., and Cheremisinoff, P.N., *Handbook of Advanced Materials Testing,* Dekker, 1994.

Davis, J.R. (ed.), *Tensile Testing,* 2nd ed., ASM International, 2004.

Dieter, G.E., *Mechanical Metallurgy,* 3rd ed., McGraw-Hill, 1986.

Dowling, N.E., *Mechanical Behavior of Materials: Engineering Methods for Deformation, Fracture, and Fatigue,* 3rd ed., Prentice Hall, 2006.

Handbook of Experimental Solid Mechanics, Society for Experimental Mechanics, 2007.

Herzberg, R.W., *Deformation and Fracture Mechanics of Engineering Materials,* 4th ed., Wiley, 1996.

Hosford, W.F., *Mechanical Behavior of Materials,* Cambridge, 2005.

Pohlandt, K., *Material Testing for the Metal Forming Industry,* Springer, 1989.

QUESTIONS

2.1 Can you calculate the percent elongation of materials based only on the information given in Fig. 2.6? Explain.

2.2 Explain if it is possible for the curves in Fig. 2.4 to reach 0% elongation as the gage length is increased further.

2.3 Explain why the difference between engineering strain and true strain becomes larger as strain increases. Is this phenomenon true for both tensile and compressive strains? Explain.

2.4 Using the same scale for stress, we note that the tensile true stress–true strain curve is higher than the engineering stress–strain curve. Explain whether this condition also holds for a compression test.

2.5 Which of the two tests, tension or compression, requires a higher-capacity testing machine than the other? Explain.

2.6 Explain how the modulus of resilience of a material changes, if at all, as it is strained: (1) for an elastic, perfectly plastic material, and (2) for an elastic, linearly strain-hardening material.

2.7 If you pull and break a tension-test specimen rapidly, where would the temperature be the highest? Explain why.

2.8 Comment on the temperature distribution if the specimen in Question 2.7 is pulled very slowly.

2.9 In a tension test, the area under the true stress–true strain curve is the work done per unit volume (the specific work). We also know that the area under the load-elongation curve represents the work done on the specimen. If you divide this latter work by the volume of the specimen between the gage marks, you will determine the work done per unit volume (assuming that all deformation is confined between the gage marks). Will this specific work be the same as the area under the true stress–true strain curve? Explain. Will your answer be the same for any value of strain? Explain.

2.10 The note at the bottom of Table 2.5 states that as temperature increases, C decreases and m increases. Explain why.

2.11 You are given the K and n values of two different materials. Is this information sufficient to determine which material is tougher? If not, what additional information do you need, and why?

2.12 Modify the curves in Fig. 2.7 to indicate the effects of temperature. Explain the reasons for your changes.

2.13 Using a specific example, show why the deformation rate, say in m/s, and the true strain rate are not the same.

2.14 It has been stated that the higher the value of m, the more diffuse the neck is, and likewise, the lower the value of m, the more localized the neck is. Explain the reason for this behavior.

2.15   Explain why materials with high m values, such as hot glass and silly putty, when stretched slowly, undergo large elongations before failure. Consider events taking place in the necked region of the specimen.

2.16   Assume that you are running four-point bending tests on a number of identical specimens of the same length and cross section, but with increasing distance between the upper points of loading. (See Fig. 2.21b.) What changes, if any, would you expect in the test results? Explain.

2.17   Would Eq. (2.10) hold true in the elastic range? Explain.

2.18   Why have different types of hardness tests been developed? How would you measure the hardness of a very large object?

2.19   Which hardness tests and scales would you use for very thin strips of material, such as aluminum foil? Why?

2.20   List and explain the factors that you would consider in selecting an appropriate hardness test and scale for a particular application.

2.21   In a Brinell hardness test, the resulting impression is found to be an ellipse. Give possible explanations for this phenomenon.

2.21   Referring to Fig. 2.22, note that the material for indenters are either steel, tungsten carbide, or diamond. Why isn't diamond used for all of the tests?

2.22   What effect, if any, does friction have in a hardness test? Explain.

2.23   Describe the difference between creep and stress-relaxation phenomena, giving two examples for each as they relate to engineering applications.

2.24   Referring to the two impact tests shown in Fig. 2.31, explain how different the results would be if the specimens were impacted from the opposite directions.

2.25   If you remove layer ad from the part shown in Fig. 2.30d, such as by machining or grinding, which way will the specimen curve? (*Hint:* Assume that the part in diagram (d) can be modeled as consisting of four horizontal springs held at the ends. Thus, from the top down, we have compression, tension, compression, and tension springs.)

2.26   Is it possible to completely remove residual stresses in a piece of material by the technique described in Fig. 2.32 if the material is elastic, linearly strain hardening? Explain.

2.27   Referring to Fig. 2.32, would it be possible to eliminate residual stresses by compression instead of tension? Assume that the piece of material will not buckle under the uniaxial compressive force.

2.28   List and explain the desirable mechanical properties for the following: (1) elevator cable, (2) bandage, (3) shoe sole, (4) fish hook, (5) automotive piston, (6) boat propeller, (7) gas-turbine blade, and (8) staple.

2.29   Make a sketch showing the nature and distribution of the residual stresses in Figs. 2.31a and b before the parts were split (cut). Assume that the split parts are free from any stresses. (*Hint:* Force these parts back to the shape they were in before they were cut.)

2.30   It is possible to calculate the work of plastic deformation by measuring the temperature rise in a workpiece, assuming that there is no heat loss and that the temperature distribution is uniform throughout. If the specific heat of the material decreases with increasing temperature, will the work of deformation calculated using the specific heat at room temperature be higher or lower than the actual work done? Explain.

2.31   Explain whether or not the volume of a metal specimen changes when the specimen is subjected to a state of (a) uniaxial compressive stress and (b) uniaxial tensile stress, all in the elastic range.

2.32   We know that it is relatively easy to subject a specimen to hydrostatic compression, such as by using a chamber filled with a liquid. Devise a means whereby the specimen (say, in the shape of a cube or a thin round disk) can be subjected to hydrostatic tension, or one approaching this state of stress. (Note that a thin-walled, internally pressurized spherical shell is not a correct answer, because it is subjected only to a state of plane stress.)

2.33   Referring to Fig. 2.19, make sketches of the state of stress for an element in the reduced section of the tube when it is subjected to (1) torsion only, (2) torsion while the tube is internally pressurized, and (3) torsion while the tube is externally pressurized. Assume that the tube is closed end.

2.34   A penny-shaped piece of soft metal is brazed to the ends of two flat, round steel rods of the same diameter as the piece. The assembly is then subjected to uniaxial tension. What is the state of stress to which the soft metal is subjected? Explain.

2.35   A circular disk of soft metal is being compressed between two flat, hardened circular steel punches having the same diameter as the disk. Assume that the disk material is perfectly plastic and that there is no friction or any temperature effects. Explain the change, if any, in the magnitude of the punch force as the disk is being compressed plastically to, say, a fraction of its original thickness.

2.36   A perfectly plastic metal is yielding under the stress state σ_1, σ_2, σ_3, where $\sigma_1 > \sigma_2 > \sigma_3$. Explain what happens if σ_1 is increased.

2.37 What is the dilatation of a material with a Poisson's ratio of 0.5? Is it possible for a material to have a Poisson's ratio of 0.7? Give a rationale for your answer.

2.38 Can a material have a negative Poisson's ratio? Explain.

2.39 As clearly as possible, define plane stress and plane strain.

2.40 What test would you use to evaluate the hardness of a coating on a metal surface? Would it matter if the coating was harder or softer than the substrate? Explain.

2.41 List the advantages and limitations of the stress–strain relationships given in Fig. 2.7.

2.42 Plot the data in Table 2.1 on a bar chart, showing the range of values, and comment on the results.

2.43 A hardness test is conducted on as-received metal as a quality check. The results indicate that the hardness is too high; thus the material may not have sufficient ductility for the intended application. The supplier is reluctant to accept the return of the material, instead claiming that the diamond cone used in the Rockwell testing was worn and blunt, and hence the test needed to be recalibrated. Is this explanation plausible? Explain.

2.44 Explain why a 0.2% offset is used to determine the yield strength in a tension test.

2.45 Referring to Question 2.44, would the offset method be necessary for a highly strained-hardened material? Explain.

PROBLEMS

2.46 A strip of metal is originally 1.5 m long. It is stretched in three steps: first to a length of 1.75 m, then to 2.0 m, and finally to 3.0 m. Show that the total true strain is the sum of the true strains in each step, that is, that the strains are additive. Show that, using engineering strains, the strain for each step cannot be added to obtain the total strain.

2.47 A paper clip is made of wire 1.20 mm in diameter. If the original material from which the wire is made from a rod 15 mm in diameter, calculate the longitudinal and diametrical engineering and true strains that the wire has undergone during processing.

2.48 A material has the following properties: UTS = 50,000 psi and $n = 0.25$. Calculate its strength coefficient K.

2.49 Based on the information given in Fig. 2.6, calculate the ultimate tensile strength of annealed 70-30 brass.

2.50 Calculate the ultimate tensile strength (engineering) of a material whose strength coefficient is 400 MPa and of a tensile-test specimen that necks at a true strain of 0.20.

2.51 A cable is made of four parallel strands of different materials, all behaving according to the equation $\sigma = K\epsilon^n$, where $n = 0.3$. The materials, strength coefficients and cross sections are as follows:

Material A: $K = 450$ MPa, $A_o = 7$ mm^2
Material B: $K = 600$ MPa, $A_o = 2.5$ mm^2
Material C: $K = 300$ MPa, $A_o = 3$ mm^2
Material D: $K = 760$ MPa, $A_o = 2$ mm^2

(a) Calculate the maximum tensile load that this cable can withstand prior to necking.
(b) Explain how you would arrive at an answer if the n values of the three strands were different from each other.

2.52 Using only Fig. 2.6, calculate the maximum load in tension testing of a 304 stainless-steel round specimen with an original diameter of 0.5 in.

2.53 Using the data given in Table 2.1, calculate the values of the shear modulus G for the metals listed in the table.

2.54 Derive an expression for the toughness of a material represented by the equation $\sigma = K(\epsilon + 0.2)^n$ and whose fracture strain is denoted as ϵ_f.

2.55 A cylindrical specimen made of a brittle material 1 in. high and with a diameter of 1 in. is subjected to a compressive force along its axis. It is found that fracture takes place at an angle of 45° under a load of 30,000 lb. Calculate the shear stress and the normal stress acting on the fracture surface.

2.56 What is the modulus of resilience of a highly cold-worked piece of steel with a hardness of 300 HB? Of a piece of highly cold-worked copper with a hardness of 150 HB?

2.57 Calculate the work done in frictionless compression of a solid cylinder 40 mm high and 15 mm in diameter to a reduction in height of 75% for the following materials: (1) 1100-O aluminum, (2) annealed copper, (3) annealed 304 stainless steel, and (4) 70-30 brass, annealed.

2.58 A material has a strength coefficient $K = 100,000$ psi Assuming that a tensile-test specimen made from this material begins to neck at a true strain of 0.17, show that the ultimate tensile strength of this material is 62,400 psi.

2.59 A tensile-test specimen is made of a material represented by the equation $\sigma = K(\epsilon + n)^n$. (a) Determine the true strain at which necking will begin. (b) Show that it is possible for an engineering material to exhibit this behavior.

2.60   Take two solid cylindrical specimens of equal diameter but different heights. Assume that both specimens are compressed (frictionless) by the same percent reduction, say 50%. Prove that the final diameters will be the same.

2.61   A horizontal rigid bar c-c is subjecting specimen a to tension and specimen b to frictionless compression such that the bar remains horizontal. (See the accompanying figure.) The force F is located at a distance ratio of 2:1. Both specimens are incompressible and have an original cross-sectional area of 1 in.^{2} and the original lengths are $a = 8$ in. and $b = 4.5$ in. The material for specimen a has a true stress–true strain curve of $\sigma = 100,000\epsilon^{0.5}$. Plot the true stress–true strain curve that the material for specimen b should have for the bar to remain horizontal during the experiment.

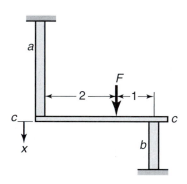

2.62   Inspect the curve that you obtained in Problem 2.61. Does a typical strain-hardening material behave in that manner? Explain.

2.63   In a disk test performed on a specimen 40 mm in diameter and 5 m thick, the specimen fractures at a stress of 500 MPa. What was the load on the disk at fracture?

2.64   In Fig. 2.32a, let the tensile and compressive residual stresses both be 10,000 psi and the modulus of elasticity of the material be 30×10^6 psi, with a modulus of resilience of 30 in.-lb/in.^{3}. If the original length in diagram (a) is 20 in., what should be the stretched length in diagram (b) so that, when unloaded, the strip will be free of residual stresses?

2.65   Show that you can take a bent bar made of an elastic, perfectly plastic material and straighten it by stretching it into the plastic range. (*Hint:* Observe the events shown in Fig. 2.32.)

2.66   A bar 1 m long is bent and then stress relieved. The radius of curvature to the neutral axis is 0.50 m. The bar is 30 mm thick and is made of an elastic, perfectly plastic material with $Y = 600$ MPa and $E = 200$ GPa. Calculate the length to which this bar should be stretched so that, after unloading, it will become and remain straight.

2.67   Assume that a material with a uniaxial yield stress Y yields under a stress state of principal stresses $\sigma_1, \sigma_2, \sigma_3$, where $\sigma_1 > \sigma_2 > \sigma_3$. Show that the superposition of a hydrostatic stress, p, on this system (such as placing the specimen in a chamber pressurized with a liquid) does not affect yielding. In other words, the material will still yield according to yield criteria.

2.68   Give two different and specific examples in which the maximum-shear-stress and the distortion-energy criteria give the same answer.

2.69   A thin-walled spherical shell with a yield stress Y is subjected to an internal pressure p. With appropriate equations, show whether or not the pressure required to yield this shell depends on the particular yield criterion used.

2.70   Show that, according to the distortion-energy criterion, the yield stress in plane strain is $1.15Y$, where Y is the uniaxial yield stress of the material.

2.71   What would be the answer to Problem 2.70 if the maximum-shear-stress criterion were used?

2.72   A closed-end, thin-walled cylinder of original length l, thickness t, and internal radius r is subjected to an internal pressure p. Using the generalized Hooke's law equations, show the change, if any, that occurs in the length of this cylinder when it is pressurized. Let $\nu = 0.33$.

2.73   A round, thin-walled tube is subjected to tension in the elastic range. Show that both the thickness and the diameter of the tube decrease as tension increases.

2.74   Take a long cylindrical balloon and, with a thin felt-tip pen, mark a small square on it. What will be the shape of this square after you blow up the balloon: (1) a larger square, (2) a rectangle, with its long axis in the circumferential directions, (3) a rectangle, with its long axis in the longitudinal direction, or (4) an ellipse? Perform this experiment and, based on your observations, explain the results, using appropriate equations. Assume that the material the balloon is made of is perfectly elastic and isotropic, and that this situation represents a thin-walled, closed-end cylinder under internal pressure.

2.75   Take a cubic piece of metal with a side length l_o and deform it plastically to the shape of a rectangular parallelepiped of dimensions l_1, l_2, and l_3. Assuming that the material is rigid and perfectly plastic, show that volume constancy requires that the following expression be satisfied: $\epsilon_1 + \epsilon_2 + \epsilon_3 = 0$.

2.76   What is the diameter of an originally 30-mm-diameter solid steel ball when it is subjected to a hydrostatic pressure of 5 GPa?

2.77 Determine the effective stress and effective strain in plane-strain compression according to the distortion-energy criterion.

2.78 (a) Calculate the work done in expanding a 2-mm-thick spherical shell from a diameter of 100 mm to 140 mm, where the shell is made of a material for which $\sigma = 200 + 50\epsilon^{0.5}$ MPa. (b) Does your answer depend on the particular yield criterion used? Explain.

2.79 A cylindrical slug that has a diameter of 1 in. and is 1 in. high is placed at the center of a 2-in.-diameter cavity in a rigid die. (See the accompanying figure.) The slug is surrounded by a compressible matrix, the pressure of which is given by the relation

$$p_m = 40,000 \frac{\Delta V}{V_{om}} \text{ psi,}$$

where m denotes the matrix and V_{om} is the original volume of the compressible matrix. Both the slug and the matrix are being compressed by a piston and without any friction. The initial pressure on the matrix is zero, and the slug material has the true stress–true strain curve of $\sigma = 15,000\epsilon^{0.4}$.

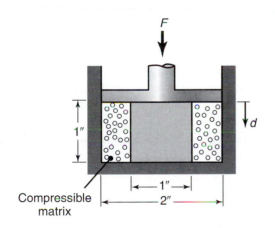

Compressible matrix

Obtain an expression for the force F versus piston travel d up to $d = 0.5$ in.

2.80 A specimen in the shape of a cube 20 mm on each side is being compressed without friction in a die cavity, as shown in Fig. 2.35d, where the width of the groove is 15 mm. Assume that the linearly strain-hardening material has the true stress–true strain curve given by $\sigma = 70 + 30\epsilon$ MPa. Calculate the compressive force required when the height of the specimen is at 3 mm, according to both yield criteria.

2.81 Obtain expressions for the specific energy for a material for each of the stress–strain curves shown in Fig. 2.7, similar to those shown in Section 2.12.

2.82 A material with a yield stress of 70 MPa is subjected to principal (normal) stresses of σ_1, $\sigma_2 = 0$, and $\sigma_3 = -\sigma_1/2$. What is the value of σ_1 when the metal yields according to the von Mises criterion? What if $\sigma_2 = \sigma_1/3$?

2.83 A steel plate has the dimensions 100 mm × 100 mm × 5 mm thick. It is subjected to biaxial tension of $\sigma_1 = \sigma_2$ with the stress in the thickness direction of $\sigma_3 = 0$. What is the largest possible change in volume at yielding, using the von Mises criterion? What would this change in volume be if the plate were made of copper?

2.84 A 50-mm-wide, 1-mm-thick strip is rolled to a final thickness of 0.5 mm. It is noted that the strip has increased in width to 52 mm. What is the strain in the rolling direction?

2.85 An aluminum alloy yields at a stress of 50 MPa in uniaxial tension. If this material is subjected to the stresses $\sigma_1 = 25$ MPa, $\sigma_2 = 15$ MPa, and $\sigma_3 = -26$ MPa, will it yield? Explain.

2.86 A cylindrical specimen 1 in. in diameter and 1-in. high is being compressed by dropping a weight of 200 lb on it from a certain height. After deformation, it is found that the temperature rise in the specimen is 300°F. Assuming no heat loss and no friction, calculate the final height of the specimen, using the following data for the material: $K = 30,000$ psi, $n = 0.5$, density = 0.1 lb/in.^{3}, and specific heat = 0.3 BTU/lb·°F.

2.87 A solid cylindrical specimen 100 mm high is compressed to a final height of 40 mm in two steps between frictionless platens; after the first step the cylinder is 70 mm high. Calculate the engineering strain and the true strain for both steps, compare them, and comment on your observations.

2.88 Assume that the specimen in Problem 2.87 has an initial diameter of 80 mm and is made of 1100-O aluminum. Determine the load required for each step.

2.89 Determine the specific energy and actual energy expended for the entire process described in the previous two problems.

2.90 A metal has a strain-hardening exponent of 0.22. At a true strain of 0.2, the true stress is 20,000 psi. (a) Determine the stress–strain relationship for this material. (b) Determine the ultimate tensile strength for this material.

2.91 The area of each face of a metal cube is 400 m^2, and the metal has a shear yield stress, k, of 140 MPa. Compressive loads of kN and 80 kN are applied at different faces (say in the x- and y-directions). What must be the compressive load applied to the z-direction to cause yielding according to the Tresca criterion? Assume a frictionless condition.

2.92 A tensile force of 9 kN is applied to the ends of a solid bar of 6.35 mm diameter. Under load, the diameter reduces to 5.00 mm. Assuming uniform deformation and volume constancy, (a) determine the engineering stress and strain, and (b) determine the true stress and strain. (c) If the original bar had been subjected to a true stress of 345 MPa and the resulting diameter was 5.60 mm, what are the engineering stress and engineering strain for this condition?

2.93 Two identical specimens 10 mm in diameter and with test sections 25 mm long are made of 1112 steel. One is in the as-received condition and the other is annealed. What will be the true strain when necking begins, and what will be the elongation of these samples at that instant? What is the ultimate tensile strength for these samples?

2.94 During the production of a part, a metal with a yield strength of 110 MPa is subjected to a stress state σ_1, $\sigma_2 = \sigma_1/3$, $\sigma_3 = 0$. Sketch the Mohr's circle diagram for this stress state. Determine the stress σ_1 necessary to cause yielding by the maximum shear stress and the von Mises criteria.

2.95 Estimate the depth of penetration in a Brinell hardness test using 500-kg load, when the sample is a cold-worked aluminum with a yield stress of 200 MPa.

2.96 The following data are taken from a stainless steel tension-test specimen:

Load, P (lb)	Extension, Δl (in.)
1600	0
2500	0.02
3000	0.08
3600	0.20
4200	0.40
4500	0.60
4600 (max)	0.86
4586 (fracture)	0.98

Also, $A_o = 0.056$ in.^{2}, $A_f = 0.016$ in.^{2}, $l_o = 2$ in. Plot the true stress–true strain curve for the material.

2.97 A metal is yielding plastically under the stress state shown in the accompanying figure.

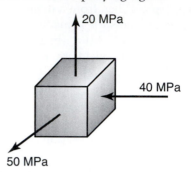

20 MPa

40 MPa

50 MPa

(a) Label the principal axes according to their proper numerical convention (1, 2, 3).
(b) What is the yield stress using the Tresca criterion?
(c) What if the von Mises criterion is used?
(d) The stress state causes measured strains of $\epsilon_1 = 0.4$ and $\epsilon_2 = 0.2$, with ϵ_3 not being measured. What is the value of ϵ_3?

2.98 It has been proposed to modify the von Mises yield criterion as

$$(\sigma_1 - \sigma_2)^a + (\sigma_2 - \sigma_3)^a + (\sigma_3 - \sigma_1)^a = C,$$

where C is a constant and a is an even integer larger than 2. Plot this yield criterion for $a = 4$ and $a = 12$, along with the Tresca and von Mises criteria, in plane stress. (*Hint:* See Fig. 2.36.)

2.99 Assume that you are asked to give a quiz to students on the contents of this chapter. Prepare three quantitative problems and three qualitative questions, and supply the answers.

Structure and Manufacturing Properties of Metals

This chapter describes the structure and properties of metals and how these properties affect metals' manufacturing characteristics:

- Crystal structures, grains and grain boundaries, plastic deformation, and thermal effects.
- The characteristics and applications of cold, warm, and hot working of metals.
- The ductile and brittle behavior of materials, modes of failure, and the effects of various factors on fracture behavior.
- Physical properties of materials and their relevance to manufacturing processes.
- General properties and engineering applications of ferrous and nonferrous metals and alloys.

3.1 | Introduction

The structure of metals, that is, the arrangement of atoms, greatly influences their *properties* and *behavior*. Understanding these structures allows us to evaluate, predict, and control their properties, as well as make appropriate selections for specific applications. For example, we will come to appreciate the reasons for the development of single-crystal turbine blades (Fig. 3.1) for use in jet engines, with properties that are better than blades made conventionally.

This chapter begins with a general review of the crystal structure of metals and the role of grain size, grain boundaries, inclusions, and imperfections as they affect plastic deformation in metalworking processes. We will then review the failure and fracture of metals, both ductile and brittle, together with factors that influence fracture, such as state of stress, temperature, strain rate, and external and internal defects in metals. The rest of the chapter is devoted to a brief discussion of the general properties and applications of ferrous and nonferrous metals and alloys, including some data to guide design and material selection.

FIGURE 3.1 Turbine blades for jet engines, manufactured by three different methods: (a) conventionally cast; (b) directionally solidified, with columnar grains, as can be seen from the vertical streaks; and (c) single crystal. Although more expensive, single-crystal blades have properties at high temperatures that are superior to those of other blades. *Source:* Courtesy of United Technologies Pratt and Whitney.

(a)   (b)   (c)

3.2 | The Crystal Structure of Metals

When metals solidify from a molten state, the atoms arrange themselves into various orderly configurations, called **crystals,** and the arrangement of the atoms in the crystal is called **crystalline structure.** The smallest group of atoms showing the characteristic **lattice structure** of a particular metal is known as a **unit cell.** It is the building block of a crystal. The three basic patterns of atomic arrangement found in most metals are (Figs. 3.2–3.4)

   a. **Body-centered cubic** (bcc).
   b. **Face-centered cubic** (fcc).
   c. **Hexagonal close packed** (hcp).

The order of magnitude of the distance between the atoms in these crystal structures is 0.1 nm (10^{-8} in.). The pattern in which the atoms are arranged in these structures has a significant influence on the properties of a particular metal. In the three basic structures illustrated in Figs. 3.2–3.4, the fcc and hcp crystals have the most densely packed configurations. In the hcp structure, the top and bottom planes are called **basal planes.**

Different crystal structures form due to the differences in the energy required to form these structures. Tungsten, for example, forms a bcc structure because it requires lower energy than would be required to form other structures. Similarly, aluminum forms an fcc structure. At different temperatures, the same metal may

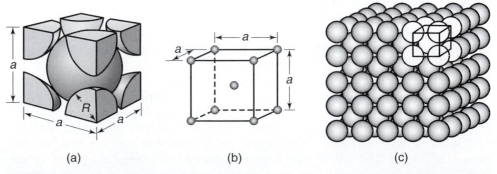

(a)   (b)   (c)

FIGURE 3.2 The body-centered cubic (bcc) crystal structure: (a) hard-ball model; (b) unit cell; and (c) single crystal with many unit cells. Common bcc metals include chromium, titanium, and tungsten. *Source:* After W.G. Moffatt.

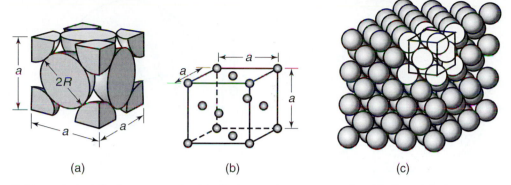

(a)   (b)   (c)

FIGURE 3.3 The face-centered cubic (fcc) crystal structure: (a) hard-ball model; (b) unit cell; and (c) single crystal with many unit cells. Common fcc metals include aluminum, copper, gold, and silver. *Source:* After W.G. Moffatt.

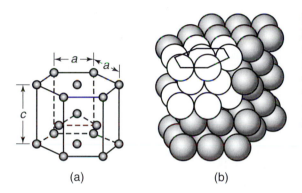

(a)   (b)

FIGURE 3.4 The hexagonal close-packed (hcp) crystal structure: (a) unit cell; and (b) single crystal with many unit cells. Common hcp metals include zinc, magnesium, and cobalt. *Source:* After W.G. Moffatt.

form different structures because of differences in the amount of energy required at each temperature. For example, iron forms a bcc structure below 912°C (1674°F) and above 1394°C (2541°F), but it forms an fcc structure between 912° and 1394°C. The appearance of more than one type of crystal structure is known as **allotropism, or polymorphism.** These structural changes are an important aspect of heat treatment of metals and alloys, as described in Section 5.2.5.

The crystal structures can be modified by adding atoms of another metal or metals. Known as **alloying,** this process typically improves the properties of the metal, as described in Section 5.2.

3.3 | Deformation and Strength of Single Crystals

When a crystal is subjected to an external force, it first undergoes **elastic deformation;** that is, it returns to its original shape when the force is removed. (See Section 2.2.) An analogy to this type of behavior is a helical spring that stretches when loaded and returns to its original shape when the load is removed. However, if the force on the crystal structure is increased sufficiently, the crystal undergoes **plastic (permanent) deformation;** that is, it does not return to its original shape when the force is removed.

There are two basic mechanisms by which plastic deformation may take place in crystal structures:

1. **Slip.** This mechanism involves the slipping of one plane of atoms over an adjacent plane (slip plane) under a shear stress, as shown schematically in

FIGURE 3.5 Permanent deformation of a single crystal under a tensile load. The highlighted grid of atoms emphasizes the motion that occurs within the lattice. (a) Deformation by slip. The *b/a* ratio influences the magnitude of the shear stress required to cause slip. Note that the slip planes tend to align themselves in the direction of pulling. (b) Deformation by twinning, involving generation of a "twin" around a line of symmetry subjected to shear. Note that the tensile load results in a shear stress in the plane illustrated.

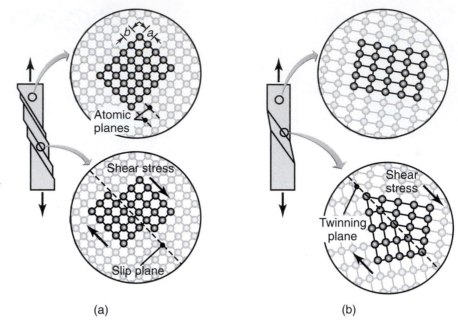

(a)   (b)

Fig. 3.5a. This mechanism is much like the sliding of playing cards against each other. Just as it takes a certain amount of force to slide playing cards against each other, so a crystal requires a certain amount of shear stress (called **critical shear stress**) to undergo permanent deformation. Thus, there must be a shear stress of sufficient magnitude within a crystal for plastic deformation to occur.

The **maximum theoretical shear stress**, τ_{max}, to cause permanent deformation in a perfect crystal is obtained as follows. When there is no stress present, the atoms in the crystal are in *equilibrium* (Fig. 3.6). Under a shear stress, the upper row of atoms moves to the right, where the position of an atom is denoted as x. Thus, when $x = 0$ or $x = b$ the shear stress is zero. Each atom of the upper row is attracted to the nearest atom of the lower row, resulting in *nonequilibrium* at positions 2 and 4; the stresses are now at a maximum, but opposite in sign. Note that at position 3, the shear stress is again zero, since this position is now symmetric. In Fig. 3.6, we assume that, as a first approximation, the shear stress varies sinusoidally. Hence, the shear stress at a displacement x is

$$\tau = \tau_{max} \sin \frac{2\pi x}{b}, \tag{3.1}$$

FIGURE 3.6 Variation of shear stress in moving a plane of atoms over another plane.

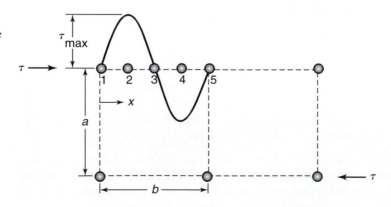

which for small values of x/b can be written as

$$\tau = \tau_{max}\frac{2\pi x}{b}.$$

From Hooke's law, we have

$$\tau = G\gamma = G\left(\frac{x}{a}\right).$$

Hence,

$$\tau_{max} = \frac{Gb}{2\pi a}. \tag{3.2}$$

If we now assume that b is approximately equal to a, then

$$\tau_{max} = \frac{G}{2\pi}. \tag{3.3}$$

It has been shown that with more refined calculations the value of τ_{max} is between $G/10$ and $G/30$.

From Eq. (3.2), it can be seen that the shear stress required to cause slip in single crystals is directly proportional to b/a. It can therefore be stated that slip in a crystal takes place along planes of maximum atomic density, or that slip takes place in closely packed planes and in closely packed directions. Because the b/a ratio is different for different directions within the crystal, a single crystal thus has different properties when tested in different directions; thus, a crystal is **anisotropic.** A common example of anisotropy is woven cloth, which stretches differently when pulled in different directions, or plywood, which is much stronger in the planar direction than along its thickness direction (where it splits easily).

2. **Twinning.** The second mechanism of plastic deformation is *twinning,* in which a portion of the crystal forms a mirror image of itself across the *plane of twinning* (Fig. 3.5b). Twins form abruptly and are the cause of the creaking sound (*tin cry*) when a tin or zinc rod is bent at room temperature. Twinning usually occurs in hcp and bcc metals by plastic deformation and in fcc metals by annealing. (See Section 5.11.4.)

3.3.1 Slip systems

The combination of a slip plane and its direction of slip is known as a **slip system.** In general, metals with five or more slip systems are ductile, whereas those with less than five are not ductile. Note that each pattern of atomic arrangement will have a different number of potential slip systems.

1. In body-centered cubic crystals, there are 48 possible slip systems; thus, the probability is high that an externally applied shear stress will operate on one of the systems and cause slip. However, because of the relatively high b/a ratio, the required shear stress is high. Metals with bcc structures (such as titanium, molybdenum, and tungsten) have good strength and moderate ductility.

2. In face-centered cubic crystals, there are 12 slip systems. The probability of slip is moderate, and the required shear stress is low. Metals with fcc structures (such as aluminum, copper, gold, and silver) have moderate strength and good ductility.

FIGURE 3.7 Schematic illustration of slip lines and slip bands in a single crystal subjected to a shear stress. A slip band consists of a number of slip planes. The crystal at the center of the upper drawing is an individual grain surrounded by other grains.

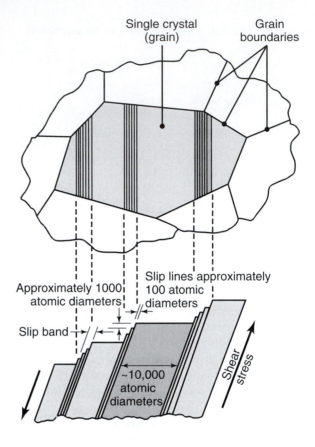

3. The hexagonal close-packed crystal has three slip systems and, thus, has a low probability of slip; however, more systems become active at elevated temperatures. Metals with hcp structures (such as beryllium, magnesium, and zinc) are generally brittle.

Note in Fig. 3.5a that (a) the portions of the single crystal that have slipped have rotated from their original angular position toward the direction of the tensile force, and (b) slip has taken place along certain planes only. With the use of electron microscopy, it has been shown that what appears to be a single slip plane is actually a **slip band,** consisting of a number of slip planes (Fig. 3.7).

3.3.2 Ideal tensile strength of metals

The ideal or theoretical tensile strength of metals can be obtained as follows: In the bar shown in Fig. 3.8, the interatomic distance is a when no external stress is applied. In order to increase this distance, a tensile force has to be applied to overcome the cohesive force between the atoms. The cohesive force is zero in the unstrained equilibrium condition. When the tensile stress reaches σ_{max} the atomic bonds between two neighboring atomic planes break; σ_{max} is known as the *ideal tensile strength*.

The ideal tensile strength can be derived by the following approach: Referring to Fig. 3.8, it can be shown that

$$\sigma_{max} = \frac{E\lambda}{2\pi a} \tag{3.4}$$

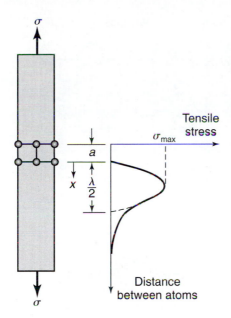

FIGURE 3.8 Variation of cohesive stress as a function of distance between a row of atoms.

and that the work done per unit area in breaking the bond between the two atomic planes (i.e., the area under the cohesive force curve) is given by

$$\text{Work} = \frac{\sigma_{max}\lambda}{\pi}. \tag{3.5}$$

This work is dissipated in creating two new fracture surfaces, thus involving surface energy of the material, γ. The total surface energy then is 2γ. Combining these equations, we find that

$$\sigma_{max} = \sqrt{\frac{E\gamma}{a}}. \tag{3.6}$$

When appropriate values are substituted into this equation, we have

$$\sigma_{max} \simeq \frac{E}{10}. \tag{3.7}$$

Thus the theoretical strength of steel, for example, would be on the order of 20 GPa (3×10^6 psi).

3.3.3 Imperfections

The actual strength of metals is approximately one to two orders of magnitude lower than the strength levels obtained from Eq. (3.7). This discrepancy has been explained in terms of **imperfections** in the crystal structure. Unlike the idealized models described previously, actual metal crystals contain a large number of imperfections and defects, which are categorized as follows:

1. **Point defects,** such as a **vacancy** (a missing atom), an **interstitial atom** (an extra atom in the lattice), or an **impurity atom** (a foreign atom that has replaced an atom of the pure metal) (Fig. 3.9).
2. **Linear,** or *one-dimensional, defects,* called **dislocations** (Fig. 3.10).
3. **Planar,** or *two-dimensional, imperfections,* such as **grain boundaries** and **phase boundaries.**
4. **Volume,** or *bulk, imperfections,* such as **voids, inclusions** (nonmetallic elements such as oxides, sulfides, and silicates), other **phases,** or **cracks.**

FIGURE 3.9 Various defects in a single-crystal lattice. *Source:* After W.G. Moffatt.

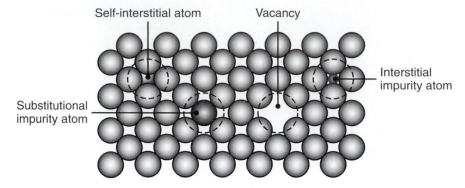

Self-interstitial atom

Vacancy

Interstitial impurity atom

Substitutional impurity atom

FIGURE 3.10 (a) Edge dislocation, a linear defect at the edge of an extra plane of atoms. (b) Screw dislocation, a helical defect in a three-dimensional lattice of atoms. Screw dislocations are so named because the atomic planes form a spiral ramp.

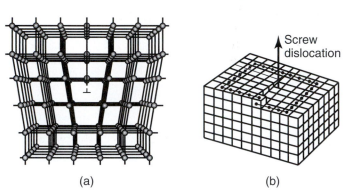

Screw dislocation

(a)   (b)

A slip plane containing a dislocation requires lower shear stress to cause slip than does a plane in a perfect lattice (Fig. 3.11). This is understandable, since slip requires far fewer atoms to move to new lattice locations when a dislocation is present. One analogy that can be used to describe the movement of an edge dislocation is the earthworm, which moves forward through a hump that starts at the tail and moves toward the head. A second analogy is that of moving a large carpet by forming a hump at one end and moving the hump forward toward the other end. In this way, the force required to move a carpet is much less than that required to slide the whole carpet along the floor.

The **density of dislocations** is the total length of dislocation lines per unit volume, (mm/mm^3 = mm^{-2}); it increases with increasing plastic deformation, by as much as 10^6 mm^{-2} at room temperature. The dislocation densities for some conditions are

a. Very pure single crystals: 0 to 10^3 mm^{-2}.
b. Annealed single crystals: 10^5 to 10^6.
c. Annealed polycrystals: 10^7 to 10^8.
d. Highly cold-worked metals: 10^{11} to 10^{12}.

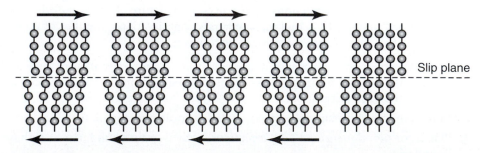

Slip plane

FIGURE 3.11 Movement of an edge dislocation across the crystal lattice under a shear stress. Dislocations help explain why the actual strength of metals is much lower than that predicted by atomic theory.

The mechanical properties of metals, such as yield and fracture strength, as well as electrical conductivity are affected by lattice defects and are known as **structure-sensitive properties.** On the other hand, physical properties, such as melting point, specific heat, coefficient of thermal expansion, and elastic constants, are not sensitive to these defects and are thus known as **structure-insensitive properties.**

3.3.4 Strain hardening (work hardening)

Although the presence of a dislocation lowers the shear stress required to cause slip, dislocations can (1) become entangled and interfere with each other and (2) be impeded by barriers, such as grain boundaries and impurities and inclusions in the material. Entanglement and impediments increase the shear stress required for slip.

The increase in the shear stress, and hence the increase in the overall strength of the metal, is known as **strain hardening,** or **work hardening.** (See Section 2.2.3.) The greater the deformation, the more the entanglements, thus increasing the metal's strength. Work hardening is used extensively to strengthen metals in metal-working processes at ambient temperature. Typical examples are strengthening wire by drawing it through a die to reduce its cross section (Section 6.5), producing the head on a bolt by forging it (Section 6.2.4), and producing sheet metal for automobile bodies and aircraft fuselages by rolling (Section 6.3). As shown by Eq. (2.11), the degree of strain hardening is indicated by the magnitude of the strain-hardening exponent, n. (See also Table 2.3.) Among the three crystal structures, hcp has the lowest n value, followed by bcc and then fcc, which has the highest.

3.4 | Grains and Grain Boundaries

Metals commonly used for manufacturing various products are composed of many individual, randomly oriented crystals (*grains*); they are **polycrystals.** When a mass of molten metal begins to solidify, crystals begin to form independently of each other, with random orientations and at various locations within the liquid mass (Fig. 3.12). Each of the crystals grows into a crystalline structure, or grain. The number and the size of the grains developed in a unit volume of the metal depend on the rate at which **nucleation** (the initial stage of formation of crystals) takes place. The *number* of different sites in which individual crystals begin to form (seven are shown in Fig. 3.12a) and the *rate* at which these crystals grow affect the size of grains developed. Generally, rapid cooling produces smaller grains, whereas slow cooling produces larger grains. (See also Section 5.3.)

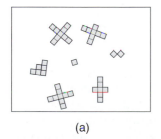   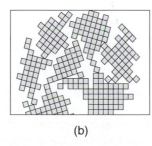   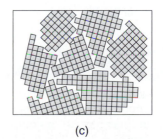   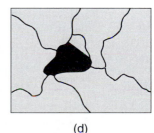

| (a) | (b) | (c) | (d) |

FIGURE 3.12 Schematic illustration of the various stages during solidification of molten metal. Each small square represents a unit cell. (a) Nucleation of crystals at random sites in the molten metal. Note that the crystallographic orientation of each site is different. (b) and (c) Growth of crystals as solidification continues. (d) Solidified metal, showing individual grains and grain boundaries. Note the different angles at which neighboring grains meet each other. *Source:* After W. Rosenhain.

FIGURE 3.13 Variation of tensile stress across a plane of polycrystalline metal specimen subjected to tension. Note that the strength exhibited by each grain depends on its orientation.

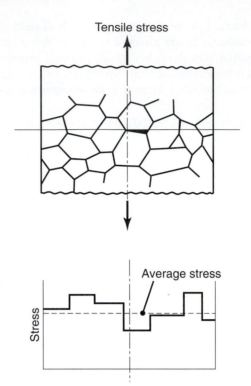

If the nucleation rate is high, the number of grains in a unit volume of metal will be greater, and consequently grain size will be small. Conversely, if the rate of growth of the crystals is high compared with their nucleation rate, there will be fewer, but larger, grains per unit volume. Note in Fig. 3.12 how, as they grow, the grains eventually interfere with and impinge upon one another. The *surfaces* that separate the individual grains are called **grain boundaries.** Each grain consists of either a single crystal (for pure metals) or a polycrystalline aggregate (for alloys).

Note that the crystallographic orientation changes abruptly from one grain to the next across the grain boundaries. Recall from Section 3.3 that the behavior of a single crystal or a single grain is *anisotropic*. The behavior of a piece of polycrystalline metal (Fig. 3.13) is therefore *isotropic* because the grains have random crystallographic orientations. Thus, properties do not vary with the direction of testing. In practice, however, perfectly isotropic metals are rare, because the crystal structure usually is not equiaxed and cold working results in preferred orientation of the crystal.

3.4.1 Grain size

Grain size significantly influences the mechanical properties of metals. Large grain size is generally associated with low strength, low hardness, and high ductility. Furthermore, large grains produce a rough surface appearance after being stretched (as in sheet metals) or compressed (as in forging and other metalworking processes). (See Section 3.6.) The yield strength, Y, is the most sensitive property and is related to grain size by the empirical formula (known as the *Hall-Petch equation*)

$$Y = Y_i + kd^{-1/2} \tag{3.8}$$

where Y_i is a basic yield stress (which can be regarded as the stress opposing the motion of dislocations), k is a constant indicating the extent to which dislocations

TABLE 3.1

Grain Sizes						
ASTM No.	−3	0	3	5	7	9	12
Grains/mm^2	1	8	64	256	1,024	4,096	32,800
Grains/mm^3	0.7	16	360	2,900	23,000	185,000	4,200,000

are piled up at barriers (such as grain boundaries), and d is the grain diameter. Equation (3.8) is valid below the recrystallization temperature of the material.

Grain size (Table 3.1) is usually measured by counting the number of grains in a given area or the number of grains that intersect a given length of a line, randomly drawn on an enlarged photograph (taken under a microscope) of the grains on a polished and etched specimen. Grain size may also be determined by referring to a standard chart. The ASTM [American Society for Testing and Materials] grain-size number, n, is related to the number of grains, N, per square inch at a magnification of 100 $\times$ (equal to 0.0645 mm^2 of actual area) by the expression

$$N = 2^{n-1}. \tag{3.9}$$

Grains of sizes between 5 and 8 are generally considered fine grains. A grain size of 7 is generally acceptable for sheet metals used to make car bodies, appliances, and kitchen utensils. Grains can also be large enough to be visible to the naked eye, such as grains of zinc on the surface of galvanized sheet steel.

3.4.2 Influence of grain boundaries

Grain boundaries have an important influence on the strength and ductility of metals. Furthermore, because they interfere with the movement of dislocations, grain boundaries also influence strain hardening. The magnitude of these effects depends on temperature, rate of deformation, and the type and amount of impurities present along the grain boundaries. Grain boundaries are more reactive (that is, they will form chemical bonds more readily) than the grains themselves, because the atoms along the grain boundaries are packed less efficiently and are more disordered than the atoms in the orderly arrangement within the grains. For this reason, corrosion is most prevalent at grain boundaries.

At elevated temperatures and in materials whose properties depend on the rate of deformation, plastic deformation also takes place by means of **grain-boundary sliding.** The **creep mechanism** (elongation under stress over a period of time, usually at elevated temperatures; see Section 2.8) results from grain-boundary sliding.

When brought into close atomic contact with certain low-melting-point metals, a normally ductile and strong metal can crack under very low stresses, a phenomenon referred to as **grain-boundary embrittlement** (Fig. 3.14). Examples include aluminum wetted with a mercury-zinc amalgam, and liquid gallium and copper at elevated temperature wetted with lead or bismuth; these elements weaken the grain boundaries of the metal by embrittlement. The term **liquid-metal embrittlement** is used to describe such phenomena, because the embrittling element is in a liquid state. However, embrittlement can also occur at temperatures well below the melting point of the embrittling element, a phenomenon known as **solid-metal embrittlement.**

Hot shortness is caused by local melting of a constituent or an impurity in the grain boundary at a temperature below the melting point of the metal itself. When such a metal is subjected to plastic deformation at elevated temperatures (*hot working;* see Section 3.7), the piece of metal crumbles and disintegrates along its grain

FIGURE 3.14
Embrittlement of copper by lead and bismuth at 350°C (660°F). Embrittlement has important effects on the strength, ductility, and toughness of materials. *Source:* After W. Rostoker.

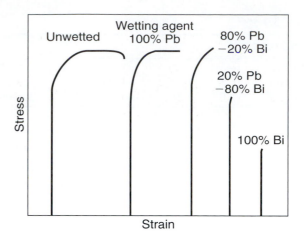

boundaries; examples include antimony in copper, and leaded steels and brass. To avoid hot shortness, the metal is usually worked at a lower temperature, which prevents softening and melting along the grain boundaries. Another form of embrittlement is **temper embrittlement** in alloy steels, which is caused by segregation (movement) of impurities to the grain boundaries.

3.5 | Plastic Deformation of Polycrystalline Metals

If a piece of polycrystalline metal with uniform equiaxed grains (i.e., having equal dimensions in all directions, as shown in the model in Fig. 3.15a) is subjected to plastic deformation at room temperature (*cold working*), the grains become permanently deformed and elongated. The deformation process may be carried out either by compressing the metal (as in forging; Section 6.2) or by subjecting it to tension (as in stretching sheet metal; Chapter 7). The deformation within each grain takes place by the mechanisms described in Section 3.3 for a single crystal.

During plastic deformation, the grain boundaries remain intact and mass continuity is maintained. The deformed metal exhibits higher strength because of the entanglement of dislocations with grain boundaries during deformation. The increase in strength depends on the amount of deformation (strain) to which the metal is subjected; the greater the deformation, the stronger the metal becomes. Because they have a larger grain-boundary surface area per unit volume of metal, the increase in strength is greater for metals with smaller grains.

Anisotropy (texture). As a result of plastic deformation, the grains in a piece of metal become elongated in one direction and contracted in the other (Fig. 3.15b). The

FIGURE 3.15 Plastic deformation of idealized (equiaxed) grains in a specimen subjected to compression, such as is done in rolling or forging of metals: (a) before deformation and (b) after deformation. Note the alignment of grain boundaries along a horizontal direction.

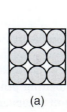

(a)

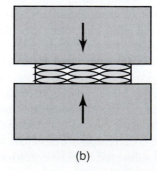

(b)

Top view

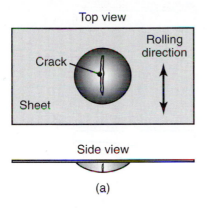

Rolling direction

Crack

Sheet

Side view

(a)

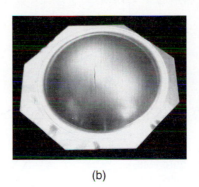

(b)

FIGURE 3.16 (a) Illustration of a crack in sheet metal subjected to bulging, such as by pushing a steel ball against the sheet. Note the orientation of the crack with respect to the rolling direction of the sheet. This material is anisotropic. (b) Aluminum sheet with a crack (vertical dark line at the center) developed in a bulge test. *Source:* Courtesy of J.S. Kallend, Illinois Institute of Technology.

behavior of the metal has become **anisotropic,** whereby its properties in the vertical direction are now different from those in the horizontal direction. The degree of anisotropy depends on how uniformly the metal is deformed. Note from the direction of the crack in Fig. 3.16, for example, that the ductility of the cold-rolled sheet in the vertical (transverse) direction is lower than in its longitudinal direction.

Anisotropy influences both the mechanical as well as the physical properties of metals. For example, sheet steel for electrical transformers is rolled in such a way that the resulting deformation imparts anisotropic magnetic properties to the sheet, thus reducing magnetic-hysteresis losses and improving the efficiency of transformers. (See also the discussion of *amorphous alloys* in Section 3.11.9.) There are two general types of anisotropy in metals:

a. **Preferred orientation.** Also called **crystallographic anisotropy,** *preferred orientation* can best be described by referring to Fig. 3.5. Note that when a metal crystal is subjected to tension, the sliding blocks rotate toward the direction of pulling; thus, slip planes and slip bands tend to align themselves with the direction of deformation. Similarly, for a polycrystalline aggregate, with grains in various orientations (Fig. 3.15), all slip directions tend to align themselves with the direction of pulling. Conversely, under compression, the slip planes tend to align themselves in a direction perpendicular to the direction of compression.

b. **Mechanical fibering.** Mechanical fibering results from the alignment of impurities, inclusions (stringers), and voids in the metal during deformation. Note that if the spherical grains in Fig. 3.15 were coated with impurities, these impurities would align themselves generally in a horizontal direction after deformation. Since impurities weaken the grain boundaries, this piece of metal would be weak and less ductile when tested in the vertical direction than in the horizontal direction. An analogy to this case would be plywood, which is strong in tension along its planar directions but peels off easily when tested in tension in its thickness direction.

3.6 | Recovery, Recrystallization, and Grain Growth

We have shown that plastic deformation at room temperature results in (a) the deformation of grains and grain boundaries, (b) a general increase in strength, and (c) a decrease in ductility. Also, plastic deformation can cause anisotropic behavior.

FIGURE 3.17 Schematic illustration of the effects of recovery, recrystallization, and grain growth on mechanical properties and shape and size of grains. Note the formation of small new grains during recrystallization. *Source:* After G. Sachs.

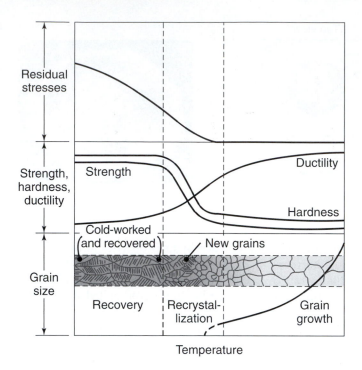

These effects can be reversed and the properties of the metal brought back to their original levels by heating the metal to within a specific temperature range for a period of time. The temperature range and the time required depend on the material and various other factors. Three events take place consecutively during the heating process:

1. **Recovery.** During recovery, which occurs at a certain temperature range below the **recrystallization temperature** of the metal (see below), the stresses in the highly deformed regions are relieved and the number of mobile dislocations is reduced. Subgrain boundaries begin to form (called **polygonization**) with no appreciable change in mechanical properties, such as hardness and strength, but with some increase in ductility (Fig. 3.17).

2. **Recrystallization.** The process in which, at a certain temperature range, new equiaxed and strain-free grains are formed, replacing the older grains, is called *recrystallization*. The temperature for recrystallization ranges approximately between $0.3T_m$ and $0.5T_m$, where T_m is the melting point of the metal on the absolute scale. The recrystallization temperature is generally defined as the temperature at which complete recrystallization occurs within approximately one hour. Recrystallization decreases the density of dislocations and lowers the strength of the metal but raises its ductility (Fig. 3.17). Metals such as lead, tin, cadmium, and zinc recrystallize at about room temperature.

   Recrystallization depends on the degree of prior cold work (work hardening); the higher the amount of cold work, the lower the temperature required for recrystallization to occur. The reason for this inverse relationship is that as the amount of cold work increases, the number of dislocations and the amount of energy stored in the dislocations (*stored energy*) also increase. The stored energy supplies some of the energy required for recrystallization. Recrystallization is a function of time, because it involves *diffusion*, that is, movement and exchange of atoms across grain boundaries.

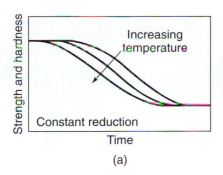

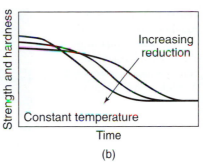

FIGURE 3.18 Variation of strength and hardness with recrystallization temperature, time, and prior cold work. Note that the more a metal is cold worked, the less time it takes to recrystallize, because of the higher stored energy from cold working due to increased dislocation density.

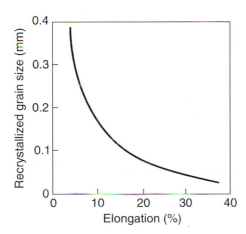

FIGURE 3.19 The effect of prior cold work on the recrystallized grain size of alpha brass. Below a critical elongation (strain), typically 5%, no recrystallization occurs.

The effects on recrystallization of temperature, time, and reduction in the thickness or height of the workpiece by cold working can be summarized as follows (Fig. 3.18):

- For a constant amount of deformation by cold working, the time required for recrystallization decreases with increasing temperature.

- The more prior cold work done, the lower the temperature required for recrystallization.

- The higher the amount of deformation, the smaller the resulting grain size during recrystallization (Fig. 3.19). Subjecting a metal to deformation is a common method of converting a coarse-grained structure to one of fine grain, with improved properties.

- Anisotropy due to preferred orientation usually persists after recrystallization. To restore isotropy, a temperature higher than that required for recrystallization may be necessary.

3. **Grain growth.** If we continue to raise the temperature of the metal, the grains begin to grow, and their size may eventually exceed the original grain size. This phenomenon is known as *grain growth,* and it has a slightly adverse effect on mechanical properties (Fig. 3.17). More importantly, however, large grains produce the **orange-peel effect,** resulting in a rough surface appearance, such as when sheet metal is stretched to form a part or when a piece of metal is subjected to compression (Fig. 3.20), such as in forging operations.

FIGURE 3.20 Surface roughness on the cylindrical surface of an aluminum specimen subjected to compression. *Source:* A. Mulc and S. Kalpakjian.

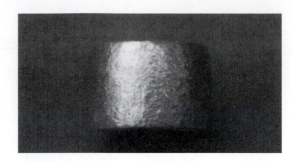

TABLE 3.2

Homologous Temperature Ranges for Various Processes	
Process	T/T_m
Cold working	<0.3
Warm working	0.3 to 0.5
Hot working	>0.6

3.7 | Cold, Warm, and Hot Working

When plastic deformation is carried out above the recrystallization temperature of the metal, it is called **hot working,** and if it is done below its recrystallization temperature, it is called **cold working.** As the name implies, **warm working** is carried out at an intermediate temperature; thus, warm working is a compromise between cold and hot working. The temperature ranges for these three categories of plastic deformation are given in Table 3.2 in terms of a ratio of T to T_m, where T is the working temperature and T_m is the melting point of the metal, both on the absolute scale. Although it is a dimensionless quantity, this ratio is known as the **homologous temperature.**

There are important technological differences in products that are processed by cold, warm, or hot working. For example, compared with cold-worked products, hot-worked products generally have (a) less dimensional accuracy, because of uneven thermal expansion and contraction during processing, and (b) a rougher surface appearance and finish, because of the oxide layer that usually develops during heating. Other important manufacturing characteristics, such as formability, machinability, and weldability, are also affected by cold, warm, and hot working to different degrees.

3.8 | Failure and Fracture

Failure is one of the most important aspects of a material's behavior because it directly influences the selection of a material for a particular application, the methods of manufacturing, and the service life of the component. Because of the many factors involved, failure and fracture of materials are complex areas of study. In this section, we consider only those aspects of failure that are of particular significance to the selection and processing of materials.

There are two general types of failure: (1) **fracture** and separation of the material, through either internal or external cracking, and (2) **buckling** (Fig. 3.21). Fracture is further divided into two general categories: ductile and brittle (Fig. 3.22).

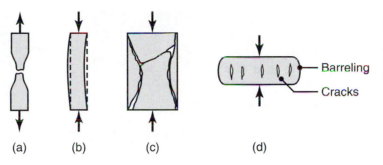

FIGURE 3.21 Schematic illustration of types of failure in materials: (a) necking and fracture of ductile materials; (b) buckling of ductile materials under a compressive load; (c) fracture of brittle materials in compression; (d) cracking on the barreled surface of ductile materials in compression. (See also Fig. 6.1b.)

It should be noted that although failure of materials is generally regarded as undesirable, certain products are indeed designed to fail. Typical examples include (a) beverage cans with tabs or entire tops of food cans that are removed by tearing the sheet metal along a prescribed path, and (b) metal or plastic screw caps for bottles.

3.8.1 Ductile fracture

Ductile fracture is characterized by plastic deformation which precedes failure of the part. In a tension test, for example, highly ductile materials such as gold and lead may neck down to a point and then fail (Fig. 3.22d). Most metals and alloys, however, neck down to a finite area and then fail. Ductile fracture generally takes place along planes on which the *shear stress is a maximum*. In torsion, for example, a ductile metal fractures along a plane perpendicular to the axis of twist, that is, the plane on which the shear stress is a maximum. Fracture in shear is a result of extensive slip along slip planes within the grains.

Upon close examination, the surface in ductile fracture (Fig. 3.23) shows a *fibrous* pattern with *dimples,* as if a number of very small tension tests have been carried out over the fracture surface. Failure is initiated with the formation of tiny voids (usually originating eventually around small inclusions or preexisting voids) which then grow and coalesce, resulting in cracks that grow in size and lead to fracture. In a tension-test specimen, fracture begins at the center of the necked region from the growth and coalescence of cavities (Fig. 3.24). The central region becomes one large crack which then propagates to the periphery of this necked region. Because of its appearance, this type of facture is called a **cup-and-cone fracture**.

Effects of inclusions. Because they are nucleation sites for voids, *inclusions* have an important influence on ductile fracture and therefore on the formability of materials. Inclusions may consist of impurities of various kinds and second-phase particles such as oxides, carbides, and sulfides. The extent of their influence depends on

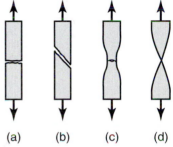

FIGURE 3.22 Schematic illustration of the types of fracture in tension: (a) brittle fracture in polycrystalline metals; (b) shear fracture in ductile single crystals (see also Fig. 3.5a); (c) ductile cup-and-cone fracture in polycrystalline metals (see also Fig. 2.2); (d) complete ductile fracture in polycrystalline metals, with 100% reduction of area.

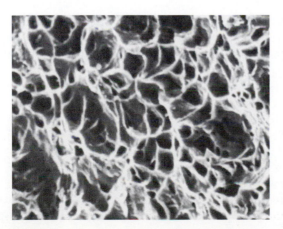

FIGURE 3.23 Surface of ductile fracture in low-carbon steel, showing dimples. Fracture is usually initiated at impurities, inclusions, or preexisting voids in the metal. *Source:* K.-H. Habig and D. Klaffke. Photo courtesy of BAM, Berlin, Germany.

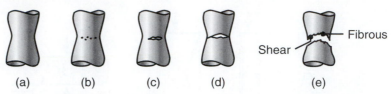

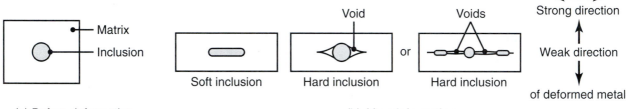

FIGURE 3.24 Sequence of events in necking and fracture of a tensile-test specimen: (a) early stage of necking; (b) small voids begin to form within the necked region; (c) voids coalesce, producing an internal crack; (d) rest of cross section begins to fail at the periphery by shearing; (e) final fracture surfaces, known as cup-(top fracture surface) and-cone (bottom surface) fracture.

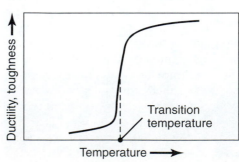

FIGURE 3.25 Schematic illustration of the deformation of soft and hard inclusions and their effect on void formation in plastic deformation. Note that hard inclusions, because they do not comply with the overall deformation of the ductile matrix, can cause voids.

factors such as their shape, hardness, distribution, and volume fraction. The greater the volume fraction of inclusions, the lower will be the ductility of the material. Voids and porosity developed during processing, such as from casting (see Section 5.12) and some metalworking operations, reduce the ductility of a material.

Two factors affect void formation:

a. the strength of the bond at the interface of an inclusion and the matrix. If the bond is strong, there is less tendency for void formation during plastic deformation.

b. The hardness of the inclusion. If the inclusion is soft, such as manganese sulfide, it will conform to the overall change in shape of the specimen or workpiece during plastic deformation. If it is hard, such as a carbide or oxide, it could lead to void formation (Fig. 3.25). Hard inclusions may also break up into smaller particles during deformation, because of their brittle nature.

The alignment of inclusions during plastic deformation leads to **mechanical fibering**. Subsequent processing of such a material must, therefore, involve considerations of the proper direction of working for maximum ductility and strength.

Transition temperature. Metals generally undergo a sharp change in ductility and toughness across a narrow temperature range called the *transition temperature* (Fig. 3.26). This phenomenon typically occurs in body-centered cubic and some

FIGURE 3.26 Schematic illustration of transition temperature. Note the narrow temperature range across which the behavior of the metal undergoes a major transition.

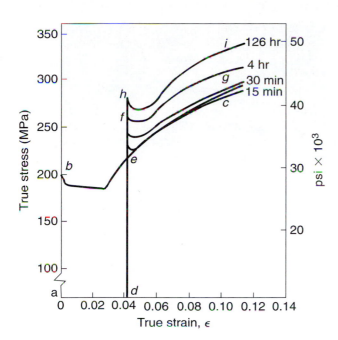

FIGURE 3.27 Strain aging and its effect on the shape of the true stress–true strain curve for 0.03% C rimmed steel at 60°C (140°F). *Source:* A.S. Keh and W.C. Leslie.

hexagonal close-packed metals; it is rarely exhibited by face-centered cubic metals. The transition temperature depends on factors such as composition, microstructure, grain size, surface finish and shape of the specimen, and rate of deformation. High rates, abrupt changes in shape, and surface notches raise the transition temperature.

Strain aging. *Strain aging* is a phenomenon in which carbon atoms in steels segregate to dislocations, thereby pinning them and thus increasing the steel's resistance to dislocation movement; the result is increased strength and reduced ductility. The effects of strain aging on the shape of the stress–strain curve in tension for low-carbon steel at room temperature are shown in Fig. 3.27. Curve *abc* is the original curve, with upper and lower yield points typical of these steels. If the tension test is stopped at point *e* and the specimen is unloaded and tested again, curve *dec* is obtained. Note that the upper and lower yield points have disappeared. However, if four hours pass before the specimen is strained again, curve *dfg* is obtained; if 126 hours pass, curve *dhi* is obtained. Instead of taking place over several days at room temperature, strain aging can occur in just a few hours at a higher temperature; it is then called **accelerated strain aging.** For steels, the phenomenon is also called **blue brittleness** because of the color of the steel.

3.8.2 Brittle fracture

Brittle fracture occurs with little or no gross plastic deformation preceding the separation of the material into two or more pieces. Figure 3.28 shows a typical example of the surface of brittle fracture. In tension, brittle fracture takes place along a crystallographic plane (called a **cleavage plane**) on which the normal tensile stress is a maximum. Body-centered cubic and some hexagonal close-packed metals fracture by cleavage, whereas face-centered cubic metals usually do not fail in brittle fracture. In general, low temperature and high rates of deformation promote brittle fracture.

In a polycrystalline metal under tension, the fracture surface has a bright granular appearance because of the changes in the direction of the cleavage planes as the crack propagates from one grain to another. Brittle fracture of a specimen in compression is more complex and in theory follows a path that is 45° to the direction of the applied force.

FIGURE 3.28 Typical fracture surface of steel that has failed in a brittle manner. The fracture path is transgranular (through the grains). Compare this surface with the ductile fracture surface shown in Fig. 3.23. Magnification: 200×. *Source:* Courtesy of Packer Engineering.

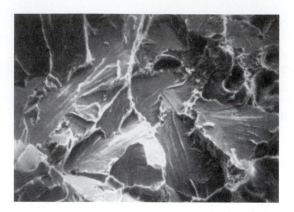

Examples of fracture along a cleavage plane include the splitting of rock salt and the peeling of layers of mica, where tensile stresses normal to the cleavage plane, caused by pulling, initiate and control the propagation of fracture through the material. Another example is the behavior of brittle materials, such as chalk, gray cast iron, and concrete, which under tension typically fail in the manner shown in Fig. 3.21a. In torsion, they fail along a plane at 45° to the axis of twist, that is, along a plane on which the tensile stress is a maximum.

Defects. An important factor in fracture is the presence of **defects** such as scratches, flaws, and external or internal cracks. Under tension, the tip of a crack is subjected to high tensile stresses which propagate the crack rapidly, because the brittle material has little capacity to dissipate energy. It can be shown that the tensile strength of a specimen with a crack perpendicular to the direction of pulling is related to the length of the crack as follows:

$$\sigma \propto \frac{1}{\sqrt{\text{Crack length}}}. \tag{3.10}$$

The presence of defects is essential in explaining why brittle materials are so weak in tension compared with their strength in compression. Under tensile stresses, cracks propagate rapidly, causing what is known as *catastrophic failure*. With polycrystalline metals, the fracture paths most commonly observed are **transgranular** (*transcrystalline,* or *intragranular*), meaning that the crack propagates *through* the grain. **Intergranular** fracture, where the crack propagates along the grain boundaries (Fig. 3.29), generally occurs when the grain boundaries (a) are soft, (b) contain a brittle phase, or (c) have been weakened by liquid- or solid-metal embrittlement (Section 3.4.2).

FIGURE 3.29 Intergranular fracture, at two different magnifications. Grains and grain boundaries are clearly visible in this micrograph. The fracture path is along the grain boundaries. Magnification: left, 100×; right, 500×. *Source:* Courtesy of Packer Engineering.

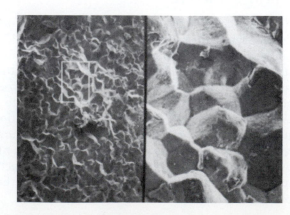

The maximum crack velocity in a brittle material is about 62% of the elastic wave-propagation (or acoustic) velocity of the material; this velocity is given by the formula $\sqrt{E/\rho}$, where E is the elastic modulus and ρ is the mass density. Thus, for steel, the maximum crack velocity is 2000 m/s (6600 ft/s).

As shown in Fig. 3.30, cracks may be subjected to stresses in different directions. Mode I is tensile stress applied perpendicular to the crack. Modes II and III are shear stresses applied in two different directions. Tearing paper, cutting sheet metal with shears, and opening pop-top cans are examples of Mode III fracture.

Fatigue fracture. In this type of fracture, minute external or internal cracks develop at flaws or defects in the material. The cracks then propagate throughout the body of the part and eventually lead to its total failure. The fracture surface in fatigue is generally characterized by the term **beach marks,** because of its appearance, as can be seen in Fig. 3.31. Under large magnification (higher than 1000×), a series of **striations** can be observed on fracture surfaces, where each beach mark consists of several striations.

Because of its sensitivity to surface defects, the fatigue life of a specimen or a part is greatly influenced by the method of preparation of its surfaces (Fig. 3.32).

The fatigue strength of manufactured products can generally be improved by the following methods:

1. Inducing compressive residual stresses on surfaces, such as by shot peening or roller burnishing (see Section 4.5.1).

2. Surface (case) hardening by various means of heat treatment (see Section 5.11.3).

3. Providing a fine surface finish on the part, thereby reducing the effects of notches and other surface imperfections.

4. Selecting appropriate materials and ensuring that they are free from significant amounts of inclusions, voids, and impurities.

Conversely, the following factors and processes can reduce fatigue strength: decarburization, surface pits due to corrosion that act as stress raisers, hydrogen embrittlement (see below), galvanizing, and electroplating (Section 4.5.1). Two of these phenomena are described below.

Stress-corrosion cracking. An otherwise ductile metal can fail in a brittle manner by stress-corrosion cracking (also called *stress cracking* or *season cracking*). After being formed, the parts may, either over a period of time or soon after the parts are made, develop cracks. Crack propagation may be intergranular or transgranular.

The susceptibility of metals to stress-corrosion cracking mainly depends on (a) the material, (b) the presence and magnitude of tensile residual stresses, and

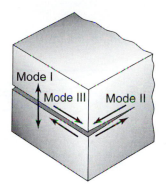

FIGURE 3.30 Three modes of fracture. Mode I has been studied extensively, because it is the most commonly observed in engineering structures and components. Mode II is rare. Mode III is the tearing process; examples include opening a pop-top can, tearing a piece of paper, and cutting materials with a pair of scissors.

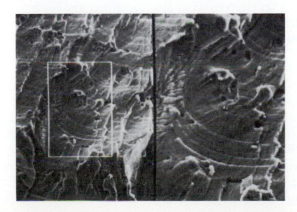

FIGURE 3.31 Typical fatigue fracture surface on metals, showing beach marks. Most components in machines and engines fail by fatigue and not by excessive static loading. Magnification: left, 500×; right, 1000×. *Source:* Courtesy of Packer Engineering.

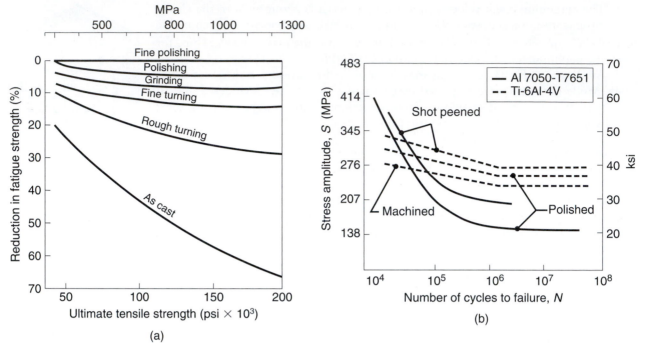

FIGURE 3.32 Reduction in fatigue strength of cast steels subjected to various surface-finishing operations. (a) Effect of surface roughness. Note that the reduction is greater as the surface roughness and strength of the steel increase. *Source:* After J.E. Shigley and L.D. Mitchell. (b) Effect of residual stress, as developed by shot peening (see Section 4.5.1). *Source:* After B.J. Hamrock, S.R. Schmid, and B.O. Jacobson.

(c) the environment. Brass and austenitic stainless steels are among the metals that are highly susceptible to stress cracking. Environmental factors such as salt water or other chemicals that could be corrosive to metals can increase stress-corrosion cracking. The usual procedure to avoid stress-corrosion cracking is to stress relieve (Section 5.11.4) the part just after it is formed. Full annealing may also be done, but this treatment reduces the strength of cold-worked parts.

Hydrogen embrittlement. The presence of hydrogen can reduce ductility and cause severe embrittlement in many metals, alloys, and nonmetallic materials, leading to premature failure. Known as hydrogen embrittlement, this phenomenon is especially severe in high-strength steels. Possible sources of hydrogen are (a) during melting of the metal; (b) during pickling, that is, removal of surface oxides by chemical or electrochemical reaction; through (c) electrolysis in electroplating; and (d) from water vapor in the atmosphere or from moist electrodes and fluxes used during welding. Oxygen also can cause embrittlement in metals, especially in copper alloys.

3.8.3 Size effect

The dependence of the properties of a material on its size is knows as *size effect*. Note from the foregoing discussions and from Eq. (3.10) that defects, cracks, imperfections, and the like are less likely to be present in a part as its size decreases. Thus, the strength and ductility of a part increase with decreasing size.

Although strength is related to the cross-sectional area of a part, its length is also important because the greater the length, the greater is the probability for defects to exist. As a common analogy, a long chain is more likely to be weaker than a shorter chain, because the probability of one of the links being weak increases with the number of links, hence the length of chain. Size effect is also demonstrated by whiskers, which, because of their small size, are either free from imperfections or do not contain the types of imperfections that affect their strength. The actual strength in such cases approaches the theoretical strength of the material. (See also Sections 8.6.10 and 10.9.2 for applications of whiskers.)

EXAMPLE 3.1 **Brittle fracture of steel plates of the hull of the R.M.S.** *Titanic*

A detailed analysis of the *Titanic* disaster in 1912 has indicated that the ship sank not so much because it hit an iceberg as because of structural weaknesses. The plates composing its hull were made of low-grade steel with a high sulfur content; they had low toughness (as determined by the Charpy test; see Section 2.9c) when chilled, as was the case in the Atlantic Ocean, and when subjected to an external impact loading. With such a material, a crack that starts in one part of a welded steel structure can propagate rapidly and completely around the hull and cause a large ship to split in two. Although the *Titanic* was built with brittle plates, as we know from the physical and photographic observations of the sunken ship, not all ships of that time were built with such low-grade steel. Furthermore, better construction techniques could have been employed, among them better welding techniques (Chapter 12) to improve the structural strength of the hull.

3.9 | Physical Properties

In addition to their mechanical properties, the **physical properties** of materials must also be considered in their selection and processing. Properties of particular interest in manufacturing are density, melting point, specific heat, thermal conductivity and expansion, electrical and magnetic properties, and resistance to oxidation and corrosion, as described next.

3.9.1 Density

The density of a metal depends on its atomic weight, atomic radius, and the packing of the atoms. Alloying elements generally have a minor effect on density, depending on the density of the alloying elements. The ranges of densities for a variety of materials at room temperature are given in Table 3.3.

Weight reduction is particularly important for aircraft and aerospace structures, automotive bodies and components, and other products for which energy consumption and power limitations are major concerns. Consequently, **specific strength** (strength-to-weight ratio) and **specific stiffness** (stiffness-to-weight ratio) of materials and structures are important considerations. (See also Section 10.9.1.) Substitution of materials for saving weight and economy is a major factor in advanced equipment and machinery as well as in consumer products such as automobiles.

Density also is an important factor in the selection of materials for high-speed equipment, such as the use of magnesium in printing and textile machinery, many components of which usually operate at very high speeds. To obtain an exposure time of 1/4000 s in cameras without sacrificing accuracy, the shutters of some high-quality 35-mm cameras are made of titanium. The resulting light weight of the components in these high-speed operations reduces inertial forces that otherwise could lead to vibrations, inaccuracies, and even part failure over time. On the other hand, there are also applications for which higher densities are desirable. Examples include

TABLE 3.3

Physical Properties of Various Materials at Room Temperature

	Density (kg/m³)	Melting point (°C)	Specific heat (J/kg K)	Thermal conductivity (W/m K)	Coefficient of thermal expansion (μm/m°C)
METAL					
Aluminum	2700	660	900	222	23.6
Aluminum alloys	2630–2820	476–654	880–920	121–239	23.0–23.6
Beryllium	1854	1278	1884	146	8.5
Copper	8970	1082	385	393	16.5
Copper alloys	7470–8940	885–1260	337–435	29–234	16.5–20
Gold	19,300	1063	129	317	19.3
Iron	7860	1537	460	74	11.5
Steels	6920–9130	1371–1532	448–502	15–52	11.7–17.3
Lead	11,350	327	130	35	29.4
Lead alloys	8850–11,350	182–326	126–188	24–46	27.1–31.1
Magnesium	1745	650	1025	154	26.0
Magnesium alloys	1770–1780	610–621	1046	75–138	26.0
Molybdenum alloys	10,210	2610	276	142	5.1
Nickel	8910	1453	440	92	13.3
Nickel alloys	7750–8850	1110–1454	381–544	12–63	12.7–18.4
Niobium (Columbium)	8580	2468	272	52	7.1
Silicon	2330	1423	712	148	7.63
Silver	10,500	961	235	429	19.3
Tantalum alloys	16,600	2996	142	54	6.5
Titanium	4510	1668	519	17	8.35
Titanium alloys	4430–4700	1549–1649	502–544	8–12	8.1–9.5
Tungsten	19,290	3410	138	166	4.5
NONMETALLIC					
Ceramics	2300–5500	—	750–950	10–17	5.5–13.5
Glasses	2400–2700	580–1540	500–850	0.6–1.7	4.6–70
Graphite	1900–2200	—	840	5–10	7.86
Plastics	900–2000	110–330	1000–2000	0.1–0.4	72–200
Wood	400–700	—	2400–2800	0.1–0.4	2–60

counterweights for various mechanisms (using lead and steel), flywheels, and components for self-winding watches (using high-density materials such as tungsten).

3.9.2 Melting point

The melting point of a metal depends on the energy required to separate its atoms. It can be seen in Table 3.3 that unlike pure metals, which have a definite melting point, the melting point of an alloy involves a wide range of temperatures, depending on the alloying elements. (See Section 5.2.) Since the recrystallization temperature of a metal is related to its melting point (Section 3.6), operations such as annealing, heat treating, and hot working require a knowledge of the melting points of the metals involved. These considerations, in turn, influence the selection of tool and die materials in manufacturing operations.

Another major influence of the melting point is in the selection of the equipment and melting practice in casting operations; the higher the melting point of the material, the more difficult the operation becomes (Section 5.5). Also, the selection of die materials, as in die casting, depends on the melting point of the workpiece material. In the electrical-discharge machining process (Section 9.13), the melting points of metals are related to the rate of material removal and tool wear.

3.9.3 Specific heat

Specific heat is the energy required to raise the temperature of a unit mass of material by one degree. Alloying elements have a relatively minor effect on the specific heat of metals. The temperature rise in a workpiece, such as resulting from forming or machining operations, is a function of the work done and the specific heat of the workpiece material (see Section 2.12.1); thus, the lower the specific heat, the higher the temperature rise in the material. If excessively high, temperature can have detrimental effects on product quality by (a) adversely affecting surface finish and dimensional accuracy, (b) causing excessive tool and die wear, and (c) resulting in adverse metallurgical changes in the material.

3.9.4 Thermal conductivity

Thermal conductivity indicates the rate at which heat flows within and through the material. Metallically bonded materials (metals) generally have high thermal conductivity, whereas ionically or covalently bonded materials (such as ceramics and plastics) have poor conductivity. Because of the large difference in their thermal conductivities, alloying elements can have a significant effect on the thermal conductivity of alloys, as can be seen in Table 3.3 by comparing the metals with their alloys.

When heat is generated by means such as plastic deformation or friction, the heat should be conducted away at a sufficiently high rate to prevent a severe rise in temperature, which can result in high thermal gradients and thus cause inhomogeneous deformation in metalworking processes. The main difficulty in machining titanium, for example, is caused by its very low thermal conductivity (Section 8.5.2).

3.9.5 Thermal expansion

Thermal expansion of materials can have several significant effects. Generally, the coefficient of thermal expansion is inversely proportional to the melting point of the material; alloying elements have a relatively minor effect on the thermal expansion of metals. Examples where relative expansion or contraction is important include electronic and computer components, glass-to-metal seals, metal-to-ceramic subassemblies (see also Section 11.8.2), struts on jet engines, and moving parts in machinery that require certain clearances for proper functioning. Shrink fits utilize thermal expansion; usually a hub is heated and fit over a shaft; when the hub cools it clamps against the shaft.

Thermal stresses result from relative expansion and contraction of components or within the material itself, leading to *cracking, warping,* or *loosening* of components in the structure during their service life. Ceramic parts and tools and dies made of relatively brittle materials are particularly sensitive to thermal stresses. Thermal conductivity, in conjunction with thermal expansion, plays the most significant role in causing thermal stresses, both in manufactured components and in tools and dies. To reduce thermal stresses, a combination of high thermal conductivity and low thermal expansion is desirable, which together reduce the temperature gradient within a body. Thermal stresses may also be caused by **anisotropy of thermal expansion** of the material, which is generally observed in hexagonal close-packed metals and in ceramics.

Thermal fatigue results from thermal cycling and causes a number of surface cracks. This phenomenon is particularly important, for example, (a) in a forging operation, when hot workpieces are placed over relatively cool dies, thus subjecting the die surfaces to thermal cycling; and (b) in interrupted cutting operations such as milling. **Thermal shock** is the term generally used to describe development of cracks after a single thermal cycle.

The influence of temperature on dimensions and the modulus of elasticity are significant in precision instruments and equipment. A spring, for example, will have a lower stiffness as its temperature increases because of reduced elastic modulus. Similarly, a tuning fork or a pendulum will have different frequencies at different temperatures. To alleviate some of the problems of thermal expansion, a family of iron-nickel alloys that have very low thermal-expansion coefficients are available and are known as **low-expansion alloys.** Consequently, these alloys also have good thermal-fatigue resistance. Typical compositions are 64% Fe-36% Ni (Invar) and 54% Fe-28% Ni-18% Co (Kovar).

3.9.6 Electrical and magnetic properties

Electrical conductivity and dielectric properties of materials are of great importance not only in electrical equipment and machinery, but also in manufacturing processes such as magnetic-pulse forming of sheet metals (Section 7.5.5) and electrical-discharge machining and electrochemical grinding of hard and brittle materials (Chapter 9).

The **electrical conductivity** of a material can be defined as a measure of how well the material conducts electric current. The units of electrical conductivity are mho/m or mho/ft, where mho is the inverse of ohm (the unit for electrical resistance). Materials with high conductivity, such as metals, are generally referred to as **conductors.** The influence of the type of atomic bonding on the electrical conductivity of materials is the same as that for thermal conductivity. Alloying elements have a major effect on the electrical conductivity of metals: The higher the conductivity of the alloying element, the higher the conductivity of the alloy.

Electrical resistivity is the inverse of conductivity, and materials with high resistivity are referred to as **dielectrics,** or **insulators. Dielectric strength** of materials is the resistivity to direct electric current and is defined as the voltage required per unit distance for electrical breakdown; the unit is V/m or V/ft.

Superconductivity is the phenomenon of almost zero electrical resistivity that occurs in some metals and alloys below a critical temperature. The highest temperature at which superconductivity has to date been exhibited, at about −123°C (−190°F), is with an alloy of lanthanum, strontium, copper, and oxygen, although other material compositions are continuously being investigated. Developments in superconductivity indicate that the efficiency of electrical components such as large high-power magnets, high-voltage power lines, and various other electronic and computer components can be markedly improved.

The electrical properties of materials such as single-crystal silicon, germanium, and gallium arsenide make them extremely sensitive to the presence and type of minute impurities as well as to temperature. Thus, by controlling the concentration and type of impurities (**dopants**), such as phosphorus and boron in silicon, electrical conductivity can be controlled as well. This property is exploited in **semiconductor** (solid-state) devices used extensively in miniaturized electronic circuitry as described in Chapter 13. Such devices are (a) very compact, (b) efficient, (c) relatively inexpensive, (d) consume little power, and (e) require no warm-up time for operation.

Ferromagnetism is the large and permanent magnetization resulting from an exchange interaction between neighboring atoms (such as iron, nickel, and cobalt) that aligns their magnetic moments in parallel. It has important applications in electric motors, generators, transformers, and microwave devices. **Ferrimagnetism** is the permanent and large magnetization exhibited by some ceramic materials, such as cubic ferrites.

The **piezoelectric** effect (*piezo* from Greek, meaning "to press"), in which there is a reversible interaction between an elastic strain and an electric field, is exhibited by some materials, such as certain ceramics and quartz crystals. This property is utilized in making *transducers*, which are devices that convert the strain from an external force to electrical energy. Typical applications of the piezoelectric effect include force or pressure transducers, strain gages, sonar detectors, and microphones.

Magnetostriction is the phenomenon of expansion and contraction of a material when subjected to a magnetic field. Pure nickel and some iron-nickel alloys, for example, exhibit this behavior. Magnetostriction is the principle behind ultrasonic machining equipment. (See Section 9.9.)

3.9.7 Resistance to corrosion

Corrosion is the deterioration of metals and ceramics, while **degradation** is a similar phenomenon in plastics. *Corrosion resistance* is an important aspect of material selection, especially for applications in the chemical, food, and petroleum industries. In addition to various possible chemical reactions from the elements and compounds present, environmental oxidation and corrosion of components and structures are also a major concern, particularly at elevated temperatures and in automobiles, aircraft, and other transportation equipment.

Corrosion resistance depends on the particular environment, as well as the composition of the material. Chemicals (acids, alkali, and salts), the environment (oxygen, pollution, and acid rain), and water (fresh or salt) may all act as corrosive media. Nonferrous metals, stainless steels, and nonmetallic materials generally have high corrosion resistance. Steels and cast irons generally have poor resistance and must therefore be protected by various means such as coatings and surface treatments (Section 4.5).

Corrosion can occur over an entire surface, or it can be localized, such as in **pitting.** It can occur along grain boundaries of metals as **intergranular corrosion** and at the interface of bolted or riveted joints as **crevice corrosion.** Two dissimilar metals may form a *galvanic cell* (two electrodes in an electrolyte in a corrosive environment, including moisture) and cause **galvanic corrosion.** Two-phase alloys (Section 5.2.3) are more susceptible to galvanic corrosion, because of the two different metals involved, than are single-phase alloys or pure metals. Thus, heat treatment can have a significant influence on corrosion resistance.

Corrosion may also act in indirect ways. **Stress-corrosion cracking** is an example of the effect of a corrosive environment on the integrity of a product that, as manufactured, contained residual stresses (Section 2.10). Likewise, metals that are cold worked are likely to contain residual stresses and thus are more susceptible to corrosion as compared with hot-worked or annealed metals.

Tool and die materials also can be susceptible to chemical attack by lubricants and coolants (Section 4.4.4). The chemical reaction alters their surface finish and adversely influences the metalworking operation. An example is tools and dies made of carbides that have cobalt as a binder (Section 8.6.4); the cobalt can be attacked by elements in the metalworking fluid (see Section 4.4.4 and 8.7), a process called **selective leaching.** Compatibility of the tool, die, and workpiece materials and the metalworking fluid is thus an important consideration in material selection.

It should be noted that chemical reactions should not always be regarded as having only adverse effects. Some advanced machining processes, such as chemical machining and electrochemical machining, are indeed based on controlled chemical reactions (Chapter 9).

Furthermore, the usefulness of some level of oxidation is exhibited by the corrosion resistance of metals such as aluminum, titanium, and stainless steel. Examples include the following:

a. Aluminum develops a thin (a few atomic layers), strong, and adherent hard oxide film that protects the surface from further environmental corrosion.

b. Titanium develops a film of titanium oxide.

c. Stainless steels, because of the chromium present in the alloy, develop a protective film on their surfaces, known as **passivation** (Section 4.2). Furthermore, when the protective film is scratched, thus exposing the metal underneath to the environment, a new oxide film forms in due time.

3.10 | General Properties and Applications of Ferrous Alloys

By virtue of their very wide range of mechanical, physical, and chemical properties, **ferrous alloys** are among the most useful of all metals. Ferrous metals and alloys contain iron as their base metal and are variously categorized as carbon and alloy steels, stainless steels, tool and die steels, cast irons, and cast steels. Ferrous alloys are produced as sheet steel for automobiles, appliances, and containers; as plates for ships, boilers, and bridges; as structural members (such as I-beams), bar products for leaf springs, gears, axles, crankshafts, and railroad rails; as stock for tools and dies; as music wire; and as fasteners such as bolts, rivets, and nuts.

A typical U.S. passenger car contains about 816 kg (1800 lb) of steel, accounting for about 55% of its weight. As an example of their widespread use, ferrous materials comprise 70 to 85% by weight of virtually all structural members and mechanical components. Carbon steels are the least expensive of all metals, but stainless steels can be costly.

3.10.1 Carbon and alloy steels

Carbon and alloy steels are among the most commonly used metals. The composition and processing of these steels are controlled in a manner that makes them suitable for a wide variety of applications. They are available in various basic product shapes: plate, sheet, strip, bar, wire, tube, castings, and forgings.

Several elements are added to steels to impart various properties such as hardenability, strength, hardness, toughness, wear resistance, workability, weldability, and machinability. Generally, the higher the percentages of these elements, the higher the particular properties that they impart to steels. Thus, for example, the higher the carbon content, the higher the hardenability of the steel and the higher its strength, hardness, and wear resistance. Conversely, ductility, weldability, and toughness are reduced with increasing carbon content.

Carbon steels. Carbon steels are generally classified as low, medium, and high (see Fig. 3.33).

1. **Low-carbon steel,** also called **mild steel,** has less than 0.30% C. It is generally used for common products such as bolts, nuts, sheet plates, tubes, and machine components that do not require high strength.

2. **Medium-carbon steel** has 0.30% to 0.60% C. It is generally used in applications requiring higher strength than those using low-carbon steels, such as machinery, automotive and agricultural equipment (gears, axles, connecting rods, crankshafts), railroad equipment, and metalworking machinery.

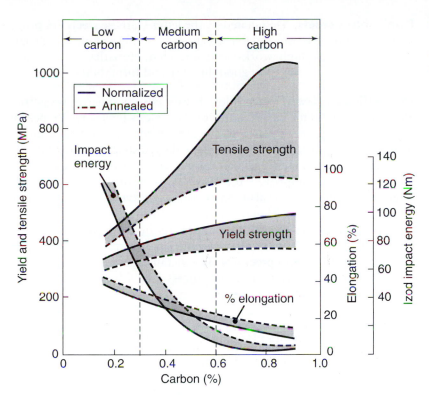

FIGURE 3.33 Effect of carbon content on the mechanical properties of carbon steel.

3. **High-carbon steel** has more than 0.60% C. It is generally used for parts requiring strength, hardness, and wear resistance; examples are springs, cutlery, cable, music wire, and rails. After being manufactured into appropriate shapes, the parts are usually heat treated and tempered. The higher the carbon content of the steel, the higher its hardness, strength, and wear resistance after heat treatment.

4. Carbon steels containing sulfur or phosphorus are known as **resulfurized** and **rephosphorized and resulfurized carbon steels,** with the major characteristic of improved machinability of the steels, as described in Section 8.5.1.

Alloy steels. Steels containing significant amounts of alloying elements are called *alloy steels*. **Structural-grade** alloy steels, as identified by ASTM specifications, are used mainly in the construction and transportation industries because of their high strength. Other types of alloy steels are also available for applications where strength, hardness, resistance to creep and fatigue, and toughness are required. These steels may also be heat treated to obtain the desired properties.

High-strength low-alloy steels. In order to improve the strength-to-weight ratio of steels, a number of *high-strength low-alloy* (HSLA) steels have been developed. These steels have a low carbon content (usually less than 0.30%) and are characterized by a microstructure consisting of fine-grain ferrite and a hard second phase of carbides, carbonitrides, or nitrides. First developed in the 1930s, HSLA steels are usually produced in sheet form by microalloying and controlled hot rolling. Plates, sheets, bars, and structural shapes are made from these steels. However, the ductility, formability, and weldability of HSLA steels are generally inferior to those of conventional low-alloy steels.

Sheet products of HSLA steels are typically used in automobile bodies to reduce weight (and thus fuel consumption), and in transportation, mining, and agricultural equipment. Plates made of HSLA steel are used in ships, bridges, and building construction, and shapes such as I-beams, channels, and angles are used in buildings and various other structures.

Dual-phase steels. Designated by the letter D, these steels are processed specially and have a mixed ferrite and martensite structure. Developed in the late 1960s, dual-phase steels have high work-hardening characteristics (i.e., a high *n* value; see Section 2.2.3) and thus have good ductility and formability.

Microalloyed steels. Microalloyed steels possess superior properties and thus can eliminate the need for heat treatment. These steels have a ferrite-pearlite microstructure with fine dispersed particles of carbonitride. A number of microalloyed steels have been produced, with a typical microalloyed steel containing 0.5% C, 0.8% Mn, and 0.1% V. When subjected to carefully controlled cooling (usually in air), these materials develop improved and uniform strength. Compared to medium-carbon steels, microalloyed steels also can provide cost savings of as much as 10%, since the additional steps of quenching, tempering, and stress relieving are not required.

Nano-alloyed steels are also under continued development. These steels have extremely small grain sizes (10–100 nm) and are produced using metallic glasses (see Section 5.10.8) as a precursor. The metallic glass is subjected to a carefully controlled vitrification (crystallization) process with a high nucleation rate, thus resulting in very fine nanoscale phases.

3.10.2 Stainless steels

Stainless steels are characterized primarily by their corrosion resistance, high strength and ductility, and high chromium content. They are called *stainless* because, in the presence of oxygen (air), they develop a thin, hard adherent film of *chromium oxide* that protects the metal from corrosion (*passivation;* see Section 4.2). This protective film builds up again if the surface is scratched. For passivation to occur, the minimum chromium content of the steel should be in the range of 10 to 12% by weight.

In addition to chromium, other alloying elements in stainless steels typically include nickel, molybdenum, copper, titanium, silicon, manganese, columbium, aluminum, nitrogen, and sulfur. The higher the carbon content, the lower the corrosion resistance of stainless steels. The reason is that the carbon combines with the chromium in the steel and forms chromium carbide, which lowers the passivity of the steel. The chromium carbides introduce a second phase in the metal, which promotes galvanic corrosion. The letter *L* is used to identify low-carbon stainless steels.

Developed in the early 1900s, stainless steels are made by techniques basically similar to those used in other types of steelmaking, using electric furnaces or the basic-oxygen process. The level of impurities is controlled by various refining techniques. Stainless steels are available in a wide variety of shapes. Typical applications are in the chemical, food-processing, and petroleum industries, and for products such as cutlery, kitchen equipment, health care and surgical equipment, and automotive trim.

Stainless steels are generally divided into five types (Table 3.4): austenitic, ferritic, martensitic, precipitation-hardening, and duplex-structure steels.

1. **Austenitic steels** (200 and 300 series) are generally composed of chromium, nickel, and manganese in iron. They have excellent corrosion resistance (but are susceptible to stress-corrosion cracking) and are nonmagnetic. These steels are the most ductile of all stainless steels and hence can be formed easily. They are hardened by cold working; however, with increasing cold work, their formability is reduced. Austenitic stainless steels are used in a wide variety of applications, such as kitchenware, fittings, lightweight transportation equipment, furnace and heat-exchanger parts, welded construction, and components subjected to severe chemical environments.

2. **Ferritic steels** (400 series) have a high chromium content: up to 27%. They are magnetic and have good corrosion resistance but have lower ductility (hence

TABLE 3.4

Room-Temperature Mechanical Properties and Typical Applications of Annealed Stainless Steels

AISI (UNS)	Ultimate tensile strength (MPa)	Yield strength (MPa)	Elongation (%)	Characteristics and typical applications
303 (S30300)	550–620	240–260	50–53	Screw-machine products, shafts, valves, bolts, bushings, and nuts; aircraft fittings; rivets; screws; studs.
304 (S30400)	565–620	240–290	55–60	Chemical and food-processing equipment, brewing equipment, cryogenic vessels, gutters, downspouts, and flashings.
316 (S31600)	550–590	210–290	55–60	High corrosion resistance and high creep strength. Chemical and pulp-handling equipment, photographic equipment, brandy vats, fertilizer parts, ketchup-cooking kettles, and yeast tubs.
410 (S41000)	480–520	240–310	25–35	Machine parts, pump shafts, bolts, bushings, coal chutes, cutlery, fishing tackle, hardware, jet engine parts, mining machinery, rifle barrels, screws, and valves.
416 (S41600)	480–520	275	20–30	Aircraft fittings, bolts, nuts, fire extinguisher inserts, rivets, and screws.

have lower formability) than austenitic stainless steels. Ferritic stainless steels are hardened by cold working and are not heat treatable. They are generally used for nonstructural applications, such as kitchen equipment and automotive trim.

3. **Martensitic steels** (400 and 500 series) generally do not contain nickel; their chromium content may be as high as 18%. These steels are magnetic and have high strength, hardness, and fatigue resistance and good ductility, but moderate corrosion resistance. They are hardenable by heat treatment. Martensitic stainless steels typically are used for cutlery, surgical tools, instruments, valves, and springs.

4. **Precipitation-hardening (PH) steels** contain chromium and nickel, along with copper, aluminum, titanium, or molybdenum. They have good corrosion resistance, good ductility, and high strength at elevated temperatures. Their main application is in aircraft and aerospace structural components.

5. **Duplex-structure steels** have a mixture of austenite and ferrite. These steels have good strength and higher resistance to corrosion (in most environments) and to stress-corrosion cracking than do the 300 series austenitic steels. Typical applications of duplex-structure steels are in water-treatment plants and heat-exchanger components.

3.10.3 Tool and die steels

Tool and die steels are specially alloyed steels (Table 3.5) and are designed for high strength, impact toughness, and wear resistance at a range of temperatures. They are commonly used in forming and machining metals. Various types of tool and die materials used for a variety of manufacturing applications are listed in Table 3.6. The main categories of these materials are described next.

1. **High-speed steels** (HSS), first developed in the early 1900s, are the most highly alloyed tool and die steels and maintain their hardness and strength at elevated operating temperatures. There are two basic types of high-speed steels: the

TABLE 3.5

Basic Types of Tool and Die Steels

Type	AISI
High speed	M (molybdenum base)
	T (tungsten base)
Hot work	H1 to H19 (chromium base)
	H20 to H39 (tungsten base)
	H40 to H59 (molybdenum base)
Cold work	D (high carbon, high chromium)
	A (medium alloy, air hardening)
	O (oil hardening)
Shock resisting	S
Mold steels	P1 to P19 (low carbon)
	P20 to P39 (others)
Special purpose	L (low alloy)
	F (carbon-tungsten)
Water hardening	W

TABLE 3.6

Typical Tool and Die Materials for Various Processes

Process	Material
Die casting	H13, P20
Powder metallurgy	
Punches	A2, S7, D2, D3, M2
Dies	WC, D2, M2
Molds for plastic and rubber	S1, O1, A2, D2, 6F5, 6F6, P6, P20, P21, H13
Hot forging	6F2, 6G, H11, H12
Hot extrusion	H11, H12, H13
Cold heading	W1, W2, M1, M2, D2, WC
Cold extrusion	
Punches	A2, D2, M2, M4
Dies	O1, W1, A2, D2
Coining	52100, W1, O1, A2, D2, D3, D4, H11, H12, H13
Drawing	
Wire	WC, diamond
Shapes	WC, D2, M2
Bar and tubing	WC, W1, D2
Rolls	
Rolling	Cast iron, cast steel, forged steel, WC
Thread rolling	A2, D2, M2
Shear spinning	A2, D2, D3
Sheet metals	
Pressworking	Zinc alloys, 4140 steel, cast iron, epoxy composites, A2, D2, O1
Deep drawing	W1, O1, cast iron, A2, D2
Shearing	
Cold	D2, A2, A9, S2, S5, S7
Hot	H11, H12, H13
Machining	Carbides, high-speed steels, ceramics, diamond, cubic boron nitride

molybdenum type (M-series) and the **tungsten type** (T-series). The M-series contain up to about 10% molybdenum, with chromium, vanadium, tungsten, and cobalt as other alloying elements. The T-series contain 12 to 18% tungsten, with chromium, vanadium, and cobalt as additional alloying elements. As compared with the T-series steels, the M-series steels generally have higher abrasion resistance, have less distortion during heat treatment, and are less expensive. The M-series steels constitute about 95% of all high-speed steels produced in the United States. High-speed steel tools can be coated with titanium nitride and titanium carbide for better resistance to wear (see Section 4.4.2).

2. **Hot-work steels** (H-series) are designed for use at elevated temperatures and have high toughness and high resistance to wear and cracking. The alloying elements are generally tungsten, molybdenum, chromium, and vanadium.

3. **Cold-work steels** (A-, D-, and O-series) are used for cold-working operations. They generally have high resistance to wear and cracking. These steels are available as oil-hardening or air-hardening types.

4. **Shock-resisting steels** (S-series) are designed for impact toughness; other properties depend on the particular composition. Typical applications for these steels are as dies, punches, and chisels.

3.11 | General Properties and Applications of Nonferrous Metals and Alloys

Nonferrous metals and **alloys** cover a very wide range of materials, from the more common metals such as aluminum, copper, and magnesium, to high-strength, high-temperature alloys such as tungsten, tantalum, and molybdenum. Although more expensive than ferrous metals, nonferrous metals and alloys have important applications because of their wide range of mechanical, physical, and chemical properties and characteristics.

A turbofan jet engine for the Boeing 757 aircraft typically contains the following nonferrous metals and alloys: 38% titanium, 37% nickel, 12% chromium, 6% cobalt, 5% aluminum, 1% niobium (columbium), and 0.02% tantalum. Without these materials, a jet engine (Fig. 3.34) could not be designed, manufactured, and operated at the required energy and efficiency levels.

Typical examples of the applications of nonferrous metals and alloys include, for example, (a) aluminum for cooking utensils and aircraft bodies, (b) copper wire for electricity and copper tubing for water in residences, (c) titanium for jet-engine turbine blades and artificial joints, and (d) tantalum for rocket engines.

3.11.1 Aluminum and aluminum alloys

Important factors in selecting *aluminum* (Al) and its alloys are their high strength-to-weight ratio, resistance to corrosion by many chemicals, high thermal and electrical conductivity, nontoxicity, reflectivity, appearance, and ease of formability and machinability; they are also nonmagnetic.

Principal uses of aluminum and its alloys, in decreasing order of consumption, include containers and packaging (aluminum beverage cans and foil), buildings and various other types of construction, transportation (aircraft and aerospace applications, buses, automobiles (see Fig. 1.5), railroad cars, and marine craft), electrical products (nonmagnetic and economical electrical conductors), consumer durables (appliances, cooking utensils, and outdoor furniture), and portable tools

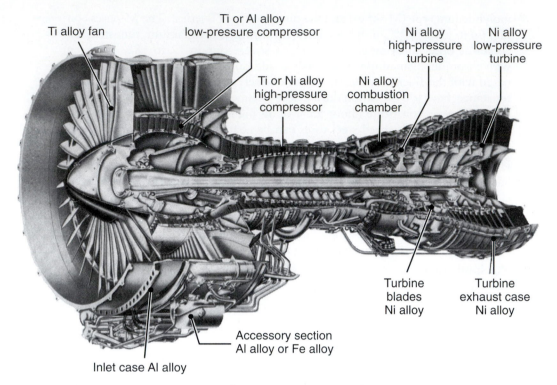

Ti alloy fan

Ti or Al alloy
low-pressure compressor

Ti or Ni alloy
high-pressure
compressor

Ni alloy
combustion
chamber

Ni alloy
high-pressure
turbine

Ni alloy
low-pressure
turbine

Turbine
blades
Ni alloy

Turbine
exhaust case
Ni alloy

Accessory section
Al alloy or Fe alloy

Inlet case Al alloy

FIGURE 3.34   Cross section of a jet engine (PW2037) showing various components and the alloys used in making them. *Source:* Courtesy of United Aircraft Pratt & Whitney.

(Tables 3.7 and 3.8). Nearly all high-voltage transmission line wiring is made of aluminum cable (such as AISI 1350-H19) with steel reinforcement (known as ACSR cable).

Because of their structural (load-bearing) components, 82% of a Boeing 747 aircraft and 70% of a Boeing 777 aircraft are made of aluminum. An important aluminum alloy family is the **aluminum-lithium** alloys, such as the 8090 series in

TABLE 3.7

Properties of Various Aluminum Alloys at Room Temperature

Alloy (UNS)	Temper	Ultimate tensile strength (MPa)	Yield strength (MPa)	Elongation in 50 mm (%)
1100 (A91100)	O	90	35	35–45
	H14	125	120	9–20
1350 (A91350)	O	85	30	23
	H19	185	165	1.5
2024 (A92024)	O	190	75	20–22
	T4	470	325	19–20
3003 (A93003)	O	110	40	30–40
	H14	150	145	8–16
5052 (A95052)	O	190	90	25–30
	H34	260	215	10–14
6061 (A96061)	O	125	55	25–30
	T6	310	275	12–17
7075 (A97075)	O	230	105	16–17
	T6	570	500	11
8090	T8X	480	400	4–5

TABLE 3.8

Manufacturing Properties and Typical Applications of Wrought Aluminum Alloys

Alloy	Characteristics*			Typical applications
	Corrosion resistance	Machinability	Weldability	
1100	A	D–C	A	Sheet-metal work, spun hollow parts, tin stock.
2014	C	C–B	C–B	Heavy-duty forgings, plate and extrusions for aircraft structural components, wheels.
3003	A	D–C	A	Cooking utensils, chemical equipment, pressure vessels, sheet-metal work, builders' hardware, storage tanks.
5054	A	D–C	A	Welded structures, pressure vessels, tube for marine uses.
6061	B	D–C	A	Trucks, canoes, furniture, structural applications.
7005	D	B–D	B	Extruded structural members, large heat exchangers, tennis racquets, softball bats.
8090	A–B	B–D	B	Aircraft frames, helicopter structural components.

*From A (excellent) to D (poor).

Table 3.7, which are lighter because of the low density of lithium. The Airbus A350 will use up to 23% of aluminum-lithium alloys in its structure, mainly in the fuselage.

As in all cases of material selection (see Section 1.5), because each aluminum alloy has its particular properties, its selection can be critical with respect to manufacturing characteristics and cost considerations. A typical aluminum beverage can, for example, consists of the following alloys: 3004 or 3104 for the can body, 5182 for the lid, and 5042 for the tab. (All sheet is in the H19 condition, which is the highest cold-worked state.) Similarly, aluminum-lithium alloys such as 8090 are advantageous for aerospace applications because of their strength-to-weight and stiffness-to-weight ratios, but their high cost restricts their use elsewhere.

Aluminum alloys are available as mill products, that is, wrought product made into various shapes by rolling, extrusion, drawing, and forging. Aluminum ingots are available for casting, as are powder metals for powder metallurgy applications (Chapter 11). There are two types of wrought alloys of aluminum: (1) alloys that can be hardened by *cold working* (designated by the letter H) and are not heat treatable, and (2) alloys that are hardenable by *heat treatment* (designated by the letter T). The letter O indicates the annealed condition. Techniques are available whereby most aluminum alloys can be machined, formed, and welded with relative ease.

3.11.2 Magnesium and magnesium alloys

Magnesium (Mg) is the lightest engineering metal available; its alloys are used in structural and nonstructural applications where weight is of primary importance. Magnesium is also an alloying element in various nonferrous metals.

Typical uses of magnesium alloys include aircraft and missile components, material-handling equipment, portable power tools (such as drills and sanders), luggage, bicycles, sporting goods, and general lightweight components. These alloys are available as either castings or wrought products, such as extruded bars and shapes, forgings, and rolled plate and sheet. Magnesium alloys are also used in printing and textile machinery to minimize inertial forces in high-speed components. Magnesium also has good vibration-damping characteristics.

TABLE 3.9

Properties and Typical Forms of Various Wrought Magnesium Alloys

Alloy	Composition (%)				Condition	Ultimate tensile strength (MPa)	Yield strength (MPa)	Elongation in 50 mm (%)	Typical forms
	Al	Zn	Mn	Zr					
AZ31B	3.0	1.0	0.2		F	260	200	15	Extrusions.
					H24	290	220	15	Sheet and plates.
AZ80A	8.5	0.5	0.2		T5	380	380	7	Extrusions and forgings.
HK31A*			0.7		H24	255	255	8	Sheet and plates.
ZK60A		5.7		0.55	T5	365	365	11	Extrusions and forgings.

*HK31A also contains 3% Th.

Because it is not sufficiently strong in its pure form, magnesium is alloyed with various elements (Table 3.9) to impart certain specific properties, particularly for high strength-to-weight ratios. A variety of magnesium alloys are available with good casting, forming, and machining characteristics. Because magnesium alloys oxidize rapidly (they are *pyrophoric*), they are a potential fire hazard, and precautions must be taken when machining, grinding, or sand casting magnesium alloys. Products made of magnesium and its alloys are, however, not a fire hazard.

3.11.3 Copper and copper alloys

First produced in about 4000 B.C., *copper* (Cu) and its alloys have properties somewhat similar to those of aluminum alloys. In addition, copper alloys are among the best conductors of electricity and heat and also have good resistance to corrosion. They can be processed easily by various forming, machining, casting, and joining techniques.

Copper alloys are often attractive for applications where combined properties, such as electrical and mechanical properties, corrosion resistance, thermal conductivity, and wear resistance, are required. Applications include electrical and electronic components, springs, cartridges for small arms, plumbing, heat exchangers, and marine hardware, as well as some consumer goods, such as cooking utensils, jewelry, and other decorative objects.

Copper alloys can acquire a wide variety of properties, and their manufacturing properties can be improved by the addition of alloying elements and by heat treatment. The most common copper alloys are brasses and bronzes; others are copper nickels and nickel silvers.

Brass, which is an alloy of copper and zinc, was one of the earliest alloys developed and has numerous applications, including decorative objects (Table 3.10). **Bronze** is an alloy of copper and tin (Table 3.11). Other bronzes include (a) *aluminum bronze*, an alloy of copper and aluminum; (b) *tin bronze;* (c) *beryllium bronze* (a beryllium copper); and (d) *phosphor bronze;* the latter two have good strength and high hardness for applications such as springs and bearings.

3.11.4 Nickel and nickel alloys

Nickel (Ni), a silver-white metal discovered in 1751, is a major alloying element that imparts strength, toughness, and corrosion resistance to metals. Nickel is used extensively in stainless steels and nickel-base alloys. These alloys are used for high-temperature applications, such as jet-engine components, rockets, and nuclear power plants, as well as in food-handling and chemical-processing equipment, coins,

TABLE 3.10

Properties and Typical Applications of Various Wrought Copper and Brasses

Type and UNS number	Nominal composition (%)	Ultimate tensile strength (MPa)	Yield strength (MPa)	Elongation in 50 mm (%)	Typical applications
Oxygen-free electronic (C10100)	99.99 Cu	220–450	70–365	55–4	Bus bars, waveguides, hollow conductors, lead in wires, coaxial cables and tubes, microwave tubes, rectifiers.
Red brass, (C23000)	85.0 Cu, 15.0 Zn	270–72	70–435	55–3	Weather stripping, conduit, sockets, fasteners, fire extinguishers, condenser and heat-exchanger tubing.
Low brass, (C24000)	80.0 Cu, 20.0 Zn	300–850	80–450	55–3	Battery caps, bellows, musical instruments, clock dials, flexible hose.
Free-cutting brass (C36000)	61.5 Cu, 3.0 Pb, 35.5 Zn	340–470	125–310	53–18	Gears, pinions, automatic high-speed screw-machine parts.
Naval brass (C46400 to C46700)	60.0 Cu, 39.25 Zn, 0.75 Sn	380–610	170–455	50–17	Aircraft turnbuckle barrels, balls, bolts, marine hardware, valve stems, condenser plates.

TABLE 3.11

Properties and Typical Applications of Various Wrought Bronzes

Type and UNS number	Nominal composition (%)	Ultimate tensile strength (MPa)	Yield strength (MPa)	Elongation in 50 mm (%)	Typical applications
Architectural bronze (C38500)	57.0 Cu, 3.0 Pb, 40.0 Zn	415	140 (as extruded)	30	Architectural extrusions, storefronts, thresholds, trim, butts, hinges.
Phosphor bronze, 5% A (C51000)	95.0 Cu, 5.0 Sn, trace P	325–960	130–550	64–2	Bellows, clutch disks, cotter pins, diaphragms, fasteners, wire brushes, chemical hardware, textile machinery.
Free-cutting phosphor bronze (C54400)	88.0 Cu, 4.0 Pb, 4.0 Zn, 4.0 Sn	300–520	130–435	50–15	Bearings, bushings, gears, pinions, shafts, thrust washers, valve parts.
Low-silicon bronze, B (C65100)	98.5 Cu, 1.5 Si	275–655	100–475	55–11	Hydraulic pressure lines, bolts, marine hardware, electrical conduits, heat-exchanger tubing.
Nickel-silver, 65–18 (C74500)	65.0 Cu, 17.0 Zn, 18.0 Ni	390–710	170–620	45–3	Rivets, screws, zippers, camera parts, base for silver plate, nameplates, etching stock.

and marine applications. Because nickel is magnetic, its alloys are also used in electromagnetic applications such as solenoids. As a metal, the principal use of nickel is in electroplating for resistance to corrosion and wear and for appearance.

Nickel alloys containing chromium, cobalt, and molybdenum have high strength and corrosion resistance at elevated temperatures. The behavior of these alloys in machining, forming, casting, and welding can be modified by various other alloying elements. A variety of nickel alloys that have a range of strengths at different

TABLE 3.12

Properties and Typical Applications of Various Nickel Alloys (All alloy names are trade names.)

Alloy (condition)	Principal alloying elements (%)	Ultimate tensile strength (MPa)	Yield strength (MPa)	Elongation in 50 mm (%)	Typical applications
Nickel 200 (annealed)	None	380–550	100–275	60–40	Chemical- and food-processing industry, aerospace equipment, electronic parts.
Duranickel 301 (age hardened)	4.4 Al, 0.6 Ti	1300	900	28	Springs, plastics-extrusion equipment, molds for glass.
Monel R-405 (hot rolled)	30 Cu	525	230	35	Screw-machine products, water-meter parts.
Monel K-500 (age hardened)	29 Cu, 3 Al	1050	750	20	Pump shafts, valve stems, springs.
Inconel 600 (annealed)	15 Cr, 8 Fe	640	210	48	Gas-turbine parts, heat-treating equipment, electronic parts, nuclear reactors.
Hastelloy C-4 (solution treated and quenched)	16 Cr, 15 Mo	785	400	54	High-temperature stability, resistance to stress-corrosion cracking.

temperatures are shown in Table 3.12. *Monel* is a nickel-copper alloy, and Inconel is a nickel-chromium alloy. *Hastelloy,* a nickel-molybdenum-chromium alloy, has good corrosion resistance and high strength at elevated temperatures. *Nichrome,* an alloy of nickel, chromium, and iron, has high oxidation and electrical resistance and is commonly used for electrical-heating elements. *Invar,* an alloy of iron and nickel, has a low coefficient of thermal expansion and has been used in precision scientific instruments and camera/optics applications. (See Section 3.9.5.)

3.11.5 Superalloys

Superalloys are important in high-temperature applications. Also known as **heat-resistant** or **high-temperature** alloys, major applications are in jet engines, gas turbines, reciprocating engines, rocket engines; tools and dies for hot working of metals; and in the nuclear, chemical, and petrochemical industries. Superalloys generally have good resistance to corrosion, mechanical and thermal fatigue, mechanical and thermal shock, creep, and erosion at elevated temperatures. Most superalloys have a maximum service temperature of about 1000°C (1800°F) for structural applications but can be as high as 1200°C (2200°F) for non-load-bearing components. Superalloys are generally identified by trade names or by special numbering systems and are available in a variety of shapes.

Superalloys are available as **iron base, cobalt base,** or **nickel base,** and they contain nickel, chromium, cobalt, and molybdenum as major alloying elements, with aluminum, tungsten, and titanium as additional elements. Iron-base superalloys generally contain 32% to 67% iron, 15% to 22% chromium, and 9% to 38% nickel. Common alloys in this group are the *Incoloy* series. Cobalt-base superalloys generally contain 35% to 65% cobalt, 19% to 30% chromium, and up to 35% nickel. Cobalt is a white-colored metal that resembles nickel. These superalloys are not as strong as nickel-base superalloys, but they retain their strength at higher temperatures better

TABLE 3.13

Properties and Typical Applications of Various Nickel-Base Superalloys at 870°C (1600°F) (All alloy names are trade names.)

Alloy	Condition	Ultimate tensile strength (MPa)	Yield strength (MPa)	Elongation in 50 mm (%)	Typical applications
Astroloy	Wrought	770	690	25	Forgings for high-temperature applications.
Hastelloy X	Wrought	255	180	50	Jet-engine sheet parts.
IN-100	Cast	885	695	6	Jet-engine blades and wheels.
IN-102	Wrought	215	200	110	Superheater and jet-engine parts.
Inconel 625	Wrought	285	275	125	Aircraft engines and structures, chemical-processing equipment.
Inconel 718	Wrought	340	330	88	Jet-engine and rocket parts.
MAR-M 200	Cast	840	760	4	Jet-engine blades.
MAR-M 432	Cast	730	605	8	Integrally cast turbine wheels.
René 41	Wrought	620	550	19	Jet-engine parts.
Udimet 700	Wrought	690	635	27	Jet-engine parts.
Waspaloy	Wrought	525	515	35	Jet-engine parts.

than do nickel-base superalloys. Nickel-base superalloys are the most common of the superalloys and are available in a wide variety of compositions (Table 3.13). The range of nickel is from 38 to 76%; nickel-base superalloys also contain up to 27% chromium and 20% cobalt. Common alloys in this group include the *Hastelloy, Inconel, Nimonic, René, Udimet, Astroloy,* and *Waspaloy* series.

3.11.6 Titanium and titanium alloys

Titanium (Ti) was discovered in 1791 but was not commercially produced until the 1950s. Although titanium is relatively expensive, its high strength-to-weight ratio and its corrosion resistance at room and elevated temperatures make it very attractive for such applications as components for aircraft, jet-engine, racing-car, and marine craft; submarine hulls; chemical and petrochemical industries; and biomaterials such as orthopedic implants (Table 3.14). Unalloyed titanium, known as commercially pure titanium, has excellent corrosion resistance for applications where strength considerations are secondary. Aluminum, vanadium, molybdenum, manganese, and other alloying elements are added to titanium alloys to impart properties such as improved workability, strength, and hardenability. Titanium alloys are available for service at 550°C (1000°F) for long periods of time, and at up to 750°C (1400°F) for shorter periods.

The properties and manufacturing characteristics of titanium alloys are extremely sensitive to small variations in both alloying as well as residual elements. Proper control of composition and processing is thus essential, including prevention of surface contamination by hydrogen, oxygen, or nitrogen during processing. These elements cause embrittlement of titanium, resulting in reduced toughness and ductility.

The body-centered cubic structure of titanium [beta-titanium, above 880°C (1600°F)] is ductile, whereas its hexagonal close-packed structure (alpha-titanium) is somewhat brittle and is very subject to stress corrosion. A variety of other titanium structures (alpha, near alpha, alpha-beta, and beta) can be obtained by alloying and heat treating, such that the properties can be optimized for specific applications. **Titanium aluminide intermetallics** (TiAl and Ti_3Al) have higher stiffness and lower density than conventional titanium alloys and can withstand higher temperatures.

TABLE 3.14

Properties and Typical Applications of Wrought Titanium Alloys

Nominal composition (%)	UNS	Condition	Temp (°C)	Ultimate tensile strength (MPa)	Yield strength (MPa)	Elongation (%)	Typical applications
99.5 Ti	R50250	Annealed	25	330	240	30	Airframes; chemical, desalination, and marine parts; plate-type heat exchangers.
			300	150	95	32	
5 Al, 2.5 Sn	R54520	Annealed	25	860	810	16	Aircraft-engine compressor blades and ducting; steam-turbine blades.
			300	565	450	18	
6 Al, 4V	R56400	Annealed	25	1000	925	14	Rocket motor cases; blades and disks for aircraft turbines and compressors; orthopedic implants; structural forgings; fasteners.
			300	725	650	14	
			425	670	570	18	
			550	530	430	35	
		Solution + age	25	1175	1100	10	
			300	980	900	10	
13 V, 11 Cr, 3Al	R58010	Solution + age	25	1275	1210	8	High-strength fasteners; aerospace components; honeycomb panels.
			425	1100	830	12	

3.11.7 Refractory metals

Molybdenum, columbium, tungsten, and tantalum are referred to as *refractory metals* because of their high melting point. These elements were discovered about two centuries ago and have been used as important alloying elements in steels and superalloys; however, their use as engineering metals and alloys did not begin until about the 1940s. More than most other metals and alloys, refractory metals and their alloys retain their strength at elevated temperatures. Consequently, they are of great importance and are used in rocket engines, gas turbines, and various other aerospace applications; in the electronics, nuclear power, and chemical industries; and as tool and die materials. The temperature range for some of these applications is on the order of 1100° to 2200°C (2000° to 4000°F), where strength and oxidation of other materials are of major concern.

1. **Molybdenum.** *Molybdenum* (Mo), a silvery white metal, has high melting point, high modulus of elasticity, good resistance to thermal shock, and good electrical and thermal conductivity. Typical applications are in solid-propellant rockets, jet engines, honeycomb structures, electronic components, heating elements, and molds for die casting. Principal alloying elements in molybdenum are titanium and zirconium. Molybdenum is used in greater amounts than any other refractory metal. It is also an important alloying element in cast and wrought alloys, such as steels and heat-resistant alloys, and imparts strength, toughness, and corrosion resistance. A major disadvantage of molybdenum alloys is their low resistance to oxidation at temperatures above about 500°C (950°F), thus necessitating the use of protective coatings.

2. **Niobium.** *Niobium* (Nb), or *columbium* (after the mineral *columbite*), possesses good ductility and formability and has greater resistance to oxidation than do other refractory metals. With various alloying elements, niobium alloys can be produced with moderate strength and good fabrication characteristics. These alloys are used in rockets; missiles; and nuclear, chemical, and superconductor applications. Niobium is also an alloying element in various alloys and superalloys.

3. **Tungsten.** *Tungsten* (W, from *wolframite*) was first identified in 1781 and is the most abundant of all refractory metals. Tungsten has the highest melting point of any metal [3410°C (6170°F)] and thus it is characterized by high strength at elevated temperatures. On the other hand, it has high density, brittleness at low temperatures, and poor resistance to oxidation.

   Tungsten and its alloys are used for applications involving temperatures above 1650°C (3000°F) such as nozzle throat liners in missiles and in the hottest parts of jet and rocket engines, circuit breakers, welding electrodes, and spark-plug electrodes. The filament wire in incandescent light bulbs is made of pure tungsten, using powder metallurgy and wire-drawing techniques. Because of its high density, tungsten is also used in balancing weights and counterbalances in mechanical systems, including self-winding watches. Tungsten is an important element in tool and die steels, imparting strength and hardness at elevated temperatures. For example, tungsten carbide (with cobalt as a binder for the carbide particles) is one of the most important tool and die materials.

4. **Tantalum.** *Tantalum* (Ta) is characterized by high melting point [3000°C (5425°F)], good ductility, and good resistance to corrosion; however, it has high density and poor resistance to chemicals at temperatures above 150°C (300°F). It is also used as an alloying element. Tantalum is used extensively in electrolytic capacitors and various components in the electrical, electronic, and chemical industries, as well as for thermal applications, such as in furnaces and acid-resistant heat exchangers. A variety of tantalum-base alloys is available in many shapes for use in missiles and aircraft.

3.11.8 Other nonferrous metals

1. **Beryllium.** Steel gray in color, *beryllium* (Be) has a high strength-to-weight ratio. Unalloyed beryllium is used in nuclear and X-ray applications, because of its low neutron absorption characteristics, and in rocket nozzles, space and missile structures, aircraft disc brakes, and precision instruments and mirrors. It is also an alloying element, and its alloys of copper and nickel are used in such applications as springs (*beryllium-copper*), electrical contacts, and nonsparking tools for use in explosive environments such as mines and in metal-powder production. Beryllium and its oxide are toxic, and thus precautions must be taken in its production and processing.

2. **Zirconium.** *Zirconium* (Zr), silvery in appearance, has good strength and ductility at elevated temperatures and has good corrosion resistance because of an adherent oxide film on its surfaces. The element is used in electronic components and nuclear power reactor applications because of its low neutron absorption characteristics.

3. **Low-melting-point metals.** The major metals in this category are lead, zinc, and tin.

   a. **Lead.** *Lead* (Pb, after *plumbum*, the root of the word "plumber") has high density, good resistance to corrosion (by virtue of the stable lead-oxide layer that forms and protects its surface), very low hardness, low strength, high ductility, and good workability. Alloying with various elements, such as antimony and tin, enhances lead's properties, making it suitable for such applications as piping, collapsible tubing, bearing alloys, cable sheathing, roofing, and lead-acid storage batteries. Lead is also used for damping sound and vibrations, radiation shielding against X-rays, printing (*type metals*), weights, and the chemical and paint industries. The oldest lead artifacts were made around 3000 B.C. Lead pipes made by the Romans and installed in the Roman baths in Bath, England, two millennia ago are still in use. Lead is also an alloying element in solders, steels, and copper alloys and promotes corrosion resistance and machinability. Because of its toxicity, however, environmental contamination by lead is now a major concern. (See, for example, *lead-free solders,* Section 12.14.3.)

b. **Zinc.** *Zinc* (Zn), which has a bluish-white color, is the fourth industrially most utilized metal, after iron, aluminum, and copper. Although known for many centuries, zinc was not investigated and developed until the eighteenth century. Zinc has two major uses: (a) for galvanizing iron, steel sheet, and wire, and (b) as an alloy base for casting. In **galvanizing,** zinc serves as the anode and protects the steel (cathode) from corrosive attack should the coating be scratched or punctured. Zinc is also used as an alloying element; brass, for example, is an alloy of copper and zinc.

Major alloying elements in zinc are aluminum, copper, and magnesium. They impart strength and provide dimensional control during casting of the metal. Zinc-base alloys are used extensively in die casting for making products such as fuel pumps and grills for automobiles, components for household appliances (such as vacuum cleaners, washing machines, and kitchen equipment), machine parts, and photoengraving plates. Another use for zinc is in superplastic alloys (see Section 2.2.7), which have good formability characteristics by virtue of their capacity to undergo large deformation without failure. Very fine grained 78% Zn-22% Al sheet is a common example of a superplastic zinc alloy that can be formed methods used commonly for forming plastics or metals.

c. **Tin.** Although used in small amounts, *tin* (Sn, after *stannum*), a silvery-white lustrous metal, is an important metal. Its most extensive use is as a protective coating on steel sheet (**tin plate**), used for making containers (tin cans) for food and various other products. Inside a sealed can, the steel is cathodic and is protected by the tin (anode), whereby the steel does not corrode. The low shear strength of the tin coatings on steel sheet also improves its performance in deep drawing and general presswork operations.

Unalloyed tin is used in such applications as lining material for water-distillation plants and as a molten layer of metal over which plate glass is made (see Section 11.11). Also called **white metals,** tin-base alloys generally contain copper, antimony, and lead. These alloying elements impart hardness, strength, and corrosion resistance. Organ pipes are made of tin alloys. Because of their low friction coefficients, which result from low shear strength and low adhesion, tin alloys are used as journal-bearing materials. These alloys are known as **babbitts** and consist of tin, copper, and antimony. **Pewter** is another alloy of this class. Developed in the fifteenth century, pewter is used for tableware, hollowware, and decorative artifacts. Tin is also an alloying element for type metals; dental alloys; and bronze (copper-tin alloy), titanium, and zirconium alloys. Tin-lead alloys are common soldering materials, with a wide range of compositions and melting points. (See Section 12.14.3.)

4. **Precious metals.** Gold, silver, and platinum are, although costly, the most important precious metals, also called *noble metals.*

a. **Gold** (Au, after *aurum*) is soft and ductile and has good corrosion resistance at any temperature. Typical applications include electric contact and terminals, jewelry, coinage, reflectors, gold leaf for decorative purposes, and dental work.

b. **Silver** (Ag, after *argentum*) is a ductile metal and has the highest electrical and thermal conductivity of any metal; however, it develops an oxide film that adversely affects its surface properties and appearance. Typical applications for silver include photographic film, electrical contacts, solders, bearings, food and chemical equipment, tableware, jewelry, and coinage. *Sterling silver* is an alloy of silver and 7.5% copper.

c. **Platinum** (Pt) is a grayish-white soft and ductile metal that has good corrosion resistance, even at elevated temperatures. Platinum alloys are used as electrical contacts, spark-plug electrodes, catalysts for automobile pollution-control devices, filaments, nozzles, dies for extruding glass fibers, and thermocouples; in the electrochemical industry; and in jewelry and dental work.

3.11.9 Special metals and alloys

1. **Shape-memory alloys.** *Shape-memory alloys*, after being plastically deformed at room temperature into various shapes, return to their original shapes upon heating. For example, a piece of straight wire made of these alloys can be wound into a helical spring; when heated with a match, the spring uncoils and returns to its original straight shape. A typical shape-memory alloy is 55% Ni and 45% Ti; other alloys include copper-aluminum-nickel, copper-zinc-aluminum, iron-manganese-silicon, and nickel-titanium. These alloys generally have good ductility, corrosion resistance, and high electrical conductivity.

    The behavior of shape-memory alloys can be reversible, that is, the shape can switch back and forth repeatedly upon periodic application and removal of heat. Typical applications include temperature sensors, clamps, connectors, fasteners, and seals that are easy to install.

2. **Amorphous alloys.** *Amorphous alloys* are a class of metal alloys that, unlike ordinary metals, do not have a long-range crystalline structure (see Section 5.10.8). These alloys have no grain boundaries, and the atoms are randomly and tightly packed. Because their structure resembles that of glasses (Section 11.10), these alloys are also called **metallic glasses**. These materials are now available in bulk quantities, as well as wire, ribbon, strip, and powder, and continue to be investigated as an important and emerging material system.

    Amorphous alloys typically consist of iron, nickel, and chromium, alloyed with carbon, phosphorus, boron, aluminum, and silicon. These alloys exhibit excellent corrosion resistance, good ductility, and high strength. They also undergo very low loss from magnetic hysteresis, making them suitable for magnetic steel cores for transformers, generators, motors, lamp ballasts, magnetic amplifiers, and linear accelerators.

    The amorphous structure was first obtained (in the late 1960s) by extremely rapid cooling of the molten alloy. One such method is called *splat cooling* or *melt spinning* (see Fig. 5.31), in which the alloy is propelled at a very high speed against a rotating metal surface. Since the rate of cooling is on the order of 10^6 K/s to 10^8 K/s, the molten alloy does not have sufficient time to crystallize. If, however, an amorphous alloy's temperature is raised and then cooled, the alloy develops a crystalline structure.

3. **Nanomaterials.** First investigated in the early 1980s, these materials have some properties that are often superior to those of traditional and commercially available materials. These characteristics include strength, hardness, ductility, wear resistance and corrosion resistance suitable for structural (load-bearing) and nonstructural applications, in combination with unique electrical, magnetic, and optical properties. Applications for nanomaterials include cutting

tools, metal powders, computer chips, flat-panel displays for laptops, sensors, and various electrical and magnetic components. (See also Sections 8.6.10, 11.8.1, and 13.18.)

Nanomaterials are available in granular form, fibers, films, and composites which contain particles that are on the order of 1 to 100 nm in size. The composition of nanomaterials may consist of any combination of chemical elements. Among the more important compositions are carbides, oxides, nitrides, metals and alloys, organic polymers, and various composites. Synthesis methods for producing nanomaterials include inert-gas condensation, plasma synthesis, electrodeposition, sol-gel synthesis, and mechanical alloying or ball milling.

4. **Metal foams.** In metal foams, usually of aluminum alloys but also of titanium or tantalum, the metal itself consists of only 5 to 20% of the structure's volume. The foams can be produced by blowing air into a molten metal and tapping the froth that forms at the surface, which solidifies into a foam. Alternative methods include (a) chemical vapor deposition onto a polymer or carbon foam lattice, (b) slip casting metal powders onto a polymer foam, and (c) doping molten or powder metals with titanium hydride, which then releases hydrogen gas at the elevated casting or sintering temperatures. Metal foams present unique combinations of strength-to-density and stiffness-to-density ratios. Although, as expected, these ratios are not as high as that of the base metal itself, the foams are very lightweight, making them an attractive material for aerospace applications; other applications include filters, lightweight beams, and orthopedic implants.

SUMMARY

- Manufacturing properties of metals and alloys depend largely on their mechanical and physical properties. These properties, in turn, are governed mainly by their crystal structure, grain boundaries, grain size, texture, and various imperfections. (Sections 3.1, 3.2)

- Dislocations are responsible for the lower shear stress required to cause slip for plastic deformation to occur. However, when dislocations become entangled with one another or are impeded by barriers such as grain boundaries, impurities, and inclusions, the shear stress required to cause slip increases; this phenomenon is called work hardening, or strain hardening. (Section 3.3)

- Grain size and the nature of grain boundaries significantly affect properties, including strength, ductility, and hardness, as well as increase the tendency for embrittlement with reduced toughness. Grain boundaries enhance strain hardening by interfering with the movement of dislocations. (Section 3.4)

- Plastic deformation at room temperature (cold working) of polycrystalline materials results in higher strength of the material as dislocations become entangled. The deformation also generally causes anisotropy, whereby the mechanical properties are different in different directions. (Section 3.5)

- The effects of cold working can be reversed by heating the metal within a specific temperature range and for a specific period of time, through the sequential processes of recovery, recrystallization, and grain growth. (Section 3.6)

- Metals and alloys can be worked at room, warm, or high temperatures. Their overall behavior, force and energy requirements during processing, and their workability depend largely on whether the working temperature is below or above their recrystallization temperature. (Section 3.7)

- Failure and fracture of a workpiece when subjected to deformation during metalworking operations is an important consideration. Two types of fracture are ductile fracture and brittle fracture. Ductile fracture is characterized by plastic deformation preceding fracture and requires a considerable amount of energy. Because it is not preceded by plastic deformation, brittle fracture can be catastrophic and requires much less energy than does ductile fracture. Impurities and inclusions, as well as factors such as the environment, strain rate, and state of stress, can play a major role in the fracture behavior of metals and alloys. (Section 3.8)

- Physical and chemical properties of metals and alloys can significantly affect design considerations, service requirements, compatibility with other materials (including tools and dies), and the behavior of the metals and alloys during processing. (Section 3.9)

- An extensive variety of metals and alloys is available, with a wide range of properties, such as strength, toughness, hardness, ductility, creep, and resistance to high temperatures and oxidation. Generally, these materials can be classified as (a) ferrous metals and alloys, (b) nonferrous metals and alloys, (c) superalloys, (d) refractory metals and their alloys, and (e) various other types, including amorphous alloys, shape-memory alloys, nanomaterials, and metal foams. (Sections 3.10, 3.11)

SUMMARY OF EQUATIONS

Theoretical shear strength of metals: $\tau_{max} = \dfrac{G}{2\pi}$

Theoretical tensile strength of metals: $\sigma_{max} = \sqrt{\dfrac{E\gamma}{a}} \simeq \dfrac{E}{10}$

Hall-Petch equation: $Y = Y_i + kd^{-1/2}$

ASTM grain-size number: $N = 2^{n-1}$

Tensile strength vs. crack length: $\sigma \propto \dfrac{1}{\sqrt{\text{Crack length}}}$

BIBLIOGRAPHY

Ashby, M.F., and Jones, D.R.H., *Engineering Materials,* Vol. 1, *An Introduction to Their Properties and Applications,* 3rd ed., Pergamon, 2005; Vol. 2, *An Introduction to Microstructures, Processing and Design,* Pergamon, 2005; Vol. 3, *Materials Failure Analysis: Case Studies and Design Implications,* Pergamon, 1993.

Ashby, M.F., *Materials Selection in Mechanical Design,* 3rd ed., Pergamon, 2005.

ASM Handbook, various volumes, ASM International.

ASM Specialty Handbooks, various volumes, ASM International.

Brandt, D.A., and Warner, J.C., *Metallurgy Fundamentals,* Goodheart-Wilcox, 2004.

Budinski, K.G., *Engineering Materials: Properties and Selection,* 8th ed., Prentice Hall, 2004.

Callister, W.D., Jr., *Materials Science and Engineering,* 7th ed., Wiley, 2006.

Davis, J.R. (ed.), *Handbook of Materials for Medical Devices,* ASM International, 2003.

Dieter, G.E., *Engineering Design: A Materials and Processing Approach,* 3rd ed., McGraw-Hill, 1999.

Farag, M.M., *Materials Selection for Engineering Design,* Prentice Hall, 1997.

Flinn, R.A., and Trojan, P.K., *Engineering Materials and Their Applications,* 4th ed., Houghton Mifflin, 1994.

Harper, C. (ed.), *Handbook of Materials for Product Design,* 3rd ed., McGraw-Hill, 2001.

Helmus, M., and Medlin, D. (eds.), *Medical Device Materials*, ASM International, 2005.

Hertzberg, R.W., *Deformation and Fracture Mechanics of Engineering Materials*, 4th ed., Wiley, 1996.

Hosford, W.F., *Physical Metallurgy*, Taylor & Francis, 2005.

Krauss, G., *Steels: Processing, Structure, and Performance*, ASM International, 2005.

Liu, A.F., *Mechanics and Mechanisms of Fracture: An Introduction*, ASM International, 2005.

Mangonon, P.C., *The Principles of Material Selection for Engineering Design*, Prentice Hall, 1999.

Material Selector, annual publication of *Materials Engineering Magazine*, Penton/IPC.

Pollock, D.D., *Physical Properties of Materials for Engineers*, 2nd ed., CRC Press, 1993.

Ratner, B.D., Hoffman, A.S., Schoen, F.J., and Lemons, J.E. (eds.), *Biomaterials Science: An Introduction to Materials in Medicine*, 2nd ed., Academic Press, 2004.

Revie, R.W. (ed.), *Uhlig's Corrosion Handbook*, Wiley-Interscience, 2000.

Roberge, P.R., and Tullmin, M., *Handbook of Corrosion Engineering*, McGraw-Hill, 1999.

Roberts, G.A., Krauss, G., Kennedy, R., and Cary, R.A., *Tool Steels*, 5th ed., ASM International, 1998.

Schaffer, J., Saxena, A., Antalovich, S., Sanders, T., and Warner, S., *The Science & Design of Engineering Materials*, 2nd ed., McGraw-Hill, 1999.

Shackelford, J.F., *Introduction to Materials Science for Engineers*, 6th ed., Macmillan, 2005.

Smith, W.F., *Principles of Materials Science and Engineering*, 3rd ed., McGraw-Hill, 1995.

Thermal Properties of Metals, ASM International, 2002.

Tool and Manufacturing Engineers Handbook, 4th ed., Vol. 3, *Materials, Finishing and Coating*, Society of Manufacturing Engineers, 1985.

Woldman's Engineering Alloys, 9th ed., ASM International, 2000.

Wroblewski, A.J., and Vanka, S., *MaterialTool: A Selection Guide of Materials and Processes for Designers*, Prentice Hall, 1997.

Wulpi, D.J., *Understanding How Components Fail*, 2nd ed., ASM International, 1999.

QUESTIONS

3.1 What is the difference between a unit cell and a single crystal?

3.2 Explain why we should study the crystal structure of metals.

3.3 What effects does recrystallization have on the properties of metals?

3.4 What is the significance of a slip system?

3.5 Explain what is meant by structure-sensitive and structure-insensitive properties of metals.

3.6 What is the relationship between nucleation rate and the number of grains per unit volume of a metal?

3.7 Explain the difference between recovery and recrystallization.

3.8 (a) Is it possible for two pieces of the same metal to have different recrystallization temperatures? Explain. (b) Is it possible for recrystallization to take place in some regions of a workpiece before other regions do in the same workpiece? Explain.

3.9 Describe why different crystal structures exhibit different strengths and ductilities.

3.10 Explain the difference between preferred orientation and mechanical fibering.

3.11 Give some analogies to mechanical fibering (such as layers of thin dough sprinkled with flour).

3.12 A cold-worked piece of metal has been recrystallized. When tested, it is found to be anisotropic. Explain the probable reason for this behavior.

3.13 Does recrystallization completely eliminate mechanical fibering in a workpiece? Explain.

3.14 Explain why we may have to be concerned with the orange-peel effect on metal surfaces.

3.15 How can you tell the difference between two parts made of the same metal, one shaped by cold working and the other by hot working? Explain the differences you might observe. Note that there are several methods that can be used to determine the differences between the two parts.

3.16 Explain why the strength of a polycrystalline metal at room temperature decreases as its grain size increases.

3.17 What is the significance of some metals, such as lead and tin, having recrystallization temperatures at about room temperature?

3.18 You are given a deck of playing cards held together with a rubber band. Which of the material-behavior phenomena described in this chapter could you demonstrate with this setup? What would be the effects of increasing the number of rubber bands holding the cards together? Explain. (*Hint:* Inspect Figs. 3.5 and 3.7.)

3.19 Using the information given in Chapters 2 and 3, list and describe the conditions that induce brittle fracture in an otherwise ductile piece of metal.

3.20 Make a list of metals that would be suitable for a (1) paper clip, (2) bicycle frame, (3) razor blade, (4) battery cable, and (5) gas-turbine blade. Explain your reasoning.

3.21 Explain the advantages and limitations of cold, warm, and hot working of metals, respectively.

3.22 Explain why parts may crack when suddenly subjected to extremes of temperature.

3.23 From your own experience and observations, list three applications each for the following metals and their alloys: (1) steel, (2) aluminum, (3) copper, (4) magnesium, and (5) gold.

3.24 List three applications that are not suitable for each of the following metals and their alloys: (1) steel, (2) aluminum, (3) copper, (4) magnesium, and (5) gold.

3.25 Name products that would not have been developed to their advanced stages, as we find them today, if alloys with high strength and corrosion and creep resistance at elevated temperatures had not been developed.

3.26 Inspect several metal products and components and make an educated guess as to what materials they are made from. Give reasons for your guess. If you list two or more possibilities, explain your reasoning.

3.27 List three engineering applications each for which the following physical properties would be desirable: (1) high density, (2) low melting point, and (3) high thermal conductivity.

3.28 Two physical properties that have a major influence on the cracking of workpieces, tools, or dies during thermal cycling are thermal conductivity and thermal expansion. Explain why.

3.29 Describe the advantages of nanomaterials over traditional materials.

3.30 Aluminum has been cited as a possible substitute material for steel in automobiles. What concerns, if any, would you have prior to purchasing an aluminum automobile?

3.31 Lead shot is popular among sportsmen for hunting, but birds commonly ingest the pellets (along with gravel) to help digest food. What substitute materials would you recommend for lead, and why?

3.32 What are metallic glasses? Why is the word *glass* used for these materials?

3.33 Which of the materials described in this chapter has the highest (a) density, (b) electrical conductivity, (c) thermal conductivity, (d) strength, and (e) cost?

3.34 What is twinning? How does it differ from slip?

PROBLEMS

3.35 Calculate the theoretical (a) shear strength and (b) tensile strength for aluminum, plain-carbon steel, and tungsten. Estimate the ratios of their theoretical strength to actual strength.

3.36 A technician determines that the grain size of a certain etched specimen is 6. Upon further checking, it is found that the magnification used was 150, instead of 100 as required by ASTM standards. What is the correct grain size?

3.37 Estimate the number of grains in a regular paper clip if its ASTM grain size is 9.

3.38 The natural frequency f of a cantilever beam is given by the expression

$$f = 0.56 \sqrt{\frac{EIg}{wL^4}},$$

where E is the modulus of elasticity, I is the moment of inertia, g is the gravitational constant, w is the weight of the beam per unit length, and L is the length of the beam. How does the natural frequency of the beam change, if any, as its temperature is increased?

3.39 A strip of metal is reduced in thickness by cold working from 25 mm to 15 mm. A similar strip is reduced from 25 mm to 10 mm. Which one of these strips will recrystallize at a lower temperature? Why?

3.40 A 1-m-long, simply supported beam with a round cross section is subjected to a load of 50 kg at its center. (a) If the shaft is made from AISI 303 steel and has a diameter of 20 mm, what is the deflection under the load? (b) For shafts made from 2024-T4 aluminum, architectural bronze, and 99.5% titanium, respectively, what must the diameter of the shaft be for the shaft to have the same deflection as in part (a)?

3.41 If the diameter of the aluminum atom is 0.5 nm, estimate the number of atoms in a grain with an ASTM size of 5.

3.42 Plot the following for the materials described in this chapter: (a) yield stress versus density, (b) modulus of elasticity versus strength, and (c) modulus of elasticity versus relative cost. *Hint:* See Table 16.4.

3.43 The following data are obtained in tension tests of brass:

Grain size (μm)	Yield stress (MPa)
15	150
20	140
50	105
75	90
100	75

Does this material follow the Hall-Petch effect? If so, what is the value of k?

3.44 It can be shown that thermal distortion in precision devices is low for high values of thermal conductivity divided by the thermal expansion coefficient. Rank the materials in Table 3.3 according to their suitability to resist thermal distortion.

3.45 Assume that you are asked to give a quiz to students on the contents of this chapter. Prepare three quantitative problems and three qualitative questions, and supply the answers.

Surfaces, Tribology, Dimensional Characteristics, Inspection, and Product Quality Assurance

Several important considerations in materials processing are described in this chapter:

- Surface structures, textures, and surface properties as they affect processing of materials.
- The role of friction, wear, and lubrication (tribology) in manufacturing processes, and the characteristics of various metalworking fluids.
- Surface treatments to enhance the appearance and performance of manufactured products.
- Engineering metrology, instrumentation, and dimensional tolerances, and their effect on product quality and performance.
- Destructive and nondestructive inspection methods for manufactured parts.
- Statistical techniques for quality assurance of products.

4.1 | Introduction

The various mechanical, physical, thermal, and chemical effects induced during processing of materials have an important influence on the **surface** of a manufactured part. A surface generally has properties and behavior that are considerably different from those of the bulk of the part. Although the **bulk material** generally determines a component's overall mechanical properties, the component's surfaces directly influence several important properties and characteristics of the manufactured part:

- Friction and wear properties of the part during subsequent processing when it comes into direct contact with tools, dies, and molds, or when it is placed in service.
- Effectiveness of lubricants during manufacturing processes and during service.

- Appearance and geometric features of the part and their role in subsequent operations such as painting, coating, welding, soldering, and adhesive bonding, as well as corrosion resistance of the part.

- Initiation of cracks due to surface defects, such as roughness, scratches, seams, and heat-affected zones, which could lead to weakening and premature failure of the part by fatigue or other fracture mechanisms.

- Thermal and electrical conductivity of contacting bodies; for example, a rough surface will have higher thermal and electrical resistance than a smooth surface because of the fewer direct contact points between the two surfaces.

Friction, wear, and lubrication (**tribology**) are *surface phenomena*. Consequently, (a) friction influences force and energy requirements and surface quality of the parts produced; (b) wear alters the surface geometry of tools and dies, which, in turn, adversely affects the quality of manufactured products and the economics of production; and (c) lubrication is an integral aspect of all manufacturing operations, as well as in the proper functioning of machinery and equipment.

The properties and characteristics of surfaces can be modified and improved by various **surface treatments**. Several mechanical, thermal, electrical, and chemical methods can be used to improve frictional behavior, effectiveness of lubricants, resistance to wear and corrosion, and surface finish and appearance.

Measurement of the relevant dimensions and features of products is an integral aspect of interchangeable manufacture, which is the basic concept of standardization and mass production. We describe the principles involved and the various instruments used in measurement. Another important aspect is **testing** and **inspection** of manufactured products, using either destructive or nondestructive methods. **Product quality** is one of the most critical aspects of manufacturing. This chapter emphasizes the technological and economic importance of *building quality into a product* rather than inspecting the product *after* it is made, and it describes the techniques used in achieving this goal.

4.2 | Surface Structure and Properties

Upon close examination of a piece of metal, we find that its surface generally consists of several layers. Referring to Fig. 4.1, the portion of the metal in the interior is called the **metal substrate** (the **bulk metal**). Its structure depends on the composition and processing history of the metal. Above this bulk metal is a layer that has usually

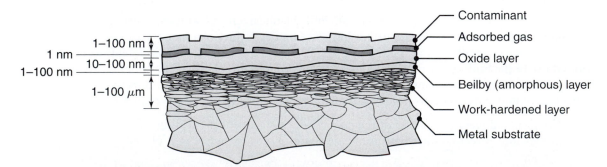

FIGURE 4.1   Schematic illustration of the cross section of the surface structure of metals. The thickness of the individual layers depends on processing conditions and the environment. *Source:* After E. Rabinowicz and B. Bhushan.

been plastically deformed and work hardened during processing. The depth and properties of the work-hardened layer (called the **surface structure**) depend on factors such as the processing method used and the extent of frictional sliding to which the surface was subjected. If the surface was produced by machining with a dull tool or under poor cutting conditions or was ground with a dull grinding wheel, this layer will be relatively thick. Furthermore, during processing, **residual stresses** in this work-hardened layer may develop due to (a) nonuniform surface deformation, or (b) severe temperature gradients.

Additionally, a **Beilby**, or **amorphous, layer** may develop on top of the work-hardened layer, consisting of a structure that is microcrystalline or amorphous. This layer develops in some machining and surface finishing operations where melting and surface flow, followed by rapid quenching, have been encountered. Unless the metal is processed and kept in an inert (oxygen-free) environment, or it is a noble metal, such as gold or platinum, an **oxide layer** usually develops on top of the work-hardened or Beilby layer. Typical examples of metals that develop an oxide layer include the following:

1. *Iron* has an oxide surface structure, with FeO adjacent to the bulk metal, followed by a layer of Fe_3O_4 and then a layer of Fe_2O_3, which is exposed to the environment.

2. *Aluminum* has a dense *amorphous* layer of Al_2O_3, with a thick, porous, hydrated aluminum-oxide layer over it.

3. *Copper* has a bright shiny surface when freshly scratched or machined. Soon after, however, it develops a Cu_2O layer, which is then covered with a layer of CuO; these layers give copper its somewhat dull color, such as seen in kitchen utensils.

4. *Stainless steels* are "stainless" because they develop a protective layer of chromium oxide, CrO (called **passivation**).

Under normal environmental conditions, surface oxide layers are generally covered with *adsorbed* layers of gas and moisture. Finally, the outermost surface of the metal may be covered with *contaminants,* such as dirt, dust, grease, lubricants, cleaning-compound residues, and pollutants from the environment.

It is thus apparent that surfaces generally have properties that are significantly different from those of the substrate. The oxide on a metal's surface, for example, is generally much harder than the base metal; hence, oxides tend to be brittle and abrasive. This surface characteristic has, in turn, several important effects on friction, wear, and lubrication in materials processing, as well as subsequent coatings on products. The factors involved in the surface structure of metals described thus far are also relevant, to a great extent, to the surface structure of plastics and ceramics. The surface texture developed on these materials depends, as with metals, on the method of production and environmental conditions.

Surface integrity. *Surface integrity* describes not only the geometric (topological) features of surfaces, but also their mechanical and metallurgical properties and characteristics. Surface integrity is an important consideration in manufacturing operations because it can influence the fatigue strength, resistance to corrosion, and service life of products.

Several **defects** caused by and produced during processing can be responsible for lack of surface integrity. These defects are usually caused by a combination of factors, such as (a) defects in the original material, (b) the method by which the surface is produced, (c) and lack of proper control of process parameters, resulting, for example, in excessive stresses and temperatures. The major surface defects encountered in practice usually consist of one or more of the following: *cracks, craters, folds, laps, seams, splatter, inclusions, intergranular attack, heat-affected zones, metallurgical transformations, plastic deformation,* and *residual stresses.*

4.3 | Surface Texture and Roughness

Regardless of the method of production, all surfaces have their own set of characteristics, referred to as **surface texture**. Although the description of surface texture as a geometrical property can be complex, certain guidelines have been established for identifying surface texture in terms of well-defined and measurable quantities (Fig. 4.2):

1. **Flaws**, or **defects**, are random irregularities, such as scratches, cracks, cavities, depressions, seams, tears, and inclusions.

2. **Lay**, or **directionality**, is the direction of the predominant surface pattern and is usually visible to the naked eye.

3. **Waviness** is a recurrent deviation from a flat surface, much like waves on the surface of water. It is described and measured in terms of (1) the space between adjacent crests of the waves (*waviness width*) and (2) the height between the crests and valleys of the waves (*waviness height*). Waviness may be caused by (a) deflections of tools, dies, and of the workpiece, (b) warping from forces or temperature, (c) uneven lubrication, and (d) vibration or any periodic mechanical or thermal variations in the system during the manufacturing operation.

4. **Roughness** consists of closely spaced irregular deviations on a scale smaller than that for waviness. Roughness, which may be superimposed on waviness, is expressed in terms of its height, its width, and the distance on the surface along which it is measured.

Surface roughness. *Surface roughness* is generally described by two methods. The **arithmetic mean value**, R_a, formerly identified as AA (for *arithmetic average*) or as CLA (for *center-line average*), is based on the schematic illustration of a rough surface, as shown in Fig. 4.3. The arithmetic mean value is defined as

$$R_a = \frac{y_a + y_b + y_c + \cdots + y_n}{n} = \frac{1}{n}\sum_{i=1}^{n} y_i = \frac{1}{l}\int_0^l |y|\, dx, \qquad (4.1)$$

where all ordinates $y_a, y_b, y_c, \ldots,$ are absolute values. The last term in Eq. (4.1) refers to the R_a of a continuous surface or wave, as is commonly encountered in analog signal processing, and l is the total measured profile length.

The **root-mean-square average**, R_q, formerly identified as RMS (for *root mean square*), is defined as

$$R_q = \sqrt{\frac{y_a^2 + y_b^2 + y_c^2 + \cdots + y_n^2}{n}} = \sqrt{\frac{1}{n}\sum_{i=1}^{n} y_i^2} = \left[\frac{1}{l}\int_0^l y^2\, dx\right]^{1/2}. \qquad (4.2)$$

The center (*datum*) line shown in Fig. 4.3 is located so that the sum of the areas above the line is equal to the sum of the areas below the line. The units generally used for surface roughness are μm (micrometer, or micron) or μin. (microinch), where 1 μm = 40 μin. and 1 μin. = 0.025 μm.

The **maximum roughness height**, R_t, may also be used as a measure of surface roughness. Defined as the vertical distance from the deepest trough to the highest peak, it indicates the amount of material that has to be removed in order to obtain a smooth surface by various means, such as polishing. Because of its simplicity, the

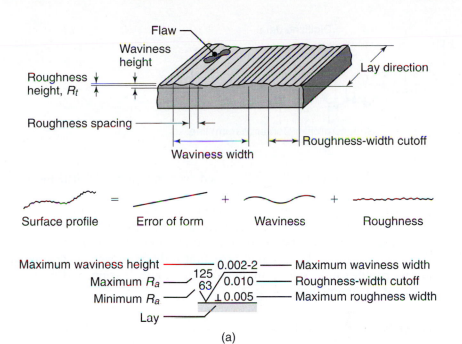

(a)

Lay symbol	Interpretation	Examples
$=$	Lay parallel to the line representing the surface to which the symbol is applied	
$\perp$	Lay perpendicular to the line representing the surface to which the symbol is applied	
X	Lay angular in both directions to line representing the surface to which symbol is applied	
P	Pitted, protuberant, porous, or particulate nondirectional lay	

(b)

arithmetic mean value was adopted internationally in the mid-1950s and is widely used in engineering practice.

It can be seen in Eqs. (4.1) and (4.2) that there is a relationship between R_a and R_q. It can be shown that for a surface roughness in the shape of a sine curve, R_q is larger than R_a by a factor of 1.11. This factor is 1.1 for most cutting processes, 1.2 for grinding, and 1.4 for lapping and honing. R_q is more sensitive to the largest asperity peaks and deepest valleys, and since these peaks and valleys are important

FIGURE 4.3 Coordinates used for measurement of surface roughness, used in Eqs. (4.1) and (4.2).

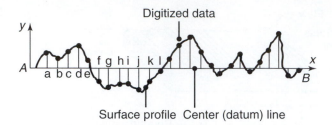

for friction and lubrication, R_q is often used even though it is a little more complicated to calculate than R_a.

In general, a surface cannot be completely described by its R_a or R_q value alone, since these values are averages. Two surfaces may have the same roughness value but their actual topography may be quite different. A few deep troughs, for example, affect the roughness values insignificantly. However, such differences in the surface profile can be significant in terms of fatigue, friction, and the wear characteristics of a manufactured product.

Symbols for surface roughness. Acceptable limits for surface roughness are specified on technical drawings by the symbols shown around the check mark in the lower portion of Fig. 4.2a, and their values are placed to the left of the check mark. Symbols used to describe a surface specify only the roughness, waviness, and lay. They do not include flaws; whenever flaws are important, a special note is included in technical drawings to describe the method to be used to inspect for surface flaws.

Measuring surface roughness. Several commercially available instruments, called **surface profilometers**, are used to measure and record surface roughness. The most commonly used instruments feature a diamond stylus traveling along a straight line over the surface (Figs. 4.4a and b). The distance that the stylus travels, which can be specified and controlled, is called the **cutoff**. (See Fig. 4.2.) To highlight the roughness, profilometer traces are recorded on an exaggerated vertical scale (a few orders of magnitude greater than the horizontal scale; Figs. 4.4c–f), called **gain** on the recording instrument. The recorded profile is therefore significantly distorted, making the surface appear much rougher than it actually is. The recording instrument compensates for any surface waviness and indicates only roughness. A record of the surface profile is made by mechanical and electronic means.

Surface roughness can be observed directly through (a) *interferometry,* and (b) optical, scanning-electron, laser, or atomic-force *microscopy.* The microscopy techniques are especially useful for imaging very smooth surfaces for which features cannot be captured by less sensitive instruments. Stereoscopic photographs are particularly useful for three-dimensional views of surfaces and can also be used to measure surface roughness. **Three-dimensional surface measurements** can be performed by three methods. An *optical-interference* microscope shines a light beam against a reflective surface and records the interference fringes that result from the incident and its reflective waves.

Laser profilometers are used to measure surfaces through either interferometric techniques or by moving an objective lens to maintain a constant focal length over a surface; the motion of the lens is then a measure of the surface. An *atomic-force microscope* can be used to measure extremely smooth surfaces and even has the capability of distinguishing atoms on atomically smooth surfaces. More commonly, it is used to scan portions of surfaces that are less than 100 μm (4000 μin.) square, with a vertical resolution of a few micrometers.

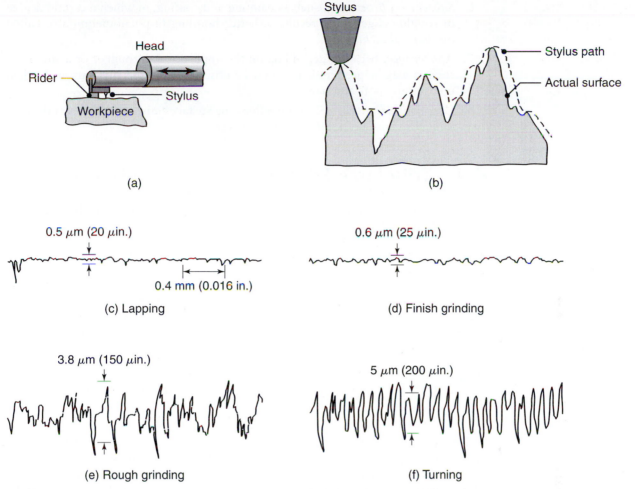

FIGURE 4.4   (a) Measuring surface roughness with a stylus. The rider supports the stylus and guards against damage. (b) Path of the stylus in measurements of surface roughness (broken line) compared with the actual roughness profile. Note that the profile of the stylus's path is smoother than the actual surface profile. Typical surface profiles produced by (c) lapping, (d) finish grinding, (e) rough grinding, and (f) turning processes. Note the difference between the vertical and horizontal scales.

Surface roughness in engineering practice.   The specified surface roughness design requirements for engineering applications can vary by as much as two orders of magnitude. The reasons and considerations for this wide range include the following:

1. *Precision required on mating surfaces,* such as seals, fittings, gaskets, tools, and dies. For example, ball bearings and gages require very smooth surfaces, whereas surfaces for gaskets and brake drums are much rougher.

2. *Tribological considerations,* that is, the effect of roughness on friction, wear, and lubrication.

3. *Fatigue and notch sensitivity,* because rougher surfaces often have shorter fatigue life.

4. *Electrical and thermal contact resistance,* because the rougher the surface, the higher the contact resistance will be.

5. *Corrosion resistance,* because the rougher the surface, the greater will be the possibility of entrapping corrosive media.

6. *Subsequent processing*, such as painting and coating, in which a certain degree of roughness generally results in better bonding (a phenomenon also called *mechanical locking*).

7. *Appearance*, because, depending on the application, a rougher or a smoother surface may be preferred. This may be observed by inspecting various kitchen utensils and pots and pans.

8. *Cost considerations*, because the finer the surface finish, the higher is the cost (a major consideration in manufacturing).

4.4 | Tribology: Friction, Wear, and Lubrication

Tribology is the science and technology of interacting surfaces, thus involving friction, wear, and lubrication.

4.4.1 Friction

Friction is defined as the resistance to relative sliding between two bodies in contact under a normal load. Metalworking processes are significantly affected by friction because of the relative motion and the forces present between tools, dies, and workpieces. Friction is an energy-dissipating (irreversible) process that results in the generation of heat. The subsequent rise in temperature can have a major detrimental effect on the overall operation (such as by causing excessive heating in grinding, leading to heat checks). Furthermore, because it impedes movement at the tool, die, and workpiece interfaces, friction significantly affects the flow and deformation of materials in metalworking processes.

We describe the role of friction, as well as other tribological considerations such as wear and lubrication, in detail while describing individual manufacturing processes. See, for example, forging (Section 6.2), rolling (6.3), extrusion (6.4), drawing (6.5), sheet-metal forming (Chapter 7), machining (Chapter 8), and abrasive processes (Chapter 9). It should be noted that friction is not always undesirable. For example, friction is necessary in the rolling of metals to make sheet and plate (Section 6.3), as otherwise it would be impossible to roll materials, just as it would be impossible to drive a car on the road. Various theories have been proposed to explain the phenomenon of friction.

For a particular theory of friction to be robust, it must explain the frictional behavior of two bodies under a variety of conditions, such as load, relative sliding speed, temperature, surface condition, and the environment. An early *Coulomb* model, for example, states that friction results from the mechanical interlocking of rough surface asperities, thereby requiring some force to be able to slide the two bodies along each other. Several more advanced models of friction have been proposed, with varying degrees of success. Because of its reasonable agreement with experimental observations, the most commonly referred to theory of friction is based on adhesion.

Adhesion theory of friction. The *adhesion theory of friction* is based on the observation that two clean, dry (that is, unlubricated) metal surfaces, regardless of how smooth they are, contact each other at only a fraction of their apparent area of contact (Fig. 4.5). The static load at the interface is thus supported by the contacting asperities. The sum of the contacting areas is known as the **real area of contact, A_r**.

Under light loads and with a large real area of contact, the normal stress at the asperity contacts (**junctions**) is low and hence it is *elastic*. As the load increases, the stresses increase and, eventually, the junctions undergo *plastic deformation*. Also, with increasing load, (a) the contact area of the asperities increases, and (b) new junctions are formed, with other asperities coming in contact with each

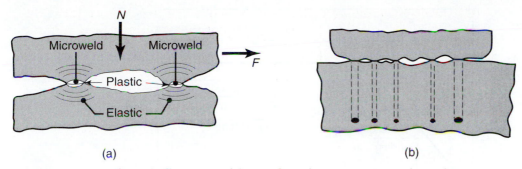

FIGURE 4.5 (a) Schematic illustration of the interface of two contacting surfaces, showing the real areas of contact and (b) sketch illustrating the proportion of the apparent area to the real area of contact. The ratio of the areas can be as high as four to five orders of magnitude.

other. Because the asperity-height distribution on the surfaces is random, some junctions will be in elastic contact, and others in plastic contact.

The intimate contact of asperities develops an **adhesive bond.** The nature of this bond involves *atomic interactions, mutual solubility,* and *diffusion.* The strength of the bond thus depends on the physical and mechanical properties of the metals in contact, the temperature, and the nature and thickness of any oxide film or other contaminants present on the surfaces. In manufacturing processes (see Chapters 6 through 9), the load at the interface is typically high; thus plastic deformation of the asperities takes place, resulting in adhesion of the junctions. In other words, the asperities form **microwelds.** Obviously, the cleaner the surfaces in contact, the stronger are the adhesive bonds.

Coefficient of friction. Sliding between two bodies under a normal load, N, is possible only by the application of a tangential force, F. According to the adhesion theory, F is the force required to *shear the junctions* (thus the friction force). The *coefficient of friction,* μ, at the interface is defined as

$$\mu = \frac{F}{N} = \frac{\tau A_r}{\sigma A_r} = \frac{\tau}{\sigma}, \tag{4.3}$$

where τ is the shear strength of the junction and σ is the normal stress, which, for a plastically deformed asperity, is equivalent to the hardness of the material (see Section 2.6). A_r is the real area of contact between the surfaces, as illustrated in Fig. 4.5b. The coefficient of friction can now be defined as

$$\mu = \frac{\tau}{\text{Hardness}}. \tag{4.4}$$

It can be seen that the nature and strength of the interface are the most significant factors in the magnitude of friction. A strong interface thus requires a high friction force for relative sliding. Equation (4.4) indicates that the coefficient of friction can be reduced either by *decreasing the shear stress* (such as by placing a thin film of low shear strength in the interface) and/or by *increasing the hardness* of the materials involved.

Two other phenomena take place in asperity interactions under high contact stresses. For a strain-hardening material, the peak of the asperity is stronger than the bulk material because of plastic deformation at the junction. Thus, when a tensile force is applied and under ideal conditions, fracture of the junction will likely follow a path either below or above the geometric interface of the two bodies. Second, a tangential motion at the junction under load will cause the contact area to increase, known as **junction growth.** The reason is that, according to yield criteria (described in Section 2.11), the *effective yield stress* of the material will decrease when it is subjected to an external shear stress. The junction (hence the real area of contact) thus has to grow in area in order to support the normal load.

FIGURE 4.6 Schematic illustration of the relation between friction force F and normal force N. Note that as the real area of contact, A_r, approaches the apparent area, A, the friction force reaches a maximum and stabilizes. At low normal forces, the friction force is proportional to normal force; most machine components operate in this region. The friction force is not linearly related to normal force in metalworking operations because of the high contact pressures involved.

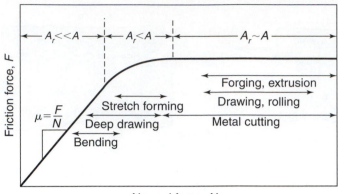

In the absence of contaminants or any fluids that may have been trapped at the interface, the real area of contact will eventually reach the apparent area of contact; this area is the maximum contact area that can be achieved and the two bodies will behave as one solid body. The friction force will now become the force required to shear the body, reaching a maximum and leveling off, as shown in Fig. 4.6. This condition is known as *sticking*. However, sticking in a sliding interface does not necessarily mean complete adhesion at the interface, as it would in welding or brazing; rather, it means that the frictional stress at the surface has reached a limiting value (which is related to the shear yield stress, k, of the material).

When two clean surfaces are pressed together with a sufficiently high normal force, *cold pressure welding* can take place; usually, however, the frictional stress is limited by contamination or by the presence of oxide layers. As the normal load or force, N, is increased further, the friction force, F, remains constant (see Fig. 4.6); hence, by definition, the coefficient of friction decreases. This situation indicates that there might be a different and more realistic means of expressing the frictional condition at an interface.

A useful approach is to define a **friction factor**, or **shear factor**, m, as

$$m = \frac{\tau_i}{k},\qquad(4.5)$$

where τ_i is the shear strength of the interface and k is the *shear yield stress* of the softer material in a sliding pair. The quantity k is equal to $Y/2$ according to the maximum-shear-stress criterion [Eq. (2.35)] and to $Y/\sqrt{3}$ according to the distortion-energy criteron (Section 2.11.2). Equation (4.5) is often referred to as the *Tresca friction model*. Note that when $m = 0$ in this definition, there is no friction, and when $m = 1$, complete sticking takes place at the interface. The magnitude of m is independent of the normal force or stress; the reason is that the shear yield stress of a thin layer of material is unaffected by the magnitude of the normal stress.

The sensitivity of friction to various factors can be illustrated by the following examples: A typical value for the coefficient of friction of steel sliding on lead is 1.0; for steel on copper, it is 0.9. However, for steel sliding on copper coated with a thin layer of lead, it is 0.2. The coefficient of friction of pure nickel sliding on nickel in a hydrogen or nitrogen atmosphere is 5; in air or oxygen, it is 3; and in the presence of water vapor, it is 1.6.

Coefficients of friction in sliding contact, as measured experimentally, vary from as low as 0.02 to as high as 100, or even higher. This range is not surprising in view of the many variables involved in the friction process. In metalworking operations that use various lubricants, the range for μ is much narrower, as shown in Table 4.1. The effects of load, temperature, speed, and environment on the coefficient of friction are difficult to generalize, as each situation must be investigated individually.

TABLE 4.1

Coefficient of Friction in Metalworking Processes		
	Coefficient of friction (μ)	
Process	Cold	Hot
Rolling	0.05–0.1	0.2–0.7
Forging	0.05–0.1	0.1–0.2
Drawing	0.03–0.1	—
Sheet-metal forming	0.05–0.1	0.1–0.2
Machining	0.5–2	—

Abrasion theory of friction. If the upper body in the model shown in Fig. 4.5a is harder than the lower body, or if its surface has protruding hard particles, then as it slides over the softer body, it will scratch and produce grooves on the lower surface (see also *Mohs hardness*, Section 2.6.6). This phenomenon, known as **plowing,** is an important aspect in abrasive friction; in fact, it can be a dominant mechanism for situations where adhesion is not particularly strong.

Plowing may involve two different mechanisms: (1) the generation of a *groove*, whereby the surface is deformed plastically, and (2) the formation of a groove because the action of the upper body generates a *chip* or *sliver* from the softer body. Both of these processes involve work supplied by a force that manifests itself as friction force. The plowing force can contribute significantly to friction and to the measured coefficient of friction at the interface.

Measuring friction. The coefficient of friction is generally determined experimentally, either by simulated tests using small-scale specimens of various shape, or during an actual manufacturing process. The techniques used generally involve measurements of either forces or dimensional changes in the specimen. One test that has gained wide acceptance, particularly for bulk deformation processes such as forging, is the **ring compression test.** In this test, a flat ring is compressed plastically between two flat platens (Fig. 4.7). As its height is reduced, the ring expands radially outward because of volume constancy. If friction at the platen-specimen interfaces is zero, both the inner and outer diameters of the ring will expand as if the ring were a solid disk.

Good lubrication   Poor lubrication

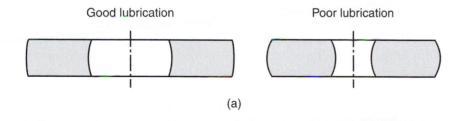

(a)

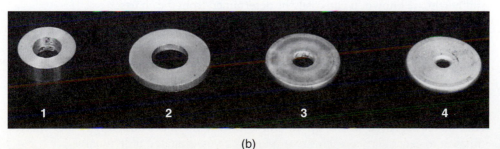

(b)

FIGURE 4.7 (a) The effects of lubrication on barreling in the ring compression test. With good lubrication, both the inner and outer diameters increase as the specimen is compressed; with poor or no lubrication, friction is high, and the inner diameter decreases. The direction of barreling depends on the relative motion of the cylindrical surfaces with respect to the flat dies. (b) Test results: (1) original specimen and (2–4) the specimen under increasing friction. *Source:* After A.T. Male and M.G. Cockcroft.

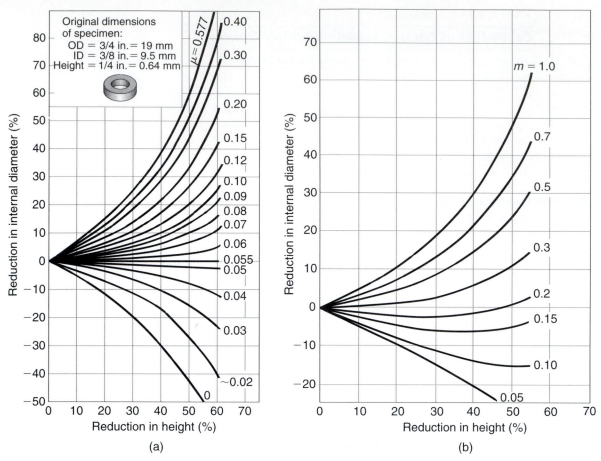

FIGURE 4.8   Charts to determine friction in ring compression tests: (a) coefficient of friction, μ, and (b) friction factor, m. Friction is determined from these charts from the percent reduction in height and by measuring the percent change in the internal diameter of the specimen after compression.

With increasing friction, the inner diameter decreases because it takes less energy to reduce the inner diameter than to expand the outer diameter of the specimen.

For a certain reduction in height, there is a critical friction value at which the internal diameter increases (from the original) if μ is low and decreases if μ is high. By measuring the change in the specimen's internal diameter and using the curves shown in Fig. 4.8 (which are developed through theoretical analyses), the coefficient of friction can be determined. The geometry of each ring has its own specific set of curves; the most common geometry has specimen proportions of outer diameter to inner diameter to height of 6:3:2. The actual size of the specimen is usually not relevant in these tests, except that when the test is conducted at elevated temperatures, smaller specimens will cool faster than large ones. Thus, once the percentage reductions in internal diameter and height are known, μ can easily be determined from the charts.

Temperature rise caused by friction.   The energy expended in overcoming friction is converted into heat, except for a small portion which remains as stored energy. Frictional heat raises the interface temperature. The magnitude of this temperature rise and its distribution throughout the workpiece depend not only on the friction force but also on speed, surface roughness, and the physical properties of the materials, especially thermal conductivity and specific heat. The temperature increases with increasing friction and speed and with low thermal conductivity and specific heat of the materials; see also Eq. (2.65). The interface temperature can be high enough to soften and melt the surface. The temperature cannot, of course, exceed the melting point of the material. Several analytical expressions have been developed to calculate the temperature rise in sliding between two bodies; Eq. (9.9) is one such expression applicable to grinding operations.

Reducing friction.   Friction can be reduced by selecting materials that exhibit low adhesion, such as relatively hard materials like carbides and ceramics, and by applying films and coatings on surfaces in contact. Lubricants such as oils or solid films such as graphite interpose an adherent film between tools, dies, and workpieces. Abrasive friction can be reduced by reducing the surface roughness of the harder of the two mating surfaces.

Note that a smooth surface inherently has fewer asperity peaks, and thus plowing will be less of a concern. On the other hand, adhesive friction is more of a concern with smooth surfaces. Friction can also be reduced significantly by subjecting the die-workpiece interface to **ultrasonic vibrations**, typically at 20 kHz. Depending on the amplitude, vibrations momentarily separate the die and the workpiece, thus allowing the lubricant to flow more freely into the interface and lubricate the interfaces more effectively.

Friction in plastics and ceramics.   Although their strength is low compared with that of metals, plastics (see Chapter 10) generally possess low frictional characteristics. This property makes polymers attractive for applications such as bearings, gears, seals, prosthetic joints, and general low-friction applications. In fact, polymers are sometimes called *self-lubricating* materials. The factors involved in metal friction described earlier are also generally applicable to polymers. However, the plowing component of friction in thermoplastics and elastomers is a significant factor because of their viscoelastic behavior (that is, they exhibit both viscous and elastic behavior) and the subsequent *hysteresis loss*.

An important factor in plastics applications is the effect of temperature rise at interfaces caused by friction. Thermoplastics lose their strength and become soft as temperature increases. Their low thermal conductivity and low melting points are thus significant. If the temperature rise is not monitored and controlled, sliding surfaces can undergo plastic deformation and thermal degradation.

The frictional behavior of ceramics (Section 11.8) has been studied extensively in view of their commercial importance. Investigations have indicated that the origin of friction in ceramics is similar to that in metals; thus, adhesion and plowing at interfaces contribute to the friction force in ceramics. Usually, adhesion is less important with ceramics because of their high hardness, so that a small real area of contact is developed.

EXAMPLE 4.1   Determining the coefficient of friction

In a ring compression test, a specimen 10 mm in height with outside diameter of 30 mm and inside diameter of 15 mm is reduced in thickness by 50%. Determine the coefficient of friction, μ, and the friction factor, m, if the outer diameter (OD) after deformation is 39 mm.

Solution.   We first have to determine the new inner diameter (ID); this diameter is obtained from volume constancy, as follows:

$$\text{Volume} = \frac{\pi}{4}(30^2 - 15^2)10 = \frac{\pi}{4}(39^2 - \text{ID}^2)5.$$

From this equation, we find that the new ID = 13 mm. Thus,

$$\text{Change in ID} = \frac{15 - 13}{15} \times 100\% = 13\% \text{ (decrease)}.$$

For a 50% reduction in height and a 13% reduction in ID, the following values are interpolated from Fig. 4.8:

$$\mu = 0.09 \quad \text{and} \quad m = 0.4.$$

4.4.2 Wear

Wear is defined as the progressive loss or undesired removal of material from a surface. Wear has important technological and economic significance, especially if it alters the shape of the workpiece, tool, and die interfaces, adversely affecting the manufacturing process and the size and quality of the parts produced. Examples of wear in manufacturing processes include dull drills, worn cutting tools, and dies in metalworking operations.

Although wear generally changes the surface topography and can also result in severe surface damage, it can have an important beneficial effect: It can reduce surface roughness by removing the peaks from asperities, as shown in Fig. 4.9. Thus, under controlled conditions, wear may be regarded as a type of smoothening or polishing process. The *running-in period* for various machines and engines produces this type of wear.

Because they are expected to wear during their normal use, especially under high loads, some components of machinery are equipped with **wear plates** (or *wear parts*). These plates can easily be replaced when worn without causing any damage

FIGURE 4.9 Changes in originally (a) wire-brushed and (b) ground-surface profiles after wear. *Source: After E. Wild and K.J. Mack.*

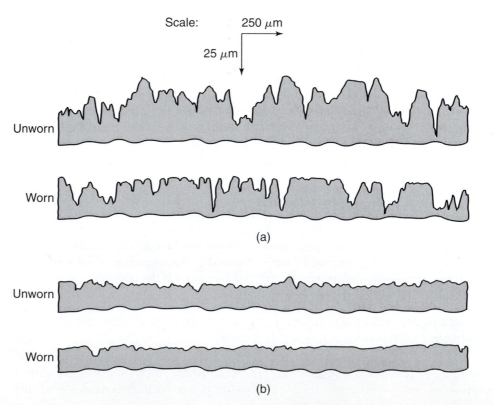

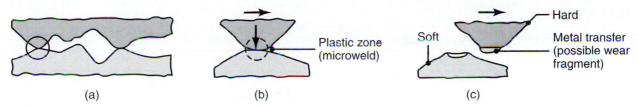

FIGURE 4.10 Schematic illustration of (a) asperities contacting, (b) adhesion between two asperities, and (c) the formation of a wear particle.

to the rest of the machinery. Examples of wear parts include cylinder liners in extrusion presses and tool inserts in forging and drawing.

Wear is generally classified into the following categories.

Adhesive wear. If a tangential force is applied to the model shown in Fig. 4.5, shearing of the junctions can take place either at the original interface of the two bodies or along a path below or above the interface (Fig. 4.10), a phenomenon known as *adhesive wear*. The fracture path depends on whether or not the strength of the adhesive bond of the asperities is higher than the cohesive strength of either of the two sliding bodies.

Because of factors such as strain hardening at the asperity contact, diffusion, and mutual solid solubility of the two bodies, the adhesive bonds are typically stronger than the base metal's. Thus, fracture at the asperity usually follows a path in the weaker or softer component because it takes less energy. A *wear fragment* is then generated. Although this fragment is attached to the harder component (see the upper member in Fig. 4.10), it is eventually detached during further rubbing at the interface, becoming a loose wear particle. This phenomenon is known as *adhesive wear*, or *sliding wear*. In severe cases, such as under high normal loads and strongly bonded asperities, adhesive wear is described as *scuffing, smearing, tearing, galling,* or *seizure*. Oxide layers on surfaces greatly influence adhesive wear. Such layers can act as a protective film, resulting in what is known as **mild wear**, consisting of small wear particles.

Based on the probability that a junction between two sliding surfaces will lead to the formation of a wear particle, the **Archard wear law** provides an expression for adhesive wear,

$$V = k\frac{LW}{3p},\qquad(4.6)$$

where V is the volume of material removed by wear from the surface, k is the *wear coefficient* (dimensionless), L is the length of travel, W is the normal load, and p is the indentation hardness of the softer body. It should be pointed out that, in some references, the factor 3 is deleted from the denominator and incorporated into the wear coefficient.

Table 4.2 lists typical values of k for a combination of materials sliding in air. Note that the wear coefficient for the same pair of materials can vary by a factor of 3, depending on whether wear is measured as a loose particle or as a transferred particle (to the mating body). Loose particles have a lower k value. Mutual solubility of the mating bodies is a significant parameter in adhesion; thus similar metal pairs have higher k values than dissimilar pairs.

Thus far, we have based the treatment of adhesive wear on the assumption that the surface layers of the two contacting bodies are clean and free from contaminants, in which case the adhesive-wear rate can be very high, leading to **severe wear**. As described in Section 4.2 and shown in Fig. 4.1, however, metal surfaces are almost always covered with contaminants and oxide layers with thicknesses generally

TABLE 4.2

Approximate Order of Magnitude for the Wear Coefficient, k, in Air			
Unlubricated	k	Lubricated	k
Mild steel on mild steel	10^{-2} to 10^{-3}	52100 steel on 52100 steel	10^{-7} to 10^{-10}
6040 brass on hardened tool steel	10^{-3}	Aluminum bronze on	10^{-8}
Hardened tool steel on hardened	10^{-4}	hardened steel	10^{-9}
tool steel		Hardened steel on	10^{-9}
Polytetrafluoroethylene (PTFE) on	10^{-5}	hardened steel	10^{-9}
tool steel			
Tungsten carbide on mild steel	10^{-6}		

ranging between 10 and 100 nm. Although this thickness may at first be regarded as insignificant, the oxide layer has a profound effect on wear behavior.

Oxide layers are usually hard and brittle. When subjected to rubbing, the oxide layer can have the following effects: If the load is light and the layer is strongly adhering to the bulk metal, the strength of the junctions between the asperities is weak and thus wear is low. In this situation, the oxide layer acts as a protective film, and the type of wear is known as *mild wear*.

The oxide layer can be broken up under high normal loads if it is brittle and is not strongly adhering to the bulk metal, or if the asperities are rubbing against each other repeatedly whereby the oxide layer breaks up by fatigue. On the other hand, if the surfaces are smooth, the oxide layers are more difficult to break up. When the oxide layer on one of the two sliding bodies is eventually broken up, a wear particle is produced. An asperity can then form a strong junction with the other mating asperity, which is now unprotected by the oxide layer. The wear rate will then be higher until a fresh oxide layer develops.

In addition to the presence of oxide layers, a surface exposed to the environment is typically covered with adsorbed layers of gas and contaminants (see Fig. 4.1). Such films, even when they are very thin, generally weaken the interfacial bond strength of contacting asperities. The magnitude of the effect of adsorbed gases and contaminants depends on many factors; for example, even small differences in environmental humidity can have a major influence on the wear rate.

Adhesive wear can be reduced by one or more of the following: (a) selecting pairs of materials that do not form a strong adhesive bond, such as one of the pair being a hard material, (b) using materials that form a thin oxide layer, (c) applying a hard coating, and (d) lubrication.

EXAMPLE 4.2 Adhesive wear in sliding

The end of a rod made of 60-40 brass is sliding over the unlubricated surface of hardened tool steel with a load of 200 lb. The hardness of brass is 120 HB. What is the distance traveled to produce a wear volume of 0.001 in^3 by adhesive wear of the brass rod?

Solution. The parameters in Eq. (4.6) for adhesive wear are as follows:

$$V = 0.001 \text{ in}^3;$$
$$k = 10^{-3} \text{ (from Table 4.2)};$$
$$W = 200 \text{ lb}$$
$$p = 120 \text{ kg/mm}^2 = 170{,}700 \text{ lb/in}^2.$$

Therefore, the distance traveled is

$$L = \frac{3Vp}{kW} = \frac{(3)(0.001)(170,700)}{(10^{-3})(200)} = 2560 \text{ in.} = 213 \text{ ft.}$$

Abrasive wear. *Abrasive wear* is caused by a hard and rough surface, or a surface with hard protruding particles, sliding against another surface. This type of wear removes particles by producing microchips or *slivers*, resulting in grooves or scratches on the softer surface (Fig. 4.11). The abrasive machining processes, described in Chapter 9, such as grinding, ultrasonic machining, and abrasive-jet machining, act in this manner, except that in these operations, the process parameters are controlled so as to produce desired shapes and surfaces, whereas abrasive wear is unintended and undesired.

The *abrasive wear resistance* of pure metals and ceramics is found to be directly proportional to their hardness. Abrasive wear can thus be reduced by increasing the hardness of materials (such as by heat treating and microstructural changes) or by reducing the normal load. Elastomers and rubbers have high abrasive wear resistance because they deform elastically and then recover after the abrasive particles cross over their surfaces. The best example is automobile tires, which have long lives (such as 70,000 km or 40,000 miles) even though they are operated on abrasive road surfaces; even hardened steels would not last long under such conditions.

In **two-body wear** the abrasive particles (such as sand) are typically carried in a jet of air or liquid, and the particles remove material from the surface by **erosion** (*erosive wear*). In **three-body wear,** the metalworking fluid may carry with it wear particles (generated over a period of time) and cause abrasive wear between the die and the workpiece. Various particles, such as from nearby machining or grinding operations or from the environment, also may contaminate the system and cause abrasive wear. Three-body wear is particularly important in forming operations. The proper filtering of the metalworking fluid is thus essential.

Corrosive wear. *Corrosive wear,* also called **oxidation** or *chemical wear,* is caused by chemical or electrochemical reactions between the surfaces and the environment. Examples of corrosive media are regular water, seawater, oxygen, acids and chemicals, and atmospheric hydrogen sulfide and sulfur dioxide. The resulting corrosive products on the mating surfaces then constitute fine wear particles. When the corrosive layer is destroyed or removed, such as by sliding or abrasion, another layer begins to form, and the process of removal and corrosive-layer formation is repeated. Corrosive wear can thus be reduced by selecting materials that will resist environmental attack, controlling the environment, and reducing operating temperatures to reduce the rate of chemical reaction.

Fatigue wear. Also called **surface fatigue** or **surface fracture wear,** *fatigue wear* is caused by surfaces being subjected to cyclic loading, such as in rolling contact in bearings. The wear particles are usually formed by a **spalling** or **pitting** mechanism. Another type of fatigue wear is **thermal fatigue.** Cracks are generated on the surface by thermal stresses from thermal cycling, such as hot workpieces repeatedly contacting

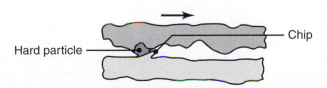

Hard particle

Chip

FIGURE 4.11   Schematic illustration of abrasive wear in sliding. Longitudinal scratches on a surface usually indicate abrasive wear.

FIGURE 4.12 Types of wear observed in a single die used for hot forging. *Source: After T.A. Dean.*

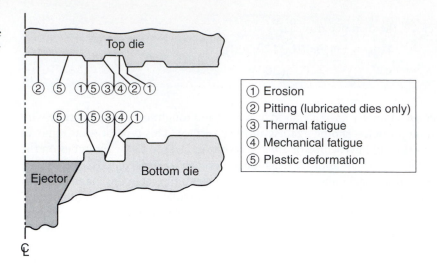

1. Erosion
2. Pitting (lubricated dies only)
3. Thermal fatigue
4. Mechanical fatigue
5. Plastic deformation

a cool die surface (*heat checking*). These cracks then join, and the surface begins to spall. This type of wear usually occurs in hot-working and die-casting dies. Fatigue wear can be reduced by (a) lowering contact stresses, (b) reducing thermal cycling, and (c) improving the quality of materials, such as by removing impurities, inclusions, and various other flaws that may act as local points for crack initiation. (See also Section 3.8.)

Other types of wear. Two other types of wear are important in manufacturing processes. **Fretting corrosion** occurs at interfaces that are subjected to very small movements, such as in vibrating machinery or collars on rotating shafts. **Impact wear** is the removal of small amounts of materials from a surface by impacting particles. Deburring by vibratory finishing and tumbling (see Section 9.8) and ultrasonic machining (see Section 9.9) are examples of impact wear.

It has been observed that, in many cases, wear of tools, dies, molds, and various machine components is the result of a combination of different types of wear. Note in Fig. 4.12, for example, that even in the same forging die, various types of wear take place in different locations. A similar situation exists in cutting tools, as shown in Fig. 8.20.

Wear of plastics and ceramics. The wear behavior of plastics is similar to that of metals. Thus, wear may occur in ways similar to those described above. In polymers, abrasive-wear behavior depends partly on the ability of the polymer to deform and recover elastically, similar to the behavior of elastomers. It also appears that the polymer's resistance to abrasive wear increases as the ratio of its hardness to elastic modulus increases. Typical polymers with good wear resistance include polyimides, nylons, polycarbonate, polypropylene, acetals, and high-density polyethylene (Chapter 10). Plastics can also be blended with internal lubricants, such as silicone, graphite, molybdenum disulfide, and Teflon (PTFE), and with rubber particles that are interspersed within the polymer matrix.

The wear resistance of reinforced plastics (Section 10.9) depends on the type, amount, and direction of reinforcement in the polymer matrix. Carbon, glass, and aramid fibers all improve wear resistance. Wear takes place when fibers are pulled out of the matrix (*fiber pullout*), and it is highest when the sliding direction is parallel to the fibers, because the fibers can be pulled out more easily in this direction. Long fibers increase the wear resistance of composites because such fibers are more difficult to pull out, and cracks in the matrix cannot propagate to the surface as easily.

When ceramics slide against metals, wear is caused by small-scale plastic deformation and by brittle surface fracture, surface chemical reactions, plowing, and some surface fatigue. Metals can be transferred to the oxide-type ceramic surfaces and forming metal oxides, whereby sliding actually takes place between the metal and the metal-oxide surface. Conventional lubricants do not appear to influence the wear of ceramics to any significant extent.

Measuring wear. Several methods can be used to observe and measure wear. Although not quantitative, the simplest method is by visual and tactile (touching) inspection. Measuring dimensional changes of components, gaging, profilometry, and weighing are more accurate methods, although the latter is not accurate for heavy parts or tools and dies. Techniques have also been developed to monitor the performance and noise level of operating machinery, as worn components emit more noise than new parts.

The lubricant can be analyzed for wear particles (*spectroscopy*). This method is precise and is widely used for critical applications, such as checking the wear of jet-engine components. *Radiography* is a method in which wear particles from an irradiated surface are transferred to the mating surface, whose amount of radiation is then measured. An example is the transfer of wear particles from irradiated cutting tools to the back side of chips (Section 8.2).

4.4.3 Lubrication

The interface between tools, dies, molds, and workpieces in manufacturing operations is subjected to a wide range of variables. The major variables are the following:

1. **Contact pressure,** ranging from elastic stresses to multiples of the yield stress of the workpiece material.

2. **Speed,** ranging from very low (such as in superplastic forming operations) to very high (such as in explosive forming, drawing of thin wire, and abrasive and high-speed machining operations).

3. **Temperature,** ranging from ambient to almost melting (such as in hot extrusion and squeeze casting).

If two surfaces slide against each other under high pressure, speed, and/or temperature, and with no protective layers at the interface, friction and wear will be high.

Lubrication regimes. Four regimes of lubrication are relevant to metalworking processes (Fig. 4.13):

1. **Thick film:** The surfaces are completely separated by a fluid film which has a thickness of about one order of magnitude greater than that of the surface roughness; consequently, there is no metal-to-metal contact between the two surfaces. In this regime, the normal load is light and is supported by the *hydrodynamic fluid film* that is formed by the wedge effect caused by the relative velocity of the two bodies and the viscosity of the fluid. The coefficient of friction is very low, typically ranging from 0.001 to 0.02, and wear is practically nonexistent.

2. **Thin film:** As the normal load increases or as the speed and viscosity of the fluid decrease (such as caused by temperature rise), the film thickness is reduced to between 3 and 10 times that of the surface roughness. There may be some metal-to-metal contact at the higher asperities. This contact increases friction and leads to slight wear.

FIGURE 4.13 Regimes of lubrication generally occurring in metalworking operations. *Source:* After W.R.D. Wilson.

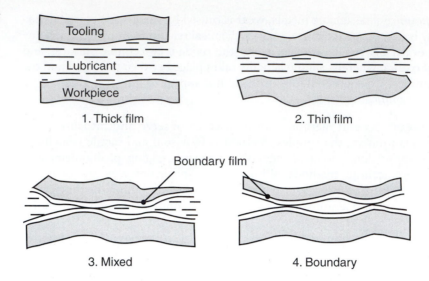

1. Thick film

2. Thin film

3. Mixed

4. Boundary

3. **Mixed:** In this regime, a significant portion of the load is carried by the metal-to-metal contact of the asperities, and the rest of the load is carried by the pressurized fluid film present in hydrodynamic pockets, such as in the valleys of asperities. The average film thickness is less than three times that of the surface roughness. With proper selection of lubricants, a strongly adhering *boundary film* a few molecules thick can develop on the surfaces. This film prevents direct metal-to-metal contact, thus reducing wear. Depending on the strength of the boundary film and various other parameters, the friction coefficient in mixed lubrication may range up to about 0.4.

4. **Boundary lubrication:** The load in this regime is supported by the contacting surfaces covered with a *boundary layer*. A lubricant may or may not be present in the surface valleys, but it is not pressurized enough to support a significant load. Depending on the boundary-film thickness and its strength, the friction coefficient ranges from about 0.1 to 0.4.

   Typical *boundary lubricants* are natural oils, fats, fatty acids, and soaps. Boundary films form rapidly on metal surfaces. As the film thickness decreases and metal-to-metal contact takes place, the chemical aspects and roughness of surfaces become significant. It should be noted that, in thick-film lubrication, the viscosity of the lubricant is the important parameter in controlling friction and wear. Chemical aspects are not particularly significant, except as they affect corrosion and staining of metal surfaces. A boundary film can *break down,* or it can be removed by being disturbed or rubbed off during sliding, or because of *desorption* due to high temperatures at the interface. Deprived of this protective layer, the clean metal surfaces then contact each other, and, as a consequence, severe wear and scoring can occur. Adherence of boundary films is thus an important aspect in lubrication.

Surface roughness and geometric effects. Surface roughness, particularly in mixed lubrication, is important, in that roughness can serve to develop local reservoirs or pockets for lubricants. The lubricant can then be trapped within the surface, and, because these fluids are incompressible, they can support a substantial portion of the normal load. The pockets also supply lubricant to regions where the boundary layer may have been destroyed. There is, however, an optimal roughness for lubricant entrapment. In metalworking operations, it is generally desirable for the workpiece, not the die, to have the rougher surface, as otherwise the workpiece surface will be damaged by the rougher and harder die surface. The recommended surface roughness on most dies is on the order of 0.40 μm (15 μin.).

In addition to surface roughness, the overall *geometry* and motion of the interacting bodies are important considerations in lubrication. The movement of the workpiece into the deformation zone must allow a supply of the lubricant to be carried into the die-workpiece interfaces. Analysis of hydrodynamic lubrication has shown that the *inlet angle* is an important parameter. As this angle decreases, more lubricant is entrained; thus, lubrication is improved, and friction is reduced. With proper selection of various parameters, a relatively thick film of lubricant can be maintained at the die-workpiece interface during metalworking operations.

The reduction of friction and wear by *entrainment* or *entrapment of lubricants* may, however, not always be desirable. The reason is that with a thick film of lubricant, the workpiece surface cannot come into full contact with the die surfaces and, as a result, it does not acquire the shiny appearance that is generally desirable on the product. A thick film of lubricant generates a grainy, dull surface on the workpiece, depending on the grain size of the material. In operations such as coining and forging (Section 6.2), trapped lubricants are especially undesirable, because they interfere with the forming operation and prevent precise shape generation. Furthermore, some operations, such as rolling, require some controlled level of friction, so that the film thickness must not be excessive.

4.4.4 Metalworking fluids

On the basis of the foregoing discussions, the functions of metalworking fluids may be summarized as follows:

- *Reduce friction*, thus reduce force and energy requirements and prevent increases in temperature.
- *Reduce wear, seizure,* and *galling.*
- *Improve material flow* in dies, tools, and molds.
- *Act as a thermal barrier* between the workpiece and tool and die surfaces, thus prevent workpiece cooling in hotworking processes.
- *Act as a release* or *parting agent* to help in the removal or ejection of parts from dies and molds.

Several types of **metalworking fluids** with a wide variety of chemistries, properties, and characteristics are available to fulfill these requirements:

1. **Oils** have high film strength, as evidenced by how difficult it can be to clean an oily surface. The sources of oils are **mineral** (petroleum or hydrocarbon), **animal,** or **vegetable** in nature. For environmental reasons, there is significant current interest in replacing mineral oils with naturally degrading vegetable oil-base stocks. Oils may be **compounded** with a variety of additives or with other oils to impart specific properties, such as lubricity, viscosity, detergents, and biocides. Although they are very effective in reducing friction and wear, oils have low thermal conductivity and specific heat and thus are not effective in conducting away the heat generated during metalworking operations. Furthermore, oils are difficult to remove from component surfaces that subsequently are to be painted or welded, and they are difficult and costly to dispose of.

2. An **emulsion** is a mixture of two immiscible liquids, usually mixtures of oil and water in various proportions, along with additives. An emulsion typically is provided as a mixture of base oil and additives and then mixed with water by

the user. Emulsions are of two types. In a *direct emulsion,* oil is dispersed in water as very small droplets, with average diameters typically ranging between 0.1 μm (4 μin.) and 30 μm (1200 μin.). In an *indirect* or *invert emulsion,* water droplets are dispersed in oil. Direct emulsions are important fluids because the presence of water (generally 95% of the emulsion by volume) gives them high cooling capacity, yet they still provide good lubrication. They are particularly effective in high-speed metal forming and cutting operations, where increases in temperature have detrimental effects on tool and die life, workpiece surface integrity, and dimensional accuracy.

3. **Synthetic solutions** are fluids that contain inorganic and other chemicals, dissolved in water. Various chemical agents are added to impart different properties. **Semisynthetic solutions** are basically synthetic solutions to which small amounts of emulsifiable oils have been added.

4. Metalworking fluids are usually blended with various **additives,** including oxidation inhibitors, rust preventatives, odor control agents, antiseptics, and foam inhibitors. Important additives in oils are sulfur, chlorine, and phosphorus. Known as **extreme-pressure additives** (EPs) and used singly or in combination, they react chemically with metal surfaces and form adherent surface films of metallic sulfides and chlorides. These films have low shear strength and good antiweld properties and thus are very effective in reducing friction and wear. While EP additives are important in boundary lubrication, these lubricants may be undesirable in some situations. For example, they may preferentially attack the cobalt binder in tungsten carbide tools and dies (Section 8.6), causing undesirable changes in surface roughness and integrity (*selective leaching*).

5. **Soaps** are generally reaction products of sodium or potassium salts with fatty acids. Alkali soaps are soluble in water, but other metal soaps are generally insoluble. Soaps are effective boundary lubricants and can also form thick-film layers at die-workpiece interfaces, particularly when applied on conversion coatings (Section 4.5) for cold metalworking applications.

6. **Greases** are solid or semisolid lubricants and generally consist of soaps, mineral oils, and various additives. They are highly viscous and adhere well to metal surfaces. Although used extensively in machinery, greases have limited use in manufacturing processes.

7. **Waxes** may be of animal or plant (*paraffin*) origin and have complex structures. Compared with greases, waxes are less "greasy" and are more brittle. They have limited use in metalworking operations, except for copper and, as chlorinated paraffin, for stainless steels and high-temperature alloys.

Solid lubricants. Because of their unique properties and characteristics, several solid materials are used as lubricants in manufacturing operations:

1. **Graphite.** The properties of *graphite* are described in Section 11.13. Graphite is weak in shear along its layers and hence has a low coefficient of friction in that direction. Thus, it can be a good solid lubricant, particularly at elevated temperatures. However, its friction is low only in the presence of air or moisture. In a vacuum or an inert-gas atmosphere, its friction is very high. In fact, graphite can be quite abrasive in these environments. Graphite may be applied either by rubbing it on surfaces or as a colloidal (dispersion of small particles) suspension in liquid carriers, such as water, oil, or alcohols.

2. **Molybdenum disulfide.** *Molybdenum disulfide* (MoS_2), another widely used lamellar solid lubricant, is somewhat similar in appearance to graphite and is used as a lubricant at room temperature. However, unlike graphite, its friction

coefficient is high in an ambient environment. Oils are common carriers for molybdenum disulfide, but it can also be rubbed onto the surfaces of a workpiece.

3. **Soft metals and polymer coatings.** Because of the their low strength, thin layers of *soft metals* and *polymer coatings* can be used as solid lubricants. Suitable metals are lead, indium, cadmium, tin, and silver, and suitable polymers include PTFE, polyethylene, and methacrylates. However, these coatings have limited applications because of their lack of strength under high stresses and at elevated temperatures. Soft metals are used to coat high-strength metals such as steels, stainless steels, and high-temperature alloys. Copper or tin, for example, is chemically deposited on the surface before the metal is processed. If the oxide of a particular metal has low friction and is sufficiently thin, the oxide layer also can serve as a solid lubricant, particularly at elevated temperatures.

4. **Glass.** Although a solid material at room temperature, *glass* becomes viscous at high temperatures and hence can serve as a liquid lubricant. Its viscosity is a function of temperature, but not of pressure, and depends on the type of glass. Its poor thermal conductivity also makes glass an attractive lubricant, acting as a thermal barrier between hot workpieces and relatively cool dies. Typical applications include hot extrusion (Section 6.4) and forging (Section 6.2).

5. **Conversion coatings.** Lubricants may not always adhere properly to workpiece surfaces, particularly under high normal and shearing stresses. This property has the greatest effects in forging, extrusion, and wire drawing of steels, stainless steels, and high-temperature alloys. For these applications, the workpiece surfaces are first transformed through chemical reaction with acids (hence the term *conversion*). The reaction leaves a somewhat rough and spongy surface, which acts as a carrier for the lubricant. After treatment, any excess acid from the surface is removed using borax or lime. A liquid lubricant, such as a soap, is then applied to the surface. The lubricant film adheres to the surface and cannot be scraped off easily. Zinc phosphate conversion coatings are often used on carbon and low-alloy steels. Oxalate coatings are used for stainless steels and high-temperature alloys. (See also Section 4.5.1.)

6. **Fullerenes** or *buckyballs* are carbon molecules in the shape of soccer balls; when placed between sliding surfaces, these molecules act like tiny ball bearings. They perform well as solid lubricants and are particularly effective in aerospace applications as bearing lubricants.

Metalworking fluid selection. Selecting a lubricant for a particular manufacturing process and workpiece material involves several factors:

- The particular manufacturing process.
- Compatibility of the lubricant with the workpiece and tool and die materials.
- The surface preparation required.
- The method of application.
- Removal of the fluid after processing.
- Contamination of the fluid by other lubricants, such as those used to lubricate machinery.
- Treatment of waste fluids.
- Storage and maintenance of the fluids.
- Biological and ecological considerations.
- Costs involved in all of the foregoing aspects.

The different functions of a metalworking fluid, whether primarily a lubricant or a coolant, must be taken into account. Water-base fluids are very effective coolants, but as lubricants they are not as effective as oils. In selecting an oil as a lubricant, the importance of its viscosity-temperature-pressure characteristics should be recognized, as low viscosity can have significant detrimental effects on friction and wear.

Metalworking fluids should not leave any harmful residues that could interfere with machinery operations. The fluids should not stain or corrode the workpieces or the equipment. The fluids should be checked periodically for deterioration caused by bacterial growth, accumulation of oxides, metal chips, and wear debris, and also for general degradation and breakdown due to temperature and time. A lubricant may carry wear particles within it and cause damage to the system; thus proper inspection and filtering of metalworking fluids are important.

After completion of manufacturing operations, metal surfaces are usually covered with lubricant residues; these should be removed prior to further workpiece processing, such as welding or painting. Various cleaning solutions and techniques can be used for this purpose (Section 4.5.2). *Biological* and *ecological considerations,* with their accompanying health and legal ramifications, are also very important. Potential health hazards may be involved in contacting or inhaling some metalworking fluids. Recycling and disposal of waste fluids are additional important factors to be considered.

4.5 | Surface Treatments, Coatings, and Cleaning

After a component is manufactured, all or sections of its surfaces may have to be processed further in order to impart desired properties and characteristics. *Surface treatments* may be necessary to

- Improve resistance to wear, erosion, and indentation (such as for slideways in machine tools, wear surfaces of machinery, and shafts, rolls, cams, gears, and bearings).
- Control friction (sliding surfaces on tools, dies, bearings, and machine ways).
- Reduce adhesion (electrical contacts).
- Improve lubrication (surface modification to retain lubricants).
- Improve resistance to corrosion and oxidation (sheet metals for automotive or other outdoor uses, gas-turbine components, and medical devices).
- Improve fatigue resistance (bearings and shafts).
- Rebuild surfaces on components (worn tools, dies, and machine components).
- Improve surface roughness (appearance, dimensional accuracy, and frictional characteristics).
- Impart decorative features, color, or special surface texture.

4.5.1 Surface treatment processes

Several processes are used for surface treatments, based on mechanical, chemical, thermal, and physical methods. Their principles and characteristics are outlined next.

1. **Shot peening, water-jet peening,** and **laser shot peening.** In *shot peening,* the surface of the workpiece is impacted repeatedly with a large number of cast-steel, glass, or ceramic shot (small balls). Using shot sizes ranging from 0.125 to 5 mm (0.005 to 0.2 in.) in diameter, they make overlapping indentations on the surface, causing plastic deformation of the surface to depths

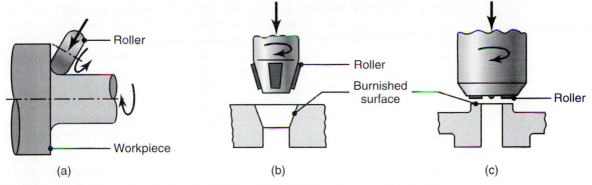

FIGURE 4.14 Examples of roller burnishing of (a) the fillet of a stepped shaft, (b) an internal conical surface, and (c) a flat surface.

up to 1.25 mm (0.05 in.). Because plastic deformation is not uniform throughout a part's thickness, the process imparts compressive residual stresses on the surface, thus improving the fatigue life of the component (see Fig. 3.32b). Shot peening is used extensively on shafts, gears, springs, oil-well drilling equipment, and jet-engine parts such as turbines and compressor blades.

In *water-jet peening*, a water jet at pressures as high as 400 MPa (60 ksi) impinges on the surface of the workpiece, inducing compressive residual stresses similar to those in shot peening. This method can be used successfully on steels and aluminum alloys.

In *laser shot peening*, the surface is subjected to laser shocks from a high-powered laser (up to 1 kW). This method can be used successfully on jet-engine fan blades and on materials such as titanium and nickel alloys, with compressive surface residual stresses deeper than 1 mm (0.04 in.).

Ultrasonic peening uses a hand tool based on a piezoelectric transducer. Operating typically at a frequency of 22 kHz, the tool can be equipped with a variety of heads for different peening applications.

2. **Roller burnishing (surface rolling).** The surface of the component is cold worked by a hard and highly polished roller or series of rollers (Fig. 4.14). Roller burnishing, which can be used on flat, cylindrical, or conical surfaces, improves the surface finish by removing scratches, tool marks, and pits. Consequently, corrosion resistance is also improved, since corrosive products and residues cannot be entrapped. Roller burnishing is typically used to improve the mechanical properties of surfaces, as well as the shape and surface finish of components. It can be used either singly or in combination with other finishing processes, such as grinding, honing, and lapping.

    Soft and ductile, as well as very hard metals can be roller burnished. Typical applications include hydraulic-system components, seals, valves, spindles, and fillets on shafts. *Internal* cylindrical surfaces are burnished by a similar process, called **ballizing** or **ball burnishing**, in which a smooth ball, slightly larger than the bore diameter, is pushed through the length of the hole.

3. **Explosive hardening.** The surface is subjected to high transient pressures by placing a layer of explosive sheet directly on the workpiece surface and detonating it. The contact pressures can be as high as 35 GPa (5×10^6 psi) and last about 2 to 3 μs. Large increases in surface hardness can be obtained by this method, with very little change (less than 5%) in the shape of the component. Railroad rail surfaces can, for example, be hardened by this method.

4. **Cladding (clad bonding).** Metals can be bonded with a thin layer of corrosion-resistant metal by applying pressure with rolls or other means. Multiple-layer cladding is also utilized in special applications. A typical application is cladding of aluminum (*Alclad*), in which a corrosion-resistant layer of aluminum alloy is clad over pure aluminum. Other applications include steels clad with stainless steel or nickel alloys. Cladding may also be performed using dies (as in cladding steel wire with copper) or using explosives.

   Laser cladding consists of the fusion of a different material over a substrate. It has been successfully applied to metals and ceramics for enhanced friction and wear behavior.

5. **Mechanical plating (mechanical coating, impact plating, peen plating).** In this process, fine metal particles are compacted over the workpiece surface by impacting it with spherical glass, ceramic, or porcelain beads. This process is typically used for hardened-steel parts for automobiles, with plating thickness usually less than 0.025 μm (0.001 in.).

6. **Case hardening (carburizing, carbonitriding, cyaniding, nitriding, flame hardening, induction hardening).** These processes are described in Section 5.11.3 and summarized in Table 5.7. In addition to the common heat sources of gas and electricity, laser beams are also used as a heat source in surface hardening of both metals and ceramics. Case hardening, as well as some of the other surface-treatment processes, induces residual stresses on surfaces. The formation of martensite in case hardening of steels induces compressive residual stresses on surfaces, which are desirable because they improve the fatigue life of components by delaying the initiation of fatigue cracks.

7. **Hard facing.** In this process, a relatively thick layer, edge, or point consisting of wear-resistant hard metal is deposited on the surface by any of the welding techniques described in Chapter 12. Several layers are usually deposited (*weld overlay*). Hard coatings of tungsten carbide, chromium, and molybdenum carbide can also be deposited using an electric arc in a process called **spark hardening**, electric spark hardening, or electrospark deposition. Hard-facing alloys are available as electrodes, rods, wire, and powder. Typical applications for hard facing include valve seats, oil-well drilling tools, and dies for hot metalworking operations. Worn parts are also hard faced for extending their use.

8. **Thermal spraying.** In *thermal spraying* (Fig. 4.15), also called **metallizing**, metal in the form of rod, wire, or powder is melted in a stream of oxyacetylene flame, electric arc, or plasma arc, and the droplets are sprayed onto a preheated surface at speeds up to 100 m/s (20,000 ft/min) using a compressed-air spray

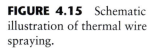

FIGURE 4.15 Schematic illustration of thermal wire spraying.

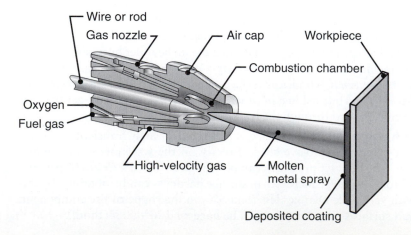

gun. The surfaces to be sprayed should be cleaned and roughened to improve bond strength. Typical applications for this process include automotive components, steel structures, storage tanks, rocket-motor nozzles, and tank cars that are sprayed with zinc or aluminum up to 0.25 mm (0.010 in.) in thickness.

There are two energy sources for thermal-spraying processes: combustion (chemical) and electrical. The most common forms of thermal spraying are:

a. Combustion spraying
- **Thermal wire spraying,** in which an oxyfuel flame melts the wire and deposits it on the surface.
- **Thermal metal-powder spraying,** in which metal powder (Section 11.2) is deposited on the surface using an oxyfuel flame.
- **Detonation gun,** in which a controlled explosion takes place using an oxyfuel gas mixture.
- **High-velocity oxyfuel gas spraying,** which has a similarly high performance as those of the foregoing processes, but can be less expensive.

b. Electrical spraying
- **Twin-wire arc,** in which an arc is formed between two consumable wire electrodes.
- **Plasma,** either conventional, high energy, or vacuum; it produces temperatures on the order of 8300°C (15,000°F) and has very good bond strength, with very low oxide content. **Low-pressure plasma spray** and **vacuum plasma spray** both produce coatings with high bond strength.

Cold spraying is a more recent process in which particles to be sprayed are not melted; thus oxidation is minimal. The spray jet is highly focused and has very high impact velocities.

9. **Surface texturing.** For technical, functional, optical, or aesthetic reasons, manufactured surfaces can be further modified by secondary operations. Texturing generally consists of (1) etching, using chemicals or sputtering techniques, (2) electric arcs, (3) laser pulses, and (4) atomic oxygen, which reacts with surfaces and produces fine, conelike surface textures. The possible adverse effects of any of these processes on the properties and performance of materials should be considered in their selection.

10. **Ceramic coating.** Techniques are available for spraying ceramic coatings for high-temperature and electrical-resistance applications, such as to withstand repeated arcing. Powders of hard metals and ceramics are used as spraying materials. Plasma-arc temperatures may reach 15,000°C (27,000°F), which are much higher than those obtained using flames. Typical applications are nozzles for rocket motors and wear-resistant parts.

11. **Vapor deposition.** In this process, the workpiece is surrounded by chemically reactive gases that contain chemical compounds of the materials to be deposited. The deposited material, typically a few μm thick, may consist of metals, alloys, carbides, nitrides, borides, ceramics, or various oxides. The substrate (workpiece) may be metal, plastic, glass, or paper. Typical applications include coatings for cutting tools, drills, reamers, milling cutters, punches, dies, and wear surfaces. (See also Section 13.5 on *semiconductor manufacturing.*)

There are two major vapor deposition process categories: *physical vapor deposition* and *chemical vapor deposition*. These techniques allow effective control of coating composition, thickness, and porosity.

a. **Physical vapor deposition** (PVD). These processes are carried out in a high vacuum and at temperatures in the range of 200°–500°C (400°–900°F).

The particles to be deposited are transported physically to the workpiece, rather than by chemical reactions, as in chemical vapor deposition.

In **vacuum deposition,** the metal to be deposited is evaporated at high temperatures in a vacuum and deposited on the substrate, which is usually at room temperature or slightly higher. Uniform coatings can be obtained on complex shapes with this method. In **arc deposition** (PV/ARC), the coating material (cathode) is evaporated by a number of arc evaporators, using localized electric arcs. The arcs produce a highly reactive plasma consisting of ionized vapor of the coating material; the vapor condenses on the substrate (anode) and coats it. Applications for this process may be functional (oxidation-resistant coatings for high-temperature applications, electronics, and optics) or decorative (hardware, appliances, and jewelry). **Pulsed-arc deposition** is a more recent related process in which the energy source is a pulsed laser.

In **sputtering,** an electric field ionizes an inert gas (usually argon). The positive ions bombard the coating material (cathode) and cause sputtering (ejecting) of its atoms. These atoms then condense on the workpiece, which is heated to improve bonding. In **reactive sputtering,** the inert gas is replaced by a reactive gas, such as oxygen, in which case the atoms are oxidized and the oxides are deposited. **Radiofrequency sputtering** (RF) is used for non-conductive materials, such as electrical insulators and semiconductor devices.

Ion plating is a generic term describing the combined processes of sputtering and vacuum evaporation. An electric field causes a glow discharge, generating a plasma; the vaporized atoms in this process are only partially ionized. **Ion-beam-enhanced (assisted) deposition** is capable of producing thin films as coatings for semiconductor, tribological, and optical applications. **Dual ion-beam-assisted deposition** is a hybrid coating technique that combines physical vapor deposition with simultaneous ion-beam bombardment and results in good adhesion on metals, ceramics, and polymers. Ceramic bearings and dental instruments are two examples of application.

b. **Chemical vapor deposition** (CVD) is a thermochemical process. In a typical application, such as for coating cutting tools with titanium nitride (TiN) (Fig. 4.16), the tools are placed on a graphite tray and heated to 950°–1050°C (1740°–1920°F) in an inert atmosphere. Titanium tetrachloride (a vapor), hydrogen, and nitrogen are then introduced into the chamber. The resulting chemical reactions form a thin coating of titanium nitride on the tool's surfaces. For coating with titanium carbide, methane is substituted for the hydrogen and nitrogen gases. Coatings obtained by chemical vapor deposition are usually thicker than those obtained via PVD. **Medium-temperature CVD** (MTCVD) produces coatings that have higher resistance to crack propagation than CVD coatings.

FIGURE 4.16 Schematic illustration of the chemical vapor deposition process.

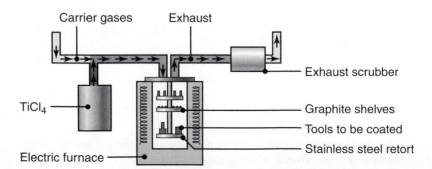
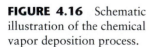

12. In **ion implantation,** ions are accelerated onto a surface (in a vacuum) to such an extent that they penetrate the substrate to a depth of a few μm. This process (not to be confused with ion plating; see below) modifies surface properties, increasing the surface hardness and improving resistance to friction, wear, and corrosion. This process can be controlled accurately, and the surface can be masked to prevent ion implantation in unwanted places. When used in specific applications, such as semiconductors (Chapter 13), this process is called **doping** (alloying with small amounts of various elements).

13. **Diffusion coating.** In this process, an alloying element is diffused into the surface, thus altering its properties. The elements can be in solid, liquid, or gaseous states. This process is given different names, depending on the diffused element. (See also carburizing, nitriding, and boronizing, in Table 5.8.)

14. **Electroplating.** The workpiece (cathode) is plated with a different metal (anode) while both are suspended in a bath containing a water-base electrolyte solution (Fig. 4.17). Although the electroplating process involves a number of reactions, the basic procedure is that the metal ions from the anode are discharged under the potential from the external source of electricity, combine with the ions in the solution, and are deposited on the cathode.

 The volume of the plated metal can be calculated from the expression

$$\text{Volume of metal plated} = cIt, \tag{4.7}$$

where I is the current in amperes, t is time, and c is a constant that depends on the plated metal, the electrolyte, and the efficiency of the system, being typically in the range of 0.03 to 0.1 mm^3/amp-s. Note that for the same volume of material deposited, the larger the workpiece surface plated, the thinner the layer will be. The time required for electroplating is typically long because the deposition rate is generally on the order of 75 μm/hr. Thin plated layers are typically on the order of 1 μm (40 μin.), and for thick layers it can be as much as 500 μm (0.02 in.).

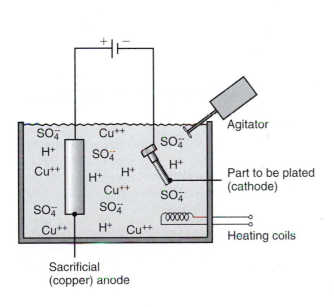

(a)

(b)

FIGURE 4.17   (a) Schematic illustration of the electroplating process and (b) examples of electroplated parts. *Source:* Courtesy of BFG Electroplating.

The plating solutions are either strong acids or cyanide solutions. As the metal is plated from the solution, it has to be periodically replenished. This is accomplished through two principal methods: (a) additional salts of metals are occasionally added to the solution, or (b) a *sacrificial anode* of the metal to be plated is used in the electroplating tank, which dissolves at the same rate that metal is deposited. There are three main methods of electroplating:

a. In **rack plating**, the parts to be plated are placed in a rack, which is then moved through a series of process tanks.

b. In **barrel plating**, small parts are placed inside a permeable barrel, which is placed inside the tank(s). Electrolytic fluid penetrates through the barrel and provides the metal for plating; electrical contact is provided through the barrel and contact with other parts. This form of electroplating is commonly performed with small parts such as bolts, nuts, gears, and fittings.

c. In **brush processing**, the electrolytic fluid is pumped through a hand-held brush with metal bristles. This process is suitable for field repair or plating of very large parts and can be used to apply coatings on large equipment without disassembly.

Common plating materials include chromium, nickel, cadmium, copper, zinc, and tin. **Chromium plating** is carried out by plating the metal first with copper, then with nickel, and finally with chromium. **Hard chromium plating** is done directly on the base metal and has a hardness of up to 70 HRC.

Typical electroplating applications include copper plating aluminum wire and phenolic boards for printed circuits, chrome plating hardware, tin plating copper electrical terminals for ease of soldering, and plating various components for enhanced appearance and resistance to wear and corrosion. Because they do not develop oxide films, noble metals (gold, silver, and platinum; see Section 3.11.8) are important electroplating materials for the electronics and jewelry industries. Plastics, such as ABS, polypropylene, polysulfone, polycarbonate, polyester, and nylon, also can have metal coatings applied through electroplating. However, because they are not electrically conductive, plastics must first be preplated by such processes as electroless nickel plating (see below). Parts to be coated may be simple or complex, and size is not a limitation. Complex shapes may have varying plating thicknesses.

15. **Electroless plating.** This process is carried out by chemical reactions and without the use of an external source of electricity. The most common application uses nickel, although copper is also used. In electroless nickel plating, nickel chloride (a metallic salt) is reduced (with sodium hypophosphite as the reducing agent) to nickel metal, which is then deposited on the workpiece. The hardness of nickel plating ranges between 425 and 575 HV, and the plating can be heat treated to a hardness of 1000 HV. The coating has excellent wear and corrosion resistance.

16. **Anodizing.** Anodizing is an oxidation process (*anodic oxidation*) in which the workpiece surfaces are converted to a hard and porous oxide layer that provides corrosion resistance and a decorative finish. The workpiece is the anode in an electrolytic cell immersed in an acid bath, resulting in chemical adsorption of oxygen from the bath. Organic dyes of various colors (typically black, red, bronze, gold, or gray) can be used to produce stable and durable surface films. Typical applications for anodizing include aluminum furniture and utensils, architectural shapes, automobile trim, picture frames, keys, and sporting goods.

Anodized surfaces also serve as a good base for painting, especially for aluminum, which otherwise would be difficult to paint.

17. **Conversion coating.** In this process, also called **chemical-reaction priming**, a coating forms on metal surfaces as a result of chemical or electrochemical reactions. Various metals, particularly steel, aluminum, and zinc, can be conversion coated. Phosphates, chromates, and oxalates are used to produce conversion coatings. These coatings are for purposes such as prepainting, decorative finishes, and protection against corrosion.

    An important application is in conversion coating of workpieces as a lubricant carrier in cold-forming operations (see Section 4.4.4), as lubricants may not always adhere properly to workpiece surfaces, particularly when subjected to high normal and shearing stresses. This condition is a problem particularly in forging, extrusion, and wire drawing of steels, stainless steels, and high-temperature alloys. For these applications, acids transform the workpiece surface by chemically reacting with it, leaving a somewhat rough and spongy surface which then acts as a carrier for the lubricant. After treatment, borax or lime is used to remove any excess acid from the surfaces. A liquid lubricant, such as a soap, is then applied to the coated surface. The lubricant film adheres strongly to the surface and cannot be scraped off easily. **Zinc phosphate** conversion coatings are often used on carbon and low-alloy steels; **oxalate** coatings are used for stainless steels and high-temperature alloys.

18. **Coloring.** As the name implies, coloring involves processes that alter the color of metals and ceramics. The change is caused by the conversion of surfaces by chemical, electrochemical, or thermal processes into chemical compounds, such as oxides, chromates, and phosphates. In **blackening**, iron and steels develop a lustrous black-oxide film, using solutions of hot caustic soda.

19. **Hot dipping.** In this process, the workpiece (usually steel or iron) is dipped into a bath of molten metal, such as zinc (for galvanized-steel sheet and plumbing supplies), tin (for tin plate and tin cans for food containers), aluminum (*aluminizing*), and *terne* (lead alloyed with 10 to 20% tin). Hot-dipped coatings on discrete parts or sheet metal provide galvanized pipe, plumbing supplies, and numerous other products with long-term resistance to corrosion. The coating thickness is usually given in terms of coating weight per unit surface area of the sheet, typically 150–900 g/m^2 (0.5–3 oz/ft^2). The service life of hot-dipped parts depends on the thickness of the zinc coating and the environment to which it is exposed. Various **precoated sheet steels** are used extensively, such as in automobile bodies and containers. Proper draining of the hot metal to remove excess coating materials is important.

20. **Porcelain enameling.** In enameling, metals are coated with a variety of glassy (vitreous) substances to provide corrosion and electrical resistance and for service at elevated temperatures. The coatings are generally classified as porcelain enamels and typically include enamels and ceramics (Section 11.8). (The word **enamel** is also used for glossy paints, indicating a smooth and relatively hard coating.) Porcelain enamels are glassy inorganic coatings consisting of various metal oxides. **Enameling** involves fusing the coating material on the substrate by heating them both to 425–1000°C (800–1800°F) to liquefy the oxides. Depending on their composition, enamels have varying resistances to alkali, acids, detergents, cleansers, and water and are available in a variety of colors.

    Typical applications for porcelain enameling include household appliances, plumbing fixtures, chemical-processing equipment, signs, cookware, and jewelry, as well as protective coatings on jet-engine components. The coating may be applied by dipping, spraying, or electrodeposition, and thicknesses are usually in the range of 0.05–0.6 mm (0.002–0.025 in.). Metals that are coated with porcelain enamel are typically steels, cast iron, and aluminum. **Glazing** is the application of glassy coatings on ceramic and earthenware to give them decorative finishes and to make them impervious to moisture. Glass coatings are used as lining for chemical resistance, with a thickness much greater than in enameling.

21. **Organic coatings.** Metal surfaces may be coated or *precoated* with a variety of organic coatings, films, and laminates to improve appearance and corrosion resistance. Coatings are applied to the coil stock on continuous lines, with thicknesses generally in the range of 0.0025–0.2 mm (0.0001–0.008 in.). These coatings have a wide range of flexibility, durability, hardness, resistance to abrasion and chemicals, color, texture, and gloss. Coated sheet metal is subsequently formed into various products, such as TV cabinets, appliance housings, paneling, shelving, siding for residential buildings, gutters, and metal furniture.

    More critical applications of organic coatings involve, for example, coatings for naval aircraft that are subjected to high humidity, seawater, rain, pollutants (such as from ship exhaust stacks), aviation fuel, deicing fluids, and battery acid, and parts that are impacted by particles such as dust, gravel, stones, and deicing salts. For aluminum structures, organic coatings typically consist of an epoxy primer and a polyurethane topcoat, with a lifetime of four to six years).

22. **Ceramic coatings.** Ceramics, such as aluminum oxide and zirconium oxide, are applied to a surface at room temperature, usually by thermal-spraying techniques. These coatings serve as a thermal barrier, especially in applications such as hot-extrusion dies, diesel-engine components, and turbine blades.

23. **Painting.** Paints are basically classified as enamels, lacquers, and water-base paints, with a wide range of characteristics and applications. They are applied by brushing, dipping, or spraying. In **electrocoating** (**electrostatic spraying**), paint particles are charged electrostatically making them attract to surfaces, producing a uniformly adherent coating.

24. **Diamond coating.** Important advances continue to be made in diamond coating of metals, glass, ceramics, and plastics, using various chemical and plasma-assisted vapor deposition processes and ion-beam enhanced deposition techniques. *Free-standing diamond films* have also been developed on the order of 1 mm (0.040 in.) thick and up to 125 mm (5 in.) in diameter, including smooth and optically clear diamond films. Combined with the important properties of diamonds, such as hardness, wear resistance, high thermal conductivity, and transparency to ultraviolet light and microwave frequencies, diamond coatings have important applications for various aerospace and electronic parts and components.

    Examples of diamond-coated products include scratchproof windows (such as for aircraft and missile sensors to protect against sandstorms), sunglasses, cutting tools (such as drills and end mills), measuring instruments, surgical knives, electronic and infrared heat seekers and sensors, light-emitting diodes, speakers for stereo systems, turbine blades, and fuel-injection nozzles. Studies are continuing on growing diamond films on crystalline copper substrates by implantation of carbon ions. An important application for such diamond films is in making computer chips (Chapter 13). Diamond can be doped to form *p*- and *n*-type ends on semiconductors to make transistors, and its high thermal conductivity allows closer packing of chips than with silicon or gallium-arsenide chips, thus significantly increasing the speed of computers.

25. **Diamond-like carbon (DLC).** Using a low-temperature, ion-beam-assisted deposition process, this material is applied as a coating of a few nm in thickness, with a hardness of about 5000 HV. Less expensive than diamond films, DLC has important applications in such areas as tools and dies, gears, bearings, microelectromechanical systems, and small (microscale) probes.

26. **Surface texturing.** Each manufacturing process produces a certain surface texture and appearance, which may be acceptable for its intended function or else may need some modification. Certainly, surfaces can be modified by grinding or polishing operations (see Chapter 9) to produce smooth surfaces. However, manufactured surfaces can further be modified by secondary operations for technical, functional, optical, or aesthetic reasons.

    Called *surface texturing*, these additional processes generally consist of the following techniques:

    a. Etching, using chemicals or sputtering techniques.

    b. Electric arcs.

    c. Lasers, using excimer lasers with pulsed beams; applications include molds for permanent-mold casting, rolls for temper mills, golf-club heads, and computer hard disks.

    d. Atomic oxygen, which reacts with surfaces and produces a fine, conelike surface texture.

4.5.2 Cleaning of surfaces

The importance of surfaces and the influence of deposited or adsorbed layers of various elements and contaminants on surfaces have been stressed throughout this chapter. A clean surface can have both beneficial as well as detrimental effects. Although a contaminated surface would reduce the tendency for adhesion and galling between mating parts, cleanliness is generally essential for more effective application of metalworking fluids, coating and painting, adhesive bonding, welding, brazing, soldering, reliable functioning of manufactured parts in machinery, manufacturing of food and beverage containers, and in assembly operations. *Contaminants* (also called soils) may consist of rust, scale, chips and other metallic and nonmetallic debris, metalworking fluids, solid lubricants, pigments, polishing and lapping compounds, and general environmental elements.

Cleaning involves the removal of solid, semisolid, or liquid contaminants from a surface. Although the word *clean,* or the degree of cleanliness of a surface, is somewhat difficult to define, two simple and common tests are based on the following simple procedures:

1. Wiping the area with a clean cloth and observing any *residues* on the cloth.

2. Observing whether water continuously coats the surface (called the *waterbreak test*). If water collects as individual droplets, the surface is not clean. (This test can easily be demonstrated by wetting dinner plates that have been cleaned to varying degrees.)

The type of cleaning process required depends on the type of contaminants to be removed. **Mechanical cleaning methods** consist of physically disturbing the contaminants, such as by wire or fiber brushing, dry or wet abrasive blasting, tumbling, steam jets, and ultrasonic cleaning. These processes are particularly effective in removing rust, scale, and other solid contaminants.

In **electrolytic cleaning,** a charge is applied to the part to be cleaned in an acqueous solution, which results in bubbles of hydrogen or oxygen. The bubbles are abrasive and aid in the removal of contaminants from the surface.

Chemical cleaning methods are effective in removing oil and grease from surfaces, including metalworking fluids. They consist of one or more of the following operations:

a. *Solution:* The soil dissolves in the cleaning solution.

b. *Saponification:* a chemical reaction that converts animal or vegetable oils into a soap that is soluble in water.

c. *Emulsification:* the cleaning solution reacts with the soil or lubricant deposits and forms an emulsion which then becomes suspended in the solution.

d. *Dispersion:* the concentration of soil on the surface is decreased by surface-active materials in the cleaning solution.

e. *Aggregation:* lubricants are removed from the surface by various agents in the cleaning fluid and collect as large dirt particles.

Cleaning fluids, including *alkaline solutions, emulsions, solvents, hot vapors, acids, salts,* and mixtures of *organic compounds,* can be used in conjunction with electrochemical processes for more effective cleaning. In **vapor degreasing,** a heated and evaporated solvent condenses on parts that are at room temperature. Dirt and lubricant are then removed and dispersed, emulsified, or dissolved in the solvent, which drips off the parts into the solvent tank. An advantage of vapor degreasing is that the solvent does not become diluted since the vapor does not contain previously removed contaminant. A drawback of this process is that the fumes can be irritating or toxic, so that environmental controls are needed.

Mechanical agitation of the surface can aid in removal of contaminants; examples of this method include wire brushing, abrasive blasting, abrasive jets, and ultrasonic vibration of a solvent bath (**ultrasonic cleaning**). In **electrolytic cleaning,** a charge is applied to the workpiece in an aqueous, often alkaline, cleaning solution. This charge results in bubbles of hydrogen or oxygen being released at the workpiece surface, depending on the polarity of the charge; these abrasive bubbles aid in the removal of contaminants.

Cleaning parts with complex shapes can be difficult. Alternative designs may thus be necessary, such as (a) avoiding deep blind holes, (b) providing appropriate drain holes in the part, or (c) producing the part by several smaller components, instead of one large component that may be difficult to clean.

4.6 | Engineering Metrology and Instrumentation

Engineering metrology is the measurement of dimensions such as length, thickness, diameter, taper, angle, flatness, and profiles. In this section we describe the characteristics of the instruments and the techniques used in engineering metrology. Numerous measuring instruments and devices are used in metrology. It is important to describe briefly the quality of an instrument, as follows:

1. **Accuracy:** The degree of agreement between the measured dimension and its true magnitude.

2. **Precision:** The degree to which the instrument gives repeated measurements.

3. **Resolution:** The smallest dimension that can be read on an instrument.

4. **Sensitivity:** The smallest difference in dimensions that the instrument can detect or distinguish.

4.6.1 Measuring instruments

1. **Line-graduated instruments.** Line-graduated instruments are used for measuring length (linear measurements) or angles (angular measurements), *graduated* meaning that it is marked to indicate a certain quantity. The simplest and most commonly used instrument for making linear measurements is a *steel rule (machinist's rule)*, bar, or tape with fractional or decimal graduations. Lengths are measured directly, to an accuracy that is limited to the nearest division, usually 1 mm or $\frac{1}{64}$ in. Rules may be rigid or flexible and they may be equipped with a hook at one end for ease of measuring from an edge. Rule-depth gages are similar to rules and slide along a special head.

   Vernier calipers have a graduated beam and a sliding jaw with a *vernier*. Also called *caliper gages,* the two jaws of the caliper contact the part being measured and the dimension is read at the matching graduated lines. The vernier improves the sensitivity of a simple rule by indicating fractions of the smallest division on the graduated beam, usually to 25 μm (0.001 in.). Vernier calipers, which can be used to measure inside or outside lengths, are also equipped with *digital readouts,* which are easier to read and less subject to human error. Vernier *height gages* are vernier calipers with setups similar to those of a depth gage and have similar sensitivity as well.

   Micrometers have a graduated, threaded spindle and are commonly used for measuring the thickness and inside or outside diameters of parts. Circumferential vernier readings to a sensitivity of 2.5 μm (0.0001 in.) can be obtained. Micrometers are also available for measuring depths (*micrometer depth gage*) and internal diameters (*inside micrometer*) with the same sensitivity. Micrometers are also available with digital readouts to reduce errors in reading. The anvils on micrometers can be equipped with conical or ball contacts; they are used to measure inside recesses, threaded rod diameters, and wall thicknesses of tubes and curved sheets.

   Diffraction gratings consist of two flat optical glasses with closely spaced parallel lines scribed on their surfaces. The grating on the shorter glass is inclined slightly; as a result, interference fringes develop when the grating is viewed over the longer glass. The position of these fringes depends on the relative position of the two sets of glasses.

2. **Indirect-reading instruments.** These instruments typically consist of calipers and dividers without any graduated scales. They are used to transfer the size measured to a direct-reading instrument, such as a graduated rule. After the legs of the instrument have been adjusted to contact the part at the desired location, the instrument is held against the rule, and the dimension is read. Because of both the experience required to use them and their dependence on graduated scales, the accuracy of this type of indirect measurement is limited. **Telescoping gages** are available for indirect measurement of holes or cavities.

   Angles are measured in degrees, radians, or minutes and seconds of arc. Because of the geometry involved, angles are usually more difficult to measure than are linear dimensions. A **bevel protractor** is a direct-reading instrument similar to a common protractor with the exception that it has a movable member. The two blades of the protractor are placed in contact with the part being measured, and the angle is read directly on the vernier scale. The sensitivity of the instrument depends on the graduations of the vernier. Another type of bevel protractor is the **combination square,** which is a steel rule equipped with devices for measuring 45° and 90° angles.

   Measuring with a **sine bar** involves placing the part on an inclined bar or plate and adjusting the angle by placing gage blocks on a surface plate. After the part is placed on the sine bar, a dial indicator (see below) is used to scan the top surface of the part. **Gage blocks** are added or removed as necessary until the top surface is parallel to the surface plate. The angle on the part is then calculated from geometric relationships. Angles can also be measured by using **angle gage blocks**. These blocks have different tapers that can be assembled in various combinations and used in a manner similar to that for sine bars. Angles on small parts can also be measured through microscopes (with graduated eyepieces) or optical projectors (see page 167).

3. **Comparative length-measuring instruments.** Unlike the instruments described thus far, instruments used for measuring *comparative lengths,* also called *deviation-type instruments*, amplify and measure variations or deviations in distance between two or more surfaces. These instruments compare dimensions, hence the word *comparative.* The common types of instruments used for making comparative measurements are described next.

   Dial indicators are simple mechanical devices that convert linear displacements of a pointer to rotation of an indicator on a circular dial. The indicator is set to zero at a certain reference surface, and the instrument or the surface to be measured (either external or internal) is brought into contact with the pointer. The movement of the indicator is read directly on the circular dial (as either plus or minus) to accuracies as high as 1 μm (40 μin.).

   Unlike mechanical systems, **electronic gages** sense the movement of the contacting pointer through changes in the electrical resistance of a strain gage or through inductance or capacitance. The electrical signals are then converted and displayed as linear dimensions. A common electronic gage is the **linear variable differential transformer** (LVDT), used extensively for measuring small displacements. Although they are more expensive than other types of gages, electronic gages have several advantages, such as ease of operation, rapid response, digital readout, less possibility of human error, versatility, flexibility, and the capability to be integrated into automated systems through microprocessors and computers.

4. **Measuring straightness, flatness, roundness, and profile.** The geometric features of straightness, flatness, roundness, and profile are important aspects of engineering design and manufacturing. For example, piston rods, instrument components, and machine-tool slideways should all meet certain specific requirements with regard to these characteristics in order to function properly. Consequently, their accurate measurement is critical.

   Straightness can be checked with straight edges or dial indicators. **Autocollimators**, resembling a telescope with a light beam that bounces back from the object, are used for accurately measuring small angular deviations on a flat surface. Optical means such as **transits** and **laser beams** are used for aligning individual machine elements in the assembly of machine components.

   Flatness can be measured by mechanical means, using a surface plate and a dial indicator. This method can also be used for measuring perpendicularity, which can be measured with the use of precision steel squares as well. Another method for measuring flatness is by **interferometry**, using an **optical flat.** The flat, a glass or fused quartz disk with parallel flat surfaces, is placed on the surface of the workpiece. When a monochromatic (one wavelength) light beam is aimed at the surface at an angle, the optical flat splits it into two beams, appearing as light and dark bands to the naked eye. The number of fringes that appear is related to the distance between the surface of the part and the bottom surface of the optical flat. Consequently, a truly flat workpiece surface (that is, when the angle between the two surfaces is zero) will not split the light beam, and no fringes will appear. When the surfaces are not flat, fringes are curved.

The interferometry method is also used for observing surface textures and scratches through microscopes for better visibility.

Roundness is generally described as deviations from true roundness (mathematically, a circle). The term **out of roundness** is actually more descriptive of the shape of the part. Roundness is essential to the proper functioning of components such as rotating shafts, bearing races, pistons, cylinders, and steel balls in bearings. The various methods of measuring roundness fall into two basic categories. In the first method, the round part is placed on a V-block or between centers and is rotated, with the pointer of a dial indicator in contact with the surface. After a full rotation of the workpiece, the difference between the maximum and minimum readings on the dial is noted; this difference is called the **total indicator reading** (TIR), or **full indicator movement**. This method is also used for measuring the straightness (*squareness*) of shaft end faces. In the second method, called **circular tracing**, the part is placed on a platform, and its roundness is measured by rotating the platform. Conversely, the probe can be rotated around a stationary part to make the measurement.

Profile may be measured by several methods. In one method, a surface is compared with a template or profile gage to check shape conformity; radii or fillets can be measured by this method. Profile may also be measured with a number of dial indicators or similar instruments. Profile-tracing instruments are the most advanced for measurement of profiles.

Threads and **gear teeth** have several important features with specific dimensions and tolerances. These dimensions must be produced accurately for the smooth operation of gears, reduction of wear and noise level, and part interchangeability. These features are measured by means of thread gages of various designs that compare the thread produced with a standard thread. The gages used include threaded plug gages, screw-pitch gages (similar to radius gages), micrometers with cone-shaped points, and snap gages with anvils in the shape of threads. Gear teeth are measured with instruments that are similar to dial indicators, with calipers and with micrometers using pins or balls of various diameters.

Special profile-measuring equipment is also available. **Optical projectors,** also called **optical comparators**, were first developed in the 1940s to check the geometry of cutting tools for machining screw threads but are now used for checking all profiles. The part is mounted on a table, or between centers, and the image is projected on a screen at magnifications up to $100\times$ or higher. Linear and angular measurements are made directly on the screen, which is equipped with reference lines and circles. The screen can be rotated to allow angular measurements as small as 1 min, using verniers.

5. **Coordinate measuring machines and layout machines.** These important machines consist of a platform on which the workpiece being measured is placed and moved linearly or rotated (Fig. 4.18); a probe, attached to a head capable of lateral and vertical movements, records all measurements. A variety of tactile and nontactile probes, such as laser-based probes, are available.

   Coordinate measuring machines (CMMs) are versatile in their capability to record measurements of complex profiles rapidly and with high sensitivity (0.25 μm [10 μin.]) Also called **measuring machines**, these machines are built rigidly and are very precise. They are equipped with digital readout or can easily be linked to computers for on-line inspection of parts. They can be placed close to machine tools for efficient inspection and rapid feedback for correction of processing parameters before the next part is made. The machines are also made more rugged to resist environmental effects in manufacturing plants, such as temperature variations, vibration, and dirt.

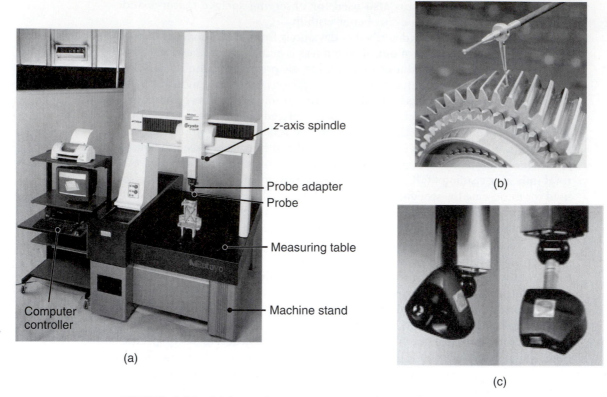

z-axis spindle

Probe adapter
Probe

Measuring table

Computer
controller

Machine stand

(a)

(b)

(c)

FIGURE 4.18   (a) A coordinate measuring machine with part being measured, (b) a touch signal probe measuring the geometry of a gear, and (c) examples of laser probes. *Source:* Courtesy of Mitutoyo America Corp.

Dimensions of large parts are measured by **layout machines,** which are equipped with digital readout. These machines are equipped with scribing tools for marking dimensions on large parts, with an accuracy of 0.04 mm (0.0016 in.).

6. **Gages.** Thus far, we have used the word *gage* (or *gauge*) to describe some types of measuring instruments, such as a caliper gage, depth gage, telescoping gage, electronic gage, strain gage, and radius gage. However, *gage* also has a variety of other meanings such as in pressure gage; gage length of a tension-test specimen; and gages for sheet metal, wire, railroad rail, and the bore of shotguns. It should also be noted that, traditionally, the words *gage* and *instrument* have been used interchangeably.

   Gage blocks are individual square, rectangular, or round metal or ceramic blocks of various sizes. Their surfaces are lapped and are flat and parallel within a range of 0.02–0.12 μm (1–5 μin.). Gage blocks are available in sets of various sizes, some sets containing almost 100 gage blocks. The blocks can be assembled in many combinations to obtain desired lengths. Dimensional accuracy can be as high as 0.05 μm (2 μin.), but environmental-temperature control is important in using gages for high-precision measurements. Although their use requires some skill, gage-block assemblies are commonly utilized in industry as an accurate reference length. Angle blocks are made similarly and are available for angular gaging.

   Fixed gages are replicas of the shapes of the parts to be measured. **Plug gages** are commonly used for holes. The *GO gage* is smaller than the *NOT GO* (or *NO GO*) *gage* and slides into any hole whose smallest dimension is less than the diameter of the gage. The NOT GO gage must not go into the hole. Two gages, one GO gage and one NOT GO gage, are required for such measurements, although both may be on the same device, either at opposite ends or in two steps at one end (*step-type gage*). Plug gages are also available for measuring internal tapers (in which deviations between the gage and the part are indicated by the looseness of the gage), splines, and threads (in which the GO gage must screw into the threaded hole).

Ring gages are used to measure shafts and similar round parts, and **ring thread gages** are used to measure external threads. The GO and NOT GO features on these gages are identified by the type of knurling on the outside diameter of the rings. **Snap gages** are commonly used to measure external dimensions. They are made with adjustable gaging surfaces for use with parts that have different dimensions. One of the gaging surfaces may be set at a different gap than that of the other, thus making a one-unit GO–NOT-GO gage. Although fixed gages are inexpensive and easy to use, they only indicate whether a part is too small or too large compared with an established standard; they do not measure actual dimensions.

There are several types of **pneumatic gages**, also called **air gages**. The gage head has holes through which pressurized air, supplied by a constant-pressure line, escapes. The smaller the gap between the gage and the hole, the more difficult it is for the air to escape, and hence the back pressure is higher. The back pressure, sensed and indicated by a pressure gage, is calibrated to read dimensional variations of holes. The air gage can be rotated during its use to observe and measure any out-of-roundness of the hole. Outside diameters of such parts as pins and shafts also can be measured, where the air plug is in the shape of a ring slipped over the part. In cases where a ring is not suitable to use, a fork-shaped gage head (with the air holes at the tips) can be used. Various shapes of air heads can be custom made for use in specialized applications on parts with different geometric features.

Air gages are easy to use and the resolution can be as fine as 0.125 μm (5 μin.). If the surface roughness of parts is too high, the readings may be unreliable. The compressed-air supply must be clean and dry for proper operation. The part being measured does not have to be free of dust, metal particles, or similar contaminants because the air will blow them away. The noncontacting nature and the low pressure of an air gage has the benefit of not distorting or damaging the part being measured, as could be the case with mechanical gages.

7. **Microscopes.** These are optical instruments used to view and measure very fine details, shapes, and dimensions on small and medium-sized tools, dies, and workpieces. Several types of microscopes are available, with various features for specialized inspection, including models with digital readout. The most common and versatile microscope used in tool rooms is the **toolmaker's microscope.** It is equipped with a stage that is movable in two principal directions and can be read to 2.5 μm (0.0001 in.). The **light-section microscope** is used to measure small surface details, such as scratches, and the thickness of deposited films and coatings. A thin light band is applied obliquely to the surface, and the reflection is viewed at 90°, showing surface roughness, contours, and other features. The **scanning electron microscope** (SEM) has excellent depth of field, and as a result, all regions of a complex part are in focus and can be viewed and photographed to show extremely fine detail. This type of microscope is particularly useful for studying surface textures and fracture patterns. Although expensive, such microscopes are capable of magnifications higher than 100,000$\times$.

4.6.2 Automated measurement

With advanced automation in all aspects of manufacturing processes and operations, a need for *automated measurement* (also called *automated inspection,* see Section 4.8.3) has developed. Flexible manufacturing systems and manufacturing cells, described in Chapter 15, have led to the adoption of advanced measurement techniques and systems. In fact, installation and use of these systems is a necessary and not an optional manufacturing strategy.

Traditionally, a batch of parts was manufactured and sent for measurement in a separate quality-control room, and if the parts passed measurement inspection, they were put into inventory. Automated inspection, on the other hand, involves various on-line sensor systems that monitor the dimensions of parts while being made and use the feedback from these measurements to correct the process whenever necessary.

To appreciate the importance of on-line monitoring of dimensions, consider the following: If a machine has been producing a certain part with acceptable dimensions, what factors contribute to subsequent deviation in the dimension of the same part produced by the same machine? The major factors could be due to

1. Static and dynamic deflections of the machine because of vibrations and fluctuating forces, caused by variations such as in the properties and dimensions of the incoming workpiece material.

2. Deformation of the machine because of thermal effects, including changes in the temperature of the metalworking fluids, machine bearings and components, or the environment.

3. Wear of the tooling, dies, and molds.

Consequently, the dimensions of the parts produced will vary, thus necessitating monitoring of dimensions during production. **In-process workpiece control** is accomplished by special gaging and is used in a variety of applications, such as high-production machining and grinding.

Traditionally, part measurements have been made after a part has been produced, a procedure known as postprocess inspection. In modern manufacturing practice, measurements are made while the part is being produced on the machine, a procedure known as *in-process, on-line,* or *real-time inspection.* Here, the term *inspection* means to check the dimensions of what is produced or being produced and to observe whether it complies with the specified dimensional accuracy for that part.

Inspection routine is particularly important for parts whose failure or malfunction has potentially serious implications, such as bodily injury or fatality. Typical examples of such failures include bolts breaking, cables snapping, switches malfunctioning, brakes failing, grinding wheels exploding, railroad wheels cracking, turbine blades failing, and pressure vessels bursting. In this section, we identify and describe the various methods that are commonly used to inspect manufactured products.

4.7 | Dimensional Tolerances

Dimensional tolerance is defined as the permissible or acceptable variation in the dimensions (height, width, depth, diameter, angles) of a part. Tolerances are unavoidable because it is virtually impossible and unnecessary to manufacture two

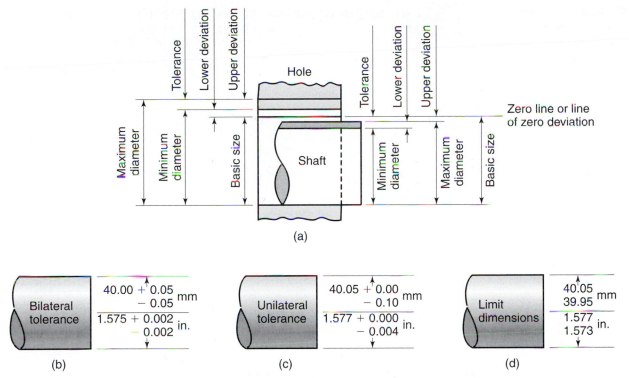

FIGURE 4.19   (a) Basic size, deviation, and tolerance on a shaft, according to the ISO system, and (b)–(d) various methods of assigning tolerances on a shaft. *Source:* After L.E. Doyle.

parts that have precisely the same dimensions. Furthermore, because close dimensional tolerances substantially increase the product cost, specifying a narrow tolerance range is undesirable economically. Tolerances become important only when a part is to be assembled or mated with another part. Surfaces that are free and not functional typically do not need close dimensional tolerance control. Thus, for example, the accuracies of the dimensions of the holes and the distance between the holes for a part such as connecting rod are far more critical than the accuracy of the rod's width and thickness at various locations along its length.

Certain terminology has been established to clearly define geometric tolerances, such as the International Standards Organization (ISO) system shown in Fig. 4.19a. Note that both the shaft and the hole have minimum and maximum diameters, respectively, the difference being the dimensional tolerance for each member. A proper engineering drawing should specify these parameters with numerical values, as shown in Fig. 4.19b.

The range of dimensional tolerances obtained in various manufacturing processes is given in Fig. 4.20. Note that there is a general relationship between tolerances and surface finish of parts manufactured by different processes. Also note the wide range of tolerances and surface finishes that can be obtained. Furthermore, the larger the part, the greater the obtainable tolerance range becomes. Experience has shown that dimensional inaccuracies of manufactured parts are approximately proportional to the cube root of the size of the part. Thus, doubling the size of a part increases the inaccuracies by $2^{1/3} = 1.26$ times, or 26%.

FIGURE 4.20 Tolerances and surface roughness obtained in various manufacturing processes. These tolerances apply to a 25-mm (1-in.) workpiece dimension. *Source:* After J.A. Schey.

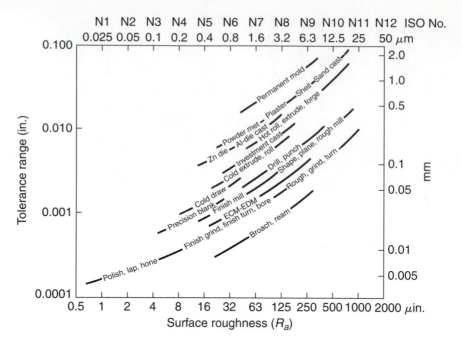

4.8 | Testing and Inspection

4.8.1 Nondestructive testing techniques

Nondestructive testing (NDT), as the name implies, is carried out in such a manner that part integrity and surface texture are preserved. The basic principles of the more commonly used nondestructive testing techniques are described below.

1. In the **liquid-penetrants technique,** fluids are applied to the surfaces of the part and allowed to penetrate into openings such as cracks, seams, and porosity. The penetrant can seep into cracks as narrow as 0.1 μm (4 μin.). Two common types of liquids are (a) *fluorescent* penetrants that fluoresce under ultraviolet light, and (b) *visible* penetrants, using dyes usually red in color, which appear as bright outlines on the surface.

   The surface to be inspected is first thoroughly cleaned and dried. The liquid is then brushed or sprayed on the surface to be inspected and allowed to remain long enough to seep into surface openings. Excess penetrant is wiped off or washed away with water or solvent. A developing agent is then applied to allow the penetrant to seep back to the surface and spread to the edges of openings, thus magnifying the size of defects. The surface is then inspected for defects, either visually (in the case of dye penetrants) or with fluorescent lighting. This method is used extensively and is capable of detecting a variety of surface defects. It is less costly to operate than other methods, but the method can detect only defects that are open to the surface, not internal defects.

2. The **magnetic-particle inspection technique** consists of placing fine ferromagnetic particles on the surface of the part. The particles can be applied either dry or in a liquid carrier such as water or oil. When the part is magnetized with a magnetic field, a discontinuity (defect) on the surface causes the particles to gather visibly around it. The collected particles generally take the shape and size of the defect. Subsurface defects can also be detected by this method, provided that they are not too deep. The ferromagnetic particles may be colored with

pigments for improved visibility. Wet particles are used for detecting fine discontinuities such as fatigue cracks.

3. In **ultrasonic inspection,** an ultrasonic beam travels through the part. An internal defect, such as a crack, interrupts the beam and reflects back a portion of the ultrasonic energy. The amplitude of the energy reflected and the time required for return indicate the presence and location of the flaws in the part. The ultrasonic waves are generated by transducers of various types and shapes, called *probes* or *search units,* They operate on the principle of *piezoelectricity* (Section 3.9.6), using materials such as quartz, lithium sulfate, and various ceramics. Most inspections are carried out at a frequency range of 1–25 MHz. *Couplants,* such as water, oil, glycerin, and grease, are used to transmit the ultrasonic waves from the transducer to the test piece.

   The ultrasonic inspection method has high penetrating power and sensitivity. It can be used to inspect flaws in large volumes of material, such as railroad wheels, pressure vessels, and die blocks, and from different directions. Its accuracy is higher than that of other nondestructive inspection methods. However, this method requires experienced personnel to interpret the results correctly.

4. The **acoustic-emission technique** detects signals (high-frequency stress waves) generated by the workpiece itself during various phenomena, such as plastic deformation, crack initiation and propagation, phase transformation, and sudden reorientation of grain boundaries. Bubble formation during boiling and friction and wear of sliding interfaces are other sources of acoustic signals. Acoustic emissions are detected by sensors consisting of piezoelectric ceramic elements. The acoustic-emission technique is typically performed by stressing elastically the part or structure, such as bending a beam, applying torque to a shaft, and pressurizing a vessel. It is particularly effective for continuous surveillance of load-bearing structures.

5. The **acoustic-impact technique** consists of tapping the surface of an object and listening to and analyzing the transmitted sound waves in order to detect discontinuities and flaws. The principle is basically similar to tapping walls, desktops, or countertops in various locations with fingers or a light hammer and listening to the sound emitted. Vitrified grinding wheels (Section 9.3.1) are tested in a similar manner (ring test) to detect cracks in the wheel that may not be visible to the naked eye. The acoustic-impact technique can be instrumented and automated and is easy to perform. However, the results depend on the geometry and mass of the part, thus requiring a reference standard to identify flaws.

6. **Radiography** involves X-ray inspection to detect internal flaws or variations in the density and thickness of the part. In **digital radiography,** a linear array of detectors is used instead of film, and the data are stored in computer memory. Computer tomography is a similar system except that the monitor produces X-ray images of thin cross sections of the workpiece. **Computer-assisted tomography (catscan)** is based on the same principle and is used widely in medical practice.

7. The **eddy-current inspection method** is based on the principle of electromagnetic induction. The part is placed in, or adjacent to, an electric coil through which alternating current (exciting current) flows at frequencies ranging from 6 to 60 MHz. This current induces eddy currents in the part. Defects existing in the part impede and change the direction of the eddy currents, causing changes in the electromagnetic field. These changes affect the exciting coil (inspection coil), whose voltage is monitored to detect the presence of flaws.

8. **Thermal inspection** involves observing temperature changes by contact- or noncontact-type heat-sensing devices, such as contact temperature probes (see Fig. 5.30c) or infrared scanners. Defects in the workpiece, such as cracks, poorly made joints, and debonded regions in laminated structures, cause a change in the temperature distribution. In **thermographic inspection**, materials such as heat-sensitive paints and papers, liquid crystals, and other coatings are applied to the surface of the part. Changes in their color or appearance indicate the presence of defects.

9. The **holography technique** produces a three-dimensional image of the part, using an optical system. This technique is generally used on simple shapes and highly polished surfaces, and the image is recorded on a photographic film. Holography has been extended to inspection of parts (**holographic interferometry**) that have various shapes and surface characteristics. Defects can be revealed in the part by using double- and multiple-exposure techniques, where the part is subjected to external forces or other changing variables (such as temperature).

    In **acoustic holography**, information on internal defects is obtained directly from the image of the interior of the part. In **liquid-surface acoustical holography**, the part and two ultrasonic transducers (one for the object beam and the other for the reference beam) are immersed in a tank filled with water. The holographic image is then obtained from the ripples in the tank. In **scanning acoustical holography**, only one transducer is used and the hologram is produced by electronic-phase detection. This system is more sensitive, the equipment is typically portable, and very large parts can be accommodated by using a water column instead of a tank.

4.8.2 Destructive testing techniques

The part tested using *destructive testing methods* no longer maintains its integrity, original shape, or surface texture. Thus, the *mechanical testing methods* described in Chapter 2 are all destructive, in that a sample (specimen) has to be removed from the part in order to test it. Other destructive tests include speed testing of grinding wheels to determine their bursting speed, high-pressure testing of pressure vessels to determine their bursting pressure, and formability tests for sheet metal (see Section 7.7.1). Hardness tests that leave large impressions (such as Brinell) also may be regarded as destructive testing, although microhardness tests are typically nondestructive because only a very small permanent indentation is made. This distinction is based on the assumption that the material is not *notch sensitive* (Section 2.9). Most glasses, highly heat-treated metals, and ceramics are notch sensitive; that is, the small indentation produced by the indenter may lower the material's strength and toughness.

4.8.3 Automated inspection

Traditionally, individual parts and assemblies of parts have been manufactured in batches, sent to inspection in quality-control rooms, and, if approved, put in inventory. If products do not pass the quality inspection, they are either reworked, recycled, scrapped, or kept in inventory on the basis of a certain acceptable deviation from the standard. Such a system (**postprocess inspection**) is obviously inefficient as it tracks the defects after they have occurred and in no way attempts to prevent defects.

In contrast, one of the most important practices in modern manufacturing is **automated inspection.** This method uses a variety of sensors that monitor the relevant parameters during the manufacturing process (**on-line inspection**). Then, using these measurements, the process automatically corrects itself to produce acceptable parts; further inspection of the part at another location in the plant is thus unnecessary. Parts may also be inspected immediately after they are produced (**in-process inspection**).

The use of appropriate sensors (see Section 14.8) and computer-control systems (Chapter 15) has enabled the integration of automated inspection into manufacturing operations. Such a system ensures that no part is moved from one manufacturing process to another (such as a turning operation on a lathe followed by cylindrical grinding) unless the part is made correctly and meets the standards set for the first operation. Automated inspection is flexible and responsive to product design changes. Furthermore, less operator skill is required, productivity is increased, and parts have higher quality, reliability, and dimensional accuracy.

Sensors for automated inspection. Rapid advances in **sensor technology** (Section 14.8) have made feasible the real-time monitoring of manufacturing operations. Using various probes and sensors, which can be tactile (touching) or nontactile, one can detect dimensional accuracy, surface roughness, temperature, force, power, vibration, tool wear, and the presence of external or internal defects. Sensors, in turn, are linked to microprocessors and computers for data storage and analysis. This capability allows rapid on-line adjustment of one or more processing parameters, in order to produce parts that are consistently within specified standards of tolerance and quality. Such systems have now become standard equipment on production machines

4.9 | Quality Assurance

Quality assurance is the total effort by a manufacturer to ensure that its products conform to a detailed set of specifications and standards. These standards cover several parameters, such as dimensions, surface finish, dimensional tolerances, composition, color, and mechanical, physical, and chemical properties of materials. The standards are usually written to ensure proper assembly using interchangeable, defect-free components and the fabrication of a product that performs as intended by its designers.

Quality assurance must be the responsibility of everyone involved with the design and manufacturing of products. The often-repeated statement that quality must be *built into a product* reflects this important concept, that quality cannot be inspected into a finished product. Every aspect of design and manufacturing operations, such as material selection, production, and assembly, must be analyzed in detail to ensure that quality is truly built into the final product.

An essential approach is to control materials and processes in such a manner that the products are made correctly in the first place. Because 100% inspection may be too costly to maintain, several methods of inspecting smaller and statistically relevant sample lots have been devised. These methods all use **statistics** to determine the probability of defects occurring in the total production batch.

Inspection involves a series of steps:

1. Inspecting incoming materials to ensure that they meet certain specific property, dimension, and surface finish and integrity requirements.

2. Inspecting individual product components to make sure that they meet specifications.

3. Inspecting the product to make sure that individual parts have been assembled properly.

4. Testing the product to make sure that it functions as designed and intended.

Inspections must be continued throughout production because there are always some variations in (a) the dimensions and properties of incoming materials, (b) the performance of tools, dies, and machines used in various stages of manufacturing, (c) possibilities of human error, (d) and errors made during assembly of the product. Consequently, no two products are made exactly alike. An important aspect of quality control is the capability to analyze defects and promptly eliminate them, or reduce them to acceptable levels. The totality of all these activities is referred to as **total quality management** (TQM).

In order to control quality, we must be able to (1) *measure quantitatively* the level of quality, and (2) *identify* all the material and process variables that can be controlled. The level of quality obtained during production can then be established by inspecting the product to determine whether it meets the specifications for dimensional tolerances, surface finish, defects, and other characteristics.

4.9.1 Statistical methods of quality control

The use of *statistical methods* is essential because of the large number of material and process variables involved in manufacturing operations. Events that occur randomly (without any particular trend or pattern) are called **chance variations.** Events that can be traced to specific causes are called **assignable variations.** For example, results from four-point bending tests have a natural range of strength predictions, because of chance variations in material strength. It was previously explained that the variations arise from the random distribution of small flaws within the material (see Section 2.5). If the bend test specimens are poorly machined so that some specimens have a notch and others are notch-free, then the resulting range of measured strengths are attributable to the production method and are assignable variations.

Although the existence of *variability* in production operations has been recognized for centuries, it was Eli Whitney (1765–1825) who first grasped its full significance when he found that interchangeable parts were indispensable to the mass production of firearms. The terms that are commonly used in **statistical quality control** (SQC) are

a. **Sample size:** The number of parts to be inspected in a sample whose properties are studied to gain information about the whole population.

b. **Random sampling:** Taking a sample from a population or lot in which each item has an equal chance of being included in the sample.

c. **Population** (also called the **universe**): The totality of individual parts of the same design from which samples are taken.

d. **Lot size:** A subset of population; a lot or several lots can be considered subsets of the population and may be treated as representative of the population.

Samples are inspected for certain characteristics and features, such as dimensional tolerances, surface finish, and defects, using the instruments and techniques described earlier in this chapter. These characteristics fall into two categories: those that can be measured *quantitatively* (method of variables) and those that can be measured *qualitatively* (method of attributes).

The **method of variables** is the quantitative *measurement* of characteristics such as dimensions, tolerances, surface finish, and physical or mechanical properties. Such measurements are made for each of the members in the group under consideration, and the results are then compared with the specifications for the part. The **method of attributes** involves observing the *presence or absence* of qualitative characteristics, such as external or internal defects in machined, formed, or welded parts, or dents in sheet-metal products, for each of the units in the group under consideration. Sample size for attributes-type data is generally larger than that for variables-type data because accurate qualitative measures are more difficult to obtain, and variance is therefore usually higher.

During inspection, measurement results typically will vary. For example, when measuring the diameter of turned shafts as they are produced on a lathe (using a micrometer), it will be found that their diameters vary, even though ideally it is desirable for all the shafts to be exactly the same size. If the measured diameters of the turned shafts in a given population are listed, it will be observed that one or more shafts have the smallest diameter, and one or more have the largest diameter. The majority of the turned shafts have diameters that lie between these extremes. The diameters can then be grouped and plotted on a *bar graph*, where each bar represents the number of parts in each diameter group (Fig. 4.21a). The bars show a **distribution** (also called a **spread** or **dispersion**) of the diameter measurements. The bell-shaped curve in Fig. 4.21a is called **frequency distribution** and represents the frequency with which parts within each diameter group are being produced.

Data from manufacturing operations often fit curves represented by a mathematically derived **normal distribution curve** (Fig. 4.21b). These curves, which are also called *Gaussian curves,* are developed on the basis of probability. The bell-shaped

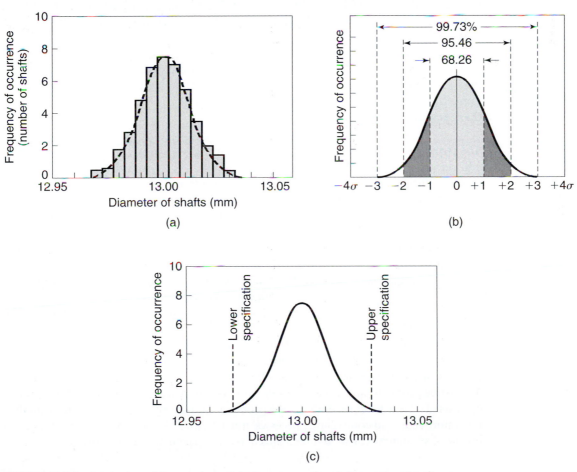

FIGURE 4.21   (a) A plot of the number of shafts measured and their respective diameters. This type of curve is called a *frequency distribution.* (b) A normal distribution curve indicating areas within each range of standard deviation. *Note:* The greater the range, the higher the percentage of parts that fall within it. (c) Frequency distribution curve, showing lower and upper specification limits.

normal distribution curve fitted to the data in Fig. 4.21b has two important features. First, it shows that the diameters of most shafts tend to cluster around an *average* value (**arithmetic mean**). This average is generally designated as $\overline{x}$ and is calculated from the expression

$$\overline{x} = \frac{x_1 + x_2 + x_3 + \cdots + x_n}{n},$$

(4.8)

where the numerator is the sum of all measured values (diameters) and n is the number of measurements (the number of shafts in this case).

The second feature of this curve is its width, indicating the *dispersion* of the diameters measured. The wider the curve, the greater is the dispersion. The difference between the largest value and smallest value is called the *range, R*:

$$R = x_{\max} - x_{\min}.$$

(4.9)

Dispersion is estimated by the **standard deviation,** which is generally denoted as σ and obtained from the expression

$$\sigma = \sqrt{\frac{(x_1 - \overline{x})^2 + (x_2 - \overline{x})^2 + (x_3 - \overline{x})^2 + \cdots + (x_n - \overline{x})^2}{n - 1}},$$

(4.10)

where x is the measured value for each part. Note from the numerator in Eq. (4.10) that (a) as the curve widens, the standard deviation becomes greater, and (b) σ has units of linear dimension.

Since the number of parts that fall within each group is known, we can calculate the percentage of the total population represented by each group. Thus, Fig. 4.21b shows that the diameters of 99.73% of the turned shafts fall within the range of $\pm 3\sigma$, 95.46% within $\pm 2\sigma$ and 68.26% within $\pm 1\sigma$; only 0.2% fall outside the $\pm 3\sigma$ range.

Six sigma. An important trend in manufacturing operations, as well as in business and service industries, is the concept of *six sigma*. Note from the preceding discussion that three sigma in manufacturing would result in 0.27%, or 2700 parts per million defective parts. In modern manufacturing this is an unacceptable rate; in fact, it has been estimated that at the three sigma level, virtually no modern computer would function properly and reliably, and in the service industries, 270 million incorrect credit-card transactions would be recorded each year in the United States alone. It has further been estimated that companies operating at three–four sigma levels lose about 10 to 15% of their total revenue due to defects. Consequently, led by major companies such as Motorola and General Electric, extensive efforts are being made to virtually eliminate defects in products and processes. The resulting savings are reported in billions of dollars.

Six sigma is a set of statistical tools, based on well-known total quality management principles, to continually measure the quality of products and services in selected projects. It includes considerations such as ensuring customer satisfaction, delivering defect-free products, and understanding process capabilities (see below). The approach consists of a clear focus on defining the problem, measuring relevant quantities, analyzing, improving, and controlling processes and activities. Because of its major impact on business, six sigma is now well recognized as a management philosophy.

4.9.2 Statistical process control

If the number of parts that do not meet set standards increases during a production run, it is essential to determine the cause (such as, for example, variability in incoming materials, machine controls, degradation of metalworking fluids, operator boredom) and take appropriate action. Although this statement at first appears to be self-evident, it was only in the early 1950s that a systematic statistical approach was developed to guide machine operators in manufacturing plants.

This approach advises the operator of the appropriate measures to take and when to take them, in order to avoid producing further defective parts. Known as *statistical process control* (SPC), this technique consists of several elements: (a) control charts and setting control limits, (b) capabilities of the particular manufacturing process, and (c) characteristics of the machinery involved.

Control charts. The frequency distribution curve given in Fig. 4.21a shows a range of shaft diameters that may fall beyond the specified design tolerance range. The same bell-shaped curve is shown in Fig. 4.21c, which now includes the specified tolerances for the diameters of the turned shafts. *Control charts* graphically represent the variations of a process over a period of time. They consist of data plotted during production, and there typically are two plots. The quantity $\bar{x}$ in Fig. 4.22a is the average for each subset of samples taken and inspected, say, a subset consisting of 5 parts. (A sample size of between 2 and 10 parts can be sufficiently accurate in manufacturing, depending on the standard deviation and provided that sample size is held constant throughout the inspection.)

The frequency of sampling depends on the nature of the process. Some processes may require continuous sampling, whereas others may require only one sample per day. Quality-control analysts are best qualified to determine this frequency for a particular situation. Since the measurements shown in Fig. 4.22a are made consecutively, the abscissa of these control charts also represents time. The solid horizontal line in this figure is the **average of averages (grand average)**, denoted as $\bar{\bar{x}}$, and represents the *population mean*. The upper and lower horizontal broken lines in these control charts indicate the **control limits** for *the process*.

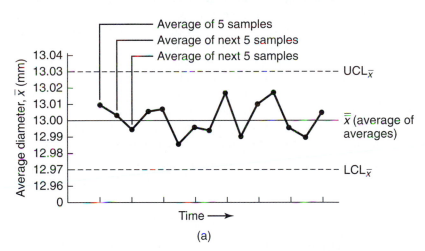

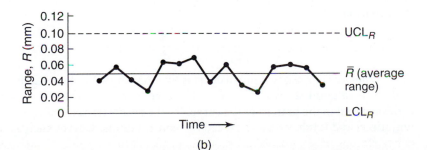

FIGURE 4.22 Control charts used in statistical quality control. The process shown is in good statistical control, because all points fall within the lower and upper control limits. In this illustration, the sample size is 5, and the number of samples is 15.

TABLE 4.3

Constants for Control Charts				
Sample size	A_2	D_4	D_3	d_2
2	1.880	3.267	0	1.128
3	1.023	2.575	0	1.693
4	0.729	2.282	0	2.059
5	0.577	2.115	0	2.326
6	0.483	2.004	0	2.534
7	0.419	1.924	0.078	2.704
8	0.373	1.864	0.136	2.847
9	0.337	1.816	0.184	2.970
10	0.308	1.777	0.223	3.078
12	0.266	1.716	0.284	3.258
15	0.223	1.652	0.348	3.472
20	0.180	1.586	0.414	3.735

The control limits are set on these charts according to statistical-control formulas designed to keep actual production within the usually acceptable $\pm 3\sigma$ range. Thus, for $\bar{x}$,

$$\text{Upper control limit (UCL}_{\bar{x}}) = \bar{x} + 3\sigma = \bar{\bar{x}} + A_2\overline{R} \tag{4.11}$$

and

$$\text{Lower control limit (LCL}_{\bar{x}}) = \bar{x} - 3\sigma = \bar{\bar{x}} - A_2\overline{R}, \tag{4.12}$$

where A_2 is obtained from Table 4.3, and $\overline{R}$ is the average of the R values.

The control limits are calculated based on the historical production capability of the equipment itself; they are generally not associated with design tolerance specifications and dimensions. They indicate the limits within which a certain percentage of measured values is normally expected to fall due to the inherent variations of the process itself. The major goal of statistical process control is to improve the manufacturing process via the aid of control charts to eliminate assignable causes.

The second control chart, given in Fig. 4.22b, shows the range R in each subset of samples. The solid horizontal line represents the average of R values, denoted as $\overline{R}$ in the lot. It is a measure of the variability in the samples. The upper and lower control limits for R are obtained from the equations

$$\text{UCL}_R = D_4\overline{R} \tag{4.13}$$

and

$$\text{LCL}_R = D_3\overline{R}, \tag{4.14}$$

where the constants D_4 and D_3 are read from Table 4.3. This table also includes values for the constant d_2 which is used in estimating the standard deviation from the equation

$$\sigma = \frac{\overline{R}}{d_2}. \tag{4.15}$$

When the curve of a control chart is like that shown in Fig. 4.22b, the process is said to be **in good statistical control.** In other words, (a) there is no clear and discernible trend in the pattern of the curve, (b) the points (measured values) are random with time, and (c) they do not exceed the control limits. Curves such as those shown

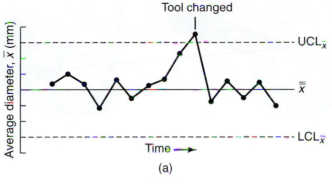

Tool changed

(a)

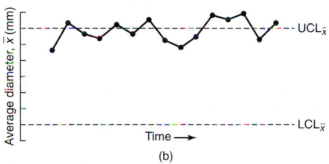

(b)

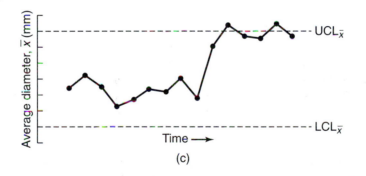

(c)

FIGURE 4.23 Control charts. (a) Process begins to become out of control because of factors such as tool wear. The tool is changed, and the process is then in good statistical control. (b) Process parameters are not set properly; thus, all parts are around the upper control limit. (c) Process becomes out of control because of factors such as a sudden change in the properties of the incoming material.

in Figs. 4.23a through c indicate certain trends. For example, about halfway in Fig. 4.23a, the diameter of the shafts increases with time, due to a change in one of the process variables such as wear of the cutting tool. If, as in Fig. 4.23b, the trend is toward consistently larger diameters (hovering around the upper control limit), it could mean that the tool settings on the lathe may not be correct and, as a result, the shafts being turned are consistently too large.

Figure 4.23c shows two distinct trends that may be caused by factors such as a change in the properties of the incoming material or a change in the performance of the cutting fluid (for example, by degradation). These situations place the process **out of control.**

Analyzing control-chart patterns and trends requires considerable experience in order to identify the specific cause(s) of an out-of-control situation. A further reason for out-of-control situations is **overcontrol** of the manufacturing process; that is, setting upper and lower control limits too close to each other, and hence setting a

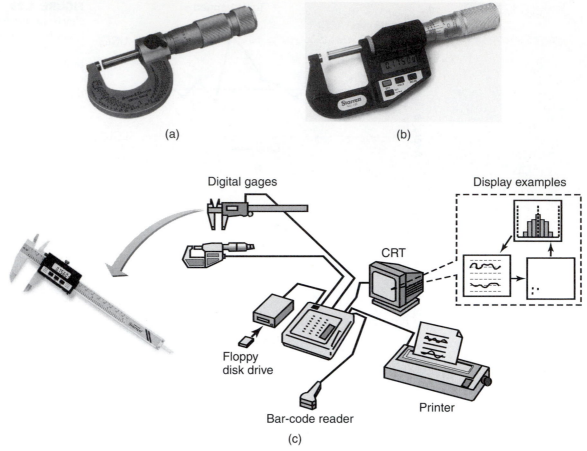

(a)   (b)

Digital gages

Display examples

CRT

Floppy
disk drive

Bar-code reader

Printer

(c)

FIGURE 4.24 (a) A vernier (analog) micrometer. (b) A digital micrometer with a range of 0 to 1 in. (0 to 25 mm) and a resolution of 50 μin. (1.25 μm). (c) Schematic illustration showing the integration of digital gages with microprocessors for real-time data acquisition for statistical process control. *Source:* (a)–(b) Courtesy of L.C. Starrett Co., (b) Courtesy of Mitutoyo Corp.

smaller standard-deviation range. In order to avoid overcontrol, control limits are set on process capability, rather than unrelated (sometimes arbitrary) ranges.

It is evident that operator training is critical for successful implementation of SPC on the shop floor. Once process-control guidelines are established, operators also should be given some authority to make adjustments in processes that are becoming out of control. This task is now made easier by a variety of available software. For example, digital readouts on electronic measuring devices are integrated directly into a computer system for real-time SPC. Figure 4.24 shows such a multi-functional computer system in which the output from a digital caliper or micrometer is analyzed in real time and displayed in several ways, such as frequency distribution curves and control charts.

Process capability. *Process capability* is defined as the limits within which individual measurement in a particular manufacturing process would normally be expected to fall when only random variation is present. It tells us that the process is capable of producing parts within certain limits of precision. Since a manufacturing process involves materials, machinery, and operators, each of these aspects can be analyzed individually to identify a problem when process capabilities do not meet part specifications.

SUMMARY

- The surface of a workpiece can significantly affect its properties and characteristics, including friction and wear characteristics, lubricant effectiveness, appearance and geometric features, and thermal and electrical conductivity of contacting bodies. (Section 4.1)

- The surface structure of a metal has usually been plastically deformed and work hardened during prior processing. Surface roughness can be quantified by various techniques. Surface integrity can be compromised by several defects. Flaws, directionality, roughness, and waviness are measurable quantities that are used to describe surface texture. (Sections 4.2, 4.3)

- Tribology is the science and technology of interacting surfaces, and encompasses friction, wear, and lubrication. Friction may be desirable or undesirable, depending on the specific manufacturing circumstances. (Section 4.4)

- Wear, defined as the progressive loss or removal of material from a surface, alters the geometry of the workpiece and tool and die interfaces and thus affects the manufacturing process, dimensional accuracy, and the quality of parts produced. Friction and wear can be reduced by using various liquid or solid lubricants, as well as by applying ultrasonic vibrations. Four regimes of lubrication are relevant to metalworking processes. (Section 4.4)

- Several surface treatments are used to impart specific physical and mechanical properties to the workpiece. The techniques employed typically include mechanical working, physical and chemical means, heat treatments, and coatings. Cleaning of a manufactured workpiece involves removal of solid, semisolid, and liquid contaminants from a surface by various means. (Section 4.5)

- Parts made are measured by a variety of instruments with specific features and characteristics. Major advances have been made in automated measurement, linking measuring equipment to microprocessors and computers for accurate in-process control of manufacturing operations. (Section 4.6)

- Dimensional tolerances and their specification are important factors in manufacturing, as they not only affect subsequent assembly of the parts made and the accuracy and operation of all types of machinery and equipment but also can significantly influence product cost. (Section 4.7)

- Several nondestructive and destructive testing techniques are available for inspection of completed parts and products. Inspection of each parts being manufactured is now possible using automated and reliable inspection techniques. (Section 4.8)

- Quality assurance is the total effort by a manufacturer to ensure that its products conform to a specific set of standards and specifications. Statistical quality-control and process-control techniques are now widely employed in defect detection and prevention. The total-quality-control philosophy focuses on prevention, rather than detection, of defects. (Section 4.9)

SUMMARY OF EQUATIONS

Arithmetic mean value: $R_a = \dfrac{y_a + y_b + y_c + \cdots + y_n}{n} = \dfrac{1}{n}\sum\limits_{i=1}^{n} y_i = \dfrac{1}{l}\int_0^l |y|\, dx$

Root-mean-square average: $R_q = \sqrt{\dfrac{y_a^2 + y_b^2 + y_c^2 + \cdots + y_n^2}{n}} = \sqrt{\dfrac{1}{n}\sum\limits_{i=1}^{n} y_i^2} = \left[\dfrac{1}{l}\int_0^l y^2\, dx\right]^{1/2}$

Coefficient of friction: $\mu = \dfrac{F}{N} = \dfrac{\tau}{\text{Hardness}}$

Friction (shear) factor: $m = \dfrac{\tau_i}{k}$

Adhesive wear: $V = k\dfrac{LW}{3p}$

Arithmetic mean: $\bar{x} = \dfrac{x_1 + x_2 + x_3 + \cdots + x_n}{n}$

Range: $R = x_{\max} - x_{\min}$

Standard deviation: $\sigma = \sqrt{\dfrac{(x_1 - \bar{x})^2 + (x_2 - \bar{x})^2 + (x_3 - \bar{x})^2 + \cdots + (x_n - \bar{x})^2}{n - 1}}$

Upper control limit: $\text{UCL}_{\bar{x}} = \bar{x} + 3\sigma = \bar{\bar{x}} + A_2\bar{R}$

Lower control limit: $\text{LCL}_{\bar{x}} = \bar{x} - 3\sigma = \bar{\bar{x}} - A_2\bar{R}$

BIBLIOGRAPHY

Aft, L.S., *Fundamentals of Industrial Quality Control,* 3d ed., Addison-Wesley, 1998.

ASM Handbook, Vol. 17: *Nondestructive Evaluation and Quality Control,* ASM International, 1989.

Bayer, R.G., *Mechanical Wear Fundamentals and Testing,* 2nd ed., Dekker, 2005.

Besterfield, D.H., *Quality Control,* 7th ed., Prentice Hall, 2004.

Bhushan, B., *Introduction to Tribology,* Wiley, 2003.

Bhushan, B. (ed.), *Modern Tribology Handbook,* CRC Press, 2001.

Booser, E.R. (ed.), *Tribology Data Handbook,* CRC Press, 1998.

Bothe, D.R., *Measuring Process Capability: Techniques and Calculations for Quality and Manufacturing Engineers,* McGraw-Hill, 1997.

Breyfogle, F., *Implementing Six Sigma: Smarter Solutions Using Statistical Methods,* 2nd ed., Wiley, 2003.

Burakowski, T., and Wiershon, T., *Surface Engineering of Metals: Principles, Equipment, Technologies,* CRC Press, 1998.

Campbell, R., *Integrated Product Design and Manufacturing Using Geometric Dimensioning and Tolerancing,* CRC Press, 2002.

Davis, J.R. (ed.), *Surface Engineering for Corrosion and Wear Resistance,* IOM Communications and ASM International, 2001.

Drake, P.J., *Dimensioning and Tolerancing Handbook,* McGraw-Hill, 1999.

Farrago, F.T., and Curtis, M.A., *Handbook of Dimensional Measurement,* 3d ed., Industrial Press, 1994.

Grant, E.L., and Leavenworth, R.S., *Statistical Quality Control,* McGraw-Hill, 1997.

Kear, F.W., *Statistical Process Control in Manufacturing Practice,* Dekker, 1998.

Krulikowski, A., *Fundamentals of Geometric Dimensioning and Tolerancing,* Delmar, 1997.

Lindsay, J.H. (ed.), *Coatings and Coating Processes for Metals,* ASM International, 1998.

Meadows, J.D., Geometric *Dimensioning and Tolerancing,* Dekker, 1995.

_____, *Measurement of Geometric Tolerances in Manufacturing,* Dekker, 1998.

Montgomery, D.C., *Introduction to Statistical Quality Control,* Wiley, 2004.

Murphy, S.D., *In-Process Measurement and Control,* Dekker, 1990.

Nachtman, E.S., and Kalpakjian, S., *Lubricants and Lubrication in Metalworking Operations,* Dekker, 1985.

Puncochar, D.E., *Interpretation of Geometric Dimensioning and Tolerancing,* 2nd ed., Industrial Press, 1997.

Rabinowicz, E., *Friction and Wear of Materials,* 2nd ed., Wiley, 1995.

Robinson, S.L., and Miller, R.K., *Automated Inspection and Quality Assurance,* Dekker, 1989.

Schey, J.A., *Tribology in Metalworking: Friction, Lubricating and Wear,* ASM International, 1983.

Stachowiak, G.W., and Batchelor, A.W., *Engineering Tribology,* Butterworth-Heinemann, 2001.

Stern, K.H. (ed.), *Metallurgical and Ceramic Protective Coatings,* Chapman & Hall, 1996.

Sudarshan, T.S. (ed.), *Surface Modification Technologies,* ASM International, 1998.

Wadsworth, H.M., *Handbook of Statistical Control Methods for Engineers and Scientists,* 2nd ed., McGraw-Hill, 1998.

Whitehouse, D.J., *Handbook of Surface Metrology,* Institute of Physics, 1994.

Williams, J.A., *Introduction to Tribology,* Cambridge University Press, 2006.

Winchell, W., *Inspection and Measurement in Manufacturing,* Society of Manufacturing Engineers, 1996.

QUESTIONS

4.1 Explain what is meant by surface integrity. Why should we be interested in it?

4.2 Why are surface-roughness design requirements in engineering so broad? Give appropriate examples.

4.3 We have seen that a surface has various layers. Describe the factors that influence the thickness of each of these layers.

4.4 What is the consequence of oxides of metals being generally much harder than the base metal? Explain.

4.5 What factors would you consider in specifying the lay of a surface?

4.6 Describe the effects of various surface defects (see Section 4.3) on the performance of engineering components in service. How would you go about determining whether or not each of these defects is important for a particular application?

4.7 Explain why the same surface roughness values do not necessarily represent the same type of surface.

4.8 In using a surface-roughness measuring instrument, how would you go about determining the cutoff value? Give appropriate examples.

4.9 What is the significance of the fact that the stylus path and the actual surface profile generally are not the same?

4.10 Give two examples each in which waviness of a surface would be (1) desirable and (2) undesirable.

4.11 Explain why surface temperature increases when two bodies are rubbed against each other. What is the significance of temperature rise due to friction?

4.12 To what factors would you attribute the fact that the coefficient of friction in hot working is higher than in cold working, as shown in Table 4.1?

4.13 In Section 4.4.1, we note that the values of the coefficient of friction can be much higher than unity. Explain why.

4.14 Describe the tribological differences between ordinary machine elements (such as meshing gears, cams in contact with followers, and ball bearings with inner and outer races) and elements of metalworking processes (such as forging, rolling, and extrusion, which involve workpieces in contact with tools and dies).

4.15 Give the reasons that an originally round specimen in a ring-compression test may become oval after deformation.

4.16 Can the temperature rise at a sliding interface exceed the melting point of the metals? Explain.

4.17 List and briefly describe the types of wear encountered in engineering practice.

4.18 Explain why each of the terms in the Archard formula for adhesive wear, Eq. (4.6), should affect the wear volume.

4.19 How can adhesive wear be reduced? How can fatigue wear be reduced?

4.20 It has been stated that as the normal load decreases, abrasive wear is reduced. Explain why this is so.

4.21 Does the presence of a lubricant affect abrasive wear? Explain.

4.22 Explain how you would estimate the magnitude of the wear coefficient for a pencil writing on paper.

4.23 Describe a test method for determining the wear coefficient k in Eq. (4.6). What would be the difficulties in applying the results from this test to a manufacturing application, such as predicting the life of tools and dies?

4.24 Why is the abrasive wear resistance of a material a function of its hardness?

4.25 We have seen that wear can have detrimental effects on engineering components, tools, dies, etc. Can you visualize situations in which wear could be beneficial? Give some examples. (*Hint:* Note that writing with a pencil is a wear process.)

4.26 On the basis of the topics discussed in this chapter, do you think there is a direct correlation between friction and wear of materials? Explain.

4.27 You have undoubtedly replaced parts in various appliances and automobiles because they were worn. Describe the methodology you would follow in determining the type(s) of wear these components have undergone.

4.28 Why is the study of lubrication regimes important?

4.29 Explain why so many different types of metal-working fluids have been developed.

4.30 Differentiate among (1) coolants and lubricants, (2) liquid and solid lubricants, (3) direct and indirect emulsions, and (4) plain and compounded oils.

4.31 Explain the role of conversion coatings. Based on Fig. 4.13, what lubrication regime is most suitable for application of conversion coatings?

4.32 Explain why surface treatment of manufactured products may be necessary. Give several examples.

4.33 Which surface treatments are functional, and which are decorative? Give several examples.

4.34 Give examples of several typical applications of mechanical surface treatment.

4.35 Explain the difference between case hardening and hard facing.

4.36 List several applications for coated sheet metal, including galvanized steel.

4.37 Explain how roller-burnishing processes induce residual stresses on the surface of workpieces.

4.38 List several products or components that could not be made properly, or function effectively in service, without implementation of the knowledge involved in Sections 4.2 through 4.5.

4.39 Explain the difference between direct- and indirect-reading linear measurements.

4.40 Why have coordinate-measuring machines become important instruments in modern manufacturing? Give some examples of applications.

4.41 Give reasons why the control of dimensional tolerances in manufacturing is important.

4.42 Give examples where it may be preferable to specify unilateral tolerances as opposed to bilateral tolerances in design.

4.43 Explain why a measuring instrument may not have sufficient precision.

4.44 Comment on the differences, if any, among (1) roundness and circularity, (2) roundness and eccentricity, and (3) roundness and cylindricity.

4.45 It has been stated that dimensional tolerances for nonmetallic stock, such as plastics, are usually wider than for metals. Explain why. Consider physical and mechanical properties of the materials involved.

4.46 Describe the basic features of nondestructive testing techniques that use electrical energy.

4.47 Identify the nondestructive techniques that are capable of detecting internal flaws and those that only detect external flaws.

4.48 Which of the nondestructive inspection techniques are suitable for nonmetallic materials? Why?

4.49 Why is automated inspection becoming an important aspect of manufacturing engineering?

4.50 Describe situations in which the use of destructive testing techniques is unavoidable.

4.51 Should products be designed and built for a certain expected life? Explain.

4.52 What are the consequences of setting lower and upper specifications closer to the peak of the curve in Fig. 4.23?

4.53 Identify factors that can cause a process to become out of control. Give several examples of such factors.

4.54 In reading this chapter, you will have noted that the specific term *dimensional tolerance* is often used, rather than just the word *tolerance*. Do you think this distinction is important? Explain.

4.55 Give an example of an assignable variation and a chance variation.

PROBLEMS

4.56 Referring to the surface profile in Fig. 4.3, give some numerical values for the vertical distances from the center line. Calculate the R_a and R_q values. Then give another set of values for the same general profile, and calculate the same quantities. Comment on your results.

4.57 Calculate the ratio of R_a/R_q for (a) a sine wave, (b) a saw-tooth profile, (c) a square wave.

4.58 Refer to Fig. 4.7b and make measurements of the external and internal diameters (in the horizontal direction in the photograph) of the four specimens shown. Remembering that in plastic deformation the volume of the rings remains constant, calculate (a) the reduction in height and (b) the coefficient of friction for each of the three compressed specimens.

4.59 Using Fig. 4.8a, make a plot of the coefficient of friction versus the change in internal diameter for a reduction in height of (1) 25%, (2) 50%, and (3) 60%.

4.60 In Example 4.1, assume that the coefficient of friction is 0.20. If all other initial parameters remain the same, what is the new internal diameter of the ring specimen?

4.61 How would you go about estimating forces required for roller burnishing? (*Hint:* Consider hardness testing.)

4.62 Estimate the plating thickness in electroplating a 50-mm solid metal ball using a current of 1 A and a plating time of two hours. Assume that $c = 0.08$.

4.63 Assume that a steel rule expands by 1% because of an increase in environmental temperature. What will be the indicated diameter of a shaft whose actual diameter is 50.00 mm?

4.64 Examine Eqs. (4.2) and (4.10). What is the relationship between R_q and σ? What would be the equation for the standard deviation of a continuous curve?

4.65 Calculate the control limits for averages and ranges for the following: number of samples = 7; $\overline{\overline{x}} = 50$; $\overline{R} = 7$.

4.66 Calculate the control limits for the following: number of samples = 7; $\overline{\overline{x}} = 40.5$; $UCL_R = 4.85$.

4.67 In an inspection with a sample size of 10 and a sample number of 40, it was found that the average range was 10 and the average of averages was 75. Calculate the control limits for averages and ranges.

4.68 Determine the control limits for the data shown in the following table:

x_1	x_2	x_3	x_4
0.65	0.75	0.67	0.65
0.69	0.73	0.70	0.68
0.65	0.68	0.65	0.61
0.64	0.65	0.60	0.60
0.68	0.72	0.70	0.66
0.70	0.74	0.65	0.71

4.69 Calculate the mean, median, and standard deviation for all of the data in Problem 4.68.

4.70 The average of averages of a number of samples of size 7 was determined to be 125. The average range was 17.82, and the standard deviation was 5.85. The following measurements were taken in a sample: 120, 132, 124, 130, 118, 132, and 121. Is the process in control?

4.71 Assume that you are asked to give a quiz to students on the contents of this chapter. Prepare three quantitative problems and three qualitative questions, and supply the answers.

CHAPTER 5

Metal-Casting Processes and Equipment; Heat Treatment

This chapter describes the fundamentals of metal casting and the characteristics of casting processes, including

- Mechanisms of the solidification of metals, characteristics of fluid flow, and the role of entrapped gases and shrinkage.
- Properties of casting alloys and their applications.
- Characteristics of expendable-mold and permanent-mold casting processes, their applications, and economic considerations.
- Design considerations and simulation techniques for casting.
- Economic considerations.

5.1 | Introduction

As described throughout this text, several methods can be used to shape materials into useful products. **Casting** is one of the oldest methods and was first used around about 4000 B.C. to make ornaments, arrowheads, and various other objects. Casting processes basically involve the introduction of molten metal into a mold cavity where, upon solidification, the metal takes the shape of the cavity. This process is capable of producing intricate shapes in a single piece, ranging in size from very large to very small, including those with internal cavities. Typical cast products are engine blocks, cylinder heads, transmission housings, pistons, turbine disks, railroad and automotive wheels, and ornamental artifacts.

All metals can be cast in, or nearly in, the final shape desired, often with only minor finishing operations required. With appropriate control of material and process parameters, parts can be cast with uniform properties throughout. As with all other manufacturing processes, a knowledge of certain fundamental aspects is essential to the production of good quality parts, with good surface finish, dimensional accuracy, strength, and few if any defects.

The important factors in casting operations are

1. **Solidification** of the metal from its molten state, and accompanying shrinkage.
2. **Flow** of the molten metal into the mold cavity.
3. **Heat transfer** during solidification and cooling of the metal in the mold.
4. **Mold material** and its influence on the casting operation.

5.2 | Solidification of Metals

An overview of the solidification of metals and alloys is presented in this section. The topics covered are essential to our understanding of the structures developed in casting and the structure-property relationships obtained in the casting processes described throughout this chapter.

Pure metals have clearly defined melting or freezing points, and solidification takes place at a constant temperature (Fig. 5.1a). When the temperature of the molten metal is reduced to the freezing point, the latent heat of fusion is given off while the temperature remains constant. At the end of this isothermal phase change, solidification is complete and the solid metal cools to room temperature. The casting contracts as it cools, due to (a) contraction from a superheated state to the metal's solidification temperature, and (b) cooling, as a solid, from the solidification temperature to room temperature (Fig. 5.1b). A significant density change also can occur as a result of phase change from liquid to solid.

Unlike pure metals, **alloys** solidify over a *range* of temperatures. Solidification begins when the temperature of the molten metal drops below the **liquidus** and is completed when the temperature reaches the **solidus**. Within this temperature range the alloy is *mushy* or *pasty*, whose composition and state are described by the particular alloy's phase diagram.

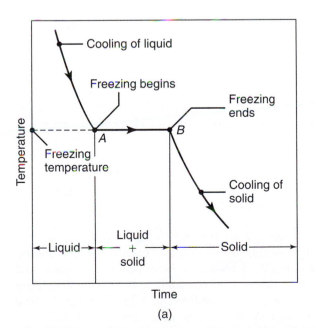

(a)

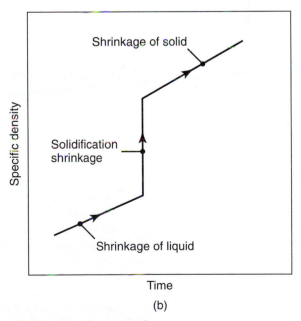

(b)

FIGURE 5.1 (a) Temperature as a function of time for the solidification of pure metals. Note that freezing takes place at a constant temperature. (b) Density as a function of time.

5.2.1 Solid solutions

Two terms are essential in describing alloys: solute and solvent. **Solute** is the minor element (such as, for example, salt or sugar) that is added to the **solvent,** which is the major element (such as water). In terms of the elements involved in a metal's crystal structure (Chapter 3), the solute is the element (solute atoms) added to the solvent (*host atoms*). When the particular crystal structure of the solvent is maintained during alloying, the alloy is called a **solid solution.**

Substitutional solid solutions. If the size of the solute atom is similar to that of the solvent atom, the solute atoms can replace solvent atoms and form a *substitutional solid solution* (see Fig. 3.9). An example of this phenomenon is brass, an alloy of zinc and copper, in which zinc (solute atom) is introduced into the lattice of copper (solvent atoms). The properties of brasses can thus be altered over a certain range by controlling the amount of zinc in copper.

Interstitial solid solutions. If the size of the solute atom is much smaller than that of the solvent atom, the solute atom occupies an interstitial position (as shown in Fig. 3.9) and forms an *interstitial solid solution*. A major example of interstitial solution is steel, an alloy of iron and carbon, in which carbon atoms are present in an interstitial position between iron atoms. As will be shown in Section 5.11, the properties of steel can thus be varied over a wide range by controlling the amount of carbon in iron. This is one reason that steel, in addition to being inexpensive, is such a versatile and important material with a wide range of properties and applications.

5.2.2 Intermetallic compounds

Intermetallic compounds are complex structures in which solute atoms are present among solvent atoms in certain specific proportions; thus, some intermetallic compounds have solid solubility. The type of atomic bonds may range from metallic to ionic. Intermetallic compounds are strong, hard, and brittle. An example is copper in aluminum, where an intermetallic compound of $CuAl_2$ can be made to precipitate from an aluminum-copper alloy; this is an example of precipitation hardening (see Section 5.11.2).

5.2.3 Two-phase alloys

A solid solution is one in which two or more elements are soluble in a solid state, forming a single homogeneous material in which the alloying elements are uniformly distributed throughout the solid. There is, however, a limit to the concentration of solute atoms in a solvent-atom lattice, just as there is a limit to the solubility of sugar in water. Most alloys consist of two or more solid phases and thus may be regarded as mechanical mixtures. Such a system with two solid phases is called a **two-phase system,** in which each phase is a homogeneous part of the total mass and has its own characteristics and properties.

A typical example of a two-phase system in metals is lead added to copper in the molten state. After the mixture solidifies, the structure consists of two phases: (a) one phase that has a small amount of lead in solid solution in copper, and (b) another phase in which lead particles, approximately spherical in shape, are *dispersed* throughout the matrix of the primary phase (Fig. 5.2a). This copper-lead alloy has properties that are different from those of either copper or lead alone.

Alloying with finely dispersed particles (*second-phase particles*) is an important method of strengthening alloys and controlling their properties. Generally, in

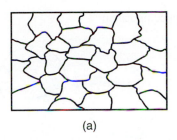

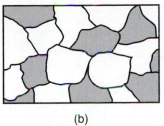

FIGURE 5.2 (a) Schematic illustration of grains, grain boundaries, and particles dispersed throughout the structure of a two-phase system, such as lead-copper alloy. The grains represent lead in solid solution of copper, and the particles are lead as a second phase. (b) Schematic illustration of a two-phase system, consisting of two sets of grains: dark and light. Dark and light grains have their own compositions and properties.

two-phase alloys, the second-phase particles present obstacles to dislocation movement, which increases the alloy's strength (Section 3.3). Another example of a two-phase alloy is the aggregate structure shown in Fig. 5.2b, which contains two sets of grains, each with its own composition and properties. The darker grains may, for example, have a different structure than the lighter grains and be brittle, whereas the lighter grains may be ductile.

5.2.4 Phase diagrams

A *phase diagram*, also called an **equilibrium diagram** or **constitutional diagram**, graphically illustrates the relationships among temperature, composition, and the phases present in a particular alloy system. *Equilibrium* means that the state of a system remains constant over an indefinite period of time. *Constitutional* indicates the relationships among structure, composition, and physical makeup of the alloy.

An example of a phase diagram is shown in Fig. 5.3 for the nickel-copper alloy. It is called a **binary phase diagram** because of the two elements (nickel and copper) present in the system. The left boundary of this phase diagram (100% Ni) indicates the melting point of nickel, and the right boundary (100% Cu) indicates the melting point of copper. (All percentages in this discussion are by weight.) Note that for a composition of, say, 50% Cu-50% Ni, the alloy begins to solidify at a temperature of 1313°C (2395°F), and solidification is complete at 1249°C (2280°F). Above 1313°C, a homogeneous liquid of 50% Cu-50% Ni exists. When cooled slowly to 1249°C, a homogeneous solid solution of 50% Cu-50% Ni results.

Between the liquidus and solidus curves, say, at a temperature of 1288°C (2350°F), is a two-phase region: a *solid phase* composed of 42% Cu-58% Ni, and a *liquid phase* of 58% Cu-42% Ni. To determine the solid composition, go left horizontally to the solidus curve and read down to obtain 42% Cu-58% Ni. The liquid composition can be determined similarly by going to the right to the liquidus curve (58% Cu-42% Ni).

The completely solidified alloy in the phase diagram shown in Fig. 5.3 is a **solid solution**, because the alloying element (Cu, solute atom) is completely dissolved in the host metal (Ni, solvent atom), and each grain has the same composition. The mechanical properties of solid solutions of Cu-Ni depend on their composition. For example, by increasing the nickel content, the properties of copper are improved. The improvement in properties is due to *pinning* (blocking) of dislocations at solute atoms of nickel, which may also be regarded as impurity atoms (see Fig. 3.9). As a result, dislocations cannot move as freely, and consequently the strength of the alloy increases.

Lever rule. The composition of various phases in a phase diagram can be determined by a procedure called the *lever rule*. As shown in the lower portion of Fig. 5.3, we first construct a lever between the solidus and liquidus lines (*tie line*), which is balanced (on the triangular support) at the nominal weight composition C_0 of the alloy.

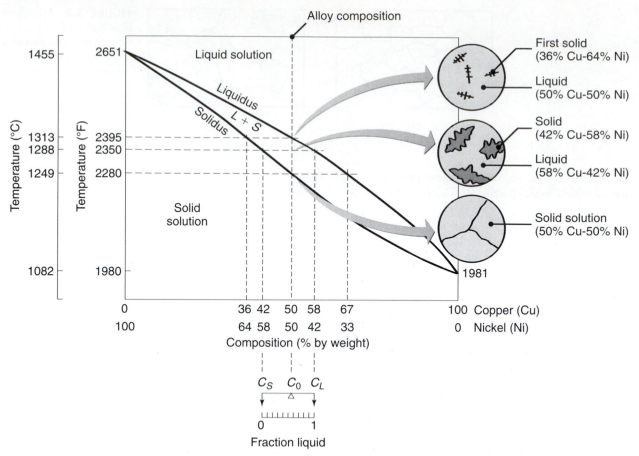

FIGURE 5.3   Phase diagram for nickel-copper alloy system obtained by a low rate of solidification. Note that pure nickel and pure copper each have one freezing or melting temperature. The top circle on the right depicts the nucleation of crystals, the second circle shows the formation of dendrites, and the bottom circle shows the solidified alloy with grain boundaries.

The left end of the lever represents the composition C_S of the solid phase and the right end of the composition C_L of the liquid phase. Note from the graduated scale in the figure that the liquid fraction is also indicated along this tie line, ranging from 0 at the left (fully solid) to 1 at the right (fully liquid).

The lever rule states that the **weight fraction of solid** is proportional to the distance between C_0 and C_L:

$$\frac{S}{S + L} = \frac{C_0 - C_L}{C_S - C_L}.$$ (5.1)

Likewise, the **weight fraction of liquid** is proportional to the distance between C_S and C_0; hence

$$\frac{L}{S + L} = \frac{C_S - C_0}{C_S - C_L}.$$ (5.2)

Note that these quantities are fractions and must be multiplied by 100 to obtain percentages.

From inspection of the tie line in Fig. 5.3 (and for a nominal alloy composition of $C_0 = 50\%$ Cu-50% Ni) it can be noted that, because C_0 is closer to C_L than it is to C_S, the solid phase contains less copper than the liquid phase. By measuring on the phase diagram and using the lever-rule equations, it can be seen that the composition of the solid phase is 42% Cu and of the liquid phase is 58% Cu, as stated in the middle circle at the right in Fig. 5.3.

Note that these calculations refer to copper. If we now reverse the phase diagram in the figure, so that the left boundary is 0% nickel (whereby nickel becomes the alloying element in copper), these calculations give us the compositions of the solid and liquid phases in terms of nickel. The lever rule is also known as the *inverse lever rule* because, as indicated by Eqs. (5.1) and (5.2), the amount of each phase is proportional to the length of the opposite end of the lever.

5.2.5 The iron-carbon system

The *iron-carbon binary system* is represented by the **iron-iron carbide phase diagram**, shown in Fig. 5.4. Note that pure iron melts at a temperature of 1538°C (2800°F), as shown at the left in Fig. 5.4; as it cools, it first forms δ-iron, then γ-iron, and finally α-iron. Commercially pure iron contains up to 0.008% C, steels up to 2.11% C, and cast irons up to 6.67% C, although most cast irons contain less than 4.5% C.

1. **Ferrite.** *Alpha ferrite,* or simply *ferrite,* is a solid solution of body-centered cubic iron and has a maximum solid solubility of 0.022% C at a temperature of 727°C (1341°F). *Delta ferrite* is stable only at very high temperatures and has no significant or practical engineering applications. Ferrite is relatively soft and ductile and is magnetic from room temperature up to 768°C (1414°F). Although very little carbon can dissolve interstitially in bcc iron, the amount of carbon significantly affects the mechanical properties of ferrite. Also, significant amounts of chromium, manganese, nickel, molybdenum, tungsten, and silicon can be contained in iron in solid solution, imparting certain desirable properties.

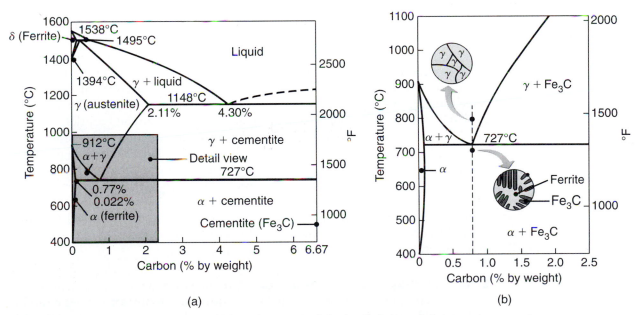

(a)

(b)

FIGURE 5.4 (a) The iron-iron carbide phase diagram and (b) detailed view of the microstructures above and below the eutectoid temperature of 727°C (1341°F). Because of the importance of steel as an engineering material, this diagram is one of the most important phase diagrams.

2. **Austenite.** Between 1394°C (2541°F) and 912°C (1674°F) iron undergoes an *allotropic transformation* (Section 3.2) from the bcc to fcc structure, becoming what is known as *gamma iron* or, more commonly, *austenite*. This structure has a solid solubility of up to 2.11% C at 1148°C (2098°F), which is about two orders of magnitude higher than that of ferrite, with the carbon occupying interstitial positions. Note that the atomic radius of Fe is 0.124 nm and for C it is 0.071 nm. Austenite is an important phase in the heat treatment of steels, described in Section 5.11. It is denser than ferrite, and its single-phase fcc structure is ductile at elevated temperatures; thus it possesses good formability. Large amounts of nickel and manganese can also be dissolved in fcc iron to impart various properties. Austenitic steel is nonmagnetic at high temperatures, and austenitic stainless steels are nonmagnetic at room temperature.

3. **Cementite.** The right boundary of Fig. 5.4a represents *cementite,* also called **carbide,** which is 100% iron carbide (Fe_3C), with a carbon content of 6.67%. (This carbide should not be confused with various carbides used for tool and die materials, described in Section 8.6.4.) Cementite is a very hard and brittle intermetallic compound (Section 5.2.2) and significantly influences the properties of steels. It can be alloyed with elements such as chromium, molybdenum, and manganese for enhanced properties.

5.2.6 The iron-iron carbide phase diagram

Various microstructures can be developed in steels depending on the carbon content and the method of heat treatment (Section 5.11). For example, consider iron with a 0.77% C content being cooled *very slowly* from a temperature of 1100°C (2012°F) in the austenite phase. The reason for the slow cooling rate is to maintain equilibrium; higher rates of cooling are used in heat treating. At 727°C (1341°F) a reaction takes place in which austenite is transformed into alpha ferrite (bcc) and cementite. Because the solid solubility of carbon in ferrite is only 0.022%, the extra carbon forms cementite.

This reaction is called a **eutectoid** (meaning *eutecticlike*) **reaction**, indicating that at a certain temperature a single solid phase (austenite) is transformed into two other solid phases, namely ferrite and cementite). The structure of the eutectoid steel is called **pearlite** because, at low magnifications, it resembles mother of pearl. The microstructure of pearlite consists of alternate layers (*lamellae*) of ferrite and cementite (Fig. 5.4b). Consequently, the mechanical properties of pearlite are intermediate between ferrite (soft and ductile) and cementite (hard and brittle).

In iron with less than 0.77% C, the microstructure formed consists of a pearlite phase (ferrite and cementite) and a ferrite phase. The ferrite in the pearlite is called *eutectoid ferrite*. The ferrite phase is called *proeutectoid ferrite* (*pro* meaning "before") because it forms at a temperature higher than the eutectoid temperature of 727°C (1341°F). If the carbon content is higher than 0.77%, the austenite transforms into pearlite and cementite. The cementite in the pearlite is called *eutectoid cementite*, and the cementite phase is called *proeutectoid cementite* because it forms at a temperature higher than the eutectoid temperature.

Effects of alloying elements in iron. Although carbon is the basic element that transforms iron into steel, other elements are also added to impart various desirable properties. The effect of these alloying elements on the iron-iron carbide phase diagram is to shift the eutectoid temperature and eutectoid composition (the percentage of carbon in steel at the eutectoid point). The eutectoid temperature may be raised or lowered from 727°C (1341°F), depending on the particular alloying element.

However, alloying elements always lower the eutectoid composition; that is, the carbon content becomes less than 0.77%. Lowering the eutectoid temperature means increasing the austenite range. Thus, an alloying element such as nickel is known as an **austenite former** because it has an fcc structure and thus tends to favor the fcc structure of austenite. Conversely, chromium and molybdenum have the bcc structure, causing these elements to favor the bcc structure of ferrite. These elements are known as **ferrite formers**.

EXAMPLE 5.1 Determining the amount of phases in carbon steel

Determine the amount of gamma and alpha phases in a 10-kg, 1040 steel casting as it is being cooled slowly to the following temperatures: (a) 900°C, (b) 728°C, and (c) 726°C.

Solution. (a) Referring to Fig. 5.4b, we draw a vertical line at 0.40% C at 900°C. We are in the single-phase austenite region, so percent gamma is 100 (10 kg) and percent alpha is zero. (b) At 728°C, the alloy is in the two-phase gamma-alpha field. When we draw the phase diagram in greater detail, we can find the weight percentages of each phase by the lever rule:

$$\text{Percent alpha} = \left(\frac{C_\gamma - C_0}{C_\gamma - C_\alpha}\right)100 = \left(\frac{0.77 - 0.40}{0.77 - 0.022}\right)100 = 50\%, \text{ or 5 kg.}$$

$$\text{Percent gamma} = \left(\frac{C_0 - C_\alpha}{C_\gamma - C_\alpha}\right)100 = \left(\frac{0.40 - 0.022}{0.77 - 0.022}\right)100 = 50\%, \text{ or 5 kg.}$$

(c) At 726°C, the alloy will be in the two-phase alpha and Fe_3C region. No gamma phase will be present. Again, we use the lever rule to find the amount of alpha present:

$$\text{Percent alpha} = \left(\frac{6.67 - 0.40}{6.67 - 0.022}\right) \times 100\% = 94\%, \text{ or 9.4 kg.}$$

5.3 | Cast Structures

The type of *cast structure* developed during solidification of metals and alloys depends on the composition of the particular alloy, the rate of heat transfer, and the flow of the liquid metal during the casting process. As described throughout this chapter, the structures developed, in turn, affect the properties of the castings.

5.3.1 Pure metals

The typical grain structure of a pure metal that has solidified in a square mold is shown in Fig. 5.5a. At the mold walls the metal cools rapidly (*chill zone*) because the walls are at ambient or slightly elevated temperature, and as a result, the casting develops a solidified **skin** (*shell*) of fine *equiaxed grains*. The grains grow in the direction opposite to the heat transfer from the mold. Grains that have a favorable orientation, called **columnar grains**, grow preferentially (see middle of Fig. 5.5). Note that grains that have substantially different orientations are blocked from further growth.

FIGURE 5.5 Schematic illustration of three cast structures of metals solidified in a square mold. (a) Pure metals, with preferred texture at the cool mold wall. Note in the middle of the figure that only favorably oriented grains grow away from the mold surface. (b) Solid-solution alloys and (c) structure obtained by heterogeneous nucleation of grains.

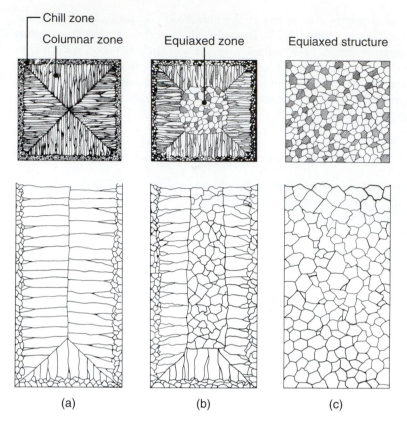

(a)   (b)   (c)

5.3.2 Alloys

Because pure metals have limited mechanical properties, they are often enhanced and modified by **alloying.** The vast majority of metals used in engineering applications are some form of an *alloy*, defined as two or more chemical elements, at least one of which is a metal.

Solidification in alloys begins when the temperature drops below the liquidus, T_L, and is complete when it reaches the solidus, T_S (Fig. 5.6). Within this temperature range, the alloy is in a mushy or pasty state with **columnar dendrites** (from the Greek *dendron* meaning "akin to," and *drys* meaning "tree"). Note in the lower right of the figure the presence of liquid metal between the dendrite arms. Dendrites have three-dimensional arms and branches (*secondary arms*), which eventually interlock, as shown in Fig. 5.7. The width of the mushy zone (where both liquid and solid phases are present) is an important factor during solidification. This zone is described in terms of a temperature difference, known as the **freezing range,** as

$$\text{Freezing range} = T_L - T_S. \tag{5.3}$$

Note in Fig. 5.6 that pure metals have a freezing range that approaches zero, and that the *solidification front* moves as a plane front, without forming a mushy zone. Eutectics solidify in a similar manner, with an approximately plane front. The type of solidification structure developed depends on the composition of the eutectic. For example, for alloys with a nearly symmetrical phase diagram, the structure is generally lamellar, with two or more solid phases present depending on the alloy system. When the volume fraction of the minor phase of the alloy is less than about 25%, the structure generally becomes *fibrous.*

For alloys, a short freezing range generally involves a temperature difference of less than 50°C (90°F), and a long freezing range greater than 110°C (200°F).

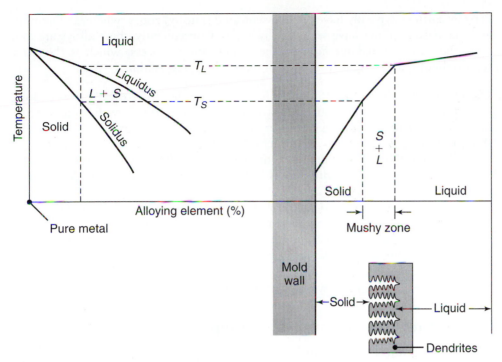

FIGURE 5.6   Schematic illustration of alloy solidification and temperature distribution in the solidifying metal. Note the formation of dendrites in the semisolid (mushy) zone.

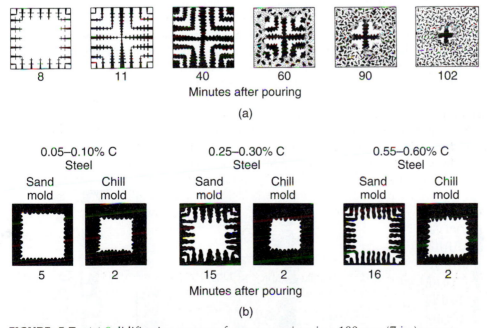

FIGURE 5.7   (a) Solidification patterns for gray cast iron in a 180-mm (7-in.) square casting. Note that after 11 min of cooling, dendrites reach each other, but the casting is still mushy throughout. It takes about two hours for this casting to solidify completely. (b) Solidification of carbon steels in sand and chill (metal) molds. Note the difference in solidification patterns as the carbon content increases. *Source:* After H.F. Bishop and W.S. Pellini.

Ferrous castings typically have narrow semisolid (mushy) zones, whereas aluminum and magnesium alloys have wide mushy zones. Consequently, these alloys are in a semisolid state throughout most of the solidification process, which is the main reason that thixocasting is feasible with these alloys (see Section 5.10.6).

Effects of cooling rate. Slow cooling rates (on the order of 10^2 K/s) or long local solidification times result in coarse dendritic structures, with large spacing between the dendrite arms. For higher cooling rates (on the order of 10^4 K/s) or short local solidification times, the structure becomes finer, with smaller dendrite arm spacing. For still higher cooling rates (on the order of 10^6 to 10^8 K/s) the structures developed are *amorphous* (without any ordered crystalline structure, as described in Section 5.10.8).

The structures developed and the resulting grain size, in turn, influence the properties of the casting. For example, as grain size decreases, (a) the strength and ductility of the cast alloy increase (see *Hall-Petch equation,* Section 3.4.1), (b) microporosity (interdendritic shrinkage voids) in the casting decreases, and (c) the tendency for the casting to crack (*hot tearing*) during solidification decreases. Also, lack of uniformity in grain size and distribution within castings results in anisotropic properties.

5.3.3 Structure-property relationships

Because all castings must possess certain specific properties to meet design and service requirements, the relationships between the properties and the structures developed during solidification are important. This section describes these relationships in terms of dendrite morphology and the concentration of alloying elements in various regions of the casting.

The compositions of dendrites and of the liquid metal in casting are given by the phase diagram of the particular alloy. When the alloy is cooled very slowly, each dendrite develops a uniform composition. Under normal cooling typically encountered in practice, however, **cored dendrites** are formed, which have a surface composition that is different from that at their centers (*concentration gradient*). The surface has a higher concentration of alloying elements than at the core of the dendrite due to solute rejection from the core toward the surface during solidification of the dendrite (known as **microsegregation**). The darker shading in the interdendritic liquid near the dendrite roots shown in Fig. 5.8 indicates that these regions have a higher solute concentration. Consequently, microsegregation in these regions is much more pronounced than in others.

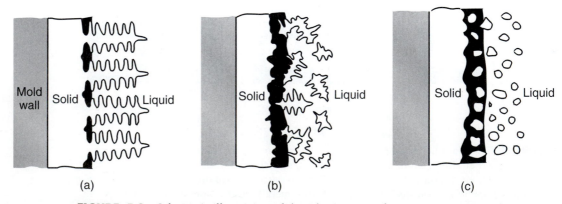

(a)   (b)   (c)

FIGURE 5.8   Schematic illustration of three basic types of cast structures: (a) columnar dendritic, (b) equiaxed dendritic, and (c) equiaxed nondendritic. *Source:* After D. Apelian.

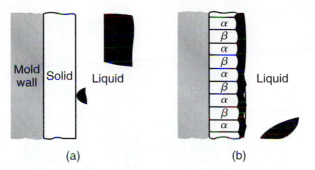

FIGURE 5.9 Schematic illustration of cast structures in (a) plane front, single phase and (b) plane front, two phase. *Source:* After D. Apelian.

In contrast to microsegregation, **macrosegregation** involves differences in composition throughout the casting itself. In situations where the solidifying front moves away from the surface of a casting as a plane front (Fig. 5.9), lower-melting-point constituents in the solidifying alloy are driven toward the center (**normal segregation**). Consequently, such a casting has a higher concentration of alloying elements at its center than at its surfaces. The opposite occurs in dendritic structures such as those for solid-solution alloys (Fig. 5.5b); that is, the center of the casting has a lower concentration of alloying elements (**inverse segregation**). The reason for this behavior is that the liquid metal (which has a higher concentration of alloying elements) enters the cavities developed from solidification shrinkage in the dendrite arms (which have solidified sooner). Another form of segregation is due to gravity (**gravity segregation**), whereby higher-density inclusions or compounds sink, and lighter elements (such as antimony in an antimony-lead alloy) float to the surface.

A typical cast structure of a solid-solution alloy, with an inner zone of equiaxed grains, is shown in Fig. 5.5b. The inner zone can be extended throughout the casting, as shown in Fig. 5.5c, by adding an **inoculant** (nucleating agent) to the alloy. The inoculant induces nucleation of grains throughout the liquid metal (**heterogeneous nucleation**). An example is the use of TiB_2 in aluminum alloys to refine grain patterns and improve mechanical properties.

Because of the presence of thermal gradients in a solidifying mass of liquid metal and because of the presence of gravity (hence density differences), *convection* has a strong influence on the cast structures developed. Convection promotes the formation of a chill zone, refines the grain size, and accelerates the transition from columnar to equiaxed grains. The structure shown in Fig. 5.8b can also be obtained by increasing convection within the liquid metal, whereby dendrite arms separate (**dendrite multiplication**). Conversely, reducing or eliminating convection results in coarser and longer columnar dendritic grains.

Convection can be enhanced by the use of mechanical or electromagnetic methods. Because the dendrite arms are not particularly strong, they can be broken up by agitation or mechanical vibration in the early stages of solidification (**rheocasting**; see Section 5.10.6). This action results in finer grain size, with equiaxed nondendritic grains that are distributed more uniformly throughout the casting (Fig. 5.8c).

5.4 | Fluid Flow and Heat Transfer

5.4.1 Fluid flow

To emphasize the importance of fluid flow in casting, let's briefly describe a basic gravity casting system, as shown in Fig. 5.10. The molten metal is poured through a **pouring basin** (*cup*). It then flows through the **sprue** to the **well**, and into **runners**

FIGURE 5.10 Schematic illustration of a typical sand mold showing various features.

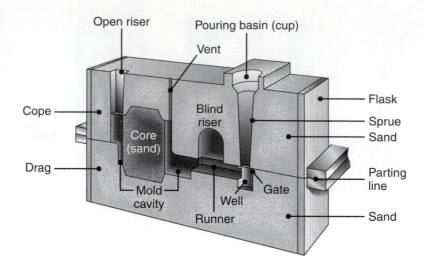

and to the mold cavity. **Risers,** also called **feeders,** serve as reservoirs of molten metal to supply the metal necessary to prevent shrinkage (which could lead to porosity). Although such a **gating system** appears to be relatively simple, successful casting requires proper design and control of the solidification process to ensure adequate fluid flow during casting. One of the most important functions of the gating system is to *trap contaminants* (such as oxides and other inclusions) in the molten metal by having the contaminants adhere to the walls of the gating system, thereby preventing their reaching the actual mold cavity. Furthermore, a properly designed gating system avoids or minimizes problems such as premature cooling, turbulence, and gas entrapment. Even before it reaches the mold cavity, the molten metal must be handled carefully to avoid forming oxides on the molten metal surfaces (from exposure to the environment) or the introduction of impurities into the molten metal.

Two basic principles of fluid flow are relevant to gating design: Bernoulli's theorem and the law of mass continuity.

Bernoulli's theorem. *Bernoulli's theorem* is based on the principle of conservation of energy and relates pressure, velocity, elevation of the fluid at any location in the system, and the frictional losses in a fluid system. Thus,

$$h + \frac{p}{\rho g} + \frac{v^2}{2g} = \text{Constant}, \tag{5.4}$$

where h is the elevation above a certain reference plane, p is the pressure at that elevation, v is the velocity of the liquid at that elevation, ρ is the density of the fluid (assuming that it is incompressible), and g is the gravitational constant. Conservation of energy requires that, at any particular location in the system, the following relationship be satisfied:

$$h_1 + \frac{p_1}{\rho g} + \frac{v_1^2}{2g} = h_2 + \frac{p_2}{\rho g} + \frac{v_2^2}{2g} + f, \tag{5.5}$$

where the subscripts 1 and 2 represent two different elevations, respectively, and f represents the frictional loss in the liquid as it travels downward through the gating system. The frictional loss includes such factors as energy loss at the liquid-mold wall interfaces and turbulence in the liquid.

Mass continuity. The mass continuity law states that for an incompressible liquid and in a system with impermeable walls, the rate of flow is constant. Thus,

$$Q = A_1 v_1 = A_2 v_2, \tag{5.6}$$

where Q is the volumetric rate of flow (such as m³/s), A is the cross-sectional area of the liquid stream, and v is the velocity of the liquid in that particular location. The subscripts 1 and 2 in the equation pertain to two different locations in the system. The permeability of the walls of the system is important because, otherwise, some liquid will permeate through (for example, in sand molds; Section 5.8.1) and the flow rate will decrease as the liquid travels through the system. Coatings are often used to inhibit such behavior in sand molds.

Sprue profile. An application of the two principles stated above is the traditional tapered design of sprues (Fig. 5.10) in which the shape of the sprue can be determined by using Eqs. (5.5) and (5.6). Assuming that the pressure at the top of the sprue is equal to the pressure at the bottom and that there are no frictional losses in the system, the relationship between height and cross-sectional area at any point in the sprue is given by the parabolic relationship

$$\frac{A_1}{A_2} = \sqrt{\frac{h_2}{h_1}}, \tag{5.7}$$

where, for example, subscript 1 denotes the top of the sprue and 2 the bottom. Recall that in a free-falling liquid, such as water from a faucet, the cross-sectional area of the stream decreases as it gains velocity downward. Thus, moving downward from the top, the cross-sectional area of the sprue must decrease. If a sprue is designed with a constant cross-sectional area, regions may develop where the molten metal loses contact with the sprue walls. As a result, **aspiration** may occur, whereby air will be sucked in or entrapped in the liquid. In many systems, however, tapered sprues are now replaced by straight-sided sprues that have a *choking* mechanism at the bottom, consisting of either a choke core or a runner choke (as shown in Fig. 5.10).

Modeling of mold filling requires the application of Eqs. (5.6) and (5.7); see also Section 5.12.5. Consider the situation shown in Fig. 5.10, where molten metal is poured into a pouring basin. It then flows through a sprue to a gate and runner and fills the mold cavity. If the pouring basin has a cross-sectional area that is much larger than the sprue bottom, the velocity of the molten metal at the top of the pouring basin will be very low. If frictional losses are due to viscous dissipation of energy, then f in Eq. (5.5) can be taken as a function of vertical distance and is often approximated as a linear function. The velocity of the molten metal leaving the gate is then obtained from Eq. (5.5) as

$$v = c\sqrt{2gh}, \tag{5.8}$$

where h is the distance from the sprue base to the liquid metal height, and c is a friction factor. This factor ranges between 0 and 1, and for frictionless flow, it is unity. The magnitude of c varies with mold material, runner layout, and channel size and can include energy losses due to turbulence as well as viscous effects.

If the liquid level has reached a height x, then the gate velocity is

$$v = c\sqrt{2g}\sqrt{h - x}. \tag{5.9}$$

The flow rate through the gate will be the product of this velocity and the gate area, according to Eq. (5.6). The shape of the casting will determine the height as a function of time. Equation (5.9) allows calculation of the mean fill time and flow rate, and dividing the casting volume by this mean flow rate then gives the mold fill time.

Simulation of mold filling helps designers specify the runner diameter and the size and number of sprues and pouring basins. To ensure that the runners don't prematurely choke, the fill time must be a small fraction of the solidification time (see Section 5.4.4). However, the velocity must not be so high as to erode the mold material (known as *mold wash*), or to result in too high of a Reynolds number (see below) because it may result in turbulence and associated air entrainment. Several computational tools (see Section 5.12.5) are now available to evaluate gating designs and help in determining the size of the mold components.

Flow characteristics. An important consideration in fluid flow in gating systems is the presence of **turbulence**, as opposed to **laminar flow** of fluids. The **Reynolds number**, Re, is used to characterize this aspect of fluid flow. This number represents the ratio of the inertia to the viscous forces in fluid flow and is expressed as

$$\text{Re} = \frac{vD\rho}{\eta}, \tag{5.10}$$

where v is the velocity of the liquid, D is the diameter of the channel, and ρ and η are the density and viscosity, respectively, of the liquid. The higher the Reynolds number, the greater is the tendency for turbulent flow. In ordinary gating systems Re ranges from 2000 to 20,000. An Re value of up to 2000 represents laminar flow. Between 2000 and 20,000 it is a mixture of laminar and turbulent flow and is generally regarded as harmless in gating systems for casting. However, Re values in excess of 20,000 represent severe turbulence, resulting in air entrainment and *dross* formation. Dross is the scum that forms on the surface of the molten metal as a result of the reaction of the liquid metal with air and other gases. Techniques for minimizing turbulence generally involve avoidance of sudden changes in flow direction and in the geometry of channel cross-sections of the gating system.

The elimination of dross or *slag* (nonmetallic products from mutual dissolution of flux and nonmetallic impurities) is another important consideration in fluid flow. This can be achieved by skimming (using *dross traps*), properly designing pouring basins and gating systems, or using filters. Filters are usually made of materials such as ceramics, mica, or fiberglass. Their proper location and placement are important for effective filtering of dross and slag.

EXAMPLE 5.2 Design and analysis of a sprue for casting

The desired volume flow rate of the molten metal into a mold is 0.01 m³/min. The top of the sprue has a diameter of 20 mm and its length is 200 mm. What diameter should be specified at the bottom of the sprue in order to prevent aspiration? What is the resultant velocity and Reynolds number at the bottom of the sprue if the metal being cast is aluminum and has a viscosity of 0.004 N-s/m²?

Solution. A pouring basin is typically provided on top of a sprue so that molten metal may be present above the sprue opening; however, this complication will be ignored in this case. Note that the metal volume flow rate is $Q = 0.01$ m³/min $= 1.667 \times 10^{-4}$ m³/s. Again, let's use the subscripts 1 for the top and 2 for the bottom of the sprue. Since $d_1 = 20$ mm $= 0.02$ m,

$$A_1 = \frac{\pi}{4}d^2 = \frac{\pi}{4}(0.002)^2 = 3.14 \times 10^{-4} \text{ m}^2.$$

Therefore,

$$v_1 = \frac{Q}{A_1} = \frac{1.667 \times 10^{-4}}{3.14 \times 10^{-4}} = 0.531 \text{ m/s}.$$

Assuming no frictional losses and recognizing that the pressure at the top and bottom of the sprue is atmospheric, Eq. (5.5) gives

$$0.2 + \frac{(0.531)^2}{2(9.81)} + \frac{p_{atm}}{\rho g} = 0 + \frac{v_2^2}{2(9.81)} + \frac{p_{atm}}{\rho g}$$

or $v_2 = 1.45$ m/s. To prevent aspiration, the sprue opening should be the same as that required by flow continuity, or

$$Q = A_2 v^2 = 1.667 \times 10^{-4} \text{ m}^3/\text{s} = A_2(1.45 \text{ m/s})$$

or $A_2 = 1.150 \times 10^{-4}$ m^2, and therefore $d = 12$ mm. The profile of the sprue will be parabolic, as suggested by Eq. (5.7). In calculating the Reynolds number, we first note from Table 3.3 that the density of aluminum is 2700 kg/m^3. The density for molten aluminum will of course be lower, but not significantly, so this value is sufficient for this problem. From Eq. (5.10),

$$\text{Re} = \frac{vD\rho}{\eta} = \frac{(1.45)(0.012)(2700)}{0.004} = 11,745.$$

As stated above, this magnitude is typical for casting molds, representing a mixture of laminar and turbulent flow.

5.4.2 Fluidity of molten metal

A term commonly used to describe the ability of the molten metal to fill mold cavities is *fluidity*. This term consists of two basic factors: (1) characteristics of the molten metal and (2) casting parameters. The following characteristics of molten metal influence fluidity:

1. **Viscosity.** Fluidity decreases as viscosity and the viscosity index (its sensitivity to temperature) increases.

2. **Surface tension.** A high surface tension of the liquid metal reduces fluidity. Oxide films developed on the surface of the molten metal have a significant adverse effect on fluidity. As an example, an oxide film on the surface of pure molten aluminum triples the surface tension.

3. **Inclusions.** As insoluble particles, inclusions can have a significant adverse effect on fluidity. This effect can be verified by observing the viscosity of a liquid such as oil with and without fine sand particles in it. It will be noted that the former will have lower viscosity.

4. **Solidification pattern of the alloy.** The manner in which solidification occurs, as described in Section 5.3, can influence fluidity. Moreover, fluidity is inversely proportional to the freezing range [see Eq. (5.3)]; thus, the shorter the range (as in pure metals and eutectics), the higher the fluidity becomes. Conversely, alloys with long freezing ranges (such as solid-solution alloys) have lower fluidity.

The following casting parameters influence fluidity and also influence the fluid flow and thermal characteristics of the system:

1. **Mold design.** The design and dimensions of components such as the sprue, runners, and risers all influence fluidity to varying degrees.

2. **Mold material and its surface characteristics.** The higher the thermal conductivity of the mold and the rougher its surfaces, the lower is the fluidity. Heating the mold improves fluidity, although it increases the solidification time, resulting in coarser grains and hence lower strength.

3. **Degree of superheat.** Defined as the increment of temperature above an alloy's melting point, superheat improves fluidity by delaying solidification.

4. **Rate of pouring.** The lower the rate of pouring into the mold, the lower the fluidity because the metal cools faster.

5. **Heat transfer.** This factor directly affects the viscosity of the liquid metal, hence its fluidity.

The term **castability** is generally used to describe the ease with which a metal can be cast to obtain a part with good quality. The interrelationships among the factors listed above are complex, particularly because this term includes the casting practices as well.

Tests for fluidity.   Several tests have been developed to quantify fluidity, although none is accepted universally. In one such common test, the molten metal is made to flow along a channel that is at room temperature; the distance the metal flows before it solidifies and stops flowing is a measure of its fluidity. Obviously, this length is a function of the thermal properties of the metal and the mold, as well as of the design of the channel. Still, such fluidity tests are useful and simulate casting situations to a reasonable degree.

5.4.3 Heat transfer

A major consideration in casting is the heat transfer during the complete cycle from pouring to solidification and cooling of the casting to room temperature. *Heat flow* at different locations in the system depends on many factors, relating to the casting material and the mold and process parameters. For instance, in casting thin sections, the metal flow rates must be high enough to avoid premature chilling and solidification. On the other hand, the flow rate must not be so high as to cause excessive turbulence, with its detrimental effects on the properties of the casting.

A typical temperature distribution in the mold-liquid metal interface is shown in Fig. 5.11. As expected, the shape of the curve will depend on the thermal properties of

FIGURE 5.11 Temperature distribution at the mold wall and liquid-metal interface during solidification of metals in casting.

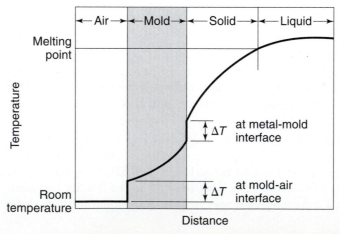

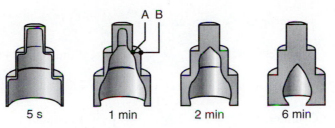

FIGURE 5.12 Solidified skin on a steel casting; the remaining molten metal is poured out at the times indicated in the figure. Hollow ornamental and decorative objects are made by a process called slush casting, which is based on this principle. *Source:* After H.F. Taylor, J. Wulff, and M.C. Flemings.

the molten metal and the mold material (such as sand, metal, or a ceramic). Heat from the liquid metal being poured is given off through the mold wall and the surrounding air. The temperature drop at the air-mold and mold-metal interfaces is caused by the presence of boundary layers and imperfect contact at these interfaces.

5.4.4 Solidification time

During the early stages of solidification, a thin solidified skin begins to form at the cool mold walls, and as time passes, the skin thickens. With flat mold walls, this thickness is proportional to the square root of time. Thus, doubling the time will make the skin $\sqrt{2} = 1.41$ times, or 41%, thicker.

The *solidification time* is a function of the volume of a casting and its surface area (**Chvorinov's rule**) and is given by

$$\text{Solidification time} = C\left(\frac{\text{Volume}}{\text{Surface area}}\right)^n, \qquad (5.11)$$

where C is a constant that reflects the mold material, the metal properties (including latent heat), and temperature. The parameter n typically is found to have a value between 1.5 and 2 and is usually taken as 2. Thus, a large sphere solidifies and cools to ambient temperature at a much lower rate than does a smaller sphere, because the volume of a sphere is proportional to the cube of its diameter and the surface area is proportional to the square of its diameter. Similarly, it can be shown that molten metal in a cube-shaped mold will solidify faster than in a spherical mold of the same volume.

The effects of mold geometry and elapsed time on skin thickness and its shape are shown in Fig. 5.12. The unsolidified molten metal has been poured out from the mold at different time intervals, ranging from 5 s to 6 min. Note that the thickness of the skin increases with elapsed time, and that it is thinner at internal angles (location A in the figure) than at external angles (location B). The metal cools slower at internal angles than at external angles. Note that this operation is very similar to making hollow chocolate shapes, and other confectionaries.

EXAMPLE 5.3 Solidification times for various solid shapes

Three pieces being cast have the same volume but different shapes. One is a sphere, one a cube, and the other a cylinder with a height equal to its diameter. Which piece will solidify the fastest and which one the slowest? Use $n = 2$.

Solution. The volume is unity, so from Eq. (5.9)

$$\text{Solidification time} \propto \frac{1}{(\text{Surface area})^2}.$$

The respective surface areas are

$$\text{Sphere: } V = \left(\frac{4}{3}\right)\pi r^3, \quad r = \left(\frac{3}{4\pi}\right)^{1/3}, \quad \text{and} \quad A = 4\pi r^2 = 4\pi\left(\frac{3}{4\pi}\right)^{2/3} = 4.84.$$

$$\text{Cube: } V = a^3, \quad a = 1, \quad \text{and} \quad A = 6a^2 = 6.$$

$$\text{Cylinder: } V = \pi r^2 h = 2\pi r^3, \quad r = \left(\frac{1}{2\pi}\right)^{1/3}, \quad \text{and}$$

$$A = 2\pi r^2 + 2\pi rh = 6\pi r^2 = 6\pi\left(\frac{1}{2\pi}\right)^{2/3} = 5.54.$$

Therefore, the respective solidification times t are

$$t_{\text{sphere}} = 0.043C, \quad t_{\text{cube}} = 0.028C, \quad \text{and} \quad t_{\text{cylinder}} = 0.033C.$$

Therefore, the cube-shaped casting will solidify the fastest, and the sphere-shaped casting will solidify the slowest.

5.4.5 Shrinkage

Metals shrink (contract) during solidification and cooling, as shown in Fig. 5.1 and Table 5.1. Note, however, that some metals expand, including gray cast iron. The reason for the latter is that graphite has a relatively high specific volume, and when it precipitates as graphite *flakes* during solidification, it causes a net expansion of the metal.

Shrinkage in a casting causes dimensional changes and sometimes cracking and is a result of the following phenomenon:

1. Contraction of the molten metal as it cools before it begins to solidify.
2. Contraction of the metal during phase change from liquid to solid (latent heat of fusion).
3. Contraction of the solidified metal (the casting) as its temperature drops to ambient temperature.

The largest amount of shrinkage occurs during cooling of the casting.

TABLE 5.1

Volumetric Solidification Contraction or Expansion for Various Cast Metals

Contraction (%)		Expansion (%)	
Aluminum	7.1	Bismuth	3.3
Zinc	6.5	Silicon	2.9
Al-4.5% Cu	6.3	Gray iron	2.5
Gold	5.5		
White iron	4–5.5		
Copper	4.9		
Brass (70-30)	4.5		
Magnesium	4.2		
90% Cu-10% Al	4		
Carbon steels	2.5–4		
Al-12% Si	3.8		
Lead	3.2		

5.5 | Melting Practice and Furnaces

Melting practice is an important aspect of casting operations because it has a direct bearing on the quality of castings. Furnaces are charged with melting stock consisting of liquid and/or solid metal, alloying elements, and various other materials such as flux and slag-forming constituents.

Fluxes are inorganic compounds that refine the molten metal by removing dissolved gases and various impurities. They have several functions, depending on the metal. For aluminum alloys, for example, there are cover fluxes (to form a barrier to oxidation), cleaning fluxes, drossing fluxes, refining fluxes, and wall-cleaning fluxes (to reduce the detrimental effect that some fluxes have on furnace linings, particularly induction furnaces). Fluxes may be added manually or can be injected automatically into the molten metal. To protect the surface of the molten metal against atmospheric reaction and contamination, as well as to refine the melt, the metal must be insulated against heat loss. Insulation is typically provided by covering the surface of the melt or mixing it with compounds that form a slag. In casting steels, the composition of the slag includes CaO, SiO_2, MnO, and FeO.

The metal *charge* may be composed of commercially pure **primary metals**, which can include remelted or recycled scrap. Clean scrapped castings, gates, and risers may also be included in the charge. If the melting points of the alloying elements are sufficiently low, pure alloying elements are added to obtain the desired composition in the melt. If the melting points are too high, the alloying elements do not mix readily with the low-melting-point metals. In this case, **master alloys (hardeners)** are often used. These usually consist of lower-melting-point alloys with higher concentrations of one or two of the needed alloying elements. Differences in specific gravity of master alloys should not be too high to cause segregation in the melt.

Melting furnaces. The melting furnaces commonly used in foundries are electric-arc, induction, crucible, and cupolas.

Electric-arc furnaces are used extensively in foundries. They have advantages such as high rate of melting (hence high production rate), much less pollution than other types of furnaces, and the ability to hold the molten metal for any length of time for alloying purposes.

Induction furnaces are especially useful in smaller foundries and produce composition-controlled smaller melts. The *coreless induction furnace* consists of a crucible, completely surrounded with a water-cooled copper coil through which high-frequency current passes. Because there is a strong electromagnetic stirring action during induction heating, this type of furnace has excellent mixing characteristics for the purposes of alloying and adding new charge of metal. The *core* or *channel furnace* uses a low frequency (as low as 60 Hz) and has a coil that surrounds only a small portion of the unit.

Crucible furnaces, which have been used extensively throughout history, are heated with various fuels, such as commercial gases, fuel oil, fossil fuel, as well as electricity. They may be stationary, tilting, or movable. Many ferrous and nonferrous metals are melted in these furnaces.

Cupolas are basically refractory-lined vertical steel vessels that are charged with alternating layers of metal, coke, and flux. Although they require major investments, cupolas operate continuously, have high melting rates, and produce large amounts of molten metal.

Levitation melting involves magnetic suspension of the molten metal. An induction coil simultaneously heats a solid billet and stirs and confines the metal (thus eliminating the need for a crucible, which could be a source of contamination of the melt with oxide inclusions). The molten metal then flows downward into an investment-casting mold (see Section 5.9.2). Investment castings made by this method are free of refractory inclusions and gas porosity and have uniform fine-grained structure.

Foundries and foundry automation. The casting operations described in the rest of this chapter are typically carried out in *foundries*. Although casting operations have traditionally involved much manual labor, modern foundries have automated and computer-integrated facilities for all aspects of their operations and produce castings at high production rates, at low cost, and with excellent quality control.

Foundry operations initially involve three separate activities: (a) pattern and mold making, which now regularly utilize computer-aided design and manufacturing and **rapid prototyping** techniques (see Section 10.12); (b) melting the metals while controlling their composition and impurities; and (c) various operations such as pouring the molten metal into molds (carried along conveyors), shake-out, cleaning, heat treatment, and inspection. All operations are automated, including the use of industrial robots (Section 14.7).

5.6 | Casting Alloys

Chapter 3 summarized the properties of wrought structures. A number of materials are useful in their cast form and don't require working to refine the microstructure before they can be successfully used in products. The general properties of various casting alloys and processes are summarized in Fig. 5.13 and Tables 5.2 through 5.5.

5.6.1 Ferrous casting alloys

The term **cast iron** refers to a family of ferrous alloys composed of iron, carbon (ranging from 2.11 to about 4.5%), and silicon (up to about 3.5%). Cast irons are usually classified according to their solidification morphology and by their structure (ferrite, pearlite, quenched and tempered, or austempered). Cast irons have lower melting temperatures than steels, which is why the casting process is so suitable for iron with high carbon content.

Cast irons represent the largest amount (by weight) of all metals cast, and they can easily be cast into complex shapes. They generally possess several desirable properties, such as strength, wear resistance, hardness, and good machinability (Chapter 8).

1. **Gray cast iron.** In this structure, graphite exists largely in the form of flakes (Fig. 5.14a). It is called *gray cast iron,* or *gray iron,* because when broken, the fracture path is along the graphite flakes and the surface has a gray, sooty appearance. The flakes act as stress raisers, thus greatly reducing ductility. Gray iron is weak in tension, although strong in compression, as are other brittle materials. On the other hand, the graphite flakes give this material the capacity to dampen vibrations by the internal friction (hence energy dissipation) caused by these flakes. Gray iron is thus a suitable and commonly used material for constructing structures and machine tool bases in which vibration damping is important (Section 8.12). Gray cast irons are specified by a two-digit ASTM designation. Class 20, for example, indicates that the material must have a minimum tensile strength of 20 ksi (140 MPa).

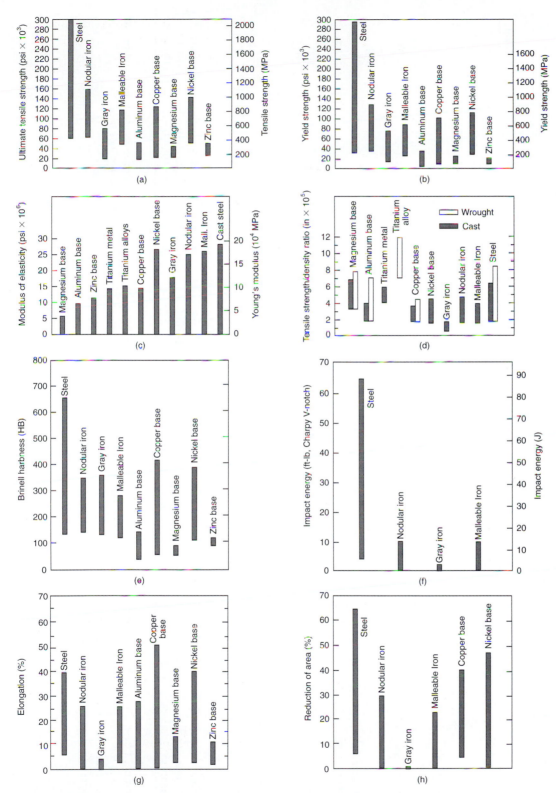

FIGURE 5.13   Mechanical properties for various groups of cast alloys. Compare with various tables of properties in Chapter 3. *Source:* Courtesy of Steel Founders' Society of America.

TABLE 5.2

General Characteristics of Casting Processes

	Sand	Shell	Evaporative pattern	Plaster	Investment	Permanent mold	Die	Centrifugal
Typical materials cast	All	All	All	Nonferrous (Al, Mg, Zn, Cu)	All	All	Nonferrous (Al, Mg, Zn, Cu)	All
Weight (kg):								
minimum	0.01	0.01	0.01	0.01	0.001	0.1	<0.01	0.01
maximum	No limit	100+	100+	50+	100+	300	50	5000+
Typ. surface finish (μm R_a)	5–25	1–3	5–25	1–2	0.3–2	2–6	1–2	2–10
Porosity[1]	3–5	4–5	3–5	4–5	5	2–3	1–3	1–2
Shape complexity[1]	1–2	2–3	1–2	1–2	1	2–3	3–4	3–4
Dimensional accuracy[1]	3	2	3	2	1	1	1	3
Section thickness (mm):								
minimum	3	2	2	1	1	2	0.5	2
maximum	No limit	—	—	—	75	50	12	100
Typ. dimensional tolerance	1.6–4 (0.25 for small)	±0.003		±0.005–0.010	±0.005	±0.015	±0.001–0.005	±0.015
Cost[1,2]								
Equipment	3–5	3	2–3	3–5	3–5	2	1	1
Pattern/die	3–5	2–3	2–3	3–5	2–3	2	1	1
Labor	1–3	3	3	1–2	1–2	3	5	5
Typical lead time[2,3]	Days	Weeks	Weeks	Days	Weeks	Weeks	Weeks–months	Months
Typical production rate[2,3]	1–20	5–50	1–20	1–10	1–1000	5–50	2–200	1–1000
Minimum quantity[2,3]	1	100	500	10	10	1000	10,000	10–10,000

Notes:
1. Relative rating, 1 best, 5 worst. For example, die casting has relatively low porosity, mid- to low-shape complexity, high dimensional accuracy, high equipment and die costs, and low labor costs. These ratings are only general; significant variations can occur depending on the manufacturing methods used.
2. Data taken from Schey, J.A., *Introduction to Manufacturing Processes*, 3rd ed., 2000.
3. Approximate values without the use of rapid prototyping technologies.

TABLE 5.3

Typical Applications for Castings and Casting Characteristics

Type of alloy	Application	Castability*	Weldability*	Machinability*
Aluminum	Pistons, clutch housings, intake manifolds, engine blocks, valve bodies, oil pans, suspension components	G–E	F*	G–E
Copper	Pumps, valves, gear blanks, marine propellers	F–G	F	G–E
Gray iron	Engine blocks, gears, brake disks and drums, machine bases	E	D	G
Magnesium	Crankcase, transmission housings, portable computer housings, toys	G–E	G	E
Malleable iron	Farm and construction machinery, heavy-duty bearings, railroad rolling stock	G	D	G
Nickel	Gas turbine blades, pump and valve components for chemical plants	F	F	F
Nodular iron	Crankshafts, heavy-duty gears	G	D	G
Steel (carbon and low alloy)	Die blocks, heavy-duty gear blanks, aircraft undercarriage members, railroad wheels	F	E	F–G
Steel (high alloy)	Gas turbine housings, pump and valve components, rock crusher jaws	F	E	F
White iron (Fe_3C)	Mill liners, shot blasting nozzles, railroad brake shoes, crushers and pulverizers	G	VP	VP
Zinc	Door handles, radiator grills	E	D	E

*E, excellent; G, good; F, fair; VP, very poor; D, difficult.

TABLE 5.4

Properties and Typical Applications of Cast Irons

Cast iron	Type	Ultimate tensile strength (MPa)	Yield strength (MPa)	Elonga-tion in 50 mm (%)	Typical applications
Gray	Ferritic	170	140	0.4	Pipe, sanitary ware
	Pearlitic	275	240	0.4	Engine blocks, machine tools
	Martensitic	550	550	0	Wear surfaces
Ductile (nodular)	Ferritic	415	275	18	Pipe, general service
	Pearlitic	550	380	6	Crankshafts, highly stressed parts
	Tempered martensite	825	620	2	High-strength machine parts, wear resistance
Malleable	Ferritic	365	240	18	Hardware, pipe fittings, general engineering service
	Pearlitic	450	310	10	Couplings
	Tempered	700	550	2	Gears, connecting rods
White	Pearlitic	275	275	0	Wear resistance, mill rolls

TABLE 5.5

Typical Properties of Nonferrous Casting Alloys

Alloy	Condition	Casting method*	UTS (MPa)	Yield strength (MPa)	Elongation in 50 mm (%)	Hardness (HB)
Aluminum						
357	T6	S	345	296	2.0	90
380	F	D	331	165	3.0	80
390	F	D	279	241	1.0	120
Magnesium						
AZ63A	T4	S, P	275	95	12	—
AZ91A	F	D	230	150	3	—
QE22A	T6	S	275	205	4	—
Copper						
Brass C83600	—	S	255	177	30	60
Bronze C86500	—	S	490	193	30	98
Bronze C93700	—	P	240	124	20	60
Zinc						
No. 3	—	D	283	—	10	82
No. 5	—	D	331	—	7	91
ZA27	—	P	425	365	1	115

*S, sand; D, die; P, permanent mold.

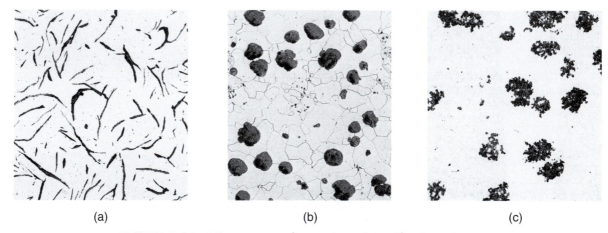

(a)   (b)   (c)

FIGURE 5.14 Microstructure for cast irons. Magnification: 100X. (a) Ferritic gray iron with graphite flakes, (b) ferritic nodular iron (ductile iron) with graphite in nodular form, and (c) ferritic malleable iron. This cast iron solidified as white cast iron, with the carbon present as cementite (Fe_3C), and was heat treated to graphitize the carbon.

The basic types of gray cast iron are called ferritic, pearlitic, and martensitic (see Section 5.11), each with different structures, properties, and applications. In **ferritic gray iron,** also known as fully gray iron, the structure consists of graphite flakes in an alpha ferrite matrix. **Pearlitic gray iron** has a structure consisting of graphite in a matrix of pearlite. Although still brittle, it is stronger than gray iron. **Martensitic gray iron** is obtained by austenitizing a pearlitic gray iron, followed by rapid quenching to produce a structure of graphite in a martensite matrix. As a result, it is very hard. Gray iron castings have relatively few shrinkage cavities and little porosity. Typical uses are for engine blocks, machine bases, electric-motor housings, pipes, and wear surfaces of machinery.

2. **Ductile iron (nodular iron).** In this structure, graphite is in *nodular* (*spheroid*) form (Fig. 5.14b), making the material more ductile and shock resistant. The shape of graphite flakes can be modified into nodules (spheres) by small additions of magnesium and/or cerium to the molten metal prior to pouring. Ductile iron can be heat treated to make it ferritic, pearlitic, or have a structure of tempered martensite. Typically used for machine parts, pipe, and crankshafts, ductile cast irons are specified by a set of two-digit numbers. Class or grade 80-55-06, for example, indicates that the material has a minimum tensile strength of 80 ksi (550 MPa), a minimum yield strength of 55 ksi (380 MPa), and a 6% elongation in 50 mm (2 in.).

3. **White cast iron.** White cast iron is very hard, wear resistant, and brittle because its structure contains large amounts of iron carbide instead of graphite. This structure is obtained either by rapid cooling of gray iron or by adjusting the material's composition by keeping the carbon and silicon content low. This cast iron is also called *white iron* because the absence of graphite gives the fracture surface a white crystalline appearance. Because of its extreme hardness and wear resistance, white cast iron typically is used for liners for machinery that processes abrasive materials, rolls for rolling mills, and railroad-car brake shoes.

4. **Malleable iron.** Malleable iron is obtained by annealing white cast iron in an atmosphere of carbon monoxide and carbon dioxide, between 800°C and 900°C (1470°F and 1650°F) for up to several hours. During this process the cementite decomposes (*dissociates*) into iron and graphite. The graphite exists as *clusters* (Fig. 5.14c) in a ferrite or pearlite matrix and thus has a structure similar to nodular iron. This structure imparts ductility, strength, and shock resistance, hence the term *malleable*. Typical uses are for railroad equipment and hardware. Malleable irons are specified by a five-digit designation; for example, 35018 indicates that the yield strength of the material is 35 ksi (240 MPa), and its elongation is 18% in 50 mm (2 in.).

5. **Compacted-graphite iron.** The compacted-graphite structure has short, thick, and interconnected flakes with undulating surfaces and rounded extremities. The mechanical and physical properties of this cast iron are between those of flake graphite and nodular graphite cast irons. Compacted-graphite iron is easy to cast and has consistent properties throughout the casting. Typical applications are automotive engine blocks, crankcases, ingot molds, cylinder heads, and brake disks; it is particularly suitable for components at elevated temperatures and resists thermal fatigue. Its machinability is better than nodular iron.

6. **Cast steels.** The high temperatures required to melt steels (see Table 3.3) require care in the selection of appropriate mold materials, particularly in view of the high reactivity of steels with oxygen. Steel castings possess properties that are more uniform (isotropic) than those made by mechanical working processes. Cast steels can be welded; however, welding alters the cast microstructure in the heat-affected zone (see Section 12.6), influencing the strength, ductility, and toughness of the base metal.

7. **Cast stainless steels.** Casting of stainless steels involves considerations similar to those for steels. Stainless steels generally have a long freezing range and high melting temperatures and develop various structures. Cast stainless steels are available in various compositions and they can be heat treated and welded. These cast products have high heat and corrosion resistance. Nickel-based casting alloys, for example, are used for severely corrosive environments and service at very high temperatures.

5.6.2 Nonferrous casting alloys

Aluminum-based alloys. Cast alloys with an aluminum base have a wide range of mechanical properties, largely because of various hardening mechanisms and heat treatments to which they can be subjected. These castings are lightweight (hence called *light-metal castings*) and have good machinability. However, except for alloys containing silicon, they generally have low resistance to wear and abrasion. Aluminum-based cast alloys have many applications, including automotive engine blocks, architectural, decorative, aerospace, and electrical.

Magnesium-based alloys. Cast alloys with a magnesium base have good corrosion resistance and moderate strength, depending on the particular heat treatment and the proper use of protective coating systems. They have very good strength-to-weight ratios and have been widely applied for aerospace and automotive structural applications.

Copper-based alloys. Cast alloys with a copper base have the advantages of good electrical and thermal conductivity, corrosion resistance, nontoxicity (unless they contain lead), good machinability, and wear properties (thus suitable as bearing materials). Mechanical properties and fluidity are influenced by the alloying elements.

Zinc-based alloys. Cast alloys with a zinc base have good fluidity and sufficient strength for structural applications. These alloys are widely used in die casting of structural shapes, electrical conduit, and corrosion-resistant parts.

High-temperature alloys. Having a wide range of properties and applications, these alloys typically require temperatures of up to 1650°C (3000°F) for casting titanium and superalloys, and higher for refractory alloys. Special techniques are used to cast high-temperature alloys for jet and rocket engine components, which would otherwise be difficult or uneconomical to produce by other means, such as forging.

5.7 | Ingot Casting and Continuous Casting

Traditionally, the first step in metal processing is the casting of the molten metal into a solid form (**ingot**) for further processing. In ingot casting, the molten metal is poured (*teemed*) from the ladle into ingot molds in which the metal solidifies; this is basically equivalent to sand casting or, more commonly, to permanent mold casting. Molten metal is poured into the permanent molds and yields the microstructure as shown in Fig. 5.5.

Gas entrapment can be reduced by bottom pouring, using an insulated collar on top of the mold, or by using an exothermic compound that produces heat when it contacts the molten metal (hot top). These techniques reduce the rate of cooling and result in a higher yield of high-quality metal. The cooled ingots are removed (*stripped*) from the molds and lowered into *soaking pits,* where they are reheated to a uniform temperature of about 1200°C (2200°F) for subsequent processing, such as rolling. Ingots may be square, rectangular, or round in cross section, and their weight ranges from a few hundred pounds to 40 tons.

5.7.1 Ferrous alloy ingots

Several reactions take place during solidification of an ingot which, in turn, have an important influence on the quality of the steel produced. For example, significant amounts of oxygen and other gases can dissolve in the molten metal during steel-making, leading to porosity defects (see Section 5.12.1). However, because the solubility limit of gases in the metal decreases sharply as its temperature decreases, much of these gases cavitate during solidification of the metal. The rejected oxygen combines with carbon, forming carbon monoxide, which causes porosity in the solidified ingot. Depending on the amount of gas evolved during solidification, three types of steel ingots can be produced.

1. **Killed steel** is fully deoxidized steel; that is, oxygen is removed and thus porosity is eliminated. In the deoxidation process the dissolved oxygen in the molten metal is made to react with elements (typically aluminum) that are added to the melt; vanadium, titanium, and zirconium are also used. These elements have an affinity for oxygen and form metallic oxides. If aluminum is used, the product is called aluminum-killed steel. The term *killed* comes from the fact that the steel lies quietly after being produced in the mold.

2. **Semi-killed steel** is partially deoxidized steel. It contains some porosity, generally in the upper central section of the ingot, but has little or no pipe, and thus scrap is reduced. Piping is less because it is compensated for by the presence of porosity in that region. Semi-killed steels are economical to produce.

3. In a **rimmed steel,** which generally has less than 0.15% carbon, the evolved gases are only partially killed or controlled by the addition of elements such as aluminum. Rimmed steels have little or no piping and have a ductile skin with good surface finish.

5.7.2 Continuous casting

Conceived in the 1860s, *continuous casting,* or **strand casting,** was first developed for casting nonferrous metal strip and is now widely used for steel production and at low cost. A continuous casting setup is shown schematically in Fig. 5.15a. The molten metal in the ladle is cleaned and equalized in temperature by blowing nitrogen gas through it for 5 to 10 min. It is then poured into a refractory-lined intermediate pouring vessel (**tundish**) where impurities are skimmed off. The tundish holds as much as three tons of metal.

Before starting the casting process, a solid *starter,* or *dummy bar,* is inserted into the bottom of the mold. The molten metal is then poured and solidifies on the starter bar (see bottom of Fig. 5.15a). The bar is withdrawn at the same rate the metal is poured, along a path supported by rollers (*pinch rolls*). The cooling rate is such that the metal develops a solidified skin (shell) to support itself during its travel downward, at speeds typically 25 mm/s (1 in./s). The shell thickness at the exit end of the mold is about 12–18 mm (0.5–0.75 in.). Additional cooling is provided by water sprays along the travel path of the solidifying metal. The molds are generally coated with graphite or similar solid lubricants to reduce friction and adhesion at the mold-metal interfaces. They are also vibrated to further reduce friction and sticking.

The continuously cast metal is then cut into desired lengths by shearing or torch cutting, or it may be fed directly into a rolling mill for further reductions in thickness and for shape rolling of products such as channels and I-beams (Section 6.3). Continuously cast metals have more uniform composition and properties than those obtained by ingot casting. Although the thickness of steel strand is typically about 250 mm (10 in.), it can be 15 mm (0.6 in.) or less. The thinner strand reduces the number of rolling operations required and improves the economy of the overall operation.

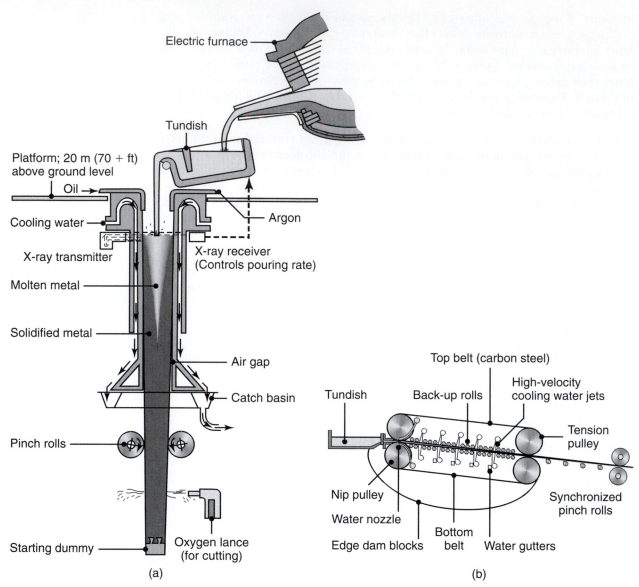

FIGURE 5.15 (a) The continuous-casting process for steel. Note that the platform is about 20 m (65 ft) above ground level. (b) Continuous strip casting of nonferrous metal strip. *Sources:* (a) American Foundrymen's Society, (b) Courtesy of Hazelett Strip-Casting Corp.

5.7.3 Strip casting

In strip casting, thin slabs or strips are produced from molten metal, which solidifies in manner similar to strand casting, but the hot solid is then rolled to form the final shape (Fig. 5.15b). The compressive stresses in rolling serve to reduce any porosity in the material and provide better properties. In effect, strip casting eliminates a hot rolling operation in the production of metal strips or slabs. In modern facilities, final thicknesses on the order of 2–6 mm (0.08–0.25 in.) can be obtained for carbon, stainless, and electrical steels (used in making the iron cores of motors, transformers, and generators) and other metals.

5.8 | Expendable-Mold, Permanent-Pattern Casting Processes

Casting processes are generally classified according to mold materials, molding processes, and methods of feeding the mold with the molten metal (see also Table 5.2). The two major categories are *expendable-mold* and *permanent-mold* casting (Section 5.10). Expendable-mold processes are further categorized as *permanent-pattern* and *expendable-pattern* processes. Expendable molds are typically made of sand, plaster, ceramics, and similar materials, which generally are mixed with various **binders** or bonding agents.

5.8.1 Sand casting

The *sand casting* process consists of (a) placing a pattern, having the shape of the desired casting, in sand to make an imprint; (b) incorporating a gating system; (c) filling the resulting cavity with molten metal; (d) allowing the metal to cool until it solidifies; (e) breaking away the sand mold; and (f) removing the casting and finishing it. Examples of parts made by sand casting are engine blocks, cylinder heads, machine-tool bases, and housings for pumps and motors. While the origins of sand casting date to ancient times (see Table. 1.1), it is still the most prevalent form of casting. In the United States alone, about 15 million tons of metal are cast by this method each year.

Sands. Sand is the product of the disintegration of rocks over extremely long periods of time. It is inexpensive and is suitable as mold material because of its resistance to high temperatures. Most sand casting operations use silica sands (SiO_2). There are two general types of sand: *naturally bonded* (*bank sand*) and *synthetic* (*lake sand*). Because its composition can be controlled more accurately, synthetic sand is preferred by most foundries.

Several factors are important in the selection of sand for molds. Sand having fine, round grains can be closely packed and forms a smooth mold surface. Good **permeability** of molds and cores allows gases and steam evolved during casting to escape easily. The mold should have good **collapsibility** (because the casting shrinks while cooling) to avoid defects in the casting. Fine sand enhances mold strength but lowers mold permeability. Sand is typically conditioned before use. *Mulling machines* are used to uniformly mull (mix thoroughly) sand with additives. Clay is typically used as a cohesive agent to bond sand particles, giving the sand better strength.

Types of sand molds. The major components of a typical sand mold are shown in Fig. 5.10. There are three basic types of sand molds: green-sand, cold-box, and no-bake molds. The most common mold material is *green molding sand,* the term *green* referring to the fact that the sand in the mold is moist or damp while the metal is being poured into it. Green molding sand is a mixture of sand, clay, and water and is the least expensive method of making molds.

In the **skin-dried** method of making molds, the mold surfaces are dried, either by storing the mold in air or drying it with torches; the molds may also be baked. These molds are generally used for large castings because of their high strength. They are stronger than green-sand molds and impart better dimensional accuracy and surface finish to the casting. However, distortion of the mold is greater, the castings are more susceptible to hot tearing because of the lower collapsibility of the mold, and the production rate is slower because of the length of drying time required.

In the **no-bake** mold process, a synthetic liquid resin is mixed with the sand, and the mixture hardens at room temperature. Because bonding of the mold takes place without heat, no-bake mold processes are also known as *cold-setting processes*. The **cold-box** mold process uses organic and inorganic binders that are blended into the sand to chemically bond the grains for greater mold strength; no heat is used. These molds are dimensionally more accurate than green-sand molds but are more expensive to produce.

Patterns. *Patterns*, which are used to mold the sand mixture into the shape of the casting, may be made of wood, plastic, or metal; see also **rapid prototyping** techniques (Section 10.12). Patterns also may be made from a combination of materials to reduce the patterns' wear in critical regions. The selection of a pattern material depends on the size and shape of the casting, the dimensional accuracy and the quantity of castings required, and the molding process to be used. Because patterns are repeatedly used to make molds, the strength and durability of the pattern material selected must reflect the number of castings the mold is expected to produce. Patterns are usually coated with a **parting agent** to facilitate their removal from the molds.

Patterns can be designed with a variety of features for specific applications as well as economic requirements.

a. **One-piece patterns** are generally used for simpler shapes and low-quantity production. They are typically made of wood and are inexpensive.

b. **Split patterns** are two-piece patterns, made so that each part forms a portion of the cavity for the casting, thus allowing casting of complicated shapes.

c. **Match-plate patterns** are a common type of mounted pattern, in which two-piece patterns are constructed by securing each half of one or more split patterns to the opposite sides of a single plate. In such constructions, the gating system can be mounted on the drag side of the pattern.

Cores. Cores are utilized for castings with internal cavities or passageways, such as in automotive engine blocks or valve bodies. They are placed in the mold cavity prior to pouring and are removed from the casting during shakeout. Like molds, cores must possess certain strength, permeability, ability to withstand heat, and collapsibility. Cores, which are typically made of sand aggregates, are anchored by **core prints**, which are recesses in the mold to support the core and to provide vents for the escape of gases. To keep the core from shifting during casting, metal supports (**chaplets**) may be used to anchor the core in place. Cores are shaped in core boxes, which are used much like patterns are used to form sand molds. The sand can be packed into the boxes with sweeps or blown into the box by compressed air from *core blowers*.

Sand-molding machines. Although the sand mixture can be compacted around the pattern by hand hammering or ramming, for high production runs, it is compacted by molding machines. These machines eliminate arduous labor and produce higher-quality castings by manipulating the mold in a controlled manner. Mechanization of the molding process can further be assisted by **jolting** the assembly, in which the flask, molding sand, and pattern are placed on a pattern plate mounted on an anvil and jolted upward by air pressure at rapid intervals. The inertial forces compact the sand around the pattern.

In **vertical flaskless molding** the halves of the pattern form a vertical chamber wall against which sand is blown and compacted. The mold halves are then packed horizontally, with the parting line oriented vertically, and moved along a pouring conveyor.

Sandslingers fill the flask uniformly with sand under a stream of high pressure. *Sandthrowers* are used to fill large flasks and are typically operated by machine. An impeller in the machine throws sand from its blades or cups at such high speeds that the machine not only places the sand but also rams it properly.

In **impact molding** the sand is compacted by controlled explosion or instantaneous release of compressed gases. This method produces molds with uniform strength and good permeability. In **vacuum molding** (also known as the *"V" process*), the pattern is covered tightly with a thin plastic sheet. A flask is then placed over the pattern and is filled with sand. A second sheet of plastic is placed on top of the sand, and a vacuum action hardens the sand to allow the pattern to be withdrawn. Both halves of the mold are made this way and then assembled. During pouring, the mold remains under a vacuum, but the casting cavity does not. When the metal has solidified, the vacuum is turned off and the sand falls away, releasing the casting.

The sand casting operation. The sequence of operations in sand casting is shown in Fig. 5.16. After the mold has been shaped and the cores have been placed in position, the two halves (*cope* and *drag*) are closed (Fig. 5.10), clamped, and weighted down (to prevent the separation of the mold sections under the pressure exerted when the molten metal is poured). The design of the *gating system* is important for proper delivery of the molten metal into the mold cavity. Turbulence must be minimized, air and gases must be allowed to escape by vents or other means, and proper temperature gradients must be established and maintained to eliminate shrinkage and porosity. The design of *risers* is also important for supplying the necessary molten metal during solidification of the casting; note that the pouring basin may also serve as a riser. After solidification, the casting is shaken out of its mold, and the sand and oxide layers adhering to the casting are removed by vibration (using a shaker) or by sand blasting. The risers and gates are removed by sawing, trimming in dies (shearing), abrasive disks, or by oxyfuel-gas cutting.

Almost all commercial alloys can be sand cast. The surface finish obtained is largely a function of the materials used in making the mold. Although dimensional accuracy is not as good as that of other casting processes, intricate shapes can be cast, such as cast-iron engine blocks and very large parts such as propellers for ocean liners. Sand casting can be economical for relatively small as well as large production runs; equipment costs are generally low.

5.8.2 Shell-mold casting

Shell-mold casting (Fig. 5.17) has grown significantly because it can produce many types of castings with close dimensional tolerances, good surface finish, and at a low cost. In this process, a mounted pattern, made of a ferrous metal or aluminum, is heated to 175°–370°C (350°–700°F), coated with a parting agent such as silicone, and clamped to a box or chamber containing a fine sand containing a 2.5 to 4% thermosetting resin binder, such as phenol-formaldehyde, which coats the sand particles. The sand mixture is then turned over the heated pattern, coating it evenly. The assembly is often placed in an oven for a short period of time to complete the curing of the resin. The shell hardens around the pattern and is removed from the pattern using built-in ejector pins. Two half-shells are made in this manner and are bonded or clamped together in preparation for pouring.

The shells are light and thin, usually 5–10 mm (0.2–0.4 in.), and hence their thermal characteristics are different from those for thicker molds. The thin shells allow gases to escape during solidification of the metal. The mold is generally used vertically and is supported by surrounding it with steel shot in a cart. The mold walls are relatively smooth, resulting in low resistance to molten metal flow and producing castings with sharper corners, thinner sections, and smaller projections than are possible in green-sand molds. Several castings can be made in a single mold

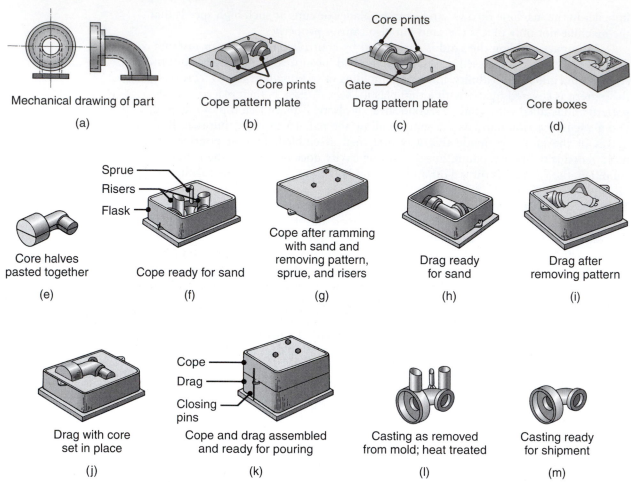

FIGURE 5.16   Schematic illustration of the sequence of operations in sand casting. (a) A mechanical drawing of the part, used to create patterns. (b–c) Patterns mounted on plates equipped with pins for alignment. Note the presence of core prints designed to hold the core in place. (d–e) Core boxes produce core halves, which are pasted together. The cores will be used to produce the hollow area of the part shown in (a). (f) The cope half of the mold is assembled by securing the cope pattern plate to the flask with aligning pins and attaching inserts to form the sprue and risers. (g) The flask is rammed with sand and the plate and inserts are removed. (h) The drag half is produced in a similar manner. (j) The core is set in place within the drag cavity. (k) The mold is closed by placing the cope on top of the drag and securing the assembly with pins. (l) After the metal solidifies, the casting is removed from the mold. (m) The sprue and risers are cut off and recycled, and the casting is cleaned, inspected, and heat treated (when necessary). *Source:* Courtesy of Steel Founders' Society of America.

using a multiple gating system. Applications include small mechanical parts requiring high precision, gear housings, cylinder heads, and connecting rods. The process is also widely used in producing high-precision molding cores, such as engine-block water jackets.

Shell-mold casting may be more economical than other casting processes, depending on various production factors, particularly energy cost. The relatively high cost of metal patterns becomes a smaller factor as the size of production run increases. The high quality of the finished casting can significantly reduce subsequent cleaning, machining, and other finishing costs. Complex shapes can be produced with less labor, and the process can be automated fairly easily.

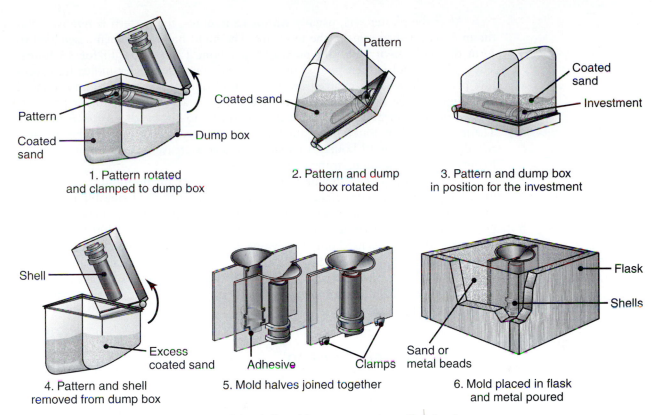

FIGURE 5.17 Schematic illustration of the shell-molding process, also called the *dump-box* technique.

Sodium silicate process. The mold material in this process is a mixture of sand and 1 to 6% sodium silicate (*waterglass*), or various other chemicals, as the binder for sand. This mixture is then packed around the pattern and hardened by blowing CO_2 gas through it. Also known as *silicate-bonded sand* or the carbon-dioxide process, this process is also used to make cores. This reduces the tendency of the casting to tear or fracture due to thermal stresses, mainly because of the core's flexibility at elevated temperatures.

Rammed graphite molding. Rammed graphite is used to make molds for casting reactive metals, such as titanium and zirconium. Sand cannot be used because these metals react vigorously with silica. The molds are packed rather like sand molds, air dried, baked at 175°C (350°F), and fired at 870°C (1600°F); they are then stored under controlled humidity and temperature. The casting procedures are similar to those for sand molds.

5.8.3 Plaster-mold casting

The *plaster-mold casting* process and the ceramic-mold and investment-casting processes are known as **precision casting** because of the high dimensional accuracy and good surface finish obtained. Typical parts made are lock components, gears, valves, fittings, tooling, and ornaments, weighing as little as 1 g (0.035 oz). In this process the mold is made of *plaster of paris* (gypsum, or calcium sulfate), with the addition of talc and silica flour to improve strength and control the time required for the plaster to set. These components are mixed with water and the resulting slurry is poured over the pattern.

After the plaster sets, usually within 15 minutes, the pattern is removed and the mold is dried to remove the moisture. The mold halves are then assembled to form the mold cavity and are preheated to about 120°C (250°F) for 16 hours. The molten metal is then poured into the mold. Because plaster molds have very low permeability, gases evolved during solidification of the metal cannot escape. Consequently, the molten metal is poured either in a vacuum or under pressure. Plaster-mold permeability can be increased substantially by the *Antioch* process. The molds are dehydrated in an autoclave (pressurized oven) for 6–12 hours, then rehydrated in air for 14 hours. Another method of increasing permeability is to use foamed plaster, containing trapped air bubbles.

Patterns for plaster molding are generally made of aluminum alloys, thermosetting plastics, brass, or zinc alloys. Wood patterns are not suitable because of the presence of water-based slurry. Because there is a limit to the maximum temperature that the plaster mold can withstand, generally about 1200°C (2200°F), plaster-mold casting is used only for casting aluminum, magnesium, zinc, and some copper-based alloys. The castings have fine detail and good surface finish. Since plaster molds have lower thermal conductivity than other types of molds, the castings cool slowly, resulting in more uniform grain structure with less warpage and better mechanical properties.

5.8.4 Ceramic-mold casting

The *ceramic-mold casting,* also called **cope-and-drag investment casting,** is another precision casting process and is similar to the plaster-mold process, with the exception that it uses refractory mold materials suitable for high-temperature applications. The slurry is a mixture of fine-grained zircon ($ZrSiO_4$), aluminum oxide, and fused silica, which are mixed with bonding agents and poured over the pattern (Fig. 5.18), which has been placed in a flask. The pattern may be made of wood or metal. After setting, the molds (ceramic facings) are removed, dried, burned off to remove volatile matter, and baked. The molds are then clamped firmly and used as all-ceramic molds. In the *Shaw* process, the ceramic facings are backed by fireclay to make the molds stronger.

The high-temperature resistance of the refractory molding materials allows these molds to be used in casting ferrous and other high-temperature alloys, stainless steels, and tool steels. Although the process is somewhat expensive, the castings have good dimensional accuracy and surface finish over a wide range of sizes and intricate shapes, where some parts weigh as much as 700 kg (1500 lb). Typical parts made are impellers, cutters for machining, dies for metalworking, and molds for making plastic or rubber components.

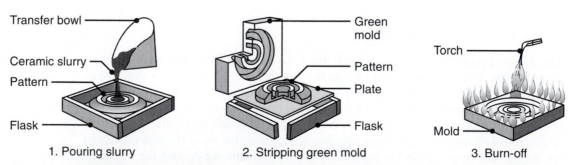

1. Pouring slurry   2. Stripping green mold   3. Burn-off

FIGURE 5.18 Sequence of operations in making a ceramic mold.

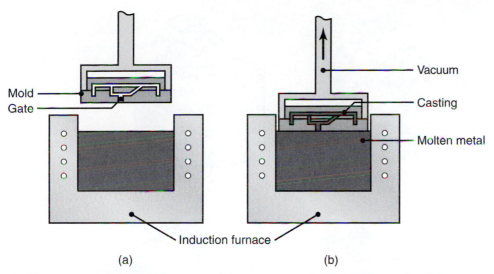

(a)   (b)

FIGURE 5.19 Schematic illustration of the vacuum-casting process. Note that the mold has a bottom gate. (a) Before and (b) after immersion of the mold into the molten metal. *Source:* After R. Blackburn.

5.8.5 Vacuum casting

The *vacuum-casting process*, also called *counter-gravity low-pressure* (CL) *process* but not to be confused with the vacuum-molding process described in Section 5.8.1, is shown in Fig. 5.19. A mixture of fine sand and urethane is molded over metal dies and cured with amine vapor. The mold is then held with a robot arm and partially immersed into molten metal in an induction furnace. The metal may be melted in air (*CLA process*) or in a vacuum (*CLV process*). The vacuum reduces the air pressure inside the mold to about two-thirds of atmospheric pressure, drawing the molten metal into the mold cavities through a gate in the bottom of the mold. After the mold is filled, it is withdrawn from the molten metal. The furnace is at a temperature usually 55°C (100°F) above the liquidus temperature; consequently, it begins to solidify within a fraction of a second. After the mold is filled, it is withdrawn from the molten metal.

This process is an alternative to investment, shell-mold, and green-sand casting and is particularly suitable for thin-walled (0.75 mm; 0.03 in.) complex shapes with uniform properties. Carbon and low- and high-alloy steel and stainless steel parts, weighing as much as 70 kg (155 lb), can be vacuum cast. These parts, which are often in the form of superalloys for gas turbines, may have walls as thin as 0.5 mm (0.02 in.). The process can be automated and production costs are similar to those for green-sand casting.

5.9 | Expendable-Mold, Expendable-Pattern Casting Processes

5.9.1 Expendable-pattern casting (lost foam)

The *expendable-pattern casting process* uses a polystyrene pattern, which evaporates upon contact with the molten metal to form a cavity for the casting, as shown in Fig. 5.20. The process is also known as *evaporative-pattern* or *lost-foam casting*

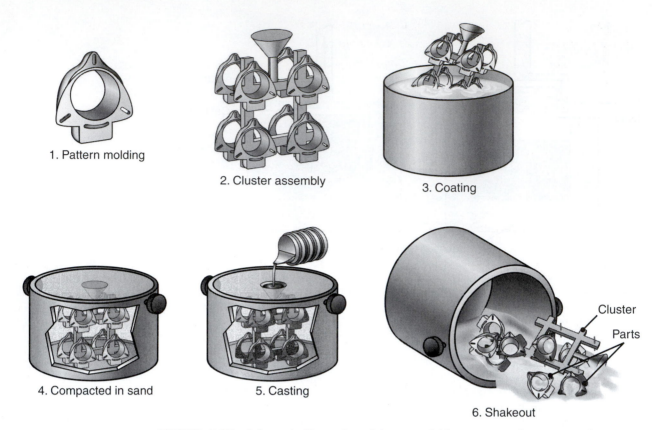

1. Pattern molding

2. Cluster assembly

3. Coating

4. Compacted in sand

5. Casting

6. Shakeout

Cluster

Parts

FIGURE 5.20 Schematic illustration of the expendable-pattern casting process, also known as lost-foam or evaporative-pattern casting.

and under the trade name *Full-Mold* process. It has become one of the more important casting processes for ferrous and nonferrous metals, particularly for the automotive industry. First, raw expandable polystyrene (EPS) beads, containing 5 to 8% pentane (a volatile hydrocarbon), are placed in a preheated die, usually made of aluminum. The polystyrene expands and takes the shape of the die cavity; additional heat is applied to fuse and bond the beads together. The die is then cooled and opened, and the pattern is removed. Complex patterns may also be made by bonding various individual pattern sections using hot-melt adhesive (Section 12.15.1).

The pattern is then coated with a water-based refractory slurry, dried, and placed in a flask. The flask is filled with loose fine sand, which surrounds and supports the pattern; it is periodically compacted by various means. Then, without removing the pattern, the molten metal is poured into the mold. This action immediately vaporizes the pattern and fills the mold cavity, completely replacing the space previously occupied by the polystyrene pattern. The heat degrades (depolymerizes) the polystyrene and the degradation products are vented into the surrounding sand. In a modification of this process, the polystyrene pattern is surrounded by a ceramic sell (*Replicast C-S process*). The pattern is burned out prior to pouring the molten metal.

The evaporative-pattern process has the following characteristics:

a. The process is relatively simple because there are no parting lines, cores, or riser systems; hence it has design flexibility.

b. Inexpensive flasks are satisfactory for this process.

c. Polystyrene is inexpensive and can easily be processed into patterns having complex shapes, various sizes, and fine surface detail.

d. The casting requires minimum cleaning and finishing operations.

e. The operation can be automated and is economical for long production runs.

f. The cost to produce the die used for expanding the polystyrene beads to make the pattern can be high.

The metal flow velocity in the mold depends on the rate of degradation of the polymer. The flow is basically laminar, with Reynolds numbers in the range of 400 to 3000 (see Section 5.4.1). The velocity of the metal-polymer pattern front is estimated to be in the range of 0.1–1 m/s and can be controlled by producing patterns with cavities or hollow sections. Because the polymer requires considerable energy to degrade, large thermal gradients are present at the metal-polymer interface. In other words, the molten metal cools faster than it would if it were poured into a cavity, which has important effects on the microstructure throughout the casting and also leads to directional solidification of the metal.

Typical parts made are aluminum engine blocks, cylinder heads, crankshafts, brake components, manifolds, and machine bases. The evaporative-pattern casting process is also used in the production of metal-matrix composites (Section 11.14). This is accomplished by embedding the foam pattern with fibers or particles throughout its volume. These then become an integral part of the casting. Further studies include modification and grain refinement of the casting by the use of grain refiners and modifier master alloys (Section 5.5) within the pattern while it is being molded.

5.9.2 Investment casting (lost-wax process)

The *investment-casting process,* also called the *lost-wax process,* was first used during the period 4000–3000 B.C. Typical parts made are mechanical components such as gears, cams, valves, and ratchets. Parts up to 1.5 m (60 in.) in diameter and weighing as much as 1140 kg (2500 lb) have successfully been cast by this process. The sequence involved in investment casting is shown in Fig. 5.21.

The pattern is made by injecting *semisolid* or *liquid* wax or plastic into a metal die in the shape of the pattern. The pattern is then removed and dipped into a slurry of refractory material, such as very fine silica and binders, ethyl silicate, and acids. After this initial coating has dried, the pattern is coated repeatedly to increase its thickness. (The term *investment* comes from investing the pattern with the refractory material.) Wax patterns require careful handling to prevent fracture. The one-piece mold is dried in air and heated to a temperature of 90°–175°C (200°–350°F) for about four hours (depending on the metal to be cast) to drive off the water of crystallization (chemically combined water) and melt out the wax. The mold may be heated to even higher temperatures to burn out the wax residue. After the mold has been preheated to the desired temperature, it is ready for molten metal. After the metal has been poured and has solidified, the mold is broken up and the casting is removed. A number of patterns can be joined in one mold, called a **tree,** to increase productivity.

The labor and materials involved can make the lost-wax process costly, but little or no finishing is required. This process is capable of producing intricate shapes from a wide variety of ferrous and nonferrous metals and alloys, with parts generally weighing from 1 g to 100 kg (0.035 oz to 75 lb) and as much as 1140 kg (2500 lb). It is suitable for casting high-melting-point alloys with a good surface finish and close dimensional tolerances.

Ceramic-shell investment casting. A variation of the investment-casting process is *ceramic-shell investment casting.* It uses the same type of wax or plastic pattern, which is (a) first dipped into a slurry with colloidal silica or ethyl silicate binder, (b) then into a fluidized bed of fine-grained fused silica or zircon flour, and (c) then

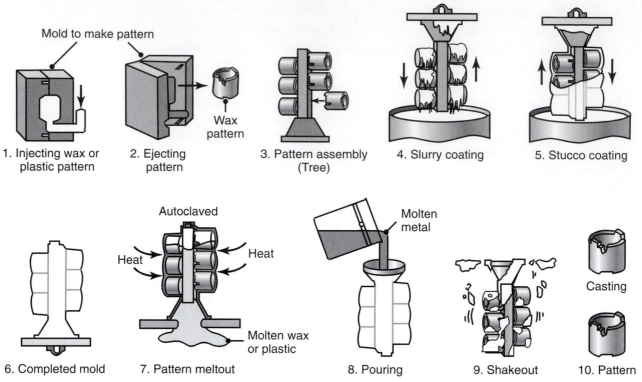

1. Injecting wax or plastic pattern
2. Ejecting pattern
3. Pattern assembly (Tree)
4. Slurry coating
5. Stucco coating
6. Completed mold
7. Pattern meltout
8. Pouring
9. Shakeout
10. Pattern

FIGURE 5.21 Schematic illustration of investment casting (lost-wax process). Castings by this method can be made with very fine detail and from a variety of metals. *Source:* Steel Founders' Society of America.

into coarse-grain silica to build up additional coatings and thickness to withstand the thermal shock of pouring. The process is economical and is used extensively for precision casting of steels, aluminum, and high-temperature alloys. If ceramic cores are used in the casting, they are removed by leaching with caustic solutions under high pressure and temperature.

The molten metal may be poured in a vacuum to extract evolved gases and reduce oxidation, further improving the quality of the casting. The castings made by this and other processes may be subjected to hot isostatic pressing to further reduce microporosity. Aluminum castings, for example, are subjected to a gas pressure of up to about 100 MPa (15 ksi) at 500°C (900°F).

EXAMPLE 5.4 Investment-cast superalloy components for gas turbines

Since the 1960s, investment-cast superalloys have been replacing wrought counterparts in high-performance gas turbines. Much development has been taking place in producing cleaner superalloys (nickel-based and cobalt-based). Improvements have been made in melting and casting techniques, such as by vacuum-induction melting, using microprocessor controls. Impurity and inclusion levels have continually been reduced, thus improving the strength, ductility, and overall reliability of these components. Such control is essential because these parts operate at a temperature only about 50°C (90°F) below the solidus.

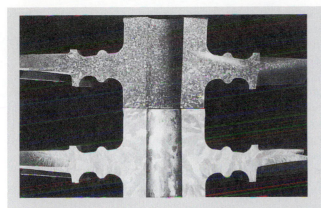

FIGURE 5.22
Microstructure of a rotor that has been investment cast (top) and conventionally cast (bottom). *Source: Advanced Materials and Processes,* October 1990, p. 25, ASM International.

The microstructure of an integrally investment-cast, gas-turbine rotor is shown in the upper portion of Fig. 5.22. Note the fine, uniform, equiaxed grains throughout the cross section. Recent techniques to obtain this result include the use of a nucleant addition to the molten metal, as well as close control of its superheat, pouring techniques, and the cooling rate of the casting. In contrast, the lower portion of Fig. 5.22 shows the same type of rotor cast conventionally; note the coarse grain structure. This rotor will have inferior properties compared to the fine-grained-rotor. Due to developments in these processes, the percentage of cast parts in aircraft engines has increased from 20% to about 45% by weight.

5.10 | Permanent-Mold Casting Processes

Permanent molds are used repeatedly and are designed so that the casting can easily be removed and the mold reused. The molds are made of metals that maintain their strength at high temperatures. Because metal molds are better heat conductors than the expendable molds described above, the solidifying casting is subjected to a higher rate of cooling. This, in turn, affects the microstructure and grain size within the casting, as described in Section 5.3.

In permanent-mold casting, two halves of a mold are made from materials such as steel, bronze, refractory metal alloys, or graphite (*semipermanent mold*). The mold cavity and the gating system are machined into the mold itself and thus become an integral part of it. If required, cores are made of metal or sand aggregate and are placed in the mold prior to casting. Typical core materials are shell or no-bake cores, gray iron, low-carbon steel, and hot-work die steel. Inserts may also used for various parts of the mold.

In order to increase the life of permanent molds, the surfaces of the mold cavity may be coated with a refractory slurry or sprayed with graphite every few castings. The coatings also serve as parting agents and thermal barriers to control the rate of cooling of the casting. Mechanical ejectors, such as pins located in various parts of the mold, may be needed for removal of complex castings. Ejectors usually leave small round impressions on castings and gating systems, as they do in injection molding of plastics (see Section 10.10.2). The molds are clamped together by mechanical means and are heated to facilitate metal flow and reduce damage (*thermal fatigue;* see Section 3.9.5) to the dies. The molten metal then flows through the gating system and fills the mold cavity, and after solidification, the molds are opened and the casting removed. Special methods of cooling the mold include water or the use of air-cooled fins, similar to those found on motorcycle or lawnmower engines that cool the engine blocks and cylinder heads.

Although the permanent-mold casting operation can be performed manually, the process is automated for large production runs. Used mostly for aluminum, magnesium, and copper alloys (because of their generally lower melting temperatures), as well as steels cast in graphite or heat-resistant metal molds, the castings have good surface finish, close dimensional tolerances, and uniform and good mechanical properties. Typical parts made are automobile pistons, cylinder heads, connecting rods, gear blanks for appliances, and kitchenware. Because of the high cost of dies, permanent-mold casting is not economical for small production runs. Furthermore, because of the difficulty in removing the casting from the mold, intricate shapes cannot be cast by this process. However, intricate internal cavities can be cast using sand cores.

5.10.1 Slush casting

It was noted in Fig. 5.12 that a solidified skin first develops in a casting and that this skin becomes thicker with time. Hollow castings with thin walls can be made by permanent-mold casting using this principle in a process called *slush casting*. The molten metal is poured into the metal mold, and after the desired thickness of solidified skin is obtained, the mold is inverted (slung) and the remaining liquid metal is poured out. The mold halves are then opened and the casting is removed. The process is suitable for small production runs and is generally used for making ornamental and decorative objects and toys from low-melting-point metals, such as zinc, tin, and lead alloys.

5.10.2 Pressure casting

In the *pressure-casting* process, also called *pressure pouring* or *low-pressure casting* (Fig. 5.23), the molten metal is forced upward into a graphite or metal mold by gas pressure, which is maintained until the metal has completely solidified in the mold. The molten metal may also be forced upward by a vacuum, which also removes dissolved gases and produces a casting with lower porosity.

FIGURE 5.23 The pressure-casting process, utilizing graphite molds for the production of steel railroad wheels. *Source:* Griffin Wheel Division of Amsted Industries Incorporated.

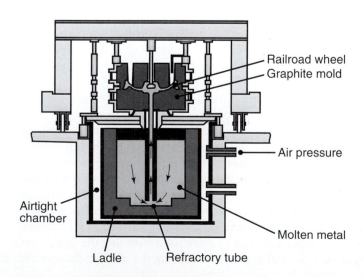

Railroad wheel
Graphite mold
Air pressure
Molten metal
Airtight chamber
Ladle   Refractory tube

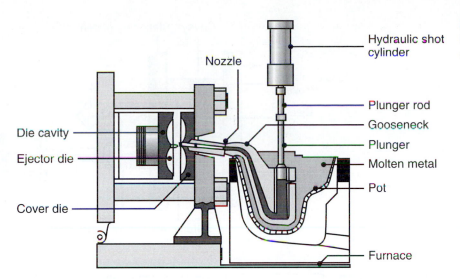

Nozzle

Hydraulic shot cylinder

Plunger rod

Gooseneck

Plunger

Molten metal

Pot

Furnace

Die cavity

Ejector die

Cover die

FIGURE 5.24 Schematic illustration of the hot-chamber die-casting process.

5.10.3 Die casting

The *die-casting* process, developed in the early 1900s, is a further example of permanent-mold casting. The molten metal is forced into the die cavity at pressures ranging from 0.7 to 700 MPa (0.1 to 100 ksi). The European term *pressure die casting,* or simply die casting, described in this section, is not to be confused with the term *pressure casting* described in Section 5.10.2. Typical parts made by die casting are transmission housings, valve bodies, motors, business machine and appliance components, hand tools, and toys. The weight of most castings ranges from less than 90 g (3 oz) to about 25 kg (55 lb).

1. **Hot-chamber process.** The *hot-chamber* process (Fig. 5.24) involves the use of a piston, which traps a specific volume of molten metal and forces it into the die cavity through a gooseneck and nozzle. The pressures range up to 35 MPa (5000 psi), with an average of about 15 MPa (2000 psi). The metal is held under pressure until it solidifies in the die. To improve die life and to aid in rapid metal cooling (thus improving productivity), dies are usually cooled by circulating water or oil through various passageways in the die block. Low-melting-point alloys such as zinc, tin, and lead are commonly cast by this process. Cycle times typically range up to 900 shots (individual injections) per hour for zinc, although very small components such as zipper teeth can be cast at 300 shots per minute.

2. **Cold-chamber process.** In the *cold-chamber* process (Fig. 5.25) molten metal is introduced into the injection cylinder (*shot chamber*). The shot chamber is not heated, hence the term *cold chamber*. The metal is forced into the die cavity at pressures usually ranging from 20 to 70 MPa (3 to 10 ksi), although they may be as high as 150 MPa (20 ksi). The machines may be horizontal or vertical. Aluminum, magnesium, and copper alloys are normally cast by this method, although other metals (including ferrous metals) are also cast. Molten metal temperatures start at about 600°C (1150°F) for aluminum and magnesium alloys and increase considerably for copper-based and iron-based alloys.

Process capabilities and machine selection. Because of the high pressures involved, the dies have a tendency to part unless clamped together tightly. Die-casting machines are rated according to the clamping force that can be exerted to keep the dies closed. The capacities of commercially available machines range from about 25 to 3000 tons. Other factors involved in the selection of die-casting machines are die size, piston stroke, shot pressure, and cost.

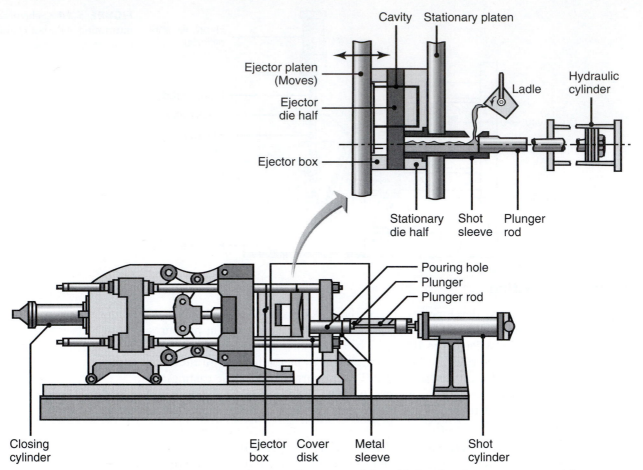

FIGURE 5.25 Schematic illustration of the cold-chamber die-casting process. These machines are large compared to the size of the casting, because high forces are required to keep the two halves of the die closed under pressure.

Die-casting dies may be made single cavity, multiple cavity (several identical cavities), combination cavity (several different cavities), or unit dies (simple small dies that can be combined in two or more units in a master holding die).

Dies are usually made of hot-work die steels or mold steels. Die wear increases with the temperature of the molten metal. **Heat-checking** of the dies (surface cracking from repeated heating and cooling) can be a problem. When die materials are selected and maintained properly, dies may last more than half a million shots before die wear becomes significant. Alloys, except magnesium alloys, generally require lubricants, which usually have a water base with graphite or other compounds in suspension. Lubricants (parting agents) are usually applied as thin coatings on die surfaces. Because of the high cooling capacity of water, these lubricants are also effective in keeping die temperatures low.

The entire die-casting and finishing process can be highly automated. Die casting has the capability for high production rates with good strength, high-quality parts with complex shapes, and good dimensional accuracy and surface detail, thus requiring little or no subsequent machining or finishing operations (*net-shape forming*). Components such as pins, shafts, and fasteners can be cast integrally (**insert molding;** see also Section 10.10.2) in a process similar to placing wooden sticks in popsicles. Ejector marks remain, as do small amounts of *flash* (thin material squeezed out between the dies; see also Section 6.2.3) at the die parting line.

TABLE 5.6

Properties and Typical Applications of Common Die-Casting Alloys

Alloy	Ultimate tensile strength (MPa)	Yield strength (MPa)	Elongation in 50 mm (%)	Applications
Aluminum 380 (3.5 Cu-8.5 Si)	320	160	2.5	Appliances, automotive components, electrical motor frames and housings, engine blocks.
Aluminum 13 (12 Si)	300	150	2.5	Complex shapes with thin walls, parts requiring strength at elevated temperatures.
Brass 858 (60 Cu)	380	200	15	Plumbing fixtures, lock hardware, bushings, ornamental castings.
Magnesium AZ91B (9 Al-0.7 Zn)	230	160	3	Power tools, automotive parts, sporting goods.
Zinc No. 3 (4 Al)	280	—	10	Automotive parts, office equipment, household utensils, building hardware, toys.
Zinc No. 5 (4 Al-1 Cu)	320	—	7	Appliances, automotive parts, building hardware, business equipment.

Source: The North American Die Casting Association.

The properties and applications of die-cast materials are given in Table 5.6. Die casting can compete favorably with other manufacturing methods, such as sheet-metal stamping and forging, or other casting processes. Because the molten metal chills rapidly at the die walls, the casting has a fine-grain, hard skin with higher strength than in the center. Consequently, the strength-to-weight ratio of die-cast parts increases with decreasing wall thickness. With good surface finish and dimensional accuracy, die casting can produce bearing surfaces that would normally be machined. Equipment costs, particularly the cost of dies, are somewhat high, but labor costs are generally low because the process usually is semi- or fully automated.

5.10.4 Centrifugal casting

The *centrifugal casting process* utilizes the inertial forces caused by rotation to force the molten metal into the mold cavities. There are three types of centrifugal casting.

1. **True centrifugal casting.** Hollow cylindrical parts, such as pipes, gun barrels, and lampposts, are produced by this technique, as shown in Fig. 5.26, in

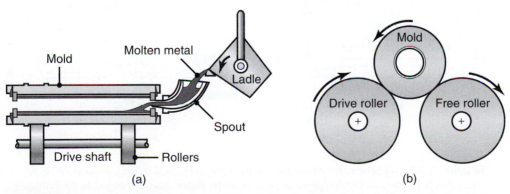

(a)   (b)

FIGURE 5.26   Schematic illustration of the centrifugal casting process. Pipes, cylinder liners, and similarly shaped hollow parts can be cast by this process.

which molten metal is poured into a rotating mold. The axis of rotation is usually horizontal but can also be vertical for short workpieces. Molds are made of steel, iron, or graphite and may be coated with a refractory lining to increase mold life. Mold surfaces can be shaped so that pipes with various outer patterns and shapes can be cast. The inner surface of the casting remains cylindrical because the molten metal is uniformly distributed by centrifugal forces. However, because of density differences in the radial direction, lighter elements such as dross, impurities, and pieces of the refractory lining tend to collect on the inner surface of the casting.

Cylindrical parts ranging from 13 mm (0.5 in.) to 3 m (10 ft) in diameter and 16 m (50 ft) long can be centrifugally cast. Wall thicknesses typically range from 6 to 125 mm (0.25 to 5 in.). The pressure generated by the centrifugal force is high (as much as 150 g's) and is necessary for casting thick-walled parts. Castings of good quality, dimensional accuracy, and external surface detail are obtained by this process. In addition to pipes, typical parts made are bushings, engine cylinder liners, street lamps, and bearing rings, with or without flanges.

2. **Semicentrifugal casting.** An example of *semicentrifugal* casting is shown in Fig. 5.27a. This method is used to cast parts with rotational symmetry, such as a wheel with spokes.

3. **Centrifuging.** Also called *centrifuge casting,* mold cavities of any shape are placed at a distance from the axis of rotation. The molten metal is poured at the center and is forced into the mold by centrifugal forces (Fig. 5.27b). The properties within the castings vary by the distance from the axis of rotation.

5.10.5 Squeeze casting

The *squeeze-casting process,* developed in the 1960s, involves solidification of the molten metal under high pressure; thus it is a combination of casting and forging (Fig. 5.28). The machinery typically includes a die, punch, and ejector pin. The pressure applied by the punch keeps the entrapped gases in solution (especially hydrogen in aluminum alloys), and the high-pressure contact at the die-metal interfaces promotes heat transfer. The higher cooling rate results in a fine microstructure with good mechanical properties and limited microporosity. The pressures in squeeze

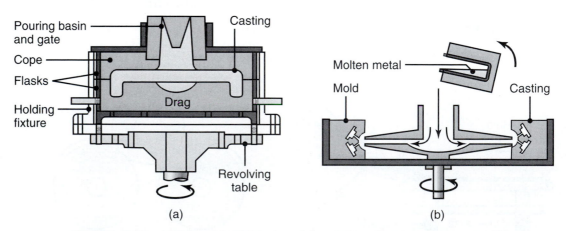

(a)   (b)

FIGURE 5.27   (a) Schematic illustration of the semicentrifugal casting process. Wheels with spokes can be cast by this process. (b) Schematic illustration of casting by centrifuging. The molds are placed at the periphery of the machine, and the molten metal is forced into the molds by centrifugal forces.

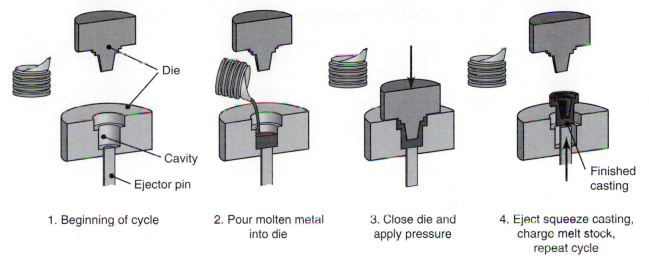

1. Beginning of cycle   2. Pour molten metal into die   3. Close die and apply pressure   4. Eject squeeze casting, charge melt stock, repeat cycle

FIGURE 5.28   Sequence of operations in the squeeze-casting process. This process combines the advantages of casting and forging.

casting are typically higher than in pressure die casting, but lower than those for hot or cold forging (Section 6.2). Parts can be made to *near-net shape* (see also Section 1.6) with complex shapes and fine surface detail, from both nonferrous and ferrous alloys. Typical products made by squeeze casting are automotive wheels, mortar bodies (a short-barreled cannon), brake drums, and valve bodies.

5.10.6 Semisolid metal forming (thixocasting) and rheocasting

Also called *semisolid metalworking,* in the *semisolid metal-forming process* (developed in the 1970s) the metal or alloy has a nondendritic, roughly spherical, and fine-grained structure when it enters the die or the mold. The alloy exhibits *thixotropic* behavior (its viscosity decreases when agitated); hence this process is also known as **thixoforming** or **thixocasting**. At rest and above its solidus temperature, for example, the alloy has the consistency of butter, but when agitated vigorously its consistency is more like machine oil. This behavior has been utilized in developing machines and technologies combining casting and forging of parts, with cast billets that are forged when 30 to 40% liquid. This technology is also used in making cast metal-matrix composites.

The main advantages of thixocasting are that the semisolid metal is just above the solidification temperature when injected into a die, resulting in reductions in solidification and cycle times, and thereby increased productivity. Furthermore, shrinkage porosity is reduced because of the lower superheat involved. Parts made include control arms, brackets, and steering components. The advantages of semi-solid metal forming over die casting are that (a) the structures developed are homogeneous, with uniform properties and high strength, (b) both thin and thick parts can be made, (c) casting alloys as well as wrought alloys can be used, and (d) parts can subsequently be heat treated. However, material and overall costs generally are higher than those for die casting.

In **rheocasting,** a slurry (a solid suspended in a liquid) is received from a melt furnace and is cooled, then magnetically stirred prior to injecting it into a mold or die. Rheocasting has been successfully applied with aluminum and magnesium and has been used to produce engine blocks, crankcases (as for motorcycles or lawn-mowers), and various marine applications.

5.10.7 Casting techniques for single-crystal components

These techniques can best be illustrated by comparing the casting of gas turbine blades, which are generally made of nickel-based superalloys. The procedures involved can also be used for other alloys and components.

Conventional casting of turbine blades. The conventional casting process involves investment casting using a ceramic mold (see Fig. 5.18). The molten metal is poured into the mold and begins solidifying at the ceramic walls. The grain structure developed is polycrystalline, and the presence of grain boundaries makes this structure susceptible to creep and cracking along those boundaries (Section 3.4) under the centrifugal forces at elevated temperatures.

Directionally solidified blades. In the *directional solidification* process (Fig. 5.29a), developed in 1960, the ceramic mold is prepared basically by investment-casting techniques and is preheated by radiant heating. The mold is supported by a water-cooled chill plate. After the metal is poured into the mold, the assembly is lowered slowly. Crystals begin to grow at the chill-plate surface. The blade is thus directionally solidified, with longitudinal (but no transverse) grain boundaries. Consequently, the blade is stronger in the direction of centrifugal forces developed in the gas turbine; that is, the longitudinal direction.

Single-crystal blades. In the process for single-crystal blades, first made in 1967, the mold is prepared basically by investment casting techniques and has a constriction in the shape of a corkscrew (Figs. 5.29b and c). This cross section allows only one crystal through. As the assembly is lowered slowly, a single crystal grows upward through the constriction and begins to grow in the mold. Strict control of the rate of movement is necessary in this process. The solidified mass in the mold is thus a single-crystal blade. Although more expensive than other blades (the largest blades are over 15 kg and can cost over $6000 each), the absence of grain boundaries makes these blades resistant to creep and thermal shock, imparting a longer and more reliable service life.

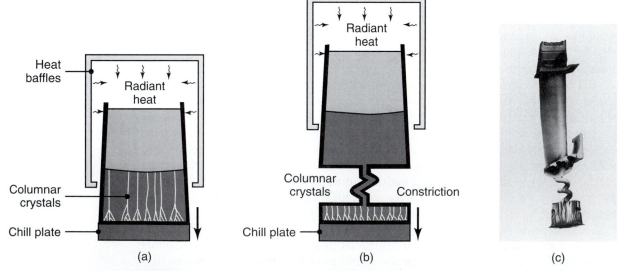

FIGURE 5.29   Methods of casting turbine blades: (a) directional solidification, (b) method to produce a single-crystal blade, and (c) a single-crystal blade with the constriction portion still attached. *Source:* (a) and (b) After B. H. Kear, (c) courtesy of ASM International.

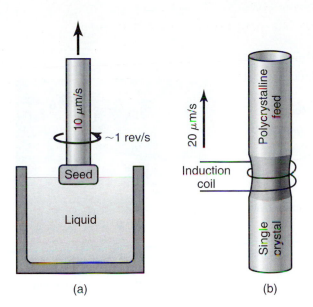

FIGURE 5.30 Two methods of crystal growing: (a) crystal pulling (Czochralski process) and (b) floating-zone method. Crystal growing is especially important in the semiconductor industry. (c) A single-crystal silicon ingot produced by the Czochralski process. *Source:* Courtesy of Intel Corp.

(a)   (b)   (c)

Single-crystal growing. With the advent of the semiconductor industry, single-crystal growing is now a major activity in the manufacture of microelectronic devices. There are basically two methods of crystal growing. In the **crystal pulling** method, known as the *Czochralski process* (Fig. 5.30a), a seed crystal is dipped into the molten metal and then pulled at a rate of about 10 μm/s while being rotated at about 1 rev/s. The liquid metal begins to solidify on the seed, and the crystal structure of the seed is continued throughout the part. *Dopants* (alloying elements) may be added to the liquid metal to impart specific electrical properties. Single crystals of silicon, germanium, and various other elements are grown by this process. Single-crystal ingots up to 400 mm (16 in.) in diameter and over 2 m (80 in.) in length have been produced by this technique, although 200- and 300-mm (8- and 12-in.) cylinders are common in the production of silicon wafers for integrated circuit manufacture (Chapter 13).

The second technique for crystal growing is the **floating-zone** method (Fig. 5.19b). Starting with a rod of polycrystalline silicon resting on a single crystal, an induction coil heats these two pieces while moving slowly upward. The single crystal grows upward while maintaining its orientation. Thin wafers are then cut from the rod, cleaned, and polished for use in microelectronic device fabrication (see Chapter 13). Because of the limited diameters that can be produced through this process, it is has largely been replaced by the Czochralski process for silicon but is still used for cylinder diameters under 150 mm. It is a cost-effective method for producing single-crystal cylinders in small quantities.

5.10.8 Rapid solidification

First developed in the 1960s, *rapid solidification* involves cooling the molten metal at rates as high as 10^6 K/s, whereby the metal does not have sufficient time to crystallize. These alloys are called **amorphous alloys** or **metallic glasses** because they do not have a long-range crystalline structure (Section 3.2). They typically contain iron, nickel, and chromium, which are alloyed with carbon, phosphorus, boron, aluminum, and silicon. Among other effects, rapid solidification results in a significant extension of solid solubility, grain refinement, and reduced microsegregation.

Amorphous alloys exhibit excellent corrosion resistance, good ductility, high strength, very little loss from magnetic hysteresis, high resistance to eddy currents, and high permeability. The last three properties are utilized in making magnetic steel

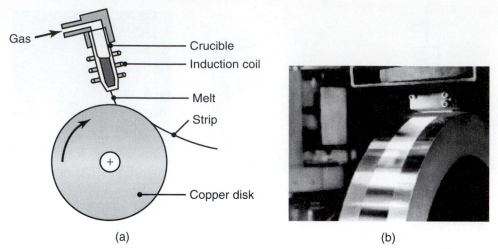

(a)   (b)

FIGURE 5.31   (a) Schematic illustration of the melt-spinning process to produce thin strips of amorphous metal, and (b) photograph of nickel-alloy production through melt spinning. *Source:* Courtesy of Siemens AG.

cores used in transformers, generators, motors, lamp ballasts, magnetic amplifiers, and linear accelerators, with greatly improved efficiency. Another major application is rapidly solidified superalloy powders, which are consolidated into near-net shapes for use in aerospace engines. Amorphous alloys are produced in the form of wire, ribbon, strip, and powder. In one process called **melt spinning** (Fig. 5.31), the alloy is melted (by induction in a ceramic crucible) and propelled under high gas pressure at very high speed against a rotating copper disk (chill block), where it chills rapidly (splat cooling).

5.11 | Heat Treatment

The various microstructures developed during metal processing can be modified by *heat treatment* techniques, involving controlled heating and cooling of the alloys at various rates (also known as **thermal treatment**). These treatments induce phase transformations that greatly influence mechanical properties such as strength, hardness, ductility, toughness, and wear resistance of the alloys. The effects of thermal treatment depend primarily on the alloy, its composition and microstructure, the degree of prior cold work, and the rates of heating and cooling during heat treatment.

5.11.1 Heat treating ferrous alloys

The microstructural changes that occur in the iron-carbon system (Section 5.2.5) are described below.

Pearlite. If the ferrite and cementite lamellae in the pearlite structure (see Section 5.2.6) of the eutectoid steel are thin and closely packed, the microstructure is called **fine pearlite**. If the lamellae are thick and widely spaced, the structure is called **coarse pearlite**. The difference between the two depends on the rate of cooling through the eutectoid temperature, a reaction in which austenite is transformed into pearlite. If the rate of cooling is relatively high (as in air), fine pearlite is produced; if slow (as in a furnace), coarse pearlite is produced.

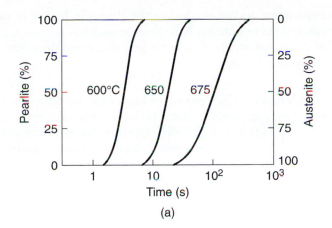

(a)

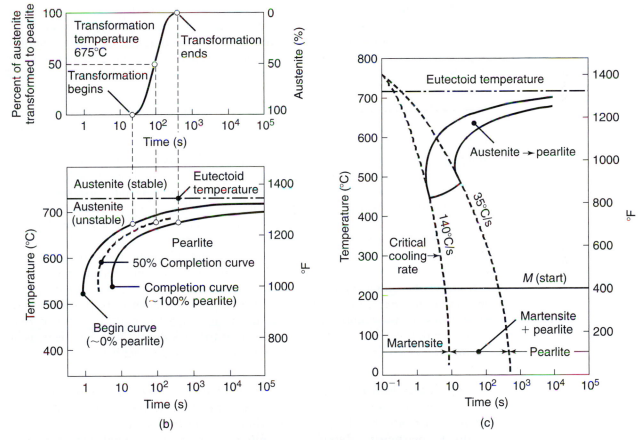

(b)   (c)

FIGURE 5.32   (a) Austenite to pearlite transformation of iron-carbon alloys as a function of time and temperature, (b) isothermal transformation diagram obtained from (a) for a transformation temperature of 675°C (1247°F), (c) microstructures obtained for a eutectoid iron-carbon alloy as a function of cooling rate. *Source:* Courtesy of ASM International.

The transformation from austenite to pearlite (and for other structures) is best illustrated by Figs. 5.32b and c. These diagrams are called **isothermal transformation (IT) diagrams** or **time-temperature-transformation (TTT) diagrams.** They are constructed from the data in Fig. 5.32a, which shows the percentage of austenite transformed into pearlite as a function of temperature and time. The higher the temperature and/or the longer the time, the higher is the percentage of austenite transformed to pearlite. Note that for each temperature, a minimum time is required for the transformation to begin and that some time later, all the austenite is transformed to pearlite.

Spheroidite. When pearlite is heated to just below the eutectoid temperature and held at that temperature for a period of time, say, for a day at 700°C (1300°F), the cementite lamellae (Fig. 5.4b) transform to *spherical* shapes (*spheroidites*). Unlike the lamellar shape of cementite, which acts as stress raisers, spheroidites are less conducive to stress concentration because of their rounded shapes. Consequently, this structure has higher toughness and lower hardness than the pearlite structure. In this form it can be cold worked, and the spheroidal particles prevent the propagation of cracks within the material during working.

Bainite. Visible only under electron microscopy, *bainite* has a very fine microstructure, consisting of ferrite and cementite. It can be produced in steels by alloying and at cooling rates that are higher than those required for transformation to pearlite. This structure, called *bainitic steel*, is generally stronger and more ductile than pearlitic steel at the same hardness level.

Martensite. When austenite is cooled rapidly (such as by quenching in water), its fcc structure is transformed to a *body-centered tetragonal* (bct) structure. This structure can be described as a body-centered rectangular prism which is slightly elongated along one of its principal axes, called *martensite*. Because it does not have as many slip systems as a bcc structure and the carbon is in interstitial positions, martensite is extremely hard and brittle, lacks toughness, and hence has limited use. Martensite transformation takes place almost instantaneously (Fig. 5.30c) because it does not involve the diffusion process (a time-dependent phenomenon that is the mechanism in other transformations).

Because of the different densities of the various phases in the structure, transformations involve volume changes. For example, when austenite transforms to martensite, its volume increases (density decreases) by as much as 4%. A similar but smaller volume expansion also occurs when austenite transforms to pearlite. These expansions and the resulting thermal gradients in a quenched part can cause internal stresses within the body, which may cause parts to crack during heat treatment, as in **quench cracking** of steels caused by rapid cooling during quenching.

Retained austenite. If the temperature at which the alloy is quenched is not sufficiently low, only a portion of the structure is transformed to martensite. The rest is *retained austenite*, which is visible as white areas in the structure along with dark needlelike martensite. Retained austenite can cause dimensional instability and cracking of the part and lowers hardness and strength.

Tempered martensite. *Tempering* is a heating process that reduces martensite's hardness and improves its toughness. The body-centered tetragonal martensite is heated to an intermediate temperature, where it transforms to a two-phase microstructure, consisting of body-centered cubic alpha ferrite and small particles of cementite. Longer tempering time and higher temperature decrease martensite's hardness. The reason is that the cementite particles coalesce and grow, and the distance between the particles in the soft ferrite matrix increases as the less stable, smaller carbide particles dissolve.

Hardenability of ferrous alloys. The capability of an alloy to be hardened by heat treatment is called its *hardenability*. It is a measure of the depth of hardness that can be obtained by heating and subsequent quenching. (The term *hardenability* should not be confused with hardness, which is the resistance of a material to indentation or scratching.) Hardenability of ferrous alloys depends on their carbon content, the grain size of the austenite, and the alloying elements. The **Jominy test** has been developed in order to determine alloy hardenability.

Quenching media. Quenching may be carried out in water, brine (saltwater), oils, molten salts, or air, as well as caustic solutions, polymer solutions, and various gases. Because of the differences in the thermal conductivity, specific heat, and heat of vaporization of these media, the rate of cooling (**severity of quench**) will also be different. In relative terms and in decreasing order, the cooling capacity of several quenching media is (a) agitated brine 5, (b) still water 1, (c) still oil 0.3, (d) cold gas 0.1, and (e) still air 0.02. Agitation is also a significant factor in the rate of cooling. In tool steels the quenching medium is specified by a letter (see Table. 3.5), such as W for water hardening, O for oil hardening, and A for air hardening. The cooling rate also depends on the surface area-to-volume ratio of the part [see Eq. (5.9)]. The higher this ratio, the higher is the cooling rate; thus, for example, a thick plate cools more slowly than a thin plate with the same surface area.

Water is a common medium for rapid cooling. However, the heated metal may form a **vapor blanket** along its surfaces from water-vapor bubbles that form when water boils at the metal-water interface. This blanket creates a barrier to heat conduction because of the lower thermal conductivity of the vapor. Agitating the fluid or the part helps to reduce or eliminate the blanket. Also, water may be sprayed on the part under high pressure. Brine is an effective quenching medium because salt helps to nucleate bubbles at the interfaces, thus improving agitation. However, brine can corrode the part. **Die quenching** is a term used to describe the process of clamping the part to be heat treated to a die, which chills selected regions of the part. In this way cooling rates and warpage can be controlled.

5.11.2   Heat treating nonferrous alloys and stainless steels

Nonferrous alloys and some stainless steels generally cannot be heat treated by the techniques used with ferrous alloys, because nonferrous alloys do not undergo phase transformations as steels do. Heat-treatable aluminum alloys (see Section 3.11.1), copper alloys, and martensitic and precipitation-hardening stainless steels are hardened and strengthened by **precipitation hardening**. This is a technique in which small particles of a different phase (called *precipitates*) are uniformly dispersed in the matrix of the original phase (see Fig. 5.2a). Precipitates form because the solid solubility of one element (one component of the alloy) in the other is exceeded.

Three stages are involved in the precipitation hardening process, which can best be described by referring to the phase diagram for the aluminum-copper system, given in Fig. 5.33. For an alloy with a composition of 95.5% Al-4.5% Cu, a single-phase (κ) substitutional solid solution of copper (solute) in aluminum (solvent) exists between 500° and 570°C (930° and 1060°F). The κ phase is aluminum rich, has an fcc structure, and is ductile. Below the lower temperature (that is, below the lower solubility curve), two phases are present: κ and θ (a hard intermetallic compound of $CuAl_2$). This alloy can be heat treated and its properties modified by solution treatment or precipitation.

Solution treatment. In *solution treatment* the alloy is heated to within the solid-solution κ phase, say, 540°C (1000°F), and cooled rapidly, such as by quenching in

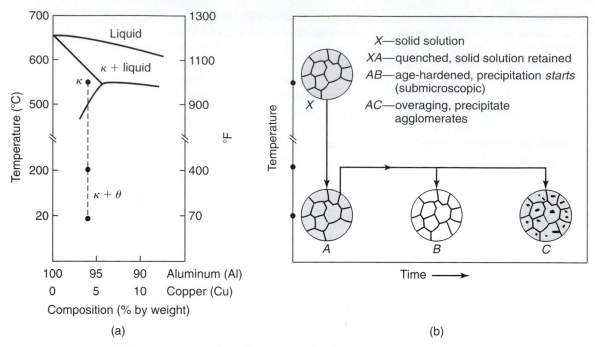

FIGURE 5.33   (a) Phase diagram for the aluminum-copper alloy system, and (b) various microstructures obtained during the age-hardening process.

water. The structure obtained soon after quenching (*A* in Fig. 5.33b) consists only of the single phase *κ*. This alloy has moderate strength and considerable ductility.

Precipitation hardening. The structure obtained in *A* in Fig. 5.33b can be strengthened by precipitation hardening. The alloy is reheated to an intermediate temperature and held there for a period of time, during which precipitation takes place. The copper atoms diffuse to nucleation sites and combine with aluminum atoms, producing the theta phase, which form as submicroscopic precipitates (shown in *B* by the small dots within the grains of the *κ* phase). This structure is stronger than that in *A*, although it is less ductile. The increase in strength is attributed to increased resistance to dislocation movement in the region of the precipitates.

Aging. Because the precipitation process is one of time and temperature, it is also called aging, and the property improvement is known as *age hardening*. If carried out above room temperature, the process is called **artificial aging**. However, several aluminum alloys harden and become stronger over a period of time at room temperature, by a process known as **natural aging**. These alloys are first quenched and then, if required, are formed at room temperature into various shapes and allowed to gain strength and hardness by natural aging. Natural aging can be slowed by refrigerating the quenched alloy.

In the precipitation process, when the reheated alloy is held at that temperature for an extended period of time, the precipitates begin to coalesce and grow. They become larger but fewer, as shown by the larger dots in *C* in Fig. 5.33b. This process is called **overaging**, which makes the alloy softer and less strong, although the part has better dimensional stability over time. Thus there is an optimal time-temperature relationship in the aging process to develop desired properties. Note that an aged alloy can be used only up to a certain maximum temperature in service; otherwise it will overage and lose its strength and hardness.

Maraging. Maraging is a precipitation-hardening treatment for a special group of high-strength iron-based alloys. The word *maraging* is derived from the words *martensite* and *aging*. In this process, one or more intermetallic compounds (Section 5.2.2) are precipitated in a matrix of low-carbon martensite. A typical maraging steel may contain 18% nickel and other elements, and aging is done at 480°C (900°F). Hardening by maraging does not depend on the cooling rate; thus, full uniform hardness can be obtained throughout large parts and with minimal distortion. Typical uses of maraging steels are for dies and tooling for casting, molding, forging, and extrusion.

5.11.3 Case hardening

The heat treatment processes described thus far involve microstructural alterations and property changes in the *bulk* of the material or component by *through hardening*. In many situations, however, alteration of only the *surface* properties of a part (hence the term *case hardening*) is desirable, particularly for improving resistance to surface indentation, fatigue, and wear. Typical applications are gear teeth, cams, shafts, bearings, fasteners, pins, automotive clutch plates, and tools and dies. Through hardening of these parts would not be desirable, because a hard part generally lacks the necessary toughness for these applications. A small surface crack can propagate rapidly through the part and cause total failure.

Several surface-hardening processes are available (Table 5.7): **carburizing** (*gas, liquid,* and *pack carburizing*), **carbonitriding, cyaniding, nitriding, boronizing,** and **flame** and **induction hardening.** Basically, the component is heated in an atmosphere containing elements (such as carbon, nitrogen, or boron) that alter the composition, microstructure, and properties of surfaces.

For steels with sufficiently high carbon content, surface hardening takes place without using any of these additional elements. Only the heat treatment processes described in Section 5.11.1 are needed to alter the microstructures, usually by flame hardening or induction hardening. Laser beams and electron beams are also used effectively to harden both small and large surfaces, and also for through hardening of relatively small parts.

Because case hardening is a localized heat treatment, case-hardened parts have a hardness gradient. Typically, the hardness is greatest at the surface and decreases below the surface, the rate of decrease depending on the composition of the metal and the process variables. Surface-hardening techniques can also be used for tempering, thus modifying the properties of surfaces that have been subjected to heat treatment. Various other processes and techniques for surface hardening, such as shot peening and surface rolling, improve wear resistance and various other characteristics, as described in Section 4.5.1.

Decarburization. This is the phenomenon in which carbon-containing alloys lose carbon from their surfaces as a result of heat treatment or hot working in a medium (usually oxygen) that reacts with the carbon. Decarburization is undesirable because it affects the hardenability of the surfaces of the part by lowering the carbon content. It also adversely affects the hardness, strength, and fatigue life of steels by significantly lowering their endurance limit. Decarburization is best avoided by processing the alloy in an inert atmosphere or a vacuum, or by using neutral salt baths during heat treatment.

TABLE 5.7

Outline of Heat Treatment Processes for Surface Hardening

Process	Metals hardened	Element added to surface	Procedure	General characteristics	Typical applications
Carburizing	Low-carbon steel (0.2% C), alloy steels (0.08–0.2% C)	C	Heat steel at 870–950° (1600–1750°F) in an atmosphere of carboaceous gases (gas carburizing) or carbon-containing solids (pack carburizing). Then quench.	A hard, high-carbon surface is produced. Hardness 55–65 HRC. Case depth <0.5–1.5 mm (<0.020 to 0.060 in.). Some distortion of part during heat treatment.	Gears, cams, shafts, bearings, piston pins, sprockets, clutch plates
Carbonitriding	Low-carbon steel	C and N	Heat steel at 700–800°C (1300–1600°F) in an atmosphere of carbonaceous gas and ammonia. Then quench in oil.	Surface hardness 55–62 HRC. Case depth 0.07–0.5 mm (0.003–0.020 in.). Less distortion than in carburizing.	Bolts, nuts, gears
Cyaniding	Low-carbon steel (0.2% C), alloy steels (0.08–0.2% C)	C and N	Heat steel at 760–845°C (1400–1550°F) in a molten bath of solutions of cyanide (e.g., 30% sodium cyanide) and other salts.	Surface hardness up to 65 HRC. Case depth 0.025–0.25 mm (0.001–0.010 in.). Some distortion.	Bolts, nuts, screws, small gears
Nitriding	Steels (1% Al, 1.5% Cr, 0.3% Mo), alloy steels (Cr, Mo), stainless steels, high-speed steels	N	Heat steel at 500–600°C (925–1100°F) in an atmosphere of ammonia gas or mixtures of molten cyanide salts. No further treatment.	Surface hardness up to 1100 HV. Case depth 0.1–0.6 mm (0.005–0.030 in.) and 0.02–0.07 mm (0.001–0.003 in.) for high speed steel.	Geards, shafts, sprockets, valves, cutters, boring bars
Boronizing	Steels	B	Part is heated using boron-containing gas or solid in contact with part.	Extremely hard and wear-resistance surface. Case depth 0.025–0.075 mm (0.001–0.003 in.).	Tool and die steels
Flame hardening	Medium-carbon steels, cast irons	None	Surface is heated with an oxyacetylene torch, then quenched with water spray or other quenching methods.	Surface hardness 50–60 HRC. Case depth 0.7–6 mm (0.030–0.25 in.). Little distortion.	Axles, crank-shafts, piston rods, lathe beds, and centers
Induction hardening	Same as above	None	Metal part is placed in copper induction coils and is heated by high-frequency current, then quenched	Same as above	Same as above

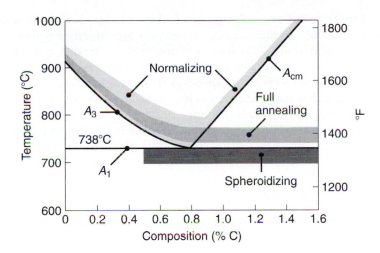

FIGURE 5.34 Temperature ranges for heat treating plain-carbon steels, as indicated on the iron-iron carbide phase diagram.

5.11.4 Annealing

Annealing is a general term used to describe the restoration of a cold-worked or heat-treated part to its original properties, so as to increase ductility (hence formability), reduce hardness and strength, or modify its microstructure. Annealing is also used to relieve residual stresses (Section 2.10) in a manufactured part for improved machinability and dimensional stability. The term *annealing* also applies to thermal treatment of glasses (Section 11.11.2) and weldments (Chapter 12).

The annealing process typically involves the following sequence: (1) heating the workpiece to a specific range of temperature, (2) holding it at that temperature for a period of time (soaking), and (3) cooling it slowly. The process may be carried out in an inert or controlled atmosphere, or is performed at low temperatures to prevent or minimize surface oxidation. Annealing temperatures may be higher than the recrystallization temperature, depending on the degree of cold work (hence *stored energy;* see Section 2.12.1). The recrystallization temperature for copper, for example, ranges between 200° and 300°C (400° and 600°F), whereas the annealing temperature required to fully recover the original properties ranges from 260° to 650°C (500° to 1200°F), depending on the degree of prior cold work.

Full annealing is a term used for annealing ferrous alloys, generally low- and medium-carbon steels. The steel is heated to above A_1 or A_3 (Fig. 5.34), and cooling takes place slowly, say, 10°C (20°F) per hour, in a furnace after it is turned off. The structure obtained in full annealing is coarse pearlite, which is soft and ductile and has small uniform grains. Excessive softness in the annealing of steels can be avoided if the entire cooling cycle is carried out in still air. This process is called **normalizing**, in which the part is heated to a temperature of A_3 or A_{cm} to transform the structure to austenite. It results in somewhat higher strength and hardness and lower ductility than in full annealing. The structure obtained is fine pearlite with small uniform grains. Normalizing is generally done to refine the grain structure, obtain uniform structure (*homogenization*), decrease residual stresses, and improve machinability.

Process annealing. In *process annealing* (also called *intermediate annealing, sub-critical annealing,* or *in-process annealing*), the workpiece is annealed to restore its ductility, a portion or all of which may have been exhausted by work hardening during prior cold working. In this way, the part can be worked further into the final desired shape. If the temperature is high and/or the time of annealing is long, grain growth may result, with adverse effects on the formability of annealed parts.

Stress-relief annealing. Residual stresses may have been induced during forming, machining, or other shaping processes or are caused by volume changes during phase transformations. To reduce or eliminate these stresses, a workpiece is generally subjected to *stress-relief annealing*, or simply **stress relieving.** The temperature and time required for this process depend on the material and the magnitude of residual stresses present. For steels, the part is heated to below A_1 in Fig. 5.34, thus avoiding phase transformations. Slow cooling, such as in still air, is generally employed. Stress relieving promotes dimensional stability in situations where subsequent relaxing of residual stresses may cause part distortion over a period of time in service. It also reduces the tendency for stress-corrosion cracking (Section 3.8.2).

5.11.5 Tempering

If steels are hardened by heat treatment, *tempering* is used in order to reduce brittleness, increase ductility and toughness, and reduce residual stresses. In tempering, the steel is heated to a specific temperature, depending on composition, and cooled at a prescribed rate. Alloy steels may undergo **temper embrittlement,** caused by the segregation of impurities along the grain boundaries at temperatures between 480° and 590°C (900° and 1100°F). The term *tempering* is also used for glasses (Section 11.11.2).

In **austempering,** the heated steel is quenched from the austenitizing temperature rapidly enough to avoid formation of ferrite or pearlite. It is held at a certain temperature until isothermal transformation from austenite to bainite is complete. It is then cooled to room temperature (usually in still air) at a moderate rate to avoid thermal gradients within the part. The quenching medium most commonly used is molten salt, and at temperatures ranging from 160° to 750°C (320° to 1380°F).

Austempering is often substituted for conventional quenching and tempering, either to (a) reduce the tendency for cracking and distortion during quenching, or (b) improve ductility and toughness while maintaining hardness. Because of the relatively short cycle time, austempering is economical for many applications. In *modified austempering*, a mixed structure of pearlite and bainite is obtained. The best example of this practice is **patenting,** which provides high ductility and moderately high strength, such as patented wire used in the wire industry (Section 6.5.3).

In **martempering** (**marquenching**), steel or cast iron is quenched from the austenitizing temperature into a hot fluid medium (such as hot oil or molten salt). It is held at that temperature until the temperature is uniform throughout the part and then cooled at a moderate rate (such as in air) to avoid temperature gradients within the part. The part is then tempered, because the structure thus obtained is primarily untempered martensite and is not suitable for most applications. Martempered steels have lower tendency to crack, distort, or develop residual stresses during heat treatment. A process that is suitable for steels with lower hardenability is *modified martempering*, in which the quenching temperature is lower and hence the cooling rate is higher.

In **ausforming,** also called **thermomechanical processing,** the steel is formed into desired shapes within controlled ranges of temperature and time to avoid formation of nonmartensitic transformation products. The part is then cooled at various rates to obtain the desired microstructures. Ausformed parts have superior mechanical properties.

5.11.6 Cryogenic treatment

In *cryogenic tempering,* the temperature of steel is lowered from room temperature to −180°C (−300°F) at a rate of as low as 2°C per minute in order to avoid thermal shock. The part is then maintained at this temperature for 24 to 36 hours. At this temperature, the conversion of austenite to martensite occurs slowly but almost completely, compared to typically only 50 to 90% in conventional quenching. As a result, additional precipitates of carbon (with chromium, tungsten, and other elements) form, residual stresses are relieved, and the material has a refined grain structure. After the 24- to 36-hour "soak" time, the parts are tempered to stabilize the martensite.

Cryogenically treated steels have higher hardness and wear resistance than untreated steels. For example, the wear resistance of D-2 tool steels (see Section 3.10.3) can increase by over 800% after cryogenic treatment, although most tool steels show a 100% to 200% increase in tool life (see Section 8.3). Applications of cryogenic treatment include tools and dies, dental instruments, aerospace materials, golf club heads, and gun barrels.

5.11.7 Design for heat treating

In addition to the metallurgical factors described above, successful heat treating involves design considerations so as to avoid problems such as cracking, warping, and development of nonuniform properties throughout the part. The cooling rate during quenching must be uniform, particularly with complex shapes of varying cross sections and thicknesses, to avoid severe temperature gradients within the part, which can lead to thermal stresses and cause cracking, residual stresses, and stress-corrosion cracking.

As a general guideline for part design for heat treating, (a) parts should have as nearly uniform thicknesses as possible, or the transition between regions of different thicknesses should be smooth; (b) internal or external sharp corners should be avoided; (c) parts with holes, grooves, keyways, splines, and unsymmetrical shapes may be difficult to heat treat because they may crack during quenching; (d) large surfaces with thin cross sections are likely to warp; and (e) hot forgings and hot-rolled products may have a *decarburized skin* and thus may not properly respond to heat treatment.

5.11.8 Cleaning, finishing, and inspecting castings

After solidification and removal from the mold or die, castings are generally subjected to several additional processes. In sand casting, the casting is shaken out of its mold and the sand and oxide layers adhering to the castings are removed by vibration or by sand blasting. Castings may also be cleaned electrochemically or by pickling with chemicals to remove surface oxides, which would adversely their machinability (Section 8.5). *Finishing operations* for castings may include straightening or forging with dies, and machining or grinding to obtain final dimensions.

Several methods are available for inspection of castings to determine quality and the presence of any defects. Castings can be inspected visually or optically for surface defects. Subsurface and internal defects are investigated using the nondestructive techniques described in Section 4.8. Test specimens are removed from various sections of a casting and tested for strength, ductility, and other mechanical properties, and to determine the presence and location of any internal defects.

Pressure tightness of cast components (such as valves, pumps, and pipes) is usually determined by sealing the openings in the casting and pressurizing it with water, oil, or air and inspecting for leaks.

5.12 | Design Considerations

As in all other engineering practice and manufacturing operations, certain guidelines and design principles pertaining to casting have been developed over many years. Although these principles were established primarily through practical experience,

TABLE 5.8

Casting Processes, and Their Advantages and Limitations		
Process	Advantages	Limitations
Sand	Almost any metal is cast; no limit to size, shape, or weight; low tooling cost.	Some finishing required; somewhat coarse finish; wide tolerances.
Shell mold	Good dimensional accuracy and surface finish; high production rate.	Part size limited; expensive patterns and equipment required.
Expendable pattern	Most metals cast with no limit to size; complex shapes.	Patterns have low strength and can be costly for low quantities.
Plaster mold	Intricate shapes; good dimensional accuracy and finish; low porosity.	Limited to nonferrous metals; limited size and volume of production; mold making time relatively long.
Ceramic mold	Intricate shapes; close tolerance parts; good surface finish.	Limited size.
Investment	Intricate shapes; excellent surface finish and accuracy; almost any metal cast.	Part size limited; expensive patterns, molds, and labor.
Permanent mold	Good surface finish and dimensional accuracy; low porosity; high production rate.	High mold cost; limited shape and intricacy; not suitable for high-melting-point metals.
Die	Excellent dimensional accuracy and surface finish; high production rate.	Die cost is high; part size limited; usually limited to nonferrous metals; long lead time.
Centrifugal	Large cylindrical parts with good quality; high production rate.	Equipment is expensive; part shape limited.

analytical methods and computer-aided design and manufacturing techniques (Chapter 15) have come into wide use, improving productivity and the quality of castings. Moreover, careful design can result in significant cost savings. Some of the advantages and limitations of casting processes that impact design are given in Table 5.8.

5.12.1 Defects in castings

Depending on casting design and the practices employed, several defects can develop in castings. Because different names have been used to describe the same defect, the International Committee of Foundry Technical Associations has developed standardized nomenclature consisting of seven basic categories of casting defects.

1. **Metallic projections,** consisting of fins, flash, rough surfaces, or massive projections such as swells.

2. **Cavities,** consisting of rounded or rough internal or exposed cavities, including blowholes, pinholes, and shrinkage cavities (see *porosity* on the next page and Fig. 5.35).

3. **Discontinuities,** such as cracks, cold or hot tearing, and cold shuts. If the solidifying metal is constrained from shrinking freely, cracking and tearing can occur. Coarse grains and the presence of low-melting segregates along the grain boundaries (intergranular) increase the tendency for hot tearing. *Cold shut* is an interface in a casting that lacks complete fusion because of the meeting of two streams of partially solidified metal.

4. **Defective surface,** such as surface folds, laps, scars, adhering sand layers, and oxide scale.

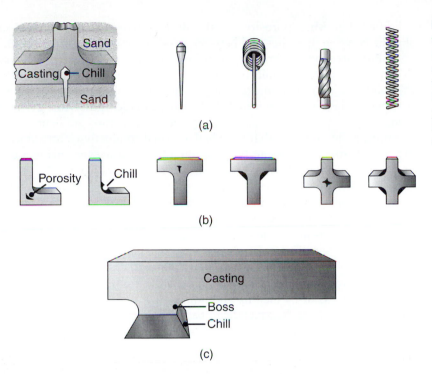

(a)

(b)

(c)

FIGURE 5.35 Various types of (a) internal and (b) external chills (dark areas at corners), used in castings to eliminate porosity caused by shrinkage. Chills are placed in regions where there is a larger volume of metal, as shown in (c).

5. **Incomplete casting,** such as misruns (due to premature solidification), insufficient volume of metal poured, molten metal being at too low a temperature or pouring the metal too slowly, and runout (due to loss of metal from the mold after pouring).

6. **Incorrect dimensions or shape,** owing to factors such as improper shrinkage allowance, pattern mounting error, uneven contraction, deformed pattern, or warped casting.

7. **Inclusions,** which form during melting, solidification, and molding. Inclusions may form (1) during melting because of reaction of the molten metal with the environment (usually oxygen) or the crucible material, (2) during chemical reactions between components in the molten metal, (3) due to slags and other foreign material entrapped in the molten metal, (4) as a result of reactions between the metal and the mold material, and (5) due to spalling of the mold and core surfaces. All of these indicate the importance of maintaining melt quality and continuously monitoring the conditions of the molds.

   Generally nonmetallic, inclusions are regarded as harmful because they act as stress raisers and thus reduce the strength of the casting. Also, for example, hard inclusions (spots) in a casting tend to chip or break tools during subsequent machining operations. Inclusions can be filtered out during processing of the molten metal.

Porosity. *Porosity* is detrimental to the ductility of a casting and its surface finish and makes it permeable, thus affecting pressure tightness of a cast pressure vessel. Porosity in a casting may be caused either by *shrinkage* or *trapped gases*, or both. Porosity due to shrinkage is explained by the fact that thin sections in a casting solidify sooner than thick sections. As a result, molten metal cannot enter into the thicker regions where the surfaces have already solidified. This leads to porosity in the thicker section because the metal has to contract but is prevented from doing so by the solidified skin. **Microporosity** can also develop when the liquid metal solidifies and shrinks between dendrites and between dendrite branches (see Fig. 5.8).

Porosity due to shrinkage can be reduced or eliminated by various means. (a) Internal or external **chills** typically are used in sand casting (Fig. 5.35) to increase the solidification rate in thicker regions. Internal chills are usually made of the same material as the castings; external chills may be made of the same material or may be iron, copper, or graphite. (b) Making the temperature gradient steep, using, for example, mold materials that have high thermal conductivity. (c) Subjecting the casting to **hot isostatic pressing** (see Section 11.3.3); however, this is a costly method and is used mainly for critical components, as in aircraft parts.

Porosity due to gases is explained by the fact that liquid metals have much greater **solubility** for gases than do solids (Fig. 5.36). When a metal begins to solidify, the dissolved gases are expelled from the solution, causing porosity. Gases may also be result of reactions of the molten metal with the mold materials. Gases either accumulate in regions of existing porosity, such as in interdendritic areas, or they cause microporosity in the casting, particularly in cast iron, aluminum, and copper.

Dissolved gases may be removed from the molten metal by flushing or purging with an inert gas or by melting and pouring the metal in a vacuum. If the dissolved gas is oxygen, the molten metal can be *deoxidized*. Steel is usually deoxidized with aluminum or silicon, and copper-based alloys with copper alloy containing 15% phosphorus.

Whether microporosity is a result of shrinkage or is caused by gases may be difficult to determine. If the porosity is spherical and has smooth walls (much like the shiny surfaces of holes in Swiss cheese), it is generally from trapped gases. If the walls are rough and angular, porosity is likely from shrinkage between dendrites. Gross porosity or *macroporosity* is from shrinkage and is generally called **shrinkage cavities.**

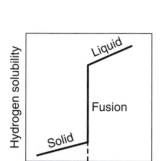

FIGURE 5.36 Solubility of hydrogen in aluminum. Note the sharp decrease in solubility as the molten metal begins to solidify.

5.12.2 General design considerations

There are two types of design issues in casting: (a) geometric features, tolerances, etc., that should be incorporated into the part; and (b) mold features that are needed to produce the desired casting. Robust design of castings usually involves the following steps:

1. Design the part so that the shape is easily cast. Several important design considerations are described throughout this chapter.

2. Select a casting process and a material suitable for the part, size, dimensional accuracy, surface texture, and mechanical properties.

3. Locate the parting line of the mold or die.

4. Locate and design the gating system to allow uniform feeding of the mold cavity with molten metal, including risers, sprue, and screens.

5. Ensure that proper controls and good practices are in place.

Design of cast parts. The following considerations are important in designing castings:

1. **Corners, angles, and section thickness.** Sharp corners, angles, and fillets should be avoided as much as possible because they act as stress raisers and may cause cracking and tearing of the metal (as well as of the dies) during solidification. Fillet radii should be selected to reduce stress concentrations and to ensure proper liquid-metal flow during pouring. The radii usually range from 3 to 25 mm ($\frac{1}{8}$ to 1 in.), although smaller radii may be permissible in small

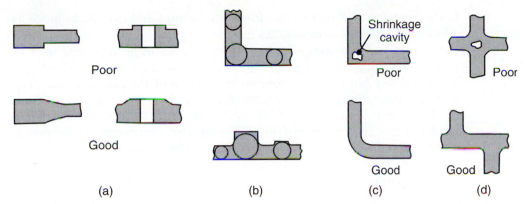

FIGURE 5.37 (a) Suggested design modifications to avoid defects in castings. Note that sharp corners are avoided to reduce stress concentrations. (b, c, d) Examples of designs showing the importance of maintaining uniform cross sections in castings to avoid hot spots and shrinkage cavities.

castings and in specific applications. On the other hand, if the fillet radii are too large, the volume of the material in those regions is also large and, consequently, the rate of cooling is lower.

Section changes in castings should be smoothly blended into each other. The location of the largest circle that can be inscribed in a particular region is critical so far as shrinkage cavities are concerned (Figs. 5.37b to d). Because the cooling rate in regions with larger circles is lower, they result in **hot spots**. These regions can then lead to **shrinkage cavities** and **porosity** (Fig. 5.37c and d). Cavities at hot spots can be eliminated using small cores. It is important to try to maintain uniform cross sections and wall thicknesses throughout the casting so as to avoid or minimize shrinkage cavities. Although they increase the cost of production, *metal chills* or *paddings* can eliminate or minimize hot spots (see Fig. 5.35).

2. **Flat areas.** Large flat areas (plain surfaces) should be avoided since they may warp during cooling or develop poor surface finish because of uneven flow of metal during pouring. One of the common techniques is to break up flat surfaces with staggered ribs and serrations.

3. **Shrinkage.** To avoid cracking of the casting during cooling, allowances should be made for shrinkage during solidification. In castings with intersecting ribs, the tensile stresses can be reduced by staggering the ribs or by modifying the intersection geometry. Pattern dimensions should also provide for shrinkage of the metal during solidification and cooling. Allowances for shrinkage, known as **patternmaker's shrinkage allowances**, typically range from about 10 to 20 mm/m ($\frac{1}{8}$ to $\frac{1}{4}$ in./ft).

4. **Draft.** A small draft (taper) is typically provided in sand-mold patterns to enable removal of the pattern without damaging the mold. Drafts generally range from 5 to 15 mm/m ($\frac{1}{16}$ to $\frac{3}{16}$ in./ft). Depending on the quality of the pattern, draft angles usually range from 0.5° to 2°. The draft angles on inside surfaces are typically twice this range; they have to be higher than those for outer surfaces because the casting shrinks inward toward the core.

5. **Dimensional tolerances.** Tolerances depend on the particular casting process, casting size, and type of pattern used. Tolerances should be as wide as possible, within the limits of good part performance, as otherwise the cost of the casting increases. In commercial practice, dimensional tolerances are usually

in the range of ± 0.8 mm ($\frac{1}{32}$ in.) for small castings and may be as much as ± 6 mm (0.25 in.) for large castings.

6. **Lettering and markings.** It is common practice to include some form of part identification in castings, such as lettering, numbers, or corporate logos. These features can be depressions on the surface of the casting or they can protrude from the surface. For example, in sand casting, a pattern plate is produced by machining on a CNC mill (Section 8.10), and it is simpler to machine letters into the pattern surface, leading to sunken letters. In die casting, on the other hand, it is simpler to machine letters into the die, leading to protruding letters.

7. **Finishing operations.** It is important to consider the subsequent machining and finishing operations that often have to take place. For example, if a hole is to be drilled in a casting, it is better to locate the hole on a flat surface than on a curved surface (to prevent the drill from wandering). An even better design would incorporate a small dimple as a starting point for the drilling operations. Castings should also include features that allow them to be easily clamped into machine tools for subsequent finishing operations.

Selecting the casting process. Casting process attributes that help in process selection are outlined in Table 5.8. As described throughout this text, process selection cannot be separated from economic considerations; they are therefore discussed in Section 5.13.

Parting line location. The location of the parting line is important because it influences mold design, ease of molding, number and shape of cores required, method of support, and the gating system, as well as the number of cores needed. A casting should be oriented in a mold such that its larger portion is relatively low and that the height of the casting is minimized. Orientation of the casting also determines the distribution of porosity, because in casting aluminum, for example, porosity can form and will float upward (due to buoyancy) and the casting will have a higher porosity on the top regions. (Recall also from Fig. 5.36 that hydrogen is soluble in liquid metal, but not in solid.) Thus, critical surfaces should be oriented so that they face downward.

A properly oriented part can then have the parting line specified (see Fig. 5.10). In general, the parting line should be (a) along a flat plane, rather than be contoured; (b) at the corners or edges of castings, rather than on flat surfaces in the middle of the casting, so that the *flash* at the parting line (material squeezing out between the two halves of the mold) will not be as visible; (c) placed as low as possible relative to the casting for less-dense metals such as aluminum alloys, and located at around the midheight for denser metals, such as steels. In sand casting, it is typical that the runners, gates, and sprue well are placed in the drag on the parting line.

Gate design and location. Gates are the connections between the runners and the part to be cast. Some of the considerations in designing gating systems are

1. Multiple gates are often preferable and are necessary for large parts. Multiple gates have the benefits of allowing lower pouring temperature and reduce the temperature gradients in the casting.

2. Gates should feed into thick sections of castings.

3. A fillet should be used where a gate meets a casting; this feature produces less turbulence than abrupt junctions.

4. The gate closest to the sprue should be placed sufficiently far away so that it can be easily removed; this distance may be as small as a few millimeters for small castings and up to 500 mm for large parts.

5. The minimum gate length should be three to five times the gate diameter, depending on the metal being cast. Its cross section should be large enough to allow filling of the mold cavity and should be smaller than the runner cross section.

6. Curved gates should be avoided, but when necessary, a straight section in the gate should be located immediately adjacent to the casting.

Runner design. The runner is a horizontal channel that receives molten metal from the sprue and delivers it to the gates. A single runner is used for simple parts, but two-runner systems can be specified for more complicated castings. Runners are also used to trap dross (a mixture of oxide and metal that forms on the surface of metals) and keep it from entering the gates and the mold cavity. Commonly, dross traps are placed at the ends of runners, and the runner projects above the gates to ensure that the metal in the gates is tapped from below the surface.

Designing other mold features. The main goal in designing a *sprue* (described in Section 5.4.1) is to achieve the required metal flow rates while preventing aspiration or excessive dross formation. Flow rates are determined such that turbulence is avoided, but the mold is filled quickly compared to the solidification time required. A *pouring basin* can be used to ensure that the metal flow into the sprue is uninterrupted. Also, if molten metal is maintained in the pouring basin during pouring, then the dross will float and will not enter the mold cavity. *Filters* are used to trap large contaminants, and they also serve to slow the metal velocity and make the flow more laminar. *Chills* can be used to speed solidification of metal in a particular region of a casting.

Establishing good practices. It has been commonly observed that a given mold design can produce acceptable parts as well as defective ones and will rarely produce only good parts or only defective ones. Quality control procedures are thus necessary to check for defective castings. Some of the common concerns are the following:

1. Starting with a high-quality molten metal is essential for producing superior castings. Pouring temperature, metal chemistry, gas entrainment, and handling procedures can all affect the quality of the metal being poured into a mold.

2. The pouring of metal should not be interrupted because it can lead to dross entrainment and turbulence. The meniscus of the molten metal in the mold cavity should experience a continuous, uninterrupted, upward advance.

3. Variation of the cooling rates within the casting can cause residual stresses. Stress relieving (Section 5.11) may thus be necessary to avoid distortion of castings in critical applications.

5.12.3 Design principles for expendable-mold casting

Expendable-mold casting processes have certain specific design considerations, which are mainly attributed to the mold material, part size, and the casting method. Important design considerations are the following.

1. **Mold layout.** The features in the mold must be placed logically and compactly, including gates when necessary. Solidification begins at one end of the mold and progresses in a uniform front across the casting, with risers solidifying last. Traditionally, mold layout has been based on experience, but more recently, commercial computer programs have become available (see also Section 5.12.5). Based on finite-difference algorithms, these techniques allow simulation of mold filling and rapid evaluation of mold layouts.

2. **Riser design.** A major concern in the design of castings is the size and placement of risers (see Fig. 5.10). Risers are essential in affecting the solidification front progression across a casting and are an important feature in mold layout described previously. Blind risers are good design features and maintain heat longer than open risers. Risers are designed according to five basic rules:

   a. The metal in the riser must not solidify before the casting does; this is usually done by avoiding the use of small risers and by using cylindrical risers with small aspect ratios (small ratios of height to cross section). Spherical risers are the most efficient shape but are difficult to work with.

   b. The riser volume must be large enough to provide sufficient liquid metal to compensate for shrinkage in the casting.

   c. Junctions between the casting and feeder should not develop a *hot spot* where shrinkage porosity can occur.

   d. Risers must be placed so that liquid metal can be delivered to locations where it is most needed.

   e. There must be sufficient pressure to drive the liquid metal into locations in the mold where it is needed; risers are therefore not as useful for metals with low density (such as aluminum alloys) as for those with a higher density (such as steel and cast irons).

3. **Machining allowance.** Because most expendable-mold castings require some additional finishing operations (such as machining and grinding), allowances should be made in casting design for these operations. Machining allowances, which are included in pattern dimensions, depend on the type of casting and increase with the size and section thickness of castings. Allowances usually range from about 2 to 5 mm (0.1 to 0.2 in.) for small castings to more than 25 mm (1 in.) for large castings.

5.12.4 Design principles for permanent-mold casting

Typical design guidelines and examples for permanent-mold casting are shown schematically for die casting in Fig. 5.38. Note that the cross sections have been reduced in order to reduce the solidification time and save material. Special considerations are also involved in designing tooling for die casting. Although designs may be modified to eliminate the draft for better dimensional accuracy, a draft angle of 0.5° or even 0.25° is usually required; otherwise galling (localized seizure or sticking of material) may occur between the part and the dies and cause part distortion.

Die-cast parts are nearly net shaped, typically requiring only the removal of gates and minor trimming to remove flashing and other minor defects. The surface finish and dimensional accuracy of die-cast parts are very good (see Table 5.2), and in general they do not require a machining allowance.

5.12.5 Computer modeling of casting processes

Because casting processes involve complex interactions among material and process variables, a quantitative study of these interactions is essential to the proper design and production of high-quality castings. In the past, such studies have presented major difficulties. Rapid advances in computers and modeling techniques have, however, led to important innovations in modeling various aspects of casting, including fluid flow, heat transfer, and microstructures developed during solidification and under various casting process conditions.

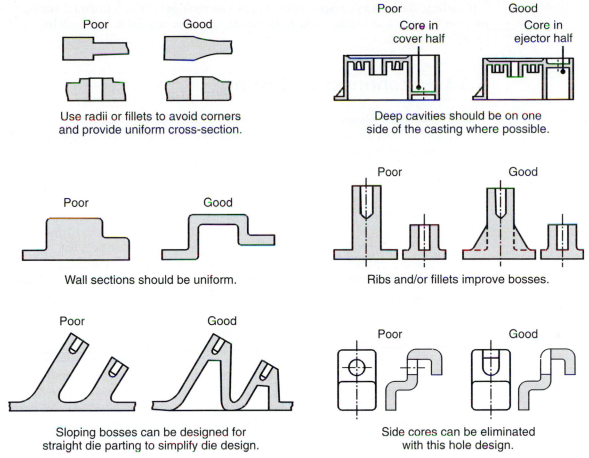

FIGURE 5.38 Suggested design modifications to avoid defects in castings.
Source: Courtesy of The North American Die Casting Association.

Modeling of *fluid flow* is based on Bernoulli's and the continuity equations (as described in Section 5.4). It predicts the behavior of the metal during pouring into the gating system and its travel into the mold cavity, as well as velocity and pressure distributions in the system. Much progress also has been made in modeling *heat transfer* in casting. These studies include investigation of the coupling of fluid flow and heat transfer and the effects of surface conditions, thermal properties of the materials involved, and natural and forced convection on cooling. Note that the surface conditions vary during solidification, as a layer of air develops between the casting and the mold wall as a result of shrinkage. Similar studies are continuing on modeling the development of *microstructures* in casting. These studies encompass heat flow, temperature gradients, nucleation and growth of crystals, formation of dendritic and equiaxed structures, impingement of grains on each other, and movement of the liquid-solid interface during solidification.

Such models are now capable of predicting, for example, the width of the mushy zone (see Fig. 5.6) during solidification and the grain size in castings. Similarly, the capability to calculate isotherms gives insight into possible hot spots and subsequent development of shrinkage cavities. With the availability of user-friendly computers and advances in computer-aided design and manufacturing (see Chapter 14), modeling techniques are becoming easier to implement. The benefits are increased productivity, improved quality, easier planning and cost estimating, and quicker response to design changes. Several commercial software programs, such as Procast® and Magmasoft, are now available on modeling and casting processes.

5.13 | Economics of Casting

When reviewing various casting operations, it was noted that some processes require more labor than others, some require expensive dies and machinery, and some take a great deal of time to complete. Each of these important factors, outlined in Table 5.1, affects to various degrees the overall cost of a casting operation. As described in greater detail in Chapter 15, the total cost of a product involves the costs of materials, labor, tooling, and equipment. Preparations for casting a product include making molds and dies that require materials, machinery, time, and effort, which all contribute to costs. Although relatively little cost is involved in molds for sand casting, die-casting dies require expensive materials and a great deal of machining and preparation. Facilities with furnaces and related equipment are also required for melting and pouring the molten metal into the molds or dies and their costs depend on the level of automation desired. Finally, costs are also involved in cleaning and inspecting castings.

The amount of labor required in casting operations can vary considerably, depending on the particular process and level of automation. Investment casting, for example, requires a great deal of labor because of the large number of steps involved in this operation. Conversely, operations such as highly automated die casting can maintain high production rates with little labor required.

The cost of equipment per casting (*unit cost*) decreases as the number of parts cast increases (Fig. 5.39); thus sustained high production rates can justify the high cost of dies and machinery. However, if demand is relatively small, the cost per casting increases rapidly; it then becomes more economical to manufacture the parts by sand casting or by other manufacturing processes. Note that Fig. 5.39 can also include other casting processes suitable for making the same part. The two processes compared (sand and die casting) produce castings with significantly different dimensional and surface-finish characteristics; thus not all manufacturing decisions should be based purely on economic considerations. In fact, parts can usually be made by more than one or two processes (see, for example, Fig. 1.6). Thus the final decision depends on both economic as well as technical considerations. The competitive aspects of manufacturing processes are discussed further in greater detail in Chapter 15.

FIGURE 5.39 Economic comparison of making a part by two different casting processes. Note that because of the high cost of equipment, die casting is economical mainly for large production runs. *Source:* The North American Die Casting Association.

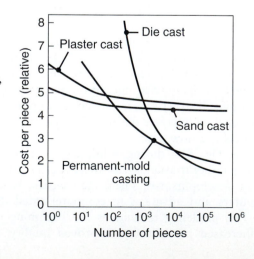

CASE STUDY | Lost-Foam Casting of Engine Blocks

One of the most important parts in an internal combustion engine is the engine block. The engine block forms the enclosure for the engine and provides the basic structure that encloses the pistons and cylinders, and encounters significant pressure during operation.

Industry trends have focused upon high-quality, low-cost lightweight designs; additional economic benefits can be attained through casting more complex geometries and by incorporating multiple components into one part. Recognizing that evaporative pattern (lost-foam) casting can simultaneously satisfy all of these requirements, Mercury Castings built a lost-foam casting line to produce aluminum engine blocks and cylinder heads.

One example of a part produced through lost-foam casting is a 60-hp three-cylinder engine block used for marine applications and illustrated in Fig. 5.40a. Previously manufactured as eight separate die castings, the block was converted to a single, 22-lb lost-foam casting with a weight and cost savings of 2 lb and $25 on each block, respectively. Lost-foam casting also allowed consolidation of the engine's cylinder head and exhaust and cooling systems into the block and eliminated the associated machining and fasteners required in sand-cast or die-cast designs. In addition, since the pattern contained holes and these could be cast without the use of cores, numerous drilling operations were eliminated.

Mercury Marine also was in the midst of developing a new V6 engine utilizing a new corrosion-resistant aluminum alloy with increased wear resistance. This engine design also required a cylinder block and head integration. It features hollow sections for water jacket cooling that could not be cored-out in die casting or semipermanent mold (the processes used for its other V6 blocks). Based on the success the foundry had with the three-cylinder lost-foam block, engineers applied lost-foam casting for the V6 die block. The new engine block involves only one casting, which is lighter and cheaper than the previous designs. Produced with an integrated cylinder head and exhaust and cooling system, this

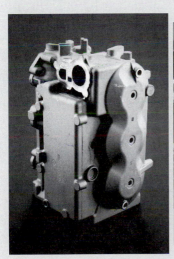

(a)

(b)

FIGURE 5.40   (a) An engine block for a 60-hp three-cylinder marine engine, produced by the lost-foam casting process, and (b) a robot pouring molten aluminum into a flask containing a polystyrene pattern. In the pressurized lost-foam process, the flask is then pressurized to 150 psi (1000 kPa). *Source:* Courtesy of Mercury Marine.

component is cast hollow to create more efficient water jacket cooling of the engine during operation.

The company has also developed a pressurized lost-foam process. A foam pattern is made, placed in a flask, and surrounded by sand. The flask is then placed in a pressure vessel, where a robot pours molten aluminum onto the polystyrene pattern (Fig. 5.40b). A lid on the pressure vessel is closed and a pressure of 150 psi is applied to the casting until it solidifies (around 15 minutes). The result is a casting with better dimensional accuracy, lower porosity, and improved strength as compared to conventional lost-foam casting.

Source: Courtesy of Mercury Marine.

SUMMARY

- Metal casting is among the oldest and most common manufacturing processes. Solidification of pure metals takes place at a clearly defined constant temperature, whereas solidification of alloys occurs over a range of temperatures, depending on their composition. Phase diagrams are important for identifying the solidification point or points for metals and alloys. (Sections 5.1 and 5.2)

- The composition and cooling rate of the melt affect the size and shape of grains and dendrites in the solidifying metal and alloy and thus influence the properties of the casting. (Section 5.3)

- Most metals shrink during solidification, although gray cast iron and a few others actually expand. The resulting loss of dimensional accuracy and sometimes cracking are difficulties that can arise during solidification and cooling. Several basic categories of casting defects have been identified. (Section 5.3)

- In casting, the molten metal or alloy may flow through a variety of passages, including pouring basins, sprues, runners, risers, and gating systems, before reaching the mold cavity. Bernoulli's theorem, the continuity law, and the Reynolds number are the analytical tools used in designing an appropriate system and in eliminating defects associated with fluid flow. Heat transfer affects fluid flow and solidification time in casting. Solidification time is a function of the volume of a casting and its surface area (Chvorinov's rule). (Section 5.4)

- Melting practices have a direct effect on the quality of castings. Factors that affect melting include: (a) inorganic compounds or fluxes that are added to the molten metal to remove dissolved gases and various impurities, (b) the type of furnace used, and (c) foundry operations, which include pattern and mold making, pouring of the melt, removal of castings from molds, cleaning, heat treatment, and inspection. (Section 5.5)

- Several ferrous and nonferrous casting alloys are available, with a wide range of properties, casting characteristics, and applications. Because castings are often designed and produced to be assembled with other mechanical components and into structures, various other considerations, such as weldability, machinability, and surface conditions, are also important. (Section 5.6)

- The traditional ingot-casting process has largely been replaced for many applications by continuous-casting methods for both ferrous and nonferrous metals. (Section 5.7)

- Casting processes are generally classified as either expendable-mold or permanent-mold casting. The most common expendable-mold methods are sand, shell-mold, plaster, ceramic-mold, and investment casting. Permanent-mold methods include slush casting, pressure casting, and die casting. Compared to permanent-mold casting, expendable-mold casting usually involves lower mold and equipment costs, but parts have lower dimensional accuracy. (Sections 5.8–5.10)

- Castings, as well as wrought parts made by other manufacturing processes, may be subjected to subsequent heat treatment operations to enhance their various properties and service life. Several transformations take place in microstructures that have widely varying characteristics and properties. Important mechanisms of hardening and strengthening involve thermal treatments, including quenching and precipitation hardening. (Section 5.11)

- General principles have been established to aid designers to produce castings that are free from defects and meet dimensional tolerance and service requirements. Because of the large number of variables involved, close control of all parameters is essential, particularly those related to the nature of liquid metal flow into the molds and dies as well as the rate of cooling in different regions of castings (Section 5.12)

- Within the limits of good performance, the economic aspects of casting are just as important as the technical considerations. Factors affecting the overall cost include the cost of materials, molds, dies, equipment, and labor, each of which varies with the particular casting operation. An important parameter is the cost per casting, which, for large production runs, can justify major expenditures that are typical of automated machinery and operations. (Section 5.13)

SUMMARY OF EQUATIONS

Bernoulli's theorem: $h + \dfrac{p}{\rho g} + \dfrac{v^2}{2g} = \text{Constant}$

Sprue contour: $\dfrac{A_1}{A_2} = \sqrt{\dfrac{h_2}{h_1}}$

Reynolds number: $\text{Re} = \dfrac{vD\rho}{\eta}$

Chvorinov's rule: $\text{Solidification time} = C\left(\dfrac{\text{Volume}}{\text{Surface area}}\right)^n$

Continuity equation: $Q = A_1 v_1 = A_2 v_2$

Metal velocity after gate: $v = c\sqrt{2gh}$

BIBLIOGRAPHY

Abrasion-Resistant Cast Iron Handbook, American Foundry Society, 2000.

Alexiades, V., *Mathematical Modeling of Melting and Freezing Processes*, Hemisphere, 1993.

Allsop, D.F., and Kennedy, D., *Pressure Die Casting, Part II: The Technology of the Casting and the Die*, Pergamon, 1983.

An Introduction to Die Casting, American Die Casting Institute, 1981.

ASM Handbook, Vol. 3: *Alloy Phase Diagrams*, ASM International, 1992.

ASM Handbook, Vol. 4: *Heat Treating*, ASM International, 1991.

ASM Handbook, Vol. 15: *Casting*, ASM International, 1988.

ASM Specialty Handbook: Cast Irons, ASM International, 1996.

Bradley, E.F., *High-Performance Castings: A Technical Guide*, Edison Welding Institute, 1989.

Campbell, J., *Castings*, 2d ed., Butterworth-Heinemann, 2003.

Case Hardening of Steel, ASM International, 1987.

Casting, in Tool and Manufacturing Engineers Handbook, Volume II: Forming, Society of Manufacturing Engineers, 1984.

Clegg, A.J., *Precision Casting Processes*, Pergamon, 1991.

Davis, J.R. (ed.), *Cast Irons*, ASM International, 1996.

Investment Casting Handbook, Investment Casting Institute, 1997.

Johns, R., *Casting Design,* American Foundrymen's Society, 1987.

Karlsson, L. (ed.), *Modeling in Welding, Hot Powder Forming and Casting,* ASM International, 1997.

Kaye, A., and Street, A.C., *Die Casting Metallurgy,* Butterworth, 1982.

Krauss, G., *Steels: Heat Treatment and Processing Principles,* ASM International, 1990.

Kurz, W., and Fisher, D.J., *Fundamentals of Solidification,* 4th ed., Trans Tech Pub., 1998.

Liebermann, H.H. (ed.), *Rapidly Solidified Alloys,* Dekker, 1993.

Powell, G.W., Cheng, S.-H., and Mobley, C.E., Jr., *A Fractography Atlas of Casting Alloys,* Battelle Press, 1992.

Rowley, M.T. (ed.), *International Atlas of Casting Defects,* American Foundrymen's Society, 1974.

Steel Castings Handbook, 6th ed., Steel Founders' Society of America, 1995.

Szekely, J., *Fluid Flow Phenomena in Metals Processing,* Academic Press, 1979.

Totten, G.E., and Howes, M.A.H., *Steel Heat Treatment,* Dekker, 1997.

Upton, B., *Pressure Die Casting,* Part 1: *Metals, Machines, Furnaces,* Pergamon, 1982.

Walton, C.F., and Opar, T.J. (eds.), *Iron Castings Handbook,* 3rd ed., Iron Castings Society, 1981.

Wieser, P.P. (ed.), *Steel Castings Handbook,* 6th ed., ASM International, 1995.

Young, K.P., *Semi-solid Processing,* Kluwer, 2000.

QUESTIONS

5.1 Describe the characteristics of (1) an alloy, (2) pearlite, (3) austenite, (4) martensite, and (5) cementite.

5.2 What are the effects of mold materials on fluid flow and heat transfer in casting operations?

5.3 How does the shape of graphite in cast iron affect its properties?

5.4 Explain the difference between short and long freezing ranges. How are they determined? Why are they important?

5.5 We know that pouring molten metal at a high rate into a mold has certain disadvantages. Are there any disadvantages to pouring it very slowly? Explain.

5.6 Why does porosity have detrimental effects on the mechanical properties of castings? Which physical properties are also affected adversely by porosity?

5.7 A spoked hand wheel is to be cast in gray iron. In order to prevent hot tearing of the spokes, would you insulate the spokes or chill them? Explain.

5.8 Which of the following considerations are important for a riser to function properly? (1) Have a surface area larger than that of the part being cast. (2) Be kept open to atmospheric pressure. (3) Solidify first. Explain.

5.9 Explain why the constant C in Eq. (5.9) depends on mold material, metal properties, and temperature.

5.10 Explain why gray iron undergoes expansion, rather than contraction, during solidification.

5.11 How can you tell whether a cavity in a casting is due to porosity or to shrinkage?

5.12 Explain the reasons for hot tearing in castings.

5.13 Would you be concerned about the fact that a portion of an internal chill is left within the casting? What materials do you think chills should be made of, and why?

5.14 Are external chills as effective as internal chills? Explain.

5.15 Do you think early formation of dendrites in a mold can impede the free flow of molten metal into the mold? Explain.

5.16 Is there any difference in the tendency for shrinkage-void formation for metals with short freezing and long freezing ranges, respectively? Explain.

5.17 It has long been observed by foundrymen that low pouring temperatures (that is, low superheat) promote equiaxed grains over columnar grains. Also, equiaxed grains become finer as the pouring temperature decreases. Explain the reasons for these phenomena.

5.18 What are the reasons for the large variety of casting processes that have been developed over the years?

5.19 Why can blind risers be smaller than open-top risers?

5.20 Would you recommend preheating the molds in permanent-mold casting? Also, would you remove the casting soon after it has solidified? Explain.

5.21 In a sand-casting operation, what factors determine the time at which you would remove the casting from the mold?

5.22 Explain why the strength-to-weight ratio of die-cast parts increases with decreasing wall thickness.

5.23 We note that the ductility of some cast alloys is very low (see Fig. 5.13). Do you think this should be a significant concern in engineering applications of castings? Explain.

5.24 The modulus of elasticity of gray iron varies significantly with its type, such as the ASTM class. Explain why.

5.25 List and explain the considerations involved in selecting pattern materials.

5.26 Why is the investment-casting process capable of producing fine surface detail on castings?

5.27 Explain why a casting may have a slightly different shape than the pattern used to make the mold.

5.28 Explain why squeeze casting produces parts with better mechanical properties, dimensional accuracy, and surface finish than expendable-mold processes.

5.29 Why are steels more difficult to cast than cast irons?

5.30 What would you recommend to improve the surface finish in expendable-mold casting processes?

5.31 You have seen that even though die casting produces thin parts, there is a limit to the minimum thickness. Why can't even thinner parts be made by this process?

5.32 What differences, if any, would you expect in the properties of castings made by permanent-mold vs. sand-casting methods?

5.33 Which of the casting processes would be suitable for making small toys in large numbers? Explain.

5.34 Why are allowances provided for in making patterns? What do they depend on?

5.35 Explain the difference in the importance of drafts in green-sand casting vs. permanent-mold casting.

5.36 Make a list of the mold and die materials used in the casting processes described in this chapter. Under each type of material, list the casting processes that are used, and explain why these processes are suitable for that particular mold or die material.

5.37 Explain why carbon is so effective in imparting strength to iron in the form of steel.

5.38 Describe the engineering significance of the existence of a eutectic point in phase diagrams.

5.39 Explain the difference between hardness and hardenability.

5.40 Explain why it may be desirable or necessary for castings to be subjected to various heat treatments.

5.41 Describe the differences between case hardening and through hardening insofar as engineering applications are concerned.

5.42 *Type metal* is a bismuth alloy used to cast type for printing. Explain why bismuth is ideal for this process.

5.43 Do you expect to see larger solidification shrinkage for a material with a bcc crystal structure or fcc? Explain.

5.44 Describe the drawbacks to having a riser that is (a) too large or (b) too small.

5.45 If you were to incorporate lettering on a sand casting, would you make the letters protrude from the surface or recess into the surface? What if the part were to be made by investment casting?

5.46 List and briefly explain the three mechanisms by which metals shrink during casting.

5.47 Explain the significance of the tree in investment casting.

5.48 Sketch the microstructure you would expect for a slab cast through (a) continuous casting, (b) strip casting, and (c) melt spinning.

5.49 The general design recommendations for a well in sand casting are that (a) its diameter should be twice the sprue exit diameter, and (b) the depth should be approximately twice the depth of the runner. Explain the consequences of deviating from these rules.

5.50 Describe the characteristics of thixocasting and rheocasting.

5.51 Sketch the temperature profile you would expect for (a) continuous casting of a billet, (b) sand casting of a cube, and (c) centrifugal casting of a pipe.

5.52 What are the benefits and drawbacks of having a pouring temperature that is much higher than the metal's melting temperature? What are the advantages and disadvantages of having the pouring temperature remain close to the melting temperature?

5.53 What are the benefits and drawbacks of heating the mold in investment casting before pouring in the molten metal?

5.54 Can a chaplet also act as a chill? Explain.

5.55 Rank the casting processes described in this chapter in terms of their solidification rate. For example, which processes extract heat the fastest from a given volume of metal and which is the slowest?

5.56 The heavy regions of parts typically are placed in the drag in sand casting and not in the cope. Explain why.

PROBLEMS

5.57 Referring to Fig. 5.3, estimate the following quantities for a 20% Cu-80% Ni alloy: (1) liquidus temperature, (2) solidus temperature, (3) percentage of nickel in the liquid at 1400°C (2550°F), (4) the major phase at 1400°C, and (5) the ratio of solid to liquid at 1400°C.

5.58 Determine the amount of gamma and alpha phases (see Fig. 5.4b) in a 10-kg, AISI 1060 steel casting as it is being cooled to the following temperatures: (1) 750°C, (2) 728°C, and (3) 726°C.

5.59 A round casting is 0.3 m in diameter and 0.5 m in length. Another casting of the same metal is elliptical in cross section, with a major-to-minor axis ratio of 3, and has the same length and cross sectional area as the round casting. Both pieces are cast under the same conditions. What is the difference in the solidification times of the two castings?

5.60 Derive Eq. (5.7).

5.61 Two halves of a mold (cope and drag) are weighted down to keep them from separating due to the pressure exerted by the molten metal (buoyancy). Consider a solid, spherical steel casting, 9 in. in diameter, that is being produced by sand casting. Each flask (see Fig. 5.10) is 20 in. by 20 in. and 15 in. deep. The parting line is at the middle of the part. Estimate the clamping force required. Assume that the molten metal has a density of 500 lb/ft^3 and that the sand has a density of 100 lb/ft^3.

5.62 Would the position of the parting line in Problem 5.61 influence your answer? Explain.

5.63 Plot the clamping force in Problem 5.61 as a function of increasing diameter of the casting, from 10 in. to 20 in.

5.64 Sketch a graph of specific volume vs. temperature for a metal that shrinks as it cools from the liquid state to room temperature. On the graph, mark the area where shrinkage is compensated for by risers.

5.65 A round casting has the same dimensions as in Problem 5.59. Another casting of the same metal is rectangular in cross section, with a width-to-thickness ratio of 3, and has the same length and cross-sectional area as the round casting. Both pieces are cast under the same conditions. What is the difference in the solidification times of the two castings?

5.66 A 75-mm-thick square plate and a right circular cylinder with a radius of 100 mm and height of 50 mm each have the same volume. If each is to be cast using a cylindrical riser, will each part require the same size riser to ensure proper feeding of the molten metal? Explain.

5.67 Assume that the top of a round sprue has a diameter of 4 in. and is at a height of 12 in. from the runner. Based on Eq. (5.7), plot the profile of the sprue diameter as a function of its height. Assume that the sprue has a diameter of 1 in. at its bottom.

5.68 Estimate the clamping force for a die-casting machine in which the casting is rectangular, with projected dimensions of 75 mm × 150 mm. Would your answer depend on whether or not it is a hot-chamber or cold-chamber process? Explain.

5.69 When designing patterns for casting, patternmakers use special rulers that automatically incorporate solid shrinkage allowances into their designs. Therefore, a 12-in. patternmaker's ruler is longer than a foot. How long should a patternmaker's ruler be for the making of patterns for (1) aluminum castings, (2) malleable cast iron, and (3) high-manganese steel?

5.70 The blank for the spool shown in the accompanying figure is to be sand cast out of A-319, an aluminum casting alloy. Make a sketch of the wooden pattern for this part. Include all necessary allowances for shrinkage and machining.

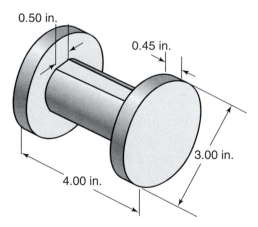

5.71 Repeat Problem 5.70, but assume that the aluminum spool is to be cast using expendable-pattern casting. Explain the important differences between the two patterns.

5.72 In sand casting, it is important that the cope mold half be held down with sufficient force to keep it from floating when the molten metal is poured in. For the casting shown in the figure on the next page, calculate the minimum amount of weight necessary to keep the cope from floating up as the molten metal is poured in. (*Hint:* The buoyancy force exerted by the molten metal on the cope is related to the effective height of the metal head above the cope.)

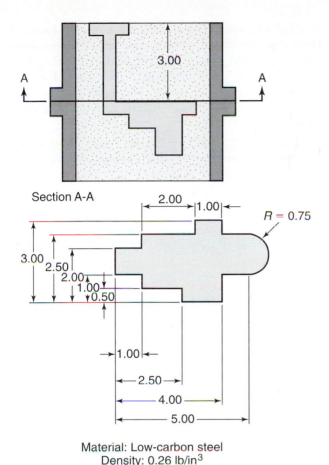

Section A-A

R = 0.75

Material: Low-carbon steel
Density: 0.26 lb/in³
All dimensions in inches

5.73 The optimum shape of a riser is spherical to en-sure that it cools more slowly than the casting it feeds. Spherically shaped risers, however, are difficult to cast. (1) Sketch the shape of a blind riser that is easy to mold but also has the smallest possible surface-area-to-volume ratio. (2) Compare the solidification time of the riser in part (1) to that of a riser shaped like a right cir-cular cylinder. Assume that the volume of each riser is the same, and that for each the height is equal to the diameter (see Example 5.2).

5.74 The part shown in the accompanying figure is a hemispherical shell used as an acetabular (mushroom-shaped) cup in a total hip replacement. Select a casting process for this part and provide a sketch of all patterns or tooling needed if it is to be produced from a cobalt-chrome alloy.

5.75 A cylinder with a height-to-diameter ratio of unity solidifies in four minutes in a sand-casting opera-tion. What is the solidification time if the cylinder height is doubled? What is the time if the diameter is doubled?

5.76 Steel piping is to be produced by centrifugal casting. The length is 12 ft, the diameter is 3 ft, and the thickness is 0.5 in. Using basic equations from dynamics and statics, determine the rotational speed needed to have the centripetal force be 70 times its weight.

5.77 A sprue is 12 in. long and has a diameter of 5 in. at the top, where the metal is poured. The molten metal level in the pouring basin is taken as 3 in. from the top of the sprue for design purposes. If a flow rate of 40 in³/s is to be achieved, what should be the diameter of the bottom of the sprue? Will the sprue aspirate? Explain.

5.78 Small amounts of slag often persist after skim-ming and are introduced into the molten metal flow in casting. Recognizing that the slag is much less dense than the metal, design mold features that will remove small amounts of slag before the metal reaches the mold cavity.

5.79 Pure aluminum is being poured into a sand mold. The metal level in the pouring basin is 10 in. above the metal level in the mold, and the runner is cir-cular with a 0.4-in. diameter. What is the velocity and rate of the flow of the metal into the mold? Is the flow turbulent or laminar?

5.80 For the sprue described in Problem 5.79, what runner diameter is needed to ensure a Reynolds num-ber of 2000? How long will a 20-in³ casting take to fill with such a runner?

5.81 How long would it take for the sprue in Problem 5.79 to feed a casting with a square cross section of 6 in. per side and a height of 4 in.? Assume the sprue is frictionless.

5.82 A rectangular mold with dimensions 100 mm × 200 mm × 400 mm is filled with aluminum with no superheat. Determine the final dimensions of the part as it cools to room temperature. Repeat the analysis for gray cast iron.

5.83 The constant C in Chvorinov's rule is given as 3 s/mm² and is used to produce a cylindrical casting with a diameter of 75 mm and a height of 125 mm. Estimate the time for the casting to fully solidify. The

Dimensions in mm

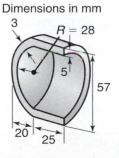

R = 28

mold can be broken safely when the solidified shell is at least 20 mm. Assuming the cylinder cools evenly, how much time must transpire after pouring the molten metal before the mold can be broken?

5.84 If an acceleration of 100 g is necessary to produce a part in true centrifugal casting and the part has an inner diameter of 10 in., a mean outer diameter of 14 in., and a length of 25 ft, what rotational speed is needed?

5.85 A jeweler wishes to produce 20 gold rings in one investment-casting operation. The wax parts are attached to a wax central sprue of a 0.5-in. diameter. The rings are located in four rows, each 0.5 in. from the other on the sprue. The rings require a 0.125-in. diameter and 0.5-in.-long runner to the sprue. Estimate the weight of gold needed to completely fill the rings, runners, and sprues. The specific gravity of gold is 19.3.

5.86 Assume that you are asked to give a quiz to students on the contents of this chapter. Prepare three quantitative problems and three qualitative questions, and supply the answers.

DESIGN

5.87 Design test methods to determine the fluidity of metals in casting (see Section 5.4.2). Make appropriate sketches and explain the important features of each design.

5.88 The accompanying figures indicate various defects and discontinuities in cast products. Review each one and offer design solutions to avoid each.

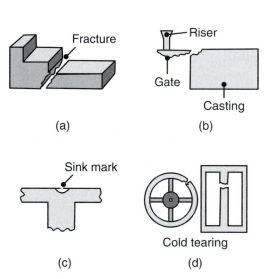

(a) (b)

(c) (d)

5.89 Utilizing the equipment and materials available in a typical kitchen, design an experiment to reproduce results similar to those shown in Fig. 5.12.

5.90 Design a test method to measure the permeability of sand for sand casting.

5.91 Describe the procedures that would be involved in making a bronze statue. Which casting process or processes would be suitable? Why?

5.92 Porosity developed in the boss of a casting is illustrated in the accompanying figure. Show that by simply repositioning the parting line of this casting, this problem can be eliminated.

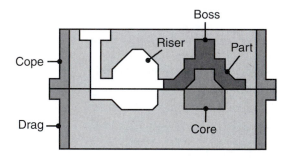

5.93 For the wheel illustrated in the accompanying figure, show how (a) riser placement, (b) core placement, (c) padding, and (d) chills may be used to help feed molten metal and eliminate porosity in the isolated hub boss.

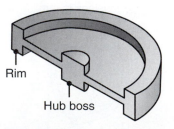

5.94 In the accompanying figure, the original casting design shown in (a) was changed to the design shown in (b). The casting is round, with a vertical axis of symmetry. As a functional part, what advantages do you think the new design has over the old one?

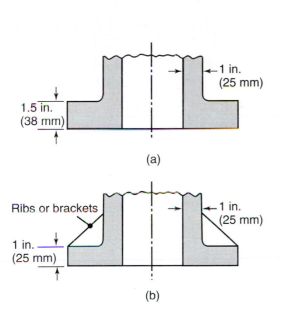

(a)

(b)

5.95 An incorrect and a correct design for casting are shown, respectively, in the accompanying figure. Review the changes made and comment on their advantages.

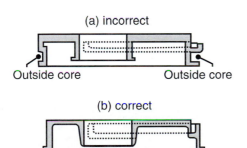

(a) incorrect

Outside core   Outside core

(b) correct

5.96 Three sets of designs for die casting are shown in the accompanying figure. Note the changes made to original die design (number 1 in each case) and comment on the reasons.

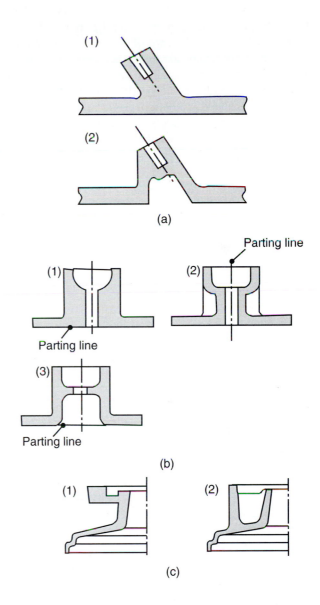

(a)

(b)

(c)

5.97 It is sometimes desirable to cool metals more slowly than they would be if the molds were maintained at room temperature. List and explain the methods you would use to slow down the cooling process.

5.98   Design an experiment to measure the constants C and n in Chvorinov's rule [Eq. (5.11)].

5.99   The part in the accompanying figure is to be cast of 10% Sn bronze at the rate of 100 parts per month. To find an appropriate casting process, consider all the processes in this chapter and reject those that are (a) technically inadmissible or (b) technically feasible but too expensive for the purpose, and (c) identify the most economical process. Write a rationale using common-sense assumptions about product cost.

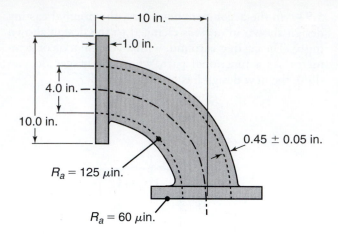

Bulk Deformation Processes

This chapter describes the bulk deformation processes of forging, rolling, extrusion, drawing, and swaging. It covers

- The fundamental principles of these processes.
- Important processing parameters such as force and power requirements, temperature, formability, tool and die materials, and metalworking fluids.
- The characteristics of machinery and equipment used.
- Die manufacturing methods and die failures.

6.1 | Introduction

In this chapter we describe metalworking processes in which the workpiece is subjected to *plastic deformation* under forces applied through various dies and tooling. Deformation processes are generally classified by type of operation as primary working or secondary working and are further divided into the categories of cold (room temperature), warm, and hot working.

Primary-working operations involve taking a solid piece of metal (generally from a cast state) and breaking it down successively into wrought materials of various shapes by the processes of forging, rolling, extrusion, and drawing. **Secondary-working** operations typically involve further processing of the products from primary working into final or semifinal products, such as bolts, gears, and sheet-metal parts.

In **bulk deformation,** which is the subject of this chapter (see Table 6.1), the parts made have a relatively small surface-area-to-volume (or to thickness) ratio; hence the term *bulk*. Typical examples are hand tools, shafts, and turbine disks. In **sheet forming,** described in Chapter 7, the surface-area-to-thickness ratio of the products made is much higher and the material is subjected to shape changes by various dies; thickness variations are usually not desirable as they can lead to part failure during manufacturing. Examples are sheet-metal hubcaps, aircraft fuselages, and beverage cans.

The processes covered in this chapter pertain only to metals. The forming and shaping processes for plastics are described in Chapter 10 and in Chapter 11 for metal powders, ceramics, glasses, composite materials, and superconductors.

TABLE 6.1

General Characteristics of Bulk Deformation Processes

Process	General characteristics
Forging	Production of discrete parts with a set of dies; some finishing operations usually necessary; similar parts can be made by casting and powder-metallurgy techniques; usually performed at elevated temperatures; dies and equipment costs are high; moderate to high labor costs; moderate to high operator skill.
Rolling	
Flat	Production of flat plate, sheet, and foil at high speeds, and with good surface finish, especially in cold rolling; requires very high capital investment; low to moderate labor cost.
Shape	Production of various structural shapes, such as I-beams and rails, at high speeds; includes thread and ring rolling; requires shaped rolls and expensive equipment; low to moderate labor cost; moderate operator skill.
Extrusion	Production of long lengths of solid or hollow products with constant cross sections, usually performed at elevated temperatures; product is then cut to desired lengths; can be competitive with roll forming; cold extrusion has similarities to forging and is used to make discrete products; moderate to high die and equipment cost; low to moderate labor cost; low to moderate operator skill.
Drawing	Production of long rod, wire, and tubing, with round or various cross sections; smaller cross sections than extrusions; good surface finish; low to moderate die, equipment, and labor costs; low to moderate operator skill.
Swaging	Radial forging of discrete or long parts with various internal and external shapes; generally carried out at room temperature; low to moderate operator skill.

6.2 | Forging

Forging denotes a family of processes to make discrete parts in which plastic deformation takes place by compressive forces applied through various dies and tooling. Forging is one of the oldest metalworking operations known, dating back to 5000 B.C. (see Table 1.1) and is used in making parts with a wide range of sizes and shapes and from a variety of metals. Simple forgings can be made with a heavy hammer and an anvil by techniques practiced by blacksmiths for centuries. Typical parts, now mostly made on modern machinery and at high production rates, are automotive engine components, turbine disks, gears, bolts, and numerous types of structural components for machinery, railroad, and other transportation equipment. With special techniques, the forging process can produce parts that are in the category of *net-shape manufacturing.*

Forging can be carried out at room temperature (*cold working*), or at elevated temperature, a process called *warm* or *hot forging,* depending on the temperature. The temperature range for these categories is given in Table 3.2 in terms of the *homologous temperature,* T/T_m, where T_m is the melting point (absolute scale) of the workpiece material. Note that the homologous recrystallization temperature for metals is about 0.5. (See also Fig. 3.17.)

6.2.1 Open-die forging

Open-die forging typically involves placing a solid cylindrical workpiece between two flat dies (platens) and reducing its height by compressing it (Fig. 6.1a), an operation that is also known as **upsetting.** The die surfaces may be flat or, more generally, have cavities of various shapes. Under ideal conditions, a solid cylinder deforms *uniformly,* as shown in Fig. 6.1a, in a process known as **homogeneous deformation.** Because in plastic deformation the volume of the cylinder remains constant

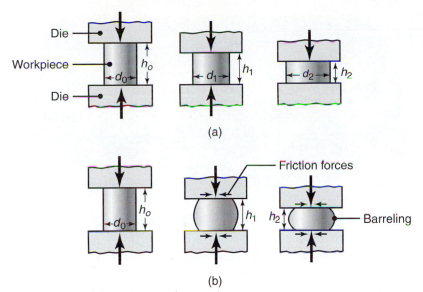

(a)

(b)

FIGURE 6.1   (a) Ideal deformation of a solid cylindrical specimen compressed between flat frictionless dies (platens), an operation known as upsetting and (b) deformation in upsetting with friction at the die–workpiece interfaces. Note barrelling of the billet caused by friction.

(see Section 2.11.5), any reduction in its height is followed by an increase in its diameter. The reduction in height is defined as

$$\text{Reduction in height} = \frac{h_o - h_1}{h_o} \times 100\%. \tag{6.1}$$

From Eqs. (2.1) and (2.9) and using absolute values (as is generally the case with bulk deformation processes), the engineering strain is

$$e_1 = \frac{h_o - h_1}{h_o} \tag{6.2}$$

and the true strain is

$$\epsilon_1 = \ln\left(\frac{h_o}{h_1}\right). \tag{6.3}$$

With a relative velocity v between the platens, the specimen is subjected to a strain rate, according to Eqs. (2.17) and (2.18), of

$$\dot{e}_1 = -\frac{v}{h_o} \tag{6.4}$$

and

$$\dot{\epsilon}_1 = -\frac{v}{h_1}. \tag{6.5}$$

If the specimen is reduced in height from h_o to h_1, the subscript 1 in the foregoing equations is replaced by subscript 2. Positions 1 and 2 in Fig. 6.1a may be regarded as instantaneous positions during a continuous upsetting operation on the workpiece. Note also that the true strain rate, $\dot{\epsilon}$, increases rapidly as the height of the specimen approaches zero.

Barreling.   Unlike the specimen shown in Fig. 6.1a, in actual practice the specimen develops a **barrel** shape during upsetting, as shown in Fig. 6.1b and in Fig. 2.14.

FIGURE 6.2 Grain flow lines in upsetting a solid, steel cylindrical specimen at elevated temperatures between two flat cool dies. Note the highly inhomogeneous deformation and barreling, and the difference in shape of the bottom and top sections of the specimen. The latter results from the hot specimen resting on the lower die before deformation starts. The lower portion of the specimen began to cool, thus exhibiting higher strength and hence deforming less than the top surface. *Source:* After J.A. Schey.

Barreling is caused primarily by *frictional forces* at the die–workpiece interfaces that oppose the outward flow of the material at these interfaces. Barreling can also occur in upsetting of hot workpieces between cool dies. The reason is that the material at and near the die–specimen interfaces cools rapidly, whereas the rest of the specimen remains relatively hot. Since the strength of the material decreases with increasing temperature (Section 2.2.6), the upper and lower portions of the specimen show a greater resistance to deformation than does the center.

As shown in Fig. 6.2, a result of barreling is that the deformation throughout the specimen becomes *nonuniform* or *inhomogeneous*. The figure shows the cross section of specimen that has been cut at midplane, then polished and etched. Note the approximately triangular stagnant zones (called *dead zones*) at the top and bottom regions. As can noted in Fig. 6.3, such inhomogeneous bulk deformation can also be observed using grid patterns (see also Section 6.4.1).

Barreling caused by friction can be minimized by applying an effective lubricant or **ultrasonic vibration** of the platens (see Section 4.4.1). In hot-working operations, barreling can be reduced by using heated dies or a thermal barrier at the interfaces, which is another role served by a lubricant.

Double barreling can also occur, especially in (a) slender specimens, with high ratios of height to cross-sectional area, and (b) when friction at the die–workpiece interfaces is very high. Under these conditions, the stagnant zone under the platens is sufficiently far from the long midsection of the workpiece, which then tends to deform more uniformly while the top and bottom portions barrel out; hence the term *double barreling*.

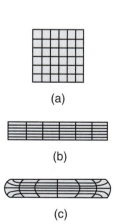

(a)

(b)

(c)

FIGURE 6.3 Schematic illustration of grid deformation in upsetting: (a) original grid pattern, (b) after deformation, without friction, and (c) after deformation, with friction. Such deformation patterns can be used to calculate the strains within a deforming body.

Forces and work of deformation under ideal conditions. If friction at the workpiece–die interfaces is zero and the material is perfectly plastic with a yield stress of Y, the normal compressive stress on the cylindrical specimen is uniform and at a level Y. The force at any height h_1 is

$$F = YA_1, \qquad (6.6)$$

where A_1 is the cross-sectional area and is obtained from volume constancy:

$$A_1 = \frac{A_o h_o}{h_1}.$$

The ideal work of deformation is the product of the specimen volume and the specific energy u [see Eq. (2.59)] and is expressed as

$$\text{Work} = \text{Volume} \int_0^{\epsilon_1} \sigma \, d\epsilon, \tag{6.7}$$

where ϵ_1 is obtained from Eq. (6.3). If the material is strain hardening, with a true-stress–true-strain curve given by the expression

$$\sigma = K\epsilon^n,$$

then the force at any stage during deformation becomes

$$F = Y_f A_1, \tag{6.8}$$

where Y_f is the flow stress of the material (see Fig. 2.37), corresponding to the true strain given by Eq. (6.3).

The expression for the work done is

$$\text{Work} = (\text{Volume})(\overline{Y})(\epsilon_1), \tag{6.9}$$

where $\overline{Y}$ is the average flow stress and is given by

$$\overline{Y} = \frac{K \int_0^{\epsilon_1} \epsilon^n \, d\epsilon}{\epsilon_1} = \frac{K\epsilon_1^n}{n+1}. \tag{6.10}$$

6.2.2 Methods of analysis

There are several methods of analysis to theoretically determine various quantities, such as stresses, strains, strain rates, forces, and local temperature rise, in deformation processing, outlined as follows.

Slab method. The *slab method* is one of the simpler methods of analyzing the stresses and loads in forging as well as other bulk deformation processes. This method requires the selection of an element in the workpiece and identification of all the normal and frictional stresses acting on that element.

1. **Forging of a rectangular workpiece in plane strain.** Consider the case of simple compression with friction (Fig. 6.4). As the flat dies compress the part and reduce its thickness, the part expands laterally. This movement at the die-workpiece interfaces causes frictional forces acting in the opposite direction to the motion. The frictional forces are indicated by the horizontal arrows shown in Fig. 6.4. For simplicity, let's also assume that the deformation is in *plane strain;* that is, the workpiece is not free to flow in the z-direction (see also Section 2.11.3).

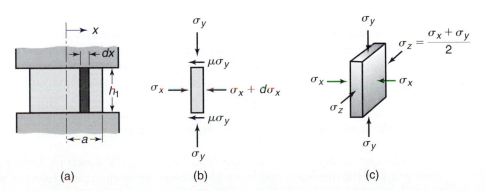

(a)   (b)   (c)

FIGURE 6.4 Stresses on an element in plane-strain compression (forging) between flat dies with friction. The horizontal stress σ_x is assumed to be uniformly distributed along the height h of the element. Identifying the stresses on an element (slab) is the first step in the slab method of analysis of metalworking processes.

Let's now take an element and indicate all the stresses acting on it, as shown in Fig. 6.4b. Note the difference between the horizontal stresses acting on the side faces of the element, which is due to the presence of frictional stresses on the element. Let's assume that the lateral stress distribution, σ_x, is uniform along the height, h, of the element.

The next step in this analysis is to balance the horizontal forces, because the element must be in static equilibrium. Assuming unit width, we have

$$(\sigma_x + d\sigma_x)h + 2\mu\sigma_y\,dx - \sigma_y h = 0,$$

or

$$d\sigma_x + \frac{2\mu\sigma_y}{h}dx = 0.$$

Note that we now have one equation but two unknowns, namely σ_x and σ_y. The necessary second equation is obtained from the yield criteria (Section 2.11), as follows: As shown in Fig. 6.4c, this element is subjected to triaxial compression. Using the distortion-energy criterion for plane strain, we obtain

$$\sigma_y - \sigma_x = \frac{2}{\sqrt{3}}Y = Y'. \tag{6.11}$$

Hence,

$$d\sigma_y = d\sigma_x.$$

Note that we have assumed σ_y and σ_x to be *principal stresses*. In the strictest sense, σ_y cannot be a principal stress because a shear stress is also acting on the same plane. However, this assumption is acceptable for low values of the coefficient of friction, μ, and is the standard practice in this method of analysis. Note also that σ_z in Fig. 6.4c is derived in a manner similar to that in Eq. (2.44). We now have two equations that can be solved by noting that

$$\frac{d\sigma_y}{\sigma_y} = -\frac{2\mu}{h}dx, \qquad \text{or} \qquad \sigma_y = Ce^{-2\mu x/h}. \tag{6.12}$$

The boundary conditions are such that at $x = a$, $\sigma_x = 0$, and hence $\sigma_y = Y'$ at the edges of the specimen. (All stresses are compressive, so that negative signs for stresses are ignored.) Thus, the value of C becomes

$$C = Y'e^{2\mu a/h},$$

and therefore,

$$p = \sigma_y = Y'e^{2\mu(a-x)/h} \tag{6.13}$$

and

$$\sigma_x = \sigma_y - Y' = Y'[e^{2\mu(a-x)/h} - 1]. \tag{6.14}$$

Equation (6.13) is plotted qualitatively and in dimensionless form in Fig. 6.5. Note that the pressure increases exponentially toward the center of the part, and also that it increases with the a/h ratio and increasing friction. For a strain-hardening material, the term Y' in Eqs. (6.13) and (6.14) is replaced by Y_f'. Because of its shape, the pressure-distribution curve in Fig. 6.5 is referred to as the *friction hill*. The pressure with friction is higher than it is without friction because the work required to overcome friction must be supplied by the upsetting force.

The area under the pressure curve in Fig. 6.5 is the upsetting **force per unit width** of the specimen, and it can be obtained by integration. However, an

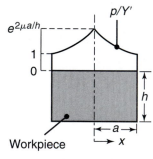

FIGURE 6.5 Distribution of die pressure, in dimensionless form of p/Y', in plane-strain compression with sliding friction. Note that the pressure at the left and right boundaries is equal to the yield stress of the material in plane strain, Y'. Sliding friction means that the frictional stress is directly proportional to the normal stress.

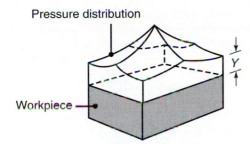

FIGURE 6.6 Die pressure distribution in compressing a rectangular workpiece with sliding friction and under conditions of plane stress, using the distortion-energy criterion. Note that the stress at the corners is equal to the uniaxial yield stress, Y, of the material.

approximate expression for the *average pressure*, p_{av}, is given by

$$p_{av} \simeq Y'\left(1 + \frac{\mu a}{h}\right). \qquad (6.15)$$

Again note the significant influence of a/h and friction on the pressure required, especially at high a/h ratios. The *forging force*, F, is the product of the average pressure and the contact area,

$$F = (p_{av})(2a)(\text{width}). \qquad (6.16)$$

Note that the expressions for pressure are in terms of an *instantaneous height*, h; hence the force at any h during a continuous upsetting operation must be calculated individually.

A rectangular specimen can be upset without being constrained on its sides (*plane stress;* see Fig. 2.35). According to the distortion-energy criterion, the normal stress distribution can be given qualitatively by the plot in Fig. 6.6. Because the elements at the corners are in uniaxial compression, the pressure there is Y, and a friction hill is present along all the edges of the specimen because of friction.

Figure 6.7 shows the lateral expansion (top view) of the edges of a rectangular specimen in actual plane-stress upsetting. Some typical results indicate that for an increase in specimen length of 40%, the increase in width (narrow dimension) is 230%. The reason for the significantly larger increase in width is that, as expected, the material flows in the direction of least resistance. Because of its smaller magnitude, the width has less cumulative frictional resistance than does the length. Likewise, after upsetting, a specimen in the shape of an equilateral cube acquires a pancake shape (with a barreled periphery) because the diagonal direction expands at a slower rate than in the other principal directions.

2. **Forging of a solid cylindrical workpiece.** The pressure distribution in forging of a solid cylindrical specimen (Fig. 6.8) also can be determined using the slab method of analysis. We first (a) isolate a segment of angle $d\theta$ in a cylinder of radius r and height h, (b) take a small element of radial length dx,

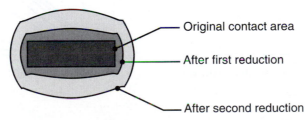

FIGURE 6.7 Increase in die–workpiece contact area of an originally rectangular specimen (viewed from the top) compressed between flat dies and with friction. Note that the length of the specimen (horizontal dimension) has increased proportionately less than its width (vertical dimension). Likewise, a specimen originally in the shape of a cube acquires the shape of a pancake after deformation with friction.

FIGURE 6.8 Stresses on an element in forging of a solid cylindrical workpiece between flat dies and with friction. Compare this figure and the stresses involved with Fig. 6.4.

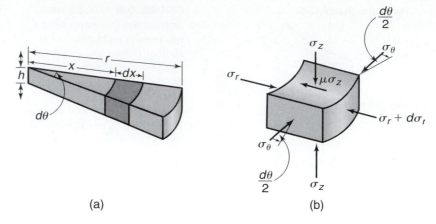

(a)   (b)

and (c) place on this element all the normal and frictional stresses acting on it. Following an approach similar to the plane-strain situation, an expression for the pressure, p, at any radius x can be obtained as

$$p = Ye^{2\mu(r-x)/h}. \tag{6.17}$$

The average pressure, p_{av}, is given approximately as

$$p_{av} \simeq Y\left(1 + \frac{2\mu r}{3h}\right), \tag{6.18}$$

and thus the forging force is

$$F = (p_{av})(\pi r^2). \tag{6.19}$$

For strain-hardening materials, Y in Eqs. (6.17) and (6.18) has to be replaced by the flow stress, Y_f. In actual forging operations, the value of the coefficient of friction, μ in Eqs. (6.17) and (6.18), can be estimated as ranging from 0.05 to 0.1 for cold forging and 0.1 to 0.2 for hot forging (Table 4.1). The selection depends on the effectiveness of the lubricant and can be higher than these values, especially in regions where the workpiece surface is devoid of lubricant. (See also Section 4.4.3.)

The effects of friction and the aspect ratio of the specimen (that is, a/h or r/h) on the average pressure, p_{av}, in upsetting are illustrated in Fig. 6.9. The

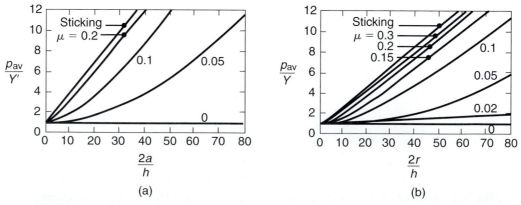

FIGURE 6.9 Ratio of average die pressure to yield stress as a function of friction and aspect ratio of the specimen: (a) plane-strain compression and (b) compression of a solid cylindrical specimen. Note that the yield stress in (b) is Y, and not Y' as it is in the plane-strain compression shown in (a). *Source:* After J.F.W. Bishop.

pressure is given in a dimensionless form and can be regarded as a pressure-multiplying factor. These curves are convenient to use and again show the significance of friction and the aspect ratio of the specimen on upsetting pressure and hence forces.

3. **Forging under sticking condition.** It can be seen that the product of μ and p is the frictional stress (surface shear stress) acting at the workpiece–die interface at any location x from the center. As p increases toward the center, μp also increases. However, the value of μp cannot be higher than the shear yield stress, k, of the material. The condition when $\mu p = k$ is known as sticking. Note that in plane strain, the value of k is $Y'/2$ (see Section 2.11.3). Sticking does not necessarily mean adhesion at the interface, but it reflects the fact that the material does not move relative to the platen surfaces.

For the sticking condition, the normal stress distribution in plane strain can be shown to be

$$p = Y'\left(1 + \frac{a - x}{h}\right). \tag{6.20}$$

Note that the pressure varies linearly with x, as also shown in Fig. 6.10. The normal stress distribution for a cylindrical specimen under sticking condition can be shown to be

$$p = Y\left(1 + \frac{r - x}{h}\right). \tag{6.21}$$

Again, the stress distribution is linear, as shown in Fig. 6.10 for plane strain.

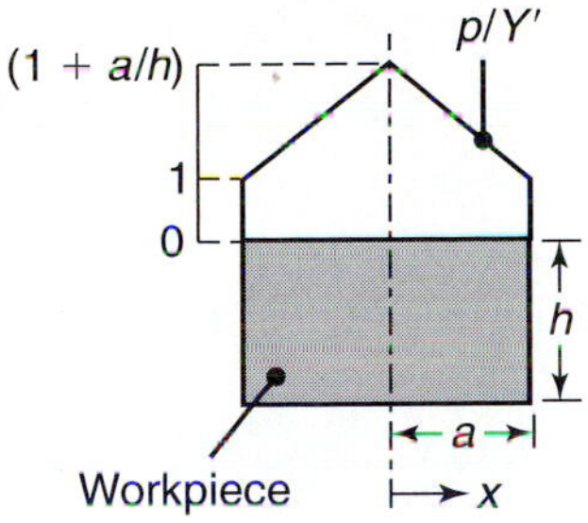

FIGURE 6.10 Distribution of dimensionless die pressure, p/Y', in compressing a rectangular specimen in plane strain and under sticking conditions. Sticking means that the frictional (shear) stress at the interface has reached the shear yield stress of the material. Note that the pressure at the edges is the uniaxial yield stress of the material in plane strain, Y'.

EXAMPLE 6.1 Calculation of upsetting force

A cylindrical specimen made of annealed 4135 steel has a diameter of 6 in. and is 4 in. high. It is upset, at room temperature, by open-die forging with flat dies to a height of 2 in. Assuming that the coefficient of friction is 0.2, calculate the upsetting force required at the end of the stroke. Use the average-pressure formula.

Solution. The average-pressure formula, from Eq. (6.18), is given by

$$p_{av} \simeq Y_f\left(1 + \frac{2\mu r}{3h}\right),$$

where Y has been replaced by Y_f because the workpiece material is strain hardening. From Table 2.3, we find that $K = 1015$ MPa (147,000 psi) and $n = 0.17$. The absolute value of the true strain is

$$\epsilon_1 = \ln\left(\frac{4}{2}\right) = 0.693,$$

and therefore

$$Y_f = K\epsilon_1^n = (147,000)(0.693)^{0.17} = 138,000 \text{ psi.}$$

The final height of the specimen, h_1, is 2 in. The radius r at the end of the stroke is found from volume constancy:

$$\frac{\pi 6^2}{4} \cdot 4 = \pi r_1^2 \cdot 2 \quad \text{and} \quad r_1 = 4.24 \text{ in.}$$

Thus

$$p_{av} \simeq 138,000 \left[1 + \frac{(2)(0.2)(4.24)}{(3)(2)} \right] = 177,000 \text{ psi.}$$

The upsetting force is

$$F = (177,000)\pi \cdot (4.24)^2 = 10^7 \text{ lb.}$$

Note that Fig. 6.9b can also be used to solve this problem.

EXAMPLE 6.2 Transition from sliding to sticking friction in upsetting

In plane-strain upsetting, the frictional stress cannot be higher than the shear yield stress, k, of the workpiece material. Thus, there may be a distance x in Fig. 6.4 where a transition occurs from sliding to sticking friction. Derive an expression for x in terms of a, h, and μ only.

Solution. The shear stress at the interface due to friction can be expressed as

$$\tau = \mu p.$$

However, the shear stress cannot exceed the yield shear stress, k, of the material, which, for plane strain, is $Y'/2$. The pressure curve in Fig. 6.5 is given by Eq. (6.13); thus, in the limit, we have

$$\mu Y' e^{2\mu(a-x)/h} = Y'/2,$$

or

$$2\mu \frac{(a-x)}{h} = \ln\left(\frac{1}{2\mu}\right).$$

Hence,

$$x = a - \left(\frac{h}{2\mu}\right) \ln\left(\frac{1}{2\mu}\right).$$

Note that, as expected, the magnitude of x decreases as μ decreases. However, the pressures must be sufficiently high to cause sticking; that is, the a/h ratio must be high. For example, let $a = 10$ mm and $h = 1$ mm. Then, for $\mu = 0.2$, $x = 7.71$ mm, and for $\mu = 0.4$, $x = 9.72$ mm.

Finite element method. In the *finite-element method,* the deformation zone in an elastic-plastic body is divided into a number of elements, interconnected at a finite number of nodal points. Next, a set of simultaneous equations is developed that represents unknown stresses and deformation increments subject to the boundary conditions. From the solution of these equations, actual velocity and stress distributions are calculated. This technique can also incorporate friction conditions at the die–workpiece interfaces as well as the actual properties of the workpiece material. Application of this technique requires inputs such as the stress–strain characteristics of the material as a function of strain rate and temperature, and frictional and heat-transfer characteristics of the die and the workpiece.

The finite-element method has been applied to relatively complex workpiece shapes in bulk deformation and sheet-metal forming. Its accuracy is influenced by

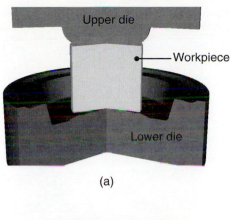

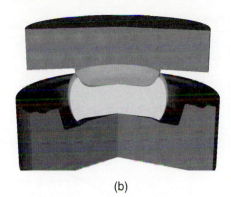

(a)

(b)

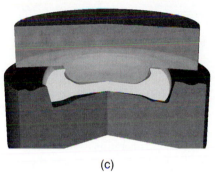

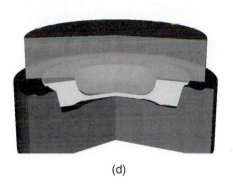

(c)

(d)

FIGURE 6.11 Deformation of a blank during forging as predicted by the software program DEFORM based on the finite-element method of analysis. *Source:* Courtesy Scientific Forming Technologies Corporation.

the number, shape, and mathematical complexity of the finite elements, the deformation increment, and the methods of calculation. The method gives a detailed picture of the actual stresses and strain distributions throughout the workpiece.

The results of applying the finite-element method of analysis on a solid cylindrical workpiece in impression-die forging are shown in Fig. 6.11. The grid pattern is known as a mesh, and its distortion is numerically predicted from theory. Note how the workpiece deforms as its height is reduced.

The finite-element method is also capable of determining the temperature distribution throughout the workpiece and predicting microstructural changes in the material during hot-working operations and the onset of defects, without using experimental data. This information is then used in modifying die design. Among other methods of analysis are the slip-line analysis and the upper-bound technique. However, these methods are largely outdated in view of the success and acceptance of the two techniques just described.

Deformation-zone geometry. The observations made concerning Fig. 2.25 are important in calculating forces in forging, as well as other bulk deformation processes. As described for hardness testing, the compressive stress required for indentation is, ideally, about three times the yield stress, Y, required for uniaxial compression (see Section 2.6.8). Recall also that (a) the deformation under the indenter is localized, making the overall deformation highly nonuniform, and (b) the deformation zone is relatively small compared with the size of the specimen. In contrast, in a simple frictionless compression test with flat overhanging dies, the top and bottom surfaces of the specimen are always in contact with the dies while the whole specimen undergoes deformation. Various situations occur between these two extreme examples of specimen-deformation geometry.

The pressures required for the frictionless condition and the deformation zones developed for various situations are shown in Fig. 6.12. Note that the important

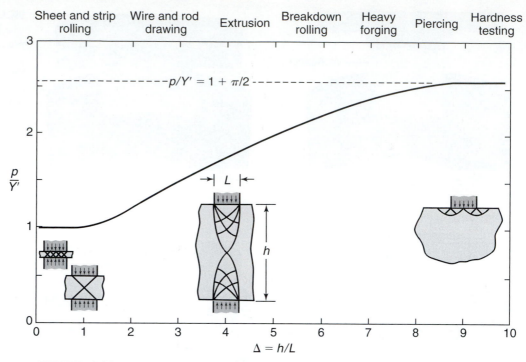

FIGURE 6.12 Die pressure required in various metalworking operations and under *frictionless* plane-strain conditions, as obtained by the slip-line analysis. Note that the magnitude of the die-workpiece contact area is an important factor in determining pressures. *Source:* After W.A. Backofen.

parameter in determining the inhomogeneity of deformation is the ratio h/L, and recall that friction significantly affects forces, particularly at small values of h/L. Deformation-zone geometry depends on the particular process and such parameters as die geometry and percent reduction of the material (Fig. 6.13).

6.2.3 Types of forging

Impression-die forging. In *impression-die forging*, the workpiece acquires the shape of the die cavity (hence the term *impression*) while it is being deformed between the closing dies. A typical example of such an operation is shown in Fig. 6.14, where it can be noted that some of the material flows radially outward, forming a **flash.** Because of its high length-to-thickness ratio (equivalent to a high a/h ratio), the flash is subjected to high pressure. This, in turn, indicates the presence of high frictional resistance as the material flows in the radially outward direction in the flash gap. Thus the flash gap is an important parameter since high friction there encourages the filling of the die cavities. Furthermore, if the forging operation is carried out at elevated temperatures (*hot forging*), the flash cools faster than does the bulk of the workpiece (because of the high surface-area-to-thickness ratio of the flash gap). As a result, the flash resists deformation more than the bulk does and thus helps filling of the die cavities.

The quality, dimensional tolerances, and surface finish of a forging depend on how well these operations are performed and controlled. Dimensional tolerances generally range between ±0.5% and ±1% of the dimensions of the forging. Tolerances for hot forging of steel are usually less than ±6 mm ($\frac{1}{4}$ in.), and in *precision forging* (see page 278) they can be as low as ±0.25 mm (0.01 in.). Factors that contribute to

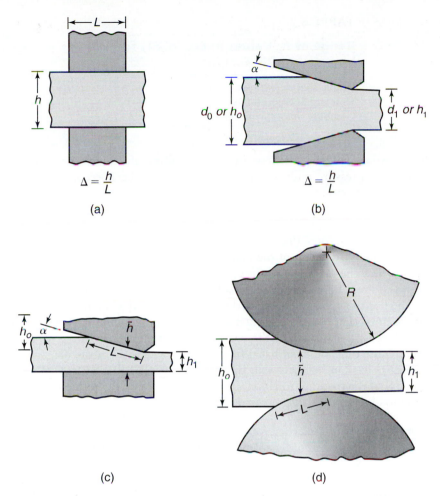

FIGURE 6.13   Examples of plastic deformation processes in plane strain, showing the h/L ratio. (a) Indenting with flat dies, an operation similar to the cogging process, as shown in Fig. 6.19, (b) drawing or extrusion of a strip with a wedge-shaped die, as described in Sections 6.4 and 6.5, (c) ironing (see also Fig. 7.53), and (d) rolling, described in Section 6.3.

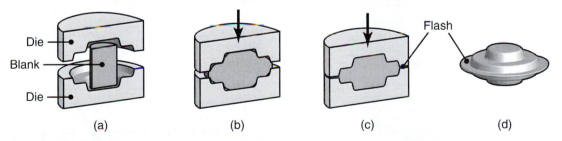

FIGURE 6.14   Schematic illustrations of stages in impression-die forging. Note the formation of a flash, or excess material, that subsequently has to be trimmed off.

dimensional inaccuracies are draft angles, radii, fillets, die wear, die closure, and mismatching of the dies. Surface finish of the forging depends on blank preparation, die surface finish, die wear, and the effectiveness of the lubricant.

   Forces in impression-die forging can be difficult to predict largely because of the generally complex shapes involved and the fact that each location within the workpiece is typically subjected to different strains, strain rates, and temperatures,

TABLE 6.2

Range of K_p Values in Eq. (6.22) for Impression-Die Forging	
Simple shapes, without flash	3–5
Simple shapes, with flash	5–8
Complex shapes, with flash	8–12

as well as variations in coefficient of friction along the die-workpiece contact lengths. Certain pressure-multiplying factors, K_p, have been recommended to be used with the expression

$$F = K_p Y_f A, \qquad (6.22)$$

where F is the forging force (load), A is the projected area of the forging, including the flash, Y_f is the flow stress of the material at the strain, the strain rate, and temperature to which the material is subjected, and K_p is to be read from Table 6.2.

A typical impression-die forging load as a function of the die stroke is shown in Fig. 6.15. For this axisymmetric workpiece, the force first increases gradually as the cavity is filled (Fig. 6.15b) and then increases rapidly as the flash forms. As the dies must close further (to shape the part), an even steeper rise in the forging load takes place. Note that the flash has a finite contact length with the die, called **land** (see Fig. 6.26). The land ensures that the flash generates sufficient resistance of the outward flow of the material to aid in die filling without contributing excessively to the forging load.

Precision forging. In precision forging, special dies are made to a higher accuracy than in ordinary impression-die forging. This operation requires higher capacity

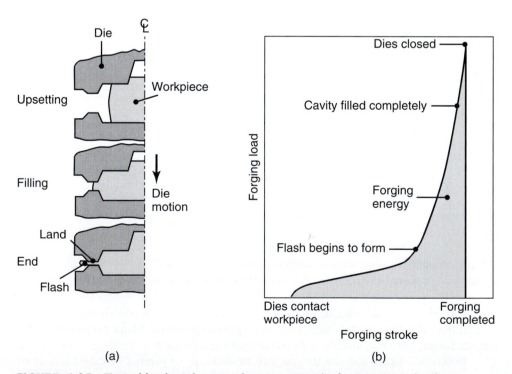

(a)   (b)

FIGURE 6.15 Typical load-stroke curve for impression-die forging. Note the sharp increase in load when the flash begins to form. *Source:* After T. Altan.

forging equipment than do other forging processes because of the higher stresses required to precisely form the part. Precision forging, as well as similar operations where the part formed is close to the final desired dimensions, such as closed-die forging and coining (see below) and powder metallurgy (Chapter 11), is also known as **near-net-shape production** (see also Section 1.6).

Aluminum and magnesium alloys are particularly suitable for precision forging, because the forging loads and temperatures required are relatively low, die wear is low, and the forgings have good surface finish. Steels and other alloys, on the other hand, are more difficult to precision forge. The choice between conventional forging and precision forging requires an economic analysis. Although precision forging requires special dies, much less machining is involved because the part is closer to or at the desired final shape.

Closed-die forging. In closed-die forging, no flash is formed and the workpiece is completely surrounded by the dies. Proper control of the volume of material vs. die-cavity volume is essential in order to produce a forging of desired shape and dimensions. Undersized blanks will prevent the complete filling of the die cavity, and oversized blanks may cause premature die failure or jamming of the dies.

Isothermal forging. Also known as **hot-die forging**, the dies in this process are heated to the same temperature as the hot blank. In this way, cooling of the workpiece is eliminated, the low flow stress of the material is maintained during forging, and material flows easier within the die cavities. The dies are generally made of nickel alloys, and complex parts with good dimensional accuracy can be forged in one stroke in hydraulic presses. Although expensive, isothermal forging can be economical for intricate forgings of expensive materials, provided that the quantity required is high enough to justify die costs.

Incremental forging. In this process the blank is forged into a shape with specially designed tooling and dies that form it in several small steps (hence the term *incremental*), similar to the *cogging* operation (Fig. 6.19), where it can be noted that the die–workpiece contact area is small. Consequently, this operation requires much lower forces as compared with conventional forging. **Orbital forging** is an example of incremental forging in which the die moves along an orbital path (Fig. 6.16) and forms the part in individual steps. The forces are lower, the operation is quieter, and part flexibility can be achieved.

6.2.4 Miscellaneous forging operations

1. **Coining.** An example of coining is the *minting of coins*, where the slug is shaped in a completely closed cavity. The pressures required can be as high as five to six times the flow stress of the material in order to produce the fine details of a coin or a medallion. Lubricants cannot be tolerated in this operation because they can be trapped in die cavities and thus prevent reproduction of fine die-surface details. This process also is used in conjunction with forging to improve surface finish and achieve the desired dimensional accuracy (**sizing**) of the products.

2. **Heading.** This is basically an upsetting operation, typically performed at the end of a rod to produce a shape with a larger cross section; examples are the heads of bolts, screws, nails, and similar type of products (Fig. 6.17). There is a tendency for the part to buckle if its length-to-diameter ratio of the part is too high. Heading is done on machines called *headers*, which are typically highly

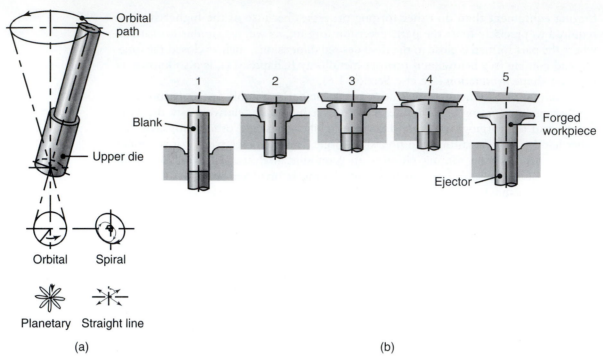

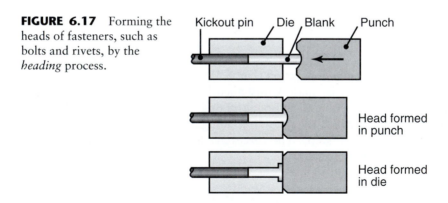

FIGURE 6.16 Schematic illustration of the orbital-forging process. Note that the die is in contact with only a portion of the workpiece surface at a time. Also called *rotary forging*, *swing forging*, and *rocking-die forging*, this process can be used for forming individual parts such as bevel gears, wheels, and bearing rings.

FIGURE 6.17 Forming the heads of fasteners, such as bolts and rivets, by the *heading* process.

automated horizontal machines with high production rates. The operation can be carried out cold, warm, or hot.

3. **Piercing.** In this process, a punch indents the workpiece surface to produce a cavity or an impression with a specific shape (Fig. 6.18). The workpiece may be confined in a die cavity, or it may be unconstrained. The piercing force depends on the punch's cross-sectional area and tip geometry, the flow stress of the material, and the friction at the interfaces. Punch pressures may be three to five times the flow stress of the material. The term *piercing* is also used to describe the process of cutting of holes with a punch and die, as described in Section 7.3.

4. **Hubbing.** In *hubbing*, a hardened punch with a particular tip geometry is pressed into the surface of a block of metal to produce a cavity (which is shallower than

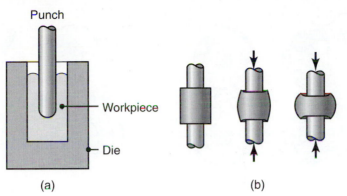

FIGURE 6.18 Examples of piercing operations.

that produced by piercing). The die cavity is then used for subsequent forming operations. The pressure required to generate a cavity is approximately three times the ultimate tensile strength (UTS) of the material of the block; thus the hubbing force required is

$$\text{Hubbing force} = 3(\text{UTS})(A), \qquad (6.23)$$

where A is the projected area of the impression. Note that the factor 3 in this formula is in agreement with the observations made with regard to hardness of materials, as described in Section 2.6.8.

5. **Cogging.** In this operation, also called *drawing out*, the thickness of a bar is reduced by successive steps at certain intervals (Fig. 6.19). This is another example of incremental forming, in which a long section of a bar, for example, can thus be reduced in thickness without the need for large dies and forces (unlike in open-die forging). The thickness rings and other parts also can be reduced in a similar manner, as illustrated in Figs. 6.19b and c.

6. **Fullering and edging.** These operations are performed, usually on bar stock, to distribute the material in certain regions prior to forging. In *edging*

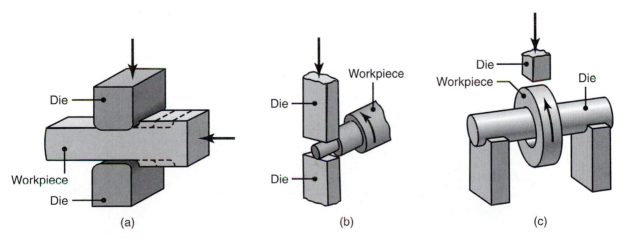

FIGURE 6.19 (a) Schematic illustration of a cogging operation on a rectangular bar. Blacksmiths use a similar procedure to reduce the thickness of parts in small increments by heating the workpiece and hammering it numerous times along the length of the part. (b) Reducing the diameter of a bar by open-die forging; note the movements of the die and the workpiece. (c) The thickness of a ring being reduced by open-die forging.

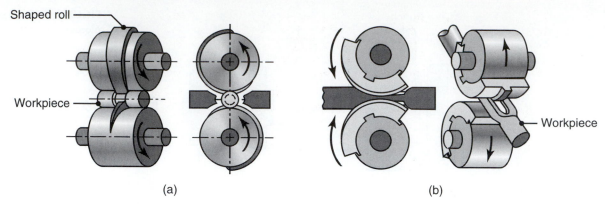

Shaped roll

Workpiece

Workpiece

(a)   (b)

FIGURE 6.20   Two illustrations of roll forging (*cross-rolling*) operations. Tapered leaf springs and knives can be made by this process using specially designed rolls. *Source: After J. Holub.*

(see Fig. 6.25), it is gathered into a localized area, whereas in *fullering,* the material is distributed away from an area. Thus, these operations are preforming a technique to make material flow easier in die cavities.

7. **Roll forging.** In *roll forging,* the cross-sectional area of a bar is reduced and altered in shape by passing it through a pair or sets of grooved rolls of various shapes (Fig. 6.20). This operation also may be used to produce parts that are basically the final product, such as tapered shafts, tapered leaf springs, table knives, and numerous tools. Roll forging is also used as a preliminary forming operation, followed by other forging and forming processes, such as in making crankshafts and various automotive components.

8. **Skew rolling.** A process similar to roll forging is *skew rolling,* which typically is used for making ball bearings. As illustrated in Fig. 6.21a, a round wire or rod stock is fed into the roll gap, and spherical blanks are formed by the continuously rotating rolls. The balls are then ground and polished in special machinery (see Section 9.6). Another method of making ball bearings is by cutting short pieces from a round bar and upsetting them between a pair of dies, as shown in Fig. 6.21b.

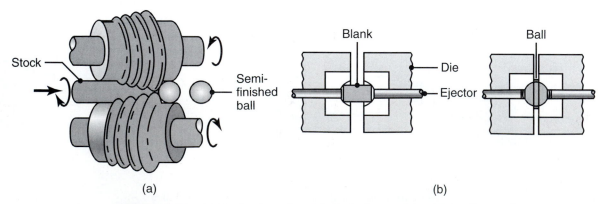

Stock

Semi-finished ball

Blank   Ball

Die

Ejector

(a)   (b)

FIGURE 6.21   (a) Production of steel balls for bearings by skew rolling and (b) production of steel balls by upsetting of a short cylindrical blank; note the formation of flash. The balls are subsequently ground and polished to be used as ball bearings and similar components.

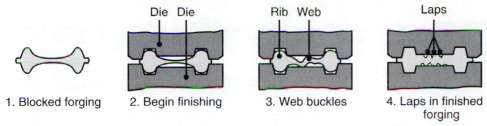

1. Blocked forging   2. Begin finishing   3. Web buckles   4. Laps in finished forging

FIGURE 6.22 Stages in lap formation in a part during forging, due to buckling of the web. Web thickness should be increased to avoid this problem.

6.2.5 Forging defects

In addition to surface cracking (see Fig. 3.21d), several other defects in forging can be caused by the material flow patterns in the die cavity.

1. As shown in Fig. 6.22, excess material in the web of a forging can buckle during forging and develop laps.

2. If the web is too thick, the excess material flows past the already forged portions and develops internal cracks (Fig. 6.23). These two examples indicate the importance of properly controlling the volume of the blank and its flow in the die cavity.

3. The die radii also can significantly affect formation of defects. For example, note in Fig. 6.24 that the material flows better around a large corner radius than it does around a small radius. With smaller radii, the material can thus fold over itself, producing a lap (**cold shut**). Such a defect can lead to fatigue failure and other problems during the service life of the forged component. The importance of inspecting forgings prior to being placed into service, particularly for critical applications, is obvious. (For inspection techniques, see Section 4.8.)

4. Another important aspect of quality in a forging is the **grain-flow pattern.** The grain-flow lines may reach a surface perpendicularly, exposing the grain boundaries directly to the environment (known as **end grains**). In service, they can be attacked by the environment (such as salt water, acid rain, or other chemically active environments), thus developing rough surfaces which act as stress raisers. End grains can be avoided by proper selection of blank orientation in the die cavity and by control of material flow during forging.

5. Because the metal flows in various directions in a forging, and there are temperature variations within the forging, the properties of a forging are generally *anisotropic.* (See Section 3.5.) Strength and ductility can vary significantly in different locations and orientations in a forged part.

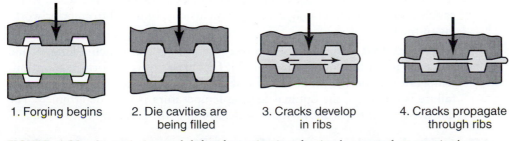

1. Forging begins   2. Die cavities are being filled   3. Cracks develop in ribs   4. Cracks propagate through ribs

FIGURE 6.23 Stages in internal defect formation in a forging because of an oversized billet. The die cavities are filled prematurely, and the material at the center of the part flows radially outward and past the filled regions as deformation continues.

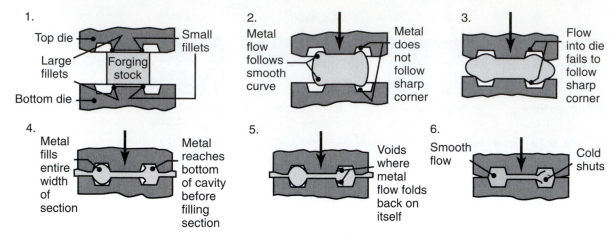

FIGURE 6.24 Effect of fillet radius on defect formation in forging. Note that small fillets (right side of the drawings) lead to defects. *Source:* Aluminum Company of America.

6.2.6 Forgeability

The forgeability of a metal may be defined as its capability to be shaped without cracking and requiring low forces. Impurities in the metal or small changes in the composition of the metal can have a significant effect on the ductility and forgeability of a metal. A number of tests have been developed to measure forgeability, but two commonly used tests are as follows:

1. **Upsetting test.** In this test, a solid cylindrical specimen is upset between two flat dies, and observations are made regarding any cracking on the barreled surfaces (see Fig. 3.21d). The higher the reduction in height prior to cracking, the greater is the forgeability of the metal. The cracks on the barreled surface are caused by *secondary tensile stresses* (so called because no external tensile stress is directly applied to the material). Friction at the die–workpiece interfaces has a marked effect on cracking; as friction increases, the specimen cracks at a lower reduction in height. Upsetting tests can be performed at various temperatures and strain rates. An optimal range for these parameters can then be specified for forging a particular material. However, such tests only serve as guidelines, since in actual forging operations, the metal is subjected to a different state of stress compared to the simple upsetting operation shown in Fig. 6.1a.

   It has been observed that the crack may be longitudinal or at a 45° angle, depending on the sign of the axial stress σ_z on the barreled surface (see Fig. 6.8b). If the stress is positive (tensile), the crack is longitudinal; if it is negative (compressive), the crack is at 45°. Surface defects will also adversely affect the results by causing premature cracking. A typical surface defect is a **seam**, which can be a longitudinal scratch, a string of inclusions, or a fold from prior working of the material.

2. **Hot-twist test.** This is a torsion test (Section 2.4), in which a long, round specimen is twisted continuously until it fails. The test is performed at various temperatures and the number of turns that each specimen undergoes before failure is observed; the optimal forging temperature is then determined. The hot-twist test is particularly useful in determining the forgeability of steels.

Effect of hydrostatic pressure on forgeability. As described in Section 2.2.8, hydrostatic pressure has a significant beneficial effect on the ductility of metals and nonmetallic materials. Experiments have indicated that cracking takes place at

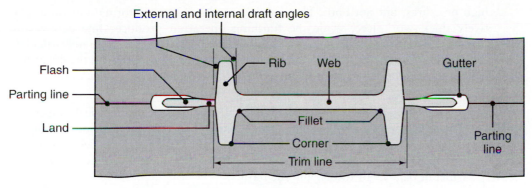

External and internal draft angles

Flash

Parting line

Land

Rib   Web   Gutter

Fillet

Corner

Trim line

Parting line

FIGURE 6.25   Standard terminology for various features of a typical forging die.

higher strain levels if the tests, described previously, are carried out in an environment of high *hydrostatic pressure*. Special techniques have been developed to forge metals in a high compressive environment in order to take advantage of this phenomenon, where the pressure-transmitting medium is usually a low-strength ductile metal. (See also *hydrostatic extrusion,* Section 6.4.3.)

Forgeability of various metals.   Based on the results of various tests and observations made during actual forging operations, the forgeability of various metals and alloys has been determined. In general, (a) aluminum, magnesium, copper and their alloys; carbon and low-alloy steels have good forgeability; and (b) high-temperature materials such as superalloys, tantalum, molybdenum, and tungsten and their alloys have poor forgeability.

6.2.7 Die design

The design of forging dies and die-material selection require knowledge of (a) the strength and ductility of the workpiece material, (b) its sensitivity to strain rate and temperature, (c) its frictional characteristics, and (d) forging temperature. *Die distortion* under high forging loads is also an important consideration, particularly if close dimensional tolerances are required. The terminology used in die design is given in Fig. 6.25.

Complex forgings require forming in a number of stages (Fig. 6.26) to ensure proper distribution of the material in the die cavities. Typically, starting with round bar stock, the bar is (1) first **preformed** (*intermediate shape*) by techniques such as fullering and edging (see Section 6.2.4), (2) forged into the final shape in two additional operations, and (c) trimmed. The reason for preforming can best be understood from Eq. (4.6), recognizing that for long die life, wear is to be minimized. In any step of a forging sequence, a die cavity is exposed to either high sliding speeds or

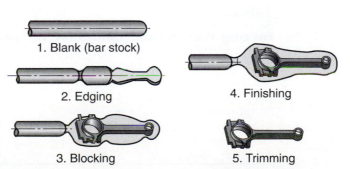

1. Blank (bar stock)

2. Edging

3. Blocking

4. Finishing

5. Trimming

FIGURE 6.26   Stages in forging a connecting rod for an internal combustion engine. Note the amount of flash developed, which is important in properly filling die cavities.

high pressures, but not both. Thus, edging produces large deformations on a relatively thick workpiece, while the finishing operation produces fine details and involves small strains, under high pressures. It is also important to accurately calculate the required volume of stock (blank) to ensure that it will properly fill the die cavity. Computer techniques are now widely available to expedite these calculations.

Some general rules of die design include the following:

1. The **parting line** is where the two dies meet. For simple symmetrical shapes, the parting line is a straight line at the center of the forging; for more complex shapes, the line may be offset and may not be in a single plane. Selection of the proper location for the parting line is based on the shape of the part, flow of the metal, balance of forces, and the flash.

2. The significance of **flash** is described in Section 6.2.3. After lateral flow has been sufficiently constrained (by the length of the land), the flash is allowed to flow into a **gutter;** thus, the extra flash does not unnecessarily increase the forging load. A general guideline for flash clearance (between the dies) is 3% of the maximum thickness of the forging. The length of the land is usually five times that of the flash clearance.

3. **Draft angles** are necessary in almost all forgings to facilitate removal of the part from the die. Draft angles usually range between 3° and 10°. Because the forging shrinks in its radial direction (as well as in other directions) as it cools, internal draft angles are made larger than external ones. Typically, internal angles are about 7° to 10° and external angles about 3° to 5°.

4. Proper selection of **die radii** for corners and fillets ensures smooth flow of the metal in the die cavity and improves die life. Small radii are generally not desirable because of their adverse effect on metal flow and their tendency to wear rapidly from stress concentration and thermal cycling. Small radii in fillets can cause fatigue cracking in dies.

Die materials. Because most forgings, particularly large ones, are performed at elevated temperatures, die materials generally must have strength and toughness at elevated temperatures, hardenability, resistance to mechanical and thermal shock, and resistance to wear (particularly to abrasive wear because of the presence of scale on heated forgings). Selection of die materials depends on the size of the die, properties of the workpiece, complexity of the workpiece shape, forging temperature, type of operation, cost of die material, number of forgings required, and heat-transfer and distortion characteristics of the die material. Common die materials are tool and die steels containing chromium, nickel, molybdenum, and vanadium (see Section 3.10.3).

Forging temperature and lubrication. The range of *temperatures* for hot forging of various metals are given in Table 6.3. *Lubrication* plays an important role in forging, in that it affects friction and wear and, consequently, the flow of metal into

TABLE 6.3

Forging Temperature Ranges for Various Metals		
Metal	°C	°F
Aluminum alloys	400–450	750–850
Copper alloys	625–950	1150–1750
Nickel alloys	870–1230	1600–2250
Alloy steels	925–1260	1700–2300
Titanium alloys	750–795	1400–1800
Refractory alloys	975–1650	1800–3000

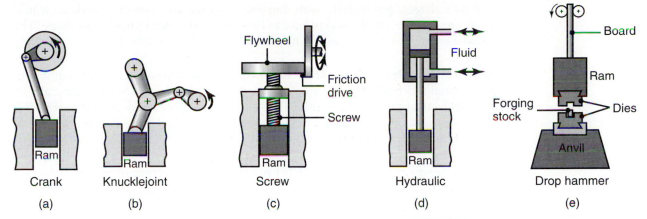

FIGURE 6.27 Schematic illustration of various types of presses used in metalworking. The choice of a press is an important consideration in the overall operation and productivity.

the die cavities. Lubricants also serve as a *thermal barrier* between the hot forging and the dies, which are typically at room temperature, thus slowing cooling of the workpiece. An additional role of a lubricant is to act as a *parting agent,* that is, to prevent the forging from sticking to the dies and enable the removal of the forging. A variety of metalworking fluids can be used in forging. For hot forging, graphite, molybdenum disulfide, and (sometimes) glass are commonly used as lubricants. For cold forging, mineral oils and soaps are common lubricants.

6.2.8 Equipment

Forging equipment of various designs, capacities, and speed-stroke characteristics is available (Fig. 6.27).

Mechanical presses. These presses are *stroke limited,* as in crank and knucklejoint presses. They are of either the crank or the eccentric type, with speeds varying from a maximum at the center of the stroke to zero at the bottom. The force available depends on the stroke position and becomes extremely large at the bottom-deadcenter position; thus, proper setup is essential to avoid breaking the dies or other equipment. The largest mechanical press has a capacity of 10^7 MN (12,000 tons).

Screw presses. Deriving their energy from a flywheel, *screw presses* transmit the forging load through a vertical screw. These presses are *energy limited* and can be used for a variety of forging operations. The presses are particularly suitable for producing small quantities, producing parts requiring precision (such as turbine blades), and for controlling the ram speed. The largest screw press has a capacity of 280 MN (31,500 tons).

Hydraulic presses. These presses have a constant low speed of operation and are *load limited*. Large amounts of energy can be transmitted to the workpiece by a constant load that is available throughout the stroke. Ram speed can be varied during the stroke. Hydraulic presses are used for both open-die and closed-die forging operations. The largest hydraulic press has a capacity of 730 MN (82,000 tons).

Hammers. *Hammers* derive their energy from the *potential energy* of the ram, which is then converted to kinetic energy; thus, hammers are *energy limited*. In *power hammers,* the ram is accelerated in the downstroke by steam or air. Hammer speeds are high, thus minimizing the cooling of the hot forgings and allowing the forging of complex shapes, particularly with thin and deep recesses (which otherwise would cool rapidly). Several blows may have to be made on the part to finalize its shape. The highest energy available in power hammers is 1150 kJ (850,000 ft-lb).

Counterblow hammers. These hammers have two rams that, as their name suggests, simultaneously approach each other to forge the part. They are generally of the mechanical-pneumatic or mechanical-hydraulic type. These machines transmit less vibration to the foundation than other hammers. The largest counterblow hammer has a capacity of 1300 kJ (1 million ft-lb).

High-energy-rate forging (HERF) machines. In this type of machine, the ram is accelerated rapidly by inert gas at high pressure, and the part is forged in one blow at a very high speed. Although there are several types of these machines, various problems associated with their operation, maintenance, die breakage, and safety considerations have greatly limited their use in industry.

Equipment selection. Selection of forging equipment depends on the size and complexity of the forging, the strength of the material and its sensitivity to strain rate, the amount of deformation involved, production rate, and cost. The number of strokes per minute ranges from a few for hydraulic presses to as many as 300 for power hammers. Generally, (a) presses are preferred for aluminum, magnesium, beryllium, bronze, and brass; and (b) hammers are preferred for copper, steels, titanium, and refractory alloys.

6.3 | Rolling

Rolling is the process of reducing the thickness or changing the cross section of a long workpiece by compressive forces applied through a set of rolls (Fig. 6.28), thus the process is similar to rolling dough with a rolling pin to reduce its thickness. Rolling, which accounts for about 90% of all metals produced by metalworking processes, was first developed in the late 1500s. The basic rolling operation is called *flat rolling,* or simply *rolling,* where the rolled products are flat plate and sheet.

Plates are generally regarded as having a thickness greater than 6 mm ($\frac{1}{4}$ in.), and are used for structural applications such as boilers, bridges, girders, ship hulls, machine structures, and nuclear vessels. Plates can be as much as 0.3 m (12 in.) thick for large boiler supports, 150 mm (6 in.) thick for reactor vessels, and 100–125 mm (4–5 in.) thick for warships and tank armor.

Sheets are generally less than 6 mm thick and are provided as flat pieces or as strip in coils to manufacturing facilities for further processing into products. They are used for automobile bodies, aircraft fuselages, office furniture, appliances, food and beverage containers, and kitchen equipment. Commercial-aircraft fuselages are typically made of about 1-mm-thick (0.040-in-thick) aluminum-alloy sheet, often of the 2000 series, and beverage cans are made of 0.15-mm-thick (0.006-in-thick) aluminum-alloy sheet such as 3104 alloy (Section 3.11.1). Aluminum foil, such as that used to wrap candy and cigarettes, has a thickness of 0.008 mm (0.0003 in.).

Traditionally, the initial form of material for rolling is an ingot; however, this practice is now rapidly being replaced by *continuous casting and rolling* (see Section 5.7), with their much higher efficiency and lower cost. Rolling is first carried out at elevated temperatures (hot rolling), wherein the coarse-grained, brittle, and porous *cast* structure of the ingot or continuously cast metal is broken down into a *wrought* structure, with finer grain size and improved properties (Fig. 6.29).

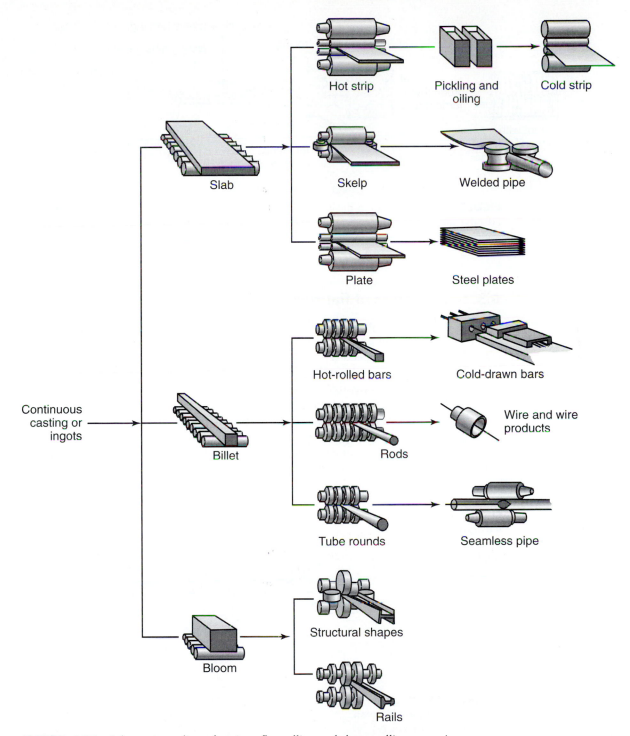

FIGURE 6.28 Schematic outline of various flat-rolling and shape-rolling operations. *Source:* American Iron and Steel Institute.

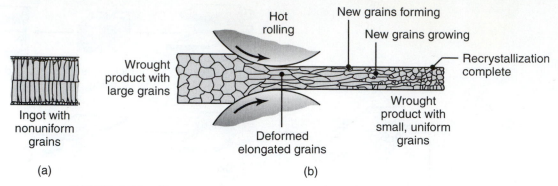

FIGURE 6.29 Changes in the grain structure of metals during hot rolling. This is an effective method to reduce grain size and refine the microstructure in metals, resulting in improved strength and good ductility. In this process cast structures of ingots or continuous castings are converted to a wrought structure.

6.3.1 Mechanics of flat rolling

The basic flat-rolling process is shown schematically in Fig. 6.30. A strip of thickness h_0 enters the roll gap and is reduced to a thickness of h_f by the powered rotating rolls at a surface speed V_r of the roll. To keep the volume rate of metal flow constant, the velocity of the strip must increase as it moves through the roll gap, as is the case with incompressible fluid flow through a converging channel. At the exit of the roll gap, the velocity of the strip is V_f (Fig. 6.31).

Since V_r is constant along the roll gap, but the strip velocity increases as it passes through the roll gap, sliding occurs between the roll and the strip. At one point along the arc of contact, however, the two velocities are the same. For this reason, this point is called **neutral point,** or **no-slip point.** To the left of this point,

FIGURE 6.30 Schematic illustration of the flat-rolling process. Note that the top roll has been removed for clarity.

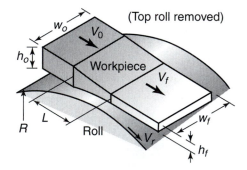

FIGURE 6.31 Relative velocity distribution between roll and strip surfaces. The arrows represent the frictional forces acting along the strip-roll interfaces. Note the difference in their direction in the left and right regions.

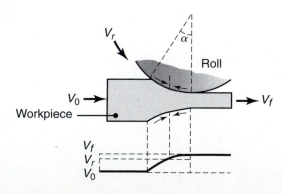

the roll moves faster than the workpiece, and to the right, the workpiece moves faster than the roll.

Because of the relative motion at the interfaces, the frictional forces (which oppose motion) act on the strip surfaces in the directions shown in Fig. 6.31. Note the basic similarity between Figs. 6.31 and 6.1b. In upsetting, the frictional forces are equal to each other, because of the symmetry of the operation. In rolling, however, the frictional force on the left of the neutral point must be greater than the frictional force on the right. This difference results in a *net frictional force* to the right, making the rolling operation possible by pulling the strip into the roll gap. Furthermore, in order to supply work to the system, the net frictional force and the surface velocity of the roll must be in the same direction. Consequently, the location of the neutral point should be toward the exit in order to satisfy these requirements.

Forward slip in rolling is defined in terms of the exit velocity of the strip, V_f, and the surface speed of the roll, V_r, as

$$\text{Forward slip} = \frac{V_f - V_r}{V_r} \tag{6.24}$$

and is a measure of the relative velocities involved.

1. **Roll pressure distribution.** It can be seen from the preceding discussion that the deformation zone in the roll gap is subjected to a state of stress similar to that in upsetting. However, the calculation of forces and stress distribution in rolling is more involved because the contact surfaces are curved. Furthermore, in cold rolling, the material at the exit is strain hardened, and thus the flow stress at the exit is higher than that at the entry.

The stresses acting on an element in the entry and exit zones, respectively, are shown in Fig. 6.32. Note that the only difference between the two elements is the *direction* of the friction force. Using the slab method of analysis for *plane strain* (described in Section 6.2.2), the stresses in rolling may be analyzed as follows.

From the equilibrium of the horizontal forces on the element shown in Fig. 6.32, we have

$$(\sigma_x + d\sigma_x)(h + dh) - 2pR\,d\phi\sin\phi - \sigma_x h \pm 2\mu p\,R\,d\phi\cos\phi = 0.$$

Simplifying and ignoring the second-order terms, this expression reduces to

$$\frac{d(\sigma_x h)}{d\phi} = 2pR(\sin\phi \mp \mu\cos\phi).$$

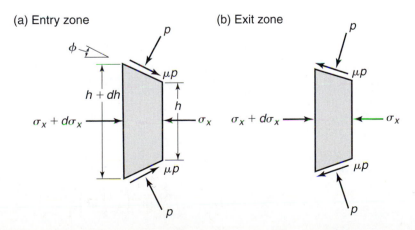

(a) Entry zone   (b) Exit zone

FIGURE 6.32 Stresses acting on an element in rolling: (a) entry zone and (b) exit zone.

In rolling practice, the angle α (see Fig. 6.31) is typically only a few degrees; hence, it can be assumed that $\sin\phi = \phi$ and $\cos\phi = 1$. Thus,

$$\frac{d(\sigma_x h)}{d\phi} = 2pR(\phi \mp \mu). \tag{6.25}$$

Because the angles involved are very small, p can be assumed to be a *principal stress*. The other principal stress is σ_x. The relationship between these two principal stresses and the flow stress, Y_f, of the material is given by Eq. (2.45) for plane strain, or

$$p - \sigma_x = \frac{2}{\sqrt{3}}Y_f = Y_f'. \tag{6.26}$$

Recall that, for a strain-hardening material, the flow stress, Y_f, in these expressions must correspond to the strain that the material has undergone at that particular location in the gap. Rewriting Eq. (6.25), we have

$$\frac{d[(p - Y_f')h]}{d\phi} = 2pR(\phi \mp \mu),$$

or

$$\frac{d}{d\phi}\left[Y_f'\left(\frac{p}{Y_f'} - 1\right)h\right] = 2pR(\phi \mp \mu),$$

which, upon differentiation, becomes

$$Y_f'h\frac{d}{d\phi}\left(\frac{p}{Y_f'}\right) + \left(\frac{p}{Y_f'} - 1\right)\frac{d}{d\phi}(Y_f'h) = 2pR(\phi \mp \mu).$$

The second term in this expression is very small, because as h decreases, Y_f' increases (due to cold working), thus making the product of Y_f' and h nearly a constant, and thus its derivative becomes zero. We now have

$$\frac{\dfrac{d}{d\phi}\left(\dfrac{p}{Y_f'}\right)}{\dfrac{p}{Y_f'}} = \frac{2R}{h}(\phi \mp \mu). \tag{6.27}$$

Letting h_f be the final thickness of the strip being rolled, we have

$$h = h_f + 2R(1 - \cos\phi),$$

or, approximately,

$$h = h_f + R\phi^2. \tag{6.28}$$

Substituting this expression for h in Eq. (6.27) and integrating, we obtain

$$\ln\frac{p}{Y_f'} = \ln\frac{h}{R} \mp 2\mu\sqrt{\frac{R}{h_f}}\tan^{-1}\sqrt{\frac{R}{h_f}}\phi + \ln C$$

or

$$p = CY_f'\frac{h}{R}e^{\mp\mu H},$$

where

$$H = 2\sqrt{\frac{R}{h_f}}\tan^{-1}\left(\sqrt{\frac{R}{h_f}}\phi\right). \tag{6.29}$$

At entry, $\phi = \alpha$; hence, $H = H_0$ with ϕ replaced by α. At exit, $\phi = 0$; hence, $H = H_f = 0$. Also, at entry and exit, $p = Y'_f$. Thus, in the **entry zone**,

$$C = \frac{R}{h_f} e^{\mu H_i}$$

and

$$p = Y'_f \frac{h}{h_0} e^{\mu(H_0 - H)}. \tag{6.30}$$

In the **exit zone**,

$$C = \frac{R}{h_f},$$

and hence

$$p = Y'_f \frac{h}{h_f} e^{\mu H}. \tag{6.31}$$

Note that the pressure p at any location in the roll gap is a function of h and its angular position ϕ along the arc of contact. These expressions also indicate that the pressure increases with increasing strength of the material, increasing coefficient of friction, and increasing R/h_f ratio. The R/h_f ratio in rolling is equivalent to the a/b ratio in upsetting, as described in Section 6.2.2.

The dimensionless theoretical pressure distribution in the roll gap is shown in Fig. 6.33. Note the similarity of this curve (friction hill) to that shown in Fig. 6.5. Also note that the neutral point shifts toward the exit as friction decreases. The reason is that when friction approaches zero, the rolls begin to slip and the relative velocity between the roll and the strip is all in one direction.

The effect of reduction in thickness of the strip on the pressure distribution is shown in Fig. 6.34. As reduction increases, the length of contact in the roll gap increases, in turn increasing the peak pressure. The curves shown in Fig. 6.34 are theoretical; actual pressure distributions, as determined experimentally, are smoother, with rounder peaks.

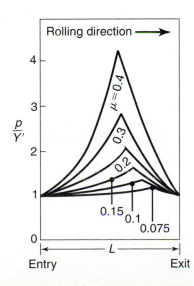

FIGURE 6.33 Pressure distribution in the roll gap as a function of the coefficient of friction. Note that as friction increases, the neutral point shifts toward the entry. Without friction, the rolls will slip, and the neutral point shifts completely to the exit. (See also Table 4.1.)

FIGURE 6.34 Pressure distribution in the roll gap as a function of reduction in thickness. Note the increase in the area under the curves with increasing reduction, thus increasing the roll force.

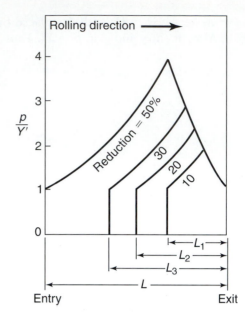

2. **Determining the location of the neutral point.** The neutral point can be determined simply by equating Eqs. (6.30) and (6.31); thus, at the neutral point,

$$\frac{h_o}{h_f} = \frac{e^{\mu H_o}}{e^{2\mu H_n}} = e^{\mu(H_o - 2H_n)},$$

or

$$H_n = \frac{1}{2}\left(H_o - \frac{1}{\mu}\ln\frac{h_o}{h_f}\right). \tag{6.32}$$

Substituting Eq. (6.32) into Eq. (6.29), we obtain

$$\phi_n = \sqrt{\frac{h_f}{R}}\tan\left(\sqrt{\frac{h_f}{R}}\cdot\frac{H_n}{2}\right). \tag{6.33}$$

3. **Front and back tension.** The roll force, F, can be reduced by various means, mainly by (a) lowering friction, (b) using rolls of smaller radii, (c) taking smaller reductions, and (d) raising workpiece temperature. In addition, a particularly effective method is to reduce the apparent compressive yield stress of the material by applying longitudinal *tension*. Recall from the discussion on yield criteria in Section 2.11 that when a tensile stress is applied to a strip (Fig. 6.35), the yield stress normal to the strip surface decreases, and thus the roll pressure decreases.

Tensile forces in rolling can be applied either at the entry (**back tension**, σ_b) or at the exit (**front tension**, σ_f) of the strip, or both. Equations (6.30) and (6.31) can now be modified to include the effect of tension for the entry and exit zones, respectively, as follows:

$$\text{Entry zone: } p = (Y_f' - \sigma_b)\frac{h}{h_o}e^{\mu(H_o - H)} \tag{6.34}$$

and

$$\text{Exit zone: } p = (Y_f' - \sigma_f)\frac{h}{h_f}e^{\mu H}. \tag{6.35}$$

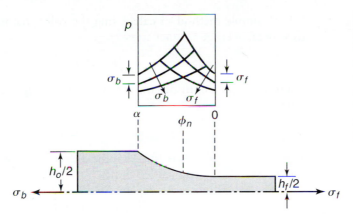

FIGURE 6.35 Pressure distribution as a function of front and back tension in rolling. Note the shifting of the neutral point and the reduction in the area under the curves (hence reduction in the roll force) as tensions increase.

Depending on the relative magnitudes of the tensile stresses applied, the neutral point may *shift*, as shown in Fig. 6.35. As expected, this shift then affects the pressure distribution, torque, and power requirements in rolling.

Tensions are particularly important in rolling thin, high-strength materials, because such materials require high roll forces. Front tension in practice is typically controlled by the torque on the coiler (**delivery reel**), around which the rolled sheet is coiled. Back tension is controlled by a braking system in the uncoiler (**payoff reel**). Special instrumentation is available for these controls.

EXAMPLE 6.3 Back tension required to cause slip in rolling

It is to be expected that in rolling a strip that if the back tension, σ_b, is too high, the rolls will begin to slip. Derive an expression for the magnitude of back tension required to make the rolls begin to slip.

Solution. Slipping of the rolls means that the neutral point has moved all the way to the exit in the roll gap. Thus, the whole contact area now becomes the entry zone, and therefore Eq. (6.34) is applicable. We also know that when $\phi = 0$, $H = 0$; hence, the pressure at the exit is given by

$$p_{\phi=0} = (Y_f' - \sigma_b)\left(\frac{h_f}{h_o}\right)e^{\mu H_o}.$$

However, the pressure at the exit is equal to Y_f'. Rearranging this equation, we have

$$\sigma_b = Y_f'\left[1 - \left(\frac{h_o}{h_f}\right)(e^{-\mu H_o})\right],$$

where H_0 is obtained from Eq. (6.29) for the condition of $\phi = \alpha$. Since all the quantities are known, the magnitude of the back tension can be calculated.

---

4. **Calculating roll forces.** It will be noted that the *roll force*, F (also called *roll-separating force*), on the strip is the product of the area under the pressure vs. contact-length curve and the strip width, w. The area can be obtained graphically or analytically, and the roll force can also be calculated from the expression

$$F = \int_0^{\phi_n} wpR\,d\phi + \int_{\phi_n}^{\alpha} wpR\,d\phi. \tag{6.36}$$

A simple method of calculating the roll force is to multiply the contact area by an average contact stress, p_{av},

$$F = Lwp_{av}, \tag{6.37}$$

where L is the length of contact and can be approximated as the projected length; thus

$$L = \sqrt{R\Delta h}, \tag{6.38}$$

where R is the roll radius and Δh is the difference between the original and final thicknesses of the strip (called **draft**).

The magnitude of p_{av} depends on the h/L ratio, where h is now the *average thickness* of the strip in the roll gap (see also Figs. 6.12 and 6.13). For large h/L ratios (such as small reductions in strip thickness and/or large roll diameters), the rolls will act in a manner similar to indenters in a hardness test. Friction is not significant, and p_{av} is obtained from Fig. 6.12, where $a = L/2$. Small h/L ratios (such as large reductions and/or large roll diameters) are equivalent to high a/h ratios; thus, friction is predominant, and p_{av} is obtained from Eq. (6.15). Note that for strain-hardening materials, the appropriate flow stresses must be determined.

As an approximation and for low frictional conditions, Eq. (6.37) can be simplified to

$$F = Lw\overline{Y}', \tag{6.39}$$

where $\overline{Y}'$ is the average flow stress in plane strain (see Fig. 2.37) of the material in the roll gap. For higher frictional conditions, an expression similar to Eq. (6.15) can be written as

$$F = Lw\overline{Y}'\left(1 + \frac{\mu L}{2h_{av}}\right). \tag{6.40}$$

5. **Roll torque and power.** The *roll torque, T,* for each roll can be calculated from the expression

$$T = \int_{\phi_n}^{\alpha} w\mu pR^2\,d\phi - \int_{0}^{\phi_n} w\mu pR^2\,d\phi.$$
$$\text{(entry zone)} \qquad \text{(exit zone)} \tag{6.41}$$

Note that the minus sign indicates the change in direction of the friction force at the neutral point. Thus, for example, if the frictional forces are equal to each other, the torque is zero.

The torque in rolling can also be estimated by assuming that the roll force, F, acts in the *middle* of the arc of contact (that is, a moment arm of $0.5L$) and that this force is perpendicular to the plane of the strip. (It has been found that whereas $0.5L$ is a good estimate for hot rolling, $0.4L$ is a better estimate for cold rolling.)

The **torque per roll** is then

$$T = \frac{FL}{2}.$$

The **power required per roll** is

$$\text{Power} = T\omega, \tag{6.42}$$

where $\omega = 2\pi N$, and N is the revolutions per minute of the roll. Consequently, the power per roll in kW is

$$\text{Power} = \frac{\pi F L N}{60,000}, \tag{6.43}$$

where F is in newtons, L is in meters, and N is the rpm of the roll. The power per roll in hp is

$$\text{Power} = \frac{\pi F L N}{33,000}, \tag{6.44}$$

where F is in lb and L is in ft.

EXAMPLE 6.4 Power required in rolling

A 9-in.-wide 6061-O aluminum strip is rolled from a thickness of 1.00 in. to 0.80 in. For a roll radius of 12 in. and roll rpm of 100, estimate the total power required for this operation.

Solution. The power needed for a set of two rolls is given by Eq. (6.44) as

$$\text{Power} = \frac{2\pi F L N}{33,000} \text{ hp,}$$

where F is obtained from Eq. (6.39) and L from Eq. (6.38). Hence,

$$F = L w \overline{Y}' \qquad \text{and} \qquad L = \sqrt{R \Delta h}.$$

Therefore,

$$L = \sqrt{(12)(1.0 - 0.8)} = 1.55 \text{ in.} = 0.13 \text{ ft} \qquad \text{and} \qquad w = 9 \text{ in.}$$

For 6061-O aluminum, we have $K = 30,000$ psi and $n = 0.2$ (from Table 2.3). The true strain in this operation is

$$\epsilon_1 = \ln\left(\frac{1.0}{0.8}\right) = 0.223.$$

Thus, from Eq. (6.10), we have

$$\overline{Y} = \frac{(30,000)(0.223)^{0.2}}{1.2} = 18,500 \text{ psi}$$

and

$$\overline{Y}' = (1.15)(18,500) = 21,275 \text{ psi.}$$

Therefore,

$$F = (1.55)(9)(21.275) = 297,000 \text{ lb,}$$

and for two rolls

$$\text{Power} = \frac{(2\pi)(297,000)(0.13)(100)}{33,000} = 735 \text{ hp.}$$

6. **Roll forces in hot rolling.** Because ingots and slabs are usually hot rolled, the calculation of forces and torque in hot rolling is important. There are, however, two difficulties: (a) the proper estimation of the coefficient of friction, μ, at elevated temperatures, which can range from 0.2 to 0.7 (see Table 4.1); and (b) the strain-rate sensitivity of materials at elevated temperatures (see Section 2.2.7).

The *average strain rate,* $\dot{\bar{\epsilon}}$, in flat rolling can be obtained by dividing the strain by the time required for an element to undergo this strain in the roll gap. The time can be approximated as L/V_r, hence

$$\dot{\bar{\epsilon}} = \frac{V_r}{L}\ln\left(\frac{h_o}{h_f}\right). \tag{6.45}$$

The flow stress, Y_f, of the material corresponding to this strain rate must first be obtained and then substituted into the proper equations. These calculations are approximate, especially because of variations in μ and temperature within the strip in hot rolling.

7. **Friction in rolling.** It is evident that without friction the rolls cannot pull the strip into the roll gap. On the other hand, as described earlier and as expected, forces and power requirements will rise as friction increases. It has been observed that in cold rolling, the coefficient of friction typically ranges between 0.02 and 0.3 (see also Table 4.1) depending on the materials involved and the lubricants used. As described in Section 4.4.1, μ will decrease with effective lubricants and in regimes approaching *hydrodynamic lubrication* (which, for instance, can occur in cold rolling of aluminum at high speeds). In hot rolling, μ may range from about 0.2 (with effective lubrication), to as high as 0.7 (indicating sticking, which typically occurs with steels, stainless steels, and high-temperature alloys). (See also Table 4.1.)

The maximum possible *draft,* that is, $h_o - h_f$, in flat rolling can be shown to be a function of friction and radius as

$$\Delta h_{\max} = \mu^2 R. \tag{6.46}$$

Thus, the higher the friction coefficient and the larger the roll radius, the greater is the maximum draft. It will be noted that the draft is zero when there is no friction. The maximum value of the angle α in Fig. 6.31 (called **angle of acceptance**) can be shown to be geometrically related to Eq. (6.46). Based on the simple model of a block sliding down an inclined plane, it can be found that

$$\alpha_{\max} = \tan^{-1}\mu. \tag{6.47}$$

If $\alpha_{\max}$ is larger than this value, the rolls begin to slip, because the friction is not high enough to pull the material through the roll gap.

8. **Roll deflections and roll flattening.** As expected, roll forces tend to bend the rolls (Fig. 6.36a), thus resulting in a strip that is thicker at its center than at its edges (**crown**). The usual technique for avoiding this problem is to grind the roll in such a way that its diameter at the center is slightly larger than at the edges. This curvature is known as **camber.** In sheet-metal rolling practice, the camber is typically less than 0.50 mm (0.02 in.) on the roll diameter. In addition, because of the heat generated during rolling, rolls can become slightly barrel-shaped. Known as **thermal camber,** this effect can be controlled by varying the location of the coolant on the rolls along their axial direction.

When properly designed, such rolls produce flat strips, as shown in Fig. 6.36b. However, a particular camber is correct only for a certain load and width of strip. In hot rolling, uneven temperature distribution in the roll also

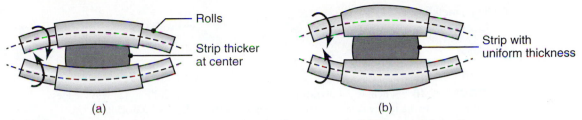

FIGURE 6.36 (a) Bending of straight cylindrical rolls (exaggerated) because of the roll force and (b) bending of rolls, ground with camber, that produce a sheet of uniform thickness during rolling.

can cause its diameter to vary along its length. Camber in hot rolling can be controlled by varying the location of the coolant (lubricant) on the rolls.

Roll forces also tend to *flatten* the rolls elastically, much like the flattening of tires on automobiles. Flattening increases the roll's radius, causing a larger contact area for the same reduction in thickness; this situation also results in an increase in the roll force, F.

It can be shown that the new (distorted) roll radius, R', is given by the expression

$$R' = R\left(1 + \frac{CF'}{h_o - h_f}\right), \tag{6.48}$$

where the value of C is 2.3×10^{-2} mm^2/kN (1.6×10^{-4} in.^{2}/klb) for steel rolls and 4.57×10^{-2} (3.15×10^{-4}) for cast-iron rolls, and F' is the **roll force per unit width** of strip, expressed in kN/mm (klb/in.). The higher the elastic modulus of the roll material, the less the roll distorts. The magnitude of R' cannot be calculated directly from Eq. (6.48) because the roll force is itself a function of the roll radius. The solution then is obtained iteratively. Note that R in all previous equations should be replaced by R' when significant roll flattening occurs.

9. **Spreading.** Rolling plates and sheets with high strip width-to-thickness ratios is essentially a process of plane strain. However, at smaller ratios, such as in rolling a square cross section, the width increases considerably during rolling, known as *spreading* (Fig. 6.37). It can be shown that spreading decreases with (a) increasing width-to-thickness ratios of the entering material, (b) decreasing friction, and (c) increasing ratios of roll radius to strip thickness. One technique of preventing spreading is by using a pair of vertical rolls that constrain the edges of the product being rolled (**edger mill**).

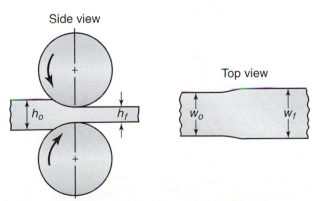

FIGURE 6.37 Increase in the width of a strip (spreading) during flat rolling. Spreading can be similarly observed when dough is rolled on a flat surface with a rolling pin.

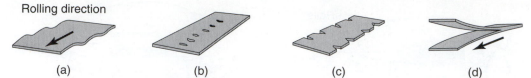

Rolling direction

(a)   (b)   (c)   (d)

FIGURE 6.38   Schematic illustration of some defects in flat rolling: (a) wavy edges, (b) zipper cracks in the center of strip, (c) edge cracks, and (d) alligatoring.

6.3.2 Defects in rolled products

Successful rolling practice requires balancing many factors, including material properties, process variables, and lubrication. There may be defects on the surfaces of the rolled plates and sheets, or there may be structural defects within the material. **Surface defects** may result from inclusions and impurities in the material, scale, rust, dirt, roll marks, and other causes related to the prior treatment and working of the material. In hot rolling of blooms, billets, and slabs, the surface is usually preconditioned by various means, such as by **scarfing** (using a torch).

Structural defects are those that distort or affect the integrity of the rolled product. Some typical defects are shown in Fig. 6.38. **Wavy edges** are caused by bending of the rolls, whereby the edges of the strip become thinner than at the center. Because the edges elongate more than the center and are restrained by the bulk of the material from expanding freely, they buckle. The cracks shown in Figs. 6.38b and c are usually caused by low material ductility and barreling of the edges. **Alligatoring** is a complex phenomenon that results from inhomogeneous deformation of the material during rolling or from defects in the original cast ingot, such as piping.

Residual stresses.   Residual stresses can develop in rolled sheets and plates because of inhomogeneous plastic deformation in the roll gap. Small-diameter rolls and/or small reductions in thickness tend to work the metal plastically at its surfaces (similarly to shot peening or roller burnishing; Section 4.5.1). This type of deformation generates compressive residual stresses on the surfaces and tensile stresses in the bulk (Fig. 6.39a). Large-diameter rolls and/or high reductions, on the other hand, tend to deform the bulk to a greater extent than the surfaces, because of frictional constraint at the surfaces along the arc of contact. This situation generates residual stresses that are opposite to those of the previous case, as shown in Fig. 6.39b.

FIGURE 6.39   The effect of roll radius on the type of residual stresses developed in flat rolling: (a) small rolls and/or small reduction in thickness and (b) large rolls and/or large reduction in thickness.

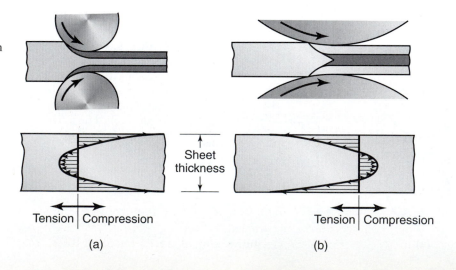

Sheet thickness

Tension | Compression   Tension | Compression

(a)   (b)

6.3.3 Vibration and chatter in rolling

Vibration and *chatter* in rolling, as well as other metalworking operations, can have significant effects on product quality and productivity. It has been estimated, for example, that modern rolling mills could operate at speeds up to 50% higher were it not for chatter. Considering the very high cost of rolling mills, this issue is a major economic concern. Chatter, generally defined as *self-excited vibration,* can occur in processes such as rolling, extrusion, drawing, machining, and grinding, as described in Sections 8.12 and 9.6.8. In rolling, it leads to random variations in the thickness of the rolled sheet and its surface finish and may lead to excessive scrap. Chatter has been found to occur predominantly in *tandem mills* (see Fig. 6.41d). Because of the many variables involved, it is a very complex phenomenon and results from interactions between the structure of the mill stand and the dynamics of the rolling operation. Rolling speed and lubrication have been found to be the two most important parameters.

The vibration modes commonly encountered in rolling are classified as torsional, third-octave, and fifth-octave chatter (*octave* meaning a frequency on the eighth degree from a given frequency).

1. **Torsional chatter** is characterized by a low resonant frequency (approximately 5 to 15 Hz) and is usually a result of forced vibration (see Section 8.12), although it can also occur simultaneously with third-octave chatter. Torsional chatter leads to minor variations in gage thickness and surface finish and is usually not significant unless it is due to factors such as malfunctioning speed controls, broken gear teeth, and misaligned shafts.

2. **Third-octave chatter** occurs in the frequency range of 125 to 240 Hz (the third musical octave is 128 to 256 Hz) and is *self-excited*; that is, energy from the rolling mill is transformed into vibratory energy, regardless of the presence of any external forces (*forced vibration*). This mode of vibration is the most serious in rolling, as it leads to significant gage variations, fluctuations in strip tension between the stands, and often results in strip breakage. Third-octave chatter is generally controlled by reducing the rolling speed. Although not always practical to implement, it has also been suggested that this type of chatter can be reduced by (a) increasing the distance between the stands of the rolling mill, (b) increasing the strip width, w, (c) incorporating *dampers* (see Section 8.12) in the roll supports, (d) decreasing the reduction per pass (draft), (e) increasing the roll radius R, and (e) increasing the strip-roll friction coefficient.

3. **Fifth-octave chatter** occurs in the frequency range of 550 to 650 Hz. It has been attributed to chatter marks that develop on work rolls, caused by (a) extended operation at certain critical speeds, (b) surface defects on a backup roll or in the incoming strip, and/or (c) improperly ground rolls. The chatter marks or striations are then imprinted onto the rolled sheet at any speed, thus adversely affecting its surface appearance. Furthermore, fifth-octave chatter is undesirable because of the objectionable noise it generates. This type of chatter can be controlled by (a) modulating the mill speed, (b) using backup rolls of progressively larger diameters in the stands of a tandem mill, (c) avoiding chatter in roll grinding, and (d) eliminating the sources of any other vibrations in the mill stand.

6.3.4 Flat-rolling practice

The initial breaking down of a cast ingot by hot rolling (at temperature similar to forging, Table 6.3) converts the coarse-grained, brittle, and porous structure into a *wrought* structure (Fig. 6.29) with finer grains and enhanced ductility. Recall that traditional methods of rolling ingots are now rapidly replaced by continuous casting and rolling (Section 5.7).

The product of the first hot-rolling operation is called a **bloom** or **slab** (see Fig. 6.28). A bloom usually has a square cross section, at least 150 mm (6 in.) on the side; a slab is usually rectangular in cross section. Blooms are further processed by *shape rolling* into structural shapes, such as I-beams and railroad rails. Slabs are rolled into plates and sheet. **Billets** are usually square in cross section and their area smaller than that of blooms. They are rolled into various shapes, such as round rods and bars, using shaped rolls. Hot-rolled round rods are used as the starting material for rod and wire drawing and are called **wire rods.**

In hot rolling of blooms, billets, and slabs, the surface of the material is usually *conditioned* (prepared for a particular operation) prior to rolling. Conditioning is done by means such as scarfing to remove heavy scale, or by rough grinding. Prior to cold rolling, the scale developed during hot rolling or any other surface defects may be removed by *pickling* with acids, by mechanical means (such as blasting with water), or by grinding.

Pack rolling is a flat-rolling operation in which two or more layers of metal are rolled together, thus improving productivity. Aluminum foil, for example, is pack rolled in two layers. It can easily be observed that one side of foil is matte, and the other side is shiny. The foil-to-foil side has a matte and satiny finish, whereas the foil-to-roll side is shiny and bright (because it has been in contact with the polished roll surfaces).

Mild steel, when stretched during sheet-forming operations, undergoes **yield-point elongation**. This phenomenon causes surface irregularities called **stretcher strains** or **Lueder's bands.** (See also Section 7.2.) To avoid this situation, the sheet metal typically is subjected to a light pass of 0.5–1.5% reduction, known as **temper rolling** (*skin pass*).

A rolled sheet may not be sufficiently flat as it leaves the roll gap because of variations in the material or in the processing parameters during rolling. To improve flatness, the strip is passed through a series of **leveling rolls.** Each roll is usually driven separately with individual electric motors. The strip is flexed in opposite directions as it passes through the sets of rollers. Several different roller arrangements are used (Fig. 6.40).

Sheet thickness is identified by a **gage number**; the smaller the number, the thicker the sheet is. There are various numbering systems for different sheet metals. Rolled sheets of copper and brass are also identified by thickness changes during rolling, such as $\frac{1}{4}$ hard, $\frac{1}{2}$ hard, and so on.

Lubrication. Ferrous alloys are usually hot rolled without a lubricant, although graphite may be used. Aqueous solutions are used to cool the rolls and break up the scale. Nonferrous alloys are hot rolled using a variety of compounded oils, emulsions, and fatty acids. Cold rolling is done with low-viscosity lubricants, including mineral oils, emulsions, paraffin, and fatty oils.

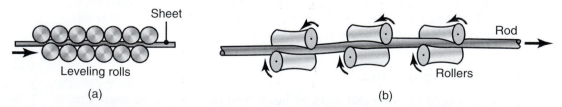

(a)   (b)

FIGURE 6.40   Schematic illustrations of roller leveling to (a) flatten rolled sheets and (b) straighten round rods.

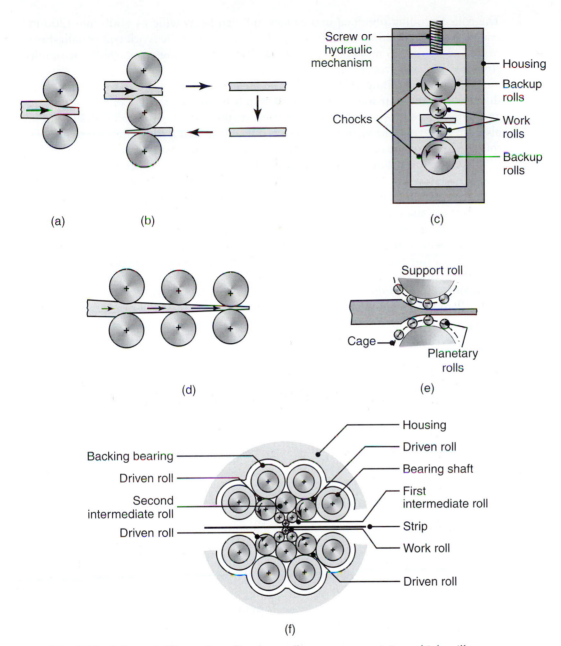

FIGURE 6.41 Schematic illustration of various roll arrangements: (a) two-high mill, (b) three-high mill, (c) four-high mill, (d) tandem rolling, with three stands, (e) planetary mill, and (f) cluster (*Sendzimir*) mill.

Equipment. A variety of rolling equipment is available having several roll arrangements, the major types of which are shown in Fig. 6.41. Small-diameter rolls are preferable because the smaller the roll radius, the lower will be the roll force. On the other hand, small rolls deflect under roll forces and have to be supported by other rolls to maintain dimensional control (Figs. 6.41c and e).

Two-high or three-high rolling mills (developed in the mid-1800s) are typically used for initial breakdown passes on cast ingots (*primary roughing*), with roll diameters ranging up to 1400 mm (55 in.). The cluster mill (Sendzimir, or Z, mill; Fig. 6.41e) is particularly suitable for cold rolling thin strips of high-strength metals.

The rolled product obtained in a cluster mill can be as wide as 5000 mm (200 in.) and as thin as 0.0025 mm (0.0001 in.). The diameter of the work roll (smallest roll) can be as small as 6 mm ($\frac{1}{4}$ in.) and is usually made of tungsten carbide for rigidity, strength, and wear resistance.

In **tandem rolling** (Fig. 6.41d), the strip is rolled continuously as its passes through a number of **stands;** a group of stands is called a **train.** Control of the gage and of the speed at which the strip travels through each roll gap is critical. Flat rolling also can be carried out with front tension only, using idling rolls (**Steckel rolling**). For this case, the torque on the roll is zero, assuming frictionless bearings.

Stiffness in rolling mills is important for controlling dimensions. Modern mills are highly automated, with rolling speeds as high as 25 m/s (5000 ft/min). The basic requirements for roll materials are mainly strength and resistance to wear. Three common roll materials are cast iron, cast steel, and forged steel. For hot rolling, roll surfaces are generally roughened and may even have notches or grooves in order to pull the metal through the roll gap at high reductions. Rolls for cold rolling are ground to a fine finish and, for special applications, are also polished.

Minimills. In *minimills,* scrap metal is melted in electric-arc furnaces, cast continuously, and rolled directly into specific lines of products. Each minimill produces essentially one kind of rolled product (such as rod, bar, or structural shapes) from one type of metal and is usually dedicated to markets within the mill's particular geographic area.

Integrated mills. *Integrated mills* are large facilities that involve complete activities, from the production of hot metal in a blast furnace to casting and rolling of finished products that are ready to be shipped to the customer. Technically, the only difference between integrated mills and minimills is the type of furnace used; however, the use of blast furnaces implies different economies of scale than electric-arc furnaces, and these are much larger, more elaborate facilities. Nearly all integrated mills specialize in flat-rolled metal or plate.

6.3.5 Miscellaneous rolling operations

1. **Shape rolling.** Straight structural shapes of various cross sections, channel sections, I-beams, and railroad rails are rolled by passing the stock through a number of pairs of specially designed rollers (Fig. 6.42). Airfoil shapes also can be produced by shape-rolling techniques. The original material shape is usually a bloom (Fig. 6.28). The design of a series of rolls (**roll-pass design**) requires special care in order to avoid defect formation and hold dimensional tolerances (although some defects may already exist in the material being rolled). For example, in shape rolling a channel, the reduction is different in various locations within the section. Thus, elongation is not uniform throughout the cross section, which can cause warping or cracking.

2. **Ring rolling.** In this process, a small-diameter, thick ring is expanded into a larger diameter, thinner ring by placing the ring between two rolls, one of which is driven (Fig. 6.43). The ring thickness is reduced by moving the rolls closer as they continue rotating. Because of volume constancy, the reduction in thickness is compensated for by an increase in the diameter of the ring. A variety of cross sections can be ring rolled with variously shaped rolls.

   This process can be carried out at room or elevated temperatures, depending on the size and strength of the product. The advantages of ring rolling, compared with other processes for making the same part, are shorter production runs, material savings, close dimensional tolerances, and favorable grain flow in the product. Typical parts made by ring rolling include large rings for rockets and

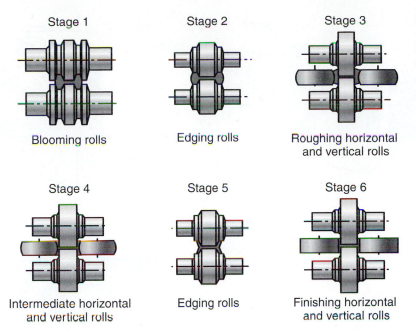

Stage 1 — Blooming rolls

Stage 2 — Edging rolls

Stage 3 — Roughing horizontal and vertical rolls

Stage 4 — Intermediate horizontal and vertical rolls

Stage 5 — Edging rolls

Stage 6 — Finishing horizontal and vertical rolls

FIGURE 6.42 Stages in shape rolling of an H-section. Several other structural sections, such as channels and rails, also are rolled by this process.

turbines, gearwheel rims, ball- and roller-bearing races, flanges and reinforcing rings for pipes, and pressure vessels.

3. **Thread and gear rolling.** This is a cold-forming process in which threads are formed on round rods or workpieces by passing them between reciprocating or rotating dies (Fig. 6.44a). Typical products made include screws, bolts, and similar threaded parts. With flat dies, the threads are formed on the rod or wire with each stroke of the reciprocating die. In all thread-rolling processes, it is essential that the material have sufficient ductility and that the rod or wire be of proper size. Lubrication is important for good surface finish and to minimize defects. Almost all externally threaded fasteners are made by this process. Two- or three-roller thread-rolling machines are also available. Threads also are formed using a rotary die (Fig. 6.44b). Production rates in thread rolling are very high, with production rates as high as 30 pieces per second. Thread rolling can also be carried out *internally* with a fluteless forming tap. The process is similar to external thread rolling and produces accurate threads with good strength.

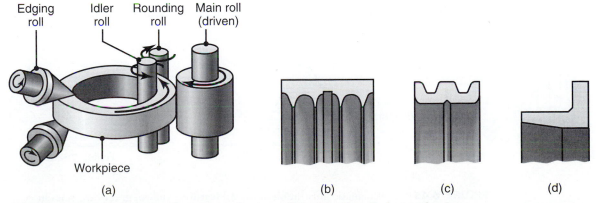

(a) (b) (c) (d)

FIGURE 6.43 (a) Schematic illustration of a ring-rolling operation. Reducing the ring thickness results in an increase in its diameter. (b)–(d) Three examples of cross sections that can be produced by ring rolling.

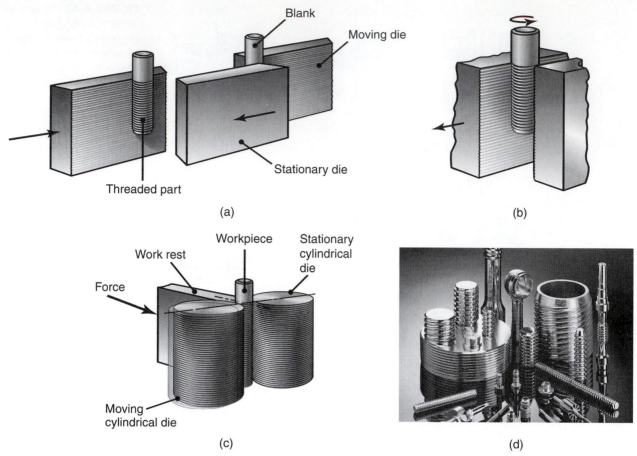

FIGURE 6.44   Thread-rolling processes: (a) and (b) reciprocating flat dies, (c) two-roller dies, and (d) thread-rolled parts, made economically and at high production rates. *Source:* (d) Courtesy of Tesker Manufacturing Corp.

The thread-rolling process produces threads without any loss of metal and with greater strength, because of the cold working involved. The surface finish is very smooth, and the process induces compressive residual stresses on part surfaces, thus improving fatigue life. Because of volume constancy in plastic deformation, a rolled thread requires a round stock of smaller diameter to produce the same major diameter as that of a machined thread (Fig. 6.45).

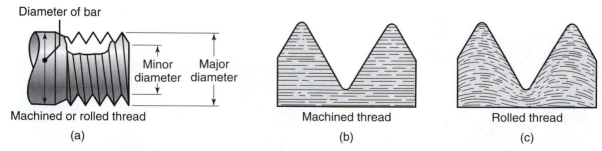

FIGURE 6.45   (a) Schematic illustration of thread features, and (b) grain-flow lines in machined and (c) rolled threads. Note that unlike machined threads, which are cut through the grains of the metal, rolled threads follow the grains and because of the cold working involved, they are stronger.

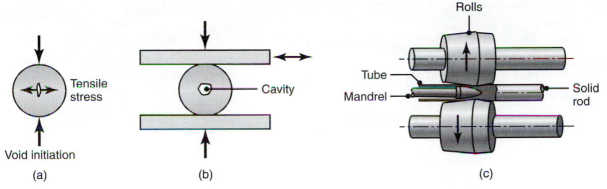

FIGURE 6.46 (a) Cracks developed in a solid round bar due to secondary tensile stresses, (b) simulation of the rotary-tube-piercing process, (c) the Mannesmann process for seamless tube making. The mandrel is held in place by a long rod, although techniques also have been developed whereby the mandrel remains in place without using a rod.

Furthermore, whereas machining removes material by cutting through the grain-flow lines of the material, rolled threads have a grain-flow pattern that improves the strength of the thread, because of the cold working involved.

Spur and helical *gears* also can be produced by processes similar to thread rolling. The process may be carried out on solid cylindrical blanks or on precut gears. Helical gears can also be made by a direct extrusion process, using specially shaped dies. Cold rolling of gears has many applications in automatic transmissions and power tools.

4. **Rotary tube piercing.** An important hot-working process, rotary tube piercing is used to make long, thick-walled seamless tubing, as shown in Fig. 6.46. This process is based on the principle that when a round bar is subjected to radial compression in the manner shown in Fig. 6.46a, tensile stresses develop at the center of the roll. When the rod is subjected to cycling compressive stresses, as shown in Fig. 6.46b, a cavity begins to form at the center of the rod.

The rotary-tube-piercing process (also known as the **Mannesmann process** and developed in the 1880s) is carried out by an arrangement of rotating rolls, as shown in Fig. 6.46c. The axes of the rolls are skewed in order to pull the round bar through the rolls by the longitudinal-force component of their rotary action. A mandrel assists the operation by expanding the hole and sizing the inside diameter of the tube. Because of the severe deformation that the metal undergoes in this process, it is important that the stock be of high quality and defect free.

5. **Tube rolling.** The diameter and thickness of tubes and pipe that are produced by various processes can be reduced by *tube rolling,* using shaped rolls, either with or without mandrels. In the **pilger mill,** the tube and an internal mandrel undergo a reciprocating motion, and the tube is advanced and rotated periodically. Steel tubing of 265 mm (10.5 in.) in diameter has been produced by this process.

6.4 | Extrusion

In the basic extrusion process (Fig. 6.47), developed in the late 1700s for producing lead pipe, a round billet is placed in a chamber and forced through a die opening by a ram. The die may be round or of various other shapes. Typical parts made are

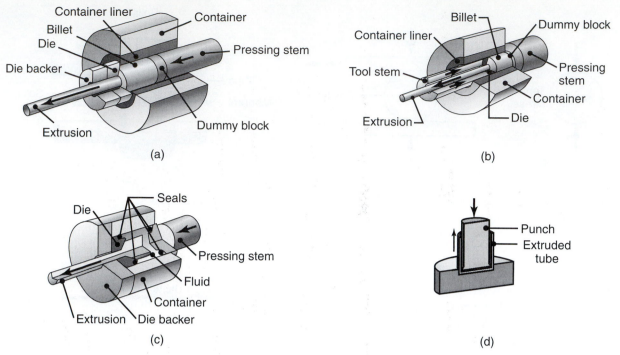

FIGURE 6.47 Types of extrusion: (a) direct, (b) indirect, (c) hydrostatic, and (d) impact.

railings for sliding doors, window frames, aluminum ladders, tubing, and structural and architectural shapes. Extrusion may be carried out cold or at elevated temperatures. Because a chamber with a constant specific volume is involved, each billet is extruded individually, and hence extrusion is basically a batch process.

There are four basic types of extrusion, as described below and illustrated in Figs. 6.47 and 6.48.

1. **Direct extrusion** (*forward extrusion*) is similar to forcing toothpaste through the opening of a tube. Note that the billet in this process slides relative to the container wall.

2. In **indirect extrusion** (*reverse, inverted,* or *backward extrusion*), the die moves toward the billet and there is no relative motion at the billet-container interface, except at the die.

3. In **hydrostatic extrusion,** the container is filled with a fluid that transmits the pressure to the billet, which is then extruded through the die. There is essentially no friction along the container walls.

4. **Impact extrusion** is a form of indirect extrusion and is particularly suitable for hollow shapes.

There are several parameters describing the extrusion process. The **extrusion ratio**, R, is defined as

$$R = \frac{A_o}{A_f}, \qquad (6.49)$$

where A_o is the cross-sectional area of the billet and A_f is the area of the extruded product (Fig. 6.47). A geometric feature that describes the shape of the extruded

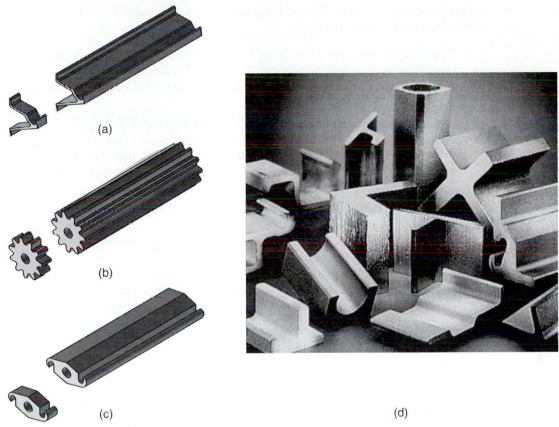

(a)

(b)

(c)

(d)

FIGURE 6.48 (a)–(c) Examples of extrusions and products made by sectioning them and (d) examples of extruded cross sections. *Source:* (a)–(c) courtesy of Kaiser Aluminum, (d) courtesy of Plymouth Extruded Shapes.

product is the **circumscribing-circle diameter** (CCD), which is the diameter of the smallest circle into which the extruded cross section will fit. The CCD for a square extruded cross section, for example, is the diagonal dimension of the cross section, or 1.44 times its side dimension. The complexity of an extrusion is described by the term **shape factor,** which is the ratio of the perimeter of the part to its cross-sectional area. Thus, for example, a solid, round extrusion is the simplest shape, having the lowest shape factor. (See also Section 11.2.2 on the shape of metal powders.)

6.4.1 Metal flow in extrusion

The *metal flow pattern* in extrusion is an important factor in the overall process, as it is in all other metalworking processes. A common technique for investigating the flow pattern is to cut a round billet lengthwise and mark one face with a square grid pattern. The two halves are then placed together in the container (they may also be *brazed* to keep the two halves intact; see Section 12.13) and are extruded as one piece. They are then taken apart and observations are made regarding the distortion of the grid lines.

Figure 6.49 shows three results typical of direct extrusion with *square dies,* where the die angle is 90°. The main factors involved in how these different flow

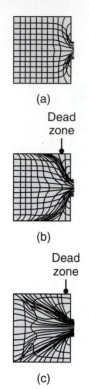

(a)

Dead zone

(b)

Dead zone

(c)

FIGURE 6.49 Schematic illustration of three different types of metal flow in direct extrusion. The die angle in these illustrations is 90°.

patterns develop are (a) friction at billet-container and billet-die interfaces and (b) thermal gradients within the billet.

1. The most homogeneous (uniform) flow pattern (Fig. 6.49a) is obtained when there is no friction at the interfaces. This type of flow occurs when the lubricant is very effective or in indirect extrusion, where there is no friction at the billet-container interfaces.

2. When friction along all interfaces is high, a **dead-metal zone** develops (Fig. 6.49b). Note the high-shear area as the material flows into the die exit (similar to liquid flowing into a funnel). This configuration may indicate that the billet surfaces (with their oxide layer and lubricant present) could enter this high-shear zone and be extruded, causing defects in the extruded product (see Section 6.4.4).

3. In the third configuration, the high-shear zone extends farther back into the billet (Fig. 6.49c). This situation can be due to high container-wall friction (which retards the flow of the billet) or from materials in which the flow stress drops rapidly with increasing temperature (such as titanium). In hot extrusion, the material near the container walls cools rapidly, whereby the material becomes stronger; as a result, the material in the central regions of the billet flows toward the die more easily than that at the outer regions. A large dead-metal zone then forms, and the flow is inhomogeneous. This flow pattern leads to a defect known as a *pipe* or *extrusion defect* (Section 6.4.4).

6.4.2 Mechanics of extrusion

In this section, we develop equations in order to determine the extrusion force under different conditions of temperature and friction, and how this force can be minimized. The ram or stem force in direct extrusion (Fig. 6.47a) for different conditions can be calculated as described below.

1. **Ideal force, no friction.** Based on the extrusion ratio, the absolute value of the true strain that the material undergoes is

$$\epsilon_1 = \ln\left(\frac{A_o}{A_f}\right) = \ln\left(\frac{L_f}{L_o}\right) = \ln R, \qquad (6.50)$$

where A_o and A_f and L_o and L_f are the areas and the lengths of the billet and the extruded product, respectively. For a perfectly plastic material with a yield stress, Y, the energy dissipated in plastic deformation per unit volume, u, is

$$u = Y\epsilon_1. \qquad (6.51)$$

Hence, the work done on the billet is

$$\text{Work} = (A_o)(L_o)(u). \qquad (6.52)$$

The work is supplied by the ram force, F, which travels a distance L_o. Therefore,

$$\text{Work} = FL_o = pA_oL_o, \qquad (6.53)$$

where p is the extrusion pressure. Equating the work of plastic deformation to the external work done, we have

$$p = u = Y\ln\left(\frac{A_o}{A_f}\right) = (Y)(\ln R), \qquad (6.54)$$

and the ideal force is $F = pA_o$.

Note that for strain-hardening materials, Y should be replaced by the *average flow stress*, $\overline{Y}$. Also note that Eq. (6.54) is equal to the area under the true-stress–true-strain curve for the material.

2. **Ideal force, with friction.** Based on the slab method of analysis (Section 6.4.4) and for small die angles, it can be shown that with friction at the die-billet interface and ignoring the container-wall friction, the extrusion pressure, p, is given by the expression

$$p = Y\left(1 + \frac{\tan \alpha}{\mu}\right)[R^{\mu \cot \alpha} - 1]. \qquad (6.55)$$

If it is assumed that the frictional stress is equal to the shear yield stress k, and that because of the dead zone formed (see Figs. 6.49b and c), the material flows along a 45° "die angle," pressure can be estimated as

$$p = Y\left(1.7 \ln R + \frac{2L}{D_o}\right). \qquad (6.56)$$

Note that as the ram travels further toward the die, L decreases, and thus the pressure, hence the force, decreases (Fig. 6.50). However, in indirect extrusion, the ram force is not a function of billet length.

3. **Actual forces.** In actual extrusion practice, as well as in all other metalworking processes, there are difficulties in estimating (1) the coefficient of friction and its variation throughout all workpiece–die contacting surfaces, (2) the flow stress of the material under the actual conditions of temperature and strain rate, and (3) the work involved in inhomogeneous deformation. A simple empirical formula has been developed in the form of

$$p = Y(a + b \ln R), \qquad (6.57)$$

where a and b are experimentally determined constants. It has been determined that an approximate value for a is 0.8, and that b ranges from 1.2 to 1.5.

4. **Optimum die angle.** The die angle has an important effect on forces in extrusion. Its effects can be summarized as follows:

a. The *ideal force* is a function of the strain that the material undergoes, thus a function of the extrusion ratio R. Consequently, it is independent of the die angle, as shown by curve (b) in Fig. 6.51.

b. The force due to *friction* increases with decreasing die angle. This is because, as can be seen in Fig. 6.47, the length of contact along the billet–die interface increases as the die angle decreases; thus the work required increases, as shown by curve (d).

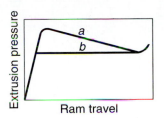

FIGURE 6.50 Schematic illustration of typical extrusion pressure as a function of ram travel: (a) direct extrusion and (b) indirect extrusion. The pressure in direct extrusion is higher because of frictional resistance at the container–billet interfaces, which decreases as the billet length decreases in the container.

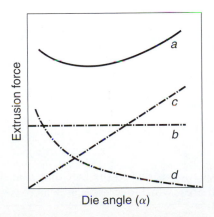

FIGURE 6.51 Schematic illustration of extrusion force as a function of die angle: (a) total force, (b) ideal force, (c) force required for redundant deformation, and (d) force required to overcome friction. Note that there is a die angle where the total extrusion force is a minimum (optimum die angle).

c. An additional force is required for *redundant work* due to the inhomogeneous deformation of the material during extrusion (see Section 2.12). This work is assumed to increase with the die angle, because the higher the angle, the more nonuniform the deformation becomes.

The *total* extrusion force is the sum of these three components. Note that as shown by curve (a) in Fig. 6.51, there is an angle at which this force is a *minimum*. Because the force is minimized at this angle, it is called the *optimum angle*.

EXAMPLE 6.5 Strain rate in extrusion

Determine the true strain rate in extruding a round billet of radius r_o as a function of distance x from the entry of a conical die.

Solution. From the geometry in the die gap and letting x be the distance from die entry, we note that

$$\tan \alpha = \frac{r_o - r}{x}, \tag{6.58}$$

or

$$r = r_o - x \tan \alpha.$$

The incremental true strain can be defined as

$$d\epsilon = \frac{dA}{A},$$

where $A = \pi r^2$. Therefore, $dA = 2\pi r \, dr$, and hence

$$d\epsilon = \frac{2dr}{r},$$

where $dr = -\tan \alpha \, dx$. We also know that

$$\dot{\epsilon} = \frac{d\epsilon}{dt} = -\left(\frac{2 \tan \alpha}{r}\right)\left(\frac{dx}{dt}\right).$$

However, $dx/dt = V$, which is the velocity of the material at any location x in the die. Hence, we have

$$\dot{\epsilon} = -\frac{2V \tan \alpha}{r}.$$

Because the flow rate is constant (volume constancy), we can write

$$V = \frac{V_o r_o^2}{r^2}$$

and therefore we have

$$\dot{\epsilon} = -\frac{2V_o r_o^2 \tan \alpha}{r^3} = -\frac{2V_o r_o^2 \tan \alpha}{(r_o - x \tan \alpha)^3}. \tag{6.59}$$

The negative sign in the equation is due to the fact that true strain is defined in terms of the cross-sectional area, which decreases as x increases.

5. **Force in hot extrusion.** Because of the strain-rate sensitivity of metals at elevated temperatures (Section 2.2.7), the force in hot extrusion is difficult to calculate accurately. It can be shown that the average true-strain rate, $\dot{\bar{\epsilon}}$, that the material undergoes is given by the expression

$$\dot{\bar{\epsilon}} = \frac{6 V_o D_o^2 \tan \alpha}{D_o^3 - D_f^3} \ln R, \qquad (6.60)$$

where V_o is the ram speed. It can be shown that (1) for high extrusion ratios, that is, $D_o \gg D_f$, and (2) a die angle of $\alpha = 45°$, as is the case with a square die and with poor lubrication (thus developing a dead zone), the true strain rate reduces to

$$\dot{\bar{\epsilon}} = \frac{6 V_o}{D_o} \ln R. \qquad (6.61)$$

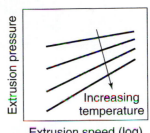

FIGURE 6.52 Schematic illustration of the effect of temperature and ram speed on extrusion pressure. Note the similarity of this figure to Fig. 2.10.

The typical effects of ram speed and temperature on extrusion pressure are shown in Fig. 6.52. As expected, the pressure increases rapidly with ram speed, especially at elevated temperatures due to increased strain-rate sensitivity (see Section 2.2.7). As speed increases, the rate of work done per unit time (hence temperature) also increases. However, the heat generated at higher speeds will not be dissipated fast enough. The rise in temperature can lead to *incipient melting* of the workpiece material and possibly cause defects. Circumferential surface cracks caused by **hot shortness** may also develop (see Section 3.4.2), a phenomenon that in extrusion is known as **speed cracking** (because of the high ram speed). Obviously, these problems can be reduced or eliminated by lowering the extrusion speed.

A parameter that is used to estimate the force in hot extrusion is an experimentally determined *extrusion constant, K_e*, which includes various factors involved in the process. Thus,

$$p = K_e \ln R. \qquad (6.62)$$

Some values of K_e for a variety of materials are given in Fig. 6.53.

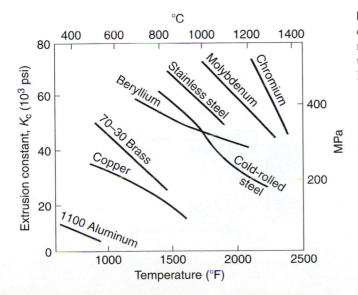

FIGURE 6.53 Extrusion constant, K_e, for various materials as a function of temperature. *Source:* After P. Loewenstein.

EXAMPLE 6.6 Force in hot extrusion

A copper billet 5 in. in diameter and 10 in. long is extruded at 1500°F at a speed of 10 in./s. Using square dies and assuming poor lubrication, estimate the force required in this operation if the extruded diameter is 2 in.

Solution. The extrusion ratio in this operation is

$$R = \frac{5^2}{2^2} = 6.25,$$

and the average true-strain rate, from Eq. (6.60), is

$$\dot{\bar{\epsilon}} = \frac{(6)(10)}{(5)} \ln 6.25 = 22 \text{ s}^{-1}.$$

From Table 2.5, let's assume that an average value for C is 19,000 psi and that $m = 0.06$. Hence,

$$\sigma = C\dot{\epsilon}^m = (19,000)(22)^{0.06} = 22,870 \text{ psi}.$$

Assuming that $\bar{Y} = \sigma$ we have, from Eq. (6.56),

$$p = \bar{Y}\left(1.7 \ln R + \frac{2L}{D_o}\right) = (22,870)\left[(1.7)(1.83) + \frac{(2)(10)}{(5)}\right] = 162,630 \text{ psi}.$$

Therefore,

$$F = (p)(A_o) = (162,630)\frac{(\pi)(5)^2}{4} = 3.2 \times 10^6 \text{ lb}.$$

6.4.3 Miscellaneous extrusion processes

1. **Cold extrusion.** *Cold extrusion* is a general term that often is used to describe a combination of processes, particularly extrusion combined with forging (Fig. 6.54). Many ductile metals can be cold extruded into various configurations, with the billet mostly at room temperature or at a temperature of a few hundred degrees. Typical parts made are automotive components and gear blanks. Cold extrusion is an important process because of its advantages such as

   a. improved mechanical properties resulting from strain hardening, provided that the heat generated by plastic deformation and friction does not recrystallize the extruded metal.

   b. good control of dimensional tolerances, thus requiring a minimum of machining and finishing operations.

FIGURE 6.54 Two examples of cold extrusion. Arrows indicate the direction of material flow. These parts may also be considered as forgings.

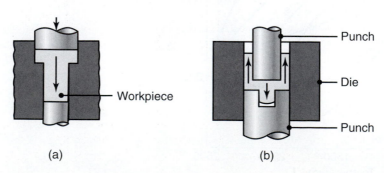

(a)   (b)

c. improved surface finish, partly due to the absence of oxide films, provided that lubrication is effective.

d. high production rates and relatively low cost.

On the other hand, the stresses on tooling and dies in cold extrusion are very high (especially with steel workpieces) and are on the order of the hardness of the material; that is, at least three times its flow stress. The design of tooling and selection of appropriate tool materials are therefore crucial to success in cold extrusion. The hardness of tooling usually ranges between 60 and 65 HRC for the punch and 58 and 62 HRC for the die. Punches are a critical component; they must have sufficient strength, toughness, and resistance to wear and fatigue.

Lubrication also is critical because new surfaces are generated during deformation, which may cause *seizure* between the workpiece and the tooling. The most effective lubrication is provided by phosphate conversion coatings on the workpiece and with soap (or wax) as the lubricant. Temperature rise in cold extrusion is an important factor, especially at high extrusion ratios. The temperature may be sufficiently high to initiate and complete the recrystallization process of the cold-worked metal, thus reducing the advantages of cold working.

2. **Impact extrusion.** A process that often is included in the category of cold extrusion is *impact extrusion* (Fig. 6.55). In this operation, the punch descends at a high speed and strikes the *blank (slug)* and extrudes it in the opposite direction. A typical example of impact extrusion is the production of collapsible tubes, such as for toothpaste. It will be noted that the thickness of the extruded tubular cross section is a function of the clearance between the punch and the die cavity.

The impact-extrusion process typically produces tubular shapes with wall thicknesses that are small in relation to their diameters, a ratio that can be as small as 0.005. The concentricity of the punch and the blank is important as otherwise the wall thickness will not be uniform. A variety of nonferrous metals are impact extruded in this manner into shapes such as those shown in Fig. 6.55b. The equipment used is typically vertical presses at production rates as high as two parts per second.

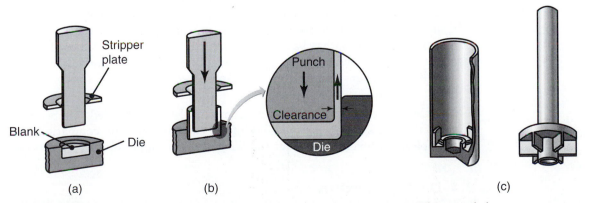

FIGURE 6.55   (a)–(b) Schematic illustration of the impact-extrusion process. The extruded parts are stripped using a stripper plate, as otherwise they may stick to the punch. (c) Two examples of products made by impact extrusion. Collapsible tubes can be produced by impact extrusion, referred to as the Hooker process.

FIGURE 6.56 Extrusion pressure as a function of the extrusion ratio for an aluminum alloy: (a) Direct extrusion, $\alpha = 90°$, (b) hydrostatic extrusion, $\alpha = 45°$, (c) hydrostatic extrusion, $\alpha = 22.5°$, and (d) ideal homogeneous deformation, calculated. *Source:* After H. Li, D. Pugh, and K. Ashcroft.

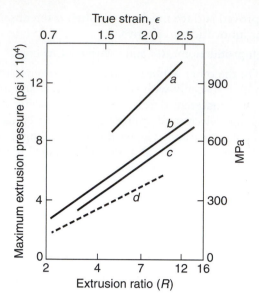

3. **Hydrostatic extrusion.** In this process, developed in the early 1950s, the pressure required for extrusion is supplied through a fluid medium that surrounds the billet, as shown in Fig. 6.47c. Thus, there is no container-wall contact and hence no friction. The high pressure in the chamber also transmits some of the fluid to the die surfaces, significantly reducing friction and forces (Fig. 6.56). Pressures in this process are typically on the order of 1400 MPa (200 ksi). Hydrostatic extrusion can also be carried out by extruding the part into a second pressurized chamber, which is under lower pressure (**fluid-to-fluid extrusion**). Because of the highly pressurized environment, this operation reduces the defects that may otherwise develop in the extruded product. A variety of metals and polymers, solid shapes, tubes and other hollow shapes, and honeycomb and clad profiles have been extruded successfully.

   Hydrostatic extrusion is usually carried out at room temperature, typically using vegetable oils as the fluid, particularly castor oil because it is a good lubricant and its viscosity is not influenced significantly by pressure. For elevated-temperature hydrostatic extrusion, waxes, polymers, and glass are used as the fluid; these materials also serve as thermal insulators and help maintain the billet temperature during extrusion. In spite of the success obtained, hydrostatic extrusion has had limited industrial applications, largely because of the somewhat complex nature of tooling, the experience required in working with high pressures and design of specialized equipment, and the long cycle times required.

4. **Coaxial extrusion.** In this process, coaxial billets are extruded together, provided that the strength and ductility of the two materials are compatible. An example is copper clad with silver. *Cladding* is another application of this process.

6.4.4 Defects in extrusion

There are three principal defects in extrusion, as described next.

1. **Surface cracking.** If the extrusion temperature, speed, and friction are too high, surface temperatures can rise significantly, leading to surface cracking and tearing (*fir-tree cracking* or *speed cracking*). These cracks are intergranular and

are usually the result of hot shortness (Section 3.4.2). They occur especially with aluminum, magnesium, and zinc alloys but are also observed with other metals, such as molybdenum alloys. This situation can typically be avoided by using lower temperatures and speeds.

Surface cracking can also occur at low temperatures and has been attributed to periodic sticking of the extruded product along the die land (**stick-slip**) during extrusion. When the product being extruded sticks to the die land, the extrusion pressure increases rapidly. Shortly thereafter, the product moves forward again, and the pressure is released. The cycle then repeats itself.

Because its appearance is similar to bamboo, this defect is known as a **bamboo defect** and is especially encountered in hydrostatic extrusion, where the pressure is sufficiently high to significantly increase the viscosity of the fluid. This situation then leads to the formation of a thick lubricant film. The billet then surges forward, which, in turn, relieves the pressure of the fluid and increases the friction. Thus, the change in the physical properties of the metalworking fluid is responsible for the stick-slip phenomenon. In such circumstances, it has been found that proper selection of a fluid is critical, and also that increasing the extrusion speed can eliminate stick-slip.

2. **Extrusion defect.** It can be noted that the type of metal flow shown in Fig. 6.49c will tend to draw the surface oxides and impurities toward the center of the billet. This defect, known as *extrusion defect*, **pipe**, **tailpipe**, and **fishtailing**, renders a considerable portion of the extruded material useless, by as much as one-third the length of the extrusion. This defect can be reduced by (a) modifying the flow pattern to a less inhomogeneous one, such as by controlling friction and minimizing temperature gradient; (b) machining the billet surface prior to extruding to eliminate scale and impurities; and (c) using a dummy block (Fig. 6.47a) that is smaller in diameter than the container, thus leaving a thin shell (*skull*) along the container wall as extrusion progresses.

3. **Internal cracking.** Cracks, variously known as **centerburst, center cracking, arrowhead fracture,** and **chevron cracking,** can develop at the center of an extruded product, as shown in Fig. 6.57. These cracks are attributed to a state of *hydrostatic tensile stress* (also called *secondary tensile stress*) at the centerline

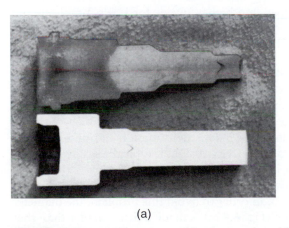

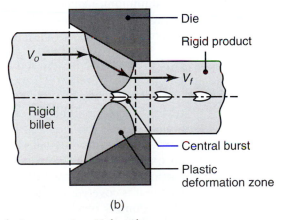

(a)   (b)

FIGURE 6.57 (a) Chevron cracking in round steel bars during extrusion. Unless the part is inspected, such internal defects may remain undetected and possibly cause failure of the part in service. (b) Deformation zone in extrusion, showing rigid and plastic zones. Note that the plastic zones do not meet, leading to chevron cracking. The same observations are also made in drawing round bars through conical dies and drawing flat sheet or plate through wedge-shaped dies. *Source:* After B. Avitzur.

of the deformation zone in the die. This situation is similar to the necked region in a uniaxial tensile-test specimen. Such cracks also have been observed in tube extrusion and in spinning of tubes (Section 7.5.4), whereby the cracks appear on the inside surfaces.

The major variables affecting hydrostatic tension are the (a) die angle; (b) extrusion ratio, hence the reduction in cross-sectional area; and (c) friction. The role of these factors can best be understood by observing the extent of inhomogeneous deformation during extrusion. Experimental results have indicated that, for the same extrusion ratio, as the die angle becomes larger, the deformation across the part becomes more inhomogeneous. Another factor is the die contact length, where it can be noted that the smaller the die angle, the longer is the contact length. The size and depth of the deformation zone increases with increasing contact length, as illustrated in Fig. 6.57b.

The h/L ratio is an important parameter, as shown in Fig. 6.12: The higher this ratio, the more inhomogeneous is the deformation. Small extrusion ratios and/or large die angles mean high h/L ratios. Inhomogeneous deformation indicates that the center of the billet is not in a fully plastic state. The reason is that the plastic deformation zones under the die-contact lengths do not reach each other (Fig. 6.57a). Likewise, small extrusion ratios and high die angles retard the flow of the material at the surfaces, while the central portions are more free to flow through the die. High h/L ratios generate hydrostatic tensile stresses in the center of the billet, causing the type of defects shown in Fig. 6.57b. These defects form more readily in materials with impurities, inclusions, and voids because these imperfections act as nucleation sites for defect formation. As for the role of friction, high friction in extrusion apparently delays formation of these cracks. These observations are also valid for the drawing of rod and wire, as described in Section 6.5.

6.4.5 Extrusion practice

A variety of materials can be extruded to various cross-sectional shapes and dimensions. Extrusion ratios in practice typically range from about 10 to over 100, and ram speeds may be up to 0.5 m/s (100 ft/min). Generally, slower speeds are preferred for aluminum, magnesium, and copper, and higher speeds for steels, titanium, and refractory alloys. Presses for hot extrusion are generally hydraulic and horizontal, and for cold extrusion they are usually vertical.

Hot extrusion. In addition to the strain-rate sensitivity of the material at elevated temperatures (Section 2.2.6), hot extrusion requires other special considerations. Temperature ranges for hot extrusion are similar to those for hot forging operations, as given in Table 6.3. Cooling of the billet in the container (which normally is not heated) can result in highly inhomogeneous deformation during extrusion. Furthermore, since the billet is heated prior to extrusion, it typically is covered with an oxide layer, unless heated in an inert atmosphere. The oxide layer affects the frictional properties, can affect the flow of the material, and can produce an extruded part that is covered with an oxide layer. In order to avoid this problem, the diameter of the dummy block ahead of the ram (Fig. 6.47a) is made a little smaller than that of the container.

Lubrication. For steels, stainless steels, and high-temperature materials, glass is an excellent lubricant. Glass maintains its viscosity at elevated temperatures, has good wetting characteristics, and acts as a thermal barrier between the billet, the container, and the die, thus minimizing cooling. In the **Séjournet process**, a circular glass pad

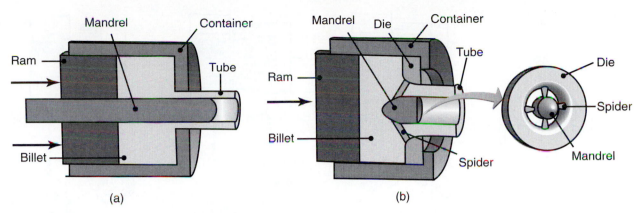

FIGURE 6.58   Extrusion of a seamless tube: (a) using an internal mandrel that moves independently of the ram. An alternative arrangement has the mandrel integral with the ram. (b) Using a spider die (see Fig. 6.59c) to produce seamless tubing.

is placed in the container at the die entrance. As extrusion progresses, this pad begins to soften and slowly melts away. In doing so, it forms an optimal die geometry in order to minimize the energy required. The viscosity–temperature index of the glass is an important factor in this application in order to control the depletion of glass at the die entrance. Solid lubricants such as graphite and molybdenum disulfide may also be used in hot extrusion. Nonferrous metals are often extruded without a lubricant (although graphite may be used) because friction stresses are moderate and surface finish is an important design concern (see *orange peel,* Section 3.6).

For materials that have a tendency to stick to the container and the die, the billet can be enclosed in a jacket (a thin-walled container) of a softer metal, such as copper or mild steel (**jacketing** or **canning**). Besides providing a low-friction interface, canning prevents contamination of the billet by the environment, or prevents the billet material from contaminating the environment (if it is toxic or radioactive). The canning technique is also used for processing metal powders (see Chapter 11).

Die design and materials.   As in all metalworking operations, die design and material selection require considerable experience. Dies with angles of 90° (**square dies,** or **shear dies**) can be successfully used for nonferrous metals, particularly aluminum. Dies for tubing typically involve the use of a ram fitted with a mandrel, as shown in Fig. 6.58a. For billets with a pierced hole, the mandrel may simply be attached to the ram, but if the billet is solid, it must first be pierced. Die materials for hot extrusion are typically hot-work die steels (Section 3.10.3). Various coatings can be applied to dies to extend their life (see Section 4.5).

Tubing and hollow extruded shapes (Fig. 6.59a) can also be produced by **welding-chamber methods,** using various special dies known as *spider, porthole,* and *bridge dies* (Fig. 6.59b–d). The metal flows around the arms of the die into strands, which then reweld themselves under the high pressures that are present at the die exit zone. Lubricants cannot be used in these operations because they penetrate interfaces and prevent rewelding in the die-exit zone. Furthermore, the welding-chamber methods are suitable only for aluminum and some of its alloys because of the methods' capacity for developing strong bonds under high pressure (*pressure welding*).

Equipment.   The basic equipment for extrusion is a hydraulic press, usually horizontal, and can be designed for a variety of extrusion operations as well as with different strokes and speeds. The largest hydraulic press for extrusion has a ram force of 160 MN (16,000 tons).

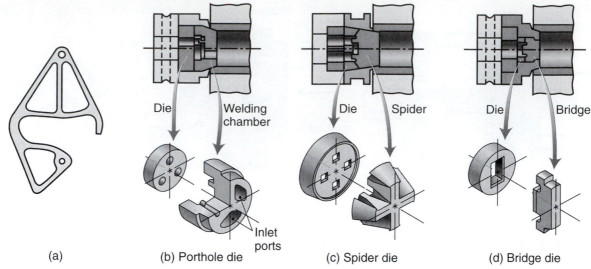

FIGURE 6.59   (a) An extruded 6063-T6 aluminum ladder lock for aluminum extension ladders. This part is 8 mm $\left(\frac{5}{16}\text{ in.}\right)$ thick and is sawed from the extrusion, as also shown in Fig. 6.48a. (b)–(d) Components of various types of dies for extruding intricate hollow shapes. *Source:* After K. Laue and H. Stenger.

6.5 | Rod, Wire, and Tube Drawing

Drawing is an operation in which the cross-sectional area of a bar or tube is reduced or changed in shape by pulling it through a converging die (Fig. 6.60). An established art by the eleventh century, the drawing process is thus somewhat similar to extrusion, except that in drawing, the bar is under tension, whereas in extrusion it is under compression.

Rod and wire drawing are generally finishing processes, and the product is either used as produced or is further processed into other shapes, such as by bending or machining. *Rods* are used for various applications, such as small pistons, structural members, shafts, and spindles, and as the raw material for making fasteners such as bolts and screws. *Wire* and wire products have a wide range of applications, such as electrical wiring, electronic equipment, cables, springs, musical instruments, paper clips, fencing, welding electrodes, and shopping carts. Wire diameters may be as small as 0.025 mm (0.001 in.).

6.5.1 Mechanics of rod and wire drawing

The major variables in the drawing process are reduction in cross-sectional area and die angle, as illustrated in Fig. 6.60. Friction also plays a major role, as will be

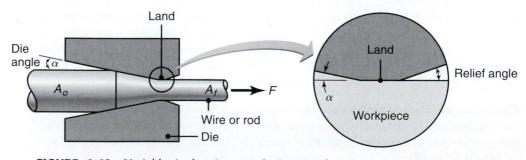

FIGURE 6.60   Variables in drawing round wire or rod.

described below. This section outlines the methods employed in estimating drawing forces for different conditions.

1. **Ideal deformation.** The drawing stress, σ_d, for the simplest case of ideal deformation (that is, no friction or redundant work) can be obtained by the same approach as that used for extrusion. Thus,

$$\sigma_d = Y \ln\left(\frac{A_o}{A_f}\right). \qquad (6.63)$$

Note that this expression is the same as Eq. (6.54), and that it also represents the energy per unit volume, u. As emphasized throughout this chapter, for strain-hardening materials, Y is replaced by an average flow stress, $\overline{Y}$, in the deformation zone. Thus, for a material that exhibits the true-stress–true-strain behavior of

$$\sigma = K\epsilon^n,$$

the quantity $\overline{Y}$ is obtained from the expression

$$\overline{Y} = \frac{K\epsilon_1^n}{n + 1}. \qquad (6.64)$$

The drawing force, F, is then

$$F = \overline{Y}A_f \ln\left(\frac{A_o}{A_f}\right). \qquad (6.65)$$

Note that, as expected, the higher the reduction in cross-sectional area and the stronger the material, the higher is the drawing force.

2. **Ideal deformation and friction.** Friction at the die–workpiece interface increases the drawing force because work has to be supplied externally to overcome friction. Using the slab method of analysis and on the basis of Fig. 6.61, the following expression for the drawing stress can be obtained:

$$\sigma_d = Y\left(1 + \frac{\tan\alpha}{\mu}\right)\left[1 - \left(\frac{A_f}{A_o}\right)^{\mu\cot\alpha}\right]. \qquad (6.66)$$

With good lubrication, the coefficient of friction, μ, in wire drawing typically ranges from about 0.03 to 0.1 (see Table 4.1). Even though Eq. (6.66) does not include the redundant work involved in the process, it is in reasonably good agreement with experimental data for small die angles and for a wide range of reductions in cross section.

3. **Redundant work.** Depending on the die angle and reduction, the material in drawing undergoes *inhomogeneous deformation*, much as it does in extrusion. It can be shown that when the redundant work of deformation is included, the expression for the drawing stress is

$$\sigma_d = \overline{Y}\left\{\left(1 + \frac{\tan\alpha}{\mu}\right)\left[1 - \left(\frac{A_f}{A_o}\right)^{\mu\cot\alpha}\right] + \frac{4}{3\sqrt{3}}\alpha^2\left(\frac{1-r}{r}\right)\right\}, \qquad (6.67)$$

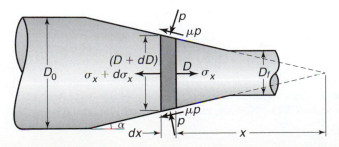

FIGURE 6.61   Stresses acting on an element in drawing of a solid cylindrical rod or wire through a converging conical die.

where r is the fractional reduction of area and α is the die angle, in radians. The first term in Eq. (6.67) represents the ideal and frictional components of work, and the second term represents the redundant-work component (which, as expected, is a function of the die angle). The larger the die angle, the greater the inhomogeneous deformation and, hence, the greater the redundant work.

For small die angles, another expression for the drawing stress that includes all three components of work is

$$\sigma_d = \overline{Y}\left[\left(1 + \frac{\mu}{\alpha}\right)\ln\left(\frac{A_o}{A_f}\right) + \frac{2}{3}\alpha\right]. \tag{6.68}$$

The last term in this expression is the redundant-work component, and it is assumed that it increases linearly with the die angle, as shown in Fig. 6.51. Because redundant deformation is a function of the h/L ratio (see Figs. 6.12 and 6.13), an **inhomogeneity factor, Φ,** for solid round cross sections has been developed and is expressed approximately as

$$\Phi = 1 + 0.12\left(\frac{h}{L}\right). \tag{6.69}$$

A simpler expression for the drawing stress is then

$$\sigma_d = \Phi\overline{Y}\left(1 + \frac{\mu}{\alpha}\right)\ln\left(\frac{A_o}{A_f}\right). \tag{6.70}$$

Equations (6.66), (6.67), (6.68), and (6.70) provide reasonable predictions for the stresses required for wire drawing.

4. **Die pressure.** Based on yield criteria and from Fig. 2.23b, and noting that the compressive stresses in the two principal directions are equal to p, the *die pressure* along the die contact length can be obtained from

$$p = Y_f - \sigma, \tag{6.71}$$

where σ is the tensile stress in the deformation zone at a particular diameter, and Y_f is the flow stress of the material at that particular diameter. Note that σ is thus equal to σ_d at the die exit and is zero at the die entry. Equation (6.71) indicates that as the tensile stress increases toward the exit, the die pressure drops toward the exit, as also shown qualitatively in Fig. 6.62.

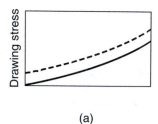

(a)

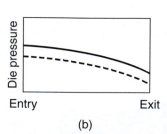

Entry   Exit

(b)

—— Without back tension
- - - - With back tension

FIGURE 6.62 Variation in the (a) drawing stress and (b) die contact pressure along the deformation zone. Note that as the drawing stress increases, the die pressure decreases (see also yield criteria, described in Section 2.11). Note the effect of back tension on the stress and pressure.

EXAMPLE 6.7 Power required and die pressure in rod drawing

A round rod of annealed 302 stainless steel is being drawn from a diameter of 10 mm to 8 mm at a speed of 0.5 m/s. Assume that the frictional and redundant work together constitute 40% of the ideal work of deformation. (a) Calculate the power required in this operation, and (b) calculate the die pressure at the exit of the die.

Solution. The true strain that the material undergoes in this operation is

$$\epsilon_1 = \ln\left(\frac{10^2}{8^2}\right) = 0.446.$$

From Table 2.3, we find that for this material and condition, $K = 1300$ MPa and $n = 0.30$. Hence,

$$\overline{Y} = \frac{K\epsilon_1^n}{n+1} = \frac{(1300)(0.446)^{0.30}}{1.30} = 785 \text{ MPa}.$$

From Eq. (6.65), the drawing force is

$$F = \overline{Y} A_f \ln\left(\frac{A_o}{A_f}\right),$$

where

$$A_f = \frac{(\pi)(0.008)^2}{4} = 5 \times 10^{-5} \text{ m}^2.$$

Hence,

$$F = (785)(5 \times 10^{-5})(0.446) = 0.0175 \text{ MN}$$

and

$$\begin{aligned} \text{Power} &= (F)(V_f) = (0.0175)(0.5) \\ &= 0.00875 \text{ MN} \cdot \text{m/s} \\ &= 0.00875 \text{ MW} = 8.75 \text{ kW}, \end{aligned}$$

and the actual power will be 40% higher, or

$$\text{Actual power} = (1.4)(8.75) = 12.25 \text{ kW}.$$

The die pressure can be obtained by noting from Eq. (6.71) that

$$p = Y_f - \sigma,$$

where Y_f represents the flow stress of the material at the exit of the die. Thus,

$$Y_f = K\epsilon_1^n = (1300)(0.446)^{0.30} = 1020 \text{ MPa}.$$

In this equation, σ is the drawing stress, σ_d. Hence, using the actual force, we have

$$\sigma_d = \frac{F}{A_f} = \frac{(1.4)(0.0175)}{0.00005} = 490 \text{ MPa}.$$

Therefore, the die pressure at the exit is

$$p = 1020 - 490 = 530 \text{ MPa}.$$

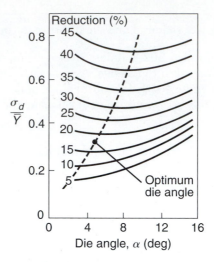

FIGURE 6.63 The effect of reduction in cross-sectional area on the optimum die angle in drawing. *Source:* After J.G. Wistreich.

5. **Drawing at elevated temperatures.** As we have seen, the flow stress of metals is a function of the strain rate. In drawing, the *average true-strain rate*, $\dot{\bar{\epsilon}}$, in the deformation zone can be given by

$$\dot{\bar{\epsilon}} = \frac{6V_o}{D_o}\ln\left(\frac{A_o}{A_f}\right). \tag{6.72}$$

which is the same as Eq. (6.61). After first calculating the average strain rate, the flow stress and the average flow stress, $\overline{Y}$, of the material can be calculated and substituted in appropriate equations.

6. **Optimum die angle.** As in extrusion, because of the various effects of the die angle on the three components of work (ideal, friction, and redundant), there is an *optimum die angle* at which the drawing force is a minimum. Figure 6.63 shows a typical example, where it can be noted that the optimum angle for the minimum drawing force increases with reduction, and that optimum angles are rather small.

7. **Maximum reduction per pass.** As reduction increases, the drawing stress increases. Obviously, there is a limit to the magnitude of the drawing stress. If it reaches the yield stress of the material at the exit, it will simply continue to yield and fail. The limiting situation can be developed based on the fact that, in the ideal case of a perfectly plastic material with a yield stress Y, the limiting condition is

$$\sigma_d = Y \ln\left(\frac{A_o}{A_f}\right) = Y, \tag{6.73}$$

or

$$\ln\left(\frac{A_o}{A_f}\right) = 1,$$

and therefore,

$$\frac{A_o}{A_f} = e.$$

Thus,

$$\text{Maximum reduction per pass} = \frac{A_o - A_f}{A_o} = 1 - \frac{1}{e} = 0.63 = 63\%. \tag{6.74}$$

It will be noted that with strain hardening, the exiting material will be stronger than the rest of the material in the die gap; consequently, the maximum reduction per pass will increase. The effects of friction and die angle on maximum reduction per pass are similar to those shown in Fig. 6.51. Because both friction and redundant deformation contribute to an increase in the drawing stress, the maximum reduction per pass will be lower than the ideal case.

EXAMPLE 6.8 Maximum reduction per pass for a strain-hardening material

Obtain an expression for the maximum reduction per pass for a material with a true-stress–true-strain curve of $\sigma = K\epsilon^n$. Ignore friction and redundant work.

Solution. From Eq. (6.63), we have

$$\sigma_d = \overline{Y} \ln \left(\frac{A_o}{A_f} \right) = \overline{Y}\epsilon_1,$$

where

$$\overline{Y} = \frac{K\epsilon_1^n}{n+1},$$

and σ_d, for this problem, can have a maximum value equal to the flow stress at ϵ_1, or

$$\sigma_d = Y_f = K\epsilon_1^n.$$

We can now write Eq. (6.73) as

$$K\epsilon_1^n = \frac{K\epsilon_1^n}{n+1}\epsilon_1,$$

or

$$\epsilon_1 = n + 1.$$

With $\epsilon_1 = \ln(A_o/A_f)$ and a maximum reduction of $(A_o - A_f)/A_o$, these expressions reduce to

$$\text{Maximum reduction per pass} = 1 - e^{-(n+1)}. \qquad (6.75)$$

Note that when $n = 0$ (that is, a perfectly plastic material), this expression reduces to Eq. (6.74), and that as n increases, the maximum reduction per pass increases.

---

8. **Drawing of flat strip.** The dies in flat strip drawing are *wedge shaped*, and there is little or no change in the width of the strip during drawing. Thus the drawing process is somewhat similar to that of rolling wide strips, and hence the process can be regarded as a **plane-strain** problem, especially at large width-to-thickness ratios. Although not of particular industrial significance in and of itself, flat drawing is the fundamental deformation mechanism in **ironing**, as described in Section 7.6.

   The approach to determining drawing force and maximum reduction is similar to that for round sections. Thus, the drawing stress for the ideal condition is

$$\sigma_d = Y' \ln \left(\frac{t_o}{t_f} \right), \qquad (6.76)$$

FIGURE 6.64 Examples of tube-drawing operations, with and without an internal mandrel. Note that a variety of diameters and wall thicknesses can be produced from the same tube stock (that has been produced by other processes, such as extrusion).

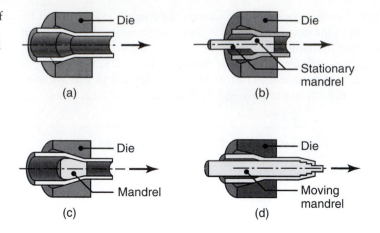

where Y' is the yield stress of the material in plane strain and t_o and t_f are the original and final thicknesses of the strip, respectively. The effects of friction and redundant deformation in strip drawing also are similar to those for round sections.

The maximum reduction per pass can be obtained by equating the drawing stress [Eq. (6.76)] to the *uniaxial yield stress* of the material (because the drawn strip is subjected only to simple tension). Thus,

$$\sigma_d = Y' \ln\left(\frac{t_o}{t_f}\right) = Y, \qquad \ln\left(\frac{t_o}{t_f}\right) = \frac{Y}{Y'} = \frac{\sqrt{3}}{2}, \qquad \text{and} \qquad \frac{t_o}{t_f} = e^{\sqrt{3}/2},$$

which reduces to

$$\text{Maximum reduction per pass} = 1 - \frac{1}{e^{\sqrt{3}/2}} = 0.58 = 58\%. \quad (6.77)$$

9. **Drawing of tubes.** Tubes produced by extrusion or other processes (such as shape rolling or the Mannesmann process) can be reduced in thickness or diameter (**tube sinking**) by the tube-drawing processes illustrated in Fig. 6.64. Shape changes can also be imparted by using dies and mandrels with various profiles. Drawing forces, die pressures, and the maximum reduction per pass in tube drawing can be calculated by methods similar to those described above.

6.5.2 Defects in drawing

Defects in drawing generally are similar to those observed in extrusion (Section 6.4.4), especially center cracking. The factors influencing center cracking are the same, namely, that the tendency for cracking increases with increasing die angle, decreasing reduction per pass, friction, and the presence of inclusions in the material.

A type of surface defect in drawing is the formation of **seams**. These are longitudinal scratches or folds in the material which can open up during subsequent forming operations, such as by upsetting, heading, thread rolling, or bending of the rod or wire.

Because of inhomogeneous deformation that the material undergoes, a cold-drawn rod, wire, or tube usually contains residual stresses. Typically, a wide range of residual stresses can be present within the rod in three principal directions, as shown in Fig. 6.65. For very light reductions, however, the surface residual stresses are compressive. Note that light reductions are equivalent to shot peening or surface rolling (which induce compressive surface stresses; see Section 4.5.1), thus

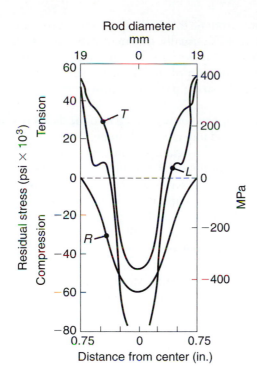

Rod diameter
mm

FIGURE 6.65   Residual stresses in cold-drawn 1045 carbon steel round rod: T = transverse direction, L = longitudinal direction, and R = radial direction. *Source:* After E. S. Nachtman.

improving fatigue life. Residual stresses also can be significant in stress-corrosion cracking over a period of time and in warping of the component (see Fig. 2.31) when a layer is subsequently removed, as by machining or grinding.

6.5.3 Drawing practice

As in all other metalworking processes, successful drawing requires proper selection of process parameters as well as other considerations. A typical die for drawing and its characteristic features are shown in Fig. 6.66. The purpose of the land is to size, that is, to set the final diameter of the product. Additionally, since the die is typically reground to extend its use, the land maintains the exit dimension of the die opening.

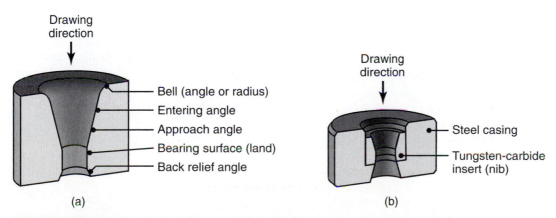

FIGURE 6.66   (a) Terminology for a typical die for drawing round rod or wire and (b) tungsten-carbide die insert in a steel casing. Diamond dies, used in drawing thin wire, are also encased in a similar manner.

Die angles usually range from 6° to 15°. Reductions in cross sectional area per pass range from about 10% to 45%; usually, the smaller the cross section, the smaller is the reduction per pass. Reductions per pass greater than 45% may result in breakdown of lubrication and deterioration of the product's surface finish. Light reductions may also be taken on rods (**sizing pass**) to improve dimensional accuracy and surface finish.

In typical drawing practice, a rod or wire is fed into the die by first **pointing** it by *swaging* (forming the tip of the rod into a conical shape, as described in Section 6.6). After it is placed in the die, the tip is clamped into the jaws of the wire-drawing machine, and the rod is drawn continuously through the die. In most wire-drawing operations, the wire passes through a series of dies (**tandem drawing**). In order to avoid excessive tension in the exiting wire, it is wound one or two turns around a *capstan* (a drum) between each pair of dies. The speed of the capstan is adjusted so that it supplies not only tension, but also a small *back tension* to the wire entering the next die. Depending on the material and its cross-sectional area, drawing speeds may be as low as for large cross sections and as high as for very fine wire.

Rods and tubes that are not sufficiently straight (or are supplied as coils) are straightened by passing them through pairs of rolls placed at different axes. The rolls subject the material to a series of bending and unbending operations, similar to the method shown in Fig. 6.40 for straightening rolled sheet or plate.

Because of strain hardening, intermediate annealing between passes may be necessary in cold drawing in order to maintain sufficient ductility of the material and avoid failure. Steel wires for springs and musical instruments are made by a heat-treatment process which can be used either prior to or after the drawing operation (**patenting**). These wires have ultimate tensile strengths as high as 4800 MPa (700,000 psi) and with tensile reduction of area of about 20%. Large cross sections may have to be drawn at elevated temperatures.

Bundle drawing. In this process, a number of wires (as many as several thousand) are drawn simultaneously as a bundle. To prevent sticking, the wires are separated from each other by a suitable material, usually a viscous lubricant. The cross section of the wires is somewhat polygonal because of the manner in which the wires are pressed together. The wires produced may be as small as 4 μm (160 μin.) in diameter, and can be made of stainless steels, titanium, and high-temperature alloys.

Dies and die materials. Die materials for drawing are generally alloy tool steels, carbides, or diamond. For drawing fine wires the die may be diamond, either a single crystal or a polycrystalline diamond in a metal matrix (similar to that shown in Fig. 6.66b). Carbide and diamond dies are made as inserts or nibs, which are then supported in a steel casing (Fig. 6.66b). A typical wear pattern on a drawing die is shown in Fig. 6.67. Note that die wear is highest at the entry. Although the die pressure is highest in this region and may be partially responsible for wear, other factors that are

FIGURE 6.67 Schematic illustration of a typical wear pattern in a wire-drawing die.

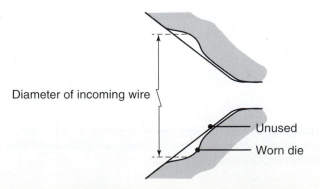

Diameter of incoming wire

Unused

Worn die

involved include (a) variations in the diameter of the entering wire; (b) vibration, which subjects the die-entry contact zone to fluctuating stresses; and (c) the presence of abrasive scale on the surface of the entering wire.

A set of idling rolls can also be used in drawing rods or bars of various shapes. This arrangement, known as **Turk's head,** is more versatile than ordinary dies because the rolls can be adjusted to various positions to produce different cross sections.

Lubrication. Proper lubrication is important in drawing operations. In **dry drawing,** the surface of the wire is coated with various lubricants, depending on the strength and frictional characteristics of the material. The rod to be drawn is first surface treated by pickling, which removes the surface scale that could lead to surface defects and considerably reduce die life (because of its abrasiveness). The bar then goes through a box (*stuffing box*) filled usually with soap powder.

With high-strength materials, such as steels, stainless steels, and high-temperature alloys, the surface of the rod may be coated either with a softer metal or with a **conversion coating** (see Section 4.5.1). Conversion coatings may consist of sulfate or oxalate coatings on the rod, which typically are then coated with soap, as a lubricant. Copper or tin can be chemically deposited as a thin layer on the surface of the metal, whereby it acts as a solid lubricant. Polymers may also be used as solid lubricants, such as in drawing titanium.

In **wet drawing,** the dies and rod are completely immersed in a lubricant. Typical lubricants include oils and emulsions (containing fatty or chlorinated additives) and various chemical compounds. The technique of **ultrasonic vibration** of the dies and mandrels is also used successfully in drawing. When carried out properly, this technique improves surface finish and die life and reduces drawing forces, thus allowing higher reductions per pass.

Equipment. Two types of equipment are generally used in drawing operations. A **draw bench,** similar to a long horizontal tensile-testing machine but with a hydraulic or chain-drive mechanism, is used for single draws of straight rods with large cross sections and for tubes with lengths up to 30 m (100 ft). Smaller cross sections are usually drawn by a **bull block,** which is basically a rotating drum around which the wire is wrapped. The tension in this setup provides the force required to draw the wire.

6.6 | Swaging

In *swaging,* also known as **rotary swaging** or **radial forging,** a rod or tube is reduced in diameter by the reciprocating radial movement of two or four dies, as shown in Fig. 6.68. The die movements are usually obtained by means of a set of rollers in a cage. The internal diameter and the thickness of the tube can be controlled with or without mandrels (Fig. 6.69). Mandrels can also be produced with longitudinal grooves (similar in appearance to a splined shaft), whereby internally shaped tubes can be swaged, such as those shown in Fig. 6.70a. The rifling in gun barrels is made by swaging a tube over a mandrel with spiral grooves. Externally shaped parts also can also be made by the swaging process (Fig. 6.70b). The swaging process is generally carried out at room temperature. Parts produced by swaging have improved mechanical properties and good dimensional accuracy.

Swaging is generally limited to part diameters of about 50 mm (2 in.), although special machines have been built to swage larger diameter parts, such as gun barrels. The length of the product is limited only by the length of the mandrel (if needed). Die angles are usually only a few degrees and may be compound; that is, the die may have more than one angle for more favorable material flow during swaging. Lubricants are used for improved surface finish and to extend die life.

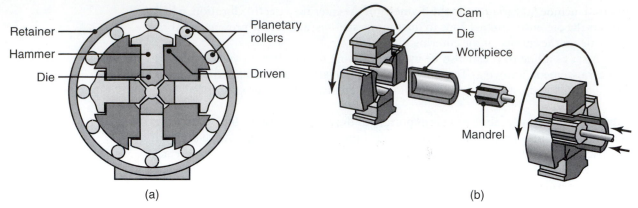

(a) (b)

FIGURE 6.68 (a) Schematic illustration of the rotary-swaging process and (b) forming internal profiles in a tubular workpiece by swaging.

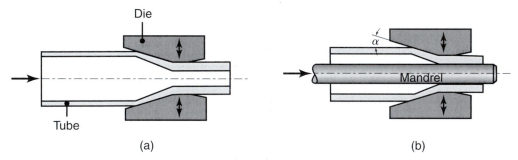

(a) (b)

FIGURE 6.69 Reduction of outer and inner diameters of tubes by swaging. (a) Free sinking without a mandrel. The ends of solid bars and wire are tapered (pointing) by this process in order to feed the material into the conical die. (b) Sinking on a mandrel. Coaxial tubes of different materials can also be swaged in one operation.

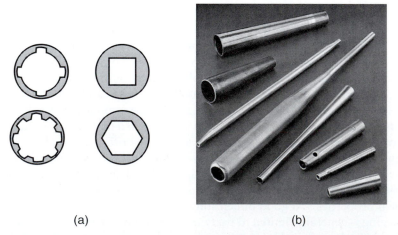

(a) (b)

FIGURE 6.70 (a) Typical cross sections produced by swaging tubular blanks with a constant wall thickness on shaped mandrels. Rifling of small gun barrels also can be made by swaging. (b) Typical parts made by swaging. *Source:* Courtesy of J. Richard Industries.

6.7 | Die Manufacturing Methods

Several manufacturing methods are used in making dies and molds for metal-working and other processes described in Chapters 5, 7, 10, and 11. These methods, which are used either singly or in various combinations, include casting, forging, machining, grinding, and electrical and electrochemical methods of die sinking. For improved surface finish and dimensional accuracy, dies are also subjected to various finishing operations, such as honing, polishing, and coating. The selection of a die manufacturing method primarily depends on the following parameters:

1. The particular process for which the die will be used.
2. Die shape and size.
3. The surface quality required.
4. Lead time required to produce the die (i.e., large dies may require months to prepare).
5. Production run (i.e., die life expected).
6. Cost.

Economic considerations often dictate the process selected because tool and die costs can be very significant in manufacturing operations. For example, the cost of a set of dies for pressworking automotive body panels (Figs. 1.5 and 7.66) may run up to $2 million; even small and simple dies can cost hundreds of dollars. On the other hand, because a large number of parts are typically produced using the same die, the die cost per piece made is generally only a small fraction of a part's manufacturing cost (Section 16.9).

The processes used for producing dies typically include *casting*, although small dies and tools also can be made by powder metallurgy techniques described in Chapter 11, as well as by *rapid tooling techniques*, described in Section 10.12. Sand casting is used for large dies weighing several tons and shell molding for small dies. Cast steels are generally preferred as die materials because of their strength and toughness and the ease with which their composition, grain size, and properties can be controlled and modified. Depending on how they solidify in the mold, cast dies, unlike those made of wrought metals, may not have directional properties; thus, they can exhibit isotropic properties on all working surfaces. However, because of shrinkage, the control of dimensional accuracy can be difficult compared with that in machined and finished dies, thus requiring additional processing.

Casting of dies is usually followed by various methods of (a) primary processing, such as forging, rolling, and extrusion; and (b) secondary processing, such as machining, grinding, polishing, and coating. Most commonly, dies are machined from cast and forged *die blocks* (called *die sinking*) by milling, turning, grinding, electrical and electrochemical machining, and polishing (Chapters 8 and 9). Conventional machining can be difficult and time consuming for high-strength, hard, tough, and wear-resistant die materials, although more recent developments include *hard machining* of dies (see *hard turning*, Section 8.9.2). Dies are commonly machined on *computer-controlled machine tools* and *machining centers* using various software packages (Fig. 1.8 and Sections 8.11 and 14.3) by which the cutter tool path is optimized for improved productivity.

Advanced machining processes, especially *electrical-discharge machining*, are also used extensively (Sections 9.9 through Section 9.15), particularly for small or medium-sized dies or for extrusion dies. These processes are generally faster and more economical than traditional machining, and the dies may not require additional finishing. However, it is important to consider any possible adverse effects that these processes may have on the properties of the die (including its fatigue life) because of possible surface damage and development of cracks.

To improve their hardness, wear resistance, and strength, die steels (Section 3.10.3) are usually *heat treated* (Section 5.11). After heat treatment, tools and dies are generally subjected to finishing operations, such as grinding and polishing, to obtain the desired surface roughness and dimensional accuracy. If not controlled properly, the grinding process can cause surface damage by generating excessive heat and inducing detrimental tensile residual stresses on die surfaces, thus reducing its fatigue life. Scratches on a die's structure can act as stress raisers as well. Dies may subsequently be subjected to various surface treatments, including *coatings* (Sections 4.5.1 and 8.6.5) for improved frictional and wear characteristics, as described in Chapter 4.

Small dies with shallow cavities also may be produced by *hubbing* (Section 6.2.4). Diamond dies for drawing fine wire are manufactured by producing holes with a thin rotating needle, coated with diamond powder suspended in oil.

6.8 | Die Failures

Failure of dies in metalworking operations, an important consideration, is generally due to one or more of the following causes:

1. Improper die design.
2. Defective die materials.
3. Improper heat treatment and finishing operations.
4. Improper installation, assembly, and alignment of die components.
5. Overheating and heat checking.
6. Excessive wear.
7. Overloading, misuse, and improper handling.

With rapid advances in *computer modeling and simulation* of processes and systems (Section 15.7), die design and optimization is now an advanced technology. Basic die-design guidelines include the following considerations:

a. Dies must have appropriate cross sections and clearances in order to withstand the forces involved.

b. Sharp corners radii, and fillets, as well as abrupt changes in cross section must be avoided as they act as stress raisers.

c. For improved strength, dies may be made in segments and may be prestressed during their assembly.

d. Dies can be designed and constructed with inserts that can be replaced when worn or cracked.

e. Die surface preparation and finishing are important, especially for die materials such as heat-treated steels, carbides, and diamond because are susceptible to cracking and chipping from *impact forces* (as occur in mechanical presses and forging hammers) or from *thermal stresses* (caused by temperature gradients within the die, as in hot-working operations).

f. Metalworking fluids can adversely affect tool and die materials, as, for example, sulfur and chlorine additives in lubricants and coolants can attack the cobalt binder in tungsten carbide and lower its strength and toughness (*leaching*, Section 3.9.7).

g. Overloading of dies can cause premature failure; a common cause of failure of cold-extrusion dies (Section 6.4.3); for example, the failure to remove an extruded part from the die before loading it with another blank.

6.9 | Economics of Bulk Forming

Several factors are involved in the cost of bulk formed parts. Depending on the complexity of the part, tool and die costs range from moderate to high. However, as in other manufacturing operations, this cost is spread over the number of parts produced with that particular tooling. Thus, even though the workpiece material cost per piece made is constant, setup and tooling costs per piece decrease as the manufacturing volume increases (Fig. 6.71).

The size of parts also has some effect on cost. Sizes range from small parts such as utensils and small automotive components, to large ones such as gears and crankshafts and connecting rods for large engines. As part size increases, the share of material cost in the total cost also increases, but at a lower rate. This occurs because (a) the incremental increase in die cost for larger dies is relatively small, (b) the machinery and operations involved are essentially the same regardless of part size, and (c) the labor involved per piece made is not that much higher.

Because they have been reduced significantly by automated and computer-controlled operations, labor costs in bulk forming are generally moderate. Also, die design and manufacturing are now performed by computer-aided design and manufacturing techniques (Chapter 15), which result in major savings in time and effort. Dies in extrusion, for example, can be simply produced through electrical discharge machining (Section 9.13) of an opening in tool steel.

The cost of bulk forming a part compared to that of making it by various casting techniques, by powder metallurgy, machining, or other methods is an important consideration in a competitive global marketplace. (Competitive aspects of manufacturing are discussed in greater detail in Chapter 16.) For example, all other factors being the same and depending on the number of pieces required, manufacturing a certain part by, say, expendable-mold casting may well be more economical than producing it by forging (Fig. 6.72). This casting method does not require expensive molds and tooling, whereas forging requires expensive dies. However, bulk forming operations are often justified by the improved mechanical properties that often result from these operations.

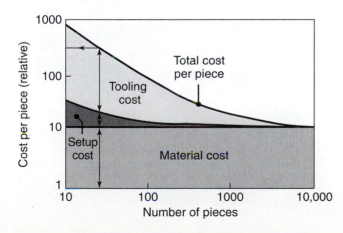

FIGURE 6.71 Typical unit cost (cost per piece) in forging. Note how the setup and the tooling costs per piece decrease as the number of pieces forged increases if all pieces use the same die.

FIGURE 6.72 Relative unit costs of a small connecting rod made by various forging and casting processes. Note that, for large quantities, forging is more economical. Sand casting is the more economical process for fewer than about 20,000 pieces.

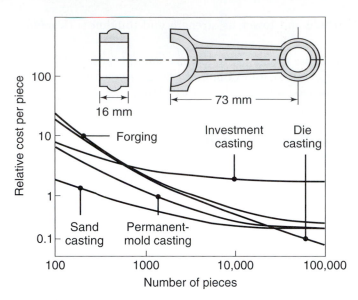

CASE STUDY | Suspension Components for the Lotus Elise Automobile

The automotive industry has been increasingly subjected to a demanding set of performance, cost, fuel efficiency, and environmental regulations. One of the main strategies in improving vehicle design with respect to all of these seemingly contradictory constraints is to optimize critical components with respect to cost or weight, while using advanced materials and manufacturing processes to maintain performance and safety. Previous design optimizations have shown that weight savings of up to 34% could be realized on suspension-system components, which is significant since suspensions make up approximately 12% of a car's mass. These weight savings could largely be achieved by (a) developing optimum designs, (b) using advanced analytical tools, and (c) by using net or near-net shape steel forgings instead of cast-iron components. In addition, significant cost savings have been demonstrated in many parts when using optimized steel forgings as compared to aluminum castings and extrusions.

The Lotus Elise (Fig. 6.73) is a high-performance sports car, designed for superior ride and handling. The Lotus group investigated the use of steel forgings and extruded aluminum instead of cast-iron suspension uprights in order to reduce cost and weight and improve reliability and performance. An aluminum-extrusion-based

FIGURE 6.73 The Lotus Elise Series 2 Sportscar. *Source:* Courtesy of Fox Valley Motorcars.

TABLE 6.4

Vertical Suspension Uprights in the Lotus Elise Series 2 Automobile

Identification	Description	Mass (kg)	Cost ($)
Benchmark	Aluminum extrusion, steel bracket, bushing, and housing	2.105	85
Phase I	Forged steel	2.685 (+28%)	27.7 (−67%)
Phase II	Forged steel	2.493 (+18%)	30.8 (−64%)

design was developed and served as a benchmark for evaluating forged designs. Further development efforts consisted of two phases, as shown in Table 6.4. The first phase involved development of a forged-steel component that could be used on the existing Elise sports car; the second phase involved production of a suspension upright for a new model.

A new design was developed using an iterative process with advanced software tools to reduce the number of components and to determine the optimum geometry. The material selected for the upright was an air-cooled hot-forged steel, which gives uniform grain size and microstructure and uniform high strength without the need for heat treatment. These materials also have approximately 20% higher fatigue strengths than traditional carbon steels such as AISI 1548-HT, used for similar applications.

The revised designs are summarized in Table 6.4. As can be seen, the optimized new forging design resulted in significant cost savings. Although it also resulted in a small weight increase compared to the aluminum-extrusion design, the weight penalty is recognized as quite small, and the use of forged steel for such components is especially advantageous in fatigue-loading conditions constantly encountered by suspension components. The new design also had certain performance advantages in that the component stiffness is now higher, which registered as improved customer satisfaction and better "feel" during driving. Furthermore, the new design reduced the number of parts required, thus satisfying another fundamental principle in design.

Source: Courtesy of Lotus Engineering and the American Iron and Steel Institute.

SUMMARY

- Bulk deformation processes generally consist of forging, rolling, extrusion, and drawing, and involve major changes in the dimensions of the workpiece. Bulk properties as well as surface characteristics of the material are important considerations. (Section 6.1)

- Forging denotes a family of processes by which deformation of the workpiece is carried out by compressive forces applied through dies. Forging is capable of producing a wide variety of parts with favorable characteristics of strength, roughness, dimensional accuracy, and reliability in service. Various types of defects can develop, depending on die geometry, quality of the billet materials, and preform shape. Several forging machines are available, with various characteristics and capabilities. (Section 6.2)

- Commonly used methods of analysis of the stresses, strains, strain rates, temperature distribution, and loads in forging and other bulk deformation processes include the slab method and the finite-element method. (Section 6.2)

- Rolling is the process of continuously reducing the thickness or changing the cross section of long stock by compressive forces applied through a set of rolls. Rolled products include plate, sheet, foil, rod, pipe, and tube, as well as shape-rolled products such as I-beams, structural shapes, rails, and bars of various cross sections. The operation involves several material and process variables, including the size of the roll relative to the thickness of the material, the amount of reduction per pass, speed, lubrication, and temperature. (Section 6.3)

- Extrusion is the process of forcing a billet through the opening of a die and it is capable of producing finite lengths of solid or hollow cross sections. Important factors include die design, extrusion ratio, lubrication, billet temperature, and extrusion speed. Cold extrusion is a combination of extrusion and forging operations and is capable of economically producing a wide variety of parts with good mechanical properties. (Section 6.4)

- Rod, wire, and tube drawing involve pulling the material through one or more dies. Proper die design and appropriate selection of materials and lubricants are essential to obtaining products of high quality and with good surface finish. The major process variables in drawing are die angle, friction, and amount of reduction per pass. (Section 6.5)

- In swaging, a solid rod or a tube is reduced in diameter by the reciprocating radial movement of two or four dies. This process is suitable for producing short or long lengths of bars or tubing with various internal or external profiles. (Section 6.6)

- Because die failure has a major economic impact, die design, material selection, and the manufacturing methods employed are of major importance. A wide variety of die materials and manufacturing methods is available, including advanced machining methods and subsequent treatment, finishing, and coating operations. (Sections 6.7, 6.8)

SUMMARY OF EQUATIONS

Forging

Pressure in plane-strain compression: $p = Y' e^{2\mu(a-x)/h}$

Average pressure in plane-strain compression: $p_{av} \simeq Y'\left(1 + \dfrac{\mu a}{h}\right)$

Pressure in axisymmetric compression: $p = Y'e^{2\mu(r-x)/h}$

Average pressure in axisymmetric compression: $p_{av} \simeq Y\left(1 + \dfrac{2\mu r}{3h}\right)$

Average pressure in plane-strain compression, sticking: $p_{av} = Y'\left(1 + \dfrac{a-x}{h}\right)$

Pressure in axisymmetric compression, sticking: $p = Y\left(1 + \dfrac{r-x}{h}\right)$

Rolling

Roll pressure in entry zone: $p = Y'_f \dfrac{h}{h_o} e^{\mu(H_0-H)}$

with back tension: $p = (Y'_f - \sigma_b)\dfrac{h}{h_o} e^{\mu(H_o-H)}$

Roll pressure in exit zone: $p = Y'_f \dfrac{h}{h_f} e^{\mu H}$

with front tension: $p = (Y'_f - \sigma_f)\dfrac{h}{h_f} e^{\mu H}$

Parameter: $H = 2\sqrt{\dfrac{R}{h_f}}\tan^{-1}\left(\sqrt{\dfrac{R}{h_f}}\phi\right)$

Roll force: $F = \displaystyle\int_0^{\phi_n} wpR\,d\phi + \int_{\phi_n}^{\alpha} wpR\,d\phi$

Roll force, approximate: $F = Lw\overline{Y}'\left(1 + \dfrac{\mu L}{2h_{av}}\right)$

Roll torque: $T = \displaystyle\int_{\phi_n}^{\alpha} w\mu pR^2\,d\phi - \int_0^{\phi_n} w\mu pR^2\,d\phi$

Roll contact length, approximate: $L = \sqrt{R\Delta h}$

Power per roll: $P = \dfrac{\pi FLN}{60{,}000}\text{ kW} = \dfrac{\pi FLN}{33{,}000}\text{ hp}$

Maximum draft: $\Delta h_{max} = \mu^2 R$

Maximum angle of acceptance: $\alpha_{max} = \tan^{-1}\mu$

Extrusion

Extrusion ratio: $R = A_o/A_f$

Extrusion pressure, ideal: $p = Y\ln R$

Extrusion pressure, with friction: $p = Y\left(1 + \dfrac{\tan\alpha}{\mu}\right)[R^{\mu\cot\alpha} - 1]$

Rod and Wire Drawing

Drawing stress, ideal: $\sigma_d = Y\ln\left(\dfrac{A_o}{A_f}\right)$

Drawing stress, with friction: $\sigma_d = Y\left(1 + \dfrac{\tan\alpha}{\mu}\right)\left[1 - \left(\dfrac{A_f}{A_0}\right)^{\mu\cot\alpha}\right]$

Die pressure: $p = Y_f - \sigma$

BIBLIOGRAPHY

Altan, T., Ngaile, G., and Shen, G. (eds.), *Cold and Hot Forging: Fundamentals and Applications,* ASM International, 2004.

Altan, T., Oh, S.-I., and Gegel, H., *Metal Forming—Fundamentals and Applications,* ASM International, 1983.

ASM Handbook, Vol. 14A: *Metalworking: Bulk Forming,* ASM International, 2005.

Blazynski, T.Z., *Plasticity and Modern Metal-Forming Technology,* Elsevier, 1989.

Davis, J.R. (ed.), *Tool Materials,* ASM International, 1995.

Dieter, G.E., *Mechanical Metallurgy,* 3rd ed., McGraw-Hill, 1986.

——— (ed.), *Workability Testing Techniques,* ASM International, 1984.

Frost, H.J., and Ashby, M.F., *Deformation-Mechanism Maps,* Pergamon, 1982.

Ginzburg, V.B., *High-Quality Steel Rolling: Theory and Practice,* Dekker, 1993.

———, *Steel-Rolling Technology: Theory and Practice,* Dekker, 1989.

Hoffman, H. (ed.), *Metal Forming Handbook,* Springer, 1998.

Hosford, W.F., and Caddell, R.M., *Metal Forming, Mechanics and Metallurgy,* 2nd ed., Prentice Hall, 1993.

Inoue, N., and Nishihara, M. (eds.), *Hydrostatic Extrusion: Theory and Applications,* Elsevier, 1985.

Kobayashi, S., Oh, S.-I., and Altan, T., *Metal Forming and the Finite-Element Method,* Oxford, 1989.

Lange, K. (ed.), *Handbook of Metal Forming,* McGraw-Hill, 1985.

Lenard, J.G., Pietrzyk, M., and Cser, L., *Mathematical and Physical Simulation of the Properties of Hot Rolled Products,* Elsevier, 1999.

Metal Forming Handbook, Schuler GmbH, 1998.

Mielnik, E.M., *Metalworking Science and Engineering,* McGraw-Hill, 1991.

Nachtman, E.S., and Kalpakjian, S., *Lubricants and Lubrication in Metalworking Operations,* Dekker, 1985.

Pietrzyk, M., and Leonard, J.G., *Thermal-Mechanical Modelling of the Flat Rolling Process,* Springer, 1991.

Prasad, Y.V.R.K., and Sasidhara, S. (eds.), *Hot Working Guide: A Compendium of Processing Maps,* ASM International, 1997.

Product Design Guide for Forging, Forging Industry Association, 1997.

Saha, P.K., *Aluminum Extrusion Technology,* ASM International, 2000.

Sheppard, T., *Extrusion of Aluminum Alloys,* Chapman & Hall, 1998.

Tool and Manufacturing Engineers Handbook, Vol. II: *Forming,* Society of Manufacturing Engineers, 1984.

Wagoner, R.H., and Chenot, J.L., *Fundamentals of Metal Forming,* Wiley, 1996.

———, *Metal Forming Analysis,* Cambridge, 2001.

QUESTIONS

Forging

6.1 How can you tell whether a certain part is forged or cast? Describe the features that you would investigate to arrive at a conclusion.

6.2 Why is the control of volume of the blank important in closed-die forging?

6.3 What are the advantages and limitations of a cogging operation? Of die inserts in forging?

6.4 Explain why there are so many different kinds of forging machines available.

6.5 Devise an experimental method whereby you can measure the force required for forging only the flash in impression-die forging. (See Fig. 6.15a.)

6.6 A manufacturer is successfully hot forging a certain part, using material supplied by Company *A.* A new supply of material is obtained from Company *B,* with the same nominal composition of the major alloying elements as that of the material from Company *A.* However, it is found that the new forgings are cracking even though the same procedure is followed as before. What is the probable reason?

6.7 Explain why there might be a change in the density of a forged product as compared to that of the cast blank.

6.8 Since glass is a good lubricant for hot extrusion, would you use glass for impression-die forging as well? Explain.

6.9 Describe and explain the factors that influence spread in cogging operations on square billets.

6.10 Why are end grains generally undesirable in forged products? Give examples of such products.

6.11 Explain why one cannot obtain a finished forging in one press stroke, starting with a blank.

6.12 List the advantages and disadvantages of using a lubricant in forging operations.

6.13 Explain the reasons why the flash assists in die filling, especially in hot forging.

6.14 By inspecting some forged products (such as a pipe wrench or coins), you can see that the lettering on them is raised rather than sunk. Offer an explanation as to why they are made that way.

Rolling

6.15 It was stated that three factors that influence spreading in rolling are (a) the width-to-thickness ratio of the strip, (b) friction, and (c) the ratio of the radius of the roll to the thickness of the strip. Explain how each of these factors affects spreading.

6.16 Explain how you would go about applying front and back tensions to sheet metals during rolling.

6.17 It was noted that rolls tend to flatten under roll forces. Which property(ies) of the roll material can be increased to reduce flattening? Why?

6.18 Describe the methods by which roll flattening can be reduced.

6.19 Explain the technical and economic reasons for taking larger rather than smaller reductions per pass in flat rolling.

6.20 List and explain the methods that can be used to reduce the roll force.

6.21 Explain the advantages and limitations of using small-diameter rolls in flat rolling.

6.22 A ring-rolling operation is being used successfully for the production of bearing races. However, when the bearing race diameter is changed, the operation results in very poor surface finish. List the possible causes, and describe the type of investigation you would conduct to identify the parameters involved and correct the problem.

6.23 Describe the importance of controlling roll speed, roll gap, temperature, and other relevant process variables in a tandem-rolling operation.

6.24 Is it possible to have a negative forward slip? Explain.

6.25 In addition to rolling, the thickness of plates and sheets can also be reduced by simply stretching. Would this process be feasible for high-volume production? Explain.

6.26 In Fig. 6.33, explain why the neutral point moves towards the roll-gap entry as friction increases.

6.27 What is typically done to make sure the product in flat rolling is not crowned?

6.28 List the possible consequences of rolling at (a) too high a speed and (b) too low a speed.

6.29 Rolling may be described as a continuous forging operation. Is this description appropriate? Explain.

6.30 Referring to appropriate equations, explain why titanium carbide is used as the work roll in Sendzimir mills, but not generally in other rolling mill configurations.

Extrusion

6.31 It was stated that the extrusion ratio, die geometry, extrusion speed, and billet temperature all affect the extrusion pressure. Explain why.

6.32 How would you go about preventing center-burst defects in extrusion? Explain why your methods would be effective.

6.33 How would you go about making a stepped extrusion that has increasingly larger cross sections along its length? Is it possible? Would your process be economical and suitable for high production runs? Explain.

6.34 Note from Eq. (6.54) that, for low values of the extrusion ratio such as $R = 2$, the ideal extrusion pressure, p, can be lower than the yield stress, Y, of the material. Explain whether or not this phenomenon is logical.

6.35 In hydrostatic extrusion, complex seals are used between the ram and the container, but not between the extrusion and the die. Explain why.

6.36 List and describe the types of defects that may occur in (a) extrusion and (b) drawing.

6.37 What is a land in a die? What is its function? What are the advantages and disadvantages to having no land?

6.38 Under what circumstances is backwards extrusion preferable to direct extrusion? When is hydrostatic extrusion preferable to direct extrusion?

6.39 What is the purpose of a container liner in direct extrusion (see Fig. 6.47a)? Why is there no container liner used in hydrostatic extrusion?

Drawing

6.40 We have seen that in rod and wire drawing, the maximum die pressure is at the die entry. Why?

6.41 Describe the conditions under which wet drawing and dry drawing, respectively, are desirable.

6.42 Name the important process variables in drawing, and explain how they affect the drawing process.

6.43 Assume that a rod drawing operation can be carried out either in one pass or in two passes in tandem. If the die angles are the same and the total reduction is the same, will the drawing forces be different? Explain.

6.44 Refer to Fig. 6.60 and assume that reduction in the cross section is taking place by pushing a rod through the die instead of pulling it. Assuming that the material is perfectly plastic, sketch the die-pressure distribution, for the following situations: (a) frictionless, (b) with friction, and (c) frictionless but with front tension. Explain your answers.

6.45 In deriving Eq. (6.74), no mention was made regarding the ductility of the original material being drawn. Explain why.

6.46 Why does the die pressure in drawing decrease toward the die exit?

6.47 What is the magnitude of the die pressure at the die exit for a drawing operation that is being carried out at the maximum reduction per pass?

6.48 Explain why the maximum reduction per pass in drawing should increase as the strain-hardening exponent, n, increases.

6.49 If, in deriving Eq. (6.74), we include friction, will the maximum reduction per pass be the same (that is, 63%), higher, or lower? Explain.

6.50 Explain what effects back tension has on the die pressure in wire or rod drawing, and discuss why these effects occur.

6.51 Explain why the inhomogeneity factor, ϕ, in rod and wire drawing depends on the ratio h/L as plotted in Fig. 6.12.

6.52 Describe the reasons for the development of the swaging process.

6.53 Occasionally, wire drawing of steel will take place within a sheath of a soft metal, such as copper or lead. Why would this procedure be effective?

6.54 Recognizing that it is very difficult to manufacture a die with a submillimeter diameter, how would you produce a 10-μm-diameter wire?

6.55 What changes would you expect in the strength, hardness, ductility, and anisotrophy of annealed metals after they have been drawn through dies? Why?

General

6.56 With respect to the topics covered in this chapter, list and explain specifically two examples each where friction (a) is desirable and (b) is not desirable.

6.57 Choose any three topics from Chapter 2 and with a specific example for each, show their relevance to the topics covered in this chapter.

6.58 Same as Question 6.57 but for Chapter 3.

6.59 List and explain the reasons that there are so many different types of die materials used for the processes described in this chapter.

6.60 Why should we be interested in residual stresses developed in parts made by the forming processes described in this chapter?

6.61 Make a summary of the types of defects found in the processes described in this chapter. For each type, specify methods of reducing or eliminating the defects.

PROBLEMS

Forging

6.62 In the free-body diagram in Fig. 6.4b, the incremental stress $d\sigma_x$ on the element was shown pointing to the left. Yet it would appear that, because of the direction of frictional stresses, μp, the incremental stress should point to the right in order to balance the horizontal forces. Show that the same answer for the forging pressure is obtained regardless of the direction of this incremental stress.

6.63 Plot the force vs. reduction in height curve in open-die forging of a solid cylindrical, annealed copper specimen 2 in. high and 1 in. in diameter, up to a reduction of 70%, for the cases of (a) no friction

between the flat dies and the specimen, (b) $\mu = 0.25$, and (c) $\mu = 0.5$. Ignore barreling and use average-pressure formulas.

6.64 Use Fig. 6.9b to provide the answers to Problem 6.63.

6.65 Calculate the work done for each case in Problem 6.63.

6.66 Determine the temperature rise in the specimen for each case in Problem 6.63, assuming that the process is adiabatic and the temperature is uniform throughout the specimen.

6.67 To determine its forgeability, a hot-twist test is performed on a round bar 25 mm in diameter and

200 mm long. It is found that the bar underwent 200 turns before it fractured. Calculate the shear strain at the outer surface of the bar at fracture.

6.68 Derive an expression for the average pressure in plane-strain compression under the condition of sticking friction.

6.69 What is the magnitude of μ when, for plane-strain compression, the forging load with sliding friction is equal to the load with sticking friction? Use average-pressure formulas.

6.70 Note that in cylindrical upsetting, the frictional stress cannot be greater than the shear yield stress, k, of the material. Thus, there may be a distance x in Fig. 6.8 where a transition occurs from sliding to sticking friction. Derive an expression for x in terms of r, h, and μ only.

6.71 Assume that the workpiece shown in the accompanying figure is being pushed to the right by a lateral force F while being compressed between flat dies. (a) Make a sketch of the die-pressure distribution for the condition for which F is not large enough to slide the workpiece to the right. (b) Make a similar sketch, except with F is now large enough so that the workpiece slides to the right while being compressed.

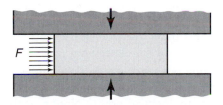

6.72 For the sticking example given in Fig. 6.10, derive an expression for the lateral force F required to slide the workpiece to the right while the workpiece is being compressed between flat dies.

6.73 Two solid cylindrical specimens, A and B, both made of a perfectly plastic material, are being forged with friction and isothermally at room temperature to a reduction in height of 25%. Originally, specimen A has a height of 2 in. and a cross-sectional area of 1 in.^{2}, and specimen B has a height of is 1 in. and a cross-sectional area of 2 in.^{2}. Will the work done be the same for the two specimens? Explain.

6.74 In Fig. 6.6, does the pressure distribution along the four edges of the workpiece depend on the particular yield criterion used? Explain.

6.75 Under what conditions would you have a normal pressure distribution in forging a solid cylindrical workpiece as shown in the accompanying figure? Explain.

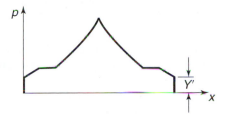

6.76 Derive the average die-pressure formula given by Eq. (6.15). (*Hint:* Obtain the volume under the friction hill over the surface by integration, and divide it by the cross-sectional area of the workpiece.)

6.77 Take two solid cylindrical specimens of equal diameter but different heights, and compress them (frictionless) to the same percent reduction in height. Show that the final diameters will be the same.

6.78 A rectangular workpiece has the following original dimensions: $2a = 100$ mm, $h = 30$ mm, and width $= 20$ mm (see Fig. 6.5). The metal has a strength coefficient of 300 MPa and a strain-hardening exponent of 0.3. It is being forged in plane strain with $\mu = 0.2$. Calculate the force required at a reduction of 20%. Do not use average-pressure formulas.

6.79 Assume that in upsetting a solid cylindrical specimen between two flat dies with friction, the dies are rotated at opposite directions to each other. How, if at all, will the forging force change from that for nonrotating dies? (*Hint:* Note that each die will now require torque but in opposite directions.)

6.80 A solid cylindrical specimen, made of a perfectly plastic material, is being upset between flat dies with no friction. The process is being carried out by a falling weight, as in a drop hammer. The downward velocity of the hammer is at a maximum when it first contacts the workpiece and becomes zero when the hammer stops at a certain height of the specimen. Establish quantitative relationships between workpiece height and velocity, and make a qualitative sketch of the velocity profile of the hammer. (*Hint:* The loss in the kinetic energy of the hammer is the plastic work of deformation; thus, there is a direct relationship between workpiece height and velocity.)

6.81 Describe how would you go about estimating the force acting on each die in a swaging operation.

6.82 A mechanical press is powered by a 30-hp motor and operates at 40 strokes per minute. It uses a flywheel, so that the rotational speed of the crankshaft does not vary appreciably during the stroke. If the stroke length is 6 in., what is the maximum contact force that can be exerted over the entire stroke length? To what height can a 5052-O aluminum cylinder with a diameter of 0.5 in. and a height of 2 in. be forged before the press stalls?

6.83 Estimate the force required to upset a 0.125-in.-diameter C74500 brass rivet in order to form a 0.25-in.-diameter head. Assume that the coefficient of friction between the brass and the tool-steel die is 0.2 and that the rivet head is 0.125 in. in thickness.

6.84 Using the slab method of analysis, derive Eq. (6.17).

Rolling

6.85 In Example 6.4, calculate the velocity of the strip leaving the rolls.

6.86 With appropriate sketches, explain the changes that occur in the roll-pressure distribution if one of the rolls is idling, that is, power is shut off to that roll.

6.87 It can be shown that it is possible to determine μ in flat rolling without measuring torque or forces. By inspecting the equations for rolling, describe an experimental procedure to do so. Note that you are allowed to measure any quantity other than torque or forces.

6.88 Derive a relationship between back tension, σ_b, and front tension, σ_f, in rolling such that when both tensions are increased, the neutral point remains in the same position.

6.89 Take an element at the center of the deformation zone in flat rolling. Assuming that all the stresses acting on this element are principal stresses, indicate the stresses qualitatively, and state whether they are tensile or compressive. Explain your reasoning. Is it possible for all three principal stresses to be equal to each other in magnitude? Explain.

6.90 It was stated that in flat rolling a strip, the roll force is reduced about twice as effectively by back tension as it is by front tension. Explain the reason for this difference, using appropriate sketches. (*Hint:* Note the shift in the position of the neutral point when tensions are applied.)

6.91 It can be seen that in rolling a strip, the rolls will begin to slip if the back tension, σ_b, is too high. Derive an analytical expression for the magnitude of the back tension in order to make the powered rolls begin to slip. Use the same terminology as applied in the text.

6.92 Derive Eq. (6.46).

6.93 In Steckel rolling, the rolls are idling, and thus there is no net torque, assuming frictionless bearings. Where, then, is the energy coming from to supply the necessary work of deformation in rolling? Explain with

appropriate sketches, and state the conditions that have to be satisfied.

6.94 Derive an expression for the tension required in Steckel rolling of a flat sheet, without friction, for a workpiece whose true-stress–true-strain curve is given by $\sigma = a + b\epsilon$.

6.95 (a) Make a neat sketch of the roll-pressure distribution in flat rolling with powered rolls. (b) Assume now that the power to both rolls is shut off and that rolling is taking place by front tension only, i.e., Steckel rolling. Superimpose on your diagram the new roll-pressure distribution, explaining your reasoning clearly. (c) After completing part (b), further assume that the roll bearings are becoming rusty and deprived of lubrication although rolling is still taking place by front tension only. Superimpose a third roll-pressure distribution diagram for this condition, explaining your reasoning.

6.96 Derive Eq. (6.28), based on the equation preceding it. Comment on how different the h values are as the angle ϕ increases.

6.97 In Fig. 6.34, assume that $L = 2L_2$. Is the roll force, F, for L now twice or more than twice the roll force for L_2? Explain.

6.98 A flat-rolling operation is being carried out where $h_0 = 0.2$ in., $h_f = 0.15$ in., $w_0 = 10$ in., $R = 8$ in., $\mu = 0.25$, and the average flow stress of the material is 40,000 psi. Estimate the roll force and the torque. Include the effects of roll flattening.

6.99 A rolling operation takes place under the conditions shown in the accompanying figure. What is the position x_n of the neutral point? Note that there are a front and back tension that have not been specified. Additional data are as follows: Material is 5052-O aluminum; hardened steel rolls; surface roughness of the rolls $= 0.02\ \mu$m; rolling temperature $= 210°$C.

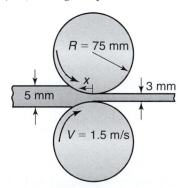

6.100 Estimate the roll force and power for an annealed low-carbon steel strip 200 mm wide and 10 mm thick, rolled to a thickness of 6 mm. The roll radius is 200 mm, and the roll rotates at 200 rpm. Let $\mu = 0.1$.

6.101 Calculate the individual drafts in each of the stands in the tandem-rolling operation shown.

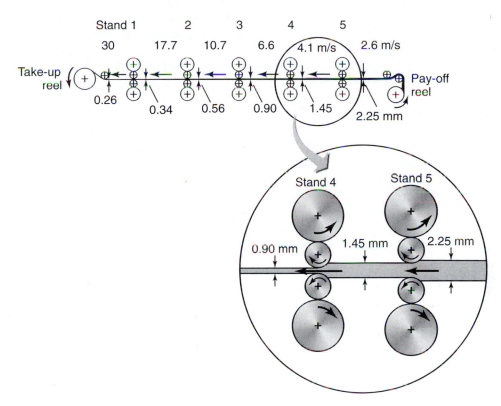

6.102 Calculate the required roll velocities for each roll in Problem 6.101 in order to maintain a forward slip of (a) zero and (b) 10%.

Extrusion

6.103 Calculate the force required in direct extrusion of 1100-O aluminum from a diameter of 6 in. to 2 in. Assume that the redundant work is 30% of the ideal work of deformation, and the friction work is 25% of the total work of deformation.

6.104 Prove Eq. (6.58).

6.105 Calculate the theoretical temperature rise in the extruded material in Example 6.6, assuming that there is no heat loss. (See Section 3.9 for information on the physical properties of the material.)

6.106 Using the same approach as that shown in Section 6.5 for wire drawing, show that the extrusion pressure is given by the expression

$$p = Y\left(1 + \frac{\tan \alpha}{\mu}\right)\left[1 - \left(\frac{A_o}{A_f}\right)^{\mu \cot \alpha}\right],$$

where A_o and A_f are the original and final workpiece diameters, respectively.

6.107 Derive Eq. (6.56).

6.108 A planned extrusion operation involves steel at 800°C, with an initial diameter of 100 mm and a final diameter of 20 mm. Two presses, one with a capacity of 20 MN and the other of 10 MN, are available for this operation. Obviously, the larger press requires greater care and more expensive tooling. Is the smaller press sufficient for this operation? If not, what recommendations would you make to allow the use of the smaller press?

6.109 Estimate the force required in extruding 70-30 brass at 700°C if the billet diameter is 125 mm and the extrusion ratio is 20.

Drawing

6.110 Calculate the power required in Example 6.7 if the workpiece material is annealed 70-30 brass.

6.111 Using Eq. (6.63), make a plot similar to Fig. 6.63 for the following conditions: $K = 100$ MPa, $n = 0.3$, and $\mu = 0.04$.

6.112 Using the same approach as that described in Section 6.5 for wire drawing, show that the drawing stress, σ_d, in plane-strain drawing of a flat sheet or plate is given by the expression

$$\sigma_d = Y' \left(1 + \frac{\tan \alpha}{\mu} \right) \left[1 - \left(\frac{h - f}{h_o} \right)^{\mu \cot \alpha} \right],$$

where h_o and h_f are the original and final thickness, respectively, of the workpiece.

6.113 Derive an analytical expression for the die pressure in wire drawing, without friction or redundant work, as a function of the instantaneous diameter in the deformation zone.

6.114 A linearly strain-hardening material with a true-stress–true-strain curve $\sigma = 5000 + 25,000\epsilon$ psi is being drawn into a wire. If the original diameter of the wire is 0.25 in., what is the minimum possible diameter at the exit of the die? Assume that there is no redundant work and that the frictional work is 15% of the ideal work of deformation. (*Hint:* The yield stress of the exiting wire is the point on the true-stress–true-strain curve that corresponds to the total strain that the material has undergone.)

6.115 In Fig. 6.65, assume that the longitudinal residual stress at the center of the rod is $-80,000$ psi. Using the distortion-energy criterion, calculate the minimum yield stress that this particular steel must have in order to sustain these levels of residual stresses.

6.116 Derive an expression for the die-separating force in frictionless wire drawing of a perfectly plastic material. Use the same terminology as in the text.

6.117 A material with a true-stress–true-strain curve $\sigma = 10,000\epsilon^{0.3}$ is used in wire drawing. Assuming that the friction and redundant work compose a total of 50% of the ideal work of deformation, calculate the maximum reduction in cross-sectional area per pass that is possible.

6.118 Derive an expression for the maximum reduction per pass for a material of $\sigma = K\epsilon^n$ assuming that the friction and redundant work contribute a total of 25% to the ideal work of deformation.

6.119 Prove that the true-strain rate $\dot{\epsilon}$, in drawing or extrusion in plane strain with a wedge-shaped die is given by the expression

$$\dot{\epsilon} = -\frac{2 \tan \alpha V_o t_o}{(t_o - 2x \tan \alpha)^2},$$

where α is the die angle, t_o is the original thickness, and x is the distance from die entry (*Hint:* Note that $d\epsilon = dA/A$.)

6.120 In drawing a strain-hardening material with $n = 0.25$ what should be the percentage of friction plus redundant work, in terms of ideal work, so that the maximum reduction per pass is 63%?

6.121 A round wire made of a perfectly plastic material with a yield stress of 30,000 psi is being drawn from a diameter of 0.1 to 0.07 in. in a draw die of 15°. Let the coefficient of friction be 0.1. Using both Eqs. (6.61) and (6.66), estimate the drawing force required. Comment on any differences in your answers.

6.122 Assume that you are asked to give a quiz to students on the contents of this chapter. Prepare three quantitative problems and three qualitative questions, and supply the answers.

DESIGN

6.123 Forging is one method of producing turbine blades for jet engines. Study the design of such blades and, referring to the relevant technical literature, prepare a step-by-step procedure for making these blades. Comment on the difficulties that may be encountered in this operation.

6.124 In comparing forged parts with cast parts, it will be noted that the same part may be made by either process. Comment on the pros and cons of each process, considering factors such as part size, shape complexity, and design flexibility in the event that a particular design has to be modified.

6.125 Referring to Fig. 6.25, sketch the intermediate steps you would recommend in the forging of a wrench.

6.126 Review the technical literature, and make a detailed list of the manufacturing steps involved in the manufacture of hypodermic needles.

6.127 Figure 6.48a shows examples of products that can be obtained by slicing long extruded sections into discrete parts. Name several other products that can be made in a similar manner.

6.128 Make an extensive list of products that either are made of or have one or more components of (a) wire, (b) very thin wire, and (c) rods of various cross sections.

6.129 Although extruded products are typically straight, it is possible to design dies whereby the extrusion is curved, with a constant radius of curvature. (a) What applications could you think of for such products?

(b) Describe your thoughts as to the shape such a die should have in order to produce curved extrusions.

6.130   Survey the technical literature and describe the design features of the various roll arrangements shown in Fig. 6.41.

6.131   The beneficial effects of using ultrasonic vibration to reduce friction in some of the processes were described in this chapter. Survey the technical literature and offer design concepts to apply such vibrations.

6.132   In the Case Study at the end of this chapter, it was stated that there was a significant cost improvement using forgings when compared to the extrusion-based design. List and explain the reasons why you think these cost savings were possible.

6.133   In the extrusion and drawing of brass tubes for ornamental architectural applications, it is important to produce very smooth surface finishes. List the relevant process parameters and make manufacturing recommendations to produce such tubes.

CHAPTER 7

Sheet-Metal Forming Processes

This chapter describes sheet-metal characteristics and the principles of forming sheets into products. Specifically, the following topics are discussed:

- Specific material properties that affect formability.
- Principles of various shearing operations.
- Fundamentals of sheet-forming operations and the important parameters involved.
- Basic design principles in sheet forming.
- Economic considerations.

7.1 | Introduction

Sheet-metal forming operations produce a wide range of consumer and industrial products, such as metal desks, appliances, aircraft fuselages, beverage cans, car bodies, and kitchen utensils. Sheet-metal forming, also called **pressworking**, **press forming**, or **stamping**, is among the most important of metalworking processes, dating back to as early as 5000 B.C., when household utensils, jewelry, and other objects were made by hammering and stamping metals such as gold, silver, and copper. Compared to those made by casting or forging, for example, sheet-metal parts offer the advantages of light weight and shape versatility.

Sheet forming, unlike bulk deformation processes, involves workpieces with a high ratio of surface area to thickness, as can be seen by inspecting simple products such as cookie sheets and hubcaps. A sheet thicker than 6 mm $\left(\frac{1}{4}\text{in.}\right)$ is generally called **plate**. Sheet metal is produced by the rolling process, as described in Section 6.3. If the sheet is thin, it is generally coiled after rolling; if thick, it is available as flat sheets or plates, which may have been decoiled and flattened prior to forming them. In a typical forming operation, a blank of suitable dimensions is first cut from a large sheet. This is usually done by a shearing process, although there are several other methods for cutting sheet and plates, as described throughout Chapters 8 and 9.

As shown in Table 7.1, each process has specific characteristics and uses different types of hard tools and dies, as well as soft tooling, such as rubber or

346

TABLE 7.1

General Characteristics of Sheet-Metal Forming Processes

Process	Characteristics
Roll forming	Long parts with constant complex cross sections; good surface finish; high production rates; high tooling costs.
Stretch forming	Large parts with shallow contours; suitable for low-quantity production; high labor costs; tooling and equipment costs depend on part size.
Drawing	Shallow or deep parts with relatively simple shapes; high production rates; high tooling and equipment costs.
Stamping	Includes a variety of operations, such as punching, blanking, embossing, bending, flanging, and coining; simple or complex shapes formed at high production rates; tooling and equipment costs can be high, but labor costs are low.
Rubber-pad forming	Drawing and embossing of simple or complex shapes; sheet surface protected by rubber membranes; flexibility of operation; low tooling costs.
Spinning	Small or large axisymmetric parts; good surface finish; low tooling costs, but labor costs can be high unless operations are automated.
Superplastic forming	Complex shapes, fine detail, and close tolerances; forming times are long, and hence production rates are low; parts not suitable for high-temperature use.
Peen forming	Shallow contours on large sheets; flexibility of operation; equipment costs can be high; process is also used for straightening parts.
Explosive forming	Very large sheets with relatively complex shapes, although usually axisymmetric; low tooling costs, but high labor costs; suitable for low-quantity production; long cycle times.
Magnetic-pulse forming	Shallow forming, bulging, and embossing operations on relatively low-strength sheets; most suitable for tubular shapes; high production rates; requires special tooling.

polyurethane. The sources of energy typically involve mechanical but also include others, such as hydraulic, magnetic, and explosive.

7.2 | Sheet-Metal Characteristics

Forming of sheet metals is generally carried out by tensile forces in the plane of the sheet, as otherwise the application of external compressive forces could lead to buckling, folding, and wrinkling of the sheet. Recall that in some bulk deformation processes (described in Chapter 6), the thickness or the lateral dimensions of the workpiece is intentionally changed to produce the part, whereas in sheet-forming processes, any change in thickness is typically due to stretching of the sheet under tensile stresses (*Poisson's effect*). Thickness decreases in sheet forming should generally be avoided as they can lead to necking and failure, as occurs in a tension test.

As described next, because the mechanics of all sheet forming basically consists of stretching and bending, certain parameters significantly influence the overall operation. These are elongation, yield-point elongation, anisotropy, grain size, residual stresses, springback, and wrinkling.

7.2.1 Elongation

Although sheet-forming operations rarely involve simple uniaxial stretching, the observations made in Section 2.2 regarding the tensile test can be useful in understanding the behavior of sheet metals in forming operations. Recall from Fig. 2.2 that a specimen subjected to tension first undergoes **uniform elongation** up to the UTS, after which it begins to neck. This elongation is then followed by further nonuniform elongation (**postuniform elongation**) until the specimen fractures. Because the sheet is being stretched during forming, high uniform elongation is thus desirable for good formability.

It was shown in Section 2.2.3 that, for a material that has a true-stress–true-strain curve represented by the equation

$$\sigma = K\epsilon^n, \tag{7.1}$$

the strain at which necking begins (**instability**) is given by

$$\epsilon = n. \tag{7.2}$$

Thus, the true uniform strain in a simple stretching operation (uniaxial tension) is numerically equal to the strain-hardening exponent, n. A high value of n indicates large uniform elongation and thus is desirable for sheet forming.

Necking of a sheet-metal specimen generally takes place at an angle ϕ to the direction of tension, as shown in Fig. 7.1a. For an isotropic sheet specimen under simple tension, the Mohr's circle is constructed as shown in Fig. 7.1b. The strain ϵ_1 is the longitudinal strain, and ϵ_2 and ϵ_3 are the two lateral strains. Since Poisson's ratio in the plastic range is 0.5, the lateral strains have the value $-\epsilon_1/2$. The narrow neck band (called **localized necking**) shown in Fig. 7.1a is in **plane strain**, because it is constrained by the material above and below the neck band.

FIGURE 7.1 (a) Localized necking in a sheet-metal specimen under tension, (b) determination of the angle of neck from the Mohr's circle for strain, (c) schematic illustrations for diffuse and localized necking, respectively, and (d) localized necking in an aluminum strip in tension; note the double neck. *Source:* S. Kalpakjian.

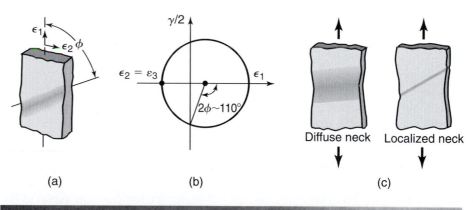

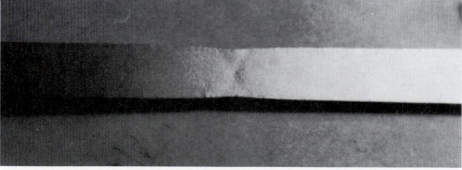

(d)

The angle ϕ can be determined from the Mohr's circle by a rotation (either clockwise or counterclockwise) of 2ϕ from the ϵ_1 position (Fig. 7.1b). For isotropic materials, this angle is about 110°, and thus the angle ϕ is about 55°. Note that although the length of the neck band essentially remains constant during the test, its thickness decreases (because of volume constancy), and the specimen eventually fractures. The angle ϕ will be different for materials that are anisotropic in the plane of the sheet (planar anisotropy).

Whether necking is *localized* or **diffuse** (Fig. 7.1c) depends on the strain-rate sensitivity, m, of the material, as given by the equation

$$\sigma = C\dot{\epsilon}^m. \tag{7.3}$$

As described in Section 2.2.7, the higher the value of m, the more diffuse the neck becomes. An example of localized necking on a strip of aluminum in tension is shown in Fig. 7.1d. Note also the double localized neck; in other words, ϕ can be in either the clockwise or counterclockwise position in Fig. 7.1a, or both (as shown in Fig. 7.1d).

The total elongation of a tension-test specimen, at a gage length of 50 mm (2 in.), is also a significant factor in formability of sheet metals, where total elongation is the sum of uniform elongation and postuniform elongation. As stated previously, uniform elongation is governed by the strain-hardening exponent, n, whereas postuniform elongation is governed by the strain-rate sensitivity index, m. The higher the m value, the more diffuse the neck is, and hence the greater the postuniform elongation becomes before fracture. Consequently, the total elongation of the material increases with increasing values of both n and m.

The following are important parameters that affect the sheet-metal forming process:

1. **Yield-point elongation.** Low-carbon steels exhibit a behavior called *yield-point elongation*, involving *upper* and *lower* yield points, as shown in Fig. 7.2a. Recall also from Section 2.2 that yield-point elongation is usually on the order of a few percent. In yield-point elongation, a material yields at a given location; subsequent yielding occurs in adjacent areas where the lower yield point is unchanged. When the overall elongation reaches the yield-point elongation, the entire specimen has been deformed uniformly. The magnitude of the yield-point elongation depends on the strain rate (with higher rates, the elongation generally increases) and the grain size of the sheet metal (as grain size decreases, yield-point elongation increases).

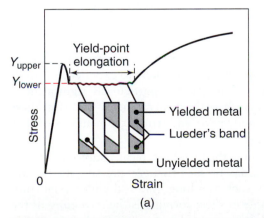

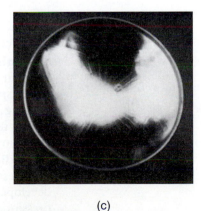

(a)   (b)   (c)

FIGURE 7.2   (a) Yield-point elongation and Lueder's bands in tensile testing, (b) Lueder's bands in annealed low-carbon steel sheet, and (c) stretcher strains at the bottom of a steel can for common household products. *Source:* (b) Courtesy of Caterpillar Inc.

Such behavior of low-carbon steels produces **Lueder's bands** (also called **stretcher strain marks** or *worms*) on the sheet, as shown in Fig. 7.2b. These bands consist of elongated depressions on the surface of the sheet and can be objectionable in the final product because of its uneven surface appearance. These bands also can cause difficulties in subsequent coating and painting operations. Stretcher strain marks can be observed on the curved bottom (dome) of steel cans for common household products, as shown in Fig. 7.2c; aluminum cans do not exhibit this behavior because aluminum alloys used for can stock do not demonstrate yield-point elongation.

The usual method of avoiding these marks is to eliminate or to reduce yield-point elongation by reducing the thickness of the sheet 0.5 to 1.5% by cold rolling, a process known as **temper rolling** or **skin rolling**. However, because of strain aging (see Section 3.8.1), yield-point elongation reappears after even a few days at room temperature, or after a few hours at higher temperatures. Thus, the sheet metal should be formed within a certain period of time (such as from one to three weeks for rimmed steel; Section 5.7.1) to avoid the reappearance of stretcher strains.

2. **Anisotropy.** Another important factor influencing sheet-metal forming is *anisotropy,* or **directionality,** of the sheet metal. Anisotropy is acquired during the thermomechanical processing history of the sheet. Recall from Section 3.5 that there are two types of anisotropy: **crystallographic anisotropy** (from preferred grain orientation) and **mechanical fibering** (from alignment of impurities, inclusions, voids, and the like throughout the thickness of the sheet during processing). Anisotropy may be present not only in the plane of the sheet (**planar anisotropy**), but also in its thickness direction (**normal** or **plastic anisotropy**). These behaviors are particularly important in deep drawing of sheet metals, as described in Section 7.6.

3. **Grain size.** *Grain size* (Section 3.4.1) of the sheet metal is important for two reasons: first, because of its effect on the mechanical properties of the material, and second, because of its effect on the surface appearance of the formed part. The coarser the grain, the rougher the surface appears (**orange peel**). An ASTM grain size of No. 7 or finer is typically preferred for general sheet-metal forming (see Section 3.4.1).

4. **Residual stresses.** *Residual stresses* can develop in sheet-metal parts because of the nonuniform deformation that the sheet undergoes during forming. When disturbed, such as by removing a portion of it, the part may distort (see Fig. 2.31). Furthermore, tensile residual stresses on surfaces can lead to **stress-corrosion cracking** of the part (Fig. 7.3) unless it is properly stress relieved. (See also Sections 3.8.2 and 5.11.4.)

5. **Springback.** Because they generally are thin and are subjected to relatively small strains during forming, sheet-metal parts are likely to experience considerable *springback* (see Section 7.4.2). This effect is particularly significant in bending and other forming operations where the bend radius-to-sheet-thickness ratio is high, such as in automotive body parts and structural members such as C-channels.

6. **Wrinkling.** Although in sheet forming the metal is typically subjected to tensile stresses, the method of forming may be such that compressive stresses are developed in the plane of the sheet, such as a long, thin rod under axial compression, which buckles. An example in sheet forming is the *wrinkling* of the flange in deep drawing (Section 7.6) because of circumferential *compressive stresses* that develop in the flange. Other terms used to describe similar phenomena are **folding** and **collapsing.** The tendency for wrinkling in sheet metals

FIGURE 7.3 Stress-corrosion cracking in a deep-drawn brass part for a light fixture. The cracks developed over a period of time. Brass and 300-series austenitic stainless steels are particularly susceptible to stress-corrosion cracking.

increases with (a) decreasing thickness, (b) nonuniformity of the thickness of the sheet, and (c) increasing length or surface area of the sheet that is not constrained or supported. Lubricants that are trapped or are not distributed evenly at the die-sheet metal interfaces can also contribute to the initiation of wrinkling.

7. **Coated sheet.** Sheet metals, especially steel, are **precoated** with a variety of organic coatings, films, and laminates; such coatings are used primarily for appearance as well as corrosion resistance. Coatings are applied to the coil stock on continuous lines, at thicknesses generally ranging from 0.0025 to 0.2 mm (0.0001 to 0.008 in.). Coatings are available with a wide range of properties, such as flexibility, durability, hardness, resistance to abrasion and chemicals, color, texture, and gloss. Coated sheet metals are subsequently formed into products such as TV cabinets, appliance housings, paneling, shelving, residential siding, and metal furniture. Zinc is used extensively as a coating on sheet steel (**galvanized** steel; Sections 3.9.7 and 4.5) to protect it from corrosion, particularly in the automotive industry. Galvanizing of steel can be done by hot dipping, electrogalvanizing, or galvannealing processes. (See also Section 4.5.1.)

7.3 | Shearing

The *shearing* process involves cutting sheet metal, as well as plates, bars, and tubing of various cross sections, into individual pieces by subjecting it to shear stresses in the thickness direction, typically using a **punch** and a **die**, similar to the action of a paper punch (Fig. 7.4). The punch and die may be of any shape, such as circular or straight blades, similar to a pair of scissors. Important variables in the shearing process are the punch force, the speed of the punch, the edge condition of the sheet, the punch and die materials, the corner radii of the punch and die, the punch-die clearance, and lubrication.

The overall features of a typical sheared edge for the two sheared surfaces are illustrated in Fig. 7.5. Note that the edges are neither smooth nor perpendicular to the plane of the sheet. The **clearance,** *c* in Fig. 7.4, is the major parameter that determines the shape and quality of the sheared edge. As shown in Fig. 7.6a, as clearance increases, the edges become rougher and the deformation zone becomes larger. Furthermore, the material is pulled into the clearance area, and the sheared edges become more and more rounded. In fact, if the clearance is too large, the sheet metal is bent and thus subjected to tensile stresses.

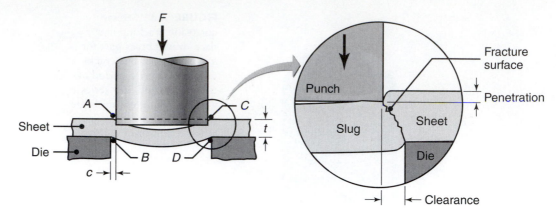

FIGURE 7.4 Schematic illustration of the shearing process with a punch and die, indicating important process variables.

In practice, clearances usually range between 2 and 8% of the sheet thickness but may be as small as 1% in fine blanking (see below). In general, clearances are smaller for softer materials, and they are higher as the sheet thickness increases. As Fig. 7.6b indicates, sheared edges can undergo severe cold working because of the high strains involved, which, in turn, can adversely affect the formability of the sheet during subsequent operations.

Observation of the shearing mechanism reveals that shearing usually starts with the formation of cracks on both the top and bottom edges of the sheet (at *A* and *B* in Fig. 7.4). These cracks eventually meet, resulting in complete separation

FIGURE 7.5 Characteristic features of (a) a punched hole and (b) the punched slug. Note that the slug has a different scale than the hole.

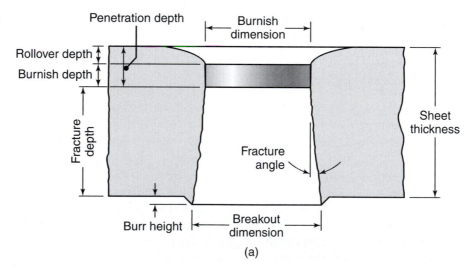

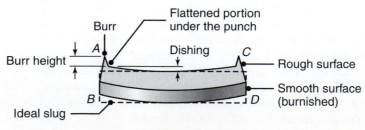

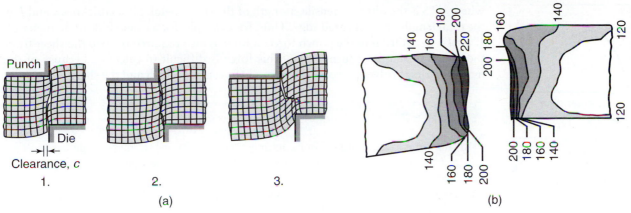

FIGURE 7.6   (a) Effect of clearance, c, on the deformation zone in shearing. Note that as clearance increases, the material tends to be pulled into the die, rather than being sheared. (b) Microhardness (HV) contours for a 6.4-mm (0.25-in.)-thick AISI 1020 hot-rolled steel in the sheared region. *Source:* After H.P. Weaver and K.J. Weinmann.

and a **rough fracture surface.** The **smooth, shiny,** and **burnished surfaces** are from the contact and rubbing of the sheared edges against the punch and the die. In the slug shown in Fig. 7.5b, the burnished surface is in the lower region, because this region is the section that rubs against the die wall. On the other hand, inspection of the sheared surface on the sheet itself reveals that the burnished surface is on the upper region in the sheared edge and results from rubbing against the punch.

The ratio of the burnished-to-rough areas on the sheared edge increases with increasing ductility of the sheet metal and decreases with increasing sheet thickness and clearance. The punch travel required to complete the shearing operation depends on the maximum shear strain that the material can undergo before fracture. Thus, a brittle or highly cold-worked material requires little travel of the punch to complete shearing.

Note from Fig. 7.6 that the deformation zone is subjected to high shear strains. The width of this zone depends on the rate of shearing, that is, the punch speed. With increasing punch speed, the heat generated by plastic deformation is confined to a smaller zone (approaching a narrow adiabatic zone), and consequently the sheared surface is smoother.

Note also the formation of a **burr** in Fig. 7.5. *Burr height* increases with increasing clearance and increasing ductility of the metal. Tooling with dull edges is also a major factor in burr formation. (See also the upcoming discussion of *slitting.*) The height, shape, and size of the burr can significantly affect subsequent forming operations, such as in flanging, where a burr can lead to subsequent cracks. Furthermore, burrs may become dislodged under external forces during their movements in mechanisms in service and interfere with the operation or contaminate lubricants. Several deburring operations are described in Section 9.8.

Punch force.   The *punch force*, F, is basically the product of the shear strength of the sheet metal and the cross-sectional area being sheared, but *friction* between the punch and the sheet can increase this force substantially. Because the sheared zone is subjected to plastic deformation, friction, and cracks, the punch-force vs. stroke curves can have various shapes. Figure 7.7 shows one typical curve for a ductile material; note that the area under the curve is the *total work* done in shearing.

An approximate empirical formula for estimating the **maximum punch force,** F_{max}, is given by

$$F_{max} = 0.7(UTS)tL \qquad (7.4)$$

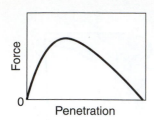

FIGURE 7.7 Typical punch force vs. penetration curve in shearing. The area under the curve is the work done in shearing. The shape of the curve depends on processing parameters and material properties.

where UTS is the ultimate tensile strength of the sheet metal, t is its thickness, and L is the total length of the sheared edge. Thus, for example, for a round hole of diameter D, $L = \pi D$. In addition to the punch force, a force is also required to strip the sheet from the punch during its return stroke. This force is difficult to calculate because of the many factors involved, especially *friction* between the punch and the sheet.

EXAMPLE 7.1 Calculation of maximum punch force

Estimate the force required in punching a 1-in. (25-mm) diameter hole through a $\frac{1}{16}$-in. (1.6-mm)-thick 5052-O aluminum sheet at room temperature.

Solution. The force is estimated from Eq. (7.4), where the UTS for this alloy is found in Table 3.7 as 190 MPa (27.5 ksi). Therefore,

$$F = 0.7\left(\frac{1}{16}\right)(\pi)(1)(27{,}500) = 3870 \text{ lb}$$

$$= 1.89 \text{ tons} = 17.3 \text{ kN}.$$

7.3.1 Shearing operations

In this section, we describe various operations that are based on the shearing process. It should first be noted that in **punching**, the sheared slug is discarded (Fig. 7.8a); in **blanking**, the slug is the part itself, and the rest is scrap. The following processes are common shearing operations.

1. **Die cutting.** *Die cutting* typically consists of the operations shown in Fig. 7.8b, where the parts produced have various uses, particularly in their assembly with other components of a product: (1) **perforating**, that is, punching a number of holes in a sheet; (2) **parting**, or shearing the sheet into two or more pieces, usually when the adjacent blanks do not have a matching contour; (3) **notching**, or removing pieces or various shapes from the edges; (4) **slitting**; and (5) **lancing**, or leaving a tab on the sheet without removing any material.

2. **Fine blanking.** Very smooth and square edges can be produced by *fine blanking* (Fig. 7.9a). One basic die design is shown in Fig. 7.9b, in which a V-shaped *stinger (impingement)* locks the sheet tightly in place and prevents the type of distortion of the material, such as that shown in Fig. 7.6. Fine blanking involves clearances on the order of 1% of the sheet thickness, as compared with as much as 8% in ordinary shearing operations. The thickness of the sheet may typically

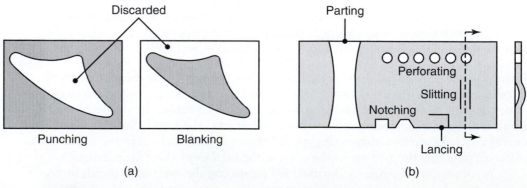

FIGURE 7.8   (a) Punching and blanking and (b) examples of shearing operations on sheet metal.

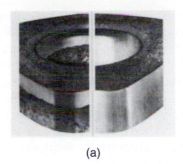

(a)

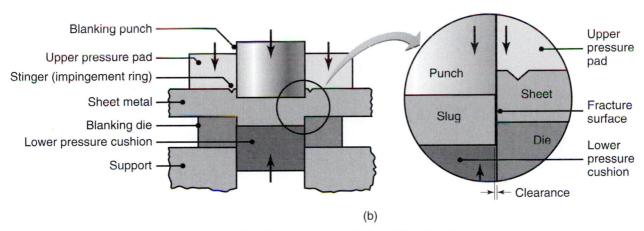

(b)

FIGURE 7.9   (a) Comparison of sheared edges by conventional (left) and fine-blanking (right) techniques and (b) schematic illustration of a setup for fine blanking. *Source:* Feintool International Holding.

range from 0.5 to 13 mm (0.02 to 0.5 in.), with a dimensional tolerance of ±0.05 mm (0.002 in.). A suitable sheet hardness is typically in the range of 50 to 90 HRB. This operation usually is carried out on triple-action hydraulic presses, triple meaning that the movements of the punch, pressure pad, and die are controlled individually. Fine blanking usually involves parts that have holes which are punched simultaneously with blanking.

3. **Slitting.** *Slitting* is a shearing operation carried out with a pair of circular blades, similar to those on a can opener (Fig. 7.10). The blades follow either a straight line, circular, or curved path. Straight slitting is commonly used in cutting wide sheet, as delivered by rolling mills, into narrower strips for further processing into individual parts. A slit edge typically has a burr, which may be

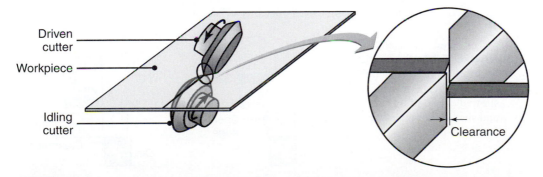

FIGURE 7.10   Slitting with rotary blades, a process similar to opening cans.

rolled over the sheet's edge using a set of rolls. There are two types of slitting equipment: (a) In the *driven* type, the blades are powered; (b) in the *pull-through* type, the strip is pulled through idling blades. Slitting operations, if not performed properly, may cause various planar distortions of the slit part or strip.

4. **Steel rules.** Sheets of soft metals, paper, thermoplastics, leather, and rubber can be blanked into various shapes with *steel-rule dies.* Such a die consists of a thin strip of hardened steel that is first bent to the shape to be sheared (similar to a cookie cutter) and then held on its edge on a flat wooden base. The die is pressed against the sheet and cuts it to the shape of the steel rule.

5. **Nibbling.** In this operation, a machine called a *nibbler* moves a straight punch up and down rapidly into a die. The sheet is fed through the punch-die gap, making a number of overlapping holes, an operation that is similar to making a large elongated hole by successively punching holes with a paper punch. Intricate slots and notches can thus be produced using standard punches. Sheets can be cut along any desired path by manual or automatic control. The process is economical for small production runs since no special dies are required.

Scrap in shearing operations. The amount of *scrap* produced (**trim loss**) in shearing operations can be significant, being as high as 30% of the original sheet for large pieces. An important factor in manufacturing costs, scrap can be reduced significantly by proper arrangement of the shapes on the sheet to be cut, called **layout** and **nesting** (see Fig. 7.65). Computer-aided design techniques are now available for minimizing scrap, particularly in large-scale operations.

7.3.2 Shearing dies

Because the formability of a sheared part can be directly influenced by the quality of its sheared edges, clearance control is important. In practice, clearances usually range between 2 and 8% of the sheet's thickness. Generally, the thicker the sheet, the larger is the clearance, to as much as 10%. Recall, however, that the smaller the clearance, the better is the quality of the sheared edge. In a process called **shaving** (Fig. 7.11), the extra material from a rough sheared edge is trimmed by cutting. Some commonly used shearing dies are described below.

1. **Punch and die shapes.** Note in Fig. 7.4 that the surfaces of the punch tip and the die are flat. Thus, the punch force builds up rapidly during shearing because the entire thickness of the sheet is sheared at the same time. The area being sheared at any instant can be controlled by beveling the punch and die surfaces, as illustrated in Fig. 7.12. The shape of a beveled punch is similar to that of a paper punch, which can easily be observed by inspecting its tip closely. The beveled geometry is particularly suitable for shearing thick sheets because it reduces the total shearing force (since it is an incremental process) and also reduces

FIGURE 7.11 Schematic illustrations of shaving on a sheared edge: (a) shaving a sheared edge and (b) shearing and shaving combined in one punch stroke.

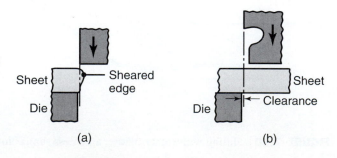

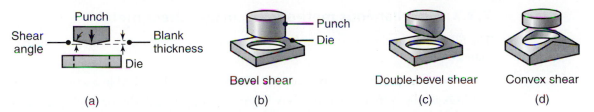

FIGURE 7.12 Examples of the use of shear angles on punches and dies. Compare these designs with that for a common paper punch.

the noise level during punching. Note also that because there will be a net lateral force in the punch shape shown in Fig. 7.12b, the punch and the press must have sufficient rigidity on the lateral direction to maintain dimensional tolerances and avoid tool breakage.

2. **Compound dies.** Several operations on the same sheet may be performed in one stroke with a *compound die* and in one station. Although these dies have higher productivity than the simple punch and die processes, these operations are usually limited to relatively simple shearing. They are somewhat slow and the dies are more expensive than those for individual shearing operations.

3. **Progressive dies.** Parts requiring multiple operations such as punching, bending, and blanking are made at high production rates using *progressive dies*. The coil strip is fed through the dies, and a different operation is performed at the same station with each stroke of a series of punches (Fig. 7.13a). An example of a part made in progressive dies is shown in Fig. 7.13b.

4. **Transfer dies.** In a *transfer die* setup, the sheet undergoes different operations at different stations, which are arranged along a straight line or a circular path. After an operation is completed, the part is transferred (hence the name transfer die) to the next station for subsequent operations.

5. **Tool and die materials.** Tool and die materials for shearing operations are generally tool steels and, for high production rates, carbides (see Table 3.6). Lubrication is important for reducing tool and die wear and for improving edge quality.

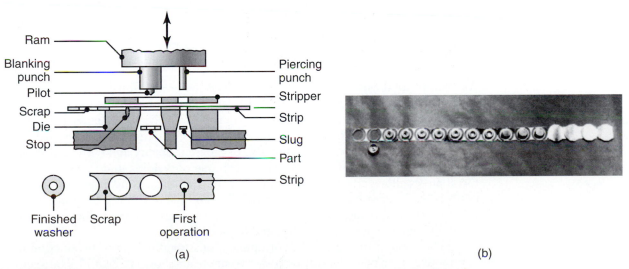

FIGURE 7.13 (a) Schematic illustration of producing a washer in a progressive die and (b) forming of the top piece of a common aerosol spray can in a progressive die. Note that the part is attached to the strip until the last operation is completed.

7.3.3 Miscellaneous methods of cutting sheet metal

The following list describes several other methods of cutting sheets and, particularly, plates.

1. The sheet or plate may be cut with a **band saw**, as described in Section 8.10.5.
2. **Oxyfuel-gas (flame) cutting** may be employed, particularly for thick plates, as widely used in shipbuilding and heavy-construction industries. (See Section 9.14.2.)
3. **Friction sawing** involves the use of a disk, or blade, that rubs against the sheet or plate at high surface speed. (See Section 8.10.5.)
4. **Water-jet cutting** and **abrasive water-jet cutting** are effective operations on sheet metals as well as on nonmetallic materials. (See Section 9.15.)
5. **Laser-beam cutting** is now widely used, with computer-controlled equipment, for high productivity, consistently cutting a variety of shapes. (See Section 9.14.1.) This process can also be combined with shearing processes.

7.3.4 Tailor-welded blanks

In sheet-metal forming operations, the blank is typically supplied in one piece, usually cut from a larger sheet, and has uniform thickness. An important technology in sheet-metal forming, particularly in the automotive industry, involves laser butt welding (see Section 12.5.2) of two or more pieces of sheet of different thicknesses and shapes (*tailor-welded blanks*, or *TWB*). The welded sheet is subsequently formed into a final shape by using any of the processes described throughout this chapter.

Laser-welding techniques are now highly developed and the weld joints are very strong. Because each welded piece can have a different thickness (as guided by design considerations such as stiffness), grade of sheet metal, coating, or other characteristics, these blanks possess the needed characteristics in the desired locations of the formed part. As a result, (a) productivity is increased, (b) the need for subsequent spot welding of the product (as in a car body) is reduced or eliminated, (c) scrap is reduced, and (d) dimensional control is improved. It should be noted, however, that because the sheet thicknesses involved are small, proper alignment of the sheets prior to welding is essential.

> **EXAMPLE 7.2** Use of tailor-welded blanks in the automotive industry
>
> An example of the use of tailor-welded blanks is the production of the automobile outer side panel, shown in Fig. 7.14a. Note that five different pieces are first blanked and then laser butt welded and stamped into the final shape. Thus, the blanks can be tailored for a particular application, including not only sheets with different shapes and thicknesses, but also of different quality and with or without coatings on one or both surfaces. This technique of welding and forming sheet-metal pieces allows significant flexibility in product design, structural stiffness and crash behavior (*crashworthiness*), formability, and the possibility of using different materials in one component, as well as weight savings and cost reduction in materials, scrap, equipment, assembly, and labor.
>
> There is a growing number of applications for this type of production in automotive companies. The various other components shown in Fig. 7.14b also utilize the advantages outlined above. For example, note that the required strength and stiffness for supporting the shock absorber is achieved by welding a round

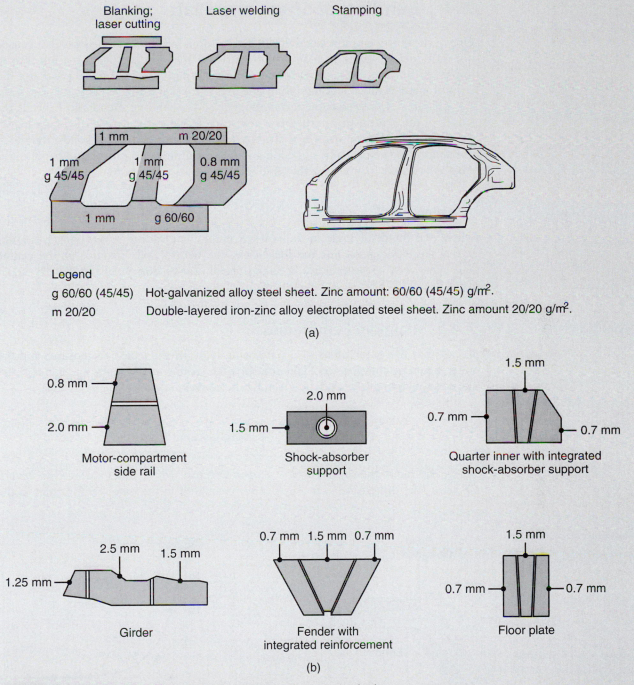

FIGURE 7.14 Examples of laser-welded and stamped automotive body components. *Source:* After M. Geiger and T. Nakagawa.

piece on the surface of the large sheet. The sheet thickness in these components varies, depending on its location and contribution to characteristics such as stiffness and strength, while allowing significant weight and cost savings.

Source: After M. Geiger and T. Nakagawa.

7.4 | Bending of Sheet and Plate

One of the most common metalworking operations is *bending,* a process that is used not only to form parts such as flanges, curls, seams, and corrugations, but also to impart stiffness (by increasing the moment of inertia). Note, for instance, that a flat strip of metal is much less rigid than one that is formed into a V cross section.

Figure 7.15a depicts the terminology used for bending. **Bend allowance** is the length of the *neutral axis* in the bend area and is used to determine the blank length for a bent part. However, as described in texts on the mechanics of solids, the radial position of the neutral axis in bending depends on the bend radius and bend angle. An approximate formula for the bend allowance, L_b, is given by

$$L_b = \alpha(R + kt),$$

where α is the bend angle (in radians), R is the bend radius, k is a constant, and t is the sheet thickness. For the ideal case, the neutral axis remains at the center, and hence $k = 0.5$. In practice, k values usually range from 0.33 for $R < 2t$ to 0.5 for $R > 2t$.

7.4.1 Minimum bend radius

Recall that the outer fibers of a part being bent are subjected to tension, and the inner fibers to compression. Theoretically, the strains at the outer and inner fibers are equal in magnitude and are given by the equation

$$e_o = e_i = \frac{1}{(2R/t) + 1}. \qquad (7.5)$$

However, due to the *shifting of the neutral axis* toward the inner surface, the *length of bend* (the dimension L in Fig. 7.15a) is smaller in the outer region than in the

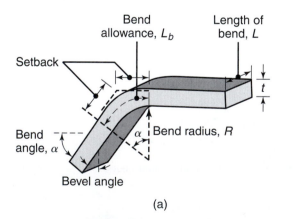

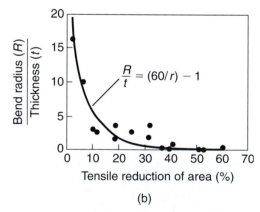

(a)   (b)

FIGURE 7.15 (a) Bending terminology. Note that the bend radius is measured to the inner surface of the bend, and that the length of the bend is the width of the sheet. (b) Relationship between the ratio of bend radius to sheet thickness and tensile reduction of area for a variety of materials. Note that sheet metal with a reduction of area of about 50% can be bent and flattened over itself without cracking, similar to folding paper. *Source:* After J. Datsko and C.T. Yang.

TABLE 7.2

Minimum Bend Radii for Various Materials at Room Temperature

Material	Material condition	
	Soft	Hard
Aluminum alloys	0	6t
Beryllium copper	0	4t
Brass, low leaded	0	2t
Magnesium	5t	13t
Steels		
austenitic stainless	0.5t	6t
low-carbon, low-alloy, and HSLA	0.5t	4t
Titanium	0.7t	3t
Titanium alloys	2.6t	4t

inner region. This phenomenon can easily be observed by bending a rectangular eraser and noting how the cross section distorts. Consequently, the outer and inner strains are different, with the difference increasing with decreasing R/t ratio.

It can be seen from Eq. (7.5) that as the R/t ratio decreases, the tensile strain at the outer fiber increases, and the material may crack after a certain strain is reached. The radius, R, at which a crack appears on the outer surface of the bend is called the *minimum bend radius*. The minimum bend radius to which a part can be bent safely is normally expressed in terms of its thickness, such as $2t$, $3t$, $4t$, and so on. Thus, for example, a bend radius of $3t$ indicates that the smallest radius to which the sheet can be bent without cracking is three times its thickness. The minimum bend radii for various materials have been determined experimentally; some typical results are given in Table 7.2. They are also available in various handbooks.

Studies have also been conducted to establish a relationship between the minimum R/t ratio and a particular mechanical property of the material. One such analysis is based on the following assumptions: (1) The true strain at cracking on the outer fiber in bending is equal to the true strain at fracture, ϵ_f, of the material in a simple tension test; (2) the material is homogeneous and isotropic; and (3) the sheet is bent in a state of plane stress, that is, its L/t ratio is small.

The true strain at fracture in tension is

$$\epsilon_f = \ln\left(\frac{A_0}{A_f}\right) = \ln\left(\frac{100}{100 - r}\right),$$

where r is the percent reduction of area of the sheet in a tension test. From Section 2.2.2, we have, for true strain,

$$\epsilon_o = \ln(1 + e_o) = \ln\left(1 + \frac{1}{(2R/t) + 1}\right) = \ln\left(\frac{R + t}{R + (t/2)}\right).$$

Equating the two expressions above and simplifying, we obtain

$$\text{Minimum}\,\frac{R}{t} = \frac{50}{r} - 1. \tag{7.6}$$

The experimental data are shown in Fig. 7.15b, where it can be noted that the curve that best fits the data is

$$\text{Minimum}\frac{R}{t} = \frac{60}{r} - 1. \tag{7.7}$$

Note that the R/t ratio approaches zero (**complete bendability**; that is, the material can be folded over itself, like a piece of paper) at a tensile reduction of area of 50%. Interestingly, this percentage is the same value obtained for spinnability of metals (described in Section 7.5.4); that is, a material with 50% reduction of area is found to be completely spinnable.

Factors affecting bendability. The bendability of a metal may be increased by increasing its tensile reduction of area, either by *heating* or by the application of *hydrostatic pressure*. Other techniques may also be employed to alter the stress environment during a bending operation, such as applying compressive forces in the plane of the sheet during bending to minimize tensile stresses in the outer fibers of the bend area.

As the length of the bend increases, the state of stress at the outer fibers changes from a uniaxial stress to a *biaxial stress*. The reason for this change is that the bend length, L, tends to become smaller due to stretching of the outer fibers (as in bending a rectangular eraser), but it is constrained by the material around the bend area. Biaxial stretching tends to reduce ductility (strain to fracture); thus, as L increases, the minimum bend radius increases (Fig. 7.16). However, at a bend length of about $10t$, the minimum bend radius increases no further, and a *plane-strain condition* is fully developed. As the R/t ratio decreases, narrow sheets (smaller length of bend) begin to crack at the edges, and wider sheets crack at the center (where the biaxial stress is the highest).

Bendability also depends on the **edge condition** of the sheet being bent. Because rough edges have various locations of stress concentration, bendability decreases as edge roughness increases. Another important factor is the amount of **cold working** that the edges undergo during shearing, as can be observed from microhardness tests in the sheared region shown in Fig. 7.6b. Removal of the cold-worked regions, such as by shaving, machining, or annealing, greatly improves the resistance to edge cracking during bending.

FIGURE 7.16 The effect of length of bend and edge condition on the ratio of bend radius to thickness for 7075-T aluminum sheet. *Source:* After G. Sachs and G. Espey.

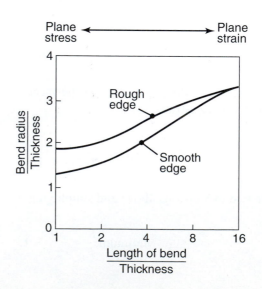

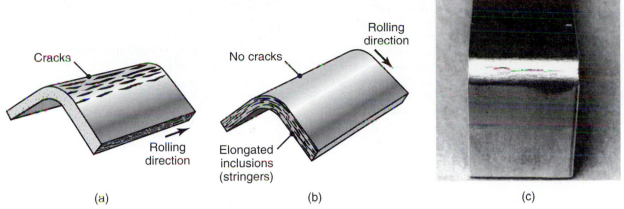

Rolling
direction

Cracks

No cracks

Rolling
direction

Rolling
direction

Elongated
inclusions
(stringers)

(a)

(b)

(c)

FIGURE 7.17 (a) and (b) The effect of elongated inclusions (stringers) on cracking in sheets as a function of the direction of bending with respect to the original rolling direction. This example shows the importance of orienting parts cut from sheet to maximize bendability. (c) Cracks on the outer radius of an aluminum strip bent to an angle of 90°; compare this part with that shown in (a).

Another significant factor in edge cracking is the amount and shape of inclusions in the sheet metal (see also Section 3.3.3). Inclusions in the form of **stringers** are more detrimental than globular-shaped inclusions; consequently, anisotropy of the sheet is also important in bendability. As depicted in Fig. 7.17, cold rolling of sheets results in **anisotropy** because of the alignment of impurities, inclusions, and voids (*mechanical fibering*). The transverse ductility is thus reduced, as shown in Fig. 7.17c (see also Fig. 3.16). In bending such a sheet, caution should be exercised in cutting or slitting the blank in the proper direction of the rolled sheet, although this may not always be possible in practice.

7.4.2 Springback

Because all materials have a finite modulus of elasticity (Table 2.1), plastic deformation is always followed by **elastic recovery** upon removal of the load. In bending, this recovery is known as *springback*. As shown in Fig. 7.18, (a) the final bend angle after springback is smaller than the angle to which it is bent and (b) the final bend radius is larger than radius to which it is bent. This phenomenon can easily be observed and verified by bending a piece of wire or a short strip metal. Note that springback will occur not only in bending flat sheets or plate, but also in bending bars, rod, and wire of any cross section.

A quantity characterizing springback is the **springback factor,** K_s, which is determined as follows. Because the bend allowance (see Fig. 7.15a) is the same before

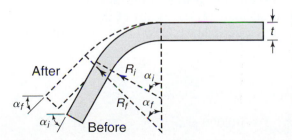

After

α_f

α_i

R_i α_i

R_f α_f

t

Before

FIGURE 7.18
Terminology for springback in bending. Note that the bend angle has become smaller. There are situations whereby the angle becomes larger, called negative springback (see Fig. 7.20).

FIGURE 7.19 Springback factor, K_s, for various materials: (a) 2024-0 and 7075-0 aluminum, (b) austenitic stainless steels, (c) 2024-T aluminum, (d) $\frac{1}{4}$-hard austenitic stainless steels, and (e) $\frac{1}{2}$-hard to full-hard austenitic stainless steels. A factor of $K_s = 1$ indicates that there is no springback. *Source:* After G. Sachs.

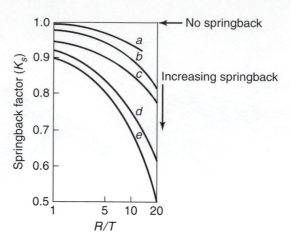

and after bending, the relationship obtained for pure bending is

$$\text{Bend allowance} = \left(R_i + \frac{t}{2} \right)\alpha_i = \left(R_f + \frac{t}{2} \right)\alpha_f. \tag{7.8}$$

From this relationship, K_s is defined as

$$K_s = \frac{\alpha_f}{\alpha_i} = \frac{(2R_i/t) + 1}{(2R_f/t) + 1}, \tag{7.9}$$

where R_i and R_f are the initial and final bend radii, respectively. Note from this expression that K_s depends only on the R/t ratio. The condition of $K_s = 1$ indicates that there is no springback, and $K_s = 0$ indicates that there is complete elastic recovery (Fig. 7.19), as in a leaf spring or any type of spring.

Recall from Fig. 2.3 that the amount of elastic recovery depends on the stress level and the modulus of elasticity, E, of the material. Thus, elastic recovery increases with the stress level and with decreasing elastic modulus. Based on this observation, an approximate formula has been developed to estimate springback, as

$$\frac{R_i}{R_f} = 4\left(\frac{R_i Y}{Et} \right)^3 - 3\left(\frac{R_i Y}{Et} \right) + 1, \tag{7.10}$$

where Y is the uniaxial yield stress of the material at 0.2% offset. (See Fig. 2.2b.)

EXAMPLE 7.3 Estimating springback

A 20-gage (0.0359-in.) steel sheet is bent to a radius of 0.5 in. Assuming that its yield stress is 40,000 psi, calculate (a) the radius of the part after it is bent, and (b) the required bend angle to achieve a 90° bend after springback has occurred.

Solution.
a. The appropriate formula is Eq. (7.10), where

$$R_i = 0.5 \text{ in.}, \qquad Y = 40,000 \text{ psi}, \qquad E = 29 \times 10^6 \text{ psi},$$

and $t = 0.0359$ in. Thus,

$$\frac{R_i Y}{Et} = \frac{(0.5)(40,000)}{(29 \times 10^6)(0.0359)} = 0.0192,$$

and

$$\frac{R_i}{R_f} = 4(0.0192)^3 - 3(0.0192) + 1 = 0.942.$$

Hence,

$$R_f = \frac{0.5}{0.942} = 0.531 \text{ in.}$$

b. To calculate the required bend angle we use Eq. (7.9), which yields

$$\frac{\alpha_f}{\alpha_i} = \frac{(2R_i/t) + 1}{(2R_f/t) + 1}$$

or $\quad \alpha_i = \alpha_f \dfrac{(2R_f/t) + 1}{(2R_i/t) + 1} = (90°)\dfrac{(2)(0.531)/(0.0359) + 1}{(2)(0.5)/(0.0359) + 1} = 95.4°.$

Negative springback. The springback commonly observed and as shown in Fig. 7.18 is called *positive springback*. Under certain conditions, however, *negative springback* also occurs, whereby the bend angle becomes larger after the bend has been completed and the load is removed. This phenomenon, which is generally associated with V-die bending, is explained by observing the sequence of deformation in Fig. 7.20. If we remove the bent piece at stage (b), it will undergo positive springback. At stage (c), the ends of the piece are touching the male punch. Note that between stages (c) and (d), the part is actually being bent in the direction opposite to that between stages (a) and (b). Note also that in both stage (b) and stage (c), there is lack of conformity between the punch radius and the inner radius of the part. In stage (d), however, the two radii are the same. Upon unloading, the part in stage (d) will spring back inwardly, because it is being *unbent* from stage (c), both at the tip of the punch and in the arms of the part. Because of the large strains that the material has undergone in the small bend area in stage (b), the amount of this inward (negative) springback can be greater than the amount of positive springback. The net result is negative springback.

Compensation for springback. In practice, springback is usually compensated for by using various techniques.

1. **Overbending** the part in the die can compensate for springback. Overbending can also be achieved by the **rotary bending** technique shown in Fig. 7.21e. The

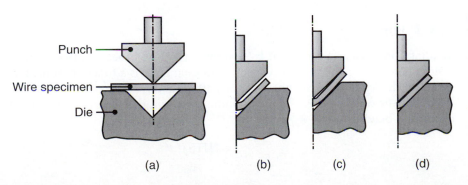

Punch — Wire specimen — Die

(a) (b) (c) (d)

FIGURE 7.20 Schematic illustration of the stages in bending round wire in a V-die. This type of bending can lead to negative springback, which does not occur in air bending (shown in Fig. 7.24a). *Source:* After K.S. Turke and S. Kalpakjian.

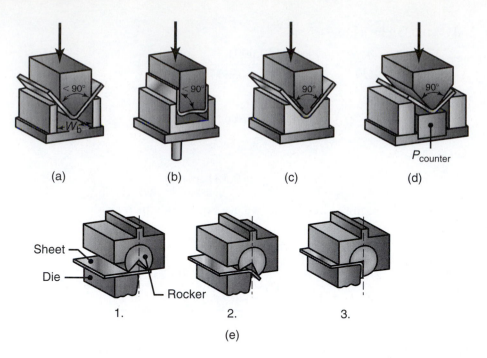

upper die has a cylindrical rocker (with an angle of <90°) and is free to rotate. As it travels downward, the sheet is clamped and bent by the rocker over the lower die (die anvil). A relief angle in the lower die allows overbending of the sheet at the end of the stroke, thus compensating for springback.

2. **Coining** the bend region by subjecting it to high localized compressive stresses between the tip of the punch and the die surface (Figs. 7.21c and d). This operation is known as **bottoming**.

3. **Stretch bending,** in which the part is subjected to tension while being bent, may also be applied. The bending moment required to make the sheet deform plastically will be reduced as the combined tension (due to bending of the outer fibers and the applied tension) in the sheet increases. As a result, springback, which is the result of nonuniform stresses due to bending, will also decrease. This technique is also used to limit springback in stretch forming of shallow automotive bodies. (See Section 7.5.1.)

4. Because springback decreases as yield stress decreases [see Eq. (7.10)], all other parameters being the same, bending may also be carried out at *elevated temperatures* to reduce springback. In practice, this is rarely done, since the temperatures required to noticeably change the yield stress are high enough to complicate handling and lubrication of the workpieces.

7.4.3 Forces

Bending forces can be estimated by assuming that the process is that of the simple bending of a rectangular beam. Thus, the bending force is a function of the material's strength, the length and thickness of the part (L and t, respectively), and the width, W, of the die opening (Fig. 7.22). Excluding friction, the general expression for the **maximum bending force**, $F_{\max}$, is

$$F_{\max} = k \frac{(\text{UTS})Lt^2}{W}, \tag{7.11}$$

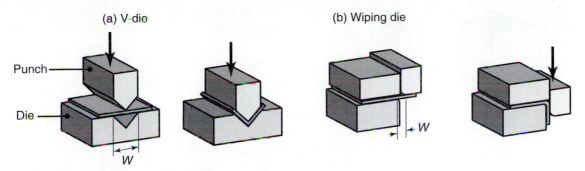

FIGURE 7.22 Common die-bending operations, showing the die-opening dimension W, used in calculating bending forces, as shown in Eq. (7.11).

where the factor k includes various factors, including friction. Its value ranges from about 1.2 to 1.33 for a V-die, 0.3–0.34 for a wiping die and 2.4–2.6 for V-dies. Equation (7.11) applies well to situations in which the punch radius and sheet thickness are small compared with the size of the die opening, W.

The bending force is also a function of punch travel. It increases from zero to a maximum and may decrease as the bend is completed; it then increases sharply as the punch bottoms in the case of die bending. In *air bending (free bending;* see Fig. 7.24a), however, the force does not increase again after it begins to decrease.

7.4.4 Common bending operations

In this section, we describe common bending operations. Some of these processes are performed on discrete sheet-metal parts; others are done continuously, as in roll forming of coiled sheet stock. The operations are described as follows.

1. **Press-brake forming.** Sheet metal or plate can be bent with simple fixtures, using a press. Parts that are long (7 m or 20 ft or more) and relatively narrow are usually bent in a *press brake*. This machine uses long dies in a mechanical or hydraulic press and is suitable for small production runs. The tooling is simple and adaptable to a wide variety of shapes (Fig. 7.23) and the process can easily be automated. Die materials for most applications are carbon steel or gray iron but may range from hardwood (for low-strength materials and small production runs) to carbides.

2. **Other bending operations.** Sheet metal may be bent by a variety of processes, as shown in Fig. 7.24. Bending of sheet metals can be carried out with two rolls (**air bending** or *free bending*), the larger one of which is flexible and typically made of polyurethane. The upper roll pushes the sheet into the flexible lower roll, imparting a curvature to the sheet, the shape of which depends on the degree of indentation into the flexible roll. Thus, by controlling the depth of penetration, it is possible to form a sheet having a variety of curvatures. Bending of relatively short pieces, such as a bushing, can also be done on highly automated **four-slide machines,** a process similar to that shown in Fig. 7.24b. Plates are bent using three rolls by the **roll bending** process, as shown in Fig. 7.24d. Note that adjusting the distance between the three roll various curvatures can be developed.

3. **Beading.** In this operation, the edge of the sheet is bent into the cavity of a die (Fig. 7.25). The bead imparts stiffness to the part by virtue of the higher moment of inertia of the edges. Beading also improves the appearance of the part and eliminates exposed sharp edges, which may be a safety hazard.

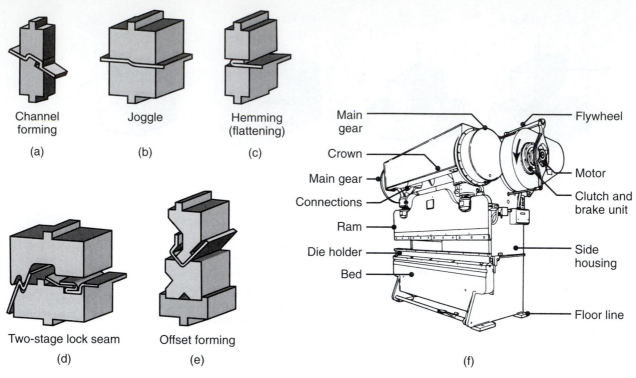

FIGURE 7.23 (a) through (e) Schematic illustrations of various bending operations in a press brake and (f) schematic illustration of a press brake. *Source:* Courtesy of Verson Allsteel Company.

FIGURE 7.24 Examples of various bending operations.

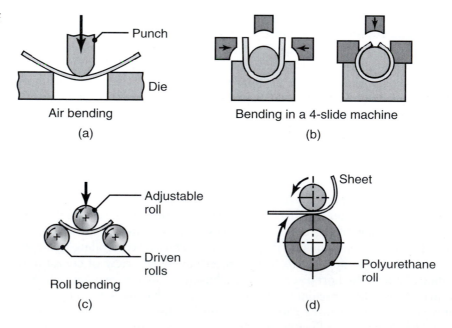

4. **Flanging.** This is a process of bending the edges of sheet metals, typically to 90°, for the purpose of imparting stiffness, appearance, or for assembly with other components. In **shrink flanging** (Fig. 7.26a), the flange periphery is subjected to compressive hoop stresses, which, if excessive, can cause it to wrinkle. The wrinkling tendency increases with decreasing radius of curvature of the

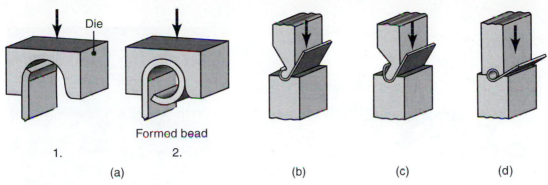

FIGURE 7.25 (a) Bead forming with a single die and (b)–(d) bead forming with two dies in a press brake.

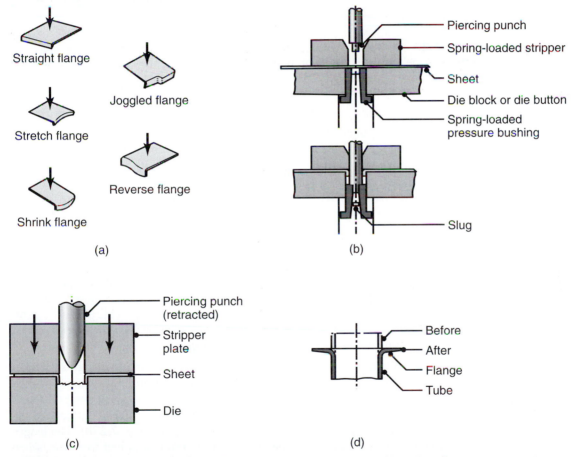

FIGURE 7.26 Illustrations of various flanging operations. (a) Flanges formed on flat sheet, (b) dimpling, and (c) piercing sheet metal with a punch to form a circular flange. In this operation, a hole does not have to be prepunched; note, however, the rough edges along the circumference of the flange. (d) Flanging of a tube; note the thinning of the periphery of the flange, due to its diametral expansion.

flange. In **stretch flanging,** on the other hand, the flange periphery is subjected to tensile stresses, which, if excessive, can lead to cracking and tearing at the edges, similar to that shown in Fig. 7.26c.

5. **Dimpling.** In this operation (Fig. 7.26b), a hole is first punched and then expanded into a flange. Flanges also may be produced by **piercing** with a

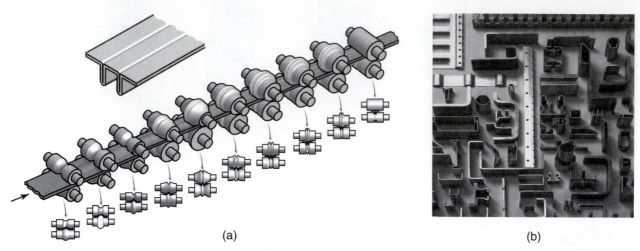

(a)   (b)

FIGURE 7.27   (a) The roll-forming operation, showing the stages in roll forming of a structural shape and (b) examples of roll-formed cross sections. *Source:* Courtesy of Sharon Custom Metal Forming, Inc.

bullet-shaped punch (Fig. 7.26c). The ends of tubes are flanged by a similar process (Fig. 7.26d). When the angle of bend is less than 90° as in fittings with conical ends, the process is called **flaring**. The condition of the edges is important in all these operations. As the ratio of flange to hole diameter increases, the strains increase proportionately. Thus, the rougher the edge, the greater will be the tendency for cracking. Sheared or punched edges may be shaved with a sharp tool (see Fig. 7.11) to improve their surface finish and thus reduce the tendency for cracking.

6. **Hemming.**   In the *hemming* process (also called **flattening**), the edge of the sheet is folded over itself (see Fig. 7.23c). Hemming increases the stiffness of the part, improves its appearance, and eliminates sharp edges (which could be a hazard). *Seaming* (Fig. 7.23d) involves joining two edges of sheet-metal pieces by hemming. Double seams are made by a similar operation, using specially shaped rollers, for watertight and airtight joints such as in food and beverage containers.

7. **Roll forming.**   This process is used for bending continuous lengths of sheet metal and for large production runs. Also called **contour roll forming** or *cold roll forming*, the metal strip in this operation is bent in stages as it passes through a series of rolls (Fig. 7.27a). Typical products include channels, gutters, siding, panels, frames, and pipes and tubing with lock seams (Fig. 7.27b). The length of the part is limited only by the amount of material supplied from the coiled stock, and the parts are usually sheared and stacked continuously. The thickness of the sheet typically ranges from about 0.125 to 20 mm (0.005 to 0.75 in.). Forming speeds are generally below 1.5 m/s (300 ft/min), although they can be much higher for special applications.

   Because dimensional tolerances, springback, tearing, and buckling of the strip are important considerations, the proper design and sequencing of the rolls require considerable experience. The rolls, which are usually mechanically driven, are generally made of carbon steel or gray iron. They may be chromium plated for better surface finish of the product and increased wear resistance. Lubricants may be used to improve roll life and the surface finish, as well as to cool the rolls and the workpiece during the operation.

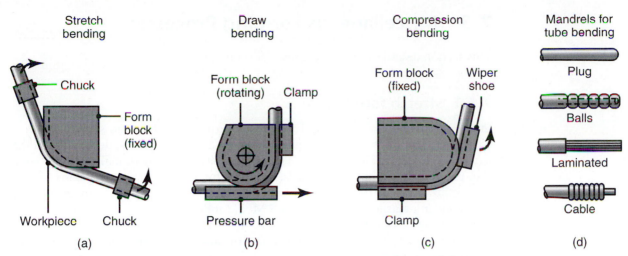

Stretch bending

Chuck

Form block (fixed)

Workpiece   Chuck

(a)

Draw bending

Form block (rotating)   Clamp

Pressure bar

(b)

Compression bending

Form block (fixed)   Wiper shoe

Clamp

(c)

Mandrels for tube bending

Plug

Balls

Laminated

Cable

(d)

FIGURE 7.28   Methods of bending tubes. Using internal mandrels, or filling tubes with particulate materials such as sand, prevents the tubes from collapsing during bending. Solid rods and structural shapes are also bent by these techniques.

7.4.5 Tube bending

Bending and forming tubes and other hollow sections is one of the most common manufacturing operations. The oldest and simplest method of bending a tube or pipe is to pack the inside with loose particles (typically sand) and bend it in a suitable fixture. The packing, which prevents the tube from buckling inward, is shaken out after the tube is bent. Tubes also can be *plugged* with various flexible internal mandrels, such as those shown in Fig. 7.28, which also illustrates various bending methods and fixtures for tubes and sections. A relatively thick tube that has a large bend radius can be bent without filling it with particulates or using plugs, since it will have less of a tendency to buckle inward.

The beneficial effect of forming metals under highly compressive stresses (see Section 2.2.8) is also demonstrated in Fig. 7.29 with respect to the bending of a tube with relatively sharp corners. Note that, in this operation, the tube is subjected to longitudinal compressive stresses, which reduce the stresses in the outer fibers in the bend area, thus improving the bendability of the material.

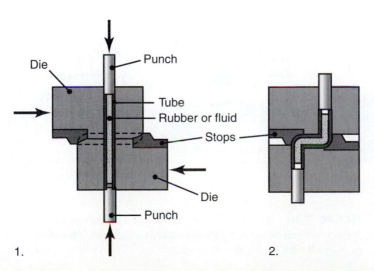

Die   Punch

Tube
Rubber or fluid

Stops

Die

Punch

1.   2.

FIGURE 7.29   A method of forming a tube with sharp angles, using an axial compressive force. Compressive stresses are beneficial in forming operations because they delay fracture. Note that the tube is supported internally with rubber or fluid to avoid collapsing during forming. *Source:* After J.L. Remmerswaal and A. Verkaik.

7.5 | Miscellaneous Forming Processes

This section describes various common forming processes.

7.5.1 Stretch forming

In this operation, the sheet metal is clamped around its edges and stretched over a die or form block which moves upward, downward, or sideways, depending on the particular machine (Fig. 7.30). This process is used primarily to make aircraft-wing skin panels, automobile door panels, and window frames. Aluminum skins for the Boeing 767 and 757 fuselages, for example, are made by stretch forming, using a blank under a tensile force as high as 9 MN (2 million lb).

In most stretch-forming operations, the blank is clamped along its narrower edges and stretched lengthwise. Controlling the amount of stretching is important to avoid tearing. Stretch forming cannot produce parts with sharp contours or reentrant corners (depressions of the surface of the die). Dies for stretch forming are generally made of zinc alloys, steel, hard plastics, or wood. Most applications require little or no lubrication. Various accessory equipment can be used in conjunction with stretch forming, including additional forming with both male and female dies while the part is under tension. Although this process is generally used for low-volume production, it is versatile and economical.

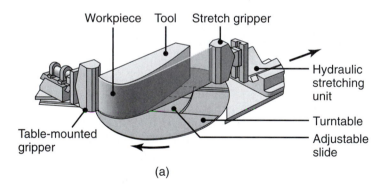

(a)

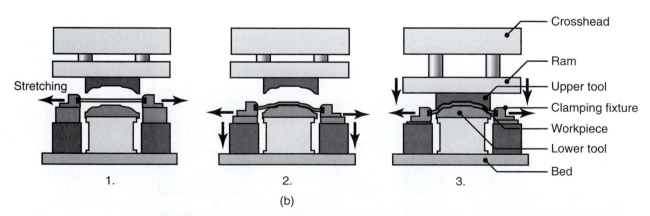

(b)

FIGURE 7.30 (a) Schematic illustration of a stretch-forming operation. Aluminum skins for aircraft can be made by this process and (b) stretch forming in a hydraulic press. *Source:* (a) Cyril Bath Co.

EXAMPLE 7.4 Work done in stretch forming

A 15-in.-long sheet with a cross sectional area of 0.5 in^2 (see Fig. 7.31) is stretched with a force, F, until $\alpha = 20°$. The material has a true-stress–true-strain curve $\sigma = 100,000\,\epsilon^{0.3}$. (a) Find the total work done, ignoring end effects and bending. (b) What is α_{max} before necking begins?

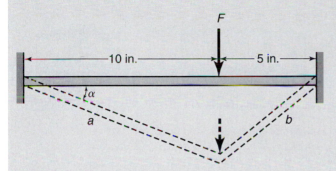

FIGURE 7.31 Workpiece considered in Example 7.4.

Solution

a. Note that because the cross section is very small compared to the length of the part, this operation is equivalent to stretching a piece of metal from 15 in. to a length of $a + b$. For $\alpha = 20°$, the final length is calculated to be $L_f = 16.8$ in. and the true strain is

$$\epsilon = \ln\left(\frac{L_f}{L_o}\right) = \ln\left(\frac{16.8}{15}\right) = 0.114.$$

The work done per unit volume (see Section 2.12) is

$$u = \int_0^{0.114} \sigma\,d\epsilon = 10^5 \int_0^{0.114} \epsilon^{0.3}\,d\epsilon = 10^5\left[\frac{\epsilon^{1.3}}{1.3}\right]_0^{0.114} = 4570 \text{ in.-lb/in}^3.$$

Since the volume of the workpiece is

$$V = (15)(0.5) = 7.5 \text{ in}^3,$$

the work done is

$$\text{Work} = (u)(V) = 34,275 \text{ in.-lb.}$$

b. The necking limit for uniaxial tension is given by Eq. (7.2). Thus,

$$L_{max} = L_o\epsilon^n = 15\epsilon^{0.3} = 20.2 \text{ in.}$$

Therefore, $a + b = 20.2$ in., and from similar triangles, we obtain

$$a^2 - 10^2 = b^2 - 5^2,$$

or

$$a^2 = b^2 + 75.$$

Hence, $a = 12$ in. and $b = 8.2$ in. Thus,

$$\cos\alpha = \frac{10}{12} = 0.833, \quad \text{or} \quad \alpha_{max} = 33.6°.$$

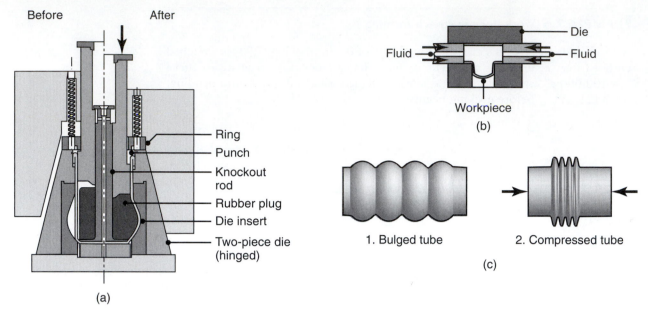

FIGURE 7.32   (a) Bulging of a tubular part with a flexible plug. Water pitchers can be made by this method. (b) Production of fittings for plumbing by expanding tubular blanks with internal pressure; the bottom of the piece is then punched out to produce a "T" section. (c) Sequence involved in manufacturing a metal bellows. *Source:* (a) and (b) After J.A. Schey.

7.5.2 Bulging

The basic process of *bulging* involves placing a tubular, conical, or curvilinear hollow part in a split female die and expanding it with a rubber or polyurethane plug (Fig. 7.32a). The punch is then retracted, the plug returns to its original shape, and the part is removed by opening the split dies. Typical products made by this process include bellows, coffee and water pitchers, barrels, and beads on tubular parts. For parts with complex shapes, the plug may be made in specific shapes in order to apply higher pressure at critical points and form the part to desired dimensions. Polyurethane plugs are very resistant to abrasion, sharp edges, wear, and lubricants, and they do not damage the surface finish of the part being formed. (See also Section 7.5.3.)

Formability in bulging operations can be enhanced by the application of compressive stresses longitudinal to the parts. *Hydraulic pressure* may also be used in bulging, although this technique requires sealing and needs hydraulic controls (Fig. 7.32b). **Segmented dies** that are expanded and retracted mechanically may also be used for bulging operations. These dies are relatively inexpensive and can be used for large production runs. *Bellows* are manufactured by a bulging process, as shown in Fig. 7.32c. After the tube is bulged at several equidistant locations, it is compressed axially to collapse the bulged regions, thus forming bellows. The tube material must be able to undergo the large strains during the collapsing process.

Embossing. This process consists of forming a number of shallow shapes, such as numbers, letters, or designs, on sheet metal, for functional as well as decorative purposes. Parts may be embossed with male and female dies, or by other means described throughout this chapter.

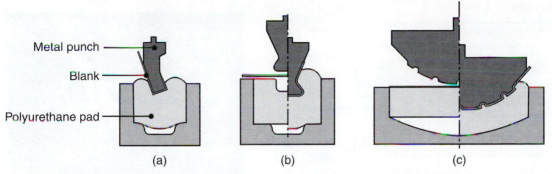

Metal punch

Blank

Polyurethane pad

(a)   (b)   (c)

FIGURE 7.33 Examples of bending and embossing sheet metal with a metal punch and a flexible pad serving as the female die. *Source:* Polyurethane Products Corporation.

7.5.3 Rubber-pad forming and hydroforming

In the processes described in the preceding sections, the dies are typically made of rigid materials. In *rubber pad forming,* one of the dies in a die set is made of a flexible material, such as a polyurethane or rubber membrane. Polyurethanes are widely used because of their resistance to abrasion, long fatigue life, and resistance to damage by burrs or sharp edges of the sheet blank. The use of the words rubber-pad forming dates back to times when polymers were not yet developed for forming operations.

In bending and embossing by the rubber-pad forming technique, as shown in Fig. 7.33, the female die is replaced with a rubber pad. The pressures applied are typically on the order of 10 MPa (1500 psi). Note that the outer surface of the sheet is now protected from damage or scratches because it is not in contact with a hard metal surface during forming. Parts can also be formed with laminated sheets of various nonmetallic materials and coatings.

In *hydroforming,* or **fluid-forming process** (Fig. 7.34), the pressure applied over the flexible membrane is controlled throughout the forming cycle, with maximum pressures reaching 100 MPa (15,000 psi). This control procedure allows proper flow of the sheet during the forming cycle to prevent wrinkling or tearing.

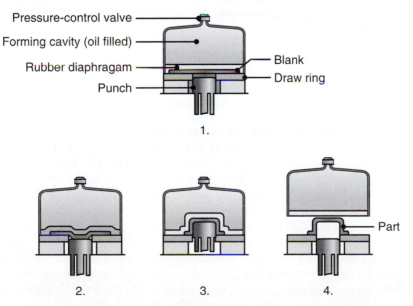

Pressure-control valve

Forming cavity (oil filled)

Rubber diaphragam

Punch

Blank

Draw ring

1.

2.   3.   4.

Part

FIGURE 7.34 The principle of the hydroforming process, also called fluid forming.

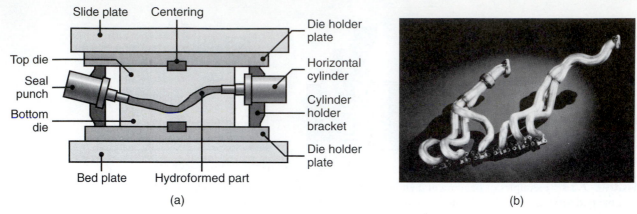

(a) (b)

FIGURE 7.35 (a) Schematic illustration of the tube hydroforming process and (b) examples of tube hydroformed parts. Automotive exhaust and structural components, bicycle frames, and hydraulic and pneumatic fittings can be produced through tube hydroforming. *Source:* Schuler GmBH.

It has been observed that deeper draws are obtained than in conventional deep drawing (described in Section 7.6), the reason being that the pressure around the rubber membrane forces the part being formed against the punch. The friction at the punch-cup interface then reduces the longitudinal tensile stresses in the part and thus delays fracture. The control of frictional conditions in rubber-pad forming as well as in other sheet-forming operations can be a critical factor in making parts successfully. Selection of proper lubricants and application methods is also important. In **tube hydroforming** (Fig. 7.35), steel or other metal tubing is bent to form a blank and is then internally pressurized by a fluid. Simple tubes (Fig. 7.35a) as well as intricate hollow tubes with varying cross sections (Fig. 7.35b) can be formed by this process. Applications of tube-hydroformed parts typically include automotive exhaust and structural components.

When selected properly, rubber-pad forming and hydroforming processes have the advantages of (a) low tooling cost, (b) flexibility and ease of operation, (c) low die wear, (d) no damage to the surface of the sheet, and (e) capability to form complex shapes, especially complex shapes that reduce weight in automotive applications.

7.5.4 Spinning

Spinning involves the forming of axisymmetric parts over a rotating mandrel, using rigid tools or rollers. The equipment used is similar to a lathe (see Section 8.9.2) with various special features and computer controls.

Conventional spinning. In *conventional spinning*, a circular blank of flat or preformed sheet metal is held against a rotating mandrel while a rigid tool deforms and shapes it over the mandrel (Fig. 7.36a). The tools may be actuated either manually or by a hydraulic mechanism. The operation involves a sequence of passes and requires considerable skill. Some typical shapes made by conventional spinning are shown in Fig. 7.37. Note that this process is particularly suitable for conical and curvilinear shapes, which would otherwise be difficult or uneconomical to form by other methods. Although most spinning is performed at room temperature, thick parts or metals with low ductility or high strength require spinning at elevated temperatures. Part diameters may range up to 6 m (20 ft). Tooling costs in spinning are relatively low; however, because the operation requires multiple passes to form the final part, it is economical for relatively small production runs only. (See Section 7.10.)

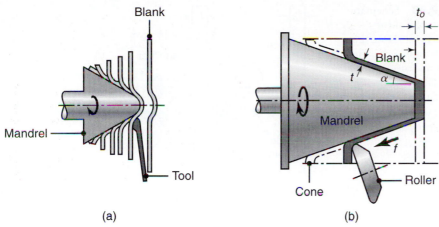

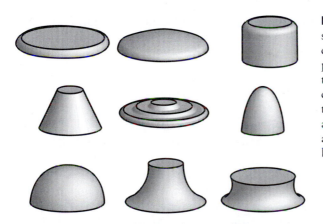

FIGURE 7.36   Schematic illustration of spinning processes: (a) conventional spinning and (b) shear spinning. Note that in shear spinning, the diameter of the spun part, unlike in conventional spinning, is the same as that of the blank. The quantity f is the feed (in mm/rev or in./rev).

FIGURE 7.37   Typical shapes produced by the conventional spinning process. Circular marks on the external surfaces of components usually indicate that the parts, such as aluminum kitchen utensils and light reflectors, have been made by spinning.

Shear spinning. In *shear spinning,* also called **power spinning, flow turning, hydrospinning,** and **spin forging,** an axisymmetric conical or curvilinear shape is generated in a manner whereby the diameter of the part remains constant (Fig. 7.36b). Parts typically made by this process include rocket-motor casings and missile nose cones. Although a single roller can be used, two rollers are preferred in order to balance the radial forces acting on the mandrel and thus avoid distortions and maintain dimensional accuracy. Little material is wasted, and the operation is completed in a relatively short time. If the operation is carried out at room temperature, the spun part has a higher yield strength than the original material, but lower ductility and toughness.

Parts typically up to about 3 m (10 ft) in diameter can be spun to close dimensional tolerances. A wide variety of shapes can be spun with relatively simple tooling, generally made of tool steel. Because of the large plastic deformation involved, the process generates considerable heat, which is usually carried away by a coolant-type fluid applied during spinning.

Referring to Fig. 7.36b, note that in shear spinning over a conical mandrel, the thickness, t, of the spun part is

$$t = t_o \sin \alpha, \tag{7.12}$$

FIGURE 7.38 Schematic illustration of a shear spinnability test. Note that as the roller advances, the spun part thickness is reduced. The reduction in thickness at fracture is called the maximum spinning reduction per pass. *Source:* After R.L. Kegg.

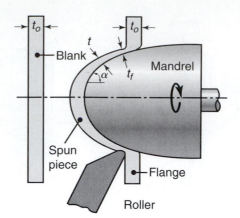

where t_o is the original sheet thickness. The force primarily responsible for supplying the energy required is the *tangential force, F_t*. For an *ideal* case in shear spinning of a cone, this force is

$$F_t = ut_o f \sin \alpha \qquad (7.13)$$

where u is the specific energy of deformation, as obtained from Eq. (2.59), and f is the feed. As described in Section 2.12, u is the area under the true-stress–true-strain curve that corresponds to a true strain related to the shear strain by the expression

$$\epsilon = \frac{\gamma}{\sqrt{3}} = \frac{\cot \alpha}{\sqrt{3}}. \qquad (7.14)$$

The actual force, however, can be as much as 50% higher than that given by Eq. (7.13) because of factors such as redundant work and friction which are difficult to calculate.

An important factor in shear spinning is the **spinnability** of the metal, defined as the maximum reduction in thickness to which a part can be spun without fracture. A simple test method has been developed to determine spinnability (Fig. 7.38). A circular blank is spun over an ellipsoid mandrel. As the thickness decreases, all materials eventually fail at some critical thickness, and the maximum spinning reduction in thickness is

$$\text{Maximum reduction} = \frac{t_o - t_f}{t_o} \times 100\%. \qquad (7.15)$$

It has been observed that ductile metals fail in tension after the reduction in thickness has taken place (as would be the case in wire and rod drawing; Section 6.5.1), whereas less ductile metals fail in the deformation zone under the roller.

The maximum reduction is then plotted vs. the tensile reduction of area of the material spun (Fig. 7.39). Note that once a material has about 50% reduction of area in a tension test, any further increase in the ductility of the original material does not improve the material's spinnability. Recall that in bending, a similar observation has also been made (see Fig. 7.15), in that maximum bendability of a sheet metal corresponds to a tensile reduction of area of about 50%.

Tube spinning. In *tube spinning*, tubes or pipes are reduced in thickness by spinning them on a cylindrical mandrel, using rollers. The operation may be carried out externally or internally (Fig. 7.40), and the part may be spun *forward* or *backward*, similar to a drawing or a backward extrusion process. (See Section 6.4.) The reduction

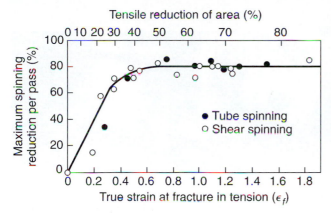

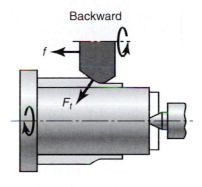

FIGURE 7.39
Experimental data showing the relationship between maximum spinning reduction per pass and the tensile reduction of area of the original material. See also Fig. 7.15.
Source: S. Kalpakjian.

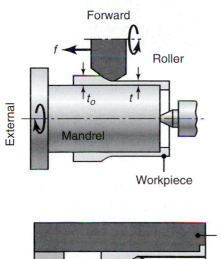

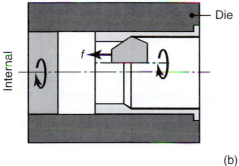

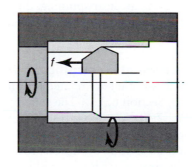

(b)

FIGURE 7.40 Examples of (a) external and (b) internal tube spinning, and the process variables involved.

in wall thickness results in a longer tube, because of volume constancy. Various external and internal profiles can be produced on tubes by controlling the path of the roller during its travel along the mandrel. This process, which can be combined with shear spinning, is used for making pressure vessels and rocket, missile, and automotive components.

Based on an approach similar to that for shear spinning, the ideal tangential force, F_t, in forward tube spinning can be expressed as

$$F_t = \overline{Y}(t_o - t)f, \tag{7.16}$$

where $\overline{Y}$ is the average flow stress of the material. Friction and redundant work of deformation make the actual force to be about twice that calculated by this formula. For backward spinning, the ideal tangential force is approximately twice that given by Eq. (7.16), the reason being that, unlike in forward spinning, the tool has to

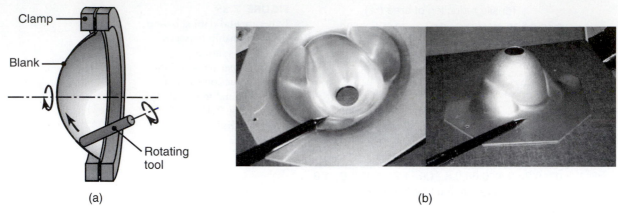

Clamp

Blank

Rotating tool

(a)

(b)

FIGURE 7.41   (a) Illustration of an incremental forming operation. Note that no mandrel is used, and that the final part shape depends on the path of the rotating tool. (b) An automotive headlight reflector produced through CNC incremental forming. Note that the part does not have to be axisymmetric. *Source:* Courtesy of J. Jesweit.

travel a shorter distance to complete the operation. (Note also the difference between direct and indirect extrusion, Section 6.4, and recall that work done is the product of force and distance travelled.)

Spinnability in tube spinning is determined by a test method similar to that for shear spinning. In the test setup, the path of the roller is inclined, whereby the thickness of the part is continually reduced, and the part eventually fails. Maximum reduction per pass in tube spinning is found to be related to the tensile reduction of area of the material (see Fig. 7.39), with results very similar to those of shear spinning. As in shear spinning, ductile metals fail in tension after the reduction in thickness has taken place, whereas less ductile metals fail in the deformation zone *under* the roller.

Incremental forming.   *Incremental forming* is a term also applied to a class of processes that are related to conventional metal spinning. (See also *incremental forging,* Section 6.2.3.) The simplest version is *incremental stretch expanding,* shown in Fig. 7.41, wherein a rotating blank is deformed by a steel rod with a smooth hemispherical tip to produce axisymmetric parts. No special tooling or mandrel is used, and the motion of the steel rod determines the final part shape, in one or more passes. The strain distribution within the workpiece depends on the tool path across the part profile, and proper lubrication is essential.

CNC incremental forming uses a computer numerical control (CNC) machine tool that is programmed to follow contours at different depths across the workpiece surface. In this arrangement, the sheet-metal blank is clamped and is stationary, and the tool rotates to assist forming. Tool paths are calculated in a manner similar to that for metal cutting, using a CAD model of the desired shape as the starting point (see Fig. 10.46). Fig. 7.41b depicts an example of a part that has been produced from CNC incremental forming. Note that the part does not have to be axisymmetric.

The main advantages of incremental forming are low tooling costs and high flexibility in the shapes that can be produced. CNC incremental forming has been used for rapid prototyping of sheet-metal parts (see Section 10.12) because the lead times associated with hard tooling are not necessary. The main drawbacks to incremental forming include low production rates and limitations on allowable materials. Most incremental forming today is performed on aluminum alloys with very high formability, so that material costs are also fairly high.

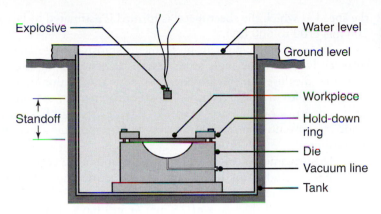

FIGURE 7.42 Schematic illustration of the explosive forming process. Although explosives are typically used for destructive purposes, their energy can be controlled and employed in forming large parts that would otherwise be difficult or expensive to produce by other methods.

7.5.5 High-Energy-Rate Forming

This section describes sheet-metal forming processes that use chemical, electrical, or magnetic sources of energy. They are called high-energy-rate processes because the energy is released in a very short period of time.

Explosive forming. The most common *explosive forming* operation is shown in Fig. 7.42. The sheet is clamped over a die, the air in the die cavity is evacuated, and the whole assembly is lowered into a tank filled with water. An explosive charge is then placed at a certain distance from the sheet surface, and detonated. The rapid conversion of the explosive into gas generates a shock wave. The pressure of this wave is sufficiently high to force the metal into the die cavity.

The peak pressure generated in water is given by the expression

$$p = K\left(\frac{\sqrt[3]{W}}{R}\right)^a,\qquad(7.17)$$

where p is the peak pressure in psi, K is a constant that depends on the type of explosive (e.g., 21,600 for TNT), W is the weight of the explosive, in pounds, R is the distance of the explosive from the workpiece (*standoff*) in feet, and a is a constant, generally taken as 1.15.

An important factor in determining peak pressure is the **compressibility** of the energy-transmitting medium (such as water) and its **acoustic impedance** (defined as the product of mass density and sound velocity in the medium). Thus, the lower the compressibility of the medium and the higher its density, the higher is the peak pressure (Fig. 7.43). Detonation speeds are typically 6700 m/s (22,000 ft/s), and

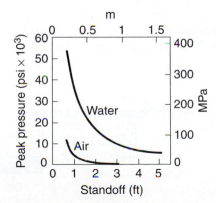

FIGURE 7.43 Effect of the standoff distance and type of energy-transmitting medium on the peak pressure obtained using 1.8 kg (4 lb) of TNT. The pressure-transmitting medium should have a high density and low compressibility. In practice, water is a commonly used medium.

the speed at which the sheet metal is formed is estimated to be on the order of 30 to 200 m/s (100 to 600 ft/s).

A variety of shapes can be formed explosively, provided that the material is sufficiently ductile at high strain rates. Depending on the number of parts to be made, dies used in this process may be made of aluminum alloys, steel, ductile iron, zinc alloys, reinforced concrete, wood, plastics, or composite materials. The final properties of the parts made by this process are basically the same as those made by conventional methods. Safety is an important aspect in explosive forming operations.

Requiring no machinery, only one die, and being versatile, explosive forming is particularly suitable for low-quantity production runs of large parts. Steel plates 25 mm (1 in.) thick and 3.6 m (12 ft) in diameter have been formed by this method. Tubes with a wall thickness of 25 mm (1 in.) have also been bulged by explosive-forming techniques. This method can also be used for much smaller parts, in which a cartridge is used as the source of energy, confined in a closed die. Parts made typically involve tubes that are bulged and expanded.

EXAMPLE 7.5 Peak pressure in explosive forming

Calculate the peak pressure in water for 0.1 lb of TNT at a standoff of 1 ft. Is this pressure sufficiently high for forming sheet metals?

Solution. Using Eq. (7.17), we find that

$$ p = (21,600)\left(\frac{\sqrt[3]{0.1}}{1}\right)^{1.15} = 9000 \text{ psi.} $$

This pressure is sufficiently high to form sheet metals. Note, for instance, that in Example 2.5, the pressure required to expand a thin-walled spherical shell of a material similar to soft aluminum alloys is only 400 psi. Also note that a process such as hydroforming (Section 7.5.3) has a maximum hydraulic pressure in the dome of about 100 MPa (15,000 psi). Other rubber-pad forming processes utilize pressures ranging from about 1500 to 7500 psi. Thus, the pressure obtained in this problem is sufficient for most sheet-forming processes.

Electrohydraulic forming. In *electrohydraulic forming,* also called **underwater-spark** or **electric-discharge forming,** the source of energy in this process is a spark from two electrodes connected with a thin wire (Fig. 7.44). The energy is stored in a bank of charged condensers; the rapid discharge of this energy through the electrodes generates a shock wave, which is strong enough to form the part. Note that this

FIGURE 7.44 Schematic illustration of the electrohydraulic forming process.

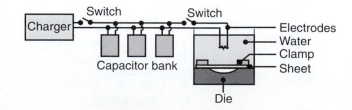

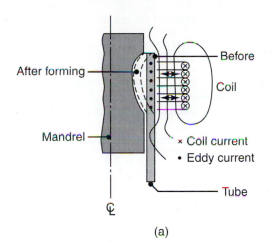

(a)   (b)

FIGURE 7.45
(a) Schematic illustration of the magnetic-pulse forming process. The part is formed without physical contact with any object. (b) Aluminum tube collapsed over a hexagonal plug by the magnetic-pulse forming process.

process is similar to explosive forming, except that it utilizes a lower level of energy and is used with smaller workpieces. It is also a safer operation than explosive forming.

Magnetic-pulse forming. In *magnetic-pulse forming,* the energy stored in a capacitor bank is discharged rapidly through a magnetic coil. In a typical application, a ring-shaped coil is placed over a tubular workpiece to be formed over another solid piece (such as a plug), thus making an integral part (Fig. 7.45). The transient magnetic field produced by the coil crosses the metal tube, generating *eddy currents* in the tube. This current, in turn, produces its own magnetic field. The forces produced by the two magnetic fields oppose each other; thus there is a repelling force between the coil and the tube. The high forces generated collapse the tube over the inner piece. Magnetic-pulse forming is used for a variety of operations, such as swaging thin-walled tubes over rods, cables, and plugs (such as placing end fittings onto torque tubes for aircraft); bulging; and flaring.

Superplastic forming. In Section 2.2.7 we described the *superplastic behavior* of some very fine grained alloys (normally less than 10 to 15 μm), where very large elongations (up to 2000%) are obtained at certain temperatures and low strain rates. These alloys, such as zinc, aluminum, and titanium, can be formed into complex shapes by employing traditional metalworking or polymer-processing techniques (such as thermoforming, vacuum forming, and blow molding, described in Chapter 10). Selection of die materials depends on the forming temperature and strength of the superplastic alloy but typically include low-alloy steels, cast tool steels, ceramics, graphite, and plaster of paris.

The high ductility and relatively low strength of superplastic alloys present the following advantages in superplastic forming:

1. Lower strength of tooling, because of the low strength of the material at forming temperatures, and, hence, lower tooling costs.

2. Ability to form complex shapes in one piece, with fine detail, close dimensional tolerances, and elimination of secondary operations.

3. Weight and material savings, because of the good formability of superplastic materials.

4. Little or no residual stresses in the formed parts.

FIGURE 7.46 Two types of structures made by combining diffusion bonding and superplastic forming of sheet metal. Such structures have a high stiffness-to-weight ratio. *Source:* Rockwell Automation, Inc.

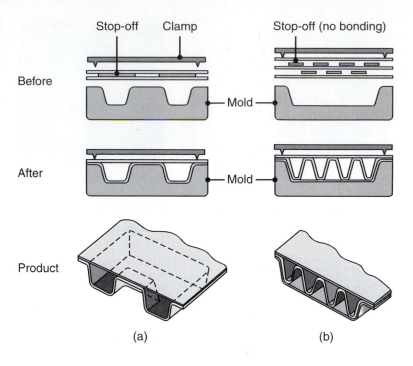

The limitations are

1. The material must not be superplastic at service temperatures.

2. Because of the extreme strain-rate sensitivity of the superplastic material, the material must be formed at sufficiently low rates (typically at strain rates of 10^{-4}/s to 10^{-2}/s).

3. Forming times range anywhere from a few seconds to several hours, and thus cycle times are much longer than in conventional forming processes. Superplastic forming is therefore a batch-forming process.

An important aspect of this process is the ability to fabricate sheet-metal structures by combining **diffusion bonding** (see Section 12.12) with superplastic forming (SPF/DB). Typical structures in which flat sheets are diffusion bonded and then formed are shown in Fig. 7.46. After bonding at selected locations of the sheets, the unbonded regions (*stop off*) are expanded into a mold by air pressure. These structures are relatively thin and have high stiffness-to-weight ratios; consequently, they are particularly important in aerospace applications.

EXAMPLE 7.6 **Applications of superplastic forming/diffusion bonding**

The majority of applications for SPF/DB produce titanium parts for military aircraft, such as the Toronado and the Mirage 2000. The components made include fuselage bulkheads, leading-edge slats, heat-exchanger ducts, and cooler outlet ducts. The nozzle fairing of the F-15 fighter aircraft is also made by this process. In civilian applications, the Airbus A340 has its water-closet, the drain, and freshwater maintenance panels (made of Ti-6Al-4V) manufactured by this process.

The superplastic forming process is usually carried out at about 900°C (1650°F) for titanium alloys and at about 500°C (930°F) for aluminum alloys; temperatures for diffusion bonding are similar. However, the presence of an oxide layer on aluminum sheets is a significant problem that degrades the bond strength in diffusion bonding. To illustrate cycle times, 718 nickel-alloy sheets 2 mm (0.080 in.) in thickness were, in one application, superplastically formed in ceramic dies at 950°C (1740°F) using argon gas at a pressure of 2 MPa (300 psi). The cycle time was four hours.

Peen forming. This process is used to produce curvatures on thin sheet metals by **shot peening** one surface of the sheet. (See Section 4.5.1.) Peening is done with cast-iron or steel shot, discharged either from a rotating wheel or with an air blast from a nozzle. In peen forming, the surface of the sheet is subjected to compressive stresses, which tend to expand the surface layer. Since the material below the peened surface remains rigid, the surface expansion causes the sheet to develop a curvature. The process also induces compressive surface residual stresses, thus improving the fatigue strength of the sheet.

The peen-forming process is used by the aircraft industry to generate curvatures on aluminum aircraft-wing skins (Fig. 7.47). Cast-steel shot about 2.5 mm (0.1 in.) in diameter at speeds of 60 m/s (200 ft/s) has been used to form wing panels 25 m (82 ft) long. For heavy sections, shot diameters as large as 6 mm $\left(\frac{1}{4} \text{ in.}\right)$ may be used. The peen-forming process is also used for straightening twisted or bent parts; out-of-round rings, for example, can be straightened by this method.

Thermal forming. This is a technology that utilizes localized heating to induce thermal-stress gradients through the thickness of the sheet. The heat source is usually a laser (*laser forming*, although a plasma torch can also be used, called *plasma forming*). The stresses developed are sufficiently high to cause localized plastic deformation of the sheet without the use of external forces, and result in, for example, a bent sheet. *Laser-assisted forming* uses lasers as a local heat source in order to reduce the flow stress of the material at specific locations and to improve its formability, thus increasing process flexibility. Applications of this process include straightening, bending, embossing, and forming of complex flat or tubular components.

Creep age forming (CAF). This is a more recent process; also called **age forming**, it involves simultaneous forming and artificial aging (see Section 5.11.2) of aluminum sheet metals. Presently, it is mainly used for the top wing-skin panels of commercial aircraft and business jets, made from 2024-T351 and the more recently developed 2022-T8 aluminum alloys. One application is in the 555-passenger Airbus A380,

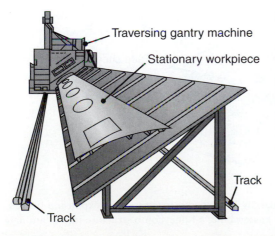

FIGURE 7.47   Schematic illustration of a peen-forming machine to shape a large sheet-metal part, such as an aircraft-skin panel. Note that the sheet is stationary and the peening head travels along its length. *Source:* Metal Improvement Company.

Traversing gantry machine

Stationary workpiece

Track

Track

which has wing skins up to 33 m (108 ft) long and 2.8 m (9.2 ft) wide, with a sheet thickness abruptly varying from 3 to 28 mm (0.12 to 1.1 in.). Springback can be a problem, since it can be up to 80% elastic recovery after forming. However, with computer modeling and simulation techniques, the process is capable of economically producing complex multiple curvatures close to net shapes.

Microforming. *Microforming* involves the use of a variety of metalworking processes to produce very small metallic parts and components (miniaturized products). Part sizes are typically in the submillimeter range, and weights are on the order of milligrams. Microforming processes are described in greater detail in Chapter 13.

Straightening. Because of the difficulties encountered in controlling all relevant parameters during production, sheet, plate, or tubular parts made may not be as straight as required. Several techniques are available to straighten these parts, including those similar to the operations and the tooling illustrated in Figs. 6.40 and 7.28; peen forming can also be used to straighten sheet metal. (Section 7.5.5.)

Manufacturing honeycomb structures. Because of their light weight and high resistance to bending forces, honeycomb structures are used for aircraft and aerospace components, as well as buildings and transportation equipment. There are two principal methods of manufacturing honeycomb structures. In the **expansion** process (Fig. 7.48a), which is the most common method, sheets are cut from a coil, and an adhesive is applied at constant intervals (node lines). The sheets are stacked and cured in an oven, whereby strong bonds develop at the adhesive joints. The block is then cut into slices of the desired dimensions and stretched to produce a honeycomb structure. This procedure is similar to expanding folded paper structures into the shape of decorative objects such as lanterns and various decorations.

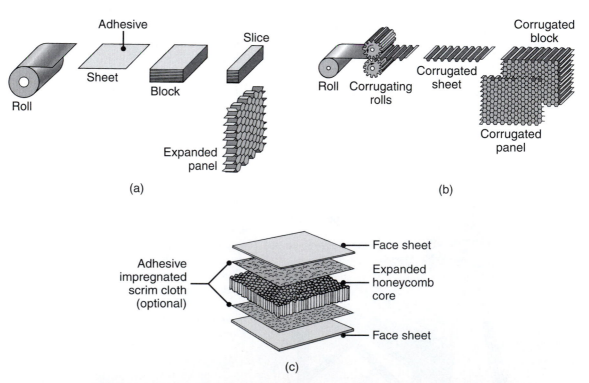

FIGURE 7.48 Methods of making honeycomb structures: (a) expansion process, (b) corrugation process, and (c) assembling a honeycomb structure into a laminate.

In the **corrugation** process (Fig. 7.48b), the sheets pass through a pair of specially designed rolls which make them into corrugated sheets. The corrugated sheets are then cut into desired lengths. Adhesive is applied to the node lines, and the block is cured. Note that no expansion of the sheets is involved in this process. The honeycomb material, made by either process, is made into a sandwich structure, as shown in Fig. 7.48c, with face sheets attached with adhesives to the top and bottom surfaces. Honeycomb structures are most commonly made from 3000-series aluminum but can also made of titanium, stainless steels, and nickel alloys.

7.6 | Deep Drawing

Deep drawing, first developed in the 1700s, is an important sheet-metal forming process. Typical parts produced by this method include beverage cans, pots and pans, containers of all shapes and sizes, kitchen sinks, and automobile panels. In this process, a flat sheet-metal blank is formed into a cylindrical or box-shaped part by means of a punch that presses the blank into the die cavity (Fig. 7.49a). Although the process is generally called deep drawing (meaning *forming deep parts*), the basic operation also produces parts with moderate depths.

The basic parameters in deep drawing a cylindrical cup are shown in Fig. 7.49b. A circular sheet blank with a diameter D_o and thickness t_o is placed over a die opening with a corner radius R_d. The blank is held in place with a **blankholder,** or **holddown ring,** under a certain force. A punch with a diameter D_p and a corner radius R_p

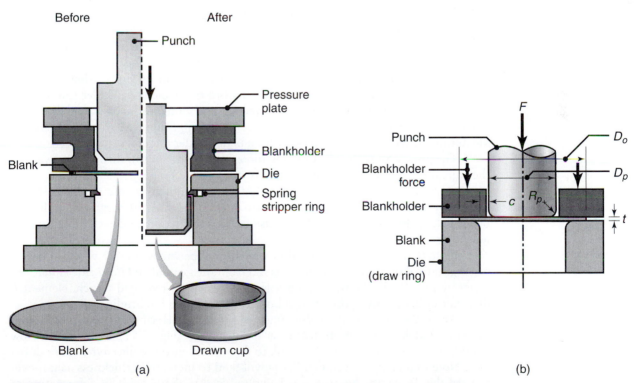

FIGURE 7.49   (a) Schematic illustration of the deep-drawing process on a circular sheet-metal blank. The stripper ring facilitates the removal of the formed cup from the punch. (b) Variables in deep drawing of a cylindrical cup. Note that only the punch force in this illustration is a dependent variable; all others are independent variables, including the blankholder force.

FIGURE 7.50 Deformation of elements in (a) the flange and (b) the cup wall in deep drawing of a cylindrical cup.

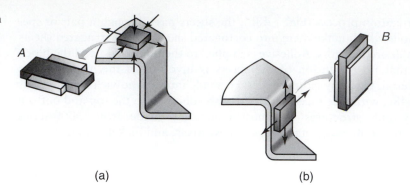

(a)   (b)

moves downward and pushes the blank into the die cavity, thus forming a cup. The significant variables in deep drawing are

1. Properties of the sheet metal.
2. Ratio of the blank diameter to the punch diameter.
3. Sheet thickness.
4. Clearance between the punch and the die.
5. Corner radii of the punch and die.
6. Blankholder force.
7. Speed of the punch.
8. Friction at the punch, die, and workpiece interfaces.

At an intermediate stage during the deep-drawing operation, the workpiece is subjected to the states of stress indicated in Fig. 7.50. On element A in the blank, the radial tensile stress is due to the blank being pulled into the cavity, and the compressive stress normal to the element is due to the pressure applied by the blankholder. With a free-body diagram of the blank along its diameter, it can be shown that the radial tensile stresses lead to compressive hoop stresses on element A. Under this state of stress, element A contracts in the hoop direction and elongates in the radial direction. It should be noted that it is the hoop stresses in the flange that tend to cause it to wrinkle during drawing, thus the necessity for a blankholder under a certain force.

The punch transmits the drawing force, F (see Fig. 7.49b), through the walls of the cup and to the flange that is being drawn into the die cavity. The cup wall, which is already formed, is subjected principally to a longitudinal tensile stress, as shown in element B in Fig. 7.50. The tensile hoop stress on element B is caused by the cup being held tightly on the punch because of its contraction under the longitudinal tensile stresses in the cup wall.

Note that the diameter of a thin-walled tube becomes smaller when subjected to longitudinal tension, as can be observed from the generalized *flow-rule equations*, given by Eq. (2.43). Thus, because it is constrained by the rigid punch, element B does not undergo any width change but elongates in the longitudinal direction.

An important aspect of drawing operations is determining how much **pure drawing** and how much **stretching** is taking place (Fig. 7.51). Note that a low blankholder force will allow the blank to flow freely into the die cavity (pure drawing). Note that element A in Fig. 7.50a will tend to increase in thickness as it moves toward the die cavity, because it is being reduced in diameter. The deformation of the sheet is mainly in the flange, and the cup wall is subjected only to elastic stresses. However, these stresses increase with an increasing D_o/D_p ratio and can eventually lead to failure when the cup wall cannot support the load required to draw the flange into the die cavity (Fig. 7.51a).

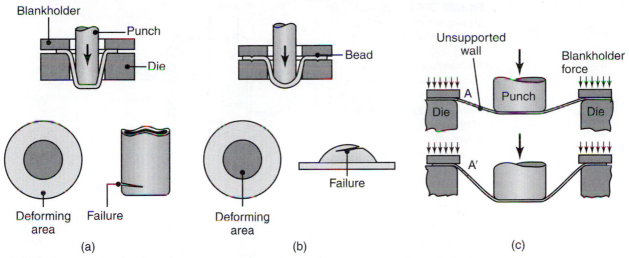

FIGURE 7.51 Examples of (a) pure drawing and (b) pure stretching; the bead prevents the sheet metal from flowing freely into the die cavity. (c) Unsupported wall and possibility of wrinkling of a sheet in drawing. *Source:* After W.F. Hosford and R.M. Caddell.

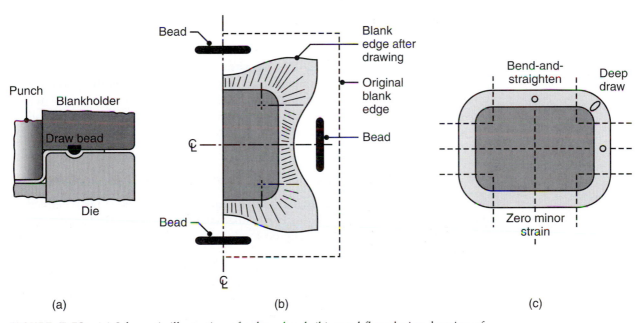

FIGURE 7.52 (a) Schematic illustration of a draw bead, (b) metal flow during drawing of a box-shaped part, using beads to control the movement of the material, and (c) deformation of circular grids in drawing. (See Section 7.7.)

Conversely, with an appropriate blankholder force or using **draw beads** (shown in Figs. 7.51b and 7.52), the blank can be prevented from flowing freely into the die cavity. The deformation of the sheet metal takes place mainly around the punch and the cup being drawn begins to stretch, eventually resulting in necking and tearing. Whether necking will be localized or diffuse depends on (a) the strain-rate sensitivity exponent, m, of the sheet metal; the higher the m value, the more diffuse the neck; (b) geometry of the punch; and (c) lubrication.

The length of the unsupported portion of the sheet (that is, the difference between the die and punch radii) is significant in that it can lead to **wrinkling.**

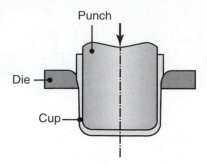

FIGURE 7.53 Schematic illustration of the ironing process. Note that the cup wall is thinner than its bottom. All beverage cans without seams (known as two-piece cans) are ironed, generally in three steps, after being deep drawn into a cup. Cans with separate tops and bottoms are known as three-piece cans.

As shown in Fig. 7.51c, element A in the sheet is being pulled into the die cavity as the punch descends. Note, however, that the blank is becoming smaller in diameter and the circumference at the element is becoming smaller as the element moves to position A'. Thus, at this position, the element is being subjected to circumferential compressive strains and is unsupported by any tooling, unlike an element between the blankholder and the die surface. Because the sheet is thin and cannot support circumferential compressive stresses to any significant extent, it tends to wrinkle in the unsupported region. This situation is particularly common in pure drawing, whereas it is less so as the process approaches pure stretching.

Ironing. If the thickness of the sheet as it enters the die cavity is more than the clearance between the punch and the die, it has to be reduced by a deformation called *ironing*. By controlling the clearance, c, ironing produces a cup with constant wall thickness (Fig. 7.53); thus, ironing can correct earing that occurs in deep drawing (as shown in Fig. 7.57). Obviously, because of volume constancy, an ironed cup will be longer than a cup produced with a large clearance.

7.6.1 Deep drawability (limiting drawing ratio)

An important parameter in drawing is the *limiting drawing ratio* (LDR), defined as the maximum ratio of blank diameter to punch diameter that can be drawn without failure, or D_o/D_p. Failure generally occurs by *thinning* of the cup wall under high longitudinal tensile stresses. Numerous attempts have been made in the past to correlate this ratio with various specific mechanical properties of the sheet metal. By observing the movement of the material into the die cavity (see Fig. 7.50), we note that the material must be capable of undergoing a reduction in width (being reduced in diameter), yet it should resist thinning under the longitudinal tensile stresses in the cup wall.

The ratio of width strain to thickness strain (Fig. 7.54) is defined as

$$R = \frac{\epsilon_w}{\epsilon_t} = \frac{\ln\left(\dfrac{w_o}{w_f}\right)}{\ln\left(\dfrac{t_o}{t_f}\right)}, \tag{7.18}$$

where R is known as the **normal anisotropy** of the sheet metal (also called **plastic anisotropy** or the **strain ratio**). Subscripts o and f refer to the original and final dimensions, respectively. An R value of unity indicates that the width and thickness strains are equal to each other; that is, the material is isotropic.

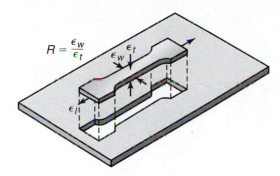

$$R = \frac{\epsilon_w}{\epsilon_t}$$

FIGURE 7.54 Definition of the normal anisotropy, R, in terms of width and thickness strains in a tensile-test specimen cut from a rolled sheet. Note that the specimen can be cut in different directions with respect to the length, or rolling direction, of the sheet.

Because errors in the measurement of small thicknesses are possible, Eq. (7.18) is generally modified, based on volume constancy, to

$$R = \frac{\ln\left(\dfrac{w_o}{w_f}\right)}{\ln\left(\dfrac{w_f l_f}{w_o l_o}\right)}, \tag{7.19}$$

where l refers to the gage length of the sheet specimen. The final length and width in a test specimen are usually measured at an elongation of 15 to 20% and for materials with lower ductility, less than the elongation at which necking begins.

Rolled sheets generally have **planar anisotropy**, and thus the R value of a specimen cut from a rolled sheet (Fig. 7.54) will depend on its orientation with respect to the rolling direction of the sheet. (See also Fig. 3.16.) An *average R* value, $\overline{R}$, is then calculated as

$$\overline{R} = \frac{R_0 + 2R_{45} + R_{90}}{4}, \tag{7.20}$$

where the subscripts 0, 45, and 90 refer, respectively, to angular orientation (in degrees) of the test specimen with respect to the rolling direction of the sheet. Note that an isotropic material has an $\overline{R}$ value of unity. Some typical values are given in Table 7.3.

TABLE 7.3

Typical Range of the Average Normal Anisotropy Ratio, $\overline{R}$, for Various Sheet Metals	
Material	$\overline{R}$
Zinc alloys	0.4–0.6
Hot-rolled steel	0.8–1.0
Cold-rolled rimmed steel	1.0–1.4
Cold-rolled aluminum-killed steel	1.4–1.8
Aluminum alloys	0.6–0.8
Copper and brass	0.6–0.9
Titanium alloys (α)	3.0–5.0
Stainless steels	0.9–1.2
High-strength low-alloy steels	0.9–1.2

FIGURE 7.55 Effect of grain size on the average normal anisotropy for various low-carbon steels. *Source:* After D.J. Blickwede.

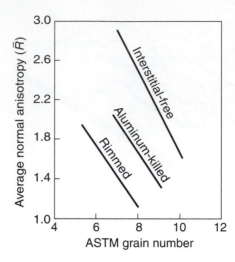

Although hexagonal close-packed metals usually have high $\bar{R}$ values, the low value for zinc given in the table is the result of its high c/a ratio in the crystal lattice. (See Fig. 3.4.) The $\bar{R}$ value also depends on the grain size and texture of the sheet metal. For cold-rolled steels, for example, $\bar{R}$ decreases as grain size decreases (Fig. 7.55). (Recall that grain size and grain number are inversely related; see Table 3.1.) For hot-rolled sheet steels, $\bar{R}$ is approximately unity because the texture developed has a random orientation.

Figure 7.56 shows the direct relationship between $\bar{R}$ and the limiting drawing ratio (LDR), as determined experimentally. In spite of its scatter, no other mechanical property of sheet metal indicates as consistent a relationship with LDR as does $\bar{R}$. Note that for an isotropic material, and based on ideal deformation, the maximum LDR is equal to $e = 2.718$.

The **planar anisotropy** of a sheet, ΔR, also can be defined in terms of directional R values as

$$\Delta R = \frac{R_0 - 2R_{45} + R_{90}}{2}, \tag{7.21}$$

which is the difference between the average of the R values in the 0° and 90° directions to rolling and the R value at 45°.

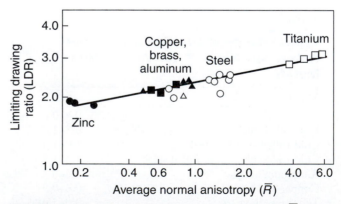

FIGURE 7.56 Effect of average normal anisotropy, $\bar{R}$, on limiting drawing ratio (LDR) for a variety of sheet metals. *Source:* After M. Atkinson.

EXAMPLE 7.7 Estimating the limiting drawing ratio

Estimate the limiting drawing ratio (LDR) that you would expect from a sheet metal that, when stretched by 23% in length, decreases in thickness by 10%.

Solution. From volume constancy of the test specimen,

$$w_o r_o l_o = w_f t_f l_f \quad \text{or} \quad \frac{w_f t_f l_f}{w_o r_o l_o} = 1.$$

From the information given, we obtain

$$\frac{l_f - l_o}{l_o} = 0.23 \quad \text{or} \quad \frac{l_f}{l_o} = 1.23$$

and

$$\frac{t_f - t_o}{t_o} = -0.10 \quad \text{or} \quad \frac{t_f}{t_o} = 0.90.$$

Hence,

$$\frac{w_f}{w_o} = 0.903.$$

From Eq. (7.18),

$$R = \frac{\ln\left(\dfrac{w_o}{w_f}\right)}{\ln\left(\dfrac{t_o}{t_f}\right)} = \frac{\ln 1.107}{\ln 1.111} = 0.965.$$

If the sheet has planar isotropy, then $R = \bar{R}$ and from Fig. 7.56 we estimate that

$$\text{LDR} = 2.4.$$

EXAMPLE 7.8 Theoretical limiting drawing ratio

Show that, assuming no thickness change of the sheet during drawing, the theoretical limiting drawing ratio is 2.718.

Solution. Reviewing Fig. 7.49b, we note that the diametral change from a blank to a cup involves a true strain of

$$\epsilon_{max} = \ln\left(\frac{\pi D_o}{\pi D_p}\right) = \ln\left(\frac{D_o}{D_p}\right).$$

The work required for plastic deformation is supplied by the radial stress on element A in Fig. 7.50a. Refer to Fig. 6.60 for wire drawing and assume that the cross section being drawn is (a) thin and rectangular, (b) with the thickness direction perpendicular to the page, and (c) a thickness that, ideally, remains constant. From Eq. (6.63), we then note that the drawing stress (the radial stress in Fig. 7.50a), which we denote as σ_d, is given by

$$\sigma_d = Y \epsilon_{max},$$

where Y is the yield stress of a perfectly plastic (ideal) material. Since, in the limit, the drawing stress is equal to the yield stress [see also the derivation of Eq. (6.74)], we have

$$Y = Y \epsilon_{max}$$

or $\epsilon_{max} = 1$. Consequently, $\ln(D_o/D_p) = 1$ and hence

$$\frac{D_o}{D_p} = 2.718.$$

FIGURE 7.57 Typical earing in a drawn steel cup, caused by the planar anisotropy of the sheet metal.

Earing. Planar anisotropy causes ears to form in drawn cups, producing a wavy edge, as shown in Fig. 7.57; the number of ears produced can be four, six, or eight, depending on the cystallographic structure of the sheet metal. The height of the ears increases with increasing ΔR. When $\Delta R = 0$, no ears form. Ears are objectionable, because they have to be trimmed off, also wasting material, which for a typical drawn aluminum beverage can is about 1–2% of its length.

It can be seen that deep drawability is thus enhanced with a high $\overline{R}$ and a low ΔR. Generally, however, sheet metals with a high $\overline{R}$ also have a high ΔR. Attempts continue to be made to develop textures in sheet metals to improve their drawability. The controlling parameters in processing metals have been found to be alloying elements, processing temperatures, annealing cycles after processing, thickness reduction in rolling, and cross (biaxial) rolling of plates to make sheets.

EXAMPLE 7.9 Estimating cup diameter and earing

A steel sheet has R values of 0.9, 1.3, and 1.9 for the 0°, 45°, and 90° directions to rolling, respectively. For a round blank 100 mm in diameter, estimate the smallest cup diameter to which it can be drawn. Will ears form during this operation?

Solution. Substituting the given values into Eq. (7.20),

$$\overline{R} = \frac{0.9 + (2)(1.3) + 1.9}{4} = 1.35.$$

The limiting drawing ratio (LDR) is defined as the maximum ratio of the diameter of the blank to the diameter of the punch that can be drawn without failure, that is, D_o/D_p. From Fig. 7.56, the LDR for this steel can be estimated to be approximately 2.5.

To determine whether or not earing will occur in this operation, we substitute the R values into Eq. (7.21). Thus,

$$\Delta R = \frac{0.9 - (2)(1.3) + 1.9}{2} = 0.1.$$

Ears will not form if $\Delta R = 0$. Since this is not the case here, ears will form in deep drawing this material.

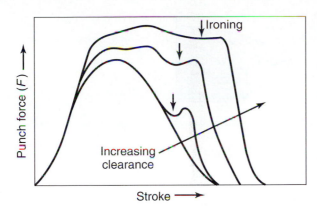

FIGURE 7.58 Schematic illustration of the variation of punch force with stroke in deep drawing. Arrows indicate the initiation of ironing. Note that ironing does not begin until after the punch has traveled a certain distance and the cup is partially formed.

Maximum punch force. The punch force, F, supplies the work required in deep drawing. As in other deformation processes, the work consists of ideal work of deformation, redundant work, frictional work, and, when present, the work required for ironing (Fig. 7.58). Because of the many variables involved in this operation and because deep drawing is not a steady-state process, accurately calculating the punch force can be difficult. Several expressions have been developed; one simple and approximate formula for the *maximum punch force* is

$$F_{\max} = \pi D_p t_o (\text{UTS}) \left(\frac{D_o}{D_p} - 0.7 \right). \tag{7.22}$$

Note that Eq. (7.22) does not specifically include parameters such as friction, the corner radii of the punch and die, or the blankholder force; however, this empirical formula makes approximate provisions for these factors.

The punch force is basically supported by the cup wall. If this force is excessive, **tearing** will occur, as shown in the lower part of Fig. 7.51a. Note that the cup was drawn to a considerable depth before failure occurred. This result can be expected by observing Fig. 7.58, where it can be seen that the punch force does not reach a maximum until after the punch has traveled a certain distance. The punch corner radius and die radius (if they are greater than 10 times the sheet thickness) do not significantly affect the maximum punch force.

7.6.2 Deep-Drawing practice

Important considerations in deep drawing are described next.

1. **Clearances and radii.** Generally, *clearances* are 7–14% greater than the original thickness of the sheet. As the clearance decreases, ironing increases, but if the clearance is too small, the blank may simply be pierced and sheared by the punch. The *corner radii* of the punch and the die are also important. If they are too small, they can cause fracture at the corners (Fig. 7.59), and if they are too large, the unsupported area wrinkles. Wrinkling in this region, as well as from the flange, causes a defect on the cup wall called puckering (wrinkling of the part between the die and the punch).

2. **Draw beads.** Draw beads (see Figs. 7.51b and 7.52) control the flow of the blank into the die cavity, and they are particularly important in drawing box-shaped or nonsymmetric parts because of the nonuniform flow of the sheet into the die. Draw beads also help in reducing the blankholder forces required, because of the stiffness imparted to the flange by the bent regions along the beads. Draw bead diameters may range from 13 to 20 mm $\left(\frac{1}{2} \text{ to } \frac{3}{4} \text{ in.} \right)$.

FIGURE 7.59 Effect of die and punch corner radii on fracture in deep drawing of a cylindrical cup. (a) Die corner radius too small; typically, it should be 5 to 10 times the sheet thickness. (b) Punch corner radius too small. Because friction between the cup and the punch aids in the drawing operation, excessive lubrication of the punch is detrimental to drawability.

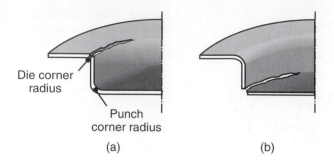

3. **Blankholder pressure.** Blankholder pressure is generally 0.7 to 1.0% of the sum of the yield and the ultimate tensile strengths of the sheet metal. Too high a blankholder force increases the punch load (because of the higher radial friction forces developed) and leads to tearing of the cup wall. On the other hand, if the blankholder force is too low, the flange may wrinkle. Since the blankholder force controls the flow of the sheet metal within the die, new presses have been designed. These presses are capable of applying a **variable blankholder force;** that is, the blankholder force can be programmed and varied throughout the punch stroke. Additional features of such modern equipment include the control of metal flow, achieved by using separate die cushions that allow the *local* blankholder force to be applied at the proper location on the sheet, thus improving drawability.

4. **Redrawing.** Parts that are too difficult to draw in one operation are generally *redrawn,* as shown in Fig. 7.60a. In **reverse redrawing** (Fig. 7.60b), the sheet is subjected to bending in the direction opposite to its original bending configuration. This reversal in bending direction results in **strain softening** and is another example of the Bauschinger effect described in Section 2.3.2. The redrawing

FIGURE 7.60 Reducing the diameter of drawn cups by redrawing operations: (a) conventional redrawing and (b) reverse redrawing. Small-diameter deep containers may undergo several redrawing operations.

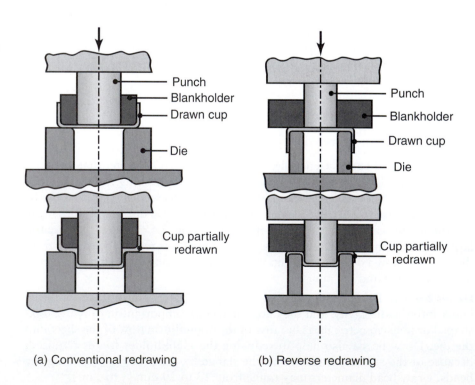

(a) Conventional redrawing   (b) Reverse redrawing

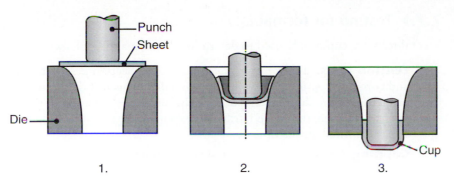

FIGURE 7.61 Stages in deep drawing without a blankholder, using a *tractrix* die profile. The tractrix is a special curve, the construction for which can be found in texts on analytical geometry or in handbooks.

operation requires lower forces than direct redrawing, and the material behaves in a more ductile manner.

5. **Drawing without a blankholder.** Deep drawing may also be carried out without a blankholder, provided that the sheet metal is sufficiently thick. The dies are specially contoured for this operation to prevent wrinkling; one example is shown in Fig. 7.61. An approximate limit for drawing without a blankholder is given by

$$D_o - D_p < 5t_o. \qquad (7.23)$$

It is apparent from Eq. (7.23) that the process can be used with thin materials for shallow draws. Although the punch stroke is longer than ordinary drawing, a major advantage of this process is the reduced cost of tooling and equipment.

6. **Tooling and equipment.** The most commonly used tool materials for deep drawing are tool steels and alloyed cast iron. Other materials, including carbides and plastics, may also be used, depending on the particular application. (See Table 3.6.) A double-action mechanical press is generally used for deep drawing (see Section 6.2.8) although hydraulic presses are also used. Punch speeds generally range between 0.1 and 0.3 m/s (20 and 60 ft/min). Speed is generally not important in drawability, although lower speeds are used for high-strength metals.

7. **Lubrication.** Lubrication in deep-drawing operations is important in order to lower forces, increase drawability, reduce tooling wear, and reduce defects in drawn parts. In general, lubrication of the punch should be minimized, as friction between the punch and the cup improves drawability.

7.7 | Formability of Sheet Metals and Modeling

The formability of sheet metals has been of great and continued interest because of its technological as well as economic significance. *Sheet-metal formability* is generally defined as the ability of a sheet to undergo the desired shape change without failure such as necking, tearing, or splitting. Three factors have a major influence on formability: (a) properties of the sheet metal, as described in Section 7.2; (b) friction and lubrication at various interfaces in the operation; and (c) characteristics of the equipment, tools, and dies used. Several techniques have been developed to test the formability of sheet metals, including the ability to predict formability by modeling the particular forming operation using several inputs, including the crystallographic texture of the sheet metal.

7.7.1 Testing for formability

Several tests are commonly used to determine the formability of sheet metal.

1. **Tension tests.** The *uniaxial tension test* (see Section 2.2) is the most basic and common test used to evaluate formability. This test determines important properties of the sheet metal, particularly, the total elongation of the sheet specimen at fracture, the strain hardening exponent, n, the planar anisotropy, ΔR, and the normal anisotropy, R.

2. **Cupping tests.** Because most sheet-forming operations basically involve biaxial stretching of the sheet, the earliest tests developed to predict formability were *cupping tests,* namely, the **Erichsen** and **Olsen** tests (*stretching*) and the **Swift** and **Fukui** tests (*drawing*). In the Erichsen test, a sheet-metal specimen is clamped over a flat die with a circular opening and a load of 1000 kg. A 20-mm-diameter steel ball is then hydraulically pressed into the sheet until a crack appears on the stretched specimen, or until the punch force reaches a maximum. The distance D, in mm, is the Erichsen number. The greater the value of D, the greater is the formability of the sheet.

   Such cupping tests measure the capability of the material to be stretched before fracturing and are relatively easy to perform. However, because the stretching under the ball is axisymmetric, they do not at all simulate the exact conditions of actual forming operations.

3. **Bulge test.** The *bulge test* has been used extensively to simulate sheet-forming operations. A circular blank is clamped at its periphery and is bulged by *hydraulic* pressure, thus replacing the punch. The operation is one of pure biaxial stretching, and no friction is involved, as would be the case in using a punch. The bulge limit (depth penetrated prior to failure) is a measure of formability, and the test is also a sensitive measure of sheet quality. Bulge tests can also be used to obtain effective-stress–effective-strain curves for biaxial loading (Section 2.11) under frictionless conditions.

4. **Forming-limit diagrams.** An aspect of sheet-metal formability is the construction of *forming-limit diagrams* (FLDs). The sheet blank is marked with a grid pattern of circles, typically 2.5 to 5 mm (0.1 to 0.2 in.) in diameter, or similar patterns, using chemical-etching or photoprinting techniques. The blank is then stretched over a punch, as shown in Fig. 7.62, and the deformation of the circles is observed and measured in regions where necking and tearing occur. Measurements are now made efficiently using specialized computer-controlled equipment. Note that the sheet is fixed by a draw bead, which prevents it from being drawn into the die. For improved accuracy of measurement, the circles on the sheet are made as small as possible and the grid lines (if used) as thin as possible.

   Because lubricants can significantly influence the test results, lubrication conditions should be stated in reporting the test results. In this manner, different lubricants can also be evaluated for the same sheet metal, or different metals can be evaluated using the same lubricant. This test can also be run with an unlubricated punch.

   In the forming-limit diagram, the **major strain** and the **minor strain** are obtained as follows: Note in Figs. 7.63b and 7.64 that after stretching, the original circle has deformed into an ellipse. The major axis of the ellipse represents the major direction and magnitude of stretching; the major strain plotted in Fig. 7.63a is the *engineering strain* (in percentage terms) in this direction. Likewise, the minor axis of the ellipse represents the magnitude of *stretching* (positive minor strain) or shrinking (negative minor strain) in the transverse

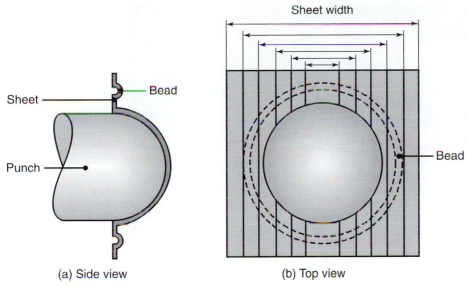

(a) Side view   (b) Top view

FIGURE 7.62 Schematic illustration of the punch-stretch test on sheet specimens with different widths, clamped along the narrower edges. Note that the narrower the specimen, the more uniaxial is the stretching. (See also Fig. 7.65.)

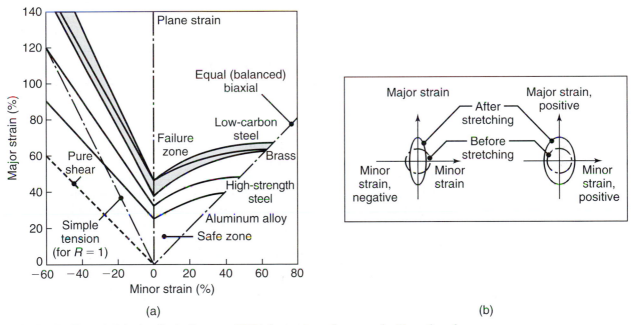

(a)   (b)

FIGURE 7.63 (a) Forming-limit diagram (FLD) for various sheet metals. Note that the major strain is always positive. The region above the curves is the failure zone; hence, the state of strain in forming must be such that it falls below the curve for a particular material; R is the normal anisotropy. (b) Illustrations of the definition of positive and negative minor strains. If the area of the deformed circle is larger than the area of the original circle, the sheet is thinner than the original thickness because the volume remains constant during plastic deformation. *Source:* After S.S. Hecker and A.K. Ghosh.

direction. Note that the major strain is always positive, because forming sheet metal takes place by stretching in at least one direction.

As an example, if we draw a circle in the center and on the surface of a sheet-metal tensile-test specimen and then stretch it plastically, the specimen

FIGURE 7.64 An example of the use of grid marks (circular and square) to determine the magnitude and direction of surface strains in sheet-metal forming. Note that the crack (tear) is generally perpendicular to the major (positive) strain. *Source:* After S.P. Keeler.

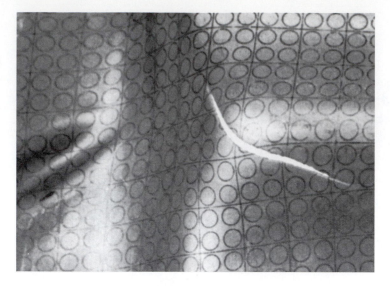

becomes narrower; thus, the minor strain is negative. Because of volume constancy in the plastic range [see Eq. (2.48)], and assuming that the normal anisotropy is $R = 1$ (that is, the width and thickness strains are equal), we have $\epsilon_w = -0.5 \, \epsilon_l$. As an experimental model, this phenomenon can easily be demonstrated by placing a circle on a wide rubber band and stretching it. On the other hand, if we place a circle on a spherical rubber balloon and inflate it, the circle simply becomes a larger circle, indicating that the minor strain is positive and equal in magnitude to the major strain. By observing the difference in surface area between the original circle and the ellipse, we can also determine whether the *thickness* of the sheet has changed during deformation. If the area of the ellipse is larger than that of the original circle, the sheet has become thinner (because volume remains constant in plastic deformation).

In order to vary the **strain path,** the specimens are prepared with varying widths, as shown in Fig. 7.62b. Thus, the center of the square specimen is subjected to a state of **equal (balanced) biaxial stretching** under the punch, whereas rectangular specimens with smaller and smaller widths approach a state of **uniaxial stretching** (simple tension). After a series of such tests is performed on a particular type of sheet metal (Fig. 7.65), the boundaries between *safe* and *failed* regions are plotted to obtain a forming-limit diagram, as shown in Fig. 7.63a). Note that this figure also includes the various strain paths, shown as straight lines: (a) The line on the right represents **equal biaxial strain;** (b) the

FIGURE 7.65 Bulge test results on steel sheets of various widths. The first specimen (farthest left) stretched farther before cracking than the last specimen. From left to right, the state of stress changes from almost uniaxial to biaxial stretching. *Source:* Courtesy of Ispat Inland, Inc.

vertical line at the center of the diagram represents **plane strain,** because the minor planar strain is zero; (c) the **simple tension** line on the left has a 2:1 slope, because the Poisson's ratio in the plastic range is 0.5 (note that the minor strain is one-half of the major strain); and (d) the **pure-shear** line has a negative 45° slope, because of the nature of deformation, as also can be seen in Fig. 2.20.

From Fig. 7.63a, it can be observed that, as expected, different materials have different forming-limit diagrams. Note that the higher the curve, the better is the formability of the material. It is also important to note that for the same minor strain, say 20%, a negative (compressive) minor strain is associated with a higher major strain before failure than is a positive (tensile) minor strain. In other words, it is desirable for the minor strain to be *negative,* because it allows shrinking in the minor direction during sheet-forming operations. Special tooling has been designed for forming sheet metals and tubing that takes advantage of the beneficial effect of negative minor strains on extending formability. (See, for example, Fig. 7.29.) The possible effect of the *rate of deformation* on forming-limit diagrams also should be assessed for each material. (See also Section 2.2.7.)

The effect of sheet-metal *thickness* on forming-limit diagrams is to raise the curves in Fig. 7.63a. Thus, the thicker the sheet, the higher its curve and the more formable the sheet is. Note, however, that a thick blank may not bend easily around small radii and may develop cracks (see Section 7.4.1).

Friction at the punch-metal interface can be significant in test results. With well-lubricated interfaces, the strains are more uniformly distributed over the punch. Also, depending on the notch sensitivity of the sheet metal, surface scratches, deep gouges, and blemishes can reduce formability and cause premature tearing and failure.

5. **Limiting dome-height test.** The *limiting dome-height* (LDH) test is performed with similarly prepared specimens as in FLD, and the *height* of the dome at failure of the sheet, or when the punch force reaches a maximum, is measured. Because the specimens are clamped, the LDH test indicates the capability of the material to stretch without failure. It has been shown that, as expected, high LDH values are related to high n and m values, as well as high total elongation of the sheet metal.

EXAMPLE 7.10 Estimating diameter of expansion

A thin-walled spherical shell made of an aluminum alloy, whose forming-limit diagram is shown in Fig. 7.63a, is being expanded by internal pressure. If the original shell diameter is 200 mm, what is the maximum diameter to which it can safely be expanded?

Solution. Because the material is being stretched in a state of equal (balanced) biaxial tension, Fig. 7.63a indicates that the maximum allowable engineering strain is about 40%. Thus,

$$e = \frac{\pi D_f - \pi D_o}{\pi D_o} = \frac{D_f - 200}{200} = 0.40.$$

Hence,

$$D_f = 280 \text{ mm.}$$

EXAMPLE 7.11 **Strains in the sheet-metal body of an automobile**

Computer programs have been developed to compute the major and minor strains and their orientations from the measured distortions of grid patterns. One such application for the various panels of an automobile body is shown in Fig. 7.66.

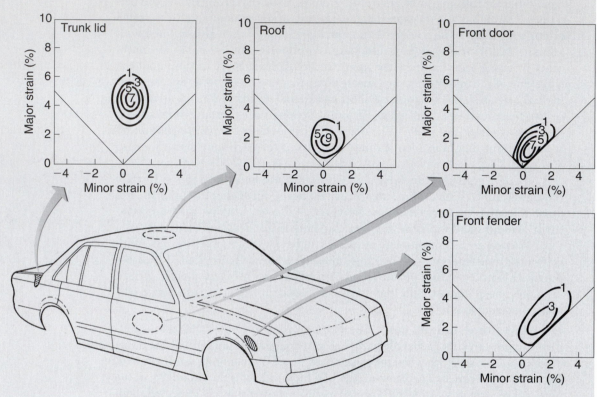

FIGURE 7.66 Major and minor strains in various regions of an automobile body.

Note that the trunk lid and the roof are subjected mainly to plane strain (that is, the vertical line in Fig. 7.63a), whereas the front door and front fender are subjected to biaxial strains. The numbers in the strain paths indicate the frequency of occurrence.

Source: After T.J. Nihill and W.R. Thorpe.

7.7.2 Dent resistance of sheet-metal parts

In some applications involving sheet-metal parts, such as automotive body panels, appliances, and office furniture, an important consideration is the dent resistance of the sheet-metal panel. A *dent* is a small, but permanent, biaxial deformation of a relatively thin material. The factors significant in dent resistance can be shown to be the yield stress, Y, the thickness, t, and the shape of the panel. *Dent resistance* is then expressed by a combination of material and geometrical parameters as

$$\text{Dent resistance} \propto \frac{Y^2 t^4}{S},$$ (7.24)

where S is the panel stiffness, which, in turn, is defined as

$$S = (E)(t^a)(\text{Shape}), \tag{7.25}$$

where the value of a ranges from 1 to 2 for most panels. As for shape, the flatter the panel, the greater is the dent resistance, because of the sheet's flexibility. Thus, dent resistance (a) increases with increasing strength and thickness of the sheet, (b) decreases with increasing elastic modulus and stiffness, and (c) decreases with decreasing curvature. Note that for $a = 2$, dent resistance is proportional to t^2. Dents are usually caused by *dynamic* forces, such as forces developed by falling objects or other objects that hit the surface of the panel from different directions. In typical automotive panels, for instance, impact velocities range up to 45 m/s (150 ft/s). Thus, the *dynamic yield stress* (that is, at high strain rates) rather than the static yield stress is the significant strength parameter. However, denting under static forces could also be important.

For materials in which yield stress increases with strain rate, denting requires higher energy levels than under static conditions. Furthermore, dynamic forces tend to cause more *localized* dents than do static forces, which tend to spread the dented area. Because a portion of the energy goes into elastic deformation, the modulus of resilience of the sheet metal [see Eq. (2.5)] is an additional factor to be considered.

7.7.3 Modeling of sheet-metal forming processes

In Section 6.2.2, we outlined the techniques for studying bulk deformation processes and described *modeling* of impression-die forging, using the finite-element method; see also Fig. 6.11. Such mathematical modeling is now also applied to sheet-forming processes. The ultimate goal is the rapid analysis of stresses, strains, flow patterns, wrinkling, and springback as functions of parameters such as material characteristics, anisotropy, friction, deformation speed, and temperature.

Through such interactive analyses, we can, for example, determine the optimum tool and die geometry to make a certain part and thus reduce or eliminate costly die tryouts. Furthermore, simulation techniques also can be used to determine the size and shape of blanks, intermediate part shapes, press characteristics, and process parameters that optimize the forming operation. The application of computer modeling, although requiring powerful computers and elaborate software, has already proven to be a cost-effective technique in sheet-metal forming operations, particularly in the automotive industry.

7.8 | Equipment for Sheet-Metal Forming

The equipment for most pressworking operations generally consists of mechanical, hydraulic, pneumatic, or pneumatic-hydraulic presses. (See also Section 6.2.8 and Fig. 6.27.) The traditional *C-frame* press design, which has an open front, has been widely used for ease of tooling and workpiece accessibility. On the other hand, the *box-type* (O-type) pillar and double-column frame structures are stiffer. Furthermore, advances in automation, industrial robots, and computer controls all have made machine accessibility less significant.

Press selection includes consideration of a number of factors, including (a) type of forming operation, (b) size and shape of the parts, (c) length of stroke of the slide(s), (d) number of strokes per minute, (e) press speed, (f) *shut height*, that is, the distance from the top of the press bed to the bottom of the slide (with the stroke down), (g) types and number of slides, such as in double-action and triple-action presses, (h) press capacity and tonnage, (i) type of controls, (j) auxiliary equipment, and (k) safety features.

Because changing dies in presses can involve a significant amount of effort and time, rapid die-changing systems have been developed. Called **single-minute exchange of dies** (SMED), such systems use automated hydraulic or pneumatic equipment to reduce die-changing times from hours down to as little as 10 minutes.

7.9 | Design Considerations

As with all other metalworking processes described throughout this text, certain design guidelines and practices for sheet-metal forming have evolved with time. Careful design, using the best established design practices, computational tools, and manufacturing techniques, is the optimum approach for achieving high-quality products and realizing cost savings. The following guidelines identify the most significant design issues of sheet-metal forming operations.

1. **Blank design.** Material scrap is the primary concern in blanking operations. Poorly designed parts will not nest and there will be considerable scrap produced between successive blanking operations (Fig. 7.67). Some restrictions on the blank shape are made by the design application, but whenever possible, blanks should be designed to reduce scrap to a minimum.

2. **Bending.** In bending operations, the main concerns are fracture of the material, wrinkling, and inability to form the required bend. As shown in Fig. 7.68, a sheet-metal part with a flange that is to be bent will force the flange to undergo compression, possibly causing buckling. This problem can be controlled by using a relief notch to limit the stresses resulting from bending, as shown in Fig. 7.68. Right-angle bends have similar difficulties, and thus relief notches can be used to avoid tearing (Fig. 7.69).

FIGURE 7.67 Examples of efficient nesting of parts for optimum material utilization in blanking. *Source:* Society of Manufacturing Engineers.

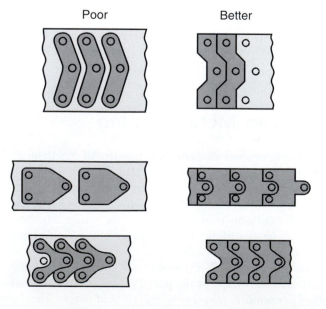

Poor   Better

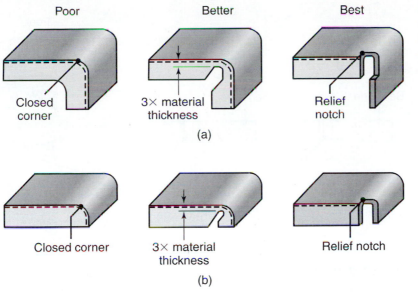

Poor   Better   Best

Closed corner   3× material thickness   Relief notch

(a)

Closed corner   3× material thickness   Relief notch

(b)

FIGURE 7.68 Control of tearing and buckling of a flange in a right-angle bend. *Source:* Society of Manufacturing Engineers.

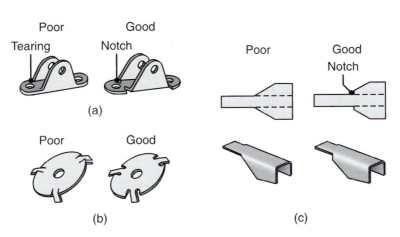

Poor   Good

Tearing   Notch

(a)

Poor   Good
            Notch

Poor   Good

(b)   (c)

FIGURE 7.69 Application of notches to avoid tearing and wrinkling in right-angle bending operations. *Source:* Society of Manufacturing Engineers.

The bend radius should be recognized as a highly stressed area, and all stress concentrations should be removed from the bend-radius location. An example is the placement of holes near bends; it is generally advantageous to move the hole away from the stress concentration, but when this is not possible, a crescent slot or ear can be used (Fig. 7.70a). Similarly, when bending flanges, tabs and notches should be avoided, since these stress concentrations will greatly reduce the part's formability. When tabs are necessary, large radii should be used to reduce the stress concentration (Fig. 7.70b).

When both bending and production of notches are to be implemented, it is important to properly orient the notches with respect to the grain direction. As shown in Fig. 7.17, bends should ideally be perpendicular to the rolling direction (or oblique, if perpendicularity is not possible) in order to avoid or minimize cracking. Bending at sharp radii can be accomplished through scoring or embossing (Fig. 7.71), but it should be recognized that this procedure can result in fracture. Burrs should not be present in the bend allowance (Fig. 7.15), since they can fracture, leading to a stress concentration that propagates the crack into the rest of the sheet.

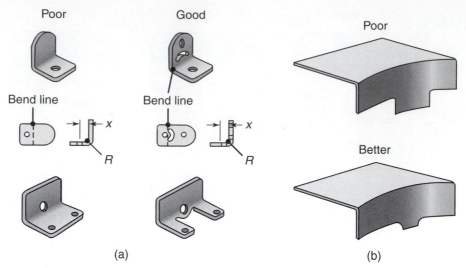

FIGURE 7.70 Stress concentrations near bends: (a) use of a crescent or ear for a hole near a bend and (b) reduction of the severity of a tab in a flange. *Source:* Society of Manufacturing Engineers.

FIGURE 7.71 Application of (a) scoring, or (b) embossing to obtain a sharp inner radius in bending. However, unless properly designed, these features can lead to fracture. *Source:* Society of Manufacturing Engineers.

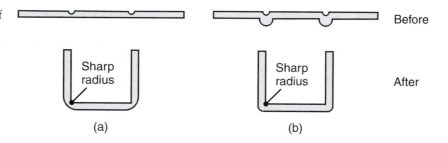

3. **Stamping and progressive-die operations.** In progressive dies (Fig. 7.13), the cost of the tooling and the number of stations are determined by the number and spacing of features on a part. It is advantageous to reduce the number of features to a minimum in order to minimize tooling costs. Closely spaced features may not provide sufficient clearance for punches, thereby possibly requiring two punches. Narrow cuts and protrusions can be challenging in forming with a single punch and die.

4. **Deep drawing.** After a deep-drawing operation, a cup will invariably spring back to some extent toward its original shape. For this reason, parts that have a vertical wall in a deep-drawn cup are difficult to form. Relief angles of at least 3° on each wall are easier to produce. Cups with sharp internal radii are difficult to produce, and deep cups often require additional ironing operations.

7.10 | Economics of Sheet-Metal Forming

Sheet-metal forming involves economic considerations similar to those for other metalworking processes described in Chapter 6. Sheet-forming operations compete with each other, as well as with other processes, more than other processes compete with each other. The forming operations are versatile, and several different processes can be used to produce the same part. For example, a cup-shaped part can be

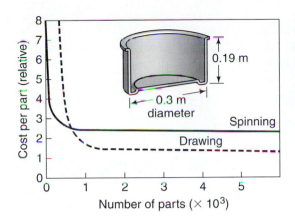

FIGURE 7.72   Cost comparison for manufacturing a cylindrical sheet-metal container by conventional spinning and deep drawing. Note that for small quantities, spinning is more economical.

formed by deep drawing, spinning, rubber-pad forming, or explosive forming. Similarly, a cup can be formed by impact extrusion or casting, or by fabrication from different pieces. As described in Section 16.5.5, almost all manufacturing processes produce some scrap, which can range up to 60% of the original material (for machining operations), and up to 25% (for hot forging). In contrast, sheet-forming operations produce scrap on the order of 10 to 25% of the original material. The scrap is now typically recycled, although scrap with residual lubricant layers is less desirable than dry scrap, because of the environmental factors involved.

As an example, note that the part shown in Fig. 7.72 can be made either by deep drawing or by conventional spinning. However, the die costs for the two processes are significantly different. Deep-drawing dies have many components and cost much more than the relatively simple mandrels and tools typically employed in spinning. Consequently, the die cost per part in drawing will be high if few parts are needed. On the other hand, this part can also be made by deep drawing in a much shorter time (seconds) than by spinning (minutes), even if the latter operation is highly automated. Furthermore, spinning requires more skilled labor. Considering all these factors, it is found that the break-even point for this particular part is at about 700 parts. Hence, deep drawing is more economical for quantities greater than that number.

CASE STUDY | Cymbal Manufacture

Cymbals (Fig. 7.73) are an essential percussion instrument for all forms of music. Modern drum-set cymbals cover a wide variety of sounds, from deep, dark, and warm, to bright, high-pitched, and cutting. Some cymbals sound musical, some are "trashy." A wide variety of sizes, shapes, weights, hammerings, and surface finishes (Fig. 7.73b) are available to achieve the desired performance.

Cymbals are produced from metals such as B20 bronze (80% copper, 20% tin, with a trace of silver), B8 bronze (92% copper, 8% tin), nickel-silver alloy, and brass, using various methods of processing. The manufacturing sequence for producing B20 bronze cymbals is shown in Fig. 7.74. The metal is first cast into mushroom-shaped ingots and then cooled to ambient temperature. It is then successively rolled, up to 14 times, water cooling the metal with each pass through the rolling mill. Special care is taken to roll the bronze at a different angle with each pass to minimize anisotropy, impart preferred grain orientation, and to develop an evenly round shape. The as-rolled blanks are then reheated and stretch formed (pressed) into the cup or bell shape that determines the cymbal's overtones. The

(a)   (b)

FIGURE 7.73   (a) A selection of common cymbals and (b) detailed view of different surface texture and finish of cymbals. *Source:* Courtesy of W. Blanchard, Sabian Ltd.

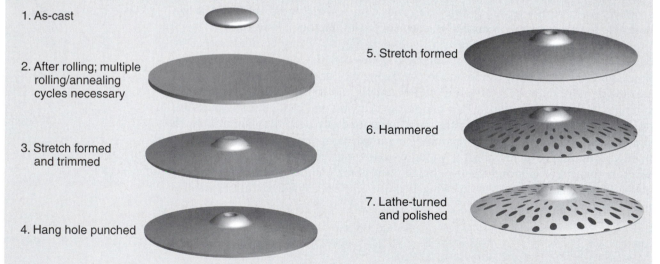

1. As-cast

2. After rolling; multiple rolling/annealing cycles necessary

3. Stretch formed and trimmed

4. Hang hole punched

5. Stretch formed

6. Hammered

7. Lathe-turned and polished

FIGURE 7.74   Manufacturing sequence for production of cymbals. *Source:* Courtesy of W. Blanchard, Sabian Ltd.

cymbals are then center-drilled or punched to create hang-holes and then trimmed on a rotary shear to approximate final diameters. This operation is followed by another stretch-forming step to achieve the characteristic Turkish dish form that controls pitch.

The cymbals are then hammered to impart a distinctive character to each instrument. Hammering can be done by hand (see Fig. 7.75a), or in automatic peen-forming machines (Fig. 7.75b). Hand hammering involves placing the bronze blank on a steel anvil, and the cymbals are then manually struck by hand hammers. Automatic peen forming is done without templates, since the cymbals have already been pressed into shape, but the pattern is controllable and uniform. The size and pattern of the peening operations depend on the desired response, such as tone, sound, and pitch of the cymbal.

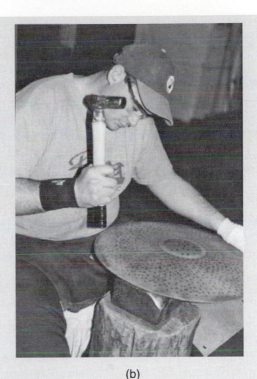

(a)   (b)

FIGURE 7.75 Hammering of cymbals: (a) automated hammering on a peening machine and (b) hand hammering of cymbals. *Source*: Courtesy of W. Blanchard, Sabian Ltd.

A number of finishing operations are then performed on the cymbals. This can merely involve cleaning and printing of identifying information, as some musicians prefer the natural surface appearance and sound of formed, hot-rolled bronze. More commonly, the cymbals are lathe-turned, without any machining fluid, in order to remove the oxide surface and reduce the thickness of the cymbal to create the desired weight and sound. As a result of this process, the surface finish becomes lustrous and, in some cases, develops a favorable surface microstructure. Some cymbals are polished to a glossy "Brilliant Finish." In many cases, the surface indentations from peening persist after finishing; this is recognized as an essential performance feature of the cymbal, and it also is an aesthetic feature that is appreciated by musicians. Various surface finishes associated with modern cymbals are shown in Fig. 7.73b.

Source: Courtesy of W. Blanchard, Sabian Ltd.

SUMMARY

- Sheet-metal forming processes involve workpieces that have a high ratio of surface area to thickness, and tensile stresses are typically applied in the plane of the sheet through various punches and dies. The sheet is generally prevented from being reduced in thickness in order to avoid necking and subsequent tearing and fracture. Material properties and behavior important to sheet-forming processes include elongation, yield-point elongation, anisotropy, and grain size. (Sections 7.1, 7.2)

- Blanks for forming operations are generally sheared from large rolled sheets, using various processes. Important process parameters include the clearance between the punch and dies, corner sharpness, and lubrication, and the quality of the sheared edge is important in subsequent forming. (Section 7.3)

- Bending of sheet, plate, and tubing is a commonly performed operation, and important parameters include the minimum bend radius (to avoid cracking in the bend area) and springback. Several techniques are available for bending sheet metal and for minimizing springback. (Section 7.4)

- Discrete sheet-metal parts can be produced by a wide variety of methods, including stretch forming, bulging, rubber-pad forming, hydroforming, spinning, and superplastic-forming methods. As in all other manufacturing processes, each forming process offers certain advantages, as well as limitations (Section 7.5).

- Deep drawing is one of the most important sheet-forming processes because it can produce container-shaped parts with high depth-to-diameter ratios. Deep drawability has been found to depend on the normal anisotropy of the sheet metal, while earing depends on its planar anisotropy. Buckling and wrinkling in drawing can be significant problems, and techniques have been developed to eliminate or minimize them. (Section 7.6)

- Tests have been developed to determine the formability of sheet metal. The most comprehensive test is the punch-stretch (bulge) test, which enables the construction of forming-limit diagrams, indicating safe and failure regions as a function of the planar strains to which the sheet is subjected. Modeling and simulation of sheet-metal forming have become important tools for predicting the behavior of sheet metal in actual forming operations. (Section 7.7)

- Several types of equipment are available for sheet-metal forming processes, and their selection depends on various factors concerning workpiece size and shape and equipment characteristics and their control features. Rapid die-changing techniques are important in improving productivity and the overall economics of forming operations. (Section 7.8)

- Design considerations in sheet-metal forming are based on factors such as shape of the blank, the type of operation to be performed, and the nature of the deformation. Several design rules and practices have been established to eliminate or minimize production problems. (Section 7.9)

- As in all other manufacturing processes, economic considerations in sheet-forming operations often determine the specific processes and practices to be employed. Because many processes are competitive, a thorough knowledge of process characteristics and capabilities is essential for economical production. (Section 7.10)

SUMMARY OF EQUATIONS

Maximum punch force in shearing: $F_{max} = 0.7(\text{UTS})tL$

Strain in bending: $e_o = e_i = \dfrac{1}{(2R/t) + 1}$

Minimum $\dfrac{R}{t} = \dfrac{50}{r} - 1$

Springback: $\dfrac{R_i}{R_f} = 4\left(\dfrac{R_i Y}{Et}\right)^3 - 3\left(\dfrac{R_i Y}{Et}\right) + 1$

Maximum bending force: $F_{max} = k\dfrac{(\text{UTS})Lt^2}{W}$

Normal anisotropy: $R = \dfrac{\epsilon_w}{\epsilon_t}$

Average: $\overline{R} = \dfrac{R_0 + 2R_{45} + R_{90}}{4}$

Planar anisotropy: $\Delta R = \dfrac{R_0 - 2R_{45} + R_{90}}{2}$

Maximum punch force in drawing: $F_{\max} = \pi D_p t_o (\text{UTS}) \left(\dfrac{D_o}{D_p} - 0.7 \right)$

Pressure in explosive forming: $p = K \left(\dfrac{\sqrt[3]{W}}{R} \right)^a$

BIBLIOGRAPHY

ASM Handbook, Vol. 14B: *Metalworking: Sheet Forming*, ASM International, 2006.

Benson, S.D., *Press Brake Technology*, Society of Manufacturing Engineers, 1997.

Buljanovic, V., *Sheet Metal Forming Processes and Die Design*, Industrial Press, 2004.

Bunge, H.J., et al. (eds.), *Formability of Metallic Materials*, Springer, 2001.

Davies, G., *Materials for Automobile Bodies*, Elsevier, 2006.

Davis, J.R. (ed.), *Tool Materials*, ASM International, 1995.

Fundamentals of Tool Design, 4th ed., Society of Manufacturing Engineers, 1998.

Gillanders, J., *Pipe and Tube Bending Manual*, FMA International, 1994.

Hosford, W.F., and Caddell, R.M., *Metal Forming, Mechanics and Metallurgy*, 2d ed., Prentice Hall, 1993.

Hu, J., Marciniak, Z., and Duncan, J., *Mechanics of Sheet Metal Forming*, Butterworth-Heinemann, 2002.

Nachtman, E.S., and Kalpakjian, S., *Lubricants and Lubrication in Metalworking Operations*, Dekker, 1985.

Pearce, R., *Sheet Metal Forming*, Springer, 2006.

Progressive Dies, Society of Manufacturing Engineers, 1994.

Rapien, B.L., *Fundamentals of Press Brake Tooling*, Hanser Gardner, 2005.

Sachs, G., *Principles and Methods of Sheet Metal Fabricating*, 2d ed., Reinhold, 1966.

Smith, D.A. (ed.), *Die Design Handbook*, 3d ed., Society of Manufacturing Engineers, 1990.

Suchy, I., *Handbook of Die Design*, McGraw-Hill, 1997.

Theis, H.E., *Handbook of Metalforming Processes*, CRC, 1999.

Tool and Manufacturing Engineers Handbook, 4th ed., Vol. 2: *Forming*, Society of Manufacturing Engineers, 1984.

Wagoner, R.H., Chan, K.S., and Keeler, S.P. (eds.), *Forming Limit Diagrams*, The Minerals, Metals and Materials Society, 1989.

QUESTIONS

7.1   Select any three topics from Chapter 2, and, with specific examples for each, show their relevance to the topics described in this chapter.

7.2   Same as for Question 7.1, but for Chapter 3.

7.3   Describe (a) the similarities and (b) the differences between the bulk deformation processes described in Chapter 6 and the sheet-metal forming processes described in this chapter.

7.4   Discuss the material and process variables that influence the shape of the curve for punch force vs. stroke for shearing (see Fig. 7.7), including its height and width.

7.5   Describe your observations concerning Figs. 7.5 and 7.6.

7.6   Inspect a common paper punch and comment on the shape of the tip of the punch as compared with those shown in Fig. 7.12.

7.7   Explain how you would estimate the temperature rise in the shear zone in a shearing operation. (*Hint:* See Section 2.12.1.)

7.8   As a practicing engineer in manufacturing, why would you be interested in the shape of the curve shown in Fig. 7.7? Explain.

7.9   Do you think the presence of burrs can be beneficial in certain applications? Give specific examples.

7.10   Explain why there are so many different types of tool and die materials used for the processes described in this chapter.

7.11 Describe the differences between compound, progressive, and transfer dies.

7.12 It has been stated that the quality of the sheared edges can influence the formability of sheet metals. Explain why.

7.13 Explain why and how various factors influence springback in bending of sheet metals.

7.14 Does the hardness of a sheet metal have any effect on its springback in bending? Explain.

7.15 As noted in Fig. 7.16, the state of stress shifts from plane stress to plane strain as the ratio of length-of-bend to sheet thickness increases. Explain why.

7.16 Describe the material properties that have an effect on the relative position of the curves shown in Fig. 7.19.

7.17 In Table 7.2, we note that hard materials have higher R/t ratios than soft ones. Explain why.

7.18 Why do tubes have a tendency to buckle when bent? Experiment with a straight soda straw, and describe your observations.

7.19 Based on Fig. 7.22, sketch and explain the shape of a U-die used to produce channel-shaped bent parts.

7.20 Explain why negative springback does not occur in air bending of sheet metals.

7.21 Give examples of products in which the presence of beads is beneficial or even necessary.

7.22 Assume that you are carrying out a sheet-forming operation and you find that the material is not sufficiently ductile. Make suggestions to improve its ductility.

7.23 In deep drawing of a cylindrical cup, is it always necessary that there be tensile circumferential stresses on the element in the cup wall, as shown in Fig. 7.50b? Explain.

7.24 When comparing hydroforming with the deep drawing process, it has been stated that deeper draws are possible in the former method. With appropriate sketches, explain why.

7.25 It will be noted in Fig. 7.50a that element A in the flange is subjected to compressive circumferential (hoop) stresses. Using a simple free-body diagram, explain why.

7.26 From the topics described in this chapter, list and explain specifically several examples where friction is (a) desirable and (b) not desirable.

7.27 Explain why increasing the normal anisotropy, R, of a sheet metal improves its deep drawability.

7.28 Explain the reason for the negative sign in the numerator of Eq. (7.21).

7.29 If you could control the state of strain in a sheet-metal forming operation, would you rather work on the left or on the right side of the forming-limit diagram? Explain.

7.30 Comment on any effect that lubrication of the punch surfaces may have on the limiting drawing ratio in deep drawing.

7.31 Comment on the role of the size of the circles on the surfaces of sheet metals in determining their formability. Are square grid patterns, as shown in Fig. 7.65, useful? Explain.

7.32 Make a list of the independent variables that influence the punch force in deep drawing of a cylindrical cup, and explain why and how these variables influence the force.

7.33 Explain why the simple-tension line in the forming-limit diagram in Fig. 7.63a states that it is for $R = 1$, where R is the normal anisotropy of the sheet metal.

7.34 What are the reasons for the development of forming-limit diagrams? Do you have any specific criticisms of such diagrams? Explain.

7.35 Explain the reasoning behind Eq. (7.20) for normal anisotropy, and Eq. (7.21) for planar anisotropy, respectively.

7.36 Describe why earing occurs. How would you avoid it? Would ears serve any useful purposes? Explain.

7.37 It was stated in Section 7.7.1 that the thicker the sheet metal, the higher is its curve in the forming-limit diagram. Explain why.

7.38 Inspect the earing shown in Fig. 7.57 and estimate the direction in which the blank was cut.

7.39 Describe the factors that should be considered in selecting the size and length of beads in sheet-metal forming operations.

7.40 It is known that the strength of metals depends on their grain size (see Section 3.4.1). Would you then expect strength to influence the magnitude of R of sheet metals? Explain.

7.41 A general rule for dimensional relationships for successful drawing without a blankholder is given by Eq. (7.23). Explain what would happen if this limit is exceeded.

7.42 Explain why the three broken lines (simple tension, plane strain, and equal biaxial stretching) in Fig. 7.63a have those particular slopes.

7.43 Select several specific parts in a typical automobile, and identify the processes described in this chapter that can be used to make those parts. Explain your reasoning.

7.44 It was stated that bendability and spinnability have a common aspect as far as properties of the workpiece materials are concerned. Describe this common aspect.

7.45   Explain the reasons that such a wide variety of sheet-forming processes has been developed and used over the years.

7.46   Prepare a summary of the types of defects found in sheet-metal forming processes and include brief comments on the reason(s) for each defect.

7.47   Which of the processes described in this chapter use only one die? What are the advantages of using only one die?

7.48   It has been suggested that deep drawability can be increased by (a) heating the flange and/or (b) chilling the punch by some suitable means. Comment on how these methods could improve drawability.

7.49   Propose designs whereby the two suggestions given in Question 7.48 can be implemented in actual production. Would production rate affect your designs? Explain.

7.50   In the manufacture of automotive-body panels from carbon-steel sheet, stretcher strains (Lueder's bands) are observed, which detrimentally affect surface finish. How can stretcher strains be eliminated? Explain.

7.51   In order to improve its ductility, a coil of sheet metal is placed in a furnace and is annealed. However, it is observed that the sheet has a lower limiting drawing ratio than it had before being annealed. Explain the reasons for this behavior.

7.52   What effects does friction have on a forming-limit diagram? Explain.

7.53   Why are lubricants generally used in sheet-metal forming? Explain, giving examples.

7.54   Through changes in clamping designs, a sheet-metal forming operation can allow the material to undergo a negative minor strain in the forming-limit diagram. Explain how this effect can be advantageous.

7.55   How would you produce the part shown in Fig. 7.35 other than by tube hydroforming?

7.56   Give three examples each of sheet-metal parts that (a) can and (b) cannot be produced by incremental forming methods.

7.57   Due to preferred orientation (see Section 3.5), materials such as iron can have higher magnetism after cold rolling. Recognizing this feature, plot your estimate of LDR vs. degree of magnetism.

7.58   Explain why a metal with a fine-grain microstructure is better suited for fine blanking than a coarse-grained metal.

7.59   What are the similarities and differences between roll forming, described in Section 7.4.4, and shape rolling, described in Section 6.3.5?

7.60   Explain how stringers can adversely affect bendability. Do they have a similar effect on formability?

7.61   In Fig. 7.56, the caption states that zinc has a high c/a ratio, whereas titanium has a low ratio. Why does this have relevance to limiting drawing ratio?

7.62   Review Eqs. (7.12) through (7.14) and explain which of these expressions can be applied to incremental forming.

PROBLEMS

7.63   Referring to Eq. (7.5), it is stated that actual values of e_o are significantly higher than the values of e_i, due to the shifting of the neutral axis during bending. With an appropriate sketch, explain this phenomenon.

7.64   Note in Eq. (7.11) that the bending force is a function of t^2. Why? (*Hint:* Consider bending-moment equations in mechanics of solids.)

7.65   Calculate the minimum tensile true fracture strain that a sheet metal should have in order to be bent to the following R/t ratios: (a) 0.5, (b) 2, and (c) 4. (See Table 7.2.)

7.66   Estimate the maximum bending force required for a $\frac{1}{8}$-in. thick and 12-in. wide Ti-5Al-2.5Sn titanium alloy in a V-die with a width of 6 in.

7.67   In Example 7.4, calculate the work done by the force-distance method, noting that work is the integral product of the vertical force, F, and the distance it moves.

7.68   What would be the answer to Example 7.4 if the tip of the mechanism applying the force, F, were fixed to the strip by some means, thus maintaining the lateral position of the force? (*Hint:* Note that the left portion of the strip will now be strained more than the right portion, as you can demonstrate with a rubber band.)

7.69   Calculate the magnitude of the force F in Example 7.4 when $\alpha = 30°$.

7.70   How would the force in Example 7.4 vary if the workpiece were made of a perfectly plastic material? Explain.

7.71   Calculate the press force required in punching 0.5-mm-thick 5052-O aluminum foil in the shape of a square hole 30 mm on each side.

7.72 A straight bead is being formed on a 1-mm-thick aluminum sheet in a 20-mm-diameter die cavity, as shown in the accompanying figure. (See also Fig. 7.25a.) Considering springback, calculate the outside diameter of the bead after it is formed and unloaded from the die. Let $Y = 150$ MPa.

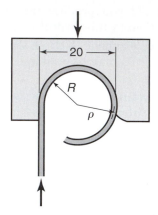

7.73 Inspect Eq. (7.10) and substituting in some numerical values, show whether the first term in the equation can be neglected without significant error in calculating springback.

7.74 In Example 7.5, calculate the amount of TNT required to develop a pressure of 10,000 psi on the surface of the workpiece. Use a standoff of 1 foot.

7.75 Estimate the limiting drawing ratio for the materials listed in Table 7.3.

7.76 For the same material and thickness as in Problem 7.66, estimate the force required for deep drawing with a blank diameter of 10 in. and a punch diameter of 9 in.

7.77 A cylindrical cup is being drawn from a sheet metal that has a normal anisotropy of 3. Estimate the maximum ratio of cup height to cup diameter that can successfully be drawn in a single draw. Assume that the thickness of the sheet throughout the cup remains the same as the original blank thickness.

7.78 Obtain an expression for the curve shown in Fig. 7.56 in terms of the LDR and the average normal anisotropy, $\overline{R}$. (*Hint:* See Fig. 2.5b.)

7.79 A steel sheet has R values of 1.0, 1.5, and 2.0 for the 0°, 45°, and 90° directions to rolling, respectively. For a round blank 150 mm in diameter, estimate the smallest cup diameter to which it can be drawn in one draw.

7.80 In Problem 7.79, explain whether ears will form and, if so, why.

7.81 A 1-mm-thick isotropic sheet metal is inscribed with a circle 4 mm in diameter. The sheet is then stretched uniaxially by 25%. Calculate (a) the final dimensions of the circle and (b) the thickness of the sheet at that location.

7.82 Conduct a literature search and obtain the equation for a tractrix curve, as used in Fig. 7.61.

7.83 In Example 7.4, assume that the stretching is done by two equal forces, F, each at 6 in. from the ends of the workpiece. (a) Calculate the magnitude of this force for $\alpha = 10°$. (b) If we want the stretching to be done up to $\alpha_{max} = 50°$ without necking, what should be the minimum value of n of the material?

7.84 Derive Eq. (7.5).

7.85 Estimate the maximum power in shear spinning a 0.5-in.-thick annealed 304 stainless steel plate that has a diameter of 12 in. on a conical mandrel of $\alpha = 30°$. The mandrel rotates at 100 rpm and the feed is $f = 0.1$ in./rev.

7.86 Obtain a common aluminum beverage can and cut it in half lengthwise with a pair of tin snips. Using a micrometer, measure the thickness of the bottom of the can and of the wall. Estimate (a) the thickness reductions in ironing of the wall and (b) the original blank diameter.

7.87 What is the force required to punch a square hole, 150-mm on each side, from a 1-mm-thick 5052-O aluminum sheet using flat dies? What would be your answer if beveled dies were used instead?

7.88 Estimate the percent scrap in producing round blanks if the clearance between blanks is to be one-tenth of the radius of the blank. Consider single- and multiple-row blanking, as shown in the accompanying figure.

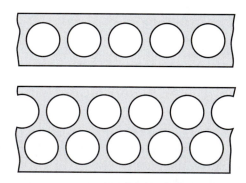

7.89 Plot the final bend radius as a function of initial bend radius in bending for (a) 5052-O aluminum, (b) 5052-H34 aluminum, (c) C24000 brass, and (d) AISI 304 stainless steel sheet.

7.90 The accompanying figure shows a parabolic profile that will define the mandrel shape in a conventional spinning operation. Determine the equation of the parabolic surface. If a spun part is to be produced from a

10-mm-thick blank, determine the minimum blank diameter required. Assume that the diameter of the profile is 6 in. at a distance of 3 in. from the open end.

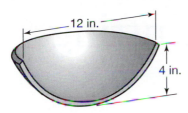

7.91 For the mandrel needed in Problem 7.90, plot the sheet-metal thickness as a function of radius if the part is to be produced by shear spinning. Comment on whether or not this operation is feasible.

7.92 Assume that you are asked to give a quiz to students on the contents of this chapter. Prepare five quantitative problems and five qualitative questions, and supply the answers.

DESIGN

7.93 Consider several shapes (such as oval, triangle, L-shaped, etc.) to be blanked from a large flat sheet by laser-beam cutting. Sketch a nesting layout to minimize scrap.

7.94 Give several structural applications in which diffusion bonding and superplastic forming can be used jointly.

7.95 On the basis of experiments, it has been suggested that concrete, either plain or reinforced, can be a suitable material for dies in sheet-metal forming operations. Describe your thoughts regarding this suggestion, considering die geometry and any other factors that may be relevant.

7.96 Metal cans are either of the two-piece variety (in which the bottom and sides are integral) or the three-piece variety (in which the sides, the bottom, and the top are each separate pieces). For a three-piece can and assuming that it can be considered as a thin-walled, internally-pressurized vessel, should the seam be (a) in the rolling direction, (b) normal to the rolling direction, or (c) oblique to the rolling direction of the sheet? Explain your answer, using equations from solid mechanics.

7.97 Investigate methods for determining optimum shapes of blanks for deep-drawing operations. Sketch the optimally shaped blanks for drawing rectangular cups, and optimize their layout on a large sheet of metal.

7.98 The design shown in the accompanying illustration is proposed for a metal tray, the main body of

which is made from flat cold-rolled sheet steel. Noting its features and that the sheet is bent in two different directions, comment on relevant manufacturing considerations. Include factors such as anisotropy of the cold-rolled sheet, its surface texture and finish, the bend directions, the nature of the sheared edges, and the method by which the handle is snapped in for assembly.

7.99 Design a box that will contain a 4 in. × 6 in. × 3 in. volume. The box is to be produced from two pieces of sheet metal and will require no tools or fasteners for its assembly.

7.100 Repeat Problem 7.99, but the box is to be made from a single piece of sheet metal.

7.101 In opening a can using an electric can opener, you will note that the lid often develops a scalloped periphery. (a) Explain why scalloping occurs. (b) What design changes for the can opener would you recommend in order to minimize or, if possible, eliminate scalloping? (c) Since lids typically are recycled or discarded, do you think it is necessary or worthwhile to make such design changes? Explain.

7.102 A recent trend in sheet-metal forming is to provide a specially-textured surface finish that develops small surface depressions (pockets) to aid lubricant entrainment. (See also Section 4.3.) Conduct a literature search on this technology, and prepare a brief technical paper on this topic.

7.103 Lay out a roll-forming line to produce any three of the cross sections shown in Fig. 7.27b.

7.104 Obtain a few pieces of cardboard and carefully cut the profiles to produce the bends shown in Fig. 7.68. Demonstrate that the designs labeled as "best" are indeed the best designs. Comment on the difference in strain states among the designs.

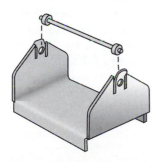

CHAPTER 8

Material-Removal Processes: Cutting

Machining is a general term describing a group of processes that consist of material removal on a workpiece and modification of its surfaces after it has been produced by the methods covered in preceding chapters. Machining processes are very versatile, having the capability of producing almost any shape with good dimensional tolerances and surface finish. The topics covered in this chapter include

- How chips are produced during machining.
- Force and power requirements.
- Mechanisms of tool wear and failure.
- Types and properties of cutting-tool materials.
- Characteristics of machine tools.
- Vibration and chatter and their implications.
- Design for machining operations.
- Economics of machining.

8.1 | Introduction

Parts manufactured by casting, forming, and various shaping processes, as described in the preceding chapters, often require further processing or finishing operations to impart specific characteristics, such as dimensional accuracy and surface finish, before the product is ready for use. This chapter and the following chapter describe in detail the various operations that are employed to obtain these characteristics. These processes are generally classified as **material-removal processes** and they are capable of producing shapes competitive with those produced by other methods. Critical choices often have to be made regarding the extent of shaping and forming a workpiece versus the extent of machining that subsequently may have to be performed to produce an acceptable part.

Although **machining** is the general term used to describe material removal, it covers several processes, which are usually divided into the following broad categories, the first of which is the subject of this chapter.

- **Cutting,** which generally involves single-point or multipoint cutting tools and processes, such as turning, boring, drilling, tapping, milling, sawing, and broaching
- **Abrasive processes,** such as grinding, honing, lapping, and ultrasonic machining (Sections 9.6 through 9.9)
- **Advanced machining processes,** which use electrical, chemical, thermal, hydrodynamic, and optical sources of energy, as well as combinations of these energy sources, to remove material from the workpiece surface (Sections 9.10 through 9.15)

As described throughout this chapter, machining processes are desirable or even necessary in manufacturing operations for the following reasons:

1. Closer **dimensional accuracy** may be required than is achievable from other metalworking processes alone. For example, in a forged crankshaft, the bearing surfaces and the holes cannot be produced with good dimensional accuracy and surface finish by forming and shaping processes alone.

2. Parts may possess external and internal **geometric features,** such as sharp corners and internal threads, that cannot be produced by forming and shaping processes.

3. Some parts are heat treated for improved hardness and wear resistance, and since heat-treated parts may undergo distortion and surface discoloration, they generally require additional **finishing operations.**

4. Special **surface characteristics** or **textures** may be required on all or selected areas of workpiece surfaces of the product that cannot be produced by other means. Engine blocks, for example, have regions that must possess far better surface finish and dimensional accuracy than other regions; also, copper mirrors with very high reflectivity, for example, are made by machining with a diamond cutting tool.

5. It may be more **economical** to machine the part than to manufacture it by other processes, particularly if the number of parts required is relatively small. Recall that metalworking processes typically require expensive dies and tooling, the cost of which can only be justified if the number of parts made is high enough.

Against these advantages, machining processes have certain limitations.

1. Machining processes inevitably **waste material** and generally require more energy and labor than other metalworking operations; thus, machining should be avoided or minimized whenever possible.

2. Removing a volume of material from a workpiece generally takes **more time** than other processes.

3. Unless carried out properly, material-removal processes can have **adverse effects** on the surface integrity of the product, including fatigue life.

In spite of these limitations, machining processes and machine tools are indispensable in manufacturing. Ever since lathes were introduced in the 1600s, these processes have developed continuously. We now have available a wide variety of computer-controlled machine tools, as well as new techniques that use various sources of energy.

As in all other manufacturing operations, it is important to view machining as a **system** consisting of *workpiece, cutting tool, tool holder, workholding devices,* and *machine tool.* Machining operations cannot be carried out efficiently and economically without a fundamental knowledge of the often complex interactions among these critical elements, as will be evident throughout this chapter. The importance of these considerations is even more evident from the fact that, in the United States alone, labor and overhead costs for machining operations are estimated to be about $300 billion per year.

8.2 | Mechanics of Chip Formation

Machining processes remove material from the workpiece surface by producing chips, as shown in Fig. 8.1. The basic mechanics of chip formation, which is essentially the same for all cutting operations, is represented by the two-dimensional model shown in Fig. 8.2. In this model, a tool moves along the workpiece at a certain velocity (cutting speed), V, and a depth of cut, t_o. A **chip** is produced just ahead of the tool by *shearing* the material continuously along the shear plane.

In the cutting process, the major *independent variables,* that is, the variables that we can change and control directly, are

- Type of cutting tool and its condition.
- Tool shape, surface finish, and its sharpness.
- Workpiece material, its condition, and the temperature at which it is machined.
- Cutting conditions, such as speed, feed, and depth of cut.
- Type of cutting fluids, if used.
- Characteristics of the machine tool, particularly its stiffness and damping.
- Tool holders, and workholding and fixturing devices.

FIGURE 8.1 Some examples of common machining processes.

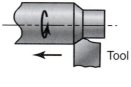

(a) Straight turning

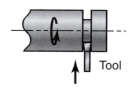

(b) Cutting off

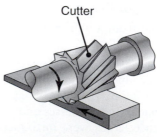

(c) Slab milling

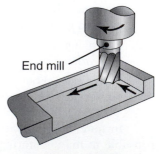

(d) End milling

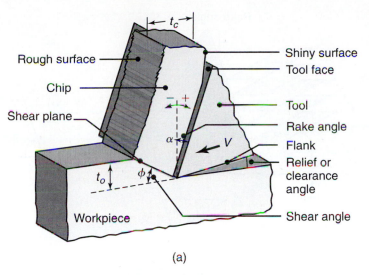

Rough surface

Chip

Shear plane

t_o

Workpiece

t_c

Shiny surface
Tool face

Tool

Rake angle

Flank

Relief or
clearance
angle

Shear angle

α V ϕ

(a)

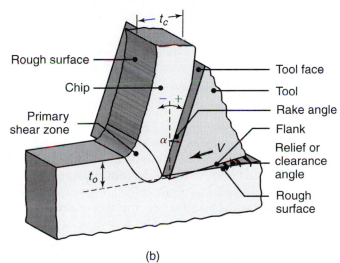

Rough surface

Chip

Primary
shear zone

t_o

t_c

Tool face

Tool

Rake angle

Flank

Relief or
clearance
angle

Rough
surface

α V

(b)

FIGURE 8.2 Schematic illustration of a two-dimensional cutting process, or orthogonal cutting: (a) orthogonal cutting with a well-defined shear plane, also known as the Merchant model and (b) orthogonal cutting without a well-defined shear plane.

The *dependent variables*, which are the variables that are influenced by changes in the independent variables, are

- Type of chip produced.
- Force required and energy dissipated in the cutting process.
- Temperature rise in the workpiece, the chip, and the tool.
- Wear, chipping, and failure of the tool.
- Surface finish and integrity of the workpiece after it is machined.

In order to appreciate the importance of studying the complex interrelationships among these variables, let's briefly ask the following questions:

a. If the surface finish of the workpiece being cut is poor and unacceptable, which of the foregoing independent variables do we modify first?

b. If the workpiece becomes too hot, thus possibly affecting its properties, what modifications should be made?

c. If the cutting tool wears rapidly and becomes dull, do we change the cutting speed, the depth of cut, the tool material itself, or some other variable?

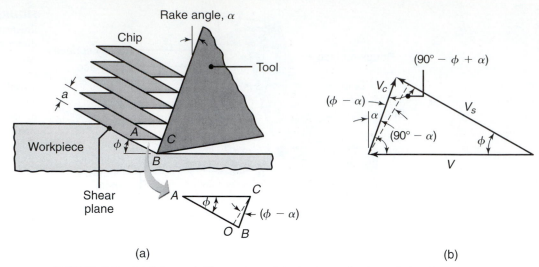

FIGURE 8.3   (a) Schematic illustration of the basic mechanism of chip formation in cutting and (b) velocity diagram in the cutting zone.

d. If the dimensional tolerance of the machined workpiece is over the specified limits, what modification should be made?

e. If the cutting tool begins to vibrate and chatter, thus affecting surface finish, what should be done to eliminate this problem?

Although almost all machining operations are three dimensional in nature, the two-dimensional model shown in Fig. 8.2 is very useful and appropriate in studying the basic mechanics of cutting process. In this model, known as **orthogonal cutting** [meaning that the cutting edge of the tool is perpendicular (orthogonal) to the cutting direction], the tool has a **rake angle,** α (positive, as shown in the figure), and a **relief,** or **clearance, angle.** Note that the sum of the rake, relief, and included angles of the tool is 90°.

Microscopic examinations reveal that chips are produced by the **shearing** mechanism shown in Fig. 8.3a. Shearing takes place along a **shear plane,** which makes an angle ϕ, called the **shear angle,** with the surface of the workpiece. Below the shear plane, the workpiece is undeformed (except for some elastic distortion), and above the shear plane, a chip is already formed and is moving up the face of the tool as cutting progresses. Because of the relative movement, there is friction between the chip and the rake face of the tool.

Note that the thickness of the chip, t_c, can be determined if t_o, α, and ϕ are known. The ratio of t_o to t_c is known as the **cutting ratio,** r, which can be expressed as

$$r = \frac{t_o}{t_c} = \frac{\sin \phi}{\cos(\phi - \alpha)}. \tag{8.1}$$

A review of this process indicates that the chip thickness is always greater than the depth of cut (also known as the **undeformed chip thickness**), hence the value of r is always less than unity. The reciprocal of r is known as the **chip compression ratio** and is a measure of how thick the chip has become compared with the depth of cut; thus, the chip compression ratio is always greater than unity.

On the basis of Fig. 8.3a, the **shear strain,** γ, that the material undergoes during cutting can be expressed as

$$\gamma = \frac{AB}{OC} = \frac{AO}{OC} + \frac{OB}{OC}, \tag{8.2}$$

or

$$\gamma = \cot \phi + \tan(\phi - \alpha). \tag{8.3}$$

Note from this equation that high shear strains are associated with low shear angles and low or negative rake angles. Shear strains of 5 or higher have been observed in actual cutting operations. Thus, compared with forming and shaping processes (see Chapter 6), the material undergoes greater deformation during cutting, as can also be seen in Table 2.4.

From Fig. 8.2, note that the undeformed chip thickness, t_o, and the depth of cut are the same parameter for orthogonal cutting; this is not true for other operations. Since chip thickness, t_c, is greater than the undeformed chip thickness, t_o, the *velocity of the chip*, V_c, must be lower than the cutting speed, V. Since mass continuity has to be maintained, we have

$$Vt_o = V_c t_c \quad \text{or} \quad V_c = Vr, \tag{8.4}$$

and therefore,

$$V_c = V \frac{\sin \phi}{\cos(\phi - \alpha)}. \tag{8.5}$$

A velocity diagram can be constructed, such as that shown in Fig. 8.3b, where, from trigonometric relationships, the following equations can be obtained:

$$\frac{V}{\cos(\phi - \alpha)} = \frac{V_s}{\cos \alpha} = \frac{V_c}{\sin \alpha}, \tag{8.6}$$

where V_s is the velocity at which shearing takes place along the shear plane. The **shear-strain rate** is the ratio of V_s to the thickness a of the sheared element (shear zone), or

$$\dot{\gamma} = \frac{V_s}{a}. \tag{8.7}$$

Experimental evidence indicates that the magnitude of a is on the order of 10^{-2} to 10^{-3} mm (10^{-3} to 10^{-4} in.). This range indicates that, even at low cutting speeds, the shear-strain rate is very high, on the order of 10^3/s to 10^6/s. A knowledge of the shear-strain rate is essential because of its effects on the strength and ductility of the material (see Section 2.2.7 and Table 2.5) and the type of chip produced.

8.2.1 Chip morphology

The type of chips produced (*chip morphology*) significantly influences surface finish and integrity, as well as the overall machining operation. When actual chips, produced under different cutting conditions, are observed under a microscope, there are significant deviations from the ideal model shown in Figs. 8.2 and 8.3a. Major types of metal chips commonly observed in practice are shown in Fig. 8.4.

Note that a chip has two surfaces: One surface has been in contact with the rake face of the tool, and the other surface is from the original surface of the workpiece. The tool side of the chip surface is shiny, or *burnished* (Fig. 8.5), caused by rubbing of the chip as it climbs up the tool face. The other surface of the chip has not come into contact with any solid body. It has a jagged, steplike appearance (as can be seen in Fig. 8.4a), which is due to the shearing mechanism of chip formation. (See also Figs. 3.5 and 3.7 for an analogy.)

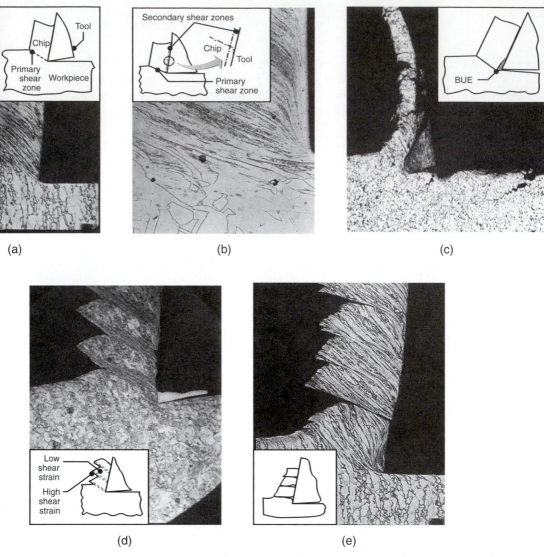

FIGURE 8.4 Basic types of chips produced in metal cutting and their micrographs: (a) continuous chip with narrow, straight primary shear zone, (b) secondary shear zone at the tool-chip interface, (c) continuous chip with built-up edge, (d) segmented or nonhomogeneous chip, and (e) discontinuous chip. *Source:* After M.C. Shaw, P.K. Wright, and S. Kalpakjian.

FIGURE 8.5 Shiny (burnished) surface on the tool side of a continuous chip produced in turning.

The basic types of chips produced in metal-cutting operations are described as follows.

1. **Continuous chips.** *Continuous chips* are typically formed at high cutting speeds and/or high rake angles (Fig. 8.4a). The deformation of the material takes place along a very narrow shear zone, called the **primary shear zone.** Such chips also may develop a **secondary shear zone** at the tool-chip interface (Fig. 8.4b), caused by friction. As expected, the secondary zone becomes thicker as tool-chip friction increases.

   In continuous chips, deformation may also take place along a wide primary shear zone with *curved boundaries* (Fig. 8.2b). Note that the lower boundary of this zone is *below* the machined surface, which subjects the machined surface to distortion and possible damage, as depicted by the vertical lines to the right of the tool tip. This situation occurs particularly often in machining soft metals at low cutting speeds and low rake angles; it can produce a poor surface finish and also induce residual surface stresses, which may be detrimental to the properties of the machined part.

   Although they generally produce good surface finish, continuous chips are not always desirable, particularly in computer-controlled machine tools, because they tend to get tangled around the tool, and the operation has to be stopped to clear away the chips. This problem can easily be alleviated with chip breaker features on cutting tools (see below).

   As a result of strain hardening (caused by the shear strains to which it is subjected), the chip usually becomes harder and stronger, and less ductile, than the original workpiece material. The increase in hardness and strength of the chip depends on the magnitude of the shear strain. As the rake angle decreases, the shear strain increases, as can be seen from Eq. (8.3), and thus the chip becomes stronger and harder.

2. **Built-up-edge chips.** A *built-up edge* (BUE) may form at the tip of the tool during cutting (Fig. 8.4c). It consists of layers of material from the workpiece that are gradually deposited on the tool (hence the term *built-up*). As it becomes larger, the BUE becomes unstable and eventually breaks up. The upper portion of the BUE is carried away on the tool side of the chip, and the lower portion is deposited randomly on the machined surface. The process of BUE formation and breakup is repeated continuously during the cutting operation.

   The built-up edge is commonly observed in practice and is one of the important factors that adversely affects surface finish and integrity in machining, as can be seen in Figs. 8.4 and 8.6. A built-up edge, in effect, changes the geometry of cutting operation. Note, for example, the large tip radius of the BUE and the rough surface finish that it produces. Because of work hardening and deposition of successive layers of material, BUE hardness increases significantly (Fig. 8.6a). Although BUE is generally undesirable, a thin but stable BUE is generally regarded as desirable because it protects the tool surface.

   Although the exact mechanism of BUE formation is not clearly understood, investigations have identified the factors that contribute to its formation: (a) Adhesion of the workpiece material to the rake face of the tool, the strength of this bond depending on the affinity of the workpiece and tool materials; see also Section 4.4. Ceramic cutting tools, for example, have much lower affinity to form BUEs than do tool steels; (b) growth of the successive layers of adhered metal on the tool; (c) tendency of the workpiece material for strain hardening; the higher the strain-hardening exponent, n, the higher is the probability for BUE formation.

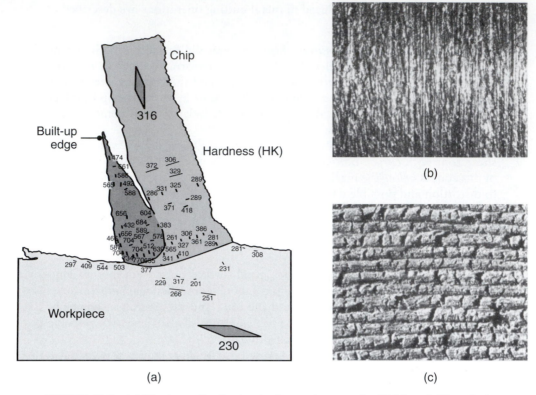

FIGURE 8.6 (a) Hardness distribution in the cutting zone for 3115 steel. Note that some regions in the built-up edge are as much as three times harder than the bulk workpiece. (b) Surface finish in turning 5130 steel with a built-up edge. (c) Surface finish on 1018 steel in face milling. Magnifications: 15×. *Source:* Courtesy of Metcut Research Associates, Inc.

Generally, the built-up edge decreases or is eliminated as (a) the cutting speed, V, increases, (b) the depth of cut, t_o, decreases, (c) the rake angle, α, increases, (d) tip radius of the tool decreases, and (e) an effective cutting fluid is applied (see Section 8.7).

3. **Serrated chips.** *Serrated chips,* also called **segmented** or **nonhomogeneous** chips, are semicontinuous chips with zones of low shear and high shear strain (Fig. 8.4e). The chips have the appearance of saw teeth (hence the term serrated). Metals, such as titanium, with low thermal conductivity and strength that decrease sharply with temperature exhibit this behavior.

4. **Discontinuous chips.** *Discontinuous chips* consist of segments that may be either firmly or loosely attached to each other (Fig. 8.4f). These chips usually form under the following conditions:

   a. The workpiece material is brittle, because it cannot undergo the high shear strains involved in cutting.

   b. The workpiece material contains hard inclusions and impurities (see Figs. 3.24 and 3.25) or has a structure such as the graphite flakes in gray cast iron (Fig. 5.14a). Impurities and hard particles act as sites for cracks, thereby producing discontinuous chips. As expected, a large depth of cut increases the probability of such defects being present in the cutting zone.

   c. The cutting speed is very low or very high.

   d. The depth of cut (undeformed chip thickness) is large or the rake angle is low.

e. The machine tool has low stiffness and poor damping.

f. There is a lack of an effective cutting fluid.

An additional factor in the formation of discontinuous chips is the magnitude of the compressive stresses on the shear plane; see also Eq. (8.17). Recall from Section 2.2.8 that the maximum shear strain at fracture increases with increasing compressive stress. Thus, if the normal stress is not sufficiently high, the material will be unable to undergo the large shear strain required to form a continuous chip.

Because of the discontinuous nature of chip formation, forces continually vary during cutting. Consequently, the stiffness of the cutting-tool holder, the workpiece holding devices, and the machine tool is important in cutting with discontinuous chips, as well as serrated-chip formation. If not sufficiently stiff, the machine tool may begin to vibrate and chatter (see Section 8.12). This effect, in turn, adversely affects the surface finish and dimensional accuracy of the machined component and may even cause damage or excessive wear of the cutting tool, as well as the machine tool itself.

Chip formation in machining nonmetallic materials. Much of the discussion thus far for metals is also generally applicable to nonmetallic materials. A variety of chips can be obtained in cutting thermoplastics, depending on the type of polymer (Chapter 10) and the process parameters, such as depth of cut, tool geometry, and cutting speed. Thermosetting plastics and ceramics, because they are generally brittle, tend to produce discontinuous chips. (See also the discussion on *ductile-regime cutting* in Section 8.9.2.)

Chip curl. *Chip curl* (Figs. 8.5 and 8.7a) is common to all machining operations with metals, as well as nonmetallic materials such as thermoplastics and wood. Although

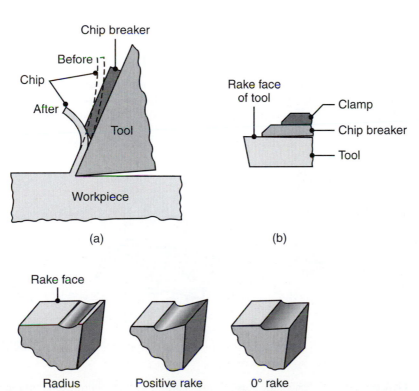

FIGURE 8.7 (a) Schematic illustration of the action of a chip breaker. Note that the chip breaker decreases the radius of curvature of the chip. (b) Chip breaker clamped on the rake face of a cutting tool. (c) Grooves on the rake face of cutting tools, acting as chip breakers. Most cutting tools now are inserts with built-in chip-breaker features.

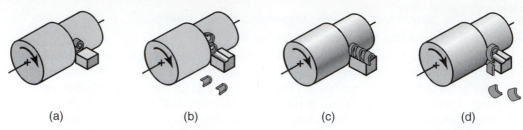

(a)   (b)   (c)   (d)

FIGURE 8.8 Various chips produced in turning: (a) tightly curled chip; (b) chip hits workpiece and breaks; (c) continuous chip moving radially outward from workpiece; and (d) chip hits tool shank and breaks off. *Source:* After G. Boothroyd.

not clearly understood, possible factors contributing to chip curl are (a) the distribution of stresses in the primary and secondary shear zones, (b) thermal gradients, (c) the work-hardening characteristics of the workpiece material, and (d) the geometry of the rake face of the tool. Process variables also affect chip curl; generally, the radius of curvature decreases (the chip becomes curlier) with decreasing depth of cut, increasing rake angle, and decreasing friction at the tool-chip interface. The use of cutting fluids and various additives in the workpiece material also influences chip curl.

Chip breakers. Long, continuous chips are undesirable because they tend to become entangled and interfere with machining operations and can also be a safety hazard to the operator. This situation is especially troublesome in high-speed automated machinery. The usual procedure to avoid the formation of continuous chips is to break the chip intermittently with a *chip breaker*. Although the chip breaker has traditionally been a piece of metal clamped to the rake face of the tool (Fig. 8.7b), it is now an integral part of the cutting tool itself (Fig. 8.7c). Chips can also be broken by modifying the tool geometry, thus controlling chip flow, as in the turning operations illustrated in Fig. 8.8.

8.2.2 Mechanics of oblique cutting

Unlike the two-dimensional situations described thus far, the majority of machining operations involve tool shapes that are three dimensional (**oblique**). The basic difference between two-dimensional and oblique cutting is illustrated in Fig. 8.9a.

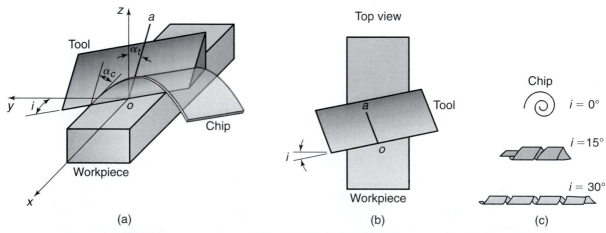

(a)   (b)   (c)

FIGURE 8.9 (a) Schematic illustration of cutting with an oblique tool, (b) top view, showing the inclination angle, i, and (c) types of chips produced with different inclination angles.

Recall that in orthogonal cutting, the tool cutting edge is perpendicular to the movement of the tool, and the chip slides directly up the rake face of the tool. In oblique cutting, on the other hand, the cutting edge is at an angle i, called the **inclination angle** (Fig. 8.9b). The chip in Fig. 8.9a flows up the rake face of the tool at an angle α_c (**chip flow angle**), measured in the plane of the tool face. (Note that this situation is similar to an angled snowplow blade, which throws the snow sideways.) The angle α_n is known as the **normal rake angle,** which is a basic geometric property of the tool. The normal rake angle is the angle between the normal oz to the workpiece surface and the line oa on the tool face.

Note that the workpiece material approaches the tool at a velocity V and leaves the workpiece surface (as a chip) with a velocity V_c. The effective rake angle, α_e, measured in the plane of these two velocities, can be calculated as follows: Assuming that the chip flow angle, α_c, is equal to the inclination angle, i (an assumption that has experimentally been found to be reasonable), the effective rake angle, α_e, is given by

$$\alpha_e = \sin^{-1}(\sin^2 i + \cos^2 i \sin \alpha_n). \tag{8.8}$$

Since i and α_n can both be measured directly, this formula can be used to calculate the magnitude of the effective rake angle. Note that as i increases, the effective rake angle increases, and the chip becomes thinner and longer. The effect of the inclination angle on chip shape is illustrated in Fig. 8.9c.

Figure 8.10 shows a typical single-point turning tool used on a lathe. Note the various angles, each of which has to be selected properly for efficient machining. Various three-dimensional cutting tools are described in greater detail in Sections 8.8 and 8.9, including those for operations such as drilling, tapping, milling, planing, shaping, broaching, sawing, and filing.

Shaving and skiving. Thin layers of material can be removed from straight or curved surfaces by a process similar to using a *plane* to shave wood. *Shaving* is particularly useful in improving the surface finish and dimensional accuracy of punched slugs or holes. (See Fig. 7.11.) Parts that are long or have a combination of shapes are shaved by *skiving* using a specially shaped cutting tool.

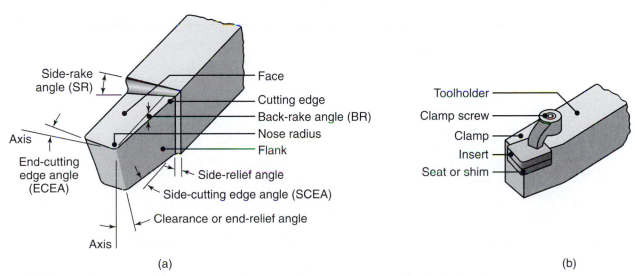

(a)

(b)

FIGURE 8.10 (a) Schematic illustration of a right-hand cutting tool for turning. Although these tools have traditionally been produced from solid tool-steel bars, they are now replaced by inserts of carbide or other tool materials of various shapes and sizes, as shown in (b).

8.2.3 Forces in orthogonal cutting

Knowledge of the cutting forces and power requirements in machining operations is essential for the following reasons.

1. **Power requirements** must be determined so that a machine tool of suitable capacity can be designed or selected for a particular application.

2. Data on cutting forces are necessary for the proper **design of machine tools** so that they are sufficiently stiff, thus avoiding excessive distortion of the machine components and vibration and chatter and maintaining desired dimensional accuracy.

3. The workpiece must be able to withstand the cutting forces without excessive **distortion** in order to maintain dimensional tolerances.

The various factors that influence the forces and power involved in orthogonal cutting are described below.

1. **Cutting forces.** The forces acting on the tool in orthogonal cutting are shown in Fig. 8.11. The **cutting force**, F_c, acts in the direction of the cutting speed, V, and supplies the energy required for the machining operation. The **thrust force,** F_t, acts in the direction normal to the cutting velocity, that is, perpendicular to the workpiece. These two forces produce the **resultant force,** F_r, which can be resolved into two components on the tool face: a **friction force,** F, along the tool-chip interface and a **normal force,** N, perpendicular to the interface. From Fig. 8.11, it can be shown that the friction force is

$$F = R \sin \beta \qquad (8.9)$$

and the normal force is

$$N = R \cos \beta. \qquad (8.10)$$

Note also that the resultant force is balanced by an equal and opposite force on the shear plane and is resolved into a **shear force,** F_s, and a **normal force,** F_n.

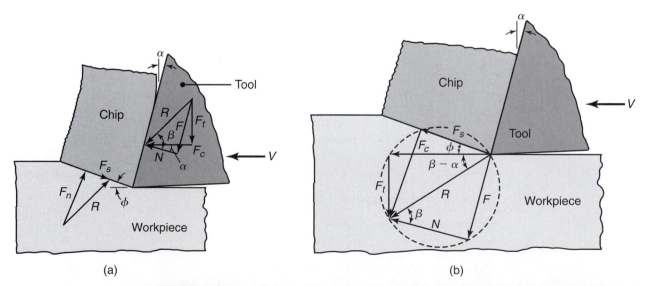

(a)   (b)

FIGURE 8.11   (a) Forces acting on a cutting tool in two-dimensional cutting. Note that the resultant forces, R, must be collinear to balance the forces. (b) Force circle to determine various forces acting in the cutting zone. *Source:* After M.E. Merchant.

From Fig. 8.11, the cutting force can be shown to be

$$F_c = R \cos(\beta - \alpha) = \frac{w t_o \tau \cos(\beta - \alpha)}{\sin \phi \cos(\phi + \beta - \alpha)}, \qquad (8.11)$$

where τ is the *average shear stress* along the shear plane.

The ratio of F to N is the **coefficient of friction**, μ, at the tool-chip interface (see also Section 4.4.1), and the angle β is known as the **friction angle.** The coefficient of friction can be expressed as

$$\mu = \tan \beta = \frac{F_t + F_c \tan \alpha}{F_c - F_t \tan \alpha}. \qquad (8.12)$$

In metal cutting, μ generally ranges from about 0.5 to 2, thus indicating that the chip encounters considerable frictional resistance while climbing up the rake face of the tool.

The magnitude of forces in typical machining operations is generally on the order of a few hundred newtons. However, the local stresses in the cutting zone and the normal stresses on the rake face of the tool are very high, because the contact areas are very small. The tool-chip contact length (Fig. 8.2), for example, is typically on the order of 1 mm (0.04 in.); thus the tool is subjected to very high stresses.

2. **Thrust force and its direction.** Although the thrust force does not contribute to the energy required in cutting, knowledge of its magnitude is important, because the tool holder, the workholding and fixturing devices, and the machine tool must be sufficiently stiff to minimize deflections caused by this particular force. For example, if the thrust force (see Fig. 8.11) is too high and the machine tool is not sufficiently stiff, the tool will be pushed away from the workpiece surface. This movement will, in turn, reduce the actual depth of cut, leading to loss of dimensional accuracy of the machined part and possibly to vibration and chatter (see Section 8.12).

Note from Fig. 8.11 that the direction of the thrust force is downward. It can be shown, however, that this force can be upward (negative), by first observing that

$$F_t = R \sin(\beta - \alpha) \qquad (8.13)$$

or

$$F_t = F_c \tan(\beta - \alpha). \qquad (8.14)$$

Because F_c is always positive (as shown in Fig. 8.11), the sign of F_t can be either positive or negative. Thus, when $\beta > \alpha$, the sign of F_t is positive (downward), and for $\beta < \alpha$ it is negative (upward). Thus, it is possible to have an upward thrust force when (a) friction at the tool-chip interface is low, and/or (b) when the rake angle is high.

This situation can be visualized by inspecting Fig. 8.11 and noting that when $\mu = 0$, $\beta = 0$, and consequently the resultant force, R, coincides with the normal force, N. In this case, then, F_t will have a thrust-force component that is upward. Also note that for the condition of $\alpha = 0$ and $\beta = 0$ the thrust force is zero. These observations have been verified experimentally, such as that shown in Fig. 8.12.

The influence of the depth of cut is obvious, because as t_o increases, R must also increase so that F_c will increase as well. This action requires additional energy to remove the extra material associated with the increased depth of cut. The change in direction and magnitude of the thrust force can play a significant role. Within a certain range of operating conditions, it can lead to *instability* in machining operations, particularly if the machine tool is not sufficiently stiff.

FIGURE 8.12 Thrust force as a function of rake angle and feed in orthogonal cutting of AISI 1112 cold-rolled steel. Note that at high rake angles, the thrust force is negative. A negative thrust force has important implications in the design of machine tools and in controlling the stability of the cutting process. *Source:* After S. Kobayashi and E.G. Thomsen.

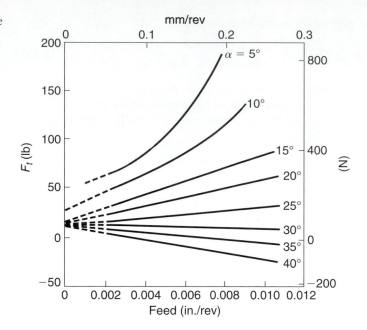

3. **Observations on cutting forces.** In addition to being a function of the strength of the workpiece material, forces in cutting are influenced by other variables. Extensive data, such as those given in Tables 8.1 and 8.2, indicate that the cutting force, F_c, increases with increasing depth of cut, decreasing rake angle, and decreasing speed. From analysis of the data in Table 8.2, the effect of cutting speed can be attributed to the fact that as speed decreases, the shear angle decreases and the coefficient of friction increases; both effects then increase the cutting force.

    The tip radius of the tool is also an important parameter: The larger the radius (and hence the duller the tool), the higher is the cutting force. Experimental evidence indicates that, for depths of cut on the order of five times the tip radius or higher, the effect of tool dullness on the cutting forces is negligible. (See also Section 8.4.)

4. **Shear and normal stresses in the cutting zone.** The stresses along the shear plane and at the tool-chip interface can be analyzed by assuming that they are uniformly distributed. The forces in the shear plane can then be resolved into shear and normal forces and stresses. Note that the area, A_s, of

TABLE 8.1

Data on Orthogonal Cutting of 4130 Steel										
α	ϕ	γ	μ	β	F_c (lb)	F_t (lb)	u_t (in.–lb/in^3 $\times 10^3$)	u_s	u_f	u_f/u_t (%)
25°	20.9°	2.55	1.46	56	380	224	320	209	111	35
35	31.6	1.56	1.53	57	254	102	214	112	102	48
40	35.7	1.32	1.54	57	232	71	195	94	101	52
45	41.9	1.06	1.83	62	232	68	195	75	120	62

$t_o = 0.0025$ in.; $w = 0.475$ in.; $V = 90$ ft/min; tool: high-speed steel.
Source: After E.G. Thomsen.

TABLE 8.2

Data on Orthogonal Cutting of 9445 Steel

α	V	ϕ	γ	μ	β	F_c	F_t	u_t	u_s	u_f	u_f/u_t (%)
+10	197	17	3.4	1.05	46	370	273	400	292	108	27
	400	19	3.1	1.11	48	360	283	390	266	124	32
	642	21.5	2.7	0.95	44	329	217	356	249	107	30
	1186	25	2.4	0.81	39	303	168	328	225	103	31
−10	400	16.5	3.9	0.64	33	416	385	450	342	108	24
	637	19	3.5	0.58	30	384	326	415	312	103	25
	1160	22	3.1	0.51	27	356	263	385	289	96	25

$t_o = 0.037$ in.; $w = 0.25$ in.; tool: cemented carbide.
Source: After M.E. Merchant.

the shear plane is

$$A_s = \frac{wt_o}{\sin \phi} \tag{8.15}$$

and therefore, the *average shear stress* in the shear plane is

$$\tau = \frac{F_s}{A_s} = \frac{F_s \sin \phi}{wt_o}, \tag{8.16}$$

and the *average normal stress* is

$$\sigma = \frac{F_n}{A_s} = \frac{F_n \sin \phi}{wt_o}. \tag{8.17}$$

Some data pertaining to these average stresses are given in Fig. 8.13. The rake angle is a parameter, and the shear plane area is increased by increasing the depth of cut during machining. The following conclusions can be drawn from these curves:

a. The shear stress on the shear plane is independent of the rake angle.

b. The normal stress on the shear plane decreases with increasing rake angle.

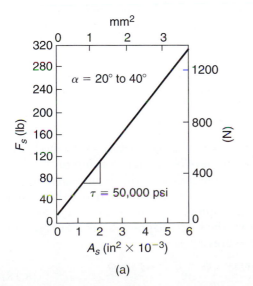

(a)

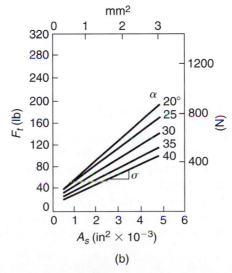

(b)

FIGURE 8.13 (a) Shear force and (b) normal force as a function of the area of the shear plane and the rake angle for 85-15 brass. Note that the shear stress in the shear plane is constant, regardless of the magnitude of the normal stress, indicating that the normal stress has no effect on the shear flow stress of the material. *Source:* After S. Kobayashi and E.G. Thomsen.

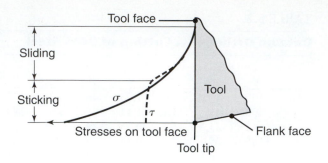

FIGURE 8.14 Schematic illustration of the distribution of normal and shear stresses at the tool-chip interface (rake face). Note that whereas the normal stress increases continuously toward the tip of the tool, the shear stress reaches a maximum and remains at that value (a phenomenon known as *sticking*; see Section 4.4.1).

c. The normal stress in the shear plane has no effect on the magnitude of the shear stress, a phenomenon that has also been verified by other mechanical tests. However, normal stress strongly influences the magnitude of the allowable shear strain in the shear zone before fracture. Recall from Section 2.2.8 that the maximum shear strain to fracture increases with the normal compressive stress. For this reason, small or negative rake angles will often be used to machine less ductile materials, in order to promote shearing without fracture.

Determining the stresses on the rake face of the tool presents considerable difficulties. One problem is accurately determining the length of contact at the tool-chip interface (Fig. 8.2). This length has been found to increase with decreasing shear angle, indicating that the contact length is a function of rake angle, cutting speed, and friction at the tool-chip interface. Another problem is that the stresses are not uniformly distributed along the rake face. *Photoelastic studies* have shown that the actual stress distribution is as shown in Fig. 8.14. Note that the normal stress at the rake face is a maximum at the tool tip and decreases rapidly toward the end of the contact length. The shear stress has a similar trend, except that it levels off about halfway along the tool-chip contact length. This behavior indicates that *sticking* is taking place (see Section 4.4.1), whereby the shear stress has reached the shear yield strength of the workpiece material. Sticking regions have been experimentally observed on some chips. Sticking regions also exist in metal-forming processes; see, for example, Section 6.2.2.

5. **Measuring cutting forces.** Cutting forces can be *measured* by using **force transducers** (such as piezoelectric crystals), as well as **force dynamometers** (with resistance-wire strain gages), mounted on the tool holder or the workholding devices on the machine tool. Forces can also be *calculated* from the amount of **power consumption** during cutting, often measured by a power monitor, provided that the mechanical efficiency of the machine tool can be determined.

8.2.4 Shear-angle relationships

Because the shear angle and the shear zone have great significance in the mechanics of cutting, much effort has been expended to develop the relationship of the shear angle to material properties and process variables. One of the earliest analyses (by M.E. Merchant, 1913–2006) is based on the assumption that (a) the shear angle adjusts itself so that the cutting force is a minimum, or (b) so that the maximum shear stress occurs in the shear plane. From the force diagrams shown in Fig. 8.11, the shear stress in the shear plane can be expressed as

$$\tau = \frac{F_s}{A_s} = \frac{F_c \sec(\beta - \alpha) \cos(\phi + \beta - \alpha) \sin \phi}{w t_o}. \tag{8.18}$$

Assuming that β is independent of ϕ, the shear angle corresponding to the maximum shear stress can be determined by differentiating Eq. (8.18) with respect to ϕ and equating to zero. Thus,

$$\frac{d\tau}{d\phi} = \cos(\phi + \beta - \alpha)\cos\phi - \sin(\phi + \beta - \alpha)\sin\phi = 0, \qquad (8.19)$$

and hence

$$\tan(\phi + \beta - \alpha) = \cot\phi = \tan(90° - \phi),$$

or

$$\phi = 45° + \frac{\alpha}{2} - \frac{\beta}{2}. \qquad (8.20)$$

Note from Eq. (8.20) that as the rake angle decreases and/or as the friction at the tool-chip interface increases, the shear angle decreases, and thus the chip becomes thicker. This result is to be expected, because decreasing α and increasing β tend to present greater resistance to the chip as it moves up the rake face of the tool, thus making the chip thicker, indicating a lower shear angle.

Another method of determining ϕ is based on slip-line analysis (by E.H. Lee and B.W. Shaffer in 1951) which gives an expression for the shear-plane angle as

$$\phi = 45° + \alpha - \beta. \qquad (8.21)$$

Note that this expression is similar to that in Eq. (8.20), indicating the same trends, although numerically it gives different values. In another study (by T. Sata and M. Mizuno in 1963), the following simple relationship has been developed:

$$\phi = \alpha \quad \text{for} \quad \alpha > 15°; \qquad (8.22)$$

$$\phi = 15° \quad \text{for} \quad \alpha < 15°. \qquad (8.23)$$

Several other expressions, based on various models and with different assumptions, have been obtained for the shear angle. Many of these expressions do not agree well with experimental data over a wide range of conditions (Fig. 8.15a), largely because shearing in chip formation rarely occurs along a thin plane. However,

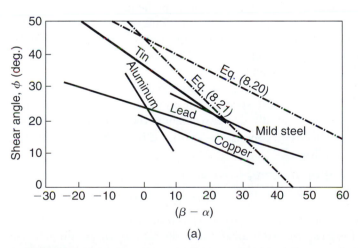

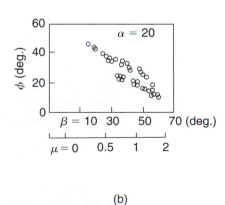

(a)   (b)

FIGURE 8.15 (a) Comparison of experimental and theoretical shear-angle relationships. More recent analytical studies have resulted in better agreement with experimental data. (b) Relation between the shear angle and the friction angle for various alloys and cutting speeds. *Source:* After S. Kobayashi.

the shear angle always decreases with increasing $\beta - \alpha$, as shown in Fig. 8.15b. More recent investigations appear to accurately predict the shear angle analytically, especially for continuous chips.

8.2.5 Specific energy

From Fig. 8.11 it can be seen that the **total power input** in cutting is

$$\text{Power} = F_c V.$$

The **total energy per unit** volume of material removed (known as **specific energy**), u_t, is then

$$u_t = \frac{F_c V}{w t_o V} = \frac{F_c}{w t_o}, \tag{8.24}$$

where w is the width of the cut. Note that u_t is simply the ratio of the cutting force to the projected area of the cut. From Figs. 8.3 and 8.11 it can be seen that the power required to overcome friction at the tool-chip interface is the product of F and V_c, or in terms of **specific energy for friction**, u_f, as

$$u_f = \frac{F V_c}{w t_o V} = \frac{F r}{w t_o} = \frac{(F_c \sin \alpha + F_t \cos \alpha) r}{w t_o}. \tag{8.25}$$

Likewise, the power required for shearing along the shear plane is the product of F_s and V_s. Thus, the **specific energy for shearing**, u_s is

$$u_s = \frac{F_s V_s}{w t_o V}. \tag{8.26}$$

The **total specific energy**, u_t, is the sum of the two energies

$$u_t = u_f + u_s. \tag{8.27}$$

Although not nearly as significant, there are two additional sources of energy in cutting. (a) One is the **surface energy** resulting from the formation of *two new surfaces* (the rake-face side of the chip and the newly machined surface) when a layer of material is removed by cutting. It can be shown that this energy represents a very small amount as compared with the shear and frictional energies involved. (b) The other source is the energy associated with the **momentum change** as the volume of metal being removed suddenly crosses the shear plane (a situation similar to the forces involved in a turbine blade from momentum changes of the liquid or gas). Although in ordinary cutting operations, the momentum energy is negligible, it can be significant at very high cutting speeds, such as above 125 m/s (25,000 ft/min) because of the higher momentum change involved.

Tables 8.1 and 8.2 provide experimental data on specific energies. Note that as the rake angle increases, the frictional specific energy remains rather constant, whereas the shear specific energy rapidly decreases. Thus, the ratio u_f/u_t increases significantly as α increases. This trend can also be predicted by obtaining an expression for the energy ratio as follows:

$$\frac{u_f}{u_t} = \frac{F V_c}{F_c V} = \frac{R \sin \beta}{R \cos(\beta - \alpha)} \cdot \frac{V r}{V} = \frac{\sin \beta}{\cos(\beta - \alpha)} \cdot \frac{\sin \phi}{\cos(\phi - \alpha)}. \tag{8.28}$$

Experimental observations have indicated that as α increases, both β and ϕ increase, and inspection of Eq. (8.28) also indicates that the ratio u_f/u_t should increase with α. Obviously, u_f and u_s are related. Although u_f is not necessary for the cutting action to take place, it affects the magnitude of u_s. The reason is that as friction increases, the shear angle decreases; a decreasing shear angle, in turn, indicates that the magnitude of u_s increases.

TABLE 8.3

Approximate Specific-Energy Requirements in Machining Operations		
	Specific energy*	
Material	W-s/mm^3	hp-min/in^3
Aluminum alloys	0.4–1.1	0.15–0.4
Cast irons	1.6–5.5	0.6–2.0
Copper alloys	1.4–3.3	0.5–1.2
High-temperature alloys	3.3–8.5	1.2–3.1
Magnesium alloys	0.4–0.6	0.15–0.2
Nickel alloys	4.9–6.8	1.8–2.5
Refractory alloys	3.8–9.6	1.1–3.5
Stainless steels	3.0–5.2	1.1–1.9
Steels	2.7–9.3	1.0–3.4
Titanium alloys	3.0–4.1	1.1–1.5

*At drive motor, corrected for 80% efficiency; multiply the energy by 1.25 for dull tools.

The calculation of all the parameters involved in these specific energies in cutting presents considerable difficulties; good theoretical computations are available, but they are difficult to perform. Consequently, the reliable prediction of cutting forces and energies is still based largely on experimental data (as shown in Tables 8.1, 8.2, and 8.3). The wide range of values in Table 8.3 can be attributed to differences in strengths within each material group and to the effects of numerous variables involved in the machining operations.

EXAMPLE 8.1 Relative energies in cutting

An orthogonal cutting operation is being carried out in which $t_o = 0.005$ in., $V = 400$ ft/min, $\alpha = 10°$, and the width of cut = 0.25 in. It is observed that $t_c = 0.009$ in., $F_c = 125$ lb, and $F_t = 50$ lb. Calculate the percentage of the total energy that is dissipated in friction at the tool-chip interface.

Solution. The percentage of energy can be expressed as

$$\frac{\text{Friction energy}}{\text{Total energy}} = \frac{FV_c}{F_c V} = \frac{Fr}{F_c},$$

where

$$r = \frac{t_o}{t_c} = \frac{5}{9} = 0.555,$$

$$F = R \sin \beta$$

$$F_c = R \cos(\beta - \alpha)$$

and

$$R = \sqrt{F_t^2 + F_c^2} = \sqrt{50^2 + 125^2} = 135 \text{ lb.}$$

Thus

$$125 = 135 \cos(\beta - 10°),$$

from which we find that

$$\beta = 32° \quad \text{and} \quad F = 135 \sin 32° = 71.5 \text{ lb.}$$

Therefore, the percentage of friction energy is calculated as

$$\text{Percentage} = \frac{(71.5)(0.555)}{125} = 0.32 = 32\%$$

and similarly, the percentage of shear energy is calculated as 68%.

EXAMPLE 8.2 Comparison of forming and machining energies

You are given two pieces of annealed 304 stainless-steel rods, each with a diameter of 0.500 in. and a length of 6 in., and you are asked to reduce their diameters to 0.480 in. (a) for one piece by *pulling* it in tension and (b) for the other by *machining* it on a lathe (see Fig. 8.8) in one pass. Calculate the respective amounts of work involved, and explain the reasons for the difference in the energies dissipated.

Solution.

(a) The work done in pulling the rod is (see Section 2.12)

$$W_{\text{tension}} = (u)(\text{volume}),$$

where

$$u = \int_0^{\epsilon_1} \sigma \, d\epsilon.$$

The true strain is found from the expression

$$\epsilon_1 = \ln\left(\frac{0.500}{0.480}\right)^2 = 0.0816.$$

From Table 2.3, we obtain the following values for K and n for this material:

$$K = 1275 \text{ MPa} = 185{,}000 \text{ psi} \qquad \text{and} \qquad n = 0.45.$$

Thus

$$u = \frac{K\epsilon_1^{n+1}}{n+1} = \frac{(185{,}000)(0.0816)^{1.45}}{1.45} = 3370 \text{ in.-lb/in}^3$$

and thus

$$W_{\text{tension}} = (3370)(\pi)(0.25)^2(6) = 3970 \text{ in.-lb.}$$

(b) From Table 8.3, let's estimate an average value for the specific energy in machining stainless steels as 1.5 hp-min/in³. The volume of material machined is

$$\text{Volume} = \frac{\pi}{4}\left[(0.5)^2 - (0.480)^2\right](6) = 0.092 \text{ in}^3.$$

The specific energy, in appropriate units, is calculated as

$$\text{Specific energy} = (15)(33{,}000)(12) = 594{,}000 \text{ in.-lb/in}^3.$$

Hence the work done in machining is

$$W_{\text{mach}} = (594{,}000)(0.092) = 54{,}650 \text{ in.-lb.}$$

Note that the work done in machining is about 14 times higher than that for tension. The reasons for the difference between the two energies are that tension involves very little strain and that there is no friction. Machining, however, involves friction, and the material removed (even though relatively small in volume) has undergone much higher strains than the bulk material undergoing tension. Assuming, from Tables 8.1 and 8.2, an average shear strain of 3, which is equivalent to an effective strain of 1.7 [see Eq. (2.58)], the material removed in machining is subjected to a strain of $1.7/0.0816 = 21$ times higher than that in tension.

These differences explain why machining consumes much more energy than reducing the diameter of this rod by stretching. It can be shown, however, that as the diameter of the rod decreases, the difference between the two energies becomes *smaller*, assuming that the same depth of material is to be removed. This result can be explained by noting the changes in the relative volumes involved in machining vs. tension as the diameter of the rod decreases.

8.2.6 Temperature

As in all other metalworking operations, the energy dissipated in machining operations is converted into heat, which, in turn, raises the temperature in the cutting zone. Knowledge of the temperature rise in cutting is important, because increases in temperature

- Adversely affect the strength, hardness, and wear resistance of the cutting tool.
- Cause dimensional changes in the part being machined, making control of dimensional accuracy difficult.
- Can induce thermal damage to the machined surface, adversely affecting its properties and service life.
- The machine tool itself may be subjected to temperature gradients, causing distortion of the machine and adversely affecting dimensional control.

Because of the work done in shearing and in overcoming friction on the rake face of the tool, the main sources of heat generation are the primary shear zone and friction on the tool-chip interface. In addition, if the tool is dull or worn, heat is also generated by the tool tip rubbing against the machined surface.

Variables affecting temperature. Various studies have been made of temperatures in cutting, based on heat transfer and dimensional analysis methods, and using experimental data. Although Fig. 8.16 shows that there are severe temperature

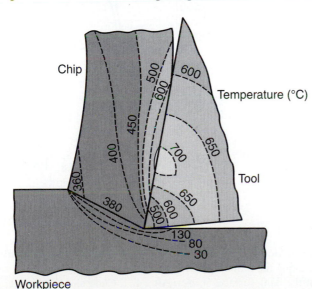

FIGURE 8.16 Typical temperature distribution in the cutting zone. Note the severe temperature gradients within the tool and the chip, and that the workpiece is relatively cool. *Source:* After G. Vieregge.

gradients in the cutting zone, a simple but approximate expression for the *mean temperature* for orthogonal cutting is

$$T = \frac{1.2Y_f}{\rho c} \sqrt[3]{\frac{Vt_o}{K}}, \qquad (8.29)$$

where T is the *mean temperature* of the tool-chip interface (°F); Y_f is the *flow stress* of the workpiece material (psi); V is the *cutting speed*; t_o is the *depth of cut* (in.); ρc is the *volumetric specific heat* of the workpiece (in.-lb/in³-°F); and K is the *thermal diffusivity* (ratio of thermal conductivity to volumetric specific heat) of the workpiece material (in²/s). Note that because the material parameters in Eq. (8.29) themselves depend on temperature, it is important to substitute appropriate values that are compatible with the *predicted* temperature range. Note that the properties in Eq. (8.29) all pertain to the workpiece. It has been shown that thermal properties of the tool material (see Section 8.6) are relatively unimportant compared with those of the workpiece.

The temperature generated in the shear plane is a function of the specific energy for shear, u_s, and the specific heat of the material. Hence, temperature rise is highest in machining materials with high strength and low specific heat, as Eq. (2.65) indicates. The temperature rise at the tool-chip interface is, of course, also a function of the coefficient of friction. Flank wear (see Section 8.3 and Fig. 8.20a) is an additional source of heat, caused by rubbing of the tool on the machined surface.

Figure 8.17 shows results from experimental measurements of temperature (using thermocouples) in turning on a lathe. Note that (a) the maximum temperature is at a location away from the tool tip and that (b) it increases with cutting speed. As the speed increases, there is little time for the heat to be dissipated, and

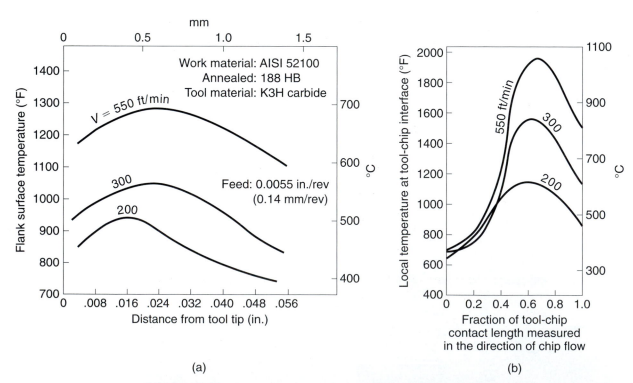

FIGURE 8.17   Temperature distribution in turning as a function of cutting speed: (a) flank temperature and (b) temperature along the tool-chip interface. Note that the rake-face temperature is higher than that at the flank surface. *Source:* After B.T. Chao and K.J. Trigger.

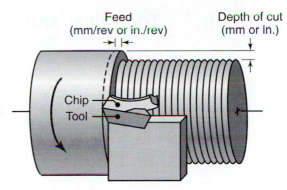

Feed (mm/rev or in./rev)   Depth of cut (mm or in.)

Chip
Tool

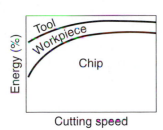

FIGURE 8.18 Proportion of the heat generated in cutting transferred to the tool, workpiece, and chip as a function of the cutting speed. Note that most of the cutting energy is carried away by the chip (in the form of heat), particularly as speed increases.

FIGURE 8.19 Terminology used in a turning operation on a lathe, where f is the feed (in mm/rev or in./rev) and d is the depth of cut. Note that feed in turning is equivalent to the depth of cut in orthogonal cutting (see Fig. 8.2), and the depth of cut in turning is equivalent to the width of cut in orthogonal cutting. See also Fig. 8.42.

hence temperature rises. Also, as cutting speed increases, a larger proportion of the heat generated is carried away by the chip, as can be seen in Fig. 8.18. The chip is a good heat sink, in that it absorbs and then carries away most of the heat generated. (This action of the chip is somewhat similar to the phenomenon of *ablation,* where layers of metal melt away from a surface, thus carrying away much of the heat.)

Note that Eq. (8.29) also indicates that the mean temperature increases with the strength of the workpiece material (because of the higher energy required) and the depth of cut (because of the ratio of surface area to chip thickness, noting also that a thin chip cools faster than a thick chip). However, for depths exceeding twice the tip radius of the tool, the depth of cut has been found to have negligible influence on the mean temperature. (See Fig. 8.28.)

Based on Eq. (8.29), another expression for the mean temperature in turning on a lathe (see Fig. 8.42) is given by

$$T \propto V^a f^b, \tag{8.30}$$

where a and b are constants, V is the cutting speed, and f is the feed (as shown in Fig. 8.19). Approximate values for a and b are given in the following table:

Tool	a	b
Carbide	0.2	0.125
High-speed steel	0.5	0.375

Techniques for measuring temperature. Temperatures and their distribution in the cutting zone may be determined by various techniques. (a) One technique is using **thermocouples** embedded in the tool or in the workpiece; this technique has been used successfully for a long time, although it involves considerable effort. (b) A simpler technique is by measuring **thermal emf** (electromotive force) at the tool-chip interface, which acts as a hot junction between two different (tool and chip) materials. (c) **Infrared radiation** from the cutting zone may also be monitored with a *radiation pyrometer;* however, this technique indicates only surface temperatures, and the accuracy of the results depends on the *emissivity* of the surfaces, which is difficult to determine precisely.

8.3 | Tool Wear and Failure

We have seen that cutting tools are subjected to high forces, elevated temperatures, and sliding; all these conditions induce wear. (See Section 4.4.2.) Because of its effects on the quality of the machined surface and the economics of machining, *tool wear* is one of the most important aspects of machining operations. A wide variety of factors are involved in tool wear, such as cutting tool and workpiece materials (including their physical, mechanical, and chemical properties), tool geometry, cutting fluids (if used), and processing parameters such as cutting speed, feed, and depth of cut. The types of wear on a tool depend on the relative roles of these variables.

The wear behavior of cutting tools is shown in Fig. 8.20, where the various regions of wear are identified as *flank wear, crater wear, nose wear,* and *chipping* of

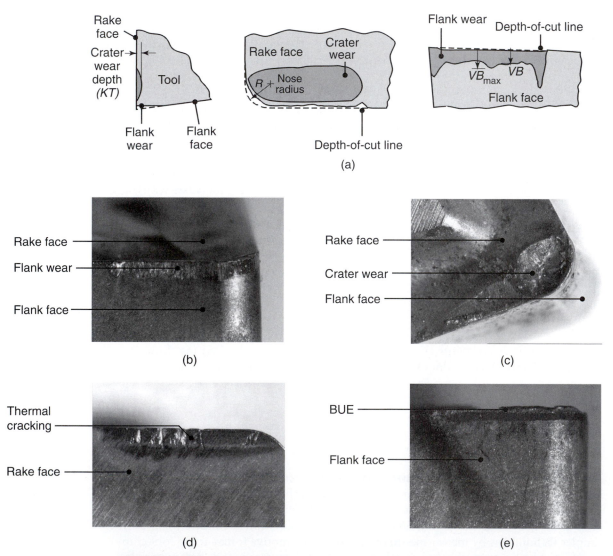

FIGURE 8.20 (a) Features of tool wear in a turning operation. The VB indicates average flank wear. (b)–(e) Examples of wear in cutting tools: (b) flank wear, (c) crater wear, (d) thermal cracking, and (e) flank wear and built-up edge. *Source:* (a) Terms and definitions reproduced with the permission of the International Organization for Standardization, ISO., copyright remains with ISO. (b)–(e) Courtesy of Kennametal, Inc.

the cutting edge. Whereas wear is generally a gradual process, chipping of the tool, especially *gross chipping*, is regarded as *catastrophic failure*. In addition to wear, *plastic deformation* of the tool also may take place, especially in tool materials that begin to lose their strength and hardness at elevated temperatures. It is obvious that the tool profile will be altered by these various wear and fracture processes, which, in turn, influence the overall machining operation.

8.3.1 Flank wear

Flank wear, which has been studied extensively, is generally attributed to

1. Sliding of the tool along the machined surface, causing adhesive and/or abrasive wear, depending on the materials involved.

2. Temperature rise, because of its adverse effects on the tool material properties.

Following an extensive study by F.W. Taylor, published in 1907, a tool-wear relationship was established for cutting various steels, as

$$VT^n = C, \qquad (8.31)$$

where V is the cutting speed, T is the time (in minutes) that it takes to develop a flank wear land (VB in Fig. 8.20a), n is an exponent that depends on cutting conditions, and C is a constant. Each combination of workpiece and tool materials and each cutting condition has its own n value and a different constant C. Equation (8.31) is a simple version of the several relationships developed by Taylor among the machining variables involved.

Tool-life curves. *Tool-life curves* are plots of experimental data obtained in machining tests (Fig. 8.21), typical in turning operations on a lathe. Note the (a) rapid decrease in tool life as cutting speed increases, (b) strong influence of the

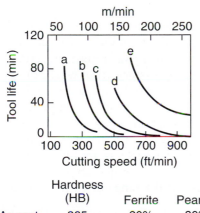

	Hardness (HB)	Ferrite	Pearlite
a. As cast	265	20%	80%
b. As cast	215	40	60
c. As cast	207	60	40
d. Annealed	183	97	3
e. Annealed	170	100	–

(a)

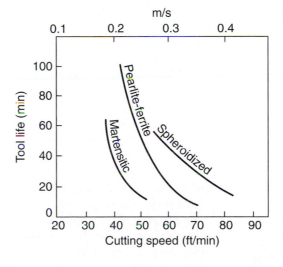

(b)

FIGURE 8.21   Effect of workpiece microstructure on tool life in turning. Tool life is given in terms of the time (in minutes) required to reach a flank wear land of a specified dimension. (a) Ductile cast iron and (b) steels, with identical hardness. Note in both figures the rapid decrease in tool life as the cutting speed increases.

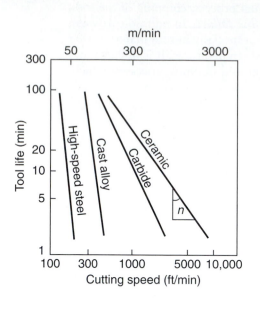

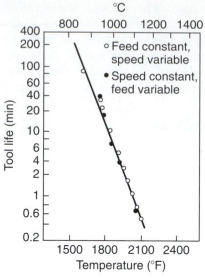

Work material: Heat-resistant alloy
Tool material: Tungsten carbide
Tool life criterion: 0.024 in. (0.6 mm) flank wear

(a)   (b)

FIGURE 8.22 (a) Tool-life curves for a variety of cutting-tool materials. The negative inverse of the slope of these curves is the exponent *n* in tool-life equations. (b) Relationship between measured temperature during cutting and tool life (flank wear). Note that high cutting temperatures severely reduce tool life. See also Eq. (8.30). *Source:* After H. Takeyama and Y. Murata.

condition of the workpiece material on tool life, and (c) large difference in tool life for different microstructures of the workpiece. Heat treatment of the workpiece is important largely because of increasing its hardness; for example, ferrite has a hardness of about 100 HB, pearlite 200 HB, and martensite 300 HB to 500 HB. (See Section 5.11.) Impurities and hard constituents in the workpiece material are also important considerations, because they reduce tool life by their abrasive action.

Tool-life curves are generally plotted on log–log paper, from which the exponent *n* can be determined (Fig. 8.22a). The range of *n* values that has been determined experimentally are given in Table 8.4. It is important to emphasize that although tool-life curves are usually linear over a certain range of cutting speeds, they are rarely so over a wide range. Moreover, it has been observed that the exponent *n* can indeed become negative at low cutting speeds. Thus, tool-life curves may actually reach a maximum and then curve downward. Consequently, caution should be exercised when using tool-life equations beyond the range of cutting speeds for which they were developed.

Because of the major influence of temperature on the physical and mechanical properties of materials, it would be expected that wear is strongly influenced by temperature. Experimental investigations have shown that there is indeed a direct relationship between flank wear and the temperature generated during machining (Fig. 8.22b). Although cutting speed has been found to be the most significant process variable in tool life, depth of cut and feed rate also are important. Thus, Eq. (8.31) can be modified as

$$VT^n d^x f^y = C, \tag{8.32}$$

where *d* is the depth of cut and *f* is the feed rate (in mm/rev or in./rev) in turning.

TABLE 8.4

Range of *n* Values for Various Cutting Tools

High-speed steels	0.08–0.2
Cast alloys	0.1–0.15
Carbides	0.2–0.5
Ceramics	0.5–0.7

The exponents x and y must be determined experimentally for each cutting condition. Taking $n = 0.15$, $x = 0.15$, and $y = 0.6$ as typical values encountered in practice, it can be seen that cutting speed, feed rate, and depth of cut are of decreasing order of importance.

Note that Eq. (8.32) can be rewritten as

$$T = C^{1/n}V^{-1/n}d^{-x/n}f^{-y/n}, \tag{8.33}$$

or

$$T \simeq C^7 V^{-7}d^{-1}f^{-4}. \tag{8.34}$$

Thus, for a constant tool life, the following observations can be made from Eq. (8.34):

1. If the feed or the depth of cut is increased, the cutting speed must be decreased, and vice versa.

2. A reduction in the cutting speed results in an increase in feed and/or depth of cut. Depending on the magnitude of the exponents, this can then result in an increase in the volume of the material removed.

Allowable wear land. Although somewhat arbitrary, typical values of the *allowable wear land* for various conditions are given in Table 8.5. For improved dimensional accuracy and surface finish, the allowable wear land may be specified smaller than the values given in the table. In practice, the recommended cutting speed for a high-speed steel tool is generally the one that gives a tool life of 60–120 min. For carbide tools, it is 30–60 min (see Table 8.9).

Optimum cutting speed. Recall that as cutting speed increases, tool life is reduced; on the other hand, if the cutting speed is low, tool life is long. Note, however, that a lower cutting speed means that the rate at which material is being removed will also be low; thus, there is an *optimum* cutting speed.

The effect of cutting speed on the volume of metal removed between tool replacements (or sharpening) can be appreciated by analyzing Fig. 8.21a. Let's assume that we are machining the material in the condition represented by curve "a." Note that if the cutting speed is 1 m/s (200 ft/min), the tool life is about 40 min. Thus, the tool travels a distance of (1 m/s)(60 s/min)(40 min) = 2400 m before it is replaced. If the cutting speed is now increased to 2 m/s, tool life will be about 5 min, and the tool travels a distance of (2)(60)(5) = 600 m. Since the volume of material removed is directly proportional to the distance the tool has traveled, we see that by *decreasing* the cutting speed, we can remove *more* material between tool changes. Note, however, that the lower the cutting speed, the longer is the time required to machine a part, thus reducing productivity, which has important economic impact (as also described in Section 8.15).

TABLE 8.5

Allowable Average Wear Land for Cutting Tools for Various Operations		
	Allowable wear land (mm)	
Operation	High-speed steels	Carbides
Turning	1.5	0.4
Face milling	1.5	0.4
End milling	0.3	0.3
Drilling	0.4	0.4
Reaming	0.15	0.15

EXAMPLE 8.3 Increasing tool life by reducing the cutting speed

Using the Taylor equation [Eq. (8.31)] for tool life and letting $n = 0.5$ and $C = 400$, calculate the percentage increase in tool life when the cutting speed is reduced by 50%.

Solution. Since $n = 0.5$, the Taylor equation can be rewritten as $V\sqrt{T} = 400$. Letting V_1 be the initial speed and V_2 the reduced speed, it can be noted that, for this problem, $V_2 = 0.5V_1$. Because C is a constant, we have the relationship

$$0.5V_1\sqrt{T_2} = V_1\sqrt{T_1}.$$

Simplifying this expression,

$$\frac{T_2}{T_1} = \frac{1}{0.25} = 4.0.$$

This relation indicates that the tool-life change is

$$\frac{T_2 - T_1}{T_1} = \left(\frac{T_2}{T_1}\right) - 1 = 4 - 1 = 3,$$

or that it is increased by 300%. Note that the reduction in cutting speed has resulted in a major increase in tool life and that, in this problem, the magnitude of C is not relevant.

8.3.2 Crater wear

The factors affecting flank wear also influence *crater wear,* but the most significant ones are temperature and the degree of chemical affinity between the tool and the workpiece. Recall that the rake face of the tool is subjected to high levels of stress and temperature, as well as sliding of the chip at relatively high speeds. As shown in Fig. 8.17b, peak temperatures can be on the order of 1100°C (2000°F). Note that, interestingly, the location of *maximum depth* of crater wear generally coincides with the location of maximum temperature at the toll-chip interface.

   Experimental evidence indicates a direct relationship between crater-wear rate and tool-chip interface temperature (Fig. 8.23). Note the sharp increase in crater wear after a certain temperature range has been reached. Figure 8.24 shows the cross section of the tool-chip interface in cutting steel at high speeds. Note the location of the crater-wear pattern and the *discoloration* of the tool (loss of temper) as a result of high temperatures to which the tool is subjected. Note also how well the discoloration profile agrees with the temperature profile shown in Fig. 8.16.

FIGURE 8.23 Relationship between crater-wear rate and average tool-chip interface temperature in turning: (a) high-speed steel tool, (b) C1 carbide, and (c) C5 carbide. Note that crater wear increases rapidly within a narrow range of temperature. *Source:* After K.J. Trigger and B.T. Chao.

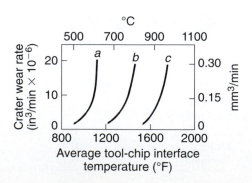

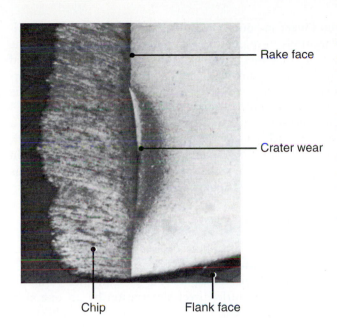

Rake face

Crater wear

Chip

Flank face

FIGURE 8.24 Interface of chip (left) and rake face of cutting tool (right) and crater wear in cutting AISI 1004 steel at 3 m/s (585 ft/min). Discoloration of the tool indicates the presence of high temperature (loss of temper). Note how the crater-wear pattern coincides with the discoloration pattern. Compare this pattern with the temperature distribution shown in Fig. 8.16. *Source:* Courtesy of P.K. Wright.

The effect of temperature on crater wear has been described in terms of a **diffusion** mechanism (that is, the movement of atoms across the tool-chip interface). Diffusion depends on the tool-workpiece material combination and on temperature, pressure, and time. As these quantities increase, the diffusion rate increases. An example of diffusion-induced crater wear is seen when a diamond cutting tool is used to machine steel. The high solubility of carbon in the steel leads to rapid crater wear and tool failure. Crater wear caused by diffusion will take place unless the factors involved are brought under control; wear may then be caused by other factors, such as abrasion and adhesion.

8.3.3 Chipping

The term *chipping* is used to describe the breaking away of a piece from the cutting edge of the tool. The chipped pieces may be very small (*microchipping* or *macrochipping*), or they may involve relatively large fragments (*gross chipping* or *fracture*). Chipping is a phenomenon that results in sudden loss of tool material. Two main causes of chipping are **mechanical shock** and **thermal fatigue** (both by interrupted cutting, as in milling operations described in Section 8.10).

Chipping by mechanical shock may occur in a region in the cutting tool where a small crack or defect already exists. High positive rake angles can also contribute to chipping, because of the small included angle of the tool tip (a phenomenon similar to chipping of a very sharp pencil). Crater wear may also contribute to chipping because it progresses toward the tool tip and weakens it, causing chipping. Thermal cracks, which are generally perpendicular to the cutting edge (as seen in Fig. 8.20d), are typically caused by the thermal cycling of the tool in interrupted cutting. Thus, chipping can be reduced by selecting tool materials with high impact and thermal-shock resistance.

8.3.4 General observations on tool wear

In addition to the wear and chipping processes already described, other phenomena also occur in tool wear (Fig. 8.20). The wear **groove** or **notch** on cutting tools has been attributed to the fact that this narrow region is the boundary [**depth-of-cut line**

(DOC)], where the chip is no longer in contact with the tool. This boundary oscillates, because of inherent variations in the cutting operation, thus accelerating the wear process. Furthermore, note that this region is in contact with the machined surface from the previous cut (see, for example, Fig. 8.19). Because a machined surface may develop a thin work-hardened layer (depending on tool sharpness and its shape), this contact would contribute to the formation of the wear groove.

Since they are hard and abrasive, scale and oxide layers on the surface of a workpiece also increase wear. In such cases, the depth of cut, d (see Fig. 8.19), should be greater than the thickness of the oxide film or the work-hardened layer. In other words, light cuts should not be taken on rusted or corroded workpieces, as otherwise the tool will travel through this thin, hard, and abrasive layer.

8.3.5 Tool-condition monitoring

With the wide use of computer-controlled machine tools and implementation of highly automated manufacturing operations, the reliable and repeatable performance of cutting tools is an important consideration. Once programmed properly, modern machine tools now operate with little direct supervision by an operator. Consequently, the failure of a cutting tool will have serious detrimental effects on the quality of the machined part as well as on the efficiency and economics of the overall machining operation. It is therefore essential to continuously and indirectly monitor the condition of the cutting tool, such as for wear, chipping, or gross failure.

Techniques for tool-condition monitoring typically fall into two general categories: direct and indirect. The **direct method** for observing the condition of a cutting tool involves *optical measurement* of wear, by periodically observing changes in the profile of the tool. This traditional method is the most commonly employed and reliable technique and is usually done using a *toolmakers' microscope*. However, because this technique requires that the machining operation be temporarily stopped, continued research on *computer vision systems* (see Section 14.8.1) for this task is being conducted.

Indirect methods of measuring wear involve correlating the tool condition with process variables such as forces, power, temperature rise, surface finish, and vibrations. The **acoustic emission** technique utilizes a piezoelectric transducer attached to a tool holder. The transducer picks up acoustic-emission signals (typically above 100 kHz) that result from the stress waves generated during machining. By analyzing the signals, tool wear and chipping can be monitored (Fig. 8.25). This technique is particularly effective in precision machining operations, where, because of the very small amounts of material removed, cutting forces are low.

A similar indirect tool-condition monitoring system consists of **transducers** that are installed in original machine tools or retrofitted on existing machines. They continually monitor parameters such as spindle *torque* and tool *forces* in various directions. A microprocessor analyzes the content of the signals (which are preamplified) and interprets it. This system is capable of differentiating the signals from tool breakage, tool wear, a missing tool, overloading of the machine, or collision of machine components (such as spindles). It can also automatically compensate for tool wear so that dimensional accuracy of the part is maintained.

The design of the transducers must be such that they are (a) nonintrusive to the machining operation, (b) accurate and repeatable in signal detection, (c) resistant to abuse and typical shop-floor environment, and (d) cost effective. Continued progress is being made in the development of such *sensors* (see Section 14.8), including the use of *infrared* and *fiber optic* techniques for temperature measurement in a variety of machining operations.

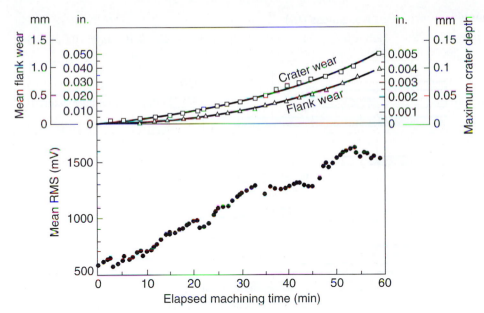

FIGURE 8.25 Relationship between mean flank wear, maximum crater wear, and acoustic emission (noise generated during cutting) as a function of machining time. This technique has been developed as a means for continuously and indirectly monitoring wear rate in various cutting processes without interrupting the operation. *Source:* After M.S. Lan and D.A. Dornfeld.

8.4 | Surface Finish and Surface Integrity

Surface finish influences not only the dimensional accuracy of machined parts, but also the properties of the parts, especially fatigue strength. Whereas **surface finish** describes the *geometric* features of surfaces, **surface integrity** pertains to *properties* such as fatigue life and corrosion resistance, which are strongly influenced by the type of surface produced. The factors that influence surface integrity are (a) temperatures generated during processing, (b) residual stresses, (c) metallurgical transformations, and (d) plastic deformation, tearing, and cracking of the surface. The ranges of surface roughness for machining and other processes are given in Fig. 8.26. As can be seen, the processes are generally organized in order of increasing surface quality, but also cost and machining time (see also Fig. 16.6).

In machining, the built-up edge, with its significant effect on modifying the tool profile, has the greatest influence on surface roughness of all the factors involved. Figure 8.27 shows surfaces obtained in two different machining operations; note the considerable damage to the surfaces from BUE. Ceramic and diamond tools generally produce better surface finish than do other tools, largely because of their much lower tendency to form BUE.

A tool that lacks sharpness has a large radius along its edges (see Fig. 8.20a), just as a dull pencil or knife does. Figure 8.28 illustrates the relationship between the radius of the cutting edge and depth of cut in orthogonal cutting. Note that at small depths of cut, the rake angle of an otherwise positive rake tool can effectively become *negative*. Thus, the tool may simply ride over the workpiece surface and not produce any chips. You can simulate this behavior by trying to scrape the surface of a stick of butter along its length with a dull knife. Note that you will not be able to remove a thin layer of butter, whereas if the knife is sharp, you will be able to do so.

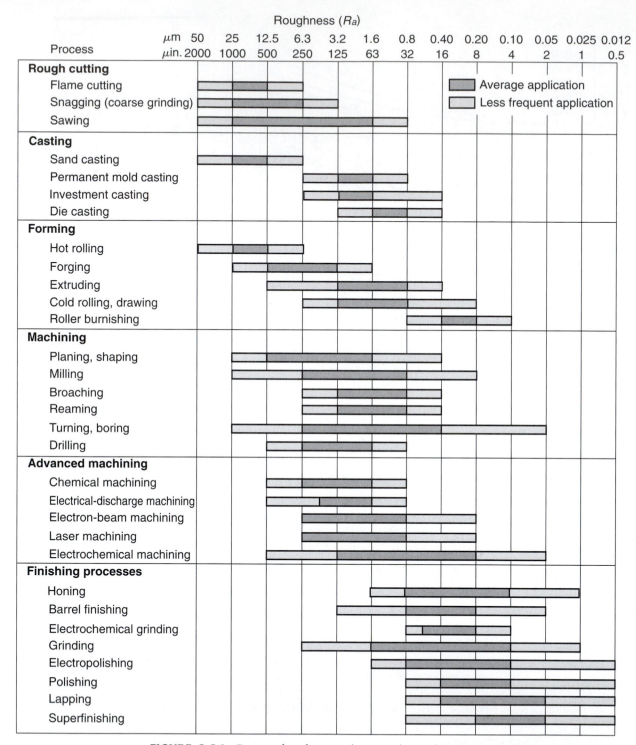

FIGURE 8.26 Range of surface roughnesses obtained in various machining processes. Note the wide range within each group, especially in turning and boring. (See also Fig. 9.27.)

If this radius is large in relation to the depth of cut, the tool will rub over the machined surface. Rubbing in machining generates frictional heat and can induce surface residual stresses, which, in turn, may cause surface damage such as tearing and cracking. In practice, the depth of cut should generally be greater than the radius on the cutting edge.

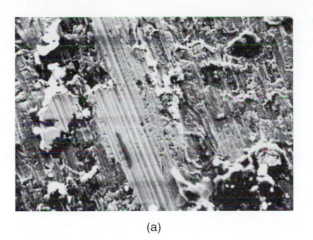
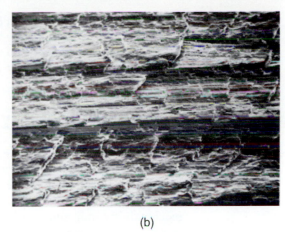

(a)   (b)

FIGURE 8.27   Surfaces produced on steel in machining, as observed with a scanning electron microscope: (a) turned surface and (b) surface produced by shaping. *Source:* After J.T. Black and S. Ramalingam.

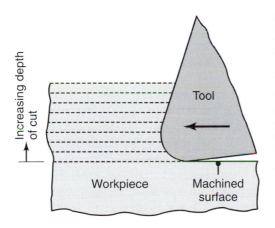

FIGURE 8.28   Schematic illustration of a dull tool in orthogonal cutting (exaggerated). Note that at small depths of cut, the rake angle can effectively become negative. In such cases, the tool may simply ride over the workpiece surface, burnishing it, instead of cutting.

Feed marks.   In turning, as in other machining operations, the tool leaves a spiral profile (*feed marks*) on the machined surface as it moves across the workpiece. (See Fig. 8.19.) As expected, the higher the feed, f, and the smaller the radius, R, the more prominent are these marks. Although not significant in rough machining operations, feed marks are important in finish machining.

In a turning operation, the peak-to-valley roughness, R_t, can be expressed as

$$R_t = \frac{f^2}{8R},$$ (8.35)

where f is the feed and R is the nose radius (see Figs. 8.10a, 8.20a and 8.41c). For R much smaller than f, the roughness is given by the expression

$$R_t = \frac{f}{\tan \alpha_s + \cot \alpha_e},$$ (8.36)

where α_s and α_e are the side and edge cutting angles, respectively (see Fig. 8.41). For face milling (see Fig. 8.57), the roughness is given by

$$R_t = \frac{f^2}{16[D \pm (2fn/\pi)]},$$ (8.37)

where D is the cutter diameter, f is the feed per tooth, and n is the number of inserts on the cutter. Equations (8.35) to (8.37) are derived considering only geometry and do not account for phenomena such as cracking of workpiece surface (see Fig. 8.6c), thermal distortion, and chatter (see Section 8.12 and Fig. 8.72).

8.5 | Machinability

Machinability is a recognizable property of a material but difficult to express quantitatively. Generally, the *machinability of a material* is defined in terms of the following four factors: (1) surface finish and integrity of the machined part, (2) tool life obtained, (3) force and power requirements, and (4) chip control. Thus, good machinability indicates *good surface finish and integrity, long tool life, low force and power requirements,* and type of *chip* that does not interfere with the machining operation and is easy to collect.

8.5.1 Machinability of steels

Because steels are among the most important engineering materials, their machinability has been studied extensively. The machinability of steels has been improved mainly by adding *lead* and *sulfur* to produce **free-machining steels**.

Leaded steels. Lead is added to molten steel and takes the form of dispersed fine lead particles. (See Fig. 5.2a.) During machining, the lead particles are sheared and smeared over the tool-chip interface, and because of their low shear strength, the lead particles act as a solid lubricant. This behavior can be verified by noting the presence of high concentrations of lead on the tool-side face of chips when machining leaded steels. In addition to this effect, it is believed that lead probably lowers the shear stress in the primary shear zone, thus reducing cutting forces and power consumption. Leaded steels are identified by the letter L between the second and third numerals, such as 10L45. (However, similar use of the letter L in identifying stainless steels refers to a low carbon content, which improves the steels' resistance to corrosion.)

Lead may be used in either nonsulfurized or resulfurized steels. However, because of its toxicity and for environmental concerns, the trend now is toward eliminating the use of leaded steels in favor of elements such as bismuth and tin (*lead-free steels*). Furthermore, leaded steels can have reduced machinability at elevated temperatures, as lead causes embrittlement of steels (see *hot shortness,* Section 3.4.2). It has also been observed that other elements can also cause solid-metal embrittlement in steels and aluminum alloys under certain conditions of stress and temperature.

Resulfurized and rephosphorized steels. *Sulfur* in steels forms *manganese-sulfide inclusions* (second-phase particles; see Fig. 8.29), which act as stress raisers in the primary shear zone. As a result, the chips produced are small and break up easily, thus improving machinability. *Phosphorus* in steels improves machinability by virtue of strengthening the ferrite, thereby increasing the hardness of the steels and producing less continuous chips.

Calcium-deoxidized steels. In these steels, oxide flakes of *calcium aluminosilicate* (CaO, SiO_2, and Al_2O_3) are formed. These flakes, in turn, lower the strength of the secondary shear zone, thus reducing tool-chip interface friction and wear and, thus, temperature. Consequently, these steels cause less crater wear of the tool, especially at high cutting speeds.

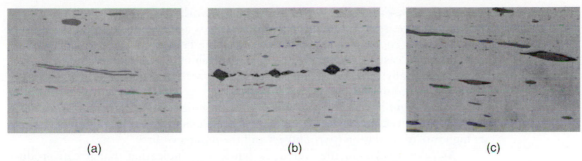

(a)   (b)   (c)

FIGURE 8.29 Photomicrographs showing various types of inclusions in low-carbon, resulfurized free-machining steels: (a) manganese-sulfide inclusions in AISI 1215 steel, (b) manganese-sulfide inclusions and glassy manganese-silicate-type oxide (dark) in AISI 1215 steel, and (c) manganese sulfide with lead particles as tails in AISI 12L14 steel. *Source:* Courtesy of Ispat Inland Inc.

Effects of other elements on machinability of steels

a. The presence of *aluminum* and *silicon* in steels is always harmful, because these elements combine with oxygen and form aluminum oxide and silicates. These compounds are hard and abrasive, thus increasing tool wear and reducing machinability.

b. *Carbon* and *manganese* have various effects on the machinability of steels, depending on their composition. As the carbon content increases, machinability decreases; however, plain low-carbon steels (less than 0.15% C) can produce poor surface finish due to forming a built-up edge. Other alloying elements, such as nickel, chromium, molybdenum, and vanadium, which improve the properties of steels, generally reduce machinability.

c. Cast steels are more abrasive, although their machinability is similar to that of wrought steels.

d. Tool and die steels are very difficult to machine and usually require annealing prior to machining. Machinability of most steels is generally improved by cold working, which reduces the tendency for built-up edge formation.

e. Austenitic (300 series) stainless steels are generally difficult to machine. Chatter could be a problem, which necessitates the use of machine tools with high stiffness and damping capacity. Ferritic stainless steels (also 300 series) have good machinability. Martensitic (400 series) stainless steels are abrasive, tend to form built-up edge, and require tool materials with high hot hardness and resistance to crater wear. Precipitation-hardening stainless steels are strong and abrasive and thus require hard and abrasion-resistant tool materials.

8.5.2 Machinability of various metals

Aluminum is generally easy to machine; however, the softer grades tend to form built-up edge, and thus develop poor surface finish. High cutting speeds, high rake angles, and high relief angles are recommended. Wrought alloys with high silicon content and cast-aluminum alloys may be abrasive and hence require harder tool materials. Dimensional control may be a problem in machining aluminum, because of its low elastic modulus and relatively high thermal coefficient of expansion.

Gray cast irons are generally machinable but are abrasive. Free carbides in castings reduce machinability and cause tool chipping or fracture, thus requiring tools with high toughness. Nodular and malleable irons are machinable using hard tool materials.

Cobalt-base alloys are abrasive and highly work hardening; they require sharp and abrasion-resistant tool materials and low feeds and speeds.

Wrought copper can be difficult to machine because of built-up edge formation, although cast-copper alloys are easy to machine. *Brasses* are easy to machine as well, especially those containing lead (*leaded free-machining brass*). *Bronzes* are more difficult to machine than brass.

Magnesium is very easy to machine, with good surface finish and prolonged tool life. However, care should be exercised when machining magnesium, because of its high rate of oxidation (*pyrophoric*) and the danger of fire.

Molybdenum is ductile and work hardening, indicating that it can produce poor surface finish, although it can be improved using sharp tools.

Nickel-base alloys are work hardening, abrasive, and strong at high temperatures; their machinability is similar to that of stainless steels.

Tantalum is very work hardening, ductile, and soft; hence, it produces a poor surface finish, and tool wear is high.

Titanium and its alloys have poor thermal conductivity (the lowest of all metals), causing significant temperature rise and formation of built-up edge; thus it can be difficult to machine.

Tungsten is brittle, strong, and very abrasive, hence its machinability is low. Its machinability improves significantly at elevated temperatures.

8.5.3 Machinability of various materials

Graphite is abrasive; thus, it requires hard, abrasion-resistant sharp tools.

Thermoplastics generally have low thermal conductivity, low elastic modulus, and low softening temperature. Consequently, machining them requires tools with a positive rake angle (to reduce cutting forces); large relief angles; small depths of cut and feed; relatively high speeds; and proper support of the workpiece, because of lack of stiffness of thermoplastics. Tools should be sharp, and external cooling of the cutting zone may be necessary to keep the chips from becoming "gummy" and sticking to the tools.

Thermosetting plastics are brittle and sensitive to thermal gradients during machining, but their machinability is generally similar to that of thermoplastics.

Because of the fibers present, *reinforced plastics* are generally very abrasive and are difficult to machine. Fiber tearing and pullout is a significant problem. Furthermore, there are environmental considerations, and machining of these materials requires proper removal of machining debris to avoid human contact with and inhalation of the loose fibers.

Metal-matrix and *ceramic-matrix composites* can be difficult to machine, depending on the properties of the individual components in the material. The reinforcing fibers can be abrasive, and the matrix material may not have sufficient ductility for good machinability.

The machinability of *ceramics* has improved steadily, especially with the development of *nanoceramics* (see Section 11.8.1) and with the selection of appropriate machining parameters, such as by *ductile-regime cutting* (Section 8.9.2).

8.5.4 Thermally assisted machining

Materials that are difficult to machine at room temperature may be machined more easily at elevated temperatures, thus lowering cutting forces and increasing tool life. In *thermally assisted machining* (**hot machining**), the source of heat is a torch, high-energy beam (such as a laser or electron beam), or plasma arc, which is focused to an area just ahead of the cutting tool. Most applications for hot machining are in turning and milling operations. The process of heating to and maintaining a uniform temperature distribution within the workpiece may be difficult to control. Furthermore, the original microstructure of the workpiece may be altered, thus adversely affecting workpiece properties. Except in isolated cases, thermally assisted machining offers no significant advantage over machining at room temperature with the use of appropriate cutting tools and fluids. Some success has been obtained in laser-assisted machining of silicon-nitride ceramics and *Stellite* (see Section 8.6.3).

8.6 | Cutting-Tool Materials

The proper selection of cutting-tool materials is among the most important considerations in machining operations, as is the selection of mold and die materials for forming and shaping processes. Recall that in machining, the tool is subjected to high temperatures, high contact stresses, rubbing on the workpiece surface, and the effects of the chip climbing up the rake face of the tool. Consequently, a cutting tool must possess the following characteristics:

- **Hardness,** particularly at elevated temperatures (**hot hardness**), so that the hardness and strength of the tool are maintained at the temperatures encountered in machining operations (Fig. 8.30).

- **Toughness,** so that impact forces on the tool in interrupted cutting operations, such as milling or turning a splined shaft, do not chip or fracture the tool.

- **Wear resistance,** so that an acceptable tool life is obtained before the tool is replaced or **indexed.**

- **Chemical stability** or **inertness** with respect to the workpiece material, so that any adverse reactions that may contribute to tool wear are avoided or minimized.

Several cutting-tool materials having a wide range of these characteristics are now available. (See Table 8.6.) Tool materials are usually divided into the following

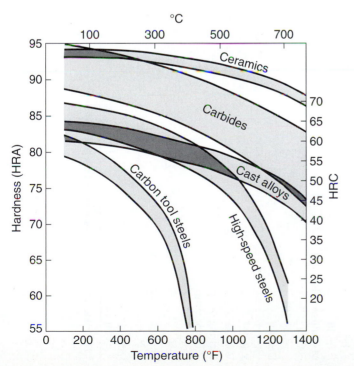

FIGURE 8.30 Hardness of various cutting-tool materials as a function of temperature (hot hardness). The wide range in each group of tool materials results from the variety of compositions and treatments available for that group.

TABLE 8.6

Typical Range of Properties of Various Tool Materials

Property	High-speed steel	Cast alloys	Carbides		Ceramics	Cubic boron nitride	Single-crystal diamond*
			WC	TiC			
Hardness	83–86 HRA	82–84 HRA	90–95 HRA	91–93 HRA	91–95 HRA	4000–5000 HK	7000–8000 HK
Compressive strength							
MPa	4100–4500	1500–2300	4100–5850	3100–3850	2750–4500	6900	6900
psi $\times 10^3$	600–650	220–335	600–850	450–560	400–650	1000	1000
Transverse rupture strength							
MPa	2400–4800	1380–2050	1050–2600	1380–1900	345–950	700	1350
psi $\times 10^3$	350–700	200–300	150–375	200–275	50–135	105–200	
Impact strength							
J	1.35–8	0.34–1.25	0.34–1.35	0.79–1.24	<0.1	<0.5	<0.2
in.-lb	12–70	3–11	3–12	7–11	<1	<5	<2
Modulus of elasticity							
GPa	200	—	520–690	310–450	310–410	850	820–1050
psi $\times 10^6$	30	—	75–100	45–65	45–60	125	120–150
Density							
kg/m^3	8600	8000–8700	10,000–15,000	5500–5800	4000–4500	3500	3500
lb/in^3	0.31	0.29–0.31	0.36–0.54	0.2–0.22	0.14–0.16	0.13	0.13
Volume of hard phase (%)	7–15	10–20	70–90	—	100	95	95
Melting or decomposition temperature							
°C	1300	—	1400	1400	2000	1300	700
°F	2370	—	2550	2550	3600	2400	1300
Thermal conductivity, W/mK	30–50	—	42–125	17	29	13	500–2000
Coefficient of thermal expansion, $\times 10^{-6}$/°C	12	—	4–6.5	7.5–9	6–8.5	4.8	1.5–4.8

*The values for polycrystalline diamond are generally lower, except impact strength, which is higher.

general categories, which are listed in the approximate chronological order in which they were developed and implemented. Note that many of these materials are also used for dies and molds, as described throughout Chapters 5, 6, 7, 10, and 11.

1. **Carbon and medium-alloy steels**
2. **High-speed steels**
3. **Cast-cobalt alloys**
4. **Carbides**
5. **Coated tools**
6. **Alumina-based ceramics**
7. **Cubic boron nitride**
8. **Silicon-nitride-based ceramics**
9. **Diamond**
10. **Whisker-reinforced and nanocrystalline tool materials**

This section describes the characteristics, applications, and limitations of these tool materials. Also discussed are their characteristics such as hot hardness, toughness, impact strength, wear resistance, resistance to thermal shock, and costs, as well as the range of cutting speeds and depth of cut for optimum performance.

8.6.1 Carbon and medium-alloy steels

Carbon steels are the oldest of tool materials and have been used widely for drills, taps, broaches, and reamers since the 1880s. Low-alloy and medium-alloy steels were developed later for similar applications, but with longer tool life. Although inexpensive and easily shaped and sharpened, these steels do not have sufficient hot hardness and wear resistance for machining at high cutting speeds, where, as we have seen, the temperature rises significantly. Note in Fig. 8.30, for example, how rapidly the hardness of carbon steels decreases as the temperature increases. Consequently, the use of these steels is limited to very low-speed cutting operations.

8.6.2 High-speed steels

High-speed-steel (HSS) tools are so named because they were developed to machine at speeds higher than previously possible. First produced in the early 1900s, high-speed steels are the most highly alloyed of the tool steels (see also Section 3.10.3). They can be hardened to various depths, have good wear resistance, and are relatively inexpensive. Because of their high toughness and resistance to fracture, high-speed steels are especially suitable for high positive-rake-angle tools (that is, small included angle), for interrupted cuts, and for use on machine tools that are subject to vibration and chatter because of their low stiffness. High-speed steels account for the largest number of tool materials used today, followed closely by various die steels and carbides. They are used in a wide variety of cutting operations that require complex tool shapes such as drills, reamers, taps, and gear cutters. Their basic limitation is the relatively low cutting speeds when compared with those of carbide tools (see below).

There are two basic types of high-speed steels: **molybdenum** (*M* series) and **tungsten** (*T* series). The *M* series contains up to about 10% molybdenum, with chromium, vanadium, tungsten, and cobalt as alloying elements. The *T* series contains 12 to 18% tungsten, with chromium, vanadium, and cobalt as alloying elements. The *M* series generally has higher abrasion resistance than the *T* series, undergoes less distortion during heat treating, and is less expensive. Consequently,

95% of all high-speed steel tools produced in the United States are made of *M*-series steels.

High-speed steel tools are available in wrought, cast, and sintered (powder-metallurgy; Chapter 11) conditions. They can also be **coated** for improved performance (see Section 8.6.5) and may also be subjected to surface treatments, such as case hardening (Section 4.5.1), for improved hardness and wear resistance.

8.6.3 Cast-cobalt alloys

Introduced in 1915, *cast-cobalt alloys* have high hardness, typically 58 HRC to 64 HRC, good wear resistance, and maintain their hardness at elevated temperatures. Their composition ranges from 38 to 53% cobalt, 30 to 33% chromium, and 10 to 20% tungsten. Commonly known as *Stellite* tools, these alloys are cast and ground into relatively simple tool shapes. They are not as tough as high-speed steels and are sensitive to impact forces; consequently, they are less suitable than high-speed steels for interrupted cutting operations. Tools made of cast-cobalt alloys are now used only for special applications that involve deep, continuous roughing operations at relatively high feeds and speeds, and as much as twice the rates possible with high-speed steels.

8.6.4 Carbides

The tool materials described thus far possess sufficient toughness, impact strength, and thermal shock resistance for many applications, but they have significant limitations on characteristics such as strength and hardness, particularly hot hardness. Consequently, they cannot be used as effectively where high cutting speeds, and hence high temperatures, are involved, and their tool life can be relatively short. *Carbides,* also known as **cemented** or **sintered carbides,** were introduced in the 1930s to meet the challenge of higher machining speeds for higher production rates.

Because of their high hardness over a wide range of temperatures (as can be seen in Fig. 8.30), high elastic modulus and thermal conductivity, and low thermal expansion, carbides are among the most important, versatile, and cost-effective tool and die materials for a wide range of applications. However, stiffness of the machine tool is important, and light feeds, low cutting speeds, and chatter can be detrimental. The two basic groups of carbides used for machining operations are *tungsten carbide* and *titanium carbide*. In order to differentiate them from coated tools (see Section 8.6.5), plain carbide tools are usually referred to as **uncoated carbides.**

1. **Tungsten carbide.** *Tungsten carbide* (WC) is a composite material consisting of tungsten-carbide particles bonded together in a *cobalt* matrix; hence, it is also called *cemented carbide.* WC is frequently compounded with carbides of titanium and niobium to impart special properties to carbide tools and dies. The amount of cobalt significantly affects the properties of carbide tools. As the cobalt content increases, strength, hardness, and wear resistance decrease, while toughness increases (Fig. 8.31). Tungsten-carbide tools are generally used for machining steels, cast irons, and abrasive nonferrous materials, and have largely replaced HSS tools in many applications for performance reasons. Tungsten-carbide tools are manufactured by powder-metallurgy techniques, described in Chapter 11.

2. **Titanium carbide.** *Titanium carbide* (TiC) has higher wear resistance than tungsten carbide but is not as tough. With a *nickel-molybdenum* alloy as the matrix, TiC is suitable for machining hard materials, mainly steels and cast irons, and for machining at higher speeds than those for tungsten carbide.

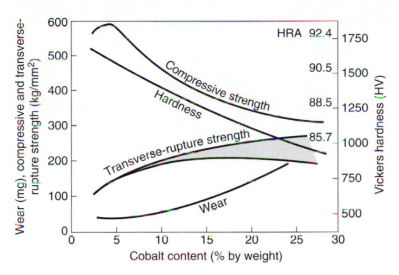

FIGURE 8.31 Effect of cobalt content in tungsten-carbide tools on mechanical properties. Note that hardness is directly related to compressive strength (see Section 2.6.8) and inversely to wear [see Eq. (4.6)].

Inserts. High-speed-steel and carbon-steel cutting tools can be shaped in one piece and ground to various geometries (see Fig. 8.10), including drills and milling cutters. However, after the cutting edge wears and becomes dull, the tool has to be removed from its holder and reground, a time-consuming process. The need for a more efficient method led to the development of *inserts*, which are individual cutting tools with a number of cutting edges and in various shapes (Fig. 8.32). Thus, a square insert, for example, has eight cutting edges, and a triangular insert has six. Inserts are available with a wide variety of **chip-breaker** features for controlling chip flow and reducing vibration and the heat generated. Optimum chip-breaker geometries are now developed using computer-aided design and finite-element techniques.

Inserts are usually clamped on the tool *shank* with various locking mechanisms (Fig. 8.32a and b). Less frequently, inserts are *brazed* to the tool shank (see Fig. 8.39); however, because of the difference in thermal expansion between the insert and the tool-shank materials, brazing must be done carefully in order to avoid cracking or warping. Clamping is the preferred method because after one cutting edge is worn, it is *indexed* (rotated in its holder) so that another edge can be used. In addition to those shown in Fig. 8.32, a wide variety of other toolholders is also available for specific applications, including toolholders with quick insert insertion and removal features for more efficient operation.

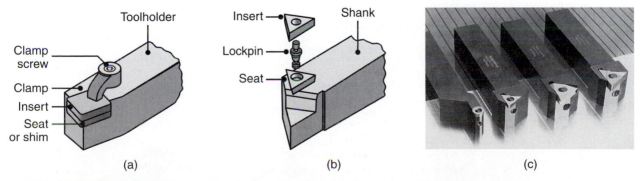

FIGURE 8.32 Methods of mounting inserts on toolholders: (a) clamping and (b) wing lockpins. (c) Examples of inserts mounted using threadless lockpins, which are secured with side screws. *Source:* Courtesy of Valenite.

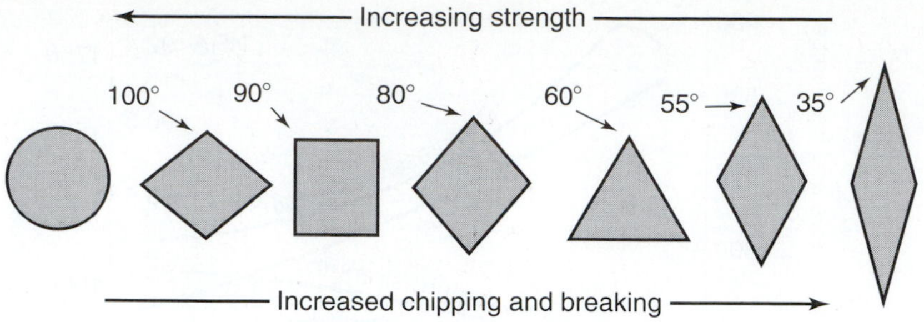

FIGURE 8.33 Relative edge strength and tendency for chipping and breaking of inserts with various shapes. Strength refers to that of the cutting edge shown by the included angles. *Source:* Courtesy of Kennametal, Inc.

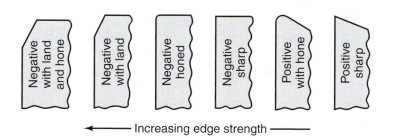

FIGURE 8.34 Edge preparations for inserts to improve edge strength. *Source:* Courtesy of Kennametal, Inc.

The strength of the cutting edge of an insert depends on its shape; the smaller the included angle of the edge (Fig. 8.33), the lower is its strength. In order to further improve edge strength and prevent chipping, insert edges are usually honed, chamfered, or are produced with a negative land (Fig. 8.34). Most inserts are honed to a radius of about 0.025 mm (0.001 in.).

8.6.5 Coated tools

A variety of materials can be used as coating materials over high-speed-steel and carbide tools (substrates) to produce *coated tools*. Because of their unique properties, coated tools can be used at high cutting speeds, thus reducing the time required for machining, hence costs. In practice, it has been observed that coated tools can improve tool life by as much as 10 times that of uncoated tools. Note in Fig. 8.35, for example, that the machining time has been reduced by a factor of more than 100 since 1900.

FIGURE 8.35 Relative time required to machine with various cutting-tool materials, with indication of the year the tool materials were introduced. Note that, within one century, machining time has been reduced by two orders of magnitude. *Source:* After Sandvik Coromant.

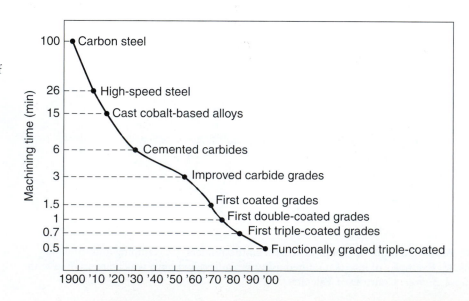

Commonly used *coating materials* include titanium nitride, titanium carbide, titanium carbonitride, and aluminum oxide (Al_2O_3), as described below. Generally in the thickness range of 2–10 μm (80–400 μin.), coatings are applied by **chemical vapor deposition** (CVD) and **physical-vapor deposition** (PVD) techniques, described in Section 4.5. The CVD process is the most commonly used coating application method for carbide tools with multiphase and ceramic coatings. The PVD-coated carbides with TiN coatings, on the other hand, have higher cutting-edge strength, less friction, lower tendency to form a built-up edge, and are smoother and more uniform in thickness, which is generally in the range of 2–4 μm (80–160 μin.). A more recent technology, particularly for multiphase coatings, is **medium-temperature chemical-vapor deposition** (MTCVD); it provides higher resistance to crack propagation than do CVD coatings.

Coatings should have the following general characteristics:

- High hardness at elevated temperatures.
- Chemical stability and inertness to the workpiece material.
- Low thermal conductivity.
- Good bonding to the substrate, to prevent flaking or spalling.
- Little or no porosity.

The effectiveness of coatings, in turn, is enhanced by hardness, toughness, and high thermal conductivity of the substrate, which may be carbide or high-speed steel. Honing (see Section 9.7) of the cutting edges is an important procedure in order to maintain the strength of the coating, and to prevent chipping at sharp edges and corners.

Some different types of coatings are discussed as follows.

1. **Titanium nitride.** *Titanium nitride* (TiN) coatings have low coefficient of friction, high hardness, good high-temperature properties, and good adhesion to the substrate. Consequently, they greatly improve the life of high-speed steel tools, as well as carbide tools, drills, and cutters. Titanium-nitride-coated tools (gold in color) perform well at higher cutting speeds and feeds; they do not perform as well as uncoated tools at low speeds because the coating can be worn off by chip adhesion. Hence, the use of appropriate cutting fluids to discourage chip-tool adhesion is important. Flank wear is significantly lower than for uncoated tools (Fig. 8.36). Flank surfaces can be reground after use, and regrinding does not remove the coating on the rake face of the tool.

2. **Titanium carbide.** *Titanium carbide* (TiC) coatings (silver-gray in color) on tungsten-carbide inserts have high resistance to flank wear, especially in machining abrasive materials.

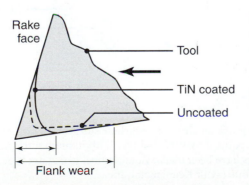

FIGURE 8.36 Wear patterns on high-speed-steel uncoated and titanium-nitride-coated cutting tools. Note that flank wear is lower for the coated tool.

3. **Titanium carbonitride.** *Titanium carbonitride* (TiCN), which is deposited by physical-vapor deposition techniques, is harder and tougher than TiN. It can be used on carbide and high-speed steel tools and is particularly effective in cutting stainless steels. TiCN coatings range from violet to mauve red in color depending on carbon content.

4. **Ceramic coatings.** Because of their good high-temperature performance, chemical inertness, low thermal conductivity, and resistance to flank and crater wear, ceramics are suitable coating materials for tools. The most commonly used ceramic coating is *aluminum oxide* (Al_2O_3). However, because ceramic coatings are very stable (not chemically reactive), oxide coatings generally bond weakly to the substrate and, thus, may have a tendency to peel off the tool or the insert.

5. **Multiphase coatings.** The desirable properties of the types of coating described above can be combined and optimized with the use of *multiphase coatings* (Fig. 8.37). Coated carbide tools are available with two or three layers of such coatings and are particularly effective in machining cast irons and steels.

    In the example shown in Fig. 8.37, the first layer over the tungsten-carbide substrate is TiC, followed by Al_2O_3 and then TiN. It is important that (a) the first layer should bond well to the substrate, (b) the outer layer should resist wear and have low thermal conductivity, and (c) the intermediate layer should bond well and be compatible with both layers.

    Typical applications of multiple-coated tools are listed as follows:

    a.   High-speed, continuous cutting: TiC/Al_2O_3.

    b.   Heavy-duty, continuous cutting: $TiC/Al_2O_3/TiN$.

    c.   Light, interrupted cutting: $TiC/TiC + TiN/TiN$.

A more recent development in coatings is **alternating multiphase layers,** with layers that are thinner than in typical multiphase coatings (Fig. 8.37). The thickness of these layers is in the range of 2–10 μm (80–400 μin.). The reason for using thinner coatings is that coating hardness increases with decreasing grain size, a phenomenon that is similar to the increase in strength of metals with

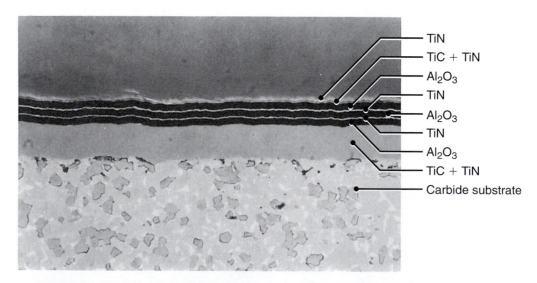

FIGURE 8.37   Multiphase coatings on a tungsten-carbide substrate. Three alternating layers of aluminum oxide are separated by very thin layers of titanium nitride. Inserts with as many as 13 layers of coatings have been made. Coating thicknesses are typically in the range of 2 to 10 μm. *Source:* Courtesy of Kennametal, Inc.

decreasing grain size (see Section 3.4.1); thus, thinner layers are harder than thicker layers.

6. **Diamond coatings.** An important development concerns the use of polycrystalline diamond as coatings, particularly for tungsten-carbide and silicon-nitride inserts. Thin-film diamond-coated inserts are available, as are thick-film diamond-brazed-tip cutting tools. Thin films are deposited on substrates by PVD and CVD techniques, whereas thick films are produced by growing a large sheet of pure diamond, which is then laser cut to shape and brazed to a carbide shank. Diamond-coated tools are particularly effective in machining abrasive materials, such as aluminum-silicon alloys, graphite, and fiber-reinforced and metal-matrix composite materials (see Section 11.14). Improvements in tool life of as much as tenfold have been obtained over that of other coated tools.

7. **Other coating materials.** Advances are continually being made in developing and testing new coating materials. **Titanium aluminum nitride** (TiAlN) is effective in machining aerospace alloys. Chromium-based coatings, such as **chromium carbide** (CrC), have been found to be effective in machining softer metals that tend to adhere to the cutting tool, such as aluminum, copper, and titanium. Other coating materials include **zirconium nitride** (ZrN) and **hafnium nitride** (HfN), **nanocoatings** with carbide, boride, nitride, oxide, or some combination, and **composite coatings**, using a variety of materials.

8.6.6 Alumina-base ceramics

Ceramic tool materials, introduced in the early 1950s, consist primarily of fine-grained, high-purity **aluminum oxide.** They are pressed into insert shapes under high pressure and at room temperature, sintered at high temperature, and are called **white,** or **cold-pressed, ceramics.** (See also Section 11.9.3.) Titanium carbide and zirconium oxide can be added to improve properties such as toughness and resistance to thermal shock.

Alumina-base ceramic tools have very high abrasion resistance and hot hardness (Fig. 8.38). Chemically, they are more stable than high-speed steels and carbides; thus they have less of a tendency to adhere to metals during machining and hence lower tendency to form a built-up edge. Consequently, good surface finish is obtained with ceramic tools, particularly in machining cast irons and steels. However, ceramics lack toughness, which can result in premature tool failure by chipping or fracture. (See Fig. 8.20.) The shape and setup of ceramic tools are also important.

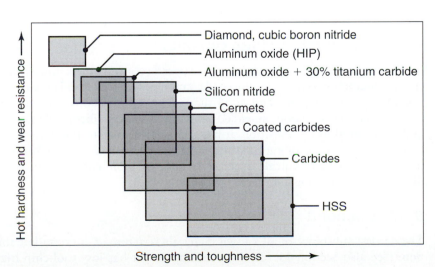

FIGURE 8.38 Ranges of properties for various groups of cutting-tool materials. (See also Tables 8.1 through 8.5.) HIP is not isostatic pressing; see Section 11.3.3.

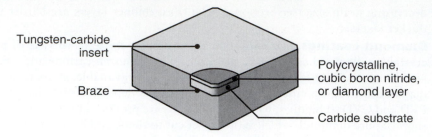

FIGURE 8.39   Construction of polycrystalline cubic-boron-nitride or diamond layer on a tungsten-carbide insert.

Negative rake angles, and hence large included angles, are generally preferred in order to avoid chipping. The occurrence of tool failure can be reduced by increasing the stiffness and damping capacity of machine tools and workholding devices, thus reducing vibration and chatter (see Section 8.12).

Cermets.   *Cermets* (from the words *cer*amic and *met*al), which are also called **black**, or **hot-pressed, ceramics** (carboxides), typically contain 70% aluminum oxide and 30% titanium carbide. Other cermets may contain molybdenum carbide, niobium carbide, or tantalum carbide. The performance of cermets is between that of ceramics and carbides (see Fig. 8.38). Although they can be coated, the benefits of coatings are somewhat controversial, as the improvement in wear resistance appears to be marginal.

8.6.7 Cubic boron nitride

Next to diamond, *cubic boron nitride* (cBN) is the hardest material presently available. The cBN cutting tools are made by bonding a 0.5- to 1-mm (0.02- to 0.04-in.) layer of *polycrystalline cubic boron nitride* to a carbide substrate by sintering under pressure (Fig. 8.39). While the carbide provides toughness, the cBN layer provides very high wear resistance and cutting-edge strength. Cubic-boron-nitride tools are also made in small sizes without a substrate. At elevated temperatures, cBN is chemically inert to iron and nickel, and its resistance to oxidation is high. It is therefore particularly suitable for machining hardened ferrous and high-temperature alloys. (See also *hard turning* in Section 8.9.2.) Because cBN tools are brittle, stiffness and damping capacity of the machine tool and fixturing devices are important to avoid vibration and chatter. Cubic boron nitride is also used as an abrasive, as described in Section 9.2.

8.6.8 Silicon-nitride-base ceramics

Silicon-nitride-base ceramic (SiN) tool materials consist of silicon nitride with additions of aluminum oxide, yttrium oxide, and titanium carbide. These tools have high toughness, hot hardness, and good thermal shock resistance. An example of this type of material is **sialon**, so called after the elements *si*licon, *al*uminum, *o*xygen, and *n*itrogen in its composition. It has higher resistance to thermal shock than silicon nitride and is recommended for machining cast irons and nickel-base superalloys at intermediate cutting speeds. Because of their chemical affinity to steels, SiN-base tools are not suitable for machining steels.

8.6.9 Diamond

The hardest substance of all known materials is *diamond*, a crystalline form of carbon. (See also Section 11.13.) As a cutting tool, it has low tool-chip friction, high wear resistance, and the ability to maintain a sharp cutting edge. It is used when very fine surface finish and dimensional accuracy are required, particularly with abrasive nonmetallic materials and soft nonferrous alloys. *Single-crystal diamond* is used for special applications, such as machining copper-front high-precision optical mirrors. Because diamond is brittle, tool shape and sharpness are important; low rake angles and large included angles are normally used to provide a strong cutting edge. Wear of diamond tools may occur by microchipping (caused by thermal stresses and oxidation) and transformation to carbon (caused by the heat generated during cutting).

Single-crystal (single-point) diamond tools have been largely replaced by **polycrystalline diamond** tools (*compacts*), which are also used as wire-drawing dies for fine wire (Section 6.5). These materials consist of very small synthetic crystals, fused by a high-pressure, high-temperature process to a thickness of about 0.5–1 mm (0.02–0.04 in.) and bonded to a carbide substrate, similar to cBN tools. (See Fig. 8.39.) The random orientation of the diamond crystals prevents the propagation of cracks through the tool, thus significantly improving its toughness.

Diamond tools can be used satisfactorily at almost any speed but are suitable mostly for light, uninterrupted finishing cuts. In order to minimize tool fracture, the single-crystal diamond must be resharpened as soon as it becomes dull. Because of its strong chemical affinity, diamond is not recommended for machining plain-carbon steels and titanium, nickel, and cobalt-base alloys. Diamond is also used as an abrasive in grinding and polishing operations (Chapter 9) and as wear-resistant coatings (Section 4.5).

8.6.10 Whisker-reinforced and nanocrystalline tool materials

In order to further improve the performance and wear resistance of cutting tools, particularly in machining abrasive and hard workpieces, new tool materials are continually being developed with enhanced properties such as (a) high fracture toughness, (b) resistance to thermal shock, (c) cutting-edge strength, and (d) hot hardness.

Whiskers (see Section 3.8.3) are being used as reinforcing fibers in composite tool materials. Examples of **whisker-reinforced** materials include silicon-nitride-base tools reinforced with silicon-carbide (SiC) whiskers, and aluminum-oxide-base tools reinforced with silicon-carbide whiskers, sometimes with the addition of *zirconium oxide* (ZrO_2). However, the affinity (low inertness) of silicon carbide to ferrous metals makes SiC-reinforced tools unsuitable for machining irons and steels.

Micrograin carbides. Progress in nanomaterials (Section 3.11.9) has led to the development of cutting tools that are made of very fine grained (*micrograin*) carbides of tungsten, titanium, and tantalum. The grain size is in the range of 0.2 to 0.8 μm (8 to 30 μin.). These materials are stronger, harder, and more wear resistant than traditional carbides. In one application, for example, drills with diameters on the order of 100 μm (0.004 in.) are being made from these materials and used in the fabrication of microelectronic circuit boards (see Chapter 13).

Functionally graded carbides. In these tools, the composition of the carbide in the insert has a gradient through its near-surface depth, instead of being uniform as is common in carbide inserts. The gradient has a smooth distribution of compositions and phases, with functions as those described regarding desirable properties of coatings on cutting tools (see Section 8.6.5). Graded mechanical properties eliminate stress concentrations and promote better tool life and performance. They are, however, more expensive and thus may not be justified for all applications.

8.6.11 Cryogenic treatment of cutting tools

Studies are continuing on the beneficial effect of cryogenic treatment of tools and other metals on their performance in machining operations. (See Section 5.11.6 for details.) In this procedure, the tool is cooled very slowly to temperatures of about $-180°$ ($-300°F$) and slowly returned to room temperature; it is then tempered. Claims have been made, depending on the combinations of tool and workpiece materials involved, that tool-life increases of up to 300% can be achieved with this procedure.

8.7 | Cutting Fluids

Also called *lubricants* and *coolants, cutting fluids* are used extensively in machining operations to

- Cool the cutting zone, thus reducing workpiece temperature and distortion, and improving tool life.
- Reduce friction and wear, hence improving tool life and surface finish.
- Reduce forces and energy consumption.
- Wash away chips.
- Protect the newly machined surfaces from environmental attack.

A cutting fluid can predominantly serve as a **coolant** and/or a **lubricant** (see Section 4.4.3). Its effectiveness in machining operations depends on a number of factors, such as the method of application, temperature, cutting speed, and type of machining operation. There are situations, however, in which the use of cutting fluids can be detrimental. For example, in interrupted cutting operations, such as milling (Section 8.10), the cooling action of the cutting fluid increases the extent of alternate heating and cooling (*thermal cycling*) to which the cutter teeth are subjected. This condition can lead to thermal cracks (*thermal fatigue* or *thermal shock*). Furthermore, cutting fluids may also cause the chip to become more curled, thus concentrating the stresses on the tool closer to its tip, thus concentrating the heat closer to the tool tip and reducing tool life.

Cutting fluids can present **biological** and **environmental hazards** (see also Section 4.4.4) that require proper recycling and disposal, thus adding to the cost of the machining operation. The use and application of cutting fluids can also be a significant item in manufacturing costs. For these reasons, **dry cutting,** or **dry machining,** has become an increasingly important approach in which no coolant or lubricant is used in the operation (see Section 8.7.2). Even though this approach would suggest that higher temperatures and more rapid tool wear would occur, some tool materials and coatings maintain a reasonable tool life. Dry cutting has been associated with high-speed machining, because of the fact that higher cutting speeds transfer a greater amount of heat from cutting to the chip (see Fig. 8.18), which is a natural strategy for reducing the need for a coolant. (See also Section 3.9.7 on possible detrimental effects of cutting fluids on some cutting tools, called *selective leaching,* such as carbide tools with cobalt binders.)

8.7.1 Types of cutting fluids and methods of application

There are four basic types of cutting fluids commonly used in machining operations: **oils, emulsions, semisynthetics,** and **synthetics,** as described in Section 4.4.4. Cutting-fluid recommendations for specific machining operations are given throughout the rest of this chapter. In selecting an appropriate cutting fluid, consideration should be given to its possible detrimental effects on the workpiece material (e.g., corrosion, stress-corrosion cracking, staining), the components of the machine tool, biological and environmental effects, and recycling and disposal.

The most common method of applying cutting fluid is **flood cooling.** Flow rates typically range from 10 L/min (3 gal/min) for single-point tools to 225 L/min (60 gal/min) per cutter for multiple-tooth cutters, such as in milling. In operations such as gun drilling and end milling, fluid pressures in the range of 700–14,000 kPa (100–2000 psi) are used to wash away the chips.

Mist cooling is another method of applying cutting fluids and is generally used with water-base fluids. Although it requires venting (to prevent inhalation of fluid particles by the machine operator and others nearby) and has limited cooling capacity, mist cooling supplies fluid to otherwise inaccessible areas and provides better visibility of the workpiece being machined. It is particularly effective in grinding operations (see Chapter 9), using air pressures in the range of 70–600 kPa (10–80 psi).

With increasing speed and power of machine tools, the heat generated in machining operations has become a significant factor (see Section 8.2.6). More recent developments include the use of **high-pressure refrigerated coolant** systems to improve the rate of heat removal from the cutting zone, as well as to reduce machining costs and avoid adverse environmental effects. High pressures on the order of 35 MPa (5000 psi) are now used to deliver the cutting fluid to the cutting zone via specially designed nozzles that aim a powerful jet of fluid to the zone. This action breaks up the chips (thus the fluid acts as a chip breaker) in situations where the chips produced would otherwise be long and continuous and thus interfere with the machining operation.

Through the cutting-tool system. We have pointed out the severity of some machining operations in terms of the difficulty of supplying cutting fluids into the cutting zone and flushing away the chips. For a more effective application, narrow passages can be produced in cutting tools, as well as in toolholders, through which cutting fluids can be applied under high pressure.

8.7.2 Near-dry and dry machining

For economic and environmental reasons, there has been a continuing worldwide trend since the mid 1990s to minimize or eliminate the use of metalworking fluids. This trend has led to the practice of *near-dry machining* (NDM), where coolant use is eliminated or reduced significantly; the significance of this approach is apparent when noting that, in the United States alone, millions of gallons of metalworking fluids are consumed each year. The major benefits of NDM include

1. Alleviating the environmental impact of using cutting fluids, improving air quality in manufacturing plants, and reducing health hazards.

2. Reducing the cost of machining operations, including the cost of maintaining, recycling, and disposing cutting fluids. This is significant because it has been estimated that metalworking fluids constitute about 7–17% of the total machining costs.

3. Further improving machined-surface quality.

The principle behind near-dry machining is the application of a fine mist of air-fluid mixture, containing a very small amount of cutting fluid, including vegetable oil. The mixture is delivered to the cutting zone, through the spindle of the machine tool, typically through a 1-mm diameter nozzle and under a pressure of 600 kPa (85 psi). The fluid is used at rates on the order of 1–100 cc/hr, which is estimated to be, at most, one ten-thousandth of that used in flood cooling; consequently, the process is also known as *minimum quantity lubrication* (MQL).

Dry machining is also a viable alternative. With major advances in cutting-tool materials, dry machining has been shown to be effective in various machining operations, especially turning, milling, and gear cutting, on steels, alloys steels, and cast irons, but generally not for aluminum alloys.

Recall that one of the functions of a cutting fluid is to flush chips away from the cutting zone. Although it first appears that this may be a problem in dry machining, tool designs have been developed that allow application of *pressurized air,* often through holes in the tool shank. The compressed air does not serve as cutting fluid and provides limited cooling, but it is very effective at clearing chips away from the cutting zone.

8.7.3 Cryogenic machining

Largely in the interest of reducing or eliminating the adverse environmental impact of using metalworking fluids, a more recent technology is the use of *liquid nitrogen* as a coolant in machining, as well as in grinding. (See Section 9.6.9.) With appropriate small-diameter nozzles, liquid nitrogen at a temperature of about $-200°C$ $(-320°F)$ is injected into the tool-workpiece interface, thus reducing its temperature. As a result, tool hardness, and hence tool life, is enhanced, thus allowing for higher cutting speeds. Furthermore, the chips become more brittle and, thus, easier to flush from the cutting zone. Because no fluids are involved and the liquid nitrogen simply evaporates, the chips can be recycled more easily, thus also improving the economics of machining operations and without any adverse effects on the environment.

8.8 | High-Speed Machining

With continuing demands for higher productivity and lower manufacturing costs, much effort has been carried out to increase the cutting speed and thus the material-removal rate in machining. Although *high-speed machining* (HSM) is a relative term, an approximate range of cutting speeds may be given as follows:

1. **High speed:** 600–1800 m/min (2000–6000 ft/min).
2. **Very high speed:** 1800–18,000 m/min (6000–60,000 ft/min).
3. **Ultrahigh speed:** >18,000 m/min.

Spindle rotational speeds in machine tools range up to 50,000 rpm, although the automotive industry, for instance, has generally limited them to 15,000 rpm for better reliability and less downtime should a failure occur during machining operations. The *spindle power* required in high-speed machining is generally on the order of 0.004 W/rpm (0.005 hp/rpm), much less than in traditional machining which is typically in the range of 0.2 to 0.4 W/rpm (0.25 to 0.5 hp/rpm). The maximum workpiece speed (see Fig. 8.54) in high-speed machining is on the order of 1 m/s (3 ft/s), and the acceleration of machine-tool components is very high.

Spindles for high rotational speeds require *high stiffness and accuracy* and generally involve an integral electric motor. The armature is built onto the shaft and the stator is placed in the wall of the spindle housing. The bearings may be rolling element or hydrostatic; the latter is more desirable because it requires less space than the former. Because of *inertia effects* during acceleration and deceleration of machine-tool components, the use of lightweight materials is an important consideration, including ceramics and composite materials. Selection of appropriate cutting-tool materials is of course a major consideration. Depending on the workpiece material, multiphase coated carbides, ceramics, cubic boron nitride, and diamond are candidate tool materials for high-speed machining.

It is also important to note that high-speed machining should be considered primarily for operations in which **cutting time** is a significant portion of the floor-to-floor time in the overall machining operation. As described in Section 16.9, **noncutting time** and various other factors (e.g., tool material costs, capital equipment costs, labor costs, etc.) are important considerations in the overall assessment of the benefits of high-speed machining for a particular application. Studies have indicated that high-speed machining is economical for certain specific applications. It has, for example, been implemented in machining (a) aluminum structural components for aircraft, (b) submarine propellers of 6 m (20 ft) diameter, made of nickel-aluminum-bronze alloy and weighing 55,000 kg (50 tons), and (c) automotive engines, with five to ten times the productivity of traditional machining. High-speed machining of complex 3- and 5-axis contours has been made possible by advances in CNC control technology (see also *machining centers*, Section 8.11).

Another major factor in the adoption of high-speed machining has been the requirement to further improve dimensional tolerances in cutting operations. As can be seen in Fig. 8.17, as the cutting speed increases, more and more of the heat generated is removed by the chip, thus the tool and, more importantly, the workpiece remain close to ambient temperature. This is beneficial because there is no thermal expansion or warping of the workpiece during machining.

The important machine-tool characteristics and special requirements in high-speed machining may be summarized as follows:

1. Spindle design for high stiffness, accuracy, and balance at very high rotational speeds, and workholding devices that can withstand high centrifugal forces.

2. Fast feed drives, bearing characteristics, and effects of inertia of the machine-tool components.

3. Selection of appropriate cutting tools and processing parameters and their computer control.

4. Effective chip removal systems for material removal at very high rates.

8.9 | Machining Processes and Machine Tools for Producing Round Shapes

This section describes the processes that produce parts which are basically *round in shape,* as outlined in Table 8.7. Typical products machined include parts as small as miniature screws for eyeglass-frame hinges and as large as pistons, cylinders, gun barrels, and turbine shafts for hydroelectric power plants. These processes are generally performed by turning the workpiece on a lathe. *Turning* means that the part is rotating while it is being machined by a stationary tool. The starting material is typically a workpiece that has been made by other processes, such as casting, shaping, forging, extrusion, and drawing. Turning processes are versatile and capable of producing a wide variety of shapes, as outlined in Fig. 8.40. The various types of turning processes are described as follows:

- **Turning** straight, conical, curved, or grooved workpieces, such as shafts, spindles, pins, handles, and various machine components.

- **Facing** to produce a flat surface at the end of the part, such as for parts that are attached to other components, or to produce grooves for O-ring seats.

- Producing various shapes by **form tools,** such as for functional purposes or for appearance.

TABLE 8.7

General Characteristics of Machining Processes

Process	Characteristics	Commercial tolerances (±mm)
Turning	Turning and facing operations are performed on all types of materials; requires skilled labor; low production rate, but medium to high rates can be achieved with turret lathes and automatic machines, requiring less skilled labor.	Fine: 0.05–0.13 Rough: 0.13 Skiving: 0.025–0.05
Boring	Internal surfaces or profiles, with characteristics similar to those produced by turning; stiffness of boring bar is important to avoid chatter.	0.025
Drilling	Round holes of various sizes and depths; requires boring and reaming for improved accuracy; high production rate; labor skill required depends on hole location and accuracy specified.	0.075
Milling	Variety of shapes involving contours, flat surfaces, and slots; wide variety of tooling; versatile; low to medium production rate; requires skilled labor.	0.13–0.25
Planing	Flat surfaces and straight contour profiles on large surfaces; suitable for low-quantity production; labor skill required depends on part shape.	0.08–0.13
Shaping	Flat surfaces and straight contour profiles on relatively small workpieces; suitable for low-quantity production; labor skill required depends on part shape.	0.05–0.13
Broaching	External and internal flat surfaces, slots, and contours with good surface finish; costly tooling; high production rate; labor skill required depends on part shape.	0.025–0.15
Sawing	Straight and contour cuts on flats or structural shapes; not suitable for hard materials unless the saw has carbide teeth or is coated with diamond; low production rate; requires only low skilled labor.	0.8

- **Boring** to enlarge a hole made by a previous process or in a tubular workpiece, or to produce internal grooves.
- **Drilling** to produce a hole, which may be followed by tapping or by boring to improve the accuracy of the hole and its surface finish.
- **Parting,** also called **cutting off,** to cut (separate) a piece from the end of a part, as in making slugs or blanks for additional processing into discrete parts.
- **Threading** to produce external or internal threads in workpieces.
- **Knurling** to produce a regularly shaped surface characteristics on cylindrical surfaces, as in making knurled knobs.

These operations may be performed at various rotational speeds of the workpiece, depths of cut, d, and feed, F, (see Fig. 8.19), depending on the workpiece and tool materials, the surface finish and dimensional accuracy required, and the capacity of the machine tool.

Roughing cuts are performed for large-scale material removal and typically involve depths of cut greater than 0.5 mm (0.02 in.) and feeds on the order of 0.2–2 mm/rev (0.008–0.08 in./rev). **Finishing cuts** usually involve lower depths of cut and feed. Most machining operations consist of roughing cuts as needed to define the shape, followed by a finishing cut to satisfy specific dimensional tolerance and surface finish requirements. Cutting speeds are usually in the range of 0.15–4 m/s (30–800 ft/min).

8.9.1 Turning parameters

The majority of turning operations involve *single-point* cutting tools. The geometry of a typical right-hand simple cutting tool for turning is shown in Fig. 8.41; such tools are identified by a standardized nomenclature. Note that the geometry shown

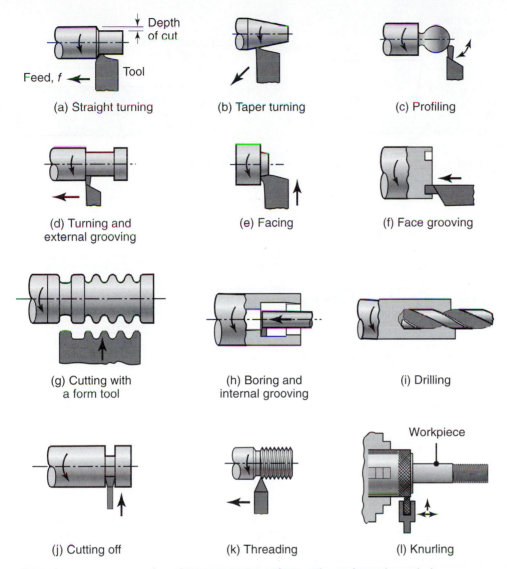

FIGURE 8.40 Variety of machining operations that can be performed on a lathe.

can be **maintained** with the proper toolholder and insert. Each group of tool and workpiece materials has an optimum set of tool angles, which have been developed over many years and largely through experience. Some data on tool geometry can be found in Table 8.8.

1. **Tool geometry.** The various angles on a cutting tool have important functions in machining operations. (a) **Rake angles** are important in controlling the direction of chip flow and in the strength of the tool tip. Positive angles improve the cutting operation by reducing forces and temperatures; however, positive angles also produce a small included angle of the tool tip (see Fig. 8.2). Depending on the toughness of the tool material, a small included angle may cause premature tool chipping and failure. (b) **Side rake angle** is more important than **back rake angle,** although the latter usually controls the direction of chip flow. (c) **Relief angles** control interference and rubbing at the tool-workpiece interface. If the relief angle is too large, the tool may chip off, and if it is too small, flank wear may be excessive. (d) **Cutting-edge angles** affect chip formation, tool

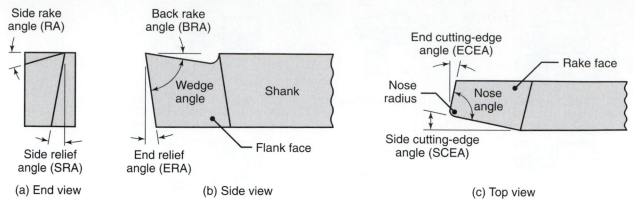

FIGURE 8.41 Designations and symbols for a right-hand cutting tool. The designation "right hand" means that the tool travels from right to left, as shown in Fig. 8.19.

TABLE 8.8

General Recommendations for Tool Angles in Turning

Material	High-speed steel					Carbide inserts				
	Back rake	Side rake	End relief	Side relief	Side and end cutting edge	Back rake	Side rake	End relief	Side relief	Side and end cutting edge
Aluminum and magnesium alloys	20	15	12	10	5	0	5	5	5	15
Copper alloys	5	10	8	8	5	0	5	5	5	15
Steels	10	12	5	5	15	−5	−5	5	5	15
Stainless steels	5	8–10	5	5	15	−5-0	−5–5	5	5	15
High-temperature alloys	0	10	5	5	15	5	0	5	5	45
Refractory alloys	0	20	5	5	5	0	0	5	5	15
Titanium alloys	0	5	5	5	15	−5	−5	5	5	5
Cast irons	5	10	5	5	15	−5	−5	5	5	15
Thermoplastics	0	0	20–30	15–20	10	0	0	20–30	15–20	10
Thermosets	0	0	20–30	15–20	10	0	15	5	5	15

strength, and cutting forces to various degrees. (e) **Nose radius** affects surface finish and tool tip strength; also, the smaller the nose radius, the rougher will be the surface finish of the workpiece, and the lower will be the strength of the tool. On the other hand, large nose radii can lead to tool chatter (see Section 8.12).

2. **Material-removal rate.** The *material-removal rate* (MRR) in machining is the volume of material removed per unit time, such as mm³/min or in³/min. Referring to Fig. 8.42a, note that for each revolution of the workpiece, a ring-shaped layer of material is removed, and its cross-sectional area is the product of the distance the tool travels in one revolution (that is, the feed f) and the depth of cut, d. The volume of this ring is the product of the cross-sectional area (that is, fd) and the average circumference of the ring [that is, πD_{avg}, where $D_{avg} = (D_o + D_f)/2$]. For light cuts on large-diameter workpieces, the average diameter can be replaced by D_o.

Thus, the material-removal rate per revolution will be $\pi D_{avg}df$. Since the rotational speed of the workpiece is N revolutions per minute, the removal rate is given by

$$MRR = \pi D_{avg}dfN. \tag{8.38}$$

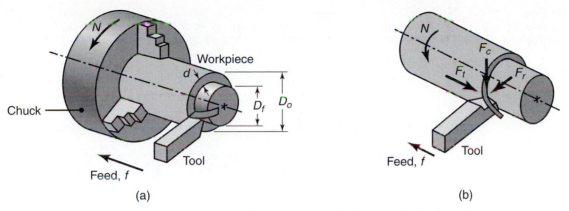

FIGURE 8.42 (a) Schematic illustration of a turning operation, showing depth of cut, d, and feed, f. The cutting speed is the surface speed of the workpiece at the tool tip. (b) Forces acting on a cutting tool in turning. F_c is the cutting force, F_t is the thrust or feed force (in the direction of feed), and F_r is the radial force that tends to push the tool away from the workpiece being machined. Compare this figure with Fig. 8.11 for a two-dimensional cutting operation.

The dimensional accuracy of this equation can be checked by substituting dimensions into the right-hand side; thus, $(mm)(mm)(mm/rev)(rev/min) = mm^3/min$, which correctly indicates the volume rate of removal. Similarly, the cutting time, T, for a workpiece of length l can be calculated by noting that the tool travels at a feed rate of $fN = (mm/rev)(rev/min) = mm/min$. Since the distance traveled is l mm, the cutting time is

$$t = \frac{l}{fN}. \tag{8.39}$$

The time does not include the time required for *tool approach* and *retraction* during the machining operation. Modern machine tools are designed and constructed to minimize this *nonproductive time,* using computer controls. A typical method employed is first to rapidly traverse the tool and then to slow its movement as the tool engages the workpiece.

3. **Forces in turning.** The three principal forces acting on a cutting tool are illustrated in Fig. 8.42b. These forces are important in the design of machine tools as well as in determining the deflection of tools for precision machining operations. The *cutting force, F_c,* acts downward on the tool tip and thus tends to deflect the tool downward. Note that the cutting force is the force that supplies the energy required for the cutting operation. As can be seen in Example 8.4, the cutting force can be calculated from the energy per unit volume, described in Section 8.2.5, and by using the data given in Table 8.3. The *thrust force, F_t,* acts in the longitudinal direction; this force is also called the *feed force,* because it is in the feed direction. The *radial force, F_r,* is in the radial direction and thus tends to push the tool away from the workpiece.

4. **Tool materials, feeds, and cutting speeds.** The general characteristics of cutting-tool materials are described in Section 8.6. Figure 8.43 gives a broad range of cutting speeds and feeds applicable for these tool materials. Specific recommendations for cutting speeds for turning various workpiece materials and for cutting tools are given in Table 8.9.

FIGURE 8.43 The range of applicable cutting speeds and feeds for a variety of cutting-tool materials.

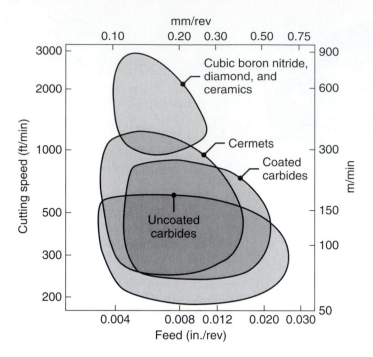

TABLE 8.9

Approximate Ranges of Recommended Cutting Speeds for Turning Operations

Workpiece material	Cutting speed	
	m/min	ft/min
Aluminum alloys	200–1000	650–3300
Cast iron, gray	60–900	200–3000
Copper alloys	50–700	160–2300
High-temperature alloys	20–400	65–1300
Steels	50–500	160–1600
Stainless steels	50–300	160–1000
Thermoplastics and thermosets	90–240	300–800
Titanium alloys	10–100	30–330
Tungsten alloys	60–150	200–500

Note: (a) The speeds given in this table are for carbides and ceramic cutting tools. Speeds for high-speed steel tools are lower than indicated. The higher ranges are for coated carbides and cermets. Speeds for diamond tools are significantly higher than any of the values indicated in the table.
(b) Depths of cut, d, are generally in the range of 0.5–12 mm (0.02–0.5 in.).
(c) Feeds, f, are generally in the range of 0.15–1 mm/rev (0.006–0.040 in./rev).

EXAMPLE 8.4 Material-removal rate and cutting force in turning

A 6-in.-long, 0.5-in.-diameter 304 stainless-steel rod is being reduced in diameter to 0.480 in. by turning on a lathe. The spindle rotates at $N = 400$ rpm, and the tool is traveling at an axial speed of 8 in./min. Calculate the cutting speed, material-removal rate, cutting time, power dissipated, and cutting force.

Solution. The cutting speed is the tangential speed of the workpiece. The maximum cutting speed is at the outer diameter, D_o, and is obtained from the expression

$$V = \pi D_o N.$$

Thus

$$V = (\pi)(0.500)(400) = 628 \text{ in./min} = 52 \text{ ft/min.}$$

The cutting speed at the machined diameter is

$$V = (\pi)(0.480)(400) = 603 \text{ in./min} = 50 \text{ ft/min.}$$

From the information given, note that the depth of cut is

$$d = \frac{0.500 - 0.480}{2} = 0.010 \text{ in.}$$

and the feed is

$$f = \frac{8}{400} = 0.02 \text{ in./rev.}$$

Thus, according to Eq. (8.38), the material-removal rate is

$$\text{MRR} = (\pi)(0.49)(0.010)(0.02)(400) = 0.123 \text{ in}^3/\text{min.}$$

The actual time to cut, according to Eq. (8.39), is

$$t = \frac{6}{(0.02)(400)} = 0.75 \text{ min.}$$

We can calculate the power required by referring to Table 8.3 and taking an average value for stainless steel as 4 W-s/mm³ = 4/2.73 = 1.47 hp-min/in³. Therefore, the power dissipated is

$$\text{Power} = (1.47)(0.123) = 0.181 \text{ hp,}$$

and since 1 hp = 396,000 in.-lb/min, the power dissipated is 71,700 in.-lb/min. The cutting force, F_c, is the tangential force exerted by the tool. Since power is the product of torque, T, and rotational speed in radians per unit time, we have

$$T = \frac{(71,700)}{(400)(2\pi)} = 29 \text{ lb-in.}$$

Since $T = (F_c)(D_{\text{avg}}/2)$, we obtain

$$F_c = \frac{(29)(2)}{(0.490)} = 118 \text{ lb.}$$

8.9.2 Lathes and lathe operations

Lathes are generally considered to be the oldest machine tools. Although woodworking lathes were first developed during the period 1000–1 B.C., metalworking lathes with lead screws were not built until the late 1700s. The most common type of lathe, shown in Fig. 8.44, was originally called an **engine lathe**, because it was powered with overhead pulleys and belts from nearby engines as the power source.

1. **Lathe components.** Lathes are typically equipped with a variety of components and accessories, as shown in Fig. 8.44. The *bed* supports all the other major components of the lathe. The *carriage*, or *carriage assembly*, which slides along the ways, consists of an assembly of the *cross slide*, *tool post*, and *apron*.

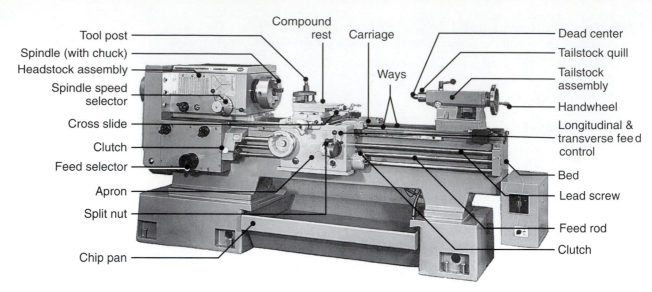

FIGURE 8.44   General view of a typical lathe, showing various major components. *Source:* Courtesy of Heidenreich & Harbeck.

The cutting tool is mounted on the *tool post*, usually with a *compound rest* that swivels to allow for tool positioning and its adjustments. The *headstock*, which is fixed to the bed, is equipped with motors, pulleys, and V-belts that supply power to the *spindle* at various rotational speeds. Headstocks have a hollow spindle to which workholding devices, such as *chucks* and *collets*, are attached. The *tailstock*, which can slide along the ways and be clamped at any position, supports the opposite end of the workpiece. The *feed rod*, which is powered by a set of gears from the headstock, rotates during operation of the lathe and provides movement to the carriage and the cross slide. The *lead screw*, which is used for accurately cutting threads, is engaged with the carriage by closing a split nut around the lead screw.

A lathe is generally specified by its *swing* (the maximum diameter of the workpiece that can be machined); by the maximum distance between the headstock and tailstock centers; and by the length of the bed. A wide variety of lathes is available, including *bench lathes, toolroom lathes, engine lathes, gap lathes,* and *special-purpose lathes.*

Tracer lathes, also called *duplicating lathes* or *contouring lathes,* are capable of turning parts with various contours, where the cutting tool follows a path that duplicates the contour of a template through a hydraulic or electrical system. *Automatic lathes,* also called *chucking machines* or *chuckers,* are used for machining individual pieces of regular or irregular shapes. *Turret lathes* are capable of performing multiple cutting operations on the same workpiece, such as turning, boring, drilling, thread cutting, and facing. Several cutting tools are mounted on the hexagonal *main turret.* The lathe usually has a *square turret* on the cross slide, with as many as four cutting tools mounted on it.

2. **Computer-controlled lathes.** In modern lathes, the movement and control of the machine tool and its components are actuated by **computer numerical controls** (CNCs); the features of such a lathe are shown in Fig. 8.45. These lathes are typically equipped with one or more turrets, each turret being equipped with a variety of cutting tools and performing several operations on different surfaces of the workpiece. These machine tools are highly automated, the operations are

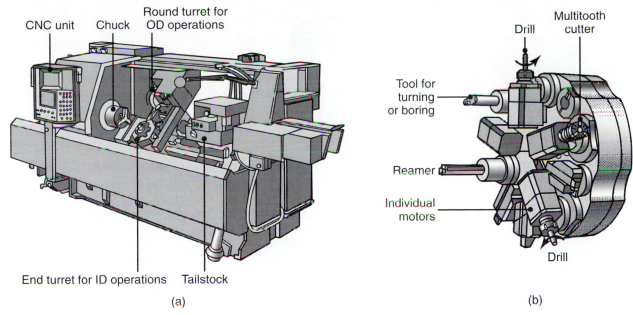

CNC unit   Chuck   Round turret for OD operations

Drill   Multitooth cutter

Tool for turning or boring

Reamer

Individual motors

Drill

End turret for ID operations   Tailstock

(a)

(b)

FIGURE 8.45   (a) A computer-numerical-control lathe, with two turrets; these machines have higher power and spindle speed than other lathes in order to take advantage of advanced cutting tools with enhanced properties. (b) A typical turret equipped with 10 cutting tools, some of which are powered.

repetitive and they maintain the desired dimensional accuracy, and less skilled labor is required (once the machine is set up) than with common lathes. They are suitable for low to medium volumes of production. The details of computer controls are presented in Chapters 14 and 15. (See also Section 8.11.)

EXAMPLE 8.5   Typical parts made on CNC turning machine tools

Figure 8.46 illustrates the capabilities of CNC turning machine tools. Workpiece materials, the number of cutting tools used, and machining times are indicated for each part. Although not as effectively or as consistently so, these parts can also be made on manual or turret lathes.

Source: Monarch Machine Tool Company.

---

3. **Turning process capabilities.**   Table 8.10 lists relative *production rates* for turning, as well as for other machining operations described in the rest of this chapter. These rates have an important bearing on productivity in machining operations. Note that there are major differences in the production rates of these processes. These differences are due not only to the inherent characteristics of the processes and machine tools, but also to various other factors, such as setup times and the types and sizes of the workpieces involved. The proper selection of a process and the machine tool for a particular product is important for minimizing production costs, as is also discussed in Section 8.15 and Chapter 16.

   The ratings given in Table 8.10 are relative, and there can be significant variations in specific applications. For example, heat-treated, high-carbon cast-steel rolls (for rolling mills; see Section 6.3) can be machined on special lathes at material-removal rates as high as 6000 cm³/min (370 in³/min), using multiple

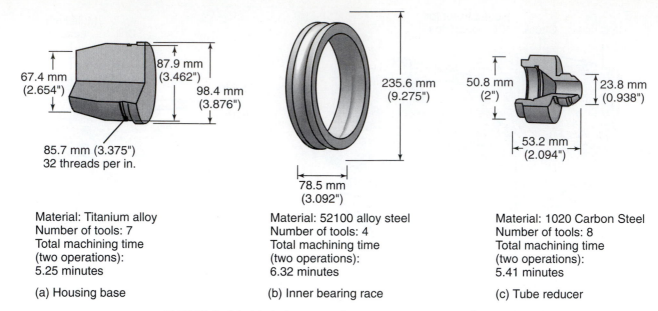

Material: Titanium alloy
Number of tools: 7
Total machining time
(two operations):
5.25 minutes

(a) Housing base

Material: 52100 alloy steel
Number of tools: 4
Total machining time
(two operations):
6.32 minutes

(b) Inner bearing race

Material: 1020 Carbon Steel
Number of tools: 8
Total machining time
(two operations):
5.41 minutes

(c) Tube reducer

FIGURE 8.46   Typical parts made on computer-numerical-control machine tools.

TABLE 8.10

Typical Production Rates for Various Cutting Operations	
Operation	Rate
Turning	
Engine lathe	Very low to low
Tracer lathe	Low to medium
Turret lathe	Low to medium
Computer-control lathe	Low to medium
Single-spindle chuckers	Medium to high
Multiple-spindle chuckers	High to very high
Boring	Very low
Drilling	Low to medium
Milling	Low to medium
Planing	Very low
Gear cutting	Low to medium
Broaching	Medium to high
Sawing	Very low to low

Note: Production rates indicated are relative: *Very low* is about one or more parts per hour; *medium* is approximately 100 parts per hour; *very high* is 1000 or more parts per hour.

cermet tools. The important factor in this operation (also called **high-removal-rate machining**) is the very high stiffness of the machine tool (to avoid tool breakage due to chatter; see Section 8.12) and the machine tool's high power, which can be up to 450 kW (600 hp).

The surface finish and dimensional accuracy obtained in turning and related operations depend on such factors as the characteristics and condition of the machine tool, stiffness, vibration and chatter, process parameters, tool geometry and wear, cutting fluids, machinability of the workpiece material, and operator skill. As a result, a wide range of surface finishes can be obtained, as shown in Fig. 8.26. (See also Fig. 9.27.)

4. **Ultraprecision machining.** There are continued and increasing demands for precision manufactured components for computer, electronics, nuclear energy, and defense applications. Examples include optical mirrors and components for optical system components, with surface finish requirements in the range of tens of nanometers [10^{-9} m or 0.001 μm (0.04 μin.)] and form accuracies in the μm and sub-μm range. The cutting tool for these *ultraprecision machining* applications is exclusively a single-crystal diamond (hence, the process is also called **diamond turning**), with a polished cutting edge that has a radius as small as in the range of tens of nanometers. Wear of the diamond can, however, be a significant problem. Recent advances include **cryogenic diamond turning**, in which the tooling system is cooled by liquid nitrogen to a temperature of about $-120°C$ ($-184°F$). (See also Section 8.7.3.)

   The workpiece materials for ultraprecision machining include copper alloys, aluminum alloys, silver, gold, electroless nickel, infrared materials, and plastics (acrylics). The depths of cut involved are in the nanometer range. In this range, hard and brittle materials produce continuous chips (known as **ductile-regime cutting**; see also the discussion of *ductile-regime grinding* in Section 9.5.3); deeper cuts tend to produce discontinuous chips.

   The machine tools for ultraprecision machining are built with very high precision and high machine, spindle, and workholding-device stiffness. These machines, some parts of which are made of structural materials with low thermal expansion and good dimensional stability, are located in a dust-free environment (*clean rooms*) where the temperature is controlled within a fraction of one degree. Vibrations from external and internal sources must be avoided as much as possible. Feed and position controls are made through laser metrology, and the machines are equipped with highly advanced computer control systems which also include thermal and geometric error-compensating features.

5. **Hard turning.** As described in Chapter 9, there are several other mechanical processes (particularly grinding) and nonmechanical methods of removing material economically from hard or hardened metals. However, it is still possible to apply traditional machining processes to hard metals and alloys by selecting an appropriate tool material and a machine tool with high stiffness. One common example is finish machining of heat-treated steel (45 HRC to 65 HRC) machine and automotive components, using polycrystalline cubic-boron-nitride (PcBN) cutting tools. Called *hard turning,* this process produces machined parts with good dimensional accuracy, surface finish, and surface integrity. It has been shown that it can compete successfully with grinding of the same components, from both technical and economic aspects. A comparative example of hard turning versus grinding is presented in Example 9.4.

6. **Cutting screw threads.** External threads are produced primarily by forming (*thread rolling;* see Fig. 6.44), which is the process by which the largest quantity of threaded parts is produced, and cutting as shown in Fig. 8.40k. When threads are produced externally or internally by cutting with a lathe-type tool, the process is called *thread cutting,* or *threading.* When the threads are cut internally with a special threaded tool (*tap*), the process is called *tapping.* External threads may also be cut with a die or by milling. Although it adds considerably to the cost of the operation, threads may also be ground for improved accuracy and surface finish.

   Automatic screw machines are designed for high-production-rate machining of screws and similar threaded parts. Because such machines are also capable of producing other components, they are generally called *automatic bar machines.* All operations on these machines are performed automatically, with tools clamped to a special turret. The bar stock is automatically fed forward through an opening in the headstock after each screw or part is machined to finished dimensions and then cut off. The machines may be equipped with single or multiple spindles, and capacities range from 3- to 150-mm ($\frac{1}{8}$- to 6-in.)-diameter bar stock.

8.9.3 Boring and boring machines

The basic boring operation is shown in Fig. 8.40h. *Boring* consists of producing circular internal profiles in hollow workpieces, or enlarging or finishing a hole made by another process, such as drilling. It is carried out using cutting tools that are similar to those used in turning. Note that the boring bar has to reach the full length of the bore, hence tool deflection and maintaining dimensional accuracy can be a problem. Therefore, the boring bar must be sufficiently stiff (made of a material with high elastic modulus, such as carbides) to minimize deflection and avoid vibration and chatter. Boring bars have been designed with capabilities for damping vibrations (see Section 8.12).

Although boring operations on relatively small workpieces can be carried out on a lathe, **boring mills** are used for large workpieces. These machine tools are either vertical or horizontal and are also capable of performing machining operations such as turning, facing, grooving, and chamfering. A *vertical boring machine* (Fig. 8.47) is similar to a lathe, but with a vertical axis of workpiece rotation. In *horizontal boring machines,* the workpiece is mounted on a table that can move horizontally, both axially and radially. The cutting tool is mounted on a spindle that rotates in the headstock and is capable of both vertical and longitudinal movements. Drills, reamers, taps, and milling cutters can also be mounted on the spindle.

8.9.4 Drilling, reaming, and tapping

One of the most common machining processes is *drilling*. **Drills** typically have a high length-to-diameter ratio (Fig. 8.48) and thus are capable of producing deep holes. However, they are somewhat flexible, depending on their length and diameter, and should be used with care in order to drill holes accurately as well as prevent the drill from breaking. Moreover, note that chips are produced within the workpiece and they move in a direction opposite to the axial movement of the drill. Consequently, chip disposal in drilling and the effectiveness of cutting fluids are important.

The most common drill is the standard-point **twist drill** (Fig. 8.48). The main features of the drill point are a *point angle*, a *lip-relief angle*, a *chisel-edge angle*, and

FIGURE 8.47 Schematic illustration of the components of a vertical boring mill.

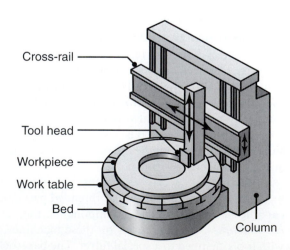

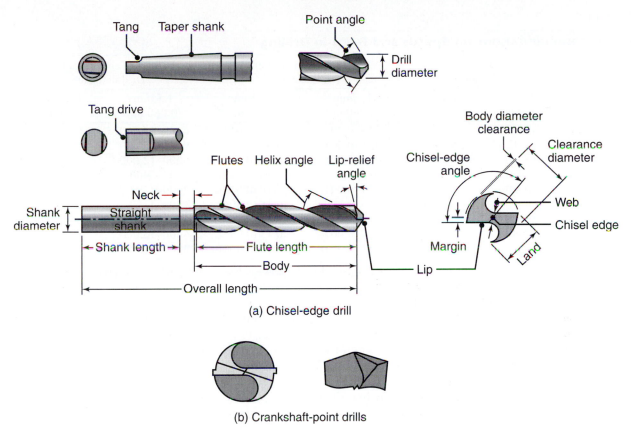

(a) Chisel-edge drill

(b) Crankshaft-point drills

FIGURE 8.48   Two common types of drills: (a) Chisel-edge drill. The function of the pair of margins is to provide a bearing surface for the drill against walls of the hole as it penetrates into the workpiece. Drills with four margins (double-margin) are available for improved drill guidance and accuracy. Drills with chip-breaker features are also available. (b) Crankshaft-point drills. These drills have good centering ability, and because chips tend to break up easily, they are suitable for producing deep holes.

a *helix angle*. The geometry of the drill tip is such that the normal rake angle and velocity of the cutting edge vary with the distance from the center of the drill. Other types of drills include the *step drill, core drill, counterboring* and *countersinking drills, center drill,* and *spade drill,* as illustrated in Fig. 8.49. *Crankshaft drills* have

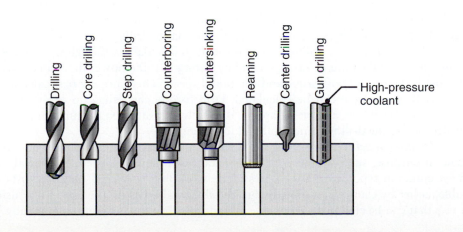

FIGURE 8.49   Various types of drills and drilling operations.

TABLE 8.11

General Recommendations for Speeds and Feeds in Drilling

Workpiece material	Surface speed m/min	Surface speed ft/min	Feed, mm/rev (in./rev) Drill diameter 1.5 mm (0.060 in.)	Feed, mm/rev (in./rev) Drill diameter 12.5 mm (0.5 in.)	Spindle speed (rpm) Drill diameter 1.5 mm (0.060 in.)	Spindle speed (rpm) Drill diameter 12.5 mm (0.5 in.)
Aluminum alloys	30–120	100–400	0.025 (0.001)	0.30 (0.012)	6400–25,000	800–3000
Magnesium alloys	45–120	150–400	0.025 (0.001)	0.30 (0.012)	9600–25,000	1100–3000
Copper alloys	15–60	50–200	0.025 (0.001)	0.25 (0.010)	3200–12,000	400–1500
Steels	20–30	60–100	0.025 (0.001)	0.30 (0.012)	4300–6400	500–800
Stainless steels	10–20	40–60	0.025 (0.001)	0.18 (0.007)	2100–4300	250–500
Titanium alloys	6–20	20–60	0.010 (0.0004)	0.15 (0.006)	1300–4300	150–500
Cast irons	20–60	60–200	0.025 (0.001)	0.30 (0.012)	4300–12,000	500–1500
Thermoplastics	30–60	100–200	0.025 (0.001)	0.13 (0.005)	6400–12,000	800–1500
Thermosets	20–60	60–200	0.025 (0.001)	0.10 (0.004)	4300–12,000	500–1500

Note: As hole depth increases, speeds and feeds should be reduced. Selection of speeds and feeds also depends on the specific surface finish required.

good centering ability, and because the chips they produce tend to break up easily, they are suitable for drilling deep holes. In gun drilling, a special drill is used for drilling deep holes; depth-to-diameter ratios of the holes produced can be 300 or higher. In the *trepanning* technique, a cutting tool produces a hole by removing a disk-shaped piece of material (*core*), usually from flat plates; thus, a hole is produced without reducing all the material to be removed to chips. The trepanning process can be used to make disks up to 150 mm (6 in.) in diameter from flat sheet or plate.

The *material-removal rate* in drilling is the ratio of volume removed to time; thus it can be expressed as

$$\text{MRR} = \frac{\pi D^2}{4} f N, \tag{8.40}$$

where D is the drill diameter, f is the feed (such as mm/revolution), and N is the rpm of the drill. See Table 8.11 for recommendations for speed and feed in drilling.

The *thrust force* in drilling is the force that acts in the direction of the hole axis. If this force is excessive, it can cause the drill to break or bend. The thrust force depends on factors such as the strength of the workpiece material, feed, rotational speed, cutting fluids, drill diameter, and drill geometry; hence accurate calculation of the thrust force on the drill has proven to be difficult. Experimental data are available as an aid in the design and use of drills and drilling equipment. Thrust forces in drilling typically range from a few newtons for small drills, to as high as 100 kN (22.5 klb) in drilling high-strength materials with large drills.

The *torque* during drilling is also difficult to calculate, although it can be estimated by using the data in Table 8.3 and by noting that the power dissipated during drilling is the product of torque and rotational speed. Therefore, by first calculating the MRR, the torque can be calculated. Drill torque can range up to 4000 N-m (3000 ft-lb). **Drill life**, as well as tap life, is usually measured by the number of holes drilled before the drill becomes dull and forces increase.

Drilling machines are used for drilling holes, tapping, reaming, and other general-purpose, small-diameter boring operations. They are generally vertical, the most common type being a **drill press**. The workpiece is placed on an adjustable table, either by clamping it directly into the slots and holes on the table or by using a vise that can be clamped to the table. The drill is then lowered, either manually by

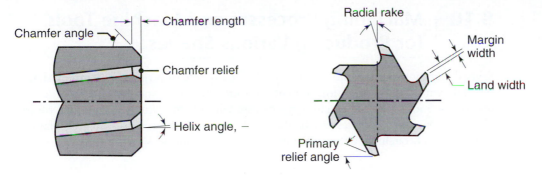

FIGURE 8.50   Terminology for a helical reamer.

hand wheel or by power feed at preset rates. In order to maintain proper cutting speeds, the spindle speed on drilling machines has to be adjustable to accommodate different sizes of drills. Drill presses are usually designated by the largest workpiece diameter that can be accommodated on the table. Sizes typically range from 150 to 1250 mm (6 to 50 in.).

Reaming and reamers.   *Reaming* is an operation that makes an existing hole dimensionally more accurate than can be achieved by drilling alone; it also improves the hole's surface finish. The most accurate holes are produced by the following sequence of operations: centering, drilling, boring, and reaming. For even better dimensional accuracy and surface finish, holes may be internally *ground* and *honed* (Chapter 9). A *reamer* (Fig. 8.50) is a multiple-cutting-edge tool with straight or helically fluted edges that removes very little material. The shanks may be straight or tapered, as in drills. The basic types of reamers are *hand* and *machine (chucking)* reamers; other types include *rose* reamers with cutting edges that have wide margins and no relief and *fluted, shell, expansion,* and *adjustable* reamers.

Tapping and taps.   Internal threads in workpieces can be produced by *tapping*. A **tap** is basically a threading tool with multiple cutting teeth (Fig. 8.51). Taps are generally available with three or four flutes; three-fluted taps are stronger because the flute is wider. *Tapered taps* are designed to reduce the torque required for tapping through holes, and bottoming taps are designed for tapping blind holes to their full depth. *Collapsible taps* are used for large-diameter holes. After tapping is completed, the tap is mechanically collapsed and removed from the hole, without having to be rotated. Tap sizes range up to 100 mm (4 in.).

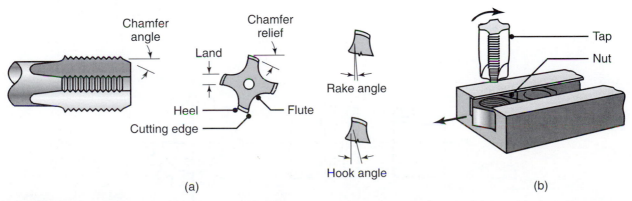

(a)   (b)

FIGURE 8.51   (a) Terminology for a tap and (b) illustration of tapping of steel nuts in high production.

8.10 | Machining Processes and Machine Tools for Producing Various Shapes

Several cutting processes and machine tools are capable of producing complex shapes typically with the use of multitooth cutting tools (Fig. 8.52 and Table 8.7). *Milling* is one of the most versatile machining processes, in which a multitooth cutter rotates along various axes with respect to the workpiece. Other processes include planing, shaping, and broaching, wherein flat as well as shaped surfaces are produced.

8.10.1 Milling operations

Milling includes a number of versatile machining operations that use a **milling cutter,** a multitooth tool that produces a number of chips per revolution, to machine a wide variety of part geometries. Parts such as the ones shown in Fig. 8.52 can be machined efficiently and repetitively using various milling cutters.

The basic types of milling operations are described below.

1. **Slab milling.** In this operation, also called **peripheral milling,** the axis of cutter rotation is parallel to the surface of the workpiece to be machined, as shown in Fig. 8.53a. The cutter, called a *plain mill,* is generally made of high-speed steel and has a number of teeth along its periphery, each tooth acting like a single-point cutting tool. Cutters used in slab milling may have *straight* or *helical* teeth, producing orthogonal or oblique cutting action, respectively. Figure 8.1c shows the helical teeth on a cutter.

   In **conventional milling,** also called *up milling,* the maximum chip thickness is at the end of the cut (Fig. 8.53b). The advantages of conventional milling are that tooth engagement is not a function of workpiece geometry, and contamination or

FIGURE 8.52 Typical parts and shapes produced by the machining processes described in Section 8.10.

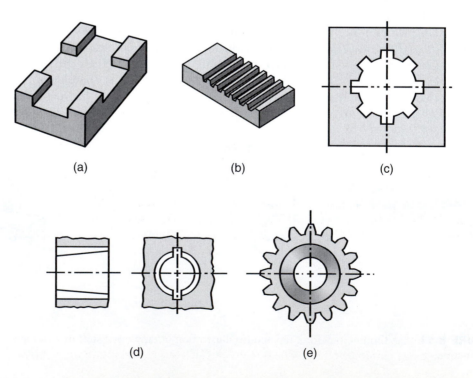

(a) (b) (c)

(d) (e)

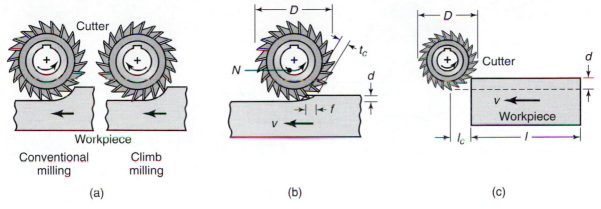

FIGURE 8.53   (a) Illustration showing the difference between conventional milling and climb milling, (b) slab-milling operation, showing depth of cut, d; feed per tooth, f; chip depth of cut, t_c; and workpiece speed, v, and (c) schematic illustration of cutter travel distance, l_c, to reach full depth of cut.

scale on the surface does not affect tool life; this process is the common method of milling. The cutting process is smooth, provided that the cutter teeth are sharp. There is, however, a tendency for the tool to chatter, and the workpiece has a tendency to be pulled upward; thus, proper clamping is important.

In **climb milling,** also called *down milling,* cutting starts at the thickest location of the chip. The advantage of climb milling is that the downward component of cutting forces holds the workpiece in place, particularly for slender parts. However, because of the resulting high-impact forces when the teeth engage the workpiece, this operation requires a rigid setup, and backlash must be eliminated in the table feed mechanism. Climb milling is not suitable for machining workpieces that have surface scale, such as hot-worked metals, forgings, and castings. The scale is hard and abrasive and thus causes excessive wear and damage to the cutter teeth, reducing tool life. Climb milling is recommended, in general, for maximum cutter life when CNC machine tools are used; a typical application is in finishing cuts on aluminum workpieces.

The cutting speed, V, in milling is the peripheral speed of the cutter, that is,

$$V = \pi D N, \tag{8.41}$$

where D is the cutter diameter and N is the rotational speed of the cutter (Fig. 8.53b). Note that the thickness of the chip in slab milling varies along its length, because of the relative longitudinal movement between cutter and workpiece. For a straight-tooth cutter, the approximate *undeformed chip thickness,* t_c (*chip depth of cut*), can be determined from the equation

$$t_c = 2f\sqrt{\frac{d}{D}}, \tag{8.42}$$

where f is the feed per tooth of the cutter (measured along the workpiece surface, that is, the distance the workpiece travels per tooth of the cutter, in mm/tooth or in./tooth), and d is the depth of cut. As t_c becomes greater, the force on the cutter tooth increases.

Feed per tooth is determined from the equation

$$f = \frac{v}{Nn}, \tag{8.43}$$

where v is the linear speed (feed rate) of the workpiece and n is the number of teeth on the cutter periphery. The dimensional accuracy of this equation can be

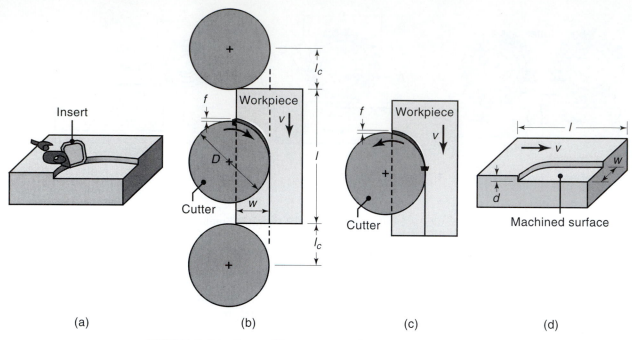

     (a)   (b)   (c)   (d)

FIGURE 8.54 Face-milling operation showing (a) action of an insert in face milling, (b) climb milling, (c) conventional milling, and (d) dimensions in face milling.

checked by substituting appropriate units for the individual terms; thus, (mm/tooth) = (m/min)(10^3 mm/m)/(rev/min)(number of teeth/rev), which is correct. The cutting time, t, is given by the expression

$$t = \frac{l + l_c}{v} \tag{8.44}$$

where l is the length of the workpiece (Fig. 8.53c) and l_c is the extent of the cutter's first contact with the workpiece. Based on the assumption that $l_c \ll l$ (although this relationship is not generally so), the *material-removal rate* is

$$\text{MRR} = \frac{lwd}{t} = wdv, \tag{8.45}$$

where w is the width of the cut, which, for a workpiece narrower than the length of the cutter, is the same as the width of the workpiece. The distance that the cutter travels in the noncutting-cycle operation is an important economic consideration and should be minimized.

2. **Face milling.** In this operation, the cutter is mounted on a spindle with an axis of rotation perpendicular to the workpiece surface and removes material in the manner shown in Fig. 8.54a. The cutter rotates at a speed of N, and the workpiece moves along a straight path at a linear speed of v. When the cutter rotates in the direction shown in Fig. 8.54b, the operation is *climb milling*; when it rotates in the opposite direction (Fig. 8.54c), it is *conventional milling*.

    Because of the relative movement between the cutting teeth and the workpiece, a face-milling cutter leaves *feed marks* on the machined surface, much as in turning operations. Surface roughness depends on insert corner geometry and feed per tooth [see also Eqs. (8.35) to (8.37)].

    The terminology for a face-milling cutter and its various angles is given in Fig. 8.55. The side view is shown in Fig. 8.56, where it can be noted that, as in turning operations, the *lead angle* of the insert in face milling has a direct

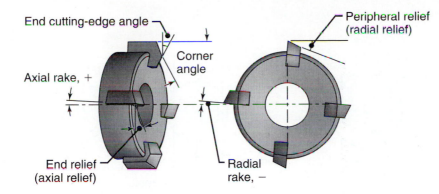

FIGURE 8.55   Terminology for a face-milling cutter.

influence on the *undeformed chip thickness*. As the lead angle (positive as shown in the figure) increases, the undeformed chip thickness (thus, also the thickness of the actual chip) decreases, and the length of contact increases. The range of lead angles for most face-milling cutters is typically from 0° to 45°. Note that the cross-sectional area of the undeformed chip remains constant. The lead angle also influences forces in milling. It can be seen that as the lead angle decreases, there is less and less of a vertical force (axial force on the cutter spindle) component.

A wide variety of milling cutters is available. The cutter diameter should be chosen so that it will not interfere with fixtures, workholding devices, and other components in the setup. In a typical face-milling operation, the ratio of the cutter diameter, D, to the width of cut, w, should be no less than 3:2. The cutting tools are usually carbide or high-speed-steel inserts and are mounted on the cutter body. (See Fig. 8.55.)

The relationship of cutter diameter and insert angles and their position relative to the surface to be milled is important, because they determine the angle at which an insert enters and exits the workpiece. Note in Fig. 8.54b for climb milling that if the insert has zero axial and radial rake angles (see Fig. 8.55), then the rake face of the insert engages the workpiece directly (thus subjecting it to a high impact force). However, as can be seen in Fig. 8.57a and b, the same insert engages the workpiece at different angles, depending on the relative positions of the cutter and the workpiece. In Fig. 8.57a, the tip of the insert makes the first contact, and hence there is potential for the cutting edge

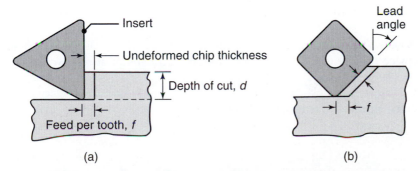

FIGURE 8.56   The effect of lead angle on the undeformed chip thickness in face milling. Note that as the lead angle increases, the undeformed chip thickness (and hence the thickness of the chip) decreases, but the length of contact (and hence the width of the chip) increases. Note that the insert must be sufficiently large to accommodate the increase in contact length.

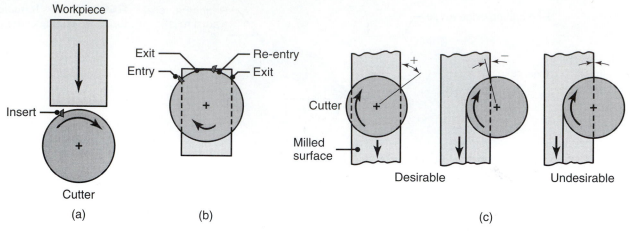

FIGURE 8.57   (a) Relative position of the cutter and the insert as it first engages the workpiece in face milling, (b) insert positions at entry and exit near the end of cut, and (c) examples of exit angles of the insert, showing desirable (positive or negative angle) and undesirable (zero angle) positions. In all figures, the cutter spindle is perpendicular to the page.

to chip off. In Fig. 8.57b, on the other hand, the contacts (at entry, reentry, and the two exits) are at a certain angle and away from the tip of the insert. There is, therefore, less of a tendency for the insert to chip off or break, because the force on the insert increases and decreases gradually. Note from Fig. 8.55 that the radial and axial rake angles also will affect insert chipping.

The exit angles for various cutter positions in face milling are shown in Fig. 8.57c. Note that in the first two examples, the insert exits the workpiece at an angle, whereby the force on the insert diminishes to zero at a slower rate (desirable) than in the third example, where the insert exits suddenly (undesirable).

EXAMPLE 8.6 Calculation of material-removal rate, power required, and cutting time in face milling

Refer to Fig. 8.54, and assume that $D = 150$ mm, $w = 60$ mm, $l = 500$ mm, $d = 3$ mm, $v = 0.6$ m/min, and $N = 100$ rpm. The cutter has 10 inserts and the workpiece material is a high-strength aluminum alloy. Calculate the material-removal rate, cutting time, and feed per tooth, and estimate the power required.

Solution. The cross section of the cut is $wd = (60)(3) = 180$ mm^2. Since the workpiece speed v is 0.6 m/min = 600 mm/min, the material-removal rate is

$$\text{MRR} = (180)(600) = 108{,}000 \text{ mm}^3/\text{min.}$$

The cutting time is given by Eq. (8.44) as

$$t = \frac{l + l_c}{v}.$$

Note from Fig. 8.54 that for this problem,

$$l_c^2 + \left(\frac{D}{2} - w\right)^2 = \left(\frac{D}{2}\right)^2$$

so that $l_c = \sqrt{Dw - w^2}$ or 73.5 mm. Thus, the cutting time is

$$t = \frac{(500 + 73.5)(60)}{600} = 57.3 \text{ s} = 0.955 \text{ min.}$$

We obtain the feed per tooth from Eq. (8.43). Noting that $N = 100$ rpm = 1.67 rev/s,

$$f = \frac{10}{(1.67)(10)} = 0.6 \text{ mm/tooth.}$$

For this material, the unit power can be taken from Table 8.3 to be 1.1 W-s/mm^3; hence, the power can be estimated as

$$\text{Power} = (1.1)(1800) = 1980 \text{ W} = 1.98 \text{ kW.}$$

3. **End milling.** The cutter in *end milling* is shown in Fig. 8.1d; it has either straight or tapered shanks for smaller and larger cutter sizes, respectively. The cutter usually rotates on an axis perpendicular to the workpiece, although it can be tilted to produce tapered surfaces as well. Flat surfaces as well as various profiles can be produced by end milling. Some end mills have cutting teeth also on their end faces; these teeth can be used as a drill to start a cavity. End mills are also available with hemispherical ends (*ball-nose end mills*) for producing curved surfaces, as in making molds and dies. *Hollow end mills* have *internal* cutting teeth and are used for machining the cylindrical surface of solid round workpieces, as in preparing stock with accurate diameters for automatic bar machines. End mills are made of high-speed steels or have carbide inserts.

   Section 8.8 has described high-speed machining and its typical applications. One of the more common applications is **high-speed milling,** using an end mill and with the same general requirements regarding, for example, the stiffness of machines and workholding devices, described earlier. A typical application is milling aluminum-alloy aerospace components and honeycomb structures, with spindle speeds on the order of 20,000 rpm. Another application is in *die sinking,* i.e., producing cavities in die blocks. Chip collection and disposal can be a significant problem in these operations, because of the high rate of material removal.

4. **Other milling operations and cutters.** Several other types of milling operations and cutters are used to machine various surfaces. In *straddle milling,* two or more cutters are mounted on an arbor and are used to machine two *parallel* surfaces on the workpiece (Fig. 8.58a). *Form milling* is used to produce curved profiles, using cutters that have specially shaped teeth (Fig. 8.58b); such cutters are also used for cutting gear teeth (see Section 8.10.7).

   Circular cutters are used for slotting and slitting operations. The teeth may be staggered, as in a saw blade (see Section 8.10.5), to provide clearance for the cutter in making deep slots. *Slitting saws* are relatively thin, usually less than 5 mm $\left(\frac{3}{16} \text{ in.}\right)$. *T-slot cutters* are used to mill T-slots, such as those in machine-tool worktables for clamping of workpieces. The slot is first milled with an end mill; a T-slot cutter then cuts the complete profile of the slot in one pass. *Key-seat cutters* are used to make the semicylindrical (Woodruff) key sets for shafts. *Angle-milling cutters,* either with a single angle or double angles, are used to produce tapered surfaces with various angles.

   Shell mills are hollow inside and are mounted on a shank, thus allowing the same size of shank to be used for different-sized cutters. The applications of shell mills is similar to that of end mills. Milling with a single cutting tooth

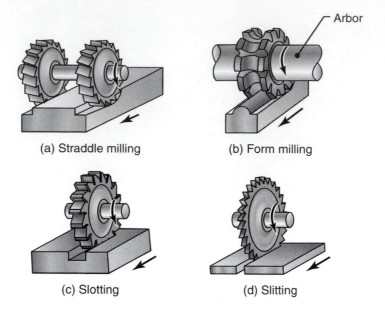

(a) Straddle milling (b) Form milling

(c) Slotting (d) Slitting

mounted on a high-speed spindle is known as *fly cutting* and can used in simple face-milling and boring operations. The cutting tool can be shaped as a single-point tool and can be clamped in various radial positions on the spindle.

5. **Toolholders.** Milling cutters are classified into two basic types. *Arbor cutters* are mounted on an arbor, such as for slab, face, straddle, and form milling. In *shank-type cutters,* the cutter and the shank are one piece. The most common examples of shank cutters are end mills. Whereas small end mills have straight shanks, larger end mills have tapered shanks for better clamping to resist the higher forces and torque involved. Cutters with straight shanks are mounted in *collet chucks* or special end-mill holders; cutters with tapered shanks are mounted in tapered toolholders. In addition to mechanical means, hydraulic tool holders and arbors also are available. The stiffness of cutters and toolholders is important for surface quality and reducing vibration and chatter in milling operations (Section 8.12). Conventional tapered toolholders have a tendency to wear and bellmouth under the radial forces in milling.

6. **Milling machines and process capabilities.** In addition to the various characteristics of milling processes described thus far, milling capabilities involve parameters such as production rate, surface finish, dimensional tolerances, and cost considerations. Table 8.12 gives the typical conventional ranges of feeds and cutting speeds for milling. Cutting speeds vary over a wide range, from 30 to 3000 m/min (90 to 10,000 ft/min) depending on workpiece material, cutting-tool material, and process parameters.

Because of their capability to perform a variety of machining operations, milling machines are among the most versatile and useful machine tools, and a wide selection of machines with numerous features is available. Used for general-purpose applications, **column-and-knee-type** machines are the most common milling machines. The spindle, to which the milling cutter is attached, may be *horizontal* (Fig. 8.59a) for slab milling or *vertical* for face and end milling, boring, and drilling operations (Fig. 8.59b). The components are moved manually or by power and with various CNC controls.

In **bed-type** machines, the worktable is mounted directly on the bed (which replaces the knee) and can move only longitudinally. These machines are not as versatile as others but have high stiffness and are used for high-production-rate

TABLE 8.12

Approximate Range of Recommended Cutting Speeds for Milling Operations

	Cutting speed	
Workpiece material	m/min	ft/min
Aluminum alloys	300–3000	1000–10,000
Cast iron, gray	90–1300	300–4200
Copper alloys	90–1000	300–3300
High-temperature alloys	30–550	100–1800
Steels	60–450	200–1500
Stainless steels	90–500	300–1600
Thermoplastics and thermosets	90–1400	300–4500
Titanium alloys	40–150	130–500

Note: (a) These speeds are for carbides, ceramic, cermets, and diamond cutting tools. Speeds for high-speed-steel tools are lower than those indicated in this table.
(b) Depths of cut, d, are generally in the range of 1–8 mm (0.04–0.3 in.).
(c) Feeds per tooth, f, are generally in the range of 0.08–0.46 mm/rev (0.003–0.018 in./rev).

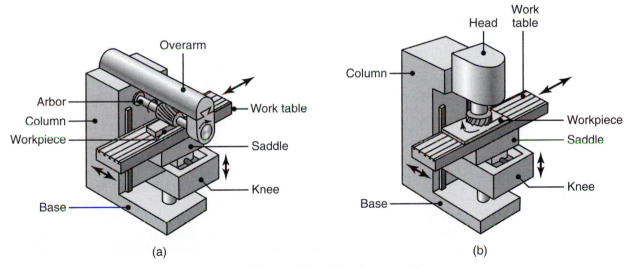

FIGURE 8.59   (a) Schematic illustration of a horizontal-spindle column-and-knee-type milling machine and (b) schematic illustration of a vertical-spindle column-and-knee-type milling machine. *Source:* After G. Boothroyd.

work. The spindles may be horizontal or vertical and of duplex or triplex types, that is, with two or three spindles for simultaneously milling two or three workpiece surfaces, respectively. **Planer-type** machines, which are similar to bed-type machines, are equipped with several heads and cutters to machine various surfaces. They are used for heavy workpieces and are more efficient than planers (see Section 8.10.2) when used for similar operations.

Rotary-table machines are similar to vertical milling machines and are equipped with one or more heads for face milling. *Profile-milling machines* have five axis movements. Using tracer fingers, *duplicating machines* (copy-milling machines) reproduce parts from a master model. They are typically used in the automotive and aerospace industries for machining complex parts and dies (die sinking), although they are replaced largely with CNC machines. These machine tools are versatile and capable of milling, drilling, boring, and tapping with consistent accuracy.

8.10.2 Planing and planers

Planing is a relatively simple machining process by which flat surfaces, as well as various cross sections with grooves and notches, are produced along the length of the workpiece. Planing is usually done on large workpieces, as large as 25 m × 15 m (75 ft × 45 ft). In a typical *planer,* the workpiece is mounted on a table that travels along a straight path. A horizontal cross rail, which can be moved vertically along the ways in the column, is equipped with one or more tool heads. The cutting tools are attached to the heads, and machining is done along a straight path. Because of the reciprocating motion of the workpiece, elapsed noncutting time during the re-turn stroke is significant, both in planing and in shaping (see Section 8.9.3). Conse-quently, planers are not efficient or economical, except for low-quantity production. Efficiency of the operation can be improved by equipping planers with toolholders and tools that cut in both directions of table travel.

8.10.3 Shaping and shapers

Machining by *shaping* is basically the same as in planing. In a *horizontal shaper,* the tool travels along a straight path, and the workpiece is stationary. The cutting tool is attached to the tool head, which is mounted on the ram. The ram has a reciprocat-ing motion, and in most machines, cutting is done during the forward movement of the ram (*push cut*). In others, it is done during the return stroke of the ram (*draw cut*). *Vertical shapers (slotters)* are used for machining notches and keyways. Shapers are also capable of producing complex shapes, such as machining helical impellers, in which the workpiece is rotated during the cut, using a master cam. Because of their low production rate, shapers are now generally used only in tool-rooms, job shops, and for repair work.

8.10.4 Broaching and broaching machines

The *broaching* operation is similar to shaping and is used to machine internal and external surfaces (Fig. 8.60), such as holes of circular, square, or irregular section; keyways; teeth of internal gears; multiple spline holes; and flat surfaces. A *broach* (Fig. 8.61) is, in effect, a long multitooth cutting tool that makes successively deeper cuts. Note that the total depth of material removed in one stroke of the broach is the

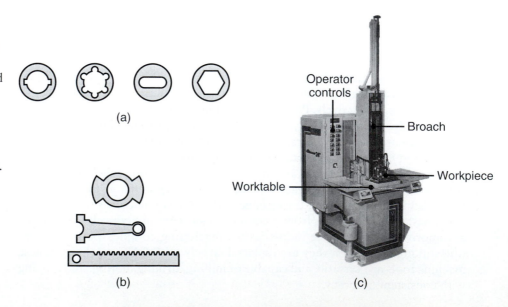

FIGURE 8.60 (a) Typical parts finished by internal broaching; (b) parts finished by surface broaching. The heavy lines indicate broached surfaces. (c) A vertical broaching machine. *Source:* (a) and (b) Courtesy of General Broach and Engineering Company; (c) Courtesy of Ty Miles, Inc.

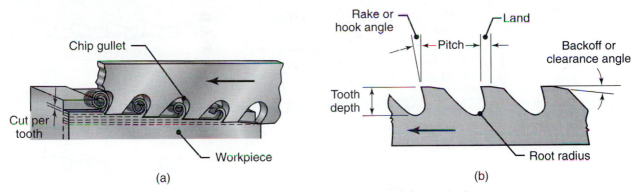

(a)   (b)

FIGURE 8.61   (a) Cutting action of a broach, showing various features and (b) terminology for a broach.

sum of the depths of cut of each tooth. A broach can remove material as deep as 6 mm (0.25 in.) in one stroke. Broaching can produce parts with good surface finish and dimensional accuracy; thus it competes favorably with other machining processes to produce similar shapes. Although broaches can be expensive, the cost is justified because of their use for high-quantity production runs.

The terminology for a broach is given in Fig. 8.61b. The rake (hook) angle depends on the material being machined and usually ranges between 0° and 20°. The clearance angle is typically 1° to 4°; finishing teeth have smaller angles. Too small a clearance angle causes rubbing of the cutter teeth against the broached surface. The pitch of the teeth depends on factors such as length of the workpiece (length of cut), tooth strength, and size and shape of chips. The tooth depth and pitch must be sufficiently large to accommodate the chips produced during broaching, particularly for long workpieces, and at least two teeth should be in contact with the workpiece at all times (similar to sawing; see Section 8.10.5).

Broaches are available with a variety of tooth profiles, including some that incorporate chip breakers (Fig. 8.62). Note that the cutting teeth on broaches have three regions: roughing, semifinishing, and finishing. Broaches are also made round with circular cutting teeth and are used to enlarge holes (Fig. 8.62). Irregular internal shapes are usually broached by starting with a round hole in the workpiece, produced by processes such as drilling or boring.

In **turn broaching** of crankshafts, the workpiece rotates between centers, the broach is equipped with multiple inserts, and it passes tangentially across the part. The operation is thus a combination of broaching and skiving. There are also machines that broach a number of crankshafts simultaneously; main bearings for engines, for example, are broached in this way.

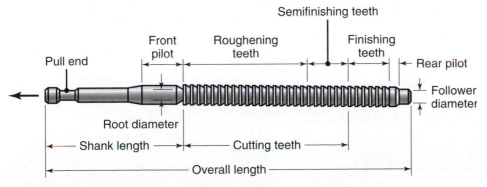

FIGURE 8.62   Terminology for a pull-type internal broach, typically used for enlarging long holes.

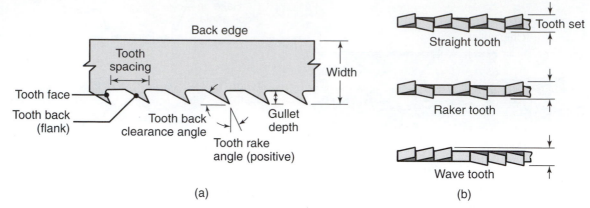

(a)

(b)

FIGURE 8.63 (a) Terminology for saw teeth and (b) types of saw teeth, staggered to provide clearance for the saw blade to prevent binding during sawing.

Broaching machines either pull or push the broaches and are available in horizontal or vertical designs. *Push broaches* are usually shorter, generally in the range of 150–350 mm (6–14 in.) long. *Pull broaches* tend to straighten the hole, whereas pushing permits the broach to follow any irregularity of the leader hole. Horizontal machines are capable of longer strokes than are vertical machines. Several types of broaching machines are available, some with multiple heads, thus allowing a variety of shapes and parts to be produced, including helical splines and rifled gun barrels. Sizes range from machines for making needlelike parts to those used for broaching gun barrels. The capacities of broaching machines are as high as 0.9 MN (100 tons) of pulling force.

8.10.5 Sawing and saws

Sawing is a cutting operation in which the tool consists of a series of small teeth, with each tooth removing a small amount of material. Saws are used for all metallic and nonmetallic machinable materials, and they are also capable of producing various shapes. The width of cut (**kerf**) in sawing is usually narrow, and hence sawing wastes little material. Typical saw-tooth and saw-blade configurations are illustrated in Fig. 8.63. Tooth spacing is usually in the range of 0.08 to 1.25 teeth per mm (2 to 32 teeth per in.).

Saw blades are generally made from carbon and high-speed steels, but steel blades with tips of carbide or high-speed steel are used for sawing harder materials (Fig. 8.64b). In order to prevent the saw from binding and rubbing during sawing, the teeth are alternately set in opposite directions so that the kerf is wider than the blade (Fig. 8.63b). At least two or three teeth should always be engaged with the workpiece in order to prevent snagging (catching of the saw tooth on the workpiece). This requirement is the reason that sawing thin materials satisfactorily is difficult or impossible. Cutting speed in sawing usually ranges up to 1.5 m/s (300 ft/min), with

FIGURE 8.64 (a) High-speed-steel teeth welded on a steel blade and (b) carbide inserts brazed to blade teeth.

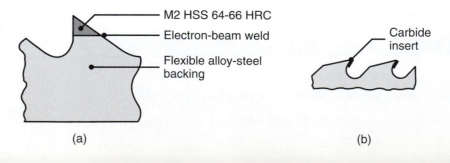

(a)

(b)

lower speeds used for high-strength metals. Cutting fluids are generally used to improve the quality of cut and the extend life of the saw.

Hacksaws have straight blades and involve reciprocating motions; they may be manually or power operated. *Circular saws (cold saws)* are generally used for high-production-rate sawing of large cross sections. Band saws have long, flexible, continuous blades and provide continuous cutting action. Blades and high-strength wire can be coated with diamond powder (*diamond-edged blades* and *diamond saws*). They are suitable for sawing hard metallic, nonmetallic, and composite materials.

Friction sawing.   *Friction sawing* is a process in which a mild-steel blade or disk rubs against the workpiece at speeds of up to 125 m/s (25,000 ft/min). The frictional energy is converted into heat, which rapidly softens a narrow zone in the workpiece. The action of the blade or disk (which is sometimes provided with teeth or notches) pulls and ejects the softened metal from the cutting zone. The heat generated in the workpiece produces a *heat-affected zone* (as in welding; see also Section 12.6) on the cut surfaces; thus, their properties can be adversely affected. Because only a small portion of the blade is engaged with the workpiece at any time, the blade cools rapidly as it passes through the air. Friction sawing is suitable for hard ferrous metals and fiber-reinforced plastics, but not for nonferrous metals because they have a tendency to stick to the blade. Disks for friction sawing as large as 1.8 m (6 ft) in diameter are used to cut off large steel sections in rolling mills. Friction sawing also can be used to remove flash from castings.

8.10.6 Filing

Filing is small-scale removal of material from a surface, corner, edge, or hole. First developed about 1000 B.C., files are usually made of hardened steels and are available in a variety of cross sections, including flat, round, half-round, square, and triangular. Files are available with various tooth forms and grades of coarseness, such as smooth cut, second cut, and bastard (intermediate) cut. Although filing is usually done by hand, various machines with automatic features are available for high production rates, with files reciprocating at up to 500 strokes/min.

Band files consist of file segments, each about 75 mm (3 in.) long, that are riveted to flexible steel bands and used in a manner similar to band saws. *Disk-type files* also are available. *Rotary files* and **burs** are used for special applications; they are usually conical, cylindrical, or spherical in shape and have various tooth profiles. The rotational speeds for burs range from 1500 rpm for cutting steel with large burs, to as high as 45,000 rpm for cutting magnesium using small burs.

8.10.7 Gear manufacturing by machining

Gears can be manufactured by casting, forging, extrusion, drawing, thread rolling, powder metallurgy, and blanking (for making thin gears such as those used in watches and small clocks). Most gears are machined and ground for better dimensional control and surface finish. Nonmetallic gears are usually made by injection molding and casting (Chapter 10).

In **form cutting** of gears, the cutting tool is similar to a form-milling cutter (see Fig. 8.58b) which is in the shape of the space between the gear teeth. The cutter travels axially along the length of the gear tooth at the appropriate depth to produce the gear tooth. After each tooth is cut, the cutter is withdrawn, the gear blank is rotated (indexed), and the cutter proceeds to cut another tooth. The process continues until all teeth are cut. Broaching also can be used to produce gear teeth and is particularly applicable to internal teeth. The broaching process is rapid and produces fine surface finish with high dimensional accuracy. However, because broaches are expensive and a separate broach is required for each size of gear, this method is suitable mainly for high-quantity production.

In **gear generating,** the tool may consist of the following:

1. *Pinion-shaped cutter.* The pinion-shaped cutter can be considered as one of the gears in a conjugate pair of gears and the other is the gear blank (Fig. 8.65a). This type of cutter is used on machines called **gear shapers** (Fig. 8.65b). The cutter has an axis parallel to that of the gear blank and rotates slowly with the blank at the same pitch-circle velocity with an axial reciprocating motion. A train of gears provides the required relative motion between the cutter shaft and the gear-blank shaft. Cutting may take place at either the downstroke or the upstroke of the machine. Because the clearance required for cutter travel is small, gear shaping is suitable for gears that are located close to obstructing surfaces such as flanges (as in the gear blank shown in Fig. 8.65b). The process can be used for low-quantity as well as high-quantity production.

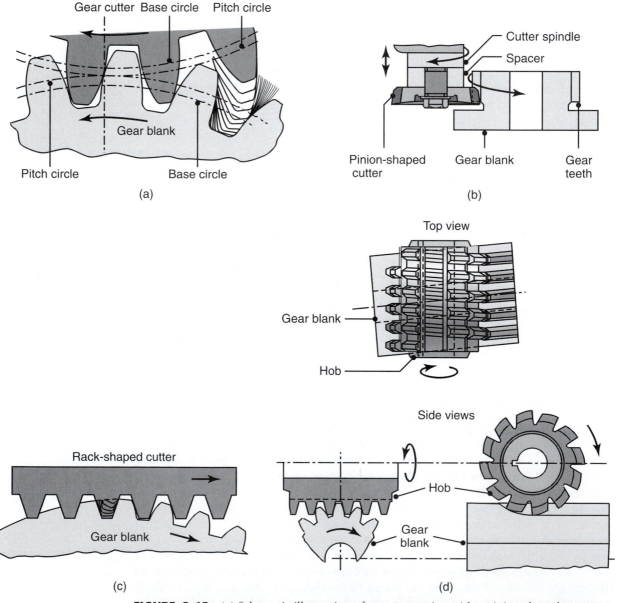

FIGURE 8.65 (a) Schematic illustration of gear generating with a pinion-shaped gear cutter, (b) schematic illustration of gear generating in a gear shaper, using a pinion-shaped cutter; note that the cutter reciprocates vertically. (c) Gear generating with a rack-shaped cutter and (d) three views of gear cutting with a hob. *Source:* After E.P. DeGarmo.

2. *Rack-shaped straight cutter.* On a **rack shaper,** the generating tool is a segment of a rack (Fig. 8.65c), which reciprocates parallel to the axis of the gear blank. Because it is not practical to have more than 6 to 12 teeth on a rack cutter, the cutter must be disengaged at suitable intervals and returned to the starting point; the gear blank, meanwhile, remains fixed.

3. *Hob.* A gear-cutting **hob** has basically the shape of a worm, or screw, that has been made into a gear-generating tool by machining a series of longitudinal slots, or gashes, into it to produce the cutting teeth (Fig. 8.65d). When hobbing a spur gear, the angle between the hob and gear-blank axes is 90° minus the lead angle at the hob threads. All motions in hobbing are rotary, and the hob and gear blank rotate continuously, as in two gears meshing, until all teeth are cut.

Gears machined by any of the three processes described may not have sufficiently high dimensional accuracy and surface finish of gear teeth for certain applications, thus requiring several *finishing* operations, including shaving, burnishing, grinding, honing, and lapping. (See Chapter 9.) Modern gear-manufacturing machines are computer controlled, and multiaxis computer-controlled machines have capabilities of machining many types of gears, using indexable cutters.

8.11 | Machining and Turning Centers

In describing individual machining processes and machine tools thus far, it will be noted that each machine, regardless of how highly it is automated, is designed to perform basically one type of operation, such as turning, milling, boring, and drilling. However, many parts have a variety of features and surfaces that require different types of machining operations to certain specific dimensional tolerances and surface finish; see, for example, Figs. 8.26 and 9.27. None of the processes and machine tools described thus far could individually produce these parts.

Traditionally, the required operations have been performed by moving the part from one machine tool to another, and then another, until all machining is completed. This method is a viable manufacturing method that can be highly automated and is the principle behind **transfer lines,** consisting of numerous machine tools arranged in a certain sequence, as described in Section 14.2.4. The workpiece, such as an engine block, moves from station to station, with a particular machining operation performed at each station; the part is then transferred automatically to the next machine for another operation, and so on. Transfer lines are commonly used in high-volume or mass production.

There are situations and products, however, for which transfer lines are not feasible or economical, particularly when the types of products to be machined change rapidly due to market demands. An important concept, developed in the late 1950s, is machining centers. A **machining center** is a computer-controlled machine tool capable of performing a variety of machining operations on different surfaces, locations, and orientations on a workpiece (Fig. 8.66). In general, the workpiece is stationary, and the cutting tools rotate, as they do in milling and drilling. The development of machining centers is intimately related to advances in **computer control of machine tools,** details of which are described in Chapter 14. Recall that, as an example of advances in modern lathes, Fig. 8.45 illustrated a numerically controlled lathe (**turning center**), with two turrets that carry several cutting tools for operations such as turning, facing, boring, and threading.

The workpiece in a machining center is placed on a **pallet** (Fig. 8.66) that can be oriented in three principal directions, as well as rotated around one or more axes on the pallet. Thus, after a particular cutting operation is completed, the workpiece does not have to be moved to another machine for subsequent operations, say,

FIGURE 8.66 A horizontal-spindle machining center, equipped with an automatic tool changer. Tool magazines in such machines can store as many as 200 cutting tools, each with its own holder. *Source:* Courtesy of Cincinnati Machine.

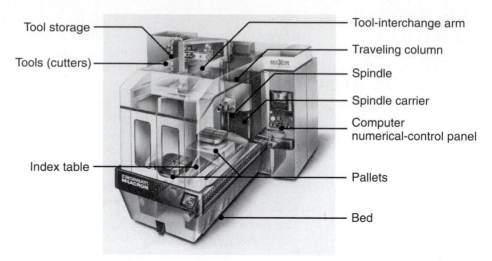

Tool storage

Tools (cutters)

Index table

Tool-interchange arm

Traveling column

Spindle

Spindle carrier

Computer numerical-control panel

Pallets

Bed

drilling, reaming, and tapping. In other words, the tools and the machine are brought to the workpiece. After all the machining operations are completed, the pallet automatically moves away, carrying the finished workpiece, and another pallet containing another workpiece to be machined is brought into position by **automatic pallet changers.** All movements in the machine are guided by computer control, with pallet-changing cycle times on the order of 10–30 s. Pallet stations are available with multiple pallets that serve the machining center. These machines are also available with various automatic components, such as loading and unloading devices.

Machining centers are equipped with a **programmable automatic tool changer.** Depending on the design, as many as 200 cutting tools can be stored in a magazine, drum, or chain (*tool storage*), and auxiliary tool storage is available on some special machining centers for complex operations. The cutting tools are automatically selected, with random access for the shortest route to the machine spindle. A common design feature is a **tool-exchange arm** that swings around to pick up a particular tool (each tool has its own toolholder) and places it in the spindle. The tools are identified by coded tags, bar codes, or memory chips applied directly to the toolholders. Tool changing times are typically on the order of a few seconds.

Machining centers also can be equipped with a **tool-checking** and/or **part-checking station** that feeds information to the computer in order to compensate for any variations in tool settings or tool wear. **Touch probes** are available that can determine reference surfaces of the workpiece, for selection of tool setting, and for on-line inspection of parts being machined. Several surfaces can be contacted, and their relative positions are determined and stored in the database of the computer software. The data are then used for programming tool paths and compensating, for example, for tool length and diameter, and tool wear.

8.11.1 Types of machining and turning centers

There are several designs for machining centers, but the two basic types are vertical spindle and horizontal spindle; many machines have the capability of using both axes. The maximum dimensions that the cutting tools can reach around a workpiece in a machining center are known as the *work envelope;* this term was first used in connection with industrial robots. (See Section 14.7.)

1. **Vertical-spindle machining centers,** or *vertical machining centers,* are suitable for performing various machining operations on flat surfaces with deep cavities, such as in mold and die making. Because the thrust forces in vertical

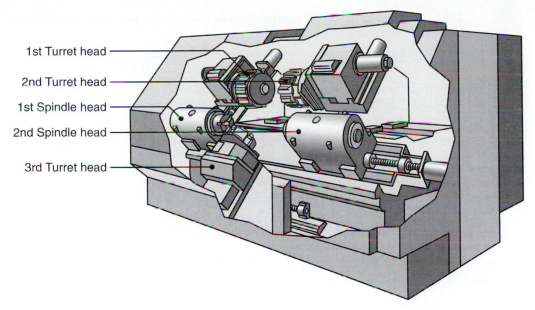

1st Turret head

2nd Turret head

1st Spindle head

2nd Spindle head

3rd Turret head

FIGURE 8.67 Schematic illustration of a computer numerical-controlled turning center. Note that the machine has two spindle heads and three turret heads, making the machine tool very flexible in its capabilities. *Source:* Courtesy of Hitachi Seiki Co., Ltd.

machining are directed downward, these machines have high stiffness and produce parts with good dimensional accuracy. Moreover, they are generally less expensive than horizontal-spindle machines.

2. **Horizontal-spindle machining centers,** or *horizontal machining centers* (see Fig. 8.66), are suitable for large or tall workpieces requiring machining on a number of their surfaces. The pallet can be rotated on different axes and to various angular positions. Another category of horizontal-spindle machines is **turning centers**, which are computer-controlled *lathes* with several features. Figure 8.67 shows a three-turret computer-numerical-control turning center. Note that this machine has two horizontal spindles and three turrets, which are equipped with a variety of cutting tools to perform several operations on a workpiece that rotates.

3. **Universal machining centers** are equipped with both vertical and horizontal spindles. They have a variety of features and are capable of machining all surfaces of a workpiece (vertical, horizontal, and diagonal), hence the term *universal*.

8.11.2 Characteristics and capabilities of machining centers

The major characteristics of machining centers are described as follows:

- They are capable of handling a variety of part sizes and shapes efficiently, economically, and with consistently high dimensional tolerances, on the order of ± 0.0025 mm (0.0001 in.).

- The machines are versatile, having as many as six axes of linear and angular movements as well as the capability of quick changeover from one type of product to another. Thus, the need for a variety of machine tools and a large amount of floor space is significantly reduced.

- The time required for loading and unloading workpieces, changing tools, gaging of workpieces being machined, and troubleshooting is reduced, thus improving productivity, reducing labor requirements (particularly the need for skilled labor), and thus minimizing overall costs.

- Machining centers are highly automated and relatively compact, so that one operator can often attend to two or more machines at the same time.
- The machines are equipped with tool-condition monitoring devices (see Section 8.3.5) for detecting tool breakage and wear, as well as probes for compensation for tool wear and for tool positioning.
- In-process and postprocess gaging and inspection of machined workpieces are now standard features of machining centers.

Machining centers are available in a wide variety of sizes and with a large selection of features, and their cost ranges from about $50,000 to $1 million and higher. Typical capacities range up to 75 kW (100 hp), and maximum spindle speeds are usually in the range of 4000–8000 rpm, but some are as high as 75,000 rpm for special applications and with small-diameter cutters. Some pallets are capable of supporting workpieces that weigh as much as 7000 kg (15,000 lb), although higher capacities are available for special applications.

8.11.3 Reconfigurable Machines and Systems

The need for flexibility of manufacturing operations has led to the more recent concept of *reconfigurable machines*, consisting of various modules that satisfy different functional requirements. The term *reconfigurable* stems from the fact that, using advanced computer hardware and reconfigurable controllers and utilizing advances in information management technologies, the machine components can be arranged and rearranged quickly in a number of configurations to meet specific production demands. Based on the typical machine-tool structure of a three-axis machining center, Fig. 8.68 shows an example of how the machine can be reconfigured to become a

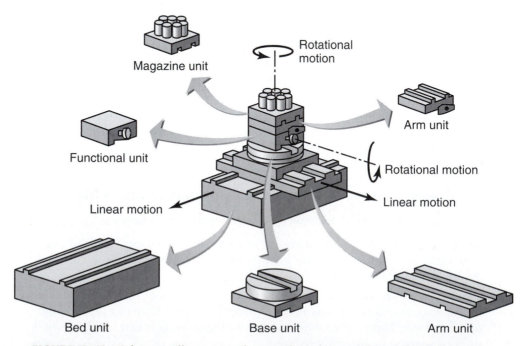

FIGURE 8.68 Schematic illustration of a reconfigurable modular machining center, capable of accommodating workpieces of different shapes and sizes, and requiring different machining operations on their various surfaces. *Source:* After Y. Koren.

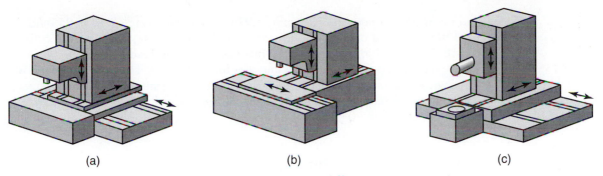

(a)   (b)   (c)

FIGURE 8.69   Three schematic illustrations of assembly of different components of a reconfigurable machining center. *Source:* After Y. Koren.

modular machining center. With the resulting flexibility, the machine can perform different machining operations while accommodating various workpiece sizes and part geometries. Another example of this flexibility is given in Fig. 8.69, where a five-axis (three linear and two rotational movements) machine can be reconfigured by assembling different modules.

EXAMPLE 8.7   Machining outer bearing races on a turning center

Outer bearing races (see Fig. 8.70) are machined on a turning center. The starting material is hot-rolled 52100 steel tube with an outside diameter of 91 mm (3.592 in.)

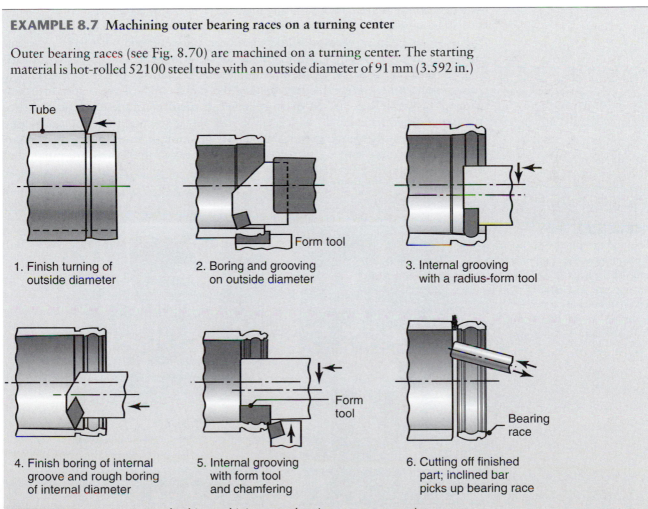

1. Finish turning of outside diameter

2. Boring and grooving on outside diameter

3. Internal grooving with a radius-form tool

4. Finish boring of internal groove and rough boring of internal diameter

5. Internal grooving with form tool and chamfering

6. Cutting off finished part; inclined bar picks up bearing race

FIGURE 8.70   Sequences involved in machining outer bearing races on a turning center.

and an inside diameter of 75.5 mm (2.976 in.). The cutting speed is 95 m/min (313 ft/min) for all operations. All cutting tools are carbide, including the cutoff tool (the last operation), which is 3.18 mm $\left(\frac{1}{8} \text{ in.}\right)$ wide instead of 4.76 mm $\left(\frac{3}{16} \text{ in.}\right)$ for the high-speed steel cutoff tool that was formerly used. The material saved by this change is significant, because the width of the race is small. The turning center was capable of machining these races at high speeds and with repeatable dimensional tolerances of ±0.025 mm (0.001 in.).

8.11.4 Hexapod machines

Developments in the design and materials used for machine-tool structures and components are continually taking place. The important goals are (a) imparting machining flexibility to machine tools, (b) increasing their machining envelope, that is, the space within which machining can be done, and (c) and making them lighter. A truly innovative machine-tool structure is a self-contained octahedral (eight-sided) machine frame. Referred to as **hexapods** (Fig. 8.71), or *parallel kinematic linked machines*, the design is based on a mechanism called the *Stewart platform* (after D. Stewart), an invention first used to position aircraft cockpit simulators. The main advantage is that the links in the hexapod are loaded axially; the bending stresses and deflections are minimal, resulting in an extremely stiff structure.

The workpiece is mounted on a stationary table. Three pairs of *telescoping tubes* (struts or legs), each with its own motor and equipped with ballscrews, are used to maneuver a rotating cutting-tool holder. During machining a part with various features and curvatures, the machine controller automatically shortens some tubes, while others are extended, so that the cutter can follow a specified path around the workpiece. Six sets of coordinates are involved in these machines (hence the term *hexapod*, meaning "six legged"): three linear sets and three rotational sets. Every motion of the cutter, even a simple linear motion, is translated into six coordinated leg lengths, moving in real time. The motions of the legs are rapid, and consequently, high accelerations and decelerations, with high inertia forces, are involved.

FIGURE 8.71 (a) A hexapod machine tool, showing its major components and (b) closeup view of the cutting tool and its head in a hexapod machining center. *Source:* National Institute of Standards and Technology.

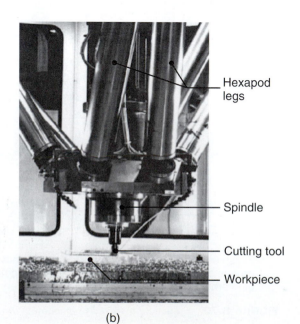

Hexapod legs

Spindle

Cutting tool

Workpiece

(a) (b)

These machines (a) have high stiffness, (b) are not as massive as machining centers, (c) have about one-third fewer parts than machining centers, (d) have a large machining envelope, hence greater access to the work zone, (e) are capable of maintaining the cutting tool perpendicular to the surface being machined, thus improving the machining operation, and (f) with six degrees of freedom, they have high flexibility in the production of parts with various geometries and sizes without the need to refixture the work in progress. Unlike most machine tools, these machines are basically portable. In fact, *hexapod attachments* are now available whereby conventional machining centers can easily be converted to a hexapod machine.

A limited number of these machines have been built, and in view of their potential as efficient machine tools, their performance is continually being evaluated regarding stiffness, thermal distortion, friction within the struts, dimensional accuracy, speed of operation, repeatability, and reliability. The cost of a hexapod machine is currently about $500,000 but is likely to come down significantly as they become more popular.

8.12 | Vibration and Chatter

In describing machining processes and machine tools throughout this chapter, it was pointed out that machine stiffness is important in controlling dimensional accuracy and surface finish of parts. In this section, we describe the adverse effects of low stiffness on product quality and machining operations, and the level of vibration and chatter in cutting tools and machines. If uncontrolled, vibration and chatter can result in the following effects:

- Poor surface finish (right central region in Fig. 8.72).
- Loss of dimensional accuracy of the workpiece.
- Premature wear, chipping, and failure of the cutting tool, which is especially critical with brittle tool materials, such as ceramics, some carbides, and diamond.
- Damage to machine-tool components from excessive vibrations.
- Objectionable noise generated, particularly if it is of high frequency, such as the squeal heard in turning brass on a lathe.

Vibration and chatter in machining are complex phenomena. Cutting operations cause two basic types of vibration: forced vibration and self-excited vibration.

1. **Forced vibration** is generally caused by some periodic force present in the machine tool, such as from gear drives, imbalance of the machine-tool components,

FIGURE 8.72 Chatter marks (right of center of photograph) on the surface of a turned part. *Source:* Courtesy of General Electric Company.

misalignment, or motors and pumps. As an example, in milling or in turning a splined shaft or a shaft with a keyway, forced vibrations are the periodic engagement of the cutter or the cutting tool with the workpiece surface, including its exit from the machined surface.

The basic solution to forced vibrations is to isolate or remove the forcing element. If the forcing frequency is at or near the natural frequency of a component of the machine-tool system, one of the frequencies may be raised or lowered. The amplitude of vibration can be reduced by increasing the stiffness or damping of the system. Although modifying the process parameters generally does not appear to greatly influence forced vibrations, changing the cutting speed and the tool geometry can be helpful.

2. Generally called **chatter**, *self-excited vibration* is caused by the interaction of the chip-removal process with the structure of the machine tool; these vibrations usually have very high amplitude. Chatter typically begins with a disturbance in the cutting zone. Such disturbances include lack of homogeneity in the workpiece material or its surface condition, changes in chip morphology during machining, or a change in frictional conditions at the tool-chip interface (as influenced by cutting fluids and their effectiveness). Self-excited vibrations can generally be controlled by (1) increasing the dynamic stiffness of the system and (2) by damping. **Dynamic stiffness** is the ratio of the amplitude of the force applied to the amplitude of vibration. Because a machine tool has different stiffnesses at different frequencies, changes in cutting parameters, such as cutting speed, also can influence chatter.

The most important type of self-excited vibration in machining is **regenerative chatter,** which is caused when a tool cuts a surface that has a roughness or disturbances left from the previous cut. Because the depth of cut thus varies, the resulting variations in the cutting force subject the tool to vibrations. The process continues repeatedly during machining, hence the term *regenerative*. This type of vibration can be observed while driving a car over a rough road (the so-called *washboard effect*).

Damping.   The term *damping* is defined as the rate at which vibrations decay, and it is an important factor in controlling machine-tool vibration and chatter.

Internal damping of structural materials.   Damping results from the energy loss within materials during vibration. Thus, for example, steel has less damping than gray cast iron, and composite materials have more damping than gray iron (Fig. 8.73). This difference in the damping capacity of materials can be observed by striking them with a gavel and listening to the sound; try, for example, striking pieces of steel, concrete, plastics, fiber-reinforced plastics, and wood.

Joints in the machine-tool structure.   Although less significant than internal damping, bolted joints in the structure of a machine tool are also a source of

FIGURE 8.73 Relative damping capacity of (a) gray cast iron and (b) epoxy-granite composite material. The vertical scale is the amplitude of vibration and the horizontal scale is time.

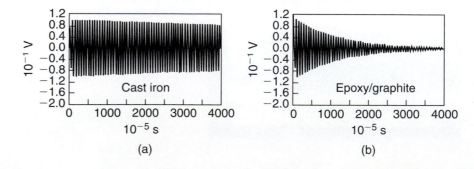

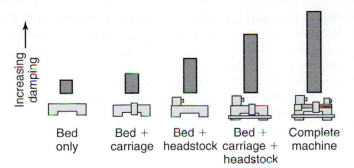

Bed only | Bed + carriage | Bed + headstock | Bed + carriage + headstock | Complete machine

damping. Because friction dissipates energy, small relative movements along dry (unlubricated) joints dissipate energy and thus improve damping. In joints where oil or grease is present, the internal friction of the lubricant layers also dissipates energy, thereby contributing to damping. In describing the machine tools for various cutting operations, we noted that all machines consist of a number of large and small components, assembled into a structure. Consequently, this type of damping is *cumulative,* owing to the presence of a number of joints in a machine tool. Note in Fig. 8.74 how damping increases as the number of components on a lathe and their contact area increase; thus, the more joints, the greater is the amount of energy dissipated, and hence the higher is the damping.

External damping. External damping is typically accomplished with the use of external dampers, which are similar to shock absorbers on automobiles. Special vibration absorbers also have been developed to be be installed on machine tools for this purpose.

Factors influencing chatter. The tendency for a particular workpiece to chatter during machining has been found to be proportional to the cutting forces and the depth and width of cut. Consequently, because cutting forces increase with strength and hardness, the tendency to chatter generally increases as the hardness of the workpiece material increases. Thus, for example, aluminum and magnesium alloys have less tendency to chatter than do martensitic and precipitation-hardening stainless steels, nickel alloys, and high-temperature and refractory alloys. An important factor in chatter is the type of chip produced during machining operations. As noted earlier, continuous chips involve steady cutting forces, and consequently, they generally do not cause chatter. Discontinuous chips and serrated chips, on the other hand, may do so because they are produced periodically, and the resulting variations in force during machining can cause chatter.

8.13 | Machine-Tool Structures

This section describes the *material* and *design* aspects of machine tools as structures with certain desired and specific characteristics. The proper design of machine-tool structures requires a knowledge of the materials available for construction, their forms and various properties, the dynamics of the particular machining process, and the cutting forces involved. *Stiffness* and *damping* are important factors in machine-tool structures. Stiffness involves the dimensions of the structural components as well as the elastic modulus of the materials used. Damping involves the type of materials used as well as the number and nature of the joints in the structure.

Materials and design.   Traditionally, the base and some of the major components of machine tools have been made of gray or nodular cast iron, which have the advantages of low cost and good damping capacity, but they are heavy. Lightweight designs are desirable because of ease of transportation, their higher natural frequencies, and also lower inertial forces of moving members. Lightweight designs and design flexibility require fabrication processes such as (a) mechanical fastening (bolts and nuts) of individual components and (b) welding. However, this approach to fabrication increases labor and material costs, because of the preparations required.

Wrought steels are the likely choice for such lightweight structures because of their low cost, availability in various section sizes and shapes (such as channels, angles, and tubes), desirable mechanical properties, and favorable characteristics, such as formability, machinability, and weldability. On the other hand, the benefit of the higher damping capacity of castings and composites is not available with steels.

In addition to stiffness, another factor that contributes to lack of precision of the machine tool is the *thermal expansion* of its components, which would lead to distortion. The sources of heat may be (a) internal, such as bearings, machine ways, and motors, as well as heat generated from the cutting zone (Section 8.2.5), or (b) external, such as nearby furnaces, heaters, sunlight, and fluctuations in cutting fluid and ambient temperatures. Equally important in machine-tool precision are *foundations,* particularly their mass and how they are installed in a plant. In one example, in installing a large grinding machine for grinding marine-propulsion gears, 2.75 m (9 ft) in diameter, with high precision, the concrete foundation is 6.7 m (22 ft deep). This large mass of concrete and the machine base greatly reduce the amplitude of vibrations and their adverse effects.

Several options exist concerning the materials used for machine-tool bases and components. For example, **acrylic concrete,** which is a mixture of concrete and polymer (polymethylmethacrylate), can easily be cast into desired shapes for machine bases and various components; several compositions of this material are available. It can also be used for *sandwich construction* with cast iron, thus combining the advantages of each type of material. **Granite-epoxy composite** has a typical composition of about 93% crushed granite and 7% epoxy binder. First used in precision grinders in the early 1980s, this composite material has several favorable properties: (1) good castability, which allows design versatility in machine tools; (2) high stiffness-to-weight ratio; (3) thermal stability; (4) resistance to environmental degradation; and (5) good damping capacity (see Fig. 8.73).

8.14 | Design Considerations

1. **General requirements for machined parts.**

   a. It is of course necessary that materials be selected to meet design requirements. However, it should also be recognized by designers that manufacturing through the processes described in this chapter is greatly simplified through selection of materials with good machinability, and it is advisable to use easy-to-machine materials whenever possible.

   b. Tolerances should be as wide as possible while leading to desired performance; surface roughness should be as high as is practicable. Often as-cast or as-formed tolerances may be sufficient and no machining is required. Excessively tight design requirements may require further costly finishing operations such as grinding, lapping, etc., and should be avoided. Figures 8.26 and 9.27 are useful guides for specifying tolerances and roughnesses and for identifying required manufacturing processes.

c. Parts should be designed so that they can be securely fixtured inside machines; this often requires incorporation of clamping features in the design. Designers should provide room for fixtures, features to facilitate clamping, and clearance for cutters. When using cast or forged blanks, the design should not require fixturing to take place on parting lines or flash.

d. Whenever possible, put all machining operation in the same plane or same diameter to reduce the number of operations. When this is not possible, try to locate them so that they can be performed from a minimum number of part setups.

e. Provide clearance at the end of a cut, such as clearance for the tool or non-critical space for a burr.

f. Burrs are unavoidable in machining. Burrs should be expected, and space should be provided for burr removal when desired. Chamfers often contain a burr, but this usually has minimal detrimental effect.

g. Design features should be such that commercially available standard cutting tools, inserts, and toolholders can be used.

2. **Design considerations for turning.**   In addition to the general considerations given above, the following considerations exist for turning operations:

a. Blanks to be machined should be as close to the final dimensions (such as by net or near-net-shape forming) as possible so as to reduce machining cycle time.

b. Thin and slender workpieces may be difficult to support properly and may deflect excessively during machining. Therefore, keep parts as short and stocky as possible.

c. Sharp corners, tapers, and major dimensional variations along the part should be avoided.

d. Radii should be large and conform to standard tool nose-radius specifications.

e. On tracer lathes, parts should be designed to minimize the number of tool changes required.

f. Side walls on parts should have a slight taper to prevent tool marks as the tool is withdrawn.

g. Knurled areas should be kept narrow; a reasonable recommendation is that the width not exceed the diameter.

3. **Design considerations for screw thread cutting.**   In addition to the general considerations, many of the design recommendations for turning operations apply to screw cutting. In addition,

a. Rolled threads are generally preferable to cut threads; whenever practical, thread cutting should be avoided.

b. Through-holes are preferable to blind holes when machining threads; internal threads in blind holes should have an unthreaded length at the bottom.

c. Designs should allow the termination of threads before the threads reach a shoulder.

d. Shallow blind-threaded holes should be avoided.

e. Burrs are unavoidable; chamfers should be specified when possible to avoid burr formation on both the internal and external threads.

f. Threaded sections should not be interrupted with holes, slots, or other discontinuities.

4. **Drilling, reaming, boring, and tapping operations** have the following additional considerations:

    a.   Holes should be specified on flat surfaces and perpendicular to the direction of drill movement for ease of drilling.

    b.   Avoid excessively small holes. Around 3-mm $\left(\frac{1}{8}\text{-in.}\right)$ diameter holes are a convenient minimum for high-production drilling.

    c.   Avoid excessively deep holes; length-to-diameter ratios of three or less should be used when possible, although 8:1 ratios are feasible.

    d.   When multiple holes are required on a part, they should use the same diameter whenever practical to avoid unnecessary tool changes.

    e.   Hole bottoms should match standard drill-point angles, and holes with interrupted surfaces should be avoided.

    f.   Through holes are usually preferred to blind holes, unless they lead to significantly larger material removal, and blind holes should be drilled deeper than subsequent reaming or tapping operations require, typically by an amount at least one-fourth the hole diameter.

5. **Milling.**   As one of the most versatile manufacturing processes, there are few features that cannot be produced in milling. However, with poor design practices, there are few features that can be produced economically in milling. The following concerns should be considered when designing a part that will be milled:

    a.   Internal corners should allow the radius to match that of the cutter whenever possible.

    b.   Milling is the rare process where sharp external corners can be manufactured without difficulty. When this is not desired, bevels should be preferred over radii, because tooling and setup costs are larger for production of radii. If the radius and shaft have the same radius, then the transition between them needs to be very accurately machined and is very difficult to produce.

    c.   Internal cavities and pockets with sharp inner corners should be avoided. When slots or key seats are required, the cutter should be allowed to define both the slot width and the end radii.

    d.   It should be recognized that using small milling cutters allows the production of any surface, but small cutters are slower, less rugged and more susceptible to chatter than large cutters. Thus, clearance should be provided in the design for the milling cutters.

6. **Broaching.**   The following considerations apply to broaching:

    a.   The use of standardized parts is especially important for broaches. Keyways, splines, gear teeth, etc., all have standard sizes, and the use of standard dimensions allows the use of common broaches.

    b.   Balanced cross sections are preferable to keep the broach from drifting and therefore to maintain tight tolerances.

    c.   Radii are difficult to achieve in broaching, and chamfers are preferred.

    d.   Inverted or dovetail splines should be avoided.

    e.   Broaching blind holes should be avoided whenever possible, but when necessary, there must be a relief at the end of the broached area.

7. **Gear machining.**   In addition to the milling considerations, the following are useful rules to bear in mind when producing gear teeth. Note that it is common to mill a bore and broach a keyway in gear teeth, so that the design rules specified for these processes (see above) apply equally to the gear product.

a. Blank design is important for proper fixturing and to ease cutting operations. Machining allowances must be present in blanks, and if machining is to be followed by further finishing operations, the part must still be oversized after machining; that is, it has a finishing allowance after machining.

b. Wide gears are more difficult to machine than narrow gears.

c. Spur gears are easier to machine than helical gears, which in turn are easier to machine than bevel gears and worm gears.

d. Dimensional tolerances and standardized gear shapers are specified by industry standards. A gear quality number should be selected so that the gear has as wide a tolerance range as possible while still meeting performance requirements.

8.15 | Economics of Machining

The advantages and limitations of machining and the technical considerations were described in detail throughout various sections in this chapter. These considerations should be viewed in light of the competitive nature of various manufacturing operations. Thus, for example, as compared with forming and shaping processes, machining requires more time and wastes more material; however, it is much more versatile and can produce parts with dimensional control and surface finish better than those provided by other processes.

This section deals with the *economics* of machining. The two most important parameters are the *minimum cost per part* and the *maximum production rate*. Typically, the **total cost per piece** consists of four items, or

$$C_p = C_m + C_s + C_l + C_t, \tag{8.46}$$

where C_p is the cost (say, in dollars) per piece; C_m is the machining cost; C_s is the cost of setting up for machining, such as mounting the tool, the cutter, and the fixtures and preparing for the particular operation; C_l is the cost of loading, unloading, and machine handling; and C_t is the tooling cost, which can take different forms depending on the application but includes factors such as tool changing, insert indexing, regrinding, and depreciation of the cutter or insert. The **machining cost** is given by

$$C_m = T_m(L_m + B_m), \tag{8.47}$$

where T_m is the machining time per piece, L_m is the labor cost of production per hour, and B_m is the burden rate, or overhead charge, of the machine, including depreciation, maintenance, indirect labor, and the like. The **setup cost** is a fixed figure, say in dollars per piece. The **loading, unloading,** and **machine-handling cost** is given by

$$C_l = T_l(L_m + B_m), \tag{8.48}$$

where T_l is the time involved in loading and unloading the part, changing speeds, changing feed rates, and so on. The **tooling cost** is expressed as

$$C_t = \frac{1}{N_i}[T_c(L_m + B_m) + D_i] + \frac{1}{N_f}[T_i(L_m + B_m)], \tag{8.49}$$

where N_i is the number of parts machined per insert, N_f is the number of parts that can be produced per insert face, T_c is the time required to change the insert, T_i is the time required to index the insert, and D_i is the depreciation of the insert in dollars. The **time** required to produce one part is

$$T_p = T_l + T_m + \frac{T_c}{N_i} + \frac{T_i}{N_f}, \tag{8.50}$$

where T_m has to be calculated for each particular operation. For example, let's consider a turning operation; the machining time (see Section 8.9.1) is

$$T_m = \frac{L}{fN} = \frac{\pi L D}{f V}, \tag{8.51}$$

where L is the length of cut, f is the feed, N is the rpm of the workpiece, D is the workpiece diameter, and V is the cutting speed. Note that appropriate units must be used in all these equations. From the Taylor tool-life equation [see Eq. (8.31)], we have

$$V T^n = C.$$

Hence,

$$T = \left(\frac{C}{V}\right)^{1/n}, \tag{8.52}$$

where T is the time, in minutes, required to reach a flank wear of certain dimension, after which the tool has to be reground or changed. The number of pieces per insert face is thus simply

$$N_f = \frac{T}{T_m}, \tag{8.53}$$

and the number of pieces per insert is given by

$$N_i = m N_f = \frac{m T}{T_m}. \tag{8.54}$$

Sometimes not all of the faces are used before the insert is discarded, so it should be recognized that m corresponds to the number of faces that are actually used, not the number of faces provided per insert. The combination of Eqs. (8.51) through (8.54) gives

$$N_p = \frac{f C^{1/n}}{\pi L D V^{(1/n)-1}}. \tag{8.55}$$

The **cost per piece**, C_p, in Eq. (8.46), can now be defined in terms of several variables. To find the optimum cutting speed and the optimum tool life for **minimum cost**, we differentiate C_p with respect to V and set it to zero. Thus, we have

$$\frac{\partial C_p}{\partial V} = 0. \tag{8.56}$$

We then find that the optimum cutting speed, V_o, is

$$V_o = \frac{C(L_m + B_m)^n}{\left(\frac{1}{n} - 1\right)^n \left\{ \frac{1}{m}[T_c(L_m + B_m) + D_i] + T_i(L_m + B_m) \right\}} \tag{8.57}$$

and the optimum tool life, T_o, is

$$T_o = \left[\left(\frac{1}{n}\right) - 1\right] \frac{\frac{1}{m}[T_c(L_m + B_m) + D_i] + T_i(L_m + B_m)}{L_m + B_m}. \tag{8.58}$$

To find the optimum cutting speed and the optimum tool life for **maximum production**, we differentiate T_p with respect to V and set the result to zero. Thus we have

$$\frac{\partial T_p}{\partial V} = 0. \tag{8.59}$$

The **optimum cutting speed** now becomes

$$V_o = \frac{C}{\left[\left(\frac{1}{n} - 1\right)\left(\frac{T_c}{m} + T_i\right)\right]^n},$$ (8.60)

and the **optimum tool life** is

$$T_o = \left(\frac{1}{n} - 1\right)\left(\frac{T_c}{m} + T_i\right).$$ (8.61)

Qualitative plots of the minimum cost per piece and the minimum time per piece, that is, the maximum production rate, are given in Fig. 8.75. The cost of a machined surface also depends on the degree of finish required (see Section 9.17); the machining cost increases rapidly with finer surface finish.

This analysis indicates the importance of identifying all relevant parameters in a machining operation, determining various cost factors, obtaining relevant tool-life

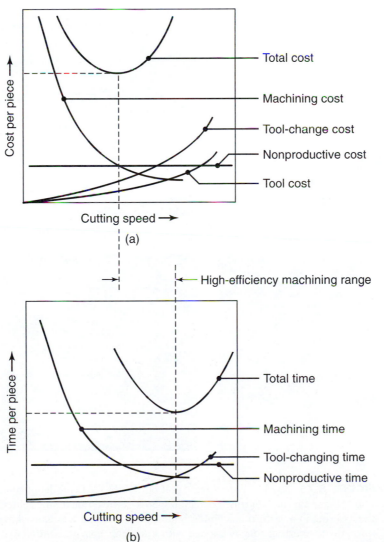

(a)

High-efficiency machining range

(b)

FIGURE 8.75   Qualitative plots showing (a) cost per piece and (b) time per piece in machining. Note that there is an optimum cutting speed for both cost and time, respectively. The range between the two optimum speeds is known as the *high-efficiency machining range.*

curves for the particular operation, and properly measuring the various time intervals involved in the overall operation. The importance of obtaining accurate data is clearly shown in Fig. 8.75, as small changes in cutting speed can have a significant effect on the minimum cost or time per piece.

CASE STUDY | Ping Golf Putters

In their efforts to develop high-end, top-performing putters, engineers at Ping Golf, Inc. in Phoenix, Arizona, recently utilized advanced machining practices in their design and production processes for a new style of putter, the Anser® series. (Fig. 8.76). Governed by a unique set of design constraints, they had the task and goal of creating putters that were both practical for production quantities and would meet specific functional and aesthetic requirements.

One of the initial decisions concerned selection of a proper material for the putter to meet its functional requirements. Four types of stainless steel (303, 304, 416, and 17-4 precipitation hardening; see Section 3.10.2) were considered for various property requirements, including machinability, durability, and the sound or feel of the particular putter material (another requirement unique

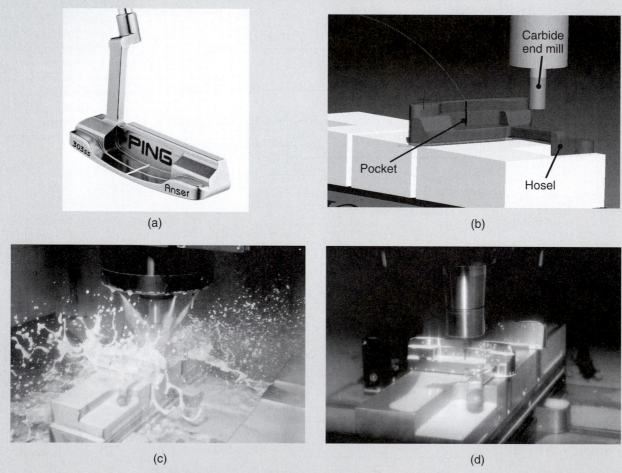

FIGURE 8.76 (a) The Ping Anser® golf putter, (b) CAD model of rough machining of the putter outer surface, (c) rough machining on a vertical machining center, (d) machining of the lettering in a vertical machining center; the operation was paused to take the photo, as normally the cutting zone is flooded with a coolant. *Source:* Courtesy of Ping Golf, Inc.

to golf equipment). Among the materials evaluated, 303 stainless steel was chosen because it is free-machining (Section 8.5.1), indicating smaller chips, lower power consumption, better surface finish, and improved tool life, thus allowing for increased machining speeds and higher productivity.

The next step of the project involved determining the optimum blank type and the sequence of operations to be performed in its production. In this case, engineers chose to develop a slightly oversized forged blank (Section 6.2). A forging was chosen because it provided a favorable internal grain structure as opposed to a casting, as castings could result in porosity and inconsistent surface finish after machining (Section 5.12.1). The blank incorporated a machining allowance, so dimensions were specified approximately 0.050–0.075 in. (1.25–1.9 mm) larger in all directions than that of the final part.

The most challenging and longest task of the project was developing the necessary programming and fixturing for each part. Beyond the common requirements of typical machined parts (including tight tolerances and repeatability), putters require an additional set of aesthetic specifications. In this case, both precise machining and the overall appearance of the finished part were imperative. A machining technique known as surfacing or contouring (commonly used in injection mold making; Section 10.10.2) was used to machine most of the finished geometry. Although this operation required additional machining time, it provided a superior finish on all surfaces and allowed machining of more complex geometries, thus adding value to the finished product.

As for all high-volume machined parts, repeatability was essential. Each forged blank was designed with a protrusion across the face of the putter, allowing for the initial locating surfaces (for fixturing). A short machining operation removed a small amount of material around the bar and produced three flat and square surfaces as a reference location for the first primary machining operation.

Each putter required six different operations in order to machine all of its surfaces, and each operation was designed to provide locating surfaces for the next step in the manufacturing process. Several operations were set up using a tombstone loading system (see Section 14.9) on a horizontal-spindle CNC milling machine. This method allowed machine operators to load and unload parts while other parts were being machined, thus significantly increasing the efficiency of the process.

Modular fixturing and tungsten-carbide cutting tools coated with TiAlN (Section 8.6.5) allowed the quick changeover between right- and left-handed parts as well as different putter models. After the initial locating operation was complete, parts were then transferred to a three-axis VMC to create the putter cavity. Since the forged blanks are near-net shape, the maximum radial depth of cut on most surfaces was 0.075 in., but the axial depth of cut of 1.5 in. inside the "cavity" of the putter was the most demanding milling operation (see Figs. 8.76b and c). The putter has small inside radii with a comparatively long depth (7× diameter or greater).

A four-axis horizontal machining center was used to reduce the number of setups in the operation. The rotary axis was used for creating the relatively complex geometry of the hosel (the socket for the shaft of the golf club). Since the hosel is relatively unsupported, chatter was the most complex challenge to overcome. Several iterations of spindle speeds were attempted in conjunction with upfront guidance from a simulation model. Modal analyses were conducted on the fixtured parts in an attempt to identify and avoid the natural frequencies of the part/fixture (see Section 8.11). The machines had spindle speeds ranging from 12,000 to 20,000 rpm, each having 30 horsepower. With the near-net-shape forging, the milling operations were designed to have low depths of cut, but high speed.

After each machining operation was completed, a small amount of hand finishing was necessary to produce a superior surface. The putters were then lightly shot blasted (with glass bead media; Section 4.5.1) for the purpose of achieving surface consistency. A black, nickel-chrome plating was then applied to all parts to enhance the aesthetic appeal and protect the stainless steel from small dings and dents and corrosion from specific chemicals that may be encountered on a golf course.

Source: Courtesy of D. Jones and D. Petersen, Ping Golf, Inc.

SUMMARY

- Machining processes are often necessary in order to impart the desired geometric features, surface finish, and dimensional accuracy characteristics to components, particularly those with complex shapes that cannot be produced economically or properly by other shaping techniques. On the other hand, machining processes inevitably waste material in the form of chips, involve longer processing times, and may have adverse effects on the surfaces produced. (Section 8.1)

- Important process variables in machining include cutting-tool shape and material; cutting conditions such as speed, feed, and depth of cut; cutting fluids; and the characteristics of the machine tool and the workpiece material. Parameters influenced by these variables include forces and power consumption, tool wear, surface finish and integrity, temperature, and dimensional accuracy of the workpiece. Commonly observed chip types are continuous, built-up edge, discontinuous, and serrated. (Section 8.2)

- Temperature rise is an important consideration, as it can have adverse effects on tool life as well as on dimensional accuracy and surface integrity of the machined part. (Section 8.2)

- Tool wear depends primarily on workpiece and tool-material characteristics, cutting speed, and cutting fluids; feeds, depth of cut, and machine-tool characteristics also have an effect. Two major types of wear are flank wear and crater wear. (Section 8.3)

- Surface finish of machined components is an important consideration, as it can adversely affect product integrity. Important variables that affect surface finish include the geometry and condition of the cutting tool, chip morphology, and process variables. (Section 8.4)

- Machinability is generally defined in terms of surface finish, tool life, force and power requirements, and chip type. Machinability of materials depends not only on their intrinsic properties and microstructure, but also on proper selection and control of process variables. (Section 8.5)

- A variety of cutting-tool materials are available, the most common being high-speed steels, carbides, ceramics, and cubic boron nitride. These materials, including their coatings, have a broad range of mechanical and physical properties, particularly hot hardness, toughness, chemical stability and inertness, and resistance to chipping and wear. (Section 8.6)

- Cutting fluids are important in machining operations, as they reduce friction, forces, and power requirements and improve tool life. Generally, slower operations with high tool pressures require a fluid with good lubricating characteristics, whereas in high-speed operations, where temperature rise can be significant, fluids with cooling capacity are preferred. (Section 8.7)

- The machining processes that produce external and internal circular profiles are turning, boring, drilling, tapping, and thread cutting. Because of the three-dimensional nature of these operations, chip movement and control are important considerations, since chips can interfere with the operation. Optimization of each process requires an understanding of the interrelationships among design and processing parameters. (Section 8.9)

- High-speed machining, ultraprecision machining, and hard turning are among more recent developments in cutting. They can help reduce machining costs, and they produce parts with exceptional surface finish and dimensional accuracy. (Sections 8.8 and 8.9)

- Complex shapes can be machined by slab, face, and end milling; broaching; and sawing. These processes use multitooth tools and cutters at various axes with respect to the workpiece. The machine tools used are now mostly computer controlled, have various features and attachments, and possess considerable flexibility in operation. (Section 8.10)

- Because of their versatility and capability of performing a variety of cutting operations, machining and turning centers are among the most important developments in machine tools. Their selection depends on factors such as part complexity required, the number and type of cutting operations to be performed, the number of cutting tools needed, the dimensional accuracy required, and the production rate required. (Section 8.11)

- Vibration and chatter in machining are important considerations for workpiece dimensional accuracy and surface finish, as well as tool life. Stiffness and damping capacity of machine tools are important factors in controlling vibration and chatter. New materials are being developed and used for constructing machine-tool structures. (Sections 8.12 and 8.13)

- Several design guidelines have been developed for parts to be produced by machining operations. (Section 8.14)

- The economics of machining processes depend on various costs. Optimum cutting speeds can be determined for minimum machining time per piece and minimum cost per piece, respectively. (Section 8.15)

SUMMARY OF EQUATIONS

Cutting ratio: $r = \dfrac{t_o}{t_c} = \dfrac{\sin \phi}{\cos(\phi - \alpha)}$

Shear strain: $\gamma = \cot \phi + \tan(\phi - \alpha)$

Velocity relationships: $\dfrac{V}{\cos(\phi - \alpha)} = \dfrac{V_s}{\cos \alpha} = \dfrac{V_c}{\sin \phi}$

Friction force: $F = R \sin \beta$

Normal force: $N = R \cos \beta$

Coefficient of friction: $\mu = \tan \beta = \dfrac{F_t + F_c \tan \alpha}{F_c - F_t \tan \alpha}$

Thrust force: $F_t = R \sin(\beta - \alpha) = F_c \tan(\beta - \alpha)$

Shear-angle relationships: $\phi = 45° + \dfrac{\alpha}{2} - \dfrac{\beta}{2}$

$$\phi = 45° + \alpha - \beta$$

$$\text{Total cutting power} = F_c V$$

$$\text{Specific energy: } u_t = \frac{F_c}{w t_o}$$

$$\text{Frictional specific energy: } u_f = \frac{Fr}{w t_o}$$

$$\text{Shear specific energy: } u_s = \frac{F_s V_s}{w t_o V}$$

$$\text{Mean temperature: } T = \frac{1.2 Y_f}{\rho c} \sqrt[3]{\frac{V t_o}{K}}$$

$$T \propto V^a f^b$$

$$\text{Tool life: } V T^n = C$$

$$\text{Material-removal rate (MRR) in turning: MRR} = \pi D_{\text{avg}} d f N$$

$$\text{in drilling: MRR} = \pi \left(\frac{D^2}{4} \right) f N$$

$$\text{in milling: MRR} = w d v$$

BIBLIOGRAPHY

Arnone, M., *High Performance Machining,* Hanser, 1998.

ASM Handbook, Vol. 16: *Machining,* ASM International, 1989.

ASM Specialty Handbook: Tool Materials, ASM International, 1995.

Astakhov, V.P., *Metal Cutting Mechanics,* CRC Press, 1998.

Boothroyd, G., and Knight, W.A., *Fundamentals of Machining and Machine Tools,* 3d ed., Dekker, 2005.

Brown, J., *Advanced Machining Technology Handbook,* McGraw-Hill, 1998.

Byers, J.P. (ed.), *Metalworking Fluids,* Dekker, 1994.

Davis, J.R. (ed.), *Tool Materials,* ASM International, 1995.

DeVries, W.R., *Analysis of Material Removal Processes,* Springer, 1992.

Dudzinski, D., Molinari, A., and Schulz, H., (eds.) *Metal Cutting and High Speed Machining,* Springer, 2002.

Erdel, B., *High-Speed Machining,* Society of Manufacturing Engineers, 2003.

Ewert, R.H., *Gears and Gear Manufacture: The Fundamentals,* Chapman & Hall, 1997.

Hoffman, E.G., *Jig and Fixture Design,* 4th ed., Industrial Press, 1996.

Kalpakjian, S. (ed.), *Tool and Die Failures: Source Book,* ASM International, 1982.

Komanduri, R., "Tool Materials," in *Kirk-Othmer Encyclopedia of Chemical Technology,* 4th ed., Vol. 24, Wiley, 1997.

Krar, S.F., and Check, A.F., *Technology of Machine Tools,* 5th ed., Glencoe Macmillan/McGraw-Hill, 1996.

Machinery's Handbook, Industrial Press, revised periodically.

Modern Metal Cutting: A Practical Handbook, Sandvik Coromant, 1996.

Nachtman, E.S., and Kalpakjian, S., *Lubricants and Lubrication in Metalworking Operations,* Dekker, 1985.

Rivin, E.I., *Stiffness and Damping in Mechanical Design,* Dekker, 1999.

Roberts, G.A., Krauss, G., and Kennedy, R., *Tool Steels,* 5th ed., ASM International, 1997.

Shaw, M.C., *Metal Cutting Principles,* 2d ed., Oxford, 2005.

Sluhan, C. (ed.), *Cutting and Grinding Fluids: Selection and Application,* Society of Manufacturing Engineers, 1992.

Stephenson, D., and Agapiou, J.S., *Metal Cutting: Theory and Practice,* 2d ed., CRC Press, 2005.

Stout, K.J., Davis, J., and Sullivan, P.J., *Atlas of Machined Surfaces,* Chapman & Hall, 1990.

Townsend, D.P., *Dudley's Gear Handbook: The Design, Manufacturing, and Application of Gears,* 2d ed., McGraw-Hill, 1991.

Trent, E.M., and Wright, P.K., *Metal Cutting,* 4th ed., Butterworth Heinemann, 2000.

Venkatesh, V.C., and Chandrasekaran, H., *Experimental Techniques in Metal Cutting,* rev. ed., Prentice Hall, 1987.

Walsh, R.A., *McGraw-Hill Machining and Metalworking Handbook,* McGraw-Hill, 1994.

Weck, M., *Handbook of Machine Tools,* 4 vols., Wiley, 1984.

QUESTIONS

8.1 Explain why the cutting force, F_c, increases with increasing depth of cut and decreasing rake angle.

8.2 What are the effects of performing a cutting operation with a dull tool tip? A very sharp tip?

8.3 Describe the trends that you observe in Tables 8.1 and 8.2.

8.4 To what factors would you attribute the large difference in the specific energies within each group of materials shown in Table 8.3?

8.5 Describe the effects of cutting fluids on chip formation. Explain why and how they influence the cutting operation.

8.6 Under what conditions would you discourage the use of cutting fluids? Explain.

8.7 Give reasons that pure aluminum and copper are generally rated as easy to machine.

8.8 Can you offer an explanation as to why the maximum temperature in cutting is located at about the middle of the tool-chip interface? (*Hint:* Note that there are two principal sources of heat: the shear plane and the tool-chip interface.)

8.9 State whether or not the following statements are true for orthogonal cutting, explaining your reasons: (a) For the same shear angle, there are two rake angles that give the same cutting ratio. (b) For the same depth of cut and rake angle, the type of cutting fluid used has no influence on chip thickness. (c) If the cutting speed, shear angle, and rake angle are known, the chip velocity can be calculated. (d) The chip becomes thinner as the rake angle increases. (e) The function of a chip breaker is to decrease the curvature of the chip.

8.10 It has been stated that it is generally undesirable to allow temperatures to rise excessively in machining operations. Explain why.

8.11 Explain the reasons that the same tool life may be obtained at two different cutting speeds.

8.12 Inspect Table 8.6 and identify tool materials that would not be particularly suitable for interrupted cutting operations, such as milling. Explain your choices.

8.13 Explain the possible disadvantages of a machining operation in a discontinuous chip is produced.

8.14 It has been noted that tool life can be almost infinite at low cutting speeds. Would you then recommend that all machining be done at low speeds? Explain.

8.15 Referring to Fig. 8.31, how would you explain the effect of cobalt content on the properties of carbides?

8.16 Explain why studying the types of chips produced is important in understanding machining operations.

8.17 How would you expect the cutting force to vary for the case of serrated-chip formation? Explain.

8.18 Wood is a highly anisotropic material; that is, it is orthotropic. Explain the effects of orthogonal cutting of wood at different angles to the grain direction on the types of chips produced.

8.19 Describe the advantages of oblique cutting. Which machining proceses involve oblique cutting? Explain.

8.20 Explain why it is possible to remove more material between tool resharpenings by lowering the cutting speed.

8.21 Explain the significance of Eq. (8.8).

8.22 How would you go about measuring the hot hardness of cutting tools? Explain any difficulties that might be involved.

8.23 Describe the reasons for making cutting tools with multiphase coatings of different materials. Describe the properties that the substrate for multiphase cutting tools should have for effective machining.

8.24 Explain the advantages and any limitations of inserts. Why were they developed?

8.25 Make a list of alloying elements in high-speed-steel cutting tools. Explain why they are used.

8.26 What are the purposes of chamfers on cutting tools? Explain.

8.27 Why does temperature have such an important effect on cutting-tool performance?

8.28 Ceramic and cermet cutting tools have certain advantages over carbide tools. Why, then, are carbide tools not replaced to a greater extent?

8.29 Why are chemical stability and inertness important in cutting tools?

8.30 What precautions would you take in machining with brittle tool materials, especially ceramics? Explain.

8.31 Why do cutting fluids have different effects at different cutting speeds? Is the control of cutting-fluid temperature important? Explain.

8.32 Which of the two materials, diamond or cubic boron nitride, is more suitable for machining steels? Why?

8.33 List and explain the considerations involved in determining whether a cutting tool should be reconditioned, recycled, or discarded after use.

8.34 List the parameters that influence the temperature in machining, and explain why and how they do so.

8.35 List and explain the factors that contribute to poor surface finish in machining operations.

8.36   Explain the functions of the different angles on a single-point lathe cutting tool. How does the chip thickness vary as the side cutting-edge angle is increased? Explain.

8.37   It will be noted that the helix angle for drills is different for different groups of workpiece materials. Why?

8.38   A turning operation is being carried out on a long, round bar at a constant depth of cut. Explain what differences, if any, there may be in the machined diameter from one end of the bar to the other. Give reasons for any changes that may occur.

8.39   Describe the relative characteristics of climb milling and up milling and their importance in machining operations.

8.40   In Fig. 8.64a, high-speed-steel cutting teeth are welded to a steel blade. Would you recommend that the whole blade be made of high-speed steel? Explain your reasons.

8.41   Describe the adverse effects of vibrations and chatter in machining.

8.42   Make a list of components of machine tools that could be made of ceramics, and explain why ceramics would be a suitable material for these components.

8.43   In Fig. 8.12, why do the thrust forces start at a finite value when the feed is zero? Explain.

8.44   Is the temperature rise in cutting related to the hardness of the workpiece material? Explain.

8.45   Describe the effects of tool wear on the workpiece and on the overall machining operation.

8.46   Explain whether or not it is desirable to have a high or low (a) n value and (b) C value in the Taylor tool-life equation.

8.47   Are there any machining operations that cannot be performed on (a) machining centers and (b) turning centers? Explain.

8.48   What is the significance of the cutting ratio in machining?

8.49   Emulsion cutting fluids typically consist of 95% water and 5% soluble oil and chemical additives. Why is the ratio so unbalanced? Is the oil needed at all? Explain.

8.50   It was stated that it is possible for the n value in the Taylor tool-life equation to be negative. Explain.

8.51   Assume that you are asked to estimate the cutting force in slab milling with a straight-tooth cutter. Describe the procedure that you would follow.

8.52   Explain the possible reasons that a knife cuts better when it is moved back and forth. Consider factors such as the material being cut, interfacial friction, and the shape and dimensions of the knife.

8.53   What are the effects of lowering the friction at the tool-chip interface (say with an effective cutting fluid) on the mechanics of cutting operations? Explain, giving several examples.

8.54   Why is it not always advisable to increase cutting speed in order to increase production rate? Explain.

8.55   It has been observed that the shear–strain rate in metal cutting is high even though the cutting speed may be relatively low. Why?

8.56   We note from the exponents in Eq. (8.30) that the cutting speed has a greater influence on temperature than does the feed. Why?

8.57   What are the consequences of exceeding the allowable wear land (see Table 8.5) for cutting tools? Explain.

8.58   Comment on and explain your observations regarding Figs. 8.34, 8.38, and 8.43.

8.59   It was noted that the tool-life curve for ceramic cutting tools in Fig. 8.22a is to the right of those for other tools. Why?

8.60   In Fig. 8.18, it can be seen that the percentage of the energy carried away by the chip increases with cutting speed. Why?

8.61   How would you go about measuring the effectiveness of cutting fluids? Explain.

8.62   Describe the conditions that are critical in benefiting from the capabilities of diamond and cubic-boron-nitride cutting tools.

8.63   The last two properties listed in Table 8.6 can be important to the life of the cutting tool. Explain why. Which of the properties listed are the least important in machining operations? Explain.

8.64   It will be noted in Fig. 8.30 that the tool materials, especially carbides, have a wide range of hardness at a particular temperature. Why?

8.65   Describe your thoughts on how would you go about recycling used cutting tools. Comment on any difficulties involved, as well as on economic considerations.

8.66   As you can see, there is a wide range of tool materials available and used successfully today, yet much research and development continues to be carried out on these materials. Why?

8.67   Drilling, boring, and reaming of large holes is generally more accurate than just drilling and reaming. Why?

8.68   A highly oxidized and uneven round bar is being turned on a lathe. Would you recommend a relatively small or large depth of cut? Explain your reasons.

8.69   Does the force or torque in drilling change as the hole depth increases? Explain.

8.70 Explain the advantages and limitations of producing threads by forming and cutting, respectively.

8.71 Describe your observations regarding the contents of Tables 8.8, 8.10, and 8.11.

8.72 The footnote to Table 8.10 states that as the depth of the hole increases, speeds and feeds should be reduced. Why?

8.73 List and explain the factors that contribute to poor surface finish in machining operations.

8.74 Make a list of the machining operations described in this chapter, according to the difficulty of the operation and the desired effectiveness of cutting fluids. (*Example:* Tapping of holes is a more difficult operation than turning straight shafts.)

8.75 Are the feed marks left on the workpiece by a face-milling cutter segments of a true circle? Explain with appropriate sketches.

8.76 What determines the selection of the number of teeth on a milling cutter? (See, for example, Figs. 8.53 and 8.55.)

8.77 Explain the technical requirements that led to the development of machining and turning centers. Why do their spindle speeds vary over a wide range?

8.78 In addition to the number of components, as shown in Fig. 8.74, what other factors influence the rate at which damping increases in a machine tool? Explain.

8.79 Why is thermal expansion of machine-tool components important? Explain, with examples.

8.80 Would using the machining processes described in this chapter be difficult on nonmetallic or rubberlike materials? Explain your thoughts, commenting on the influence of various physical and mechanical properties of workpiece materials, the cutting forces involved, the part geometries, and the fixturing required.

8.81 The accompanying illustration shows a part that is to be machined from a rectangular blank. Suggest the type of operations required and their sequence, and specify the machine tools that are needed.

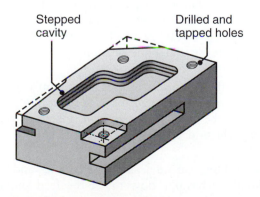

8.82 Select a specific cutting-tool material and estimate the machining time for the parts shown in the accompanying three figures: (a) pump shaft, stainless steel; (b) ductile (nodular) iron crankshaft; (c) 304 stainless-steel tube with internal rope thread.

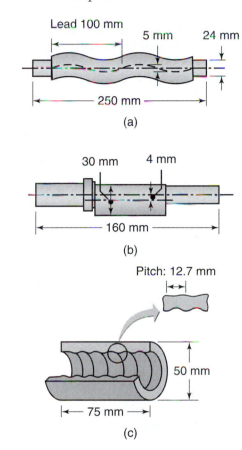

8.83 Why is the machinability of alloys generally difficult to assess?

8.84 What are the advantages and disadvantages of dry machining?

8.85 Can high-speed machining be performed without the use of cutting fluids? Explain.

8.86 If the rake angle is 0°, then the frictional force is perpendicular to the cutting direction and, therefore, does not contribute to machining power requirements. Why, then, is there an increase in the power dissipated when machining with a rake angle of, say, 20°?

8.87 Would you recommend broaching a keyway on a gear blank before or after the teeth are machined? Explain.

8.88 Given your understanding of the basic metal-cutting process, describe the important physical and chemical properties of a cutting tool.

8.89 Negative rake angles are generally preferred for ceramic, diamond, and cubic boron nitride tools. Why?

8.90 If a drill bit is intended for woodworking applications, what material is it most likely to be made from? (*Hint:* Temperatures rarely rise to 400°C in woodworking.) Are there any reasons why such a drill bit cannot be used to drill a few holes in a piece of metal? Explain.

8.91 What are the consequences of a coating on a cutting tool that has a different coefficient of thermal expansion than does the substrate? Explain.

8.92 Discuss the relative advantages and limitations of near-dry machining. Consider all relevant technical and economic aspects.

8.93 In modern manufacturing with computer-controlled machine tools, which types of metal chips are undesirable and why?

8.94 Explain why hacksaws are not as productive as band saws.

8.95 Describe workpieces and conditions under which broaching would be the preferred method of machining.

8.96 With appropriate sketches, explain the differences between and similarities among the following processes: (a) shaving, (b) broaching, and (c) turn broaching.

8.97 Why is it difficult to use friction sawing on non-ferrous metals? Explain.

8.98 Review Fig. 8.68 on modular machining centers, and explain workpieces and operations that would be suitable on such machines.

8.99 Describe types of workpieces that would not be suitable for machining on a machining center. Give specific examples.

8.100 Give examples of (a) forced vibration and (b) self-excited vibration in general engineering practice.

8.101 Tool temperatures are low at low cutting speeds and high at high cutting speeds, but low again at even higher cutting speeds. Explain why.

8.102 Explain the technical innovations that have made high-speed machining advances possible, and the economic motivations for high-speed machining.

PROBLEMS

8.103 Assume that in orthogonal cutting the rake angle is 15° and the coefficient of friction is 0.2. Using Eq. (8.20), determine the percentage increase in chip thickness when friction is doubled.

8.104 Prove Eq. (8.1).

8.105 With a simple analytical expression prove the validity of the statement in the last paragraph in Example 8.2.

8.106 Using Eq. (8.3), make a plot of the shear strain, γ, vs. the shear angle, ϕ, with the rake angle, α, as a parameter. Describe your observations.

8.107 Assume that in orthogonal cutting, the rake angle is 10°. Plot the shear plane angle and cutting ratio as a function of the friction coefficient.

8.108 Derive Eq. (8.12).

8.109 Determine the shear angle in Example 8.1. Is this calculation exact or an estimate? Explain.

8.110 The following data are available from orthogonal cutting experiments. In both cases, depth of cut (feed) $t_o = 0.13$ mm, width of cut $w = 2.5$ mm, rake angle $\alpha = -5°$, and cutting speed $V = 2$ m/s.

	Workpiece material	
	Aluminum	Steel
Chip thickness, t_c, mm	0.23	0.58
Cutting force, F_c, N	430	890
Thrust force, F_t, N	280	800

Determine the (a) shear angle ϕ [do not use Eq. (8.20)], (b) friction coefficient μ, (c) shear stress τ and shear strain γ on the shear plane, (d) chip velocity V_c and shear velocity V_s, and (e) energies u_f, u_s, and u_t.

8.111 Estimate the temperatures for the conditions of Problem 8.110 for the following workpiece properties:

	Workpiece material	
	Aluminum	Steel
Flow strength, Y_f, MPa	120	325
Thermal diffusivity, K, mm^2/s	97	14
Volumetric specific heat, ρc, N/mm^2 °C	2.6	3.3

8.112 In a dry cutting operation using a $-5°$ rake angle, the measured forces were $F_c = 1330$ N and $F_t = 740$ N. When a cutting fluid was used, these forces were $F_c = 1200$ N and $F_t = 710$ N. What is the change in the friction angle resulting from the use of a cutting fluid?

8.113 In the dry machining of aluminum with a 10° rake angle tool, it is found that the shear angle is 25°. Determine the new shear angle if a cutting fluid is applied which decreases the friction coefficient by 15%.

8.114 Taking carbide as an example and using Eq. (8.30), determine how much the feed should be changed in order to keep the mean temperature constant when the cutting speed is tripled.

8.115 With appropriate diagrams, show how the use of a cutting fluid can affect the magnitude of the thrust force, F_t, in orthogonal cutting.

8.116 An 8-in-diameter stainless-steel bar is being turned on a lathe at 600 rpm and at a depth of cut $d = 0.1$ in. If the power of the motor is 5 hp and has a mechanical efficiency of 80%, what is the maximum feed that you can have at a spindle speed of 500 rpm before the motor stalls?

8.117 Using the Taylor equation for tool wear and letting $n = 0.3$, calculate the percentage increase in tool life if the cutting speed is reduced by (a) 30% and (b) 60%.

8.118 The following flank wear data were collected in a series of machining tests using C6 carbide tools on 1045 steel (HB = 192). The feed rate was 0.015 in./rev and the width of cut was 0.030 in. (a) Plot flank wear as a function of cutting time. Using a 0.015 in. wear land as the criterion of tool failure, determine the lives for the four cutting speeds shown. (b) Plot the results on a log–log plot and determine the values of n and C in the Taylor tool-life equation. (Assume a straight line relationship.) (c) Using these results, calculate the tool life for a cutting speed of 300 ft/min.

Cutting speed V, ft/min	Cutting time min	Flank wear in.
400	0.5	0.0014
	2.0	0.0023
	4.0	0.0030
	8.0	0.0055
	16.0	0.0082
	24.0	0.0112
	54.0	0.0150
600	0.5	0.0018
	2.0	0.0035
	4.0	0.0060
	8.0	0.0100
	13.0	0.0145
	14	0.0160
800	0.5	0.0050
	2.0	0.0100
	4.0	0.0140
	5.0	0.0160
1000	0.5	0.0100
	1.0	0.0130
	1.8	0.0150
	2.0	0.0160

8.119 Determine the n and C values for the four tool materials shown in Fig. 8.22a.

8.120 Using Eq. (8.30) and referring to Fig. 8.18a, estimate the magnitude of the coefficient a.

8.121 (a) Estimate the machining time required in rough turning a 1.5-m-long, annealed aluminum-alloy round bar 75-mm in diameter, using a high-speed-steel tool. (b) Estimate the time for a carbide tool. Let feed = 2 mm/rev.

8.122 A 150-mm-long, 75-mm-diameter titanium-alloy rod is being reduced in diameter to 65 mm by turning on a lathe in one pass. The spindle rotates at 400 rpm and the tool is traveling at an axial velocity of 200 mm/min. Calculate the cutting speed, material removal rate, time of cut, power required, and the cutting force.

8.123 Calculate the same quantities as in Example 8.4 but for high-strength cast iron and at $N = 500$ rpm.

8.124 A 0.75-in-diameter drill is being used on a drill press operating at 300 rpm. If the feed is 0.005 in./rev, what is the material removal rate? What is the MRR if the drill diameter is tripled?

8.125 A hole is being drilled in a block of magnesium alloy with a 15-mm drill at a feed of 0.1 mm/rev. The spindle is running at 500 rpm. Calculate the material removal rate and estimate the torque on the drill.

8.126 Show that the distance l_c in slab milling is approximately equal to $\sqrt{Dd}$ for situations where $D \gg d$.

8.127 Calculate the chip depth of cut in Example 8.6.

8.128 In Example 8.6, which of the quantities will be affected when the spindle speed is increased to 200 rpm?

8.129 A slab-milling operation is being carried out on a 20-in.-long, 6-in.-wide high-strength-steel block at a feed of 0.01 in./tooth and a depth of cut of 0.15 in. The cutter has a diameter of 2.5 in., has six straight cutting teeth, and rotates at 150 rpm. Calculate the material removal rate and the cutting time, and estimate the power required.

8.130 Referring to Fig. 8.54, assume that $D = 200$ mm, $w = 30$ mm, $l = 600$ mm, $d = 2$ mm, $v = 1$ mm/s, and $N = 200$ rpm. The cutter has 10 inserts and the workpiece material is 304 stainless steel. Calculate the material removal rate, cutting time, and feed per tooth, and estimate the power required.

8.131 Estimate the time required for face milling an 8-in.-long, 3-in.-wide brass block using an 8-in-diameter cutter with 12 HSS teeth.

8.132 A 12-in.-long, 2-in.-thick plate is being cut on a band saw at 150 ft/min. The saw has 12 teeth per in. If the feed per tooth is 0.003 in., how long will it take to saw the plate along its length?

8.133 A single-thread hob is used to cut 40 teeth on a spur gear. The cutting speed is 200 ft/min and the hob has a diameter of 4 in. Calculate the rotational speed of the spur gear.

8.134 In deriving Eq. (8.20) it was assumed that the friction angle, β, was independent of the shear angle, ϕ. Is this assumption valid? Explain.

8.135 An orthogonal cutting operation is being carried out under the following conditions: depth of cut = 0.10 mm, width of cut = 5 mm, chip thickness = 0.2 mm, cutting speed = 2 m/s, rake angle = 15°, cutting force = 500 N, and thrust force = 200 N. Calculate the percentage of the total energy that is dissipated in the shear plane during cutting.

8.136 An orthogonal cutting operation is being carried out under the following conditions: depth of cut = 0.020 in., width of cut = 0.1 in., cutting ratio = 0.3, cutting speed = 300 ft/min, rake angle = 0°, cutting force = 200 lb, thrust force = 150 lb, workpiece density = 0.26 lb/in³, and workpiece specific heat = 0.12 BTU/lb°F. Assume that (a) the sources of heat are the shear plane and the tool-chip interface; (b) the thermal conductivity of the tool is zero and there is no heat loss to the environment; (c) the temperature of the chip is uniform throughout. If the temperature rise in the chip is 155°F, calculate the percentage of the energy dissipated in the shear plane that goes into the workpiece.

8.137 It can be shown that the angle ψ between the shear plane and the direction of maximum grain elongation (see Fig. 8.4a) is given by the expression

$$\psi = 0.5 \cot^{-1}\left(\frac{\gamma}{2}\right),$$

where γ is the shear strain, as given by Eq. (8.3). Assume that you are given a piece of the chip obtained from orthogonal cutting of an annealed metal. The rake angle and cutting speed are also given but you have not seen the setup on which the chip was produced. Outline the procedure that you would follow to estimate the power required in producing this chip. Assume that you have access to a fully equipped laboratory and a technical library.

8.138 A lathe is set up to machine a taper on a bar stock 120-mm in diameter; the taper is 1 mm per 10 mm. A cut is made with an initial depth of cut of 4 mm at a feed rate of 0.250 mm/rev and at a spindle speed of 150 rpm. Calculate the average metal removal rate.

8.139 Develop an expression for optimum feed rate that minimizes the cost per piece if the tool life is as described by Eq. (8.34).

8.140 Assuming that the coefficient of friction is 0.25, calculate the maximum depth of cut for turning a hard aluminum alloy on a 20-hp lathe (with a mechanical efficiency of 80%) at a width of cut of 0.25 in., rake angle of 0°, and a cutting speed of 300 ft/min. What is your estimate of the material's shear strength?

8.141 Assume that, using a carbide cutting tool, you measure the temperature in a cutting operation at a speed of 250 ft/min and feed of 0.0025 in./rev as 1200°F. What would be the approximate temperature if the cutting speed is increased by 50%? What should the speed be to lower the maximum temperature to 800°F?

8.142 A 3-in-diameter gray cast-iron cylindrical part is to be turned on a lathe at 500 rpm. The depth of cut is 0.25 in. and a feed is 0.02 in./rev. What should be the minimum horsepower of the lathe?

8.143 (a) A 6-in.-diameter aluminum bar with a length of 12 in. is to have its diameter reduced to 5 in. by turning. Estimate the machining time if an uncoated carbide tool is used. (b) What is the time for a TiN-coated tool?

8.144 Calculate the power required for the cases given in Problem 8.143.

8.145 Using trigonometric relationships, derive an expression for the ratio of shear energy to frictional energy in orthogonal cutting, in terms of angles α, β, and ϕ only.

8.146 For a turning operation using a ceramic cutting tool, if the cutting speed is increased by 50%, by what factor must the feed rate be modified to obtain a constant tool life? Let $n = 0.5$ and $y = 0.6$.

8.147 Using Eq. (8.35), select an appropriate feed for $R = 1$ mm and a desired roughness of 1 μm. How would you adjust this feed to allow for nose wear of the tool during extended cuts? Explain your reasoning.

8.148 In a drilling operation, a 0.5-in. drill bit is being used in a low-carbon steel workpiece. The hole is a blind hole which will then be tapped to a depth of 1 in. The drilling operation takes place with a feed of 0.010 in./rev and a spindle speed of 700 rpm. Estimate the time required to drill the hole prior to tapping.

8.149 Assume that in the face-milling operation shown in Fig. 8.54, the workpiece dimensions are 5 in. by 10 in. The cutter is 6 in. in diameter, has 8 teeth, and rotates at 300 rpm. The depth of cut is 0.125 in. and the feed is 0.005 in./tooth. Assume that the specific energy required for this material is 2 hp-min/in³ and that only 75% of the cutter diameter is engaged during cutting. Calculate (a) the power required and (b) the material removal rate.

8.150 Calculate the ranges of typical machining times for face milling a 10-in.-long, 2-in.-wide cutter and at a depth of cut of 0.1 in. for the following workpiece materials: (a) low-carbon steel, (b) titanium alloys, (c) aluminum alloys, and (d) thermoplastics.

8.151 A machining-center spindle and tool extend 12 in. from its machine-tool frame. What temperature change can be tolerated to maintain a tolerance of 0.0001 in. in machining? A tolerance of 0.001 in.? Assume that the spindle is made of steel.

8.152 In the production of a machined valve, the labor rate is $19.00 per hour and the general overhead rate is $15.00 per hour. The tool is a square ceramic insert and costs $25.00; it takes five minutes to change and one minute to index. Estimate the optimum cutting speed from a cost perspective. Let $C = 100$ for V_o in m/min.

8.153 Estimate the optimum cutting speed in Problem 8.152 for maximum production.

8.154 Develop an equation for optimum cutting speed in a face milling operation using a milling cutter with inserts.

8.155 Develop an equation for optimum cutting speed in turning where the tool is a high-speed-steel tool that can be reground periodically.

8.156 Assume that you are an instructor covering the topics in this chapter and you are giving a quiz on the quantitative aspects to test the understanding of the students. Prepare several numerical problems, and supply the answers to them.

DESIGN

8.157 Tool life could be greatly increased if an effective means of cooling and lubrication were developed. Design methods of delivering a cutting fluid to the cutting zone and discuss the advantages and shortcomings of your design.

8.158 Devise an experimental setup whereby you can perform an orthogonal cutting operation on a lathe using a short round tubular workpiece.

8.159 Cutting tools are sometimes designed so that the chip-tool contact length is controlled by recessing the rake face some distance away from the tool tip (see the leftmost design in Fig. 8.7c). Explain the possible advantages of such a tool.

8.160 The accompanying illustration shows drawings for a cast-steel valve body before (left) and after (right) machining. Identify the surfaces that are to be machined (noting that not all surfaces are to be machined). What type of machine tool would be suitable to machine this part? What type of machining operations are involved, and what should be the sequence of these operations?

8.161 Make a comprehensive table of the process capabilities of the machining operations described in this chapter. Use several columns describe the (a) machines involved, (b) type of tools and tool materials used, (c) shapes of blanks and parts produced, (d) typical maximum and minimum sizes produced, (e) surface finish produced, (f) dimensional tolerances produced, and (g) production rates achieved.

8.162 A large bolt is to be produced from hexagonal bar stock by placing the hex stock into a chuck and machining the cylindrical shank of the bolt by turning on a lathe. List and explain the difficulties that may be involved in this operation.

8.163 Design appropriate fixtures and describe the machining operations required to produce the piston shown in Fig. 12.62.

8.164 In Figs. 8.16 and 8.17b, we note that the maximum temperature is about halfway up the face of the tool. We have also described the adverse effects of temperature on various tool materials. Considering the mechanics of cutting operations, describe your thoughts on the technical and economic merits of embedding a small insert, made of materials such as ceramic or carbide, halfway up the rake face of a tool made of a material with lower resistance to temperature than ceramic or carbide.

8.165 Describe your thoughts on whether chips produced during machining can be used to make useful products. Give some examples of possible products and comment on their characteristics and differences as compared to the same products made by other manufacturing processes. Which types of chips would be desirable for this purpose? Explain.

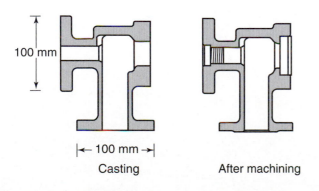

Casting   After machining

8.166 Experiments have shown that it is possible to produce thin, wide chips, such as 0.08 mm (0.003 in.) thick and 10 mm (4 in.) wide, which would be similar to rolled sheet. Materials used have been aluminum, magnesium, and stainless steel. A typical setup would be similar to orthogonal cutting, by machining the periphery of a solid round bar with a straight tool moving radially inward (plunge). Describe your thoughts on producing thin metal sheet by this method, its surface characteristics, and its properties.

8.167 One of the principal concerns with coolants is degradation due to biological attack by bacteria. To prolong their life, chemical biocides are often added, but these biocides greatly complicate the disposal of coolants. Conduct a literature search regarding the latest developments in the use of environmentally benign biocides in cutting fluids.

8.168 If expanded honeycomb panels (see Section 7.5.5) were to be machined in a form milling operation (see Fig. 8.58b), what precautions would you take to keep the sheet metal from buckling due to cutting forces? Think of as many solutions as you can.

8.169 The part shown in the accompanying figure is a power-transmitting shaft; it is to be produced on a lathe. List the operations that are appropriate to make this part and estimate the machining time.

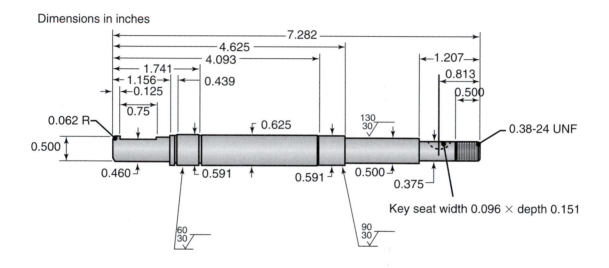

Material-Removal Processes: Abrasive, Chemical, Electrical, and High-Energy Beams

This chapter describes the essential features of finishing operations, commonly performed to improve dimensional tolerances and surface finish of products. Among the topics covered are

- The technology of grinding wheels and mechanics of grinding operations.
- Types of grinding machines and advanced abrasive machining processes.
- Various abrasive machining operations, including lapping, honing, polishing, chemical mechanical polishing, and the use of coated abrasives.
- Processes based on nonmechanical means of material removal, including chemical and electrochemical machining, electrical discharge machining, laser and electron beam machining, and abrasive jet machining.
- Deburring operations.
- Design and economic considerations for the processes described.

9.1 | Introduction

In all the machining processes described in detail in Chapter 8, the cutting tool is made of a certain material and has a clearly defined geometry. Furthermore, the machining process is carried out by chip removal, the mechanics of which are reasonably well understood. There are, however, many situations in manufacturing where the workpiece material is either *too hard* or *too brittle*, or its *shape* is difficult to produce with sufficient dimensional accuracy by any of the machining methods described previously. One of the best methods for producing such parts is using **abrasives**. An abrasive is a small, hard particle that has sharp edges and an irregular shape, unlike typical cutting tools. Abrasives are capable of removing small amounts of material from a surface by a cutting process that produces tiny chips.

Abrasive machining processes are generally among the last operations performed on manufactured products, although they are not necessarily confined to fine or small-scale material removal from workpieces, and they can indeed compete economically with some machining processes, such as milling and turning. Because they are hard, abrasives are also used in finishing very hard or heat-treated parts; shaping hard nonmetallic materials, such as ceramics and glasses; removing unwanted weld beads; cutting off lengths of bars, structural shapes, masonry, and concrete; and cleaning surfaces, for which jets of air or water containing abrasive particles are used.

In addition to abrasive machining options, several **advanced machining processes** have been developed, beginning in the 1940s. Also called *nontraditional* or *unconventional* machining, these processes are based on electrical, chemical, fluid, and thermal principles, and are attractive when one or more of the following occur:

1. The hardness and strength of the workpiece material is very high, typically above 400 HB.

2. The part is too flexible or slender to support the machining or grinding forces, or parts are difficult to clamp in workholding devices.

3. The shape of the part is complex, such as internal and external profiles or small-diameter deep holes.

4. Surface finish and dimensional accuracy required are better than those obtainable by other processes.

5. Temperature rise or residual stresses in the workpiece are undesirable or unacceptable.

When selected and applied properly, advanced machining processes offer significant economic and technical advantages over the traditional machining methods described in the preceding chapter.

9.2 | Abrasives

The abrasives commonly used in manufacturing are as follows:

1. **Conventional abrasives**

   - *Aluminum oxide* (Al_2O_3)
   - *Silicon carbide* (SiC)

2. **Superabrasives**

   - *Cubic boron nitride* (cBN)
   - *Diamond*

Abrasives are significantly harder than conventional cutting-tool materials, as can be seen by comparing Tables 8.6 and 9.1. The last two of the four abrasives listed above are the two hardest materials known, hence the term *superabrasives*. In addition to hardness, an important characteristic of an abrasive is **friability**, that is,

TABLE 9.1

Knoop Hardness Range for Various Materials and Abrasives			
Common glass	350–500	Titanium nitride	2000
Flint, quartz	800–1100	Titanium carbide	1800–3200
Zirconium oxide	1000	Silicon carbide	2100–3000
Hardened steels	700–1300	Boron carbide	2800
Tungsten carbide	1800–2400	Cubic boron nitride	4000–5000
Aluminum oxide	2000–3000	Diamond	7000–8000

the ability of an abrasive grain to fracture (break down) into smaller pieces. Friability gives abrasives *self-sharpening characteristics,* which are important in maintaining the sharpness of the abrasives during use. High friability indicates low strength or low fracture resistance of the abrasive; thus, a highly friable abrasive grain fragments more rapidly under grinding forces than an abrasive grain with low friability. Aluminum oxide, for example, has lower friability than silicon carbide.

The **shape** and **size** of an abrasive grain also affect its friability. Blocky grains, which may be analogous to negative-rake-angle cutting tools (see Fig. 8.28), are, for example, less friable than platelike grains. As for size, because the probability of defects in smaller grains is lower (due to the *size effect;* Section 3.8.3), they are stronger and less friable than larger grains. The importance of friability in abrasive processes is described further in Section 9.5.

Types of abrasives. Abrasives found in nature include *emery, corundum (alumina), quartz, garnet*, and *diamond*. However, natural abrasives contain unknown amounts of impurities and possess nonuniform properties; consequently, their performance is inconsistent and unreliable. As a result, aluminum oxides and silicon carbides used as abrasives are made synthetically to control impurities.

1. Synthetic **aluminum oxide** (Al_2O_3), first made in 1893, is obtained by fusing bauxite, iron filings, and coke. Aluminum oxides are divided into two groups: fused and unfused. **Fused aluminum oxides** are categorized as *white* (very friable), *dark* (less friable), and *monocrystalline* (single crystal). **Unfused alumina,** also known as *ceramic aluminum oxides,* can be harder than fused alumina. The purest form of fused alumina is **seeded gel.** First introduced in 1987, seeded gel has a particle size on the order of 0.2 μm (8 μin.), which is much smaller than abrasive grains commonly used in industry. Seeded gels are sintered (see also Section 11.4) to form larger sizes. Because of their hardness and relatively high friability, seeded gels maintain their sharpness and hence are used for difficult-to-grind materials.

2. **Silicon carbide** (SiC), first discovered in 1891, is made with silica sand, petroleum coke, and small amounts of sodium chloride. Silicon carbides are available in *green* (more friable) and *black* (less friable) types and generally have higher friability than aluminum oxides; hence, they have a higher tendency to fracture and remain sharp.

3. **Cubic boron nitride** (cBN) was first produced in the 1970s. Its properties and characteristics are described in Sections 9.6.7 and 11.8.1.

4. **Diamond** was first used as an abrasive in 1955. It is also produced synthetically, in which case it is known as *synthetic* or *industrial diamond*. Its properties and characteristics are described in Sections 8.6.9 and 11.13.2.

Grain size. As used in manufacturing operations, abrasives are generally very small compared with the size of the cutting tools and inserts described in Chapter 8. Also, abrasives have sharp edges, thus allowing the removal of very small quantities of material from the workpiece surface; thus very fine surface finish and dimensional accuracy can be obtained. (See Figs. 6.26 and 9.27.) The size of an abrasive grain is identified by a **grit number,** which is a function of sieve size; the smaller the sieve size, the larger is the grit number. For example, grit number 10 is rated as very coarse, 100 as fine, and 500 as very fine. As commonly observed, sandpaper and emery cloth also are identified in this manner, with the grit number printed on the back of the abrasive paper or cloth.

9.3 | Bonded Abrasives

Because each abrasive grain typically removes only a very small amount of material at a time, high rates of material removal can be obtained only if a large number of grains act together. This is accomplished by using *bonded abrasives,* typically in the form of a **grinding wheel** (Fig. 9.1). The abrasive grains are held together by a **bonding material** (various types of which are described in Section 9.3.1), which acts as supporting posts or braces between the grains. Some porosity is essential in bonded abrasives to provide clearance for the minute chips being produced as well as to provide cooling; otherwise, the chips would interfere with the grinding operation. Note that it would be impossible to use a grinding wheel that is fully dense (solid) with no porosity. Porosity can easily be observed simply by looking at the surface of any grinding wheel. Other features of the grinding wheel shown in Fig. 9.1 are described in Sections 9.4 and 9.5.

Some of the more commonly used types of grinding wheels are shown in Fig. 9.2 for conventional abrasives and in Fig. 9.3 for superabrasives. Note that, due to the high cost of the latter, only a small portion of the periphery of the wheels consists of a layer of superabrasives. Bonded abrasives are marked with a standardized system of letters and numbers, indicating the type of abrasive, grain size, grade, structure, and bond type. Figure 9.4 shows the marking system for aluminum-oxide and silicon-carbide bonded abrasives, and Fig. 9.5 shows the marking system for diamond and cubic-boron-nitride bonded abrasives.

9.3.1 Bond types

The common bond types for bonded abrasives are vitrified, resinoid, rubber, and metal and are used for conventional abrasives as well as for superabrasives (except rubber bonds).

1. **Vitrified.** Essentially a glass, a *vitrified bond* is also called a *ceramic bond,* particularly outside the United States; it is the most common and widely used bond. The raw materials in the bond consist of feldspar (a crystalline mineral) and various clays. These materials are first mixed with the abrasives, moistened, and then molded under pressure into the shape of grinding wheels. These "green" products (similar to powder-metallurgy parts; Section 11.3) are then slowly fired, up to a temperature of about 1250°C (2300°F), to fuse the glass to develop structural strength. The wheels are then cooled slowly to prevent

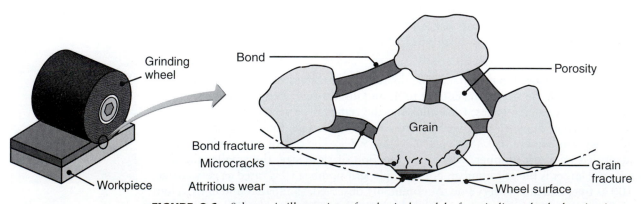

FIGURE 9.1 Schematic illustration of a physical model of a grinding wheel, showing its structure and grain wear and fracture patterns.

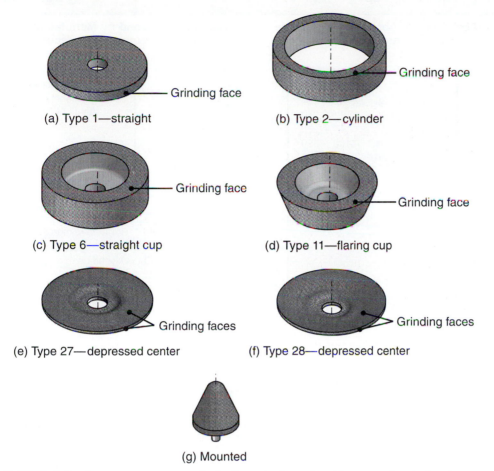

FIGURE 9.2 Some common types of grinding wheels made with conventional abrasives (aluminum oxide and silicon carbide). Note that each wheel has a specific grinding face; grinding on other surfaces is improper and unsafe.

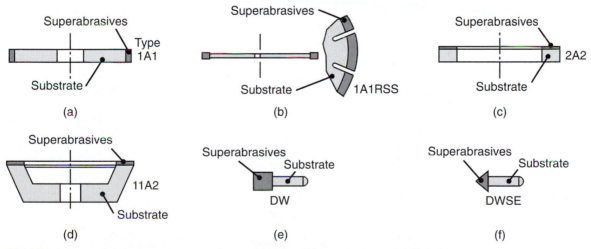

FIGURE 9.3 Examples of superabrasive wheel configurations. The rim consists of superabrasives and the wheel itself (core) is generally made of metal or composites. Note that the basic numbering of wheel types (such as 1, 2, and 11) is the same as that shown in Fig. 9.2. The bonding materials for the superabrasives are (a), (d), and (e) resinoid, metal, or vitrified, (b) metal, (c) vitrified, and (f) resinoid.

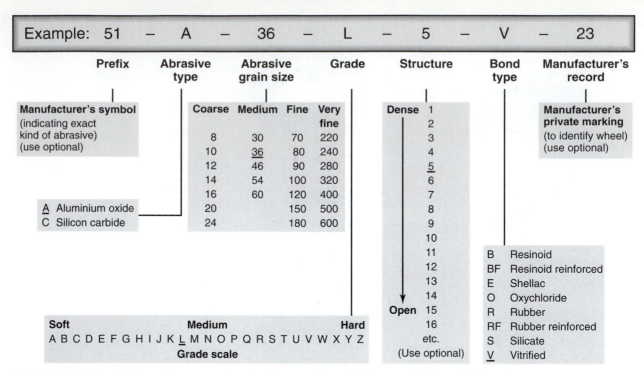

FIGURE 9.4 Standard marking system for aluminum-oxide and silicon-carbide bonded abrasives.

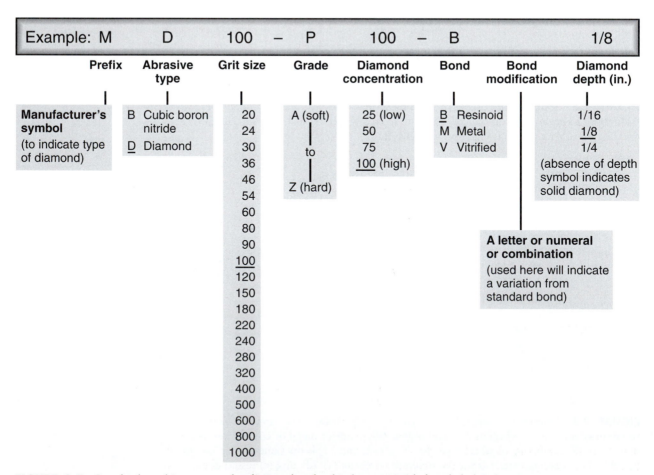

FIGURE 9.5 Standard marking system for diamond and cubic-boron-nitride bonded abrasives.

thermal cracking, finished to size, inspected for quality and dimensional accuracy, and tested for defects.

Vitrified bonds produce wheels that are strong, stiff, porous, and resistant to oils, acids, and water; however, because the wheels are brittle, they lack resistance to mechanical and thermal shock. Vitrified wheels are also available with steel backing plates or cups for better structural support during their use.

2. **Resinoid.** *Resinoid* bonding materials are *thermosets* (see Section 10.4) and are available in a wide range of compositions and properties. Because the bond is an organic compound, wheels with resinoid bonds are also called **organic wheels.** The basic manufacturing procedure consists of mixing the abrasive with liquid or powdered phenolic resins and additives, pressing the mixture into the shape of a grinding wheel, and curing it at temperatures of about 175°C (350°F) Because the elastic modulus of thermosetting resins is lower than that of glasses, resinoid wheels are more flexible than vitrified wheels. A more recent development is the use of *polyimide* (see Section 10.6) as a substitute for the phenolic in resinoid wheels. Polyimide is tough and has good resistance to high temperatures.

    Reinforced resinoid wheels, in which one or more layers of generally fiberglass mats of various mesh sizes provide the reinforcement, are widely used. The main purpose of the reinforcement is to retard the disintegration of the wheel should it break for some reason, rather than to improve its strength. Large-diameter wheels can additionally be supported with one or more internal rings (made of round steel bar) which are inserted during molding of the wheel.

3. **Rubber.** The most flexible bond used in abrasive wheels is *rubber.* Such wheels are manufactured by mixing crude rubber, sulfur, and abrasive grains together, rolling the mixture into sheets, cutting out circles, and heating them under pressure to vulcanize the rubber. Thin wheels can be made in this manner and are used like saws for cutting-off operations (cut-off blades).

4. **Metal bonds.** Abrasive grains, usually diamond or cubic boron nitride, are bonded in a metal matrix to the periphery of a metal wheel, typically to depths of 6 mm (0.25 in.) or less. (See Fig. 9.3.) The bonding is carried out under high pressure and temperature. The wheel itself (core) may be made of aluminum, bronze, steel, ceramics, or composite materials, depending on special requirements for the wheel, such as strength, stiffness, and dimensional stability. Superabrasive wheels may be layered so that a single abrasive layer is plated or brazed to a metal wheel.

5. **Other bonds.** In addition to those described above, other types of bonds include *silicate, shellac,* and *oxychloride* bonds. Because these bonds have limited uses they are not discussed further here.

9.3.2 Wheel grade and structure

The *grade* of a bonded abrasive is a measure of the bond's strength, and it includes both the *type* and the *amount* of bond in the wheel. Because strength and hardness are directly related, the grade is also referred to as the *hardness* of a bonded abrasive. Thus, a hard wheel has a stronger bond and/or a larger amount of bonding material than a soft wheel. The *structure* is a measure of the *porosity* (the spacing between the grains, as can be seen in Fig. 9.1). of the bonded abrasive. Some porosity is essential to provide clearance for the grinding chips, as otherwise they would interfere with the grinding operation. The structure of bonded abrasives ranges from dense to open (see Fig. 9.4).

9.4 | Mechanics of Grinding

Grinding is basically a *chip removal process* in which the cutting tool is an individual abrasive grain. The following are major factors that differentiate the action of a single grain from that of a single-point cutting tool (see Fig. 8.2):

1. The individual grain has an irregular geometry and is spaced randomly along the periphery of the wheel (Fig. 9.6).

2. The average rake angle of the grains is highly negative, typically $-60°$ or even lower; consequently, the shear angles are very low (see Section 8.2.4).

3. The grains in the periphery of a grinding wheel have different radial positions.

4. The cutting speeds of grinding wheels are very high (Table 9.2), typically on the order of 30 m/s (6000 ft/min).

An example of chip formation by an abrasive grain is shown in Fig. 9.7. Note the negative rake angle, the low shear angle, and the very small size of the chip (see also Example 9.1). Grinding chips are easily collected on a piece of adhesive tape held against the sparks of a grinding wheel. From direct observation it will be noted that a variety of metal chips can be obtained in grinding.

The mechanics of grinding and the variables involved can best be studied by analyzing the *surface-grinding* operation, shown in Fig. 9.8. In this figure, a grinding wheel of diameter D is removing a layer of metal at a depth d, known as the **wheel depth of cut.** An individual grain on the periphery of the wheel is moving at a tangential velocity V (*up*, or *conventional, grinding,* as shown in Fig. 9.8. see also *milling,* Section 8.10.1), and the workpiece is moving at a velocity v. The grain is removing a chip with an *undeformed thickness* (**grain depth of cut**), t, and an undeformed length, l. For the condition of $v \ll V$, the *undeformed-chip length, l* is approximately

$$l \simeq \sqrt{Dd}. \tag{9.1}$$

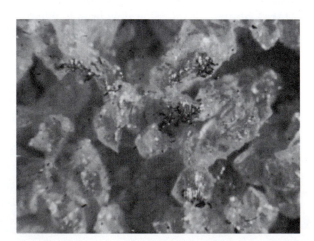

FIGURE 9.6 The grinding surface of an abrasive wheel (A46-J8V), showing grains, porosity, wear flats on grains (see also Fig. 9.7b), and metal chips from the workpiece adhering to the grains. Note the random distribution and shape of the abrasive grains. Magnification: 50×.

TABLE 9.2

Typical Ranges of Speeds and Feeds for Abrasive Processes				
Process variable	Conventional grinding	Creep-feed grinding	Buffing	Polishing
Wheel speed (m/min)	1500–3000	1500–3000	1800–3600	1500–2400
Work speed (m/min)	10–60	0.1–1	—	—
Feed (mm/pass)	0.01–0.05	1–6	—	—

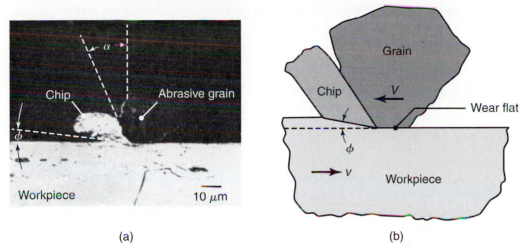

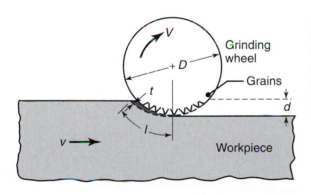

FIGURE 9.7   (a) Grinding chip being produced by a single abrasive grain. Note the large negative rake angle of the grain. (b) Schematic illustration of chip formation by an abrasive grain. Note the negative rake angle, the small shear angle, and the wear flat on the grain. *Source:* (a) After M.E. Merchant.

FIGURE 9.8   Basic variables in surface grinding. In actual grinding operations, the wheel depth of cut, d, and contact length, l, are much smaller than the wheel diameter, D. The dimension t is called the *grain depth of cut.*

For *external (cylindrical) grinding* (see Section 9.6),

$$l = \sqrt{\frac{Dd}{1 + (D/D_w)}},$$ (9.2)

and for *internal grinding,*

$$l = \sqrt{\frac{Dd}{1 - (D/D_w)}},$$ (9.3)

where D_w is the diameter of the workpiece.

The relationship between t and other process variables can be derived as follows: Let C be the number of cutting points per unit area of wheel surface, and v and V the surface speeds of the workpiece and the wheel, respectively (Fig. 9.8). Assuming the width of the workpiece to be unity, the number of grinding chips produced per unit time is VC, and the volume of material removed per unit time is vd.

Letting r be the ratio of the chip width, w, to the average chip thickness, the volume of a chip with a rectangular cross-sectional area and constant width along its length is

$$\text{Vol}_{\text{chip}} = \frac{wtl}{2} = \frac{rt^2l}{4}.$$ (9.4)

The volume of material removed per unit time is the product of the volume of each chip and the number of chips produced per unit time. Thus,

$$VC\frac{rt^2l}{4} = vd,$$

and because $l = \sqrt{Dd}$, the undeformed chip thickness in surface grinding is

$$t = \sqrt{\frac{4v}{VCr}\sqrt{\frac{d}{D}}}. \tag{9.5}$$

Experimental observations indicate the value of C to be approximately on the order of 0.1 to 10 per mm^2 (10^2 to 10^3 per in^2); the finer the grain size of the wheel, the larger is this number. The magnitude of r is between 10 and 20 for most grinding operations. Substituting typical values for a grinding operation into Eqs. (9.1) through (9.5), it can be noted that l and t are very small quantities. For example, typical values for t are in the range of 0.3–0.4 μm (12–160 μin.).

EXAMPLE 9.1 Chip dimensions in grinding

Estimate the undeformed chip length and the undeformed chip thickness for a typical surface-grinding operation. Let $D = 200$ mm, $d = 0.05$ mm, $C = 2$ per mm^2, and $r = 15$.

Solution. The formulas for undeformed length and thickness, respectively, are

$$l = \sqrt{Dd} \quad \text{and} \quad t = \sqrt{\frac{4v}{VCr}\sqrt{\frac{d}{D}}}.$$

From Table 9.2 the following values are selected:

$$v = 0.5 \text{ m/s} \quad \text{and} \quad V = 30 \text{ m/s}.$$

Therefore,

$$l = \sqrt{(200)(0.05)} = 3.2 \text{ mm} = 0.126 \text{ in.}$$

and

$$t = \sqrt{\frac{(4)(0.5)}{(30)(2)(15)}\sqrt{\frac{0.05}{200}}} = 0.006 \text{ mm} = 2.3 \times 10^{-4} \text{ in.}$$

Note that because of plastic deformation, the actual length of the chip is shorter and the thickness greater than these values. (See Fig. 9.7.)

9.4.1 Grinding forces

A knowledge of forces is essential not only in the design of grinding machines and workholding devices, but also in determining the deflections that the workpiece and the machine will undergo. Deflections, in turn, adversely affect dimensional accuracy of the workpiece, which is especially critical in precision grinding.

In the *force* on the grain (see the discussion of cutting force, F_c, in Section 8.2.3) is proportional to the cross-sectional area of the undeformed chip, it can be shown that the **relative grain force** is given by the expression

$$\text{Relative grain force} \propto \frac{v}{VC}\sqrt{\frac{d}{D}}. \tag{9.6}$$

The actual force is then the product of the relative grain force and the strength of the metal being ground.

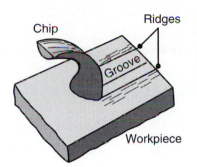

FIGURE 9.9 Chip formation and plowing (plastic deformation without chip removal) of the workpiece surface by an abrasive grain.

The **specific energy** consumed in producing a grinding chip consists of three components:

$$u = u_{\text{chip}} + u_{\text{plowing}} + u_{\text{sliding}}. \qquad (9.7)$$

The quantity u_{chip} is the specific energy required for chip formation by plastic deformation, u_{plowing} is the specific energy required for plowing, which is plastic deformation without chip removal (Fig. 9.9), and the last term, u_{sliding}, can best be understood by observing the grain in Fig. 9.7b. The grain develops a **wear flat** as a result of the grinding operation (similar to flank wear in cutting tools; see Section 8.3). The wear flat is in contact with the surface being ground and, because of friction, requires energy for sliding. The larger the wear flat, the higher is the grinding force.

Typical specific-energy requirements in grinding are given in Table 9.1. Note that these energy levels are much higher than those in cutting operations with single-point tools, as given in Table 8.3. This difference has been attributed to the following factors:

1. **Size effect.** As previously stated, the size of grinding chips is very small, as compared with chips produced in other cutting operations, by about two orders of magnitude. As described in Section 3.8.3, the smaller the size of a piece of metal, the higher is its strength; consequently, grinding involves higher specific energy than machining operations. Studies have indicated that extremely high dislocation densities (see Section 3.3.3) occur in the shear zone during chip formation, thus influencing the grinding energies by virtue of increased strength.

2. **Wear flat.** A wear flat (see Fig. 9.7b) requires frictional energy for sliding; this energy contributes significantly to the total energy consumed. The size of the wear flat in grinding is much larger than the grinding chip, unlike in metal cutting by a single-point tool, where flank wear land is small compared with the size of the chip (see Section 8.3).

3. **Chip morphology.** Because the average rake angle of a grain is highly negative (see Fig. 9.7), the shear strains in grinding are very large. This indicates that the energy required for plastic deformation to produce a grinding chip is higher than in other machining processes. Furthermore, note that plowing consumes energy without contributing to chip formation (see Fig. 9.9).

EXAMPLE 9.2 Forces in surface grinding

Assume that you are performing a surface-grinding operation on a low-carbon steel workpiece using a wheel of diameter $D = 10$ in. that rotates at $N = 4000$ rpm. The width of cut is $w = 1$ in., depth of cut is $d = 0.002$ in., and the feed rate of the workpiece is $v = 60$ in./min. Calculate the cutting force, F_c (the force tangential to the wheel), and the thrust force, F_n (the force normal to the workpiece).

TABLE 9.3

Approximate Specific-Energy Requirements for Surface Grinding

Workpiece material	Hardness	Specific energy	
		W-s/mm^3	hp-min/in^3
Aluminum	150 HB	7–27	2.5–10
Cast iron (class 40)	215 HB	12–60	4.5–22
Low-carbon steel (1020)	110 HB	14–68	5–25
Titanium alloy	300 HB	16–55	6–20
Tool steel (T15)	67 HRC	18–82	6.5–30

Solution. We first determine the material removal rate as follows:

$$\text{MRR} = dwv = (0.002)(1)(60) = 0.12 \text{ in}^3/\text{min.}$$

The power consumed is given by

$$\text{Power} = (u)(\text{MRR}),$$

where u is the specific energy, as obtained from Table 9.3. For low-carbon steel, let's estimate u to be 15 hp-min/in^3. Hence,

$$\text{Power} = (15)(0.12) = 1.8 \text{ hp.}$$

By noting that 1 hp = 33,000 ft-lb/min = 396,000 in.-lb/min, we obtain

$$\text{Power} = (1.8)(396,000) = 712,800 \text{ in.-lb/min.}$$

Since power is defined as

$$\text{Power} = T\omega,$$

where T is the torque and is equal to $(F_c)(D/2)$ and ω is the rotational speed of the wheel in radians per minute, we also have $\omega = 2\pi N$. Thus,

$$712,800 = (F_c)\left(\frac{10}{2}\right)(2\pi)(4000),$$

and therefore, $F_c = 57$ lb. The thrust force, F_n, can be calculated by noting from experimental data in the technical literature that it is about 30% higher than the cutting force, F_c. Consequently,

$$F_n = (1.3)(57) = 74 \text{ lb.}$$

9.4.2 Temperature

Temperature rise in grinding is an important consideration because it can adversely affect surface properties and cause residual stresses on the workpiece. Furthermore, temperature gradients in the workpiece cause distortions due to thermal expansion and contraction. When some of the heat generated during grinding is conducted into the workpiece, the heat expands the part being ground, thus making it difficult to control dimensional accuracy. The work expended in grinding is mainly converted into heat. The *surface temperature rise*, ΔT, has been found to be a function of the ratio of the total energy input to the surface area ground. Thus, in *surface grinding*, if w is the width and L is the length of the surface area ground, then

$$\Delta T \propto \frac{uwLd}{wL} \propto ud. \tag{9.8}$$

If we introduce size effect and assume that u varies inversely with the undeformed chip thickness t, then the temperature rise is

$$\text{Temperature rise} \propto D^{1/4} d^{3/4} \left(\frac{V}{v} \right)^{1/2}. \qquad (9.9)$$

The *peak temperatures* in chip generation during grinding can be as high as 1650°C (3000°F). However, the time involved in producing a chip is extremely short (on the order of microseconds); hence melting of the chip may or may not occur. Because, as in machining, the chips carry away much of the heat generated (see Fig. 8.18), only a small fraction of the heat generated is conducted to the workpiece. Experiments indicate that as much as one-half the energy dissipated in grinding is conducted to the chip, a percentage that is higher than that in machining (see Section 8.2). On the other hand, the heat generated by sliding and plowing is conducted mostly into the workpiece.

Sparks. The sparks observed in grinding metals are actually glowing chips. The glowing occurs because of the *exothermic reaction* of the hot chips with oxygen in the atmosphere; sparks have not been observed with any metal ground in an oxygen-free environment. The color, intensity, and shape of the sparks depend on the composition of the metal being ground. If the heat generated by the exothermic reaction is sufficiently high, the chip may melt and, because of surface tension, solidify as a shiny spherical particle. Observation of these particles under scanning electron microscopy has revealed that they are hollow and have a fine dendritic structure (see Fig. 5.8), indicating that they were once molten (by exothermic oxidation of hot chips in air) and they resolidified rapidly. It has been suggested that some of the spherical particles may also be produced by plastic deformation and rolling of chips at the grain-workpiece interface during grinding.

9.4.3 Effects of temperature

The major effects of temperature in grinding are

1. **Tempering.** Excessive temperature rise caused by grinding can temper (Section 5.11.5) and soften the surfaces of steel components, which are often ground in the heat-treated and hardened state. Grinding process parameters must therefore be chosen carefully to avoid excessive temperature rise. Grinding fluids (Section 9.6.9) can effectively control temperatures.

2. **Burning.** If the temperature rise is excessive, the workpiece surface may burn. Burning produces a bluish color on steels, which indicates oxidation at high temperatures. A burn may not be objectionable in itself; however, the surface layers may undergo metallurgical transformations, with martensite formation in high-carbon steels from reaustenization followed by rapid cooling (see Section 5.11). This effect is known as *metallurgical burn,* which is an especially serious concern with nickel-base alloys.

3. **Heat checking.** High temperatures in grinding lead to thermal stresses and may cause thermal cracking of the workpiece surface, known as *heat checking*. (See also Section 5.10.3.) Cracks are usually perpendicular to the grinding direction. Under severe grinding conditions, however, parallel cracks may also develop. Heat checking is detrimental from fatigue as well as aesthetic standpoints.

4. **Residual stresses.** Temperature change and gradients within the workpiece are mainly responsible for residual stresses in grinding. Other contributing factors are the physical interactions of the abrasive grain in chip formation and the

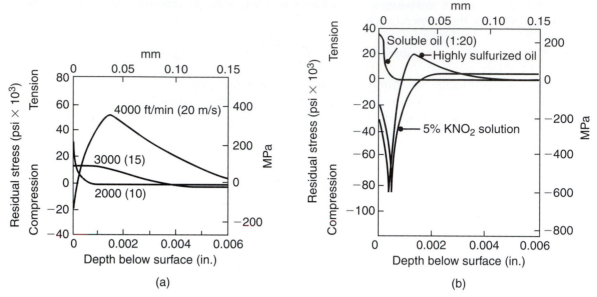

FIGURE 9.10 Residual stresses developed on the workpiece surface in grinding tungsten:
(a) effect of wheel speed and (b) effect of type of grinding fluid. Tensile residual stresses on a
surface are detrimental to the fatigue life of ground components. The variables in grinding
can be controlled to minimize residual stresses, a process known as *low-stress grinding*.
Source: After N. Zlatin.

sliding of the wear flat along the workpiece surface, causing plastic deformation
of the surface. Two examples of residual stresses in grinding are shown in
Fig. 9.10, demonstrating the effects of wheel speed and the type of grinding
fluid used. The method and direction of the application of grinding fluid also
can have a significant effect on residual stresses. Because of the deleterious effect
of tensile residual stresses on fatigue strength (Section 3.8.2), process parameters should be chosen properly. Residual stresses can usually be lowered by
using softer grade wheels (*free-cutting wheels*), lower wheel speeds, and higher
work speeds, a procedure known as **low-stress**, or **gentle, grinding**.

9.5 Grinding Wheel Wear

Grinding wheel wear is an important consideration because it adversely affects the
shape and accuracy of ground surfaces, a situation similar to wear of cutting tools
(Section 8.3). Grinding wheels wear by three different mechanisms as described next.

1. **Attritious wear.** The cutting edges of a sharp grain become dull by attrition
   (known as *attricious wear*), developing a *wear flat* (see Fig. 9.7b) that is similar to flank wear in cutting tools. Wear is caused by the interaction of the grain
   with the workpiece material, resulting in complex physical and chemical reactions. These reactions involve diffusion, chemical degradation or decomposition
   of the grain, fracture at a microscopic scale, plastic deformation, and melting.

   Attritious wear is low when the two materials are chemically inert with
   respect to each other, much like with the use of cutting tools. The more inert
   the materials, the lower will be the tendency for reaction and adhesion to occur
   between the grain and the workpiece being ground. For example, because aluminum oxide is relatively inert to iron, its rate of attritious wear when it is used

to grind steels is much lower than that for silicon carbide and diamond. On the other hand, silicon carbide can dissolve in iron, and hence it is not suitable for grinding steels. Cubic boron nitride has a higher inertness to steels, and hence it is suitable for use as an abrasive. Consequently, the selection of the type of abrasive for low attritious wear should be based on the reactivity of the grain and the workpiece, as well as their relative mechanical properties, such as hardness and toughness. The environment and the type of grinding fluid used also have an influence on grain-workpiece material interactions.

2. **Grain fracture.** Because abrasive grains are brittle, their fracture characteristics in grinding are important. If the wear flat caused by attritious wear is excessive, the grain becomes dull, and the grinding operation becomes inefficient and produces high temperatures. Optimally, the grain should fracture or fragment at a moderate rate so that new sharp cutting edges are produced continuously during grinding. Note that this process is equivalent to breaking a piece of dull chalk into two or more pieces in order to expose new sharp edges. Recall that Section 9.2 has described *friability* of abrasives, which gives them their self-sharpening characteristics, an important consideration in effective grinding.

   The selection of grain type and size for a particular application also depends on the attritious-wear rate. Note that a grain-workpiece material combination with high attritious wear and low friability causes dulling of grains and development of a large wear flat. Grinding then becomes inefficient, and surface damage, such as burning, is likely to occur.

   The following combinations are generally recommended for grain selection:

   a. *Aluminum oxide:* for steels, ferrous alloys, and alloy steels.
   b. *Silicon carbide:* for cast iron, nonferrous metals, and hard and brittle materials (such as carbides, ceramics, marble, and glass).
   c. *Diamond:* for ceramics, cemented-carbide ceramics, and some hardened steels.
   d. *Cubic boron nitride:* for steels and cast irons at 50 HRC (such as hardened tool steels) or above, and for high-temperature superalloys.

3. **Bond fracture.** The strength of the bond (*grade*) is a significant parameter in grinding. Note that if the bond is too strong, dull grains cannot be dislodged so that other, sharp grains along the circumference of the grinding wheel can begin to contact the workpiece and remove chips. Thus, the grinding process becomes inefficient. On the other hand, if the bond is too weak, the grains are easily dislodged, and the wear rate of the wheel increases. Consequently, maintaining dimensional accuracy of the workpiece becomes difficult. In general, softer bonds are recommended for harder materials and for reducing residual stresses and thermal damage to the workpiece. Hard-grade wheels are used for softer materials and for removing large amounts of material at high rates (see also Section 9.5.3).

9.5.1 Dressing, truing, and shaping of grinding wheels

Dressing is the process of conditioning worn grains on the surface of a grinding wheel in order to produce sharp new grains and for truing an out-of-round wheel. Dressing is necessary when excessive attritious wear dulls the wheel (a phenomenon called **glazing**, because of the shiny appearance of the wheel surface) or when the wheel becomes loaded. **Loading** occurs when the porosities on the grinding surfaces of the wheel (see Fig. 9.6) become filled or clogged with chips. Loading can occur

(a) while grinding soft workpiece materials, (b) by improper selection of the grinding wheel, such as a wheel with low porosity, and (c) by improper selection of processing parameters. A loaded wheel obviously grinds very inefficiently, generating much frictional heat and causing surface damage and loss of dimensional accuracy.

Dressing is done by any of the following techniques:

1. A specially shaped diamond-point tool or diamond *cluster* is moved across the width of the grinding face of a rotating wheel, removing a very small layer from the wheel surface with each pass. This method can be used either dry or wet (using grinding fluids), depending on whether the wheel is to be used dry or wet, respectively, during grinding.

2. A set of star-shaped steel disks is manually pressed against the rotating grinding wheel, and material is removed from the wheel surface by crushing the grains. This method produces a coarse grinding surface on the wheel and is used only for rough grinding operations on bench or pedestal grinders (Section 9.6.5).

3. *Abrasive sticks* may be held against the grinding surface of the wheel. This is common when grinding with softer wheels but is not appropriate for precision grinding operations.

4. More recent developments in dressing techniques, for metal-bonded diamond wheels, involve the use of electrical-discharge and electrochemical machining techniques, which erode away very small layers of the metal bond, thus exposing new diamond cutting edges.

5. Dressing for *form grinding* involves **crush dressing**, or *crush forming*, and consists of pressing a metal roll on the surface of the grinding wheel, which is usually a vitrified wheel. The roll (made of high-speed steel, tungsten carbide, or boron carbide) has a machined or ground profile and thus reproduces a replica of this profile on the surface of the grinding wheel being dressed (see Fig. 9.11).

Dressing techniques and the frequency with which the wheel surface is dressed are significant factors affecting grinding forces and workpiece surface finish. Modern computer-controlled grinders (Section 9.6) are equipped with automatic dressing features, which dress the wheel during the grinding operation. For a typical aluminum-oxide wheel, the depth removed during dressing is on the order of 5 to 15 μm (200 to 600 μin.), but for a cBN wheel, it would be 2 to 10 μm (80 to 400 μin.). Modern dressing systems have a resolution as low as 0.25 to 1 μm (10 to 40 μin.).

Dressing can also be performed to generate a certain shape or form on a grinding wheel for the purpose of grinding specific profiles on workpieces. (See Section 9.6.2.) *Truing* is an operation by which a wheel is restored to its original shape. A round wheel is *shaped* to make its circumference a true circle, hence the word *truing*. Grinding wheels can also be shaped to the form to be ground on the workpiece. The grinding face on the Type 1 straight wheel shown in Fig. 9.2a is cylindrical and thus produces a flat surface; however, this surface can be shaped by dressing the wheel to various forms. Modern grinders are equipped with computer-controlled shaping features, whereby the diamond dressing tool automatically traverses the wheel face along a certain prescribed path (Fig. 9.11). Note in this figure that the axis of the diamond dressing tool remains normal to the wheel face at the point of contact.

9.5.2 Grinding ratio

Grinding wheel wear is generally correlated with the amount of material ground by a parameter called the *grinding ratio*, G, which is defined as

$$G = \frac{\text{Volume of material removed}}{\text{Volume of wheel wear}}. \tag{9.10}$$

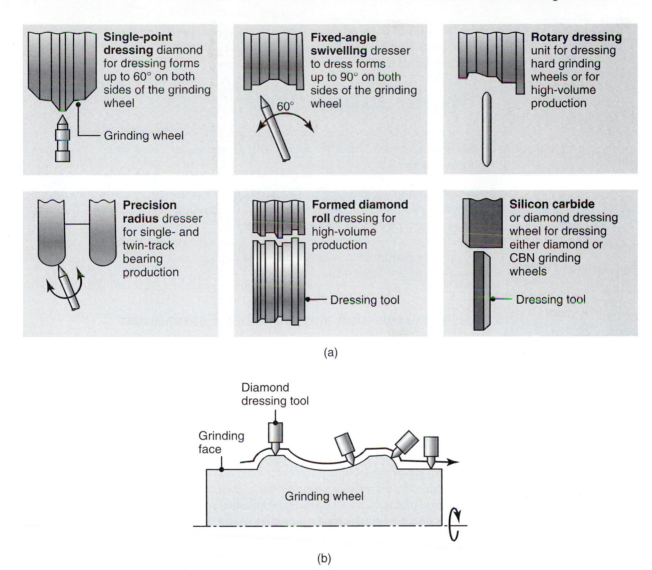

FIGURE 9.11 (a) Methods of grinding wheel dressing and (b) shaping the grinding face of a wheel by dressing it with computer-controlled shaping features. Note that the diamond dressing tool is normal to the wheel surface at point of contact. *Source:* OKUMA America Corporation.

In practice, grinding ratios vary widely, ranging from 2 to 200 and higher, depending on the type of wheel, the workpiece material, the grinding fluid, and process parameters such as depth of cut and speeds of the wheel and workpiece. It should be noted that attempting to obtain a high grinding ratio in practice isn't necessarily desirable, because high ratios may indicate grain dulling and possible surface damage. A lower ratio may well be acceptable if an overall economic analysis justifies it.

Soft-acting or hard-acting wheels. During a grinding operation, a particular wheel may *act soft* (meaning wear rate is high) or *hard* (wear rate is low), regardless of its grade. Note, for example, that an ordinary pencil acts soft when you write on rough paper and acts hard on soft paper. This behavior is a function of the force on the grain. The higher the force, the greater is the tendency for the grains to fracture or be dislodged from the wheel surface, and hence the higher is the wheel wear and the lower is the grinding ratio. Equation (9.6) indicates that the grain force increases with the strength of the workpiece material, the work speed, and the depth of cut and decreases with increasing wheel speed and wheel diameter. Thus, a wheel acts soft when v and d increase, or when V and D decrease.

EXAMPLE 9.3 Action of a grinding wheel

A surface-grinding operation is being carried out with the wheel rotating at a constant spindle speed. Will the wheel act soft or act hard as it wears down over a period of time?

Solution. Referring to Eq. (9.6), it will be noted that the parameters that change with time in this operation are the wheel surface speed, V, and the wheel diameter, D. As both become smaller with time, the relative grain force increases, and therefore the wheel acts softer. Some grinding machines are equipped with variable-speed spindle motors to accommodate these changes and to make provisions for wheels of different diameter.

9.5.3 Wheel selection and grindability of materials

Proper selection of a grinding wheel for a given application greatly influences the quality of surfaces produced, as well as the economics of the operation. The selection involves not only the shape of the wheel with respect to the shape of the part, but the characteristics of the workpiece material as well. The *grindability* of materials, like machinability (Section 8.5) or forgeability (Section 6.2.6), is difficult to define precisely. It is a general indication of how easy it is to grind a material, and includes considerations such as surface finish, surface integrity, wheel wear, cycle time, and overall economics. As with machinability, grindability of a material can be greatly enhanced by proper selection of process parameters, the type of wheel, grinding fluids, machine characteristics, and fixturing devices.

Grinding practices have been well established over the years for a wide variety of metallic and nonmetallic materials, including newly developed materials for aerospace applications. Specific recommendations for selecting wheels and process parameters can be found in various handbooks. Some examples of such recommendations are C60-L6V for cast irons, A60-M6V for steels, C60-I9V or D150-R75B for carbides, and A60-K8V for titanium. Ceramics can be ground with relative ease by using diamond wheels, as well as by carefully selecting process parameters; a typical wheel for ceramics is D150-N50M.

Ductile regime grinding. It has been shown that with light passes and rigid machines with good damping capacity, it is possible to obtain continuous chips in grinding ceramics. (See Figs. 9.9 and 9.7b.) This procedure is known as *ductile regime grinding* and produces good workpiece surface integrity. However, ceramic chips are typically 1–10 μm (40–400 μin.) in size and hence are more difficult to remove from grinding fluids than metal chips, thus requiring fine filtration.

9.6 | Grinding Operations and Machines

Grinding operations are carried out with a wide variety of wheel-workpiece configurations. The selection of a grinding process for a particular application depends on part shape, part size, ease of fixturing, and the production rate required. The basic

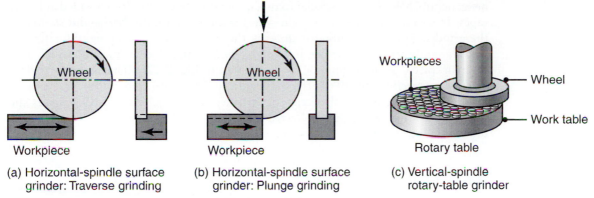

(a) Horizontal-spindle surface grinder: Traverse grinding

(b) Horizontal-spindle surface grinder: Plunge grinding

(c) Vertical-spindle rotary-table grinder

FIGURE 9.12 Schematic illustrations of surface-grinding operations: (a) traverse grinding with a horizontal-spindle surface grinder, (b) plunge grinding with a horizontal-spindle surface grinder, producing a groove in the workpiece, and (c) vertical-spindle rotary-table grinder (also known as the *Blanchard-type* grinder).

types of grinding operations are surface, cylindrical, internal, and centerless grinding. The relative movement of the wheel in these operations may be along the surface of the workpiece (*traverse grinding, through feed grinding,* or *cross-feeding*), or it may be radially *into* the workpiece (*plunge grinding*). Surface grinders constitute the largest percentage of grinders in use, followed by bench grinders (usually with two grinding wheels), cylindrical grinders, and tool and cutter grinders.

Modern grinding machines are computer controlled, with features such as automatic part loading and unloading, clamping, cycling, gaging, dressing, and wheel shaping. Grinders can also be equipped with probes and gages for determining the relative position of the wheel and workpiece surfaces, as well as with tactile sensing features whereby diamond dressing-tool breakage, if any, can be detected during the dressing cycle. Because grinding wheels are brittle and are operated at high speeds, certain safety procedures must be followed carefully in their handling, storage, and use.

9.6.1 Surface grinding

Surface grinding is one of the most common grinding operations (Fig. 9.12) and involves grinding flat surfaces. Typically, the workpiece is secured on a magnetic chuck mounted the worktable of a **surface grinder** (Fig. 9.13). Nonmagnetic materials

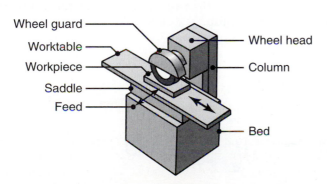

FIGURE 9.13 Schematic illustration of a horizontal-spindle surface grinder.

are generally held by vises, special fixtures, vacuum chucks, or double-sided adhesive tapes. In surface grinding, a straight wheel is mounted on the *horizontal spindle* of the grinder. Traverse grinding is done as the table reciprocates longitudinally and feeds laterally after each stroke. In *plunge grinding*, the wheel is moved radially into the workpiece, as in grinding a groove shown in Fig. 9.12b. The size of a surface grinder is specified by the surface dimensions of length and width that can be ground on the machine. Other types of surface grinders include *vertical spindles* and *rotary tables* (Fig. 9.12c), also referred to as *Blanchard-type* grinders. These configurations allow a number of parts to be ground in one setup.

9.6.2 Cylindrical grinding

In *cylindrical grinding*, also called *center-type grinding*, the workpiece's external cylindrical surfaces and shoulders are ground, such as crankshaft bearings, spindles, pins, bearing rings, and rolls for rolling mills. The rotating cylindrical workpiece reciprocates laterally along its axis, although in grinders used for large and long workpieces, the grinding wheel reciprocates. The latter design is called a *roll grinder* and is capable of grinding rolls as large as 1.8 m (72 in.) in diameter used in metal rolling (see Fig. 6.29).

The workpiece in cylindrical grinding is held between centers, held in a chuck, or mounted on a faceplate in the headstock of the grinder. For straight cylindrical surfaces, the axes of rotation of the wheel and workpiece are parallel. Separate motors drive the wheel and workpiece at different speeds. Long workpieces with two or more diameters are ground on cylindrical grinders. Cylindrical grinding also can produce shapes in which the wheel is dressed to the form to be ground on the workpiece (*form grinding* and *plunge grinding*). Cylindrical grinders are identified by the maximum diameter and length of the workpiece that can be ground, similar to lathes (Section 8.9.2).

In *universal grinders*, both the workpiece and the wheel axes can be moved and swiveled around a horizontal plane, thus permitting the grinding of tapers and other shapes. These machines are equipped with computer controls, thereby reducing labor and producing parts accurately and repetitively. Cylindrical grinders can also be equipped with computer-controlled features so that *noncylindrical* parts (such as cams) can be ground on rotating workpieces. The workpiece spindle speed is synchronized such that the distance between the workpiece and wheel axes is varied continuously to produce a particular shape.

Thread grinding is done on cylindrical grinders, as well as on centerless grinders (see Section 9.6.4), with specially dressed wheels that match the shape of the threads (Fig. 9.14). The workpiece and wheel movements are synchronized to produce the pitch of the thread, usually in about six passes. Although this operation is costly, it produces threads more accurately than any other manufacturing process, and the threads have a very fine surface finish.

FIGURE 9.14 Threads produced by (a) traverse and (b) plunge grinding.

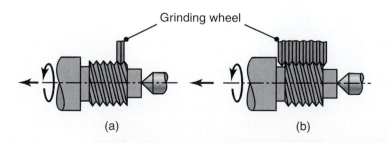

(a)   (b)

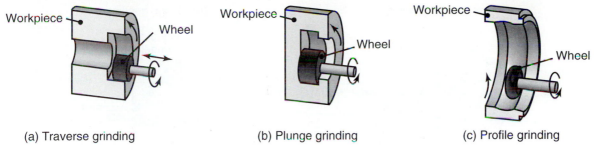

(a) Traverse grinding   (b) Plunge grinding   (c) Profile grinding

FIGURE 9.15   Schematic illustrations of internal-grinding operations.

9.6.3 Internal grinding

In *internal grinding* (Fig. 9.15), a small wheel is used to grind the inside diameter of parts, such as bushings and bearing races. The workpiece is held in a rotating chuck, and the wheel rotates at 30,000 rpm or higher. Internal profiles also can be ground with profile-dressed wheels that move radially into the workpiece. The headstock of internal grinders can be swiveled on a horizontal plane to grind tapered holes.

9.6.4 Centerless grinding

Centerless grinding is a high-production process for continuously grinding cylindrical surfaces in which the workpiece is supported not by centers (hence the term *centerless*) or chucks, but by a blade, as shown in Fig. 9.16. Typical parts made by this operation include cylindrical roller bearings, piston pins, engine valves, camshafts, and similar components. Parts with diameters as small as 0.1 mm (0.004 in.) can be ground using this process. Centerless grinders are now capable of wheel surface speeds on the order of 10,000 m/min (35,000 ft/min), using cubic boron nitride abrasive wheels. This continuous-production process requires little operator skill.

In *through-feed grinding* (Fig. 9.16a), the workpiece is supported on a work-rest blade and is ground between two wheels. Grinding is done by the larger wheel, while the smaller wheel regulates the axial movement of the workpiece. The *regulating wheel,* which is rubber bonded, is tilted and runs at a speed of only about 5% of that of the grinding wheel.

Parts with variable diameters, such as bolts, valve tappets, and distributor shafts, can be ground by centerless grinding. Called *infeed grinding* or *plunge grinding* (Fig. 9.16b), the process is similar to plunge or form grinding with cylindrical grinders. Tapered pieces are centerless ground by *end-feed grinding*. High-production-rate thread grinding can be done with centerless grinders, using specially dressed wheels. In *internal centerless grinding,* the workpiece is supported between three rolls and is internally ground. Typical applications include sleeve-shaped parts and rings.

9.6.5 Other types of grinders

Several special-purpose grinders are available for various applications. *Bench grinders* are used for routine offhand grinding of tools and small parts. They are usually equipped with two wheels mounted on the two ends of the shaft of an electric motor. One wheel is usually coarse for rough grinding, and the other is fine for finish grinding. *Pedestal,* or *stand, grinders* are placed on the floor and are used similarly to bench grinders.

Universal tool and cutter grinders are used for grinding single-point or multi-point cutting tools and cutters. They are equipped with special workholding devices

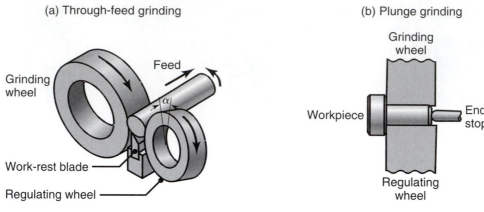

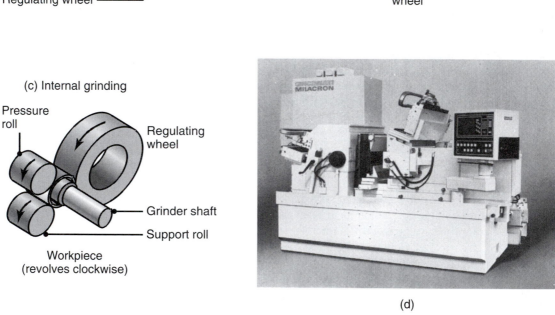

(d)

FIGURE 9.16 (a–c) Schematic illustrations of centerless-grinding operations and (d) a computer-numerical-control centerless grinding machine. *Source:* Cincinnati Milacron, Inc.

for accurate positioning of the tools to be ground. *Tool-post grinders* are self-contained units and are usually mounted on the tool post of a lathe (see Fig. 8.44). The workpiece is mounted on the headstock and is ground by moving the tool post. These grinders are versatile, but the lathe surfaces should be protected from abrasive debris.

Swing-frame grinders are typically used in foundries for grinding large castings. Rough grinding of castings is called *snagging* and is usually done on floorstand grinders, using wheels as large as 0.9 m (36 in.) in diameter. *Portable grinders*, either air or electrically driven, or with a flexible shaft connected to the shaft of an electric motor or gasoline engine, are available for operations such as grinding off weld beads (see Fig. 12.5) and *cutting-off* operations using thin abrasive disks.

9.6.6 Creep-feed grinding

Although grinding has traditionally been associated with small rates of material removal and fine finishing operations; grinding can also be used for large-scale metal removal operations, similar to milling, broaching, and planing (Sections 8.9.1 to 8.9.4). In *creep-feed grinding*, developed in the late 1950s, the wheel depth of cut,

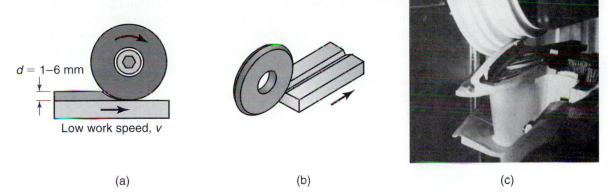

FIGURE 9.17 (a) Schematic illustration of the creep-feed grinding process. Note the large wheel depth of cut. (b) A groove produced on a flat surface in one pass by creep-feed grinding using a shaped wheel. Groove depth can be on the order of a few mm. (c) An example of creep-feed grinding with a shaped wheel. *Source:* Courtesy of Blohm, Inc. and Society of Manufacturing Engineers.

d, is as much as 6 mm (0.25 in.), and the workpiece speed is low (Fig. 9.17). The wheels are mostly softer-grade resin bonded with open structure to keep temperatures low and improve surface finish. Grinders with capabilities for continuously dressing the grinding wheel with a diamond roll are also available. The machines used for creep-feed grinding have special features, such as (a) high power of up to 225 kW (300 hp), (b) high stiffness, because of the high forces due to the larger depth of material removed, (c) high damping capacity, (d) variable and well-controlled spindle and worktable speeds, and (e) ample capacity for grinding fluids.

In view of its overall economics and competitive position with respect to other machining processes, creep-feed grinding has been shown to be economical for specific applications, such as in grinding shaped punches, key seats, twist-drill flutes, the roots of turbine blades (Fig. 9.17c), and various complex superalloy parts. The wheel is dressed to the shape of the workpiece to be produced. Consequently, the workpiece does not have to be previously milled, shaped, or broached; thus, *near-net-shape* castings and forgings are particularly suitable parts for creep-feed grinding. Although a single pass is generally sufficient, a second pass may be necessary for improved surface finish.

9.6.7 Heavy stock removal by grinding

Grinding processes can be used for heavy stock removal, thus competing with machining operations, particularly milling, turning, and broaching, and can be economical for specific applications. Because this is a rough grinding operation, it can have detrimental effects on the workpiece surface and its integrity. In this operation, surface finish is of secondary importance, and the grinding wheel (or abrasive belt; see Section 9.8) is used to its maximum capacity for minimum cost per piece. The geometric tolerances in this process are on the same order as those obtained by other machining processes. (See also Fig. 9.27.)

9.6.8 Grinding chatter

Chatter is particularly significant in grinding because it adversely affects surface finish and wheel performance. Vibrations during grinding may be caused by bearings, spindles, and unbalanced grinding wheels, as well as external sources, such as from nearby machinery. The grinding process itself can also cause regenerative chatter.

The analysis of chatter in grinding is similar to that for machining operations (see Section 8.12 and involves *self-excited vibration* and *regenerative chatter*. Thus, the important variables are (a) stiffness of the grinder and of the workholding devices and (b) damping of the system. Additional factors that are unique to grinding chatter include (a) nonuniformities in the grinding wheel itself, (b) the dressing techniques used, and (c) uneven wheel wear.

Because the foregoing variables produce characteristic **chatter marks** on ground surfaces, a study of these marks often can lead to the source of the vibration problem. General guidelines have been established to reduce the tendency for chatter in grinding. These are (a) using a soft-grade wheel, (b) dressing the wheel frequently, (c) changing dressing techniques, (d) reducing the material removal rate, and (e) supporting the workpiece rigidly. Advanced techniques include the use of sensors to monitor grinding chatter and its control by chatter suppression systems.

9.6.9 Grinding fluids

The functions of grinding fluids are similar to those for cutting fluids, as described in Section 8.7. Although grinding and other abrasive machining processes can be performed dry, the use of a fluid is usually preferred to (a) prevent excessive temperature rise in the workpiece, (b) improve the part's surface finish and dimensional accuracy, and (c) improve the efficiency of the operation by reducing wheel wear and loading and lowering power consumption.

Grinding fluids typically are *water-base emulsions* and chemicals and synthetics; oils may be used for thread grinding. The fluids may be applied as a stream (flood) or as mist, which is a mixture of fluid and air. Because of the high surface speeds involved in grinding, an *air stream* or *air blanket* develops around the periphery of the wheel, preventing the fluid from reaching the grinding zone. Special *nozzles* that conform to the shape of the wheel's grinding surface can be designed for effective application of the grinding fluid under high pressure.

The temperature of water-base grinding fluids can rise significantly during their use as they remove heat from the grinding zone, as otherwise the workpiece would expand, making it difficult to control dimensional tolerances. The common method employed to maintain even temperature is to use a refrigerating system (chiller) through which the fluid is circulated. As also described for cutting fluids in Section 8.7, the biological and ecological aspects, treatment, recycling, and disposal of grinding fluids are among important considerations in their selection and use. The practices employed must comply with federal, state, and local laws and regulations.

Cryogenic grinding. A more recent technology is the use of liquid nitrogen as a coolant in grinding operations, largely in the interest of reducing or eliminating the adverse environmental impact of metalworking fluids (see also Section 8.7.2). With small-diameter nozzles, liquid nitrogen at around $-200°C$ ($-320°F$) is injected into the wheel-workpiece grinding zone, thus reducing its temperature. As stated previously, high temperatures can have significant adverse effects on the part, such as poor surface finish and integrity. Experimental studies have indicated that, as compared with the use of traditional grinding fluids, cryogenic grinding is associated with a reduced level of surface (metallurgical) burn and oxidation, improved surface finish, lower tensile residual stress levels, and less loading of the grinding wheel (hence, less need for dressing). Cryogenic grinding appears to be particularly suitable for materials such as titanium, which has very low thermal conductivity (see Table 3.3), low specific heat, and high reactivity. Other effects, such as on fatigue life of the ground pieces, as well as the economic advantages of cryogenic grinding, continue to be investigated.

EXAMPLE 9.4 Grinding vs. hard turning

Hard turning is described in Section 8.9.2, an example of which is the machining of heat-treated steels (usually above 45 HRC), using a single-point, polycrystalline cBN cutting tool. In view of the discussions presented thus far in this chapter, it is evident that grinding and hard turning can be competitive in specific applications. Hard turning continues to be increasingly competitive with grinding, and dimensional tolerances and surface finish produced are approaching those for grinding. Furthermore, (a) turning requires much less energy than grinding (compare, for example, Tables 8.3 and 9.3), (b) thermal and other damage to the workpiece surface is less likely to occur, (c) cutting fluids may not be necessary, and (d) the machine tools are less expensive.

In addition, finishing the part while it is still chucked in the lathe or turning center eliminates the need for material handling and setting the part in the grinder. On the other hand, workholding devices for large and slender workpieces for hard turning can present difficulties because the machining forces are higher than those in grinding and the workpiece will tend to deflect. Furthermore, tool wear and its control in hard turning can be a significant problem as compared with the automatic dressing of grinding wheels in modern grinders. It is thus evident that the competitive position of hard turning vs. grinding must be evaluated individually for each application and in terms of product surface integrity, quality, as well as overall economics.

9.7 | Finishing Operations

In addition to the abrasive machining processes described thus far, several processes are generally used on workpieces as the final finishing operation, using mainly abrasive grains. However, finishing operations can contribute significantly to production time and product cost; hence they should be specified with due consideration to their costs and benefits.

Commonly used finishing operations are described next.

1. **Coated abrasives.** Typical examples of *coated abrasives* are sandpaper and emery cloth. The grains in coated abrasives are more pointed than those used for grinding wheels. They are electrostatically deposited on flexible backing materials, such as paper or cloth (Fig. 9.18), with their long axes perpendicular to the plane of the backing. The matrix (coating) is made of resins. Coated abrasives are available as sheets, belts, and disks, and usually have a much more open structure than the abrasives used in grinding wheels. Coated abrasives are used extensively in finishing flat or curved surfaces of metallic and nonmetallic parts, in finishing metallographic specimens, and in woodworking. The surface finish obtained depends primarily on grain size.

   Coated abrasives are also used as *belts* for high-rate material removal. **Belt grinding** is an important production process and in some cases has replaced

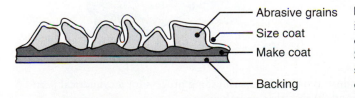

Abrasive grains
Size coat
Make coat
Backing

FIGURE 9.18 Schematic illustration of the structure of a coated abrasive. Sandpaper, developed in the sixteenth century, and emery cloth are common examples of coated abrasives.

FIGURE 9.19 Schematic illustration of a honing tool to improve the surface finish of bored or ground holes.

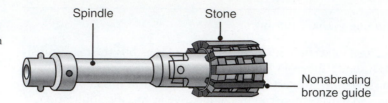

conventional grinding operations such as grinding of camshafts, with 8 to 16 lobes (of approximately elliptic shapes) per shaft. Belt speeds are usually in the range of 700–1800 m/min (2500–6000 ft/min). Machines for abrasive-belt operations require proper belt support and rigid construction to minimize vibrations.

A more recent development is **microreplication**, in which aluminum-oxide abrasives in the shape of tiny pyramids are placed in a predetermined orderly arrangement on the belt surface. When used on stainless steels and superalloys, their performance is more consistent, and the temperature rise is lower than when using other coated abrasives. Typical applications of microreplication include surgical implants, turbine blades, and medical and dental instruments.

2. **Wire brushing.** In this process, the workpiece is held against a circular wire brush that rotates at high speed. The tips of the wire produce longitudinal scratches on the workpiece surface. Wire brushing is used to produce a fine surface texture but can also serve as a light material-removal process.

3. **Honing.** *Honing* is an operation used primarily to give holes a fine surface finish. The honing tool (Fig. 9.19) consists of a set of aluminum-oxide or silicon-carbide bonded abrasives called *stones*. The stones are mounted on a mandrel that rotates in the hole, applying a radial force with a reciprocating axial motion, thus producing a crosshatched pattern. The stones can be adjusted radially for different hole sizes. The surface finish can be controlled by the type and size of abrasive used, the speed of rotation, and the pressure applied. A fluid is used to remove chips and to keep temperatures low. If not implemented properly, honing can produce holes that may not be straight and cylindrical, but rather bell-mouthed, wavy, barrel shaped, or tapered. Honing is also used on external cylindrical or flat surfaces, and to remove sharp edges on cutting tools and inserts. (See Fig. 8.32.)

In **superfinishing**, the pressure applied is very light, and the motion of the honing stone has a short stroke. The operation is controlled so that the grains do not travel along the same path across the surface of the workpiece. Figure 9.20 shows examples of external superfinishing of a round part.

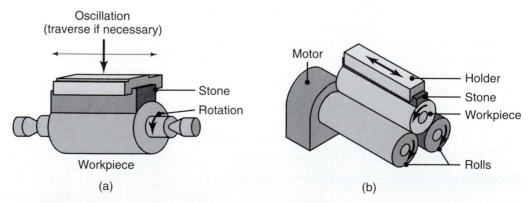

(a)   (b)

FIGURE 9.20 Schematic illustration of the superfinishing process for a cylindrical part: (a) cylindrical microhoning and (b) centerless microhoning.

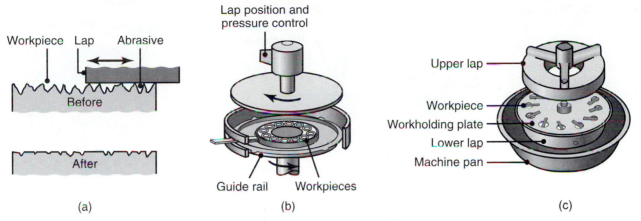

FIGURE 9.21 (a) Schematic illustration of the lapping process (see also Fig. 4.9) and (b) production lapping on flat surfaces, and (c) production lapping on cylindrical surfaces.

4. **Electrochemical honing.** This process combines the fine abrasive action of honing with electrochemical action. Although the equipment is costly, the process is as much as five times faster than conventional honing, and the tool lasts up to ten times longer. Electrochemical honing is used primarily for finishing internal cylindrical surfaces.

5. **Lapping.** *Lapping* is a finishing operation on flat or cylindrical surfaces. The *lap* (Fig. 9.21a) is usually made of cast iron, copper, leather, or cloth. The abrasive particles are embedded in the lap, or they may be carried through a slurry. Depending on the hardness of the workpiece, lapping pressures range from 7–140 kPa (1–20 psi). Dimensional tolerances on the order of ±0.4 mm (±16 min.) can be obtained with the use of fine abrasives, up to size 900. Surface finish can be as smooth as 0.025–0.1 μm (1–4 μin.). Production lapping on flat or cylindrical workpieces is done on machines such as those shown in Figs. 9.21b and c. Lapping is also performed on curved surfaces, such as spherical objects and glass lenses, using specially shaped laps. *Running-in* of mating gears also can be done by lapping.

6. **Polishing.** *Polishing* is a process that produces a smooth, lustrous surface finish. Two basic mechanisms are involved in the polishing process: (1) fine-scale abrasive removal and (2) softening and smearing of surface layers by frictional heating. The shiny appearance of polished surfaces results from the smearing action. Polishing is done with disks or belts, made of fabric, leather, or felt, and coated with fine powders of aluminum oxide or diamond. Parts with irregular shapes, sharp corners, deep recesses, and sharp projections are difficult to polish.

7. **Laser polishing.** This method involves rapid melting and resolidification of a surface (at depths of submicrons) using short laser pulses, in the range of micro- or nanoseconds. The melting action smoothens the surface by reducing asperity heights (see Fig. 4.4) on ferrous and nonferrous metallic workpieces as well as on glass and diamond. The surface developed by laser polishing is suitable for optical use and also has lower friction, thus making it attractive for applications such as automotive cylinder liners, including those for Audi and Volkswagen engines. Unlike traditional polishing operations, this process is much faster and is also applicable on workpieces with uneven surfaces, using programmable controls.

FIGURE 9.22 Schematic illustration of the chemical-mechanical polishing process. This process is widely used in the manufacture of silicon wafers and integrated circuits, where it is known as *chemical-mechanical planarization*. Additional carriers and more disks per carrier also are possible.

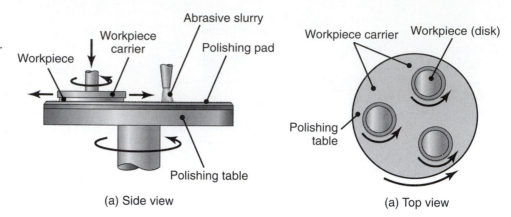

(a) Side view   (a) Top view

8. **Buffing.** *Buffing* is similar to polishing, with the exception that very fine abrasives are used on soft disks typically made of cloth. The abrasive is supplied externally from a stick of abrasive compound. Buffing may be done on polished parts to obtain an even finer surface finish.

9. **Electropolishing.** Mirrorlike finishes can be obtained on metal surfaces by *electropolishing*, a process that is the reverse of electroplating (see Section 4.5.1). Because there is no mechanical contact with the workpiece, this process is particularly suitable for polishing irregular shapes. The electrolyte attacks projections and peaks on the workpiece surface at a higher rate than for the rest of the surface, thus producing a smooth surface. Electropolishing is also used for deburring operations (Section 9.9).

10. **Chemical-mechanical polishing.** Chemical-mechanical polishing (CMP) is very important in the semiconductor industry. The process, shown in Fig. 9.22, uses a suspension of abrasive particles in a water-base solution with a chemistry selected to produce controlled corrosion. This process removes material from the workpiece surfaces through combined actions of abrasion and corrosion. The result is exceptionally fine surface finish and a workpiece that is essentially flat (plane). For this reason, the process is often referred to as **chemical-mechanical planarization.**

     A major application of this process is the polishing of silicon wafers (Section 13.4), in which case the primary function of CMP is to polish a wafer at the microlevel. To evenly remove material across the whole wafer, the wafer is held on a rotating carrier face down and is pressed against a polishing pad attached to a rotating disk, as shown in Fig. 9.22. Both the carrier and the pad rotate in order to avoid development of a linear lay (see Section 4.3). The pad contains grooves in order to help supply slurry to all wafers.

     Specific abrasive and solution chemistry combinations have been developed for the polishing of copper, silicon, silicon dioxide, aluminum, tungsten, and other metals. For silicon dioxide or silicon polishing, for example, an alkaline slurry of colloidal silica (SiO_2) particles in a KOH solution or NH_4OH is continuously fed to the pad/wafer interface.

11. **Polishing processes using magnetic fields.** In this technique, abrasive slurries are supported with and made more effective by magnetic fields. There are two basic methods:

     a. **Magnetic float polishing** of ceramic balls is illustrated schematically in Fig. 9.23a. In this process, a magnetic fluid, containing abrasive grains and extremely fine ferromagnetic particles in a carrier fluid such as water or

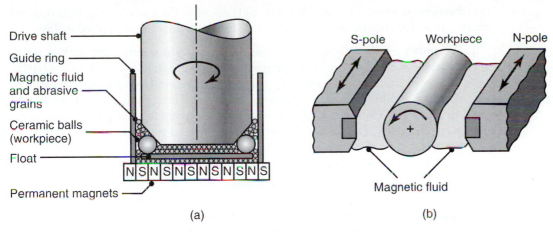

FIGURE 9.23 Schematic illustration of the use of magnetic fields to polish balls and rollers: (a) magnetic float polishing of ceramic balls and (b) magnetic-field-assisted polishing of rollers. *Source:* After R. Komanduri, M. Doc, and M. Fox.

kerosene, fills the chamber within a guide ring. The ceramic balls are located between a drive shaft and a float. The abrasive grains, ceramic balls, and the float (which is made of a nonmagnetic material) are all suspended by magnetic forces. The balls are pressed against the rotating drive shaft and are polished by abrasive action. The forces applied by the abrasive particles on the balls are extremely small and controllable, and hence the polishing action is very fine. Polishing times are much lower than those for other polishing methods; thus, the process is very economical and the surfaces produced have few or no defects.

b. **Magnetic-field-assisted polishing** of ceramic rollers is illustrated in Fig. 9.23b. In this process, a ceramic or steel roller (the workpiece) is clamped and rotated on a spindle. The magnetic poles are oscillated, introducing a vibratory motion to the magnetic-abrasive conglomerate. This action polishes the cylindrical roller surface. Bearing steels of 63 HRC have been mirror finished in 30 seconds by this process.

9.8 | Deburring

Burrs are thin ridges, usually triangular in shape, that develop along the edges of a workpiece from processes such as shearing of sheet materials (see Fig. 7.5), trimming forgings and castings, and machining. Burrs may interfere with the assembly of parts and can cause jamming of parts, misalignment, and short circuiting in electrical components. Also, burrs may reduce fatigue life (see Section 7.3). Because they are usually sharp, burrs also can be a safety hazard to personnel. The need for deburring may be reduced by adding chamfers to sharp edges on parts. On the other hand, burrs on thin drilled or tapped components, such as tiny parts in watches, can provide extra thickness and thus improve the holding torque of very small screws.

Several *deburring processes* are available: (1) manually with files (see Section 8.10.6); (2) mechanically by cutting; (3) wire brushing (Section 9.7); (4) flexible abrasive finishing, using rotary nylon brushes with the nylon filaments embedded with abrasive grits; (5) abrasive belts (Section 9.7); (6) ultrasonics (Section 9.9); (7) electropolishing (Section 9.7); (8) electrochemical machining (Section 9.11);

(9) vibratory finishing; (10) shot blasting; (11) abrasive-flow machining; (12) thermal energy, and (13) robotic deburring. The last five processes are described next.

1. **Vibratory** and **barrel-finishing** processes are used to improve the surface finish and remove burrs from large numbers of relatively small parts. This is a batch-type operation in which specially shaped *abrasive pellets* or *media* are placed in a container along with the parts to be deburred. The container is either vibrated or tumbled. The impact of individual abrasives and metal particles removes sharp edges and burrs from the parts. Depending on the application, this process is performed dry or wet, and liquid compounds may be added for other requirements, such as degreasing and providing resistance to corrosion.

2. In **shot blasting** (also called *grit blasting*), abrasive particles (usually sand) are propelled by a high-velocity jet of air or by a rotating wheel onto the surface of the workpiece. Shot blasting is particularly useful in deburring both metallic and nonmetallic materials and stripping, cleaning, and removing surface oxides from workpieces. The surface produced by shot blasting has a matte finish. Small-scale polishing and etching also can be done by this process on bench-type units (*microabrasive blasting*).

3. In **abrasive-flow machining,** abrasive grains, such as silicon carbide or diamond, are mixed in a puttylike matrix, which is then forced back and forth through the openings and passageways in the workpiece. The movement of the abrasive matrix under pressure erodes away burrs and sharp corners and polishes the part. The process is particularly suitable for workpieces with internal cavities that are inaccessible by other means. Pressures applied range from 0.7 to 22 MPa (100 to 3200 psi). External surfaces also can be deburred using this process by containing the workpiece within a fixture that directs the abrasive media to the edges and areas to be deburred.

4. The **thermal-energy method** of deburring consists of placing the part in a chamber that is then injected with a mixture of natural gas and oxygen. This mixture is ignited, producing a heat wave with a temperature of 3300°C (6000°F). The burrs instantly heat up and are melted away, while the temperature of the part itself rises only to about 150°C (300°F). The thermal energy method is effective on a wide range of materials including zinc, aluminum, brass, steel, stainless steel, cast iron, and thermoplastics. However, larger burrs or flash from forgings and castings tend to form beads after melting and the process can distort thin and slender parts. Also, this method does not polish or buff the workpiece surfaces as do many of the other deburring processes.

5. **Robotic deburring.** Deburring and flash removal are now being performed increasingly by programmable robots (see Section 14.7 and the case study for Chapter 14), using a force-feedback system for control of the process. The main advantage of robotic deburring is flexibility, both in terms of the geometries that can be deburred as well as the media that can be applied. This method eliminates tedious and costly manual labor and results in more consistent deburring. The main drawback is the capital equipment cost involved—robotic deburring systems with multiple spindles can cost over $250,000.

9.9 | Ultrasonic Machining

In *ultrasonic machining* (UM), material is removed from a workpiece surface by the mechanism of microchipping or erosion with abrasive particles. The tip of the tool (Fig. 9.24a), called a *sonotrode,* vibrates at amplitudes of 0.05 to 0.125 mm (0.002 to 0.005 in.) and at a frequency of 20 kHz. This vibration, in turn, transmits a high

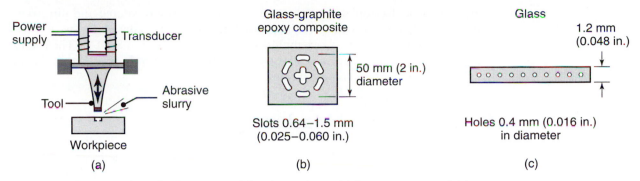

FIGURE 9.24 (a) Schematic illustration of the ultrasonic-machining process; material is removed through microchipping and erosion. (b) and (c) Typical examples of cavities produced by ultrasonic machining. Note the dimensions of cut and the types of workpiece materials.

velocity to fine abrasive grains between the tool and the surface of the workpiece. The grains are in a water slurry with concentrations ranging from 20 to 60% by volume; the slurry also carries away the debris from the cutting area. Although the grains are usually boron carbide, aluminum oxide and silicon carbide are also used, with grain sizes ranging from 100 (for roughing) to 1000 (for finishing).

Ultrasonic machining is best suited for hard and brittle materials, such as ceramics, carbides, glass, precious stones, and hardened steels. The tip of the tool is usually made of low-carbon steel and is attached to a transducer through the tool-holder. With fine abrasives, dimensional tolerances of 0.0125 mm (0.0005 in.) or better can be held in this process. Figures 9.24b and c show two applications of ultrasonic machining.

Microchipping in ultrasonic machining is possible because of the high stresses produced by particles striking a solid surface. The contact time between the particle and the surface is very short (10 to 100 μs), and the area of contact is very small. The contact time, t_o, can be expressed as

$$t_o \simeq \frac{5r}{c_o}\left(\frac{c_o}{v}\right)^{1/5},\qquad (9.11)$$

where r is the radius of a spherical particle, c_o is the elastic wave velocity in the workpiece $\left(c_o = \sqrt{E/\rho}\right)$, and v is the velocity with which the particle strikes the surface. The force F of the particle on the surface is obtained from the rate of change of momentum; that is,

$$F = \frac{d(mv)}{dt},\qquad (9.12)$$

where m is the mass of the particle. The *average force, F_{ave}*, of a particle striking the surface and rebounding is

$$F_{\text{ave}} = \frac{2mv}{t_o}.\qquad (9.13)$$

Substitution of numerical values into Eq. (9.13) indicates that even small particles can exert significant forces and, because of the very small contact area, the contact stresses are very high. In brittle materials, these stresses are sufficiently high to cause microchipping and surface erosion. (See also the discussion of *abrasive-jet machining* in Section 9.15.)

Rotary ultrasonic machining (RUM). In this process, the abrasive slurry is replaced by a tool with metal-bonded diamond abrasives that have been either impregnated or electroplated on the tool surface. The tool is rotated and ultrasonically vibrated, and the workpiece is pressed against it at a constant pressure; the operation is thus similar to a face milling operation. (See Section 8.10.1.) The rotary ultrasonic machining process is particularly effective in producing deep holes in ceramics at high material removal rates.

9.10 | Chemical Machining

The *chemical machining* (CM) process was developed based on the fact that certain chemicals attack metals and etch them, thereby removing small amounts of material from the surface (Table 9.4). The material is removed from a surface by chemical dissolution, using **reagents,** or **etchants,** such as acids and alkaline solutions. Chemical machining is the oldest of the nontraditional machining processes and has been used for many years for engraving metals and hard stones and in the production of printed-circuit boards and microprocessor chips. Parts can also be deburred by chemical means.

TABLE 9.4

General Characteristics of Advanced Machining Processes

Process	Characteristics	Process parameters and typical material removal rate or cutting speed
Chemical machining (CM)	Shallow removal (up to 12 mm) on large flat or curved surfaces; blanking of thin sheets; low tooling and equipment cost; suitable for low production runs.	0.025–0.1 mm/min
Electrochemical machining (ECM)	Complex shapes with deep cavities; highest rate of material removal; expensive tooling and equipment; high power consumption; medium to high production quantity.	V: 5–25 DC; A: 2.5–12 mm/min, depending on current density
Electrochemical grinding (ECG)	Cutting off and sharpening hard materials, such as tungsten-carbide tools; also used as a honing process; higher material removal rate than grinding.	A: 1–3 A/mm^2; typically 1500 mm^3/min per 1000 A
Electrical-discharge machining (EDM)	Shaping and cutting complex parts made of hard materials; some surface damage may result; also used for grinding and cutting; versatile; expensive tooling and equipment.	V: 50–380; A: 0.1–500; typically 300 mm^3/min
Wire EDM	Contour cutting of flat or curved surfaces; expensive equipment.	Varies with workpiece material and its thickness
Laser-beam machining (LBM)	Cutting and hole making on thin materials; heat-affected zone; does not require a vacuum; expensive equipment; consumes much energy; extreme caution required in use.	0.50–7.5 m/min
Electron-beam machining (EBM)	Cutting and hole making on thin materials; very small holes and slots; heat-affected zone; requires a vacuum; expensive equipment.	1–2 mm^3/min
Water-jet machining (WJM)	Cutting all types of nonmetallic materials to 25 mm (1 in.) and greater in thickness; suitable for contour cutting of flexible materials; no thermal damage; environmentally safe process.	Varies considerably with workpiece material
Abrasive water-jet machining (AWJM)	Single or multilayer cutting of metallic and nonmetallic materials.	Up to 7.5 m/min
Abrasive-jet machining (AJM)	Cutting, slotting, deburring, flash removal, etching, and cleaning of metallic and nonmetallic materials; tends to round off sharp edges; some hazard because of airborne particulates.	Varies considerably with workpiece material

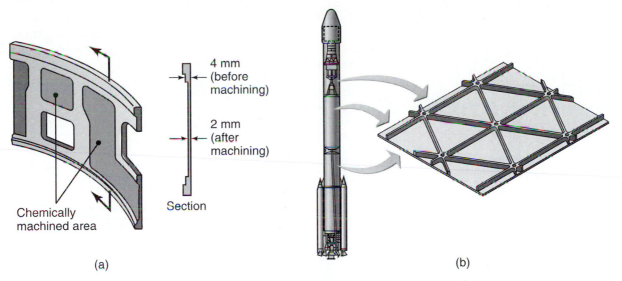

FIGURE 9.25 Examples of chemical milling: (a) missile skin-panel section contoured by chemical milling to improve the stiffness-to-weight ratio of the part and (b) weight reduction of space launch vehicles by chemical milling of aluminum-alloy plates. These panels are chemically milled after the plates have first been formed into shape, such as by roll forming or stretch forming. *Source:* ASM International.

9.10.1 Chemical milling

In *chemical milling,* shallow cavities are produced on sheets, plates, forgings, and extrusions, either for design requirements or for weight reduction in parts (Fig. 9.25). Chemical milling has been used on a wide variety of metals, with depths of material removal to as much as 12 mm (0.5 in.). Selective attack by the chemical reagent on different areas of the workpiece surfaces is controlled by removable layers of a **maskant** (Fig. 9.26) or by partial immersion in the reagent.

Chemical milling is used in the aerospace industry, particularly for removing shallow layers of material from large aircraft, missile skin panels, and extruded parts for airframes. The process is also used to fabricate microelectronic devices (see Sections 13.8 and Section 13.14), often referred to as **wet etching.** Tank capacities for reagents are as large as 3.7×15 m (12×50 ft). Figure 9.27 shows the range of surface finish and dimensional tolerances obtained by chemical machining and other machining processes. Some surface damage may result from chemical

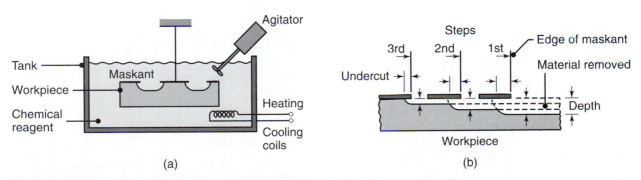

FIGURE 9.26 (a) Schematic illustration of the chemical machining process. Note that no forces are involved in this process. (b) Stages in producing a profiled cavity by chemical machining.

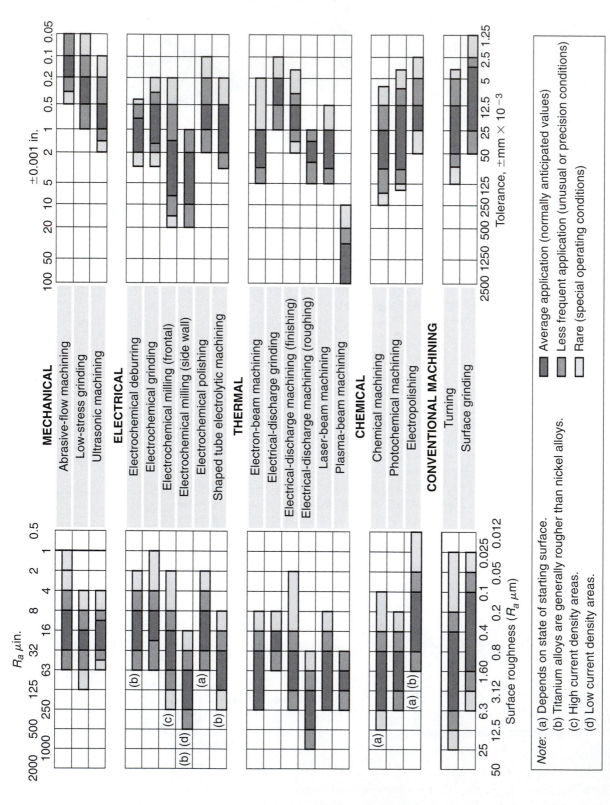

FIGURE 9.27 Surface roughness and dimensional tolerance capabilities of various machining processes. Note the wide range within each process. (See also Fig. 8.26.) *Source: Machining Data Handbook*, 3rd ed., © 1980. Used by permission of Metcut Research Associates, Inc.

milling, because of preferential etching and intergranular attack by the reagents (see Section 3.4), which adversely affect surface properties. Chemical milling of welded and brazed structures may produce uneven material removal because the filler metal or the structure material may machine preferentially. Also, chemical milling of castings may result in uneven surfaces caused by porosity and nonuniformity of the structure.

9.10.2 Chemical blanking

Chemical blanking is similar to blanking of sheet metal (see Fig. 7.8) in that it is used to produce features that penetrate through the thickness of the material, with the exception that material is removed by chemical dissolution rather than by shearing. Typical applications for chemical blanking include burr-free etching of printed-circuit boards (Fig. 13.31) and the production of decorative panels, thin sheet-metal stampings, and small or complex shapes.

9.10.3 Photochemical blanking

Also called *photoetching, photochemical blanking* is a modification of chemical milling. Material is removed by photographic techniques, usually from flat thin sheet. Complex burr-free shapes can be blanked (Fig. 9.28) on metals as thin as 0.0025 mm (0.0001 in.); the process is also used for etching (see Section 13.8). Typical applications for photochemical blanking include fine screens, printed-circuit cards, electric-motor laminations, flat springs, and masks for color television. Photochemical blanking is capable of forming very small parts for which traditional blanking dies (Section 7.3) are difficult to make, and the process is effective for blanking fragile workpieces and materials. Handling of chemical reagents requires special safety precautions to protect personnel against exposure to both liquid chemicals and volatile chemicals. Disposal of chemical by-products from this process is also a major consideration, although some by-products can be recycled. Although skilled labor is required for photochemical blanking, tooling costs are low, the process can be automated, and it is economical for medium to high production volume.

FIGURE 9.28   Typical parts made by chemical blanking; note the fine detail. *Source:* Courtesy of Buckabee-Mears St. Paul.

9.11 | Electrochemical Machining

Electrochemical machining (ECM) is basically the reverse of electroplating. An **electrolyte** (Fig. 9.29) acts as current carrier, and the high rate of electrolyte movement in the tool-workpiece gap washes metal ions away from the workpiece (*anode*) before they have a chance to plate onto the tool (*cathode*). Note that the cavity produced is the female mating image of the tool. Modifications of this process are used for operations such as turning, facing, slotting, trepanning, and profiling in which the electrode becomes the cutting tool.

The shaped tool is generally made of brass, copper, bronze, or stainless steel. The electrolyte is a highly conductive inorganic salt solution, such as sodium chloride mixed in water or sodium nitrate, and is pumped at a high rate through the passages in the tool. A DC power supply in the range of 5–25 V maintains current densities, which for most applications are 1.5–8 A/mm^2 (1000–5000 A/in^2) of active machined surface. Machines that have current capacities as high as 40,000 A and as small as 5 A are available. Because the metal removal rate is only a function of ion exchange rate, it is not affected by the strength, hardness, or toughness of the workpiece (which must be electrically conductive).

The material removal rate in electrochemical machining can be calculated from the equation

$$MRR = CI\eta, \tag{9.14}$$

where MRR = mm^3/min; I = current in amperes; and η = current efficiency, which typically ranges from 90 to 100%. C is a material constant in mm^3/A-min and, for pure metals, depends on valence; the higher the valence, the lower is its value. For most metals, the value of C typically ranges between about 1 and 2. If a cavity of uniform cross-sectional area A_o (in mm^2) is being electrochemically machined, the feed rate, f, in mm/min would be

$$f = \frac{MRR}{A_o}. \tag{9.15}$$

Note that the feed rate is the speed at which the electrode is penetrating the workpiece.

Electrochemical machining is generally used for machining complex cavities in high-strength materials, particularly in the aerospace industry for mass production of turbine blades, jet-engine parts, and nozzles (Fig. 9.30). It is also used for machining forging-die cavities (*die sinking*) and producing small holes. The ECM process leaves a burr-free surface; in fact, it can also be used as a deburring process. It does not cause any thermal damage to the part, and the lack of tool forces prevents distortion of the part, as would be the case with typical machining operations.

FIGURE 9.29 Schematic illustration of the electrochemical-machining process. This process is the reverse of electroplating, described in Section 4.5.1.

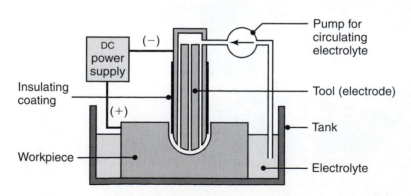

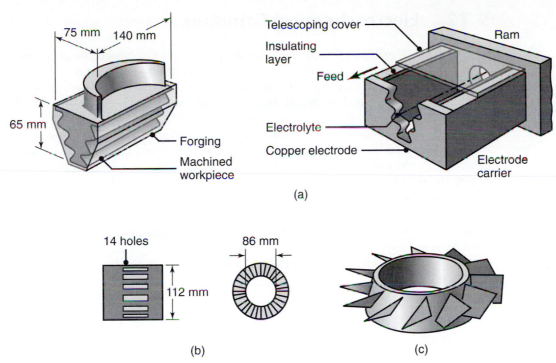

FIGURE 9.30 Typical parts made by electrochemical machining. (a) Turbine blade made of a nickel alloy, 360 HB; the part on the right is the shaped electrode. (b) Thin slots on a 4340-steel roller-bearing cage and (c) integral airfoils on a compressor disk. *Source:* (a) ASM International.

Furthermore, there is no tool wear, and the process is capable of producing complex shapes as well as machining hard materials. However, the mechanical properties of components made by ECM should be compared with those produced by other machining methods. Electrochemical-machining systems are now available as numerically controlled machining centers, with the capability for high production rates, high flexibility, and maintenance of close dimensional tolerances.

A modification of ECM is the **shaped-tube electrolytic machining** (STEM) process, typically used for drilling small-diameter deep holes, as in turbine blades. The tool is a titanium tube, coated with an electrically insulating resin to prevent removal of material from other regions. Holes as small as 0.5 mm (0.02 in.) can be drilled, at depth-to-diameter ratios of as high as 300:1.

Electrochemical machining systems are now available as *numerically controlled machining centers,* with the capability of high production rates, high flexibility, and the maintenance of close dimensional tolerances. The ECM process can also be combined with electrical-discharge machining (EDM; see Section 9.13) on the same machine. Further developments include *hybrid machining systems,* in which ECM and other advanced machining processes (as described throughout this chapter) are combined to take advantage of the capabilities of each process.

Pulsed electrochemical machining. *Pulsed electrochemical machining* (PECM) is a refinement of ECM. It uses very high current densities (on the order of 100 A/cm^2), but the current is pulsed rather than direct. The purpose of pulsing is to eliminate the need for high electrolyte flow rates, which limits the usefulness of ECM in die and moldmaking. Investigations have indicated that PECM improves fatigue life over ECM, and the process can be used to eliminate the recast layer left on die and mold surfaces by electrical discharge machining. (See Section 9.13.)

9.12 | Electrochemical Grinding

Electrochemical grinding (ECG) combines the processes of electrochemical machining and conventional grinding (Table 9.4) on equipment that is similar to a conventional grinder, except that the wheel is a rotating cathode with abrasive particles (Fig. 9.31a). The wheel is metal bonded, with diamond or aluminum-oxide abrasives, and rotates at a surface speed of 1200–2000 m/min (4000–7000 ft/min). Current densities range from 1–3 A/mm^2 (500–2000 A/in^2). The abrasives serve as insulators between the grinding wheel and the workpiece and mechanically remove electrolytic products from the working area. A flow of electrolyte, usually sodium nitrate, is provided for the electrochemical-machining phase of the operation. The majority of metal removal in ECG is by electrolytic action, and typically less than 5% of metal is removed by the abrasive action of the wheel; consequently, wheel wear is very low. Finishing cuts are usually made by the grinding action, but only to produce a surface with good finish and dimensional accuracy.

The material removal rate in electrochemical grinding can be calculated from the equation

$$MRR = \frac{GI}{\rho F}, \tag{9.16}$$

where MRR is in mm^3/min, G = mass in grams, I = current in amperes, ρ = density in g/mm^3, and F = Faraday's constant (96,485 coulombs/mole). The speed of penetration, V_s, of the grinding wheel into the workpiece is given by the equation

$$V_s = \left(\frac{G}{\rho F}\right)\left(\frac{E}{gK_p}\right)K, \tag{9.17}$$

where V_s is in mm^3/min; E = cell voltage in volts; g = wheel-workpiece gap in mm; K_p = coefficient of loss, which is in the range of 1.5 to 3; and K = electrolyte conductivity in Ω^{-1} mm^{-1}.

Electrochemical grinding is suitable for applications similar to those for milling, grinding, and sawing (Fig. 9.31b) and has been successfully applied to carbides and high-strength alloys. It is not readily adaptable to cavity-sinking operations, such as

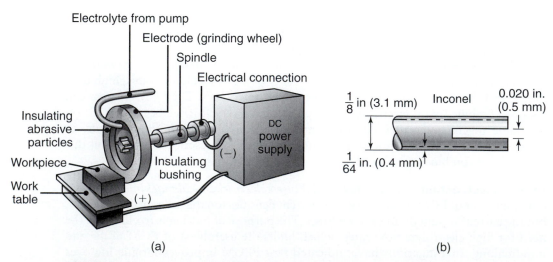

(a)   (b)

FIGURE 9.31   (a) Schematic illustration of the electrochemical grinding process and (b) thin slot produced on a round nickel-alloy tube by this process.

die making, because deep cavities are not accessible for grinding. The process offers a distinct advantage over traditional diamond-wheel grinding when processing very hard materials, for which wheel wear can be high. ECG machines are available with numerical controls, thus improving dimensional accuracy and providing repeatability and increased productivity.

EXAMPLE 9.5 Machining time in electrochemical machining vs. drilling

A round hole 12.5 mm (0.5 in.) in diameter is being produced in a titanium-alloy block by electrochemical machining. Using a current density of 6 A/mm², estimate the time required for machining a 20-mm-deep hole. Assume that the efficiency is 90%. Compare this time with that required for ordinary drilling.

Solution. Note from Eqs. (9.13) and (9.14) that the feed rate can be expressed by the equation

$$f = \frac{CI\eta}{A_o}.$$

Letting $C = 1.6$ mm³/A-min and $I/A_o = 6$ A/mm², the feed rate is $f = (1.6)(6)(0.9) = 8.64$ mm/min. Since the hole is 20 mm deep,

$$\text{Machining time} = \frac{20}{8.64} = 2.3 \text{ min.}$$

To determine the drilling time, refer to Table 8.12 and note the data for titanium alloys. Selecting the following values for a 12.5-mm drill—rpm = 300 and feed = 0.15 mm/rev—it can be seen that the feed rate is $(300 \text{ rev/min})(0.15 \text{ mm/rev}) = 45$ mm/min. Since the hole is 20 mm deep,

$$\text{Drilling time} = \frac{20}{45} = 0.45 \text{ min,}$$

which is about one-fifth of the time required for ECM.

9.13 | Electrical-Discharge Machining

The principle of *electrical-discharge machining* (EDM), also called *electrodischarge* or *spark-erosion machining*, is based on erosion of metals by spark discharges (Table 9.4). We know that when two current-conducting wires are allowed to touch each other, an arc is produced. When the point of contact between the two wires is closely examined, it will be noted that a small portion of the metal has been eroded away, leaving a small crater. Although this phenomenon has been known since the discovery of electricity, it was not until the 1940s that a machining process based on this principle was developed. It has become one of the most important and widely used production technologies in manufacturing industries.

The EDM system (Fig. 9.32) consists of a shaped tool (**electrode**) and the workpiece, which are connected to a DC power supply and placed in a **dielectric** (electrically nonconducting) **fluid**. When a voltage is applied to the tool, a magnetic field causes suspended particles in the dielectric fluid to concentrate between the electrode and workpiece, eventually forming a bridge for current to flow to the workpiece. An intense electrical arc is then generated, causing sufficient heating to melt a portion of the workpiece and, usually, some of the tooling material as well. In addition, the dielectric fluid is heated rapidly, causing evaporation of the fluid

FIGURE 9.32 Schematic illustration of the electrical-discharge machining process.

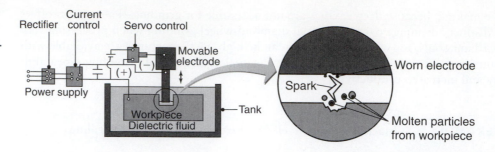

in the arc gap. This evaporation, in turn, increases the resistance of the interface, until the arc can no longer be maintained. Once the arc is interrupted, heat is removed from the gas bubble by the surrounding dielectric fluid, and the bubble collapses (cavitates). The associated shock wave and flow of dielectric fluid flush debris from the workpiece surface and entrain any molten workpiece material into the dielectric fluid. The capacitor discharge is repeated at rates of between 50 and 500 kHz, with voltages usually ranging between 50 and 380 V and currents from 0.1 to 500 A.

The dielectric fluid (a) acts as an insulator until the potential is sufficiently high, (b) acts as a flushing medium and carries away the debris in the gap, and (c) provides a cooling medium. The gap between the tool and the workpiece (*overcut*) is critical; thus, the downward feed of the tool is controlled by a servomechanism, which automatically maintains a constant gap. The most common dielectric fluids are mineral oils, although kerosene and distilled and deionized water may be used in specialized applications. The workpiece is fixtured within the tank containing the dielectric fluid, and its movements are controlled by numerically controlled systems. The machines are equipped with a pump and filtering system for the dielectric fluid.

The EDM process can be used on any material that is an electrical conductor. The melting point and latent heat of melting are important physical properties that determine the volume of metal removed per discharge. As these values increase, the rate of material removal slows. The volume of material removed per discharge is typically in the range of 10^{-6} to 10^{-4} mm³ (10^{-10} to 10^{-8} in³). Since the process does not involve mechanical energy, the hardness, strength, and toughness of the workpiece material do not necessarily influence the removal rate. The frequency of discharge or the energy per discharge is usually varied to control the removal rate, as are the voltage and current. The rate and surface roughness increase with increasing current density and decreasing frequency of sparks.

Electrodes for EDM are usually made of graphite, although brass, copper, or copper-tungsten alloy may be used. The tools are shaped by forming, casting, powder metallurgy, or machining. Electrodes as small as 0.1 mm (0.005 in.) in diameter have been used. Tool wear is an important factor because it adversely affects dimensional accuracy and the shape produced. *Tool wear* can be minimized by reversing the polarity and using copper tools, a process called **no-wear EDM.**

Electrical-discharge machining has numerous applications, such as producing die cavities for large automotive-body components (die-sinking machining centers); narrow slots; turbine blades; various intricate shapes (Fig. 9.33a and b); and small-diameter deep holes (Fig. 9.33c), using tungsten wire as the electrode. Stepped cavities can be produced by controlling the relative movements of the workpiece in relation to the electrode (Fig. 9.34).

The material removal rate in electrical-discharge machining is basically a function of the current and the melting point of the workpiece material, although other process variables, such as temperature and frequency, also have an effect. The following

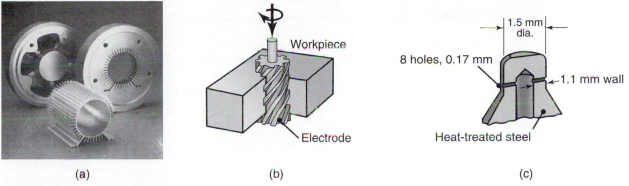

(a)   (b)   (c)

FIGURE 9.33   (a) Examples of shapes produced by the electrical-discharge machining process, using shaped electrodes. The two round parts in the rear are a set of dies for extruding the aluminum piece shown in front; see also Section 6.4. (b) A spiral cavity produced using a shaped rotating electrode. (c) Holes in a fuel-injection nozzle produced by electrical-discharge machining. *Source:* (a) Courtesy of AGIE USA Ltd. (b) *American Machinist.*

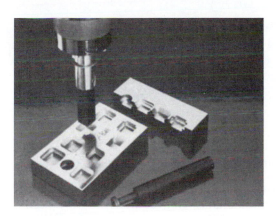

FIGURE 9.34   Stepped cavities produced with a square electrode by EDM. In this operation, the workpiece moves in the two principal horizontal directions, and its motion is synchronized with the downward movement of the electrode to produce these cavities. Also shown is a round electrode capable of producing round or elliptical cavities. *Source:* Courtesy of AGIE USA Ltd.

approximate empirical relationship can be used as a guide to estimate the metal removal rate in EDM:

$$\text{MRR} = 4 \times 10^4 \, I T_w^{-1.23}, \tag{9.18}$$

where MRR is in mm³/min, I is the current in amperes, and T_w is the melting point of the workpiece in °C.

The wear rate of the electrode, W_t, can be estimated from the empirical equation

$$W_t = (1.1 \times 10^{11}) I T_t^{-2.38} \tag{9.19}$$

where W_t is mm³/min and T_t is the melting point of the electrode material in °C. The wear ratio of the workpiece to electrode, R, can be estimated from the expression

$$R = 2.25 T_r^{-2.38} \tag{9.20}$$

where T_r is the ratio of workpiece to electrode melting points. In practice, the value of R varies over a wide range, typically between 0.2 and 100. (Note that in abrasive processing, grinding ratios are also in the same range; see Section 9.5.2.)

Metal removal rates in EDM usually range from 2 to 400 mm³/min. High rates produce a very rough finish, with a molten and resolidified (recast) structure with poor surface integrity and low fatigue properties. Thus, finishing cuts are made at low metal removal rates, or the recast layer is later removed by finishing operations such as grinding. A more recent technique is to use an oscillating electrode, which provides very fine surface finish, requiring significantly less benchwork to improve the luster of cavities.

EXAMPLE 9.6 Machining time in electrical-discharge machining vs. drilling

Calculate the machining time for producing the hole in Example 9.5 by EDM, and compare the time with that for drilling and for ECM. Assume that the titanium alloy has a melting point of 1600°C (see Table 3.3) and that the current is 100 A. Calculate the wear rate of the electrode, assuming that the melting point of the electrode is 1100°C.

Solution.

1. Using Eq. (9.18),

$$\text{MRR} = (4 \times 10^4)(100)(1600^{-1.23}) = 458 \text{ mm}^3/\text{min}.$$

The volume of the hole is

$$V = \pi \left[\frac{(12.5)^2}{4} \right](20) = 2452 \text{ mm}^3.$$

Hence, the machining time for EDM is 2454/458 = 5.4 min; this time is 2.35 times that for ECM and 11.3 times that for drilling. Note, however, that if the current is increased to 300 A, the machining time for EDM will be only 1.8 minutes, which is less than the time for ECM.

2. The wear rate of the electrode is calculated using Eq. (9.19). Thus,

$$W_t = (11 \times 10^3)(100)(1100^{-2.38}) = 0.064 \text{ mm}^3/\text{min}.$$

9.13.1 Electrical-discharge grinding

The grinding wheel in *electrical-discharge grinding* (EDG) is made of graphite or brass and contains no abrasives. Material is removed from the workpiece surface by repetitive spark discharges between the rotating wheel and the workpiece. The material removal rate can be estimated from the equation

$$\text{MRR} = KI, \tag{9.21}$$

where MRR is in mm^3/min, I = current in amperes, and K = workpiece material factor in $\text{mm}^3/\text{A-min}$; for example, $K = 16$ for steel and 4 for tungsten carbide.

The EDG process can be combined with electrochemical grinding, a process called **electrochemical-discharge grinding** (ECDG). Material is removed by chemical action, with the electrical discharges from the graphite wheel breaking up the oxide film, which is washed away by the electrolyte flow. The process is used primarily for grinding carbide tools and dies but can also be used for grinding fragile parts, such as surgical needles, thin-walled tubes, and honeycomb structures. The ECDG process is faster than EDG, but power consumption for ECDG is higher.

In **sawing** with EDM, a setup similar to a band or circular saw (but without any teeth) is used with the same electrical circuit as in EDM. Narrow slits can be made at high rates of metal removal. Because the cutting forces are negligible, the process can be used on slender components.

9.13.2 Wire EDM

An important variation of EDM is *wire EDM* (Fig. 9.35), or *electrical-discharge wire cutting*. In this process, which is similar to contour cutting with a band saw

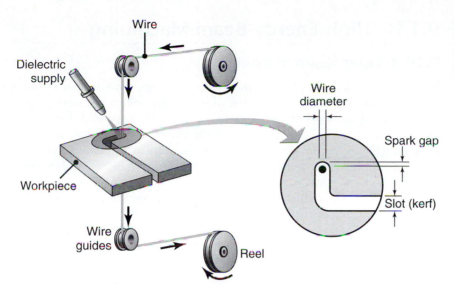

FIGURE 9.35 Schematic illustration of the wire EDM process. As many as 50 hours of machining can be performed with one reel of wire, which is then recycled.

(Section 8.10.5), a slowly moving wire travels along a prescribed path, cutting the workpiece with the discharge sparks acting like cutting teeth. This process is used to cut plates as thick as 300 mm (12 in.) and for making punches, tools, and dies from hard metals; it can also cut intricate components for the electronics industry (Table 9.4).

The wire is usually made of brass, copper, or tungsten and is typically about 0.25 mm (0.01 in.) in diameter, making narrow cuts possible. Zinc- or brass-coated, as well as multicoated, wires are also used. The wire should have sufficient tensile strength and fracture toughness, and high electrical conductivity. The wire is generally used only once, as it is relatively inexpensive, and is then recycled. It travels at a constant velocity in the range of 0.15 to 9 m/min (6 to 360 in./min), and a constant gap (*kerf*; Fig. 9.35) is maintained during the cut.

The cutting speed is generally given in terms of the cross-sectional area cut per unit time. Some typical examples are 18,000 mm²/hr (28 in²/hr) for 50-mm thick (2-in.) D2 tool steel, and 45,000 mm²/hr (70 in²/hr) for 150-mm thick (6-in.) aluminum. These material removal rates indicate a linear cutting speed of 18,500/50 = 360 mm/hr = 6 mm/min, and 45,000/150 = 300 mm/hr = 5 mm/min, respectively.

The material removal rate for wire EDM can be obtained from the expression

$$\text{MRR} = V_f h b, \tag{9.22}$$

where MRR is in mm³/min; V_f is the feed rate of the wire into the workpiece in mm/min; h is the thickness or height in mm; and the kerf, b (see Fig. 9.35), is denoted as $b = d_w + 2s$, where d_w = wire diameter in mm and s = gap, in mm, between wire and workpiece during machining.

Modern wire-EDM machines (**multiaxis EDM wire-cutting machining centers**) are equipped with (a) computer controls to control the cutting path of the wire and its angle with respect to the workpiece plane, (b) automatic self-threading features in case of wire breakage, (c) multiheads for cutting two parts at the same time, (d) automatic controls that prevent wire breakage, and (e) programmed machining strategies. Two-axis computer-controlled machines can produce cylindrical shapes in a manner similar to that of a turning operation or cylindrical grinding.

9.14 | High-Energy-Beam Machining

9.14.1 Laser-beam machining

In *laser-beam machining* (LBM), the source of energy is a laser (an acronym for *light amplification by stimulated emission of radiation*), which focuses optical energy on the surface of the workpiece (Fig. 9.36a). The highly focused, high-density energy melts and evaporates small portions of the workpiece in a controlled manner. This process, which does not require a vacuum, is used to machine a variety of metallic and nonmetallic workpieces. There are several types of lasers used in manufacturing operations: CO_2 (pulsed or continuous wave), Nd:YAG (neodymium: yttrium-aluminum-garnet), Nd:glass, ruby, and excimer (from the words *ex*cited and *dimer*, meaning two mers, or two molecules of the same chemical composition). The typical applications of laser-beam machining are outlined in Tables 9.4 and 9.5.

Important physical parameters in LBM are the *reflectivity* and *thermal conductivity* of the workpiece surface and its specific heat and latent heats of melting and evaporation. The lower these quantities, the more efficient is the process. The cutting depth, t, may be expressed as

$$t \propto \frac{P}{vd}, \tag{9.23}$$

where P is the power input, v is the cutting speed, and d is the laser-beam spot diameter. The surface produced by LBM is usually rough and has a heat-affected zone (see Fig. 12.15) that, for critical applications, may have to be removed or heat treated. Kerf width is an important consideration, as it is in other cutting processes such as sawing, wire EDM, and electron-beam machining.

Laser beams may be used in combination with a gas stream, such as oxygen, nitrogen, or argon (*laser-beam torch*), for cutting thin sheet materials. High-pressure inert-gas-assisted (e.g., nitrogen) laser cutting is used for stainless steel and aluminum as it leaves an oxide-free edge that can improve weldability. Gas streams also have the important function of blowing away molten and vaporized material from the workpiece surface.

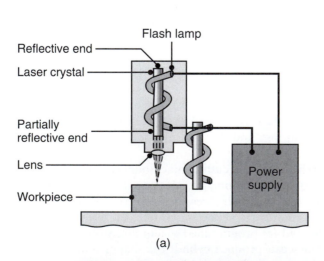

(a)

Reflective end
Flash lamp
Laser crystal
Partially reflective end
Lens
Workpiece
Power supply

(b)

FIGURE 9.36 (a) Schematic illustration of the laser-beam machining process and (b) cutting sheet metal with a laser beam. *Source:* (b) Courtesy of Rofin-Sinat, Inc.

TABLE 9.5

General Applications of Lasers in Manufacturing	
Application	Laser type
Cutting	
Metals	PCO_2; $CWCO_2$; Nd:YAG; ruby
Plastics	$CWCO_2$
Ceramics	PCO_2
Drilling	
Metals	PCO_2; Nd:YAG; Nd:glass; ruby
Plastics	Excimer
Marking	
Metals	PCO_2; Nd:YAG
Plastics	Excimer
Ceramics	Excimer
Surface treatment (metals)	$CWCO_2$
Welding (metals)	PCO_2; $CWCO_2$; Nd:YAG; Nd:glass; ruby

Note: P = pulsed; CW = continuous wave.

Laser-beam machining is used widely in cutting and drilling metals, nonmetals, and composite materials (Fig. 9.36). The abrasive nature of composite materials and the cleanliness of the operation have made laser-beam machining an attractive alternative to traditional machining methods. Holes as small as 0.005 mm (0.0002 in.) and with hole depth-to-diameter ratios of 50 to 1 have been produced in various materials, although a more practical minimum is 0.025 mm (0.001 in.). Extreme caution should be exercised with lasers, as even low-power lasers can cause damage to the retina of the eye if proper safety precautions are not observed.

Laser beams are also used for (a) *welding* (see Section 12.5), (b) small-scale *heat treating* of metals and ceramics to modify their surface mechanical and tribological properties (see Section 4.4), and (c) *marking* of parts with letters, numbers, codes, etc. Marking also can be done by devices such as punches, pins, styli, scroll rolls, or stamps by etching. Although the equipment is more expensive than for other methods, marking and engraving with lasers have become increasingly common, due to the accuracy, reproducibility, flexibility, ease of automation, and on-line application in a wide variety of manufacturing operations.

The inherent flexibility of the laser-beam cutting process with fiber-optic-beam delivery, simple fixturing, low setup times, and the availability of multi-kW machines and 2D and 3D computer-controlled systems are attractive features. Thus, laser cutting can compete successfully with cutting of sheet metal using the traditional punching processes, described in Section 7.3, although the two processes can be combined for improved overall efficiency.

9.14.2 Electron-beam machining and plasma-arc cutting

The source of energy in *electron-beam machining* (EBM) is high-velocity electrons, which strike the surface of the workpiece and generate heat (Fig. 9.37). The applications of this process are similar to those of laser-beam machining, except that EBM requires a vacuum. The machines use voltages in the range of 50–200 kV to accelerate the electrons to speeds of 50 to 80% of the speed of light. Electron-beam machining is used for very accurate cutting of a wide variety of metals. Surface finish is better and kerf width is narrower than that for other thermal-cutting processes. (See also the discussion in Section 12.5 on electron-beam welding.) The interaction of the

FIGURE 9.37 Schematic illustration of the electron-beam machining process. Unlike LBM, this process requires a vacuum, and hence workpiece size is limited by the chamber size.

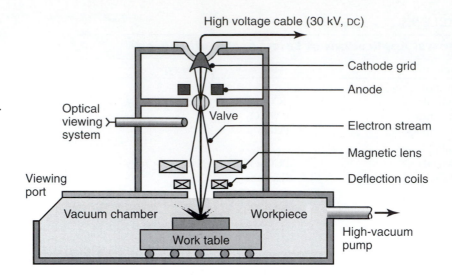

electron beam with the workpiece surface produces hazardous X-rays; consequently, the equipment should be used only by highly trained personnel.

In **plasma-arc cutting** (PAC), *plasma beams* (ionized gas) are used for rapid cutting of nonferrous and stainless-steel plates. The temperatures generated are very high (9400°C, or 17,000°F, in the torch for oxygen as a plasma gas); consequently, the process is fast, kerf width is small, and the surface finish is good. Material removal rates are much higher than in the EDM and LBM processes, and parts can be machined with good reproducibility. Plasma-arc cutting is highly automated through the use of programmable controls.

A more traditional method is **oxyfuel-gas cutting** (OFC), using a torch as in welding (Chapter 12). The operation is particularly useful for cutting steels, cast irons, and cast steels. Cutting occurs mainly by oxidation and burning of the steel, with some melting taking place as well. Kerf widths are usually in the range of 1.5–10 mm (0.06–0.4 in.).

9.15 | Water-Jet, Abrasive Water-Jet, and Abrasive-Jet Machining

We know that when we put our hand across a jet of water or air, there is considerable concentrated force acting on our hand. This force is a result of momentum change of the stream and, in fact, is the principle on which the operation of water or gas turbines is based. This principle is also used in water-jet, abrasive water-jet, and abrasive-jet machining, described next.

1. **Water-jet machining** (WJM). In this process, also called *hydrodynamic machining* (Fig. 9.38a), the force from the jet is utilized in cutting and deburring operations. The water jet acts like a saw and cuts a narrow groove in the material. Although pressures as high as 1400 MPa (200 ksi) can be generated, a pressure level of about 400 MPa (60 ksi) is typically used for efficient operation. Jet-nozzle diameters usually range between 0.05 and 1 mm (0.002 and 0.040 in.). A water-jet cutting machine and typical parts produced are shown in Fig. 9.38.

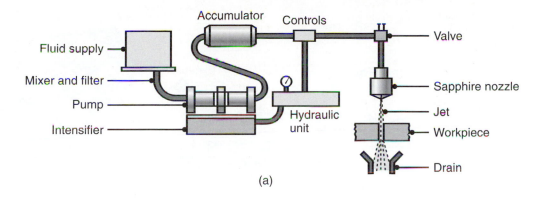

(a)

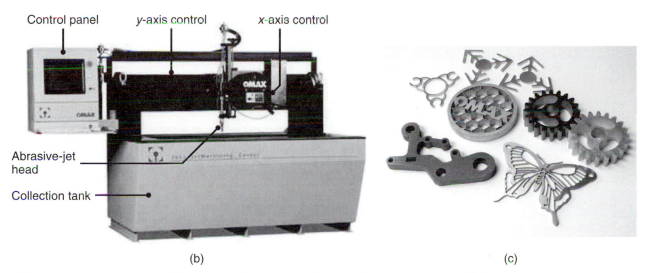

(b)   (c)

FIGURE 9.38 (a) Schematic illustration of water-jet machining, (b) a computer-controlled water-jet cutting machine, and (c) examples of various nonmetallic parts machined by the water-jet cutting process. *Source:* Courtesy of OMAX Corporation.

A variety of nonmetallic materials can be cut with this technique, including plastics, fabrics, rubber, wood products, paper, leather, insulating materials, brick, and composite materials (Fig. 9.38b). Thicknesses range to 25 mm (1 in.) or even higher. Vinyl and foam coverings for some automobile dashboards are, for example, cut using multiple-axis, robot-guided water-jet machining equipment. Because it is an efficient and clean operation compared with other cutting processes, WJM also is used in the food-processing industry for cutting and slicing food products (Table 9.4).

The advantages of this process are that (a) cuts can be started at any location without the need for predrilled holes, (b) no heat is produced, (c) no deflection of the rest of the workpiece takes place (hence the process is suitable for flexible materials), (d) little wetting of the workpiece takes place, (e) the burr produced is minimal, and (f) it is environmentally friendly and safe.

2. **Abrasive water-jet machining.** In *abrasive water-jet machining* (AWJM), the water jet contains abrasive particles such as silicon carbide or aluminum oxide, thus increasing the material removal rate over that of water-jet machining. Metallic, nonmetallic, and composite materials of various thicknesses can be cut in single or multiple layers, particularly heat-sensitive materials that

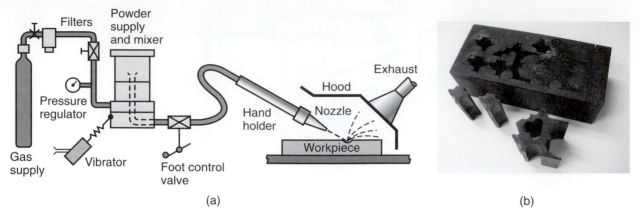

FIGURE 9.39   (a) Schematic illustration of the abrasive-jet machining process and (b) examples of parts produced by abrasive-jet machining; the parts are 50 mm (2 in.) thick and are made of 304 stainless steel. *Source:* Courtesy of OMAX Corporation.

cannot be machined by processes in which heat is produced. The minimum hole size that can be produced satisfactorily to date is about 3 mm (0.12 in.), and maximum hole depth is on the order of 25 mm (1 in.). Cutting speeds can be as high as 7.5 m/min (2.5 ft/min) for reinforced plastics but are much lower for metals. With multiple-axis and robotic-control machines, complex three-dimensional parts can be machined to finish dimensions by this process. Nozzle life is improved by making it from rubies, sapphires, and carbide-base composite materials.

3. **Abrasive-jet machining.**   In *abrasive-jet machining* (AJM), a high-velocity jet of dry air, nitrogen, or carbon dioxide, containing abrasive particles, is aimed at the workpiece surface under controlled conditions (Fig. 9.39). The impact of the particles develops sufficiently high concentrated force (see also Section 9.7) for operations such as (a) cutting small holes, slots, or intricate patterns in very hard or brittle metallic and nonmetallic materials, (b) deburring or removing small flash from parts, (c) trimming and beveling, (d) removing oxides and other surface films, and (e) general cleaning of components with irregular surfaces.

The gas supply pressure is on the order of 850 kPa (125 psi), and the abrasive-jet velocity can be as high as 300 m/s (100 ft/s) and is controlled by a valve. The handheld nozzles are usually made of tungsten carbide or sapphire. The abrasive size is in the range of 10–50 μm (400–2000 μin.). Because the flow of the free abrasives tends to round off corners, designs for abrasive-jet machining should avoid sharp corners; holes made in metal parts tend to be tapered. Because of airborne particulates, there is some safety hazard involved in using this process.

9.16 | Design Considerations

The important design considerations for the processes and operations described in this chapter are outlined below.

9.16.1 Grinding and abrasive machining processes

Typical grinding operations are performed on workpieces produced to close to their final shape (near-net shape), often through machining operations. They are generally not economical for removing large volumes of material, except in creep-feed grinding,

Hence, design considerations for grinding operations are somewhat similar to those for machining, described in Section 8.14. In addition, specific attention should be given to the following considerations:

1. Parts to be ground should be designed so that they can be mounted securely in suitable fixtures and workholding devices. Workpieces that are thin, straight, or tubular may distort during grinding, thus requiring special attention.

2. If high dimensional accuracy is required, interrupted surfaces, such as holes and keyways, should be avoided, as they can cause vibrations and chatter.

3. Parts for cylindrical grinding should be balanced, and long and slender designs should be avoided to minimize deflections. Fillets and corner radii should be as large as possible.

4. In centerless grinding, short pieces may be difficult to grind accurately because of their lack of support on the blade. In through-feed grinding, only the largest diameter on the parts can be ground.

5. Designs requiring precise form grinding should be kept simple to avoid frequent form dressing of the wheel.

6. Deep and small-diameter holes, and blind holes requiring internal grinding, should be avoided, or they should include a relief (see Fig. 9.40).

7. In order to maintain good dimensional accuracy, designs should preferably allow for all grinding operations to be done without having to reposition the workpiece. (Note that this guideline is also applicable to all manufacturing processes and operations.)

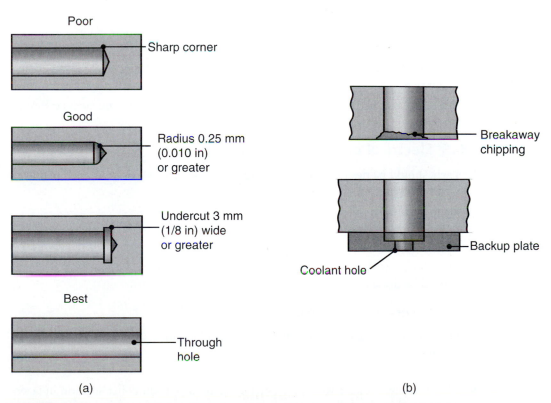

FIGURE 9.40 Design guidelines for internal features, especially as applied to holes. (a) Guidelines for grinding the internal surfaces of holes. These guidelines generally hold for honing as well. (b) The use of a backing plate for producing high-quality through holes by ultrasonic machining. *Source: After J. Bralla.*

9.16.2 Ultrasonic machining

1. Avoid sharp profiles, corners, and radii, because these are eroded by the abrasive slurry.
2. Holes produced will have some taper because of oblique collisions between abrasives and vertical sidewalls.
3. Because of the tendency for chipping of brittle materials at the exit end of holes, the bottom of the parts should have a backup plate (Fig. 9.40b).

9.16.3 Chemical machining

1. Because the etchant attacks all exposed surfaces continuously, designs involving sharp corners, deep and narrow cavities, large tapers, folded seams, or porous workpiece materials should be avoided.
2. Because the etchant attacks the material in both vertical and horizontal directions, *undercuts* may develop (as shown by the areas under the edges of the maskant in Fig. 9.26). Typically, tolerances of $\pm 10\%$ of the material thickness can be maintained in chemical blanking.
3. In order to improve the production rate, the bulk of the workpiece should be preshaped by other processes (such as by machining) prior to chemical machining.

9.16.4 Electrochemical machining and grinding

1. Because of the tendency for the electrolyte to erode away sharp profiles, electrochemical machining is not suited for producing sharp square corners or flat bottoms.
2. Controlling the electrolyte flow may be difficult, so irregular cavities may not be produced to the desired shape with acceptable dimensional accuracy.
3. Designs should make provision for a small taper for holes and cavities to be machined.
4. If flat surfaces are to be produced, the electrochemically ground surface should be narrower than the width of the grinding wheel.

9.16.5 Electrical discharge machining

1. Parts should be designed so that the required electrodes can be shaped properly and economically.
2. Deep and narrow slots should be avoided.
3. For economic production, the surface finish specified should not be too fine, as should be the case for all manufacturing processes.
4. In order to achieve a high production rate, the majority of material removal should be done by conventional processes (roughing out).

9.16.6 Laser- and electron-beam machining

1. Designs with sharp corners should be avoided since they can be difficult to produce.
2. Deep cuts will produce tapered walls.
3. The reflectivity of the workpiece surface is an important consideration in laser-beam machining; dull and unpolished surfaces are preferable.
4. Any adverse effects on the properties of the machined parts caused by high local temperatures and heat-affected zone should be investigated.

5. Because vacuum chambers have limited capacity, part sizes should closely match the size of the vacuum chamber for a high-production rate per cycle for electron-beam machining.

6. If a part requires electron-beam machining on only a small portion of the workpiece, consideration should be given to manufacturing it as several smaller components and assembling them after electron-beam machining.

9.17 | Process Economics

This chapter has indicated that the grinding process can be employed both as a finishing operation and for large-scale removal operations. The use of grinding as a finishing operation is often necessary because forming and machining processes alone usually cannot produce parts with the desired dimensional accuracy and surface finish (see, for example, Fig. 9.27). However, because it is an additional operation, grinding can contribute significantly to product cost. Creep-feed grinding, on the other hand, has proven to be an economical alternative to machining operations such as milling, even though wheel wear is high.

All finishing operations contribute to product cost. On the basis of the discussion thus far, it can be seen that as the surface finish improves, more operations typically are required, and hence the cost increases. Note in Fig. 9.41, for example, how rapidly cost increases as surface finish is improved by processes such as grinding and honing.

Much progress has been made in automating the equipment involved in finishing operations, including advanced computer controls, the increased use of robotics (see Section 14.7), and the availability of powerful and user-friendly software. Consequently, production times and labor costs have been reduced, even though modern machine tools would require significant capital investment. If finishing costs are likely to be an important factor in manufacturing a particular product, the conceptual and design stages should involve an analysis of the degree of surface finish and

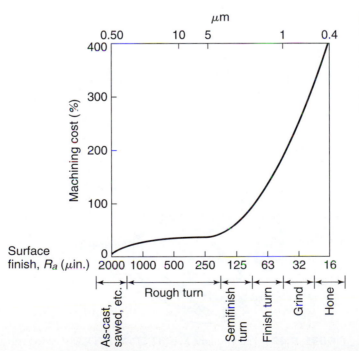

FIGURE 9.41   Increase in the cost of machining and finishing operations as a function of the surface finish required. Note the rapid increase associated with finishing operations.

dimensional accuracy required (see Chapter 1). Furthermore, all processes that precede finishing operations should be analyzed for their capability to produce a more acceptable surface finish and dimensional accuracy (net-shape manufacturing). This task can be accomplished through proper selection of tools and process parameters, as well as the characteristics of the machine tools involved.

The economic production run (see Chapter 16) for a particular machining process depends on the cost of tooling and equipment, the material removal rate, operating costs, and the level of operator skill required, as well as the cost of secondary and finishing operations that may be necessary. In chemical machining, the costs of reagents, maskants, and disposal, together with the cost of cleaning the parts, are important factors. In electrical-discharge machining, the cost of electrodes and the need to replace them periodically can be significant.

The rate of material removal, and hence the production rate, can also vary significantly in these processes. The cost of tooling and equipment also varies significantly, as does the operator skill required. If possible, the high capital investment for machine tools such as for electrical and high-energy-beam machining should be justified in terms of the production runs they are able to generate versus the difficulty and cost of manufacturing the same part by other methods.

CASE STUDY | Manufacture of Stents

Heart attacks, strokes, and other cardiovascular diseases claim one life every 33 seconds in the United States alone. These illnesses are attributed mostly to coronary artery disease, which is the gradual buildup of fat (cholesterol) within the artery wall causing the coronary arteries to become narrowed or blocked. This condition reduces the blood flow to the heart muscle and eventually leads to a heart attack, stroke, or other cardiovascular diseases. One of the most popular methods today for keeping blocked arteries open is to implant a stent into the artery. The MULTI-LINK TETRA™, shown in Fig. 9.42, consists of a tiny mesh tube that is expanded using a balloon dilatation catheter and implanted into a blocked or partially blocked coronary artery. A stent serves as a scaffolding or mechanical brace to keep the artery open. A stent offers the patient a minimally invasive method for treating coronary

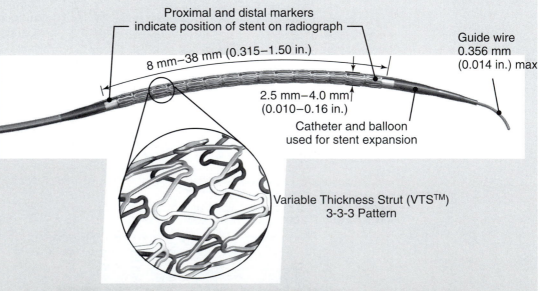

FIGURE 9.42   The Guidant MULTI-LINK TETRA™ coronary stent system.

heart disease for which the alternative is usually open-heart bypass surgery, a procedure with greater risk, pain, rehabilitation time, and cost to the patient.

Stent manufacturing is extremely demanding because the accuracy and precision of its design are paramount for proper performance. The manufacture of the stent must allow satisfaction of all of the design constraints and provide an extremely reliable device; thus, strict quality control procedures must be in place throughout the manufacturing process. When designing a stent there are many material-selection factors to consider, such as radial strength, corrosion resistance, endurance limit, flexibility, and biocompatibility. Radial strength is important because the stent must withstand the pressure the artery exerts on the stent upon expansion. Because the stent is implantable in the body it must be corrosion resistant; it must also be able to withstand the stress undulations caused by heart-beats, and consequently, the material must also have high resistance to fatigue. In addition, the stent must have the appropriate wall thickness in order to be flexible enough to negotiate through the tortuous anatomy of the heart. Above all, the material must be biocompatible with the body, because it remains there for the rest of the patient's life. A standard material for stents, one that satisfies all of these requirements, is 316L stainless steel (Section 3.10.2).

The pattern of a stent also affects its performance. A finite-element analysis of the stent's design is first performed to determine how the stent will perform under the stresses induced from the beating heart. The strut pattern, tube thickness, and strut-leg width all affect the performance of the stent. Many different patterns are possible, such as the MULTI-LINK TETRA™ stent, which is shown in Figure 9.43, including some of the critical dimensions.

A stent starts out as a drawn stainless-steel tube with outer diameter that matches the final stent dimension, and a wall thickness selected to provide proper strength when expanded. The drawn tube is then laser machined (Section 9.14.1) to achieve the desired pattern (Fig. 9.44a). This method has proven to be very effective because of the small, intricate pattern of the stent and the tight dimensional tolerances that it must have. As the laser cuts the pattern of the stent, it leaves small metal slugs that need to be removed; therefore, it is very important that the laser cut all the way through the tube wall. A thick oxide layer or slag inevitably develops on the stainless steel due to the thermal and chemical attack from the air; in addition, weld splatter, burrs, and other surface defects result from laser machining. Finishing operations are therefore required to remove the splatter and oxide layer.

The first finishing operation after laser machining involves chemical etching to remove as much of the slag on the stent as possible; this operation is typically done in an acid solution, and the resulting surface is shown in Fig. 9.44b. Once the

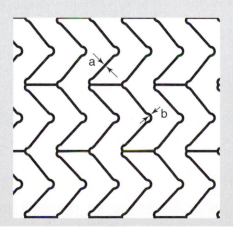

Notes:
a. 0.12 mm (0.0049 in.) section thickness to provide radiopacity
b. 0.091 mm (0.0036 in.) thickness for flexibility

FIGURE 9.43 Detail of the 3-3-3 MULTI-LINK TETRA™ pattern.

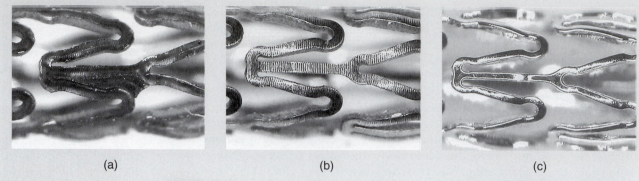

(a)   (b)   (c)

FIGURE 9.44 Evolution of the stent surface. (a) MULTI-LINK TETRA™ after lasing. Note that a metal slug is still attached. (b) After removal of slag. (c) After electropolishing.

slag has been sufficiently removed, any residual burrs from the laser machining process need to be removed and the stent surface must be finished. It is very important that the surface finish of the stent be smooth to prevent possible formation of clots (thrombi) on the stent. To remove the sharp points, the stent is electropolished (Section 9.8) by passing an electrical current through an electrochemical solution to ensure proper surface finish (Fig. 9.44c) that is both shiny and smooth. The stent is then placed onto a balloon catheter assembly, and sterilized and packaged for delivery to the surgeon.

Source: Courtesy of K.L. Graham, Guidant Corporation.

SUMMARY

- Abrasive machining and advanced machining processes are often necessary and economical when workpiece hardness is high, the materials are brittle or flexible, part shapes are complex, and surface finish and dimensional tolerance requirements are high. (Section 9.1)

- Conventional abrasives consist basically of aluminum oxide and silicon carbide, and superabrasives consist of cubic boron nitride and diamond. Friability of abrasive grains is an important factor, as are their shape and size. (Section 9.2)

- Grinding wheels typically consist of a combination of abrasive grains and bonding agents. Important characteristics of wheels include the types of abrasive grain and bond, grade, and hardness. Wheels may be reinforced with metallic or fibers to maintain wheel integrity in case of fracture. (Section 9.3)

- The mechanics of grinding processes help establish quantitative relationships regarding chip dimensions, grinding forces and energy requirements, temperature rise, residual stresses, and any adverse effects on surface integrity of ground components. (Section 9.4)

- Wheel wear is an important consideration in surface quality and product integrity. Wear is usually monitored in terms of the grinding ratio, defined as the ratio of material removed to the volume of wheel wear. Dressing and truing of wheels are done by various techniques, with the aid of computer controls. (Section 9.5)

- A variety of abrasive machining processes and equipment is available for surface, external, and internal grinding, as well as for large-scale material removal processes. The proper selection of abrasives, process variables, and grinding fluids is important for obtaining the desired surface finish and dimensional accuracy and for avoiding damage such as burning, heat checking, harmful residual stresses, and chatter. (Section 9.6)

- Ultrasonic machining removes material by microchipping and is best suited for hard and brittle materials. (Section 9.7)
- Several finishing operations are available for improving surface finish. Because finishing processes can contribute significantly to product cost, proper selection and implementation of finishing operations are important. (Section 9.8)
- Deburring may be necessary for certain finished components. Commonly used deburring methods include vibratory and barrel finishing and shot blasting, although thermal-energy processes and other methods also can be used. (Section 9.9)
- Advanced machining processes include chemical and electrical means and high-energy beams and are particularly suitable for hard materials and complex part shapes. The effects of these processes on surface integrity must be investigated as they can damage surfaces, thus reducing fatigue life. (Sections 9.10–9.14)
- Water-jet, abrasive water-jet, and abrasive-jet machining processes can be used effectively for cutting as well as deburring operations. Because they do not use hard tooling, they have inherent flexibility of operation. (Section 9.15)
- As in all other manufacturing processes, certain design guidelines should be followed for effective use of advanced machining processes. (Section 9.16)
- The processes described in this chapter have unique capabilities. However, because they involve different types of machinery, controls, process variables, and cycle times, the competitive aspects of each process with respect to a particular product must be studied individually. (Section 9.17)

SUMMARY OF EQUATIONS

Undeformed chip length, l:

surface grinding: $l = \sqrt{Dd}$

external grinding: $l = \sqrt{\dfrac{Dd}{1 + D/D_w}}$

internal grinding: $l = \sqrt{\dfrac{Dd}{1 - D/D_w}}$

Undeformed chip thickness, surface grinding: $t = \sqrt{\dfrac{4v}{VCr}\sqrt{\dfrac{d}{D}}}$

Relative grain force: $\propto \dfrac{v}{VC}\sqrt{\dfrac{d}{D}}$

Temperature rise in surface grinding: $\Delta T \propto D^{1/4}d^{3/4}\left(\dfrac{V}{v}\right)^{1/2}$

Grinding ratio: $G = \dfrac{\text{Volume of material removed}}{\text{Volume of wheel wear}}$

Average force of particle in ultrasonic machining: $F_{\text{ave}} = \dfrac{2mv}{t_o}$; where $t_o = \dfrac{5r}{c_o}\left(\dfrac{c_o}{v}\right)^{1/5}$

Material removal rate in

electrochemical machining: $MRR = CI\eta$

electrochemical grinding: $MRR = \dfrac{GI}{\rho F}$

electrical-discharge machining: $MRR = 4 \times 10^4 \, IT_w^{-1.23}$

electrical-discharge grinding: $MRR = KI$

wire EDM: $MRR = V_f \, hb$, where $b = d_w + 2s$

Penetration rate in electrochemical grinding: $V_s = \left(\dfrac{G}{dF}\right)\left(\dfrac{E}{gK_p}\right)K$

Wear rate of electrode in electrical-discharge machining: $W_t = (1.1 \times 10^{11})IT_t^{-2.38}$

Laser-beam cutting time: $t = \dfrac{P}{vd}$

BIBLIOGRAPHY

ASM Handbook, Vol. 16: *Machining,* ASM International, 1989.

Borkowski, J., and Szymanski, A., *Uses of Abrasives and Abrasive Tools,* Ellis Horwood, 1992.

Brown, J., *Advanced Machining Technology Handbook,* McGraw-Hill, 1998.

Chryssolouris, G., and Sheng, P., *Laser Machining, Theory & Practice,* Springer, 1991.

Crafer, R.C., and Oakley, P.J., *Laser Processing in Manufacturing,* Chapman & Hall, 1993.

El-Hofy, H.A.-G., *Advanced Machining Processes,* McGraw-Hill, 2005.

Gillespie, L.K., *Deburring and Edge Finishing Handbook,* Society of Manufacturing Engineers/American Society of Mechanical Engineers, 2000.

Guitran, E.B., *The EDM Handbook,* Hanser-Gardner, 1997.

Hwa, L.S., *Chemical Mechanical Polishing in Silicon Processing,* Academic Press, 1999.

Jain, V.K., and Pandey, P.C., *Theory and Practice of Electrochemical Machining,* Wiley, 1993.

Jameson, E.C., *Electrical Discharge Machining,* Society of Manufacturing Engineers, 2001.

Krar, S., and Ratterman, E., *Superabrasives: Grinding and Machining with CBN and Diamond,* McGraw-Hill, 1990.

Krar, S., *Grinding Technology,* 2nd ed., Delmar, 1995.

Malkin, S., *Grinding Technology: Theory and Applications of Machining with Abrasives,* Wiley, 1989.

Marinescu, I.D. (ed.), *Handbook of Advanced Ceramics Machining,* CRC Press, 2006.

Marinescu, I.D., Hitchiner, M., Uhlmann, E., and Rowe, W.B. (eds.), *Handbook of Machining with Grinding Wheels,* CRC Press, 2006.

Marinescu, I.D., Tonshoff, H.K., and Inasaki, I. (eds.), *Handbook of Ceramics Grinding and Polishing,* Noyes, 1999.

Maroney, M.L., *A Guide to Metal and Plastic Finishing,* Industrial Press, 1991.

McGeough, J.A., *Advanced Methods of Machining,* Chapman & Hall, 1988.

Momber, A.W., and Kovacevic, R., *Principles of Abrasive Water Jet Machining,* Springer, 1998.

Nachtman, E.S., and Kalpakjian, S., *Lubricants and Lubrication in Metalworking Operations,* Dekker, 1985.

Powell, J., *Laser Cutting,* Springer, 1991.

Salmon, S.C., *Modern Grinding Process Technology,* McGraw-Hill, 1992.

Schneider, A.F., *Mechanical Deburring and Surface Finishing Technology,* Dekker, 1990.

Shaw, M.C., *Principles of Abrasive Processing,* Oxford, 1996.

Sluhan, C. (ed.), *Cutting and Grinding Fluids: Selection and Application,* Society of Manufacturing Engineers, 1992.

Sommer, C., and Sommer, S., *Wire EDM Handbook,* Technical Advanced Publishing Co., 1997.

————, *Non-Traditional Machining Handbook,* Advance Publishing, 1999.

Steen, W.M., *Laser Material Processing,* Springer, 1991.

Szymanski, A., and Borkowski, J., *Technology of Abrasives and Abrasive Tools,* Ellis Horwood, 1992.

Taniguchi, N. (ed.), *Nanotechnology,* Oxford, 1996.

Tool and Manufacturing Engineers Handbook, 4th ed., Vol. 1: *Machining,* Society of Manufacturing Engineers, 1983.

Webster, J.A., Marinescu, I.D., and Trevor, T.D., *Abrasive Processes: Theory, Technology, and Practice,* Dekker, 1996.

QUESTIONS

9.1 Why may grinding operations be necessary for parts that have been machined by other processes?

9.2 Explain why there are so many different types and sizes of grinding wheels.

9.3 Why are there large differences between the specific energies involved in grinding (Table 9.3) and in machining (Table 8.3)? Explain.

9.4 Describe the advantages of superabrasives over conventional abrasives. Are there any limitations? Explain.

9.5 Give examples of applications for the grinding wheels shown in Fig. 9.2.

9.6 Explain why the same grinding wheel may act soft or act hard.

9.7 Describe your understanding of the role of friability of abrasive grains on the performance of grinding wheels.

9.8 Explain the factors involved in selecting the appropriate type of abrasive for a particular grinding operation.

9.9 What are the effects of wear flat on the grinding operation? Are there similarities to the effects of flank wear in metal cutting? Explain.

9.10 It was stated that the grinding ratio, G, depends on the following factors: (1) type of grinding wheel, (2) workpiece hardness, (3) wheel depth of cut, (4) wheel and workpiece speeds, and (5) type of grinding fluid. Explain why.

9.11 List and explain the precautions you should take when grinding with high precision. Comment on the role of the machine tool, processing parameters, the type of grinding wheel, and use of grinding fluids.

9.12 Describe the methods you would use to determine the number of active cutting points per unit surface area on the periphery of a straight (Type 1; see Fig. 9.2a) grinding wheel. What is the significance of this number?

9.13 Describe and explain the difficulties involved in grinding parts made of (a) thermoplastics, (b) thermosets, (c) ceramics.

9.14 Explain why ultrasonic machining is not suitable for soft and ductile metals.

9.15 It is generally recommended that a soft-grade wheel be used for grinding hardened steels. Explain why.

9.16 Explain the reasons that the processes described in this chapter may adversely affect the fatigue strength of materials.

9.17 Describe the factors that may cause chatter in grinding operations and give the reasons they cause chatter.

9.18 Outline the methods that are generally available for deburring manufactured parts. Discuss the advantages and limitations of each method.

9.19 In which of the processes described in this chapter are the physical properties of the workpiece material important? Explain.

9.20 Give all possible technical and economic reasons that the material removal processes described in this chapter may be preferred, or even required, over those described in Chapter 8.

9.21 What processes would you recommend for die sinking in a die block, such as that used for forging? Explain. (See also Section 6.7.)

9.22 The proper grinding surfaces for each type of wheel are shown in Fig. 9.2. Explain why grinding on other surfaces of the wheel is improper and/or unsafe.

9.23 Note that wheel (b) in Fig. 9.3 has serrations along its periphery. Explain the reason for such a design.

9.24 In Fig. 9.10, it will be noted that wheel speed and grinding fluids can have a major effect on the type and magnitude of residual stresses developed in grinding. Explain the possible reasons for these phenomena.

9.25 Explain the consequences of allowing the workpiece temperature to rise excessively in grinding operations.

9.26 Comment on any observations you have regarding the contents of Table 9.4.

9.27 Why has creep-feed grinding become an important manufacturing process? Explain.

9.28 There has been a trend in manufacturing industries to increase the spindle speed of grinding wheels. Explain the possible advantages and limitations of such an increase in speed.

9.29 Why is preshaping or premachining of parts generally desirable in the advanced machining processes described in this chapter? Explain.

9.30 Why are finishing operations sometimes necessary? How could they be minimized to reduce product costs? Explain, with examples.

9.31 Why has the wire-EDM process become so widely used in industry, especially in tool and die manufacturing? Explain.

9.32 Make a list of the material removal processes described in this chapter that may be suitable for the following workpiece materials: (1) ceramics, (2) cast iron, (3) thermoplastics, (4) thermosets, (5) diamond, and (6) annealed copper. Explain.

9.33 Explain why producing sharp corners and profiles using some of the processes described in this chapter can be difficult.

9.34 How do you think specific energy, u, varies with respect to wheel depth of cut and hardness of the workpiece material? Explain.

9.35 It is stated is Example 9.2 that the thrust force in grinding is about 30% higher than the cutting force. Why is it higher?

9.36 Why should we be interested in the magnitude of the thrust force in grinding? Explain.

9.37 Why is the material removal rate in electrical-discharge machining a function of the melting point of the workpiece material? Explain.

9.38 Inspect Table 9.4 and, for each process, list and describe the role of various mechanical, physical, and chemical properties of the workpiece material on performance.

9.39 Which of the processes listed in Table 9.4 would not be applicable to nonmetallic materials? Explain.

9.40 Why does the machining cost increase rapidly as surface finish requirements become finer?

9.41 Which of the processes described in this chapter are particularly suitable for workpieces made of (a) ceramics, (b) thermoplastics, and (c) thermosets? Explain.

9.42 Other than cost, is there a reason that a grinding wheel intended for a hard workpiece cannot be used for a softer workpiece? Explain.

9.43 How would you grind the facets on a diamond, such as for a ring, since diamond is the hardest material known?

9.44 Define dressing and truing, and describe the difference between them.

9.45 What is heat checking in grinding? What is its significance? Does heat checking occur in other manufacturing processes? Explain.

9.46 Explain why parts with irregular shapes, sharp corners, deep recesses, and sharp projections can be difficult to polish.

9.47 Explain the reasons why so many different deburring operations have been developed over the years.

9.48 Note from Equation (9.8) that the grinding temperature decreases with increasing work speed. Does this mean that for a work speed of zero, the temperature is infinite? Explain.

9.49 Describe the similarities and differences in the action of metalworking fluids in machining vs. grinding operations.

9.50 Are there any similarities among grinding, honing, polishing, and buffing? Explain.

9.51 Is the grinding ratio an important factor in evaluating the economics of a grinding operation? Explain.

9.52 Although grinding can produce a very fine surface finish on a workpiece, is this necessarily an indication of the quality of a part? Explain.

9.53 If not performed properly, honing can produce holes that are bellmouthed, wavy, barrel-shaped, or tapered. Explain how this is possible.

9.54 Which of the advanced machining processes described in this chapter causes thermal damage to workpieces? List and explain the possible consequences of such damage.

9.55 Describe your thoughts regarding laser-beam machining of nonmetallic materials. Give several possible applications and include their advantages as compared to other processes.

9.56 It was stated that graphite is the generally preferred material for EDM tooling. Would graphite also be appropriate for wire EDM? Explain.

9.57 What is the purpose of the abrasives in electrochemical grinding? Explain.

PROBLEMS

9.58 In a surface-grinding operation, calculate the chip dimensions for the following process variables: $D = 8$ in., $d = 0.001$ in., $v = 30$ ft/min, $V = 5000$ ft/min, $C = 500$ per in^2, and $r = 20$.

9.59 If the workpiece strength in grinding is increased by 50%, what should be the percentage decreases in the wheel depth of cut, d, in order to maintain the same grain force, all other variables being the same?

9.60 Taking a thin, Type 1 grinding wheel as an example, and referring to texts on stresses in rotating bodies, plot the tangential stress, σ_t, and radial stress,

σ_r, as a function of radial distance (from the hole to the periphery of the wheel). Note that because the wheel is thin, this situation can be regarded as a plane-stress problem. How would you determine the maximum combined stress and its location in the wheel? Explain.

9.61 Derive a formula for the material removal rate in surface grinding in terms of process parameters. Use the same terminology as in the text.

9.62 Assume that a surface-grinding operation is being carried out under the following conditions: $D = 250$ mm, $d = 0.1$ mm, $v = 0.5$ m/s, and $V = 50$ m/s.

These conditions are then changed to the following: $D = 150$ mm, $d = 0.1$ mm, $v = 0.3$ m/s, and $V = 25$ m/s. What is the difference in the temperature rise from the initial condition?

9.63 For a surface-grinding operation, derive an expression for the power dissipated in imparting kinetic energy to the chips. Comment on the magnitude of this energy. Use the same terminology as in the text.

9.64 The shaft of a Type 1 grinding wheel is attached to a flywheel only, which is rotating at a certain initial rpm. With this setup, a surface-grinding operation is being carried out on a long workpiece and at a constant workpiece speed, v. Obtain an expression for estimating the linear distance ground on the workpiece before the wheel comes to a stop. Ignore wheel wear.

9.65 Calculate the average impact force on a steel plate by a 1-mm spherical aluminum-oxide abrasive grain, dropped from heights of (a) 1 m, (b) 2 m, and (c) 10 m. Plot the results and comment on your observations.

9.66 A 50-mm-deep hole, 25 mm in diameter, is being produced by electrochemical machining. Assuming that a high production rate is more important than the quality of the machined surface, estimate the maximum current and the time required to perform this operation.

9.67 If the operation in Problem 9.66 were performed on an electrical-discharge machine, what would be the estimated machining time?

9.68 A cutting-off operation is being performed with a laser beam. The workpiece being cut is $\frac{1}{4}$ in. thick and 4 in. long. If the kerf is $\frac{1}{6}$ in. wide, estimate the time required to perform this operation.

9.69 Referring to Table 3.3, identify two metals or metal alloys that, when used as workpiece and electrode, respectively, in EDM would give the (1) lowest and (2) highest wear ratios, R. Calculate these quantities.

9.70 It was stated in Section 9.5.2 that, in practice, grinding ratios typically range from 2 to 200. Based on the information given in Section 9.13, estimate the range of wear ratios in electrical-discharge machining, and then compare them with grinding ratios.

9.71 It is known that heat checking occurs when grinding under the following conditions: spindle speed of 4000 rpm, wheel diameter of 10 in., and depth of cut of 0.0015 in., and a feed rate of 50 ft/min. For this reason, the spindle speed is to be kept at 3500 rpm. If a new, 8-in.-diameter wheel is now used, what should be the spindle speed before heat checking occurs? What spindle speed should be used to maintain the same grinding temperatures as those encountered with the existing operating conditions?

9.72 A hard aerospace aluminum alloy is to be ground. A depth of 0.003 in. is to be removed from a cylindrical section 8 in. long and with a 3-in. diameter. If each part is to be ground in not more than one minute, what is the approximate power requirement for the grinder? What if the material is changed to a hard titanium alloy?

9.73 A grinding operation is taking place with a 10-in. grinding wheel at a spindle rotational speed of 4000 rpm. The workpiece feed rate is 50 ft/min and the depth of cut is 0.002 in. Contact thermometers record an approximate maximum temperature of 1800°F. If the workpiece is steel, what is the temperature if the spindle speed is increased to 5000 rpm? What if it is increased to 10,000 rpm?

9.74 The regulating wheel of a centerless grinder is rotating at a surface speed of 25 ft/min and is inclined at an angle of 5°. What is the feed rate of material past the grinding wheel?

9.75 Using some typical values, explain what changes, if any, take place in the magnitude of the impact force of a particle in ultrasonic machining of a hardened-steel workpiece as its temperature is increased?

9.76 Estimate the percent increase in the cost of the grinding operation if the specification for the surface finish of a part is changed from 63 μin. to 16 μin.

9.77 Assume that the energy cost for grinding an aluminum part is $0.90 per piece. Letting the specific energy requirement for this material be 8 Ws/mm^3, what would be the energy cost if the workpiece material is changed to T15 tool steel?

9.78 Derive an expression for the angular velocity of the wafer as a function of the radius and angular velocity of the pad in chemical mechanical polishing.

9.79 A 25-mm-thick copper plate is being machined by wire EDM. The wire moves at a speed of 1.5 m/min and the kerf width is 1.5 mm. Calculate the power required. (Assume that it takes 1550 J to melt 1 gram of copper.)

9.80 An 8-in.-diameter grinding wheel, 1 in. wide, is used in a surface grinding operation performed on a flat piece of heat-treated 4340 steel. The wheel is rotating with a surface speed $V = 5,000$ fpm, depth of cut $d = 0.002$ in./pass, and cross feed $w = 0.15$ in. The reciprocating speed of the work is $v = 20$ ft/min., and the operation is performed dry. (a) What is the length of contact between the wheel and the work? (b) What is the volume rate of metal removed? (c) Letting $C = 300$, estimate the number of chips formed per time. (d) What is the average volumer per chip? (e) If the tangential cutting force on the workpiece is

$F_c = 10$ lb, what is the specific energy for the operation?

9.81   A 150-mm-diameter tool steel ($u = 60$ W-s/mm^3) work roll for a metal rolling operation is being ground using a 250-mm-diameter, 75-mm-wide Type 1 grinding wheel. The work roll rotates at 10 rpm. Estimate the chip dimensions and grinding force if $d = 0.04$ mm, $r = 12$, $C = 5$ grains per mm^2, and the wheel rotates at $N = 3000$ rpm.

9.82   Estimate the contact time and average force for the following particles striking a steel workpiece at 1 m/s.

Use Eqs. (9.11) and (9.13) and comment on your findings. (a) 5-mm-diameter steel shot; (b) 0.1-mm-diameter cubic boron nitride particles; (c) 3-mm-diameter tungsten sphere; (d) 75-mm-diameter rubber ball; (e) 3-mm-diameter glass beads. (*Hint:* See Tables 2.1, 3.3, and 8.6.)

9.83   Assume that you are an instructor covering the topics in this chapter, and you are giving a quiz on the quantitative aspects to test the understanding of the students. Prepare three quantitative problems, and supply the answers.

DESIGN

9.84   Would you consider designing a machine tool that combines, in one machine, two or more of the processes described in this chapter? Explain. For what types of parts could such a machine be useful? Make preliminary sketches for such machines.

9.85   With appropriate sketches, describe the principles of various fixturing methods and devices that can be used for each of the processes described in this chapter.

9.86   As also described in Section 9.4, surface finish can be an important consideration in the design of products. Describe as many parameters as you can that could affect the final surface finish in grinding, including the role of process parameters as well as the setup and the equipment used.

9.87   Size effect in grinding was described in Section 9.4.1. Design a setup and suggest a series of experiments whereby size effect can be investigated.

9.88   Describe how the design and geometry of the workpiece affect the selection of an appropriate shape and type of a grinding wheel.

9.89   Prepare a comprehensive table of the capabilities of abrasive machining processes, including the shapes of parts ground, types of machines involved, typical maximum and minimum workpiece dimensions, and production rates.

9.90   How would you produce a thin circular disk with a thickness that decreases linearly from the center outward?

9.91   Marking surfaces of manufactured parts with letters and numbers can be done not only with labels and stickers but also by various mechanical and non-mechanical means (see also Section 9.14.1). Make a list of some of these methods, explaining their advantages and limitations.

9.92   On the basis of the information given in Chapters 8 and 9, comment on the feasibility of producing a 10-mm diameter, 100-mm deep through hole in a copper alloy (a) by conventional drilling and (b) other methods.

9.93   Conduct a literature search and explain how observing the color, brightness, and shape of sparks produced in grinding can be a useful guide to identifying the type of material being ground and its condition.

9.94   Visit a large hardware store and inspect the various grinding wheels on display. Make a note of the markings on the wheels and, based on the marking system shown in Figs. 9.4 and 9.5, comment on your observations, including the most commonly found types and sizes of wheels in the store.

9.95   Obtain a small grinding wheel and observe its surfaces using a magnifier or a microscope, and compare with Fig. 9.6. Rub the periphery of the wheel while pressing it hard against a variety of flat metallic and nonmetallic materials. Describe your observations regarding (a) the type of chips produced, (b) the surfaces developed, and (c) the changes, if any, to the grinding wheel surface.

9.96   In reviewing the abrasive machining processes in this chapter it will noted that some processes use bonded abrasives while others involve loose abrasives. Make two separate lists for these two types and comment on your observations.

9.97   Based on the topics covered in Chapters 6 through 9, make a comprehensive table of holemaking processes. (a) Describe the advantages and limitations of each method, (b) comment on the quality and surface integrity of the holes produced, and (c) give examples of specific applications.

9.98 Precision engineering is a term used to describe manufacturing high-quality parts with close dimensional tolerances and good surface finish. Based on their process capabilities, make a list of advanced machining processes (in decreasing order of quality of parts produced). Include a brief commentary on each method.

9.99 It can be seen that several of the processes described in this chapter can be employed, either singly or in combination, to produce or finish tools and dies for metalworking operations. Prepare a brief technical paper on these methods, describing their advantages and limitations, and giving typical applications.

9.100 List the processes described in this chapter that would be difficult to apply to a variety of nonmetallic or rubberlike materials. Explain your thoughts, commenting on such topics as part geometries and the influence of various physical and mechanical properties of workpiece materials.

9.101 Make a list of the processes described in this chapter in which the following properties are relevant or significant: (a) mechanical, (b) chemical, (c) thermal, and (d) electrical. Are there processes in which two or more of these properties are important? Explain.

Properties and Processing of Polymers and Reinforced Plastics; Rapid Prototyping and Rapid Tooling

This chapter covers the characteristics and processing of polymers and reinforced plastics (composite materials), including

*   The structure, properties, and behavior of thermoplastics, thermosets, elastomers, and reinforced plastics, and their manufacturing characteristics.
*   The processing methods employed to produce parts, beginning with extrusion and followed by various forming and molding processes, and the processing parameters and the equipment involved.
*   Fundamentals of various processes that are unique to producing reinforced plastic parts and the equipment involved.
*   The principles of various rapid prototyping operations, where short runs of predominantly plastic parts can be produced quickly.
*   Design considerations in forming polymers and reinforced plastics.
*   The economics of processing these materials into products.

10.1 | Introduction

The word **plastics** was first used around 1909 and is a commonly used synonym for polymers. *Plastics*, from the Greek word *plastikos*, means "it can be molded and shaped." Plastics are one of numerous polymeric materials having extremely large molecules (*macromolecules*). Consumer and industrial products made of polymers include food and beverage containers, packaging, signs, housewares, textiles, medical devices, foams, paints, safety shields, and toys. Compared with metals, polymers are

TABLE 10.1

Approximate Range of Mechanical Properties for Various Engineering Plastics at Room Temperature

Material	UTS (MPa)	E (GPa)	Elongation in 50 mm (%)	Poisson's ratio (ν)
ABS	28–55	1.4–2.8	75–5	—
ABS (reinforced)	100	7.5	—	0.35
Acetals	55–70	1.4–3.5	75–25	—
Acetals (reinforced)	135	10	—	0.35–0.40
Acrylics	40–75	1.4–3.5	50–5	—
Cellulosics	10–48	0.4–1.4	100–5	—
Epoxies	35–140	3.5–17	10–1	—
Epoxies (reinforced)	70–1400	21–52	4–2	—
Fluorocarbons	7–48	0.7–2	300–100	0.46–0.48
Nylon	55–83	1.4–2.8	200–60	0.32–0.40
Nylon (reinforced)	70–210	2–10	10–1	—
Phenolics	28–70	2.8–21	2–0	—
Polycarbonates	55–70	2.5–3	125–10	0.38
Polycarbonates (reinforced)	110	6	6–4	—
Polyesters	55	2	300–5	0.38
Polyesters (reinforced)	110–160	8.3–12	3–1	—
Polyethylenes	7–40	0.1–0.14	1000–15	0.46
Polypropylenes	20–35	0.7–1.2	500–10	—
Polypropylenes (reinforced)	40–100	3.6–6	4–2	—
Polystyrenes	14–83	1.4–4	60–1	0.35
Polyvinyl chloride	7–55	0.014–4	450–40	—

generally characterized by low density, low strength and stiffness (Table 10.1), low electrical and thermal conductivity, good resistance to chemicals, and a high coefficient of thermal expansion.

On the other hand, the useful temperature range for most polymers is generally low, being up to about 350°C (660°F), and they are not as dimensionally stable in service, over a period of time, as metals. Plastics can be machined, cast, formed, and joined into a wide variety of complex shapes with relative ease. Few if any additional surface-finishing operations are required, an important advantage over metals. Plastics are commercially available as sheet, plate, film, rods, and tubing of various cross sections.

The word *polymer* was first used in 1866. The earliest polymers were made of **natural organic materials** from animal and vegetable products, cellulose being the most common example. With various chemical reactions, cellulose is modified into *cellulose acetate,* used in making photographic films (celluloid), sheets for packaging, and textile fibers; it is also modified into cellulose nitrate for plastics, explosives, rayon (a cellulose textile fiber), and varnishes. The earliest **synthetic polymer** was *phenol formaldehyde,* a thermoset developed in 1906 and called *Bakelite* (a trade name, after L.H. Baekeland, 1863–1944).

The development of modern plastics technology began in the 1920s, when raw materials necessary for making polymers were extracted from coal and petroleum products. *Ethylene* was the first example of such a raw material, as it became the building block for *polyethylene*. It is the product of the reaction between acetylene and hydrogen, and acetylene itself is the product of the reaction between coke and

FIGURE 10.1 Basic structure of some polymer molecules: (a) ethylene molecule, (b) polyethylene, a linear chain of many ethylene molecules, and (c) molecular structure of various polymers. These molecules are examples of the basic building blocks for plastics.

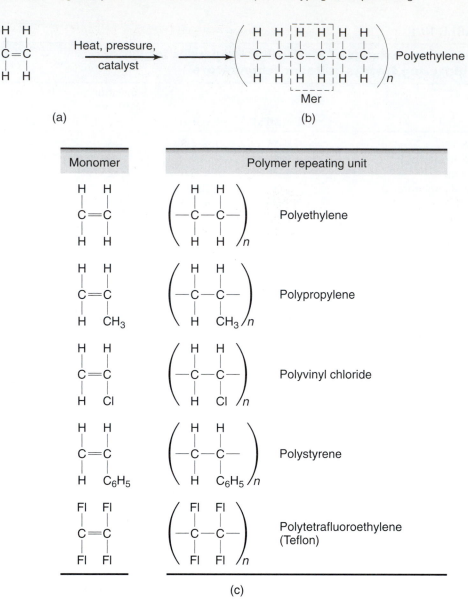

methane. Likewise, commercial polymers including polypropylene, polyvinyl chloride, polymethylmethacrylate, and polycarbonate are all made in a similar manner; these materials are known as *synthetic organic polymers*. An outline of the basic process of making various synthetic polymers is given in Fig. 10.1.

An important group of materials is **polymer-matrix reinforced plastics** (a type of **composite material**), which exhibit a wide range of properties such as stiffness, strength, resistance to creep, and high strength-to-weight and stiffness-to-weight ratios. Their applications include numerous consumer and industrial products, as well as products in the automotive and aerospace industries. Another important advance in manufacturing is **rapid prototyping and tooling**, also called *desktop manufacturing* or *free-form fabrication*. This method is a family of processes by which a solid physical model of a part is made directly from a three-dimensional CAD drawing, involving a time span that is much shorter than that for traditional prototype manufacturing.

10.2 | The Structure of Polymers

A polymer's properties basically depend on (a) the structure of individual polymer molecules, (b) the shape and size of the molecules, and (c) how the molecules are arranged to form a polymer structure. Polymer molecules are characterized by their extraordinary size, a feature that distinguishes them from other organic chemical compositions. A **monomer** is the basic building block of polymers. The word **mer**, from the Greek *meros*, meaning "part," indicates the smallest repeating unit, similar to the term *unit cell* used in connection with crystal structures of metals (described in Section 3.2). Most monomers are organic materials in which carbon atoms are joined in *covalent* (electron-sharing) bonds with other atoms, such as hydrogen, oxygen, nitrogen, fluorine, chlorine, silicon, and sulfur. An ethylene molecule is a simple monomer consisting of carbon and hydrogen atoms (Fig. 10.1a).

The term **polymer** means many mers or units, generally repeated hundreds or thousands of times in a chainlike structure. Polymers are *long-chain molecules*, also called **macromolecules** or **giant molecules** that are formed by *polymerization*, that is, by linking and cross-linking different monomers.

10.2.1 Polymerization

Monomers can be linked in repeating units to make longer and larger molecules by a chemical reaction called *polymerization reaction*. Polymerization processes are complex and are described only briefly here. Although there are several variations, two basic polymerization processes are condensation polymerization and addition polymerization:

a. In **condensation polymerization**, polymers are produced by the formation of bonds between two types of reacting mers. A characteristic of this reaction is that reaction by-products such as water are condensed out, hence the term *condensation*. This process is also known as *step-growth* or *step-reaction polymerization,* because the polymer molecule grows step by step until all of one reactant is consumed.

b. In **addition polymerization**, also known as *chain-growth* or *chain-reaction polymerization*, bonding takes place without reaction by-products. It is called *chain-reaction polymerization* because of the high rate at which long molecules form simultaneously, usually within a few seconds, a rate much higher than that for condensation polymerization. In this reaction, an initiator is added to open the double bond between the carbon atoms and begins the linking process by adding more monomers to a growing chain. For example, ethylene monomers (Fig. 10.1a) link to produce polyethylene (Fig. 10.1b); other examples of addition-formed polymers are shown in Fig. 10.1c.

Some basic characteristics and types of polymers are described below.

1. **Molecular weight.** The sum of the molecular weight of the mers in the polymer chain is the *molecular weight* of the polymer. The higher the molecular weight in a given polymer, the greater is the length of the chain. Because polymerization is a random event, the polymer chains produced are not all of equal length, and the chain lengths produced fall into a traditional distribution curve (see Fig. 11.4). The *average* molecular weight of a polymer is determined and expressed on a statistical basis by averaging. The spread observed in the weight distribution is referred to as the **molecular weight distribution** (MWD). Molecular weight and molecular weight distribution have a strong influence on the properties of the polymer. For example, tensile and impact strength,

FIGURE 10.2 Effect of molecular weight and degree of polymerization on the strength and viscosity of polymers.

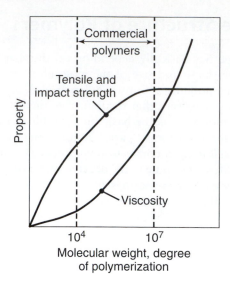

resistance to cracking, and viscosity in the molten state of a polymer all increase with increasing molecular weight (Fig. 10.2). Most commercial polymers have a molecular weight between 10,000 and 10,000,000.

2. **Degree of polymerization.** It is sometimes more convenient to express the size of a polymer chain in terms of the *degree of polymerization* (DP), defined as the ratio of the molecular weight of the polymer to the molecular weight of the repeating unit. In polymer processing (described in Section 10.10), the higher the DP, the higher is the polymer's viscosity, or resistance to flow (Fig. 10.2), which affects ease of shaping and overall cost of processing.

3. **Bonding.** During polymerization, the monomers are linked together in a *covalent bond*, forming a polymer chain. Because of their strength, covalent bonds are also called **primary bonds.** The polymer chains are, in turn, held together by the secondary bonds, such as van der Waals bonds, hydrogen bonds, and ionic bonds. **Secondary bonds** are much weaker than primary bonds, by one to two orders of magnitude. In a given polymer, the increase in strength and viscosity with molecular weight comes in part from the fact that the longer the polymer chain, the greater is the energy required to overcome the strength of the secondary bonds. For example, ethylene mers with a DP of 1, 6, 35, 140, and 1350 are, respectively, in the form of gas, liquid, grease, wax, and hard plastic at room temperature.

4. **Linear polymers.** The chainlike polymers shown in Fig. 10.1 are called *linear polymers* because of their linear structure (Fig. 10.3a). A linear molecule does not mean that it is straight in shape. In addition to those shown in Fig. 10.3, other linear polymers include polyamides (nylon 6,6) and polyvinyl fluoride. Generally, a polymer consists of more than one type of structure; thus, a linear polymer may contain some branched and cross-linked chains. (See parts 5 and 6 of this list.) As a result of branching and cross-linking, the polymer's properties change.

5. **Branched polymers.** The properties of a polymer depend not only on the type of monomers in the polymer, but also on their arrangement in the molecular structure. As shown in Fig. 10.3b, *branched polymers* have side-branch chains that are attached to the main chain during the synthesis of the polymer. Branching interferes with the relative movement of the molecular chains; as a result, resistance to deformation increases and stress cracking resistance is

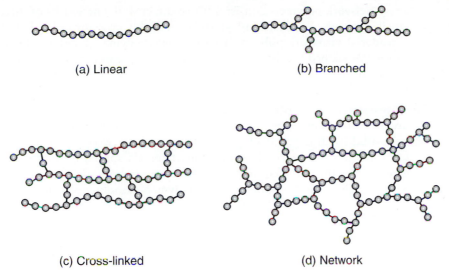

FIGURE 10.3 Schematic illustration of polymer chains. (a) Linear structure; thermoplastics such as acrylics, nylons, polyethylene, and polyvinyl chloride have linear structures. (b) Branched structure, such as polyethylene, and (c) cross-linked structure; many rubbers and elastomers have this structure. Vulcanization of rubber produces this structure. (d) Network structure, which is basically highly cross-linked; examples include thermosetting plastics such as epoxies and phenolics.

affected. Also, the density of branched polymers is lower than that of linear-chain polymers because branches interfere with the packing efficiency of polymer chains. The behavior of branched polymers can be compared with that of linear-chain polymers by making an analogy with a pile of tree branches (branched polymers) and a bundle of straight logs (linear-chain polymers). Note that it is more difficult to move a branch within the pile of branches than to move a log in its bundle. The three-dimensional entanglements of branches make movements more difficult, a phenomenon akin to increased strength.

6. **Cross-linked polymers.** Generally three-dimensional (spatial) in structure, *cross-linked polymers* have adjacent chains linked by covalent bonds (Fig. 10.3c). Polymers with a cross-linked chain structure are called **thermosets** or **thermosetting plastics,** such as epoxies, phenolics, and silicones (Section 10.6). Cross-linking has a major influence on the properties of polymers (generally imparting hardness, strength, stiffness, brittleness, and better dimensional stability; see Fig. 10.4), as well as in the **vulcanization** of rubber (Section 10.8).

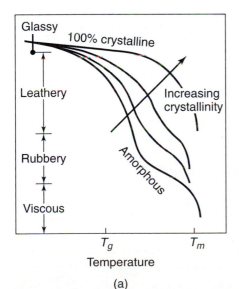

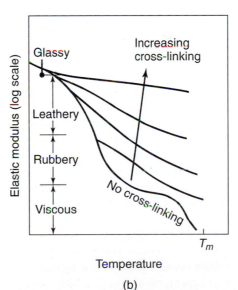

FIGURE 10.4 Behavior of polymers as a function of temperature and (a) degree of crystallinity and (b) cross-linking. The combined elastic and viscous behavior of polymers is known as viscoelasticity.

Network polymers consist of three-dimensional networks of three active covalent bonds, as shown in Fig. 10.3d. A highly cross-linked polymer is also considered a network polymer. Thermoplastic polymers that have already been formed or shaped can be cross-linked to obtain greater strength by being subjected to high-energy radiation, such as ultraviolet light, X-rays, or electron beams. However, excessive radiation can cause *degradation* of the polymer.

7. **Copolymers and terpolymers.** If the repeating units in a polymer chain are all of the same type, the molecule is called a **homopolymer.** However, as with solid-solution metal alloys (see Section 5.2.1), two or three different types of monomers can be combined to impart certain special properties and characteristics to the polymer, such as improvement of both the strength and toughness as well as the formability of the polymer. *Copolymers* contain two types of polymers, such as styrene-butadiene, used widely for automobile tires. *Terpolymers* contain three types, such as ABS (acrylonitrile-butadiene-styrene), used for helmets, telephones, and refrigerator liners.

EXAMPLE 10.1 Degree of polymerization in polyvinyl chloride

Determine the molecular weight of a polyvinyl chloride (PVC) mer. If a PVC polymer has an average molecular weight of 50,000, what is the degree of polymerization?

Solution. From Fig. 10.1c, note that each PVC mer has three hydrogen atoms, two carbon atoms, and one chlorine atom. Since the atomic numbers of these elements are 1, 12, and 35.5, respectively, the weight of a PVC mer is $(3)(1) + (2)(12) + (1)(35.5) = 62.5$; and the degree of polymerization is $50,000/62.5 = 800$.

10.2.2 Crystallinity

Polymers such as polymethylmethacrylate, polycarbonate, and polystyrene are generally *amorphous;* that is, the polymer chains exist without long-range order. (See also the discussion of *amorphous alloys* in Section 3.11.9.) The amorphous arrangement of polymer chains is often described as similar to a bowl of spaghetti, or worms in a bucket, all intertwined with each other. In some polymers, however, it is possible to impart some crystallinity and thereby modify their characteristics. This may be done either during the synthesis of the polymer or by deformation (shaping) during its subsequent processing.

The crystalline regions in polymers are called **crystallites** (Fig. 10.5). The crystals are formed when the long molecules arrange themselves in an orderly manner, similar to folding a fire hose in a cabinet or facial tissue in a box. Thus, a partially crystalline (semicrystalline) polymer can be regarded as a two-phase material (see Section 5.2.3), one phase being crystalline and the other amorphous. By controlling the rate of solidification during cooling, it is possible to impart different **degrees of crystallinity** to polymers, although a polymer can never be 100% crystalline.

Crystallinity ranges from an almost complete crystal (up to about 95% by volume in the case of polyethylene) to slightly crystallized, but mostly amorphous polymers. The degree of crystallinity is also affected by branching. A linear polymer can become highly crystalline, but a highly branched polymer cannot. Although the latter may develop some low degree of crystallinity, it will never acquire a high crystalline content because the branches interfere with the alignment of the chains into a regular crystal array.

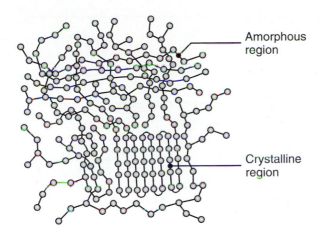

FIGURE 10.5 Amorphous and crystalline regions in a polymer. Note that the crystalline region (crystallite) has an orderly arrangement of molecules. The higher the crystallinity, the harder, stiffer, and less ductile is the polymer.

Amorphous region

Crystalline region

Effects of crystallinity. The mechanical and physical properties of polymers are greatly influenced by the degree of crystallinity. As crystallinity increases, polymers become stiffer, harder, less ductile, denser, less rubbery, and more resistant to solvents and heat (Fig. 10.4). The increase in density with increasing crystallinity is caused by crystallization shrinkage and a more efficient packing of the molecules in the crystal lattice. For example, the highly crystalline form of polyethylene, known as *high-density polyethylene* (HDPE), has a specific gravity in the range of 0.941 to 0.970 (80 to 95% crystalline) and is stronger, stiffer, tougher, and less ductile than low-density polyethylene (LDPE), which is about 60 to 70% crystalline and has a specific gravity of about 0.910 to 0.925.

Optical properties of polymers also are affected by the degree of crystallinity. The reflection of light from the boundaries between the crystalline and the amorphous regions in the polymer (see Fig. 10.5) causes opaqueness. Furthermore, because the index of refraction is proportional to density, the higher the density difference between the amorphous and the crystalline phases, the greater is the opaqueness of the polymer. Polymers that are completely amorphous can be transparent, such as polycarbonate and acrylics.

10.2.3 Glass-transition temperature

Amorphous polymers do not have a specific melting point, but they undergo a distinct change in their mechanical behavior across a narrow range of temperature. At low temperatures, amorphous polymers are hard, rigid, brittle, and glassy, and at high temperatures, they are rubbery or leathery. The temperature at which this transition occurs is called the *glass-transition temperature,* T_g, or the *glass point* or *glass temperature.* The term *glass* is included in this definition because glasses, which are amorphous solids (see Section 3.11.9), behave in the same manner (as you can see by holding a glass rod over a flame and observing the change in its behavior). Although most amorphous polymers exhibit this behavior, there are some exceptions, such as polycarbonate, which is (a) not rigid or brittle below its glass-transition temperature and (b) tough at ambient temperature and is, for example, used for safety helmets and shields.

To determine T_g, the specific volume of the polymer is measured and plotted against temperature to observe the sharp change in the slope of the curve, as can be seen in Fig. 10.6. However, in the case of highly cross-linked polymers, the slope of the curve changes *gradually* near T_g, thus making it difficult to determine T_g for these polymers. The glass-transition temperature varies for different polymers (Table 10.2); for example, room temperature is above T_g for some polymers and below it for others. Unlike amorphous polymers, partly crystalline polymers have a distinct

FIGURE 10.6 Specific volume of polymers as a function of temperature. Amorphous polymers, such as acrylic and polycarbonate, have a glass-transition temperature, T_g, but do not have a specific melting point, T_m. Partly crystalline polymers, such as polyethylene and nylons, contract sharply at their melting points during cooling.

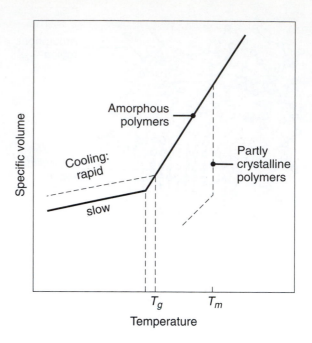

TABLE 10.2

Glass-Transition and Melting Temperatures of Selected Polymers		
Material	T_g (°C)	T_m (°C)
Nylon 6,6	57	265
Polycarbonate	150	265
Polyester	73	265
Polyethylene		
high density	−90	137
low density	−110	115
Polymethylmethacrylate	105	—
Polypropylene	−14	176
Polystyrene	100	239
Polytetrafluoroethylene (Teflon)	−90	327
Polyvinyl chloride	87	212
Rubber	−73	—

melting point, T_m (Fig. 10.6; see also Table 10.2). Because of the structural changes (known as first-order changes) that occur, the specific volume of the polymers drops rapidly as their temperature is reduced.

10.2.4 Polymer blends

To improve the brittle behavior of amorphous polymers below their glass-transition temperature, they can be blended, usually with small quantities of an elastomer (see Section 10.8), known as **rubber-modified polymers.** These tiny particles are dispersed throughout the amorphous polymer, enhancing its toughness and impact strength by improving its resistance to crack propagation. Blending may involve several components (**polyblends**) to obtain the favorable properties of different polymers. **Miscible blends** involve mixing without separation of two phases (a process similar to alloying of metals) and enable polymer blends to become more ductile. Polymer blends account for about 20% of all polymers produced.

10.2.5 Additives in polymers

In order to impart certain specific properties, polymers are usually compounded with *additives*. The additives modify and improve the characteristics of the polymers such as their stiffness, strength, color, weatherability, flammability, arc resistance for electrical applications, and ease of subsequent processing.

1. **Fillers** are generally wood flour (fine sawdust), silica flour (fine silica powder), clay, powdered mica, and short fibers of cellulose, glass, and asbestos. Depending on their type, fillers improve the strength, hardness, toughness, abrasion resistance, dimensional stability, and/or stiffness of plastics. These properties are greatest at various percentages of different types of polymer-filler combinations. As in reinforced plastics (Section 10.9), a filler's effectiveness depends on the nature and strength of the bond between the filler material and the polymer chains. Because of their low cost, fillers are important in reducing the overall cost of polymers.

2. **Plasticizers** are added to some polymers to impart flexibility and softness by lowering their glass-transition temperature. Plasticizers are low-molecular-weight solvents with high boiling points (that is, they are nonvolatile). They reduce the strength of the secondary bonds between the long-chain molecules, thus making the polymer soft and flexible. The most common use of plasticizers is in polyvinyl chloride (PVC), which remains flexible during its many uses. Other applications of plasticizers include thin sheet, film, tubing, shower curtains, and clothing materials.

3. Most polymers are adversely affected by **ultraviolet radiation** (sunlight) and oxygen, which weaken and break the primary bonds, resulting in the scission (splitting) of the long-chain molecules. The polymer then degrades and becomes brittle and stiff. On the other hand, degradation may be beneficial, as in the disposal of plastic objects by subjecting them to environmental attack. (See also *biodegradable plastics,* Section 10.7.3.) A typical example of imparting protection against ultraviolet radiation is the compounding of some plastics and rubbers with *carbon black* (soot) which absorbs a high percentage of the ultraviolet radiation. Protection against degradation by oxidation, particularly at elevated temperatures, is achieved by adding *antioxidants* to the polymer. Applying various *coatings* is another means of protecting polymers against environmental attack.

4. The wide variety of colors available in plastics is obtained by adding **colorants**; these materials are either organic (*dyes*) or inorganic (*pigments*). Pigments are dispersed particles and generally have greater resistance than dyes to temperature and light. The selection of a colorant depends on the polymer's service temperature and the length of exposure to light.

5. Most polymers will ignite if the temperature is sufficiently high. The **flammability** of polymers (defined as the ability to *support* combustion) varies considerably, depending on their composition (such as the chlorine and fluorine content). Flammability can be reduced either by making the polymer from less flammable raw materials or by adding **flame retardants,** such as compounds of chlorine, bromine, and phosphorus. Examples of polymers with different burning characteristics include (a) fluorocarbons (e.g., *Teflon*) which do not burn, (b) carbonate, nylon, and vinyl chloride which burn but are *self-extinguishing;* and (c) acetal, acrylic, ABS, polyester, polypropylene, and styrene which burn and are not self-extinguishing.

6. **Lubricants** may be added to polymers to reduce friction during their subsequent processing into useful products and to prevent parts from sticking to the molds. Lubrication is also important in preventing thin polymer films from sticking to each other (see also Chapter 4).

10.3 | Thermoplastics: Behavior and Properties

It was noted earlier that the bonds between adjacent long-chain molecules (secondary bonds) are much weaker than the covalent bonds (primary bonds) within each molecule, and that it is the strength of the secondary bonds that determines the overall strength of the polymer. Linear and branched polymers have weak secondary bonds. As the temperature of the polymer is raised above its glass-transition temperature or melting point, certain polymers become easier to form or mold into desired shapes. The increased temperature weakens the secondary bonds (due to thermal-induced vibration of the long molecules), and the adjacent chains can thus move more easily under the shaping forces. If the polymer is then cooled, it returns to its original hardness and strength; in other words, the effects of the process are *reversible*. However, the repeated heating and cooling of thermoplastics can cause **degradation** (**thermal aging**).

Polymers that exhibit this behavior are known as **thermoplastics**; typical examples include acrylics, cellulosics, nylons, polyethylenes, and polyvinyl chloride. The behavior of thermoplastics depends primarily on temperature and rate of deformation. Below the glass-transition temperature, most polymers are *glassy* (described as rigid, brittle, or hard), and they behave like an elastic solid. If the load exceeds a certain critical value, the polymer fractures, just as a piece of glass does at room temperature. In the glassy region, the relationship between stress and strain is linear, or

$$\sigma = E\epsilon. \tag{10.1}$$

If the polymer is tested in torsion, then

$$\tau = G\gamma. \tag{10.2}$$

The glassy behavior can be represented by a spring that has a stiffness equivalent to the modulus of elasticity of the polymer (Fig. 10.7a). Note that the strain is completely recovered when the load is removed at time t_1. When the applied stress is increased, the polymer eventually fractures, just as a piece of glass does at room temperature. The mechanical properties of several polymers listed in Table 10.1 indicate that thermoplastics are about two orders of magnitude less stiff than metals, and their ultimate tensile strength is about one order of magnitude lower than that of metals. (See Table 2.1.) Typical stress–strain curves for some thermoplastics and thermosets at room temperature are shown in Fig. 10.8. Note that the plastics in the figure exhibit different behaviors, which can be described as rigid, soft, brittle, flexible, and so on. Plastics, like metals, also undergo fatigue and creep phenomena. Some of the major characteristics of thermoplastics are described next.

1. **Effects of temperature and deformation rate.** The typical effects of temperature on the strength and elastic modulus of thermoplastics are similar to those for metals (see Sections 2.2.6 and 2.2.7). Thus, with increasing temperature, the strength and modulus of elasticity decrease, and the toughness increases (Fig. 10.9). The effect of temperature on impact strength is shown in Fig. 10.10; note the large difference in the impact behavior of various polymers.

   If the temperature of a thermoplastic polymer is raised above its T_g, it first becomes leathery and then rubbery with increasing temperature (Fig. 10.4). Finally, at higher temperatures, e.g., above T_m for crystalline thermoplastics, it becomes a viscous fluid, with viscosity decreasing as temperature and strain rate are increased. Thermoplastics display **viscoelastic** behavior, as demonstrated by the spring and dashpot models shown in Fig. 10.7c and d, known as the *Maxwell* and *Kelvin* (or *Voigt*) models, respectively. When a constant load is applied, the polymer first stretches at a high strain rate and then continues to elongate over a period of time (*creep*, see Section 2.8), because of its viscous

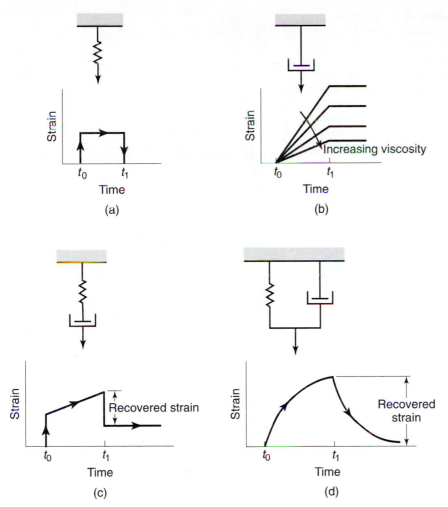

FIGURE 10.7   Various deformation modes for polymers: (a) elastic, (b) viscous, (c) viscoelastic (Maxwell model), and (d) viscoelastic (Voigt or Kelvin model). In all cases, an instantaneously applied load occurs at time t_o, resulting in the strain paths shown.

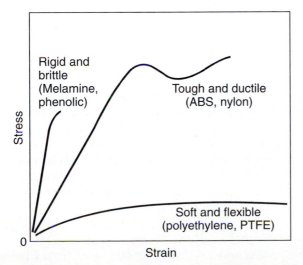

FIGURE 10.8   General terminology describing the behavior of three types of plastics. PTFE is polytetrafluoroethylene (Teflon, a trade name). *Source:* After R.L.E. Brown.

FIGURE 10.9 Effect of temperature on the stress–strain curve for cellulose acetate, a thermoplastic. Note the large drop in strength and increase in ductility with a relatively small increase in temperature. *Source:* After T.S. Carswell and H.K. Nason.

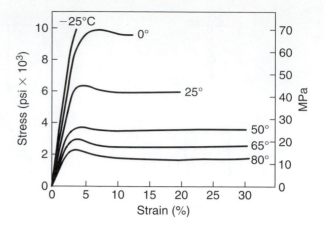

FIGURE 10.10 Effect of temperature on the impact strength of various plastics. Note that small changes in temperature can have a significant effect on impact strength. *Source:* After P.C. Powell.

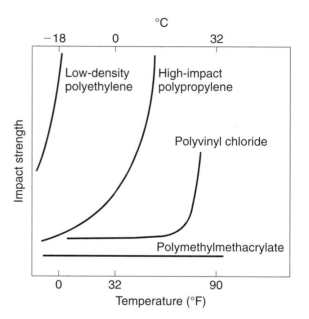

behavior. Note in the models shown in Fig. 10.7 that the elastic portion of the elongation is reversible (elastic recovery), but the viscous portion is not.

The viscous behavior is expressed by

$$\tau = \eta\left(\frac{dv}{dy}\right) = \eta\dot{\gamma}, \tag{10.3}$$

where η is the *viscosity* and $dv/dy = \dot{\gamma}$ is the *shear-strain rate*, $\dot{\gamma}$, as shown in Fig. 10.11. The viscosities of some polymers are given in Fig. 10.12. When the shear stress, τ, is directly proportional to the shear-strain rate, the behavior of the thermoplastic polymer is known as *Newtonian*. For many polymers, however, Newtonian behavior is not a good approximation, and Eq. (10.3) will give poor predictions of process performance. For example, polyvinyl chloride, polyethylene (low density and high density), and polypropylene have viscosities that decrease markedly with increasing strain rate (**pseudoplastic behavior**). Their viscosity, as a function of strain rate, can be expressed as

$$\eta = A\dot{\gamma}^{1-n}, \tag{10.4}$$

where A is the *consistency index* and n is the *power-law index* for the polymer.

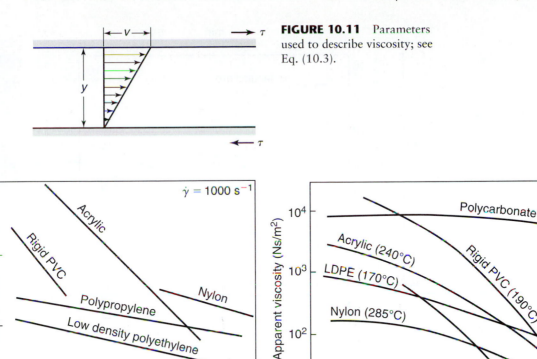

FIGURE 10.11 Parameters used to describe viscosity; see Eq. (10.3).

FIGURE 10.12 Viscosity of some thermoplastics as a function of (a) temperature and (b) shear rate. *Source:* After D.H. Morton-Jones.

Note from the foregoing equations that the viscous behavior of thermoplastics is similar to the strain-rate sensitivity of metals (Section 2.2.7), given by the expression

$$\sigma = C\dot{\epsilon}^{m}, \qquad (10.5)$$

where, for Newtonian behavior, $m = 1$.

Thermoplastics have high m values, indicating that they can undergo large *uniform* deformations in tension before fracture. Note in Fig. 10.13 how, unlike with ordinary metals, the necked region elongates considerably. This characteristic, which is the same as in superplastic metals (described in Section 2.2.7), enables thermoplastics to be formed into complex shapes, such as bottles for soft drinks, cookie and meat trays, and lighted signs, as described in Section 10.10. Note also that the effect of increasing the rate of load application is that the strength of the polymer is increased.

Between T_g and T_m, thermoplastics exhibit leathery and rubbery behavior, depending on their structure and degree of crystallinity, as shown in Fig. 10.4. A term combining the strains caused by elastic behavior (e_e) and viscous flow (e_v) is called the *viscoelastic modulus*, E_r, and is expressed as

$$E_r = \frac{\sigma}{(e_e + e_v)}. \qquad (10.6)$$

This modulus essentially represents a time-dependent elastic modulus.

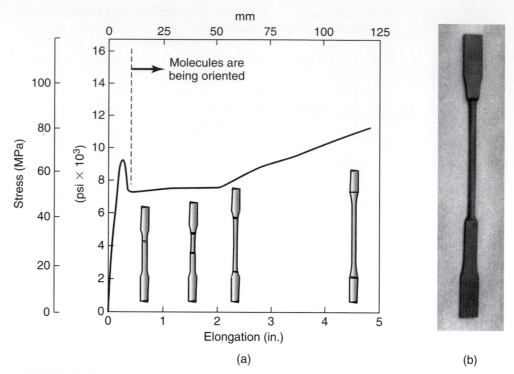

FIGURE 10.13   (a) Load-elongation curve for polycarbonate, a thermoplastic, and (b) high-density polyethylene tension-test specimen, showing uniform elongation (the long, narrow region in the specimen). *Source:* (a) After R.P. Kambour and R.E. Robertson.

The viscosity η of polymers, a measure of the resistance of their molecules in sliding along each other, depends on temperature, pressure, and on the polymer's structure and molecular weight. The effect of temperature can be represented by

$$\eta = \eta_o e^{E/kT}, \tag{10.7}$$

where η_o is a material constant, E is the activation energy (the energy required to initiate a reaction), k is Boltzmann's constant (the thermal-energy constant, or 13.8×10^{-24} J/K), and T is the temperature (in K).

Thus, as the temperature of the polymer is increased, η decreases, because of the higher mobility of the molecules. Note that viscosity may be regarded as the inverse of fluidity of molten metals in a casting process (Section 5.4.2). Increasing the molecular weight (and hence increasing the length of the chain) increases η, because of the greater number of secondary bonds present. As the molecular-weight distribution widens, there are more shorter chains, and η decreases. The effect of increasing pressure is that viscosity increases, because of reduced *free volume*, or *free space* (defined as the volume in excess of the true volume of the crystal in the crystalline regions of the polymer).

Based on experimental observations that at the glass-transition temperature, T_g, polymers have a viscosity η of about 10^{12} Pa-s, an empirical relationship between viscosity and temperature has been developed for linear thermoplastics:

$$\log \eta = 12 - \frac{17.5 \, \Delta T}{52 + \Delta T}. \tag{10.8}$$

In this equation, $\Delta T = T - T_g$, in K or °C; thus, we can estimate the viscosity of the polymer at any temperature.

2. **Creep and stress relaxation.** Because of their viscoelastic behavior, thermoplastics are particularly susceptible to *creep* and *stress relaxation* phenomena, described in Section 2.8. Referring back to the dashpot models in Fig. 10.7b, c, and d, note that under a constant load, the polymer undergoes further strain, and thus it creeps. The recovered strain depends on the stiffness of the spring and, hence, the modulus of elasticity.

According to the Maxwell model of viscoelastic behavior (Fig. 10.7c), the strain caused by an applied force is called the *creep function* and is given by

$$\epsilon(t) = \left(\frac{1}{k} + \frac{1}{\eta}t \right)F, \tag{10.9}$$

where k is the spring stiffness (AE/l for a linear spring), η is the coefficient of viscosity for the dashpot, and F is the applied force. For the Voigt model (Fig. 10.7d), the creep function is

$$\epsilon(t) = \frac{1}{k}\left[1 - e^{-(\mu/\eta)t}\right]F. \tag{10.10}$$

Stress relaxation in polymers occurs over a period of time. Recall that stress relaxation refers to a gradual reduction in stress under an applied strain. For the Maxwell model (Fig. 10.7c), the *relaxation function* is

$$\sigma(t) = \frac{k\Delta l}{A}e^{-(k/\eta)t} = \frac{k\Delta l}{A}e^{-t/\lambda} \tag{10.11}$$

where Δl is the elongation and A is the instantaneous cross-sectional area. The factor $\lambda = \eta/k$ has the units of time and characterizes the rate of decay of the stress; hence, it is referred to as the *relaxation time*. Because both the stiffness and the viscosity of polymers depend on temperature to varying degrees, the relaxation time also depends on temperature.

3. **Orientation.** When thermoplastics are permanently deformed or shaped, say by stretching, the long-chain molecules align in the general direction of elongation. This process is called *orientation*, and just as with metals, the polymer becomes *anisotropic* (see Section 3.5). Also, the specimen becomes stronger and stiffer in the elongated (stretched) direction than in the transverse direction. This process is an important technique to enhance the strength and toughness of polymers; however, orientation weakens the polymer in the transverse direction.

4. **Crazing.** When subjected to tensile stresses or to bending, some thermoplastics such as polystyrene and polymethylmethacrylate, develop localized, wedge-shaped narrow regions of highly deformed material, a phenomenon called *crazing*. Although they may appear to be like cracks, crazes are areas of spongy material, typically containing about 50% voids. With increasing tensile load on the specimen, the voids coalesce to form a crack, eventually leading to fracture. Crazing has been observed in transparent glassy polymers as well as in other types of polymers. The environment and the presence of solvents, lubricants, and water vapor enhance the formation of crazes (**environmental stress cracking** and **solvent crazing**). Residual stresses in the material also contribute to crazing and cracking of the polymer, as do radiation and ultraviolet radiation.

A related phenomenon is **stress whitening.** When subjected to tensile stresses, such as by folding or bending, the plastic becomes lighter in color, a phenomenon that is usually attributed to the formation of microvoids in the material. As a result, the material becomes less translucent (transmits less light), or more opaque. This behavior can easily be demonstrated by bending thin plastic components commonly found in household products and toys.

5. **Water absorption.** An important limitation of some polymers, such as nylons, is their ability to absorb water (*hygroscopy*). Water acts as a plasticizing agent (it makes the polymer more plastic); thus, in a essence, it lubricates the chains in the amorphous region. Typically, with increasing moisture absorption, the glass-transition temperature, yield stress, and elastic modulus of the polymer are lowered severely. Dimensional changes also occur because of water absorption, such as in a humid environment.

6. **Thermal and electrical properties.** Compared with metals, plastics generally are characterized by low thermal and electrical conductivity, low specific gravity (ranging from 0.90 to 2.2), and a relatively high coefficient of thermal expansion (about one order of magnitude higher; see Table 3.3). Because polymers generally have low electrical conductivity, they are commonly used for numerous electrical and electronic components. Polymers can, however, be made to be electrically conductive, as described in Section 10.7.2.

EXAMPLE 10.2 Lowering the viscosity of a polymer

In processing a batch of polycarbonate at 170°C to make a certain part, it is found that the polycarbonate's viscosity is twice that desired. Determine the temperature at which this polymer should be processed.

Solution. From Table 10.2, the T_g for polycarbonate is 150°C. Its viscosity at 170°C is determined by Eq. (10.8):

$$\log \eta = 12 - \frac{17.5(20)}{52 + 20} = 7.14.$$

Hence, $\eta = 13.8$ MPa-s, and because this magnitude is twice what is desired, the new viscosity should be 6.9 MPa-s. We substitute this new value to find the new temperature:

$$\log (6.9 \times 10^6) = 12 - \frac{17.5(\Delta T)}{52 + \Delta T}.$$

Therefore, $\Delta T = 21.7$, or the new temperature is $150 + 21.7 = 171.7°C$. Note that in this solution the numbers for temperature have been rounded, and that viscosity is very sensitive to temperature.

EXAMPLE 10.3 Stress relaxation in a thermoplastic member under tension

A long piece of thermoplastic is stretched between two rigid supports at a stress level of 5 MPa. After 30 minutes, the stress level is found to have decayed to one-half of the original level. How long will it take for the stress level to reach one-tenth of the original value?

Solution. Substituting these data into Eq. (10.11) as modified for normal stress, we have, noting that at $t = 0$, $\sigma = 5$ MPa,

$$5 = \frac{k\Delta l}{A} e^0 = \frac{k\Delta l}{A}.$$

Therefore, Eq. (10.11) becomes

$$\sigma(t) = 5e^{-t\lambda}.$$

Also, at $t = 30$ minutes, $\sigma = 2.5$ MPa, so that

$$2.5 = 5e^{-(30)/\lambda}; \qquad -\frac{30}{\lambda} = \ln\left(\frac{2.5}{5}\right); \qquad \lambda = 43.3 \text{ min.}$$

Thus, the time needed for $\sigma = 0.5$ MPa is

$$\ln\left(\frac{0.5}{5}\right) = -\frac{t}{43.3}, \text{ or } t = 99.7 \text{ min.}$$

10.4 | Thermosets: Behavior and Properties

When the long-chain molecules in a polymer are cross-linked in a three-dimensional (spatial) arrangement, the structure in effect becomes one giant molecule with strong covalent bonds. Such polymers are called **thermosetting polymers**, or *thermosets,* because during polymerization, the network is completed, and the shape of the part is permanently set. An important behavior is that this **curing** (*cross-linking*) reaction, unlike that of thermoplastics, is irreversible. The response of a thermosetting plastic to temperature thus can be likened to baking a cake or boiling an egg; once the cake is baked and cooled, or the egg is boiled and cooled, reheating it will not alter its shape.

Some thermosets, such as epoxy, polyester, and urethane, cure at room (ambient) temperature, whereby the heat of the exothermic reaction cures the plastic. Thermosetting polymers do not have a sharply defined glass-transition temperature. The polymerization process for thermosets generally takes place in two stages: (1) The first stage occurs at the chemical plant, where the molecules are partially polymerized into linear chains, and (2) the second stage occurs at the parts-producing (manufacturing) plant, where cross-linking is completed under heat and pressure during the molding and shaping of the part (Section 10.10).

Because of the nature of their bonds, the strength and hardness of thermosets, unlike those of thermoplastics, are not affected by temperature or rate of deformation. A typical thermoset is *phenolic,* which is a product of the reaction between phenol and formaldehyde. Common products made from this polymer are the handles and knobs on cooking pots and pans, and components of light switches and outlets. Thermosetting plastics (Section 10.6) generally possess better mechanical, thermal, and chemical properties, electrical resistance, and dimensional stability than do thermoplastics. However, if the temperature is increased sufficiently, the thermosetting polymer begins to burn up, degrade, and char.

10.5 | Thermoplastics: General Characteristics and Applications

This section outlines the general characteristics and typical applications of major thermoplastics, particularly as they relate to manufacturing of products and their service life. General recommendations for various plastics applications are given in Table 10.3. The major thermoplastics are described as follows.

1. **Acetals** (from *ace*tic and *al*cohol) have good strength; stiffness; and resistance to creep, abrasion, moisture, heat, and chemicals. Typical applications include mechanical parts and components where high performance is required over a

TABLE 10.3

General Recommendations for Plastic Products

Design requirement	Typical applications	Plastics
Mechanical strength	Gears, cams, rollers, valves, fan blades, impellers, pistons.	Acetals, nylon, phenolics, polycarbonates, polyesters, polypropylenes, epoxies, polyimides.
Wear resistance	Gears, wear strips and liners, bearings, bushings, roller-skate wheels.	Acetals, nylon, phenolics, polyimides, polyurethane, ultrahigh-molecular-weight polyethylene.
Frictional properties high	Tires, nonskid surfaces, footware, flooring.	Elastomers, rubbers.
low	Sliding surfaces, artificial joints.	Fluorocarbons, polyesters, polyethylene, polyimides.
Electrical resistance	All types of electrical components and equipment, appliances, electrical fixtures.	Polymethylmethacrylate, ABS, fluorocarbons, nylon, polycarbonate, polyester, polypropylenes, ureas, phenolics, silicones, rubbers.
Chemical resistance	Containers for chemicals, laboratory equipment, components for chemical industry, food and beverage containers.	Acetals, ABS, epoxies, polymethylmethacrylate, fluorocarbons, nylon, polycarbonate, polyester, polypropylene, ureas, silicones.
Heat resistance	Appliances, cookware, electrical components.	Fluorocarbons, polyimides, silicones, acetals, polysulfones, phenolics, epoxies.
Functional and decorative features	Handles, knobs, camera and battery cases, trim moldings, pipe fittings.	ABS, acrylics, cellulosics, phenolics, polyethylenes, polpropylenes, polystyrenes, polyvinyl chloride.
Functional and transparent features	Lenses, goggles, safety glazing, signs, food-processing equipment	Acrylics, polycarbonates, polystyrenes, polysulfones. laboratory hardware.
Housings and hollow shapes	Power tools, housings, sport helmets, telephone cases.	ABS, cellulosics, phenolics, polycarbonates, polyethylenes, polypropylene, polystyrenes.

long period: bearings, cams, gears, bushings, rollers, impellers, wear surfaces, pipes, valves, showerheads, and housings. Common trade name: *Delrin*.

2. **Acrylics** (such as **polymethylmethacrylate**, or PMMA) possess moderate strength, good optical properties, and weather resistance. They are transparent but can be made opaque, and generally are resistant to chemicals and have good electrical resistance. Typical applications include lenses, lighted signs, displays, window glazing, skylights, bubble tops, automotive lenses, windshields, lighting fixtures, and furniture. Common trade names: *Orlon, Plexiglas,* and *Lucite*.

3. **Acrylonitrile-butadiene-styrene** (ABS) is dimensionally stable and rigid and has good resistance to impact, abrasion, chemicals, and electricity, strength and toughness, and low-temperature properties. Typical applications include pipes,

fittings, chrome-plated plumbing supplies, helmets, tool handles, automotive components, boat hulls, telephones, luggage, housing, appliances, refrigerator liners, and decorative panels.

4. **Cellulosics** have a wide range of mechanical properties, depending on their composition. They can be made rigid, strong, and tough. However, they weather poorly and are affected by heat and chemicals. Typical applications include tool handles, pens, knobs, frames for eyeglasses, safety goggles, machine guards, helmets, tubing and pipes, lighting fixtures, rigid containers, steering wheels, packaging film, signs, billiard balls, toys, and decorative parts.

5. **Fluorocarbons** possess good resistance to temperature, chemicals, weather, and electricity; they also have unique nonadhesive properties and low friction. Typical applications include linings for chemical-processing equipment, nonstick coatings for cookware, electrical insulation for high-temperature wire and cable, gaskets, low-friction surfaces, bearings, and seals. Common trade name: *Teflon*.

6. **Polyamides** (from the words *poly*, *am*ine, and carboxyl *acid*) are available in two main types.

   a. **Nylons** (a coined word) have good mechanical properties and abrasion resistance. They are self-lubricating and resistant to most chemicals. All nylons are hygroscopic (they absorb water). Moisture absorption reduces mechanical properties and increases part dimensions. Typical applications include gears, bearings, bushings, rollers, fasteners, zippers, electrical parts, combs, tubing, wear-resistant surfaces, guides, and surgical equipment.

   b. **Aramids** (*a*romatic poly*amides*) have very high tensile strength and stiffness. Typical applications include fibers for reinforced plastics (composite materials), bulletproof vests, cables, and radial tires. Common trade name: *Kevlar*.

7. **Polycarbonates** are versatile and have good mechanical and electrical properties; they also have high impact resistance and can be made resistant to chemicals. Typical applications include safety helmets, optical lenses, bullet-resistant window glazing, signs, bottles, food-processing equipment, windshields, load-bearing electrical components, electrical insulators, medical instruments, business-machine components, guards for machinery, and parts requiring dimensional stability. Common trade name: *Lexan*.

8. **Polyesters** (thermoplastics; see also Section 10.6) have good mechanical, electrical, and chemical properties, good abrasion resistance, and low friction. Typical applications include gears, cams, rollers, load-bearing members, pumps, and electromechanical components. Common trade names: *Dacron*, *Mylar*, and *Kodel*.

9. **Polyethylenes** possess good electrical and chemical properties. Their mechanical properties depend on their composition and structure. The three major classes of polyethylenes are low density (LDPE), high density (HDPE), and ultrahigh molecular weight (UHMWPE). Typical applications for LDPE and HDPE include housewares, bottles, garbage cans, ducts, bumpers, luggage, toys, tubing, bottles, and packaging material. UHMWPE is used in parts requiring high-impact toughness and abrasive wear resistance; examples include artificial knee and hip joints.

10. **Polyimides** have the structure of a thermoplastic, but the nonmelting characteristic of a thermoset (see Section 10.6). Common trade name: *Torlon*.

11. **Polypropylenes** have good mechanical, electrical, and chemical properties and good resistance to tearing. Typical applications include automotive trim and

components, medical devices, appliance parts, wire insulation, TV cabinets, pipes, fittings, drinking cups, dairy-product and juice containers, luggage, ropes, and weather stripping.

12. **Polystyrenes** are inexpensive, have generally average properties, and are somewhat brittle. Typical applications include disposable containers, packaging, foam insulation, appliances, automotive, radio, and TV components, housewares, and toys and furniture parts (as a wood substitute).

13. **Polysulfones** have excellent resistance to heat, water, and steam and are highly resistant to some chemicals, but are attacked by organic solvents. Typical applications include steam irons, coffeemakers, hot-water containers, medical equipment that requires sterilization, power-tool and appliance housings, aircraft cabin interiors, and electrical insulators.

14. **Polyvinyl chloride** (PVC) has a wide range of properties, is inexpensive and water resistant, and can be made rigid or flexible. It is not suitable for applications that require strength and heat resistance. *Rigid PVC* is tough and hard and is used for signs and in the construction industry, such as for pipes and conduits. *Flexible PVC* is used in wire and cable coatings, low-pressure flexible tubing and hose, footwear, imitation leather, upholstery, records, gaskets, seals, trim, film, sheet, and coatings. Common trade names: *Saran* and *Tygon*.

10.6 | Thermosets: General Characteristics and Applications

This section outlines the general characteristics and typical applications of major thermosetting plastics:

1. **Alkyds** (from *alkyl*, meaning "alcohol," and *acid*) possess good electrical insulating properties, impact resistance, and dimensional stability and have low water absorption. Typical applications include electrical and electronic components.

2. **Aminos** (**urea** and **melamine**) have properties that depend on composition. Generally, aminos are hard and rigid and are resistant to abrasion, creep, and electrical arcing. Typical applications include small appliance housings, countertops, toilet seats, handles, and distributor caps. Urea is used for electrical and electronic components, and melamine is used for dinnerware.

3. **Epoxies** have excellent mechanical and electrical properties, dimensional stability, strong adhesive properties, and good resistance to heat and chemicals. Typical applications include electrical components that require mechanical strength and high insulation, tools and dies, and adhesives. *Fiber-reinforced epoxies* (Section 10.9) have excellent mechanical properties and are used in pressure vessels, rocket motor casings, tanks, and similar structural components.

4. **Phenolics** are rigid, but brittle; they are dimensionally stable and have high resistance to heat, water, electricity, and chemicals. Typical applications include knobs, handles, laminated panels, telephones, bond material to hold abrasive grains together in grinding wheels, and electrical components, such as wiring devices, connectors, and insulators.

5. **Polyesters** (thermosetting plastics; see also Section 10.5) have good mechanical, chemical, and electrical properties. Polyesters are generally reinforced with glass or other fibers. Typical applications include boats, luggage, chairs, automotive

bodies, swimming pools, and material for impregnating cloth and paper. Polyesters are also available as casting resins.

6. **Polyimides** possess good mechanical, physical, and electrical properties at elevated temperatures; they also have creep resistance and low friction and wear characteristics. Polyimides have the nonmelting characteristics of a thermoset, but the structure of a thermoplastic. Typical applications include pump components (bearings, seals, valve seats, retainer rings, and piston rings); electrical connectors for high-temperature use; aerospace parts; high-strength, impact-resistant structures; sports equipment; and safety vests.

7. **Silicones** have properties that depend on composition. Generally, they weather well, possess excellent electrical properties over a wide range of humidity and temperatures, and resist chemicals and heat. (See also Section 10.8.) Typical applications include electrical components that require strength at elevated temperatures, oven gaskets, heat seals, and waterproof materials.

10.7 | High-Temperature Polymers, Electrically Conducting Polymers, and Biodegradable Plastics

This section describes three important trends in the development of polymers.

10.7.1 High-temperature polymers

Polymers and polymer blends for high-temperature applications are available, particularly for the aerospace industry. High-temperature resistance may be short term at relatively high temperatures, or long term at lower temperatures. (See, for example, Section 3.11.6 for a similar consideration regarding titanium alloys.) Short-term exposure to high temperatures generally requires *ablative* materials, or materials that wear, melt, or vaporize to dissipate the heat, such as phenolic-silicone copolymers, used for rocket and missile components at temperatures of thousands of degrees. Long-term exposure for polymers is presently confined to temperatures on the order of 260°C (500°F). High-temperature thermoplastic polymers include fluorine-containing thermoplastics, polyketones, and polyimides. High-temperature thermosetting polymers include phenolic resins, epoxy resins, silicone-based thermosetting polymers, and phenolic-fiberglass systems.

10.7.2 Electrically conducting polymers

The electrical conductivity of some polymers can be increased by **doping**, that is, introducing certain impurities in the polymer, such as metal powder, salts, and iodides. Also, the conductivity of polymers increases with moisture absorption, and the electronic properties of polymers can be altered by irradiation as well. Applications of electrically conducting polymers include microelectronic devices, rechargeable batteries, capacitors, fuel cells, catalysts, fuel-level sensors, de-icer panels, antistatic coatings, and as conducting adhesives for surface-mount technologies (see also Section 12.14.4).

10.7.3 Biodegradable plastics

About one-third of plastics produced today are in the disposable-products sector, such as beverage bottles, packaging, and garbage bags. These plastics contribute about 10% of municipal solid waste, and on a volume basis, they contribute between two and three times their weight. With the growing use of plastics and increasing concern over environmental issues regarding disposal of plastic products and limited landfills, major efforts are continuing to develop biodegradable plastics.

Most plastic products have traditionally been made from synthetic polymers that are (a) derived from nonrenewable natural resources, (b) are not biodegradable, and (c) are difficult to recycle. **Biodegradability** means that microbial species in the environment (e.g., microorganisms in soil and water) will degrade a portion of (or even the entire) polymeric material, under the proper environmental conditions and without producing toxic by-products (see also *biological cycle,* Section 1.4). The end products of the degradation of the biodegradable portion of the material are carbon dioxide and water. Because of the variety of constituents present in the materials, biodegradable plastics can be regarded as composite materials; consequently, only a portion of these plastics may be truly biodegradable.

Three different *biodegradable plastics* have thus far been developed. They have different degradability characteristics, and they degrade over different periods of time, ranging from a few months to a few years.

1. The **starch-based system** is the farthest along of the three types of bioplastics in terms of production capacity. Starch may be extracted from potatoes, wheat, rice, and corn. In this system, starch granules are processed into a powder, which is heated and becomes a sticky liquid; various binders and additives are also blended in the starch to impart special characteristics to the bioplastic materials. The liquid is then cooled, formed into pellets, and processed in conventional plastic-processing equipment (Section 10.10).

2. In the **lactic-based system,** fermenting corn or other feedstock produce lactic acid; it is then polymerized to form a polyester resin.

3. In the third system, **organic acids** are added to a **sugar** feedstock. Through a specially developed process, the resulting reaction produces a highly crystalline and very stiff polymer that, after further processing, behaves in a manner similar to polymers developed from petroleum.

Numerous attempts continue to be made to produce fully biodegradable plastics, by using various agricultural waste (*agrowastes*), plant carbohydrates, plant proteins, and vegetable oils. Typical applications include the following:

- Disposable tableware made from a cereal substitute, such as rice grains or wheat flour.
- Plastics made almost entirely from starch extracted from potatoes, wheat, rice, and corn.
- Plastic parts made from coffee beans and rice hulls, dehydrated and molded under high pressure and temperature.
- Water-soluble and compostable polymers, for medical and surgical applications.
- Food and beverage containers (made from potato starch, limestone, cellulose, and water) that can dissolve in storm sewers and oceans without affecting marine life or wildlife.

The long-range performance of biodegradable plastics, both during their useful life cycle as products and in landfills, has not yet been fully assessed. There is also concern that emphasis on biodegradability may divert attention from the issue of *recyclability* of plastics and the efforts for *conservation* of materials and energy. A major consideration is the fact that the cost of today's biodegradable polymers is substantially higher than that of synthetic polymers. Consequently, a mixture of agricultural waste such as hulls from corn, wheat, rice, and soy (as the major component) and biodegradable polymers (as the minor component) is an attractive alternative.

Recycling. Recycled plastics are being used increasingly for a variety of products, including automotive-body components, packaging materials, and architectural structural shapes. Plastic products carry the following numerals within a triangular mark with arrows, meaning that the product is recyclable: 1-PETE (polyethylene), 2-HDPE (high-density polyethylene), 3-V (vinyl), 4-LDPE (low-density polyethylene), 5-PP (polypropylene), 6-PS (polystyrene), and 7-others.

10.8 | Elastomers (Rubbers): General Characteristics and Applications

The terms *elastomer* and *rubber* are often used interchangeably. Generally, an **elastomer** is defined as being capable of recovering substantially in shape and size after a load has been removed; **rubber** is defined as being capable of quickly recovering from large deformations. Elastomers (from the words *elastic* and *mer*) comprise a large family of amorphous polymers that have (a) a low glass-transition temperature, (b) the characteristic ability to undergo large elastic deformations without rupture, (c) are soft, and (d) have a low elastic modulus. The structure of these polymers is highly *kinked* (tightly twisted or curled); they stretch but then return to their original shape after the load is removed (Fig. 10.14). They can be cross-linked, the best known example of which is the elevated-temperature **vulcanization** (after Vulcan, the Roman god of fire) of rubber with sulfur, discovered by Charles Goodyear in 1839. Note that an automobile tire, which is one giant molecule, cannot be softened and reshaped.

The hardness of elastomers, which is measured with a durometer (see Section 2.6.7), increases with increasing cross-linking of the molecular chains. A variety of additives can be blended with elastomers to impart specific properties, as is done with polymers. Elastomers have a wide range of applications, such as high-friction and nonskid surfaces, protection against corrosion and abrasion, electrical insulation, and shock and vibration insulation. Specific examples include tires, hoses, weather stripping, footwear, linings, gaskets, seals, printing rolls, and flooring.

A characteristic of elastomers is their hysteresis loss in stretching or compression (Fig. 10.14). The clockwise loop in Fig. 10.14 indicates energy loss, whereby mechanical energy is converted into heat; this property is desirable for absorbing vibrational energy (damping) and sound deadening.

The major types of elastomers are:

1. **Natural rubber.** The base for natural rubber is **latex**, a milklike sap obtained from the inner bark of a tropical tree. It has good resistance to abrasion and fatigue and high frictional properties, but low resistance to oil, heat, ozone, and sunlight. Typical applications include tires, seals, shoe heels, couplings, and engine mounts.

2. **Synthetic rubbers.** Compared with natural rubbers, synthetic rubbers have improved resistance to heat, gasoline, and chemicals and a higher useful-temperature range. Synthetic rubbers include synthetic natural rubber, butyl, styrene butadiene, polybutadiene, and ethylene propylene. Examples of synthetic rubbers that are resistant to oil are neoprene, nitrile, urethane, and silicone.

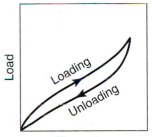

FIGURE 10.14 Typical load-elongation curve for elastomers. The area within the clockwise loop, indicating loading and unloading paths, is the hysteresis loss. Hysteresis gives rubbers the capacity to dissipate energy, damp vibration, and absorb shock loading, as in automobile tires and vibration dampeners for machinery.

Typical applications of synthetic rubbers include tires, shock absorbers, seals, and belts.

3. **Silicones.** Silicones (see also Section 10.6) have the highest useful temperature range of all elastomers, up to 315°C (600°F), but their other properties such as strength and resistance to wear and oils are generally inferior to those of other elastomers. Typical applications include seals, gaskets, thermal insulation, high-temperature electrical switches, and various electronic components.

4. **Polyurethane.** Polyurethane has very good overall properties of high strength, stiffness, hardness, and exceptional resistance to abrasion, cutting, and tearing. Typical applications include seals, gaskets, cushioning, diaphragms for rubber-pad forming of sheet metals (see Section 7.5.3), and auto-body parts such as bumpers.

10.9 | Reinforced Plastics

Among the major group of important materials are *reinforced plastics* (**composite materials**), as shown in Fig. 10.15. These **engineered materials** are defined as a combination of two or more chemically distinct and insoluble phases whose properties and structural performance are superior to those of the constituents acting independently. Although plastics possess mechanical properties (particularly strength, stiffness, and creep resistance) that are generally inferior to those of metals and alloys, these properties can be improved by embedding reinforcements of various types to produce reinforced plastics. As shown in Table 10.1, reinforcements improve the strength, stiffness, and creep resistance of plastics, and particularly their strength-to-weight and stiffness-to-weight ratios. Reinforced plastics now have a wide variety of applications in aircraft, space vehicles, offshore structures, piping, electronics, automobiles, boats, ladders, and sporting goods.

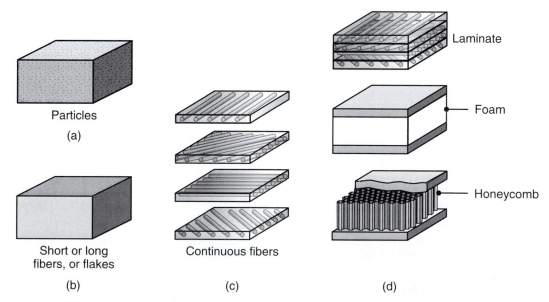

FIGURE 10.15   Schematic illustration of types of reinforcing plastics: (a) matrix with particles, (b) matrix with short or long fibers or flakes, (c) continuous fibers, and (d) laminate or sandwich composite structures using a foam or honeycomb core (see also Fig. 7.48 on making of honeycombs).

The oldest example of composites is the addition of straw to clay for making mud huts and bricks for structural use, dating back to 4000 B.C. The straws act as the reinforcing fibers, and the clay is the matrix. Another example of a composite material is the reinforcing of masonry and concrete with iron rods, beginning in the 1800s. In fact, concrete itself is a composite material, consisting of cement, sand, and gravel. In *reinforced concrete*, steel rods impart the necessary tensile strength to the composite, since concrete is brittle and generally has little or no useful tensile strength.

10.9.1 Structure of polymer-matrix-reinforced plastics

Reinforced plastics consist of *fibers* (the *discontinuous* or *dispersed* phase) in a poly-mer **matrix** (the *continuous* phase), as can be seen in Fig. 10.15. Commonly used fibers include glass, graphite, aramids, and boron (Table 10.4). These fibers are strong and stiff and have a high specific strength (strength-to-weight ratio) and spe-cific modulus (stiffness-to-weight ratio), as shown in Fig. 10.16. They are, however, generally brittle and abrasive and lack toughness. Thus, the fibers, by themselves, have little value as structural members. The polymer matrix is less strong and less stiff, but tougher than the fibers; reinforced plastics thus combine the advantages of each of the two constituents (Table 10.5).

In addition to having high specific strength and specific modulus, reinforced plastic structures also have higher fatigue resistance, toughness, and creep resistance than unreinforced plastics. The percentage of fibers (by volume) in reinforced plas-tics usually ranges between 10 and 60%. The highest practical fiber content is 65%; higher percentages generally result in diminished structural properties. When a com-posite material has more than one type of fiber, it is called a **hybrid,** which generally has even better properties than a reinforced plastic with only one type of fiber.

TABLE 10.4

Typical Properties of Reinforcing Fibers

Type	Tensile strength (MPa)	Elastic modulus (GPa)	Density (kg/m³)	Relative cost
Boron	3500	380	2600	Highest
Carbon				
high strength	3000	275	1900	Low
high modulus	2000	415	1900	Low
Glass				
E type	3500	73	2480	Lowest
S type	4600	85	2540	Lowest
Kevlar				
29	2800	62	1440	High
49	2800	117	1440	High
129	3200	85	1440	High
Nextel				
312	1630	135	2700	High
610	2770	328	3960	High
Spectra				
900	2270	64	970	High
1000	2670	90	970	High

Note: These properties vary significantly, depending on the material and method of preparation. Strain to failure for these fibers is typically in the range of 1.5 to 5.5%.

FIGURE 10.16 Specific tensile strength (ratio of tensile strength to density) and specific tensile modulus (ratio of modulus of elasticity to density) for various fibers used in reinforced plastics. Note the wide range of specific strength and stiffness available.

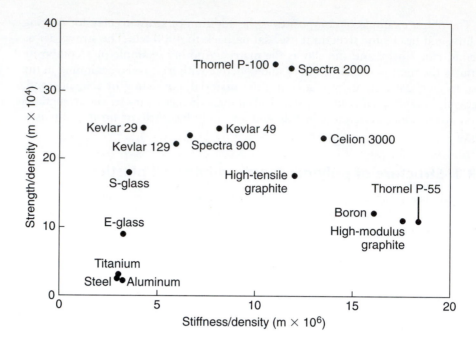

TABLE 10.5

Types and General Characteristics of Reinforced Plastics and Metal-Matrix and Ceramic-Matrix Composites

Material	Characteristics
Fiber	
Glass	High strength, low stiffness, high density; E (calcium aluminoborosilicate) and S (magnesiaaluminosilicate) types are commonly used; lowest cost.
Graphite	Available typically as high modulus or high strength; less dense than glass; low cost.
Boron	High strength and stiffness; has tungsten filament at its center (coaxial); highest density; highest cost.
Aramids (Kevlar)	Highest strength-to-weight ratio of all fibers; high cost.
Other	Nylon, silicon carbide, silicon nitride, aluminum oxide, boron carbide, boron nitride, tantalum carbide, steel, tungsten, and molybdenum; see Chapters 3, 8, 9, and 10.
Matrix	
Thermosets	Epoxy and polyester, with the former most commonly used; others are phenolics, fluorocarbons, polyethersulfone, silicon, and polyimides.
Thermoplastics	Polyetheretherketone; tougher than thermosets, but lower resistance to temperature.
Metals	Aluminum, aluminumlithium alloy, magnesium, and titanium; fibers used are graphite, aluminum oxide, silicon carbide, and boron.
Ceramics	Silicon carbide, silicon nitride, aluminum oxide, and mullite; fibers used are various ceramics.

10.9.2 Reinforcing fibers: characteristics and manufacture

The major types of reinforcing fibers and their characteristics are described next.

1. **Polymer fibers.** The most commonly used reinforcing fibers are **aramids** (such as *Kevlar*). They are among the toughest fibers and have very high specific strength (Fig. 10.16 and Table 10.4); they can undergo some plastic deformation before fracture and, consequently, have higher toughness than that of brittle fibers. However, aramids absorb moisture (as do nylons), which reduces their properties and complicates their application, as hygrothermal stresses must be considered. Other polymer reinforcements include rayon, nylon, and acrylics.

   A high-performance polyethylene fiber is *Spectra* (a trade name), which has ultrahigh molecular weight and high molecular-chain orientation. Compared to aramid fibers, the fiber has better abrasion resistance and flexural fatigue and at a comparable cost, and also, because of its lower density, it has higher specific strength and specific stiffness. However, it has a low melting point and poor adhesion characteristics with the matrix as compared with other fibers. New polymer fibers are continually being introduced, including *Nextel*.

   Most synthetic fibers used in reinforced plastics are extruded through the tiny holes of a device called a spinneret (resembling a shower head) to form continuous semisolid filaments. The extruder forces the polymer through the spinneret, which may have from one to several hundred holes. If the polymers are thermoplastics, they are first melted in the extruder, as described in Section 10.10.1. Thermosetting polymers also can be formed into fibers by first dissolving or chemically treating them, so that they can be extruded. These operations are performed at high production rates and with very high product reliability. As the filaments emerge from the holes in the spinneret, the liquid polymer is converted first to a rubbery state and then solidified.

   This process of extrusion and solidification of continuous filaments is called **spinning**, a term which is also used for the production of natural textiles (such as cotton or wool), where short pieces of fiber are twisted into yarn. There are four methods of spinning fibers, as described next.

   a. In **melt spinning**, shown in Fig. 10.17, the polymer is melted for extrusion through the spinneret, and then directly solidified by cooling. A typical spinneret for this operation has about 50 holes of around 0.25 mm (0.01 in.) in diameter, and is about 5 mm (0.2 in.) thick. The fibers that emerge from the spinneret are cooled by forced-air convection and are simultaneously pulled, so that their final diameter becomes much smaller than the spinneret opening. Polymers such as nylon, olefin, polyester, and PVC are produced in this matter. Because of the important applications of nylon and polyester fibers, melt spinning is the most important fiber-manufacturing process.

      Melt-spun fibers also can be extruded from the spinneret into various other cross sections, such as trilobal (triangle with curved sides), pentagonal, octagonal, as well as hollow shapes. While various cross sections have specific applications, hollow fibers trap air and thus provide additional thermal insulation.

   b. **Wet spinning** is the oldest process for fiber production and is used for polymers that have been dissolved in a solvent. The spinnerets are submerged in a chemical bath, and as the filaments emerge, they precipitate in the chemical bath, producing a fiber that is then wound onto a bobbin. The term wet spinning refers to the use of a precipitating liquid bath, resulting in wet fibers that require drying before they can be used. Acrylic, rayon, and aramid fibers can be produced by this process.

FIGURE 10.17 The melt spinning process for producing polymer fibers. The fibers are used in a variety of applications, including fabrics and as reinforcements for composite materials.

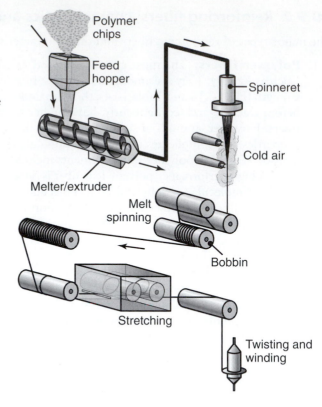

c. **Dry spinning** is used for thermosets, carried by a solvent. However, instead of precipitating the polymer by dilution, as in wet spinning, solidification is achieved by evaporating the solvent in a stream of air or inert gas. The filaments do not come in contact with a precipitating liquid, thus eliminating the need for drying. Dry spinning may be used for the production of acetate, triacetate, polyether-based elastane, and acrylic fibers.

d. **Gel spinning** is a special process used to obtain fibers with high strength or special properties. The polymer is not completely melted or dissolved in liquid, but the molecules are bonded together at various points in liquid-crystal form. This operation produces strong interchain forces in the resulting filaments, thus significantly increasing the tensile strength of the fibers. In addition, the liquid crystals are aligned along the fiber axis by the strain encountered during extrusion. The filaments emerge from the spinneret with an unusually high degree of orientation relative to each other, further enhancing fiber strength. This process is also called dry-wet spinning, because the filaments first pass through air and then are cooled further in a liquid bath. Some high-strength polyethylene and aramid fibers are produced by gel spinning.

   A necessary step in the production of most fibers is the application of significant stretching to induce orientation of the polymer molecules in the fiber direction. This orientation is the main reason for the high strength of the fibers compared to the polymer in bulk form. The stretching can be done while the polymer is still pliable just after extrusion from the spinneret, or it can be performed as a cold-drawing operation. The strain induced can be as high as 800%.

2. **Glass fibers.** These fibers are the most widely used and least expensive of all fibers. (See also *glasses*, Section 11.10.2.) The composite material containing

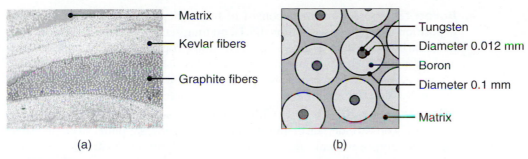

Matrix
Kevlar fibers
Graphite fibers

Tungsten
Diameter 0.012 mm
Boron
Diameter 0.1 mm
Matrix

(a)   (b)

FIGURE 10.18   (a) Cross section of a tennis racket, showing graphite and aramid (Kevlar) reinforcing fibers and (b) cross section of boron-fiber-reinforced composite material. *Source:* (a) After J. Dvorak and F. Garrett.

glass fibers is called *glass-fiber reinforced plastic* (GFRP) and may contain between 30 and 60% fibers by volume. Glass fibers are made by drawing molten glass through small openings in a platinum die and then mechanically elongated, cooled, and wound on a roll. A protective coating or sizing may be applied to facilitate the passage of the glass fibers through machinery. The fibers are treated with **silane** (a silicon hydride) for improved wetting and bonding between the fiber and the matrix.

The principal types of glass fibers are (a) the **E type,** a calcium aluminoborosilicate glass, which is used most often; (b) the **S type,** a magnesia-aluminosilicate glass, which has higher strength and stiffness but is more expensive than the other two types, and (c) the **E-CR type,** which offers higher resistance to elevated temperature and acid corrosion than the other two types.

3. **Graphite fibers.** Graphite fibers (Fig. 10.18a), although more expensive than glass fibers, have the desirable combination of low density, high strength, and high stiffness; the composite is called *carbon-fiber reinforced plastic* (CFRP). Graphite fibers are made by **pyrolysis** of organic **precursors,** commonly polyacrylonitrile (PAN) because of its lower cost. Rayon and pitch (the residue from catalytic crackers in petroleum refining) also can be used as precursors. *Pyrolysis* is the process of inducing chemical changes by heat, such as burning a length of yarn, which becomes carbon and black in color.

With PAN, the fibers are partially cross-linked, at a moderate temperature, in order to prevent melting during subsequent processing steps, and the fibers are simultaneously elongated. At this point, the fibers are *carburized,* that is, they are exposed to elevated temperature to expel the hydrogen (dehydrogenation) and nitrogen (denitrogenation) from the PAN. The temperature range is up to about 1500° (2730°F) for carbonizing and to 3000°C (5400°F) for graphitizing.

The difference between carbon and graphite, although the terms are often used interchangeably, depends on the temperature of pyrolysis and the purity of the material. Carbon fibers are generally 80 to 95% carbon, and graphite fibers are usually more than 99% carbon; the rest is graphite and carbon, respectively. The fibers are classified by the magnitude of their elastic modulus, which typically ranges from 35 to 800 GPa; tensile strengths typically range from 250 to 2600 MPa.

Conductive graphite fibers are also available, making it possible to impart electrical and thermal conductivity to reinforced plastics, such as in electromagnetic and radio-frequency shielding and lightning protection. These fibers are coated with a metal, usually nickel, typically to a thickness of 0.5 μm (20 μin.) on a graphite-fiber core 7 μm (280 μin.) in diameter, using a continuous electroplating process.

4. **Boron fibers.** Boron fibers consist of boron deposited by chemical vapor-deposition techniques (see Section 4.5.1), on tungsten fibers (Fig. 10.18b), as well as on carbon fibers. These fibers have favorable properties, such as high strength and stiffness in tension and compression and resistance to high temperatures. However, because of the tungsten, they have high density and are expensive, thus increasing the weight and cost of the reinforced plastic component.

5. **Miscellaneous fibers.** Other fibers that are used for composite materials include silicon carbide, silicon nitride, aluminum oxide, sapphire, steel, tungsten, molybdenum, boron carbide, boron nitride, and tantalum carbide. (See also Chapter 11.) Metallic fibers are drawn as described in Section 6.5, although at the smaller diameters, the wires are drawn in bundles (see also Section 6.5.3). **Whiskers**, described in Section 3.8.3, also are used as reinforcing fibers. Whiskers are tiny needlelike single crystals that grow to 1–10 μm (40–400 μin.) in diameter and have aspect ratios (defined as the ratio of fiber length to diameter) ranging from 100 to 15,000. Because of their small size, either the whiskers are free of imperfections or the imperfections they contain do not significantly affect their strength, which approaches the theoretical strength of the material (Section 3.3.2).

10.9.3 Fiber size and length

The mean diameter of fibers used in reinforced plastics is usually less than 0.01 mm (0.0004 in.). The fibers are very strong and rigid in tension, because the molecules in the fibers are *oriented* in the longitudinal direction, and their cross sections are so small that the probability is low that any defects exist in the fiber. Glass fibers, for example, can have tensile strengths as high as 4600 MPa (650 ksi), whereas the strength of glass in bulk form is much lower; thus, glass fibers are stronger than steel.

Fibers are classified as **short** or **long** fibers, also called **discontinuous fibers** or **continuous fibers**, respectively. Short fibers typically have an *aspect ratio* between 20 and 60, and long fibers between 200 and 500. The short- and long-fiber designations are, in general, based on the following observations: In a given fiber, if the mechanical properties of the composite improve as a result of increasing the fiber length, then the fiber is denoted as a short fiber. When no additional improvement in properties occurs, the fiber is denoted as a long fiber. In addition to the discrete fibers, reinforcements in composites may also be in the form of (a) continuous *roving* (slightly twisted strand of fibers), (b) *woven* fabric (similar to cloth), (c) *yarn* (twisted strand), and (d) *mats* of various combinations. As shown in Fig. 10.15, reinforcing elements may also be in the form of particles and flakes.

10.9.4 Matrix materials

The matrix in reinforced plastics has three important functions:

1. Support and transfer the stresses to the fibers, which carry most of the load.
2. Protect the fibers against physical damage and the environment.
3. Reduce propagation of cracks in the composite by virtue of the ductility and toughness of the plastic matrix.

Matrix materials are usually epoxy, polyester, phenolic, fluorocarbon, polyethersulfone, or silicon. The most common materials are epoxies (80% of all reinforced plastics) and polyesters, which are less expensive than epoxies. Polyimides, which resist exposure to temperatures in excess of 300°C (575°F), are used with graphite fibers. A thermoplastic such as polyetheretherketone (PEEK) is also used as

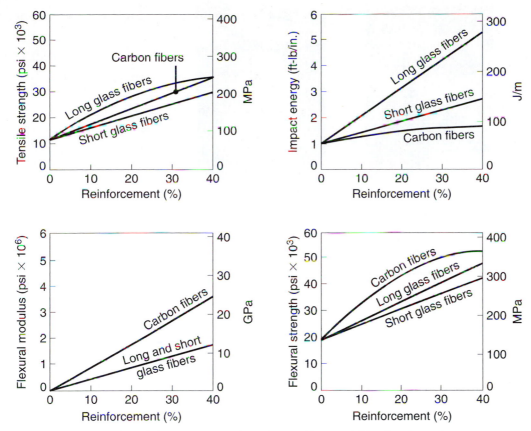

FIGURE 10.19   Effect of the percentage of reinforcing fibers and fiber length on the mechanical properties of reinforced nylon. Note the significant improvement with increasing percentage of fiber reinforcement. *Source:* Courtesy of Wilson Fiberfill International.

a matrix material; it has higher toughness than thermosets, but its resistance to temperature is lower, being limited to the range of 100–200°C (200–400°F).

10.9.5 Properties of reinforced plastics

The overall properties of reinforced plastics depend on (a) the type, shape, and orientation of the reinforcing material, (b) the length of the fibers, and (c) the volume fraction (percentage) of the reinforcing material. Short fibers are less effective than long fibers (Fig. 10.19), and their properties are strongly influenced by time and temperature. Long fibers transmit the load through the matrix better and, thus, are commonly used in critical applications, particularly at elevated temperatures.

A critical factor in reinforced plastics is the *strength of the bond* between the fiber and the polymer matrix, since the load is transmitted through the fiber-matrix interface. The importance of proper bonding can be appreciated by inspecting Fig. 10.20, which shows the fracture surfaces of a reinforced plastic. Weak bonding in the composite causes **fiber pullout** and **delamination**, particularly under adverse environmental conditions, including temperature and humidity. Adhesion at the interface can be improved by special surface treatments, such as coatings and coupling agents. Glass fibers, for example, are treated with *silane* (see Section 10.9.2) for improved wetting and bonding between the fiber and the matrix.

Generally, the highest stiffness and strength in reinforced plastics is obtained when the fibers are aligned in the direction of the tension force, which makes the

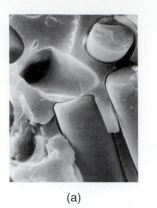

(a)   (b)

FIGURE 10.20   (a) Fracture surface of glass-fiber-reinforced epoxy composite. The fibers are 10 μm (400 μin.) in diameter and have random orientation. (b) Fracture surface of a graphite-fiber-reinforced epoxy composite. The fibers are 9–11 μm (360–440 μin) in diameter. Note that the fibers are in bundles and are all aligned in the same direction. *Source:* Courtesy of L.J. Broutman.

FIGURE 10.21   Tensile strength of glass-reinforced polyester as a function of fiber content and fiber direction in the matrix. *Source:* After R.M. Ogorkiewicz.

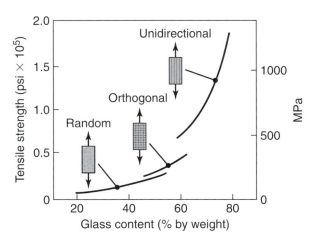

composite highly anisotropic (Fig. 10.21). As a result, other properties of the composite, such as stiffness, creep resistance, thermal expansion, and thermal and electrical conductivity also are anisotropic. The transverse properties of such a unidirectionally reinforced structure are much lower than the longitudinal properties. Note, for example, how easily you can split fiber-reinforced packaging tape, yet how strong it is when you pull on it (tension).

For a specific service condition, a reinforced-plastic part can be given an optimal configuration. For example, if the reinforced-plastic part is to be subjected to forces in biaxial directions (such as a thin-walled, pressurized vessel), the fibers are crisscrossed in the matrix. (See also the discussion of *filament winding* in Section 10.11.2.) Reinforced plastics may also be made with various other materials and shapes of the polymer matrix in order to impart specific properties (such as permeability and dimensional stability), to make processing easier, and to reduce costs.

Strength and elastic modulus of reinforced plastics. The strength of a reinforced plastic with longitudinal fibers can be determined in terms of the strength of the fibers and matrix and the volume fraction of fibers in the composite. In the following equations, c refers to the composite, f to the fiber, and m to the matrix.

The total load F_c on the composite is shared by the fiber load F_f and the matrix load F_m; thus,

$$F_c = F_f + F_m, \tag{10.12}$$

which can be rewritten as

$$\sigma_c A_c = \sigma_f A_f + \sigma_m A_m, \tag{10.13}$$

where A_c, A_f, and A_m are the cross-sectional areas of the composite, fiber, and matrix, respectively, and $A_c = A_f + A_m$. Let's now denote x as the area fraction of the fibers in the composite. (Note that x also represents the volume fraction because the fibers are uniformly longitudinal in the matrix.) Equation (10.13) can now be rewritten as

$$\sigma_c = x\sigma_f + (1 - x)\sigma_m. \tag{10.14}$$

The fraction of the total load carried by the fibers can now be calculated as follows: First, note that in the composite under a tensile load, the strains sustained by the fibers and the matrix are the same (that is, $e_c = e_f = e_m$), and then recall from Section 2.2 that

$$e = \frac{\sigma}{E} = \frac{F}{AE}.$$

Consequently,

$$\frac{F_f}{F_m} = \frac{A_f E_f}{A_m E_m}. \tag{10.15}$$

Since we know the relevant quantities for a specific case, Eq. (10.12) can be used to determine the fraction F_f/F_c. Then, using the foregoing relationships, the elastic modulus E_c of the composite can be calculated by replacing σ in Eq. (10.14) with E. Hence,

$$E_c = xE_f + (1 - x)E_m. \tag{10.16}$$

EXAMPLE 10.4 Properties of a graphite-epoxy-reinforced plastic

Assume that a graphite-epoxy-reinforced plastic with longitudinal fibers contains 20% graphite fibers, which have a strength of 2500 MPa and elastic modulus of 300 GPa. The strength of the epoxy matrix is 120 MPa, and it has an elastic modulus of 100 GPa. Calculate (a) the elastic modulus of the composite and (b) the fraction of the load supported by the fibers.

Solution

(a) The data given are $x = 0.2$, $E_f = 300$ GPa, $E_m = 100$ GPa, $E_m = 100$ GPa, $\sigma_f = 2500$ MPa, and $\sigma_m = 120$ MPa. From Eq. (10.16),

$$E_c = 0.2(300) + (1 - 0.2)100 = 60 + 80 = 140 \text{ GPa}.$$

(b) The load fraction F_f/F_m is obtained from Eq. (10.15):

$$\frac{F_f}{F_m} = \frac{(0.2)(300)}{(0.8)(100)} = 0.75.$$

Since

$$F_c = F_f + F_m$$

and

$$F_m = \frac{F_f}{0.75},$$

we obtain

$$F_c = F_f + \frac{F_f}{0.75} = 2.33 F_f,$$

or

$$F_f = 0.43 F_c.$$

Therefore, the fibers support 43% of the load, even though they occupy only 20% of the cross-sectional area (and hence volume) of the composite.

10.9.6 Applications of reinforced plastics

The first application of reinforced plastics (in 1907) was for an acid-resistant storage tank made of a phenolic resin (as the matrix) with asbestos fibers. *Formica*, commonly used for countertops, was developed in the 1920s. Epoxies were first used as a matrix material in the 1930s. Beginning in the 1940s, boats were made with fiberglass, and reinforced plastics were used for aircraft, electrical equipment, and sporting goods. Major developments in composites began in the 1970s, and these materials are now called **advanced composites.** Glass-fiber or carbon-fiber reinforced hybrid plastics are available for high-temperature applications, with continuous use ranging up to about 300°C (550°F).

Reinforced plastics are typically used in military and commercial aircraft and rocket components, helicopter blades, automotive bodies, leaf springs, drive shafts, pipes, ladders, pressure vessels, sporting goods, sports and military helmets, boat hulls, and various other structures. Examples of specific applications include components in the DC-10 aircraft, the L-1011 aircraft, and the Boeing 727, 757, 767, and 777 aircraft. The Boeing 777 is made of about 9% composites by total weight (which is triple the composite content of prior Boeing transport aircraft). The floor beams and panels and most of the vertical and horizontal tail are made of composite materials.

By virtue of the resulting weight savings, reinforced plastics also have reduced aircraft fuel consumption by about 2%. The newly designed Airbus jumbo jet A380, with a capacity of 550 to 700 passengers, has horizontal stabilizers, ailerons, wing boxes and leading edges, secondary mounting brackets of the fuselage, and the deck structure made of composites with carbon fibers, thermosetting resins, and thermoplastics. The upper fuselage will be made of alternating layers of aluminum and glass-fiber-reinforced epoxy prepregs.

The structure of the Lear Fan 2100 passenger aircraft is almost totally made of graphite-epoxy reinforced plastic. Nearly 90% of the structure of the lightweight Voyager aircraft, which circled the earth in 1986 without refueling, was made of carbon-reinforced plastic. The contoured frame of the Stealth bomber is made of composites consisting of carbon and glass fibers, epoxy-resin matrices, high-temperature polyimides, and other advanced materials. Boron-fiber reinforced composites are used in military aircraft, golf-club shafts, tennis rackets, fishing rods, and sailboards. A more recent example is the development of a small, all-composite ship (twin-hull catamaran design) for the U.S. Navy, capable of speeds of 50 knots (58 mph).

The processing of polymer-matrix-reinforced plastics, described Section 10.11, presents significant challenges. Several innovative techniques have been developed to manufacture both large and small parts by a combination of processes such as molding, forming, cutting, and assembly. Careful inspection and testing of reinforced plastics is essential in critical applications in order to ensure that good bonding between the reinforcing fiber and the matrix has been obtained throughout the structure. It has been shown that, in some instances, the cost of inspection can be as high as one quarter of the total cost of the composite product.

10.10 | Processing of Plastics

The processing of plastics involves operations similar to those used in forming and shaping metals, as described in Chapters 6 and 7. Plastics can be molded, cast, shaped, formed, machined, and joined into many shapes with relative ease and with few or no additional operations required (Table 10.6). As indicated earlier, plastics melt (thermoplastics) or cure (thermosets) at relatively low temperatures (Table 10.2) and thus, unlike metals, are easy to handle and require less energy to process. However, because the properties of plastic parts and components are greatly influenced by the method of manufacture and the processing parameters, their proper control is essential for good part quality. Plastics are usually shipped to manufacturing plants as *pellets* or *powders* and are melted just before the shaping process. Plastics are also available as sheet, plate, rod, and tubing, which may then be formed into a variety of products. *Liquid* plastics are often used to make reinforced-plastic parts.

TABLE 10.6

Characteristics of Processing Plastics and Reinforced Plastics

Process	Characteristics
Extrusion	Long, uniform, solid or hollow, simple or complex cross sections; wide range of dimensional tolerances; high production rates; low tooling cost.
Injection molding	Complex shapes of various sizes and with fine detail; good dimensional accuracy; high production rates; high tooling cost.
Structural foam molding	Large parts with high stiffness-to-weight ratio; low production rates; less expensive tooling than in injection molding.
Blow molding	Hollow thin-walled parts of various sizes; high production rates and low cost for making beverage and food containers.
Rotational molding	Large hollow shapes of relatively simple design; low production rates; low tooling cost.
Thermoforming	Shallow or deep cavities; medium production rates; low tooling costs.
Compression molding	Parts similar to impression-die forging; medium production rates; relatively inexpensive tooling.
Transfer molding	More complex parts than in compression molding, and higher production rates; some scrap loss; medium tooling cost.
Casting	Simple or intricate shapes, made with flexible molds; low production rates.
Processing of reinforced plastics	Long cycle times; dimensional tolerances and tooling costs depend on the specific process.

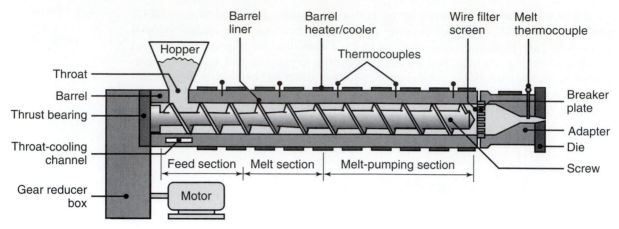

FIGURE 10.22 Schematic illustration of a typical extruder.

10.10.1 Extrusion

In *extrusion*, raw thermoplastic materials in the form of pellets, granules, or powder are placed into a hopper and fed into the extruder barrel (Fig. 10.22). The barrel is equipped with a *screw* that blends and conveys the pellets down the barrel. The internal friction and shear stresses from the mechanical action of the screw, along with heaters around the extruder's barrel, heats the pellets and liquefies them. The screw action also builds up pressure in the barrel.

Screws have three distinct sections: (1) a *feed section*, which conveys the material from the hopper area into the central region of the barrel; (2) a *melt*, or *transition*, *section*, where the heat generated from shearing of the plastic causes melting to begin; and (3) a *pumping section*, where additional shearing and melting occurs, with pressure buildup at the die. The lengths of these sections can be modified to accommodate the melting characteristics of individual plastics.

1. **Mechanics of polymer extrusion.** The pumping section of the screw determines the rate of polymer flow through the extruder. Consider a uniform screw geometry with narrow *flights* and small clearance with the barrel (Fig. 10.23). At any point in time, the molten plastic is in the shape of a helical ribbon, and this ribbon is being conveyed toward the extruder outlet by the screw flights. If the pressure is constant along the pumping zone, then the volume flow rate of plastic out of the extruder, or *drag flow*, is given by

$$Q_d = \frac{vHW}{2},$$ (10.17)

FIGURE 10.23 Geometry of the pumping section of an extruder screw.

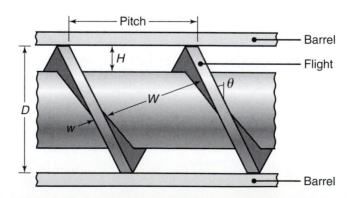

where v is the velocity of the flight in the extrusion direction, H is the channel depth, and W is the width of the polymer ribbon. From the geometry defined in Fig. 10.23, we can write

$$V = \omega \cos \theta = \pi DN \cos \theta \qquad (10.18)$$

and

$$W = \pi D \sin \theta - w, \qquad (10.19)$$

where ω is the angular velocity of the screw, D is the screw diameter, N is the shaft speed (usually in rev/min), θ is the flight angle, and w is the flight width. If the flight width, w, can be considered to be negligibly small, then the drag flow can be simplified as

$$Q_d = \frac{\pi^2 HD^2 \, N \sin \theta \cos \theta}{2}. \qquad (10.20)$$

The actual flow rate can be larger than this value if there is significant pressure buildup in the feed or melt sections of the screw, but it is usually smaller because of the high pressure at the die end of the barrel. Therefore, the flow rate through the extruder can be taken as

$$Q = Q_d - Q_p, \qquad (10.21)$$

where Q_p is a flow correction due to pressure. For Newtonian fluids (see Section 10.3), Q_p can be taken as

$$Q_p = \frac{WH^3 p}{12\eta(l/\sin \theta)} = \frac{p\pi DH^3 \sin^2 \theta}{12\eta l}, \qquad (10.22)$$

where l is the length of the pumping section. Hence, Eq. (10.21) becomes

$$Q = \frac{\pi^2 HD^2 \, N \sin \theta \cos \theta}{2} - \frac{p\pi DH^3 \sin^2 \theta}{12\eta l}. \qquad (10.23)$$

Equation (10.23) is known as the **extruder characteristic**. If the power-law index n from Eq. (10.4) is known, then the extruder characteristic is given by the following approximate relationship (after Rauwendaal, 1984):

$$Q = \left(\frac{4+n}{10}\right)\left(\pi^2 HD^2 \, N \sin \theta \cos \theta\right) - \frac{p\pi DH^3 \sin^2 \theta}{(1+2n)4\eta}. \qquad (10.24)$$

The die plays a major role in determining the output of the extruder. The **die characteristic** is the expression relating flow to the pressure drop across the die and in general form is written as

$$Q_{\text{die}} = Kp \qquad (10.25)$$

where Q_{die} is the flow through the die, p is the pressure at the die inlet, and K is a function of die geometry. The determination of K is usually complicated and difficult to obtain analytically, although computer-based tools are becoming increasingly available for predictions. More commonly, K is determined experimentally. However, one closed-form solution for extruding solid round cross sections is given by

$$K = \frac{\pi D_d^4}{128\eta l_d}, \qquad (10.26)$$

where D_d is the die opening diameter and l_d is the die land. If both the extruder characteristic and the die characteristic are known, we have two simultaneous algebraic equations that can be solved for the pressure and flow rate during the operation.

2. **Process characteristics.** Once the extruded product exits the die, it is cooled, either by air or by passing it through a water-filled channel. Controlling the rate and uniformity of cooling is important for minimizing product shrinkage and distortion. The extruded product can also be drawn (*sized*) by a puller after it has cooled; the extruded product is then coiled or cut into desired lengths. Complex shapes with constant cross section can be extruded with relatively inexpensive tooling. This process is also used to extrude elastomers.

   Because the material being extruded is still soft as it leaves the die and the pressure is relieved, the cross section of the extruded product is different than the shape of the die opening; this effect is known as **die swell**. (See Fig. 10.57b.) Thus, the diameter of a round extruded part, for example, is larger than the die opening, the difference between the two diameters depending on the type of polymer. Considerable experience is required to design proper dies for extruding complex cross sections of specific shape and dimensions. Modern software can assist in the design of extrusion dies, such as POLYFLOW, which numerically simulates polymer extrusion processes of all kinds, including reverse and coextrusion.

   Extruders are generally rated by the diameter of the barrel and by the length-to-diameter (L/D) ratio of the barrel. Typical commercial units are 25 to 200 mm (1 to 8 in.) in diameter, with L/D ratios of 5 to 30. Production-size extrusion equipment costs between \$30,000 and \$100,000, with an additional \$30,000 for equipment for downstream cooling and winding of the extruded product. Large production runs are therefore generally required to justify such an expenditure.

EXAMPLE 10.5 Analysis of a plastic extruder

An extruder screw with a thread angle of 20° has a melt-pumping zone that is 1 m long, with a channel depth of 7 mm for a 50-mm-diameter barrel. The screw is used to extrude 5-mm-diameter circular nylon bars through a die with a 20-mm land length at 300°C. If the extruder is operated at 50 rpm = 0.833 rev/s, determine the extruder and die characteristics and obtain the operating flow rate. What is the speed of material leaving the extruder? Ignore die swell.

Solution. Note from Fig. 10.12 that the viscosity of nylon at 300°C is around 300 Ns/m², and that the final bar cross-sectional area is $A = \pi D_d^2/4 = 1.96 \times 10^{-5}$ m². The extruder characteristic is given by Eq. (10.23) as

$$Q = \frac{\pi^2 \, HD^2 \, N \sin \theta \cos \theta}{2} - \frac{p\pi DH^3 \sin^2 \theta}{12 \, \eta l}$$

$$= \frac{\pi^2 (0.007)(0.050)^2 (0.833) \sin 20° \cos 20°}{2} - \frac{\pi (0.050)(0.007)^3 \sin^2 20°}{12(300)(1)} p$$

$$= 2.31 \times 10^{-5} - (1.75 \times 10^{-12}) \, p$$

where Q is in m³/s and p is in N/m². From Eq. (10.26),

$$K = \frac{\pi D_d^4}{128 \, \eta l_d} = \frac{\pi (0.005)^4}{128(300)(0.020)} = 2.56 \times 10^{-12},$$

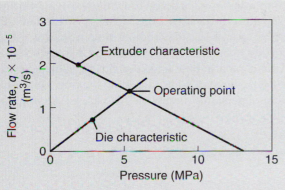

FIGURE 10.24 Extruder and die characteristics for Example 10.5.

so that the die characteristic is given by Eq. (10.25) as

$$Q = Kp = (2.56 \times 10^{-12})p.$$

The extruder and die characteristics are plotted in Fig. 10.24. Note that for both to be valid, the extruder must operate at the intersection of the two lines. This can be obtained from the graph or directly from the foregoing equations as $p = 5.4$ MPa. At this pressure, the flow rate can be calculated as $Q = 1.37 \times 10^{-5}$ m³/s. Based on the cross-sectional area of the product, the final velocity is

$$Q = vA; \qquad v = \frac{Q}{A} = \frac{1.37 \times 10^{-5}}{1.96 \times 10^{-5}} = 0.70 \text{ m/s}.$$

3. **Sheet and film extrusion.** Polymer sheet and film can be produced using a flat extrusion die with a thin, rectangular opening. The polymer is extruded by forcing it through a specially designed die, after which the extruded sheet is taken up first on water-cooled rolls and then by a pair of rubber-coated pull-off rolls.

　　Thin polymer films are typically made from a tube produced by an extruder (Fig. 10.25). In this process, a thin-walled tube is extruded vertically

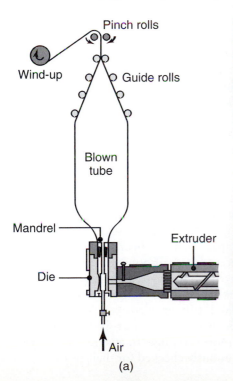

FIGURE 10.25
(a) Schematic illustration of production of thin film and plastic bags from a tube produced by an extruder and then blown by air and (b) a blown-film operation. *Source:* Courtesy of Windmoeller & Hoelscher Corp.

(a)

(b)

and expanded (blown) into a balloon shape by blowing warm air through the center of the extrusion die until the desired film thickness is reached. The balloon is usually cooled by air from holes in a cooling ring around it, which can also act as a barrier to further expansion of the balloon. The **blown film** is sold as wrapping film (after the cooled bubble has been slit) or as bags (where the bubble is pinched and cut off). Film is also produced by shaving solid round billets of plastics, especially polytetrafluoroethylene (PTFE), by *skiving* (shaving with specially designed knives).

EXAMPLE 10.6 Blown film

Assume that a plastic shopping bag, made from blown film, has a lateral (width) dimension of 400 mm. (a) What should be the extrusion die diameter? (b) These bags are relatively strong. How is this strength achieved?

Solution.

(a) The perimeter of the bag is $(2)(400) = 800$ mm. Since the original cross section of the film was round, the blown diameter can be calculated from $\pi D = 800$, or $D = 255$ mm. Recall that in this process a tube is expanded 1.5 to 2.5 times the extrusion die diameter. Selecting the maximum value of 2.5, the die diameter is calculated as $255/2.5 = 100$ mm.

(b) Note in Fig. 10.25 that, after being extruded, the bubble is being pulled upward by the pinch rolls. Thus, in addition to diametral stretching and the resulting molecular orientation, the film is stretched and oriented in the longitudinal direction as well. The biaxial orientation of the polymer molecules significantly improves the strength and toughness of the blown film.

---

4. **Miscellaneous extrusion processes.** Pellets, which are used for other plastic-processing methods described throughout the rest of this chapter, are also made by extrusion. In this case, the extruded product is a small-diameter rod which is then chopped into short lengths, or *pellets*, as it is extruded. With some modifications, extruders can also be used as simple melters for other shaping processes, such as injection molding and blow molding. Extrusion of tubes is also a necessary first step for related processes such as extrusion blow molding and blown film (see above).

Plastic tubes and pipes are produced in an extruder equipped with a *spider die*, as shown in Fig. 10.26a (see also Fig. 6.58 for details). For the production of *reinforced hoses* that can withstand higher pressures, woven fiber or wire reinforcements are fed through the extruder using specially designed dies.

Coextrusion, shown in Fig. 10.26b, involves simultaneous extrusion of two or more polymers through a single die. The product cross section thus contains different polymers, each with its own characteristics and function. Coextrusion is commonly used for shapes such as sheet, film, and tubes, especially in food packaging where different layers of polymers have different functions. These functions include (a) inertness for food and liquids, (b) serving as barriers to fluids such as water or oil, and (c) labeling of the product.

Plastic-coated electrical wire, cable, and strips are also extruded and coated by coextrusion. The wire is fed into the die opening at a controlled rate with the extruded plastic in order to produce a uniform coating. To ensure proper insulation, extruded electrical wires are continuously checked for their resistance as

Spider die

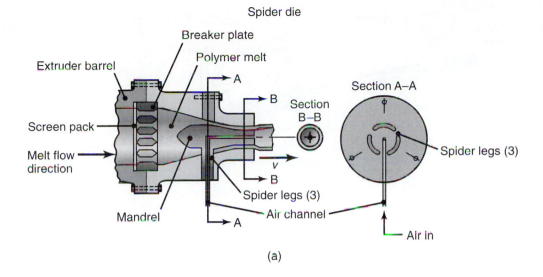

(a)

Co-extrusion blow molding

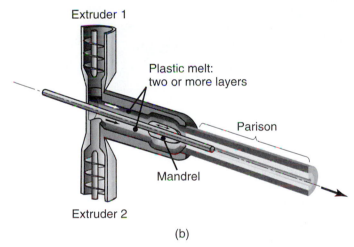

(b)

FIGURE 10.26 Extrusion of plastic tubes: (a) extrusion using a spider die (see also Fig. 6.59) and pressurized air and (b) coextrusion of tube for producing a bottle.

they exit the die; they are also marked automatically with a roller to identify the specific type of wire. *Plastic-coated paper clips* are also made by coextrusion.

10.10.2 Injection molding

Injection molding is very similar to the same process as hot-chamber die casting shown in Fig. 5.24. The pellets, or granules, are fed into a heated cylinder, where they are melted. The melt is then forced into a split-die chamber (Fig. 10.27a), either by a hydraulic plunger or by the rotating-screw of an extruder. Most modern equipment is of the *reciprocating-screw* type (Fig. 10.27b). As the pressure builds up at the mold entrance, the rotating screw starts to move backward, under pressure, to a predetermined distance, thus controlling the volume of material to be injected. The screw then stops rotating and is pushed forward hydraulically, forcing the molten plastic into the mold cavity. Injection-molding pressures usually range from 70 to 200 MPa (10,000 to 30,000 psi).

FIGURE 10.27 Injection molding with (a) a plunger and (b) a reciprocating rotating screw. Telephone receivers, plumbing fittings, tool handles, and housings are examples of parts made by injection molding.

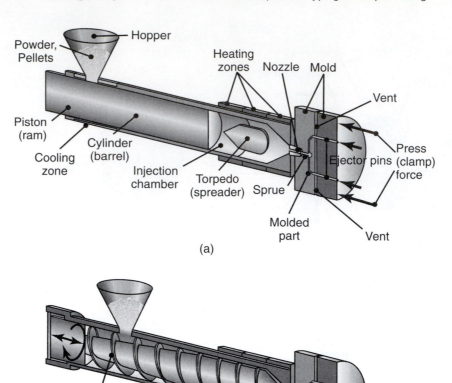

(a)

(b)

Typical injection-molded products include cups, containers, housings, tool handles, knobs, electrical and communication components (such as cell phones), toys, and plumbing fittings. Although the molds are relatively cool for thermoplastics, thermosets are molded in heated molds, where *polymerization* and *cross-linking* take place. In either case, after the part is sufficiently cooled (for thermoplastics) or set or cured (for thermosets), the molds are opened, and the part is ejected. The molds are then closed, and the process is repeated automatically. Elastomers are also injection molded. Molds with moving and unscrewing mandrels are also used; they allow the molding of parts with multiple cavities and internal and external threads.

Because the material is molten when injected into the mold, complex shapes and good dimensional accuracy can be achieved. However, as in metal casting (Chapter 5), the molded part shrinks during cooling. Note also that plastics have higher thermal expansion coefficient than metals (see Table 3.3). Linear shrinkage for plastics typically ranges between 0.005 and 0.025 mm/mm or in./in., and the volumetric shrinkage is typically in the range of 1.5 to 7%. (See also Table 5.1 and Section 11.4.) Shrinkage in injection molding is compensated for by oversizing the molds.

Various aspects of injection molding are discussed next.

1. **Molds.** Injection molds have several components, depending on part design, such as runners, cores, cavities, cooling channels, inserts, knockout pins, and ejectors. There are three basic types of molds:

   a. *Cold-runner two-plate* mold (Fig. 10.28 and Fig. 10.29a), which is the basic and simplest mold design.

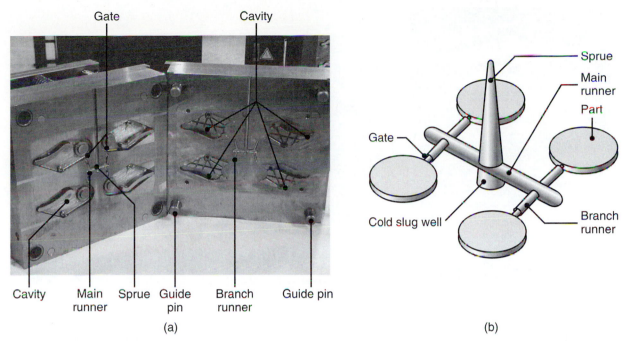

FIGURE 10.28   Illustration of mold features for injection molding: (a) two-plate mold, with important features identified and (b) injection molding of four parts, showing details and the volume of material involved. *Source:* Courtesy of Tooling Molds West, Inc.

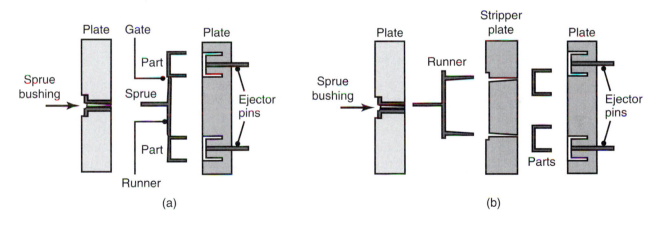

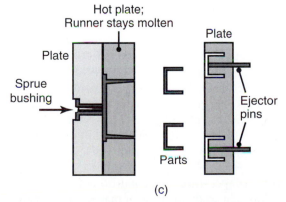

FIGURE 10.29   Types of molds used in injection molding: (a) two-plate mold, (b) three-plate mold, and (c) hot-runner mold.

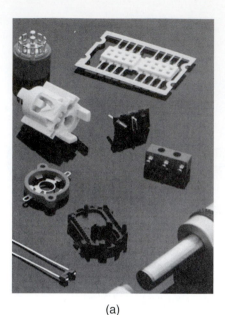

(a)

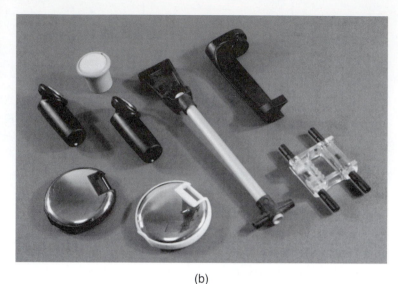

(b)

FIGURE 10.30   Products made by insert injection molding. Metallic components are embedded in these parts during molding. *Source:* (a) Courtesy of Plainfield Molding, Inc. and (b) Courtesy of Rayco Mold and Mfg. LLC.

   b.   *Cold-runner three-plate* mold (Fig 10.29b), in which the runner system is separated from the part after the mold is opened.

   c.   *Hot-runner* mold, also called *runnerless* mold (Fig. 10.29c), in which the molten plastic is kept hot in a heated runner plate.

In cold-runner molds, the solidified plastic in the channels that connect the mold cavity to the end of the barrel must be removed, usually by trimming. This scrap is usually chopped and recycled. In hot-runner molds, which are more expensive, there are no gates, runners, or sprues attached to the part. Cycle times are shorter, because only the injection-molded part must be cooled and ejected.

   Metallic components, such as screws, pins, and strips, can also be placed in the mold cavity, thereby becoming an integral part of the injection-molded product (**insert molding**; see Fig. 10.30). The most common examples are electrical components and hand tools such as plastic-handled screwdrivers. *Multicomponent* injection molding, also called *coinjection* or *sandwich* molding, allows the forming of parts with a combination of colors and shapes. Examples include multicolor molding of rear-light covers for automobiles and ball joints made of different materials. Printed film can also be placed in the mold cavity; thus, parts need not be decorated or labeled after molding.

   Injection molding is a high-rate production process, with good dimensional control. Typical cycle times range from 5 to 60 s but can be several minutes for thermosetting materials. The molds, generally made of tool steels or beryllium-copper alloy, may have *multiple cavities* so that more than one part can be made in one cycle (as in die casting; Section 5.10.3). Mold design and control of material flow in the die cavities are important factors in the quality of the product. Other factors that also affect part quality include injection pressure, temperature, and condition of the resin. *Computer models* have been developed to study the flow of material in dies and thereby improve die design and establish appropriate process parameters.

2. **Machines.** Injection-molding machines are usually horizontal, and the clamping force on the dies is supplied generally by hydraulic means, although electrical types are also available. Electrically driven models weigh less and are quieter than hydraulic machines. Vertical machines are used for making small, close-tolerance parts and for insert molding. Injection-molding machines are rated according to the capacity of the mold and the clamping force. Although in most machines, this force generally ranges from 0.9 to 2.2 MN (100 to 250 tons), the largest machine in operation has a capacity of 45 MN (5000 tons) and can produce parts weighing up to 25 kg (55 lb); however, parts typically weigh 100 to 600 g (3 to 20 oz). Modern machines used in injection molding are equipped with microprocessors and microcomputers in a control panel, and monitor all aspects of the operation. Because of the high cost of dies, typically ranging from $20,000 to $200,000, high-volume production is required to justify such an expenditure.

3. **Overmolding** and **ice-cold molding.** *Overmolding* is a technique for producing, in one operation, hinged joints and ball-and-socket joints. Two different plastics are used to ensure that no bond will form between the two parts, which may interfere with the free movement of the components. *Ice-cold molding* uses the same kind of plastic to form both components, such as in a hinge. The process takes place in one cycle and involves a two-cavity mold, using cooling inserts positioned such that no bond forms between the two pieces.

4. **Reaction-injection molding.** In the *reaction-injection molding* (RIM) process, a mixture of two or more reactive fluids is forced into the mold cavity (Fig. 10.31). Chemical reactions then take place rapidly in the mold, and the polymer solidifies, producing a thermoset part. Major applications include automotive bumpers and fenders, thermal insulation for refrigerators and freezers, and stiffeners for structural components. Various reinforcing fibers, such as glass or graphite, may also be used to improve the product's strength and stiffness.

5. **Structural foam molding.** The *structural-foam-molding* process is used to make plastic products that have a solid skin and a cellular inner structure. Typical products include furniture components, TV cabinets, business-machine housings, and storage-battery cases. Although there are several foam-molding

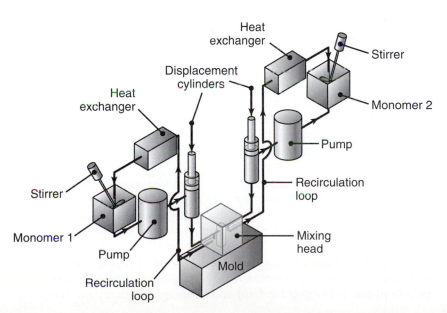

FIGURE 10.31   Schematic illustration of the reaction-injection molding process.

processes, they are basically similar to injection molding or extrusion. Both thermoplastics and thermosets can be used for foam molding, but thermosets are in the liquid-processing form, similar to the polymers used for reaction-injection molding.

6. **Injection foam molding.** In *injection foam molding*, also called *gas-assist molding*, thermoplastics are mixed with a *blowing agent* (usually an inert gas, such as nitrogen, or a chemical agent that produces gas during molding), which expands the polymer. The core of the part is cellular, and the skin is rigid. The thickness of the skin can be as much as 2 mm (0.08 in.), and part densities are as low as 40% of the density of the solid plastic. Consequently, parts have a high stiffness-to-weight ratio and can weigh as much as 55 kg (120 lb).

EXAMPLE 10.7 Injection molding of gears

A 250-ton injection-molding machine is used to make 4.5-in.-diameter spur gears with a thickness of 0.5 in. The gears have a fine-tooth profile. How many gears can be injection molded in one set of molds? Does the thickness of the gears influence the answer?

Solution. Because of the fine-tooth detail involved, the pressures required in the mold cavity will probably be on the order of 100 MPa (15 ksi). The cross-sectional (projected) area of the gear is $\pi(4.5)^2/4 = 15.9$ in^2. Assuming that the parting plane of the two halves of the mold is in the midplane of the gear, the force required is $(15.9)(15,000) = 238,500$ lb. The capacity of the machine is 250 tons; hence the available clamping force is $(250)(2000) = 500,000$ lb. Therefore, the mold can accommodate two cavities, producing two gears per cycle. Because it does not influence the cross-sectional area of the gear, the thickness of the gear does not directly influence the pressures involved, and hence the answer is the same for different gear thicknesses.

10.10.3 Blow molding

Blow molding is a modified combination of extrusion and injection-molding processes. In **extrusion blow molding**, a tube is (a) extruded (usually vertically), (b) clamped into a mold with a cavity much larger than the tube's diameter, and (c) then blown outward to fill the mold cavity (Fig. 10.32a). Blowing is usually done with an air blast, at a pressure of 350 to 700 kPa (50 to 100 psi). In some operations the extrusion may be continuous, and the molds move with the tubing. The molds close around the tubing, closing off both ends (thereby sectioning the tube), and then move away as air is injected into the tubular piece. The part is then cooled and ejected. *Corrugated pipe* and tubing are made by continuous blow molding, whereby the pipe or tube is extruded horizontally and blown into moving molds. (See also Section 10.10.1.)

In **injection blow molding**, a short tubular piece (**parison**) is first injection molded (Fig. 10.32b). The dies are then opened, and the parison is transferred to a blow-molding die. Hot air is injected into the parison, which expands and fills the mold cavity. Typical products made by this process include plastic beverage bottles and hollow containers.

Multilayer blow molding involves the use of coextruded tubes, or parisons, thus allowing the use of multilayer structures. Typical examples of multilayer structures include plastic packaging for food and beverages, with characteristics such as odor

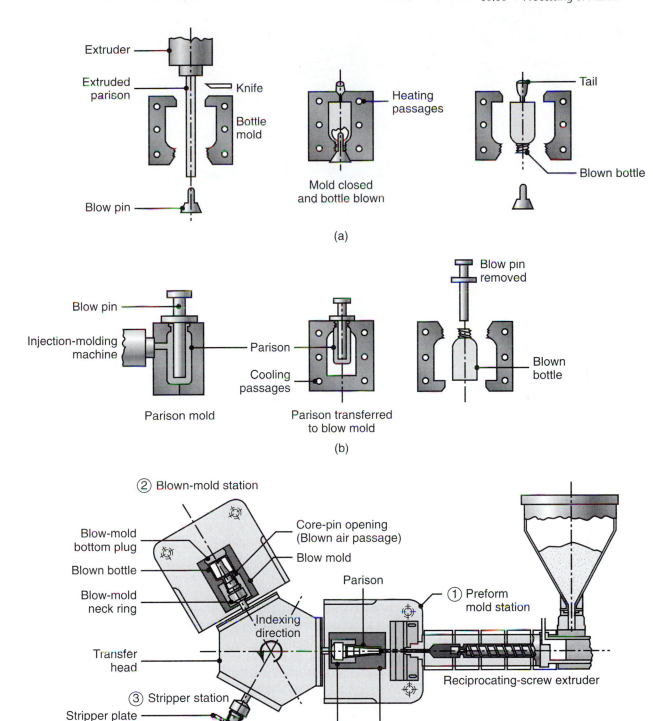

FIGURE 10.32 Schematic illustrations of (a) the blow-molding process for making plastic beverage bottles, (b) the injection-blow-molding process, and (c) a three-station injection-blow-molding machine for making plastic bottles.

FIGURE 10.33 The rotational molding (rotomolding or rotocasting) process. Trash cans, buckets, carousel horses and plastic footballs can be made by this process.

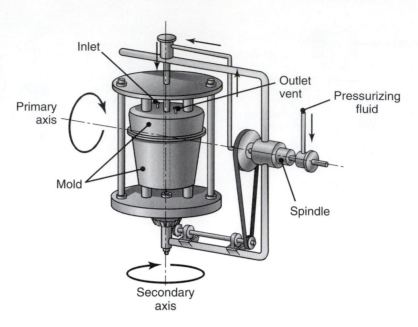

and permeation barrier, taste and aroma protection, resistance to scuffing, printing capability, and ability to be filled with hot fluids. Other applications include the cosmetics and pharmaceutical industries.

10.10.4 Rotational molding

Most thermoplastics and some thermosets can be formed into large hollow parts by *rotational molding*. Typical parts made include trash cans, boat hulls, buckets, housings, toys, carrying cases, and footballs. The thin-walled metal mold is made of two pieces (*split female mold*), designed to be rotated about two perpendicular axes (Fig. 10.33). A premeasured quantity of powdered plastic is placed inside a warm mold. The powder is produced from a polymerization process that precipitates the powder from a liquid. The mold is then heated, usually in a large oven, while it rotates about the two axes. This action tumbles the powder against the mold, where the heat fuses the powder without melting it. In making some parts, a chemical agent is added to the powder, and cross-linking occurs after the part is formed in the mold by continued heating. Various metallic or plastic *inserts* may also be molded into the parts made by this process, as in insert molding, described above.

Liquid polymers, called **plastisols** (vinyl plastisols being the most common), can also be used in a process called **slush molding**. The mold is simultaneously heated and rotated, and the tumbling action forces the particles of plastic against the inside walls of the heated mold. Upon contact, the material melts and coats the walls of the mold. The part is cooled while still rotating, and is then removed by opening the mold.

Rotational molding can produce parts with complex hollow shapes, with wall thicknesses as small as 0.4 mm (0.016 in.). Parts as large as $1.8 \times 1.8 \times 3.6$ m ($6 \times 6 \times 12$ ft) have been formed using this method. The outer surface finish of the part is a replica of the surface finish of the mold walls. Although cycle times are longer than in other processes, equipment costs are low. Quality-control considerations generally involve the weight of powder placed in the mold, the rotation of the mold, and the temperature-time relationship during the oven cycle.

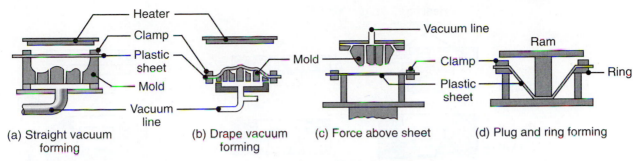

(a) Straight vacuum forming (b) Drape vacuum forming (c) Force above sheet (d) Plug and ring forming

FIGURE 10.34   Various thermoforming processes for thermoplastic sheet. These processes are commonly used in making advertising signs, cookie and candy trays, panels for shower stalls, and packaging.

10.10.5 Thermoforming

Thermoforming is a family of processes for forming thermoplastic sheet or film over a mold with the application of heat and pressure or vacuum (Fig. 10.34). A sheet (made by sheet extrusion; see above) is first heated in an oven to the *sag* (softening) *point*, but not to the melting point. It is then removed from the oven, placed over a mold, and forced against the mold by the application of a vacuum. Since the mold is usually at room temperature, the shape of the plastic is set upon contacting the mold. Because of the low strength of the materials formed, the pressure differential caused by the vacuum is usually sufficient for forming, although air pressure or mechanical means are also applied for some parts.

Typical parts made by this method include packaging, advertising signs, refrigerator liners, appliance housings, and panels for shower stalls. Parts with openings or holes cannot be formed by this method because the necessary pressure differential cannot be developed during forming. Because thermoforming is a combination of drawing and stretching operations, much like sheet-metal forming (see Chapter 7), the material should exhibit high uniform elongation, as otherwise it will neck and fail. Thermoplastics have a high capacity for uniform elongation by virtue of their high strain-rate sensitivity exponent, m. (See Section 2.2.7.)

Hollow parts can be produced by using twin sheets. In this operation, the two mold halves are brought together with the heated sheets between them. The sheets are drawn to the mold halves through a combination of vacuum from the mold and the compressed air that is introduced between the sheets. The molds also join the sheet around the mold cavity. (See the discussion on *hot-plate welding* in Section 12.17.1.)

Molds for thermoforming are usually made of aluminum, since high strength is not a requirement. The holes in the molds are generally less than 0.5 mm (0.02 in.) in diameter in order not to leave any marks on the formed sheets. Quality considerations include tears, nonuniform wall thickness, improperly filled molds, and poor surface details (part definition).

10.10.6 Compression molding

In *compression molding,* a preshaped charge of material, a premeasured volume of powder, or a viscous mixture of liquid resin and filler material is placed directly in a heated mold cavity. Molding is done under pressure with a plug or the upper half of the die, as shown in Fig. 10.35. The flash is removed by trimming or some other means. Typical parts made by this method include dishes, handles, container caps, fittings, electrical and electronic components, washing-machine agitators, and housings. Elastomers and fiber-reinforced parts with long chopped fibers are also formed exclusively by this process.

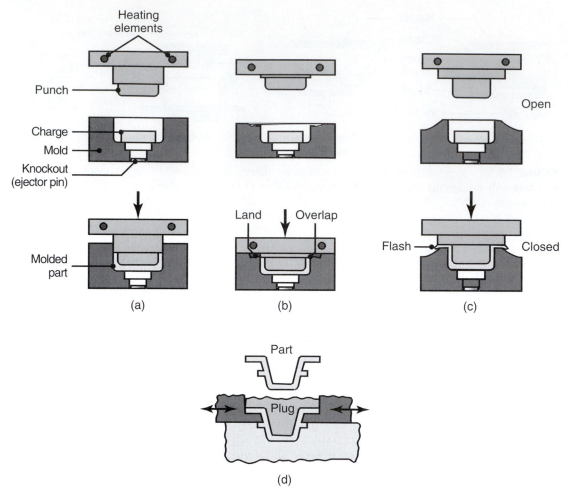

FIGURE 10.35   Types of compression molding, a process similar to forging: (a) positive, (b) semipositive, and (c) flash. The flash in part (c) is trimmed off. (d) Die design for making a compression-molded part with undercuts. Such designs are also used in other molding and shaping operations.

Compression molding is used mainly with thermosetting plastics, with the original material in a partially polymerized state. Cross-linking is completed in the heated die, with curing times typically ranging from 0.5 to 5 min, depending on the polymer, and part geometry and its thickness. The thicker the part, the longer it will take to cure. Because of their relative simplicity, dies for compression molding generally cost less than dies for injection molding. Three types of compression molds are available: (1) **flash type** for shallow or flat parts, (2) **positive** for high-density parts, and (3) **semipositive** for high-quality production. Undercuts in parts are not recommended; however, dies can be designed to be opened sideways (Fig. 10.35d) to allow easy removal of the part. In general, the part complexity in this process is less than with injection molding, and the dimensional control is better.

10.10.7 Transfer molding

Transfer molding represents a further development of the compression-molding process. The uncured thermosetting material is placed in a heated transfer pot or chamber (Fig. 10.36). After the material reaches the proper temperature, it is injected

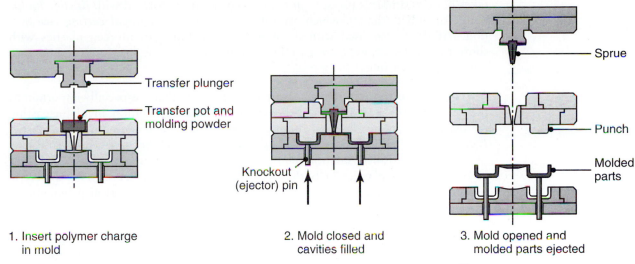

Transfer plunger

Transfer pot and molding powder

Sprue

Knockout (ejector) pin

Punch

Molded parts

1. Insert polymer charge in mold

2. Mold closed and cavities filled

3. Mold opened and molded parts ejected

FIGURE 10.36 Sequence of operations in transfer molding of thermosetting plastics. This process is particularly suitable for making intricate parts with varying wall thicknesses.

into heated, closed molds. Depending on the type of machine used, a ram, a plunger, or a rotating screw feeder forces the material to flow through the narrow channels into the mold cavity. Because of internal friction this flow generates internal heat, which raises the temperature of the material and homogenizes it. Curing then takes place by cross-linking. Because the polymer is molten as it enters the molds, the complexity of the part and dimensional control approach those for injection molding.

The process is particularly suitable for intricate shapes that have varying wall thicknesses. Typical parts made by transfer molding include electrical and electronic components and rubber and silicone parts. The molds for this process tend to be more expensive than those for compression molding, and material is wasted in the channels of the mold during filling. (See also the discussion of *resin transfer molding* in Section 10.11.1.)

10.10.8 Casting

Some thermoplastics, such as nylons and acrylics, and thermosetting plastics, such as epoxies, phenolics, polyurethanes, and polyester, can be *cast* in rigid or flexible molds into a variety of shapes (Fig. 10.37a). Typical parts cast include large gears, bearings, wheels, thick sheets, and components that require resistance to abrasive wear. In casting thermoplastics, a mixture of monomer, catalyst, and various additives is heated and poured into the mold. The part is formed after polymerization

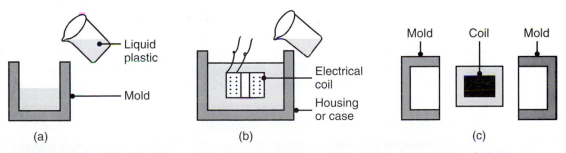

Liquid plastic

Mold

Electrical coil

Housing or case

Mold   Coil   Mold

(a)   (b)   (c)

FIGURE 10.37 Schematic illustration of (a) casting, (b) potting, and (c) encapsulation of plastics.

takes place at ambient pressure. Intricate shapes can be formed with *flexible molds* (made of polyurethane), which are then peeled off. *Centrifugal casting* (see Section 5.10.4) is also used with thermoplastics, including reinforced plastics with short fibers. Thermosets are cast in a similar manner, and parts are similar to thermoplastic castings.

Potting and encapsulation. A variation of casting that is especially important to the electrical and electronics industry is potting and encapsulation, involving casting the plastic around an electrical component, thus totally embedding it in the plastic. *Potting* (Fig. 10.37b) is done in a housing or case, which becomes an integral part of the product. In *encapsulation* (Fig. 10.37c), the component is covered with a layer of the solidified plastic. In both applications, the plastic serves as a dielectric (nonconductor). Structural members, such as hooks and studs, may also be made by partial encapsulation.

Foam molding. Products such as *styrofoam* cups and food containers, insulating blocks, and shaped packaging materials (such as for cameras, computers, appliances, and electronics) are made by *foam molding*. The material is made of *expandable polystyrene,* produced by placing **polystyrene beads** (obtained by polymerization of styrene monomer) containing a blowing agent in a mold, and exposing them to heat, usually by steam. As a result of exposure to heat, the beads expand (to as much as 50 times their original size) and acquire the shape of the mold. The amount of expansion can be controlled through temperature and time. A common method of molding involves the use of *preexpanded* beads, in which the beads are expanded by steam (or hot air, hot water, or an oven) in an open-top chamber. The beads are then placed in a storage bin and allowed to stabilize for a period of 3 to 12 hours. They can then be molded into shapes as described previously.

   Polystyrene beads are available in three sizes: (a) small, for products such as cups, (b) medium, for molded shapes such as containers, and (c) large, for molding of insulating blocks (which can then be cut to size). The bead size chosen depends on the minimum wall thickness of the product. Beads can also be colored prior to expansion, or integrally colored beads can be used.

Polyurethane foam processing. This process is used for making products such as cushions and insulating blocks. It involves several processes that basically consist of mixing two or more chemical components. The reaction produces a cellular structure, which solidifies in the mold. Various low-pressure and high-pressure machines, with computer controls for proper mixing, are available for this operation.

10.10.9 Cold forming and solid-phase forming

Cold-working processes described in Chapters 6 and 7, such as rolling, deep drawing, extrusion, closed-die forging, coining, and rubber forming, can also be used to form thermoplastics at room temperature (*cold forming*). Typical polymers commonly formed include polypropylene, polycarbonate, ABS, and rigid PVC. The major considerations are that (a) the material must be sufficiently ductile at room temperature, hence, polystyrenes, acrylics, and thermosets cannot be formed; and (b) the material's deformation must be nonrecoverable, in order to minimize springback and creep.

   The advantages of cold forming of plastics over other methods of shaping are listed as follows:

1. Strength, toughness, and uniform elongation of the material are increased.

2. Polymers with high molecular weight can be used to make parts with superior properties.

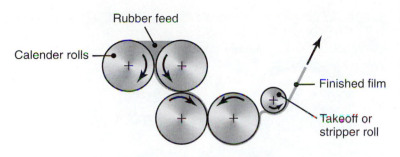

FIGURE 10.38   Schematic illustration of calendering. Sheets produced by this process are subsequently used in processes such as thermoforming.

3. Forming speeds are not affected by the thickness of the part, since there is no heating or cooling involved.

4. Typical cycle times are shorter than those for molding processes.

Solid-phase forming is carried out at a temperature about to 10°–20°C (20°–40°F) below the melting temperature of the plastic (see Table 10.2), if it is a crystalline polymer and it is formed while still in a solid state. The advantages of solid-phase forming over cold forming are that forming forces and springback are lower for the former. These processes are not as widely used as hot-processing methods and are generally restricted to special applications.

10.10.10 Processing elastomers

Elastomers can be formed by a variety of processes used for shaping thermoplastics. Thermoplastic elastomers are commonly shaped by extrusion and injection molding, extrusion being the most economical and the fastest process. In terms of its processing characteristics, a thermoplastic elastomer is a polymer; in terms of its function and performance, it is a rubber (see Section 10.8). These polymers can also be formed by blow molding and thermoforming. Thermoplastic polyurethane can be shaped by all conventional methods, and it can also be blended with thermoplastic rubbers, polyvinyl chloride compounds, ABS, and nylon. For extrusion, the temperatures are in the range of 170° to 230°C (340° to 440°F), and for molding, they range up to 60°C (140°F). Typical extruded products include hoses, moldings, and inner tubes. Injection-molded products cover a broad range of applications, such as children's toys, wiring harnesses and other components for automobiles, and housings, footings, and controls for appliances.

Rubber and some thermoplastic sheets are formed by the **calendering** process (Fig. 10.38), wherein a warm mass of the compound is fed through a series of rolls (*masticated*) and is then stripped off in the form of a sheet. The rubber also may be formed between two surfaces of a fabric liner. Discrete rubber products, such as gloves and balloons, are made by repeatedly dipping a rigid form into a liquid compound of rubber dissolved in solvent that adheres to the form. The solvent is then allowed to evaporate and the rubber is then *vulcanized*, usually in steam, and stripped from the form.

10.11 | Processing of Polymer-Matrix-Reinforced Plastics

As described in Section 10.9, reinforced plastics are among the most important materials and can be engineered to meet specific design requirements such as high strength-to-weight and stiffness-to-weight ratios and creep resistance. Because of

FIGURE 10.39
Reinforced-plastic
components for a Honda
motorcycle. The parts shown
are front and rear forks, a
rear swing arm, a wheel, and
brake disks.

their unique structure and the characteristics of their individual components, reinforced plastics require special methods to shape them into useful products (Fig. 10.39).

The care required and the numerous steps involved in manufacturing reinforced plastics make processing costs substantial. Thus, careful assessment and integration of the design and manufacturing processes (*concurrent engineering;* see Section 1.2) is essential in order to minimize costs while maintaining product integrity and production rate. An important environmental concern with respect to reinforced plastics is the dust generated during processing, such as airborne carbon fibers, which are known to remain in the work area long after fabrication of parts has been completed.

Reinforced plastics can usually be fabricated by the methods described in this chapter, with consideration for the presence of more than one type of material in the composite. As described in Section 10.9, the reinforcement may consist of chopped fibers, woven fabric or mat, roving or yarn (slightly twisted fiber), or continuous lengths of fiber. Short fibers are commonly added to thermoplastics in injection molding; milled fibers can be used in reaction-injection molding; and longer chopped fibers are used primarily in compression molding of reinforced plastics. In order to obtain good bonding between the fibers and the polymer matrix, as well as to protect the fibers during subsequent processing steps, the fibers are first surface treated by *impregnation* (*sizing*).

When the impregnation is done as a separate step, the resulting partially cured sheets are referred to by various terms:

1. **Prepregs.** The continuous fibers are aligned (Fig. 10.40a) and subjected to surface treatment to enhance their adhesion to the polymer matrix. They are then coated by being dipped in a resin bath and made into a *sheet* or *tape* (Fig. 10.40b). Finally, individual pieces of the sheet are assembled into laminated structures, such as the horizontal stabilizer for the F-14 fighter aircraft. Special computer-controlled tape-laying machines have been developed for this purpose. Typical products made include flat or corrugated architectural paneling, panels for construction and electric insulation, and structural components of aircraft that require good retention of properties and fatigue strength under various environmental conditions.

2. **Sheet-molding compound (SMC).** Continuous strands of reinforcing fiber are chopped into short fibers (Fig. 10.41) and deposited over a layer of resin paste, usually a polyester mixture, carried on a polymer film such as polyethylene. A second layer of resin paste is deposited on top, and the sheet is

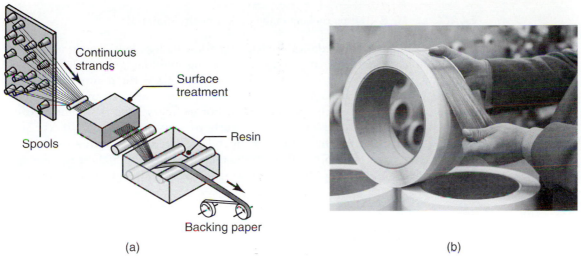

(a)   (b)

FIGURE 10.40   (a) Manufacturing process for polymer-matrix composite and (b) boron-epoxy prepreg tape. *Source:* After T.-W. Chou, R.L. McCullough, and R.B. Pipes and (b) Textron Systems.

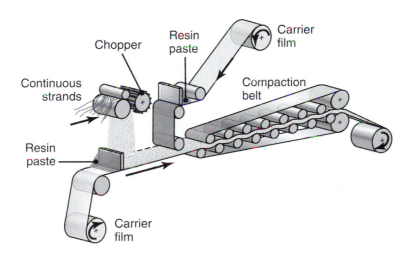

FIGURE 10.41
Manufacturing process for producing reinforced-plastic sheets. The sheet is still viscous at this stage and can later be shaped into various products. *Source:* After T.-W. Chou, R.L. McCullough, and R.B. Pipes.

pressed by passing it through rollers. The product is then gathered into rolls, or placed into containers in layers, and stored until it undergoes a *maturation period,* reaching the desired molding viscosity. The maturing process involves controlled temperature and humidity and usually takes one day. The matured SMC, which has a leatherlike feel, has a shelf life of about 30 days and must be processed within this period. Alternatively, the resin and the fibers can be mixed together only at the time they are placed in the mold.

3. **Bulk-molding compound (BMC).** These compounds are in the shape of billets, generally up to 50 mm (2 in.) in diameter; they are made in the same manner as SMCs, that is, by extrusion. When processed into products, BMCs have flow characteristics similar to those of dough; hence, they are called *dough-molding compounds* (DMCs).

4. **Thick-molding compound (TMC).** This compound combines the characteristics of BMCs (lower cost) and SMCs (higher strength); it is usually injection molded, using chopped fibers of various lengths. One of its applications is for electrical components, because of the high dielectric strength of TMCs.

EXAMPLE 10.8 Tennis rackets made of composite materials

In order to impart certain desirable characteristics, such as light weight and stiffness, composite-material tennis rackets are being manufactured with graphite, fiberglass, boron, ceramic (silicon carbide), and Kevlar as the reinforcing fibers. Rackets have a foam core; some have unidirectional reinforcement, and others have braided reinforcement. Rackets with boron fibers have the highest stiffness, followed by rackets with graphite (carbon), glass, and Kevlar fibers. The racket with the lowest stiffness has 80% fiberglass, whereas the stiffest racket has 95% graphite and 5% boron fibers; thus, the latter has the highest percentage of inexpensive reinforcing fiber and the smallest percentage of the most expensive fiber.

10.11.1 Molding

There are five basic methods for molding reinforced plastics, as described next.

1. **Compression molding.** In this operation, the material is placed between two molds, and pressure is applied. Depending on the material, the molds are either at room temperature or are heated to accelerate hardening. The material may be in bulk form (*bulk-molding compound*, or BMC), which is a viscous, sticky mixture of polymers, fibers, and additives. It is generally shaped into a log, which is cut into the desired size before being placed in the mold. Fiber lengths generally range from 3 to 50 mm (0.125 to 2 in.), although fibers 75 mm long (3 in.) may also be used. Sheet-molding compounds (SMC) can also be used in molding; these compounds are similar to BMCs, except that the resin-fiber mixture is laid between plastic sheets to make a sandwich that can easily be handled. The sheets are removed after the SMC is placed in the mold.

2. **Vacuum-bag molding.** As shown in Fig. 10.42, prepregs are laid in a mold, and the pressure required to form the shape and to develop good bonding is obtained by covering the layup with a plastic bag and developing a vacuum. If additional heat and pressure are necessary, the entire assembly is placed in an *autoclave*. Care should be exercised to maintain fiber orientation, if specific

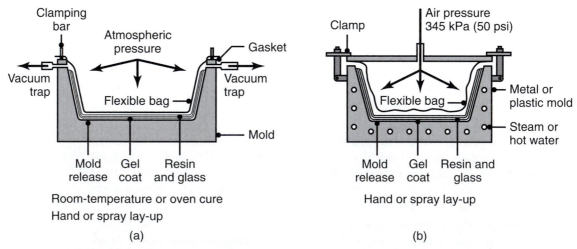

FIGURE 10.42 (a) Vacuum-bag forming and (b) pressure-bag forming. *Source:* After T.H. Meister.

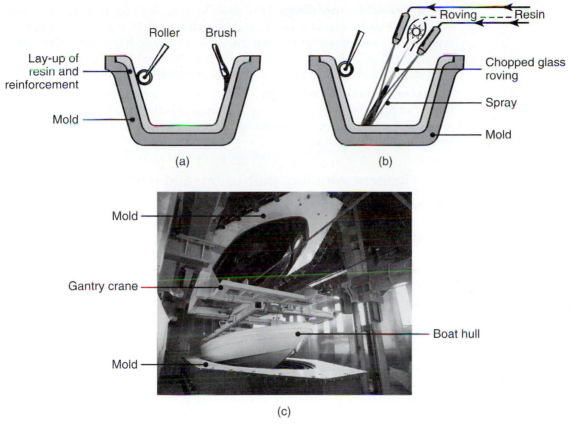

FIGURE 10.43 Manual methods of processing reinforced plastics: (a) hand lay-up and (b) spray-up. These methods are also called *open-mold processing*. (c) A boat hull made by these processes. *Source:* Courtesy of Genmar Holdings, Inc.

fiber orientations are desired. In materials with chopped fibers, no specific orientation is intended. In order to prevent the resin from sticking to the vacuum bag and to facilitate removal of any excess, several sheets of various materials (*release cloth* or *bleeder cloth*) are placed on top of the prepreg sheets. Although the molds can be made of metal, usually aluminum, more often they are made from the same resin (with reinforcement) as the material to be cured. This condition eliminates any problems resulting from differences in thermal expansion between the mold and the part.

3. **Contact-molding.** This process is used in making products with high surface-area-to-thickness ratios, such as swimming pools, boats, tub and shower units, and housings. It uses a single male or female mold (Fig. 10.43), made of materials such as reinforced plastics, wood, or plaster. The contact molding process is a wet method, in that the reinforcement is impregnated with the resin at the time of molding. The simplest method is called **hand lay-up.** The materials are placed and formed in the mold by hand (Fig. 10.43a), and the squeezing action expels any trapped air and compacts the part.

   Molding may also be done by spraying (*spray-up;* see Fig. 10.43b). Although the spraying operation can be automated, these processes are relatively slow, and labor costs are high. They are simple, however, and the tooling is inexpensive. Only the mold-side surface of the part is smooth, and the choice of materials is limited. Many types of boat hulls can be made by this process, as shown in Fig. 10.43c.

4. **Resin transfer molding.** This process is based on transfer molding (Section 10.10.7), whereby a resin, mixed with a catalyst, is forced by a piston-type, positive-displacement pump into a mold cavity filled with fiber reinforcement. The process is a viable alternative to hand lay-up, spray-up, and compression molding, particularly for low- or intermediate-volume production.

5. **Transfer/injection molding.** This is an automated operation that combines the processes of compression molding, injection molding, and transfer molding. Consequently, it takes advantage of each process and produces parts with enhanced properties.

10.11.2 Filament winding, pultrusion, and pulforming

Filament winding. *Filament winding* is an important process whereby the resin and fibers are combined at the time of curing. Symmetric parts, such as pipes and storage tanks, as well as axisymmetric parts are produced by this method. The reinforcing filament, tape, or roving is wrapped continuously around a rotating mandrel or form. The reinforcements are impregnated by passing through a polymer bath (Fig. 10.44a). The process can be modified by wrapping the mandrel with prepreg material.

The products made by filament winding are very strong because of their highly reinforced structure. The process also can be used for strengthening cylindrical or spherical pressure vessels (Fig. 10.44b) made of materials such as aluminum or titanium. The presence of a metal inner lining thus makes the part impermeable. Filament winding can be used directly over solid-rocket propellant forms. Seven-axis computer-controlled machines (see Chapter 14) have been developed for making asymmetric parts that automatically dispense several unidirectional prepregs. Typical asymmetric parts made include aircraft engine ducts, fuselages, propellers, blades, and struts.

Pultrusion. Parts with high length-to-cross-sectional area ratios and various constant profiles, such as rods, structural profiles, and tubing (similar to drawn metal products; Section 6.5), are made by the *pultrusion* process. Typical products made by this process include golf club shafts, drive shafts, and structural members such as ladders, walkways, and handrails. In this process, developed in the early 1950s, the continuous reinforcement (roving or fabric) is pulled through a thermosetting-polymer bath, and then through a long heated steel die (Fig. 10.45). The product is cured during its travel through the die and then cut into desired lengths. The most common material used in pultrusion is polyester with glass reinforcements.

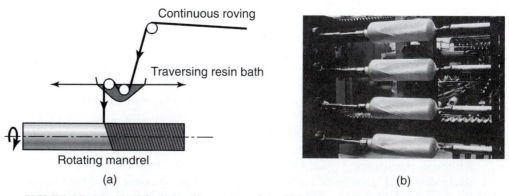

Continuous roving

Traversing resin bath

Rotating mandrel

(a)   (b)

FIGURE 10.44   (a) Schematic illustration of the filament-winding process and (b) fiberglass being wound over aluminum liners for slide-raft inflation vessels for the Boeing 767 aircraft. *Source:* Advanced Technical Products Group, Inc., Lincoln Composites.

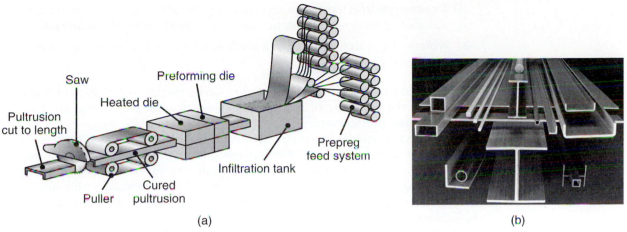

(a)

(b)

FIGURE 10.45   (a) Schematic illustration of the pultrusion process and (b) examples of parts made by pultrusion. *Source:* Courtesy of Strongwell Corporation.

Pulforming. Continuously reinforced products with a constant cross section are made by *pulforming*. After being pulled through the polymer bath, the composite is clamped between the two halves of a die and cured into a finished product. The dies recirculate and shape the products successively. Common examples of products made by this method include glass-fiber-reinforced hammer handles and curved automotive leaf springs.

10.11.3 Product quality

The major quality considerations for the processes described thus far principally involve internal voids and gaps between successive layers of material. Volatile gases that develop during processing must be allowed to escape from the lay-up through the vacuum bag to avoid porosity due to trapped gases within the lay-up. Micro-cracks also may develop due to improper curing or during transportation and handling. These defects can be detected using ultrasonic scanning and other techniques, described in Section 4.8.1.

10.12 | Rapid Prototyping and Rapid Tooling

Making a *prototype,* that is, a first full-scale model of a product (see Fig. 1.3), has traditionally involved flexible manufacturing processes such as machining operations described in Chapter 8, using a variety of tooling and machines, and usually taking several weeks or months. An important advance in manufacturing operations is *rapid prototyping,* also called **desktop manufacturing** or **free-form fabrication,** a process by which a solid physical model of a part is made directly from a three-dimensional CAD drawing. Developed in the 1980s, rapid prototyping entails several different consolidation techniques, such as resin curing, deposition, solidification, and sintering, as described in this section.

    The importance and economic impact of rapid prototyping can best be appreciated by noting the following:

1. The conceptual product design is viewed in its entirety and from different angles on a monitor through a three-dimensional CAD system, as described in Section 15.4.

2. A prototype from various nonmetallic and metallic materials is manufactured and studied thoroughly from functional, technical, and esthetic aspects.

3. Prototyping is accomplished in a much shorter time and at lower cost than by traditional methods.

It should be noted, however, that modern CNC machine tools (Chapters 8 and 15) also provide the capability of producing complex shapes quickly and are a viable option for rapid prototyping.

This section emphasizes use of **additive manufacturing**, that is, processes that *build* parts in layers. Prototypes also can be produced through **subtractive processes** (basically involving computer-controlled machining operations) or **virtual prototyping** (involving advanced graphics and software). Additive manufacturing inherently involves integrated computer-driven hardware and software. In order to visualize the methodology used, it is helpful to visualize constructing a loaf of bread by stacking and bonding individual slices on top of each other. All of the processes described in this section *build parts slice by slice* in a similar manner. As an example, Fig. 10.46 shows the computational steps involved in producing a part. The main difference between the various additive processes lies in the method of producing the individual slices, which are typically 0.1 to 0.5 mm (0.004 to 0.020 in.) thick but can be higher for some systems.

Part production in rapid prototyping typically takes a few hours for most parts. Consequently, these processes are not viable for large production runs, especially since the workpiece material in rapid prototyping is often an expensive polymer or a laminate. The characteristics of the rapid-prototyping processes are summarized in Table 10.7, and Table 10.8 lists typical material properties that are attainable by these methods.

10.12.1 Stereolithography

The *stereolithography* (STL) process is based on the principle of curing (hardening) a *liquid photopolymer* into a specific shape. Consider what happens when a laser beam is focused on and translated across the surface of a liquid *photopolymer*. The laser serves to cure the photopolymer by providing the energy necessary for polymerization. Laser energy is absorbed by the polymer; according to the Beer-Lambert law, the exposure decreases exponentially with depth according to the rule

$$E(z) = E_o^{-z/D_p}, \tag{10.27}$$

where E is the exposure in energy per area, E_o is the exposure at the resin surface ($z = 0$), and D_p is the penetration depth (in the z-direction) at the laser wavelength and is a property of the resin. At the cure depth, the polymer is exposed to enough energy per area for for it to gel, or

$$E_c = E_o^{-C_d/D_p}, \tag{10.28}$$

where E_c is the exposure necessary to transform the working liquid to a gel and C_d is the cure depth. Thus, solving for the cure depth,

$$C_d = D_p \ln\left(\frac{E_o}{E_c}\right), \tag{10.29}$$

which represents the thickness at which the resin has polymerized into a gel, although this state does not have particularly high strength. Recognizing this condition, the controlling software slightly overlaps the cured volumes, although additional curing under fluorescent lamps is often necessary as a finishing operation.

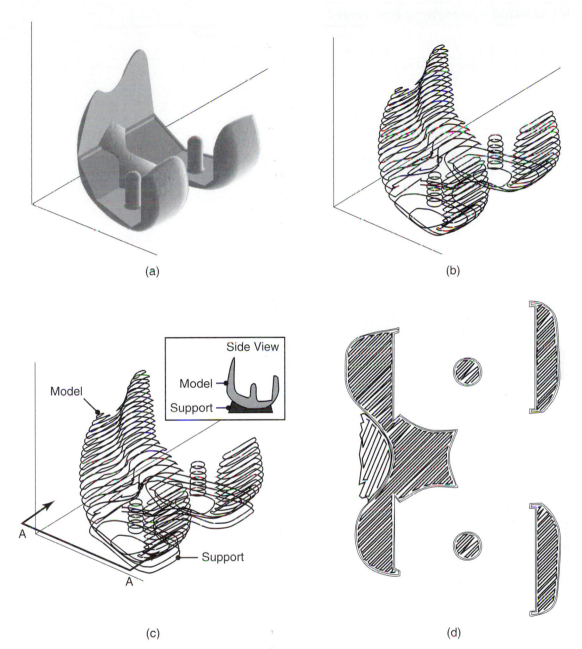

FIGURE 10.46   The computational steps involved in producing a stereolithography file:
(a) three-dimensional description of the part and (b) the part is divided into slices.
(Only 1 in 10 is shown.) (c) Support material is planned and (d) a set of tool directions is
determined for manufacturing each slice. Shown is the extruder path at section A-A from
(c), for a fused-deposition modeling operation.

The polymer at the periphery of the laser spot does not receive sufficient exposure to polymerize. It can be shown that the cured line width, L_w, at the surface is
given by

$$L_w = B\sqrt{\frac{C_d}{2D_p}}, \qquad (10.30)$$

where B is the diameter of the laser-beam spot.

TABLE 10.7

Characteristics of Rapid-Prototyping Processes

Supply phase	Process	Layer creation technique	Phase-change type	Materials
Liquid	Stereolithography	Liquid-layer curing	Photopolymerization	Photopolymers (acrylates, epoxies, colorable resins, and filled resins).
	Polyjet	Liquid-layer curing	Photopolymerization	Photopolymers
	Fused-deposition modeling	Extrusion of melted plastic	Solidification by cooling	Thermoplastics (ABS, polycarbonate, and polysulfone).
Powder	Three-dimensional printing	Binder-droplet deposition onto powder layer	No phase change	Polymer, ceramic, and metal powder with binder.
Selective laser sintering	Layer of powder	Laser-driven	Sintering or melting	Polymers, metals with binder, metals, ceramics, and sand with binder.
Electron-beam melting	Layer of powder	Electron-beam	Melting	Titanium and titanium alloys, cobalt chrome.

TABLE 10.8

Mechanical Properties of Selected Materials for Rapid Prototyping

Process	Material	Tensile strength (MPa)	Elastic modulus (GPa)	Elongation in 50 mm (%)	Notes
Stereo-lithography	Somos 7120a	63	2.59	2.3–4.1	Transparent amber; good general-purpose material for rapid prototyping.
	Somos 9120a	32	1.14–1.55	15–25	Transparent amber; good chemical resistance; good fatigue properties; used for producing patterns in rubber molding.
	WaterShed 11120	47.1–53.6	2.65–2.88	3.3–3.5	Optically clear with a slight green tinge; similar mechanical properties as ABS; used for rapid tooling.
	Prototool 20Lb	72–79	10.1–11.2	1.2–1.3	Opaque beige; higher strength polymer suitable for automotive components, housings, and injection molds.
Polyjet	FC 700	42.3	2.0	15–25	Transparent amber; good impact strength, good paint absorption and machinability.
	FC800	49.9–55.1	2.5–2.7	15–25	White, blue, or black; good humidity resistance; suitable for general-purpose applications.
	FC900	2.0–4.6	—	47	Gray or black; very flexible material, simulates the feel of rubber or silicone.
Fused-deposition modeling	Polycarbonate	52	2.0	3	White; high-strength polymer suitable for rapid prototyping and general use.
	ABS	22	1.63	6	Available in multiple colors, most commonly white; a strong and durable material suitable for general use.
	PC-ABS	34.8	1.83	4.3	Black; good combination of mechanical properties and heat resistance.
Selective laser sintering	Duraform PA	44	1.6	9	White; produces durable heat- and chemical-resistant parts; suitable for snap-fit assemblies and sandcasting or silicone tooling.
	Duraform GF	38.1	5.9	2	White; glass-filled form of Duraform PA, has increased stiffness and is suitable for higher temperature applications.
	SOMOS 201	17.3	14	130	Multiple colors available; mimics rubber mechanical properties.
	ST-100c	305	137	10	Bronze-infiltrated steel powder.
Electron-beam melting	Ti-6Al-4V	970-1030	120	12-16	Can be treated by HIP to obtain up to 600 MPa fatigue strength.

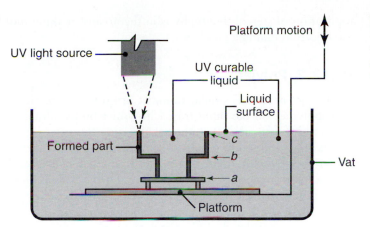

FIGURE 10.47   Schematic illustration of the stereolithography process. *Source:* Courtesy of 3D Systems.

Stereolithography (Fig. 10.47) equipment involves a vat filled with a photocurable liquid acrylate photopolymer and contains a mechanism whereby a platform can be lowered and raised vertically. The liquid is a mixture of acrylic monomers, oligomers (polymer intermediates), and a photoinitiator that initiates polymerization when exposed to laser light. When the platform is at its highest position, the layer of liquid above it is shallow. A *laser,* generating an ultraviolet beam, is then focused along a selected surface area of the photopolymer at surface *a* and moved in the *x-y*-direction.

The beam cures that portion of the photopolymer (say, a ring-shaped portion), producing a thin solid body. The platform is then lowered sufficiently to cover the cured polymer with another layer of liquid polymer, and the sequence is repeated. In Fig. 10.47, the process is repeated until level *b* is reached. Note that, thus far, the platform is now lowered by a vertical distance *ab*, and that we have a cylindrical part with a constant wall thickness.

At level *b*, the *x-y* movements of the beam are wider, so that we now have a flange-shaped piece that is being built over the previously formed part below it. After the desired thickness of the liquid has been cured, the process is repeated, producing another cylindrical section between levels *b* and *c*. Note that the surrounding liquid polymer is still fluid, because it has not been exposed to the ultraviolet beam, and that the part has been produced from the *bottom up* in individual "slices." The unused portion of the liquid polymer can be used again to make another part or prototype.

Note that the word *stereolithography,* as used to describe this process, comes from the fact that the movements are three-dimensional and that the process is similar to lithography (in which the image to be printed on a flat surface is ink receptive and the blank areas are ink repellent). After the operation is completed, the part is removed from the platform, blotted, and cleaned ultrasonically and with an alcohol bath. Finally, the part is exposed to UV radiation, for up to a few hours, to fully cure and harden the polymer part.

By controlling the movements of the beam and the platform through a servo-control system, a variety of parts can be formed by this process. Total cycle times range from a few hours to a day. Progress continues to be made toward improvements in (a) accuracy and dimensional stability of the prototypes produced, (b) less expensive liquid modeling materials, (c) CAD interfaces to transfer geometrical data to model-making systems, and (d) strength so that the prototypes made by this process can truly be considered prototypes and models in the traditional sense. Depending on capacity, the cost of machines for stereolithography ranges from $100,000 to $500,000, while the cost of the liquid polymer is on the order of $300 per gallon. Maximum part size is $0.5 \times 0.5 \times 0.6$ m ($20 \times 20 \times 24$ in.). One major application of stereolithography is in the area of making molds and dies for casting and injection molding (see below).

10.12.2 Polyjet

The polyjet process is similar to inkjet printing, where eight print heads deposit the photopolymer on the build tray. Ultraviolet bulbs, alongside the jets, immediately cure and harden each layer, thus eliminating the need for any postmodeling curing that exists in stereolithography. The result is a smooth surface of thin layers as small as 16 μm (0.0006 in.) that can be handled immediately after the process is completed. Two different materials are used for rapid prototyping: One material is used for the actual model, while a second, gel-like resin is used for support, such as those shown in Fig. 10.46. Each material is simultaneously jetted and cured, layer by layer. When the model is completed, the support material is removed with an aqueous solution. Build sizes are fairly large, with an envelope of up to $500 \times 400 \times 200$ mm (20 $\times$ 16 $\times$ 8 in.).

The polyjet process has capabilities similar to stereolithography and uses similar resins (Table 10.8). The main advantages are the ability to avoid the necessity for part cleanup, lengthy postprocess curing operations, and the much smaller layer thickness, thus allowing for better resolution.

10.12.3 Fused-deposition modeling

In the *fused-deposition modeling* (FDM) process (Fig. 10.48), a gantry-robot-controlled extruder head moves in two principal directions over a table. The table can be raised and lowered as needed. A thermoplastic filament is extruded through the small orifice of a heated die. The initial layer is placed on a foam foundation by

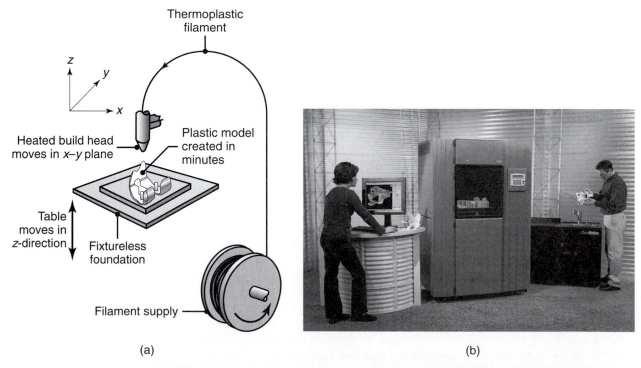

(a)   (b)

FIGURE 10.48   (a) Schematic illustration of the fused-deposition modeling process and (b) the FDM Vantage X rapid prototyping machine. *Source:* Courtesy of Stratasys, Inc.

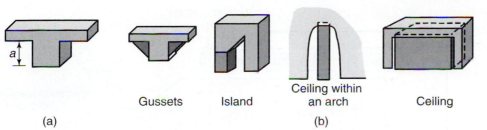

FIGURE 10.49 (a) A part with a protruding section that requires support material and (b) common support structures used in rapid-prototyping machines. *Source:* After P.F. Jacobs.

extruding the filament at a constant rate while the extruder head follows a predetermined path (Fig. 10.46d). When the first layer is completed, the table is lowered so that subsequent layers can be superposed.

Occasionally, complicated parts may be required, such as that shown in Fig. 10.49a. This is a difficult part to manufacture directly, because once the part has been constructed up to height a, the next slice would require the filament to be placed in a location where no material exists below to support it. The solution to this problem is to extrude a support material separately from the modeling material, so that a filament can be placed safely in the center of the part. The support material is extruded with a less dense spacing of filament on a layer, so that it is weaker than the model material and hence it can be broken off or dissolved after the part is completed.

In the FDM process, the extruded layer's thickness is determined by the extruder-die diameter, the diameters typically being in the range of 0.33 to 0.12 mm (0.013 to 0.005 in.). This thickness represents the best achievable dimensional tolerance in the vertical direction. In the x-y plane, however, dimensional accuracy can be as fine as 0.025 mm (0.001 in.), as long as a filament can be extruded into the feature. Close examination of an FDM-produced part will indicate that a *stepped* surface exists on oblique exterior planes. If the roughness of this surface is unacceptable, chemical-vapor polishing or a heated tool can be used to smoothen the surface; also, a coating can be applied, often in the form of a polishing wax. The overall dimensional tolerances may be compromised unless care is taken in applying these finishing operations.

10.12.4 Selective laser sintering

Selective laser sintering (SLS) is a process based on the sintering of polymer (or, less commonly, metallic) powders selectively into an individual object. (See also Chapter 11.) The basic components in this process are illustrated in Fig. 10.50. The bottom of the processing chamber is equipped with two cylinders: a *part-build cylinder*, which is lowered incrementally to where the sintered part is being formed, and a *powder-feed cylinder*, which is raised incrementally to supply powder to the part-build cylinder through a roller mechanism.

A thin layer of powder is first deposited in the part-build cylinder. A laser beam, guided by a process-control computer (using instructions generated by the 3D CAD program of the desired part), is then focused on that layer, tracing and melting (or, for metals, *sintering*; see Section 11.4) a particular cross section, which then rapidly resolidifies into a solid mass (after the laser beam is moved to another section). The powder in other areas remains loose, yet supports the solid portion. Another layer of powder is then deposited, and this cycle is repeated continuously until the entire three-dimensional part has been produced. The loose particles are then shaken off and recovered.

A variety of materials can be used in this process, including polymers (ABS, PVC, nylon, polyester, polystyrene, and epoxy), wax, and metals, as well as ceramics with appropriate binders. It is most common to use polymers because of the smaller, less expensive, and less complicated lasers required for sintering. With ceramics and

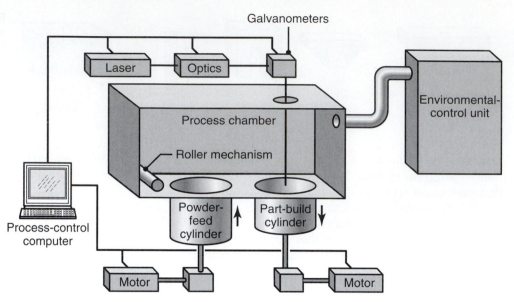

FIGURE 10.50 Schematic illustration of the selective-laser-sintering process. *Source:* After C. Deckard and P.F. McClure.

metals, it is common to sinter only a polymer binder that has been blended with the ceramic or metal powders; sintering is then completed in a furnace.

Electron-beam melting. A process similar to SLS and electron beam welding (see Section 12.5.1) uses the energy source associated with an electron-beam to melt titanium or cobalt chrome powder and make metal prototypes. The workpiece is produced in a vacuum, and the part build size is limited to around $200 \times 200 \times 180$ mm ($8 \times 8 \times 6.3$ in.). (See also Section 12.4.1.) *Electron-beam melting* (EBM) is up to 95% efficient from an energy standpoint (compared to 10–20% efficiency for laser sintering), so that the titanium powder is actually melted and fully-dense parts can be produced. A volume build rate of up to 60 cm^3/hr (3.7 in^3/hr) can be obtained, with individual layer thicknesses of 0.050–0.200 mm (0.002–0.008 in.). Hot isostatic pressing (Section 11.3.3) can be performed on parts to improve fatigue strength. Although mainly applied to titanium and cobalt chrome to date, the process is being developed for stainless steels, aluminum and copper alloys.

10.12.5 Three-Dimensional Printing

In the *three-dimensional printing* (3DP) process, a print head deposits an inorganic binder material onto a layer of nonmetallic or metallic powder, as shown in Fig. 10.51. A piston, supporting the powder bed, is lowered incrementally and, with each step, a layer is deposited and then fused by the binder.

Three-dimensional printing allows considerable flexibility in the materials and binders used. Commonly used powder materials are blends of polymers and fibers, foundry sand, and even metals. Furthermore, since multiple binder print heads can be incorporated into one machine, it is possible to produce full-color prototypes by having different color binders (Fig. 10.52). The effect is a three-dimensional analog to printing photographs using three ink colors on an inkjet printer.

The parts produced through the 3DP process are somewhat porous and therefore may lack strength. Three-dimensional printing of metal powders can be combined with sintering and metal infiltration (see Section 11.4) to produce fully dense parts using the sequence shown in Fig. 10.53. Here, the part is produced as previously, by directing binder onto powders to produce the part. However, the build sequence is then followed by sintering to burn off the binder and partially fuse the metal powders,

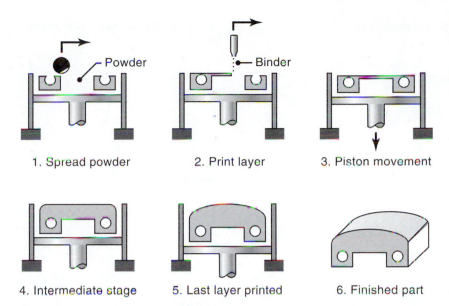

FIGURE 10.51 Schematic illustration of the three-dimensional-printing process. *Source:* After E. Sachs and M. Cima.

1. Spread powder

2. Print layer

3. Piston movement

4. Intermediate stage

5. Last layer printed

6. Finished part

just as in metal injection molding described in Section 11.3.4. Common metals used in 3DP are stainless steels, aluminum, and titanium. Infiltration materials typically are copper and bronze, which provide good heat transfer capabilities as well as wear resistance. This approach represents an efficient strategy for *rapid tooling* (see below).

10.12.6 Direct (rapid) manufacturing and rapid tooling

Parts produced by various rapid-prototyping operations are useful not only for design evaluation and troubleshooting, but also occasionally for direct application into products, or to aid in the direct manufacture of marketable products. Also, it is often desirable, for functional reasons, to use metallic parts, while the best developed and most available rapid-prototyping operations involve polymeric workpieces. Although prototypes can be produced from metal stock using the machining operations described in Chapter 8, there may be a cost advantage to using other manufacturing approaches. Rapid prototyping techniques are often incorporated into conventional processes to streamline them and make them more economically competitive.

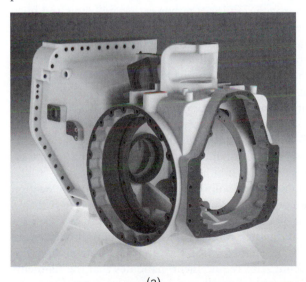

(a)

(b)

FIGURE 10.52 (a) and (b) Examples of parts produced through three-dimensional printing. Full color parts also are possible, and the colors can be blended throughout the volume. *Source:* Courtesy of ZCorp, Inc.

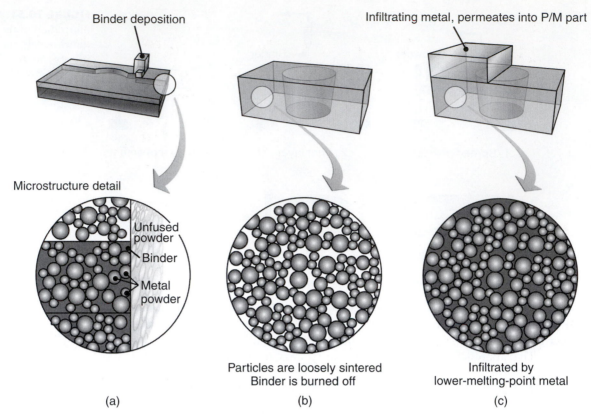

FIGURE 10.53   The three-dimensional printing process: (a) part build, (b) sintering, and (c) infiltration steps to produce metal parts. *Source:* Courtesy of the ProMetal Division of Ex One Corporation.

The simplest method to apply rapid prototyping operations to other manufacturing processes is in the direct production of patterns or molds (see Section 5.8). As an example, Fig. 10.54 shows an approach for investment casting. Here, the individual patterns are made in a rapid-prototyping operation (in this case, stereolithography) and then used as the patterns in assembling a tree for investment casting. It should be noted that this approach requires a polymer that will melt and burn from the ceramic mold completely; such polymers are available for all forms of polymer rapid-prototyping operations. Furthermore, the parts as drawn in CAD programs are usually software modified to account for shrinkage, and it is the modified part that is produced in the rapid-prototyping machinery.

As another example, 3DP can easily produce a ceramic-casting shell (Section 5.8.4), in which an aluminum-oxide or aluminum-silica powder is fused with a silica binder. The molds have to be postprocessed in two steps: curing at around 150°C (300°F) and then firing at 1000°–1500°C (1840°–2740°F). Such parts are suitable for shell-casting operations (see Section 5.8.4), and similar methods allow for the direct production of sand molds and injection molding dies.

Another common application of rapid tooling is injection molding (see Section 10.10.2), where the mold (or, more typically, a *mold insert*) is manufactured by rapid prototyping. Molds for slip casting of ceramics (see Section 11.9.1) can also be produced in this manner. To produce individual molds, rapid-prototyping processes are used directly, but the molds will be shaped with the desired permeability. For example, in fused-deposition modeling, this requirement mandates that the filaments be placed onto the individual slices with a small gap between adjacent filaments. These filaments are then positioned at right angles in adjacent layers.

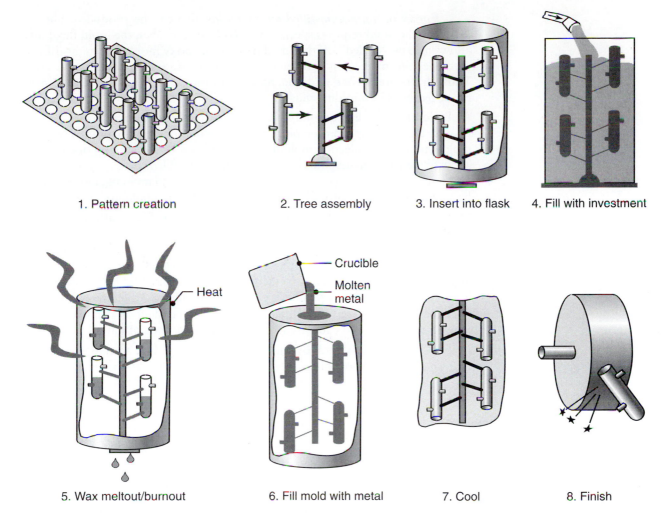

1. Pattern creation   2. Tree assembly   3. Insert into flask   4. Fill with investment

5. Wax meltout/burnout   6. Fill mold with metal   7. Cool   8. Finish

FIGURE 10.54   Manufacturing steps for investment casting that uses rapid-prototyped wax parts as patterns. This approach uses a flask for the investment, but a shell method can also be used. *Source*: 3D Systems, Inc.

The advantage of rapid tooling is the capability to produce a mold or a mold insert that can be used to manufacture components without the time lag (typically several months) traditionally required for the procurement of tooling. Furthermore, the design is simplified, because the designer need only analyze a CAD file of the desired part; software then produces the tool geometry and automatically compensates for shrinkage.

Other rapid tooling approaches based on rapid-prototyping technologies are described next.

1. **RTV (room-temperature vulcanizing) molding/urethane casting** can be performed by preparing a pattern of a part by any rapid prototyping operation. The pattern is coated with a parting agent and may or may not be modified to define mold parting lines. Liquid RTV rubber is poured over the pattern and cures (usually within a few hours) to produce mold halves. The mold is then used in injection-molding operations using liquid urethanes in injection molding or reaction-injection molding operations (see Section 10.10.2). One main limitation of this approach is mold life; curing of the polyurethane in the mold causes progressive damage and the mold may only be suitable for as few as 25 parts.

Epoxy or *aluminum-filled epoxy molds* also can be produced, but the design of the mold requires special care. With RTV rubber, the mold flexibility allows it to be "peeled" off the cured part. With epoxy molds, the high stiffness precludes this method of part removal, and mold design is more complicated. Thus, drafts are needed and undercuts and other design features that can be produced by RTV molding must be avoided.

2. **ACES (acetal clear epoxy solid) injection molding** (also known as *direct AIM*) refers to the use of rapid prototyping (usually stereolithography) to directly produce molds suitable for injection molding. The molds are shells with an open end to allow filling with a material such as epoxy, aluminum-filled epoxy, or low-melting-point metals. Depending on the polymer used in injection molding, mold life may be as few as 10 parts, although a few hundred parts per mold are possible.

3. In the **sprayed metal tooling** process, shown in Fig. 10.55, a pattern is created through rapid prototyping. A metal spray operation (see Section 4.5.1) then coats the pattern surface with a zinc-aluminum alloy. The metal coating is placed in a flask and potted with an epoxy or aluminum-filled epoxy material. In some applications, cooling lines can be incorporated into the mold before the epoxy is applied. The pattern is removed; two such mold halves then are suitable for injection molding operations. Mold life is highly dependent on the material and temperatures used and can vary from a few to thousands of parts.

4. In the **Keltool process,** an RTV mold is produced based on a rapid-prototyped pattern as above. The mold is then filled with a mixture of powdered A6 tool

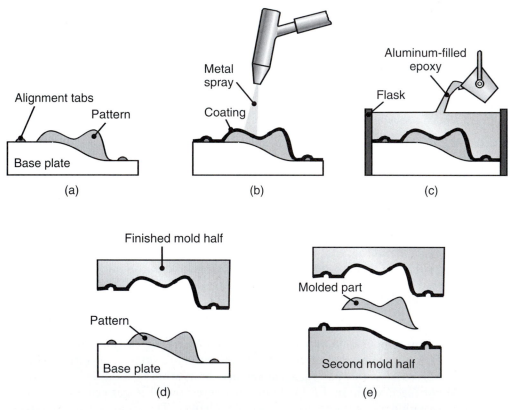

FIGURE 10.55   Production of tooling for injection molding by the sprayed-metal tooling process: (a) a pattern and base plate are prepared through a rapid-prototyping operation, (b) a zinc-aluminum alloy is sprayed onto the pattern (see Section 4.5.1), (c) the coated base plate and pattern assembly is placed in a flask and back-filled with aluminum-impregnated epoxy, (d) after curing, the base plate is removed from the finished mold, and (e) after a second mold half is prepared, a part is injection molded.

steel (Section 3.10.3), tungsten carbide, and polymer binder, and allowed to cure. The so-called *green* tool (see Section 11.3) is then fired to burn off the polymer and fuse the steel and tungsten-carbide powders. The tool is then infiltrated with copper in a furnace to produce the final mold. The mold can subsequently be machined or polished to attain superior surface finish and dimensional tolerances. Keltool molds are limited to around $150 \times 150 \times 150$ mm ($6 \times 6 \times 6$ in.), so typically a mold insert is produced, suitable for high-volume molding operations. Depending on the material and processing conditions, mold life can range from 100,000 to 10 million parts.

EXAMPLE 10.9   Rapid prototyping of an injection manifold

The Rover Group, a subsidiary of British Aerospace, is well known for four-wheel-drive vehicles and high-performance automobiles. An innovative new injection-manifold design (Fig. 10.56), based on a plenum chamber for efficient movement of air into the cylinders, was designed for a new Rover power unit. However, the chamber is very intricate and has large trapped volumes, and consequently, a prototype had to be constructed from one piece to provide precise flow characteristics.

Normally, Rover would have relied on traditional patternmaking, followed by complex core-making and casting operations. However, this operation is expensive and has a typical turnaround time of about 16 weeks. Using stereolithography, Rover produced a prototype multitrack injection manifold from a CAD file in only 39 hours, and at less than 10% of the cost of the conventional prototype. More importantly, due to the prototype's strength and accuracy, it could be attached to an engine test bed, and its volumetric efficiency (ability to allow the maximum air volume into the cylinder during one stroke under different conditions) could be directly measured. This procedure allowed optimization of the complex inlet tracts and plenum chamber, and allowed Rover to incorporate design iterations into the product before committing to the purchase of expensive tooling and machinery.

Source: Courtesy of 3D Systems, Inc. and The Rover Group.

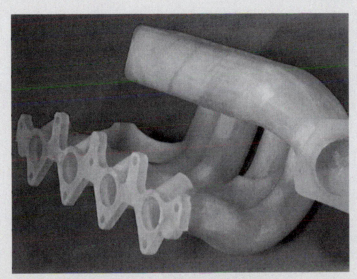

FIGURE 10.56   Rapid prototyped model of an injection-manifold design, produced through stereolithography.
Source: 3D Systems.

10.13 | Design Considerations

Design considerations for forming and shaping plastics are, to a large extent, similar to those for processing metals. Selection of an appropriate material and available processes from an extensive list requires consideration of (a) mechanical and physical properties, (b) service requirements, (c) possible long-range effects of processing on properties and behavior, such as dimensional stability and degradation, (d) manufacturability, (e) economics, and (f) ultimate disposal and recycling at the end of the product's life cycle. The following list summarizes the major considerations involved in design for plastics and for reinforced plastics.

1. Compared with metals, plastics have lower strength and stiffness, although the strength-to-weight and stiffness-to-weight ratios for reinforced plastics are higher than for many metals. Thus, section sizes should be selected accordingly, with a view to maintaining a sufficiently high section modulus (the ratio of moment of inertia to the distance from the neutral axis to the surface of the part) for the required stiffness. Improper part design or assembly can lead to warping and shrinking (Fig. 10.57a).

2. The overall part shape often determines the choice of the particular forming or molding process. Even after a particular process is selected, the design of the part and die or mold should be such that it will not cause problems concerning shape generation (Fig. 10.57b), dimensional control, and surface finish. As in casting metals and alloys, material flow in the mold cavities should be properly controlled. The effects of molecular orientation during processing should also be considered, especially in extrusion, thermoforming, and blow molding.

3. Polymers are capable of acquiring complex shapes in molding operations. In fact, one advantage of thermoplastics is that one molded part can replace what would otherwise be an assembly of several parts. They are available in a wide

FIGURE 10.57 Examples of design modifications to eliminate or minimize distortion of plastic parts: (a) suggested design changes to minimize distortion, (b) die design (exaggerated) for extrusion of square sections. Without this design modification, product cross sections would not have the desired shape because of the recovery of the material, known as die swell. (c) Design change in a rib to minimize pull-in caused by shrinkage during cooling and (d) stiffening of the bottom of thin plastic containers by doming, similar to the process used to make the bottoms of aluminum beverage cans and similar containers.
Source: (a) After F. Strasser.

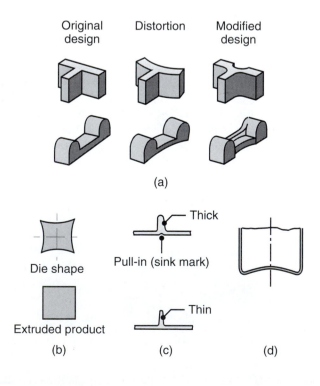

variety of material properties and colors but usually require fairly large production quantities to justify tooling expenditures (see Section 10.14).

4. Large variations in cross section (Fig. 10.57c) and abrupt changes in geometry should be avoided in order to achieve better product quality and increased mold life. Uniform wall thicknesses should be maintained whenever possible, and if changes are required, they should be made gradually. Reinforcing ribs are often used to increase stiffness; the ribs should be thinner than the section they reinforce, and they should not be higher than around three times the wall thickness. Sink marks (Fig. 10.57c) can be disguised with surface texture or grooves. Ribs should also incorporate a draft of 0.5°–1.5°.

5. Contraction in large cross sections can cause porosity in plastic parts. Conversely, because of a lack of stiffness, removing thin sections from molds may be difficult. The low elastic modulus of plastics further requires that shapes be properly selected for improved stiffness of the component (Fig. 10.57d), particularly when material saving is important. These considerations are similar to those in designing metal castings and forgings.

6. Holes in thermoplastic parts can be produced but represent a complication in mold design. Flashing will often develop at the edge of a hole; through holes are preferable to blind holes when using core pins because the pin can be supported at both ends.

7. Parts should be designed to facilitate ejection from molds so that side features, such as flanges or holes that would interfere with ejection, should be avoided. Such features can be produced with retractable side cores, although they complicate mold design.

8. Screw threads can be molded, usually using one of three basic approaches: (a) A core containing the thread geometry is used and is rotated after molding to remove the part; (b) the part is placed with its axis on the parting line. This approach leads to a small amount of flash on the threads but eliminates the need for a core. (c) For flexible polymers, a rounded thread can be molded and stripped from the mold, but the thread depth must be small and the thread angle must be large for this approach to be successful.

9. Lettering, numbering, and other surface features can be incorporated. Whether these features are raised or depressed depends on the mold-making approach. If the mold is machined, it is easier to cut letters into a mold than to remove the surrounding material; thus, letters should be raised in the part and recessed in the mold. On the other hand, if a mold is produced through room-temperature vulcanization (RTV) from a machined blank, it is easier to have raised letters on the mold and depressed letters on the part.

10.14 Economics of Processing Plastics

As in all processes, design and manufacturing decisions are ultimately based on performance and cost, including the costs of equipment, tooling, and production. The final selection of a process or processes also depends greatly on production volume. High equipment and tooling costs in plastics processing can be acceptable only if the production run is large, as is also the case in casting and forging. However, the use of rapid-prototyping operations makes some processes economical for limited production runs, as described above, but the tools and molds have limited life. Rapid prototyping is suitable for prototypes and even limited production runs but require expensive consumables and are therefore unsuitable for moderate to high production

TABLE 10.9

Comparative Costs and Production Volumes for Processing of Plastics

Process	Equipment capital cost	Production rate	Tooling cost	Typical production volume, number of parts						
				10	10^2	10^3	10^4	10^5	10^6	10^7
Machining	Med	Med	Low	←	→					
Compression molding	High	Med	High			←			→	
Transfer molding	High	Med	High			←			→	
Injection molding	High	High	High				←			→
Extrusion	Med	High	Low		*					
Rotational molding	Low	Low	Low	←		→				
Blow molding	Med	Med	Med				←			→
Thermoforming	Low	Low	Low	←		→				
Casting	Low	Very low	Low	←	→					
Forging	High	Low	Med	←	→					
Foam molding	High	Med	Med				←			→

*Continuous process.

Source: After R.L.E. Brown, *Design and Manufacture of Plastic Parts.* Copyright © 1980 by John Wiley & Sons, Inc. Reprinted by permission of John Wiley & Sons, Inc.

runs. A general guide to economical processing of plastics and composite materials is given in Table 10.9.

Several types of equipment are used in plastics forming and shaping processes; the most expensive is injection-molding machines, with costs being directly proportional to the clamping force. A machine with a 2000-kN (225-ton) clamping force costs about $100,000, and one with a 20,000-kN (2250-ton) clamping force costs about $450,000. (See also Table 16.8.) For composite materials, equipment and tooling costs for most molding operations are generally high, and production rates and economic production quantities vary widely.

The optimum number of cavities in the die for making the product in one cycle is an important consideration, as in die casting (Section 5.10.3). For small parts, a number of cavities can be made in a mold, with runners to each cavity. If the part is large, then only one cavity may be accommodated. As the number of cavities increases, so does the cost of the die. Larger dies may be considered for larger numbers of cavities, thus increasing die cost even further. On the other hand, more parts will be produced per machine cycle on a larger die, thereby increasing the production rate. Hence, a detailed analysis has to be made to determine the optimum number of cavities, die size, and machine capacity.

CASE STUDY | Invisalign Orthodontic Aligners

Orthodontic braces have been available to straighten teeth for over 50 years. These braces involve metal, ceramic, or plastic brackets that are adhesively bonded to teeth, with fixtures for attachment to a wire which then forces compliance on the teeth and straightens them to the desired shape within a few years. Conventional orthodontic braces are a well-known and successful technique for long-term dental health. There are, however, several drawbacks to conventional braces: (a) They are aesthetically unappealing, (b) the sharp wires and brackets can be painful, (c) they trap food, leading to premature tooth decay, (d) brushing and flossing of teeth are much more difficult and less effective with braces in place, and (e) certain foods must be avoided because they can damage the braces.

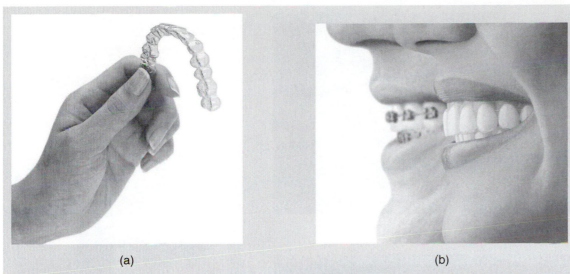

(a)   (b)

FIGURE 10.58   (a) An aligner for orthodontic use, manufactured using a combination of rapid tooling and thermoforming and (b) comparison of conventional orthodontic braces to the use of transparent aligners. *Source:* Courtesy of Align Technologies, Inc.

One solution is the Invisalign product, made by Align Technologies. It consists of a series of aligners, each of which is worn by the person for approximately two weeks. Each aligner (see Fig. 10.58) consists of a precise geometry which incrementally moves the teeth to their desired positions. Because they are inserts that can be removed for eating, brushing, and flossing, most of the drawbacks of conventional braces are eliminated. Furthermore, since they are produced from a transparent plastic, they do not seriously affect the person's appearance.

The Invisalign product uses a combination of advanced technologies in a production process shown in Fig. 10.59. The treatment begins with an orthodontist making a polymer impression of the patient's teeth (Fig. 10.59a). These impressions are used to create a three-dimensional CAD representation of the teeth, as shown in Fig. 10.59b. Proprietary computer-aided design software then assists in the development of a treatment strategy for moving the teeth in an optimal manner.

Once a treatment plan has been developed, the computer-based information is used to produce the aligners. This is done through a novel application of stereolithography. Although a number of materials are available for stereolithography, they have a characteristic yellow-brown shade to them, and are therefore unsuitable for direct application as an orthodontic product. Instead, the stereolithography machine produces patterns of the desired incremental positions of the teeth (Fig. 10.59c). A sheet of clear polymer is then thermoformed (see Section 10.10.5) over these patterns to produce the aligners. These are sent to the treating orthodontist, and new aligners are given to the patient as needed, usually about every two weeks.

The Invisalign product has proven to be very popular with patients who wish to promote dental health and to preserve their teeth long into their lives. The use of stereolithography to produce accurate thermoforming molds quickly and inexpensively allows this orthodontic treatment to be economically viable.

Source: Courtesy Align Technologies, Inc.

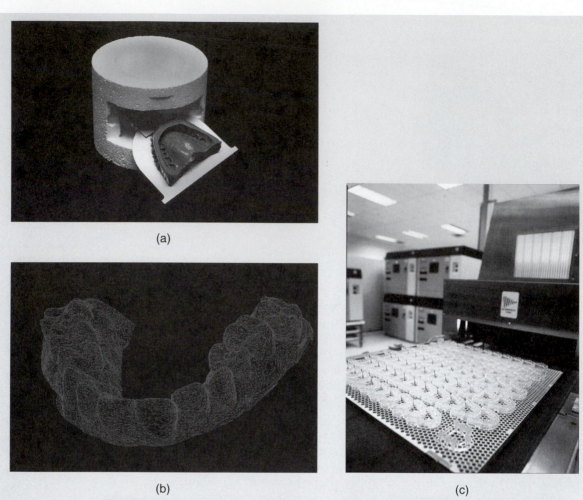

(a)

(b)   (c)

FIGURE 10.59 Manufacturing sequence for Invisalign orthodontic aligners: (a) creation of a polymer impression of the patient's teeth, (b) computer modeling to produce CAD representations of desired tooth profiles, and (c) production of incremental models of desired tooth movement. An aligner is produced by thermoforming a transparent plastic sheet against this model. *Source:* Courtesy of Align Technologies, Inc.

SUMMARY

- Polymers are an important class of materials because they possess a very wide range of mechanical, physical, chemical, and optical properties. Compared with metals, plastics are generally characterized by lower density, strength, elastic modulus, and thermal and electrical conductivity, and higher coefficient of thermal expansion. (Section 10.1)

- Plastics are composed of polymer molecules and various additives. The smallest repetitive unit in a polymer chain is called a mer. Monomers are linked by polymerization processes to form larger molecules. The properties of the polymer depend on the molecular structure, the degree of crystallinity, and additives. The glass-transition temperature indicates the brittle and ductile regions of polymers. (Section 10.2)

- Thermoplastics become soft and easy to form at elevated temperatures. Their mechanical behavior can be characterized by various spring and dashpot models and includes phenomena such as creep and stress relaxation, crazing, and water absorption. (Section 10.3)

- Thermosets are obtained by cross-linking polymer chains; they do not become soft to any significant extent with increasing temperature. (Section 10.4)
- Thermoplastics and thermosets have a very wide range of consumer and industrial applications with a variety of characteristics. (Sections 10.5, 10.6)
- Among important aspects of polymers are biodegradable plastics; several formulations have been developed. (Section 10.7)
- Elastomers have the characteristic ability to undergo large elastic deformations and return to their original shape when unloaded. Consequently, they have important applications as tires, seals, footwear, hoses, belts, and shock absorbers. (Section 10.8)
- Reinforced plastics, also called composite materials, are an important class of engineered materials that have superior mechanical properties and are lightweight. The reinforcing fibers are usually glass, graphite, aramids, and boron, and epoxies commonly serve as a matrix material. The properties of reinforced plastics primarily depend on their composition and the orientation of the fibers. (Section 10.9)
- Plastics can be formed and shaped by a variety of processes, such as extrusion, molding, casting, and thermoforming. Thermosets are generally molded or cast. (Section 10.10)
- Reinforced plastics are processed into structural components, using liquid plastics, prepregs, and bulk- and sheet-molding compounds. Fabricating techniques include various molding methods, filament winding, and pultrusion. (Section 10.11)
- Rapid prototyping has become an increasingly important technology because of its inherent flexibility, low cost, and the much shorter times required for making prototypes. Typical techniques include stereolithography, cyberjet, fused-deposition modeling, selective laser sintering, and three-dimensional printing. Rapid production of tooling is a further development of these processes. (Section 10.12)
- The design of plastic parts includes considerations of their relatively low strength and stiffness, high thermal expansion, and generally low resistance to temperature. (Section 10.13)
- Because of the variety of low-cost materials and processing techniques available, the economics of processing plastics and reinforced plastics is an important consideration, particularly when compared with metal components, and includes die and mold costs, cycle times, and production volume as important parameters. (Section 10.14)

SUMMARY OF EQUATIONS

Linearly elastic behavior:

in tension: $\sigma = E\epsilon$

in shear: $\tau = G\gamma$

Viscous behavior: $\tau = \eta\left(\dfrac{dv}{dy}\right) = \eta\dot{\gamma}$

Strain-rate sensitivity: $\sigma = C\dot{\epsilon}^m$

Viscoelastic modulus: $E_r = \dfrac{\sigma}{(e_e + e_v)}$

Viscosity:

$$\eta = \eta_o e^{E/kT}$$

$$\log \eta = 12 - \frac{17.5\ \Delta T}{52 + \Delta T}$$

Stress relaxation: $\tau = \tau_o e^{-t/\lambda}$

Relaxation time: $\lambda = \dfrac{\eta}{G}$

Strength of composite: $\sigma_c = x\sigma_f + (1 - x)\sigma_m$

Elastic modulus of composite: $E_c = xE_f + (1 - x)E_m$

Extruder characteristic: $Q = \dfrac{\pi^2 H D^2 N \sin\theta \cos\theta}{2} - \dfrac{p\pi D H^3 \sin^2\theta}{12\eta l}$

Die characteristic: $Q_{\text{die}} = Kp$

for round die: $K = \dfrac{\pi D_d^4}{128\eta l_d}$

Cure depth in stereolithography: $C_d = D_p \ln\left(\dfrac{E_o}{E_c}\right)$

Line width in stereolithography: $L_w = B\sqrt{\dfrac{C_d}{2D_p}}$

BIBLIOGRAPHY

Agarwal, B.D., Broutman, L.J., and Chandrashekhara, K., *Analysis and Performance of Fiber Composites*, 3rd ed., Wiley, 2006.

ASM Handbook, Vol. 21: *Composites*, ASM International, 2001.

Baird, D.G., and Collias, D.I., *Polymer Processing: Principles and Design*, Wiley, 1998.

Beaman, J.J., Barlow, J.W., Bourell, D.L., and Crawford, R., *Solid Freeform Fabrication*, Kluwer, 1997.

Bertholet, J.-M., *Composite Materials: Mechanical Behavior and Structural Analysis*, Springer, 1999.

Bhowmick, A.K., and Stephens, H.L., *Handbook of Elastomers*, 2nd ed., CRC, 2000.

Buckley, C.P., Bucknall, C.B., and McCrum, N.G., *Principles of Polymer Engineering*, 2nd ed., Oxford, 1997.

Campbell, F. (ed.), *Manufacturing Processes for Advanced Composites*, Elsevier, 2003.

Campbell, P., *Plastic Component Design*, Industrial Press, 1996.

Chanda, M., and Roy, S.K., *Plastics Technology Handbook*, 3rd ed., Dekker, 1998.

Chawla, K.K., *Composite Materials: Science and Engineering*, 2nd ed., Springer, 1998.

Chua, C.K., and Leong, L.K.F., *Rapid Prototyping: Principles and Applications in Manufacturing*, World Scientific Co., 2000.

Cooper, K.G., *Rapid Prototyping Technology*, Dekker, 2001.

Daniel, I.M., and Ishai, O., *Engineering Mechanics of Composite Materials*, 2nd ed., Oxford, 2005.

Dimov, S.S., and Pham, D.T., *Rapid Manufacturing: The Technologies and Applications of Rapid Prototyping and Rapid Tooling*, Springer, 2001.

Erhard, G., *Designing with Plastics*, Hanser Gardner, 2006.

Gastrow, H., *Injection Molds: 130 Proven Designs*, Hanser Gardner, 2002.

Griskey, R.G., *Polymer Process Engineering*, Chapman & Hall, 1995.

Gutowski, T.G., *Advanced Composites Manufacturing*, Wiley, 1997.

Harper, C.A., *Handbook of Plastics, Elastomers, and Composites*, 3rd ed., McGraw-Hill, 1996.

————, *Handbook of Plastic Processes*, Wiley-Interscience, 2006.

Hilton, P.D., and Jacobs, P.F., *Rapid Tooling: Technologies and Industrial Applications*, Marcel Dekker, 2000.

Johnson, P.S., *Rubber Processing: An Introduction*, Hanser Gardner, 2001.

Lu, L., Fuh, J.Y.H., and Wong, Y.S., *Laser-Induced Materials and Processes for Rapid Prototyping*, Kluwer, 2001.

MacDermott, C.P., and Shenoy, A.V., *Selecting Thermoplastics for Engineering Applications*, 2nd ed., Dekker, 1997.

Mallick, P.K. (ed.), *Composites Engineering Handbook*, Dekker, 1997.

Malloy, R.A., *Plastic Part Design for Injection Molding: An Introduction*, Hanser Gardner, 1994.

Mazumdar, S.K., *Composites Manufacturing: Materials, Products and Process Engineering*, CRC Press, 2001.

Miller, E., *Introduction to Plastics and Composites: Mechanical Properties and Engineering Applications*, Dekker, 1995.

Nielsen, L.E., and Landel, R.F., *Mechanical Properties of Polymers and Composites*, 2nd ed., Dekker, 1994.

Potter, K., *Introduction to Composite Products: Design, Development and Manufacture*, Chapman & Hall, 1997.

———, *Resin Transfer Molding*, Chapman & Hall, 1997.

Rauwendaal, C., *Polymer Extrusion*, 4th ed., Hanser Gardner, 2001.

Rosato, D.V., *Plastics Processing Data Handbook*, 2nd ed., Chapman & Hall, 1997.

———, and Rosato, M.G., *Injection Molding Handbook*, 3rd ed., Kluwer Academic Publishers, 2000.

———, and Rosato, M.G., *Plastics Design Handbook*, Kluwer Academic Publishers, 2001.

Rosen, S.R., *Thermoforming: Improving Process Performance*, Society of Manufacturing Engineers, 2002.

Rudin, A., *Elements of Polymer Science and Engineering*, 2nd ed., Academic Press, 1999.

Shastri, R., *Plastics Product Design*, Dekker, 1996.

Shenoy, A, *Thermoplastic Melt Rheology and Processing*, CRC Press, 1996.

Skotheim, T.A., *Handbook of Conducting Polymers*, 2 vols., Dekker, 1986.

Starr, T.F., *Pultrusion for Engineers*, CRC Press, 2000.

———, *Plastics: Materials and Processing*, 2nd ed., Prentice Hall, 1999.

Tadmor, Z., and Goqos, C., *Principles of Polymer Processing*, 2nd ed., Wiley, 2006.

Tool and Manufacturing Engineers Handbook, 4th ed., Vol. 8: *Plastic Part Manufacturing*, Society of Manufacturing Engineers, 1996.

Vollrath, L., and Haldenwanger, H.G., *Plastics in Automotive Engineering: Materials, Components, Systems*, Hanser-Gardner, 1994.

Zachariades, A.E., and Porter, R.S., *High-Modulus Polymers: Approaches to Design and Development*, Dekker, 1995.

QUESTIONS

10.1 Summarize the most important mechanical and physical properties of plastics in engineering applications.

10.2 What are the major differences between the properties of plastics and of metals?

10.3 What properties are influenced by the degree of polymerization? Explain.

10.4 Give applications for which flammability of plastics would be a major concern in their various applications.

10.5 What properties do elastomers have that thermoplastics, in general, do not have?

10.6 Is it possible for a material to have a hysteresis behavior that is the opposite of that shown in Fig. 10.14, whereby the arrows are counterclockwise? Explain.

10.7 Observe the behavior of the tension-test specimen shown in Fig. 10.13 and state whether the material has a high or low m value. (See Section 2.2.7.) Explain.

10.8 Why would we want to synthesize a polymer with a high degree of crystallinity? Explain.

10.9 Add more items to the applications column in Table 10.3.

10.10 Discuss the significance of the glass-transition temperature, T_g, in engineering applications of polymers.

10.11 Why does cross-linking improve the strength of polymers?

10.12 Describe the methods by which optical properties of polymers can be altered.

10.13 Explain the reasons that elastomers were developed. Based on the contents of this chapter, are there any substitutes for elastomers? Explain.

10.14 Give several examples of plastic products or components for which creep and stress relaxation would be important considerations.

10.15 Describe your opinions regarding recycling of plastics versus developing plastics that are biodegradable.

10.16 Explain how you would go about determining the hardness of the plastics described in this chapter. Would there be any significant difficulties involved? Explain.

10.17 Distinguish between composites and alloys. Give several examples.

10.18 Describe the functions of the matrix and the reinforcing fibers in reinforced plastics. What fundamental differences are there in the characteristics of the two materials?

10.19   What products have you personally seen that are made of reinforced plastics? How can you tell that they are reinforced?

10.20   Referring to earlier chapters, identify metals and alloys that have strengths comparable to those of reinforced plastics.

10.21   Compare the relative advantages and limitations of (1) metal-matrix composites, (2) reinforced plastics, and (3) ceramic-matrix composites.

10.22   This chapter has described the numerous advantages of composite materials. What limitations or disadvantages do these materials have? What suggestions would you make to overcome these limitations?

10.23   A hybrid composite is defined as a composite material containing two or more different types of reinforcing fibers. What advantages would such a composite have over other composites?

10.24   Why are fibers capable of supporting a major portion of the load in composite materials? Explain.

10.25   Assume that you are manufacturing a product in which all the gears are made of metal. A salesperson visits you and asks you to consider replacing some of the metal gears with plastic ones. Prepare a list of the questions that you would raise before making such a decision. Explain.

10.26   Review the three curves in Fig. 10.8 and describe some applications for each type of behavior. Explain your choices.

10.27   Repeat Question 10.26 for the curves given in Fig. 10.10.

10.28   Do you think that honeycomb structures could be used in any of the components of passenger cars? If so, which components? Explain.

10.29   Other than those described in this chapter, what materials can you think of that can be regarded as composite materials? Explain.

10.30   What applications for composite materials can you think of in which high thermal conductivity would be desirable? Explain.

10.31   Conduct a survey of a variety of sports equipment and identify the components that are made of composite materials. Explain the reasons for and advantages of using composites for these specific applications.

10.32   Several material combinations and structures have been described in this chapter. In relative terms, identify those that would be suitable for applications involving one of the following: (a) very low temperatures, (b) very high temperatures, (c) vibrations, and (d) high humidity.

10.33   Explain how you would go about determining the hardness of the reinforced plastics and composite materials described in this chapter. What type of tests would you use? Are hardness measurements for these types of materials meaningful? Does the size of the indentation make a difference in your answer? Explain.

10.34   Describe the advantages of applying traditional metalworking techniques covered in the preceding chapters to the forming and shaping of plastics.

10.35   Describe the advantages of cold forming of plastics over other methods of processing.

10.36   Explain the reasons that some forming and shaping processes are more suitable for certain plastics than for others.

10.37   Would you use thermosetting plastics in injection molding? Explain.

10.38   By inspecting plastic containers, such as for baby powder, you can see that the lettering on them is raised and not sunk. Offer an explanation as to why they are molded in that way.

10.39   Give examples of several parts that are suitable for insert molding. How would you manufacture these parts if insert molding were not available?

10.40   What manufacturing considerations are involved in making a metal beverage container versus a plastic one? Explain.

10.41   Inspect several electrical components, such as light switches, outlets, and circuit breakers, and describe the process or processes used in making them.

10.42   Inspect several similar products that are made of metals and plastics, such as a metal bucket and a plastic bucket of similar shape and size. Comment, for example, on their respective thicknesses, and explain the reasons for their differences, if any.

10.43   Make a list of processing methods used for reinforced plastics. Identify which of the following fiber orientation and arrangement capabilities each has: (1) uniaxial, (2) cross-ply, (3) in-plane random, and (4) three-dimensional random.

10.44   As you may have observed, some plastic products have lids with integral hinges; that is, no other material or part is used at the junction of the two parts. Identify such products and describe a method for making them.

10.45   Explain why some shaping operations, such as blow molding and film-bag making, are done vertically and why buildings that house equipment for these operations have ceilings 10 to 15 m (35 to 50 ft) high.

10.46   Consider the case of a coffee mug being produced by rapid prototyping. Describe how the top of the handle can be manufactured, since there is no material directly beneath the arch of the handle.

10.47 Make a list of the advantages and disadvantages of each of the rapid-prototyping operations.

10.48 Explain why finishing operations are generally needed for rapid-prototyping operations. If you are making a prototype of a toy car, list the finishing operations you would want to perform.

10.49 A continuing topic of research involves producing parts from rapid-prototyping operations and then using them in experimental stress analysis in order to infer the strength of the final parts produced by conventional manufacturing operations. List the concerns that you may have with this approach and outline means of addressing these concerns.

10.50 Because of relief of residual stresses during curing, long unsupported overhangs in parts made by stereolithography will tend to curl. Suggest methods of controlling or eliminating this problem.

10.51 One of the major advantages of stereolithography and cyberjet is that semi- and fully-transparent polymers can be used, so that internal details of parts can readily be discerned. List parts or products for which this feature is valuable.

10.52 Based on the processes used to make fibers, as described in this chapter, explain how you would produce carbon foam. How would you produce metal foam?

10.53 Die swell in extrusion is radially uniform for circular cross sections but is not uniform for other cross sections. Recognizing this fact, make a qualitative sketch of a die profile that will produce (a) square and (b) triangular cross sections of extruded polymer, respectively.

10.54 What are the advantages of using whiskers as a reinforcing material? Are there any limitations?

10.55 By incorporating small amounts of blowing agent, it is possible to produce polymer fibers with gas cores. List some applications for such fibers.

10.56 With injection-molding operations, it is common practice to remove the part from its runner and then to place the runner into a shredder and recycle into pellets. List the concerns you would have in using such recycled pellets as opposed to so-called virgin pellets.

10.57 What characteristics make polymers attractive for applications such as gears? What characteristics would be drawbacks for such applications?

10.58 Can polymers be made to conduct electricity? Explain, giving several examples.

10.59 Why is there so much variation in the stiffness of various polymers? What is its engineering significance? Explain, with examples.

10.60 Explain why thermoplastics are easier to recycle than thermosets.

10.61 Describe the principle of shrink-wrap.

10.62 List the characteristics required of a polymer for the following applications: (a) a total hip replacement insert, (b) a golf ball, (c) an automotive dashboard, (d) clothing, and (e) a child's doll.

10.63 How can you tell whether a part is made of a thermoplastic or a thermoset? Explain.

10.64 Describe the features of an extruder screw and comment on their specific functions.

10.65 An injection-molded nylon gear is found to contain small pores. It is recommended that the material be dried before molding it. Explain why drying will solve this problem.

10.66 What determines the cycle time for (a) injection molding, (b) thermoforming, and (c) compression molding? Explain.

10.67 Does the pull-in defect (sink marks) shown in Fig. 10.57 also occur in metal-forming and casting processes? Explain.

10.68 List the differences between the barrel section of an extruder and that on an injection-molding machine.

10.69 Identify processes that are suitable for making small production runs of plastic parts, such as quantities of 100 or fewer. Explain.

10.70 Review the Case Study at the end of this chapter and explain why aligners cannot be made directly by rapid prototyping operations.

10.71 Explain why rapid prototyping approaches are not suitable for large production runs.

10.72 List and explain methods for quickly manufacturing tooling for injection molding.

10.73 Careful analysis of a rapid-prototyped part indicates that it is made up of layers, with a white filament outline visible on each layer. Is the material a thermoset or a thermoplastic? Explain.

10.74 List the advantages of using a room-temperature vulcanized (RTV) rubber mold in injection molding.

10.75 What are the similarities and differences between stereolithography and cyberjet?

10.76 Explain how color can be incorporated into rapid-prototyped components.

PROBLEMS

10.77 Calculate the areas under the stress–strain curve (toughness) for the material in Fig. 10.9. Plot them as a function of temperature and describe your observations.

10.78 Note in Fig. 10.9 that, as expected, the elastic modulus of the polymer decreases as temperature increases. Using the stress–strain curves given in the figure, make a plot of the modulus of elasticity versus temperature.

10.79 Calculate the percentage increase in the mechanical properties of reinforced nylon from the data given in Fig. 10.19.

10.80 A rectangular cantilever beam 75 mm high, 25 mm wide, and 1 m long is subjected to a concentrated force of 100 N at its end. Select two different unreinforced and reinforced materials from Table 10.1 and calculate the maximum deflection of the beam. Then select aluminum and steel, and for the same beam dimensions, calculate the maximum deflection. Compare and comment on the results.

10.81 Sections 10.5 and 10.6 described several plastics and their applications. Rearrange this information by making a table of products and the type of plastics that can be used to make the products.

10.82 Determine the dimensions of a tubular steel drive shaft for a typical automobile. If you now replace this shaft with shafts made of unreinforced and reinforced plastic, respectively, what should be the shaft's new dimensions to transmit the same torque for each case? Choose the materials from Table 10.1 and assume a Poisson's ratio of 0.4.

10.83 Calculate the average increase in the properties of the plastics listed in Table 10.1 as a result of their reinforcement, and describe your observations.

10.84 In Example 10.4, what would be the percentage of the load supported by the fibers if their strength is 1250 MPa and the matrix strength is 240 MPa? What if the strength is unaffected but the elastic modulus of the fiber is 600 GPa while that of the matrix is 50 GPa?

10.85 Estimate the die clamping force required for injection molding 10 identical 1.5-in.-diameter disks in one die. Include the runners of appropriate length and diameter.

10.86 A 2-liter plastic beverage bottle is made from a parison which has the same diameter as the threaded neck of the bottle and has a length of 5 in. Assuming uniform deformation during blow molding, estimate the wall thickness of the tubular section.

10.87 Estimate the consistency index and power-law index for the polymers in Fig. 10.12.

10.88 An extruder has a barrel diameter of 100 mm. The screw rotates at 100 rpm, has a channel depth of 6 mm, and a flight angle of 17.5°. What is the highest flow rate of polypropylene that can be achieved?

10.89 The extruder in Problem 10.88 has a pumping section that is 2.5 m long and is used to extrude round polyethylene solid rod. The die has a land of 1 mm and a diameter of 5 mm. If the polyethylene is at a mean temperature of 250°C, what is the flow rate through the die? What is the flow rate for a die diameter of 10 mm?

10.90 An extruder has a barrel diameter of 4 in., a channel depth of 0.25 in., a flight angle of 18°, and a pumping zone that is 6 ft long. It is used to pump a plastic with a viscosity of 100×10^{-4} lb-s/in². If the die characteristic is experimentally determined as $Q_x = (0.00210 \text{ in}^5/\text{lb-s})p$, what should be screw speed be to achieve a flow rate of 7 in³/s from the extruder?

10.91 What flight angle should be used on a screw so that a flight translates a distance equal to the barrel diameter with each revolution?

10.92 For a laser providing 10 kJ of energy to a spot with diameter of 0.25 mm, determine the cure depth and the cured line width in stereolithography. Let $E_c = 6.36 \times 10^{10}$ J/m² and $D_p = 100 \ \mu m$.

10.93 For the stereolithography system described in Problem 10.92, estimate the time required to cure a layer defined by a 40-mm circle if adjacent lines overlap each other by 10% and the power available is 10 MW.

10.94 The extruder head in a fused-deposition-modeling setup has a diameter of 1 mm and produces layers that are 0.25 mm thick. If the velocities of the extruder head and polymer extrudate are both 50 mm/s, estimate the production time for generating a 50-mm solid cube. Assume that there is a 15-second delay between layers as the extruder head is moved over a wire brush (for cleaning).

10.95 Using the data given in Problem 10.94 and assuming that the porosity of the support material is 50%, calculate the production rate for making a 100-mm high cup with an outside diameter of 88 mm and wall thickness of 6 mm. Consider both the case with the closed end (a) down and (b) up.

10.96 What would the answer to Example 10.5 be if the nylon has a power law viscosity with $n = 0.5$? What if $n = 0.2$?

10.97 Referring to Fig. 10.7, plot the relaxation curves (i.e., the stress as a function of time) if a unit strain is applied at time $t = t_o$.

10.98 Derive a general expression for the coefficient of thermal expansion in the fiber direction for a continuous fiber-reinforced composite.

10.99 Estimate the number of molecules in a typical automobile tire. Estimate the number of atoms.

10.100 Calculate the elastic modulus and percentage of load supported by fibers in a composite with an epoxy matrix ($E = 100$ GPa), with 20% fibers made of (a) high-modulus carbon and (b) Kevlar 29.

10.101 Calculate the stress in the fibers and in the matrix for Problem 10.100. Assume that the cross-sectional area is 50 mm^2 and $F_c = 2000$ N.

10.102 Consider a composite consisting of reinforcing fibers with $E_f = 300$ GPa. If the allowable fiber stress is 200 MPa and the matrix strength is 50 MPa, what should be the matrix stiffness so that the fibers and matrix fail simultaneously?

10.103 Assume that you are asked to give a quiz to students on the contents of this chapter. Prepare five quantitative problems and five qualitative questions, and supply the answers.

DESIGN

10.104 Conduct a survey of the recent technical literature and present data indicating the effects of fiber length on such mechanical properties as the strength, the elastic modulus, and impact energy of reinforced plastics.

10.105 Discuss the design considerations involved in replacing a metal beverage container with one made of plastic.

10.106 Using specific examples, discuss the design issues involved in various products made of plastics versus reinforced plastics.

10.107 Make a list of products, parts, or components that are not currently made of plastics, and offer reasons why they are not.

10.108 In order to use a steel or aluminum container to hold an acidic material, such as tomato juice or sauce, the inside of the container is typically coated with a polymeric barrier. Describe the methods of producing such a container. (See also Chapter 7.)

10.109 Using the information given in this chapter, develop special designs and shapes for possible new applications of composite materials.

10.110 Would a composite material with a strong and stiff matrix and soft and flexible reinforcement have any practical uses? Explain.

10.111 Make a list of products for which the use of composite materials could be advantageous because of their anisotropic properties.

10.112 Name several product designs in which both specific strength and specific stiffness are important.

10.113 Describe designs and applications in which strength in the thickness direction of a composite is important.

10.114 Design and describe a test method to determine the mechanical properties of reinforced plastics in their thickness direction.

10.115 We have seen that reinforced plastics can be adversely affected by environmental factors such as humidity, chemicals, and temperature variations. Design and describe test methods to determine the mechanical properties of composite materials under these conditions.

10.116 As with other materials, the mechanical properties of composites are obtained by preparing appropriate specimens and testing them. Explain what difficulties you might encounter in preparing specimens for testing and in the actual testing process itself.

10.117 Add a column to Table 10.1 describing the appearance of these plastics, including available colors and opaqueness.

10.118 It is possible to weave fibers in three dimensions and then to impregnate the weave with a curable resin. Describe the property differences that such materials would have as compared to laminated composite materials.

10.119 Instead of having a constant cross section, it may be possible to make fibers or whiskers with a varying cross section or a fiber with a wavy surface. What advantages would such fibers have? Explain.

10.120 Polymers (either plain or reinforced) can be a suitable material for dies in sheet-metal forming operations described in Chapter 7. Describe your thoughts, considering die geometry and any other factors that may be relevant.

10.121 For ease of sorting for recycling, all plastic products are now identified with a triangular symbol with a single-digit number at its center and two or more letters under it. Explain what these numbers indicate and why such symbols are used.

10.122 Obtain different kinds of toothpaste tubes and carefully cut them across with a sharp razor blade. Comment on your observations regarding the type of materials used and how the tube could be produced.

10.123 Design a machine that uses rapid-prototyping technologies to produce ice sculptures. Describe its basic features, commenting on the effect of size and shape complexity on your design.

10.124 A manufacturing technique is being proposed that uses a variation of fused-deposition modeling: There are two polymer filaments that are melted and mixed before being extruded to produce the part. What advantages could this method have? Explain.

Properties and Processing of Metal Powders, Ceramics, Glasses, and Superconductors

This chapter explores the manufacturing processes and technologies for producing net-shape parts from metal and ceramic powders, as well as glass, diamond and graphite. Specifically, the following are described:

- Methods of producing metal and ceramic powders
- Processes used to shape compacts of powers, including the mechanics of powder compaction.
- Finishing operations to improve dimensional tolerances and surface properties, as well as part aesthetics.
- Methods of producing superconductor materials and commercially important forms of carbon, such as graphite and diamond.
- Design and competitive aspects of these processes.

11.1 | Introduction

In the manufacturing processes described in the preceding chapters, the raw materials used were either in a molten state or in solid form. This chapter describes groups of operations whereby metal powders, ceramics, and glasses are processed into products, as well as those involved in processing metal and ceramic matrix composite materials and superconductors. Design considerations in each process are also described.

The **powder metallurgy** (P/M) process is capable of making complex parts by compacting metal powders in dies and sintering them (heating without melting) to net- or near-net-shape products. The availability of a wide range of powder compositions, the capability to produce parts to net dimensions and the economics of the overall operation make this process attractive for many applications.

This chapter next describes the structure, properties, and processing of **ceramics, glasses, graphite,** and **diamond.** These materials have significantly different mechanical and physical properties compared to metals, including hardness, high-temperature resistance, and electrical and optical properties. Consequently, they have important and unique applications.

The chapter then describes the structure, properties, and processing of **metal-matrix** and **ceramic-matrix composites.** Recall that polymer-matrix composites are described in Chapter 10. Because of the large number of possible combinations of elements, a great variety of composite materials is now available for a wide range of consumer and industrial applications, particularly in the aircraft and aerospace industries.

The chapter ends with a description of how superconductors are processed into products, such as magnetic coils for magnetic resonance imaging and other devices that monitor magnetic fields.

11.2 | Powder Metallurgy

One of the first uses of powder metallurgy was in making the tungsten filaments for incandescent light bulbs in the early 1900s. Applications now include gears, cams, bushings, cutting tools, porous products such as filters and oil-impregnated bearings, and automotive components such as piston rings, valve guides, connecting rods, and hydraulic pistons (Fig. 11.1). Recent advances in P/M technology also permit structural parts of aircraft, such as landing gears, engine-mount supports, engine disks, impellers, and engine nacelle frames, to be made from metal powders.

Powder metallurgy has become competitive with processes such as casting, forging, and machining, particularly for relatively complex parts made of high-strength and hard alloys. Nearly 70% of P/M part production is for automotive applications. Parts made by this process have good dimensional accuracy, and their sizes range from tiny balls for ball-point pens to parts weighing about 50 kg (110 lb), although most parts made by P/M weigh less than 2.5 kg (5.5 lb). A typical family car now contains, on average, 13 kg (28.6 lb) of precision metal parts made by powder metallurgy, an amount that has been increasing by about 10% annually.

The powder-metallurgy operation consists of the following steps, in sequence:

1. **Powder production.**
2. **Blending.**
3. **Compaction.**
4. **Sintering.**
5. **Finishing operations.**

FIGURE 11.1   (a) Examples of typical parts made by powder-metallurgy processes and (b) upper trip lever for a commercial irrigation sprinkler, made by P/M. Made of unleaded brass alloy, it replaces a die-cast part, at a 60% cost savings. *Source:* Courtesy of Metal Powder Industries Federation.

(a)

(b)

The finishing operations may include processes such as coining, sizing, machining, and infiltration for improved quality, dimensional accuracy, and part strength.

11.2.1 Production of metal powders

Metal powders can be produced by several methods, the choice of which depends on the particular requirements of the end product. Sources for metals generally are their bulk form, ores, salts, and other compounds. These forms are then produced, by various methods, into powders. The shape, size distribution, porosity, chemical purity, and bulk and surface characteristics of the powder particles depend on the particular process used (Fig. 11.2). These characteristics are important, because they significantly affect flow during powder compaction and reactivity in subsequent sintering operations. Particle sizes produced range from 0.1 to 1000 μm (0.04 to 4 μin.).

The methods of powder production are

1. **Atomization.** *Atomization* produces a liquid-metal stream by injecting molten metal through a small orifice (Fig. 11.3a). The stream is broken up by jets of inert gas, air, or water. The size of the particles formed depends on the temperature of the metal, the rate of flow, nozzle size, and jet characteristics. Melt-atomization methods are widely used for production of powders for P/M. In one variation of this method, a consumable electrode is rotated rapidly in a helium-filled chamber (Fig. 11.3d); the centrifugal force breaks up the molten tip of the electrode, producing metal particles.

2. **Reduction.** *Reduction* of metal oxides (removal of oxygen) involves gases, such as hydrogen and carbon monoxide, as reducing agents, whereby very fine metallic oxides are reduced to the metallic state. The powders produced by this method are spongy and porous and have uniformly sized spherical or angular shapes.

Acicular (chemical decomposition)	Irregular rodlike (chemical decomposition, mechanical comminution)

(a) One-dimensional

Flake (mechanical comminution)	Dendritic (electrolytic)

(b) Two-dimensional

Spherical [atomization, carbonyl (Fe), precipitation from a liquid]	Irregular (atomization, chemical decomposition)	Rounded (atomization, chemical decomposition)	Porous (reduction of oxides)	Angular [mechanical disintegration, carbonyl (Ni)]

(c) Three-dimensional

FIGURE 11.2   Particle shapes and characteristics of metal powders and the processes by which they are produced.

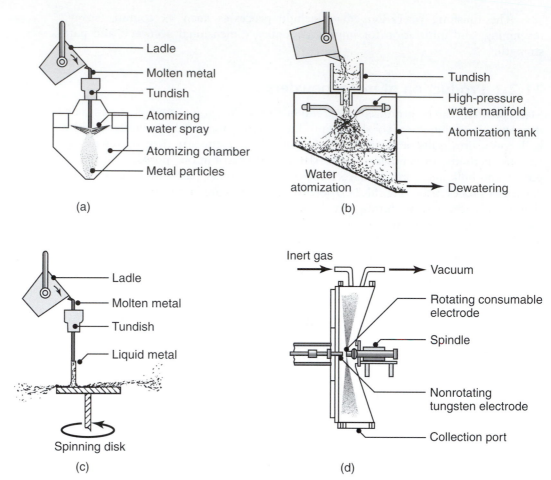

FIGURE 11.3 Methods of metal-powder production by atomization: (a) gas atomization, (b) water atomization, (c) atomization with a rotating consumable electrode, and (d) centrifugal atomization with a spinning disk or cup.

3. **Electrolytic deposition.** This operation uses either aqueous solutions or fused salts. The powders produced are among the purest of all metal powders.

4. **Carbonyls.** *Metal carbonyls*, such as iron carbonyl $[Fe(CO)_5]$ and nickel carbonyl $[Ni(CO)_4]$, are formed by letting iron or nickel react with carbon monoxide. The reaction products are then decomposed to iron and nickel, producing small, dense, and uniform spherical particles of high purity.

5. **Comminution.** Mechanical *comminution* (*pulverization*) involves crushing, milling in a *ball mill*, or grinding brittle or less ductile metals into small particles. A ball mill is a machine (see Fig. 11.26) with a rotating hollow cylinder that is partly filled with steel or white cast-iron balls. When made from brittle metals, the powder particles have angular shapes, and when made from ductile metals, they are flaky and are not particularly suitable for powder-metallurgy applications.

6. **Mechanical alloying.** In this process, powders of two or more pure metals are mixed in a ball mill. Under the impact of the hard balls, the powders repeatedly fracture and bond together by diffusion, producing an alloy.

7. **Other methods.** Less commonly used methods of powder production include

   a. **precipitation** from a chemical solution.

   b. production of fine metal chips by **machining**.

   c. **vapor condensation**.

   d. high-temperature **extractive metallurgy**.

   e. reaction of volatile halides (a compound of halogen and an electropositive element) with liquid metals.

   f. controlled reduction and reduction/carburization of solid oxides.

Nanopowders. More recent developments include the production of *nanopowders* of various metals, including copper, aluminum, iron, and titanium. (See also the discussion of *nanomaterials* in Section 3.11.9.) When the metals are subjected to large plastic deformation by compression and shear at stress levels of 5500 MPa (800 ksi) during processing of these powders, their particle size is reduced, and the material becomes pore free, thus possessing enhanced properties.

Microencapsulated powders. These powders are completely coated with a binder and are compacted by warm pressing. (See also the discussion of *metal injection molding* in Section 11.3.4.) In electrical applications, the binder acts like an insulator, preventing electricity from flowing and thus reducing eddy-current losses.

11.2.2 Particle size, distribution, and shape

Particle size is usually measured and controlled by screening, that is, by passing the metal powder through screens (*sieves*) of various mesh sizes. The operation involves using a vertical stack of screens with increasing mesh size as the powder flows downward through the screens. The larger the mesh size, the smaller is the opening in the screen. For example, a mesh size of 30 has openings of 600 μm, size 100 has 150 μm, and size 400 has 38 μm. (This method is similar to the technique used for numbering abrasive grains; the higher the number, the smaller is the size of the abrasive particle; see Section 9.2.)

In addition to screen analysis, several other methods are also used for particle-size analysis, particularly for powders finer than 45 μm (0.0018 in.):

1. **Sedimentation,** which involves measuring the rate at which particles settle in a fluid.

2. **Microscopic analysis,** including the use of transmission and scanning electron microscopy.

3. **Light scattering** from a *laser* that illuminates a sample consisting of particles suspended in a liquid medium; the particles cause the light to be scattered, which is then focused on a detector that digitizes the signals and computes the particle-size distribution.

4. **Optical means,** such as having particles blocking a beam of light, which is then sensed by a *photocell*.

5. **Suspension of particles** in a liquid and subsequent detection of particle size and distribution by electrical *sensors*.

The *size distribution* of particles is important because it affects the processing characteristics of the powder and is given in terms of a *frequency-distribution* plot, as shown in Fig. 11.4a. (For a detailed description of this type of plot, see Section 4.9.1 and Fig. 4.21a.) Note in this plot that the highest percentage of the

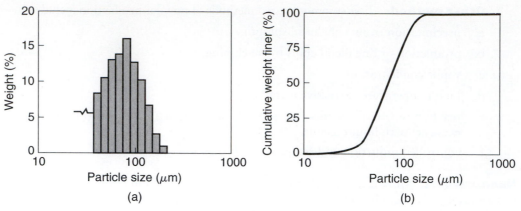

FIGURE 11.4   (a) Distribution of particle size, given as weight percentage; note that the highest percentage of particles have a size between 75 and 90 μm. (b) Cumulative particle-size distribution as a function of weight. *Source:* After R.M. German.

particles (by weight) has a size in the range of 75 to 90 μm; this maximum is called the *mode size*. The same particle-size data are also plotted in the form of a *cumulative distribution* (Fig. 11.4b). Thus, for example, about 75% of the particles (by weight) have a size finer than 100 μm. Note that the cumulative weight reaches 100% at a particle size of about 200 μm.

The *particle shape* has a major influence on its processing characteristics. The shape is generally described in terms of aspect ratio or shape index. *Aspect ratio* is the ratio of the largest dimension to the smallest dimension of the particle and ranges from 1 for a spherical particle to about 10 for flakelike or needlelike particles. *Shape index*, or **shape factor** (SF), is a measure of the surface area to the volume of the particle with reference to a spherical particle of equivalent diameter. Thus, the shape factor for a flake is higher than that for a sphere.

EXAMPLE 11.1 Particle shape-factor determination

Determine the shape factor for a (a) spherical particle, (b) cubic particle, and (c) cylindrical particle with a length-to-diameter ratio of 2.

Solution

a. The ratio of surface area A to volume V can be expressed as

$$\frac{A}{V} = \frac{k}{D_{eq}},$$

where k is the shape factor and D_{eq} is the diameter of a sphere that has the same volume as the particle being considered. In the case of a spherical particle, its diameter is $D = D_{eq}$, its area $A = \pi D^2$, and its volume $V = \pi D^3/6$. Thus,

$$k = \left(\frac{A}{V}\right)D_{eq} = \left[\frac{(\pi D^2)}{(\pi D^3/6)}\right]D = 6.$$

b. The surface area of a cube of length L is $6L^2$, and its volume is L^3. Hence,

$$\frac{A}{V} = \frac{6L^2}{L^3} = \frac{6}{L}.$$

The particle's equivalent diameter then is

$$D_{eq} = \left(\frac{6V}{\pi}\right)^{1/3} = \left(\frac{6L^3}{\pi}\right)^{1/3} = 1.24L.$$

Therefore, the shape factor is

$$k = \left(\frac{A}{V}\right)D_{eq} = \left(\frac{6}{L}\right)(1.24\ L) = 7.44.$$

c. The surface area of a cylindrical particle with length-to-diameter ratio of 2 is

$$A = \left(\frac{2\pi D^2}{4}\right) + \pi DL = \left(\frac{\pi D^2}{2}\right) + 2\pi D^2 = 2.5\pi D^2.$$

The particle's volume is

$$V = \left(\frac{\pi D^2}{4}\right)(L) = \left(\frac{\pi D^2}{4}\right)(2D) = \frac{\pi D^3}{2},$$

and its equivalent diameter is

$$D_{eq} = \left(\frac{6V}{\pi}\right)^{1/3} = \left(\frac{6\pi D^3}{2\pi}\right)^{1/3} = 1.442D.$$

Hence, its shape factor is

$$k = \left(\frac{A}{V}\right)D_{eq} = \left[\frac{(2.5\pi D^2)}{(\pi D^3/2)}\right](1.442D) = 7.21.$$

11.2.3 Blending metal powders

Blending (mixing) powders is the second step in powder-metallurgy processing and is carried out for the following purposes:

1. Powders of different metals and materials can be blended in order to impart specific physical and mechanical properties and characteristics to the P/M part. Mixtures of powders can be made from metal alloys, or else blends of different metal powders can be produced. Proper mixing is essential to ensure uniformity of properties throughout the part.

2. Even when a single metal is used, the powders may significantly vary in size and shape; hence they must be blended to obtain uniformity from part to part. The ideal mix is one in which all the particles of each material, and of each size and morphology, are distributed uniformly.

3. *Lubricants* can be mixed with the powders to improve their flow characteristics during processing. They reduce friction between the metal particles, improve flow of the powder metals into the dies, and improve die life. Lubricants are typically stearic acid or zinc stearate, in a proportion of from 0.25 to 5% by weight.

4. Various additives, such as *binders* as in sand molds, are used to develop sufficient *green strength* (see below) and facilitate sintering.

Powder mixing must be carried out under controlled conditions in order to avoid contamination and deterioration. *Deterioration* is caused by excessive mixing, which may alter the shape of the particles and work-harden them, thus making the subsequent compaction more difficult. Powders can be mixed in air, in inert atmospheres (to avoid oxidation), or in liquids, which act as lubricants and make the mix more uniform. Several types of blending equipment are available. These operations are now controlled by microprocessors to improve and maintain quality.

Because of their high surface-area-to-volume ratio, metal powders are *explosive*, particularly aluminum, magnesium, titanium, zirconium, and thorium powders. Great care must be exercised both during blending and during storage and handling. Among precautions to be taken are (a) grounding equipment, (b) preventing the generation of sparks, by using nonsparking tools and avoiding the use of friction as a source of heat, and (c) avoiding dust clouds, open flames, and chemical reactions.

11.3 | Compaction of Metal Powders

Compaction is the step in which the blended powders are pressed into shapes (Fig. 11.5a and b), using dies and presses that are either hydraulically or mechanically actuated (see also Fig. 6.27). Pressing is generally carried out at room temperature, although it can be done at elevated temperatures as well. The purposes of compaction are to obtain the required shape, density, and particle-to-particle contact and to make the part sufficiently strong to be handled and processed further. The as-pressed powder is known as a **green compact.**

Density is relevant at three different stages in P/M processing: (1) as loose powder, (2) as a green compact, and (3) after sintering. The particle shape, average size, and size distribution all affect the packing density of loose powder. Spherical powders with a wide size distribution result in a high packing density. However, compacts made from these powders have poor green strength and hence are unsuitable for P/M parts compacted by the die-pressing method (which is the most commonly used method of fabricating P/M parts). Consequently, powders with some irregularity of

FIGURE 11.5 (a) Compaction of metal powder to produce a bushing and (b) a typical tool and die set for compacting a spur gear. *Source:* Courtesy of Metal Powder Industries Federation.

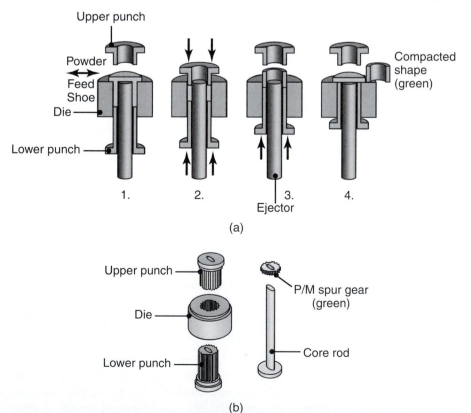

shape (for example, water-atomized ferrous powders; see Fig. 11.3) are preferred, even though the fill density of powder in the die is generally lower than that of spherical powders. For reproducibility of part dimensions, the fill density of the powder should be consistent from one powder lot to the next. Spherical powders are preferred for *hot isostatic pressing*, a technique that is described in Section 11.3.3.

The density after compaction (**green density**) depends primarily on the (a) compaction pressure, (b) powder composition, and (c) hardness of the powder (Fig. 11.6a). The higher the compacting pressure and the softer powder, the higher is the green density. The density and its uniformity within a compact can be improved with the addition of a small quantity of admixed (blended-in) lubricant.

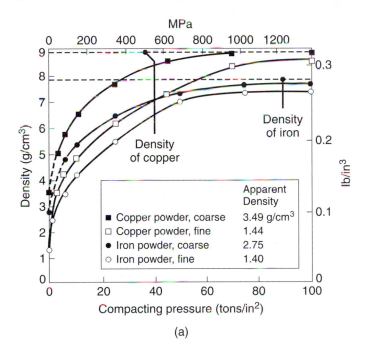

(a)

FIGURE 11.6 (a) Density of copper- and iron-powder compacts as a function of compacting pressure. Density greatly influences the mechanical and physical properties of P/M parts. (b) Effect of density on tensile strength, elongation, and electrical conductivity of copper powder. (IACS is International Annealed Copper Standard for electrical conductivity.) *Source:* (a) After F.V. Lenel.

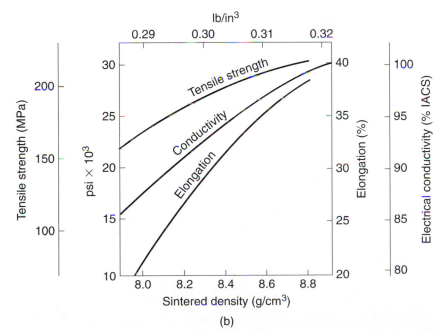

(b)

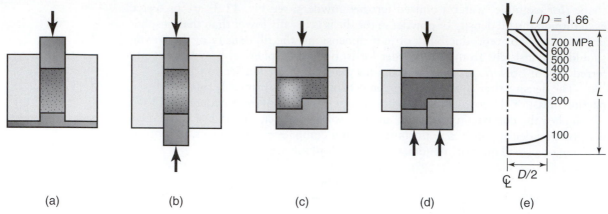

(a)   (b)   (c)   (d)   (e)

FIGURE 11.7 Density variation in compacting metal powders in different dies: (a) and (c) single-action press and (b) and (d) double-action press, where the punches have separate movements. Note the greater uniformity of density in (d) as compared with (c). Generally, uniformity of density is preferred, although there are situations in which density variation, and hence variation of properties, within a part may be desirable. (e) Pressure contours in compacted copper powder in a single-action press. *Source:* After P. Duwez and L. Zwell.

The effect of particle shape on green density can best be understood by considering two powder grades with the same chemical composition and hardness: one with a spherical particle shape and the other with an irregular shape. The spherical grade will have a higher apparent density (fill density), but after compaction under a higher pressure, compacts from both grades will have similar green densities. When comparing two similar powders pressed under some standard conditions, the powder that gives a higher green density is said to have a higher *compressibility*. (See also the discussion of *sintered density* in Section 11.4.)

The higher the density, the higher will be the strength and elastic modulus of the part (Fig. 11.6b). The reason is that the higher the density, the higher will be the amount of solid metal in the same volume, hence the greater will be the part's resistance to external forces. Because of the friction present between the metal particles in the powder and between the punches and the die walls, the density can vary considerably *within* the part. This variation can be minimized by proper punch and die design and by controlling friction. Thus, for example, it may be necessary to use *multiple punches*, with separate movements in order to ensure that the density is more uniform throughout the part (Fig. 11.7).

EXAMPLE 11.2 Density of metal-powder-lubricant mix

Zinc stearate is a lubricant that commonly is mixed with metal powders prior to compaction, in proportions up to 2% by weight. Calculate the theoretical and apparent densities of an iron powder-zinc-stearate mix, assuming that (a) 1000 g of iron powder is mixed with 20 g of lubricant, (b) the density of the lubricant is 1.10 g/cm^3, (c) the theoretical density of the iron powder is 7.86 g/cm^3 (from Table 3.3), and (d) the apparent density of the iron powder is 2.75 g/cm^3 (from Fig. 11.6a).

Solution. The volume of the mixture is

$$V = \left(\frac{1000}{7.86}\right) + \left(\frac{20}{1.10}\right) = 145.4 \text{ cm}^3.$$

The combined weight of the mixture is $1000 + 20 = 1020$ g, and thus its theoretical density is $1020/145.41 = 7.01$ g/cm³. The apparent density of the iron powder is given as 2.75 g/cm³; hence its density is $(2.75/7.86)100 = 35\%$ of the theoretical density. Assuming a similar percentage for the mixture, we can estimate the apparent density of the mix as $(0.35)(7.01) = 2.45$ g/cm³. (*Note:* Although g and g/cm³ are not SI units, they continue to be commonly used in the powder-metallurgy industry.)

11.3.1 Pressure distribution in powder compaction

As can be seen in Fig. 11.7c, in single-action pressing (see Fig. 11.7a and c), the pressure decays rapidly toward the bottom of the compact. The pressure distribution along the length of the compact can be determined by using the slab method of analysis, described in Section 6.2.2. As also done in Fig. 6.4, we first describe the operation in terms of its coordinate system, as shown in Fig. 11.8, where D is the diameter of the compact, L is the compact's length, and p_0 is the pressure applied by the punch. An element dx thick is then taken, with all the relevant stresses indicated on it, namely, the compacting pressure, p_x, the die-wall pressure, σ_r, and the frictional stress along the die wall, $\mu\sigma_r$. Note that the frictional stresses act upward on the element, because the punch movement is downward.

Balancing the vertical forces acting on this element,

$$\left(\frac{\pi D^2}{4}\right)p_x - \left(\frac{\pi D^2}{4}\right)(p_x + dp_x) - (\pi D)(\mu\sigma_r)\,dx = 0, \qquad (11.1)$$

which can be simplified to

$$D\,dp_x + 4\mu\sigma_r\,dx = 0.$$

We now have one equation, but two unknowns (p_x and σ_r). Let's now introduce a factor k, which is a measure of the interparticle friction during compaction:

$$\sigma_r = kp_x.$$

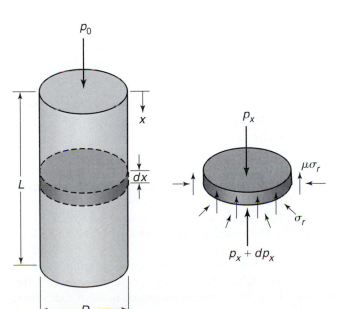

FIGURE 11.8 Coordinate system and stresses acting on an element in compaction of powders. The pressure is assumed to be uniform across the cross section. (See also Fig. 6.4.)

If there is no friction between the particles, $k = 1$, the powder behaves like a fluid, and thus $\sigma_r = p_x$, signifying a state of hydrostatic pressure. We now have the expression

$$dp_x + \frac{4\mu k p_x \, dx}{D} = 0,$$

or

$$\frac{dp_x}{p_x} = -\frac{4\mu k \, dx}{D}.$$

This expression is similar to that given in Section 6.2.2 for upsetting. In the same manner, this expression can be integrated, noting that the boundary condition in this case is $p_x = p_0$ when $x = 0$;

$$p_x = p_0 e^{-4\mu kx/D}. \tag{11.2}$$

It can be seen from this expression that the pressure within the compact decays as the coefficient of friction, the parameter k, and the length-to-diameter ratio increase.

EXAMPLE 11.3 Pressure decay in compaction

Assume that a powder mix has the values of $k = 0.5$ and $\mu = 0.3$. At what depth will the pressure in a straight cylindrical compact 10 mm in diameter become (a) zero and (b) one-half the pressure at the punch?

Solution. For case (a), $p_x = 0$. Consequently, from Eq. (11.2)

$$0 = p_0 e^{-(4)(0.3)(0.5)x/10}, \quad \text{or} \quad e^{-0.06x} = 0.$$

Note that the value of x must approach ∞ for the pressure to decay to 0. For case (b), we have $p_x/p_0 = 0.5$. Therefore,

$$e^{-0.06x} = 0.5, \quad \text{or} \quad x = 11.55 \text{ mm}.$$

In practice, a pressure drop of 50% is considered severe as the compact density will then be unacceptably low. This example shows that, under the conditions assumed, uniaxial compaction of a cylinder of even a length-to-diameter ratio of about 1.2 will be unsatisfactory.

11.3.2 Equipment

The compacting pressure required for pressing metal powders ranges from 70 MPa (10 ksi) for aluminum to 800 MPa (120 ksi) for high-density iron parts (Table 11.1). The pressure required depends on the characteristics and shape of the particles, the method of blending, and the lubrication.

Press capacities are on the order of 1.8–2.7 MN (200–300 tons), although presses with much higher capacities are available for special applications. However, most applications require less than 100 tons. For small capacities, crank or eccentric-type mechanical presses are used; for higher capacities, toggle or knucklejoint presses are employed. (See Fig. 6.27.) Hydraulic presses with capacities as high as 45 MN (5000 tons) can be used for compacting large parts.

The selection of a press depends on part size and configuration, density requirements, and production rate. The higher the pressing speed, the greater is the tendency to trap air in the die cavity. Proper die design, including provision of vents, is thus important so that trapped air does not hamper compaction.

TABLE 11.1

Compacting Pressures for Various Metal Powders

	Pressure	
	MPa	psi $\times$ 10^3
Metal		
Aluminum	70–275	10–40
Brass	400–700	60–100
Bronze	200–275	30–40
Iron	350–800	50–120
Tantalum	70–140	10–20
Tungsten	70–140	10–20
Other materials		
Aluminum oxide	110–140	16–20
Carbon	140–165	20–24
Cemented carbides	140–400	20–60
Ferrites	110–165	16–24

11.3.3 Isostatic pressing

For improved compaction, parts may be subjected to a number of additional operations, such as *rolling, forging,* and *isostatic pressing.* Because, as we have seen, the density of die-compacted powders can vary significantly, powders are subjected to *hydrostatic pressure* in order to achieve more uniform compaction, a process similar to pressing cupped hands together when making snowballs.

In **cold isostatic pressing** (CIP), the metal powder is placed in a flexible rubber mold made of neoprene rubber, urethane, polyvinyl chloride, or other elastomers (Fig. 11.9). The assembly is then pressurized hydrostatically in a chamber, filled usually with water. The most common pressure is 400 MPa (60 ksi), although pressures

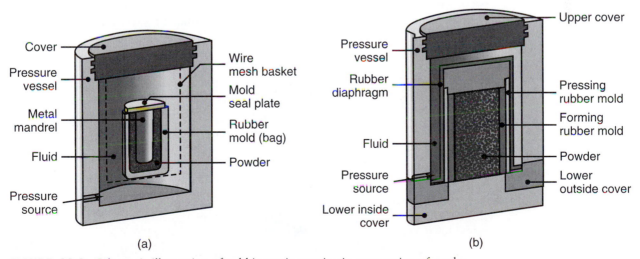

(a)   (b)

FIGURE 11.9 Schematic illustration of cold isostatic pressing in compaction of a tube. (a) The *wet-bag process,* where the rubber mold is inserted into a fluid that is subsequently pressurized. In the arrangement shown, the powder is enclosed in a flexible container around a solid core rod or mandrel. (b) The *dry bag process,* where the rubber mold does not contact the fluid, but instead is pressurized through a diaphragm.

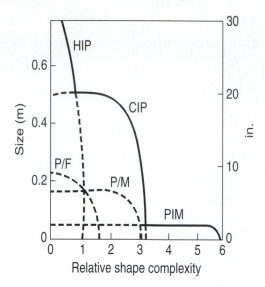

FIGURE 11.10 Process capabilities of part size and shape complexity for various P/M operations; P/F is powder forging. *Source:* Metal Powder Industries Federation.

of up to 1000 MPa (150 ksi) may be used. The applications of CIP and other compacting methods, in terms of size and part complexity, are shown in Fig. 11.10.

In **hot isostatic pressing** (HIP), the container is usually made of a high-melting-point sheet metal, and the pressurizing medium is inert gas or vitreous (glasslike) fluid (Fig. 11.11). A typical operating condition for HIP is 100 MPa (15 ksi) at 1100°C (2000°F), although the trend is toward higher pressures and temperatures. The main advantage of HIP is its ability to produce compacts with essentially 100% density, strong bonding among the particles, and good mechanical properties. Although relatively expensive, typical applications include making superalloy components for the aerospace industry, final densification for tungsten-carbide cutting tools and P/M tool steels, and closing internal porosity and improving properties in superalloy and titanium-alloy castings.

The main advantage of isostatic pressing is that, because of the uniformity of pressure from all directions and the absence of die-wall friction, it produces compacts with almost uniform grain structure and density, irrespective of shape. Parts with high length-to-diameter ratios have been produced by this method, with very

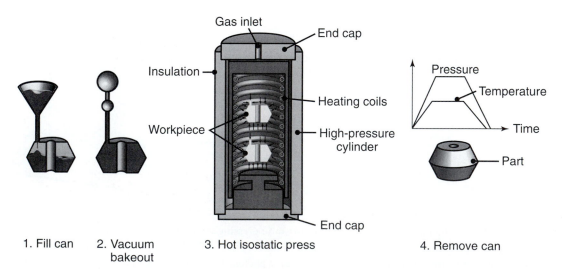

FIGURE 11.11 Schematic illustration of the sequence of steps in hot isostatic pressing. Diagram (4.) shows the pressure and temperature variation versus time.

uniform density, high strength, high toughness, and good surface details. The limitations of this method include wider dimensional tolerances produced, higher time required, relatively small production quantities output, and higher cost per piece.

11.3.4 Miscellaneous compacting and shaping processes

1. **Metal injection molding.** In *metal injection molding* (MIM), very fine metal powders (generally <45 μm, and often <10 μm) are blended with a 25–45% polymer or a wax-based binder. The mixture then undergoes a process similar to injection molding of plastics (Section 10.10.2); it is injected into a mold at a temperature of 135°–200°C (275°–400°F). The molded green parts are then placed in a low-temperature oven to burn off the plastic (*debinding*), or else the binder is partially removed with a solvent. Some binder is purposely left to maintain some strength for handling; the green parts are finally sintered in a furnace at temperatures as high as 1375°C (2500°F). The term *powder injection molding* (PIM) is used with a broader reference to molding of both metal and ceramic powders.

   Generally, metals that are suitable for powder injection molding are those that melt at temperatures above 1000°C (1830°F); examples are carbon and stainless steels, tool steels, copper, bronze, and titanium. Typical parts made by this method include components for guns, surgical instruments, automobiles, and watches.

   The major advantages of powder injection molding over conventional compaction are

   a. Complex shapes, having wall thicknesses as small as 5 mm (0.2 in.), can be molded and then easily removed from the dies. Most commercially produced parts fabricated by this method weigh from a fraction of a gram to about 250 g (0.55 lb) each.
   b. Mechanical properties are nearly equal to those of wrought parts.
   c. Dimensional tolerances are good.
   d. High production rates can be achieved by using multicavity dies.
   e. Parts produced by the MIM process compete well against small investment-cast parts (Section 5.9.2), small forgings, and complex machined parts. It does not, however, compete well with zinc and aluminum die casting (Section 11.12) and with screw machining (Section 8.9.2).

   The major limitation of PIM is the high cost and availability of fine metal powders.

2. **Rolling.** In *powder rolling*, also called *roll compaction*, the powder is fed into the roll gap in a two-high rolling mill (see Fig. 6.41a) and is compacted into a continuous strip at speeds of up to 0.5 m/s (100 ft/min), as shown in Fig. 11.12. The process can be carried out at room or at elevated temperatures, depending on the powder metal. Sheet metal for electrical and electronic components and for coins can be made by powder rolling.

3. **Extrusion.** Powders may be compacted by *hot extrusion*, in which the powder is encased in a metal container and extruded. (See also Section 6.4.) Superalloy powders, for example, are hot extruded for improved properties. Preforms made from the extrusions may be reheated and forged in a closed die to their final shape.

4. **Pressureless compaction.** In this process, the die is filled with metal powder by gravity, and the powder is sintered directly in the die. Because of the resulting

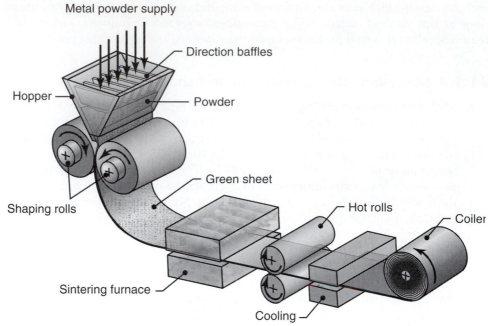

FIGURE 11.12   An example of powder rolling. The purpose of direction baffles in the hopper is to ensure uniform distribution of powder across the width of the strip.

low density, pressureless compaction is used principally for porous parts, such as filters.

5. **Ceramic molds.** In this process, molds for shaping metal powders are made by the technique used in investment casting. After the ceramic mold is made, it is filled with metal powder, placed in a steel container, and the space between the mold and the container is filled with particulate material. The container is then evacuated, sealed, and subjected to hot isostatic pressing. Titanium-alloy compressor rotors for missile engines, for example, have been made by this process.

6. **Spray deposition.** This is a shape-generation method in which the basic components are an atomizer, a spray chamber with inert atmosphere, and a mold for producing performs. Although there are several variations, the best known is the *Osprey* process (Fig. 11.13). After the metal is atomized, it is deposited on a cooled preform mold (usually made of copper or ceramic) where it solidifies.

FIGURE 11.13   Spray casting (Osprey process) in which molten metal is sprayed over a rotating mandrel to produce seamless tubing and pipe.

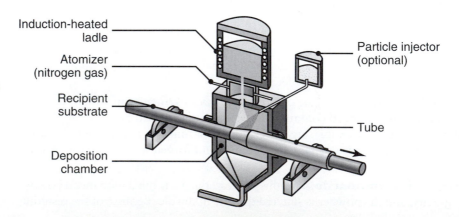

The metal particles weld together, developing a density that is normally above 99% of the solid metal density. The mold may have various shapes, such as billets, tubes, disks, and cylinders. The spray-deposited parts may be subjected to additional shaping and consolidation processes, such as forging, rolling, and extrusion. The grain size of the part is fine, and its mechanical properties are comparable with those of wrought products of the same alloy.

11.3.5 Punch and die materials

The selection of punch and die materials for P/M depends on the abrasiveness of the powder metal and the number of parts to be made. The most commonly used die materials are air- or oil-hardening tool steels, such as D2 or D3, with a hardness range of 60–64 HRC (Section 3.10.3). Because of their greater hardness and wear resistance, tungsten-carbide dies are used for more severe applications. Punches are generally made of similar materials as dies. Close control of die and punch dimensions is essential for proper compaction and longer die life. Too large a clearance between the punch and the die will allow the metal powder to enter the gap, interfere with the operation, and produce eccentric parts. Diametral clearances are generally less than 25 μm (0.001 in.). Die and punch surfaces must be lapped or polished in the direction of tool movements for improved die life and overall performance.

11.4 | Sintering

Sintering is the process whereby compacted metal powder is heated in a controlled-atmosphere furnace to a temperature below its melting point, but sufficiently high to allow bonding (fusion) of the individual metal particles. Prior to sintering, the compact is brittle and its strength, known as **green strength,** is low. The nature and strength of the bond between the particles, and hence of the sintered compact, depend on the mechanisms of diffusion, plastic flow, evaporation of volatile materials in the compact, recrystallization, grain growth, and pore shrinkage.

The density of a sintered part mainly depends on the part's green density and on the sintering conditions, in terms of temperature, time, and furnace atmosphere. The sintered density increases with increasing temperature and time and usually with the use of a more deoxidizing type of furnace atmosphere. For structural parts, a higher sintered density is very desirable, as it leads to better mechanical properties. On the other hand, it is often preferable to minimize the increase in density during sintering for better dimensional accuracy.

Better properties and dimensional accuracy can be achieved by using a powder with a high compressibility, i.e., a powder that will give a high green density and maintain a moderate sintering temperature. Such a powder also has another important benefit in that larger parts can be produced with a specific press capacity. A considerable amount of development work by powder producers is thus aimed at producing powder grades with higher compressibility.

Sintering temperatures (Table 11.2) are generally within 70 to 90% of the melting point of the metal or alloy (see Table 3.2). *Sintering times* range from a minimum of about 10 minutes for iron and copper alloys to as long as 8 hours for tungsten and tantalum. Continuous-sintering furnaces are used for most production today. These furnaces have three chambers: (1) a burn-off chamber, to volatilize the lubricants in the green compact in order to improve bond strength and prevent cracking; (2) a high-temperature chamber, for sintering; and (3) a cooling chamber.

Proper control of the furnace atmosphere is essential for successful sintering and to obtain optimum properties. An oxygen-free atmosphere is necessary in order

TABLE 11.2

Sintering Temperature and Time for Various Metal Powders

Material	Temperature (°C)	Time (min)
Copper, brass, and bronze	760–900	10–45
Iron and iron graphite	1000–1150	8–45
Nickel	1000–1150	30–45
Stainless steels	1100–1290	30–60
Alnico alloys (for permanent magnets)	1200–1300	120–150
Ferrites	1200–1500	10–600
Tungsten carbide	1430–1500	20–30
Molybdenum	2050	120
Tungsten	2350	480
Tantalum	2400	480

to control the carburization and decarburization of iron and iron-base compacts and to prevent oxidation of powders. Oxide inclusions have a detrimental effect on mechanical properties (see Section 3.8). For the same volume of inclusions, the smaller inclusions have a larger effect because there are more of them per unit volume of the part. The gases most commonly used for sintering a variety of other metals are hydrogen, dissociated or burned ammonia, partially combusted hydrocarbon gases, and nitrogen. A vacuum is generally used for sintering refractory metal alloys and stainless steels.

Sintering mechanisms. Sintering mechanisms are complex and depend on the composition of the metal particles as well as the processing parameters. As the temperature increases, two adjacent particles begin to form a bond by **diffusion** (*solid-state bonding*), as shown in Fig. 11.14a. As a result, the strength, density, ductility, and thermal and electrical conductivities of the compact increase (Fig. 11.15). At the

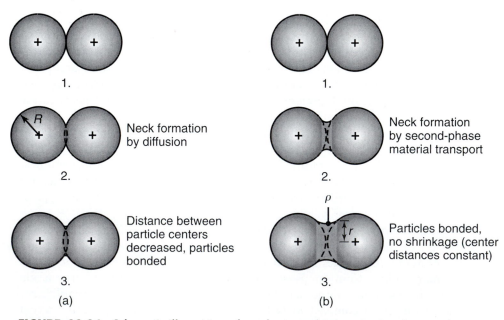

FIGURE 11.14   Schematic illustration of two basic mechanisms in sintering metal powders: (a) solid-state material transport and (b) liquid-phase material transport. R = particle radius, r = neck radius, and ρ = neck profile radius.

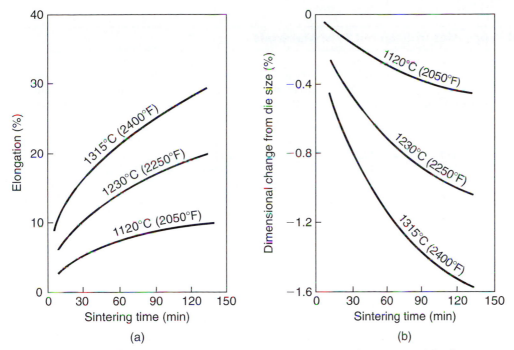

FIGURE 11.15 Effect of sintering temperature and time on (a) elongation and (b) dimensional change during sintering of type 316L stainless steel. *Source:* ASM International.

same time, however, the compact shrinks; hence allowances should be made for shrinkage, as is done in casting. (See Section 5.12.2.)

A second sintering mechanism is **vapor-phase transport**. Because the material is heated very close to its melting temperature, metal atoms will release to the vapor phase. At convergent geometries (the interface of two particles), the melting temperature is locally higher, and the vapor resolidifies. Thus, the interface grows and strengthens, while each particle shrinks as a whole. If two adjacent particles are of different metals, *alloying* can take place at the interface of the two particles. One of the particles may have a lower melting point than the other, in which case one particle may melt and, because of surface tension, surround the particle that has not melted (**liquid-phase sintering**), as shown in Fig. 11.14b. An example is cobalt in tungsten-carbide tools and dies (Section 8.6.4). Stronger and denser parts can be obtained in this way.

Depending on temperature, time, and processing history, different structures and porosities can be obtained in a sintered compact. However, porosity cannot be completely eliminated, because some voids remain after compaction and gases evolve during sintering and are trapped. Porosities can consist of either a network of interconnected pores or closed holes. Their presence is important in making P/M parts such as filters and bearings (with some porosity to entrap lubricants).

Spark sintering, also called **spark plasma sintering** (SPS), is another technique. In this process, loose metal powders are placed in a graphite mold, heated by electric current, subjected to a high-energy discharge, and compacted, all in one step. The rapid discharge of electricity strips any contaminants or oxide coating (such as that found on aluminum) from the surfaces of the particles, thus encouraging good bonding during compaction at elevated temperatures. A promising application for SPS is the densification of ceramic and hard metal powders, although it has also been effective for aluminum alloys. The SPS process has recently been applied to processing of nanoscale powders.

TABLE 11.3

Typical Mechanical Properties of Selected P/M Materials

Designation	MPIF type	Condition	Ultimate tensile strength (MPa)	Yield strength (MPa)	Hardness	Elongation in 25 mm (%)	Elastic modulus (GPa)
Ferrous							
FC-0208	N	AS	225	205	45 HRB	<0.5	70
		HT	295	—	95 HRB	<0.5	70
	R	AS	415	330	70 HRB	1	110
		HT	550	—	35 HRC	<0.5	110
	S	AS	550	395	80 HRB	1.5	130
		HT	690	655	40 HRC	<0.5	130
FN-0405	S	AS	425	240	72 HRB	4.5	145
		HT	1060	880	39 HRC	1	145
	T	AS	510	295	80 HRB	6	160
		HT	1240	1060	44 HRC	1.5	160
Aluminum							
601 AB		AS	110	48	60 HRH	6	—
pressed bar		HT	252	241	75 HRH	2	—
Brass							
CZP-0220	T	—	165	76	55 HRH	13	—
	U	—	193	89	68 HRH	19	—
	W	—	221	103	75 HRH	23	—
Titanium							
Ti-6Al-4V		HIP	917	827	—	13	—
Superalloys							
Stellite 19		—	1035	—	49 HRC	<1	—

Note: MPIF = Metal Powder Industries Federation; AS = as sintered; HT = heat treated; HIP = hot isostatically pressed.

Typical mechanical properties for several sintered P/M alloys are given in Table 11.3. Note the effect of heat treating on the properties of metals. To evaluate the differences between the properties of P/M wrought vs. cast metals and alloys, compare this table with tables in Chapter 3 and others. Table 11.4 shows the effects of various processing methods on the mechanical properties of a particular titanium alloy. Note that hot isostatically pressed (HIP) titanium has properties that are similar to those for cast and forged titanium. Recall, however, that forged components

TABLE 11.4

Mechanical Property Comparisons for Ti-6Al-4V Titanium Alloy

Process	Density (%)	Yield stress (MPa)	Ultimate tensile strength (MPa)	Elongation (%)	Reduction of area (%)
Cast	100	840	930	7	15
Cast and forged	100	875	965	14	40
Powder metallurgy					
Blended elemental (P + S)*	98	786	875	8	14
Blended elemental (HIP)*	>99		875	9	17
Realloyed (HIP)	100	880	975	14	26
Electron-beam melting	100	910	970	16	—

*P + S = pressed and sintered; HIP = hot isostatically pressed.
Source: After R.M. German.

are likely to require additional machining processes (unless the components are precision forged to net shape), while P/M components may not. Consequently, powder metallurgy can be a competitive alternative to casting and forging operations.

EXAMPLE 11.4 Shrinkage in sintering

In solid-state bonding during sintering of a powder-metal green compact, the linear shrinkage is 4%. If the desired sintered density is 95% of the theoretical density of the metal, what should be the density of the green compact? Ignore the small changes in mass that occur during sintering.

Solution. Linear shrinkage is defined as $\Delta L/L_o$, where L_o is the original length of the part. The volume shrinkage during sintering can then be expressed as

$$V_{sint} = V_{green}\left(1 - \frac{\Delta L}{L_o}\right)^3. \tag{11.3}$$

The volume of the green compact must be larger than that of the sintered part. However, the mass does not change during sintering, so we can rewrite this expression in terms of the density, ρ, as

$$\rho_{green} = \rho_{sint}\left(1 - \frac{\Delta L}{L_o}\right)^3. \tag{11.4}$$

Therefore,

$$\rho_{green} = 0.95(1 - 0.04)^3 = 0.84 \quad \text{or} \quad 84\%.$$

EXAMPLE 11.5 Production of automotive-engine main bearing caps by powder metallurgy

Main bearing caps (MBCs), as shown in Fig. 11.16, are essential structural elements in internal combustion engines. Most MBCs are made from gray cast iron (see Section 5.6), which is cast, followed by extensive machining operations. Grey iron is a very versatile material that exhibits excellent castability and good machinability (due to its low hardness and flake graphite in the microstructure), compressive strength, and vibration-damping properties. However, it has low fatigue strength and is very brittle, which limits its use for highly stressed applications.

Powder-metallurgy MBCs were used for the first time in 1993 in General Motors 3100 and 3800 V6 engines. P/M parts were chosen to replace the formerly used cast-iron parts because of P/M's design flexibility, functional advantages, and its net-shape capability. P/M eliminated the need for major investments in machining lines required for finishing the cast parts to final shape and dimensions. A high-strength, low-cost P/M alloy steel was developed for this application, involving liquid-phase sintering and using an optimized low-alloy combination in a carbon-steel matrix, designated as Zenith Material 833 (ZM833).

In the P/M process, individual MBCs are molded into net shape, except for the new side bolt holes, which are drilled and tapped. Critical to the success of the P/M approach was the molding of the long bolt holes, which would be costly and difficult to machine. This molding step requires "upright" compaction (vertical bolt holes), which presents significant powder-compaction challenges due to the irregular shape because of the bearing arch. The advantages of upright compaction,

FIGURE 11.16   Powder-metal main bearing caps for 3.8- and 3.1-liter General Motors engines. *Source:* Courtesy of Zenith Sintered Products, Inc., Milwaukee, WI.

however, include the ability to press-in the bearing notch and the features of the top face, which provide identification and orientation for assembly. Upright compaction also permits use of a much smaller compacting press than for "flat" pressing (horizontal bolt holes), with associated cost savings. In addition, the 3800 engine's rear MBC has several features that would be impossible to compact in flat orientation; thus, the use of upright compaction was inevitable. Principal features of the MBCs produced by this method include recessed bolt-hole bosses, a rectangular oil-drain hole, vertical side grooves, a double arch to incorporate an oil seal, and a raised-arch feature for improved strength.

Once the desired fatigue strength was achieved, the ZM833 material was used to produce actual MBCs that were then subjected to severe engine-endurance testing in both engines. It was found that the cast-iron engine block generally fractured before the P/M cap cracked. Often, several engine blocks had to be sacrificed to obtain a fatigue-test point, but eventually an *S-N* curve (see Fig. 2.26) was generated for the P/M material so that a safety margin could be determined. The P/M was proven to be a robust design by repeated successful fatigue testing, simulating 1.8 times the most severe running conditions predicted for the engine.

Material ductility is a functional benefit, since it is possible to induce cracks in a very brittle material during press-fit installation. The ZM833 P/M material had the following properties: (a) density of 6.6 g/cm^3, (b) tensile strength of 450 MPa (65,000 psi), (c) hardness of 70 HRB, (d) modulus of elasticity of 107 GPa (15.5×10^6 psi); the P/M material developed for this application has a lower modulus of elasticity than that of cast iron, resulting in significantly lower initial tensile stresses in highly stressed areas of the MBC, (e) tensile elongation of 3 to 4%, which compares with that for gray iron, at 0.5%, and that for ductile iron, typically at 3%, and (f) it was found that the P/M alloy's fatigue endurance limit, as determined by flexural bending fatigue tests, was intermediate between those of the gray and ductile irons, being double the gray iron's level.

Source: Adapted from "Powder Metallurgy Main Bearing Caps for Automotive Engines," by T.M. Cadle, M.A. Jarrett, and W.L. Miller, *Int. Jr. of Powder Metallurgy,* 30(3): 275–282, 1994.

11.5 | Secondary and Finishing Operations

In order to further improve the properties of sintered powder-metallurgy products or to impart certain specific characteristics, several additional operations may be carried out after the sintering process:

1. **Re-pressing**, also called **coining** and **sizing**, is an additional compaction operation, performed under high pressure in presses. The purposes of this operation are to improve dimensional accuracy and surface finish of the sintered part and to increase the part's strength.

2. **Forging** involves the use of unsintered or sintered alloyed-powder preforms that are subsequently hot forged in heated, confined dies to the desired final shapes; the preforms may also be shaped by impact forging (see also *hammers*, Section 6.2.8). The forging process is generally referred to as *powder-metallurgy forging* (P/F), and when the unsintered perform is later sintered, the process is usually referred to as *sinter forging*. Ferrous and nonferrous powders can be processed in this manner, resulting in *full-density* products (on the order of 99.9% of the theoretical density of the material).

   The deformation of the preform may consist of the following two modes: *upsetting* and *re-pressing*. In upsetting, the material flows laterally outward, as also shown in Fig. 6.1. The preform is thus subjected to compressive and shear stresses during deformation. As a result, any interparticle residual oxide films present are broken up, and the part has better toughness, ductility, and fatigue strength. In *re-pressing*, the material flow is mainly in the direction of the movement of the punch; as a result, the mechanical properties of the part are not as high as those made by upsetting.

   P/M forged products have good surface finish and dimensional tolerances, little or no flash, and uniform and fine grain size throughout the part. The superior properties obtained make this technology particularly suitable for applications such as highly stressed components for use in automobiles (such as connecting rods), jet engines, military hardware, and off-road equipment.

3. The inherent porosity of powder-metallurgy components can be taken advantage of by **impregnating** the components with a fluid. A typical application is to impregnate the sintered part with oil, which is usually done by immersing the part in heated oil. Bearings and bushings that are internally lubricated, with up to 30% oil by volume, are made by this method. Internally lubricated components thus have a continuous supply of lubricant during their service life. Universal joints are now made with grease-impregnated P/M techniques, as a result of which these parts no longer require grease fittings.

4. **Infiltration** is a process whereby a metal slug, with a lower melting point than that of the part, is placed on top of the sintered part, and the assembly is heated to a temperature sufficient to melt the slug. The molten metal infiltrates the pores of the other part by *capillary action*, resulting in a relatively pore-free part with good density and strength. The most common application is the infiltration of iron-base compacts with copper.

   The advantages of this technique are that hardness and tensile strength are improved and the pores of the part are filled, thus preventing moisture penetration, which otherwise could cause corrosion. Infiltration may also be done using a lead slug, whereby, because of the low shear strength of lead, the infiltrated part has lower frictional characteristics than the uninfiltrated one (see Section 4.4.4). Some bearing materials are formed in this manner.

5. Powder-metal parts may be subjected to various other operations:

   a. *Heat treating* (quenching and tempering, and steam treating; Section 5.11) for improved strength, hardness, and resistance to wear.

   b. *Machining* (turning, milling, drilling, tapping, and grinding; Chapters 8 and 9) for making undercuts and slots, improving surface finish and dimensional accuracy, and producing threaded holes and other surface features.

   c. *Finishing* (deburring, burnishing, mechanical finishing, plating, and coating) for improved surface characteristics, corrosion, fatigue resistance, and appearance.

EXAMPLE 11.6 Production of tungsten carbide for tools and dies

Tungsten carbide is an important tool and die material, mainly because of its hardness, strength, and wear resistance over a wide range of temperatures (see Section 8.6.4). Powder-metallurgy techniques are used in making these carbides. First, powders of tungsten and carbon are blended together in a ball mill or a rotating mixer. The mixture (typically 94% tungsten and 6% carbon, by weight) is heated to approximately 1500°C (2800°F) in a vacuum-induction furnace. As a result, the tungsten is carburized, forming tungsten carbide in a fine powder form. A binding agent (usually cobalt) is then added to the tungsten carbide (with an organic fluid such as hexane), and the mixture is ball milled to produce a uniform and homogeneous mix. This process can take several hours, or even days.

   The mixture is then dried and consolidated, usually by cold compaction, at pressures in the range of 200 MPa (30,000 psi). Finally, it is sintered in a hydrogen-atmosphere or vacuum furnace at a temperature of 1350° to 1600°C (2500° to 2900°F) depending on its composition. At this temperature, the cobalt is in a liquid phase and acts as a binder for the carbide particles. (Powders may also be hot pressed at the sintering temperature, using graphite dies.) During sintering, the tungsten carbide undergoes a linear shrinkage of about 16%, corresponding to a volume shrinkage of about 40%; thus, control of size and shape is important for producing tools with accurate dimensions. A combination of other carbides, such as titanium carbide and tantalum carbide, can likewise be produced using mixtures made by the method described in this example.

11.6 | Design Considerations for Powder Metallurgy

Because of the unique properties of metal powders and their flow characteristics in the die, and the brittleness of green compacts, there are certain design principles that should be followed, as also illustrated in Figs. 11.17 through 11.19.

1. The shape of the compact must be kept as simple and uniform as possible. Features such as sharp changes in contour, thin sections, variations in thickness, and high length-to-diameter ratios should be avoided.

2. Provision must be made for ease of ejection of the green compact from the die without damaging the compact; thus, holes or recesses should be parallel to the axis of punch travel. Chamfers also should be provided.

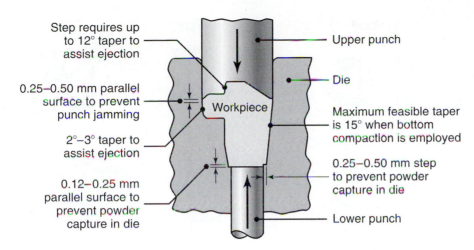

Step requires up
to 12° taper to
assist ejection

Upper punch

Die

0.25–0.50 mm parallel
surface to prevent
punch jamming

Workpiece

2°–3° taper to
assist ejection

Maximum feasible taper
is 15° when bottom
compaction is employed

0.12–0.25 mm
parallel surface to
prevent powder
capture in die

0.25–0.50 mm step
to prevent powder
capture in die

Lower punch

FIGURE 11.17 Die geometry and design features for powder-metal compaction. *Source:* Metal Powder Industries Federation.

3. As is the case for parts made by most other processes, P/M parts should be made with the widest dimensional tolerances, consistent with their intended applications, in order to increase tool and die life and reduce production costs.

4. Generally, part walls should not be less than 1.5 mm (0.060 in.) thick, although walls as thin as 0.34 mm (0.0135 in.) have been pressed successfully on components 1 mm (0.04 in.) in length. Walls with length-to-thickness ratios greater than 8:1 can be difficult to press, and density variations are virtually unavoidable.

5. Simple steps in parts can be produced if their size does not exceed 15% of the overall part length. Larger steps can be pressed, but they require more complex, multiple-action tooling (see also Fig. 11.7).

6. Letters and numbers can be pressed into parts if they are oriented perpendicular to the direction of pressing; letters can also be raised or recessed. However, raised letters are more susceptible to damage in the green stage and interfere with proper stacking during sintering.

7. Flanges and overhangs can be produced by a step in the die; however, such protrusions may be broken during ejection of the green part and may require more elaborate tooling. Also, a long flange should incorporate a draft around the flange, a radius at the bottom edge, and a radius at the juncture of the flange and/or component body to reduce stress concentrations and the likelihood of fracture.

8. A true radius cannot be pressed into the edge of a part, because it would require the punch to be feathered (that is, a smooth transition) to a zero thickness (Fig. 11.18e). Chamfers or flats are preferred for pressing; a common design approach is to use a 45° angle and 0.25-mm (0.010-in.) flat (Fig. 11.18d).

9. Features such as keys, keyways, and holes that are used for transmitting torques on gears and pulleys can be shaped during compaction. Bosses (see, for example, Fig. 5.35c) can be produced if proper drafts are used and their length is small compared with the overall size of the pressed component.

10. Notches and grooves can be made if they are oriented perpendicular to the pressing direction (Fig. 11.18c). It is recommended that circular grooves not exceed a depth of 20% of the overall depth of the component. Rectangular grooves should not exceed a depth of 15% of the overall depth of the component; see Fig. 11.19b.

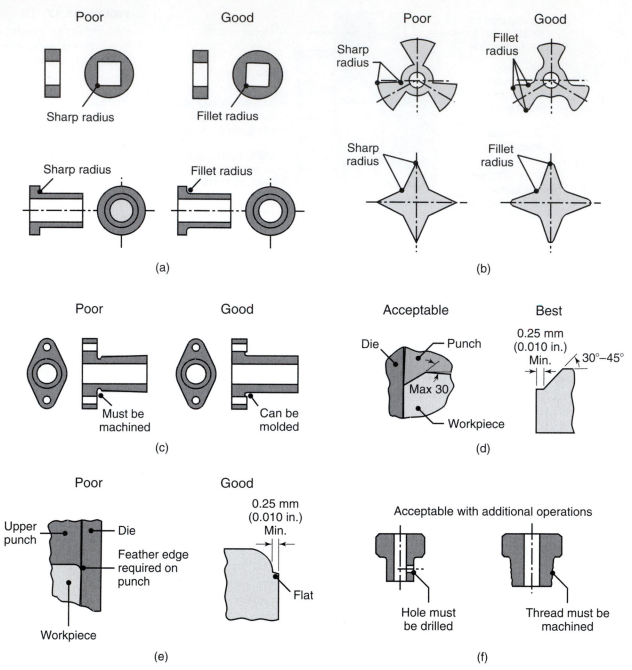

FIGURE 11.18 Examples of P/M parts, showing various poor and good designs. Note that sharp radii and reentry corners should be avoided, and that threads and transverse holes have to be produced separately by additional operations such as machining or grinding. *Source:* Metal Powder Industries Federation.

11. Parts produced through metal-injection molding have design constraints similar to injection molding. Wall thicknesses should be as uniform as possible in order to minimize distortion during sintering. Molds and dies should be designed with smooth transitions to prevent powder accumulation and to allow uniform distribution of metal powder (Fig. 11.20).

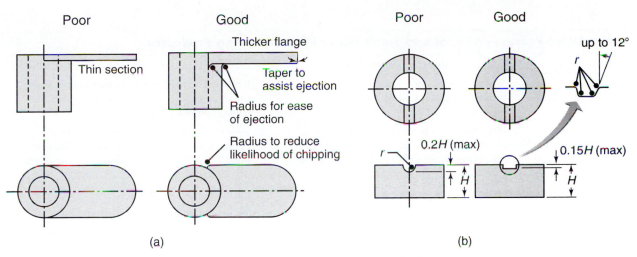

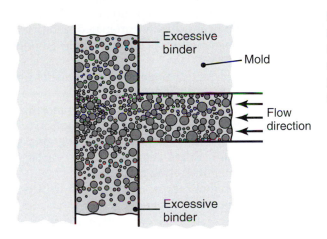

FIGURE 11.19   (a) Design features for use with unsupported flanges and (b) design features for use with grooves. *Source:* Metal Powder Industries Federation.

FIGURE 11.20   The use of abrupt transitions in molds for powder injection molding causing nonuniform metal-powder distribution within a part.

12. Dimensional tolerances of sintered parts are usually on the order of $\pm0.05-0.1$ mm ($\pm0.002-0.004$ in.). Tolerances improve significantly with additional operations such as sizing, machining, and grinding. (See Fig. 9.27.)

11.7 | Economics of Powder Metallurgy

Major cost elements in P/M parts production are associated with the powders, compacting dies, and equipment for compacting and sintering. The cost of metal powders per unit weight is much higher (by a factor of about 1.5 to 7 or even more) than that for molten metal for casting or for wrought bar stock for machining and forming. The cost also depends on the method of powder production, its quality, and the quantity purchased. The least expensive powder metal is iron, followed by, in increasing order, aluminum, zinc, copper, chromium, stainless steel, molybdenum, tungsten, cobalt, niobium, zirconium, and tantalum.

The net- or near-net-shape capability of P/M is a highly significant factor because of its cost-effectiveness in numerous applications. However, due to the high

TABLE 11.5

Competitive Features of P/M and Some Other Manufacturing Processes

Process	Advantages over P/M	Limitations as compared with P/M
Casting	Wide range of part shapes and sizes produced; generally low mold and setup cost.	Some waste of material in processing; some finishing required; may not be feasible for some high-temperature alloys.
Forging (hot)	High production rate of a wide range of part sizes and shapes; high mechanical properties through control of grain flow.	Some finishing required; some waste of material in processing; die wear; relatively poor surface finish and dimensional control.
Extrusion (hot)	High production rate of long parts; complex cross sections may be produced.	Only a constant cross-sectional shape can be produced; die wear; poor dimensional control.
Machining	Wide range of part shapes and sizes; short lead time; flexibility; good dimensional control and surface finish; simple tooling.	Waste of material in the form of chips; relatively low productivity.

cost of punches, dies, and various equipment for P/M processing, production volume must be sufficiently high to warrant this major expenditure, typically upwards of 50,000 parts per year. Modern P/M equipment is highly automated, and production is ideal for many automotive parts made in millions per year, with the labor cost per part being very low.

Powder metallurgy has become increasingly competitive with cast, forged, extruded, or machined parts (Table 11.5). Many die-pressed and sintered parts are used without the need for any machining, although some may require minor finishing operations. Even in highly complex parts, finish machining of P/M parts involves relatively simple operations, such as drilling and tapping holes or grinding some surfaces. Furthermore, an important P/M feature is that a single part may replace an assembly of several parts made by various manufacturing processes.

P/M forging is used mainly for critical applications in which full density and the accompanying superior fatigue resistance are essential. For specific applications, P/M forging competes with processes such as conventional forging and casting, in terms of both product properties and production cost. Automotive connecting rods can be produced by both P/M forging and casting. Metal injection molding (MIM) uses finer, and therefore more expensive, powder, and also requires more production steps, such as feedstock preparation, molding, debinding, and sintering. Because of these higher cost elements, MIM is cost-effective mainly for small but highly complex parts (generally weighing less than 100 g) required in large quantities.

Relatively large aerospace parts can be produced in small quantities when there are critical property requirements or special metallurgical considerations. For example, some nickel-base superalloys are so highly alloyed that they are prone to segregation during casting (see Section 5.3.3) and can be processed only by P/M methods. The powders may first be consolidated by hot extrusion (Section 6.4) and then hot forged (Section 6.2) in dies maintained at a high temperature. All beryllium processing is also based on powder metallurgy, and involves either cold isostatic pressing and sintering or hot isostatic pressing. Many of these materials are difficult to process and may require much finish machining even when the parts are made by P/M methods. This is in sharp contrast to the numerous ferrous press-and-sinter P/M parts that require practically no finish machining. To a large extent, the special and considerably expensive P/M methods are used for aerospace applications because no competing processes exist.

11.8 | Ceramics: Structure, Properties, and Applications

Ceramics are compounds of metallic and nonmetallic elements. The term ceramics refers both to the material and to the ceramic product itself. In Greek, the word *keramos* means "potter's clay," and *keramikos* means "clay products." Because of the large number of possible combinations of elements, a wide variety of ceramics is now available for widely different consumer and industrial applications.

The earliest use of ceramics was in pottery and bricks, dating back to before 4000 B.C. Ceramics have been used for many years in automotive spark plugs as an electrical insulator and for high-temperature strength; they have become increasingly important materials in heat engines and various other applications (Table 11.6), as

TABLE 11.6

Types and General Characteristics of Ceramics and Glasses

Type	General characteristics
Oxide ceramics	
Alumina	High hot hardness and abrasion resistance, moderate strength and toughness; most widely used ceramic; used for cutting tools, abrasives, and electrical and thermal insulation.
Zirconia	High strength and toughness; resistance to thermal shock, wear, and corrosion; partially stabilized zirconia and transformation-toughened zirconia have better properties; suitable for heat-engine components.
Carbides	
Tungsten carbide	High hardness, strength, toughness, and wear resistance, depending on cobalt binder content; commonly used for dies and cutting tools.
Titanium carbide	Not as tough as tungsten carbide but has a higher wear resistance; has nickel and molybdenum as the binder; used as cutting tools.
Silicon carbide	High-temperature strength and wear resistance; used for engines components and as abrasives.
Nitrides	
Cubic boron nitride	Second hardest substance known, after diamond; high resistance to oxidation; used as abrasives and cutting tools.
Titanium nitride	Used as coatings on tools, because of its low friction characteristics.
Silicon nitride	High resistance to creep and thermal shock; high toughness and hot hardness; used in heat engines.
Sialon	Consists of silicon nitrides and other oxides and carbides; used as cutting tools.
Cermets	Consist of oxides, carbides, and nitrides; high chemical resistance but is somewhat brittle and costly; used in high-temperature applications.
Nanophase ceramics	Stronger and easier to fabricate and machine than conventional ceramics; used in automotive and jet-engine applications.
Silica	High temperature resistance; quartz exhibits piezoelectric effects; silicates containing various oxides are used in high-temperature, nonstructural applications.
Glasses	Contain at least 50% silica; amorphous structure; several types available, with a wide range of mechanical, physical, and optical properties.
Glass ceramics	High crystalline component to their structure; stronger than glass; good thermal-shock resistance; used for cookware, heat exchangers, and electronics.
Graphite	Crystalline form of carbon; high electrical and thermal conductivity; good thermal-shock resistance; also available as fibers, foam, and buckyballs for solid lubrication; used for molds and high-temperature components.
Diamond	Hardest substance known; available as single-crystal or polycrystalline form; used as cutting tools and abrasives and as die insert for fine wire drawing; also used as coatings.

well as in tools, dies, and molds. Modern applications of ceramics include cutting tools (see Section 8.6), whiteware and tiles for architectural applications and automotive components, such as exhaust-port liners, coated pistons, and cylinder liners, with the desirable properties of strength and corrosion resistance at high operating temperatures (for improved engine efficiency).

Some properties of ceramics are significantly better than those of metals, particularly their hardness and thermal and electrical resistance. Ceramics are available as a single crystal or in polycrystalline form. Grain size has a major influence on the strength and properties of ceramics. The finer the grain size, the higher are the strength and toughness, hence the term **fine ceramics.** Ceramics are generally divided into the categories of **traditional ceramics** (whiteware, tiles, brick, pottery, and abrasive wheels) and **industrial ceramics,** also called **engineering** or **high-tech ceramics** (heat exchangers, cutting tools, semiconductors, and prosthetics).

11.8.1 Structure and types of ceramics

The structure of ceramic crystals is among the most complex of all materials, containing various elements of different sizes. The bonding between the atoms is generally *covalent* (electron sharing, hence strong bonds) and *ionic* (primary bonding between oppositely charged ions, hence strong bonds).

Among the oldest raw materials for ceramics is **clay,** a fine-grained sheetlike structure. The most common example is *kaolinite* (from Kaoling, a hill in China), a white clay, consisting of silicate of aluminum with alternating weakly bonded layers of silicon and aluminum ions (Fig. 11.21). When added to kaolinite, water attaches itself to the layers (adsorption), makes them slippery, and gives wet clay its well-known softness and plastic properties (*hydroplasticity*) that make it formable. Other raw materials for ceramics that are found in nature are *flint* (rock of very fine grained silica, SiO_2) and *feldspar* (a group of crystalline minerals consisting of aluminum silicates, potassium, calcium, or sodium). In their natural state, these raw materials generally contain various impurities which, for reliable performance, have to be removed prior to further processing of the materials into useful products. Highly refined raw materials produce ceramics with enhanced properties.

The various types of ceramics are

1. **Oxide ceramics**

   Alumina. Also called *corundum* or *emery, alumina* (aluminum oxide, Al_2O_3) is the most widely used *oxide ceramic,* either in pure form or as a raw material to be mixed (blended) with other oxides. It has high hardness and moderate strength. Although alumina exists in nature, it contains varying amounts of impurities and thus possesses nonuniform properties in its natural form. As a result, its behavior is unreliable. Aluminum oxide, as well as silicon carbide and many other ceramics, is now almost totally manufactured synthetically, so that its quality can be controlled (see also Section 9.2).

FIGURE 11.21 The crystal structure of kaolinite, commonly known as clay; compare with Figs. 3.2–3.4 for metals.

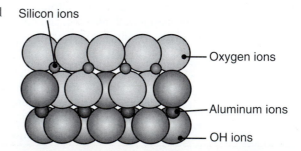

Silicon ions

Oxygen ions

Aluminum ions

OH ions

First made in 1893, *synthetic aluminum oxide* is obtained by the fusion of molten bauxite (an aluminum-oxide ore, the principal source of aluminum), iron filings, and coke in electric furnaces. The material is then crushed and graded by size by passing the particles through standard screens. Parts made of aluminum oxide are then *cold pressed and sintered* (*white ceramics*). Their properties are further improved by minor additions of other ceramics, such as titanium oxide and titanium carbide. Structures containing various alumina and other oxides are known as *mullite* and *spinel* and are used as refractory materials for high-temperature applications. The mechanical and physical properties of alumina are particularly suitable for applications such as electrical and thermal insulation, cutting tools, and abrasives.

Zirconia. *Zirconia* (zirconium oxide, ZrO_2) is white in color and has good toughness, resistance to wear, thermal shock, and corrosion; it also has low thermal conductivity and low friction coefficient. **Partially stabilized zirconia** (PSZ) has high strength and toughness and performs more reliably than zirconia. It is obtained by doping the zirconia with oxides of calcium, yttrium, or magnesium, a process which forms a material with fine particles of tetragonal zirconia in a cubic lattice. Typical applications of PSZ include dies for hot extrusion of metals and as zirconia beads for grinding and dispersion media for aerospace coatings, automotive primers and topcoats, and fine glossy print on flexible food packaging.

Another important characteristic of PSZ is the fact that its coefficient of thermal expansion is only about 20% lower than that of cast iron, and its thermal conductivity is about one-third that of other ceramics. Consequently, it is very suitable for heat-engine components, such as cylinder liners and valve bushings, to keep the cast-iron engine assembly intact during service (that is, no loosening of components or excessive tightness). Further improvements upon properties include **transformation-toughened zirconia** (TTZ) which, because of dispersed tough phases in the ceramic matrix, has higher toughness than that of PSZ.

2. **Other ceramics**   (see also Section 8.6)

Carbides. Typical examples of *carbides* are tungsten and titanium, used as cutting tools and die materials, and silicon carbide, used as abrasives (e.g., in grinding wheels):

a. **Tungsten carbide** (WC) consists of tungsten-carbide particles with cobalt as a binder. The amount of binder has a major influence on the material's properties. Toughness increases with the cobalt content (see Fig. 8.31), whereas hardness, strength, and resistance decrease.

b. **Titanium carbide** (TiC) has nickel and molybdenum as the binder and is not as tough as tungsten carbide.

c. **Silicon carbide** (SiC) has good resistance to wear, thermal shock, and corrosion. It also has a low friction coefficient and retains its strength at elevated temperatures. It is suitable for high-temperature components in heat engines and is also used as an abrasive (Section 9.2). Synthetic silicon carbide is made from silica sand, coke, and small amounts of sodium chloride and sawdust, in a process similar to making synthetic aluminum oxide (see above).

Nitrides. Another important class of ceramics is *nitrides*:

a. **Cubic boron nitride** (cBN) is the second hardest known substance, after diamond, and has special applications such as abrasives in grinding wheels and as cutting tools. It does not exist in nature and was first made synthetically in the 1970s, with techniques similar to those used for making synthetic diamond (Section 11.13.2).

b. **Titanium nitride** (TiN) is used extensively as coatings on cutting tools, improving tool life because of its low friction characteristics.

c. **Silicon nitride** (Si_3N_4) has high resistance to creep at elevated temperatures, low thermal expansion, and high thermal conductivity, and hence it resists thermal shock. It is suitable for high-temperature structural applications, such as in automotive engine and gas-turbine components, cam-follower rollers, bearings, sand-blast nozzles, and components for the paper industry.

Sialon. *Sialon* consists of silicon nitride with various additions of aluminum oxide, yttrium oxide, and titanium carbide. (See Section 8.6.8.) The word *sialon* is derived from the words *si*licon, *al*uminum, *o*xygen, and *n*itrogen. Sialon has higher strength and thermal-shock resistance than silicon nitride and is used primarily as a cutting-tool material.

Cermets. *Cermets* are combinations of *cer*amics bonded with a *met*allic phase. Introduced in the 1960s and also called **black ceramics** or *hot-pressed ceramics*, they combine the high-temperature oxidation resistance of ceramics and the toughness, thermal-shock resistance, and ductility of metals. A common application of cermets is cutting tools, a typical composition being 70% Al_2O_3 and 30% TiC. Other cermets contain various oxides, carbides, and nitrides, and are developed for high-temperature applications such as nozzles for jet engines and aircraft brakes. Cermets, which can be regarded as composite materials, are used in various combinations of ceramics and metals, bonded by powder-metallurgy techniques. (See also Section 11.14.)

3. **Silica.** Abundant in nature, *silica* is a polymorphic material; that is, it can have different crystal structures. The cubic structure is found in refractory bricks used for high-temperature furnace applications. Most glasses contain more than 50% silica. The most common form of silica is **quartz**, a hard, abrasive hexagonal crystal used extensively as oscillating crystals of fixed frequency in communications applications, because it exhibits the *piezoelectric* effect. (See Section 3.9.6.)

   Silicates are products of the reaction of silica with oxides of aluminum, magnesium, calcium, potassium, sodium, and iron. Examples include clay, asbestos, mica, and silicate glasses. **Lithium aluminum silicate** has very low thermal expansion and thermal conductivity and good thermal-shock resistance. However, it has very low strength and fatigue life, and thus it is suitable only for nonstructural applications, such as catalytic converters, regenerators, and heat-exchanger components.

4. **Nanoceramics (nanophase ceramics).** *Nanoceramics* consist of atomic clusters containing a few thousand atoms. They exhibit ductility at significantly lower temperatures than conventional ceramics and are stronger and easier to fabricate, with fewer flaws. Control of particle size, its distribution, and contamination during processing is important. Applications of nanophase ceramics in the automotive industry include valves, rocker arms, turbocharger rotors, and cylinder liners; they are also used in jet-engine components. Other applications in various stages include coatings, microbatteries, optical filters, very thin capacitors, nanoabrasives for lapping, solar cells, and valves for artificial hearts. Nanocrystalline second-phase particles, on the order of 100 nm or less, and fibers also are used as reinforcement in composites (see Sections 11.14). They have enhanced properties such as tensile strength and creep resistance. (See also *nanomaterials,* in Sections 3.11.9 and 13.18.)

11.8.2 General properties and applications of ceramics

Compared with metals, ceramics have the following relative characteristics: brittleness, high compressive strength and hardness at elevated temperatures, high elastic modulus, low toughness, low density, low thermal expansion, and low thermal and electrical conductivity. However, because of the wide variety of compositions and grain size, the mechanical and physical properties of ceramics vary significantly. For example, the electrical conductivity of ceramics can be modified from poor to good, which is the principle behind semiconductors (Section 13.3). Ceramics have a wide range of properties depending on the presence of defects, cracks (surface or internal), impurities, and different methods of production.

1. **Mechanical properties.** The mechanical properties of several engineering ceramics are given in Table 11.7. Note that, because of their sensitivity to cracks, impurities, and porosity, their strength in tension (transverse rupture strength) is approximately one order of magnitude lower than their compressive strength. Such defects lead to the initiation and propagation of cracks under tensile stresses, severely reducing tensile strength (see also Section 3.8). Consequently, reproducibility of their properties and reliability (acceptable performance over a specified period of time) of ceramic components are critical aspects in their service life.

   The tensile strength of polycrystalline ceramic parts increases with decreasing grain size and increasing porosity. Common earthenware has a porosity ranging between 10 and 15%, whereas the porosity of hard *porcelain* (a white ceramic composed of kaolin, quartz, and feldspar) is about 3%. An empirical relationship is given by

$$\text{UTS} \simeq \text{UTS}_o e^{-nP}, \tag{11.5}$$

   where P is the volume fraction of pores in the solid, UTS_o is the tensile strength at zero porosity, and the exponent n ranges between 4 and 7.

   The modulus of elasticity is likewise affected by porosity, as given by

$$E \simeq E_o(1 - 1.9P + 0.9P^2), \tag{11.6}$$

   where E_o is the modulus at zero porosity. Equation (11.6) is valid up to 50% porosity.

TABLE 11.7

Approximate Range of Properties of Various Ceramics at Room Temperature

Material	Symbol	Transverse rupture strength (MPa)	Compressive strength (MPa)	Elastic modulus (GPa)	Hardness (HK)	Poisson's ratio (v)	Density (kg/m^3)
Aluminum oxide	Al_2O_3	140–240	1000–2900	310–410	2000–3000	0.26	4000–4500
Cubic boron nitride	cBN	725	7000	850	4000–5000	—	3480
Diamond	—	1400	7000	830–1000	7000–8000	—	3500
Silica, fused	SiO_2	—	1300	70	550	0.25	—
Silicon carbide	SiC	100–750	700–3500	240–480	2100–3000	0.14	3100
Silicon nitride	Si_3N_4	480–600	—	300–310	2000–2500	0.24	3300
Titanium carbide	TiC	1400–1900	3100–3850	310–410	1800–3200	—	5500–5800
Tungsten carbide	WC	1030–2600	4100–5900	520–700	1800–2400	—	10,000–15,000
Partially stabilized zirconia	PSZ	620	—	200	1100	0.3	5800

Note: These properties vary widely, depending on the condition of the material.

Ceramics, unlike most metals and thermoplastics, generally lack impact toughness and thermal-shock resistance because of their inherent lack of ductility. In addition to fatigue failure under cyclic loading, ceramics (particularly glasses) exhibit a phenomenon called **static fatigue**: When subjected to a static tensile load over a period of time, these materials may suddenly fail. This phenomenon occurs in environments where water vapor is present. Static fatigue, which does not occur in dry air or a vacuum, has been attributed to a mechanism similar to stress-corrosion cracking of metals (Section 3.8.2).

Ceramic components that are to be subjected to tensile stresses in service may be **prestressed**, much like prestressed concrete. Prestressing shaped ceramic components subjects them to compressive stresses. Methods used for prestressing include (1) heat treatment and chemical tempering; Sections 5.11 and 11.11.2); (2) laser treatment of surfaces; (3) coating with ceramics with different degrees of thermal expansion; and (4) surface-finishing operations, such as grinding, in which compressive residual stresses are induced on the surfaces. Significant advances have been made in improving the toughness and other properties of ceramics, including **machinable ceramics** (see also Section 8.5.3).

2. **Physical properties.** Ceramics generally have relatively low specific gravity, ranging from about 3 to 5.8 for oxide ceramics, as compared with 7.86 for iron (Table 3.3). They have very high melting or decomposition temperatures. The thermal conductivity of ceramics varies by as much as three orders of magnitude, depending on their composition, whereas the thermal conductivity of metals varies by one order. The thermal conductivity of ceramics, as well as of other materials, decreases with increasing temperature and porosity because air is a very poor thermal conductor.

The thermal conductivity, k, is related to porosity by

$$k = k_o(1 - P),\qquad(11.7)$$

where k_o is the conductivity at zero porosity.

The thermal-expansion characteristics of ceramics are shown in Fig. 11.22. Thermal expansion and thermal conductivity induce thermal stresses that can lead to thermal shock or thermal fatigue (see Section 3.9.5). The tendency for *thermal cracking* (also called *spalling* when a piece or a layer from the surface breaks off), is lower with low thermal expansion and high thermal conductivity. Fused silica, for example, has high thermal shock resistance because of its virtually zero thermal expansion.

A common example that illustrates the importance of low thermal expansion is heat-resistant ceramics for cookware and stove tops. These ceramics can sustain high thermal gradients, from hot to cold and vice versa. The relative thermal expansion of ceramics and metals is important also for the use of ceramic components in heat engines. The fact that the thermal conductivity of partially stabilized zirconia components is close to that of the cast iron in engine blocks (see Fig. 11.22) is a further advantage in the use of PSZ in heat engines.

An additional characteristic, typically exhibited by oxide ceramics, is the **anisotropy of thermal expansion** whereby thermal expansion varies in different directions of the ceramic. This behavior causes thermal stresses that can lead to cracking of the ceramic component.

The *optical properties* of ceramics can be controlled by various formulations and control of structure, thereby imparting different degrees of transparency and colors. Single-crystal sapphire, for example, is completely transparent,

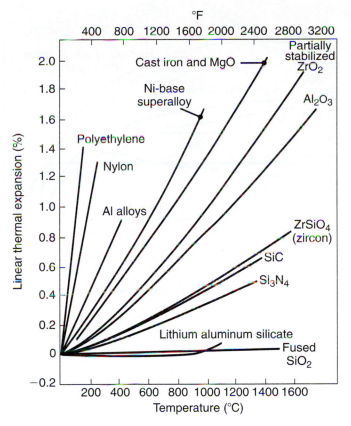

FIGURE 11.22 Effect of temperature on thermal expansion for several ceramics, metals, and plastics. Note that the expansions for cast iron and for partially stabilized zirconia (PSZ) are within about 20%.

zirconia is white, and fine-grained polycrystalline aluminum oxide is a translucent gray. Porosity also influences the optical properties of ceramics, much like trapped air in ice cubes, which makes ice less transparent and gives it a white appearance.

Although ceramics basically are resistors, they can be made electrically conductive by alloying them with certain elements (*doping*), thus making them act like a semiconductor or even a superconductor (see Section 11.15).

3. **Applications.** As shown in Table 11.6, ceramics have numerous consumer and industrial applications. Several types of ceramics are used in the electrical and electronics industry, because of their high electrical resistivity, dielectric strength (the voltage required for electrical breakdown per unit thickness), and magnetic properties suitable for applications such as magnets for speakers. An example of such a ceramic is *porcelain*. Certain ceramics such as lead zirconate titanate (PZT) and barium titanate ($BaTiO_3$) also have good *piezoelectric* properties.

The capability of ceramics to maintain their strength and stiffness at elevated temperatures (Figs. 11.23 and 11.24) makes them very attractive for high-temperature applications. The higher operating temperatures made possible by the use of ceramic components enable more efficient fuel burning and reduced emissions. Internal combustion engines are only about 30% efficient, but with the use of ceramic components, their operating performance can be improved by at least 30%. Ceramics that have been used successfully, especially in gasoline and diesel-engine components and as rotors, are silicon nitride, silicon carbide, and partially stabilized zirconia. Their high resistance to wear also makes them suitable for applications such as cylinder liners, bushings, seals,

FIGURE 11.23 Effect of temperature on the strength of various engineering ceramics. Note that much of the strength is maintained at high temperatures; compare with Figs. 2.9 and 8.30.

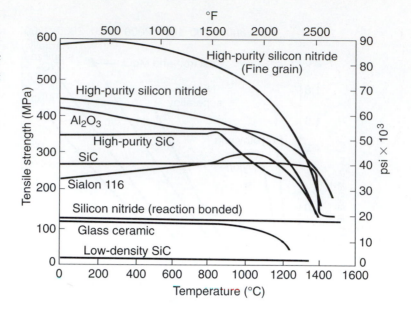

FIGURE 11.24 Effect of temperature on the modulus of elasticity for various ceramics; compare with Fig. 2.9. *Source:* After D.W. Richerson.

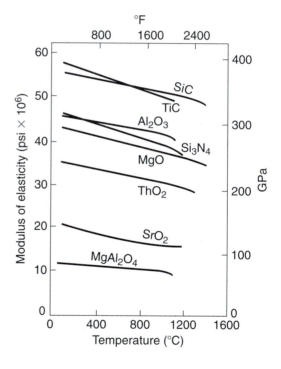

and bearings. Ceramics are also used to coat metal, which may be done to reduce wear, prevent corrosion, and/or provide a thermal barrier.

The low density and high elastic modulus of ceramics make it possible to reduce engine weight and also reduce the inertial forces generated by moving parts. High-speed components for machine tools are also candidates for ceramics, replacing metals. Their higher elastic modulus makes them attractive for improving their stiffness as well, thus reducing the tendency for vibration and chatter (see Section 8.12), as well as improving the dimensional accuracy of parts being machined. Silicon-nitride ceramics are used as ball bearings and rollers in machines.

Because of their strength and inertness, ceramics are also used as *biomaterials* to replace joints in the human body, as prosthetic devices, and for dental work. Commonly used *bioceramics* include aluminum oxide, silicon nitride, and various compounds of silica. Furthermore, ceramics can be made to be porous, thus allowing bone to grow into the porous surface and develop a strong mechanical bond.

EXAMPLE 11.7 Effect of porosity on properties

If a fully dense ceramic has the properties of UTS_o = 100 MPa, E_o = 400 GPa, and K_o = 0.5 W/m-K, what are these properties at 10% porosity? Assume that $n = 5$.

Solution. Using Eqs. (11.5) through (11.7), we have the following:

$$UTS = 100e^{-(5)(0.1)} = 61 \text{ MPa}$$

$$E = 400[1 - (1.9)(0.1) + (0.9)(0.1)^2] = 328 \text{ GPa}$$

and

$$k = 0.5 (1 - 0.1) = 0.45 \text{ W/m-K}.$$

EXAMPLE 11.8 Ceramic Ball and Roller Bearings

Silicon-nitride ceramic ball and roller bearings are used when high temperature, high speed, or marginally lubricated conditions occur. The bearings can be made entirely from ceramics or just the ball and rollers, in which case they are referred to as hybrid bearings (Fig. 11.25). Examples of machines utilizing ceramic and hybrid bearings include high-performance machine-tool spindles (see Section 8.13), metal-can seaming heads, high-speed flow meters, and the Space Shuttle main booster rocket's liquid oxygen and hydrogen pumps.

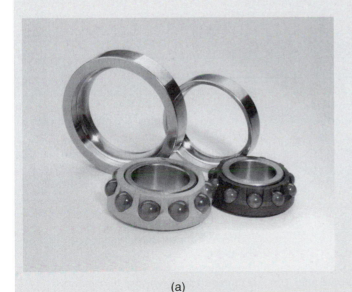

(a)

(b)

FIGURE 11.25 A selection of ceramic bearings and races. *Source:* Courtesy of Timken, Inc.

Ceramic balls have high wear resistance, high fracture toughness, perform well with little or no lubrication, and have low density. The balls have a coefficient of thermal expansion one-fourth that of steel, and they can withstand temperatures of up to 1400°C (2550°F). The ceramic balls have a diametral tolerance of 0.13 μm (5 μin.) and a surface roughness of 0.02 μm (0.8 μin.). Produced from titanium and carbon nitride by powder-metallurgy techniques, the full-density titanium carbonitride (TiCN) or silicon nitride (Si_3N_4) bearing-grade material can be twice as hard as chromium steel and 40% lighter. Components up to 300 mm (12 in.) in diameter can be produced.

11.9 | Shaping Ceramics

Several techniques can be used for shaping ceramics into useful products (Table 11.8). Generally, the procedure involves the following steps: (a) crushing or grinding the raw materials into very fine particles, (b) mixing the particles with additives to impart certain desirable characteristics, and (c) shaping, drying, and firing the material. *Crushing* (also called *comminution* or *milling*) of the raw materials is usually done in a ball mill (Fig. 11.26b), either dry or wet. Wet crushing is more effective because it keeps the particles together and prevents the suspension of fine particles in air. The ground particles are then mixed with *additives*, the function of which is one or more of the following:

1. *Binder,* for the ceramic particles.
2. *Lubricant,* for mold release and to reduce internal friction between particles during molding.
3. *Wetting agent,* to improve mixing.
4. *Plasticizer,* to make the mix more plastic and formable.
5. *Deflocculent,* to make the ceramic-water suspension uniform. Typical defloc-culents include Na_2CO_3 and Na_2SiO_3 in amounts of less than 1%; they change

TABLE 11.8

General Characteristics of Ceramics Processing Methods

Process	Advantages	Limitations
Slip casting	Large parts; complex shapes; low equipment cost.	Low production rate; limited dimensional accuracy.
Extrusion	Hollow shapes and small diameters; high production rate.	Parts have constant cross section; limited thickness.
Dry pressing	Close tolerances; high production rate with automation.	Density variation in parts with high length-to-diameter ratios; dies require high abrasive-wear resistance; equipment can be costly.
Wet pressing	Complex shapes; high production rate.	Limited part size and dimensional accuracy; tooling costs can be high.
Hot pressing	Strong, high-density parts.	Protective atmospheres required; die life can be short.
Isostatic pressing	Uniform density distribution.	Equipment can be costly.
Jiggering	High production rate with automation; low tooling cost.	Limited to axisymmetric parts; limited dimensional accuracy.
Injection molding	Complex shapes; high production rate.	Tooling costs can be high.

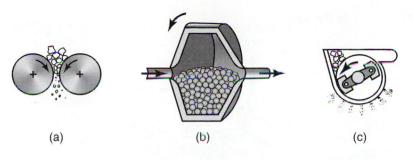

FIGURE 11.26 Methods of crushing ceramics to obtain very fine particles: (a) roll crushing, (b) ball milling, and (c) hammer milling.

(a)   (b)   (c)

the electrical charges on the particles of clay so that they repel instead of attract each other. Water is added to make the mixture more pourable and less viscous.

6. Various agents to control *foaming* and *sintering*.

The three basic shaping processes for ceramics are casting, plastic forming, and pressing, and are described below.

11.9.1 Casting

The most common casting process is **slip casting**, also called *drain casting* (Fig. 11.27). A *slip* is a suspension of ceramic particles in a liquid, generally water. In this process, the slip is poured into a porous mold made of plaster of paris. The slip must have sufficient fluidity and low viscosity to flow easily into the mold, much like the fluidity of molten metals. After the mold has absorbed some of the water from the outer layers of the suspension, it is inverted, and the remaining suspension is poured out (for making hollow objects, as in slush casting of metals, illustrated in Fig. 5.12). The top of the part is then trimmed, the mold is opened, and the part is removed.

Large and complex parts, such as plumbing ware, art objects, and dinnerware, can be made by slip casting. Although dimensional control is limited and the production rate is low, mold and equipment costs are also low. In some applications, components of the product (such as handles for cups and pitchers) are made separately and then joined, using the slip as an adhesive. Molds for slip casting may consist of several components.

For solid ceramic parts, the slip is supplied continuously into the mold to replenish the absorbed water, as otherwise the part will shrink. At this stage, the part is described as a soft solid or semirigid. The higher the concentration of solids in the slip, the less water has to be removed. The part, called *green* (as in powder metallurgy), is then fired.

While the ceramic parts are still green, they may be machined to produce certain features or dimensional accuracy to the parts. Because of the delicate nature of the green compacts, however, machining is usually done manually or with simple

1.   2.   3.   4.   5.

FIGURE 11.27 Sequence of operations in slip casting a ceramic part. After the slip has been poured, the part is dried and fired in an oven to give it strength and hardness. The step in (4.) is a trimming operation. *Source:* After F.H. Norton.

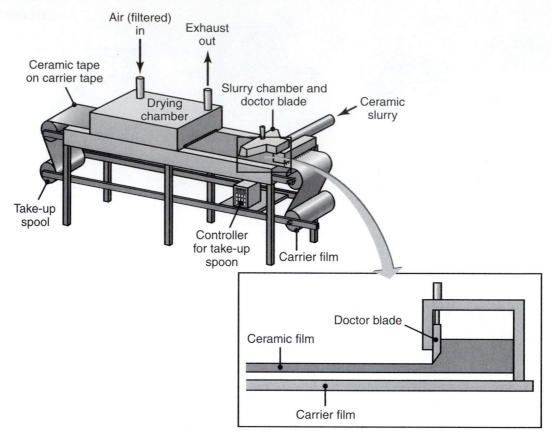

FIGURE 11.28 Production of ceramic sheets through the doctor-blade process.

tools. The flashing (similar to a flash in forging, shown in Fig. 6.14c) in a slip casting, for example, may be removed gently, with a fine wire brush, or holes can be drilled. Detailed work, such as tapping of threads, is generally not done on green compacts because warpage due to firing makes such machining not viable.

Thin ceramic sheets, less than 1.5 mm (0.06 in.) thick, can be made by a casting technique called the **doctor-blade process.** In this operation, the slip is cast over a moving plastic belt, and its thickness is controlled by a blade, as shown in Fig. 11.28. Other processes include *rolling* the slip between pairs of rolls and casting the slip over a paper tape, which is then burned off during firing.

11.9.2 Plastic forming

Plastic forming (also called *soft, wet,* or *hydroplastic forming*) can be done by various methods, such as extrusion, injection molding, or molding and jiggering (as done on a potter's wheel). Plastic forming tends to orient the layered structure of clays along the direction of material flow, as is the case in metal forming. This orientation leads to anisotropic behavior of the material, both in subsequent processing and in the final properties of the ceramic product.

In **extrusion,** the clay mixture, containing 20 to 30% water, is forced through a die opening by screw-type equipment (see, for example, Fig. 10.22). The cross section of the extruded product is constant, but there are limitations on the wall thickness for hollow extrusions. Tooling costs are low and production rates are high.

11.9.3 Pressing

The various methods of pressing are

1. **Dry pressing.** Similar to powder-metal compaction, *dry pressing* is used for relatively simple shapes such as whiteware, refractories, and abrasive products. The process has the same high production rates and close control of dimensional tolerances as in powder metallurgy. Although the moisture content of the mixture is generally below 4%, it may be as high as 12%. Organic and inorganic binders, such as stearic acid, wax, starch, and polyvinyl alcohol, are usually added to the mixture; the binders also act as lubricants. The pressing pressure is between 35 and 200 MPa (5 and 30 ksi). Modern presses for dry pressing are highly automated. Dies are usually made of carbides or hardened steel and must have high wear resistance to withstand the abrasive ceramic particles; thus they can be expensive.

   The density of dry-pressed ceramics can vary significantly (Fig. 11.29) because of friction between particles and at the mold walls, as in powder-metal compaction (see Fig. 11.7). Several methods may be used to minimize density variations. Vibratory pressing and impact forming are used, particularly for nuclear-reactor fuel elements. Isostatic pressing also reduces density variations. Density variations cause warping during firing. Warping is particularly severe for parts that have high length-to-diameter ratios; the recommended maximum ratio is 2:1.

2. **Wet pressing.** In this operation, the part is formed in a mold while under high pressure in a hydraulic or mechanical press. Wet pressing is generally used to make intricate shapes. The moisture content of the part usually ranges from 10 to 15%. Production rates are high, but part size is limited. Dimensional control is difficult because of shrinkage during drying, and tooling costs can be high.

3. **Isostatic pressing.** Used extensively in powder metallurgy, *isostatic pressing* is also used for ceramics in order to obtain uniform density throughout the part. Among products made by this process are (a) automotive spark-plug insulators made of porcelain and (b) silicon-nitride vanes for high-temperature use.

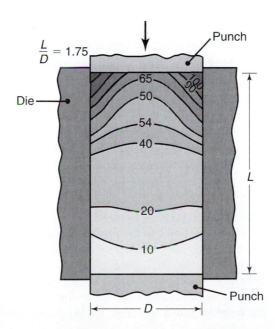

FIGURE 11.29 Density variation in pressed compacts in a single-action press. Note that the variation increases with increasing *L/D* ratio; see also Fig. 11.7e. *Source:* After W.D. Kingery.

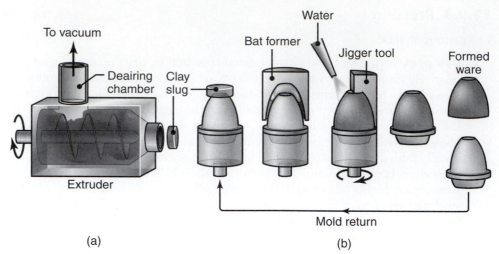

FIGURE 11.30   (a) Extruding and (b) jiggering operations in shaping ceramics. *Source:* After R.F. Stoops.

4. **Jiggering.** A combination of processes is used to make axisymmetric parts, such as ceramic plates. Clay slugs are first extruded, then formed into a bat over a plaster mold, and finally jiggered on a rotating mold (Fig. 11.30). *Jiggering* is an operation in which the clay bat is formed with templates or rollers. The part is then dried and fired. The process has limited dimensional accuracy.

5. **Injection molding.** The advantages of *injection molding* of plastics and powder metals were indicated earlier in Sections 10.10.2 and 11.3.4. *Ceramic injection molding* (CIM) is now used extensively for *precision forming* of ceramics for high-technology applications such as in rocket-engine components, piezoelectric scanners, and medical devices such as ultrasonic scalpels. The raw material is mixed with a binder, such as a thermoplastic polymer (polypropylene, low-density polyethylene, ethylene vinyl acetate, or wax). The binder is usually removed by pyrolysis, and the part is sintered by firing. Injection molding can produce thin sections, typically less than 10–15 mm (0.4–0.6 in.), using most engineering ceramics, such as alumina, zirconia, silicon nitride, and silicon carbide. Thicker sections require careful control of the materials used and of processing parameters in order to avoid internal voids and cracks, such as those due to shrinkage.

6. **Hot pressing.** In this operation, also called *pressure sintering*, pressure and temperature are applied simultaneously, thus reducing porosity and making the part denser and stronger. *Hot isostatic pressing* (see Section 11.3.3) also may be used in this process, particularly to improve the quality of high-technology ceramics. Because of the presence of both pressure and temperature, die life in hot pressing can be shorter than in others processes. Protective atmospheres are usually employed, and graphite is a commonly used punch and die material.

11.9.4 Drying and firing

After the ceramic has been shaped by any of the methods described above, the next step is to dry and fire the part to give it strength. *Drying* is a critical stage because of the tendency for the part to warp or crack from variations in moisture content throughout the part, particularly for complex shapes. Thus, the control of atmospheric humidity and temperature is important.

Loss of moisture results in shrinkage of the part by as much as 15 to 20% of the original moist size (Fig. 11.31). In a humid environment, the evaporation rate is

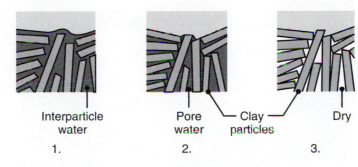

FIGURE 11.31 Shrinkage of wet clay, caused by removal of water during drying; shrinkage may be as much as 20% by volume. *Source:* After F.H. Norton.

low, and consequently the moisture gradient across the thickness of the part is lower than that in a dry environment. The low moisture gradient, in turn, prevents a large and uneven shrinkage from the surface to the interior during drying. The dried part (called *green*) can be machined relatively easily at this stage to bring it closer to its final shape, although it must be handled carefully.

Firing, also called *sintering*, involves heating the part to an elevated temperature in a controlled environment, similar to the sintering process in powder metallurgy, and some shrinkage occurs during firing. Firing gives the ceramic part its strength and hardness. The improvement in properties results from (1) development of a strong bond between the complex oxide particles in the ceramic and (2) reduced porosity. A more recent technology, although not yet commercialized, involves **microwave sintering** of ceramics, in furnaces operating at more than 2 GHz.

Nanophase ceramics (see Sections 11.8.1 and 13.18) can be sintered at lower temperatures than those for conventional ceramics. Nanophase ceramics are also easier to fabricate and can be compacted at room temperature to high densities, hot pressed to theoretical density, and formed into net-shape parts without binders or sintering aids.

EXAMPLE 11.9 Dimensional changes during shaping of ceramic components

A solid cylindrical ceramic part is to be made whose final length must be $L = 20$ mm. It has been established that for this material, linear shrinkages during drying and firing are 7% and 6%, respectively, based on the dried dimension L_d. Calculate (a) the initial length L_o of the part and (b) the dried porosity P_d if the porosity of the fired part, P_f, is 3%.

Solution

(a) On the basis of the information given, and remembering that firing is preceded by drying, we can write

$$\frac{(L_d - L)}{L_d} = 0.06,$$

or

$$L = (1 - 0.06)L_d.$$

Hence,

$$L_d = \frac{20}{0.94} = 21.28 \text{ mm}$$

and

$$L_o = (1 + 0.07)L_d = (1.07)(21.28) = 22.77 \text{ mm}.$$

(b) Since the final porosity is 3%, the actual volume V_a of the ceramic material is

$$V_a = (1 - 0.03)V_f = 0.97V_f,$$

where V_f is the fired volume of the part. Since the linear shrinkage during firing is 6%, the dried volume V_d of the part can be determined as

$$V_d = \frac{V_f}{(1 - 0.06)^3} = 1.2V_f.$$

Hence,

$$\frac{V_a}{V_d} = \frac{0.97}{1.2} = 0.81 = 81\%.$$

Therefore, the porosity P_d of the dried part is 19%.

11.9.5 Finishing operations

After firing, additional operations may be performed to give the part its final shape, remove surface flaws, and improve surface finish and dimensional accuracy. The processes involved are (a) grinding, (b) lapping, (c) ultrasonic machining, (d) electrical-discharge machining, (e) laser-beam machining, (f) abrasive water-jet machining, and (g) tumbling to remove sharp edges and grinding marks. The choice of the process is important in view of the brittle nature of most ceramics and the additional costs involved in these processes. The effect of the finishing operation on the properties of the product also must be considered. Because of notch sensitivity of ceramics, the finer the finish, the higher is the part's strength. To improve appearance and strength, and to make them impermeable, ceramic products are often coated with a glaze material (see also Section 4.5.1) which forms a glassy coating after firing.

11.10 | Glasses: Structure, Properties, and Applications

Glass is an *amorphous solid* with the structure of a liquid; in other words, it has been *supercooled* (cooled at a rate too high for crystals to form). Generally, glass is defined as an inorganic product of fusion that has cooled to a rigid condition without crystallizing. Glass has no distinct melting or freezing point; thus its behavior is similar to that of amorphous polymers (see Section 10.2.2). It has been estimated that there are some 750 different types of commercially available glasses today. The uses of glass range from windows, bottles, and cookware to glasses with special mechanical, electrical, high-temperature, chemical-resistant, corrosion-resistant, and optical characteristics. Special glasses are used in *fiber optics* for communication, with little loss in signal power, and in *glass fibers* with very high strength for reinforced plastics (Section 10.9.2).

All glasses contain at least 50% silica (known as a *glass former*). The composition and properties of glasses, except strength, can be modified greatly by the addition of oxides of aluminum, sodium, calcium, barium, boron, magnesium, titanium, lithium, lead, and potassium. Depending on their function, these oxides are known as *intermediates* or *modifiers*. Glasses are generally resistant to chemical attack and are also ranked by their resistance to acid, alkali, or water corrosion.

TABLE 11.9

General Characteristics of Various Types of Glasses				
	Soda-lime glass	Lead glass	Borosilicate glass	Fused	96% Silica glass
Density	High	Highest	Medium	Low	Lowest
Strength	Low	Low	Moderate	High	Highest
Resistance to thermal shock	Low	Low	Good	Better	Best
Electrical resistivity	Moderate	Best	Good	Good	Good
Hot workability	Good	Best	Fair	Poor	Poorest
Heat treatability	Good	Good	Poor	None	None
Chemicals resistance	Poor	Fair	Good	Better	Best
Impact abrasion resistance	Fair	Poor	Good	Good	Best
Ultraviolet-light transmission	Poor	Poor	Fair	Good	Good
Relative cost	Lowest	Low	Medium	High	Highest

11.10.1 Types of glasses

Almost all commercial glasses are categorized by type, as shown in Table 11.9:

1. **Soda-lime glass** (the most common type).
2. **Lead-alkali glass.**
3. **Borosilicate glass.**
4. **Aluminosilicate glass.**
5. **96% silica glass.**
6. **Fused silica.**

Glasses are also classified as colored, opaque (white and translucent), multiform (variety of shapes), optical, photochromatic (darkens when exposed to light, as in some types of sunglasses), photosensitive (changes from clear to opal), fibrous (drawn into long fibers, as in fiberglass), and foam or cellular glass (containing air bubbles, hence a good thermal insulator). Glasses also are referred to as **hard** or **soft**, usually in the sense of a thermal property rather than a mechanical property (as in hardness). Thus, a soft glass softens at a lower temperature than does a hard glass. Soda-lime and lead-alkali glasses are considered soft, and the rest are considered hard.

11.10.2 Mechanical properties

For all practical purposes, the behavior of glass, as for most ceramics, can be considered as linearly elastic and brittle. The range of elastic modulus for most commercial glasses is 55–90 GPa (8–13 million psi), and the Poisson's ratio ranges from 0.16–0.28. The hardness of glasses, as a measure of resistance to scratching, ranges from 5–7 on the Mohs scale (Section 2.6.6), equivalent to a range of approximately 350–500 HK.

In *bulk* form glass has a strength of less than 140 MPa (20 ksi). The relatively low strength of bulk glass is attributed to the presence of small flaws and microcracks on the surface of the glass, some or all of which may be introduced during normal handling of the glass by inadvertent abrading. These defects reduce the strength of glass by two to three orders of magnitude, compared with its ideal (defect-free) strength (see also Section 3.3.2). Glasses can be strengthened by thermal or chemical treatments to obtain high strength and toughness, as described in Section 11.11.2.

The strength of glass can theoretically reach as high as 35 GPa (5 million psi). When molten glass is freshly drawn into fibers (*fiberglass*), its tensile strength ranges from 0.2 to 7 GPa (30 to 1000 ksi), with an average value of about 2 GPa (300 ksi). Thus, glass fibers are stronger than steel and are used to reinforce plastics in applications such as boats, automobile bodies, furniture, and sports equipment. The strength of glass is generally measured by bending it (see also Section 2.5). The surface of the glass is first thoroughly abraded (roughened) to ensure that the test gives a reliable strength level in actual service under adverse conditions. The phenomenon of *static fatigue* observed in ceramics (see Section 11.8.2) is also exhibited by glasses. As a general rule, if a glass item (such as a glass shelf) must withstand a load for 1000 hours or longer, the maximum stress that can be applied to it is approximately one-third the maximum stress that the same item can withstand during the first second of loading.

11.10.3 Physical properties

Glasses have low thermal conductivity and high dielectric strength. Their thermal expansion coefficient is lower than those for metals and plastics (see Table 3.3) and may even approach zero. For example, titanium-silicate glass, a clear, synthetic high-silica glass, and fused silica, a clear, synthetic amorphous silicon dioxide of very high purity, both have near-zero coefficient of expansion. (See Fig. 11.22.) Optical properties of glasses, such as reflection, absorption, transmission, and refraction, can be modified by varying their composition and treatment.

11.10.4 Glass ceramics

Glass ceramics (such as *Pyroceram*, a trade name) contain large proportions of several oxides, and hence their properties are a combination of those for glass and ceramics. They have a high crystalline component to their microstructure and most are stronger than glass. First developed in 1957, glass ceramics are first shaped and then heat treated to cause **devitrification** (recrystallization) of the glass. Unlike most glasses, which are clear, glass ceramics generally are white or gray in color.

The hardness of glass ceramics ranges approximately from 520 to 650 HK. They have a near-zero coefficient of thermal expansion; hence they have good thermal shock resistance. They also have high strength because of the absence of porosity usually found in conventional ceramics. The properties of glass ceramics can be improved by modifying their composition and by heat-treatment techniques. Glass ceramics are suitable for cookware, heat exchangers for gas-turbine engines, radomes (housings for radar antenna), and electrical and electronics applications.

11.11 | Forming and Shaping Glass

All glass forming and shaping processes begin with molten glass, which has the appearance of red-hot viscous syrup, supplied from a melting furnace or tank. Glass products generally can be categorized as

- **Flat sheet** or **plate**, ranging in thickness from about 0.8 to 10 mm (0.03 to 0.4 in.), such as window glass, glass doors, and glass tabletops.
- **Rods** and **tubing** used for chemicals, laboratory glassware, neon lights, and decorative artifacts.

- **Discrete products**, such as bottles, vases, headlights, and television tubes.
- **Glass fibers** to reinforce composite materials and for fiber optics.

Flat-sheet glass is made by the *float method*, although it has traditionally been made by *drawing* or *rolling* from the molten state. All of these processes are continuous operations. In the **drawing** process, the molten glass passes through a pair of rolls. The solidifying glass is squeezed between the rolls, forming a sheet, which is then moved forward over a set of smaller rolls. In the **rolling** process, the molten glass is squeezed between rollers, shaping it into a sheet. The surfaces of the glass also can be embossed with a pattern by shaping the roller surfaces accordingly. Glass sheet produced by these two processes has a usually rough surface appearance. Thus, in making plate glass, for example, both surfaces have to be ground parallel and polished for a smooth appearance.

In the **float method** (Fig. 11.32), developed in the 1950s, molten glass from the furnace is fed into a bath of molten tin, under controlled atmosphere. The glass floats on the tin bath and then moves over rollers into another chamber (*lehr*) and solidifies. **Float glass** has a smooth (*fire-polished*) surface and needs no further finishing operations.

Glass tubing is manufactured by the process shown in Fig. 11.33. In this operation, molten glass is wrapped around a rotating hollow cone-shaped or cylindrical *mandrel* and is drawn out by a set of rollers. Air is blown through the hollow mandrel to keep the glass tube from collapsing. The machines used in this process may be horizontal, vertical, or slanted downward. **Glass rods** are made in a similar manner.

Continuous **fibers** are drawn through multiple (200 to 400) orifices in heated platinum plates, at speeds as high as 500 m/s (1700 ft/s). Fibers as small as 2 μm (80 μin.) in diameter can be produced by this method. In order to protect their surfaces, the fibers are subsequently coated with chemicals such as silane. Short glass fibers, used as thermal insulating material (**glass wool**) or for acoustic insulation, are made by a *centrifugal spraying process* in which molten glass is ejected (spun) from a rotating head. The diameter of the fibers is typically in the range of 20 to 30 μm (800 to 1200 μin.).

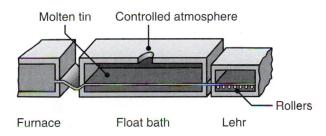

Molten tin   Controlled atmosphere

Furnace   Float bath   Lehr   Rollers

FIGURE 11.32 The float method of forming sheet glass. *Source:* Corning Glass Works.

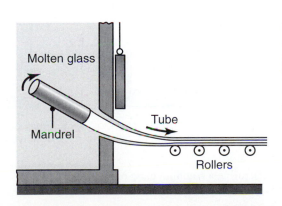

Molten glass

Mandrel   Tube   Rollers

FIGURE 11.33 Continuous manufacturing process for glass tubing. Air is blown through the mandrel to keep the tube from collapsing. *Source:* Corning Glass Works.

11.11.1 Manufacture of discrete glass products

Several processes are used for making discrete glass objects, including blowing, pressing, centrifugal casting, and sagging.

Blowing. The *blowing* process is used to make hollow, thin-walled glass items, such as bottles and flasks, and is similar to blow molding of thermoplastics, described in Section 10.10.3. The steps involved in the production of an ordinary glass bottle by the blowing process are illustrated in Fig. 11.34. Blown air expands a *gob* of heated glass against the inner walls of a mold, which are usually coated with a parting agent (such as oil or emulsion) to prevent the glass part from sticking to it. After shaping, the two halves of the mold are opened and the product is removed. The surface finish of products made by this process is acceptable for most applications. Although it is difficult to control the wall thickness of the product, this process is used for high rates of production. Light bulbs are made in automatic blowing machines, at a rate of 2000 bulbs per minute.

Pressing. In this operation, a gob of molten glass is placed in a mold and is pressed into shape with the use of a shaped plunger. The mold may be made in one piece (Fig. 11.35), or it may be a split mold (Fig. 11.36). After pressing, the solidifying glass acquires the shape of the cavity between the mold and the plunger. Because of the confined environment, the product has higher dimensional accuracy than can be

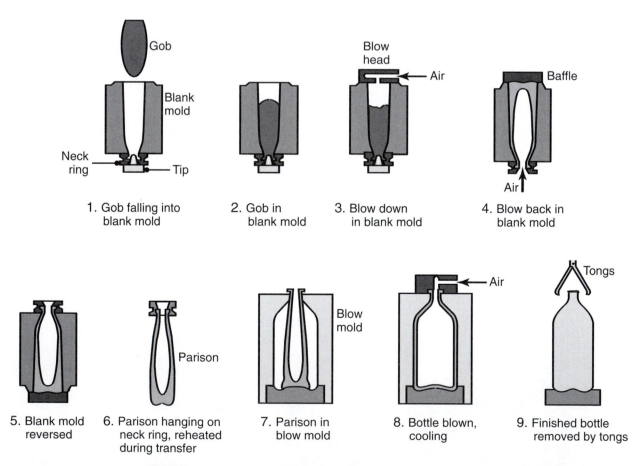

FIGURE 11.34 Stages in manufacturing a common glass bottle. *Source:* After F.H. Norton.

1. Empty mold   2. Loaded mold   3. Glass pressed   4. Finished piece

FIGURE 11.35 Manufacturing steps for a glass item by pressing in a mold. *Source:* Corning Glass Works.

1. Empty mold   2. Loaded mold   3. Glass pressed   4. Finished product

FIGURE 11.36 Pressing glass in a split mold. Note that the use of a split mold is essential to be able to remove the part; see also Figs. 10.34, 10.35, and 10.36. *Source:* After E.B. Shand.

obtained with the blowing method. However, pressing cannot be used on thin-walled items or for parts such as bottles from which the plunger cannot be retracted.

Centrifugal casting. Also known as *spinning* in the glass industry, the *centrifugal-casting* process is similar to that for metals (see Section 5.10.4). In this operation, the centrifugal force pushes the molten glass against the cool mold wall, where it solidifies. Typical products include picture tubes for televisions and missile nose cones.

Sagging. Shallow dish-shaped or lightly embossed glass parts can be made by the *sagging* process. A sheet of glass is placed over the mold and is heated, whereby the glass sags by its own weight and takes the shape of the mold cavity. The operation is similar to thermoforming of thermoplastic sheets (see Fig. 10.34), but without applying pressure or a vacuum. Typical parts made include dishes, lenses for sunglasses, mirrors for telescopes, and lighting panels.

11.11.2 Techniques for treating glass

Glass can be strengthened by thermal tempering, chemical tempering, and laminating, as described below. Glass products may also be subjected to annealing and other finishing operations.

1. **Thermal tempering.** Also called *physical tempering* or *chill tempering,* the surfaces of the hot glass are cooled rapidly, as illustrated in Fig. 11.37. As a result, the cooler surfaces shrink, and because the bulk is still hot, tensile stresses develop on the surfaces. As the bulk of the glass then begins to cool, it contracts. The solidified surfaces are now forced to contract, thus developing residual compressive surface stresses and interior tensile stresses. Compressive surface stresses improve the strength of the glass, as they do in other materials. Recall that the higher the coefficient of thermal expansion of the glass and the lower its thermal conductivity, the higher is the level of residual stresses developed and, hence, the stronger the glass becomes. Thermal tempering takes a relatively short time (minutes) and can be applied to most glasses. Because of the large amount of energy stored from residual stresses, **tempered glass** shatters into a large number of pieces when broken.

FIGURE 11.37 Stages in the development of residual stresses in tempered glass plate.

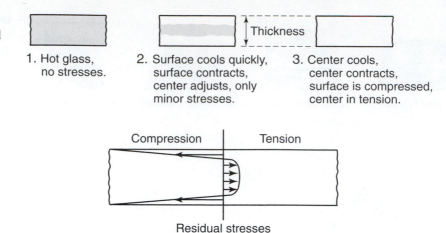

1. Hot glass, no stresses.

2. Surface cools quickly, surface contracts, center adjusts, only minor stresses.

3. Center cools, center contracts, surface is compressed, center in tension.

Compression | Tension

Residual stresses

2. **Chemical tempering.** In this operation, the glass is heated in a bath of molten KNO_3, K_2SO_4, or $NaNO_3$, depending on the type of glass, whereby ion exchange takes place, with larger atoms replacing the smaller atoms on the surface of the glass. As a result, residual compressive stresses develop on the surface, a situation similar to forcing a wedge between two bricks in a wall. The time required for chemical tempering is longer (about one hour) than for thermal tempering. Chemical tempering may be performed at various temperatures; at low temperatures, distortion of the part is minimal, and complex shapes can be treated. At elevated temperatures, there may be some distortion of the part, but the product can then be used at higher temperatures without loss of strength.

3. **Laminated glass.** This is a strengthening method in which two pieces of flat glass are assembled with a thin sheet of tough plastic, such as polyvinyl butyral (PVB), between them; hence the process is also called **laminate strengthening**. When laminated glass is broken, its pieces are held together by the plastic sheet, a phenomenon that can be observed on a shattered automobile windshield.

Finishing operations. Glass products may be subjected to further operations, such as cutting, drilling, grinding, and polishing, to meet specific product requirements. Sharp edges and corners can be smoothened by (a) grinding, as can be seen in glass tops for desks and shelves or (b) holding a torch against the edges (**fire polishing**) to cause localized softening, which then results in edge rounding due to surface tension.

As in metal products, residual stresses also can develop in glass products if they are not cooled at a sufficiently low rate. In order to ensure that the product is free from these stresses, it is *annealed* by a process similar to stress-relief annealing of metals. In this process, the glass is heated to a certain temperature and cooled gradually. Depending on the size, thickness, and type of glass, annealing times may range from a few minutes to as long as 10 months, as in the case of a 600-mm (24-in.) mirror for a telescope.

11.12 | Design Considerations for Ceramic and Glass Products

Ceramic and glass products require careful selection of composition, processing methods, finishing operations, and methods of assembly into other metallic or non-metallic components. Limitations such as general lack of tensile strength, sensitivity to external and internal defects, and low impact toughness are important considerations. These limitations have to be balanced against desirable characteristics, such as their hardness, resistance to abrasion, compressive strength at room and elevated temperatures, and diverse physical properties. Control of processing parameters and the type and level of impurities in the raw materials are important. As in all other design decisions, various factors should also be considered, including the number of parts required and the costs of tooling, equipment, and labor.

The possibilities of dimensional changes, warping, and cracking during processing are significant factors in selecting methods for shaping these materials. When a ceramic or a glass component is part of a larger assembly, compatibility with other components is another important consideration. Particularly important in such cases are thermal expansion (as in seals) and the type of loading. The potential consequences of part failure are always a significant factor in designing ceramic products.

11.13 | Graphite and Diamond

11.13.1 Graphite

Graphite is a crystalline form of carbon with a *layered structure* of basal planes or sheets of close-packed carbon atoms. Although brittle, graphite has high electrical and thermal conductivity and resistance to thermal shock and high temperature, although it begins to oxidize at 500°C (930°F). It is thus an important material for applications such as electrodes, brushes for motors, heating elements, high-temperature fixtures and furnace parts, crucibles for melting metals, molds for casting of metals (see, for example, Fig. 5.23), and seals (because of its low friction and high wear resistance). A common use of graphite is in ordinary pencils, in combination with clay.

Its low thermal-neutron-absorption cross section and high scattering cross section make graphite also suitable for nuclear applications. A characteristic of graphite is its resistance to chemicals; hence it is also used as filters for corrosive fluids. An important use of graphite is as **fibers** in composite materials and reinforced plastics, as described in Section 10.9.2.

Because of its layered structure, graphite is weak when sheared along the layers (see also Fig. 3.4). This characteristic, in turn, gives graphite its low frictional properties as a solid lubricant (Section 4.4.4), although its frictional properties are low only in an environment of air or moisture. Graphite is abrasive and a poor lubricant in a vacuum. Unlike other materials, the strength and stiffness of graphite increase with temperature.

Another form of graphite is soccer-ball-shaped carbon molecules, called **buckyballs** (after Buckminster Fuller, 1895–1983, inventor of the geodesic dome) or **fullerenes** (after Fuller). These chemically inert spherical molecules are produced from soot and act much like solid-lubricant particles. Fullerenes become superconductors when mixed with metals. Another development is **microcellular carbon foam,** which has uniform porosity and isotropic-strength characteristics. It is used, for example, as reinforcing components in aerospace structures that can be shaped directly.

Graphite is generally graded in terms of decreasing order of grain size: *industrial, fine grain,* and *micrograin.* As with ceramics, the mechanical properties of graphite improve with decreasing grain size. Micrograin graphite can be impregnated with copper and is used as electrodes for electrical-discharge machining and for furnace fixtures. *Amorphous graphite* is known as *lampblack* (black soot) and is used as a pigment. Graphite is usually processed by molding or forming, oven baking, and then machining to the final shape. It is available commercially in square, rectangular, or round shapes of various sizes.

Carbon nanotubes.   Carbon nanotubes have a similar geometric structure as graphite sheets; one can picture a nanotube as a rolled-up sheet of graphite. These nanotubes are a few nanometers in diameter and typically a few micrometers in length. While they have been suggested for a number of applications, only a few nanotube applications have been commercialized to date. Presently, nanotubes are most often cited as a natural building material for new microelectromechanical systems (see Section 13.18).

11.13.2 Diamond

A principal form of carbon is *diamond,* which has a covalently bonded structure and is the hardest substance known (7000 to 8000 HK). Diamond is brittle and it begins to decompose in air at about 700°C (1300°F), but in nonoxidizing environments it resists higher temperatures. The high hardness of diamond makes it an important material (a) for cutting tools (Section 8.6.9), either as a single crystal or in polycrystalline form, (b) as an abrasive in grinding wheels (Section 9.2) and for dressing of grinding wheels (sharpening of abrasive grains; Section 9.5.1), and (c) as a die material for drawing thin wire (Section 6.5.3), with a diameter of less than 0.06 mm (0.0025 in.).

Synthetic or **industrial diamond** was first made in 1955. A principal method of manufacturing it is to subject graphite to a hydrostatic pressure of 14 GPa (2 million psi) and a temperature of 3000°C (5400°F). Synthetic diamond is identical to natural diamond for industrial applications and has superior properties because of its lack of impurities. It is available in various sizes and shapes, the most common abrasive grain size being 0.01 mm (0.004 in.) in diameter. *Gem-quality synthetic diamond* is now made with electrical conductivity 50 times higher than that for natural diamond and 10 times more resistance to laser damage for optics applications. Potential applications of this quality of diamond include heat sinks for computers, in the telecommunications and integrated-circuit industries, and windows for high-power lasers. Diamond particles can also be *coated* (with nickel, copper, or titanium) for improved performance in grinding operations.

Diamondlike carbon (DLC), used as a coating, is described in Section 4.5.1. DLC is widely applied for its high wear resistance; applications include cutting tools, razors for shaving, and high-performance automotive engine components.

11.14 | Processing Metal-Matrix and Ceramic-Matrix Composites

Developments in composite materials are continually taking place, with a wide range and form of polymeric, metallic, and ceramic materials being used both as fibers and as matrix materials with the aim of improving strength, toughness, stiffness, resistance to high temperatures, and reliability in service, particularly under adverse environmental conditions.

11.14.1 Metal-matrix composites

The advantage of a metal matrix over a polymer matrix is its higher resistance to elevated temperatures and higher ductility and toughness. The limitations of metal-matrix composites are higher density and greater difficulty in processing. The matrix

TABLE 11.10

Metal-Matrix Composite Materials and Typical Applications

Fiber	Matrix	Typical applications
Graphite	Aluminum	Satellite, missile, and helicopter structures.
	Magnesium	Space and satellite structures.
	Lead	Storage-battery plates.
	Copper	Electrical contacts and bearings.
Boron	Aluminum	Compressor blades and structural supports.
	Magnesium	Antenna structures.
	Titanium	Jet-engine fan blades.
Alumina	Aluminum	Superconductor restraints in fusion power reactors.
	Lead	Storage-battery plates.
	Magnesium	Helicopter transmission structures.
Silicon carbide	Aluminum, titanium	High-temperature structures.
	Superalloy (cobalt base)	High-temperature engine components.
Molybdenum, tungsten	Superalloy	High-temperature engine components.

materials in these composites are usually aluminum, aluminum-lithium, magnesium, and titanium, although other metals are also being investigated. The fiber materials typically are graphite, aluminum oxide, silicon carbide, and boron, with beryllium and tungsten as other possibilities.

Because of their high specific stiffness, light weight, and high thermal conductivity, boron fibers in an aluminum matrix have, for example, been used for structural tubular supports in the Space Shuttle Orbiter. Other applications include bicycle frames, sporting goods, and stabilizers for aircraft and helicopters. An important consideration is the proper bonding of fibers to the metal matrix. Applications of metal-matrix composites are in gas turbines, electrical components, and various structural components (Table 11.10).

There are three principal methods of manufacturing metal-matrix composites into near-net-shape parts, as described below.

1. **Liquid-phase processing** basically consists of casting the liquid matrix and the solid reinforcement, using either conventional casting processes or pressure-infiltration casting techniques. In the latter process, pressurized gas is used to force the liquid matrix metal into a preform (usually as sheet or wire) made of the reinforcing fibers.

2. **Solid-phase processing** basically involves powder-metallurgy techniques, including cold and hot isostatic pressing. Proper mixing of the constituents for homogeneous distribution of the fibers throughout the matrix is important. A use of this technique, employed in tungsten-carbide tool and die manufacturing with cobalt as the matrix material, is described in Example 11.6. In making complex MMC parts with whisker or fiber reinforcement, die geometry and control of process variables are very important in ensuring proper distribution and orientation of the fibers within the part. MMC parts made by powder-metallurgy processes are subsequently heat treated for optimum properties.

3. **Two-phase processing** consists of the processes of *rheocasting*, described in Section 5.10.6, and *spray atomization and deposition*. In the latter processes, the reinforcing fibers are mixed with a matrix that contains both liquid and solid phases.

EXAMPLE 11.10 **Aluminum matrix composite brake calipers**

One of the trends in automobile design and manufacture is the increased use of lighter weight designs in order to realize improved performance and/or fuel economy. This can be seen in the development of metal-matrix composite brake calipers. Traditional brake calipers are made from cast iron, and can weigh about 3 kg (6.6 lb) each in a small car, and up to 14 kg (30 lb) in a truck. The cast-iron caliper could be completely redesigned using aluminum to achieve weight savings, but this would require a larger volume of material, and the space available between the wheel and rotor is highly constrained.

A new brake caliper was designed (Fig. 11.38) using an aluminum alloy reinforced with precast composite inserts using continuous ceramic fiber. The fiber

FIGURE 11.38
Aluminum-matrix composite brake caliper, using nanocrystalline alumina-fiber reinforcement. *Source:* Courtesy of 3M Specialty Materials Division.

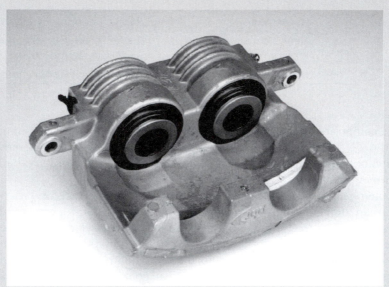

is a nanocrystalline alumina fiber [UTS = 3100 MPa (450 ksi), density = 3.9 g/cm^3], with diameter of 10–12 μm, and fiber volume fraction of 65%. The resulting metal-matrix composite had a tensile strength of 1500 MPa, and a density of 3.48 g/cm^3. Finite-element analysis confirmed that the design exceeded minimum design requirements and matched deflections of cast-iron calipers in a volume-constrained environment. The new brake caliper resulted in a weight savings of 50% and brings additional benefits of corrosion resistance and ease of recyclability.

11.14.2 Ceramic-matrix composites

Ceramic-matrix composites are another important class of engineered materials. Recall that ceramics are strong and stiff and resist high temperatures, but generally lack toughness. On the other hand, matrix materials such as silicon carbide, silicon nitride, aluminum oxide, and mullite (a compound of aluminum, silicon, and oxygen) retain their strength to 1700°C (3100°F). Also, carbon-carbon-matrix composites retain much of their strength up to 2500°C (4500°F), although they lack oxidation resistance at high temperatures. Applications for ceramic-matrix composites are in jet and automotive engines, equipment for deep-sea mining, pressure vessels, and various structural components.

There are several processes to make ceramic-matrix composites; three common ones are briefly described next.

1. **Slurry infiltration,** the most common process, involves the preparation of a fiber preform that is hot pressed and then impregnated with a slurry that contains the

matrix powder, a carrier liquid, and an organic binder. High strength, toughness, and uniform structure are obtained by the slurry-infiltration process, but the product has limited high-temperature properties, due to the low melting temperature of the matrix materials used. A further improvement of this process is *reaction bonding* or *reaction sintering* of the slurry.

2. **Chemical synthesis** processes involve the solgel and polymer-precursor techniques. In the *solgel* process, a sol (a colloidal fluid with the liquid as the continuous phase) containing fibers is converted to a gel, which is then subjected to heat treatment to produce a ceramic-matrix composite. The *polymer-precursor* method is analogous to the process used in making ceramic fibers.

3. **Chemical vapor infiltration** involves a porous fiber preform which is infiltrated with the matrix phase, using the chemical vapor deposition technique (described in Section 4.5.1). The product has very good high-temperature properties, but the process is time consuming and costly.

In addition to the processes described above, techniques such as melt infiltration, controlled oxidation, and hot-press sintering (still largely in the experimental stage) are at various stages of development for improving the properties and performance of these composites.

11.14.3 Miscellaneous composites

Described below are several other composites.

1. Composites may consist of *coatings* of various kinds on base metals or substrates. Examples include plating of aluminum and other metals over plastics for decorative purposes and enamels, dating to prior to 1000 B.C., or applying similar vitreous (glasslike) coatings on metal surfaces for various functional or ornamental purposes (see Section 4.5).

2. Another example of a composite is *glass-reinforced fiber-metal laminate* (GLARE), used extensively in aircraft wings, fuselage sections, tail surfaces, and doors for the Airbus A380 aircraft. GLARE consists of many layers of aluminum interspersed with layers of glass-fiber reinforced epoxy matrix (prepreg). GLARE has advantages of weight savings and improved impact and fatigue properties over conventional aerospace materials.

3. Composites such as *cemented carbides*, usually tungsten carbide and titanium carbide, with cobalt and nickel, respectively, as a binder, are made into tools and dies (see Section 8.6).

4. *Grinding wheels* are typically made of aluminum-oxide, silicon-carbide, diamond, or cubic-boron-nitride grains. The abrasive particles are held together with various organic, inorganic, or metallic binders (see Chapter 9).

5. Another composite material is granite particles embedded in an epoxy matrix. (See Section 8.13.) It has high strength and good frictional characteristics and vibration-damping capacity (better than that of gray cast iron). This composite is used as machine-tool beds for some precision grinders.

11.15 | Processing Superconductors

Although superconductors (Section 3.9.6) have major energy-saving potential in the generation, storage, and distribution of electrical power, their processing into useful shapes and sizes for practical applications presents significant difficulties. Two basic

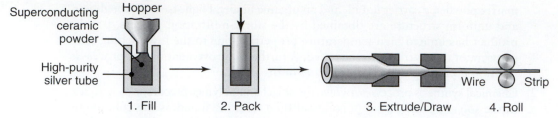

FIGURE 11.39 Schematic illustration of the steps involved in the powder-in-tube process. *Source:* Courtesy of Concurrent Technologies Corporation.

types of superconductors are *metals* (**low-temperature superconductors,** LTSCs), including combinations of niobium, tin, and titanium, and *ceramics* (**high-temperature superconductors,** HTSCs), including various copper oxides. In this application, *high temperature* means closer to *ambient temperature;* consequently, the HTSCs are of more practical use.

Ceramic superconducting materials are available in powder form. The major difficulty in manufacturing them is their inherent brittleness and anisotropy, which make it difficult to align the grains in the proper direction for high efficiency. The smaller the grain size, the more difficult it is to align the grains.

The basic manufacturing process for superconductors consists of the following steps:

1. Preparing the powder, mixing it, and grinding it in a ball mill to a grain size of 0.5 to 10 mm.
2. Shaping.
3. Heat treating, to improve grain alignment.

The most common shaping process is **oxide powder in tube** (OPIT), as shown in Fig. 11.39. In this operation, the powder is packed into silver tubes (because silver has the highest electrical conductivity of all metals) and sealed at both ends. The tubes are then mechanically worked (by such processes as swaging, drawing, extrusion, isostatic pressing, and rolling) into final shapes, which may be wire, tape, coil, or bulk.

Other principal superconductor-shaping processes include (a) coating of silver wire with superconducting material, (b) deposition of superconductor films by laser ablation (ablation refers to laser heating causing layers of material to melt away from a surface, thus carrying away much of the heat), (c) the doctor-blade process (see Section 11.9.1), (d) explosive cladding (see Section 12.11), and (e) chemical spraying.

CASE STUDY | Hot Isostatic Pressing of Valve Lifter

A HIP clad valve lifter, used in a full range of medium- to heavy-duty truck diesel engines, is shown in Fig. 11.40. The 0.2 kg (0.45 lb) valve lifter rides on the camshaft, and opens and closes the engine valves. As such, it is desired to have a tungsten-carbide face for wear resistance, with a steel shaft for fatigue resistance. Before the HIP valve lifter was developed, parts were produced through furnace brazing, but resulted in occasional field failures and relatively high scrap rates.

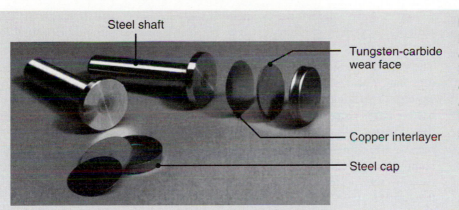

Steel shaft

Tungsten-carbide
wear face

Copper interlayer

Steel cap

FIGURE 11.40 A valve lifter for heavy-duty diesel engines, produced from a hot-isostatically-pressed carbide cap on a steel shaft. *Source:* Courtesy of Metal Powder Industries Federation and Bodycote, Inc.

The required annual production of these parts is over 400,000; hence high scrap rates are particularly objectionable.

The HIP clad product consists of a 9% Co bonded tungsten carbide face, made from powder and pressed and sintered, a steel-sheet metal cap fitted over the WC disk, a copper alloy foil interlayer, and a steel shaft. The steel cap is electron-beam welded (Section 12.5) to the steel shaft and the assembly is then hot isostatically pressed to provide a very strong bond. The HIP operation takes place at 1010°C (1850°F) and at a pressure of 100 MPa (15,000 psi). The tungsten carbide surface has a density of 14.52–14.72 g/cm^3, a hardness of 90.8 ± 5 HRA, and a minimum transverse rupture strength of 2450 MPa (355,000 psi).

Secondary operations consist of grinding the face to remove any protruding sheet-metal cap and to expose the wear-resistant tungsten carbide face. The high reliability of the HIP bond drastically reduced scrap rates to under 0.2%. No field failures have been experienced in over four years of full production. Production costs were also substantially reduced because of the hot isostatic pressing operation.

Source: Metal Powder Industries Federation and Bodycote, Inc.

SUMMARY

- The powder-metallurgy process is capable of economically producing relatively complex parts in net- or near-net shape to close dimensional tolerances, from a wide variety of metal and alloy powders. (Section 11.1)

- The steps in powder-metallurgy processing are powder production, blending, compaction, sintering, and additional processing to improve dimensional accuracy, surface finish, mechanical or physical properties, or appearance. (Sections 11.2–11.5)

- Design considerations in powder metallurgy include the shape of the compact, the ejection of the green compact from the die without fracture, and the acceptable dimensional tolerances of the application. (Section 11.6)

- The powder-metallurgy process is suitable for medium- to high-volume production runs and for relatively small parts and has competitive advantages over other processing methods. (Sections 11.7)

- Ceramics have characteristics such as high hardness and strength at elevated temperatures, high modulus of elasticity, brittleness, low toughness, low density, low thermal expansion, and low thermal and electrical conductivity. (Section 11.8)

- Three basic shaping processes for ceramics are casting, plastic forming, and pressing. The resulting product is then dried and fired to give it the proper strength. Finishing operations, such as machining and grinding, may be performed to give the part its final shape, or the part may be subjected to surface treatments to improve specific properties. (Section 11.9)

- Almost all commercial glasses are categorized as one of six types that indicate their composition. Glass in bulk form has relatively low strength, but it can be strengthened by thermal or chemical treatments to obtain high strength and toughness. (Section 11.10)

- Continuous methods of glass processing are floating, drawing, and rolling. Discrete glass products can be manufactured by blowing, pressing, centrifugal casting, and sagging. After initial processing, glasses can be strengthened by thermal or chemical tempering or by laminating. (Section 11.11)

- Design considerations for ceramics and glasses are guided by factors such as general lack of tensile strength and toughness, and sensitivity to external and internal defects. Warping and cracking are important considerations, as are the methods employed for production and assembly. (Section 11.12)

- Graphite, buckyballs (fullerenes), and diamond are forms of carbon that display unusual combinations of properties. These materials have unique applications. (Section 11.13)

- Metal-matrix and ceramic-matrix composites possess unique combinations of properties and have increasingly broad applications. Metal-matrix composites are processed through liquid-phase, solid-phase, and two-phase processes; ceramic-matrix composites can be processed by slurry infiltration, chemical synthesis, and chemical-vapor infiltration. (Section 11.14)

- Manufacture of superconductors into useful products is a challenging area, because of the anisotropy and inherent brittleness of the materials involved. Although other processes are being developed, the basic and common practice involves packing the powder into a silver tube and deforming it plastically into desired shapes. (Section 11.15)

SUMMARY OF EQUATIONS

Shape factor of particles: $k = \left(\dfrac{A}{V}\right) D_{eq}$

Pressure distribution in compaction: $p_x = p_o e^{-4\mu kx/D}$

Volume: $V_{sint} = V_{green}\left(1 - \dfrac{\Delta L}{L_o}\right)^3$

Density: $\rho_{green} = \rho_{sint}\left(1 - \dfrac{\Delta L}{L_o}\right)^3$

Tensile strength: $UTS \simeq UTS_o e^{-nP}$

Elastic modulus: $E \simeq E_o(1 - 1.9P + 0.9P^2)$

Thermal conductivity: $k = k_o(1 - P)$

BIBLIOGRAPHY

Powder Metallurgy

ASM Handbook, Vol. 7: *Powder Metal Technologies and Applications*, ASM International, 1998.

German, R.M., *A-Z of Powder Metallurgy*, Elsevier, 2006.

——, *Powder Metallurgy and Particulate Materials Processing*, Metal Powder Industries Federation, 2005.

——, and Bose, A., *Injection Molding of Metals and Ceramics*, Metal Powder Industries Federation, 1997.

Karlsson, L. (ed.), *Modeling in Welding, Hot Powder Forming and Casting*, ASM International, 1997.

Pease III, L.F., and West, W.G., *Fundamentals of Powder Metallurgy*, Metal Powder Industries Federation, 2002.

Powder Metallurgy Design Guidebook, American Powder Metallurgy Institute, revised periodically.

Powder Metallurgy Design Manual, 3rd ed., Metal Powder Industries Federation, 1998.

Upadhyaya, G.S., *Sintering Metallic and Ceramic Materials: Preparation, Properties and Applications*, Wiley, 2000.

Ceramics and Other Materials

Barsoum, M.W., *Fundamentals of Ceramics*, McGraw-Hill, 1996.

Buchanan, R.C., *Ceramic Materials for Electronics: Processing, Properties, and Applications*, 3rd ed., Dekker, 2004.

Cranmer, D.C., and Richerson, D.W., *Mechanical Testing Methodology for Ceramic Design and Reliability*, Dekker, 1998.

Harper, C.A. (ed.), *Handbook of Ceramics, Glasses, and Diamonds*, McGraw-Hill, 2001.

Holand, W., and Beall, G.H., *Design and Properties of Glass-Ceramics*, American Chemical Society, 2001.

Jahanmir, S., *Friction and Wear of Ceramics*, Dekker, 1994.

King, A.G., *Ceramics Processing and Technology*, Noyes Pub., 2001.

Lu, H.Y., *Introduction to Ceramic Science*, Dekker, 1996.

Prelas, M.A., Popovici, G., and Bigelow, L.K. (eds.), *Handbook of Industrial Diamonds and Diamond Films*, Dekker, 1997.

Rahaman, M.N., *Ceramic Processing Technology and Sintering*, Dekker, 1996.

Reed, J.S., *Principles of Ceramics Processing*, 2nd ed., Wiley, 1995.

Richerson, D.W., *Modern Ceramic Engineering: Properties, Processing, and Use in Design*, 3rd ed., Dekker, 2005.

Wilks, J., and Wilks, E., *Properties and Applications of Diamond*, Butterworth-Heinemann, 1991.

Composites

ASM Engineered Materials Handbook, Desk Edition, ASM International, 1995.

ASM Handbook, Vol. 21: *Composites*, ASM International, 2001.

Belitskus, D.L., *Fiber and Whisker Reinforced Ceramics for Structural Applications*, Dekker, 2004.

Chawla, K.K., *Composite Materials*, 2nd ed., Springer, 1998.

——, *Ceramic Matrix Composites*, 2nd ed., Springer, 2003.

Gutowski, T.G. (ed.), *Advanced Composites Manufacturing*, Wiley, 1997.

Hoa, S.V., *Computer-Aided Design for Composite Structures*, Dekker, 1996.

Mallick, P.K. (ed.), *Composites Engineering Handbook*, Dekker, 1997.

Ochiai, S., *Mechanical Properties of Metallic Composites*, Dekker, 1994.

QUESTIONS

Powder Metallurgy

11.1 Explain the advantages of blending different metal powders in making P/M products.

11.2 Green strength can be important in powder-metal processing. Explain why.

11.3 Give the reasons that injection molding of metal powders has become an important process.

11.4 Describe the events that occur during sintering.

11.5 What is mechanical alloying and what are its advantages over conventional alloying of metals, described in Section 5.2?

11.6 It is possible to infiltrate P/M parts with various resins, as can be done with metals. What possible benefits would result from infiltration? Give some examples.

11.7 What concerns would you have when electroplating P/M parts? Explain.

11.8 Describe the effects of different shapes and sizes of metal powders in P/M processing, commenting on the magnitude and significance of the shape factor of the particles.

11.9 Comment on the shapes of the curves and their relative positions shown in Fig. 11.6.

11.10 Should green compacts be brought up to the sintering temperature slowly or rapidly? Explain.

11.11 What are the effects of using fine vs. coarse powders in making P/M parts? Explain.

11.12 Are the requirements for punch and die materials in powder metallurgy different than those for forging and extrusion, described in Chapter 6? Explain.

11.13 Describe the relative advantages and limitations of cold and hot isostatic pressing, respectively.

11.14 Why do mechanical and physical properties depend on the density of P/M parts? Explain.

11.15 Comment on the type of press required to compact powders using the set of punches shown in Fig. 11.7d. (See also Chapters 6 and 7.)

11.16 Explain the difference between impregnation and infiltration. Give some applications for each.

11.17 What are the advantages of making tool steels by P/M techniques over traditional methods, such as casting and subsequent metalworking techniques? Explain.

11.18 Why do compacting pressure and sintering temperature depend on the type of powder metal used? Explain.

11.19 Name various methods of powder production and describe the morphology of powders produced by each method.

11.20 It was indicated that there are hazards involved in P/M processing. Describe their causes.

11.21 What is screening of metal powders? Why is it done?

11.22 Why are there density variations in compacted metal powders? How can they be reduced?

11.23 It has been stated that P/M can be competitive with other processes, such as casting and forging. Explain why this is so, commenting on technical and economic advantages.

11.24 Selective laser sintering was described in Section 10.12.4 as a rapid-prototyping technique. What similarities does this process have with the processes described in this chapter? Explain.

11.25 Prepare an illustration similar to Fig. 6.28, showing the variety of P/M manufacturing options.

Ceramics and Other Materials

11.26 Describe the major differences between ceramics, metals, thermoplastics, and thermosets.

11.27 Explain why ceramics are weaker in tension than in compression.

11.28 Why do the mechanical and physical properties of ceramics decrease with increasing porosity? Explain.

11.29 What engineering applications could benefit from the fact that, unlike metals, ceramics generally maintain their modulus of elasticity at elevated temperatures?

11.30 Explain why the mechanical-property data given in Table 11.7 have such a broad range. What is the significance of this wide range in engineering applications? Explain.

11.31 List and explain the factors that you would consider when replacing a metal component with a ceramic component. Give examples of such possible substitutions, also commenting on their shape and size.

11.32 How are ceramics made tougher? Explain.

11.33 Describe situations and applications in which static fatigue can be important.

11.34 Explain the difficulties involved in making large ceramic components. What recommendations would you make to overcome these difficulties?

11.35 Explain why ceramics are effective cutting-tool materials, as described in Section 8.6. Would ceramics also be suitable as die materials for metal forming? Explain.

11.36 Describe applications in which the use of a ceramic material with a zero coefficient of thermal expansion would be desirable.

11.37 Give reasons for the development of ceramic-matrix components. Name some present and future possible applications.

11.38 List the factors that are important in drying ceramic components and explain why they are important.

11.39 It has been stated that the higher the coefficient of thermal expansion of glass and the lower its thermal conductivity, the higher is the level of residual stresses developed during processing. Explain why.

11.40 What types of finishing operations are typically performed on ceramics? Why are they done?

11.41 What should be the property requirements for the metal balls used in a ball mill (see Fig. 11.26b)? Explain why these particular properties are important.

11.42 Which properties of glasses allow them to be shaped into bottles by blowing? Are there any similarities to some of the sheet-metal forming operations, described in Chapter 7? Explain.

11.43 What properties should a plastic sheet possess when used in laminated glass, as in automotive windshields? Explain.

11.44 Consider some ceramic products that you are familiar with and outline a sequence of processes involved in manufacturing each of them.

11.45 Explain the difference between physical and chemical tempering of glass.

11.46 What do you think is the purpose of the operation shown in Fig. 11.27d? Explain.

11.47 Injection molding is a process that is used for plastics, powder metals, and ceramics. Why is it suitable for all these different types of materials?

11.48 Are there any similarities between the strengthening mechanisms for glass and those for other metallic and nonmetallic materials described throughout this text? Explain, giving specific examples.

11.49 Describe and explain the differences in the manner in which each of the following flat surfaces would fracture when struck with a large piece of rock: (a) ordinary window glass, (b) tempered glass, and (c) laminated glass.

11.50 Describe the similarities and the differences between the processes described in this chapter and those in Chapters 5 through 10.

11.51 What is the doctor-blade process? Why was it developed?

11.52 Describe the methods by which glass sheet is manufactured.

11.53 Describe the differences and similarities in producing metal and ceramic powders. Which of these processes would be suitable for producing glass powder? Explain.

11.54 How are glass fibers made? List and explain the various applications of these fibers.

11.55 Would you consider diamond a ceramic? Explain.

11.56 What are the similarities and differences between injection molding, metal injection molding, and ceramic injection molding?

11.57 Aluminum oxide and partially stabilized zirconia are normally white in appearance. Can they be colored? If so, how?

11.58 It was stated that ceramics have a wider range of strengths in tension than do metals. List the reasons why this is so.

PROBLEMS

11.59 Estimate the number of particles in a 500-g sample of iron powder, if the particle size is 50 μm.

11.60 Assume that the surface of a copper particle is covered with a 0.1-μm-thick oxide layer. What is the volume occupied by this layer if the copper particle itself is 75 μm in diameter? What could be the role of this oxide layer in subsequent processing of the powders? Explain.

11.61 Determine the shape factor for a flakelike particle with a ratio of surface area to thickness of $12 \times 12 \times 1$, for a cylinder with dimensional ratios $1:1:1$, and for an ellipsoid with an axial ratio of $5 \times 2 \times 1$.

11.62 It was stated in Section 3.3 that the energy in brittle fracture is dissipated as surface energy. It is also noted that the comminution process for powder preparation generally involves brittle fracture. What are the relative energies involved in making spherical powders of diameters 1, 10, and 100 μm, respectively?

11.63 Referring to Fig. 11.6a, what should be the volume of loose, fine iron powder in order to make a solid cylindrical compact 25 mm in diameter and 15 mm high?

11.64 In Fig. 11.7e note that the pressure is not uniform across the diameter of the compact. Explain the reasons for this variation.

11.65 Plot the family of pressure-ratio p_x/p_o curves as a function of x for the following ranges of process parameters: $\mu = 0$ to 1, $k = 0$ to 1, and $D = 5$ to 50 mm.

11.66 Derive an expression, similar to Eq. (11.2), for compaction in a square die with dimensions a by a.

11.67 For the ceramic material described in Example 11.7, calculate (a) the porosity of the dried part if the porosity of the fired part is to be 9% and (b) the initial length, L_o, of the part if the linear shrinkages during drying and firing are 8% and 7%, respectively.

11.68 What would be the answers to Problem 11.67 if the quantities given were halved?

11.69 Plot the UTS, E, and k values for ceramics as a function of porosity, P, and describe and explain the trends that you observe in their behavior.

11.70 Plot the total surface area of a 1-g sample of aluminum powder as a function of the natural log of particle size.

11.71 Conduct a literature search and determine the largest size of metal powders that can be produced in atomization chambers.

11.72 Coarse copper powder is compacted in a mechanical press at a pressure of 20 tons/in^2. During subsequent sintering, the green part shrinks an additional 8%. What is the final density of the part?

11.73 A gear is to be manufactured from iron powder, with a final density that is 90% of that of cast iron. It is known that the shrinkage in sintering will be approximately 5%. For a gear 2.5 in. in diameter and with a 0.75-in. hub, what is the required press force?

11.74 What volume of powder is needed to make the gear in Problem 11.73 if its thickness is to be 0.5 in.?

11.75 The axisymmetric part shown in the accompanying figure is to be produced from fine copper powder and is to have a tensile strength of 200 MPa. Determine the compacting pressure and the initial mass of powder needed if the compacted volume is 8 cm^3.

Dimensions in mm

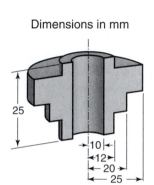

11.76 What techniques, other than the powder-in-tube process, could be used to produce superconducting monofilaments? Explain.

11.77 Describe other methods that can be used to manufacture the parts shown in Fig. 11.1a. Comment on the advantages and limitations of each of these methods over P/M.

11.78 If a fully-dense ceramic has the properties of $UTS_o = 180$ MPa and $E_o = 300$ GPa, what are these properties at 20% porosity for values of $n = 4, 5, 6$, and 7, respectively?

11.79 Calculate the thermal conductivities for ceramics at porosities of 1%, 5%, 10%, 20%, and 30% for $k_o = 0.7$ W/m-K.

11.80 A ceramic has $k_o = 0.65$ W/m-K. If this ceramic is shaped into a cylinder with a porosity distribution of $P = 0.1(x/L)(1 - x/L)$, where x is the distance from one end of the cylinder and L is the total cylinder length, estimate the average thermal conductivity of the cylinder.

11.81 Assume that you are asked to give a quiz to students on the contents of this chapter. Prepare three quantitative problems and three qualitative questions, and supply the answers.

DESIGN

11.82 Make sketches of several P/M products in which density variations would be desirable. Explain why, in terms of the function of these parts.

11.83 Compare the design considerations for P/M products with those for products made by (a) casting and (b) forging. Describe your observations.

11.84 It is known that in the design of P/M gears, the distance between the outside diameter of the hub and the gear root should be as large as possible. Explain the reasons for this design consideration.

11.85 How are the design considerations for ceramics different, if at all, than those for the other materials described in this chapter? Explain.

11.86 Are there any shapes or design features that are not suitable for production by powder metallurgy? By ceramics processing? Explain.

11.87 What design modifications would you recommend for the part shown in Problem 11.75? Explain your reasoning.

11.88 The axisymmetric parts shown in the accompanying figure are to be produced through P/M. Describe the design changes that you would recommend.

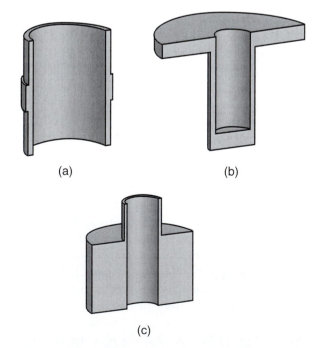

(a)   (b)

(c)

11.89 Assume that in a particular design, a metal beam is to be replaced with one made of ceramics.

Discuss the differences in the behavior of the two beams, such as with respect to its strength, stiffness, deflection, and resistance to temperature and to the environment.

11.90 Describe your thoughts regarding designs of internal combustion engines that have ceramic pistons.

11.91 Assume that you are employed in technical sales in powder-metallurgy products. What applications currently using non-P/M parts would you try to develop? What would you advise your potential customers during your sales visits? What kind of concerns and questions do you think they would ask?

11.92 Pyrex cookware displays a unique phenomenon: It functions well for a large number of cycles and may then shatter into many pieces. Investigate this phenomenon, list the probable causes, and discuss the manufacturing considerations that may alleviate or contribute to such failures.

11.93 It has been noted that the strength of brittle materials, such as ceramics and glasses, is very sensitive to surface defects such as scratches (notch sensitivity).

Obtain some pieces of these materials, make scratches on them, and test them by carefully clamping in a vise and bending them. Comment on your observations.

11.94 Make a survey of the technical literature and describe the differences, if any, between the quality of glass fibers made for use in reinforced plastics and those made for use in fiber-optic communication. Comment on your observations.

11.95 Describe your thoughts on the processes that can be used to make (a) small ceramic statues, (b) whiteware for bathrooms, (c) common brick, and (d) floor tile.

11.96 As described in this chapter, one method of producing superconducting wire or strip is by compacting powders of these materials, placing them into a tube, and drawing them through dies, or rolling them. Describe your thoughts concerning the possible difficulties involved in each step of this production.

11.97 Review Fig. 11.18 and prepare a similar figure for constant-thickness parts, as opposed to the axisymmetric parts shown in the figure.

Joining and Fastening Processes

This chapter describes the principles, characteristics, and applications of major joining processes and operations. The topics covered include

- Oxyfuel gas welding and arc welding processes, using nonconsumable and consumable electrodes.
- Solid-state welding processes.
- Laser-beam and electron-beam welding.
- The nature and characteristics of the weld joint and factors involved in weldability of metals.
- Adhesive bonding and applications.
- Mechanical fastening methods and fasteners.
- Economic considerations in joining operations.
- Good joint design practices and process selection.

12.1 | Introduction

Joining is a general, all-inclusive term covering numerous processes that are important aspects of manufacturing operations. The necessity of joining and assembling components can be appreciated simply by inspecting various products such as automobiles, bicycles, printed circuit boards, machinery, and appliances. Joining may be preferred or necessary for one or more of the following reasons:

- The product is impossible or uneconomical to manufacture as a single piece.
- The product is easier to manufacture in individual components, which are then assembled, than as a single piece.
- The product may have to be taken apart for repair or maintenance during its service life.
- Different properties may be desirable for functional purposes of the product; for example, surfaces subjected to friction and wear, or corrosion and environmental attack, typically require characteristics different than those of the component's bulk.
- Transportation of the product in individual components and subsequent assembly may be easier and more economical than transporting it as a single piece.

Although there are different ways of categorizing the wide variety of joining and fastening processes, this chapter will follow the sequence shown in the section headings. Joining methods can be categorized in terms of their common principle of operation. The joints produced often undergo metallurgical and physical changes, which, in turn, have major effects on the properties and performance of the welded components.

Fusion welding involves melting and coalescing materials by means of heat, usually supplied by electrical or high-energy means. The processes consist of *oxyfuel gas welding*, consumable- and nonconsumable-electrode *arc welding*, and *high-energy-beam welding*. As described in Section 9.14, the high-energy processes are also used for cutting and machining applications.

Solid-state welding involves joining without fusion; that is, there is no liquid (molten) phase in the joint. The basic categories of solid state welding are *cold, ultrasonic, friction, resistance, explosion welding*, and *diffusion bonding*.

Brazing and **soldering** use filler metals and involve lower temperatures than used in welding. The heat required is supplied externally.

Adhesive bonding is an important technology because of its unique advantages for applications requiring strength, sealing, insulation, vibration damping, and resistance to corrosion between dissimilar or similar metals. Included in this category are *electrically conducting adhesives* for surface-mount technologies.

Mechanical fastening processes use a wide variety of fasteners, bolts, nuts, screws, and rivets. *Joining nonmetallic materials* can be accomplished by such means as mechanical fastening, adhesive bonding, fusion by various external or internal heat sources, diffusion, and preplating with metal.

The individual groups of joining processes have, as in all other manufacturing processes, several important characteristics, such as joint design (Fig. 12.1), size and shape of the parts to be joined, strength and reliability of the joint, cost and maintenance of equipment, and skill level of labor (Tables 12.1 and 12.2).

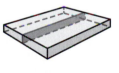 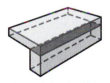  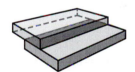

| (a) Butt joint | (b) Corner joint | (c) T joint | (d) Lap joint | (e) Edge joint |

FIGURE 12.1   Examples of welded joints.

TABLE 12.1

Comparison of Various Joining Methods

Method	Strength	Design variability	Small parts	Large parts	Tolerances	Relibility	Ease of maintenance	Visual inspection	Cost
Arc welding	1	2	3	1	3	1	2	2	2
Resistance welding	1	2	1	1	3	3	3	3	1
Brazing	1	1	1	1	3	1	3	2	3
Bolts and nuts	1	2	3	1	2	1	1	1	3
Riveting	1	2	3	1	1	1	3	1	2
Fasteners	2	3	3	1	2	2	2	1	3
Seaming, crimping	2	2	1	3	3	1	3	1	1
Adhesive bonding	3	1	1	2	3	2	3	3	2

Note: 1, very good; 2, good; 3, poor.

TABLE 12.2

General Characteristics of Joining Processes

Joining process	Operation	Advantage	Skill level required	Welding position	Current type	Distortion*	Cost of equipment
Shielded metal arc	Manual	Portable and flexible	High	All	AC, DC	1 to 2	Low
Submerged arc	Automatic	High deposition	Low to medium	Flat and horizontal	AC, DC	1 to 2	Medium
Gas metal arc	Semiautomatic or automatic	Most metals	Low to high	All	DC	2 to 3	Medium to high
Gas tungsten arc	Manual or automatic	Most metals	Low to high	All	AC, DC	2 to 3	Medium
Flux-cored arc	Semiautomatic or automatic	High deposition	Low to high	All	DC	1 to 3	Medium
Oxyfuel	Manual	Portable and flexible	High	All	—	2 to 4	Low
Electron beam, laser beam	Semiautomatic or automatic	Most metals	Medium to high	All	—	3 to 5	High

*1, highest; 5, lowest.

12.2 | Oxyfuel Gas Welding

Developed in the early 1900s, *oxyfuel gas welding* (OFW) is a general term used to describe any welding process that uses a **fuel gas**, combined with *oxygen* to produce a flame, as the source of the heat required to melt the metals at the joint. The most common gas welding process uses *acetylene;* the process is known as oxyacetylene gas welding (OAW) and is typically used for structural sheet-metal fabrication, automotive bodies, and various other repair work.

The heat generated is a result of the combustion of acetylene gas (C_2H_2) in a mixture with oxygen, in accordance with a pair of chemical reactions. The primary combustion process, which occurs in the inner core of the flame (Fig. 12.2), involves the reaction

$$C_2H_2 + O_2 \rightarrow 2CO + H_2 + \text{Heat}. \tag{12.1}$$

This reaction dissociates the acetylene into carbon monoxide and hydrogen and produces about one-third of the total heat generated in the flame. The secondary combustion process is

$$2CO + H_2 + 1.5O_2 \rightarrow 2CO_2 + H_2O + \text{Heat}. \tag{12.2}$$

This reaction consists of the further burning of the hydrogen and the carbon monoxide and produces about two-thirds of the total heat. Note that the reaction also produces water vapor. The temperatures developed in the flame can reach 3300°C (6000°F).

Flame types. The proportion of acetylene and oxygen in the gas mixture is an important factor in oxyfuel gas welding. At a ratio of 1:1, that is, when there is no excess oxygen, the flame is considered to be *neutral* (Fig. 12.2a). With a greater oxygen supply, the flame can be harmful, especially for steels, because it oxidizes the metal; hence it is known as an **oxidizing flame** (Fig. 12.2b). Only in welding copper

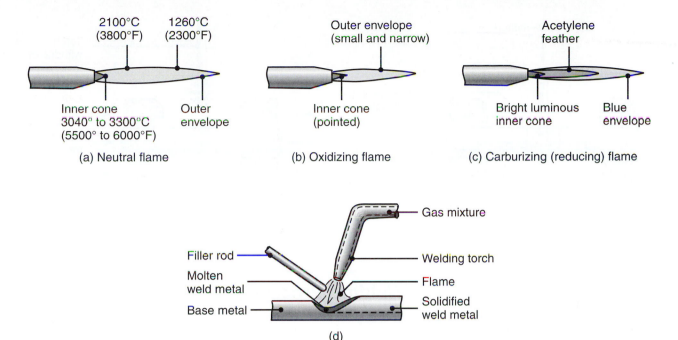

FIGURE 12.2 Three basic types of oxyacetylene flames used in oxyfuel gas welding and cutting operations: (a) neutral flame, (b) oxidizing flame, and (c) carburizing, or reducing, flame. (d) Schematic illustration of the oxyfuel gas welding operation.

and copper-base alloys is an oxidizing flame desirable, because it forms a thin protective layer of *slag* (compounds of oxides) over the molten metal. If the oxygen is insufficient for full combustion, the flame is known as a **reducing** (having excess acetylene) or **carburizing flame** (Fig. 12.2c). The temperature of a reducing flame is lower, hence suitable for applications requiring low heat, such as brazing, soldering, and flame hardening operations.

Other fuel gases, such as hydrogen and methylacetylene propadiene, can also be used in oxyfuel gas welding. However, the temperatures developed by these gases are low; hence they are used for welding metals with low melting points, such as lead, and for parts that are thin and small. The flame with pure hydrogen gas is colorless; hence it is difficult to adjust the flame by eyesight, as is the case with other gases.

Filler metals. Filler metals are used to supply additional metal to the weld zone during welding. They are available as **filler rods** or **wire** (Fig. 12.2d) and may be bare or coated with flux. The purpose of the flux is to retard oxidation of the surfaces being welded, by generating a *gaseous shield* around the weld zone. The flux also helps to dissolve and remove oxides and other substances from the weld zone, thus enhancing the development of a stronger joint. The *slag* developed (compounds of oxides, fluxes, and electrode coating materials) protects the molten puddle of metal against oxidation as it cools.

Pressure gas welding. In this method, the two components to be welded are heated at their interface by means of a torch using, typically, an oxyacetylene gas mixture (Fig. 12.3a). After the interface begins to melt, the torch is withdrawn. An axial force is then applied to press the two components together (Fig. 12.3b) and is maintained until the interface solidifies. Note the formation of a flash due to the upsetting of the joined ends of the two components.

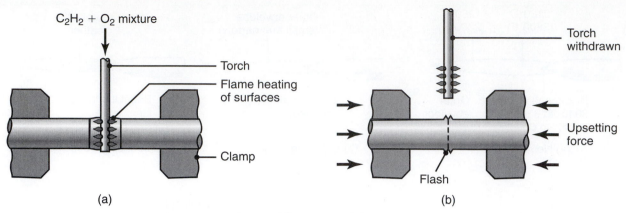

FIGURE 12.3 Schematic illustration of the pressure gas welding process (a) before and (b) after. Note the formation of a flash at the joint, which can later be trimmed off.

12.3 | Arc Welding Processes: Consumable Electrode

In arc welding, developed in the mid-1800s, the heat required is obtained through electrical energy. Using either a consumable or a nonconsumable electrode (rod or wire), an arc is produced between the tip of the electrode and the parts to be welded, using AC or DC power supplies. Arc welding includes several processes (Table 12.2), as described throughout this section.

12.3.1 Heat transfer in arc welding

The heat input in arc welding can be calculated from the equation

$$\frac{H}{l} = e\frac{VI}{v} \tag{12.3}$$

where H is the heat input (J or BTU), l is the weld length, V is the voltage applied, I is the current, in amperes, and v is the welding speed. The term e is the efficiency of the process, and varies from around 75% for shielded metal arc welding to 90% for gas metal arc welding and submerged arc welding. The efficiency is an indication that not all of the available energy is beneficially used to melt material, as some of the available heat is conducted through the workpiece, some is lost by radiation, and still more by convection to the surrounding environment.

The heat input given by Eq. (12.3) melts a certain volume of material, usually the electrode or filler metal, and can also be expressed as

$$H = u(\text{Volume}) = uAl \tag{12.4}$$

where u is the specific energy required for melting, and A is the cross section of the weld. Some typical values of u are given in Table 12.3. Equations (12.3) and (12.4) allow an expression of the welding speed:

$$v = e\frac{VI}{uA}. \tag{12.5}$$

Although these equations have been developed for arc welding, similar expressions can be obtained for other fusion-welding operations as well, taking into account differences in weld geometry and process efficiency.

TABLE 12.3

Approximate Specific Energies Required to Melt a Unit Volume of Commonly Welded Metals		
	Specific energy, u	
Material	J/mm^3	BTU/in^3
Aluminum and its alloys	2.9	41
Cast irons	7.8	112
Copper	6.1	87
Bronze (90Cu-10Sn)	4.2	59
Magnesium	2.9	42
Nickel	9.8	142
Steels	9.1–10.3	128–146
Stainless steels	9.3–9.6	133–137
Titanium	14.3	204

Note: 1 BTU = 1055 J = 780 ft-lb

EXAMPLE 12.1 Estimating the welding speed for different materials

Consider the situation where a welding operation is being performed with $V = 20$ volts, $I = 200$ A, and the cross-sectional area of the weld bead is 30 mm^2. Estimate the welding speed if the workpiece and electrode are made of (a) aluminum, (b) carbon steel, and (c) titanium. Use an efficiency of 75%.

Solution. For aluminum, we note from Table 12.3 that the specific energy required is $u = 2.9$ J/mm^3. Therefore, from Eq. (12.5),

$$v = e\frac{VI}{uA} = (0.75)\frac{(20)(200)}{(2.9)(30)} = 34.5 \text{ mm/s.}$$

Similarly, for carbon steel, we estimate u as 12.3 J/mm^3 (average of extreme values in Table 12.3), leading to $v = 8.1$ mm/s. For titanium, $u = 14.3$ J/mm^3, so that $v = 7.0$ mm/s.

12.3.2 Shielded metal arc welding

Shielded metal arc welding (SMAW) is one of the oldest, simplest, and most versatile joining processes; currently, about one-half of all industrial and maintenance welding is performed by this process. The electric arc (hence also the term *electric arc welding*) is generated by touching the tip of a coated electrode to the workpiece and then withdrawing it quickly to a distance sufficient to maintain the arc (Fig. 12.4a). The electrodes are in the shape of thin, long sticks (see Section 12.3.8); hence this process is also known as **stick welding.** The heat generated melts a portion of the electrode tip, its coating, and the base metal in the immediate area of the arc. A weld forms after the molten metal (a mixture of the base metal, electrode metal, and substances from the coating on the electrode) solidifies in the weld area. The electrode coating deoxidizes and provides a shielding gas in the weld area to protect it from the oxygen in the environment.

In this operation, the bare section at the end of the electrode is clamped to one terminal of the power source, while the other terminal is connected to the part being welded (Fig. 12.4b). The current usually ranges between 50 and 300 A, with power requirements generally less than 10 kW; too low a current causes incomplete fusion,

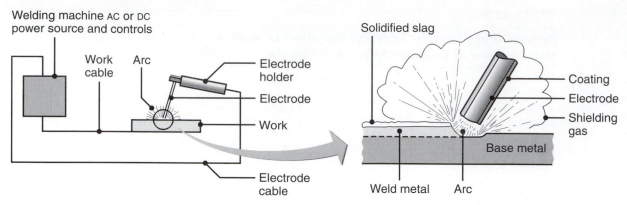

FIGURE 12.4   Schematic illustration of the shielded metal arc welding process. About one-half of all large-scale industrial welding operations use this process.

and too high a current can damage the electrode coating and reduce its effectiveness. The current may be AC or DC; for sheet-metal welding, DC is preferred, because of the steady arc produced.

The **polarity** (direction of current flow) of the DC current can be important, depending on factors such as the type of electrode, the metals to be welded, and the depth of the heated zone. In **straight polarity,** the workpiece is positive and the electrode negative. It is preferred for sheet metals, because of its shallow penetration, and for joints with very wide gaps. In **reverse polarity,** where the electrode is positive and the workpiece negative, deeper weld penetration is possible. When the current is AC, the arc pulsates rapidly, a condition that is suitable for welding thick sections and using large diameter electrodes at maximum currents.

The SMAW process requires a relatively small variety of electrodes, and the equipment consists of a power supply, power cables, and an electrode holder. This process is commonly used in general construction, shipbuilding, and for pipelines, as well as for maintenance work, since the equipment is portable and can be easily maintained. It is especially useful for work in remote areas, where portable fuel-powered generators can be used as the power supply. The SMAW process is best suited for workpiece thicknesses of 3–19 mm (0.12–0.75 in.), although this range can easily be extended by using multiple-pass techniques (Fig. 12.5). This process requires that slag (compounds of oxides, fluxes, and electrode coating materials) be cleaned after each weld bead. Unless removed completely, the solidified slag can cause severe corrosion of the weld area and can subsequently lead to failure of the weld. Slag should also be completely removed, such as by wire brushing, before another weld is applied for multiple-pass welding; thus labor costs are high, as are the material costs.

FIGURE 12.5   A weld zone showing the build-up sequence of individual weld beads in deep welds.

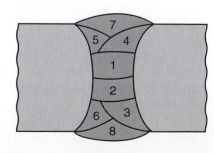

EXAMPLE 12.2 Current in shielded metal arc welding

A shielded metal arc welding operation takes place on a steel workpiece (with a steel electrode) with a 20-volt power supply. If a weld with a triangular cross section with a 10-mm leg length is to be produced, estimate the current needed for a welding speed of 10 mm/s. Use an efficiency of 75%.

Solution. The cross-sectional area of the weld is calculated from the given geometry as

$$A = \frac{1}{2}bh = \frac{1}{2}(10)(10) = 50 \text{ mm}^2.$$

The specific energy needed to melt the steel electrode is taken from Table 12.3 as 10.3 J/mm². Therefore, Eq. (12.5) yields

$$v = e\frac{VI}{uA}; \qquad I = \frac{vuA}{eV} = \frac{(10)(10.3)(50)}{(0.75)(20)} = 343 \text{ A}.$$

12.3.3 Submerged arc welding

In *submerged arc welding* (SAW), the weld arc is shielded by *granular flux* (consisting of lime, silica, manganese oxide, calcium fluoride, and other elements) which is fed into the weld zone by gravity flow through a nozzle (Fig. 12.6). The thick layer of flux completely covers the molten metal and prevents weld spatter and sparks and suppresses the intense ultraviolet radiation and fumes. The flux also acts as a thermal insulator, allowing deep penetration of heat into the workpiece. The unfused flux is recovered (using a *recovery tube*), treated, and reused.

The consumable electrode is a coil of bare wire 1.5–10 mm $\left(\frac{1}{16}-\frac{3}{8} \text{ in.}\right)$ in diameter and is fed automatically through a tube (**welding gun**). Electric currents usually range between 300 and 2000 A; the power supplies are usually connected to standard single- or three-phase power lines with a primary rating up to 440 V. Because the flux is fed by gravity, the SAW process is somewhat limited to welds in a horizontal or flat position, with a backup piece. Circular welds also can be made on pipes, provided that they are rotated during welding.

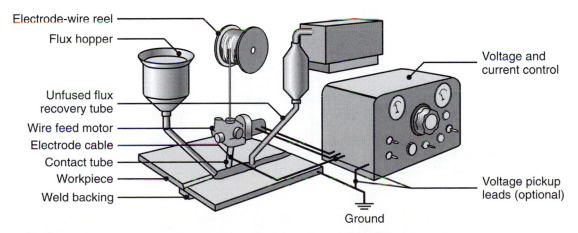

FIGURE 12.6 Schematic illustration of the submerged arc welding process and equipment. Unfused flux is recovered and reused.

The SAW process can be automated for greater economy and is used to weld a variety of carbon- and alloy-steel and stainless-steel sheet or plate, at speeds as high as 5 m/min (16 ft/min). Typical applications include thick-plate welding for shipbuilding and fabrication of pressure vessels. The quality of the weld is very high, with good toughness, ductility, and uniformity of properties. The process provides very high welding productivity, depositing 4 to 10 times the amount of weld metal per hour as deposited by the SMAW process.

12.3.4 Gas metal arc welding

In *gas metal arc welding* (GMAW), the weld area is shielded by an external source of gas, such as argon, helium, carbon dioxide, or various other gas mixtures (Fig. 12.7a). In addition, deoxidizers are usually present in the electrode metal itself, in order to prevent oxidation of the molten weld puddle. The consumable bare wire is fed automatically through a nozzle into the weld arc (Fig. 12.7b), and multiple weld layers can be deposited at the joint.

Developed in the 1950s, the GMAW process is suitable for a variety of ferrous and nonferrous metals and is used extensively in the metal-fabrication industry. The process is rapid, versatile, and economical; its welding productivity is double that of

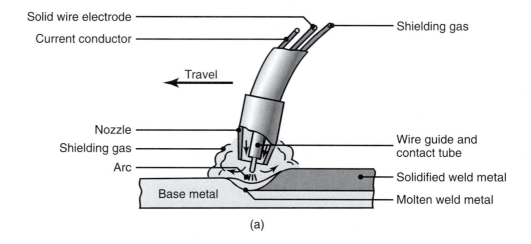

(a)

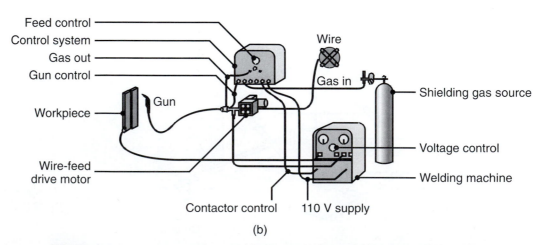

(b)

FIGURE 12.7   (a) Gas metal arc welding process, formerly known as MIG welding (for metal inert gas) and (b) basic equipment used in gas metal arc welding operations.

the SMAW process, and it can easily be automated and lends itself readily to robotics and flexible manufacturing systems (see Chapters 14 and 15).

Formerly called *MIG welding* (for *metal inert gas*), the metal in this process is transferred in one of three ways:

1. **Spray transfer.** Small droplets of molten metal from the electrode are transferred to the weld area, at rates of several hundred droplets per second. The transfer is spatter free and very stable. High DC current, high voltages, and large-diameter electrodes are used, with argon or argon-rich gas mixtures used as the shielding gas. The average current required in this process can be reduced by using *pulsed arcs,* which are high-amplitude pulses superimposed over a low, steady current. The process can be used for all welding positions.

2. **Globular transfer.** Carbon dioxide rich gases are used, and globules of molten metal are propelled by the forces of the electric arc that transfer the metal. This operation thus results in considerable spatter. The welding currents are high, with greater weld penetration and welding speed than in spray transfer. Heavier sections are commonly welded by this method.

3. **Short circuiting.** The metal is transferred in individual droplets, at rates of more than 50/s, as the electrode tip touches the molten weld pool and short circuits. Low currents and voltages are used, with carbon-dioxide rich gases and electrodes made of small-diameter wire. The power required is about 2 kW. The temperatures involved are relatively low; hence this method is suitable only for thin sheets and sections [less than 6 mm (0.25 in.) thick], although with thicker materials, incomplete fusion may occur (see Fig. 12.20). This process is very easy to use and is a common method for welding ferrous metals in thin sections; however, pulsed arc systems are gaining wider usage for thin ferrous and nonferrous metals.

12.3.5 Flux-cored arc welding

The *flux-cored arc welding* (FCAW) process (Fig. 12.8) is similar to gas metal arc welding, with the exception that the electrode is tubular in shape and is filled with flux

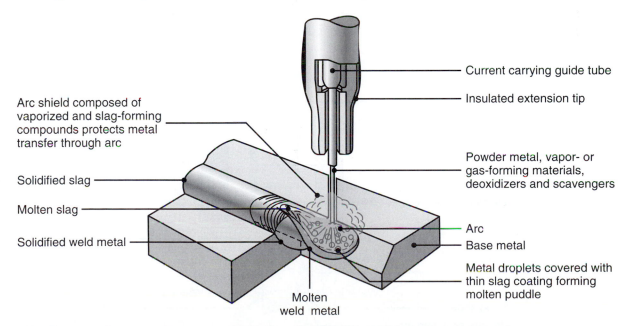

Arc shield composed of vaporized and slag-forming compounds protects metal transfer through arc

Solidified slag

Molten slag

Solidified weld metal

Molten weld metal

Current carrying guide tube

Insulated extension tip

Powder metal, vapor- or gas-forming materials, deoxidizers and scavengers

Arc

Base metal

Metal droplets covered with thin slag coating forming molten puddle

FIGURE 12.8 Schematic illustration of the flux-cored arc welding process. This operation is similar to gas metal arc welding.

(hence the term *flux cored*). Cored electrodes produce a more stable arc, improve weld contour, and improve the mechanical properties of the weld metal. The flux is much more flexible than the brittle coating used on SMAW electrodes. The tubular electrode can be provided in long coiled lengths. The power required is about 20 kW.

The electrodes are usually 0.5–4 mm (0.020–0.15 in.) in diameter. Small diameter electrodes are used in welding thinner materials, as well as making it relatively easy to weld parts out of position. The flux chemistry enables welding of many different base metals. *Self-shielded cored electrodes* are also available; these electrodes contain emissive fluxes that shield the weld area against the surrounding atmosphere and thus do not require external gas shielding.

The flux-cored arc welding process combines the versatility of SMAW with the continuous and automatic electrode feeding feature of GMAW. It is economical and is used for welding a variety of joints with different thicknesses, mainly with steels, stainless steels, and nickel alloys. A major advantage of FCAW is the ease with which specific weld metal chemistries can be developed. By adding alloys to the flux core, virtually any alloy composition can be developed. This process is easy to automate and is readily adaptable to flexible manufacturing systems and robotics.

12.3.6 Electrogas welding

Electrogas welding (EGW) is primarily used for welding the vertical edges of sections in one pass with the pieces placed edge to edge (*butt welding;* see Fig. 12.1a). It is classified as a machine welding process because it requires special equipment (Fig. 12.9). The weld metal is deposited into a weld cavity between the two pieces to be joined. The space is enclosed by two water-cooled copper dams (*shoes*) to prevent the molten slag from running off. Mechanical drives move the shoes upward. Circumferential welds such as on pipes are also possible, with the workpiece rotating.

Single or multiple electrodes are fed through a conduit, and a continuous arc is maintained, using flux-cored electrodes at currents up to 750 A, or solid electrodes at 400 A; power requirements are about 20 kW. Shielding is provided by an inert gas, such as carbon dioxide, argon, or helium, depending on the type of material being welded. The gas may be provided from an external source, or it may be produced from a flux-cored electrode, or both. Weld thickness ranges from 12 to 75 mm (0.5 to

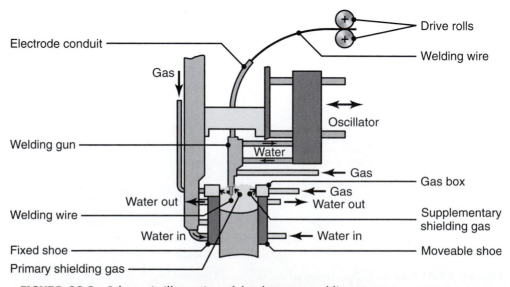

FIGURE 12.9   Schematic illustration of the electrogas welding process.

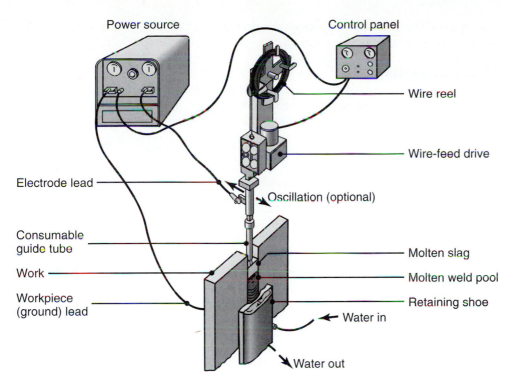

Power source

Control panel

Wire reel

Wire-feed drive

Electrode lead

Oscillation (optional)

Consumable
guide tube

Molten slag

Work

Molten weld pool

Workpiece
(ground) lead

Retaining shoe

Water in

Water out

FIGURE 12.10 Equipment used for electroslag welding operations.

3 in.) on steels, titanium, and aluminum alloys. Typical applications are in the
construction of bridges, ships, pressure vessels, storage tanks, and thick-walled and
large-diameter pipes.

12.3.7 Electroslag welding

In *electroslag welding* (ESW) the arc is started between the electrode tip and the bottom of the part to be welded (Fig. 12.10). Flux is added and melted by the heat of
the arc. After the molten slag reaches the tip of the electrode, the arc is extinguished;
energy is supplied continuously through the electrical resistance of the molten slag.
Single or multiple solid as well as flux-cored electrodes may be used, and the guide
may be nonconsumable (conventional method) or consumable.

Electroslag welding has applications that are similar to those for electrogas
welding. The process is capable of welding plates with thicknesses ranging from
50 mm to more than 900 mm (2 in. to more than 36 in.). Welding is done in one pass.
The current required is about 600 A at 40–50 V, although higher currents are used
for thick plates. The travel speed of the weld is 12–36 mm/min (0.5–1.5 in./min). The
weld quality is good, and the process is used for heavy structural steel sections, such
as heavy machinery and nuclear reactor vessels.

12.3.8 Electrodes for arc welding

The *electrodes* for the consumable electrode arc welding processes described thus far
are classified according to the strength of the deposited weld metal, the current (AC or
DC), and the type of coating. Electrodes are identified by numbers and letters or by
color code, particularly if they are too small to imprint with identification. Typical
coated electrodes are 150 to 460 mm (6 to 18 in.) long and 1.5 to 8 mm ($\frac{1}{16}$ to $\frac{5}{16}$ in.)

in diameter. The diameter of the electrode decreases with the thickness of the sections to be welded and the current required.

Electrode coatings. Electrodes are *coated* with claylike materials that include silicate binders and powdered materials such as oxides, carbonates, fluorides, metal alloys, and cellulose (cotton cellulose and wood flour). The coating, which typically is brittle and has complex interactions during welding, has the following basic functions:

1. Stabilize the arc.
2. Generate gases to act as a shield against the surrounding atmosphere (the gases produced are carbon dioxide and water vapor, and carbon monoxide and hydrogen in small amounts).
3. Control the rate at which the electrode melts.
4. Act as a flux to protect the weld against the formation of oxides, nitrides, and other inclusions and, with the slag produced, protect the molten weld pool.
5. Add alloying elements to the weld zone to enhance the properties of the weld, including deoxidizers to prevent the weld from becoming brittle.

The electrode coating or slag must be removed from the weld surfaces after each pass (see also Fig. 12.5) in order to ensure weld quality. A manual or powered wire brush can be used for this purpose. Bare electrodes and wires made of stainless steels and aluminum alloys also are available, and are used as filler metals in various welding operations.

12.4 | Arc Welding Processes: Nonconsumable Electrode

Nonconsumable electrode arc welding processes typically use a **tungsten electrode**. As one pole of the arc, the electrode generates the heat required for welding; a shielding gas is supplied from an external source. The three basic processes are described in the upcoming subsections.

12.4.1 Gas tungsten arc welding

In *gas tungsten arc welding* (GTAW), formerly known as *TIG welding* (for *tungsten inert gas*), a filler metal is typically supplied from a **filler wire** (Fig. 12.11a). However, welding also may be done without filler metals, such as in welding close-fit joints. The composition of filler metals must be similar to that of the metals to be welded. Flux is not used and the shielding gas is usually argon or helium, or a mixture of the two. Because the tungsten electrode is not consumed in this operation, a constant and stable arc gap is maintained at a constant level of current.

The power supply (Fig. 12.11b) ranges from 8 to 20 kW and is either DC at 200 A or AC at 500 A, depending on the metals to be welded. In general, AC is preferred for aluminum and magnesium, because the cleaning action of AC removes oxides and improves weld quality. Thorium or zirconium may be used in the tungsten electrodes to improve their electron-emission characteristics. Contamination of the tungsten electrode by the molten metal can be a significant problem in this process, particularly in critical applications, as contamination can cause discontinuities in the weld; therefore, contact of the electrode with the molten metal pool must be avoided.

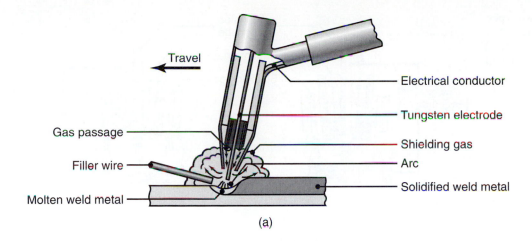

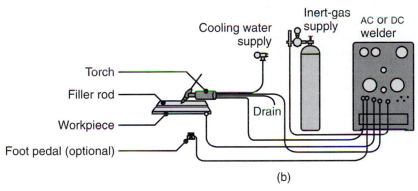

FIGURE 12.11   (a) Gas tungsten arc welding process, formerly known as TIG welding (for tungsten inert gas) and (b) equipment for gas tungsten arc welding operations.

The GTAW process is used for a wide variety of applications and metals, particularly aluminum, magnesium, titanium, and refractory metals, and it is especially suitable for thin metals. The cost of the inert gas makes this process more expensive than SMAW, but it provides welds with very high quality and good surface finish.

12.4.2 Atomic hydrogen welding

In *atomic hydrogen welding* (AHW), an arc is generated between two tungsten electrodes in a shielding atmosphere of flowing hydrogen gas. The hydrogen gas is normally diatomic (H_2), but near the arc, where the temperatures are over 6000°C (11,000°F), the hydrogen breaks down into its atomic form, simultaneously absorbing a large amount of heat from the arc. When the hydrogen strikes a relatively cold surface, i.e., the weld zone, it recombines into its diatomic form and releases the stored heat very rapidly. The energy in AHW can easily be varied by changing the distance between the arc stream and the workpiece surface. This process is not used as often because shielded arc welding processes are often preferable to atomic hydrogen welding, mainly because of the availability of inexpensive inert gases.

12.4.3 Plasma arc welding

In *plasma arc welding* (PAW), developed in the 1960s, a concentrated plasma arc is produced and directed toward the weld area. The arc is stable and reaches

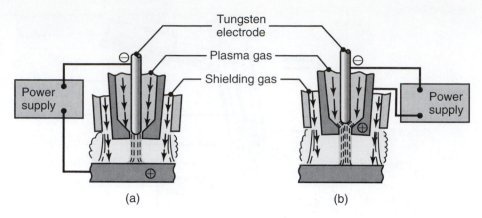

temperatures as high as 33,000°C (60,000°F). Plasma is ionized hot gas, composed
of nearly equal numbers of electrons and ions. The plasma is initiated between the
tungsten electrode and the orifice, using a low-current pilot arc. Unlike the arc in
other processes, the plasma arc is concentrated, because it is forced through a relatively small orifice. Operating currents are usually below 100 A, but they can be
higher for special applications. When a filler metal is used, it is fed into the arc, as in
GTAW. Arc and weld-zone shielding is supplied through an outer shielding ring,
using gases such as argon, helium, or mixtures of these gases.

   There are two methods of plasma arc welding. In the **transferred arc** method
(Fig. 12.12a), the part being welded is part of the electrical circuit. The arc thus
transfers from the electrode to the workpiece, hence the term transferred. In the
nontransferred method (Fig. 12.12b), the arc is between the electrode and the
nozzle, and the heat is carried to the workpiece by the plasma gas; this technique is
also used for *thermal spraying* (described in Section 4.5.1).

   Compared with other arc welding processes, plasma arc welding has (a) a
greater energy concentration, hence deeper and narrower welds can be made, (b) better arc stability, (c) less thermal distortion, and (d) higher welding speeds, such as
120–1000 mm/min (5–40 in./min). A variety of metals can be welded, generally with
part thicknesses less than 6 mm (0.25 in.). The high heat concentration can completely
penetrate through the joint (**keyhole technique**), to thicknesses as much as 20 mm
(0.75 in.) for some titanium and aluminum alloys. In the keyhole technique, the force
of the plasma arc displaces the molten metal and produces a hole at the leading edge
of the weld pool. Plasma arc welding is often used for butt and lap joints, because of
high energy concentration, better arc stability, and higher welding speeds.

12.5 | High-Energy-Beam Welding

Joining with high-energy beams, principally laser-beam and electron-beam welding,
has important applications in modern manufacturing, because of high quality and
technical and economic advantages. The general characteristics of high-energy
beams and their unique applications in machining operations are also described in
Section 9.14.

12.5.1 Electron-beam welding

In *electron-beam welding* (EBW), heat is generated by high-velocity, narrow-beam
electrons. The kinetic energy of the electrons is converted into heat as the electrons
strike the workpiece. This process requires special equipment to focus the beam on
the workpiece and requires a vacuum. The higher the vacuum, the more the beam

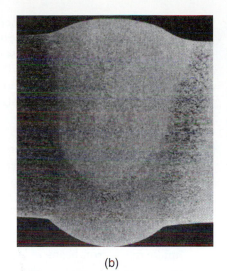

(a)   (b)

FIGURE 12.13

Comparison of the size of weld beads in (a) electron-beam or laser-beam welding with that in (b) conventional (tungsten arc) welding. *Source:* American Welding Society, *Welding Handbook,* 8th ed., 1991.

penetrates into the part and the greater is the depth-to-width ratio. The level of vacuum is specified as HV (high vacuum) or MV (medium vacuum), both of which are used commonly. NV (no vacuum) can be effective on some materials. Almost any similar or dissimilar metals can be butt or lap welded using this process, with thicknesses ranging from foil to plates as thick as 150 mm (6 in.). The intense energy is also capable of producing holes in the workpiece (see Section 9.14). Generally, no shielding gas, flux, or filler metal is required. Capacities of electron-beam guns range to 100 kW.

Developed in the 1960s, the EBW process has the capability of producing high-quality, deep and narrow welds that are almost parallel sided and have small heat-affected zones. (See Section 12.6.) Depth-to-width ratios range between 10 and 30 (Fig. 12.13). Welding parameters can be precisely controlled, with speeds as high as 12 m/min (40 ft/min). Distortion and shrinkage in the weld area are minimal, and the weld quality is good, with very high purity. Typical applications of EBW include welding of aircraft, missile, nuclear, and electronic components, as well as gears and shafts in the automotive industry. However, electron-beam welding equipment generates X-rays, and hence proper monitoring and periodic maintenance are important. A related process is electron-beam melting, discussed in Section 10.12.4.

12.5.2 Laser-beam welding

Laser-beam welding (LBW) uses a high-power laser beam as the source of heat (see Fig. 9.36 and Table 9.5) to produce a fusion weld. Because the laser beam can be highly focused to as small a diameter as 10 μm (400 μin.), it has high energy density and therefore deep penetrating capability. Consequently, this process is particularly suitable for welding deep and narrow joints (Fig. 12.13a), with depth-to-width ratios typically ranging from 4 to 10. The laser beam may be **pulsed** (milliseconds) for applications such as spot welding of thin materials, with power levels up to 100 kW. **Continuous** multi-kW laser systems are used for deep welds on thick sections. The efficiency of this process decreases with increasing *reflectivity* of the workpiece materials. To further improve the process' performance, oxygen may be used for welding steels, and inert gases for nonferrous metals.

The LBW process can be automated and used on a variety of materials, with thicknesses of up to 25 mm (1 in.); it is particularly effective on thin workpieces. Typical metals and alloys welded include aluminum, titanium, ferrous metals, copper, superalloys, and refractory metals. Welding speeds range from 2.5 m/min (8 ft/min) to as high as 80 m/min (250 ft/min) for thin metals. Because of the nature of the process, welding can be done in otherwise inaccessible locations. Safety is a particularly important consideration in laser-beam welding, because of the extreme hazards to the eye as well as the skin. Solid-state (YAG) lasers are particularly harmful.

Laser-beam welding produces welds of good quality, with minimum shrinkage and distortion. Laser welds have good strength and are generally ductile and free of porosity. In the automotive industry, welding of transmission components is a widespread application of this process; among numerous other applications is the welding of thin parts for electronic components, hermetic welding of pace makers, and transmission shaft components (where it competes with *friction welding* as described in Section 12.9). As shown in Fig. 7.14, another important application of laser welding is the forming of butt-welded sheet metal, particularly for automotive body panels. (See the discussion of *tailor-welded blanks* in Section 7.3.4.)

The major advantages of LBW over electron-beam welding are

- The laser beam can be transmitted through air; hence a vacuum is not required, unlike in electron-beam welding.
- Because laser beams can be shaped, manipulated, and focused optically (using fiber optics), the process can easily be automated.
- The laser beams do not generate X-rays, unlike electron beams.
- The quality of the weld is better with laser-beam welding, with less tendency for incomplete fusion, spatter, porosity, and distortion.

EXAMPLE 12.3 Laser-beam welding of razor blades

Figure 12.14 shows a close-up of one model of a Gillette Sensor razor cartridge. Each of the two narrow, high-strength blades has 13 pinpoint welds, 11 of which can be seen, as darker spots about 0.5 mm in diameter, on each blade in the

FIGURE 12.14 Gillette Sensor razor cartridge, with laser-beam welds.

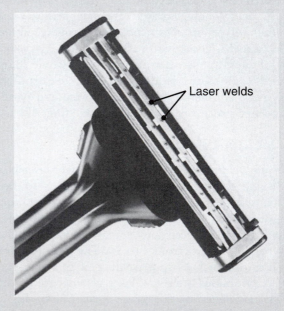

Laser welds

photograph. The welds are made with an Nd:YAG laser equipped with fiber-optic delivery, thus providing flexible beam manipulation to the exact locations along the length of the blade. With a set of these machines, production rate is 3 million welds per hour, with accurate and consistent weld quality.

Source: Courtesy of Lumonics Corporation, Industrial Products Division.

12.6 | The Fusion Welded Joint

This section describes the following aspects of fusion welding processes:

- The nature, properties, and quality of the *welded joint*.
- *Weldability* of metals.
- *Testing* welds.

A typical fusion welded joint is shown in Fig. 12.15, where three distinct zones can be identified:

1. The **base metal**, that is, the metal to be joined.
2. The **heat-affected zone** (HAZ).
3. The **fusion zone**, that is, the region that has melted during welding.

The metallurgy and properties of the second and third zones depend greatly on the metals being joined (single phase, two phase, dissimilar), the particular welding process, the filler metals used (if any), and the process variables. A joint produced without a filler metal is called **autogenous**, and the weld zone consists of the *resolidified* base metal. A joint made with a filler metal has a central zone called the **weld metal**, which is composed of a mixture of the base and filler metals.

The mechanical properties of a welded joint depend on several factors: (a) the rate of heat application and the thermal properties of metals, because they control the magnitude and distribution of temperature in a joint during welding; (b) the microstructure and grain size in the joint, which in turn are affected by the heat applied, temperature rise, degree of prior cold work of the metals, and the rate of cooling after the weld is made, and (c) various other factors, such as the geometry of the weld bead and the presence of cracks, residual stresses, inclusions, and oxide films.

Solidification of the weld metal. After heat has been applied and filler metal (if any) has been introduced into the weld area, the molten weld joint begins to cool to ambient temperature. The *solidification process* is similar to that in casting and begins with the formation of columnar (dendritic) grains. These grains are relatively long and form parallel to the heat flow. (See Fig. 5.5.) Because metals are much better

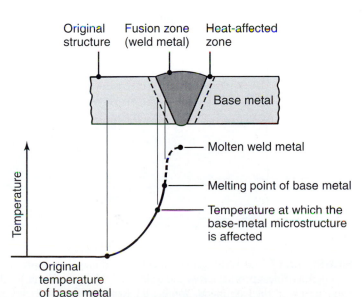

FIGURE 12.15

Characteristics of a typical fusion weld zone in oxyfuel gas welding and arc welding processes.

FIGURE 12.16 Grain structure in (a) a deep weld and (b) a shallow weld. Note that the grains in the solidified weld metal are perpendicular to their interface with the base metal.

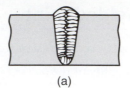
(a)

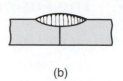
(b)

heat conductors than the surrounding air, the grains lie parallel to the plane of the two plates or sheets being welded (Fig. 12.16a). The grains developed in a *shallow* weld are shown in Fig. 12.16b. Grain structure developed and size depend on the specific alloy, the particular welding process, and the filler metal used.

The weld metal has basically an *alloyed cast structure,* and because it has cooled somewhat slowly, it typically has coarse grains. Consequently, this structure has relatively low strength, hardness, toughness, and ductility (Fig. 12.17). However, the mechanical properties of the joint can be improved with proper selection of filler metal composition or with subsequent heat treatments; the results depend on the particular alloy, its composition, and the thermal cycling to which the joint is subjected. For example, cooling rates may be controlled and reduced by *preheating* the weld area prior to welding. Preheating is particularly important for metals with high thermal conductivity, such as aluminum and copper, as otherwise the heat rapidly dissipates during welding.

Heat-affected zone. The *heat-affected zone* is within the base metal itself, as can be seen in Fig. 12.15. It has a microstructure different than that of the base metal prior to welding, because it has been subjected to elevated temperatures for a period of time during welding. The regions of the base metal that are far away from the heat source do not undergo any microstructural changes during welding. The properties and microstructure of the HAZ depend on (a) the rate of heat input and cooling and (b) the temperature to which this zone has been raised during welding. Metallurgical factors such as original grain size, grain orientation, and degree of prior cold work, the specific heat and thermal conductivity of the metals also influence the size and characteristics of HAZ.

The strength and hardness of the heat-affected zone depend partly on how the original strength and hardness of the particular alloy had been developed prior to welding. As described in Chapters 3 and 5, they may, for example, have been

(a)

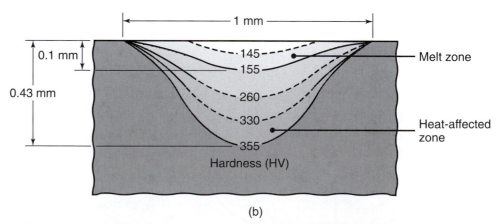
(b)

FIGURE 12.17 (a) Weld bead on a cold-rolled nickel strip produced by a laser beam and (b) microhardness profile across the weld bead. Note the lower hardness of the weld bead as compared with the base metal. *Source:* IIT Research Institute.

FIGURE 12.18
Intergranular corrosion of a weld joint in ferritic stainless-steel welded tube, after exposure to a caustic solution. The weld line is at the center of the photograph. *Source:* Courtesy of Allegheny Ludlum Corp.

developed by cold working, solid solution strengthening, precipitation hardening, or various other heat treatments. Among metals strengthened by these various methods, the simplest to analyze is base metal that has been cold worked, say, by cold rolling or forging (Chapter 6).

The heat applied during welding *recrystallizes* the elongated grains (preferred orientation) of the cold-worked base metal. Grains that are away from the weld metal will recrystallize into fine equiaxed grains. However, grains close to the weld metal, having been subjected to elevated temperatures for a longer period of time, will grow. (See also Section 3.6.) The grain growth will result in a region that is softer and has lower strength, and such a joint will be weakest in its heat-affected zone. The grain structure of such a weld that has been exposed to corrosion by chemical reaction is shown in Fig. 12.18. The central vertical line is the juncture of the two workpieces. The effects of heat during welding on the HAZ for joints made with *dissimilar* metals, and for alloys strengthened by other mechanisms, are complex, and a discussion is beyond the scope of this text.

12.6.1 Weld quality

Because of its past history of thermal cycling and attendant microstructural changes, a welded joint may develop *discontinuities*. Welding discontinuities can also be caused by inadequate or careless application of established welding techniques or substandard operator training. The major discontinuities that affect weld quality are described below:

1. **Porosity.** *Porosity* in welds may be caused by (a) *trapped gases* that are released during melting of the weld zone but trapped during solidification, (b) *chemical reactions* that occur during welding, or (c) *contaminants* present. Most welded joints contain some porosity, which is generally spherical in shape or in the form of elongated pockets. (See also Section 5.12.1.) The *distribution* of porosity in the weld zone may be random, or it may be concentrated in a certain region. Particularly important in porosity is the presence of *hydrogen,* which may be due to the use of damp fluxes or to environmental humidity. The result is, for example, porosity in aluminum-alloy welds and hydrogen embrittlement in steels. (See also Section 3.8.2.)

     Porosity in welds can be reduced by the following methods:

   a. Proper selection of electrodes and filler metals.

   b. Improvement of the welding operation, such as by preheating the weld area or increasing the rate of heat input.

   c. Proper cleaning of the weld zone and prevention of contaminants from entering the weld zone.

   d. Lowering the welding speed to allow time for gases to escape.

   e. Shot peening the weld bead (see Section 4.5.1).

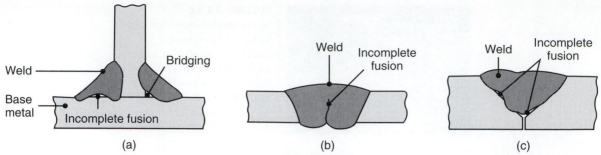

FIGURE 12.19   Examples of various incomplete fusion in welds. (a) Incomplete fusion in fillet welds, (b) incomplete fusion from oxide or dross at the center of a joint, especially in aluminum, and (c) incomplete fusion in a groove weld.

2. **Slag inclusions.** *Slag inclusions* are compounds, such as oxides, fluxes, and electrode coating materials that are trapped in the weld zone. If the shielding gases used are not effective, contamination from the environment may also contribute to slag inclusions. Maintenance of proper welding conditions is also important. For example, when proper techniques are used, the molten slag will float to the surface of the molten weld metal and not be entrapped. Slag inclusions, or *slag entrapment,* may be prevented by the following methods:

   a. Cleaning the weld-bead surface before the next layer is deposited (see Fig. 12.5), typically using a hand or power wire brush or chisel.

   b. Providing adequate shielding gas.

   c. Redesigning the joint (see Section 12.17) to permit sufficient space for proper manipulation of the puddle of molten weld metal.

3. **Incomplete fusion and incomplete penetration.** *Incomplete fusion* (*lack of fusion*) produces poor weld beads, such as those shown in Fig. 12.19. A better weld can be obtained by the following methods:

   a. Raising the temperature of the base metal.

   b. Cleaning the weld area prior to welding.

   c. Modifying the joint design.

   d. Changing the type of electrode.

   e. Providing adequate shielding gas.

   Incomplete penetration occurs when the depth of the welded joint is insufficient. Penetration can be improved by the following methods:

   a. Increasing the heat input.

   b. Lowering travel speed during welding.

   c. Modifying the design of the joint.

   d. Ensuring that the surfaces to be joined fit properly.

4. **Weld profile.** *Weld profile* is important not only because of its effects on the strength and appearance of the weld, but also because it can indicate incomplete fusion or the presence of slag inclusions in multiple layer welds.

   Underfilling, shown in Fig. 12.20a, results when the joint is not filled with the proper amount of weld metal.

   Undercutting results from melting away of the base metal and subsequent development of a groove in the shape of a sharp recess or notch. If deep or sharp, an undercut can act as a stress raiser and thus reduce the fatigue strength of the joint.

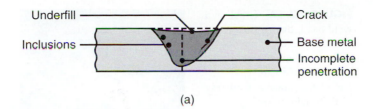

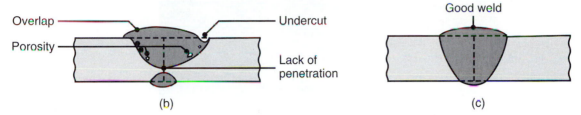

FIGURE 12.20   Examples of various defects in fusion welds.

Overlap (Fig. 12.20b) is a surface discontinuity, generally caused by poor welding practice and improper selection of the materials. A proper weld bead without these defects is shown in Fig. 12.20c.

5. **Cracks.** *Cracks* may develop in various locations and directions in the weld area. The typical types of cracks are longitudinal, transverse, crater, underbead, and toe cracks (Fig. 12.21). Cracks generally result from a combination of the following factors:

a. Temperature gradients, causing thermal stresses in the weld zone.

b. Variations in the composition of the weld zone, causing different contractions in the zone.

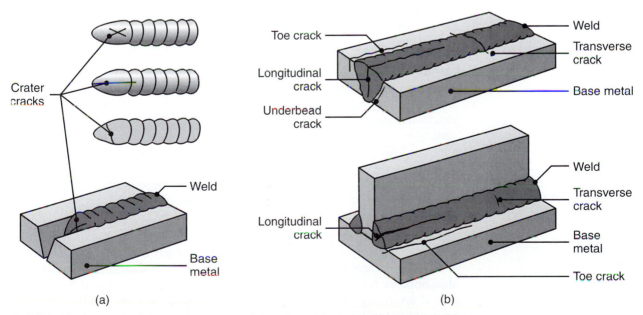

FIGURE 12.21   Types of cracks in welded joints. The cracks are caused by thermal stresses that develop during solidification and contraction of the weld bead and the welded structure: (a) crater cracks and (b) various types of cracks in butt and T joints.

FIGURE 12.22 Crack in a weld bead, due to the fact that the two components were not allowed to contract after the weld was completed. *Source:* Courtesy of Packer Engineering.

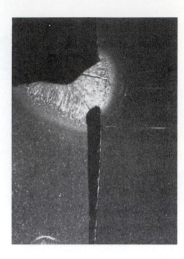

    c. Embrittlement of grain boundaries by segregation of elements, such as sulfur, to the grain boundaries (see Section 3.4.2) as the solid-liquid boundary moves when the weld metal begins to solidify.

    d. Hydrogen embrittlement (see Section 3.8.2).

    e. Inability of the weld metal to contract during cooling (Fig. 12.22), a situation similar to the development of *hot tears* in castings, due to excessive constraint of the workpiece (see also Section 5.12.1).

    Cracks are classified as **hot cracks** (developing while the joint is still at elevated temperatures) and **cold cracks** (that develop after the weld metal has solidified). Basic crack-prevention measures are

    a. Modifying the joint design to minimize thermal stresses from shrinkage during cooling.

    b. Changing welding process parameters, procedures, and sequence.

    c. Preheating the components being welded.

    d. Avoiding rapid cooling of the components after welding.

6. **Lamellar tears.** In describing the anisotropy of plastically deformed metals in Section 3.5, it was stated that because of the alignment of nonmetallic impurities and inclusions (stringers), the workpiece is weaker in its thickness direction than in other directions. This condition, called *mechanical fibering,* is particularly evident in rolled plates and structural shapes. In welding such components, *lamellar tears* may develop, because of shrinkage of the restrained members in the structure during cooling. Such tears can be avoided by providing for shrinkage of the members or by modifying the joint design to make the weld bead penetrate the weaker member to a greater extent.

7. **Surface damage.** During welding, some of the metal may spatter and be deposited as small droplets on adjacent surfaces. Also, in arc welding, the electrode may inadvertently be allowed to contact the parts being welded at places outside of the weld zone (**arc strikes**). Such surface discontinuities may be objectionable for reasons of appearance or the subsequent use of the welded part. If severe, they may adversely affect the properties of the welded structure, particularly for notch-sensitive metals.

8. **Residual stresses.** Because of localized heating and cooling during welding, expansion and contraction of the weld area cause *residual stresses* in the

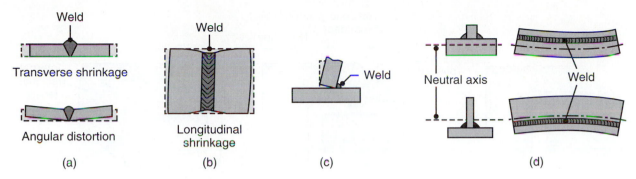

Weld

Transverse shrinkage

Angular distortion

(a)

Weld

Longitudinal shrinkage

(b)

Weld

(c)

Neutral axis

Weld

(d)

FIGURE 12.23   Distortion and warping of parts after welding, caused by differential thermal expansion and contraction of different regions of the welded assembly. Warping can be reduced or eliminated by proper weld design and fixturing prior to welding.

workpiece. (See also Section 2.10.) Residual stresses can cause detrimental effects such as

a.   Distortion, warping, and buckling of the welded parts (Fig. 12.23).

b.   Stress-corrosion cracking (see Section 3.8.2).

c.   Further distortion if a portion of the welded structure is subsequently removed, say, by machining or sawing.

d.   Reduced fatigue life.

The type and distribution of residual stresses in welds are best described by reference to Fig. 12.24a. When two plates are being welded, a long, narrow region is subjected to elevated temperatures, whereas the plates as a whole are essentially at ambient temperature. As the weld is completed and time elapses, the heat from the weld area dissipates laterally into the plates. The plates thus begin to expand longitudinally while the welded length begins to contract. These two opposing effects cause residual stresses that typically are distributed as illustrated in Fig. 12.24b. Note that, as expected, the magnitude of compressive residual stresses in the plates diminishes to zero at a point away from the weld area. Because no external forces are acting on the welded plates, the tensile and compressive forces represented by these residual stresses must balance each other.

   In complex welded structures, residual stress distributions are three dimensional and thus difficult to analyze. The preceding example involves only two plates that are not restrained from movement; in other words, the plates are

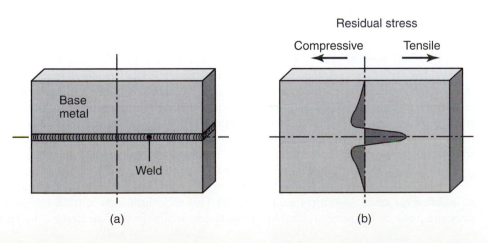

Residual stress

Compressive   Tensile

Base metal

Weld

(a)

(b)

FIGURE 12.24   Residual stresses developed in a straight butt joint. *Source:* Courtesy of the American Welding Society.

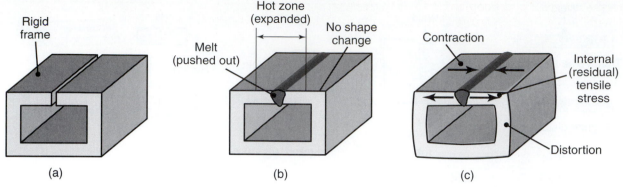

FIGURE 12.25 Distortion of a welded structure. (a) before welding, (b) during welding, with weld bead placed in joint, and (c) after welding, showing distortion in the structure. *Source:* After J.A. Schey.

not an integral part of a larger structure. If, however, the plates are restrained, reaction stresses will develop because the plates are not free to expand or contract. This situation is particularly present in structures with high stiffness.

Events leading to the *distortion* of a welded structure is shown in Fig. 12.25. Prior to welding, the structure is stress-free, as shown in Fig. 12.25a. The shape may be fairly rigid, and fixturing may also be present in order to support the structure. When the weld bead is placed, the molten metal fills the gap between the surfaces to be joined, and flows outwards to form the weld bead. At this point, the weld is not under any stress. The weld bead then solidifies, and both the weld bead and the surrounding material cool to room temperature. As these materials cool, they contract but are constrained by the bulk of the surrounding structure. As a result, the structure distorts (Fig. 12.25c) and residual stresses develop.

Stress relieving of welds. The effects of residual stresses, such as distortion, buckling, and cracking, can be reduced by *preheating* the base metal or the parts to be welded. Preheating reduces distortion by lowering the cooling rate and the magnitude of thermal stresses (by reducing the elastic modulus of the metals being welded). This method also reduces shrinkage and the tendency for cracking of the joint. Preheating is also effective in welding hardenable steels (Section 5.11) for which rapid cooling of the weld area would otherwise be harmful. The pieces may be heated in a furnace or electrically or inductively; thin sections may be heated by radiant lamps or a hot-air blast. For optimum results, preheating temperatures and cooling rates must be controlled in order to attain acceptable strength and toughness in the welded structure.

Residual stresses can be reduced by *stress relieving* or *stress-relief annealing* the welded structure (Section 5.11.4). The temperature and the time required for stress relieving depend on the type of material and magnitude of the residual stresses developed. Other methods of stress relieving include peening, hammering, and surface rolling the weld bead area. (See Section 4.5.1.) These processes induce compressive residual stresses at the surface, thus reducing or eliminating tensile residual stresses in the weld. For multilayer welds, the first and last layers should not be peened, because peening may cause damage to those layers.

Residual stresses can also be relieved or reduced by plastically deforming the structure by a small amount. (See Fig. 2.32.) This technique can be used in welded pressure vessels by pressurizing the vessels internally (*proof stressing*), in turn producing axial and hoop stresses (see also Section 2.11). In order to reduce the possibility of sudden fracture under high internal pressure, the weld must be free from significant notches or discontinuities, which could act as points of stress concentration.

In addition to stress relieving, welds may also be *heat treated* in order to modify and improve properties such as strength and toughness. The techniques employed include (a) annealing, normalizing, or quenching and tempering of steels and (b) solution treatment and aging of precipitation-hardenable alloys (see Section 5.11).

12.6.2 Weldability

Weldability is generally defined as (a) a metal's capability to be welded into a specific structure with certain specific properties and characteristics and (b) that the welded structure will satisfactorily meet service requirements. As can be expected from the discussions thus far in this chapter, weldability involves a large number of variables, thus making generalizations difficult. Material characteristics, such as alloying elements, impurities, inclusions, grain structure, and processing history of the base metal and filler metal, are all important. Other factors that influence weldability are properties such as strength, toughness, ductility, notch sensitivity, elastic modulus, specific heat, melting point, thermal expansion, surface-tension characteristics of the molten metal, and corrosion.

Preparation of surfaces for welding is important, as are the nature and properties of surface oxide films and adsorbed gases. The welding process employed significantly affects the temperatures developed and their distribution in the weld zone. Additional factors that influence weldability are shielding gases, fluxes, moisture content of the coatings on electrodes, welding speed, welding position, cooling rate, preheating, and postwelding techniques (such as stress relieving and heat treating).

The following list briefly summarizes the general weldability of specific groups of metals. The weldability of these materials can vary significantly, with some requiring special welding techniques and proper control of all processing parameters.

1. *Aluminum alloys:* weldable at a high rate of heat input; alloys containing zinc or copper generally are considered unweldable.
2. *Cast irons:* generally weldable.
3. *Copper alloys:* similar to that of aluminum alloys.
4. *Lead:* weldable.
5. *Magnesium alloys:* weldable with the use of protective shielding gas and fluxes.
6. *Molybdenum:* weldable under well-controlled conditions.
7. *Nickel alloys:* weldable.
8. *Niobium (columbium):* weldable under well-controlled conditions.
9. *Stainless steels:* weldable.
10. *Steels, galvanized and prelubricated:* weldability adversely affected by the presence of zinc coating and lubricant layer.
11. *Steels, high-alloy:* generally good weldability under well-controlled conditions.
12. *Steels, low-alloy:* fair to good weldability.
13. *Steels, plain-carbon:* (a) excellent weldability for low-carbon steels, (b) fair to good weldability for medium-carbon steels, and (c) poor weldability for high-carbon steels.
14. *Tantalum:* weldable under well-controlled conditions.

15. *Tin:* weldable.
16. *Titanium alloys:* weldable with the use of proper shielding gases.
17. *Tungsten:* weldable under well-controlled conditions.
18. *Zinc:* difficult to weld; soldering preferred (see Section 12.13.3).
19. *Zirconium:* weldable with the use of proper shielding gases.

12.6.3 Testing welded joints

The *quality* of a welded joint is typically established by testing the joint. Several standardized tests and procedures have been developed and established and are available from organizations such as the American Society for Testing and Materials (ASTM), the American Welding Society (AWS), the American Society of Mechanical Engineers (ASME), the American Society of Civil Engineers (ASCE), and various federal agencies. Welded joints may be tested either *destructively* or *nondestructively* (see Section 4.8). Each technique has its capabilities, limitations, reliability, and need for special equipment and operator skill.

1. **Destructive testing techniques.** Five methods of destructively testing welded joints are commonly used.

   a. **Tension test.** Longitudinal and transverse tension tests are performed on specimens removed from actual welded joints and from the weld-metal area. Stress-strain curves are then obtained by the procedures described in Section 2.2. These curves indicate the yield strength (Y), ultimate tensile strength (UTS), and ductility of the welded joint in different locations and directions. Ductility is measured in terms of percentage elongation and percentage reduction of area. Tests of weld hardness may also be used to indicate weld strength and microstructural changes in the weld zone.

   b. **Tension-shear test.** The specimens in the tension-shear test (Fig. 12.26a) are specially prepared to simulate actual welded joints and procedures.

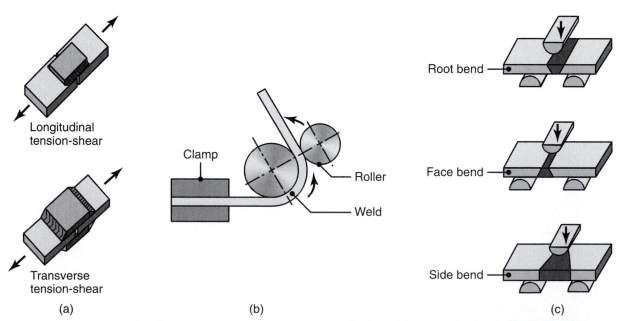

FIGURE 12.26 (a) Types of specimens for tension-shear testing of welds, (b) wraparound bend test method, and (c) three-point bending of welded specimens. (See also Fig. 2.21.)

The specimens are then subjected to tension, and the shear strength of the weld metal and the location of fracture are determined.

c. **Bend test.** Several bend tests can be used to determine the ductility and strength of welded joints. In one test, the welded specimen is bent around a fixture (*wraparound bend test*; Fig. 12.26b). In another test, the specimens are tested in *three-point transverse bending* (Fig. 12.26c; see also Fig. 2.21).

d. **Fracture toughness test.** Fracture toughness tests commonly use impact-testing techniques, described in Section 2.9. Charpy V-notch specimens are prepared and tested for toughness; other toughness tests include the drop-weight test, in which the energy is supplied by a falling weight.

e. **Corrosion and creep tests.** Because of the difference in the composition and microstructure of the materials in the weld zone, preferential corrosion may take place in this zone. (See Fig. 12.18.) Creep tests are essential in determining the behavior of welded joints at elevated temperatures.

f. **Testing of spot welds.** Spot-welded joints may be tested for weld-nugget strength (Fig. 12.27), using the (a) *tension-shear,* (b) *cross-tension,* (c) *twist,* and (d) *peel tests.* Because they are inexpensive and easy to perform, tension-shear tests are commonly used in sheet-metal fabricating facilities. The cross-tension and twist tests are capable of revealing flaws, cracks, and porosity in the weld area. The peel test is commonly used for thin sheets. After the joint has been bent and peeled, the shape and size of the torn-out weld nugget are evaluated.

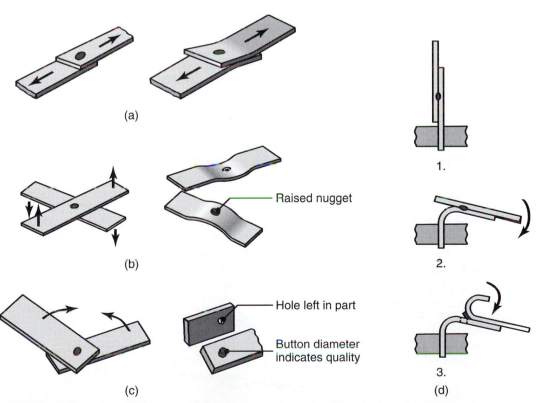

FIGURE 12.27 (a) Tension shear test for spot welds, (b) cross-tension test, (c) twist test, and (d) peel test.

2. Nondestructive testing techniques. Welded structures often have to be tested nondestructively, particularly for critical applications where weld failure can be catastrophic, such as in pressure vessels, pipelines, load-bearing structural members, and power plants. Nondestructive testing techniques for welded joints usually consist of visual, radiographic, magnetic-particle, liquid-penetrant, and ultrasonic testing methods, as described in detail in Section 4.8.1.

12.6.4 Welding process selection

In addition to the role of material characteristics described thus far, selection of a welding process for a particular application involves the following considerations:

1. Configuration of the parts or structure to be welded and their shape, thickness, and size.
2. The methods used to manufacture component parts.
3. Service requirements, such as the type of loading and stresses generated.
4. Location, accessibility, and ease of welding.
5. Effects of distortion and discoloration.
6. Joint appearance.
7. Costs involved in various operations, such as edge preparation, welding, and postprocessing of the weld, including machining and finishing operations.

12.7 | Cold Welding

In *cold welding* (CW), pressure is applied to the mating faces of the parts, through either dies or rolls. Because of the resulting plastic deformation, it is important and necessary that at least one, but preferably both, of the mating parts be sufficiently ductile. The interface is usually cleaned by wire or power brushing prior to welding. However, in joining two dissimilar metals that are mutually soluble, brittle *intermetallic compounds* may form (see Section 5.2.2), resulting in a weak and brittle joint. An example is the bonding of aluminum and steel, where a brittle intermetallic compound is formed at the interface. The best bond strength and ductility are obtained with two similar materials. Cold welding can be used to join small workpieces made of soft, ductile metals; applications include electrical connections, wire stock, and sealing of heat sensitive containers (such as those containing explosives, such as detonators).

Roll bonding. In *roll bonding,* or *roll welding* (ROW), the pressure required for cold welding of long pieces or continuous strips is applied through a pair of rolls (Fig. 12.28). A common application is in manufacturing some U.S. coins, as described in Example 12.4. As in other similar processes, surface preparation is important for improved interfacial strength. Another common application of roll bonding is the

FIGURE 12.28 Schematic illustration of the roll-bonding, or cladding, process.

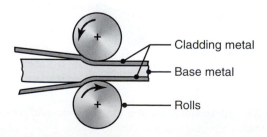

Cladding metal

Base metal

Rolls

production of bimetallic strips for thermostats and similar controls, using two layers of materials with different coefficients of thermal expansion. (See also Section 12.8.) Bonding in only selected regions in the interface can be achieved by depositing a parting agent such as graphite or ceramic (called *stop-off*). The use of this technique in *superplastic forming* of sheet-metal structures is shown in Fig. 7.46.

Roll bonding can also be carried out at elevated temperatures (*hot roll bonding*) for improved interfacial strength. Typical examples include *cladding* (Section 4.5.1) (a) pure aluminum over precipitation-hardened aluminum-alloy sheet (Section 5.11.2) and (b) stainless steel over mild steel for corrosion resistance.

EXAMPLE 12.4 Roll bonding of the U.S. quarter

The technique used for manufacturing composite U.S. quarters is roll bonding. Two outer layers of the coin are 75% Cu-25% Ni (cupronickel), each 1.2 mm (0.048 in.) thick, with an inner layer of pure copper 5.1 mm (0.20 in.) thick. For good bond strength, the mating surfaces (*faying surfaces*) are first chemically cleaned and wire brushed. The strips are then rolled to a thickness of 2.29 mm (0.090 in.). A second rolling operation reduces the final thickness to 1.36 mm (0.0535 in.). The strips thus undergo a total reduction in thickness of 82%.

Because volume constancy is maintained in plastic deformation (Section 2.11.5), there is a major increase in the surface area between the layers, thus generating clean interfacial surfaces and hence better bonding. This extension in surface area under the high pressure of the rolls, combined with the solid solubility of nickel in copper (see Section 5.2.1), produces a strong bond.

12.8 | Ultrasonic Welding

In *ultrasonic welding* (USW), the faying surfaces of the two members are subjected to a static normal force and oscillating shearing (tangential) stresses. The shearing stresses are applied by the tip of a **transducer** (Fig. 12.29a), similar to that used for

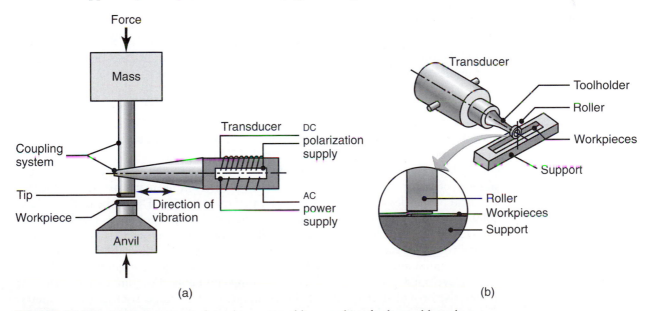

(a)   (b)

FIGURE 12.29   (a) Components of an ultrasonic welding machine for lap welds and (b) ultrasonic seam welding using a roller.

ultrasonic machining (see Fig. 9.24a). The frequency of oscillation generally ranges from 10 to 75 kHz. The energy required in this operation increases with the thickness and hardness of the materials being joined. Proper coupling between the transducer and the tip (called *sonotrode*, from the words *son*ic and el*ectrode*) is important for effective bonding. The welding tip can be replaced with rotating disks (Fig. 12.29b) for seam welding structures (similar to those shown in Fig. 12.35 for *resistance seam welding*), one component of which can be a sheet or foil.

The shearing stresses cause small-scale plastic deformation at the workpiece interfaces, breaking up oxide films and contaminants, thus allowing good contact and producing a strong solid state bond. Temperatures generated in the weld zone are usually in the range of one-third to one-half the melting point (absolute scale) of the metals joined (Table 3.2); therefore, no melting and fusion take place. In certain situations, however, temperatures can be sufficiently high to cause significant metallurgical changes in the weld zone.

The ultrasonic welding process is reliable and versatile, and can be used with a combination of a wide variety of metallic and nonmetallic materials, including dissimilar metals (such as in making *bimetallic strips*). It is used extensively in joining plastics and in the automotive and consumer electronics industries for lap welding of sheet, foil, and thin wire and in packaging with foils. The mechanism of joining thermoplastics (see Section 12.17.1) by ultrasonic welding is different than that for metals, and interfacial melting takes place, due to the much lower melting temperature of plastics (see Table 10.2).

12.9 | Friction Welding

Note that in the joining processes described thus far, the energy required for welding is supplied externally. In *friction welding* (FRW), the heat required for welding is, as the name implies, generated through friction at the interface of the two members being joined; thus, the source of energy is mechanical. The significant rise in temperature from friction can be demonstrated simply by quickly rubbing one's hands together or by rapidly sliding down a rope.

In the friction welding process, one of the members remains stationary while the other is placed in a chuck, collet, or a similar fixture (see Fig. 8.42) and rotated at a high constant speed. The two members to be joined are then brought into contact under an axial force (Fig. 12.30). The surface speed of rotation may be as high as 900 m/min (3000 ft/min). The rotating member must be clamped securely to resist both torque and axial forces without slipping in fixtures. After sufficient interfacial contact is established, (a) the rotating member is brought to a sudden stop, so that the weld is not destroyed by subsequent shearing action, and (b) the axial force is increased.

The weld zone is usually confined to a narrow region, depending on (a) the amount of heat generated, (b) the thermal conductivity of the materials, and (c) the mechanical properties of the materials at elevated temperature. The shape of the welded joint depends on the rotational speed and the axial force applied, as can be seen in Fig. 12.31. These factors must be controlled to obtain a uniformly strong joint. The radially outward movement of the hot metal at the interface (flash) pushes oxides and other contaminants out of the interface, further increasing joint strength.

Developed in the 1940s, friction welding can be used to join a wide variety of materials, provided that one of the components has some rotational symmetry. Solid as well as tubular parts can be joined by this method, with good joint strength. Solid steel bars up to 100 mm (4 in.) in diameter and pipes up to 250 mm

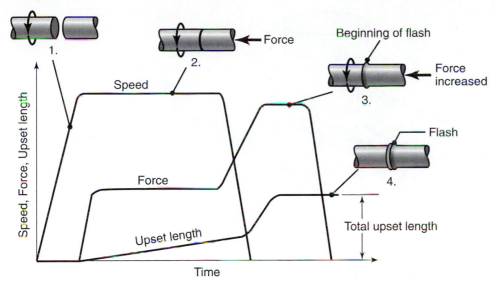

FIGURE 12.30   Sequence of operations in the friction welding process. (1) The part on the left is rotated at high speed, (2) the part on the right is brought into contact under an axial force, (3) the axial force is increased, and the part on the left stops rotating; flash begins to form, and (4) after a specified upset length or distance is achieved, the weld is completed. The upset length is the distance the two pieces move inward during welding after their initial contact; thus, the total length after welding is less than the sum of the lengths of the two pieces. If necessary, the flash can be removed by secondary operations, such as machining or grinding.

(a) High pressure or low speed

(b) Low pressure or high speed

(c) Optimum

FIGURE 12.31   Shapes of the fusion zone in friction welding as a function of the force applied and the rotational speed.

(10 in.) outside diameter have been welded successfully using this process. Because of the combined heat and pressure, the interface in FRW develops a flash by plastic deformation (upsetting) of the heated zone; if objectionable, this flash can easily be removed by subsequent machining or grinding. (See also the Case Study at the end of this chapter.)

There are several variations on the friction welding process, as described below.

1. **Inertia friction welding.** *Inertia friction welding* (IFW) is a modification of FRW, although the term has been used interchangeably with *friction welding*. The energy required in this process is supplied through the kinetic energy of a flywheel. In this operation, (a) the flywheel is first accelerated to the proper speed, (b) the two members are brought into contact, (c) an axial force is applied, and (d) as the friction at the interface begins to slow down the flywheel, the axial force is increased. The weld is completed when the flywheel comes to a stop. The timing of this sequence is important for weld quality. The rotating mass in inertia friction welding machines, hence the energy level, can be adjusted depending on workpiece cross section and material properties.

FIGURE 12.32 Schematic illustration of the friction stir welding process. Aluminum-alloy plates up to 75 mm (3 in.) thick have been welded by this process. *Source:* TWI, Cambridge, United Kingdom.

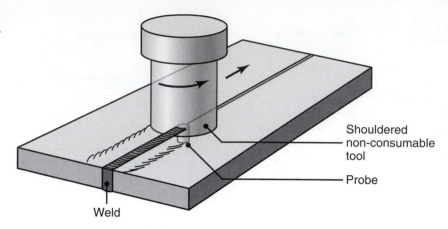

Shouldered non-consumable tool

Probe

Weld

2. **Linear friction welding.** *Linear friction welding* (LFW) involves subjecting the interface of the two parts to be joined to a linear reciprocating motion of at least one of the components to be joined. Thus, the parts do not have to be circular or tubular in cross section. In this process, one part is moved across the face of the other part, using a balanced reciprocating mechanism. Linear friction welding is capable of joining square or rectangular components, as well as round parts, made of metals or plastics. In one application, a rectangular titanium-alloy part is friction welded at a linear frequency of 25 Hz, an amplitude of ±2 mm (0.08 in.), and a pressure of 100 MPa (15,000 psi), acting on a 240-mm^2 (0.38-in^2) interface. Various other metal parts have been welded successfully, with rectangular cross sections as large as 50×20 mm (2×0.8 in.).

3. **Friction stir welding.** Whereas in conventional friction welding, heating of interfaces is achieved through friction by rubbing two contacting surfaces, in the *friction stir welding* (FSW) process, a third body is rubbed against the two surfaces to be joined. The rotating tool is a small (5 to 6 mm in diameter, 5 mm in height) rotating member that is plunged into the joint (Fig. 12.32). The contact pressures cause frictional heating, raising the temperature to the range of 230° to 260°C (450° to 500°F). The probe at the tip of the rotating tool forces heating and mixing or stirring of the material in the joint.

   This process was originally intended for welding of aerospace alloys, especially aluminum extrusions, although current applications include polymers and composite materials. The thickness of the welded parts can be as small as 1 mm (0.04 in.) and as much as 30 mm (1.2 in.) or so. The weld quality is high, with minimal pores and with uniform material structure. The welds are produced with low heat input and, therefore, low distortion and little microstructural changes; furthermore, there are no fumes or spatter produced. The welding equipment may be a simple conventional, vertical-spindle milling machine, such as that shown in Fig. 8.59b, and the process is relatively easy to implement and can easily be automated.

12.10 | Resistance Welding

Resistance welding (RW) operations cover a number of processes in which the heat required for welding is produced by means of the *electrical resistance* between the two members to be joined; hence, the major advantages include the fact that they

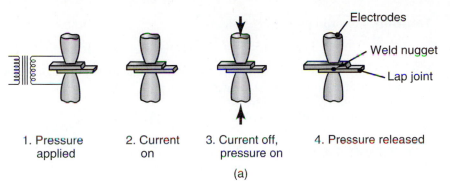

1. Pressure   2. Current   3. Current off,   4. Pressure released
   applied   on   pressure on

(a)

FIGURE 12.33
(a) Sequence in the resistance spot welding operation, and (b) cross section of a spot weld, showing weld nugget and light indentation by the electrode on sheet surfaces.

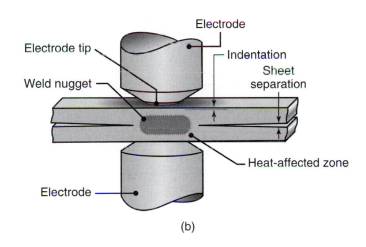

(b)

don't require consumable electrodes, shielding gases, or flux. The heat generated in resistance welding is given by the general expression

$$H = I^2Rt, \tag{12.6}$$

where H = heat generated, in joules (watt-seconds); I = current, in amperes; R = resistance, in ohms; and t = flow time of the current in seconds. This equation is generally modified to represent the actual heat energy available in the weld by including a factor K that represents the energy losses through radiation and conduction. Thus, Eq. (12.6) becomes $H = I^2RtK$, where the value of K is less than unity.

The total resistance in these processes, such as in the *resistance spot welding* shown in Fig. 12.33, is the sum of the following components:

1. Resistance of the electrodes.
2. Electrode-workpiece contact resistances.
3. Resistances of the individual parts to be welded.
4. Workpiece-workpiece contact resistances (faying surfaces).

EXAMPLE 12.5 Heat generated in resistance spot welding

Assume that two 1-mm (0.04-in.) thick steel sheets are being spot welded at a current of 5000 A and current-flow time $t = 0.1$s. Using electrodes 5 mm (0.2 in.) in diameter, estimate the amount of heat generated and its distribution in the weld zone. Use an effective resistance of 200 $\mu\Omega$.

Solution. According to Eq. (12.6),

$$\text{Heat} = (5000)^2(0.0002)(0.1) = 500 \text{ J.}$$

If we assume that the material below the electrode is heated enough to melt and fuse, we can calculate the weld nugget volume as

$$V = \left(\frac{\pi}{4}d^2\right)(t) = \left(\frac{\pi}{4}(5)^2\right)(2) = 39.3 \text{ mm}^3.$$

From Table 12.3, u for steel is 9.7 J/mm^3 (average of extreme values). Therefore, the heat required to melt the weld nugget is, from Eq. (12.4),

$$H = u(\text{Volume}) = (9.7)(39.3) = 381 \text{ J.}$$

Consequently, the remaining heat (119 J), or 24%, is dissipated into the metal surrounding the nugget.

The magnitude of the current in resistance welding operations may be as high as 100,000 A, although the voltage is typically only 0.5–10 V. The actual temperature rise at the joint depends on the specific heat and thermal conductivity of the metals to be joined. Consequently, because they have high thermal conductivity, metals such as aluminum and copper require high heat concentrations. Electrode materials should have high thermal conductivity and strength at elevated temperatures and are typically made of copper alloys. Similar and dissimilar metals can be joined by resistance welding.

Developed in the early 1900s, resistance welding processes require specialized machinery, and most are now operated with programmable computer control. The machinery is generally not portable, and the process is most suitable for use in manufacturing plants and machine shops.

There are five basic methods of resistance welding: spot, seam, projection, flash, and upset welding. Lap joints are used in the first three processes and butt joints in the last two.

12.10.1 Resistance spot welding

In *resistance spot welding* (RSW), the tips of two opposing solid, round electrodes contact the surfaces of the lap joint of two sheet metals, and resistance heating produces a spot weld (Fig. 12.33a). In order to obtain a good bond in the **weld nugget** (Fig. 12.33b), pressure is continually applied until the current is turned off. Accurate control and timing of the electric current and pressure are thus essential. Currents usually range from 3000 to 40,000 A, depending on the materials being welded and their thickness.

The strength of the weld depends on the surface roughness and cleanliness of the mating surfaces; thus oil, paint, and thick oxide layers should be removed before welding. However, the presence of uniform, thin oxide layers and other contaminants is not critical. The weld nugget is generally up to 10 mm (0.375 in.) in diameter. The surface of the weld spot has a slightly discolored indentation.

Spot welding is the simplest and most commonly used resistance welding process. Welding may be performed by means of single or multiple electrodes, and the pressure required is supplied through mechanical or pneumatic means. The *rocker-arm type* of spot welding machines is generally used for smaller parts, whereas the *press type* machine is used for larger workpieces. The shape and surface condition of the electrode tip and accessibility of the area to be welded are important

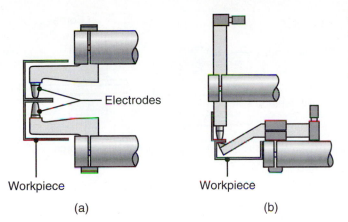

Electrodes

Workpiece   Workpiece

(a)   (b)

factors in spot welding. A variety of electrode shapes is used to spot weld areas that are difficult to reach (Fig. 12.34).

Modern equipment for spot welding is computer controlled for optimum timing of current and pressure, and the spot welding guns are manipulated by programmable robots (Section 14.7). Spot welding is widely used for fabricating sheet-metal products. Applications range from attaching handles to cookware to rapid spot welding of automobile bodies, using multiple electrodes. A typical automobile may have as many as 2,000 spot welds.

12.10.2 Resistance seam welding

Resistance seam welding (RSEW) is a modification of resistance spot welding wherein the electrodes are replaced by wheels or rollers (Fig. 12.35). With a continuous AC power supply, the electrically conducting rotating electrodes produce continuous spot welds whenever the current reaches a sufficiently high level in the AC cycle. The welds produced are actually overlapping spot welds and can produce a joint that is liquid and gas tight (Fig. 12.35b). The typical welding speed is 1.5 m/min (60 in./min) for thin sheet. With *intermittent* application of current, a series of spot welds at various intervals can be made along the length of the seam (Fig. 12.35c), a procedure called **roll spot welding**. The RSEW process is used in

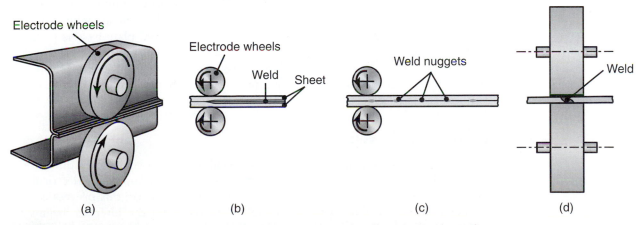

Electrode wheels

Electrode wheels

Weld   Sheet

Weld nuggets

Weld

(a)   (b)   (c)   (d)

FIGURE 12.35   (a) Illustration of the seam welding process, with rolls acting as electrodes, (b) overlapping spots in a seam weld, (c) cross section of a roll spot weld, and (d) mash seam welding.

FIGURE 12.36 Schematic illustration of resistance projection welding: (a) before and (b) after. The projections on sheet metal are produced by embossing operations, as described in Section 7.5.2.

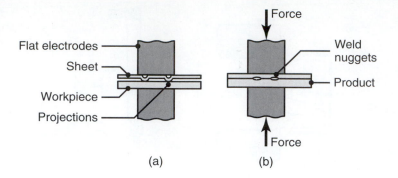

making the longitudinal (side) seam of cans for household products, mufflers, gasoline tanks, and other containers. In *mash seam welding,* the overlapping welds are about one to two times the sheet thickness. Tailor-welded blanks, described in Section 7.3.4, may also be made by this process.

High-frequency resistance welding (HFRW) is similar to resistance seam welding, except that a high-frequency (up to 450 kHz) current is employed. Applications include butt-welded tubing, structural sections such as I-beams, spiral pipe and tubing, finned tubes for heat exchangers, and wheel rims. In **high-frequency induction welding** (HFIW), the tube or pipe to be welded is subjected to high-frequency induction heating, using an induction coil ahead of the rollers.

12.10.3 Resistance projection welding

In *resistance projection welding* (RPW), high electrical resistance at the joint is developed by embossing (Section 7.5.2) one or more projections (dimples) on one of the surfaces to be welded (Fig. 12.36). The projections may be round or oval for specific design or part strength purposes. Because of the small contact areas involved, high localized temperatures are generated at the projections, which are in contact with the flat mating part. Weld nuggets are similar to those produced in spot welding, and are formed as the electrodes exert pressure to compress the projections. The electrodes are generally made of copper-base alloys; they have flat tips and are water cooled to keep their temperature low.

Spot welding equipment can be used by modifying the electrodes. Although embossing workpieces is an added expense, RPW produces a number of welds in one stroke, extends electrode life, and is capable of welding metals with different thicknesses. Nuts and bolts can also be welded to sheet and plate by this process, with projections that may be produced by operations such as machining or forging. Joining a network of wires, such as metal baskets, grills, oven racks, and shopping carts, is also considered as resistance projection welding, because of the small contact area between crossing wires (*grids*).

12.10.4 Flash welding

In *flash welding* (FW), also called **flash butt welding,** heat is generated from the arc as the ends of the two members begin to make contact, developing an electrical resistance at the joint (Fig. 12.37). Because of the arc's presence, this process is also classified as arc welding. After the temperature rises and the interface begins to soften, an axial force is applied at a controlled rate, and a weld is formed by plastic deformation of the joint. Note that this is a process of *hot upsetting* (see Fig. 6.1), hence the term **upset welding** (UW). A significant amount of metal may be expelled

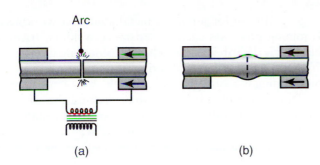

FIGURE 12.37 Flash welding process for end-to-end welding of solid rods or tubular parts. (a) Before and (b) after.

from the joint as a shower of sparks during the flashing process. The joint may later be machined to improve its appearance. Because impurities and contaminants are also squeezed out during this operation, the quality of the weld is good. The machines for FW are usually automated and large, with a variety of power supplies ranging from 10 to 1500 kVA.

The flash welding process is suitable for end-to-end or edge-to-edge joining of similar or dissimilar metals 1–75 mm (0.05–3 in.) in diameter, and sheet and bars 0.2–25 mm (0.01–1 in.) thick. Thinner sections have a tendency to buckle under the axial force applied during the process. Rings made by forming processes such as those shown in Figs. 7.24b and c also can be flash butt welded. This technique is also used to repair broken band saw blades (Section 8.10.5), with fixtures that are attached to the band saw frame. Flash welding can be automated for reproducible welding operations. Typical applications include (a) joining pipe and tubular shapes for metal furniture and windows, (b) welding high-speed steels to steel shanks, and (c) welding the ends of coils of sheet or wire for continuous operation of rolling mills and wire-drawing equipment.

12.10.5 Stud arc welding

Stud arc welding (SW) is similar to the flash welding process. The stud, such as a bolt or threaded rod, hook, or hanger, serves as one of the electrodes while being joined to another member, which is typically a flat plate (Fig. 12.38). In order to properly concentrate the heat generated, prevent oxidation, and retain the molten metal in the weld zone, a disposable ceramic ring (*ferrule*) is placed around the joint. The equipment for stud welding can be automated, with various controls for arcing and applying pressure; portable equipment is also available. The process has numerous applications in the automotive, construction, appliance, electrical, and shipbuilding industries.

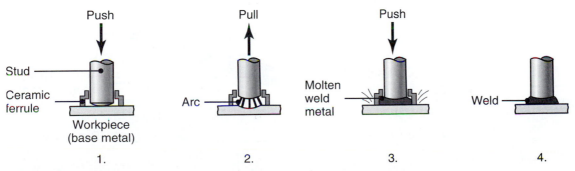

FIGURE 12.38 Sequence of operations in stud arc welding, used for welding bars, threaded rods, and various fasteners on metal plates.

Stud welding is a more general term for joining a metal stud to a workpiece and can be done by various joining processes such as resistance welding, friction welding, or any other suitable means described in this chapter. Shielding gases may or may not be used, depending on the particular application.

Capacitor discharge stud welding is a similar process in which a DC arc is produced from a capacitor bank. No ferrule or flux is required, because the welding time is very short (on the order of 1 to 6 milliseconds). The process is capable of stud welding on thin metal sheets with coated or painted surfaces. The choice between this process and stud arc welding depends on factors such as the type of metals to be joined, part thickness, stud diameter, and shape of the joint.

12.10.6 Percussion welding

The resistance welding processes described thus far typically require a transformer to supply power requirements. The electrical energy for welding may also be stored in a capacitor, as is done in capacitor discharge stud welding described above. *Percussion welding* (PEW) uses this technique, in which the power is discharged in a very short period of time (1–10 milliseconds), thus developing high localized heat at the joint. This process is useful where heating of the components adjacent to the joint is to be avoided, such as in electronic components.

12.11 | Explosion Welding

In *explosion welding* (EXW), the necessary contact pressure is applied by detonating a layer of explosive placed over one of the members being joined (called the *flyer plate*; Fig. 12.39). The pressures developed are extremely high (on the order of 10^{10} Pa), and the kinetic energy of the flyer plate striking the mating member produces a turbulent, wavy interface (vortexes). This impact mechanically interlocks the two mating surfaces (Fig. 12.40). In addition, cold pressure welding by plastic deformation also takes place (Section 4.4). The flyer plate is placed at an angle, whereby any oxide films present at the interface are broken up and propelled from the interface. As a result, bond strength in explosion welding is very high.

The explosive used in this operation may be in the form of flexible plastic sheet, cord, granular solid, or a liquid that is cast or pressed onto the flyer plate. Detonation speeds are typically 2400–3600 m/s (8000–12,000 ft/s), depending on the type of explosive, thickness of the explosive layer, and its packing density. Detonation is carried out using a standard commercial blasting cap.

Explosion welding is particularly suitable for cladding plates and slabs with dissimilar metals, particularly for the chemical industry. Plates as large as 6 × 2 m

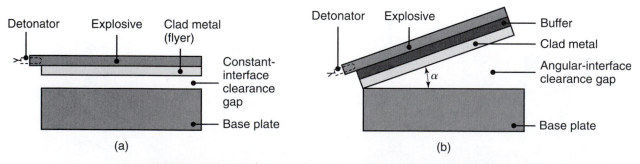

FIGURE 12.39 Schematic illustration of the explosion welding process: (a) constant interface clearance gap and (b) angular interface clearance gap.

(a)   (b)

FIGURE 12.40   Cross sections of explosion welded joints: (a) titanium (top) on low-carbon steel (bottom) and (b) incoloy 800 (iron-nickel-base alloy) on low-carbon steel. The wavy interfaces shown improve the shear strength of the joint. Some combinations of metals, such as tantalum and vanadium, produce a much less wavy interface. If the two metals have little metallurgical compatibility, an interlayer may be added that has compatibility with both metals. *Source:* Courtesy of DuPont Company.

$(20 \times 7 \text{ ft})$ have been explosively clad. The resulting material may then be rolled into thinner sections. Tube and pipe are often joined to the holes in header plates of boilers and tubular heat exchangers by this method. The explosive is placed inside the tube, and the explosion expands the tube and seals it tightly against the plate.

12.12 | Diffusion Bonding

Diffusion bonding, also called **diffusion welding** (DFW), is a solid-state joining process in which the strength of the joint results primarily from diffusion (movement of atoms across the interface) and, to a lesser extent, from some plastic deformation of the faying surfaces. This process requires temperatures of about $0.5T_m$ (where T_m is the melting point of the metal on the absolute scale) in order to have a sufficiently high diffusion rate between the parts to be joined. The bonded interface in DFW has essentially the same physical and mechanical properties as those of the base metal. Bond strength depends on pressure, temperature, time of contact, and the cleanliness of the faying surfaces. These requirements can be lowered by using filler metal at the interfaces.

The practice of diffusion bonding dates back centuries, when goldsmiths bonded gold over copper. To produce this material, called **filled gold,** a thin layer of gold foil is first made by hammering. The foil is placed over copper, and a weight is placed on top of it. The assembly is then placed in a furnace and left there until a good bond is obtained (a process also called *hot pressure welding,* HPW).

In diffusion bonding, the two parts are usually heated in a furnace or by electrical resistance. The pressure required may be applied by (a) using dead weights, (b) a press, (c) using differential gas pressure, or (d) from the relative thermal expansion of the parts to be joined. High-pressure autoclaves are also used for bonding complex parts. The process is generally most suitable for dissimilar metal pairs, although it is also used for reactive metals, such as titanium, beryllium, zirconium, and refractory metal alloys. Diffusion bonding is also important in sintering in powder metallurgy (Section 11.4) and for processing composite materials (Section 11.15).

FIGURE 12.41 Sequence of operations in diffusion bonding and superplastic forming of a structure with three flat sheets. See also Fig. 7.46. *Source:* After D. Stephen and S.J. Swadling.

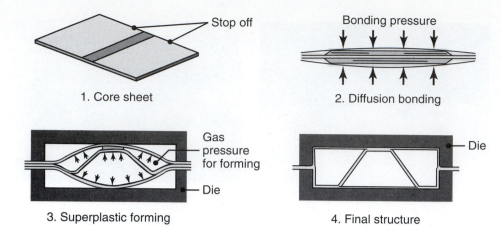

Because diffusion involves migration of the atoms across the joint, the process is slower than other welding methods. Although DFW is used for fabricating complex parts in small quantities for the aerospace, nuclear, and electronics industries, it has been automated to make it suitable and economical for moderate volume production, as in bonding sections of orthopedic implants and sensors (see Fig. 13.48).

Diffusion bonding/superplastic forming. An important development, beginning in the 1970s, is the ability to fabricate complex sheet-metal structures by combining the processes of *diffusion bonding* and *superplastic forming* (DB/SPF; see also Section 7.5.5). After selected locations of the sheets are diffusion bonded, the unbonded regions (stop-off) are expanded into a mold by air pressure. Typical structures made by this process are shown in Fig. 12.41. The structures are thin and have high stiffness-to-weight and stiffness-to-weight ratios (see also Section 3.9.1); thus they are particularly important in aircraft and aerospace applications. This process improves productivity by (a) eliminating mechanical fasteners, (b) reducing the number of parts required, (c) producing parts with good dimensional accuracy and low residual stresses, and (d) reducing labor costs and lead times. This technology is well advanced for titanium structures (typically Ti-6Al-4V alloy) for aerospace applications, as well as for aluminum alloys and various other alloys.

12.13 | Brazing and Soldering

Two joining processes that involve lower temperatures than those required for welding are *brazing* and *soldering*. Brazing and soldering are arbitrarily distinguished by temperature, although brazing temperatures are higher than those for soldering.

12.13.1 Brazing

In *brazing,* a process first used as far back as 3000–2000 B.C., a filler metal is placed at or between the faying surfaces to be joined, and the temperature is raised to melt the filler metal, but not the workpieces (Fig. 12.42a). The molten metal fills the closely fitting space by *capillary action.* Upon cooling and solidification of the filler metal, a strong joint is developed. In **braze welding**, the filler metal is deposited at the joint, as illustrated in Fig. 12.42b.

Filler metals for brazing generally melt above 450°C (840°F), but below the melting point (*solidus temperature*) of the metals to be joined (see Fig. 5.3). Thus,

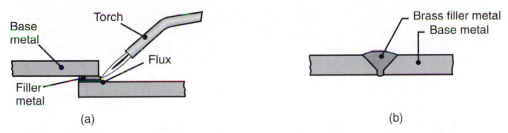

FIGURE 12.42 (a) Brazing and (b) braze welding operations.

this process is unlike liquid-state welding processes, in which the workpieces must melt in the weld area for fusion to occur. Any difficulties associated with heat-affected zones (Section 12.6), warping, and residual stresses are, therefore, reduced in the brazing process. The strength of the brazed joint depends on joint design and the bond at the interfaces of the workpiece and filler metal. Consequently, the surfaces to be brazed should be chemically or mechanically cleaned to ensure full capillary action; thus, the use of a flux is important.

The *clearance* between mating surfaces is an important parameter, as it directly affects the strength of the brazed joint (Fig. 12.43). Note that the smaller the gap, the higher is the *shear strength* of the joint. Also, there is an optimum gap to achieve maximum *tensile strength*. The typical joint clearance ranges from 0.025 to 0.2 mm (0.001 to 0.008 in.), and because clearances are very small, surface roughness of the mating surfaces is also important (see also Section 4.3).

Filler metals. Several *filler metals* (**braze metals**) are available, with a range of brazing temperatures (Table 12.4) and in a variety of shapes such as wire, strip, rings, shims, preforms, and filings or powder. Note that filler metals for brazing, unlike those for other welding operations, generally have significantly different compositions than those of the metals to be joined. The choice of filler metal and its composition is important in order to avoid *embrittlement* of the joint (by grain boundary penetration of liquid metal; see Section 3.4.2), formation of brittle intermetallic compounds at the joint, and galvanic corrosion in the joint.

Because of diffusion between the filler metal and the base metal, mechanical and metallurgical properties of joints can change during the service life of brazed components or in subsequent processing of the brazed parts. For example, when titanium is brazed with pure tin filler metal, it is possible for the tin to completely diffuse into the titanium base metal by subsequent aging or heat treatment. When that happens, a well-defined interface no longer exists.

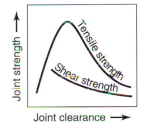

FIGURE 12.43 The effect of joint clearance on tensile and shear strength of brazed joints. Note that unlike tensile strength, shear strength continually decreases as clearance increases.

TABLE 12.4

Typical Filler Metals for Brazing Various Metals and Alloys

Base metal	Filler metal	Brazing temperature (°C)
Aluminum and its alloys	Aluminum-silicon	570–620
Magnesium alloys	Magnesium-aluminum	580–625
Copper and its alloys	Copper-phosphorus	700–925
Ferrous and nonferrous alloys (except aluminum and magnesium)	Silver and copper alloys, copper-phosphorus	620–1150
Iron-, nickel-, and cobalt-base alloys	Gold	900–1100
Stainless steels, nickel- and cobalt-base alloys	Nickel-silver	925–1200

Fluxes. The use of a *flux* is essential in brazing in order to prevent oxidation and to remove oxide films from workpiece surfaces to be joined. Brazing fluxes are generally made of borax, boric acid, borates, fluorides, and chlorides, and are available as paste, slurry, or powder. *Wetting agents* may also be added to improve both the wetting characteristics of the molten filler metal and its capillary action. Because they are corrosive, fluxes should be removed after brazing, especially in hidden crevices, usually by washing vigorously with hot water.

Surfaces to be brazed must be clean and free from rust, oil, lubricants, and other contaminants. Clean surfaces (see Section 4.5.2) are essential to obtain the proper wetting and spreading characteristics of the molten filler metal in the joint, as well as in developing maximum bond strength. Sand blasting and other processes (see also Section 9.9) also may be used to improve surface finish of the faying surfaces.

12.13.2 Brazing methods

The *heating methods* used in brazing also identify the various brazing processes described below. A variety of special fixtures may be used to hold the parts together during brazing, some with provision for allowing thermal expansion and contraction.

1. **Torch brazing.** The source of heat in *torch brazing* (TB) is oxyfuel gas with a carburizing flame (see Fig. 12.2c). Brazing is performed by first heating the joint with the torch, and then depositing the brazing rod or wire in the joint. More than one torch may be used in this operation. Part thicknesses are usually in the range of 0.25–6 mm (0.01–0.25 in.). Although it can be automated as a production process, torch brazing is difficult to control and requires skilled labor.

2. **Furnace brazing.** In *furnace brazing* (FB), the parts are precleaned and preloaded with brazing metal in appropriate configurations and then placed in the furnace (Fig. 12.44). The whole assembly is heated uniformly in the furnace. Furnaces (Section 5.5) may be batch type for complex shapes, or continuous type for high production runs, especially for small parts with simple joint designs. **Vacuum furnaces** or *neutral atmospheres* are used for metals that react with the environment, such as stainless steels where the passivation layer of chromium oxide (see Section 3.10.2) can compromise weld strength.

3. **Induction brazing.** The source of heat in *induction brazing* (IB) is induction heating by high-frequency AC current. (See Section 5.5.) Parts are first preloaded with filler metal and then placed near the induction coils for rapid heating. Unless a protective atmosphere is used in the induction furnace, fluxes are generally needed. Part thicknesses are usually less than 3 mm (0.125 in.). Induction brazing is particularly suitable for continuous brazing of parts.

4. **Resistance brazing.** In *resistance brazing* (RB), the source of heat is through electrical resistance of the components to be brazed. Electrodes are used for this purpose, as in resistance welding. Parts are either preloaded with filler metal or

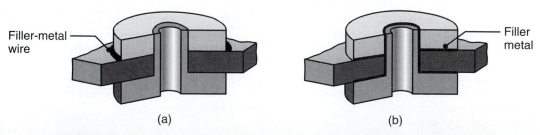

Filler-metal wire

Filler metal

(a)   (b)

FIGURE 12.44   An example of furnace brazing: (a) before and (b) after. Note that the filler metal is a shaped wire.

the filler metal is supplied externally during brazing. Parts commonly brazed by this process have a thickness of 0.1–12 mm (0.004–0.5 in.). As in induction brazing, the process is rapid, heating zones can be confined to very small areas, and the process can be automated to produce uniform quality.

5. **Dip brazing.** *Dip brazing* (DB) is carried out by dipping the assemblies to be brazed into either a molten filler metal bath or a molten salt bath (at a temperature just above the melting point of the filler metal), which serves as the heat source. All workpiece surfaces are thus coated with the filler metal. Dip brazing in metal baths is used only for small parts, such as fittings, usually of less than 5 mm (0.2 in.) in thickness or diameter. Molten salt baths, which also act as fluxes, are used for complex assemblies of parts with various thicknesses. Depending on the size of the parts and the bath, as many as 1000 joints can be made at one time by dip brazing.

6. **Infrared brazing.** The heat source in *infrared brazing* (IRB) is a high-intensity quartz lamp. This process is particularly suitable for brazing very thin components, usually less than 1 mm (0.04 in.) thick, including honeycomb structures (see Fig. 7.48). The radiant energy of the lamp is focused on the joint, and the process can be carried out in a vacuum. *Microwave heating* may also be used.

7. **Diffusion brazing.** *Diffusion brazing* (DFB) is carried out in a furnace in which, with proper control of temperature and time, the filler metal diffuses into the faying surfaces of the components to be joined. The brazing time required may range from 30 min to 24 hrs. Diffusion brazing is used for strong lap or butt joints and for difficult joining operations. Because the rate of diffusion at the interface does not depend on the thickness of the components, part thicknesses may range from foil to as much as 50 mm (2 in.).

8. **High energy beams.** For specialized high-precision applications and with high-temperature metals and alloys, *electron-beam* and *laser-beam* heating may also be used (see also Section 12.5).

Braze welding. The joint in *braze welding* is prepared as in fusion welding. Using an oxyacetylene torch with an oxidizing flame, filler metal is deposited at the joint. Thus, considerably more filler metal is used compared to brazing. However, temperatures in braze welding are generally lower than in fusion welding, and thus part distortion is minimal. The use of a flux is essential in this process for proper joint strength. Although the principal use of braze welding is in maintenance and in repair of parts, such as ferrous castings and steel components, it can be automated and used for mass production as well.

Examples of typical brazed joints are given in Fig. 12.45. In general, dissimilar metals can be assembled with good joint strength, including carbide drill bits (*masonry*

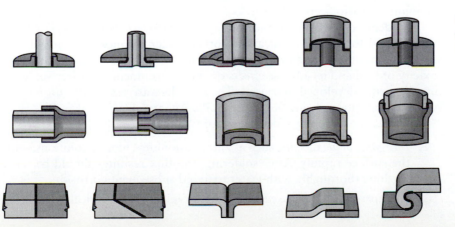

FIGURE 12.45 Joint designs commonly used in brazing operations.

FIGURE 12.46 Joint designs commonly used for soldering.

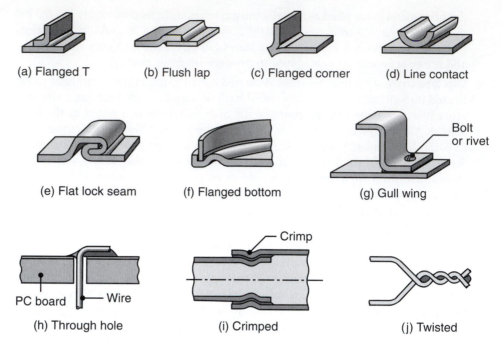

(a) Flanged T   (b) Flush lap   (c) Flanged corner   (d) Line contact

(e) Flat lock seam   (f) Flanged bottom   (g) Gull wing

(h) Through hole   (i) Crimped   (j) Twisted

drill) and carbide inserts on steel shanks (see Fig. 8.32). The shear strength of brazed joints can reach 800 MPa (120 ksi), using brazing alloys containing silver (*silver solder*). Intricate, lightweight shapes can be joined rapidly and with little distortion.

12.13.3 Soldering

In soldering, the filler metal (**solder**), which typically melts below 450°C (840°F), fills the joint by capillary action (as in brazing) between closely fitting or closely placed components (Fig. 12.46). Soldering with copper-gold and tin-lead alloys was first practiced as long ago as 4000–3000 B.C. Heat sources for soldering are typically soldering irons, torches, or ovens. Soldering can be used to join various metals and part thicknesses and is used extensively in the electronics industry. Although manual soldering operations require skill and are time consuming, soldering speeds can be high with automated equipment.

Unlike in brazing, temperatures in soldering are relatively low; consequently, a soldered joint has very limited use at elevated temperatures. Moreover, because solders do not generally have much strength, they are not used for load-bearing structural members. Because of the small faying surfaces, butt joints are rarely made with solders. In other situations, joint strength is improved by *mechanical interlocking* of the joint. (See also Section 12.17.)

Solders are generally tin-lead alloys in various proportions. For higher joint strength and for special applications, other solder compositions are tin-zinc, lead-silver, cadmium-silver, and zinc-aluminum alloys (Table 12.5). Because of the toxicity of lead and its adverse effects on the environment, **lead-free solders** are continually being developed and used. These solders are essentially *tin-based* solders, typical compositions being 96.5% Sn-3.5% Ag, and 42% Sn-58% Bi.

Fluxes for soldering generally are of two types:

1. *Inorganic acids* or *salts*, such as zinc ammonium chloride solutions, which clean the surface rapidly. After soldering, the flux residues should be removed by washing thoroughly with water to avoid subsequent corrosion.

2. *Noncorrosive resin-based fluxes,* used typically in electrical applications.

TABLE 12.5

Types of Solders and Their Applications	
Solder	Typical application
Tin-lead	General purpose
Tin-zinc	Aluminum
Lead-silver	Strength at higher than room temperature
Cadmium-silver	Strength at high temperatures
Zinc-aluminum	Aluminum; corrosion resistance
Tin-silver	Electronics
Tin-bismuth	Electronics

Solderability. *Solderability* may be defined in a manner similar to weldability, described in Section 12.6.2. Briefly, (a) copper and precious metals, such as silver and gold, are easy to solder; (b) iron and nickel are more difficult to solder; (c) aluminum and stainless steels are difficult to solder, because of their strong, thin oxide film (see Section 4.2); these and other metals can be soldered by using special fluxes that modify surfaces; and (d) other materials, such as cast irons, magnesium, and titanium, as well as nonmetallic materials such as graphite and ceramics, may be soldered by first plating the parts with metallic elements. (See also Section 12.17.3 for a discussion of the use of a similar technique for joining ceramics.) Solderability can be improved by coating metals; a common example is **tinplate**, which is *steel* coated with *tin* and widely used for food containers.

Soldering methods. There are several soldering methods, which are similar to the brazing methods described in Section 12.14.2:

1. Reflow (paste) soldering (RS).
2. Wave soldering (WS), used for automated soldering of printed circuit boards.
3. Torch soldering (TS).
4. Furnace soldering (FS).
5. Iron soldering (INS), using a soldering iron.
6. Induction soldering (IS).
7. Resistance soldering (RS).
8. Dip soldering (DS).
9. Infrared soldering (IRS).
10. Ultrasonic soldering, in which a transducer subjects the molten solder to ultrasonic cavitation, which removes the oxide films from the surfaces to be joined; the need for a flux is thus eliminated.

Most of these processes are closely related to associated welding and brazing operations described earlier in this chapter. The first two techniques, which are significantly different from the other soldering methods, are described in more detail below.

1. **Reflow (paste) soldering.** Semisolid in consistency, solder pastes consist of solder-metal particles bound together by flux and wetting agents. They have high viscosity but maintain a solid shape for relatively long periods of time (similar to the behavior of greases and cake frostings). In this operation, the paste is placed directly onto the joint (or on flat objects for finer detail), applied using a *screening* or stenciling process (Fig. 12.47). This is a commonly used operation

FIGURE 12.47 Screening solder paste onto a printed circuit board in reflow soldering. *Source:* After V. Solberg.

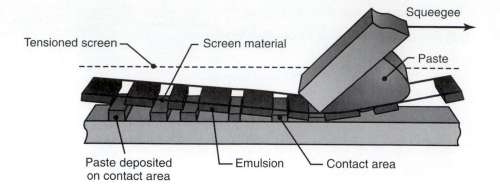

for mounting of electrical components to printed circuit boards (Section 13.13). An additional benefit of this method is that the surface tension of the paste helps keep surface-mount packages aligned on their pads, a feature that improves the reliability of such solder joints.

Once the paste and the components are placed, the assembly is heated in a furnace, and reflow soldering takes place. In this operation, the product must be heated in a controlled manner, so that the following sequence of events occurs:

1. Solvents present in the paste are evaporated.
2. The flux in the paste is activated, and fluxing action occurs.
3. The components are carefully preheated.
4. The solder particles are melted and wet the joint.
5. The assembly is cooled at a low rate to prevent thermal shock to and fracture of the solder joint.

While this process appears to be straightforward, there are several process variables for each stage, and proper control over temperatures and exposure time must be maintained at each stage to ensure good joint strength. Regardless, reflow soldering is the most common soldering method in circuit board manufacture.

2. **Wave soldering.** *Wave soldering* is a very common method for soldering circuit components to their boards (Section 13.13). To understand wave soldering, it is important to recognize the fact that molten solder does not wet all surfaces. Indeed, solder will not bond to most polymer surfaces and thus can easily be removed while in a molten state. Also, as can be observed with a simple hand-held soldering iron, the solder wets metal surfaces and forms a good bond only when the metal is preheated to a certain temperature. Therefore, wave soldering requires separate fluxing and preheating operations before it can be applied successfully.

Wave soldering is schematically illustrated in Fig. 12.48a. A standing laminar wave of molten solder is first generated by a pump. Preheated and prefluxed circuit boards are then conveyed over the wave. The solder wets the exposed metal surfaces, but it does not stay attached to the polymer package for the integrated circuits, and it does not bond to the polymer-coated circuit boards. An *air knife* (basically a high-velocity jet of hot air) blows excess solder from the joint, thus preventing bridging between adjacent leads.

When surface-mount packages are to be wave soldered, they must first be adhesively bonded (see Section 12.14) to the circuit board before the soldering can commence. This operation is usually accomplished by screening or stenciling

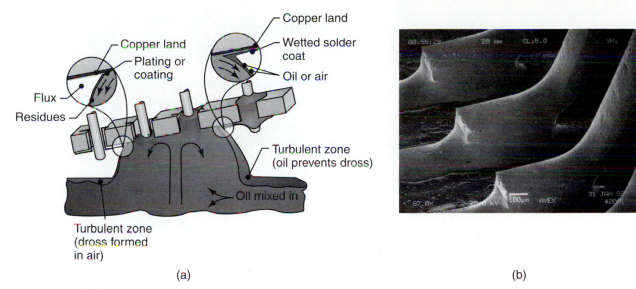

Copper land
Copper land
Plating or coating
Wetted solder coat
Flux
Oil or air
Residues
Turbulent zone (oil prevents dross)
Oil mixed in
Turbulent zone (dross formed in air)

(a)

(b)

FIGURE 12.48   (a) Schematic illustration of the wave soldering process, and (b) SEM image of a wave soldered joint on a surface-mount device. (See also Section 13.13.)

epoxy onto the boards, placing the components in their proper locations, curing the epoxy, inverting the board, and performing wave soldering. Figure 12.48b shows a scanning-electron microscopy photograph of a typical surface-mount joint.

EXAMPLE 12.6   Soldering of components onto a printed circuit board

The computer and consumer electronics industries place extremely high demands on electronic components (see also Chapter 13). It is expected that the integrated circuits and other electronic devices will function reliably for extended periods of time, during which they can be subjected to significant temperature variations and to vibration. In recognition of this requirement, it is essential that the solder joints used to attach such devices to circuit boards be sufficiently strong and reliable, and also that the solder joints be applied extremely rapidly with automated equipment.

A continuing trend in these industries is toward continual reduction of chip sizes and increasing compactness of circuit boards. Further space savings are achieved by mounting integrated circuits into surface-mount packages, which allow tighter packing on a circuit board and, more importantly, the mounting of components on both sides of a circuit board.

A challenging problem arises when a printed circuit board has both surface-mount and in-line circuits on the same board, and it is desired to solder all the joints via a reliable automated process. A subtle point should be recognized: All of the in-line circuits should be restricted to insertion from one side of the board to ease assembly.

The basic steps in soldering the connections on such a board are as follows:

1. Apply solder paste to one side.
2. Place the surface-mount packages onto the board; also, insert in-line packages through the primary side of the board.
3. Reflow the solder.

4. Apply adhesive to the secondary side of the board.

5. Attach the surface mount devices on the secondary side using the adhesive.

6. Cure the adhesive.

7. Perform a wave soldering operation on the secondary side to electrically bond the surface mounts and the in-line circuits to the board.

Applying solder paste is done with chemically etched stencils or screens, so that the paste is placed only onto the designated areas of a circuit board. (Stencils are more widely used for fine-pitch devices, and produce a more uniform paste thickness.) Surface-mount circuit components are then placed on the board, and the board is heated in a furnace to around 200°C (400°F) to reflow the solder and thus produce strong connections between the surface mount and the circuit board.

At this point the components, with leads, are inserted into the primary side of the board, their leads are crimped, and the board is flipped over. An adhesive pattern is printed onto the board, using a dot of epoxy at the center of a surface-mount component location. The surface-mount packages are then placed onto the adhesive, usually by high-speed automated, computer controlled systems. The adhesive is then cured, the board is flipped, and wave soldering is done. The wave soldering operation simultaneously joins the surface-mount components to the secondary side and solders the leads of the in-line components from the board's primary side. The board is then cleaned and inspected prior to performing electronic quality checks.

12.14 | Adhesive Bonding

Numerous components and products can be joined and assembled with an **adhesive**, instead of using any of the joining methods described thus far. **Adhesive bonding** has been a common method of joining and assembly for applications such as bookbinding, labeling, packaging, home furnishings, and footwear. Plywood, developed in 1905, is a typical example of adhesive bonding of several layers of wood with glue. Adhesive bonding has been gaining increased acceptance in manufacturing ever since its first use on a large scale in assembling load-bearing components in military aircraft manufactured during World War II (1939–1945) and became a common industrial technique beginning in the 1960s. Examples of adhesively bonded joints are given in Fig. 12.49.

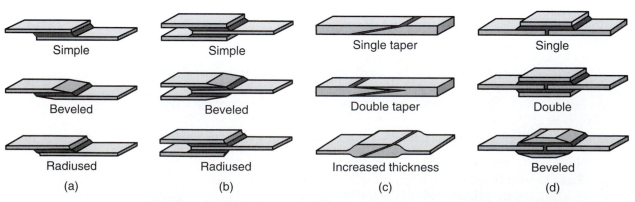

FIGURE 12.49 Various configurations for adhesively bonded joints: (a) single lap, (b) double lap, (c) scarf, and (d) strap.

Major industries that now use adhesive bonding extensively are aerospace, automotive, appliances, and building products. Applications include attaching rearview mirrors to windshields, automotive brake lining assemblies, laminated windshield glass (see Section 11.11.2), appliances, helicopter blades, honeycomb structures, and aircraft bodies and control surfaces.

Adhesives are available in various forms such as liquids, pastes, solutions, emulsions, powder, tape, and film. When applied, adhesives typically are on the order of 0.1 mm (0.004 in.) thick.

12.14.1 Types of adhesives

Numerous types of adhesives are available, and continue to be developed, that provide adequate joint strength, including fatigue strength and resistance to environmental attack. The three basic types of adhesives are listed as follows:

1. **Natural adhesives,** such as starch, dextrin (a gummy substance derived from starch), soya flour, and animal products.
2. **Inorganic adhesives,** such as sodium silicate and magnesium oxychloride.
3. **Synthetic organic adhesives,** which may be thermoplastics (for nonstructural and some structural bonding) or thermosetting polymers (primarily for structural bonding).

Because of their cohesive strength, synthetic organic adhesives are the most important adhesives in manufacturing processes, particularly for load-bearing applications (*structural adhesives*). They are generally classified as follows:

1. **Chemically reactive,** such as polyurethanes, silicones, epoxies, cyanoacrylates, modified acrylics, phenolics, and polyimides; also included are anaerobics, which cure in the absence of oxygen (such as Loctite® for threaded fasteners).
2. **Pressure sensitive,** such as natural rubber, styrene-butadiene rubber, butyl rubber, nitrile rubber, and polyacrylates.
3. **Hot melt,** which are thermoplastics such as ethylene-vinyl-acetate copolymers, polyolefins, polyamides, polyester, and thermoplastic elastomers.
4. **Reactive hot-melt,** which have a thermoset portion based on urethane's chemistry, with improved properties.
5. **Evaporative** or **diffusion,** including vinyls, acrylics, phenolics, polyurethanes, synthetic rubbers, and natural rubbers.
6. **Film** and **tape,** such as nylon-epoxies, elastomer-epoxies, nitrile-phenolics, vinyl-phenolics, and polyimides.
7. **Delayed tack,** including styrene-butadiene copolymers, polyvinyl acetates, polystyrenes, and polyamides.
8. **Electrically** and **thermally conductive,** including epoxies, polyurethanes, silicones, and polyimides (see Section 12.15.4).

Adhesive systems may be classified based on their specific chemistries:

1. **Epoxy based systems.** These systems have high strength and high-temperature properties, to as high as 200°C (400°F); typical applications include automotive brake linings and as bonding agents for sand molds for casting. (See Section 5.8.1.)
2. **Acrylics.** Suitable for applications on substrates that are not clean or are contaminated.

3. **Anaerobic systems**. Curing of these adhesives is by oxygen deprivation, and the bond is usually hard and brittle; curing times can be reduced by external heating or ultraviolet (UV) radiation.

4. **Cyanoacrylate**. The bond lines are thin, and the bond sets within 5 to 40 s.

5. **Urethanes**. High toughness and flexibility at room temperature; widely used as sealants.

6. **Silicones**. Highly resistant to moisture and solvents, silicones have high impact and peel strength; however, curing times are typically in the range of one to five days.

Many of these adhesives can be combined to optimize their properties; examples include epoxy-silicon, nitrile-phenolic, and epoxy-phenolic. The least expensive adhesives are epoxies and phenolics, followed by polyurethanes, acrylics, silicones, and cyanoacrylates. Adhesives for high-temperature applications at up to about 260°C (500°F), such as polyimides and polybenzimidazoles, are generally the most expensive adhesives.

Depending on the particular application, an adhesive generally must have one or more of the following properties (Table 12.6):

• Strength (shear and peel).

• Toughness.

• Resistance to various fluids and chemicals.

• Resistance to environmental degradation, including heat and moisture.

• Capability to wet the surfaces to be bonded.

Adhesive joints are designed to withstand shear, compressive, and tensile forces, but they should not be subjected to peeling forces (Fig. 12.50). Note, for example, how easily an adhesive tape can be peeled from a surface. During peeling, the behavior of an adhesive may be brittle, or it may be ductile and tough (thus requiring large forces to peel it).

TABLE 12.6

Typical Properties and Characteristics of Chemically Reactive Structural Adhesives

	Epoxy	Polyurethane	Modified acrylic	Cyanocrylate	Anaerobic
Impact resistance	Poor	Excellent	Good	Poor	Fair
Tension-shear strength, MPa (10^3 psi)	15–22 (2.2–3.2)	12–20 (1.7–2.9)	20–30 (2.9–4.3)	18.9 (2.7)	17.5 (2.5)
Peel strength,* N/m (lb/in.)	<523 (3)	14,000 (80)	5250 (30)	<525 (3)	1750 (10)
Substrates bonded	Most	Most smooth, nonporous	Most smooth, nonporous	Most nonporous metals or plastics	Metals, glass, thermosets
Service temperature range, °C (°F)	−55 to 120 (−70 to 250)	−40 to 90 (−250 to 175)	−70 to 120 (−100 to 250)	−55 to 80 (−70 to 175)	−55 to 150 (−70 to 300)
Heat cure or mixing required	Yes	Yes	No	No	No
Solvent resistance	Excellent	Good	Good	Good	Excellent
Moisture resistance	Good-Excellent	Fair	Good	Poor	Good
Gap limitation, mm (in.)	None	None	0.5 (0.02)	0.25 (0.01)	0.60 (0.025)
Odor	Mild	Mild	Strong	Moderate	Mild
Toxicity	Moderate	Moderate	Moderate	Low	Low
Flammability	Low	Low	High	Low	Low

*Peel strength varies widely depending on surface preparation and quality.

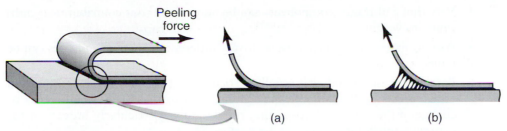

FIGURE 12.50 Characteristic behavior of (a) brittle and (b) tough and ductile adhesives in a peeling test. This test is similar to peeling adhesive tape from a solid surface.

12.14.2 Surface preparation and application

Surface preparation is very important in adhesive bonding, as joint strength depends greatly on the absence of dirt, dust, oil, and various other contaminants. Note, for example, that it is virtually impossible to apply an adhesive tape over dusty or oily surfaces. Contaminants also affect the wetting ability of the adhesive, and they can prevent spreading of the adhesive evenly over the interface. Thick, weak, or loose oxide films on workpieces are also detrimental to adhesive bonding. On the other hand, a porous or thin and strong oxide film may be desirable, particularly one with some surface roughness that would improve adhesion at the interfaces. However, the roughness must not be too high, as air may be trapped and joint strength will thus be reduced. Various compounds and primers are available to modify surfaces to increase the strength of adhesive bonds.

12.14.3 Process capabilities

A wide variety of similar and dissimilar metallic and nonmetallic materials and components with different shapes, sizes, and thicknesses can be bonded to each other by adhesives. Adhesive bonds for structural applications are rarely suitable for service above 250°C (500°F). Adhesive bonding can also be combined with mechanical fastening methods (see Section 12.15) to further improve the strength of the bond. An important consideration in the use of adhesives in production is curing time, which can range from a few seconds at high temperatures to several hours at room temperature, particularly for thermosetting adhesives, thus lowering production rate. Joint design and bonding methods require care and skill, and also special equipment, such as fixtures, presses, tooling, and autoclaves and ovens for curing, is also usually required.

Nondestructive inspection of the quality and strength of adhesively bonded components can be difficult, although some of the techniques described in Section 4.8.1, such as acoustic impact (tapping), holography, infrared detection, and ultrasonic testing, can be effective.

The *advantages* of adhesive bonding may be summarized as follows.

1. It provides a bond at the interface, either for structural strength or for non-structural applications such as sealing, insulating, preventing electrochemical corrosion between dissimilar metals, and reducing vibration and noise through internal damping at the joints.

2. It distributes the load at an interface, thus eliminating localized stresses that typically result from joining the components with welds or mechanical fasteners such as bolts and screws. Moreover, since no holes are required, structural integrity of the components is maintained.

3. The external appearance of the joined components is unaffected.

4. Very thin and fragile components can be bonded without contributing significantly to weight.

5. Porous materials and materials with very different properties and sizes can be joined.

6. Because adhesive bonding is usually carried out between room temperature and about 200°C (400°F), there is no significant distortion of the components or change in their original properties; this factor is particularly important for materials and components that are heat sensitive.

The *limitations* of adhesive bonding are

1. The service temperatures are relatively low.

2. The bonding time can be long.

3. Surface preparation is essential.

4. It is difficult to test adhesively bonded joints nondestructively, particularly with large structures.

5. The reliability of adhesively bonded structures during their service life and under hostile environmental conditions (*degradation* by temperature, oxidation, radiation, stress corrosion, and dissolution) may be a significant concern, particularly for critical applications.

12.14.4 Electrically conducting adhesives

Although the majority of adhesive bonding is for mechanical strength and structural integrity, an important advance is the development and application of *electrically conducting adhesives,* particularly to replace lead-base solders. Applications of electrically conducting adhesives include calculators, remote controls, and control panels, as well as high-density use in electronic assemblies, liquid-crystal displays, pocket TVs, and electronic games.

These adhesives require curing or setting temperatures that are lower than those for soldering. Electrical conductivity in adhesives is obtained by the addition of *fillers,* such as silver, copper, aluminum, nickel, gold, and graphite. (See also Section 10.7.2.) Silver is the most commonly used metal, because of its very high electrical conductivity, with up to 85% silver content. There is a minimum volume concentration of fillers in order to make the adhesive electrically conducting, being typically in the range of 40–70%. Fillers generally are in the form of flakes or particles and include polymeric particles, such as polystyrene, that are coated with thin films of silver or gold.

The size, shape, and distribution of the particles and the nature of contact among the individual particles, as well as how heat and pressure are applied, can be controlled so as to impart isotropic or anisotropic electrical conductivity to the adhesive. Note that fillers that improve electrical conductivity also improve thermal conductivity of the adhesively bonded joints. Matrix materials are generally epoxies, although various thermoplastics are also used.

12.15 | Mechanical Fastening

Countless products, including mechanical pencils, watches, computers, bicycles, engines, and aircraft have components that are fastened *mechanically,* whereby two or more components are assembled in such a way that they can be taken apart during the product's service life or life cycle (see also Section 14.10). Mechanical fastening may be preferred over other methods because of its

1. Ease of manufacturing.
2. Ease of assembly, disassembly, and transportation.
3. Ease of parts replacement, maintenance, and repair.
4. Ease in creating designs that require movable joints, such as hinges, sliding mechanisms for doors, and adjustable components and fixtures.
5. Lower overall cost of manufacturing the product.

The most common method of mechanical fastening is by using bolts, nuts, screws, rivets, pins, and similar **fasteners.** Most mechanical fastening typically requires that the components have *holes* through which fasteners are inserted and secured.

12.15.1 Hole preparation

Hole preparation is an important aspect of mechanical fastening. Depending on the type of material, its properties, and its thickness, a hole in a solid body can be produced by various processes such as punching, drilling, chemical and electrical means, and high energy beams (see Chapters 7, 8, and 9). Recall also from Chapters 5, 6, and 11 that holes also may be produced as an *integral part* of the product during casting, forging, extrusion, and powder metallurgy. In this way, additional operations are avoided and costs are reduced. For improved dimensional accuracy and surface finish, many of these holemaking operations may be followed by finishing operations, such as shaving, deburring, reaming, and honing.

Because of their fundamental differences, each type of holemaking operation produces a hole with different surface finish and properties and dimensional characteristics. For example, a punched hole will have an axial lay (see Sections 4.3 and 7.3), whereas a drilled hole will have a circumferential lay. The most significant influence of a hole in a solid body is to act as a stress concentration, and thus its presence can reduce the component's fatigue life. Fatigue life can best be improved by inducing compressive residual stresses on the cylindrical surface of the hole. A common technique is to push a round rod (*drift pin*) through the hole and expand it by a very small amount. This is a process that plastically deforms the surface layers of the hole in a manner similar to shot peening or roller burnishing (see Section 4.5.1).

12.15.2 Threaded fasteners

Bolts, screws, and nuts are among the most commonly used *threaded fasteners.* References on machine design describe in detail numerous standards and specifications, including thread dimensions, tolerances, pitch, strength, and the quality of materials used to make these fasteners. Bolts and screws may be secured with nuts or they may be *self-tapping,* whereby the screw either cuts or forms the thread into the part to be fastened. The latter method is particularly effective and economical for plastic products.

If the joint is to be subjected to vibration, such as in aircraft and various types of machinery and engines, several specially designed nuts and lock washers are commercially available. They increase the frictional resistance in the torsional direction, thus preventing vibrational loosening of the fasteners.

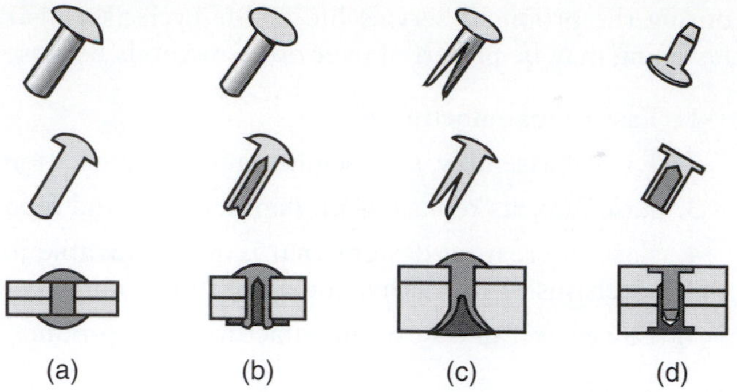

FIGURE 12.51 Examples
of rivets: (a) solid,
(b) tubular, (c) split,
or bifurcated, and
(d) compression.

(a)   (b)   (c)   (d)

12.15.3 Rivets

The most common method of permanent or semipermanent mechanical joining or
assembly is by riveting (Fig. 12.51). Thousands of rivets are used in the assembly of
each large commercial aircraft. The basic types of rivets are solid, hollow, and blind
(which are inserted from one side only). Installing a rivet basically consists of placing
the rivet in the hole and deforming the end of its shank by upsetting (similar to the
heading process shown in Fig. 6.17), performed by hand or by mechanized means,
including the use of robots. Riveting may be done either at room temperature or hot,
using special tools, or by using explosives placed in the cavity of the rivet.

12.15.4 Various methods of fastening

Several other fastening methods may be used in joining and assembly applications;
the most common types are described below.

1. **Stitching, stapling, and clinching.** The process of *metal stitching* or
   stapling (Fig. 12.52) is much like that of ordinary stapling of papers. This
   operation is fast and particularly suitable for joining thin metallic and nonmetallic materials, and it does not require any holes. A common example is
   the stapling of cardboard boxes and containers. In *clinching*, illustrated in
   Fig. 12.52b, the fastening material should be sufficiently thin and ductile to
   withstand the large localized deformations during sharp bending.

2. **Seaming.** *Seaming* is based on the simple principle of folding two thin pieces
   of material together, much like folding two or more pieces of paper. Common
   examples of seaming are the lids of beverage cans and containers for food and
   household products (*lock seams*, Fig. 12.53). In seaming, the materials should be
   capable of undergoing bending and folding at very small radii (see Section 7.4.1
   and Fig. 7.15); otherwise, they may crack, and the seams will not be airtight or

FIGURE 12.52 Examples
of various fastening methods.
(a) Standard loop staple,
(b) flat clinch staple,
(c) channel strap, and
(d) pin strap.

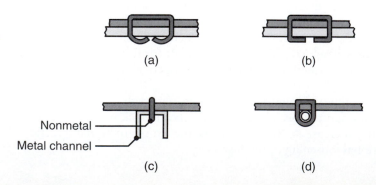

(a)   (b)

Nonmetal ——
Metal channel ——

(c)   (d)

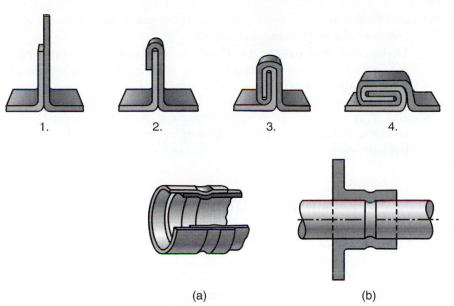

FIGURE 12.53 Stages in forming a double-lock seam. See also Fig. 7.23.

1. 2. 3. 4.

FIGURE 12.54 Two examples of mechanical joining by crimping.

(a)   (b)

watertight. The performance and airtightness of lock seams may be improved with adhesives, coatings, and polymeric materials at internal interfaces, or by soldering, as is done in some steel cans.

3. **Crimping.** The *crimping* process is a method of joining without using fasteners. Caps on glass bottles, for example, are assembled by crimping, as are some connectors for electrical wiring. Crimping can be done with beads or dimples (Fig. 12.54), which can be produced by *shrinking* (Section 7.4.4) or by swaging operations (Section 6.6). Crimping can be used on both tubular and flat parts.

4. **Snap-in fasteners.** Figure 12.55 shows a selection of typical *spring* and *snap-in fasteners*. Snap fits are widely used in the assembly of automobile bodies

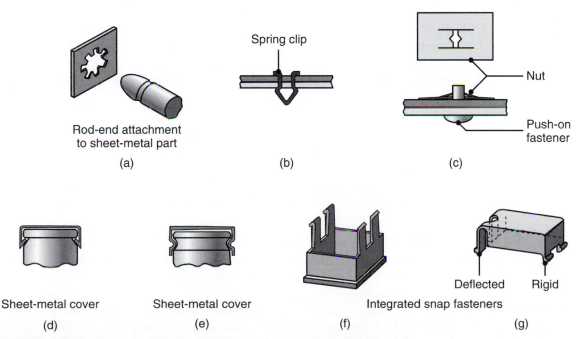

Rod-end attachment to sheet-metal part

(a)

Spring clip

(b)

Nut

Push-on fastener

(c)

Sheet-metal cover

(d)

Sheet-metal cover

(e)

Integrated snap fasteners

(f)

Deflected   Rigid

(g)

FIGURE 12.55 Examples of spring and snap-in fasteners to facilitate assembly.

and household appliances. They are economical and permit easy and rapid component assembly and disassembly.

5. **Shrink fits and press fits.** Components also may be assembled by shrink fitting and press fitting. *Shrink fitting* is based on the principle of the differential thermal expansion and contraction of two components. Typical applications include the assembly of die components and mounting gears and cams on shafts. In *press fitting,* one component is forced over another, resulting in high joint strength. Examples include shrink fit toolholders and hubs on shafts.

12.16 | Joining Nonmetallic Materials

12.16.1 Joining thermoplastics

As described in Section 10.3, thermoplastics soften and melt as temperature increases. Consequently, they can be joined by means whereby heat is generated at the interface of the components to be joined, either through external means or internally. The heat softens the thermoplastic at the interface to a viscous or molten state; the application of pressure then allows *fusion* to take place, ensuring a good bond. Filler materials of the same type of polymer also may be used for stronger bonds.

Oxidation can be a problem in joining some polymers, such as polyethylene, because of degradation; thus an inert shielding gas (such as nitrogen) is typically used to prevent oxidation. Because of the low thermal conductivity of thermoplastics (see Table 3.3), the heat source may burn or char the surfaces of the components to be joined if applied at too high a rate. This would cause difficulties in developing a sufficiently deep fusion for good joint strength.

External heat sources may consist of the following, depending on the compatibility of the polymers to be joined:

1. Hot air or gases, or infrared radiation using high-intensity focused quartz heat lamps.

2. Heated tools and dies that contact and heat the surfaces to be joined; this process, known as *hot-tool* or *hot-plate welding,* is commonly used to butt-weld pipe and tubing.

3. Radio-frequency or dielectric heating, particularly useful for thin films.

4. Lasers, using defocused beams at low power to prevent degradation of the polymer.

5. Electrical resistance wire or braids, or carbon-based tapes, sheet, and ropes; these are placed at the interface and allow passing of electrical current (a process known as *resistive implant welding*). These elements at the interface, which may also be subjected to a radio-frequency field (*induction welding*), must be compatible with the intended use of the joined product, since they remain in the weld zone.

Internal heat sources are developed through the following means:

1. Ultrasonic welding (Section 12.8), the most commonly used process for thermoplastics, particularly amorphous polymers such as ABS and high-impact polystyrene.

2. Friction welding (Section 12.9), also called *spin welding* for polymers; it includes linear friction welding, also called *vibration welding,* which is particularly useful

for joining polymers with a high degree of crystallinity (see Section 10.2.2), such as acetal, polyethylene, nylon, and polypropylene.

3. *Orbital welding*, which is similar to friction welding, except that the rotary motion of one part is in an orbital path (see also Fig. 6.16a).

Other joining methods.   *Adhesive bonding* of polymers is a versatile process, commonly used for joining sections of PVC or ABS pipe. The liquid adhesive is applied to the connecting sleeve and pipe surfaces, sometimes using a primer to improve adhesion. Adhesive bonding of polyethylene, polypropylene, and polytetrafluoroethylene (Teflon) can be difficult, because these polymers have low surface energy, and hence adhesives do not readily bond to their surfaces. The surfaces generally have to be treated chemically to improve bond strength. The use of adhesive *primers* or *double-sided adhesive tapes* can also be effective. Bonding may also be achieved by using solvents (*solvent bonding*).

Coextruded multilayer food wrappings consist of different types of films bonded by the *heat* generated during *extrusion* (see Section 10.10.1). Each film has a specific function, such as to keep out moisture and oxygen or to facilitate heat sealing during packaging. Some wrappings have as many as seven layers, all bonded together during production of the film (*cocuring*).

A method of producing well-defined polymer weld joints is the **Clearweld**® process. In this operation, a toner is first applied to an interface between two polymers, or one or both of the polymers are doped with the toner during polymer manufacturing. A laser is then used that does not heat the polymer, but does heat the toner. Thus, the heating can take place along well-defined interfaces, leading to fusion and strong polymer welds.

Thermoplastics may also be joined by *mechanical* means, including the use of fasteners and self-tapping screws. The strength of the joint depends on the particular method used and the inherent toughness and resilience of the plastic to resist tearing at the holes present for mechanical fastening. *Integrated snap fasteners* (Fig. 12.55f and g) are gaining wide acceptance for their simplified and cost effective assembly operations.

Bonding may also be achieved by *magnetic* means, by embedding tiny particles in the polymer. The dimensions of the particles are on the order of 1 μm (40 μin.). A high-frequency field then causes induction heating of the polymer and melts it at the interfaces to be joined (*electromagnetic bonding*).

12.16.2 Joining thermosets

Because they do not soften or melt with increasing temperature, thermosetting plastics, such as epoxy and phenolics, are usually joined by using (1) threaded or other molded-in inserts (see Fig. 10.30), (2) mechanical fasteners, and (3) solvent bonding. Bonding with solvents typically involves the following sequence: (a) roughening the surfaces of the thermoset parts with an abrasive cloth or paper, (b) wiping the surfaces with a solvent, and (c) pressing the surfaces together and maintaining the pressure until sufficient bond strength is developed.

12.16.3 Joining ceramics and glasses

As with any other material used in products, ceramics and glasses have to be assembled into components, either with the same type of material or with different materials. Parts can be joined using adhesive bonding, mechanical means, and shrink or press fitting. The *relative thermal expansion* of the two materials should be taken into consideration to avoid damage if the assembly is subjected to elevated temperature during its service life. (See Section 3.9.5.)

Ceramics. A commonly used technique that is effective in joining difficult-to-bond material combinations consists of first applying a coating of a material that bonds itself easily to one or both parts, thus acting as a bonding agent, just as an adhesive does between a piece of wood and a metal that otherwise would not bond to each other. Thus, the surface of alumina ceramics, for example, can be *metallized* (see Section 4.5.1). In this technique, known as the *Mo-Mn process,* the ceramic part is first coated with a slurry that, after being fired, forms a glassy layer. This layer is then plated with nickel, and, because the part now has a metallic surface, it can be brazed to a metal surface. Depending on their particular structure, ceramics can be joined unto metals by diffusion bonding (Section 12.12). It may be necessary to place a metallic layer at the joint to make it stronger.

Ceramic parts also can be joined or assembled together during their shaping process (see Section 11.9), a common example of which is attaching the handle of a coffee mug. Thus, in a sense, the shaping of the product is done integrally, rather than as an additional operation after the part is already made.

As described in Section 8.6.4, tungsten carbide has a matrix (binder) of cobalt, and titanium carbide has matrix of nickel-molybdenum alloy. Consequently, with both binders being metal, carbides can easily be brazed to other metals. Common applications include brazing carbide cutting tools to steel toolholders (see Fig. 8.32d), brazing carbide tips to masonry drills, and brazing cubic boron nitride or diamond tips to carbide inserts (see Fig. 8.39).

Glasses. As evidenced by the availability of numerous glass objects, glasses can easily be bonded to each other. This is accomplished by first softening the glass surfaces to be joined and then pressing the two pieces together while the assembly cools. Bonding of glass to metals is also possible because of the diffusion of metal ions into the amorphous surface structure of glass.

12.17 | Design Considerations in Joining

12.17.1 Design for welding

As in all other manufacturing processes, the optimum choice in welding is the one that satisfies all design and service requirements at minimum cost. Some examples of weld characteristics are given in Fig. 12.56. General design guidelines for welding may be summarized as follows:

1. Product design should minimize the number of welds, as welding can be costly when it is not automated.
2. The weld location should be selected to avoid excessive stresses or stress concentrations in the welded structure, as well as for appearance.
3. The weld location should be selected so as not to interfere with subsequent processing of the part or with its intended use and appearance.
4. Parts should fit properly before welding; the method employed to produce the edges (sawing, machining, shearing, flame cutting, etc.) can affect weld quality.
5. Modification of the design may avoid the need for edge preparation.
6. Weld bead size should be kept to a minimum to conserve weld metal as well as for appearance.
7. Mating surfaces for some processes may require uniform cross sections at the joint, such as is shown for flash welding in Fig. 12.57.

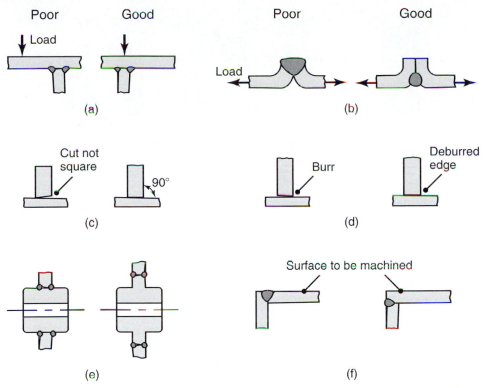

FIGURE 12.56 Design guidelines for welding. *Source:* After Bralla, J.G. (ed.), *Handbook of Product Design for Manufacturing*, 2nd ed., McGraw-Hill, 1999.

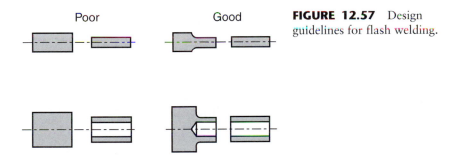

FIGURE 12.57 Design guidelines for flash welding.

EXAMPLE 12.7 Weld design selection

Three different types of weld designs are illustrated in Fig. 12.58. As shown in Fig. 12.58a, the two vertical joints can be welded either externally or internally. Note that full-length external welding takes considerable time and requires more weld material than the alternative design, which consists of intermittent internal welds. Moreover, in the alternative method, the appearance of the structure is improved, and distortion is reduced because of the lower energy content placed in the structure (see Fig. 12.25).

Although both designs require the same amount of weld material and welding time (Fig. 12.58b), further analysis will show that the design on the right can carry three times the moment M than can the design on the left. In Fig. 12.58c, the weld

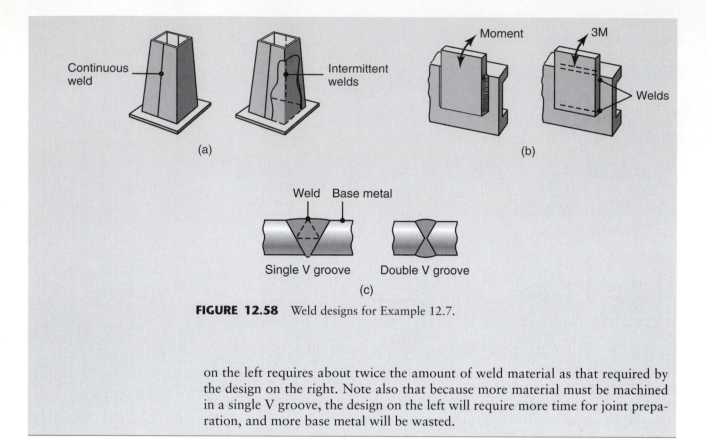

FIGURE 12.58 Weld designs for Example 12.7.

on the left requires about twice the amount of weld material as that required by the design on the right. Note also that because more material must be machined in a single V groove, the design on the left will require more time for joint preparation, and more base metal will be wasted.

12.17.2 Design for brazing and joining

Some general design guidelines for brazing are given in Fig. 12.59. Note that strong joints require a greater contact area for brazing than for welding. Design guidelines for soldering are similar to those for brazing. Figures 12.45 and 12.46 show some examples of frequently used joint designs. Note again the importance of large contact surfaces in order to the develop sufficient joint strength.

FIGURE 12.59 Examples of good and poor designs for brazing.

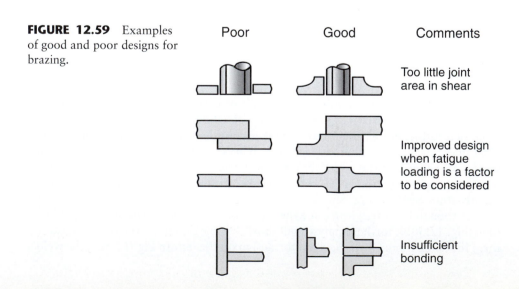

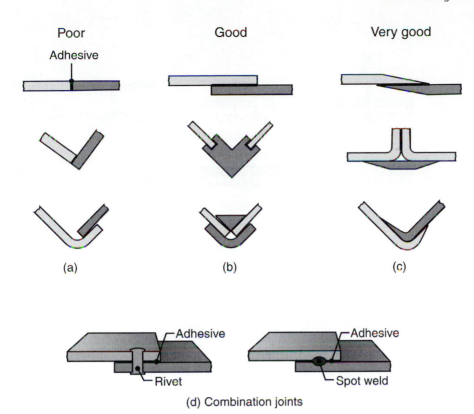

FIGURE 12.60   Various joint designs in adhesive bonding. Note that good designs require large contact areas for better joint strength.

12.17.3 Design for adhesive bonding

Designs for adhesive bonding should ensure that joints are subjected to compressive, tensile, or shear forces, but not to peeling or cleavage (see Fig. 12.50). Several joint designs for adhesive bonding are given in Fig. 12.60. Note that they vary considerably in their strength. Consequently, selection of an appropriate design is important and should include considerations such as the type of loading and the environment to which the bonded structure will be subjected, particularly over its service life.

Butt joints require large bonding surfaces, and because of the force couple developed at the joint, lap joints tend to distort under a tensile force. The coefficients of thermal expansion of the components to be bonded should preferably be close so as to avoid internal stresses during adhesive bonding. Note also that thermal cycling can cause differential movement across the joint, possibly reducing joint strength.

12.17.4 Design for mechanical fastening

The design of mechanical joints requires consideration of (a) the type of loading to which the structure will be subjected, (b) the size and spacing of the holes, and (c) compatibility of the fastener material with the components to be joined. Incompatibility may lead to galvanic corrosion, also known as *crevice corrosion*. For example, in using a steel bolt or rivet to fasten copper sheets, the bolt is anodic and the copper plate cathodic, thus resulting in rapid corrosion and loss of joint strength. Similar reactions take place when aluminum or zinc fasteners are used on copper products. Some design guidelines for riveting are given in Fig. 12.61.

Other general design guidelines for mechanical joining include the following (see also Section 14.10).

1. It is generally less costly to use fewer, but larger, fasteners than a large number of smaller fasteners.
2. Part assembly should be accomplished with a minimum number of fasteners.

FIGURE 12.61 Design guidelines for riveting. *Source*: Bralla, J.G. (ed.), *Handbook of Product Design for Manufacturing*, 2nd ed., McGraw-Hill, 1999.

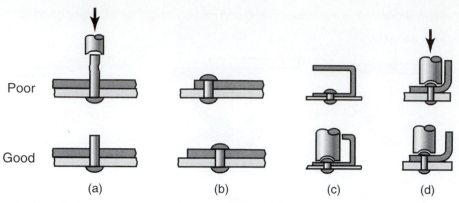

3. The fit between parts to be joined should be as loose as permissible to reduce costs and facilitate the assembly process.

4. Fasteners of standard size should be used whenever possible.

5. Holes should not be too close to edges or corners, as the material may tear when subjected to external forces.

12.18 | Economic Considerations

The basic economic considerations for particular joining processes have been described individually. It can be seen that because of the wide variety of considerations, it is difficult to make generalizations regarding costs. The overall relative costs for these processes have been included in the last columns in Tables 12.1 and 12.2. Note in Table 12.1, for example, that because of the preparations involved, brazing and mechanical fastening can be the most costly methods. On the other hand, because of their highly automated nature, resistance welding and seaming and crimping processes are the least costly of the methods listed.

Although the cost of adhesive bonding depends on the specific application, the overall economics of the process makes adhesive bonding an attractive choice, and sometimes it is the only process that is feasible or practical for a particular application.

Table 12.2 lists the cost of equipment involved for major categories of joining processes. Note that because of the specialized equipment involved, laser-beam welding can be the most expensive process, whereas the traditional oxyfuel gas welding and shielded metal arc welding are the least expensive processes. Typical ranges of equipment costs for some processes are listed below (in alphabetical order). The cost of some equipment can, however, significantly exceed these values, depending on the size of the equipment and the level of automation and controls implemented.

Electron-beam welding: $90,000 to over $1 million
Electroslag welding: $15,000 to $25,000
Flash welding: $5000 to $1 million
Friction welding: $75,000 to over $1 million
Furnace brazing: $2000 to $300,000
Gas metal arc and flux-cored arc welding: $1000 to $3000
Gas tungsten arc welding: $1000 to $5000
Laser-beam welding: $30,000 to $1 million
Plasma arc welding: $1500 to $6000
Resistance welding: $20,000 to $50,000
Shielded metal arc welding: $300 to $2000

As in all other manufacturing operations, an important aspect of the economics of joining is in process automation and optimization. (See Chapters 14 and 15.)

Particularly important is the extensive and effective use of industrial robots, programmed in such a manner that they can accurately track complex weld paths (such as in *tailor-welded blanks;* Section 7.3.4), using machine vision systems and closed-loop controls. As a result, the repeatability and accuracy of welding processes have greatly been improved.

CASE STUDY | Friction Welding of Monosteel® Pistons

There has been a sustained effort among heavy truck manufacturers to design and manufacture diesel engines with reduced emissions. For this reason, a number of technologies have become more prevalent since the 1980s, reflecting the need for *green design* (see Section 16.4). Exhaust-gas recirculation (the reintroduction of a portion of the spent exhaust gases into the intake stream of the engine) has become standard and is known to reduce nitrous oxide emission. Unfortunately, this strategy leads to less efficient combustion and less component durability because of the abrasive wear particles (see Section 4.4.2) and acids that are recirculated into the engine. To maintain and even improve engine efficiency, engine manufacturers have increased cylinder pressures and operating temperatures, which lead to an even more demanding environment for engine components.

In the U.S. market, the traditional aluminum pistons in diesel engines were found to be unable to function reliably in modern engine designs. The problems identified with pistons were a tendency to "mushroom" and potentially fracture under the high firing pressures in the cylinder, inadequate cooling of the piston and scuffing (wear) at the pin that joins the piston to the connecting rod. A solution is a Monosteel® piston, shown in Fig. 12.62, which has the following design attributes:

- The piston is produced from steel, which has higher strength and high-temperature mechanical properties than the aluminum alloys previously used (see Section 3.10).

- A two-piece design allows incorporation of an oil gallery, allowing circulation of cooling oil in the piston. One of the main advantages of the Monosteel® design is the use of a very large gallery, resulting in effective heat transfer

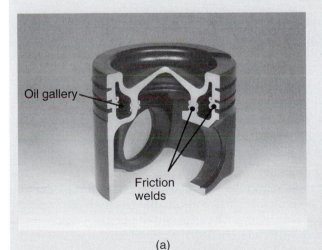
(a)

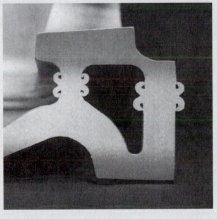
(b)

FIGURE 12.62   The Monosteel® piston. (a) Cutaway view of the piston, showing the oil gallery and friction welded sections, and (b) detail of the friction welds before the external flash is removed by machining. Note that the piston has been rotated in this photo.

from the piston. This design has been shown to reduce piston temperatures in the rim by around 40°C compared with earlier piston designs.

- The piston's steel skirt is much more rigid than the aluminum skirt, resulting in smaller deformation and allowing for designs with tighter clearances. This results in a more stable piston with less oil consumption (thus further reducing harmful exhaust emissions).

Monosteel® pistons are produced from two forged components, which are machined prior to welding. The process used to join these components is inertia friction welding (Section 12.9) which has the following advantages in this application:

- The process leads to well-controlled, reliable, and repeatable high-quality welds.
- Friction welds are continuous and do not involve any porosity, thereby producing a high-strength weld that seals the oil gallery.
- The welding process is fairly straightforward to optimize, the main process variables being energy (or spindle speed for a given flywheel) and contact pressure.

The friction welding process does not require any operator intervention or expertise, as it is entirely machine controlled. Although the capital investment is significant compared to other welding technologies, weld quality (and the ability to weld this application) is significantly more favorable.

The Monosteel® piston shown in Fig. 12.62 was produced on a 250-ton force capacity inertia friction welder, using a peripheral velocity of 7.5 m/s (1500 ft/min), and a contact pressure of 140 MPa (20 ksi) (see Fig. 12.30). As can be seen, the weld zone contains the optimum flash shape (see Fig. 12.31), which is removed from the exterior piston surface by a turning operation (Section 8.9.1), after which the piston skirt is ground (Section 9.6). Production takes place at relatively high rates; 40–60 seconds cycle times are typical but can be higher or lower depending on piston size.

The Monosteel® piston has been applied to multiple engine platforms and has been in high-volume production since 2003.

Source: Courtesy of D. Adams, Manufacturing Technology, Inc., and K. Westbrooke, Federal Mogul, Inc.

SUMMARY

- Because almost all products are an assemblage of many components, joining and fastening processes are an important aspect of manufacturing, as well as of servicing and transportation of parts. (Section 12.1)

- A major category of joining processes is fusion welding, in which the two pieces to be joined melt together and coalesce by means of heat; filler metals may or may not be used. Two common methods are consumable electrode arc welding and nonconsumable electrode arc welding. The selection of the particular method depends on factors such as workpiece material; shape complexity, size, and thickness; and type of joint. (Sections 12.3 and 12.4)

- Electron-beam and laser-beam welding are major categories of joining using high-energy beams. They produce small and high-quality weld zones and thus have important and unique applications in manufacturing. (Section 12.5)

- Because the joint undergoes important metallurgical and physical changes, the nature, properties, and quality of the weld joint are important factors to consider. Weldability of metals, weld design, and process selection are factors that also should be considered. (Section 12.6)

- In solid-state welding, joining takes place without fusion, and pressure is applied either mechanically or by explosives. Surface preparation and cleanliness can be important. Ultrasonic welding and resistance welding are two major examples of solid-state welding, particularly important for sheet metals and foil. (Sections 12.7–12.11)

- Diffusion bonding, especially combined with superplastic forming, is an effective means of fabricating complex sheet-metal structures with high strength-to-weight and stiffness-to-weight ratios. (Section 12.12)

- Brazing and soldering involve the use of filler metals at the interfaces to be joined. These processes require lower temperatures than those in welding and are capable of joining dissimilar metals with intricate shapes and thicknesses. (Section 12.13)

- Adhesive bonded joints have favorable characteristics such as strength, sealing, insulating, vibration damping, and resistance to corrosion between dissimilar metals. An important development is electrically conducting adhesives for surface-mount technologies. (Section 12.14)

- Mechanical fastening is among the oldest and most commonly used techniques. A wide variety of shapes and sizes of fasteners and fastening techniques have been developed for numerous permanent and semipermanent applications. (Section 12.15)

- Several joining techniques are available for joining thermoplastics and thermosetting plastics and various types of ceramics and glasses. (Section 12.16)

- General design guidelines have been established for joining, some of which are applicable to a variety of processes, whereas others require special considerations, depending on the particular application. (Section 12.17)

- The economics of joining operations involves factors such as equipment cost, labor cost, and skill level required, as well as processing parameters such as time required, joint quality, and the need to meet specific requirements. The implementation of automation, process optimization, and computer controls has a major impact on costs involved. (Section 12.18)

SUMMARY OF EQUATIONS

Heat input in welding: $\dfrac{H}{l} = e\dfrac{VI}{v}$

Welding speed: $v = e\dfrac{VI}{uA}$

Heat input in resistance welding: $H = I^2Rt$

BIBLIOGRAPHY

Adams, R.D., (ed.), *Adhesive Bonding*, CRC Press, 2005.

Baghdachi, J., *Adhesive Bonding Technology*, Dekker, 1996.

Bickford, J.H., and Nassar, S. (eds.), *Handbook of Bolts and Bolted Joints*, Dekker, 1998.

Bowditch, M.A., and Baird, R.J., *Oxyfuel Gas Welding*, Goodheart-Wilcox, 2003.

Cary, H.B., and Helzer, S., *Modern Welding Technology*, 6th ed., Prentice Hall, 2004.

Duley, W.W., *Laser Welding*, Wiley, 1999.

Evans, G.M., and Bailey, N., *Metallurgy of Basic Weld Metal*, Wooodhead, 1997.

Grong, O., *Metallurgical Modeling of Welding*, The Institute of Metals, 1994.

Hicks, J.G., *Welded Joint Design*, 2nd ed., Abington, 1997.

Houldcroft, P.T., *Welding and Cutting: A Guide to Fusion Welding and Associated Cutting Processes*, Industrial Press, 2nd ed., 2001.

Humpston, G., and Jacobson, D.M. *Principles of Soldering*, ASM International, 2004.

Hwang, J.S., *Modern Solder Technology for Competitive Electronics Manufacturing*, McGraw-Hill, 1996.

Introduction to the Nondestructive Testing of Welded Joints, 2nd ed., American Society of Mechanical Engineers, 1996.

Jacobson, D.M., and Humpston, G., *Principles of Brazing*, ASM International, 2005.

Jeffus, L.F., *Welding: Principles and Applications*, 5th ed., Delmar, 2002.

Judd, M., and Brindley, K., *Soldering in Electronics Assembly*, 2nd ed., Newnes, 1999.

Kou, S., *Welding Metallurgy*, 2nd ed., Wiley, 2002.

Lancaster, J.F., *The Metallurgy of Welding*, 6th ed., Chapman & Hall, 1999.

Lippold, J.C., and Kotecki, D.J., *Welding Metallurgy and Weldability of Steels*, Wiley, 2005.

Mandal, N.R., *Aluminum Welding*, ASM International, 2002.

Manko, H.H., *Soldering Handbook for Printed Circuits and Surface Mounting*, Van Nostrand Reinhold, 1995.

Minnick, W.H., *Gas Metal Arc Welding Handbook*, Goodheart-Wilcox, 1999.

Mouser, J.D., *Welding Codes, Standards, and Specifications*, McGraw-Hill, 1997.

Nicholas, M.G., *Joining Processes: Introduction to Brazing and Diffusion Bonding*, Chapman & Hall, 1998.

Parmley, R.O. (ed.), *Standard Handbook of Fastening and Joining*, 3rd ed., McGraw-Hill, 1997.

Pecht, M.G., *Soldering Processes and Equipment*, Wiley, 1993.

Petrie, E.M., *Handbook of Adhesives and Sealants*, 2nd ed., McGraw-Hill, 2006.

Powell, J., CO_2 *Laser Cutting*, 2nd ed., Springer, 1998.

Rotheiser, J., *Joining of Plastics: Handbook for Designers and Engineers*, Hanser Gardner, 2004.

Satas, D., *Handbook of Pressure-Sensitive Adhesive Technology*, 3rd ed., Satas & Associates, 1999.

Schultz, H., *Electron Beam Welding*, Woodhead, 1994.

Schwartz, M.M., *Brazing*, 2nd ed., ASM International, 2003.

————, *Ceramic Joining*, ASM International, 1990.

————, *Joining of Composite-Matrix Materials*, ASM International, 1994.

Speck, J.A., *Mechanical Fastening, Joining, and Assembly*, Dekker, 1997.

Steen, W.M., *Laser Material Processing*, 2nd ed., Springer, 1998.

Swenson, L.-E., *Control of Microstructures and Properties in Steel Arc Welds*, CRC Press, 1994.

Tres, P.A., *Designing Plastic Parts for Assembly*, 3rd ed., Hanser-Gardner, 1998.

Weld Integrity and Performance, ASM International, 1997.

Woodgate, R.W., *Handbook of Machine Soldering*, Wiley, 1996.

QUESTIONS

12.1   Explain the reasons that so many different welding processes have been developed.

12.2   List the advantages and limitations of mechanical fastening as compared with adhesive bonding.

12.3   Describe the similarities and differences between consumable and nonconsumable electrodes.

12.4   What determines whether a certain welding process can be used for workpieces in a horizontal, a vertical, or an upside-down position, or for all types of positions? Explain, giving appropriate examples.

12.5   Comment on your observations regarding Fig. 12.5.

12.6   Discuss the need for and role of fixtures in holding workpieces in proper positions for the welding operations described in this chapter.

12.7   Describe the factors that influence the size of the two weld beads shown in Fig. 12.13.

12.8   Why is the quality of welds produced by submerged arc welding very good? Explain.

12.9   Explain the factors involved in electrode selection in arc welding processes.

12.10   Explain why the electroslag welding process is particularly suitable for thick plates and heavy structural sections.

12.11   What are the similarities and differences between consumable and nonconsumable electrode arc welding processes? Explain.

12.12   In Table 12.2, there is a column on the distortion of welded components, ranging from lowest to highest. Explain why the degree of distortion varies among different welding processes.

12.13 Explain why the grains shown in Fig. 12.16 grow in the particular directions shown.

12.14 Prepare a table listing the processes described in this chapter and provide, for each process, the range of welding speeds as a function of workpiece material and its thickness.

12.15 Explain what is meant by *solid-state welding*.

12.16 Describe your observations concerning Figs. 12.19, 12.20, and 12.21.

12.17 What advantages does friction welding have over the other joining methods described in this chapter? Explain.

12.18 Why is diffusion bonding, when combined with superplastic forming of sheet metals, an attractive fabrication process? Does it have any limitations? Explain.

12.19 Can roll bonding be applied to various part configurations? Explain.

12.20 Comment on your observations concerning Fig. 12.41.

12.21 If electrical components are to be mounted to both sides of a circuit board, what soldering process(es) would you use? Explain.

12.22 Discuss the factors that influence the strength of (a) a diffusion bonded component and (b) a cold welded component.

12.23 Describe the difficulties you might encounter in applying explosion welding in a factory environment in urban areas.

12.24 Inspect the periphery of a U.S. quarter and comment on your observations. Is the cross section, i.e., the thickness of individual layers, symmetrical? Explain.

12.25 What advantages do resistance welding processes have over others described in this chapter? Explain.

12.26 What does the strength of a weld nugget in resistance spot welding depend on? Why?

12.27 Explain the significance of the magnitude of the pressure applied through the electrodes during resistance welding operations.

12.28 Which materials can be friction stir welded, and which cannot? Explain.

12.29 List the joining methods that would be suitable for a joint that will encounter high stresses and will have to be disassembled several times during the product life, and rank the methods.

12.30 Inspect Fig. 12.31 and explain why those particular fusion-zone shapes are developed as a function of pressure and speed. Comment on the influence of the properties of the material.

12.31 Which applications could be suitable for the roll spot welding process shown in Fig. 12.35c? Give specific examples.

12.32 Give several examples concerning the bulleted items listed at the beginning of Section 12.1.

12.33 Could the projection welded parts shown in Fig. 12.36 also be made by any of the processes described in other parts of this text? Explain.

12.34 Describe the factors that influence the flattening of the interface after resistance projection welding takes place (see Fig. 12.36).

12.35 What factors influence the shape of the upset joint in flash welding, as shown in Fig. 12.37b? Why?

12.36 Explain how you would fabricate the structures shown in Fig. 12.41b with methods other than diffusion bonding and superplastic forming.

12.37 Make a survey of metal containers for household products and foods and beverages. Identify those that have utilized any of the processes described in this chapter. Describe your observations.

12.38 Which process uses a solder paste? What are the advantages to this process?

12.39 Explain why some joints may have to be preheated prior to welding.

12.40 What are the similarities and differences between fusion welding and casting of metals, described in Chapter 5?

12.41 Explain the role of the excessive restraint (stiffness) of various components to be welded on possible weld defects.

12.42 Discuss the weldability of several metals and explain why some metals are easier to weld than others.

12.43 Does the filler metal have to be of the same composition as that of the base metal to be welded? Explain.

12.44 Describe the factors that contribute to the difference in properties across a welded joint.

12.45 How does the weldability of steel vary as its carbon content increases? Why?

12.46 Are there common factors among the weldability, solderability, castability, formability, and machinability of metals? Explain giving appropriate examples.

12.47 Describe the procedure you would follow in inspecting a weld for a critical application. If you find a flaw during your inspection, how would you go about determining whether or not this flaw is important for the particular application?

12.48 Do you think it is acceptable to differentiate brazing and soldering arbitrarily by temperature of application? Explain.

12.49 Loctite® is an adhesive used to keep metal bolts from vibrating loose; it basically glues the bolt to the nut after the nut is inserted in the bolt. Explain how this adhesive works.

12.50 List the joining methods that would be suitable for a joint that will encounter high stresses and cyclic (fatigue) loading, and rank the methods in order of preference.

12.51 Why is surface preparation important in adhesive bonding? Explain.

12.52 Why have mechanical joining and fastening methods been developed? Give several specific examples of their applications.

12.53 Explain why hole preparation may be important in mechanical joining of components.

12.54 What precautions should be taken in mechanical joining of dissimilar metals? Why?

12.55 What difficulties are involved in joining plastics? What about in joining ceramics? Explain.

12.56 Comment on your observations concerning the numerous joints shown in various figures in Section 12.17.

12.57 How different is adhesive bonding from other joining methods? What limitations does it have? Explain.

12.58 It was noted that soldering is generally applied to thinner components. Why?

12.59 Explain why adhesively bonded joints tend to be weak in peeling.

12.60 Inspect various household products and describe how they are joined and assembled. Explain why those particular processes were used.

12.61 Name several products that have been assembled by (a) seaming, (b) stitching, and (c) soldering.

12.62 Suggest methods of attaching a round bar (made of thermosetting plastic) perpendicularly to a flat metal plate.

12.63 Describe the tooling and equipment that are essential to perform the double-lock seaming operation shown in Fig. 12.53, starting with flat sheet. (See also Fig. 7.23.)

12.64 What joining methods would be suitable in assembling a thermoplastic cover over a metal container? Note that the cover has to be removed periodically.

12.65 Repeat Question 12.64, but for a cover made of (a) a thermosetting plastic, (b) metal, and (c) ceramic. Describe the factors involved in your selection of assembly methods.

12.66 Do you think the strength of an adhesively bonded structure is as high as that obtained by diffusion bonding? Explain.

12.67 Comment on workpiece size limitations, if any, for each of the processes described in this chapter.

12.68 Describe part shapes that would be difficult, or impossible, to join by the processes described in this chapter. Gives specific examples.

12.69 Give several applications of electrically conducting adhesives.

12.70 Give several applications for fasteners used in various household products and explain why other joining methods have not been used instead.

12.71 Comment on workpiece shape limitations, if any, for each of the processes described in this chapter.

12.72 List and explain the guidelines that must be followed to avoid cracks in welded joints, such as hot tearing, hydrogen-induced cracking, and lamellar tearing.

12.73 If a built-up weld is to be constructed (see Fig. 12.5), should all of it be done at once or should it be done a little at a time? Assume that sufficient time is allowed for cooling between beads.

12.74 Describe the reasons that fatigue failure generally occurs in the heat-affected zone of welds and not through the weld bead itself.

12.75 If the parts to be welded are preheated, is the likelihood that porosity will form increased or decreased? Explain.

12.76 What is the advantage of electron-beam and laser-beam welding processes as compared to arc welding? Explain.

12.77 Describe the common types of discontinuities in welds and explain the methods by which they can be avoided.

12.78 What are the sources of weld spatter? How can spatter be controlled? Explain.

12.79 Describe the functions and characteristics of electrodes. What functions do coatings have? How are electrodes classified? Explain.

12.80 Describe the advantages and limitations of explosion welding.

12.81 Explain the difference between resistance seam welding and resistance spot welding.

12.82 Could any of the processes described in this chapter be used to make a large bolt by welding the head to the shank? (See Fig. 6.17.) Explain the advantages and limitations of this approach.

12.83 Describe wave soldering. What are the advantages and disadvantages of this process?

12.84 What are the similarities and differences between a bolt and a rivet? Explain.

12.85 It is common practice to tin plate electrical terminals to facilitate soldering. Why is tin a suitable material?

12.86 Review the contents of Table 12.3 and explain why some materials require more heat than others to melt a given volume.

PROBLEMS

12.87 Two flat copper sheets (each 1.5 mm thick) are being spot welded with a current of 7000 A and a current flow time of 0.3 s. The electrodes are 5 mm in diameter. Estimate the heat generated in the weld zone. Assume that the resistance is 200 $\mu\Omega$.

12.88 Calculate the temperature rise in Problem 12.87 assuming that the heat generated is confined to the volume of material directly between the two round electrodes and that the temperature distribution is uniform.

12.89 Calculate the range of allowable currents for Problem 12.87, if the temperature must be between 0.7 and 0.85 times the melting temperature of copper. Repeat this problem for carbon steel.

12.90 In Fig. 12.24 assume that most of the top portion of the top piece is cut horizontally with a sharp saw. Thus, the residual stresses will be disturbed and, as described in Section 2.10, the part will undergo shape change. For this case, how will the part distort? Explain.

12.91 The accompanying figure shows a metal sheave that consists of two matching pieces of hot-rolled, low-carbon-steel sheets. These two pieces can be joined either by spot welding or by V-groove welding. Discuss the advantages and limitations of each process for this application.

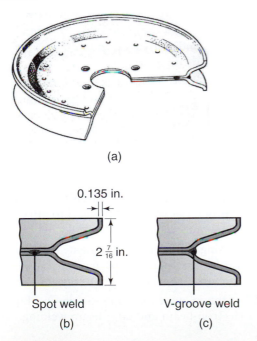

(a)

0.135 in.

$2\frac{7}{16}$ in.

Spot weld

V-groove weld

(b)   (c)

12.92 A welding operation takes place on an aluminum-alloy plate. A pipe 50 mm in diameter with a 4-mm wall thickness and a 60-mm length is butt-welded onto a section of $15 \times 15 \times 5$ mm angle iron. The angle iron is of an L-shape and has a length of 0.3 m. If the weld zone in a gas tungsten arc welding process is approximately 8 mm wide, what would be the temperature increase of the entire structure due to the heat input from welding only? What if the process were an electron-beam welding operation with a bead width of 6 mm? Assume that the electrode requires 1500 J and the aluminum alloy requires 1200 J to melt 1 gram.

12.93 A shielded metal arc welding operation is taking place on carbon steel to produce a fillet weld (see Fig. 12.21b). The desired welding speed is about 25 mm/s. If the power supply is 10 V, what current is needed if the weld width is to be 7 mm?

12.94 The energy applied in friction welding is given by the formula $E = IS^2/C$, where I is the moment of inertia of the flywheel, S is the spindle speed in rpm, and C is a constant of proportionality (5873, when the moment of inertia is given in lb-ft^2). For a spindle speed of 600 rpm and an operation in which a steel tube (3.5 in. OD and 0.25 in. wall thickness) is welded to a flat frame, what is the required moment of inertia of the flywheel if all of the energy is used to heat the weld zone (approximated as a material 0.25 in. deep and directly below the tube)? Assume that 1.4 ft-lbm is required to melt the electrode.

12.95 In oxyacetylene, arc, and laser-beam cutting, the processes basically involve a localized melting of the workpiece. If an 80-mm diameter hole is to be cut from a 250-mm diameter, 12-mm thick plate, plot the mean temperature rise in the plate as a function of kerf width. Assume that one-half of the energy goes into the plate and the other half goes into the blank.

12.96 Refer to the simple butt and lap joints shown in Fig. 12.1. (a) Assuming the area of the butt joint is 3×20 mm and referring to the adhesive properties given in Table 12.6, estimate the minimum and maximum tensile force that this joint can withstand. (b) Estimate these forces for the lap joint assuming its area is 15×15 mm.

12.97 A rivet can buckle if it is too long, as shown in Fig. 12.61. Using information from solid mechanics, determine the length-to-diameter ratio of a rivet that will not buckle during riveting.

12.98   Repeat Example 12.2 if the workpiece is (a) magnesium, (b) copper, and (c) nickel.

12.99   A submerged arc welding operation takes place on 10 mm thick stainless steel plate, producing a butt weld as shown in Fig. 12.20c. The weld geometry can be approximated as a trapezoid with 15 mm and 10 mm as the top and bottom dimensions, respectively. If the voltage applied is 40 V at 400 A, estimate the welding speed if a stainless steel filler wire is used.

12.100   Assume that you are asked to give a quiz to students on the contents of this chapter. Prepare three quantitative problems and three qualitative questions, and supply the answers.

DESIGN

12.101   Design a machine that can perform friction welding of two cylindrical pieces and be able to remove the flash from the welded joint. (See Fig. 12.30.)

12.102   How would you modify your design in Problem 12.101 if one of the pieces to be welded has a noncircular cross section?

12.103   Describe part shapes that cannot be joined by any of the friction welding processes.

12.104   Make a comprehensive outline of joint designs relating to the processes described in this chapter. Give specific examples of engineering applications for each type of joint.

12.105   Review the two weld designs shown in Fig. 12.58a and based on the topics covered in courses on the strength of materials, show that the design on the right is capable of supporting a larger moment.

12.106   In the building of large ships, it is necessary to weld large sections of steel plates together to form a hull. For this application, consider each of the welding operations described in this chapter and list the benefits and drawbacks of that particular operation for this product. Which welding process would you particularly recommend? Why?

12.107   Examine various household products and describe how they are joined and assembled. Explain why those particular processes are used for these particular applications.

12.108   A major cause of erratic behavior (hardware bugs) and failures of computer equipment is fatigue failure of the soldered joints, especially in surface-mount devices and those with bond wires. (See Fig. 12.48.) Design a test fixture for cyclic loading of surface-mount joints for fatigue testing.

12.109   Using two strips of steel 1 in. wide and 8 in. long, design and fabricate a joint that gives the highest strength in a tension test in the longitudinal direction.

12.110   Make an outline of the general guidelines for safety in welding operations. For each of the operations described in this chapter, prepare a poster which concisely and effectively gives specific instructions for safe practices in welding and for cutting. (Review the various publications of the National Safety Council and other similar organizations.)

12.111   A common practice for repairing expensive broken or worn parts, such as may occur when, for example, a fragment is broken from a forging, is to fill the area with layers of weld bead and then to machine the part back to its original dimensions. Prepare a list of the precautions that you would suggest to someone who uses this approach.

12.112   In the roll bonding process shown in Fig. 12.28, how would you go about ensuring that the interfaces are clean and free of contaminants, so that a good bond is developed? Explain.

12.113   Alclad stock is made from 5182 aluminum alloy, both sides of which are coated with a thin layer of pure aluminum. The 5182 provides high strength, while the two outside layers of pure aluminum provide good corrosion resistance because of their stable oxide film. For these reasons, Alclad is commonly used in aerospace structural applications. Investigate other common roll-bonded materials and their uses, and prepare a summary table.

12.114   Obtain a soldering iron and attempt to solder two wires together. First, try to apply the solder at the same time as you first put the soldering iron tip to the wires. Second, preheat the wires before applying the solder. Repeat the same procedure for a cool surface and a heated surface, respectively. Record your results and explain your findings.

12.115   Perform a literature search to determine the types and properties of adhesives used to affix artificial hips onto the human femur.

12.116   Using the Internet, investigate the geometry of the heads of screws that are permanent fasteners, that is, ones that can be screwed in but not out.

12.117   Obtain an expression similar to Eq. (12.6) but for electron-beam and laser-beam welding.

Fabrication of Microelectronic, Micromechanical, and Microelectromechanical Devices; Nanomanufacturing

This chapter presents the science and technology involved in the production of microscopic devices, particularly microelectronic and microelectromechanical systems, as well as the materials commonly used with these products. We describe

- The unique properties of silicon that make it ideal for producing oxide and dopants and complimentary metal-on-oxide semiconductor devices.
- Treatment of a cast ingot and machining operations to produce a wafer.
- Lithography, etching, and doping processes.
- Wet and dry etching for circuit manufacture.
- Electrical connections at all levels, from a transistor to a computer.
- Packages for integrated circuits and manufacturing methods for printed circuit boards.
- Specialized processes for manufacturing MEMS devices.
- Nanomanufacturing.

13.1 | Introduction

Micromanufacturing, by definition, refers to manufacturing on a microscopic scale, thus not visible to the naked eye (see Fig 1.7). The terms micromanufacture, microelectronics, and microelectromechanical systems (MEMS) are not strictly limited to such small length scales but, instead, suggest a material and manufacturing strategy. Generally, this type of manufacturing relies heavily on lithography, wet and dry etching, and coating techniques. In addition, micromanufacturing of semiconductors exploits the unique ability of silicon to form oxides and **complimentary metal-on-oxide**

semiconductors (CMOS). Examples of products that rely upon micromanufacturing technologies include a wide variety of sensors and probes, inkjet printing heads, microactuators and associated devices, magnetic hard-drive heads, and microelectronic devices such as computer processors and memory chips.

Although semiconducting materials have been used in electronics for several decades, it was the invention of the transistor in 1947 that set the stage for what would become one of the greatest technological advancements in all history. Microelectronics has played an ever-increasing role since **integrated circuit** (IC) technology (Fig. 13.1) became the foundation for personal computers, cellular telephones, information systems, automotive control, and telecommunications.

The basic building block of a complex IC is the transistor. A **transistor** (Fig. 13.2) is a three-terminal device that acts as a simple on-off switch: If a positive voltage is applied to the "gate" terminal, a conducting channel is formed between

(a)   (b)

(c)   (d)

FIGURE 13.1   (a) A 300-mm (11.8-in.) wafer with a large number of dies fabricated onto its surface, (b) detail view of an Intel 45-nm chip including a 153 Mbit SRAM (static random access memory) and logic test circuits, (c) image of the Intel® Itanium®2 processor, and (d) Pentium processor motherboard. *Source:* Courtesy of Intel Corporation.

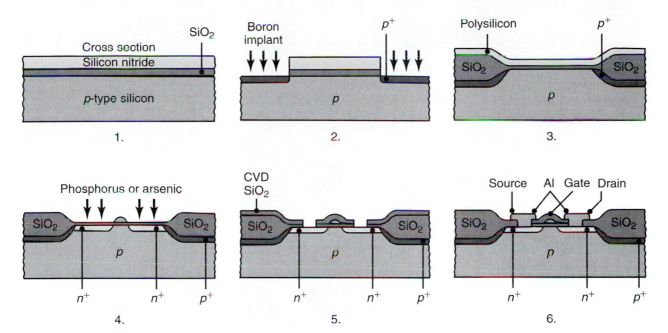

FIGURE 13.2   Cross-sectional views of the fabrication of a metal oxide semiconductor (MOS) transistor. *Source:* After R.C. Jaeger.

the "source" and "drain" terminals, allowing current to flow between those two terminals (switch closed). If no voltage is applied to the gate, no channel is formed, and the source and drain are isolated from each other (switch opened). Figure 13.3 shows how the basic processing steps, described in detail in this chapter, are combined to form a **metal-oxide-semiconductor field effect transistor (MOSFET)**.

In addition to the metal-oxide semiconductor structure, the **bipolar junction transistor (BJT)** also is used, but to a lesser extent. While the actual fabrication steps for these transistors are very similar to those for both the MOSFET and MOS technologies, their circuit applications are different. Memory circuits, such as **random access memory (RAM)**, and microprocessors consist primarily of MOS devices, whereas linear circuits, such as amplifiers and filters, are more likely to contain bipolar transistors. Other differences between these two types of devices include the higher operating speeds, breakdown voltage, and current needed (with a lower current for the MOSFET).

The major advantage of today's ICs is their high degree of complexity, reduced size, and low cost. As fabrication technologies becomes more advanced, the size of devices decreases, and, consequently, more components can be placed onto a **chip** (a small slice of semiconducting material on which the circuit is fabricated). In addition, mass processing and automation have greatly helped reduce the cost of each completed circuit. The components fabricated include transistors, diodes, resistors, and capacitors. Typical chip sizes produced today range from 0.5×0.5 mm $(0.02 \times 0.02$ in.) to more than 50×50 mm (2×2 in.) Modern technology now allows densities in the tens of millions of devices per chip, referred to as **very large-scale integration (VLSI)**. Processors used for cellular phones are common examples of VLSI. When ICs contain more than 100 million devices, the term **ultra large-scale integration (ULSI)** is applied. The Intel dual-core Itanium® 2 processor, for example, contains over 1.7 billion transistors.

This chapter begins with a description of the properties of common semiconductors, such as silicon, gallium arsenide, and polysilicon. It then describes, in detail,

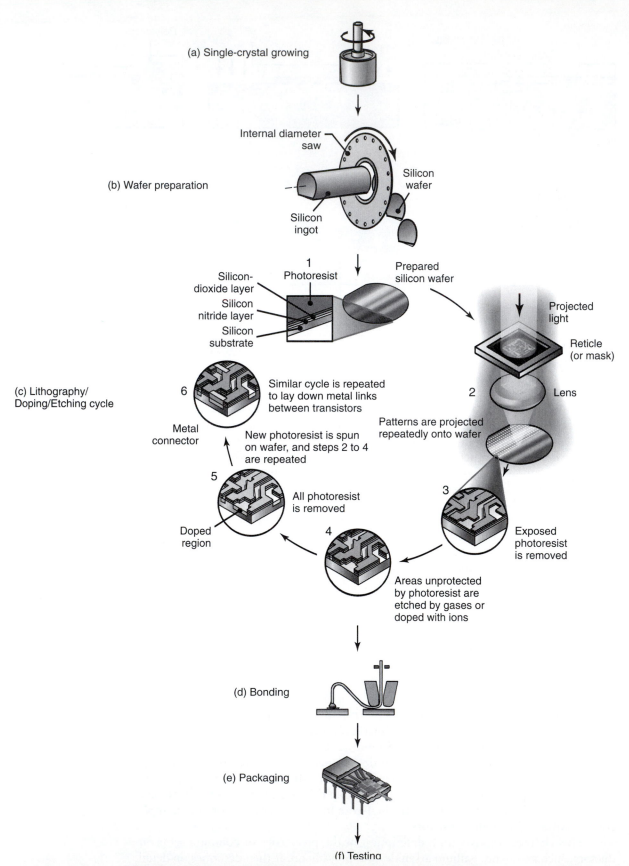

(a) Single-crystal growing

(b) Wafer preparation

Internal diameter saw

Silicon wafer

Silicon ingot

1 Photoresist

Silicon-dioxide layer

Silicon nitride layer

Silicon substrate

Prepared silicon wafer

Projected light

Reticle (or mask)

2   Lens

Patterns are projected repeatedly onto wafer

(c) Lithography/Doping/Etching cycle

6

Similar cycle is repeated to lay down metal links between transistors

Metal connector

New photoresist is spun on wafer, and steps 2 to 4 are repeated

5

All photoresist is removed

Doped region

4

3

Exposed photoresist is removed

Areas unprotected by photoresist are etched by gases or doped with ions

(d) Bonding

(e) Packaging

(f) Testing

FIGURE 13.3   General fabrication sequence for integrated circuits.

the current processes employed in fabricating microelectronic devices and integrated circuits (Fig. 13.3), including IC testing, packaging, and reliability. The chapter also describes a potentially more important development concerning the manufacture of **microelectromechanical systems** (MEMS). These are combinations of electrical and mechanical systems with characteristic lengths of less than 1 mm (0.040 in.). These devices utilize many of the batch-processing technologies used for the manufacture of electronic devices, although several other unique processes have been developed as well.

Although MEMS is a term that came into use around 1987, it has been applied to a wide variety of applications, including precise and rapid sensors, microrobots for nanofabrication, medical delivery systems, and artificial organs. Currently, there are relatively few microelectromechanical systems in use, such as accelerometers and pressure sensors, with on-chip electronics. Often, MEMS is a label applied to micro-mechanical and microelectromechanical devices, such as pressure sensors, valves, and micromirrors. This is not strictly accurate, since MEMS require an integrated microelectronic controlling circuitry; even so, MEMS sales worldwide were around $5 billion in 2005 and are projected to grow by 15% annually.

13.2 | Clean Rooms

Clean rooms are essential for the production of integrated circuits, a fact that can be appreciated by noting the scale of manufacturing that is to be performed. Integrated circuits are typically a few millimeters in length, and the smallest features in a transistor on the circuit may be as small as a few tens of nanometers (nano = 10^{-9}). This size range is smaller than particles that we normally don't consider as harmful, such as dust, smoke, perfume, and bacteria. However, if these contaminants are present on a silicon wafer during processing, they can severely compromise the performance of the whole device. It is thus essential that all potentially harmful particles be eliminated from the integrated circuit manufacturing environment.

There are various levels of cleanliness, defined by the **class** of the clean rooms. The classification system refers to the number of 0.5 μm or larger particles within a cubic foot of air. Thus, a class 10 clean room has 10 or fewer such particles per cubic foot. The size and the number of particles are important to define the class of a clean room, as shown in Fig. 13.4. Most clean rooms for microelectronics manufacturing

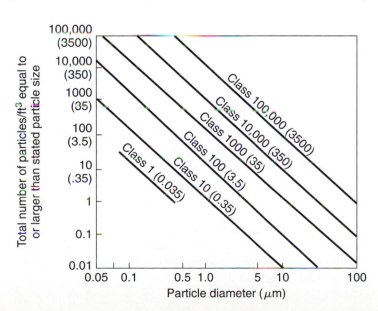

FIGURE 13.4 Allowable particle size concentrations for various clean room classes.

range from class 1 to class 10. In comparison, the contamination level in modern hospitals is on the order of 10,000 particles per cubic foot of air.

To obtain controlled atmospheres that are free from particulate contamination, all ventilating air is passed through a *high-performance particulate air* (HEPA) filter. In addition, the air is usually conditioned so that it is at 21°C (70°F) and 45% relative humidity. Clean rooms are designed such that the cleanliness at critical processing locations is greater than in the clean room in general. This is accomplished by directing the filtered ventilating air so that it displaces ambient air and directs dust particles away from the operation. This can be facilitated by laminar-flow hooded work areas.

The largest source of contaminants in a clean room is people. Skin particles, hair, perfume and makeup, clothing, bacteria, and viruses are given off naturally by people, and in sufficiently large numbers to quickly compromise a class 100 clean room. For these reasons, most clean rooms require special coverings, such as white laboratory coats, gloves and hair nets, as well as avoidance of perfumes and make-up. The most stringent clean rooms require full-body coverings, called bunny suits. There are other stringent precautions, as well. For example, the use of a pencil instead of a ball point pen can produce objectionable graphite particles, and the paper used is a special clean room paper.

13.3 | Semiconductors and Silicon

Semiconductor materials have electrical properties that lie between those of conductors and insulators; they exhibit resistivities between 10^{-3} and 10^8 Ω-cm. Semiconductors have become the foundation for electronic devices, because their electrical properties can be altered by adding controlled amounts of selected impurity atoms into their crystal structures. These impurity atoms, known as **dopants,** either have one more valence electron (*n*-type, or *negative dopant*) or one fewer valence electron (*p*-type, or *positive dopant*) than the atoms in the semiconductor lattice. For silicon, a Group IV element in the periodic table, typical *n*-type and *p*-type dopants include phosphorus and arsenic (Group V) and boron (Group III), respectively. The electrical operation of semiconductor devices is controlled by creating regions of different doping types and concentrations.

Although the earliest electronic devices were fabricated on *germanium,* **silicon** is the industry standard. The abundance of silicon in its alternative forms in the Earth's crust is second only to that of oxygen, thus making it economically attractive. Silicon's main advantage over germanium is its larger energy gap (1.1 eV) compared with that of germanium (0.66 eV). This larger energy gap allows siliconbased devices to operate at temperatures about 150°C (270°F) higher than the operating temperatures of devices fabricated on germanium [about 100°C (180°F)]. Furthermore, the oxidized form of silicon (silicon dioxide, SiO_2) enables the production of metal-oxide-semiconductor (MOS) devices, which are the basis for MOS-transistors. These materials are used in memory devices and processors and the like and account for by far the largest volume of semiconductor material produced worldwide.

The *crystallographic structure* of silicon is a diamond-type fcc structure, as shown in Fig. 13.5, along with the Miller indices of an fcc material. (Miller indices are a notation for identifying planes and directions within a unit cell.) A crystallographic plane is defined by the reciprocal of its intercepts on the three axes. Since anisotropic etchants (described in Section 13.8.1) preferentially remove material in specific crystallographic planes, the orientation of the silicon crystal in a *wafer* (see Section 13.5) is important.

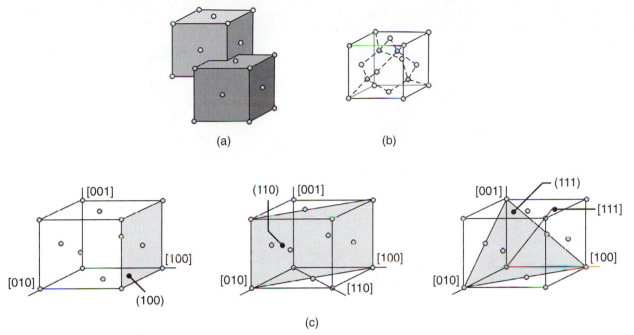

FIGURE 13.5 Crystallographic structure and Miller indices for silicon. (a) Construction of a diamond-type lattice from interpenetrating face-centered cubic cells (one of eight penetrating cells shown) and (b) the diamond-type lattice of silicon. The interior atoms have been shaded darker than the surface atoms. (c) Miller indices for a cubic lattice.

An important processing advantage of silicon is that its oxide (silicon dioxide) is an excellent insulator and is thus used for isolation and passivation purposes; conversely, germanium oxide is water soluble and unsuitable for electronic devices. However, silicon has some limitations, which has encouraged the development of compound semiconductors, specifically **gallium arsenide** (GaAs). Its major advantage over silicon is its capability for light emission, thus allowing fabrication of devices such as lasers and light-emitting diodes (LEDs). Additionally, its larger energy gap (1.43 eV) allows for higher maximum operating temperature, to about 200°C (400°F).

Devices fabricated on gallium arsenide also have much higher operating speeds than those fabricated on silicon. Some disadvantages of gallium arsenide include its considerably higher cost, greater processing complications, and the difficulty of growing high-quality oxide layers, the need for which is emphasized throughout this chapter.

13.4 | Crystal Growing and Wafer Preparation

Silicon, which occurs naturally in the forms of silicon dioxide and various silicates, must undergo a series of purification steps in order to become a high-quality, defect-free, and single-crystal material required for semiconductor-device fabrication. The purification process begins by heating silica and carbon together in an electric furnace, resulting in 95- to 98%-pure polycrystalline silicon. This material is converted to an alternative form, commonly trichlorosilane, which in turn is purified and decomposed in a high-temperature hydrogen atmosphere. The result is an extremely high-quality **electronic-grade silicon** (EGS).

Single-crystal silicon is usually obtained by using the **Czochralski, or CZ, process** (see Fig. 5.30). This method uses a seed crystal that is dipped into a silicon

FIGURE 13.6 Finishing operations on a silicon ingot to produce wafers: (a) and (b) grinding of the end and cylindrical surfaces of a silicon ingot, (c) machining of a notch or flat, (d) slicing of wafers, (e) end grinding of wafers, and (f) chemical-mechanical polishing of wafers.

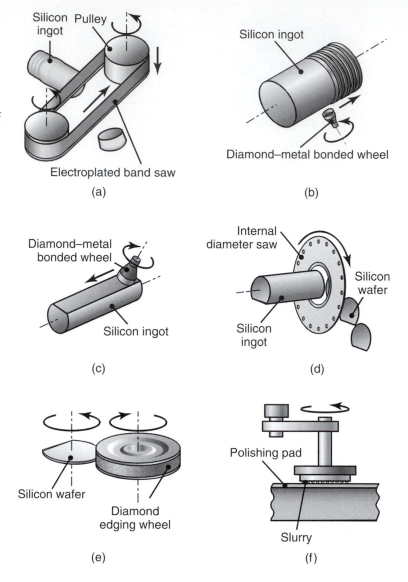

melt and then slowly pulled out while being rotated. At this point, controlled amounts of impurities can be added to the system to obtain a uniformly doped crystal. Typical pull rates are on the order of 20 μm/s (800 μin./s). This growing technique produces a cylindrical single-crystal ingot, typically over 1 m (40 in.) long and 100–300 mm (4–12 in.) in diameter. However, because this technique does not allow precise control of the diameter, ingots are commonly grown to a diameter a few millimeters larger than required and then ground to a precise diameter.

Silicon wafers are produced from ingots by a sequence of machining and finishing operations, as illustrated in Fig. 13.6. A notch or flat is often machined into the silicon cylinder to identify its crystal orientation (Fig. 13.6c). The crystal is then sliced into individual **wafers** by using an inner-diameter blade (Fig. 13.6d). This method uses a rotating blade with its cutting edge on the inner ring. While the substrate depth required for most electronic devices is no more than several microns, wafers are typically cut to a thickness of about 0.5 mm (0.02 in.), thus providing the necessary mass to withstand temperature variations and provide mechanical support during subsequent fabrication steps.

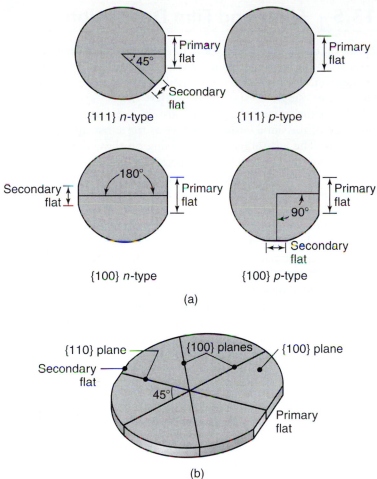

FIGURE 13.7
Identification of single-crystal wafers of silicon. This identification scheme is common for 150-mm (6-in.) diameter wafers, but notches are more common for larger wafers.

{111} *n*-type {111} *p*-type

{100} *n*-type {100} *p*-type

(a)

{110} plane {100} planes {100} plane

Secondary flat Primary flat

(b)

The wafer is then ground along its edges with a diamond wheel. This operation gives the wafer a rounded profile that is more resistant to chipping. Finally, the wafers are polished and cleaned to remove surface damage caused by the sawing process. This operation is commonly performed by *chemical-mechanical polishing* (also referred to as *chemical-mechanical planarization* or CMP), as described in Section 9.7.

In order to properly control the manufacturing process, it is important to determine the orientation of the crystal in a wafer. Wafers, therefore, have notches or flats machined into them for identification, as described above and shown in Fig. 13.7. Most commonly, the (100) or (111) plane of the crystal defines the wafer surface, although (110) surfaces also can be used for micromachining applications. Wafers are also identified by a laser scribe marking produced by the manufacturer. Laser scribing of information may take place on the front or on the back side of the wafer. The front side of some wafers has an exclusion edge area, 3 to 10 mm in size, reserved for the scribe information, including lot numbers, orientation, and a unique wafer identification code.

Device fabrication takes place over the entire wafer surface. Wafers are typically processed in lots of 25 or 50 wafers with 150–200 mm (6–8 in.) diameters each, or lots of 12–25 wafers with 300-mm (12-in.) diameters each. In this way, they can easily be handled and transferred during processing. Because of the small device size and large wafer diameter, thousands of individual circuits can be placed on one wafer. Once processing is completed, the wafer is sliced into individual **chips**, each containing one complete integrated circuit.

13.5 | Films and Film Deposition

Films of many different types, particularly insulating and conducting, are used extensively in microelectronic-device processing. Common deposition films include polysilicon, silicon nitride, silicon dioxide, tungsten, titanium, and aluminum. In some cases, single-crystal silicon wafers merely serve as a mechanical support on which custom *epitaxial layers* are grown (see Section 13.6). These silicon epitaxial films are of the same lattice structure as the substrate, so they are also single-crystal materials. The advantages of processing on these deposited films, instead of on the actual wafer surface, include the presence of fewer impurities (notably carbon and oxygen), improved device performance, and the production of tailored material properties not obtainable on the wafers themselves.

Some of the major functions of deposited films include **masking** for diffusion or implantation and protection of the semiconductor's surface. In masking applications, the film must effectively inhibit the passage of dopants while also being able to be etched into patterns of high resolution. Upon completion of device fabrication, films are also applied to protect the underlying circuitry. Films used for masking and protection include silicon dioxide, phosphosilicate glass (PSG), borophosphosilicate glass (BPSG), and silicon nitride. Each of these materials has distinct advantages, and they often are used in combination.

Other films contain dopant impurities and are used as doping sources for the underlying substrate. Conductive films are used primarily for device interconnection. These films must have a low resistivity, be capable of carrying large currents, and be suitable for connecting to terminal-packaging leads with wire bonds. Aluminum and copper are generally used for this purpose. To date, aluminum is the more common material, because it can be dry etched more easily. However, aluminum is difficult to use for small structures and high current densities, and for this reason, copper, with its low resistivity, has been of significant research interest. Increasing circuit complexity has required up to six levels of conductive layers, which must all be separated by insulating films.

Films may be deposited by a number of techniques, involving a variety of pressures, temperatures, and vacuum systems (see also Section 4.5.1):

1. **Evaporation.** One of the simplest and oldest methods of film deposition is *evaporation*, used primarily for depositing metal films. In this process, the metal is heated in a vacuum until vaporization; upon evaporation, the metal forms a thin layer on the surface. The heat of evaporation is usually provided by a heating filament or electron beam.

2. **Sputtering.** Another method of metal deposition is *sputtering*, which involves bombarding a target with high-energy ions, usually argon (Ar^+), in a vacuum. Sputtering systems usually include a DC power source to produce the energized ions. As the ions impinge on the target, atoms are knocked off and are subsequently deposited on wafers mounted within the system. Although some argon may be trapped within the film, this technique provides very uniform coverage. Advanced sputtering techniques include use of a radio-frequency power source (**RF sputtering**) and introduction of magnetic fields (**magnetron sputtering**).

3. **Chemical vapor deposition.** In one of the most common techniques, *chemical vapor deposition* (CVD), film deposition is achieved by the reaction and/or decomposition of gaseous compounds (see also Section 4.5.1). In this technique, silicon dioxide is routinely deposited by the oxidation of silane or a chlorosilane. Figure 13.8a shows a continuous CVD reactor that operates at atmospheric pressure.

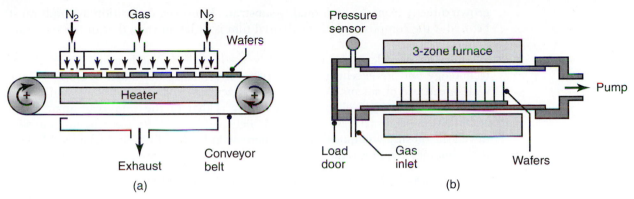

FIGURE 13.8   Schematic diagrams of a (a) continuous, atmospheric-pressure CVD reactor and (b) low-pressure CVD reactor. *Source:* After S.M. Sze.

4. **Low-pressure chemical vapor deposition.** Figure 13.8b shows the apparatus for a method similar to CVD that operates at lower pressures, referred to as *low-pressure chemical vapor deposition* (LPCVD). Capable of coating hundreds of wafers at a time, this method has a much higher production rate that than of atmospheric-pressure CVD and provides superior film uniformity with less consumption of carrier gases. This technique is commonly used for depositing polysilicon, silicon nitride, and silicon dioxide.

5. **Plasma-enhanced chemical vapor deposition.** The *plasma-enhanced chemical vapor deposition* (PECVD) process involves placing wafers in radio-frequency (RF) plasma containing the source gases and offers the advantage of maintaining low wafer temperature during deposition. However, the films deposited generally include hydrogen and are of lower quality than those deposited by other methods.

Silicon **epitaxy** layers, in which the crystalline layer is formed using the substrate as a seed crystal, can be grown by a variety of methods. If the silicon is deposited from the gaseous phase, the process is known as **vapor-phase epitaxy** (VPE). In another variation, called **liquid-phase epitaxy** (LPE), the heated substrate is brought into contact with a liquid solution containing the material to be deposited.

Another high-vacuum process, called **molecular-beam epitaxy** (MBE), uses evaporation to produce a thermal beam of molecules that deposit on the heated substrate. This process offers a very high degree of purity. In addition, since the films are grown one atomic layer at a time, excellent control of doping profiles is achieved, which is especially important in gallium-arsenide technology. However, MBE suffers from relatively low growth rates compared with those of other conventional film-deposition techniques.

13.6 | Oxidation

The term *oxidation* refers to the growth of an oxide layer by the reaction of oxygen with the substrate material. Oxide films can also be formed by the deposition techniques described in Section 13.5. Thermally grown oxides, described in this section, display a higher level of purity than deposited oxides, because the former are grown directly from the high-quality substrate. However, deposition methods must be used if the composition of the desired film is different than that of the substrate material.

Silicon dioxide is the most widely used oxide in modern integrated-circuit technology and its excellent characteristics are one of the major reasons for the widespread use of silicon. Aside from its functions of dopant masking and device isolation, silicon dioxide's most critical role is that of the *gate-oxide* material in MOSFETs. Silicon surfaces have an extremely high affinity for oxygen, and a freshly sawed silicon slice will quickly grow a native oxide of 3–4 nm in thickness. Modern IC technology requires oxide thicknesses from a few to a few hundred nanometers.

1. **Dry oxidation.** Dry oxidation is a relatively simple process and is accomplished by elevating the substrate temperature, typically to 750°–1100°C (1380°–2020°F), in an oxygen-rich environment with variable pressure. Silicon dioxide is produced according to the chemical reaction

$$Si + O_2 \rightarrow SiO_2. \tag{13.1}$$

Most oxidation is carried out in a batch process where up to 150 wafers are placed in a furnace. **Rapid thermal processing** (RTP) is a related process combining **rapid thermal oxidation** (RTO) with an anneal step, called **rapid thermal annealing** (RTA), and is used to produce thin oxides on a single wafer.

As a layer of oxide forms, the oxidizing agents must be able to pass through the oxide layer and reach the silicon surface where the actual reaction takes place. Thus, an oxide layer does not continue to grow on top of itself, but rather it grows from the silicon-silicon-dioxide interface outward. Some of the silicon substrate is consumed in the oxidation process (Fig. 13.9). The ratio of oxide thickness to the amount of silicon consumed is found to be 1:0.44. Thus, for example, to obtain an oxide layer 100 nm thick, approximately 44 nm of silicon will be consumed. This condition does not present a problem, as substrates are always grown sufficiently thick.

One important effect of this consumption of silicon is the rearrangement of dopants in the substrate near the interface. Some dopants deplete away from the oxide interface while others pile up, hence processing parameters have to be adjusted to compensate for this effect.

2. **Wet oxidation.** Another oxidizing technique uses a water-vapor atmosphere as the agent and is appropriately called *wet oxidation*. The chemical reaction involved in wet oxidation is

$$Si + 2H_2O \rightarrow SiO_2 + 2H_2. \tag{13.2}$$

This method offers a considerably higher growth rate than that of dry oxidation, but it suffers from a lower oxide density and, therefore, a lower dielectric strength. The common practice in industry is to combine both dry and wet oxidation methods, growing an oxide in a three-part layer: dry, wet, and dry.

FIGURE 13.9 Growth of silicon dioxide, showing consumption of silicon.

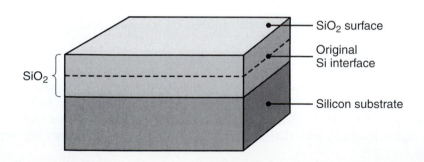

This approach combines the advantages of the much higher growth rate in wet oxidation and the high quality obtained in dry oxidation.

3. **Selective oxidation.**   The two oxidation methods described above are useful primarily for coating the entire silicon surface with oxide, but it is also necessary to oxidize only certain portions of the surface. The procedure used for this task, called *selective oxidation,* uses silicon nitride (Sections 8.6 and 11.8.1), which inhibits the passage of oxygen and water vapor. Thus, by masking certain areas with silicon nitride, the silicon under these areas remains unaffected while the uncovered areas are oxidized.

13.7 | Lithography

Lithography is the process by which the geometric patterns that define devices are transferred from a **reticle** (also called a **photomask,** or **mask**) to the substrate surface. A summary of lithographic techniques is given in Table 13.1 and they are compared in Fig. 13.10. There are several forms of lithography, but the most common form used today is *photolithography*. Electron-beam and X-ray lithography are of great interest because of their ability to transfer patterns of higher resolution, which is a necessary feature for increased miniaturization of integrated circuits. However, most integrated circuit applications can be successfully manufactured with photolithography.

A reticle is a glass or quartz plate with a pattern of the chip deposited onto it, usually with a chromium film although iron oxide and emulsion also are used. The reticle image can have the same size as the desired structure on the chip, but it is often an enlarged image (occasionally 5× to 20× larger, although 10× magnification is most common). The enlarged images are then focused onto a wafer through a lens, in a process known as *reduction lithography*.

In current practice, the lithographic process is applied to microelectronic circuits several times (modern IC devices can require up to 40 lithography steps), each time using a different reticle to define the different areas of the devices and interconnect. Typically designed at several thousand times their final size, reticle patterns undergo a series of reductions before being applied permanently to a defect-free quartz plate. Computer-aided design (Section 15.4) has had a major impact on reticle design and generation. Cleanliness is especially important in lithography, and robots and specialized wafer-handling apparatus are now used in order to minimize dust and dirt contamination.

Once the film deposition process is completed and the desired reticle patterns have been generated, the wafer is cleaned and coated with an organic **photoresist** (PR). A photoresist consists of three principal components: (a) a polymer that

TABLE 13.1

General Characteristics of Lithography Techniques

Method	Wavelength (nm)	Finest feature size (nm)
Ultraviolet (photolithography)	365	350
Deep UV	193	190
Extreme UV	10–20	30–100
X-ray	0.01–1	20–100
Electron beam	—	80

Source: After P.K. Wright.

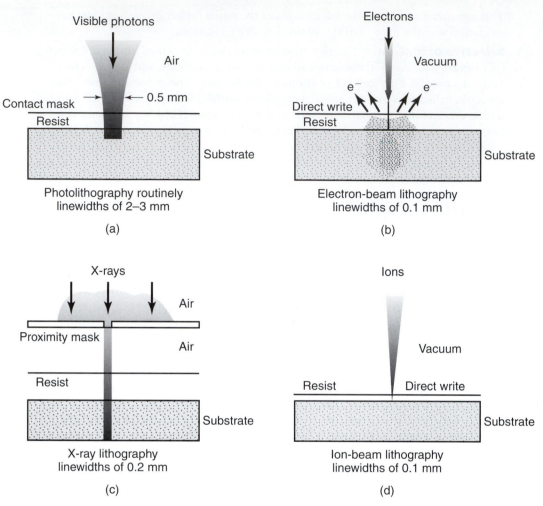

FIGURE 13.10 Comparison of four different lithography techniques: (a) Photolithography, (b) electron-beam lithography, (c) X-ray lithography, and (d) ion-beam lithography.

changes its structure when exposed to radiation, (b) a sensitizer that controls the reactions in the polymer, and (c) a solvent, necessary to deliver the polymer in liquid form.

The wafer is then placed inside a resist spinner, and the photoresist is applied as a viscous liquid onto the wafer (Fig. 13.11). Photoresist layers of 0.5–2.5 μm (20–100 μin.) thick are obtained by spinning at several thousand rpm for 30 to 60 seconds to give uniform coverage. Proper control of the resist layer is needed to ensure proper performance of subsequent lithography operations (see Section 13.7). The thickness of the resist is given by the expression

$$t = \frac{kC^{\beta}\eta^{\gamma}}{\omega^{\alpha}}, \tag{13.3}$$

where t is the resist thickness, C is the polymer concentration in mass per volume, η is the viscosity of the polymer, ω is the angular velocity during spinning, and k, α, γ, and β are constants for the particular spinning system. Where masking levels are considered critical, a **barrier antireflective layer** (BARL) or **barrier antireflective coating** (BARC) is applied either beneath or on top of the photoresist to provide line width control, especially over aluminum.

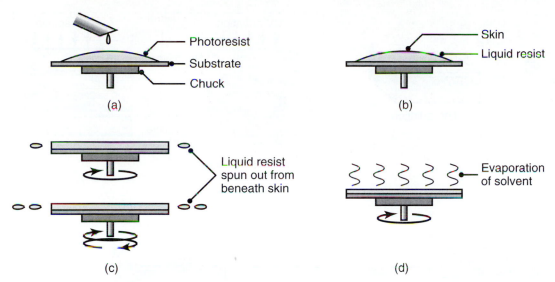

FIGURE 13.11 Spinning of an organic coating on a wafer: (a) Liquid is dispensed, (b) liquid is spread over the wafer surface by spinning at low speed, (c) speed is increased, developing a uniform coating thickness and expelling excess liquid, and (d) evaporation of solvent at final spin speed to obtain organic coating.

The next step in lithography is **prebaking** the wafer to remove the solvent from the photoresist and harden it. This step is carried out in a convection oven or hot plate at around 100°C (273°F) for a period of 10 to 30 min. The pattern is then transferred to the wafer through **stepper** or **step-and-scan** systems. With wafer steppers (Fig. 13.12a), the full image is first exposed in one flash, and then the reticle pattern is refocused onto another adjacent section of the wafer. With step-and-scan systems

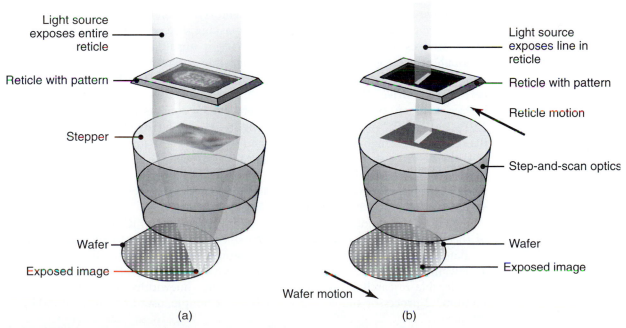

FIGURE 13.12 Schematic illustration of (a) wafer stepper technique for pattern transfer and (b) step-and-scan technique.

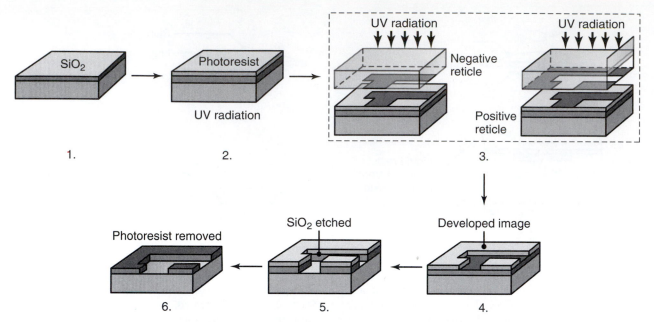

FIGURE 13.13 Pattern transfer by lithography. Note that the mask in Step 3 can be either a positive or a negative image of the pattern. *Source:* After W.C. Till and J.T. Luxon.

(Fig. 13.12b), the exposing light source is focused into a line, and the reticle and wafer are translated simultaneously in opposite directions to transfer the pattern.

The wafer must be carefully aligned under the desired reticle. In this crucial step, called **registration**, the reticle must be aligned correctly with the previous layer on the wafer. Upon development and removal of the exposed photoresist, a duplicate of the reticle pattern will appear in the PR layer. As can be seen in Fig. 13.13, the reticle can be either a negative or a positive image of the desired pattern. A positive reticle uses the ultraviolet (UV) radiation to break down the chains in the organic film, so that these chains are preferentially removed by the developer. Positive masking is more commonly used than negative masking, because with negative masking the photoresist can swell and distort, making it unsuitable for small geometries, although newer negative photoresist materials do not have this problem.

Following the exposure and development sequence, **postbaking** the wafer is performed to drive off solvent and toughen and improve the adhesion of the remaining resist. In addition, a deep UV treatment, which consists of baking the wafer to 150°–200°C (300°–400°F) in ultraviolet light, can also be used to further strengthen the resist against high-energy implants and dry etches. The underlying film not covered by the PR is then implanted or etched away (Sections 13.8 and 13.9). Finally, the photoresist is stripped by exposing it to a wet stripper or an oxygen plasma, a technique referred to as **ashing**. The lithography process may be repeated as many as 40 times in the fabrication of the most advanced ICs.

One of the major issues in lithography is **linewidth**, which refers to the width of the smallest feature obtainable on the silicon surface. As circuit densities have escalated over the years, device sizes and features have become smaller and smaller. Today, minimum commercially feasible linewidths are between 0.04 and 1 μm with considerable research being conducted at smaller linewidths. The Intel dual-core Itanium® 2 processor shown in Fig. 13.1c, for example, uses linewidths of 0.04 μm (40 nm).

As pattern resolution, and therefore device miniaturization, is limited by the wavelength of the radiation source used, the need has arisen to move to wavelengths shorter than those in the ultraviolet range, such as "deep" UV wavelengths, "extreme" UV wavelengths, electron beams, and X-rays (see Table 13.1). In these technologies, the photoresist is replaced by a similar resist that is sensitive to a specific range of shorter wavelengths.

Extreme-ultraviolet lithography. The pattern resolution in photolithography is limited by light diffraction. One method of reducing the effects of diffraction is to use ever-shorter wavelengths. *Extreme-ultraviolet* (EUV) *lithography* uses light at a wavelength of 13 nm in order to obtain features around 30 to 100 nm in size. The waves are focused by highly reflective molybdenum/silicon mirrors (instead of glass lenses that absorb EUV light) through the mask to the wafer surface.

X-ray lithography. Although photolithography is the most widely used lithography technique, it has fundamental resolution limitations associated with light diffraction. *X-ray lithography* is superior to photolithography, because of the shorter wavelength of the radiation and the very large depth of focus involved. This characteristic allows much finer patterns to be resolved, and X-ray lithography is far less susceptible to dust. Furthermore, the aspect ratio (defined as the ratio of depth to lateral dimension) can be more than 100 with X-ray lithography but is limited to around 10 with photolithography.

However, in order to achieve this benefit, synchrotron radiation is needed, which is expensive and available at only a few research laboratories. Given the large capital investment required for a manufacturing facility, industry has preferred to refine and improve optical lithography instead of investing new capital in X-ray based production. X-ray lithography is currently not widespread, although the LIGA process (see Section 13.15) fully exploits the benefits of X-ray lithography.

Electron-beam and ion-beam lithography. Like X-ray lithography, *electron-beam* (e-beam) and *ion-beam* (i-beam) *lithography* are superior to photolithography with respect to the resolutions attainable. These methods involve high current density in narrow electron or ion beams (*pencil sources*) that scan a pattern, one pixel at a time, onto a wafer. The masking is done by controlling the point-by-point transfer of the stored pattern, and is therefore performed by software.

These techniques have the advantages of accurate control of exposure over small areas of the wafer, large depth of focus, and low defect densities. Resolutions are limited to around 10 nm, because of electron scatter, although 2-nm resolutions have been reported for some materials. It should be noted, however, that the scan time increases significantly as the resolution increases, because more highly focused beams are required. The main drawback of these techniques is that electron and ion beams need to be maintained in a vacuum, which significantly increases equipment complexity and production cost. Furthermore, the scan time for a wafer for these techniques is much slower than that for other lithographic methods.

SCALPEL. In the SCALPEL process, from *Scattering with Angular Limitation Projection Electron-Beam Lithography* (Fig. 13.14), a mask is produced from a roughly 0.1-μm thick membrane of silicon nitride and patterned with an approximately 50-nm thick coating of tungsten. High-energy electrons pass through both the silicon nitride and the tungsten, but the tungsten scatters the electrons widely, whereas the silicon nitride results in very little scattering. An aperture blocks the scattered electrons, resulting in a high-quality image at the wafer.

The limitation to this process is the small-sized masks that are currently in use, but the process has high potential. Its most significant advantage is the fact that energy does not need to be absorbed by the reticle, but instead, it is blocked by the aperture, which is not as fragile or expensive as the reticle.

FIGURE 13.14 Schematic illustration of the SCALPEL process.

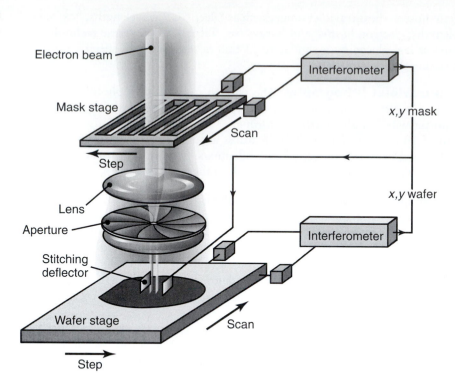

Soft lithography. *Soft lithography* refers to a number of processes for pattern transfer. All processes require that a master mold be created by standard lithography techniques described above. The master mold is then used to produce an elastomeric pattern or stamp as shown in Fig. 13.15. An elastomer that has been commonly used for the stamp is silicone rubber, or polydimethylsiloxane (PDMS), because it is chemically inert and hygroscopic (it does not swell with humidity), has good thermal stability, strength, durability, and surface properties.

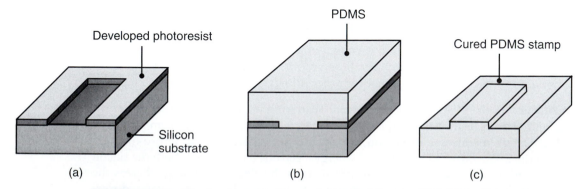

FIGURE 13.15 Production steps for a polydimethylsiloxane (PDMS) mold for soft lithography: (a) A developed photoresist is produced through standard lithography (see Fig. 13.13), (b) a PDMS stamp is cast over the photoresist, (c) the PDMS stamp is peeled off the substrate to produce a stamp. The stamp shown has been rotated to emphasize replication of surface features; the master pattern can be used several times. *Source:* After Y. Xia and G.M. Whitesides.

Several PDMS stamps can be produced from the same pattern, and each stamp can be used many times. Some of the common soft lithography processes are the following:

- **Microcontact printing** (μCP). In *microcontact printing*, the PDMS stamp is coated with an "ink" and then pressed against a surface. The peaks of the pattern are in contact with the opposing surface and a thin layer of the ink is transferred, often only 1 molecule thick (self-assembled monolayer or boundary film, see Section 4.4.3). This thin film can serve as a mask for selective wet etching described below or can be used to impart a desired chemistry onto the surface.

- **Microtransfer molding** (μTM). In this process, shown in Fig. 13.16a, the recesses of the PDMS mold are filled with a liquid polymer precursor and then pushed against a surface. After the polymer has cured, the mold is peeled off, leaving behind a pattern suitable for further processing.

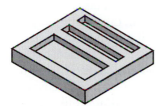

1. Prepare PDMS stamp

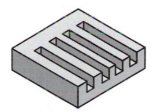

1. Prepare PDMS stamp

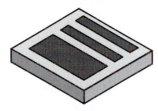

2. Fill cavities with polymer precursor

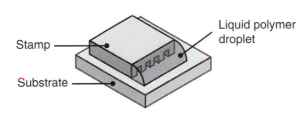

2. Press stamp against surface; apply drop of liquid polymer to end of stamp.

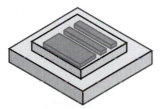

3. Press stamp against surface; allow precursor to cure

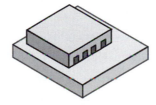

3. Remove excess liquid; allow polymer to cure

4. Peel off stamp

(a)

4. Peel off stamp

(b)

FIGURE 13.16 Soft lithography techniques: (a) microtransfer molding (μTM) and (b) micromolding in capillaries (MIMIC). *Source:* After Y. Xia and G.M. Whitesides.

- **Micromolding in capillaries** (MIMIC). In the MIMIC technique, shown in Fig. 13.16b, the PDMS stamp pattern consists of channels that use capillary action to wick a liquid into the stamp, either from the side of the stamp or from reservoirs within the stamp. The liquid can be a thermosetting polymer, a ceramic sol-gel, or suspensions of solids within liquid solvents. Good pattern replication can occur, as long as the channel aspect ratio is moderate, and the actual channel dimensions allowed depending on the liquid used. The MIMIC process has been used to produce all-polymer field effect transistors and diodes and has various applications in sensors.

13.8 | Etching

Etching is the process by which entire films, or particular sections of films or the substrate, are removed. One of the most important criteria in this important process is **selectivity,** which refers to the ability to etch one material without etching another. Tables 13.2 and 13.3 provide a summary of etching processes. In silicon technology, an etching process must effectively etch the silicon-dioxide layer with minimal removal of the underlying silicon or the resist material. In addition, polysilicon and metals must be etched into high-resolution lines with vertical wall profiles and with minimal removal of the underlying insulating film. Typical etch rates range from tens to several thousands of nm/min, and *selectivities* (defined as the ratio of the etch rates of the two films) can range from 1:1 to 100:1.

13.8.1 Wet etching

Wet etching involves immersing the wafers in a liquid, usually acidic, solution. The main drawback to most wet-etching operations is that they are *isotropic,* that is, they etch in all directions of the workpiece at the same rate. This condition results in undercuts beneath the mask material (see, for example, Fig. 13.17a) and thus limits the resolution of geometric features in the substrate.

Effective etching requires the following conditions:

1. Transport of etchant to the surface.
2. A chemical reaction to remove material.
3. Transporting reaction products away from the surface.
4. Ability to rapidly stop the etching process in order to obtain superior pattern transfer (etch stop), usually using an underlying layer with high selectivity.

If the first or third steps limit the speed of the process, agitation or stirring of the solution can increase etching rates (see also Fig. 9.26a). If the second step limits the speed of the process, the etching rate will strongly depend on temperature, etching material, and solution composition. Reliable etching therefore requires both good temperature control and repeatable stirring capability.

Isotropic etchants are widely used for the following procedures:

1. Removal of damaged surfaces.
2. Rounding of sharp etched corners to avoid stress concentrations.
3. Reduction of roughness after anisotropic etching.
4. Creation of structures in single-crystal slices.
5. Evaluation of defects.

TABLE 13.2

Comparison of Etch Rates

		Etch rate (nm/min)[a]							
Etchant	Target material	Polysilicon n^+	Polysilicon, undoped	SiO_2	SiN	Phospho-silicate glass, annealed	Aluminum	Titanium	Photoresist (OCG-820PR)
Wet etchants									
Concentrated HF (49%)	Silicon oxides	0	—	2300	14	3600	4.2	>1000	0
25:1 HF:H$_2$O	Silicon oxides	0	0	9.7	0.6	150	—	—	0
5:1 BHF[b]	Silicon oxides	9	2	100	0.9	440	140	>1000	0
Silicon etchant (126 HNO$_3$:60H$_2$O:5NH$_4$F)	Silicon	310	100	9	0.2	170	400	300	0
Aluminum etchant (16H$_3$PO$_4$:1HNO$_3$:1HAc:2H$_2$O)	Aluminum	<1	<1	0	0	<1	660	0	0
Titanium etchant (20 H$_2$O:1 H$_2$O$_2$:1HF)	Titanium	1.2	—	12	0.8	210	>10	880	0
Piranha (50 H$_2$SO$_4$:1H$_2$O$_2$)	Cleaning off metals and organics	0	0	0	0	0	180	240	>10
Acetone (CH$_3$COOH)	Photoresist	0	0	0	0	0	0	0	>4000
Dry etchants									
CF$_4$ + CHF$_3$ + He, 450W	Silicon oxides	190	210	470	180	620	—	>1000	220
SF$_6$ + He, 100W	Silicon nitride	73	67	31	82	61*	>1000	69	—
SF$_6$m 125 W	Thin silicon nitrides	170	280	110	280	140	—	>1000	310
O$_2$, 400W	Ashing Photoresist	0	0	0	0	0	0	0	340

Source: After K. Williams and R. Muller, J. *Microelectromechanical Systems*, Vol. 5, 1996, pp. 256–269.

Notes:

[a] Results are for fresh solutions at room temperature unless otherwise noted. Actual etch rates will vary with temperature and prior use of solution, area of exposure of film, other materials present, and film impurities and microstructure.

[b] Buffered hydrofluoric acid, 33% NH4F and 8.3% HF by weight.

TABLE 13.3

General Characteristics of Silicon Etching Operations

	Temperature (°C)	Etch rate (μm/min)	{111}/{100} selectivity	Nitride etch rate (nm/min)	SiO_2 etch rate (nm/min)	p^{++} etch stop
Wet etching						
HF:HNO_3:CH_3COOH	25	1–20	—	Low	10–30	No
KOH	70–90	0.5–2	100:1	<1	10	Yes
Ethylene-diamine pyrochatechol (EDP)	115	0.75	35:1	0.1	0.2	Yes
$N(CH_3)_4OH$ (TMAH)	90	0.5–1.5	50:1	<0.1	<0.1	Yes
Dry (plasma) etching						
SF_6	0–100	0.1–0.5	—	200	10	No
SF_6/C_4F_8 (DRIE)	20–80	1–3	—	200	10	No

Source: Adapted from N. Maluf, *An Introduction to Microelectromechanical Systems Engineering*, Artech House, 2000.

Microelectronic devices and MEMS (see Sections 13.14 through 13.16) require accurate machining of structures; this task is done through masking, which is a challenge with isotropic etchants. The strong acids used will (a) etch aggressively, [at a rate of up to 50 μm/min with an etchant of 66% nitric acid (HNO_3) and 34% hydrofluoric acid (HF), although etch rates of 0.1–1 μm/s are more typical], and (b) produce rounded cavities. Furthermore, the etch rate is very sensitive to agitation and, therefore, lateral and vertical features are difficult to control.

The size of the features in an integrated circuit determines its performance, and for this reason, there is a strong need to produce well-defined, extremely small structures. Such small features cannot be attained through isotropic etching, because of the poor definition that results from undercutting of masks.

Anisotropic etching takes place when etching is strongly dependent on compositional or structural variations in the material. There are two basic kinds of anisotropic etching: *orientation-dependent etching* and *vertical etching*. Most vertical

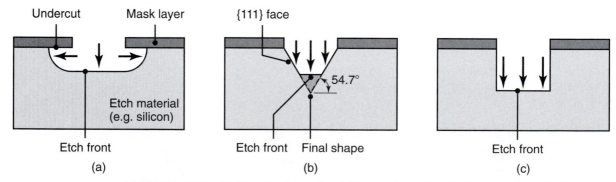

FIGURE 13.17   Etching directionality. (a) Isotropic etching: Etch proceeds vertically and horizontally at approximately the same rate; note the significant mask undercut. (b) Orientation-dependent etching (ODE): Etch proceeds vertically, terminating on {111} crystal planes, with little mask undercut. (c) Vertical etching: Etch proceeds vertically, with little mask undercut. *Source:* Courtesy of K.R. Williams.

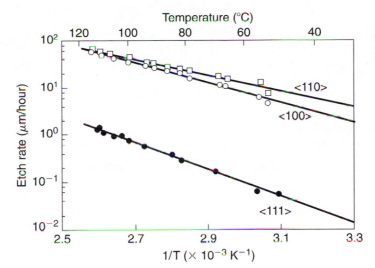

FIGURE 13.18 Etch rates of silicon at different crystallographic orientations, using ethylene-diamine/pyrocatechol-in-water as the solution. *Source:* After H. Seidel et al., *J. Electrochemical Society,* 1990, pp. 3612–3626.

etching is done with dry plasmas, discussed in Section 13.8.2. Orientation-dependent etching commonly occurs in a single crystal when etching takes place at different rates at different directions, as shown in Fig. 13.17b. When orientation-dependent etching is performed properly, the etchants produce geometric shapes with walls defined by the crystallographic planes that resist the etchants. Figure 13.18, for example, shows the vertical etch rate for silicon as a function of temperature. As can be seen, etching rate is more than one order of magnitude lower in the [111] crystal direction than in other directions; hence well-defined walls can be obtained along the [111] crystal direction.

The **anisotropy ratio** (AR) for etching is defined by

$$AR = \frac{E_1}{E_2},$$ (13.4)

where E is the etch rate and the subscripts refer to two crystallographic directions of interest. Recall that selectivity refers to the etch rates between materials of interest. The anisotropy ratio is unity for isotropic etchants and can be as high as 400:200:1 for (110)/(100)/(111) silicon. The {111} planes always etch the slowest, but the {100} and {110} plane etch rates can be controlled through etchant chemistry.

Masking is also a concern in anisotropic etching, but silicon oxide is less valuable as a mask material for different reasons than in isotropic etching. Anisotropic etching is slower than isotropic etching (typically 3 μm/min; 120 μin./min), and thus anisotropic etching through a wafer may take several hours. Silicon oxide may etch too rapidly to use as a mask, and hence a high-density silicon-nitride mask may be needed.

Often, it is important to rapidly halt the etching process (*etch stop*). This situation is typically the case when thin membranes are to be manufactured or when features with very precise thicknesses are needed. Conceptually, this task can be accomplished by removing the wafer from the etching solution. However, etching depends to a great extent on the ability to circulate fresh etchants to the desired locations. Since the circulation varies across the wafer surface, this strategy for halting the etching process would lead to large variations in etch depth.

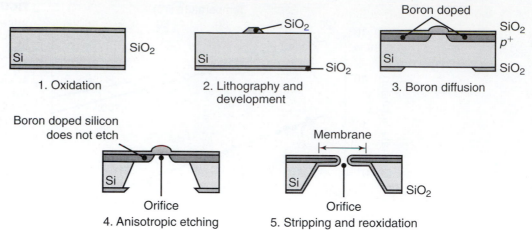

FIGURE 13.19   Application of a boron etch stop and back etching to form a membrane and orifice. *Source:* After I. Brodie and J.J. Murray.

The most common approach for producing uniform feature sizes across a wafer is to use a boron etch stop, where a boron layer is diffused or implanted into silicon. Examples of common etch stops include the placement of a boron-doped layer beneath silicon and the placement of silicon oxide (SiO_2) beneath silicon nitride (Si_3N_4). Since anisotropic etchants do not attack boron-doped silicon as aggressively as they do undoped silicon, surface features or membranes can be created by **back etching.** Figure 13.19 shows an example of the boron etch-stop approach.

Numerous etchant formulations have been developed. Some of the more common wet etchants are summarized next.

1. Silicon dioxide is commonly etched with hydrofluoric (HF) acid solutions. The driving chemical reaction in pure HF etching is

$$SiO_2 + 6HF \rightarrow H_2SiF_6 + 2H_2O. \qquad (13.5)$$

It is, however, rare that silicon dioxide is etched purely through the reaction in Eq. (13.5). Hydrofluoric acid is a weak acid, and it does not completely dissociate into hydrogen and fluorine ions in water. HF_2^- is an additional ion that exists in hydrofluoric acid, and HF_2^- attacks silicon oxide about 4.5 times faster than does HF alone. The reaction involving the HF_2^- ion is

$$SiO_2 + 2HF_2^- + H^+ \rightarrow SiF_6^{2-} + 2H_2O. \qquad (13.6)$$

The pH value of the etching solution is critical, because acidic solutions have sufficient hydrogen ions to dissociate the HF_2^- ions into HF ions. As HF and HF_2^- are consumed, the etch rate decreases. For that reason, a buffer of ammonium fluoride (NH_4F) is used to maintain the pH and, thus, keep the concentrations of HF and HF_2^- constant, stabilizing the etch rate. Such an etching solution is referred to as a *buffered hydrofluoric acid* (BHF) or *buffered oxide etch* (BOE) and involves the reaction

$$SiO_2 + 4HF + 2NH_4F \rightarrow (NH_4)_2SiF_6 + 2H_2O. \qquad (13.7)$$

2. Silicon nitride is etched with phosphoric acid (H_3PO_4), usually at an elevated temperature, typically 160°C (320°F). The etch rate of phosphoric acid decreases with water content, so a *reflux system* is used to return condensed water vapor to the solution to maintain a constant etch rate.

3. Etching silicon often involves mixtures of nitric acid (HNO_3) and hydrofluoric acid (HF). Water can be used to dilute these acids. The preferred buffer is acetic acid because it preserves the oxidizing power of HNO_3; this system is referred to as an *HNA etching system*. A simplified description of this etching process is that the nitric acid oxidizes the silicon and then the hydrofluoric acid removes the silicon oxide. This two-step process is a common approach for chemical machining of metals. (See Section 9.10.) The overall reaction is

$$18HF + 4HNO_3 + 3Si \rightarrow 3H_2SiF_6 + 4NO + 8H_2O. \tag{13.8}$$

The etch rate is limited by the silicon-oxide removal; a buffer of ammonium fluoride is therefore usually used to maintain etch rates.

4. Anisotropic etching, or orientation-dependent etching, of single-crystal silicon can be done with solutions of potassium hydroxide, although other etchants have also be used. The reaction is

$$Si + 2OH^- + 2H_2O \rightarrow SiO_2(OH)_2^{2-} + 2H_2. \tag{13.9}$$

Note that this reaction does not require potassium to be the source of the OH ions; KOH attacks {111}-type planes much more slowly than other planes. Isopropyl alcohol is sometimes added to KOH solutions to reduce etch rates and increase the uniformity of etching. KOH is also extremely valuable in that it stops etching when it contacts a very heavily doped *p*-type material (*boron etch stop;* see Fig. 13.19).

5. Aluminum is etched through a solution typically consisting of 80% phosphoric acid (H_3PO_4), 5% nitric acid (HNO_3), 5% acetic acid (CH_3COOH), and 10% water. The nitric acid first oxidizes the aluminum, and the oxide is then removed by the phosphoric acid and water. This solution can be masked with a photoresist.

6. Wafer cleaning is accomplished through *Piranha solutions*, which have been in use for decades. These solutions consist of hot mixtures of sulfuric acid (H_2SO_4) and peroxide (H_2O_2). Piranha solutions strip photoresist and other organic coatings and remove metals on the surface, but do not affect silicon dioxide or silicon nitride, thus making it an ideal cleaning solution. Bare silicon forms a thin layer of hydrous silicon oxide, which is removed through a short dip in hydrofluoric acid.

7. Although photoresist can be removed through Piranha solutions, acetone is commonly used for this purpose instead. Acetone dissolves the photoresist, but it should be noted that if the photoresist is excessively heated during a process step, it will be significantly more difficult to remove with acetone. In such a case, the photoresist can be removed through an ashing plasma.

EXAMPLE 13.1 Processing of a *p*-type region in *n*-type silicon

Assume that we want to create a *p*-type region within a sample of *n*-type silicon. Draw cross sections of the sample at each processing step in order to accomplish this task.

Solution.   See Fig. 13.20. This simple device is known as a *pn junction diode,* and the physics of its operation is the foundation for most semiconductor devices.

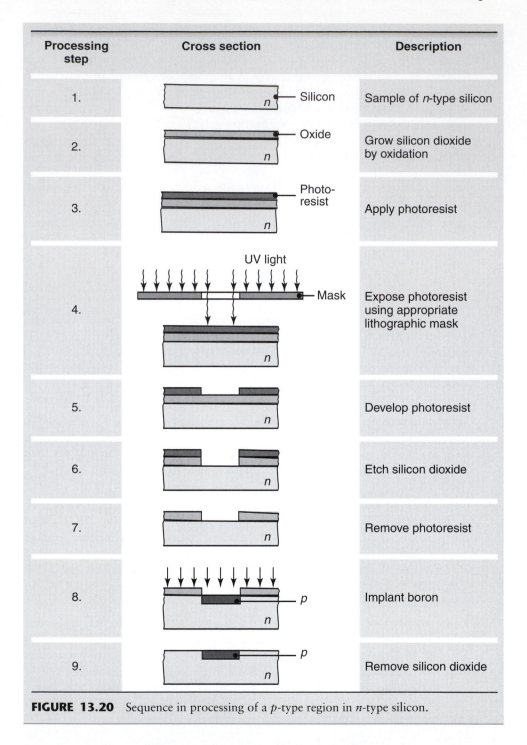

Processing step	Cross section	Description
1.	*n* — Silicon	Sample of *n*-type silicon
2.	— Oxide *n*	Grow silicon dioxide by oxidation
3.	Photo-resist *n*	Apply photoresist
4.	UV light — Mask *n*	Expose photoresist using appropriate lithographic mask
5.	*n*	Develop photoresist
6.	*n*	Etch silicon dioxide
7.	*n*	Remove photoresist
8.	*p* *n*	Implant boron
9.	*p* *n*	Remove silicon dioxide

FIGURE 13.20   Sequence in processing of a *p*-type region in *n*-type silicon.

13.8.2 Dry etching

Modern integrated circuits are etched exclusively through *dry etching*, which involves the use of chemical reactants in a low-pressure system. In contrast to the wet process, described above, dry etching can have a high degree of directionality, resulting in highly anisotropic etch profiles (Fig. 13.17c). Furthermore, the dry etching process requires only small amounts of the reactant gases, whereas the solutions used in the wet etching process have to be refreshed periodically. Dry etching usually involves a

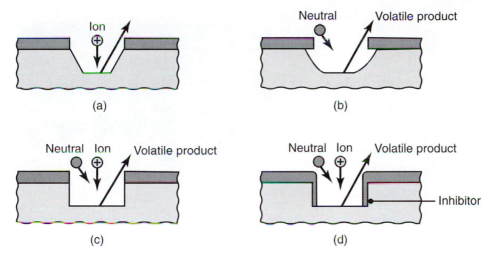

FIGURE 13.21   Machining profiles associated with different dry-etching techniques: (a) sputtering, (b) chemical, (c) ion-enhanced energetic, and (d) ion-enhanced inhibitor. *Source:* After M. Madou.

plasma or discharge in areas of high electric and magnetic fields; any gases that are present are dissociated to form ions, electrons, or highly reactive molecules.

There are several specialized dry-etching techniques.

1. **Sputter etching.** *Sputter etching* removes material by bombarding it with noble-gas ions, usually Ar^+. The gas is ionized in the presence of a cathode and anode (Fig. 13.21). If a silicon wafer is the target, the momentum transfer associated with bombardment of atoms causes bond breakage and material to be ejected (sputtered). If the silicon chip is the substrate, the material in the target is deposited onto the silicon after it has been sputtered by the ionized gas. Some of the concerns with sputter etching are

   a. The ejected material can be redeposited onto the target, especially when the aspect ratios are large.

   b. Sputter etching is not material selective; most materials sputter at about the same rate, and therefore masking is difficult.

   c. Sputter etching is slow, with etch rates limited to tens of nm/min.

   d. Sputtering can cause damage to or excessive erosion of the material.

   e. The photoresist is difficult to remove.

2. **Reactive plasma etching.** Also referred to as *dry chemical etching, reactive plasma etching* involves chlorine or fluorine ions (generated by RF excitation) and other molecular species that diffuse to and chemically react with the substrate. The volatile compound formed is removed by a vacuum system. The mechanism of reactive plasma etching is illustrated in Fig. 13.22. Here, a reactive species is produced, such as CF_4, dissociating upon impact with energetic electrons to produce fluorine atoms (step 1). The reactive species then diffuse to the surface (step 2), become adsorbed (step 3), and chemically react to form a volatile compound (step 4). The reactant then desorbs from the surface (step 5) and diffuses into the bulk gas, where it is removed by the vacuum system.

   Some reactants polymerize on the surface and require additional removal, either with oxygen in the plasma reactor or by an external ashing operation. The electrical charge of the reactive species is not high enough to cause damage

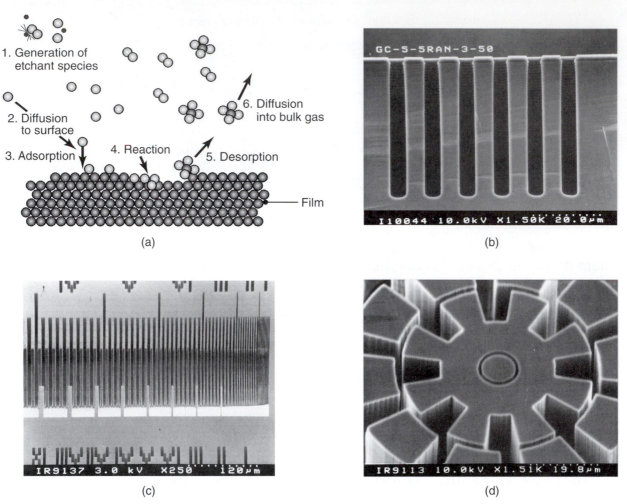

FIGURE 13.22 (a) Schematic illustration of reactive plasma etching, (b) example of deep reactive ion etched trench; note the periodic undercuts, or scalloping. (c) Near vertical sidewalls produced through DRIE with an anisotropic etching process, and (d) an example of cryogenic dry etching, showing a 145-μm deep structure etched into Si using a 2.0-μm thick oxide masking layer. The substrate temperature was $-140°$C during etching. *Source:* (a) After M. Madou. (b) to (d) Courtesy of R. Kassing and I.W. Rangelow, University of Kassel, Germany.

through impact on the surface, so that no sputtering occurs. Thus, the etching is isotropic, and undercutting of the mask takes place (Fig. 13.17a). Table 13.2 lists some of the more commonly used dry etchants, their target materials, and their typical etch rates.

3. **Physical-chemical etching.** Processes such as *reactive ion-beam etching* (RIBE) and *chemically assisted ion-beam etching* (CAIBE) combine the advantages of physical and chemical etching. Although these processes use chemically reactive species to remove material, these procedures are physically assisted by the impact of ions onto the surface. In RIBE, also known as *deep reactive ion etching* (DRIE), vertical trenches hundreds of micrometers deep can be produced by periodically interrupting the etch process and depositing a polymer layer. When performed with an isotropic dry etching process, this operation results in scalloped sidewalls, as shown in Fig. 13.22b. Anisotropic DRIE can produce near-vertical sidewalls (Fig. 13.22c).

In CAIBE, ion bombardment can assist dry chemical etching by

a. Making the surface more reactive.

b. Clearing the surface of reaction products and allowing the chemically reactive species access to the cleared areas.

c. Providing the energy to drive surface chemical reactions; however, the neutral species do most of the etching.

Physical-chemical etching is extremely useful, because the ion bombardment is directional, so that etching is anisotropic. Also, the ion bombardment energy is low and does not contribute significantly to mask removal. This factor allows generation of near-vertical walls with very large aspect ratios. Since the ion bombardment does not directly remove material, masks can be used.

4. **Cryogenic dry etching.** This procedure is an approach used to obtain very deep features with vertical walls. The workpiece is lowered to cryogenic temperatures, and chemically assisted ion-beam etching takes place. The low temperatures involved ensure that sufficient energy is not available for a surface chemical reaction to take place, unless the ion bombardment direction is normal to the surface. Oblique impacts, such as occur on side walls in deep crevices, cannot drive the chemical reactions; therefore, very smooth vertical walls can be produced, as shown in Fig. 13.22d.

Because dry etching is not selective, etch stops cannot be directly applied. Dry-etch reactions must be terminated when the target film is removed. Optical emission spectroscopy is often used to determine the "end point" of a reaction. Filters can be used to capture the wavelength of light emitted during a particular reaction. A noticeable change in light intensity at the point of etching will be detected.

EXAMPLE 13.2 Comparison of wet and dry etching

Consider a case where a $\langle 100 \rangle$ wafer has an oxide mask placed on it in order to produce square or rectangular holes. The sides of the square mask are precisely oriented with the $\langle 110 \rangle$ direction (see Fig. 13.7) of the wafer surface, as shown in Fig. 13.23. Describe the possible hole geometries that can be produced by etching.

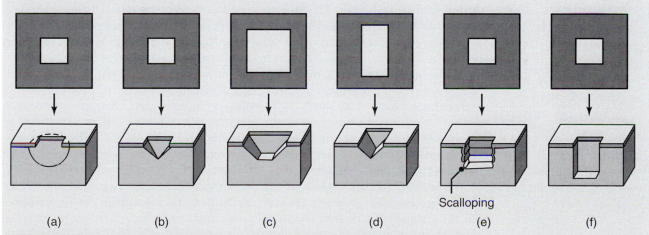

Scalloping

(a)   (b)   (c)   (d)   (e)   (f)

FIGURE 13.23 Various types of holes generated from a square mask in (a) isotropic (wet) etching, (b) orientation-dependent etching (ODE), (c) ODE with a larger hole, (d) ODE of a rectangular hole, (e) deep reactive ion etching, and (f) vertical etching. *Source:* After M. Madou.

Solution. Isotropic etching results in the cavity shown in Fig. 13.23a. Since etching occurs at constant rates in all directions, a rounded cavity is produced that undercuts the mask. An orientation-dependent etchant produces the cavity shown in Fig. 13.23b. Since etching is much faster in the $\langle 100 \rangle$ and $\langle 110 \rangle$ directions than in the $\langle 111 \rangle$ direction, sidewalls defined by the $\{111\}$ plane are generated. For silicon, for example, these sidewalls are at an angle of 54.74° to the surface.

The effect of a larger mask or shorter etch time is shown in Fig. 13.23c. The resultant pit is defined by $\langle 111 \rangle$ sidewalls and a bottom in the $\langle 100 \rangle$ direction parallel to the surface. A rectangular mask and the resultant pit are shown in Fig. 13.23d. Deep reactive ion etching is depicted in Fig. 13.23d. Note that a polymer layer is periodically deposited onto the sidewalls of the hole to allow for deep pockets, but scalloping (greatly exaggerated in the figure) is unavoidable. A hole produced from chemically reactive ion etching is shown in Fig. 13.23f.

13.9 | Diffusion and Ion Implantation

As stated in Section 13.3, the operation of microelectronic devices often depends on regions of different doping types and concentrations. The electrical character of these regions can be altered by introducing dopants into the substrate, accomplished by *diffusion* and *ion implantation* processes. Since many different regions of microelectronic devices must be defined, this step in the fabrication sequence is repeated several times.

In the diffusion process, the movement of atoms results from thermal excitation. Dopants can be introduced to the substrate in the form of a deposited film, or the substrate can be exposed to a vapor containing the dopant source. This operation takes place at elevated temperatures, usually 800°–1200°C (1500°–2200°F). Dopant movement within the substrate is a function of temperature, time, and the diffusion coefficient (or diffusivity) of the dopant species, as well as the type and quality of the substrate material. Because of the nature of diffusion, dopant concentration is very high at the surface and drops off sharply away from the surface.

To obtain a more uniform concentration within the substrate, the wafer is heated further to drive in the dopants, a process called **drive-in diffusion**. The fact that diffusion, desired or undesired, will always occur at high temperatures is invariably taken into account during subsequent processing steps. Although the diffusion process is relatively inexpensive, it is highly isotropic.

Ion implantation is a much more extensive process and requires specialized equipment (Fig. 13.8c). Implantation is accomplished by accelerating ions through a high-voltage beam of as much as 1 million volts and then choosing the desired dopant by means of a magnetic mass separator. In a manner similar to that used in cathode-ray tubes, the beam is swept across the wafer by sets of deflection plates, thus ensuring uniform coverage of the substrate. The complete implantation system must be operated in a vacuum.

The high-velocity impact of ions on the silicon surface damages the lattice structure, resulting in lower electron mobilities. This condition is undesirable, but the damage can be repaired by an annealing step, which involves heating the substrate to relatively low temperatures, usually 400°–800°C (750°–1500°F), for 15–30 min. This process provides the energy that the silicon lattice requires to rearrange and mend itself. Another important function of annealing is to allow the dopant to move from the interstitial to substitutional sites (see Fig. 3.9), where they are electrically active.

13.10 | Metallization and Testing

The generation of a complete and functional integrated circuit requires that devices be *interconnected*; this must take place on a number of levels, as illustrated in Fig. 13.24. **Interconnections** are made by metals that exhibit low electrical resistance and good adhesion to dielectric (insulator) surfaces. Aluminum and aluminum-copper alloys remain the most commonly used materials for this purpose in VLSI (very large scale integration) technology today. However, as device dimensions continue to shrink, electromigration has become more of a concern with aluminum interconnects.

Electromigration is the process by which metal atoms are physically moved by the impact of drifting electrons under high-current conditions. Low-melting-point metals such as aluminum are especially prone to electromigration. In extreme cases, this condition can lead to severed and/or shorted metal lines. Solutions to this problem include the addition of sandwiched metal layers such as tungsten and titanium, and more recently, the use of pure copper, which displays lower resistivity and significantly better antielectromigration performance than does aluminum.

Metals are deposited by standard deposition techniques, and interconnection patterns are generated by lithographic and etching processes. Modern ICs typically have one to four layers of metallization, each layer of metal being insulated by a

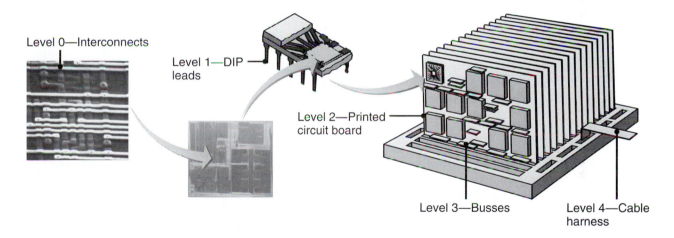

Level	Element example	Interconnection method
Level 0	Transistor within an IC	IC metallization
Level 1	ICs, other discrete components	Package leads or module interconnections
Level 2	IC packages	Printed circuit board
Level 3	Printed circuit boards	Connectors (busses)
Level 4	Chassis or box	Connectors/cable harnesses
Level 5	System, e.g., computer	

FIGURE 13.24   Connections between elements in the hierarchy for integrated circuits.

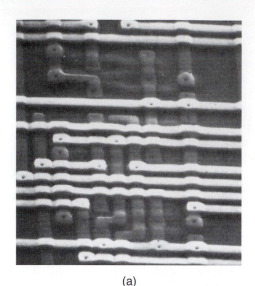

(a)

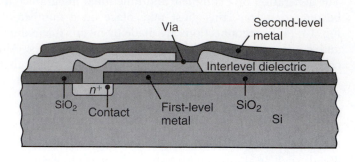

(b)

FIGURE 13.25   (a) Scanning electron microscope photograph of a two-level metal interconnect; note the varying surface topography. (b) Cross section of a two-level metal interconnect structure. *Source:* After R.C. Jaeger.

dielectric, either silicon oxide or borophosphosilicate glass. **Planarization** (producing a planar surface) of these interlayer dielectrics is critical to reducing metal shorts and linewidth variation of the interconnect. A common method for achieving a planar surface has been a uniform oxide etch process that smoothens out the peaks and valleys of the dielectric layer.

However, today's standard for planarizing high-density interconnects is **chemical-mechanical polishing** (see Section 9.7), also called *chemical-mechanical planarization*. This process involves physically polishing the wafer surface, similar to a disc or belt sander polishing the ridges on a piece of wood. A typical CMP process combines an abrasive medium with a polishing compound or slurry and can polish a wafer to within 0.03 μm (1.2 μin.) of being perfectly flat, with a R_q roughness (see Section 4.4) on the order of 0.1 nm for a new, bare silicon wafer.

Different layers of metal are connected together by **vias**, and access to the devices on the substrate is achieved through **contacts** (Fig. 13.25). As devices have become smaller and faster, the size and speed of some chips have become dominated by the resistance of the metallization process itself and the capacitance of the dielectric and the transistor gate. Wafer processing is completed upon application of a *passivation layer*, usually silicon nitride (Si_3N_4), which acts as an ion barrier for sodium ions and also provides excellent scratch resistance.

The next step in production is to test each of the individual circuits on the wafer (Fig. 13.26). Each chip, also referred to as a **die**, is tested with a computer-controlled platform which contains needlelike probes to access the bonding pads on the die. The probes are of two forms:

1. **Test patterns or structures.** The probe measures test structures, often outside of the active die, placed in the so-called *scribe line* (the vacant space between dies). These structures consist of transistors and interconnect structures that measure various quantities such as resistivity, contact resistance, and electromigration.

2. **Direct probe.** This approach uses 100% testing on the bond pads of each die.

The platform indexes across the wafer, testing whether each circuit functions properly, using computer-generated timing wave forms. If a defective chip is

FIGURE 13.26   A probe (top center) checking for defects in a wafer; an ink mark is placed on each defective die. *Source:* Courtesy of Intel Corp.

encountered, it is marked with a drop of ink. Up to one-third of the cost of a micro-machined part can be incurred during this testing.

After the wafer-level testing is complete, back grinding may be done to remove a large amount of the original substrate on the opposite side of the wafer from the fabricated integrated circuit. Final die thickness depends on the packaging require-ment, but anywhere from 25 to 75% of the wafer thickness can be removed. After back grinding, each die is separated from the wafer. Diamond sawing is a commonly used separation technique and results in very straight edges, with minimal chipping and cracking damage. The chips are then sorted, with the passing packaged and the inked dice discarded.

13.11 | Wire Bonding and Packaging

The working dice must be attached to a more rugged foundation to ensure reliability. One simple method is to *bond* a die to its packaging material with an *epoxy cement* (see Section 12.14). Another method uses a *eutectic bond*, made by heating metal-alloy systems. One widely used mixture is 96.4% gold and 3.6% silicon, which has a eutectic point (see Fig. 5.4) at 370°C (700°F). Once the chip has been bonded to its substrate, it must be electrically connected to the package leads. This task is accom-plished by *wire bonding* very thin [25-μm (0.001-μin.) diameter] gold wires from the package leads to bonding pads located around the perimeter or down the center of the die (Fig. 13.27a). The bonding pads on the die are typically 50 μm (2000 μin.) or more per side, and the bond wires are attached using thermocompression, ultra-sonic, or thermosonic techniques (Fig. 13.28).

The connected circuit is now ready for final *packaging*. The packaging opera-tion largely determines the overall cost of each completed IC, since the circuits are mass produced on the wafer but are then packaged individually. Packages are

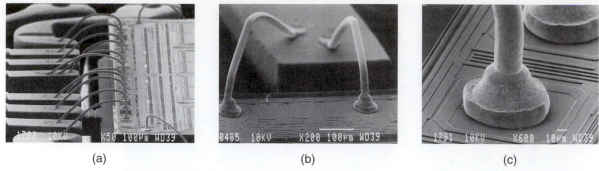

(a)   (b)   (c)

FIGURE 13.27   (a) SEM photograph of wire bonds connecting package leads (left-hand side) to die bonding pads. (b) and (c) Detailed views of (a). *Source:* Courtesy of Micron Technology, Inc.

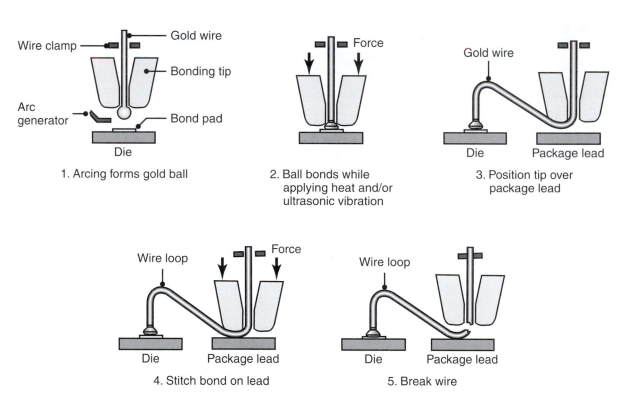

FIGURE 13.28   Schematic illustration of the thermosonic ball and stitch process. *Source:* After N. Maluf.

available in a variety of styles (Table 13.4), and selection of the appropriate one must take into account operating requirements. Consideration of a circuit's package includes consideration of chip size, the number of external leads, operating environment, heat dissipation, and power requirements. ICs that are used for military and industrial applications, for example, require packages with particularly high strength, toughness, hermicity, and high-temperature resistance.

Packages are produced from polymers, metals, or ceramics. Metal containers are produced from alloys such as Kovar (an iron-cobalt-nickel alloy with a low coefficient of thermal expansion; Section 3.9.5) and provide a hermetic seal and good thermal conductivity but are limited in the number of leads that can be accommodated. Ceramic packages are usually produced from Al_2O_3, are hermetic, and

TABLE 13.4

Summary of Molded-Plastic IC Packages

Package	Abbreviation	Number of pins		Description
		min.	max.	
Through-hole mount				
Dual in-line	DIP	8	64	Two in-line rows of leads.
Single in-line	SIP	11	40	One in-line row of leads.
Zigzag in-line	ZIP	16	40	Two rows with staggered leads.
Quad in-line package	QUIP	16	64	Four in-line rows of staggered leads.
Surface mount				
Small-outline IC	SOIC	8	28	Small package with leads on two sides.
Thin small-outline package	TSOP	26	70	Thin version of SOIC.
Small-outline J-lead	SOJ	24	32	Same as SOIC, with leads in a J-shape.
Plastic leaded chip carrier	PLCC	18	84	J-shaped leads on four sides.
Thin quad flat pack	TQFP	32	256	Wide but thin package with leads on four sides.

have good thermal conductivity, with higher lead counts than metal packages. They are, however, more expensive than metal packages. Plastic packages are inexpensive, with high lead counts, but they have high thermal resistance and are not hermetic.

An older but still common style of packaging is the **dual in-line package** (DIP), shown schematically in Fig. 13.29a. Characterized by low cost (if plastic) and ease of handling, DIP packages are made of thermoplastic, epoxy, or ceramic, and they can have from 2 to 500 external leads. Ceramic packages are designed for use over a broader temperature range for high-performance and military applications and cost considerably more than plastic packages. A flat ceramic package is shown in Fig. 13.29b, in which the package and all the leads are in the same plane. This package style does not offer the ease of handling or the modular design of the DIP package; therefore, it is usually affixed permanently to a multiple-level circuit board in which the low profile of the flat pack is essential.

Surface-mount packages have become the standard for modern integrated circuits. Some common examples are shown in Fig. 13.29c, where it can be seen that the main difference among them is in the shape of the connectors. The DIP connection to the surface board is by way of prongs that are inserted into corresponding holes, whereas a surface mount is soldered onto specially fabricated pad or land designs. Package size and land layouts are selected from standard patterns and usually require adhesive bonding of the package to the board, followed by wave soldering of the connections (see Section 12.14.3).

Faster and more versatile chips require increasingly closely spaced connections. **Pin-grid arrays** (PGAs) use tightly packed pins that connect by way of through-holes onto printed circuit boards. However, PGAs, and indeed other in-line and surface-mount packages, are extremely susceptible to plastic deformation of the wires and legs, especially with small-diameter closely spaced wires. One way to achieve tight packing of connections and avoid the difficulties of slender connections is through **ball-grid arrays** (BGAs), shown in Fig. 13.29d. Such arrays have a plated solder coating on a number of closely spaced metal balls on the underside of the package. The spacing between the balls can be as small as 50 μm (2000 μin.), but more commonly, the spacing is standardized as 1.0 mm (0.040 in.), 1.27 mm (0.050 in.), or 1.5 mm (0.060 in.).

BGAs can be designed with over 1000 connections, but such high numbers of connections are extremely rare; usually, 200–300 connections are sufficient for demanding applications. With reflow soldering (see Section 12.14.3), the solder

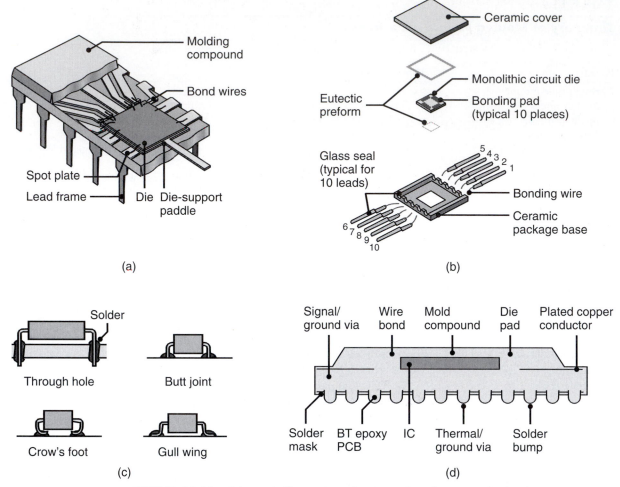

FIGURE 13.29 Schematic illustration of various IC packages: (a) dual in-line (DIP), (b) ceramic flat pack, (c) common surface-mount configurations, and (d) ball-grid array (BGA). *Source:* After R.C. Jaeger, A.B. Glaser, and G.E. Subak-Sharpe.

serves to center the BGAs by surface tension, resulting in well-defined electrical connections for each ball.

Flip-chip technology (FCT) refers to the attachment procedure of ball-grid arrays and is depicted in Fig. 13.30. The final encapsulation with an epoxy is necessary, not only to more securely attach the IC package to the printed circuit board, but also to evenly distribute thermal stresses during its operation.

FIGURE 13.30
Illustration of flip-chip technology: (a) flip-chip package with solder-plated metal balls and pads on the printed circuit board, (b) flux application and placement, (c) reflow soldering, and (d) encapsulation.
Source: After P.K. Wright.

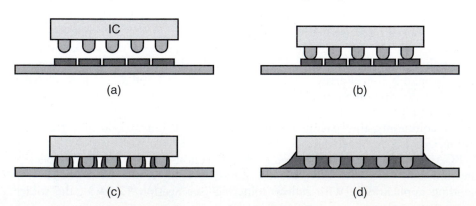

After the chip has been sealed in the package, it undergoes final testing. Because one of the main purposes of packaging is isolation from the environment, testing at this stage usually encompasses heat, humidity, mechanical shock, corrosion, and vibration. Destructive tests (Section 4.8.2) are also performed to investigate the effectiveness of sealing.

13.12 | Yield and Reliability of Chips

Yield is defined as the ratio of functional chips to the total number of chips produced. The overall yield of the total IC manufacturing process is the product of the wafer yield, bonding yield, packaging yield, and test yield. This value can range from only a few percent for new processes to above 90% for mature manufacturing lines, with most loss of yield occurring during wafer processing due to its complex nature. In this stage, wafers are commonly separated into regions of good and bad chips. Failures at this stage can arise from point defects (such as oxide pinholes), film contamination, or metal particles, as well as from area defects, such as uneven film deposition or etch nonuniformity.

A major concern about completed ICs is their **reliability** and **failure rate.** Since no device has an infinite lifetime, statistical methods are used to characterize the expected lifetimes and failure rates of microelectronic devices (see also *Six Sigma,* Section 4.9.1). The unit for failure rate is the *failure in time* (FIT), defined as the number of failures per one billion device-hours. However, complete systems may have millions of devices, so the overall failure rate in entire systems is correspondingly higher. Failure rates higher than 100 FIT are generally unacceptable.

Equally important in failure analysis is determination of the failure mechanism, that is, the actual incident or component that causes the device to fail. Common failures due to processing involve (a) diffusion regions, leading to nonuniform current flow and junction breakdown, (b) oxide layers (dielectric breakdown and accumulation of surface charge), (c) lithography (uneven definition of features and mask misalignment), and (d) metal layers (poor contact and electromigration resulting from high current densities). Other failures can originate from improper chip mounting, degradation of wire bonds, and loss of package hermeticity. Wire-bonding and metallization failures account for over one-half of all integrated-circuit failures.

Because device lifetimes are very long (10 years or more), it is impractical to study device failure under normal operating conditions. One method of studying failures efficiently is by **accelerated life testing,** which involves accelerating the conditions whose effects are known to cause device breakdown. Cyclic variations in temperature, humidity, voltage, and current are used to stress the components. Statistical data taken from these tests are then used to predict device failure modes and device life under normal operating conditions. Chip mounting and packaging are strained by cyclical temperature variations.

13.13 | Printed Circuit Boards

Packaged integrated circuits are seldom used alone; they are usually combined with other ICs to serve as building blocks of a yet larger system. A *printed circuit board* (PCB) is the *substrate* for the final interconnections among all completed chips and serves as the communication link between the busses to other PCBs and the microelectronic circuitry within each packaged IC (see Fig. 13.24). In addition to the ICs,

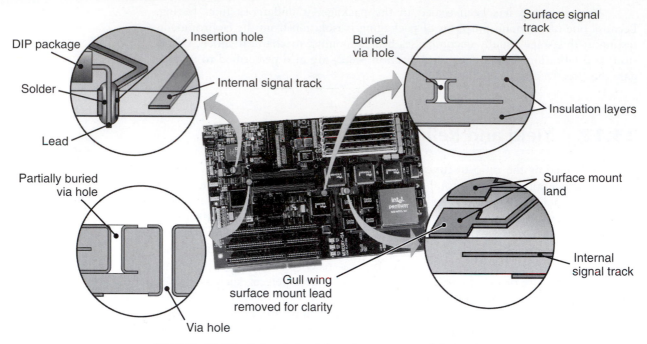

FIGURE 13.31   Printed circuit board structures and design features.

circuit boards also usually contain discrete circuit components, such as resistors and capacitors, that would (a) take up too much "real estate" on the limited silicon surface, (b) have special power dissipation requirements, or (c) cannot be implemented on a chip. Other common discrete components include inductors, which cannot be integrated onto the silicon surface, and high-performance transistors, large capacitors, precision resistors, and crystals for frequency control.

A printed circuit board is basically a plastic resin material containing several layers of copper foil (Fig. 13.31). *Single-sided* PCBs have copper tracks on only one side of an insulating substrate, while *double-sided* boards have copper tracks on both sides. Multilayer boards can also be constructed from alternating copper and insulator layers. Single-sided boards are the simplest form of circuit board. Double-sided boards usually must have locations where electrical connectivity is established between the features on both sides of the board. This structure is accomplished with **vias,** as shown in Fig. 13.31. Multilayer boards can have partial, buried, or through-hole vias to allow for flexible PCBs. Double and multilayer boards are beneficial in that IC packages can be bonded to both sides of the board, allowing more compact designs.

The insulating material of the board is usually an epoxy resin 0.25 to 3 mm (0.01 to 0.12 in.) thick, reinforced with an epoxy/glass fiber, referred to as E-glass (see Section 10.9.2). They are produced by impregnating sheets of glass fiber with epoxy and then pressing the layers together between hot plates or rolls. The heat and pressure cure the board, resulting in a stiff and strong basis for printed circuit boards. Boards are sheared to a desired size, and approximately 3-mm-diameter locating holes are then drilled or punched into the board corners to permit alignment and proper location of the board within chip-insertion machines. Holes for vias and connections are punched or produced through CNC drilling (Section 14.3); stacks of boards can be drilled simultaneously to increase production rates.

The conductive patterns on circuit boards are defined by lithography, although originally they were produced through screen-printing technologies, hence the terms *printed circuit board* and *printed wiring board* (PWB). In the *subtractive method*, a copper foil is bonded to the circuit board. The desired pattern on the board is then defined by a positive mask developed through photolithography, and the remaining copper is removed through wet etching. In the *additive method,* a negative mask is placed directly onto an insulator substrate to define the desired shape. Electroless plating and electroplating of copper (see Section 4.5.1) serve to define the connections, tracks, and lands on the circuit board.

The ICs and other discrete components are then fastened to the board by soldering. This procedure is the final step in making the integrated circuits (and the microelectronic devices they contain) accessible. *Wave soldering* and *reflow paste soldering* (Section 12.14.3) are the preferred methods of soldering ICs onto circuit boards.

Basic design considerations in laying out PCBs are

1. Wave soldering should be used only on one side of the board; thus all through-hole mounted components should be inserted from the same side of the board. Surface-mount devices placed on the insertion side of the board must be reflow soldered in place because this side is not exposed to the solder. Surface-mount devices on the leg side can be wave soldered.

2. To allow good solder flow in wave soldering, IC packages should carefully be laid out on the printed circuit board. Inserting the packages in the same direction is advantageous for automated placing, whereas random orientations can cause difficulties in flow of solder across all of the connections.

3. The spacing of ICs is determined mainly by the need to remove heat during operation. Sufficient clearance between packages and adjacent boards is required to allow forced air flow and heat convection.

4. There should be sufficient space around each IC package to allow for rework and repair without disturbing adjacent devices.

13.14 | Micromachining of MEMS Devices

The topics described thus far have dealt with the manufacture of integrated circuits and products that operate based purely on electrical or electronic principals. These processes also are suitable for manufacturing devices that incorporate mechanical elements or features as well.

The following types of devices can be made through the approach described in Fig. 13.3:

1. **Microelectronic devices.** These semiconductor-based devices often have the common characteristic of extreme miniaturization and use electrical principles in their design.

2. **Micromechanical devices.** This term refers to a product that is purely mechanical in nature and has dimensions between atomic length scales and a few millimeters, such as some very small gears and hinges.

3. **Microelectromechanical devices.** These products combine mechanical and electrical or electronic elements at very small length scales. Most sensors are examples of microelectromechanical devices.

4. **Microelectromechanical systems** (MEMS). These microelectromechanical devices also incorporate an integrated electrical system in one product. Microelectromechanical systems are rare compared to microelectronic, micromechanical, or microelectromechanical devices, typical examples being air-bag sensors and digital micromirror devices, as discussed in the Case Study for this chapter.

It should be noted that microelectronic devices are semiconductor-based, whereas microelectromechanical devices and portions of MEMS do not have this material restriction. This allows the use of many more materials and the development of processes suitable for these materials. Regardless, silicon is often used because several highly advanced and reliable manufacturing processes have been developed for microelectronic applications.

The production of features from micrometers to millimeters in size is called micromachining. MEMS devices have been constructed from **polycrystalline silicon (polysilicon)** and **single-crystal silicon** because the technologies for integrated-circuit manufacture, described earlier in this chapter, are well developed and exploited for these devices, and other, new processes have been developed that are compatible with the existing processing steps. The use of anisotropic etching techniques allows the fabrication of devices with well-defined walls and high aspect ratios, and for this reason, some single-crystal silicon MEMS devices have been fabricated.

One of the recognized difficulties associated with using silicon for MEMS devices is the high adhesion encountered at small length scales and the associated rapid wear. Most commercial devices are designed to avoid friction by, for example, using flexing springs instead of bushings. However, this approach complicates designs and makes some MEMS devices unfeasible. Consequently, significant research is being conducted to identify materials and lubricants that provide reasonable life and performance.

Silicon carbide, diamond, and metals such as aluminum, tungsten, and nickel have been investigated as potential MEMS materials. Lubricants have also been investigated. It is known that surrounding the MEMS device in a silicone oil, for example, practically eliminates adhesive wear (see Section 4.4.2), but it also limits the performance of the device. Self-assembling layers of polymers are also being investigated, as well as novel and new materials with self-lubricating characteristics. However, the tribology of MEMS devices remains a main technological barrier to expansion of their already widespread use.

The area of MEMS and MEMS devices is rapidly advancing, and new processes or variations on existing processes are continually being developed. It has been suggested that MEMS technology can have widespread industrial applications; however, only a few industries, such as the computer, medical, and automotive industries have exploited MEMS thus far. Many of the processes described in this chapter have not yet become widely applied, but are of interest to researchers and practitioners in MEMS.

EXAMPLE 13.3 Accelerometer and control system for airbags

An example of a commercial MEMS-based product is the accelerometer and control system for airbag deployment in an automobile, as illustrated in Fig. 13.32. Such accelerometers, based on the principle of lateral resonators, represent one of the largest commercial applications of MEMS today and are widely used as sensors for automotive air bag deployment systems.

A central mass is suspended over the substrate but is anchored through four slender beams which act as springs to center the mass under static-equilibrium conditions. An acceleration causes the mass to deflect, reducing or increasing the clearance between the fins on the mass and the stationary fingers on the substrate. By measuring the electrical capacitance between the mass and fins, the deflection of the mass, and therefore the acceleration or deceleration of the system, can be directly measured. Figure 13.32 shows an arrangement for the measurement of

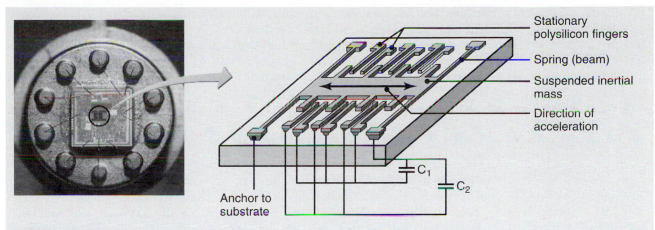

FIGURE 13.32   The Analog Devices ADXL-50 accelerometer. This MEMS-based product contains a surface-micromachined sensor, on-chip excitation, self-test, and signal-controlling circuitry. The schematic illustration shows the structure of the suspended mass. The entire chip measures 0.500×0.625 mm. *Source:* Courtesy of Analog Devices, Inc.

acceleration in one direction, but commercial sensors employ several masses so that accelerations can be measured in multiple directions simultaneously.

Figure 13.32 shows the 50-g surface micromachined accelerometer (ADXL-50) with on-board signal-conditioning and self-diagnostic electronics. The polysilicon sensing element (visible in the center of the die) occupies only 5% of the total die area, and the whole chip measures 0.500 by 0.625 mm (0.020 by 0.025 in.). The mass is approximately $0.3 \, \mu$g, and the sensor has a measurement accuracy of 5% over the ±50-g range.

13.14.1   Bulk micromachining

Until the early 1980s, bulk micromachining was the most common form of machining at micrometer scales. This machining process uses orientation-dependent etches on single-crystal silicon. (See Fig. 13.17b.) The approach is based on etching down into a surface, stopping on certain crystal faces, doped regions, and etchable films to form the required structure. As an example of this process, consider the fabrication of the silicon cantilever shown in Fig. 13.33. By using the masking techniques described in Section 13.7, a rectangular patch of the *n*-type silicon substrate is changed to *p*-type silicon through boron doping. Recall that etchants for orientation-dependent

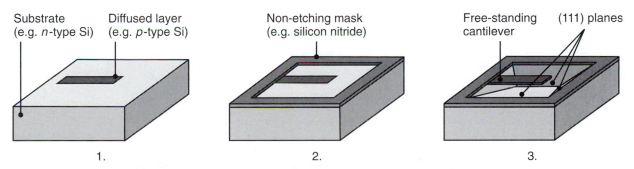

FIGURE 13.33   Schematic illustration of the steps in bulk micromachining: (1) diffuse dopant in desired pattern, (2) deposit and pattern masking film, and (3) orientation-dependent etch, leaving behind a freestanding structure. *Source:* After K.R. Williams.

etching, such as potassium hydroxide, will not be able to etch heavily boron-doped silicon, hence this patch will not be etched.

A mask is then produced, such as with silicon nitride on silicon. When etched with potassium hydroxide, the undoped silicon will be removed rapidly, while the mask and the doped patch will essentially be unaffected. Etching progresses until the (111) planes are exposed in the *n*-type silicon substrate, and they undercut the patch, leaving a suspended cantilever, as shown in the figure.

13.14.2 Surface micromachining

Although bulk micromachining is useful for producing very simple shapes, it is restricted to single-crystal materials, since polycrystalline materials will not machine at different rates in different directions when using wet etchants. Many MEMS applications require the use of other materials, so that alternatives to bulk micromachining are needed. One such method is *surface micromachining*, the basic steps of which are illustrated in Fig. 13.34 for silicon devices. A spacer or sacrificial layer is deposited onto a silicon substrate coated with a thin dielectric layer (called an *isolation* or *buffer layer*).

Phosphosilicate glass, deposited by chemical vapor deposition, is the most common material for a spacer layer, because it etches very rapidly in hydrofluoric acid. Figure 13.34b shows the spacer layer after the application of masking and etching. At this stage, a structural thin film is deposited onto the spacer layer; the film can be polysilicon, metal, metal alloy, or a dielectric (Fig. 13.34c). The structural film is then patterned, usually through dry etching in order to maintain vertical walls and tight dimensional tolerances. Finally, wet etching of the sacrificial layer leaves a free-standing, three-dimensional structure, as shown in Fig. 13.34e. It should be noted that the wafer must be annealed to remove the residual stresses in

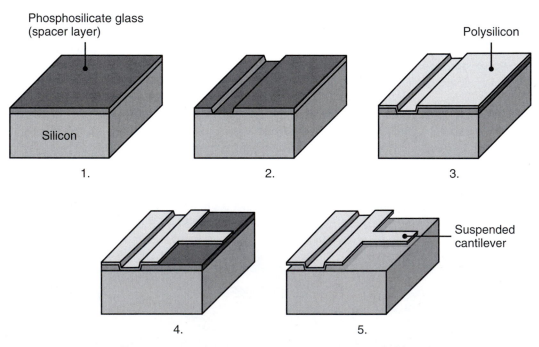

FIGURE 13.34 Schematic illustration of the steps in surface micromachining:
(1) deposition of a phosphosilicate glass (PSG) spacer layer, (2) etching of spacer layer,
(3) deposition of polysilicon, (4) etching of polysilicon, and (5) selective wet etching of PSG,
leaving the silicon substrate and the deposited polysilicon unaffected.

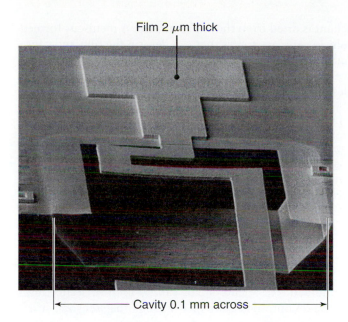

Film 2 μm thick

Cavity 0.1 mm across

FIGURE 13.35 A microlamp produced by a combination of bulk and surface micromachining processes. *Source:* After K.R. Williams, Agilent Technologies.

the deposited metal before it is patterned, as otherwise the structural film will severely warp once the spacer layer is removed.

Figure 13.35 shows a microlamp that emits a white light when current is passed through it. This part was produced through a combination of surface and bulk micromachining. The top patterned layer is a 2.2-μm thick layer of plasma-etched tungsten, forming a meandering filament and bond pad. The rectangular overhang is dry-etched silicon nitride. The steeply sloped layer is wet-HF-etched phosphosilicate glass. The substrate is silicon, which is ODE etched.

The etchant used to remove the spacer layer must be carefully chosen. It must preferentially dissolve the spacer layer while leaving the dielectric, the silicon, and the structural film as intact as possible. With large features and narrow spacer layers, this task becomes very difficult to do, and etching can take several hours. To reduce the etch time, additional etch holes can be designed into the microstructures to increase access of the etchant to the spacer layer.

Another difficulty that must be overcome in this operation is **stiction** after wet etching. Consider the situation illustrated in Fig. 13.36. After the spacer layer has

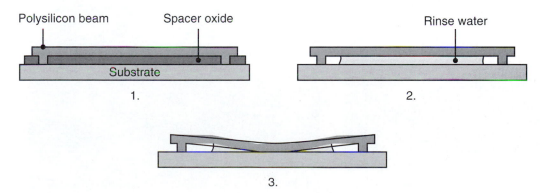

Polysilicon beam   Spacer oxide   Rinse water

Substrate

1.

2.

3.

FIGURE 13.36 Stiction after wet etching: (1) unreleased beam, (2) released beam before drying, and (3) released beam pulled to the surface by capillary forces during drying. Once contact is made, adhesive forces prevent the beam from returning to its original shape. *Source:* After B. Bhushan.

been removed, the liquid etchant is dried from the wafer surface. A meniscus formed between the layers then results in capillary forces that can deform the film and cause it to contact the substrate as the liquid evaporates. Since adhesion forces are more significant at small length scales, it is possible that the film may permanently stick to the surface, and thus the desired three-dimensional features will not be produced.

EXAMPLE 13.4 Surface micromachining of a hinge for a mirror actuation system

Surface micromachining is a very common technology for the production of microelectromechanical systems. Applications include accelerometers, pressure sensors, micropumps, micromotors, actuators, and microscopic locking mechanisms. Often these devices require very large vertical walls, which cannot be directly manufactured because the high vertical structure is difficult to deposit. This difficulty is overcome by machining large flat structures horizontally, and then rotating or folding them into an upright position, as shown in Fig. 13.37.

Figure 13.37a shows a micromirror that has been inclined with respect to the surface on which it was manufactured. Such systems can be used for reflecting light (that is oblique to a surface) onto detectors or towards other sensors. It is apparent that a device which has such depth, and has the aspect ratio of the deployed mirror, is very difficult to machine directly. Instead, it is easier to surface micromachine the mirror along with a linear actuator, and then fold the mirror into a deployed position. In order to do so, special hinges, as shown in Fig. 13.37b, are integrated into the design.

Figure 13.38 shows the hinge during its manufacture, requiring the following steps:

1. A 2-μm-thick layer of phosphosilicate glass (PSG) is deposited onto the substrate material.

2. A 2-μm-thick layer of polysilicon (Poly1 in Fig. 13.38a) is deposited onto the PSG, patterned by photolithography, and is dry etched to form the desired structural elements, including the hinge pins.

3. A second layer of sacrificial PSG, with a thickness of 0.5 μm, is deposited (Fig. 13.38b).

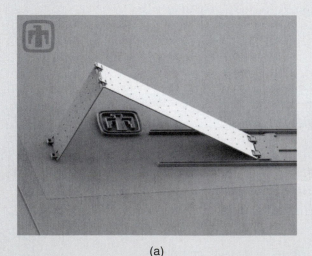

(a)

(b)

FIGURE 13.37   SEM image of a deployed micromirror and (b) detail of the micromirror hinge. *Source:* Sandia National Laboratories.

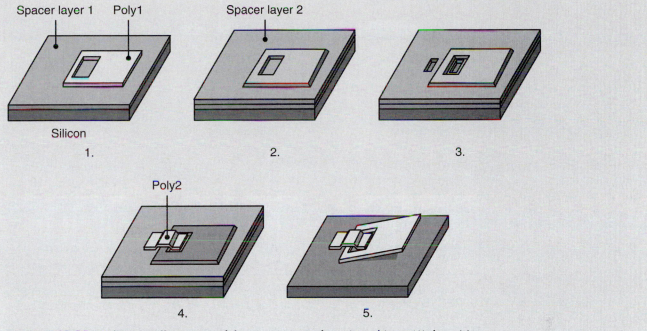

FIGURE 13.38 Schematic illustration of the steps in manufacturing a hinge: (1) deposition of a phosphosilicate glass (PSG) spacer layer and polysilicon layer (see Fig. 13.34), (2) deposition of a second spacer layer, (3) selective etching of the PSG, (4) deposition of polysilicon to form a staple for the hinge, and (5) rotation of the hinge after selective wet etching of the PSG.

4. The connection locations are etched through both layers of PSG (Fig. 13.38c).

5. A second layer of polysilicon (Poly2 in Fig. 13.38d) is deposited, patterned, and etched.

6. The sacrificial layers of PSG are then removed through wet etching.

Hinges such as these have very high friction. Consequently, if mirrors as shown are manually and carefully manipulated with probe needles, they will remain in position. Often, such mirrors are combined with linear actuators to precisely control their deployment.

SCREAM. Another approach for making very deep MEMS structures is the *single-crystal silicon reactive etching and metallization* (SCREAM) process, depicted in Fig. 13.39. In this technique, standard lithography and etching processes produce trenches 10–50 μm (400–2000 μin.) deep, which are then protected by a layer of chemically vapor-deposited silicon oxide. An anisotropic etch step removes the oxide only at the bottom of the trench, and the trench is then extended through dry etching. An isotropic etch, using sulfur hexafluoride (SF_6), laterally etches the exposed sidewalls at the bottom of the trench. This undercut, when it overlaps adjacent undercuts, releases the machined structures.

SIMPLE. An alternative to SCREAM is the *silicon micromachining by single-step plasma etching* (SIMPLE) technique, as depicted in Fig. 13.40. This technique uses a chlorine-gas-based plasma etch process that machines *p*-doped or lightly doped silicon anisotropically, but heavily *n*-doped silicon isotropically. A suspended MEMS device can thus be produced in one plasma etching step, as shown in the figure.

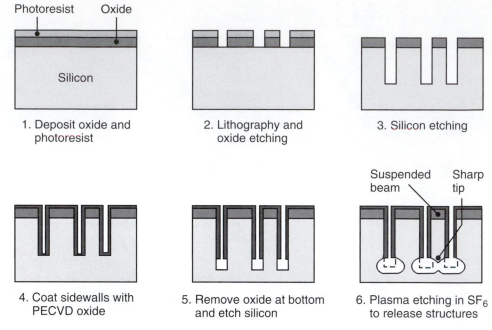

1. Deposit oxide and photoresist

2. Lithography and oxide etching

3. Silicon etching

4. Coat sidewalls with PECVD oxide

5. Remove oxide at bottom and etch silicon

6. Plasma etching in SF_6 to release structures

FIGURE 13.39 Steps involved in the SCREAM process. *Source:* After N. Maluf.

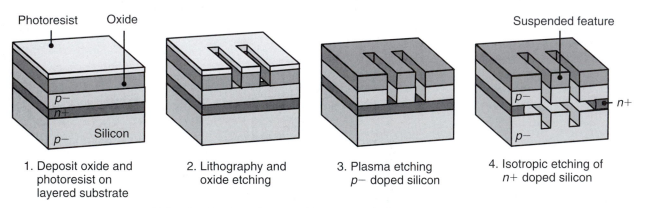

1. Deposit oxide and photoresist on layered substrate

2. Lithography and oxide etching

3. Plasma etching $p-$ doped silicon

4. Isotropic etching of $n+$ doped silicon

FIGURE 13.40 Schematic illustration of silicon micromachining by the single-step plasma etching (SIMPLE) process.

Some of the concerns regarding the SIMPLE process are

1. The oxide mask is machined, although at a slower rate, by the chlorine-gas plasma; therefore, relatively thick oxide masks are required.

2. The isotropic etch rate is low, typically 50 nm/min; consequently, this process is very slow.

3. The layer beneath the structures will contain deep trenches, which may affect the motion of free-hanging structures.

Etching combined with fusion bonding. The terms fusion bonding and diffusion bonding (see Section 12.12) are interchangeable, but fusion bonding is the preferred term for MEMS applications. Very tall structures can be produced in single crystal silicon through a combination of *silicon fusion bonding and deep reactive ion etching* (SFB-DRIE), as illustrated in Fig. 13.41. First, a silicon wafer is prepared with

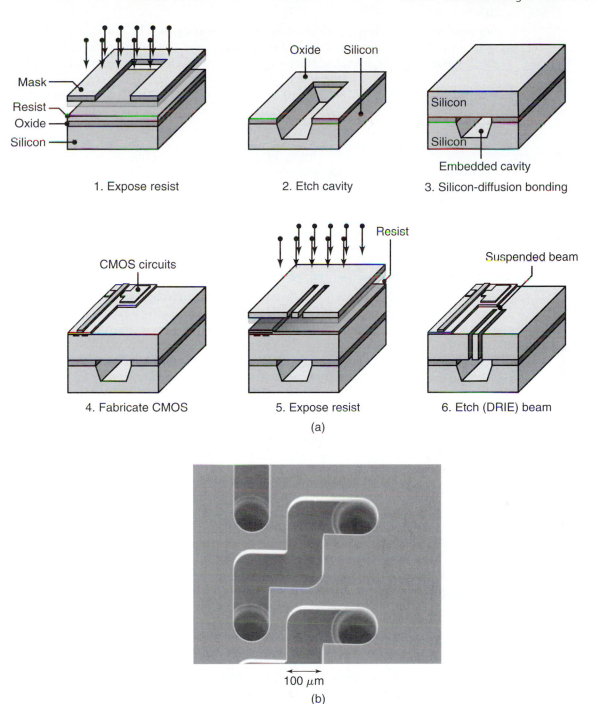

FIGURE 13.41 (a) Schematic illustration of silicon fusion bonding combined with deep reactive ion etching to produce suspended cantilever beams. (b) A micro-fluid-flow device manufactured by applying the DRIE process to two separate wafers and then aligning and silicon fusion bonding them together. Afterward, a Pyrex layer (not shown) is anodically bonded over the top to provide a window for observing fluid flow. *Source:* (a) After N. Maluf, (b) after K.R. Williams.

an insulating oxide layer, with the deep trench areas defined by a standard lithography procedure. This step is followed by conventional wet or dry etching to form a large cavity. A second layer of silicon is fusion bonded to this layer, which can then be ground and polished to the desired thickness, if necessary. At this stage, integrated circuitry is manufactured through the steps outlined in Fig. 13.3. A protective resist is applied and exposed, and the desired trenches are etched by deep reactive ion etching through to the cavity in the first layer of silicon.

EXAMPLE 13.5 Operation and fabrication sequence for a thermal ink-jet printer

Thermal ink-jet printers are among the most successful applications of MEMS to date. These printers operate by ejecting nano- or picoliters (10^{-12} liters) of ink from a nozzle towards paper. Ink-jet printers use a variety of designs, but silicon machining technology is most applicable to high-resolution printers. It should be noted that a resolution of 1200 dots per inch (dpi) requires a nozzle pitch of approximately 20 μm (800 μin.).

The operation of an ink-jet printer is illustrated in Fig. 13.42. When an ink droplet is to be generated and expelled, a tantalum resistor below a nozzle is heated. This resistor heats a thin film of ink so that a bubble forms within 5 microseconds. The bubble expands rapidly, with internal pressures reaching 1.4 MPa (200 psi), and as a result, fluid is rapidly forced out of the nozzle. Within 24 microseconds, the tail of the ink droplet separates because of surface tension. The heat source is then removed (turned off), and the bubble collapses inside the nozzle. Within 50 microseconds, sufficient ink has been drawn from a reservoir into the nozzle to form the desired meniscus for the next droplet.

Traditional ink-jet printer heads were produced with electroformed nickel nozzles, fabricated separately from the integrated circuitry, and required a bonding

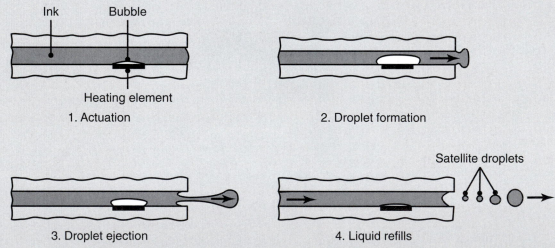

1. Actuation

2. Droplet formation

3. Droplet ejection

4. Liquid refills

FIGURE 13.42   Sequence of operations of a thermal ink-jet printer. (1) Resistive heating element is turned on, rapidly vaporizing the ink and forming a bubble. (2) Within 5 microseconds, the bubble has expanded and displaced liquid ink from the nozzle. (3) Surface tension breaks the ink stream into a bubble, which is discharged at high velocity. The heating element is turned off at this time, so that the bubble collapses as heat is transferred to the surrounding ink. (4) Within 24 microseconds, an ink droplet (and undesirable satellite droplets) are ejected, and surface tension of the ink draws more liquid from the reservoir. *Source:* From F.G. Tseng, "Microdroplet Generators," in M. Gad-el-hak (ed.), *The MEMS Handbook*, CRC Press, 2002.

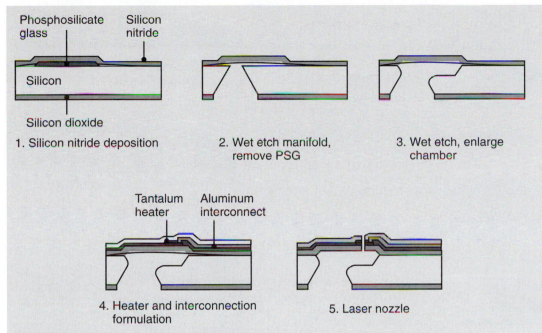

FIGURE 13.43   The manufacturing sequence for producing thermal ink-jet printer heads. *Source:* From F.G. Tseng, "Microdroplet Generators," in M. Gad-el-hak (ed.), *The MEMS Handbook,* CRC Press, 2002.

operation to attach these two components. With increasing printer resolution, however, it is more difficult to bond the components with a tolerance under a few micrometers. For this reason, single-component, or monolithic, fabrication is of interest.

The fabrication sequence for a monolithic ink-jet printer head is shown in Fig. 13.43. A silicon wafer is prepared and coated with a phosphosilicate-glass (PSG) pattern and low-stress silicon-nitride coating. The ink reservoir is obtained by isotropically etching the back side of the wafer, followed by PSG removal and then enlargement of the reservoir. The required CMOS (complimentary metal-oxide-semiconductor) controlling circuitry is then produced (this step is not shown in Fig. 13.43), and a tantalum heater pad is deposited. The aluminum interconnection between the tantalum pad and the CMOS circuit is formed, and the nozzle is produced through laser machining. An array of such nozzles can be placed inside an ink-jet printing head, and resolutions of 2400 dpi or higher can be achieved.

13.15 | LIGA and Related Microfabrication Processes

LIGA is a German acronym for the combined processes of X-ray lithography, electrodeposition, and molding (**X-ray Lithographie, Galvanoformung und Abformung**). The LIGA process, illustrated in Fig. 13.44, involves the following steps:

1. A very thick (up to hundreds of microns) resist layer of polymethylmethacrylate (PMMA) is deposited onto a primary substrate.

2. The PMMA is exposed to collimated X-rays and developed.

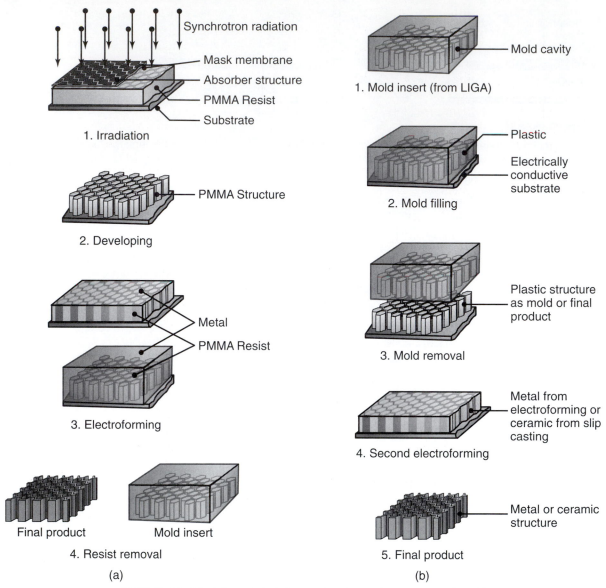

FIGURE 13.44 The LIGA (lithography, electrodeposition, and molding) technique: (a) primary production of a metal product or mold insert and (b) use of the primary part for secondary operations, or replication. *Source:* Courtesy of IMM Institute für Mikrotechnik.

   3. Metal is electrodeposited onto the primary substrate.

   4. The PMMA is removed or stripped, resulting in a freestanding metal structure.

   5. Plastic injection molding is done in the metal structure, which acts as a mold.

Depending on the application, the final product from a LIGA process may be

   1. A freestanding metal structure, resulting from the electrodeposition process.

   2. A plastic injection-molded structure.

   3. An investment-cast metal part, where the injection molded structure was used as a blank.

   4. A slip-cast ceramic part, produced with the injection-molded parts as the molds.

The substrate used in LIGA is an electrical conductor or a conductor-coated insulator. Examples of primary substrate materials include austenitic steel plate; silicon wafers with a titanium layer; and copper plated with gold, titanium, or nickel. Metal-plated ceramic and glass have also been used. The surface may be roughened by grit blasting to encourage good adhesion of the resist material.

Resist materials must have high X-ray sensitivity, dry- and wet-etching resistance when unexposed, and thermal stability. The most common resist material is polymethylmethacrylate, which has a very high molecular weight (more than 10^6 g per mole; see Section 10.2.1). The X-rays break the chemical bonds, leading to the generation of free radicals and significantly reduced molecular weight in the exposed region. Organic solvents then preferentially dissolve the exposed PMMA in a wet etching process. After development, the remaining three-dimensional structure is rinsed and dried, or it is spun and blasted with dry nitrogen.

Electrodeposition of metal usually involves nickel (see Section 4.5.1 and Fig. 4.17). The nickel is deposited onto exposed areas of the substrate; it fills the PMMA structure and can even coat the resist (Fig. 13.44a). Nickel is the material of choice because of the relative ease in electroplating with well-controlled deposition rates and residual stress control. Electroless plating of nickel (Section 4.5.1) is also possible, and the nickel can be deposited directly onto electrically insulating substrates. However, because nickel displays high wear rates in MEMS, significant research continues to be directed toward the use of other materials or coatings.

After the metal structure has been deposited, precision grinding removes either the substrate material or a layer of the deposited nickel, a process referred to as *planarization* (see also Section 13.10). The need for planarization is obvious when it is recognized that three-dimensional MEMS devices require micrometer tolerances on layers several hundreds of micrometers thick. Planarization is difficult to achieve, and conventional lapping (Section 9.7) leads to preferential removal of the soft PMMA and smearing of the metal. Planarization is usually accomplished with a diamond lapping procedure, referred to as *nanogrinding*. Here, a diamond slurry loaded, soft-metal plate is used to remove material in order to maintain flatness within 1 μm (40 μin.) over a 75-mm (3-in.) diameter substrate.

If cross-linked (Section 10.2.1), the PMMA resist is then exposed to synchrotron X-ray radiation and removed by exposure to an oxygen plasma or through solvent extraction. The result is a metal structure, which may be used for further processing. Examples of freestanding metal structures produced through electrodeposition of nickel are shown in Fig. 13.45.

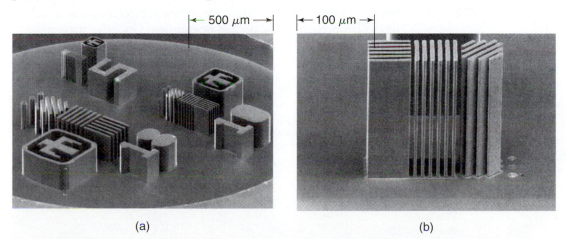

(a)   (b)

FIGURE 13.45   (a) Electroformed nickel structures and (b) detail of nickel lines and spaces. *Source:* After T. Christenson, *The MEMS Handbook*, CRC Press, 2002.

TABLE 13.5

Comparison of Micromold Manufacturing Techniques

	Production technique		
	LIGA	Laser machining	EDM
Aspect ratio	10–50	10	up to 100
Surface roughness	<50 nm	100 nm	0.3–1 μm
Accuracy	<1 μm	1–3 μm	1–5 μm
Mask required?	Yes	No	No
Maximum height	1–500 μm	200–500 μm	μm to mm

Source: L. Weber, W. Ehrfeld, H. Freimuth, M. Lacher, M. Lehr, and P. Pech, *SPIE Micromachining and Microfabrication Process Technology II*, Austin, TX, 1996.

Although the processing steps used to make freestanding metal structures are extremely time consuming and expensive, the main advantage of LIGA is that these structures serve as molds for the rapid replication of submicron features through molding operations. Table 13.5 lists and compares the processes that can be used for producing micromolds, where it can be seen that LIGA provides some distinct advantages. Reaction injection molding, injection molding, and compression molding processes (Section 10.10) have also been used to make these micromolds.

EXAMPLE 13.6 Production of rare-earth magnets

A number of scaling issues in electromagnetic devices indicate that there is an advantage in using rare-earth magnets from the samarium cobalt (SmCo) and neodymium iron boron (NdFeB) families. These materials are available in powder form and are of interest because they can produce magnets that are an order of magnitude more powerful than conventional magnets (Table 13.6). Thus, these materials can be used when effective miniature electromagnetic transducers are to be produced.

The processing steps used to manufacture these magnets are shown in Fig. 13.46. The polymethylmethacrylate mold is produced by exposure to X-ray radiation and solvent extraction. The rare-earth powders are mixed with an epoxy binder and applied to the PMMA mold through a combination of calendering (see Fig. 10.38) and pressing. After curing in a press at a pressure around 70 MPa (10 ksi), the substrate is planarized. The substrate is then subjected to a magnetizing field of at least 35 kilooersteds in the desired orientation. Once the material has been magnetized, the PMMA substrate is dissolved, leaving behind the rare-earth magnets, as shown in Fig. 13.47.

Source: After T. Christenson, Sandia National Laboratories.

TABLE 13.6

Comparison of Properties of Permanent-Magnet Materials

Material	Energy product (Gauss-Oersted $\times 10^{-6}$)
Carbon steel	0.20
36% cobalt steel	0.65
Alnico I	1.4
Vicalloy I	1.0
Platinum-cobalt	6.5
$Nd_2Fe_{14}B$, fully dense	40
$Nd_2Fe_{14}B$, bonded	9

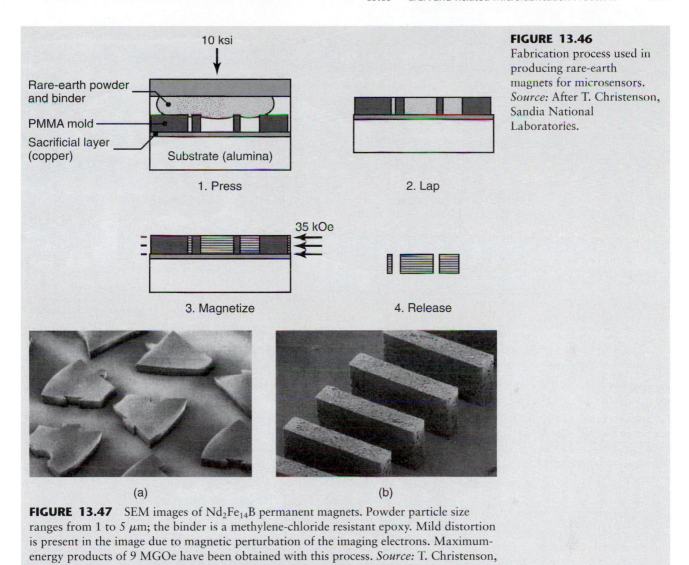

FIGURE 13.46
Fabrication process used in producing rare-earth magnets for microsensors. *Source:* After T. Christenson, Sandia National Laboratories.

FIGURE 13.47 SEM images of $Nd_2Fe_{14}B$ permanent magnets. Powder particle size ranges from 1 to 5 μm; the binder is a methylene-chloride resistant epoxy. Mild distortion is present in the image due to magnetic perturbation of the imaging electrons. Maximum-energy products of 9 MGOe have been obtained with this process. *Source:* T. Christenson, in Gad-el-Hak (ed.), *The MEMS Handbook,* CRC Press, 2002.

Multilayer X-ray lithography. The LIGA technique is very suitable for producing MEMS devices with large aspect ratios and reproducible shapes. Often, however, a multilayer stepped structure that cannot be made directly through LIGA is required. For nonoverhanging geometries, direct plating can be applied. In this technique, a layer of electrodeposited metal with surrounding PMMA is produced as described for LIGA above. A second layer of PMMA resist is then bonded to this structure and X-ray exposed with an aligned X-ray mask.

It is often useful to have overhanging geometries within complex MEMS devices. A batch diffusion bonding and release procedure has been developed for this purpose, schematically illustrated in Fig. 13.48a. This process involves the preparation of two PMMA patterned and electroformed layers, with the PMMA being subsequently removed. The wafers are then aligned, face to face, with guide pins that press-fit into complimentary structures on the opposite surface. Next, the substrates are joined in a hot press, and a sacrificial layer on one substrate is etched away, leaving one layer bonded to the other. An example of such a structure is shown in Fig. 13.48b.

HEXSIL. The HEXSIL process, illustrated in Fig. 13.49, combines HEXagonal honeycomb structures, SILicon micromachining, and thin-film deposition to produce

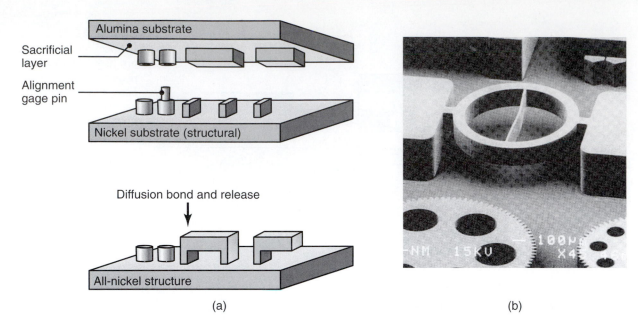

Sacrificial layer

Alignment gage pin

(a)   (b)

FIGURE 13.48   (a) Multilevel MEMS fabrication through wafer-scale diffusion bonding and (b) a suspended ring structure, for measurement of tensile strain, formed by two-layer wafer-scale diffusion bonding. *Source:* After T. Christenson, Sandia National Laboratories.

FIGURE 13.49
Illustrations of the HEXagonal honeycomb structure, SILicon micromachining, and thin-film deposition (HEXSIL process).

1. Etch deep in silicon wafer

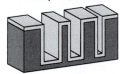

2. Deposit sacrificial oxide

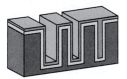

3. Deposit undoped poly

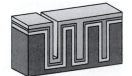

4. Deposit in-situ doped poly

5. Blanket etch planar surface layer to oxide

6. Deposit electroless nickel

7. Lap and polish to oxide layer

8. HF etch release and mold ejection

9. Go to step 2: Repeat mold cycle

Wafer
Sacrificial oxide
Undoped poly
Doped poly
Electroless nickel

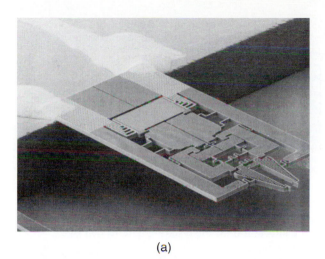
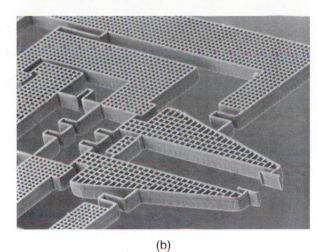

(a)   (b)

FIGURE 13.50   (a) SEM image of micro-scale tweezers, used in microassembly and microsurgery and (b) detailed view of the tweezers. *Source:* Courtesy of MEMS Precision Instruments.

high aspect-ratio, free-standing structures. HEXSIL can produce tall structures with a shape definition that rivals LIGA's capabilities.

In HEXSIL, a deep trench is first produced in single-crystal silicon, using dry etching, and followed by a shallow wet etch to make the trench walls smoother. The depth of the trench matches the desired structure height and is practically limited to around 100 μm. An oxide layer is then grown or deposited onto the silicon, followed by an undoped polycrystalline silicon layer, which leads to good mold filling and good shape definition. A doped silicon layer is then deposited, to provide for a resistive portion of the microdevice. Electroplated or electroless nickel plate is then deposited. Fig. 13.49 shows various trench widths to demonstrate the different structures that can be produced in HEXSIL.

Figure 13.50 shows microtweezers produced through the HEXSIL process. A thermally activated bar is used to activate the tweezers, which have been used for microassembly and microsurgery applications.

13.16 | Solid Freeform Fabrication of Devices

Solid freeform fabrication is another term for rapid prototyping, described in Section 10.12. This method is unique in that complex three-dimensional structures are produced through additive manufacturing, as opposed to material removal. Many of the advances in rapid prototyping are also applicable to MEMS manufacture.

Microstereolithography.   Recall that *stereolithography* involves curing of a liquid thermosetting polymer, using a photoinitiator and a highly focused light source. Conventional stereolithography uses layers between 75 and 500 μm (0.003 in. and 0.02 in.) in thickness, with a laser dot focused to a 0.25-mm (0.01 in.) diameter.

The **microstereolithography** process uses the same approach, but the laser is more highly focused, to a diameter as small as 1 μm (40 μin.), and layer thicknesses are about 10 μm (400 μin.). This technique has a number of economic advantages, but the MEMS devices are difficult to integrate with the controlling circuitry because stereolithography produces nonconducting polymer structures.

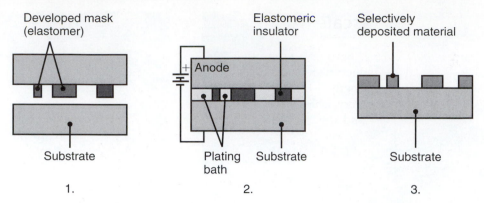

FIGURE 13.51 The instant masking process: (1) bare substrate, (2) during deposition, with the substrate and instant mask in contact, and (3) the resulting pattern deposit. *Source:* After A. Cohen, MEMGen Corporation.

Instant masking is another technique for producing MEMS devices (Fig. 13.51). The solid freeform fabrication of MEMS devices by instant masking is also known as **electrochemical fabrication** (EFAB). A mask of elastomeric material is first produced through the conventional photolithography techniques described in Section 13.7. The mask is pressed against the substrate in an electrodeposition bath, so that the elastomer conforms to the substrate and excludes plating solution in contact areas. Electrodeposition takes place in areas that are not masked, eventually producing a mirror image of the mask. By using a sacrificial filler, made of a second material, complex three-dimensional shapes can be produced, complete with overhangs, arches, and other features, using instant masking technology.

13.17 | Mesoscale Manufacturing

Conventional manufacturing processes, such as those described in Chapters 5 through 12, typically produce parts that can be described as visible to the naked eye. Such parts are generally referred to as *macroscale*, the word macro being derived from *makros* in Greek, meaning "long." The processing of such parts is known as *macromanufacturing* and is the most developed and best understood size range from a design and manufacturing perspective.

In contrast, examples of *mesomanufacturing* are components for miniature devices such as hearing aides, medical devices such as stents and valves, mechanical watches, and extremely small motors and bearings. Note that mesomanufacturing overlaps both macro- and micromanufacturing in Fig. 1.7.

Two general approaches have been used for mesomanufacturing: scaling macromanufacturing processes *down*, and scaling micromanufacturing processes *up*. Examples of the former are a lathe with a 1.5-W motor that measures $32 \times 25 \times 30.5$ mm and weighs 100 g. Such a lathe can machine brass to a diameter as small as 60 μm and with a surface roughness of 1.5 μm (see Fig. 9.27). Similarly miniaturized versions are milling machines, mechanical presses, and various other machine tools.

Scaling up of micromanufacturing processes is similarly performed; LIGA can produce microscale devices, but the largest parts that can be made by LIGA are mesoscale. For this reason, mesomanufacturing processes are often indistinguishable from their micromanufacturing implementations.

13.18 | Nanoscale Manufacturing

In *nanomanufacturing*, parts are produced at nanometer length scales. The term usually refers to manufacturing strategies below the micrometer scale, or between 10^{-6} and 10^{-9} m in length. Many of the features in integrated circuits are at this length scale, but very little else with manufacturing relevance. Molecularly engineered medicines and other forms of biomanufacturing are the only present commercial examples. However, it has been recognized that many physical and biological processes act at this length scale, and that this approach holds much promise for future innovations.

Nanomanufacturing takes two basic approaches: *top-down* and *bottom-up*. Top-down approaches use large building blocks (such as a silicon wafer) and various manufacturing processes (such as lithography and wet and plasma etching) to construct ever-smaller features and products (microprocessors, sensors, and probes). At the other extreme, bottom-up approaches use small building blocks (such as atoms, molecules, or clusters of atoms and molecules) to build up a structure. In theory, this is similar to additive manufacturing technologies described in Section 10.12. However, when placed in the context of nanomanufacturing, bottom-up approaches suggest the manipulation and construction of products on an atomistic or molecular scale.

It has been pointed out that bottom-up approaches are widely used in nature (building cells is a fundamental bottom-up approach), whereas human manufacturing has, for the most part, consisted of top-down approaches. In fact, there are presently no nanomanufactured products that have demonstrated commercial viability. Some prospects that are most applicable to engineering materials are described below.

1. **Carbon nanotube devices.** Carbon nanotubes (see also Section 11.13.1) can be thought of as tubular forms of graphite and are of interest for the development of nanoscale devices. These nanotubes are produced by laser ablation of graphite, carbon-arc discharge, and, most often, chemical vapor deposition (Section 4.5.1). Carbon nanotubes can be single-walled (SWNT) or multiwalled (MWNT) and can be doped with various elements.

   Carbon nanotubes have exceptional strength, which makes them attractive as reinforcing fibers for composite materials (Sections 10.9 and 11.4). However, carbon nanotubes have very low adhesion with most materials, so that delamination with a matrix limits their effectiveness in these applications. Also, it is difficult to disperse the nanotubes properly throughout the structure, and their effectiveness as a reinforcement is limited if the nanotubes are clumped. A few products contain carbon nanotubes, such as a bicycle frame for the 2006 Tour de France and specialty baseball bats and tennis racquets. As of today, nanotubes provide only a fraction of the reinforcement by volume or effectiveness in these products, with graphite filling the major role.

   Another material characteristic of carbon nanotubes is their very high electrical current-carrying capability. Carbon nanotubes can be made as semiconductors or conductors, depending on the orientation of the graphite in the nanotube. Armchair nanotubes (Fig. 13.52) are theoretically able to carry a current density more than 1000 times higher than silver or copper, which makes them attractive for electrical connections in nanodevices.

   Carbon nanotubes also have been incorporated into polymers to improve their static electricity discharge capability, especially in fuel lines for automotive and aerospace applications. Additional proposed uses for carbon nanotubes include storage of hydrogen for use in hydrogen-powered vehicles, flat panel displays, X-ray and microwave generators, and nanoscale sensors.

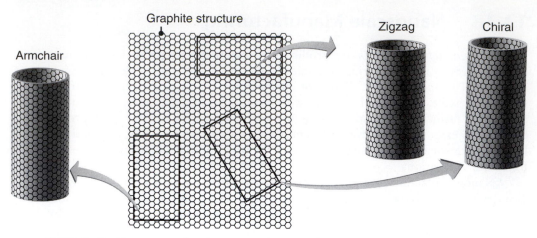

FIGURE 13.52 Forms of carbon nanotubes: armchair, zigzag, and chiral. Armchair nanotubes are noteworthy for their high electrical conductivity, whereas zigzag and chiral nanotubes are semiconductors.

2. **Nanophase ceramics.** Nanophase ceramics (see also Section 11.8.1) have received interest because by using nanoscale particles in producing the ceramic material, there can be a simultaneous and major increase in both the strength and ductility of the ceramic. Also, nanophase ceramics are utilized for catalysis because of their high surface-to-area ratios. Nanophase particles can be used as a reinforcement, such as with SiC particles in an aluminum matrix (Section 11.14).

CASE STUDY | Digital Micromirror Device

An example of a commercial MEMS-based product is the Digital Pixel Technology (DPT™) device, illustrated in Fig. 13.53. This device uses an array of *digital micromirror devices* (DMD™) to project a digital image, such as in computer-driven projection systems. The aluminum mirrors can be tilted so that light is directed into or away from the optics that focus light onto a screen; thus each mirror can represent a pixel of an image's resolution. The mirror allows light or dark pixels to be projected, but levels of gray can also be accommodated. Since the switching time is about 15 microseconds, which is much faster than the human eye can respond to, the mirror will switch between the on and off states in order to reflect the proper light dose to the optics.

The fabrication steps for producing the DMD device are shown in Fig. 13.54. Note that the sequence is similar to other surface micromachining examples (see Example 13.5) but has the following important differences:

- All micromachining steps take place at temperatures below 400°C, sufficiently low to ensure that no damage occurs to the electronic circuit.
- A thick silicon dioxide layer is deposited and is chemical-mechanical polished to provide an adequate foundation for the MEMS device.
- The landing pads and electrodes are produced from aluminum, deposited through sputtering (Section 13.5).
- High device reliability requires low stresses and high strength in the torsional hinge, produced from a proprietary aluminum alloy.

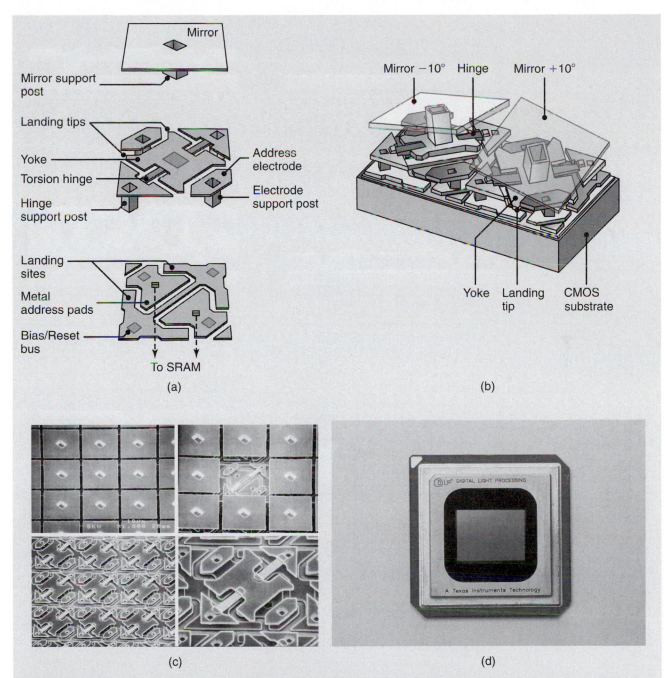

FIGURE 13.53 The Texas Instruments Digital Pixel Technology (DPT™) Device:
(a) exploded view of a single digital micromirror device (DMD™), (b) view of two adjacent
DMD pixels and (c) images of DMD arrays, with some mirrors removed for clarity; each
mirror measures approximately 17 μm (670 μin.) on a side. (d) A typical DPT device, used
for digital projection systems, high definition televisions, and other image display systems.
The device shown contains 1,310,720 micromirrors and measures less than 50 mm (2 in.)
per side. *Source:* Courtesy of Texas Instruments Corp.

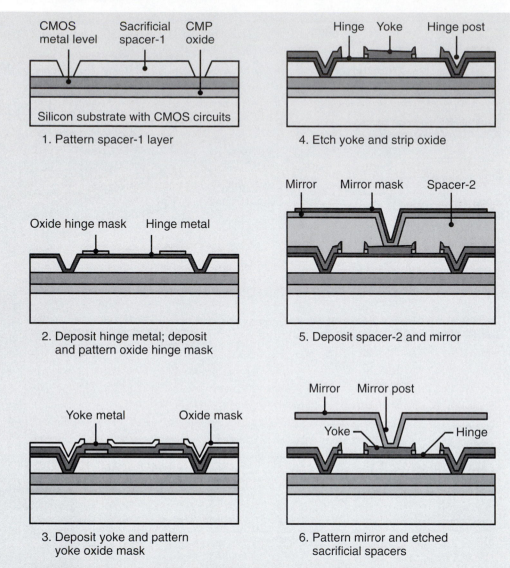

FIGURE 13.54 Manufacturing sequence for the Texas Instruments DMD device.

- The MEMS portion of the DMD is very delicate, and hence special care must be taken in separating the dice. When completed, a wafer saw (see Fig. 13.8c) cuts a trench along the edges of the DMD, which allows the individual dice to be broken apart at a later stage.

- A special step deposits a layer that prevents adhesion between the yoke and landing pads.

- The DMD is placed in a hermetically sealed ceramic package (Fig. 13.55) with an optical window.

An array of such mirrors represents a grayscale screen. Using three mirrors (one each for red, green, and blue light) for each pixel, the color image with millions of discrete colors. Digital pixel technology is widely applied in digital projection systems, high-definition television, and various other optical applications.

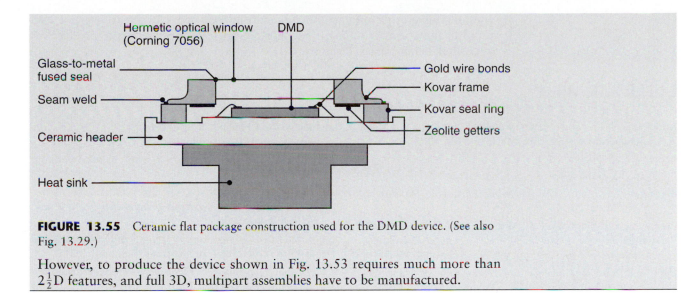

FIGURE 13.55 Ceramic flat package construction used for the DMD device. (See also Fig. 13.29.)

However, to produce the device shown in Fig. 13.53 requires much more than $2\frac{1}{2}$D features, and full 3D, multipart assemblies have to be manufactured.

SUMMARY

- The microelectronics industry continues changing rapidly, with the possibilities for new device concepts and circuit designs appearing to be endless. MEMS devices have become important technologies. (Section 13.1)

- Semiconductor materials, such as silicon, gallium arsenide, and polysilicon, have unique properties, where their electrical properties lie between those of conductors and insulators. (Section 13.3)

- Techniques are well developed for single-crystal growing and wafer preparation; after fabrication, the wafer is sliced into individual chips, each containing one complete integrated circuit. (Section 13.4)

- The fabrication of microelectronic devices and integrated circuits involves several different types of processes. After bare wafers have been prepared, they undergo repeated oxidation, film deposition, and lithographic and etching steps to open windows in the oxide layer in order to provide access to the silicon substrate. (Section 13.5–13.9)

- Dopants are introduced into various regions of the silicon structure to alter their electrical characteristics, a task done by diffusion and ion implantation. (Section 13.10)

- Microelectronic devices are interconnected by multiple metal layers, and the completed circuit is packaged and made accessible through electrical connections. (Section 13.11)

- Yield in devices is significant for economic considerations, and reliability has become increasingly important, because of the long expected life of these devices. (Section 13.12)

- Requirements for more flexible and faster integrated circuits require packaging that places many closely spaced contacts on a printed circuit board. (Section 13.13)

- MEMS devices are manufactured through techniques and with materials that, for the most part, have been pioneered in the microelectronics industry. Bulk and surface micromachining, LIGA, SCREAM, HEXSIL, and fusion bonding of multiple layers are the most prevalent methods. (Sections 13.14–13.16)

- Mesomanufacturing and nanomanufacturing are areas of great promise and rapid change, using top-down and bottom-up techniques to make products at very small length scales. (Sections 13.17–13.18)

BIBLIOGRAPHY

Anderson, B.L., and Anderson, R.L., *Fundamentals of Semiconductor Devices*, McGraw-Hill, 2004.

Berger, L.I., *Semiconductor Materials*, CRC Press, 1997.

Bhushan, B., *Handbook of Nanotechnology*, Springer, 2004.

Blackwell, G.R. (ed.), *The Electronic Packaging Handbook*, CRC Press, 2000.

Brar, A.S., and Narayan, P.B., *Materials and Processing Failures in the Electronics and Computer Industries: Analysis and Prevention*, ASM International, 1993.

Campbell, S.A., *The Science and Engineering of Microelectronic Fabrication*, 2nd ed., Oxford, 2001.

Chandrakasan, A., and Brodersen, R. (eds.), *Low Power CMOS Design*, IEEE, 1998.

Chandrakasan, A., and Brodersen, R., *Low Power Digital CMOS Design*, Kluwer, 1995.

Chang, C.-Y., and Sze, S.M. (eds.), *ULSI Devices*, Wiley-Interscience, 2000.

Davis, J.A., and Meindl, J.D. (eds.), *Interconnect Technology and Design for Gigascale Integration*, Springer, 2003.

Elwenspoek, M., and Jansen, H., *Silicon Micromachining*, Cambridge University Press, 2004.

Elwenspoek, M., and Wiegerink, R., *Mechanical Microsensors*, Springer, 2001.

Gad-el-Hak, M. (ed.), *The MEMS Handbook*, 2nd ed., CRC Press, 2005.

Gardner, J.W., Varadan, V., and Awadelkarim, O.O., *Microsensors, MEMS and Smart Devices*, Wiley, 2001.

Griffin, P.B., Plummer, J.D., and Deal, M.D., *Silicon VLSI Technology: Fundamentals, Practice and Modeling*, Prentice Hall, 2000.

Harper, C.A. (ed.), *Electronic Packaging and Interconnection Handbook*, 4th ed., McGraw-Hill, 2004.

———, *High-Performance Printed Circuit Boards*, McGraw-Hill, 2000.

Hwang, J.S., *Modern Solder Technology for Competitive Electronics Manufacturing*, McGraw-Hill, 1996.

Javits, M.W. (ed.), *Printed Circuit Board Materials Handbook*, McGraw-Hill, 1997.

Judd, M., and Brindley, K., *Soldering in Electronics Assembly*, 2nd ed., Newnes, 1999.

Khandour, R.S., *Printed Circuit Boards*, McGraw-Hill, 2005.

Kovacs, G.T.A., *Micromachined Transducers Sourcebook*, McGraw-Hill, 1998.

Liu, C., *Foundations of MEMS*, Prentice Hall, 2005.

Madou, M.J., *Fundamentals of Microfabrication*, 2nd ed., CRC Press, 2002.

Mahajan, S., and Harsha, K.S.S., *Principles of Growth and Processing of Semiconductors*, McGraw-Hill, 1998.

Mahalik, N., *Micromanufacturing and Nanotechnology*, Springer, 2005.

Maluf, N., and Williams, K., *An Introduction to Microelectromechanical Systems Engineering*, 2nd ed., Artech House, 2004.

Manko, H.H., *Soldering Handbook for Printed Circuits and Surface Mounting*, Van Nostrand Reinhold, 1995.

Matisoff, B.S., *Handbook of Electronics Manufacturing*, 3rd ed., Chapman & Hall, 1996.

May, G.S., and Spanos, C.J., *Fundamentals of Semiconductor Manufacturing and Process Control*, Wiley, 2006.

Nishi, Y., and Doering, R. (eds.), *Handbook of Semiconductor Manufacturing Technologies*, Dekker, 2000.

Ohring, M., *Reliability & Failure of Electronic Materials and Devices*, Academic Press, 1998.

Pierret, R.F., *Advanced Semiconductor Fundamentals*, 2nd ed., Prentice Hall, 2002.

Poole, C.P., and Owens, F.J., *Introduction to Nanotechnology*, Wiley, 2003.

Quirk, M., and Serda, J., *Semiconductor Manufacturing Technology*, Prentice Hall, 2000.

Rizvi, S., *Handbook of Photomask Manufacturing Technology*, CRC Press, 2005.

Robertson, C., *Printed Circuit Board Designer's Reference*, CRC Press, 2003.

Schroeder, D.K., *Semiconductor Material and Device Characterization*, 3rd ed., Wiley, 2006.

Sze, S.M. (ed.), *Semiconductor Devices: Physics and Technology*, 2nd ed., Wiley, 2001.

Taur, Y., and Ning, T.H., *Fundamentals of Modern VLSI Devices*, Cambridge, 1998.

Ulrich, R.K., and Brown, W.D. (eds.), *Advanced Electronic Packaging*, Wiley, 2006.

Van Zandt, P., *Microchip Fabrication: A Practical Guide to Semiconductor Processing*, McGraw-Hill, 2000.

Wolf, S., *Microchip Manufacturing*, Lattice Press, 2003.

Wolf, S., and Tauber, R.N., *Silicon Processing for the VSLI Era: Process Technology*, 2nd ed., Lattice Press, 1999.

van Zant, P., *Microchip Fabrication: A Practical Guide to Semiconductor Processing*, 5th ed., McGraw-Hill, 2004.

Varadan, V.K., Jiang, X., and Varadan, V.V., *Microstereolithography and Other Fabrication Techniques for 3D MEMS*, Wiley, 2001.

QUESTIONS

13.1 Define the terms *wafer, chip, device, integrated circuit*, and *surface mount*.

13.2 Why is silicon the most commonly used semiconductor in IC technology? Explain.

13.3 What do the terms VLSI, IC, CVD, CMP, and DIP stand for?

13.4 How do *n*-type and *p*-type dopants differ? Explain.

13.5 How is epitaxy different than other forms of film deposition?

13.6 Comment on the differences between wet and dry etching.

13.7 How is silicon nitride used in oxidation?

13.8 What are the purposes of prebaking and postbaking in lithography?

13.9 Define selectivity and isotropy and their importance in relation to etching.

13.10 What do the terms *linewidth* and *registration* refer to?

13.11 Compare diffusion and ion implantation.

13.12 What is the difference between evaporation and sputtering?

13.13 What is the definition of yield? How important is yield? Comment on its economic significance.

13.14 What is accelerated life testing? Why is it practiced?

13.15 What do BJT and MOSFET stand for?

13.16 Explain the basic processes of (a) surface micromachining and (b) bulk micromachining.

13.17 What is LIGA? What are its advantages over other processes?

13.18 What is the difference between isotropic and anisotropic etching?

13.19 What is a mask? What is its composition?

13.20 What is the difference between chemically assisted ion etching and dry plasma etching?

13.21 Which process(es) in this chapter allow(s) fabrication of products from polymers? (See also Chapter 10.)

13.22 What is a PCB?

13.23 With an appropriate sketch, describe the thermosonic stitching process.

13.24 Explain the difference between a die, a chip, and a wafer.

13.25 Why are flats or notches machined onto silicon wafers? Explain.

13.26 What is a via? What is its function?

13.27 What is a flip chip? Describe its advantages over a surface-mount device.

13.28 Explain how IC packages are attached to a printed circuit board if both sides will contain ICs.

13.29 In a horizontal epitaxial reactor (see the accompanying figure), the wafers are placed on a stage (susceptor) that is tilted by a small amount, usually 1°–3°. Why is this procedure done?

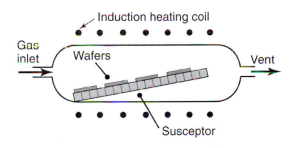

13.30 The accompanying table describes three changes in the manufacture of a wafer: increase of the wafer diameter, reduction of the chip size, and increase of the process complexity. Complete the table by filling in the words "increase," "decrease," or "no change" to indicate the effect that each change would have on wafer yield and on the overall number of functional chips.

Effects of Manufacturing Changes

Change	Wafer yield	Number of functional chips
Increase wafer diameter		
Reduce chip size		
Increase process complexity		

13.31 The speed of a transistor is directly proportional to the width of its polysilicon gate, with a narrower gate resulting in a faster transistor and a wider gate resulting in a slower transistor. Knowing that the manufacturing process has a certain variation for the gate width, say $\pm0.1\ \mu$m, how might a designer alter the gate size of a critical circuit in order to minimize its speed variation? Are there any penalties for making this change? Explain.

13.32 A common problem in ion implantation is channeling, in which the high-velocity ions travel deep into the material through channels along the crystallographic planes before finally being stopped. What is one simple way to stop this effect?

13.33   The MEMS devices discussed in this chapter apply macroscale machine elements, such as spur gears, hinges, and beams. Which of the following machine elements can and cannot be applied to MEMS, and why?

a. ball bearing
b. helical springs
c. bevel gears
d. rivets
e. worm gears
f. bolts
g. cams

13.34   Figure 13.7b shows the Miller indices on a wafer of (100) silicon. Referring to Fig. 13.5, identify the important planes for the other wafer types illustrated in Fig. 13.7a.

13.35   Referring to Fig. 13.23, sketch the holes generated from a circular mask.

13.36   Explain how you would produce a spur gear if its thickness were one-tenth its diameter and its diameter were (a) 10 μm, (b) 100 μm, (c) 1 mm, (d) 10 mm, and (e) 100 mm.

13.37   Which clean room is cleaner, a Class 10 or a Class 1?

13.38   Describe the difference between a microelectronic device, a micromechanical device and MEMS.

13.39   Why is silicon often used with MEMS and MEMS devices?

13.40   Explain the purpose of a spacer layer in surface micromachining.

13.41   What do the terms SIMPLE and SCREAM stand for?

13.42   Which process(es) in this chapter allow the fabrication of products from ceramics? (See also Chapter 11.)

13.43   What is HEXSIL?

13.44   Describe the differences between stereolithography and microstereolithography.

13.45   Lithography produces projected shapes; consequently, true three-dimensional shapes are more difficult to produce. Which of the processes described in this chapter are best able to produce three-dimensional shapes, such as lenses?

13.46   List and explain the advantages and limitations of surface micromachining as compared to bulk micromachining.

13.47   What are the main limitations to the LIGA process?

13.48   Describe the process(es) that can be used to make the microtweezers shown in Fig. 13.49 other than HEXSIL.

PROBLEMS

13.49   A certain wafer manufacturer produces two equal-sized wafers, one containing 500 chips and the other containing 300 chips. After testing, it is observed that 50 chips on each type of wafer are defective. What are the yields of the two wafers? Can any relationship be established between chip size and yield?

13.50   A chlorine-based polysilicon etch process displays a polysilicon:resist selectivity of 4:1 and a polysilicon:oxide selectivity of 50:1. How much resist and exposed oxide will be consumed in etching 350 nm of polysilicon? What should the polysilicon:oxide selectivity be in order to remove only 4 nm of exposed oxide?

13.51   During a processing sequence, four silicon-dioxide layers are grown by oxidation: 400 nm, 150 nm, 40 nm, and 15 nm. How much of the silicon substrate is consumed?

13.52   A certain design rule calls for metal lines to be no less than 2 μm wide. If a 1 μm-thick metal layer is to be wet etched, what is the minimum photoresist width allowed? (Assume that the wet etch is perfectly isotropic.) What would be the minimum photoresist width if a perfectly anisotropic dry-etch process were used?

13.53   Using Fig. 13.18, obtain mathematical expressions for the etch rate as a function of temperature.

13.54   If a square mask of side length 100 μm is placed on a {100} plane and oriented with a side in the $\langle 110 \rangle$ direction, how long will it take to etch a hole 4 μm deep at 80°C using thylene-diamine/pyrocatechol? Sketch the resulting profile.

13.55   Obtain an expression for the width of the trench bottom as a function of time for the mask shown in Fig. 13.17b.

13.56   Estimate the time of contact and average force when a fluorine atom strikes a silicon surface with a velocity of 1 mm/s. *Hint:* See Eqs. (9.11) and (9.13).

13.57   Calculate the undercut in etching a 10-μm-deep trench if the anisotropy ratio is (a) 200, (b) 2, and (c) 0.5. Calculate the sidewall slope for these three cases.

13.58   Calculate the undercut in etching a 10-μm-deep trench for the wet etchants listed in Table 13.3. What would the undercut be if the mask were made of silicon oxide?

13.59 Estimate the time required to etch a spur-gear blank from a 75-mm-thick slug of silicon.

13.60 A resist is applied in a resist spinner spun operating at 2000 rpm, using a polymer resist with viscosity of 0.05 N-s/m. The measured resist thickness is 1.5 μm. What is the expected resist thickness at 6000 rpm? Let $\alpha = 1.0$ in Eq. (13.3).

13.61 Examine the hole profiles in the accompanying figure and explain how they might be produced.

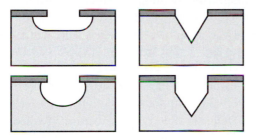

13.62 A polyimide photoresist requires 100 mJ/cm^2 per μm of thickness in order to develop properly. How long does a 150 μm film need to develop when exposed by a 1000 W/m^2 light source?

13.63 How many levels are required to produce the micromotor shown in Fig. 13.22d?

13.64 It is desired to produce a 500 by 500 μm diaphragm, 25 μm thick, in a silicon wafer 250 μm thick. Given that you will use a wet etching technique with KOH in water with an etch rate of 1 μm/min, calculate the etching time and the dimensions of the mask opening that you would use on a (100) silicon wafer.

13.65 If the Reynolds number for water flow through a pipe is 2000, calculate the water velocity if the pipe diameter is (a) 10 mm; (b) 100 μm. Do you expect flow in MEMS devices to be turbulent or laminar? Explain.

DESIGN

13.66 The accompanying figure shows the cross section of a simple *npn* bipolar transistor. Develop a process flow chart to fabricate this device.

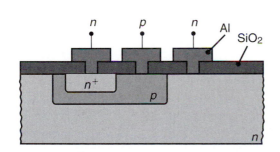

13.67 Referring to the MOS transistor cross section in the accompanying figure and the given table of design rules, what is the smallest transistor size W obtainable? Which design rules, if any, have no impact on the magnitude of W? Explain.

Rule No.	Rule name	Value (μm)
R1	Minimum polysilicon width	0.50
R2	Minimum poly-to-contact spacing	0.15
R3	Minimum enclosure of contact by diffusion	0.10
R4	Minimum contact width	0.60
R5	Minimum enclosure of contact by metal	0.10
R6	Minimum metal-to-metal spacing	0.80

13.68 The accompanying figure shows a mirror that is suspended on a torsional beam; it can be inclined through electrostatic attraction by applying a voltage on either side of the mirror at the bottom of the trench. Make a flow chart of the manufacturing operations required to produce this device.

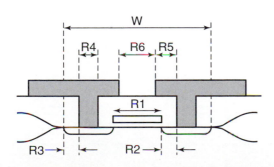

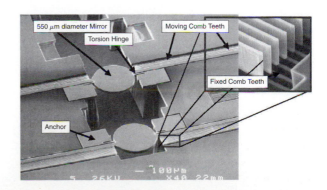

13.69   Referring to Fig. 13.36, design an experiment to find the critical dimensions of an overhanging cantilever that will not stick to the substrate.

13.70   Explain how you would manufacture the device shown in Fig. 13.32.

13.71   Inspect various electronic and computer equipment, take them apart as much as you can, and identify components that may have been manufactured by the techniques described in this chapter.

13.72   Do any aspects of this chapter's contents and the processes described bear any similarity to the processes described throughout previous chapters in this book? Explain and describe what they are.

13.73   Describe your understanding of the important features of clean rooms and how they are maintained.

13.74   Describe products that would not exist without the knowledge and techniques described in this chapter. Explain.

13.75   Review the technical literature and give more details regarding the type and shape of the abrasive wheel used in the wafer-cutting operation shown in Step 2 in Fig. 13.6.

13.76   It is well known that microelectronic devices may be subjected to hostile environments (such as high temperature, humidity, and vibration) as well as physical abuse (such as being dropped on a hard surface). Describe your thoughts on how you would go about testing these devices for their endurance under these conditions.

13.77   Conduct a literature search and determine the smallest diameter hole that can be produced by (a) drilling; (b) punching; (c) water-jet cutting; (d) laser machining; (e) chemical etching; and (f) EDM.

13.78   Design an accelerometer similar to the one shown in Fig. 13.32 using the (a) SCREAM process and (b) HEXSIL process, respectively.

13.79   Conduct a literature search and write a one-page summary of applications in biomems.

13.80   Describe the crystal structure of silicon. How does it differ from the structure of FCC? What is the atomic packing factor?

Automation of Manufacturing Processes and Operations

This chapter describes the art and science of automation in manufacturing operations, including the use of robots, sensors, and fixturing. The topics described include

- The use of hard automation for very large production quantities.
- Numerical control of machines for improved productivity and increased flexibility.
- Control strategies, including open-loop, closed-loop, and adaptive control.
- Industrial robots in various phases of manufacturing operations.
- Sensors for monitoring and controlling manufacturing processes.
- Design considerations for product assembly and disassembly.
- The impact of automation on product design and process economics.

14.1 | Introduction

Until the early 1950s, most manufacturing operations were carried out on traditional machinery that required skilled labor and lacked flexibility. Each time a different product was manufactured, the machinery had to be retooled and the movement of materials had to be rearranged. Furthermore, new products and parts with complex shapes required numerous trial-and-error attempts by the operator to set the proper processing parameters. Also, because of the human involvement, making parts that were exactly alike was difficult and time consuming. These circumstances meant that processing methods were generally inefficient and that labor costs were a significant portion of the overall production costs. The necessity for reducing the labor share of product cost gradually became apparent, as did the need to improve the efficiency and flexibility of manufacturing operations.

Productivity also became a major concern. Defined as the optimum use of all resources, such as materials, energy, capital, labor, and technology, or as output per employee per hour, productivity basically measures operating efficiency. **Mechanization** of machinery and operations had, by and large, reached its peak by the 1940s, although with rapid advances in the science and technology of manufacturing, the efficiency of manufacturing operations began to improve, and the percentage of total cost represented by labor costs began to decline. The next step in

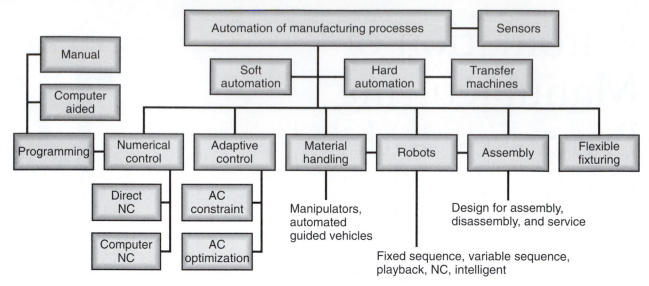

FIGURE 14.1   Outline of topics described in this chapter.

improving productivity was **automation,** from the Greek word *automatos,* meaning "self-acting." This term was coined in the mid-1940s by the U.S. automobile industry to indicate *automatic handling and processing* of parts in production machines. During the past six decades, major advances and breakthroughs in the types and in the extent of automation have occurred, made possible largely through rapid advances in the capacity and sophistication of **computers** and **control systems.**

This chapter follows the outline shown in Fig. 14.1. First, it reviews the history and principles of automation and how they have helped integrate various operations and activities in manufacturing plants. It then introduces the important concept of control of machines and systems through **numerical control** and **adaptive-control** techniques. Next, an essential aspect of manufacturing is discussed, namely material handling, which has been developed into various systems, particularly those that include the use of **industrial robots. Sensor technology** is then described; this technology is an essential element in the control and optimization of machinery, processes, and systems. Other developments discussed include **flexible fixturing** and **assembly operations,** taking full advantage of advanced manufacturing technologies, particularly **flexible manufacturing systems.** Finally, the chapter presents major developments in **computer-integrated manufacturing systems** and their impact on all aspects of manufacturing operations.

14.2 | Automation

Automation is generally defined as the process of having machines follow a predetermined sequence of operations with little or no human involvement, using specialized equipment and devices that perform and control manufacturing processes. As described in Section 14.8 and in Chapter 15, automation is achieved through the use of a variety of devices, sensors, actuators, and specialized equipment that are capable of (a) observing all aspects of the manufacturing operation, (b) making decisions concerning the changes that should be made, and (c) controlling all aspects of the operation.

Automation is an *evolutionary* rather than a revolutionary concept, and it has been implemented successfully in the following basic areas of activity:

- **Manufacturing processes and operations.** Machining, grinding, forging, cold extrusion, casting, and plastics molding are typical examples of processes that have been extensively automated.

- **Material handling.** Materials and parts in various stages of completion are moved throughout a plant by computer-controlled equipment, including the use of robots, with little or no human guidance.

- **Inspection.** Parts are automatically, especially while they are being made, inspected for defects, dimensional accuracy, and surface finish.

- **Assembly.** Individually manufactured parts are automatically assembled into subassemblies and, finally, into products.

- **Packaging.** Products are packaged automatically.

14.2.1 Evolution of automation

Some metalworking processes were developed as early as 4000 B.C. (see Table 1.1); and it was at the beginning of the Industrial Revolution, in the 1750s, that automation began to be introduced. Machine tools began to be developed starting in the late 1890s. Mass production and transfer machines were developed in the 1920s, particularly in the automobile industry; however, these systems had *fixed* automatic mechanisms and were designed to produce specific products. The major breakthrough in automation began with *numerical control* (NC) of machine tools, in the early 1950s. Since this historic development, rapid progress has been made in automating almost all aspects of manufacturing Table 14.1).

Manufacturing involves various levels of automation, depending on the processes involved, the type of product, and the quantity or volume to be produced. Manufacturing systems (see Fig. 14.2) include the following classifications, in order of increasing automation:

- **Job shops** typically use general-purpose machines and machining centers (Section 8.11) with high levels of human labor involvement.

- **Rapid-prototyping facilities** (Section 10.12) have less human involvement than job shops and exploit the unique capabilities of computer-driven rapid-prototyping machines.

- **Stand-alone NC production** use **numerically controlled machines** (Section 14.3) but still with significant operator/machine interaction.

- **Manufacturing cells** (Section 15.9) consist of cluster of machines with integrated computer control and flexible material handling, often with industrial robots (Section 14.7).

- **Flexible manufacturing cells** (Section 15.10) use computer control of all aspects of manufacturing, the simultaneous incorporation of several manufacturing cells, and automated material handling systems.

- **Flexible manufacturing** lines involve computer-controlled machinery in production lines instead of cells; part transfer is through hard automation and product flow is more limited than in flexible manufacturing systems. However, the throughput is higher, resulting in higher production quantities.

- **Flowlines** and **transfer lines** consist of organized groupings of machinery with automated material handling between machines; the manufacturing line is designed for limited or no flexibility, because the goal is to produce a single part.

TABLE 14.1

Developments in the History of Automation and Control of Manufacturing Processes. (See also Table 1.1)

Date	Development
1500–1600	Water power for metalworking; rolling mills for coinage strips.
1600–1700	Hand lathe for wood; mechanical calculator.
1700–1800	Boring, turning, and screw cutting lathe, drill press.
1800–1900	Copying lathe, turret lathe, universal milling machine; advanced mechanical calculators.
1808	Sheet-metal cards with punched holes for automatic control of weaving patterns in looms.
1863	Automatic piano player (Pianola).
1900–1920	Geared lathe; automatic screw machine; automatic bottle-making machine.
1920	First use of the word "robot."
1920–1940	Transfer machines; mass production.
1940	First electronic computing machine.
1943	First digital electronic computer.
1945	First use of the word "automation."
1947	Invention of the transistor.
1952	First prototype numerical control machine tool.
1954	Development of the symbolic language APT (Automatically Programmed Tools); adaptive control.
1957	Commercially available NC machine tools.
1959	Integrated circuits; first use of the term "group technology."
1960	Industrial robots.
1965	Large-scale integrated circuits.
1968	Programmable logic controllers.
1970s	First integrated manufacturing system; spot welding of automobile bodies with robots; microprocessors; minicomputer-controlled robot; flexible manufacturing system; group technology.
1980s	Artificial intelligence; intelligent robots; smart sensors; untended manufacturing cells.
1990–2000s	Integrated manufacturing systems; intelligent and sensor-based machines; telecommunications and global manufacturing networks; fuzzy-logic devices; artificial neural networks; Internet tools; virtual environments; high-speed information systems.

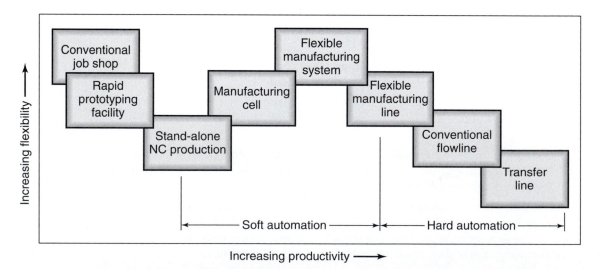

FIGURE 14.2 Flexibility and productivity of various manufacturing systems. Note the overlap between the systems, which is due to the various levels of automation and computer control that are applicable in each group. (See also Chapter 15 for more details.)

14.2.2 Goals of automation

Automation has the following primary goals:

1. **Integrate** various aspects of manufacturing operations so as to improve product quality and uniformity, minimize cycle times and effort involved, and reduce labor costs.

2. **Improve productivity** by reducing manufacturing costs through better control of production: Raw materials and parts are loaded, fed, and unloaded on machines faster and more efficiently; machines are used more effectively; and production is organized more efficiently.

3. **Improve quality** by improving the repeatability of manufacturing processes.

4. **Reduce human involvement,** boredom, and the possibility of human error.

5. **Reduce workpiece damage** caused by manual handling of parts.

6. **Economize on floor space** by arranging machines, material handling and movement, and auxiliary equipment more efficiently.

7. **Raise the level of safety** for personnel, especially under hazardous working conditions.

Automation and production quantity. Production quantity or volume is crucial in determining the type of machinery and the level of automation required to produce parts economically. **Total production quantity** is defined as the total number of parts to be produced, in various **lot sizes.** Lot size greatly influences the economics of production, as described in Chapter 16. **Production rate** is defined as the number of parts produced per unit time. The approximate and generally accepted ranges of production quantities for some typical applications are shown in Table 14.2. Note that, as expected, **experimental** and **prototype** products represent the lowest volume. (See also Section 10.12.)

Small quantities can be manufactured in *job shops* (Fig. 14.2). These operations have high part variety, meaning that different parts can be produced in a short time without extensive changes in tooling and in operations. On the other hand, machinery in job shops generally requires skilled labor, and the production quantity and rate are low; as a result, the cost per part can be high (Fig. 14.3). When products involve a large labor component, their production is referred to as **labor intensive.** (See also Section 14.12.)

Piece-part production usually involves very small quantities and is suitable for job shops. The majority of piece-part production is in lot sizes of 50 or less. Quantities for **small-batch production** typically range from 10 to 100, using general-purpose machines and machining centers. **Batch production** usually involves lot sizes between 100 and 5000 and utilizes machinery similar to that used for small-batch production, but with specially designed fixtures for higher production rates.

TABLE 14.2

Approximate Annual Quantity of Production		
Type of production	Number produced	Typical products
Experimental or prototype	1–10	All types
Piece or small batch	<5000	Aircraft, machine tools, dies
Batch or high quantity	5000–100,000	Trucks, agricultural machinery, jet engines, diesel engines, orthopedic devices
Mass production	100,000+	Automobiles, appliances, fasteners, bottles, food and beverage containers

FIGURE 14.3 General
characteristics of three types
of production methods: job
shop, batch production, and
mass production.

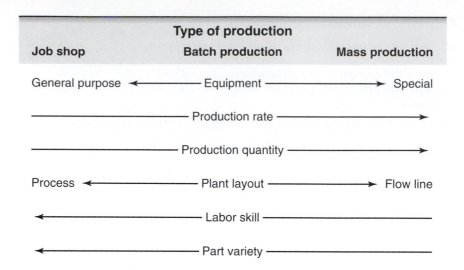

Mass production generally involves quantities over 100,000, and requires special-purpose machinery (**dedicated machines**) and automated equipment for transferring materials and parts. Although the machinery, equipment, and specialized tooling are expensive, both the labor skills required and the labor costs are relatively low, because of the high level of automation. However, these production systems are organized for a specific type of product and, hence, lack flexibility. Most manufacturing facilities operate with a variety of machines in combination and with various levels of automation and computer controls.

14.2.3 Applications of automation

Automation can be applied to the manufacture of all types of goods, from raw materials to finished products, and in all types of production, from job shops to large manufacturing facilities. The decision to automate a new or existing facility requires that the following additional considerations be taken into account:

- Type of product manufactured.
- Quantity and the rate of production required.
- Particular phase of the manufacturing operation to be automated.
- Level of skill in the available workforce.
- Reliability and maintenance problems that may be associated with automated systems.
- Economics.

Because automation generally involves a high initial cost of equipment and requires knowledge of the principles of operation and maintenance, a decision about the implementation of even low levels of automation must involve a study of the true needs of an organization. It is not unusual for a company to begin the implementation of automation with great enthusiasm and with high across-the-board goals, only to discover that the economic benefits of automation were largely illusory rather than real and that, in the final assessment, automation was not cost effective. Thus, **selective automation**, rather than *total* automation, of a facility may then be desirable. Generally, the higher the level of skill available in the workforce, the lower is the need for automation, provided that labor costs are justified and that there are a sufficient number of workers available. Conversely, if a manufacturing facility is already automated, the skill level required is lower.

14.2.4 Hard automation

In *hard automation,* also called **fixed-position automation,** the production machines are designed to produce a standardized product, such as an engine block, a valve, a gear, or a spindle. Although product size and processing parameters can be changed, these machines are specialized and thus lack flexibility. They cannot be modified to any significant extent to accommodate a variety of products with different shapes and dimensions. (See also the discussion of *group technology* in Section 15.8.) Because these machines are expensive to design and construct, their economical use requires the production of parts in very large quantities.

The machines used in hard-automation applications are usually fabricated on the **building-block (modular) principle.** They are generally called **transfer machines** and consist of two major components: power-head production units and transfer mechanisms. **Power-head production units** consist of a frame or bed, electric drive motors, gearboxes, and tool spindles and are self-contained. Because of their intrinsic modularity, these units can easily be regrouped to produce a different part. Transfer machines consisting of two or more power-head units can be arranged on the shop floor in *linear, circular,* or *U patterns.* **Buffer storage** features are often incorporated in these systems to permit continued operation in the case of tool failure or individual machine breakdown.

Transfer mechanisms and **transfer lines** are used to move the workpiece from one station to another in the machine, or from one machine to another, usually controlled by sensors and other devices. Workpieces are transferred by several methods: (1) *rails* along which the parts, usually placed on pallets, are pushed or pulled by various mechanisms (Fig. 14.4a); (2) *rotary indexing tables* (Fig. 14.4b); and (3) *overhead conveyors.* Tools on transfer machines can easily be changed in toolholders with quick-change features (see, for example, Fig. 8.66). These machines are also equipped with various automatic gaging and inspection systems to ensure that a part produced in one station is within acceptable dimensional tolerances before it is transferred to the next station. (As described in Section 14.10, transfer machines are also used extensively in automatic assembly.)

Figure 14.5 shows the **transfer lines,** or **flow lines,** in a very large system, consisting of a number of transfer machines for producing cylinder heads for engine blocks. This system is capable of producing 100 cylinder heads per hour. Note the various machining operations performed: milling, drilling, reaming, boring, tapping, and honing, as well as cleaning, and gaging.

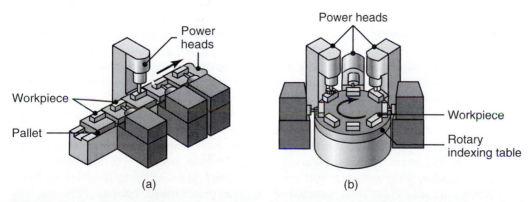

(a)   (b)

FIGURE 14.4   Two types of transfer mechanisms: (a) straight and (b) circular pattern.

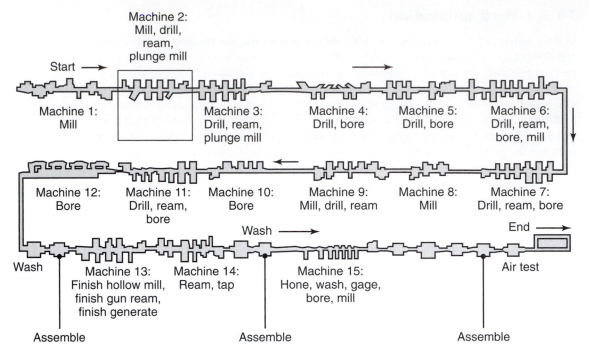

FIGURE 14.5   A traditional transfer line for producing engine blocks and cylinder heads. *Source:* Courtesy of Ford Motor Company.

14.2.5 Soft automation

In contrast to hard automation, *soft automation,* also called *flexible automation* and *programmable automation,* has greater flexibility through computer control of the machine and its functions by various software programs (see Section 14.3 and 14.4 for details). The machines can easily be reprogrammed to produce a part that has a different shape or dimensions. Flexible automation is the principle behind **flexible manufacturing systems** (see Section 15.10), which have high levels of efficiency and productivity.

14.2.6 Programmable controllers

The control of a manufacturing operation has traditionally been performed by devices such as timers, switches, relays, counters, and similar hard-wired devices that are based on mechanical, electromechanical, and pneumatic principles. Beginning in 1968, **programmable logic controllers** (PLCs) were introduced to replace these devices. Because PLCs eliminate the need for relay control panels and because they can easily be reprogrammed and take up less space, they have been widely adopted in manufacturing systems and operations. Their basic functions are (a) on-off motion, (b) sequential operations, and (c) feedback control.

PLCs are also used in system control, with high-speed digital-processing and communication capabilities, improving the overall efficiency of the operation; also, they perform reliably in industrial environments. However, because of advances in numerical control machines (Section 14.3), they have become less common in new installations, although they still represent a very large existing installation base. Microcomputers are used extensively in new installations because they are less expensive than PLCs and are easier to program and to network.

14.2.7 Total productive maintenance

The management and maintenance of a wide variety of machines, equipment, and systems are among the significant aspects that affect productivity in a manufacturing organization. Consequently, the concepts of *total productive maintenance* (TPM) and *total productive equipment management* (TPEM) have become important. These activities include continuous analysis of such factors as (a) equipment breakdown and equipment problems; (b) monitoring and improving equipment productivity; (c) implementation of preventive and predictive maintenance; (d) reduction of setup time, idle time, and cycle time; (d) full utilization of machinery and equipment and improvement of their effectiveness; and (e) reduction of defective products. Teamwork is an important aspect of this activity and involves the full cooperation of machine operators, maintenance personnel, engineers, and management of the organization.

14.3 | Numerical Control

Numerical control (NC) is a method of controlling the movements of machine components by directly inserting coded instructions, in the form of numbers and letters, into the system. The system automatically interprets these data and converts them to output signals. These signals, in turn, control various machine components, such as, for example, (a) turning spindles on and off, (b) changing tools, (c) moving the workpiece or the tools along specific paths, and (d) turning cutting fluids on and off.

The importance of numerical control can be illustrated by the following example: Assume that several holes are to be drilled on a part in the positions shown in Fig. 14.6. In the traditional manual method of machining this part, the operator positions the drill bit with respect to the workpiece, using reference points given by any of the three methods shown in the figure. The operator then proceeds to drill the holes. Let's first assume that 100 parts, all having exactly the same shape and dimensional accuracy, are to be drilled. Obviously, this operation is going to be tedious, because the operator has to repeatedly go through the same procedure. Moreover, the probability is high that, for a variety of reasons (see Section 4.9), not all drilled holes will be the same. Let's now assume that during this production run, the last 10 parts require holes in different positions. The machinist now has to reposition the worktable, an operation that will be time consuming.

Numerical control machines are capable of producing such parts repeatedly and accurately by simply loading different part programs, as described in Section 14.4. Data concerning all aspects of the machining operation can be stored on hard disks.

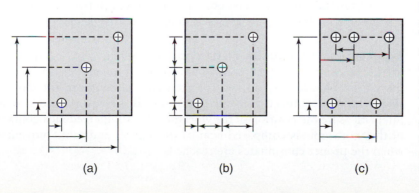

(a)   (b)   (c)

FIGURE 14.6 Positions of drilled holes in a workpiece. Three methods of measurements are shown: (a) absolute dimensioning, referenced from one point at the lower left of the part, (b) incremental dimensioning, made sequentially from one hole to another, and (c) mixed dimensioning, a combination of both methods.

Specific information is then relayed to the machine tool's control panel. On the basis of input information, various devices on the machine are actuated to obtain the necessary machine setup. Complex operations, such as turning a part that has various contours or die sinking in a milling machine, can easily be carried out.

14.3.1 Computer numerical control

In *computer numerical control* (CNC), the control hardware (mounted on the NC machine) follows directions received from *local* computer software. There are two types of computerized systems: direct numerical control and computer numerical control.

In **direct numerical control** (DNC), several machines are directly controlled, step by step, by a central computer. In this system, the operator has access to the central computer, and the status of all machines in a manufacturing facility can be monitored and assessed from the central computer. However, DNC has a significant disadvantage: If, for some reason, the computer shuts down, all the machines become inoperative. **Distributed numerical control** involves the use of a central computer serving as the control system over a number of individual CNC machines with dedicated microcomputers. This system provides large memory and computational capabilities and offers flexibility while overcoming the disadvantage of direct numerical control.

Computer numerical control is a system in which a control microcomputer (*onboard computer*) is an integral part of a machine or piece of equipment. The part program may be prepared at a remote site by the programmer, and it may incorporate information obtained from drafting software packages and from machining simulations (in order to ensure that the part program is bug free). The machine operator can easily and manually program the onboard computer, can modify the programs directly, prepare programs for different parts, and store the programs. CNC systems are widely used today because of the availability of small computers with large memory, microprocessors, and program-editing capabilities, as well as the increased flexibility, accuracy, and versatility of such computers.

14.3.2 Principles of numerical control machines

Fig. 14.7 depicts the basic elements and operation of a typical NC machine. The functional elements involved are

1. **Data input:** The numerical information is read and stored in computer memory.
2. **Data processing:** The programs are read into the machine control unit for processing.
3. **Data output:** This information is translated into commands (typically pulsed commands) to the servomotor (Fig. 14.8), which then moves the machine table to specific positions by means of stepping motors, lead screws, and other devices.

Types of control circuits. In the **open-loop** system (Fig. 14.8a), the signals are sent to the servomotor by the controller, but the movements and final positions of the worktable are not checked for accuracy. In contrast, the **closed-loop** system (Fig. 14.8b) is equipped with various transducers, sensors, and counters that accurately measure the position of the worktable. Through **feedback control,** the position of the worktable is compared against the signal, and table movements terminate when the proper coordinates are reached.

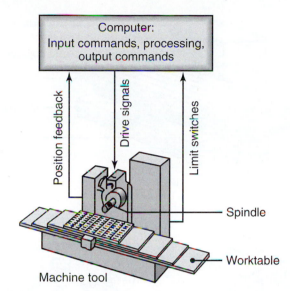

FIGURE 14.7 Schematic illustration of the major components of a numerical control machine tool.

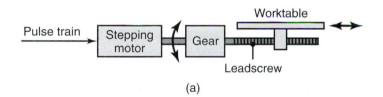

(a)

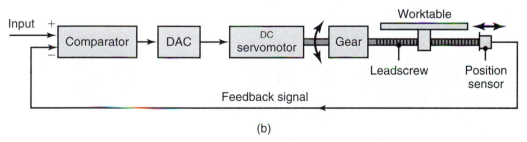

(b)

FIGURE 14.8 Schematic illustration of the components of (a) an open loop and (b) a closed-loop control system for a numerical control machine. (DAC is digital-to-analog converter.)

Position measurement in NC machines is accomplished by two methods. (a) In the *direct measuring system*, a sensing device reads a graduated scale on the machine table or slide for linear movement (Fig. 14.9a). This system is more accurate than indirect methods, because the scale is built into the machine, and *backlash* (the play between two adjacent mating gear teeth) in the mechanisms is not significant. (b) In the *indirect measuring system*, **rotary encoders** or **resolvers** (Fig. 14.9b and c) convert rotary movement to translation movement. In this system, backlash can significantly affect measurement accuracy. Position feedback mechanisms use various sensors that are based mainly on magnetic and photoelectric principles.

FIGURE 14.9 (a) Direct measurement of the linear displacement of a machine-tool worktable and (b) and (c) indirect measurement methods.

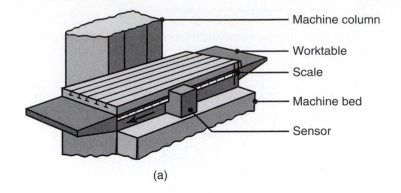

Machine column

Worktable

Scale

Machine bed

Sensor

(a)

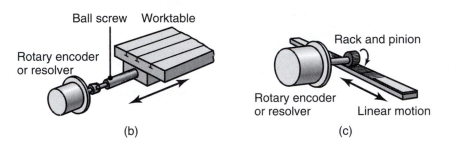

Ball screw   Worktable

Rotary encoder or resolver

(b)

Rack and pinion

Rotary encoder or resolver

Linear motion

(c)

EXAMPLE 14.1 Comparison of open-loop and closed-loop control

Consider the simple, one-dimensional open-loop and closed-loop control systems illustrated in Fig. 14.8. Assume that the worktable mass, m, is known and that control is achieved by changing the force applied to the worktable through a torque to the gear. If the system is frictionless, develop equations for the force needed to obtain a new position for (a) open-loop control and (b) closed-loop control, where both position and velocity are measured.

Solution. For the open-loop control system, the new position is determined by the equations of motion:

$$x - x_0 = \frac{1}{2}at^2$$

where it is assumed that the initial velocity is zero (at time $t = 0$) and the initial position is x_0. By applying $F = ma$, but realizing that the mass needs to accelerate for the first half of the desired motion and decelerate for the second half, the force as a function of time is given by

$$F = ma \qquad \text{for} \qquad 0 < t < \frac{x - x_0}{a}$$

and

$$F = -ma \qquad \text{for} \qquad \frac{x - x_0}{a} < t < \frac{2(x - x_0)}{a}.$$

In theory, the acceleration selected for this system could be defined by the maximum torque that can be developed by the motor to minimize the time required to attain the new position. However, motors cannot attain their maximum torque instantaneously and, furthermore, the frictionless situation is not realistic. Note that when external forces act upon the worktable, it will move; there is no provision in an open-loop control system to ensure that the desired position will be achieved or maintained.

For the closed-loop control system, the force is given by

$$F = -k_v \dot{x} + k_p(x - x_0)$$

where $\dot{x}$ is the instantaneous velocity, x is the desired position, and k_v and k_p are the velocity and position *gains* of the worktable, respectively. If the gains are selected properly, the closed-loop control system can maintain the desired position and be stable even if external forces are applied.

14.3.3 Types of control systems

There are two basic types of control systems in numerical control: point-to-point and contouring.

1. In the **point-to-point system**, also called the **positioning system**, each axis of the machine is driven separately by lead screws and, depending on the type of operation, at different speeds. The machine initially moves at maximum velocity (in order to reduce nonproductive time) but decelerates as the tool approaches its numerically defined position. Thus, in an operation such as drilling, the positioning and drilling take place *sequentially* (Fig. 14.10a). After the hole is drilled, the tool retracts upward and moves rapidly to another position, and the operation is repeated. Punching is another application of this system.

2. In the **contouring system**, also known as the **continuous-path system**, the positioning and the operation are both performed along controlled paths, as in milling (Fig. 14.10b). Because the tool operates as it travels along a prescribed path, accurate control and synchronization of velocities and movements are important. Other applications of the contouring system include lathes, grinders, welding machinery, and machining centers.

Interpolation. Movement along a path (*interpolation*) occurs incrementally by one of several basic methods (Fig. 14.11). Figure 14.12 illustrates examples of actual paths in drilling, boring, and milling operations. In all interpolations, the path controlled is that of the *center of rotation* of the tool. Compensation for different types

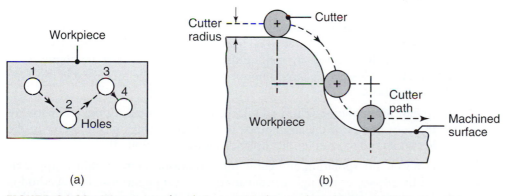

(a)   (b)

FIGURE 14.10   Movement of tools in numerical control machining. (a) Point-to-point system: The drill bit drills a hole at position 1, is then retracted and moved to position 2, and so on. (b) Continuous path by a milling cutter; note that the cutter path is compensated for by the cutter radius. This path can also compensate for cutter wear.

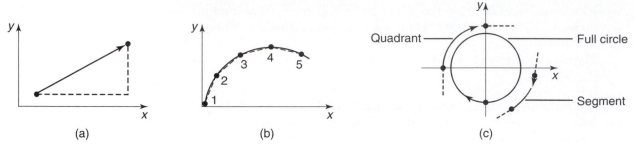

FIGURE 14.11   Types of interpolation in numerical control: (a) linear, (b) continuous path approximated by incremental straight lines, and (c) circular.

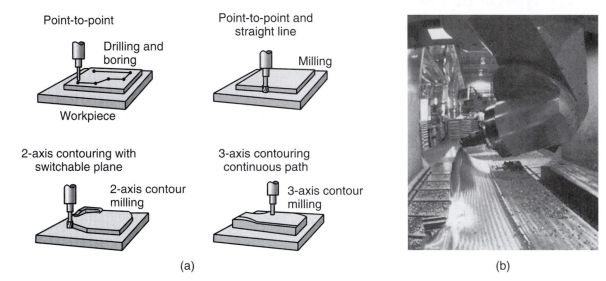

FIGURE 14.12   (a) Schematic illustration of drilling, boring, and milling operations with various cutter paths and (b) machining a sculptured surface on a five-axis numerical control machine. *Source:* Courtesy of the Ingersoll Milling Machine Co.

of tools, different tool diameters, or for tool wear during machining can be made in the NC program.

1. In **linear interpolation,** the tool moves in a *straight* line, from start to end (Fig. 14.11a), on two or three axes. Theoretically, all types of profiles can be produced by this method, by making the increments between the points small (Fig. 14.11b). However, a large amount of data has to be processed in this type of interpolation.

2. In **circular interpolation** (Fig. 14.11c), the inputs required for the tool path are (a) the coordinates of the end points, (b) the coordinates of the center of the circle and its radius, and (c) the direction of the tool along the arc.

3. In **parabolic interpolation** and **cubic interpolation,** the tool path is approximated by *curves,* using higher order mathematical equations. This method is effective in five-axis machines (see Section 8.11.2) and is particularly useful in die-sinking operations, such as for sheet forming of automotive bodies. These interpolations are also used for the movements of industrial robots. (See Section 14.7.)

14.3.4 Positioning accuracy of numerical control machines

Positioning accuracy is defined by how accurately the machine can be positioned to a certain coordinate system. An NC machine generally has a positioning accuracy of at least ± 3 μm (0.0001 in.). **Repeatability** is defined as the closeness of agreement of repeated position movements under the same operating conditions of the machine; it is usually about ± 8 μm (0.0003 in.). **Resolution** is defined as the smallest increment of motion of the machine components; it is usually about 2.5 μm (0.0001 in.).

The *stiffness* of the machine tool (see Section 8.12) and elimination of the *backlash* in its gear drives and lead screws are important for dimensional accuracy. Backlash in modern machines is eliminated by such means as using preloaded ball screws. Rapid response of the machine to command signals requires that friction and inertia effects be minimized, such as, for example, by reducing the mass of moving components of the machine.

14.3.5 Advantages and limitations of numerical control

Numerical control has the following *advantages* over conventional methods of machine control:

1. Flexibility of operation is improved, as well as the ability to produce complex shapes with good dimensional accuracy, good repeatability, reduced scrap, high production rates, high productivity, and high product quality.
2. Tooling costs are reduced, because templates and similar fixtures are not required.
3. Machine adjustments are easy to make.
4. More operations can be performed with each setup, the lead time for setup and machining required is less, design changes are easier to accommodate, and inventory is reduced.
5. Programs can be prepared rapidly and they can be recalled at any time.
6. Faster prototype production is possible.
7. Required operator skill is not as high, and the operator has more time to attend to other tasks in the work area.

The major *limitations* of numerical control are (a) the relatively high initial equipment cost, (b) the need for and cost of programming and computer time, and (c) the special maintenance that requires trained personnel. Because NC machines are complex systems, breakdowns can be costly; thus preventive maintenance is essential. These limitations are often easily outweighed by the overall economic advantages of NC.

14.4 | Programming for Numerical Control

A program for numerical control consists of a sequence of directions that instructs the machine to carry out specific operations. *Programming for NC* may be accomplished with the same software that is used for CAD, be done on the shop floor, or be purchased from an outside source. The program contains instructions and commands:

a. *Geometric instructions* pertain to relative movements between the tool and the workpiece.
b. *Processing instructions* concern parameters such as spindle speeds, feeds, cutting tools, and cutting fluids.

c. *Travel instructions* pertain to the type of interpolation and to the speed of movement of the tool or the worktable.

d. *Switching instructions* relate, for example, to the on-off position for coolant supplies, direction or lack of spindle rotation, tool changes, workpiece feeding, clamping, and so on.

Manual part programming consists first of calculating the dimensional relationships of the tool, workpiece, and worktable, on the basis of the engineering drawings of the part (including CAD; see Section 15.4), the operations to be performed, and the sequence of the operations. A program is then prepared, detailing the information necessary for carrying out the particular operation. Because they are familiar with machine tools and process capabilities, skilled machinists can, with some training in programming, do manual programming. The work involved is tedious, time consuming, and uneconomical; consequently, manual programming is used mostly in simple point-to-point applications.

Computer-aided part programming involves special symbolic **programming languages** that determine the coordinate points of corners, edges, and surfaces of the part. A programming language is a means of communicating with the computer and involves the use of symbolic characters. The programmer describes the component to be processed in this language, and the computer converts that description to commands for the NC machine. Several programming languages, with various features and applications, are commercially available. The first language that used English-like statements (called **APT**, for **Automatically Programmed Tools**) was developed in the late 1950s; this language, in its various expanded forms, is still occasionally used for both point-to-point and continuous-path programming.

Complex parts are machined using graphics-based, computer-aided machining programs. A tool path is created in a largely graphic environment that is similar to a CAD program (see Section 15.4). The machine code (**G-Code**) is created automatically by the program. This code is valuable in communicating machining instructions to the CNC hardware, but it is difficult to edit and troubleshoot it without software interpreters. When minor difficulties are encountered, such as (a) using different end-mill diameters than originally programmed or (b) changing the cutting speed to avoid chatter, the machine operator needs to modify the program; this is difficult using G-Code.

In *shop-floor programming*, CNC programming software is used directly on the machine tool controller. This allows higher-level geometry and processing information to be sent to the CNC controller instead of G-code. G-code is then developed by the dedicated computer under the control of the machine operator. The advantage of this method is that any changes that are made to the machining program are sent back to the programming group and stored as a shop-proven design iteration and can be re-used or standardized across a family of parts. Before production begins, the programs must be verified, either by viewing a simulation of the process on a monitor or by first making the part from an inexpensive material such as aluminum, wood, wax, or plastic.

14.5 | Adaptive Control

In *adaptive control* (AC), the operating parameters automatically adapt themselves to conform to new circumstances, such as changes in the dynamics of the particular process and any disturbances that may arise. This approach is basically a feedback system. It will be recognized that human reactions to occurrences in everyday life already contain dynamic feedback control. For example, driving your car on a smooth road is relatively easy, and you need to make few, if any, adjustments. However, on a rough road, you may have to steer to avoid potholes by observing the condition of the road. Also, your body feels the car's rough movements and vibrations; you then react by changing the direction and the speed of the car to minimize the effects of the rough road and to increase the comfort of your ride. Similarly, an **adaptive controller** would check these conditions, adapt an appropriate desired braking profile (for example, antilock brake system and traction control), and then use feedback to implement it.

Several adaptive-control systems are commercially available for applications such as ship steering, chemical-reactor control, rolling mills, and medical technology. Although adaptive control has, for some time, been widely used in continuous processing in the chemical industry and in oil refineries, its successful application to manufacturing processes is relatively recent. In manufacturing operations adaptive control helps (a) optimize *production rate,* (b) optimize *product quality,* and (c) minimize *production costs.* Application of AC in manufacturing is particularly important in situations where workpiece dimensions and quality are not uniform, such as a poor casting or an improperly heat-treated part.

Gain scheduling is perhaps the simplest form of adaptive control. In gain scheduling, a different gain is selected depending on the measured operating conditions. A different gain is assigned to each region of the system's operating space. With advanced adaptive controllers, the gain may vary continuously with changes in operating conditions.

Adaptive control is a logical extension of CNC systems. As described in Section 14.4, the part programmer sets the processing parameters, based on the existing knowledge of the workpiece material and various data on the particular manufacturing process. Whereas in CNC machine these parameters are held constant during a particular processing cycle, in adaptive control the system is capable of automatic adjustments *during* processing through closed-loop feedback control (Fig. 14.13).

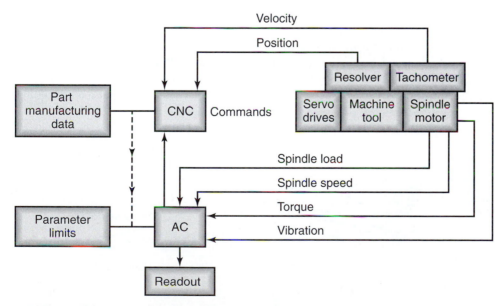

FIGURE 14.13 Schematic illustration of the application of adaptive control (AC) for a turning operation. The system monitors such parameters as cutting force, torque, and vibrations; if they are excessive, AC modifies process variables, such as feed and depth of cut, to bring them back to acceptable levels.

Principles and applications of adaptive control.   The basic functions common to adaptive-control systems are

1. Determining the operating conditions of the process, including measures of performance. This information is typically obtained by using sensors that measure process parameters, such as force, torque, vibration, and temperature.

2. Configuring the process control in response to the operating conditions. Large changes in the operating conditions may provoke a decision to make a major switch in control strategy. More modest alterations may include the modification of process parameters, such as the speed of operation or of the feed in machining.

3. Continue to monitor the operation, making further changes in the controller as needed.

In an operation such as turning on a lathe (Section 8.9.2), for example, the adaptive-control system would sense, in real time, cutting forces, torque, temperature, tool-wear rate, tool chipping or tool fracture and surface finish of the workpiece. The system then converts this information into commands that modify the process parameters on the machine tool to hold the parameters constant (or within certain limits) or to optimize the machining operation.

Systems that place a constraint on a process variable are called *adaptive-control constraint* (ACC) systems. Thus, for example, if the thrust force and the cutting force (and hence the torque) increase excessively (because, for instance, of the presence of a hard region in a casting), the AC system changes the cutting speed or the feed to lower the cutting force to an acceptable level (Fig. 14.14). Note that without AC or the direct intervention of the operator, high cutting forces may cause the tools to chip or break or cause the workpiece to deflect or distort excessively. As a result, the dimensional accuracy and surface finish of the part deteriorate.

Systems that optimize an operation are called **adaptive-control optimization** (ACO) systems. Optimization may, for example, involve maximizing material-removal rate between tool changes or indexing, or improving surface finish. Currently, most systems are based on ACC, because the development and the proper implementation of ACO are more complex.

Response time during the operation must be short for AC to be effective, particularly in high-speed machining operations (see Section 8.8). Assume, for example, that a turning operation is being performed on a lathe at a spindle speed of 1000 rpm, and the tool suddenly chips off, adversely affecting the surface finish and dimensional

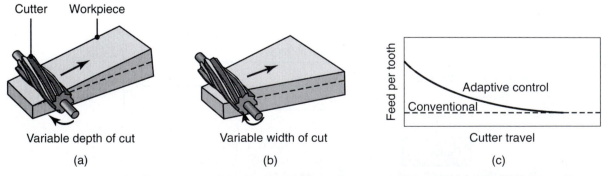

FIGURE 14.14   An example of adaptive control in slab milling. As the depth of cut or the width of cut increases, the cutting forces and the torque increase; the system senses this increase and automatically reduces the feed to avoid excessive forces or tool breakage. *Source:* After Y. Koren.

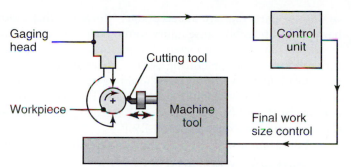

FIGURE 14.15 In-process inspection of workpiece diameter in a turning operation. The system automatically adjusts the radial position of the cutting tool in order to machine the correct diameter.

accuracy of the part being machined. In order for the AC system to be effective, the sensing system must respond within a very short time, as otherwise the damage to the workpiece will be extensive.

For adaptive control to be effective, *quantitative relationships* must be established and stored in the computer memory as mathematical models. For example, if the tool-wear rate in a machining operation is excessive, the computer must be able to determine how much of a change (increase or decrease) in cutting speed or feed is necessary in order to reduce the wear rate to an acceptable level. The system must also be able to compensate for dimensional changes in the workpiece due to such causes as tool wear and temperature rise (Fig. 14.15). If the operation is, for example, grinding (Section 9.6), the computer software must contain the desired quantitative relationships among process variables and such parameters as wheel wear, dulling of abrasive grains, grinding forces, temperature, surface finish, and part deflections. Likewise, in bending of a sheet in a V-die (Section 7.4), data on the dependence of springback on punch travel and on other material and process variables must be stored in the computer memory.

It is apparent that, because of the numerous factors involved, mathematical equations for such quantitative relationships in manufacturing processes are difficult to establish. Compared to the various other parameters involved, forces and torque in machining have been found to be the easiest to monitor by AC. Various solid-state power controls are commercially available, in which power is displayed or interfaced with data-acquisition systems.

14.6 | Material Handling and Movement

Material handling is defined as the functions and systems associated with the transportation, storage, and control of materials and parts in the total manufacturing cycle of a product. During this cycle, raw materials and parts (*work-in-progress*) are typically moved from storage to machines, from machine to machine, from inspection to assembly and to inventory (see also *just-in-time*, Section 15.12), and finally to shipment. For example, (a) a forging is mounted on a milling-machine bed (Fig. 8.66) for machining for better dimensional accuracy; (b) the machined forging may subsequently be ground for even better surface finish and dimensional accuracy; and (c) the part is inspected prior to being assembled into a finished product. Similarly, cutting tools have to be mounted on lathes, dies have to be placed in presses or hammers, grinding wheels will be mounted on spindles, and parts will be placed in special fixtures for dimensional measurement and inspection.

Material handling operations must be repeatable and reliable. Consider, for example, what would happen if a part is loaded improperly into a chuck or collet on a lathe, or a die or a mold. The consequences of such action may well be broken tools or dies, or parts that are unacceptable. Moreover, this action can also present safety hazards and possibly cause injury to the operator and nearby personnel.

Plant layout is important for the orderly flow of materials and components throughout the manufacturing cycle. Obviously, distances required for moving raw materials and parts in progress should be minimized, and storage areas and service centers should be organized accordingly. For parts requiring multiple operations, equipment should be grouped around the operator or the industrial robot. (See also *cellular manufacturing,* Section 15.9.)

Important aspects of material handling are summarized below

1. **Methods of material handling.** Several factors must be considered in selecting an appropriate material-handling method for a particular manufacturing operation:

   a. Shape, size, weight, and characteristics of the parts.

   b. Distances involved and the position and orientation of the parts during movement and at their final destination.

   c. Conditions of the path along which the parts are to be transported.

   d. Level of automation and control desired, and integration with other equipment in the system.

   e. Operator skill required.

   f. Economic considerations.

   For small batch operations, raw materials and parts can be handled and transported by hand, but this method can be time consuming and thus costly. Also, because it involves manual operations, this practice can be unpredictable and unreliable; it can even be unsafe to the operator, for reasons such as the weight and size of the parts to be moved and such environmental factors as heat and smoke in older foundries and forging plants.

2. **Equipment.** Several types of equipment can be used to move materials: conveyors, rollers, self-powered monorails, carts, forklift trucks, automated guided vehicles, and various mechanical, electrical, magnetic, pneumatic, and hydraulic devices and manipulators. **Manipulators** are designed to be controlled directly by the operator, or they can be automated for repetitive movements (such as the loading and unloading of parts from machine tools, presses, dies, and furnaces). They are capable of gripping and moving heavy parts and orienting them as required. Machinery combinations that have the capability of conveying parts without the use of additional material-handling equipment are called **integral transfer devices.**

   Warehouse space can be used efficiently, thus reducing labor costs, by using **automated storage/retrieval systems** (AS/RS). There are several such systems available, with varying degrees of automation and complexity. However, these systems are not desirable because of the importance of *minimal or zero inventory* and on *just-in-time production* methods, described in Section 15.12.

3. **Automated guided vehicles.** Flexible material handling and movement, with real-time control, is an integral part of modern manufacturing. First developed in the 1950s, *automated guided vehicles* (AGVs) are used extensively in flexible manufacturing (Fig. 14.16). This transport system has high flexibility and is capable of efficient delivery to different workstations. The movements of AGVs can be planned so that they also interface with automated storage/retrieval systems.

   There are several types of AGVs commonly in use, with a variety of designs, load-carrying capacities, and features for specific applications. Among

(a)   (b)

FIGURE 14.16 (a) A self-guided vehicle (Tugger type). This vehicle can be arranged in a variety of configurations to pull caster-mounted cars; it has a laser sensor to ensure that the vehicle operates safely around people and various obstructions. (b) A self-guided vehicle configured with forks for use in a warehouse. *Source:* Courtesy of Egemin, Inc.

these are (a) unit-load vehicles with pallets that can be handled from both sides; (b) light-load vehicles, equipped with bins or trays for light manufacturing; (c) tow vehicles (see Fig. 14.16a); (d) forklift-style vehicles (Fig. 14.16b); and (e) assembly-line vehicles to carry subassemblies to final assembly. Pallet sizes vary, and loading and unloading are accomplished either manually or automatically, by using various transfer mechanisms. The load capacity of AGVs can range up to 7000 lbs.

AGVs are guided automatically along pathways with in-floor wiring (for *magnetic guidance*) or tapes or fluorescent-painted strips (for *optical guidance*, called *chemical guide path*). Some systems may require additional operator guidance. *Autonomous guidance* involves no wiring or tapes and uses various optical, ultrasonic, and inertial techniques with onboard controllers. *Routing* of the AGV can be controlled and monitored from a central computer, such that the system optimizes the movement of materials and parts in case of congestion around workstations, machine breakdown, or the failure of one section of the manufacturing system. Proper traffic management on the factory floor is important. Sensors and various controls on the vehicles are designed to avoid collisions with other AGVs or machinery on the plant floor.

4. **Coding systems.** Various *coding systems* have been developed to locate and identify parts and subassemblies throughout the manufacturing system and to correctly transfer them to their appropriate stations:

   a. **Bar coding** is the most widely used and least costly system. The codes are printed on labels, which are attached to the parts themselves and read by fixed or portable bar code readers or handheld scanners.

   b. **Magnetic strips,** such as those on the back of a credit card, constitute the second most common coding system.

   c. **Radio frequency** (RF) **tags** constitute the third system. Although expensive, the tags do not require the clear line of sight necessary for the first two systems. In addition, they have a long range, from hundreds of meters for conventional RF tags to around 10–30 meters for bluetooth systems (see Section 15.14), and are rewritable.

   d. **Acoustic waves, optical character recognition,** and **machine vision** are other identification methods. (See Section 14.8.1.)

14.7 | Industrial Robots

The word *robot* was coined in 1920 by the Czech author K. Čapek in his play *R.U.R.* (Rossum's Universal Robots). It is derived from the Czech word *robota*, meaning "worker." An *industrial robot* has been defined as a reprogrammable multifunctional manipulator designed to move materials, parts, tools, or other devices by means of variable programmed motions and to perform a variety of other tasks. In a broader context, the term *robot* also includes manipulators that are activated directly by an operator. Introduced in the early 1960s, the first industrial robots were used in hazardous operations, such as handling toxic and radioactive materials and the loading and unloading of hot workpieces from furnaces and in foundries.

In its earliest applications a rule-of-thumb for the need for robots was the three D's (*dull, dirty,* and *dangerous*) and the three H's (*hot, heavy,* and *hazardous*). Today, however, industrial robots have much broader applications and are very important components in manufacturing processes and operations and have helped improve productivity and product quality and significantly reduce labor costs.

14.7.1 Robot components

To appreciate the functions of robot components and their capabilities, one merely needs to observe the flexibility and capability of the diverse movements of the human arm, wrist, hand, and fingers in, for example, reaching for and gripping an object from a shelf, using a hand tool or a pair of scissors, or operating a machine or an automobile. The basic components of an industrial robot are described next (Fig. 14.17a).

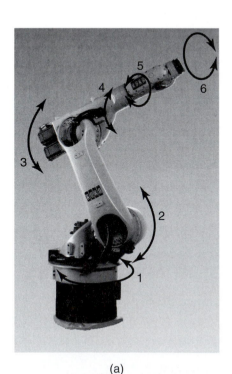

(a)

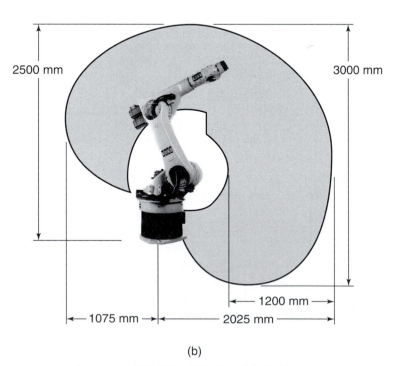

(b)

FIGURE 14.17   (a) Schematic of a six-axis KR-30 KUKA robot; the payload at the wrist is 30 kg and repeatability is ±0.15 mm (±0.006 in.). The robot has mechanical brakes on all its axes. (b) The work envelope of the KUKA robot, as viewed from the side. *Source:* Courtesy of KUKA Robotics.

1. **Manipulator.** Also called **arm and wrist,** the *manipulator* is a mechanical unit that provides motions (trajectories) similar to those of a human arm and hand, using various devices such as linkages, gears, and joints. The end of the wrist can reach a point in space with a specific set of coordinates and in a specific orientation. Most robots have six rotational joints. (See Fig. 14.17a.) There are also four-degrees-of-freedom (d.o.f.) and five-d.o.f. robots, but these kinds are not fully dexterous, because full dexterity, by definition, requires six d.o.f. Seven-d.o.f. (or *redundant*) robots for special applications also are available.

2. **End effector.** The end of the wrist in a robot is equipped with an *end effector.* Also called **end-of-arm tooling,** end effectors can be custom made to meet special handling requirements. *Mechanical grippers* are the most commonly used end effectors and are equipped with two or more fingers. The selection of an appropriate end effector for a specific application depends on such factors as the *payload* (weight of the object to be lifted and moved; load-carrying capacity), environment, reliability, and cost. Depending on the type of operation, end effectors may be equipped with any of the following:

   a. Grippers, hooks, scoops, electromagnets, vacuum cups, and adhesive fingers for material handling (Fig. 14.18a).

   b. Spray guns for painting.

   c. Various attachments, such as for spot and arc welding and for arc cutting.

   d. Power tools, such as drills, nut drivers, burrs, and sanding belts.

   e. Measuring instruments, such as dial indicators and laser or contact probes.

   Compliant end effectors are used to handle fragile materials or to facilitate assembly. They can use elastic mechanisms to limit the force that can be applied to a workpiece or part, and they can be designed with a specific required stiffness. For example, end effectors can be designed to be stiff in the axial direction, but compliant in lateral directions. This arrangement prevents damage to parts in those assembly operations in which slight misalignments can occur.

3. **Power supply.** Each motion of the manipulator, in linear and rotational axes, is controlled and regulated by independent actuators that use an electrical, pneumatic, or hydraulic power supply; each has its own characteristics, advantages, and limitations.

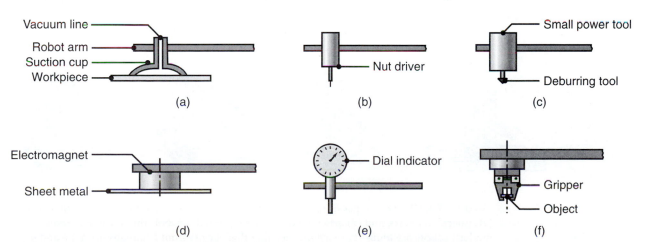

FIGURE 14.18 Various devices and tools that can be attached to end effectors to perform a variety of operations.

4. **Control system.** Whereas manipulators and end effectors are the robot's arms and hands, the control system is the *brain* of a robot. Also known as the **controller,** the control system is the communications and information-processing system that gives commands for the movements of the robot; it stores data to initiate and terminate movements of the manipulator. The controller acts as the *nerves* of the robot; it interfaces with computers and other equipment such as manufacturing cells or assembly operations. **Feedback devices** are an important part of a robot's control system. Robots with a *fixed set of motions* have **open-loop control.** In this system, commands are given, and the robot arm goes through its motions. However, accuracy of the movements is not monitored and the system does not have a self-correcting capability (see also Section 14.3.1 and Fig. 14.8). In **closed-loop systems,** positioning feedback leads to better accuracy.

As in numerical control machines, the types of control in industrial robots are *point-to-point* and *continuous-path* (Section 14.3.3). Depending on the particular task, the *positioning repeatability* required for an industrial robot may be as small as 0.050 mm (0.002 in.), as in assembly operations for electronic printed circuitry (see Section 13.13). Accuracy and repeatability vary greatly with payload and with position within the *work envelope* (see Section 14.7.2) and, as such, are very difficult to quantify for most robots.

14.7.2 Classification of robots

Robots may be classified by basic types, as illustrated in Fig. 14.19:

1. **Cartesian,** or rectilinear.
2. **Cylindrical.**
3. **Spherical,** or polar.
4. **Articulated,** or revolute, jointed, or anthropomorphic.

Robots may be attached permanently to the plant floor, move along overhead rails (**gantry robots**), or they may be equipped with wheels and move along the factory floor (**mobile robots**). Robots can also be classified according to their operation and programming as follows:

1. **Fixed-sequence and variable-sequence robots.** Also called a **pick-and-place robot,** a fixed-sequence robot is programmed for a specific sequence of

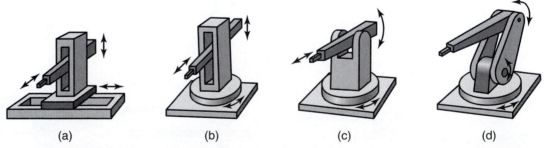

(a)   (b)   (c)   (d)

FIGURE 14.19 Four types of industrial robots: (a) Cartesian (rectilinear), (b) cylindrical, (c) spherical (polar), and (d) articulated (revolute, jointed, or anthropomorphic). Some modern robots are *anthropomorphic,* meaning that they resemble humans in shape and in movement. These complex mechanisms are made possible by powerful computer processors and fast motors that can maintain a robot's balance and accurate movement control.

point-to-point movements; the cycle is repeated continuously. These robots are simple and relatively inexpensive. The *variable-sequence robot* can be programmed for multiple specific operation sequences, any of which it can execute when given the proper cue, such as a signal from a controlling or scheduling computer, a bar code, or a signal from an inspection station.

2. **Playback robot.** An operator leads or walks the playback robot and its end effector through the desired path; in other words, the operator teaches the robot by showing it what to do. The robot memorizes and records the path and sequence of motions; it can then repeat them continually without any further action or guidance by the operator. The robot has a *teach pendant*, which uses handheld button boxes that are connected to the control panel. These button boxes are used to control and guide the robot through the work to be performed. These movements are registered in the memory of the controller and are automatically reenacted by the robot whenever required.

3. **Numerically controlled robot.** This type of robot is programmed and operated much like a numerically controlled machine. The robot is servocontrolled by digital data, and its sequence of movements can be changed with relative ease. As in NC machines, there are two basic types of controls: point to point and continuous path. *Point-to-point* robots are easy to program and have a higher payload and a larger **work envelope** (also called the **working envelope,** the maximum extent or reach of the robot hand or working tool in all directions), as illustrated in Figs. 14.17b and 14.20. *Continuous-path* robots have a higher accuracy than point-to-point robots, but they have a lower payload. More advanced robots have a complex system of path control, enabling high-speed movements with high accuracy.

4. **Intelligent (sensory) robot.** The *intelligent robot* is capable of performing some of the functions and tasks carried out by human beings. It is equipped with a variety of sensors with visual (*computer vision;* see Section 14.8) and *tactile* (touching) capabilities. Much like humans, the robot observes and evaluates the immediate environment and its proximity to other objects, especially machinery, in its path, by *perception* and *pattern recognition.* The robot then makes appropriate decisions for the next movement and proceeds accordingly. Because its operation is complex, powerful computers are required to control this type of robot.

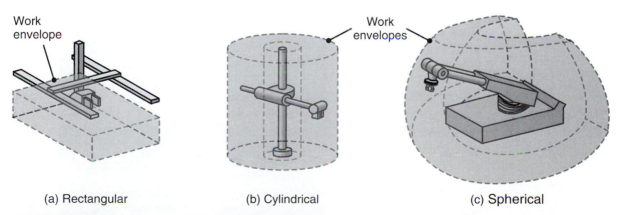

(a) Rectangular   (b) Cylindrical   (c) Spherical

FIGURE 14.20 Work envelopes for three types of robots. The selection depends on the particular application (See also Fig. 14.17b.)

14.7.3 Applications and selection of robots

Major applications of industrial robots include the following:

1. *Material handling,* some examples being (a) casting and molding operations, in which molten metal, raw materials, and parts in various stages of completion are handled without operator interference; (b) heat treating, in which parts are loaded and unloaded from furnaces and quench baths; and (c) forming and shaping operations, in which parts are loaded and unloaded from presses and various other types of machinery.

2. *Spot welding* by robots, especially for automobile and truck bodies, produces reliable welds of high quality (Fig. 14.21). Robots also perform other similar operations, such as arc welding, arc cutting, and riveting (see Chapter 12).

3. *Finishing operations,* such as grinding, deburring, and polishing (Chapter 9) can be done by using appropriate tools attached to end effectors.

4. *Applying adhesives and sealants,* such as in the automobile frame shown in Fig. 14.22.

FIGURE 14.21   Spot welding automobile bodies with industrial robots. *Source:* Courtesy of Ford Motor Co.

FIGURE 14.22   Sealing joints of an automobile body with an industrial robot. *Source:* Courtesy of Cincinnati Milacron, Inc.

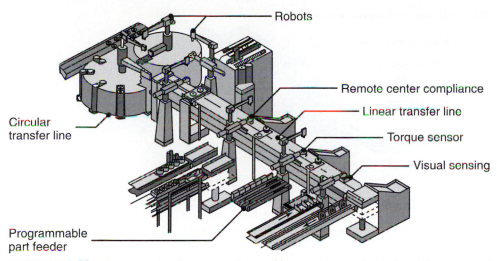

Robots

Remote center compliance

Linear transfer line

Torque sensor

Visual sensing

Circular
transfer line

Programmable
part feeder

FIGURE 14.23   An example of automated assembly operations using industrial robots and circular and linear transfer lines.

5. *Spray painting,* particularly of parts with complex shapes, and cleaning operations; the motions for one piece are repeated accurately for the next piece.

6. *Automated assembly,* performing very repetitive operations (Fig. 14.23); see also Section 14.10.

7. *Inspection and gaging* in various stages of manufacture, at speeds much higher than those that can be achieved by humans.

Robot selection. Factors that influence the selection of robots in manufacturing operations are (a) cost, (b) payload, (c) speed of movement, (d) reliability, (e) repeatability, (f) arm configuration, (g) number of degrees of freedom, (h) control system, (i) program memory, and (j) work envelope. Consequently, robots are rarely off-the-shelf items; instead they must be integrated with controllers, end effectors, and their environment.

14.8 | Sensor Technology

A *sensor* is a device that produces a signal in response to detection or measurement of a specific quantity or a property, such as position, force, torque, pressure, temperature, humidity, speed, acceleration, or vibration. Traditionally, sensors, actuators, and switches have been used to set limits on the performance and movements of machines, such as stops on machine-tool slideways to restrict worktable movements, pressure and temperature gages with automatic shutoff features, and governors on engines to prevent excessive speed of operation. Sensor technology is an important aspect of manufacturing processes and systems and is essential for data acquisition, monitoring, communication, and computer control of machines and systems (Fig. 14.24).

Because they convert one quantity to another, sensors also are often referred to as **transducers. Analog sensors** produce a signal, such as voltage, that is proportional to the measured quantity. **Digital sensors** have digital (numeric) outputs that can directly be transferred to computers. **Analog-to-digital converters** (ADCs) are used for interfacing analog sensors with computers.

FIGURE 14.24
A toolholder equipped with thrust-force and torque sensors (*smart tool holder*), capable of continuously monitoring the machining operation. (See Section 14.5.) *Source:* Courtesy of Cincinnati Milacron, Inc.

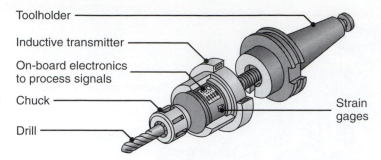

Toolholder
Inductive transmitter
On-board electronics to process signals
Chuck
Drill
Strain gages

14.8.1 Sensor classification

Sensors that are of interest in manufacturing processes and operations may be classified as follows:

1. *Mechanical* sensors, which measure such quantities as position, shape, velocity, force, torque, pressure, vibration, strain, and mass.
2. *Electrical* sensors, which measure voltage, current, charge, and conductivity.
3. *Magnetic* sensors, which measure magnetic field, flux, and permeability.
4. *Thermal* sensors, which measure temperature, flux, conductivity, and specific heat.
5. *Acoustic, ultrasonic, chemical, optical, radiation, laser,* and *fiber-optic* sensors.

Sensors are essential to the control of intelligent robots; they continue to be developed with capabilities that resemble those of the senses of human beings (*smart sensors;* see below). Depending on its application, a sensor may be made of metallic, nonmetallic, organic, or inorganic materials as well as fluids, gases, plasmas, or semiconductors. Using the special characteristics of these materials, sensors convert the quantity or property measured to analog or digital output.

Actuators physically contact the object and take appropriate action, usually by electromechanical means. The operation of a common mercury thermometer is based on the difference between the thermal expansion of mercury and that of glass. Similarly, the presence of a physical body or barrier can be detected by a *break of a beam of light,* sensed by a photoelectric cell. A *proximity sensor,* which senses and measures the distance between it and an object or a moving member of a machine, can be based on acoustics, magnetism, capacitance, or optics.

Sensors are also classified as follows.

1. **Tactile sensing** involves the continuous sensing of varying contact forces, commonly by an *array of sensors.* Such a system is capable of performing within an arbitrary three-dimensional space. Fragile parts, such as glass bottles, electronic devices, or eggs, can be handled by robots with *compliant (smart) end effectors.* These effectors can sense the force applied to the object being handled, using piezoelectric devices, strain gages, magnetic induction, ultrasonics, and optical systems of fiber optics and light-emitting diodes. Tactile sensors that are capable of measuring and controlling gripping forces and moments in three axes have been designed (Fig. 14.25).

    The *gripping force* of an end effector is sensed, monitored, and controlled through closed-loop feedback devices. Compliant grippers that have force-feedback capabilities and sensory perception can, however, be complicated and require powerful computers, and thus can be costly. *Anthropomorphic end effectors* are designed to simulate the human hand and fingers and to have the

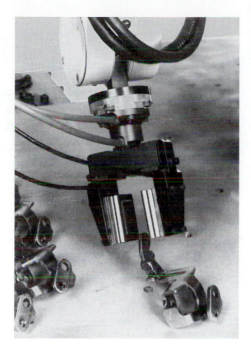

FIGURE 14.25 A robot gripper with tactile sensors. In spite of their capabilities, tactile sensors are now being used less frequently, because of their high cost and low durability (lack of robustness) in industrial applications. *Source:* Courtesy of Lord Corporation.

capability of sensing touch, force, movement, and pattern. The ideal tactile sensor must also sense slip (such as a heavy object slipping between one's fingers), a motion that human fingers sense readily.

2. **Visual sensing (machine vision; computer vision)** involves cameras that optically sense the presence (and hence also the absence) and shape of an object (Fig. 14.26). A microprocessor processes the image, usually in less than one second. The image is then measured, and the measurements are digitized (**image recognition**). There are two basic systems of machine vision: **Linear arrays** sense only one dimension, such as the presence of an object or some feature on its surface. **Matrix arrays** sense two or three dimensions and are capable of detecting, for example, a properly inserted component in a printed circuit or a properly made solder joint, an operation known as **assembly verification**. When used in automated inspection systems (Section 4.8.3), these sensors can also detect cracks and flaws.

   Machine vision is particularly suitable for parts that have inaccessible features, in hostile manufacturing environments, for measuring a large number of small features, and in situations where physical contact with the part may cause damage to the part. Examples of applications of machine vision include (a) on-line, real-time inspection in sheet-metal stamping lines; and (b) sensors for machine tools that can sense tool offset and tool breakage, verify part placement and fixturing, and monitor surface finish; see also Fig. 14.26. Machine vision is capable of in-line identification and inspection of parts and of rejection of defective parts. With visual sensing capabilities, end effectors can pick up parts and grip them in the proper orientation and location.

3. **Smart sensors** have the capability to perform a logic function, conduct two-way communication, and make decisions and take appropriate actions. The necessary input, and the knowledge required to make decisions, can be built into a smart sensor. For example, a computer chip with sensors can be programmed to turn a machine tool off in the event that a cutting tool fails. Likewise, a smart sensor can stop a mobile robot or a robot arm from accidentally coming in contact with a machine or people by sensing quantities such as distance, heat, and noise.

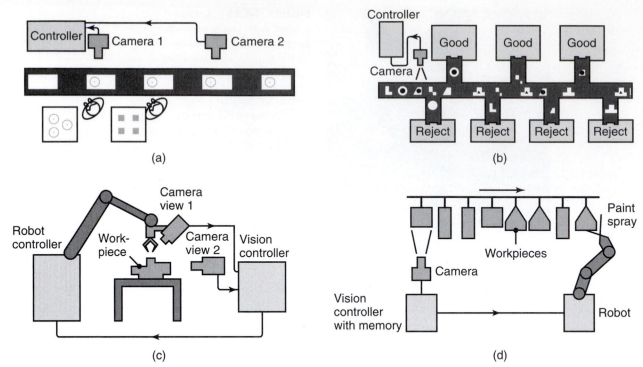

FIGURE 14.26 Examples of machine vision applications: (a) in-line inspection of parts, (b) identifying parts with various shapes, and inspection and rejection of defective parts, (c) use of cameras to provide positional input to a robot relative to the workpiece, and (d) painting of parts with different shapes by means of input from a camera; the system's memory allows the robot to identify the particular shape to be painted and to proceed with the correct movements of a paint spray attached to the end effector.

Sensor selection. The selection of a sensor depends on such factors as (a) the particular quantity to be measured or sensed, (b) its interaction with other components in the system, (c) its expected service life, (d) the required level of performance, (e) difficulties associated with use of the sensor, and (f) its cost. An important consideration is the type of environment in which the sensor is to be used. Rugged (*robust*) sensors are designed to withstand, for instance, extremes of temperature, shock and vibration, humidity, corrosion, dust and various contaminants, fluids, electromagnetic radiation, and other interferences present in industrial environments. (See also the discussion of *robust design* in Section 16.2.3.)

14.8.2 Sensor fusion

Sensor fusion involves the integration of several different sensors in such a manner that the individual data from each of the sensors (such as force, vibration, temperature, and dimensions) are combined to provide a higher level of information and reliability. A simple and common example of sensor fusion occurs when we drink from a cup of hot tea or coffee. Although such a practice is taken for granted, it can readily be seen that this process involves simultaneous data input from the person's eyes, lips, tongue, fingers, and hands. Through our five senses (sight, hearing, smell, taste, and touch), there is, in this example, real-time monitoring of temperature, movements, and positions. Thus, for example, if the coffee is too hot, the movement of the cup toward the lip is controlled accordingly.

The earliest applications of sensor fusion were in robot movement control and in missile-flight tracking and similar military applications, primarily because these activities involve movements that mimic human behavior. An example of sensor fusion in manufacturing is a machining operation in which a number of different, but integrated, sensors continuously monitor quantities such as (a) dimensions and surface finish of the workpiece being machined; (b) cutting forces, vibration, and tool wear and fracture; (c) temperatures in various regions of the tool-workpiece system; and (d) spindle power.

An essential aspect in sensor fusion is **sensor validation,** in which the failure of one particular sensor is detected so that the control system retains high reliability. In validation, the receipt of redundant data from different sensors is essential. Although complex and relatively expensive, sensor fusion and validation is now possible due to the advances made in sensor size, quality, and technology, and continued developments in control systems, artificial intelligence, expert systems, and artificial neural networks (described in Chapter 15).

14.9 | Flexible Fixturing

In workholding devices for manufacturing operations the words *fixture, clamp,* and *jig* are often used interchangeably and sometimes in pairs, such as in *jigs and fixtures.* Common workholding devices include chucks, collets, and mandrels. Although many of these devices are operated manually, particularly in job shops, other workholding devices are designed and operated at various levels of mechanization and automation, such as *power chucks,* which are driven by mechanical, hydraulic, or electrical means.

Fixtures are generally designed for specific purposes; **clamps** are simple multi-functional devices; **jigs** have various reference surfaces and points for accurate alignment of parts and tools and are widely used in mass production (see also the discussion of *pallets* in Section 8.11). These devices may be used for actual manufacturing operations (in which case the forces exerted on the part must maintain the part's position in the machine without slipping or distortion), or they may be used to hold workpieces for purposes of measurement and inspection, where the part is not subjected to any forces.

Workholding devices have certain ranges of capacity; for example, (a) a specific collet can accommodate rods or bars only within a certain range of diameters; (b) four-jaw chucks can accommodate square or prismatic workpieces of various sizes; and (c) other devices and fixtures are designed and made for specific workpiece shapes and dimensions and for specific tasks, called **dedicated fixtures.** It is very simple to reliably fixture a workpiece in the shape of a rectangular bar, such as by clamping it between the parallel jaws of a vise. If the part has curved surfaces, it is possible to shape the contacting surfaces of the jaws themselves by machining them (called *machinable jaws*) to conform to the workpiece surfaces.

The emergence of flexible manufacturing systems (Section 15.10) has necessitated the design and use of workholding devices and fixtures that have **built-in flexibility.** There are several methods of *flexible fixturing,* based on different principles (also called **intelligent fixturing systems**), and the term itself has been defined somewhat loosely. Basically, however, these devices are capable of quickly accommodating a range of part shapes and dimensions without the necessity of making extensive changes and adjustments or requiring operator intervention, both of which would adversely affect productivity.

FIGURE 14.27
Components of a modular workholding system. *Source: Courtesy of Carr Lane Manufacturing Co.*

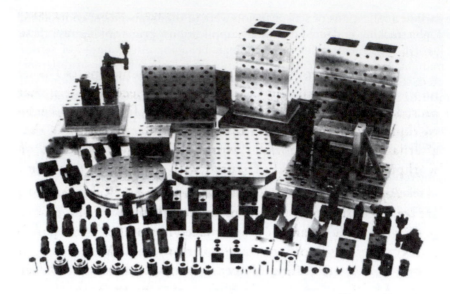

In addition to the use of some of the methods outlined previously, including the use of robots and computer controls for manipulating part orientation for proper clamping, other techniques for fixturing are described as follows.

1. **Modular fixturing.** *Modular fixturing* is often used for small or moderate lot sizes (Fig. 14.27), especially when the cost of dedicated fixtures and the time required to make them are difficult to justify. Complex workpieces can be located within machines through fixtures produced quickly from standard components and can be disassembled when a production run is completed. These modular fixtures are usually based on tooling plates or blocks configured with grid holes or T-slots upon which a fixture is constructed.

    A number of other standard components, such as locating pins, adjustable stops, workpiece supports, V-blocks, clamps, and springs, can be mounted onto the base plate or block to quickly produce a fixture. By computer-aided fixture planning for specific situations, such fixtures can be assembled and modified using robots. As compared with dedicated fixturing, modular fixturing has been shown to be low in cost, have a shorter lead time, provide greater ease of repair of damaged components, and offer a more intrinsic flexibility of application.

2. **Tombstone fixtures.** Also referred to as *pedestal-type fixtures*, tombstone fixtures have between two and six vertical faces (hence resembling tombstones) onto which parts can be mounted. Tombstone fixtures are typically used in automated or robot-assisted manufacturing; the machine tool performs the desired operations on the part or parts on one face, then flips or rotates the tombstone to begin work on other parts. These fixtures allow feeding more than one part into a machine but are not as flexible as other fixturing approaches. Thus, tombstone fixtures are commonly used for higher volume production, typically as in the automotive industry (see also the Case Study in Chapter 8).

3. **Bed-of-nails device.** This fixture consists of a series of air-actuated pins that conform to the shape of the external part surfaces. Each pin moves as necessary to conform to the shape at its point of contact with the part; the pins are then mechanically locked against the part. The fixture is compact and has high stiffness and is reconfigurable.

4. **Adjustable-force clamping.** Figure 14.28 shows a schematic illustration of another flexible fixturing system. In this system, referred to as an *adjustable-force*

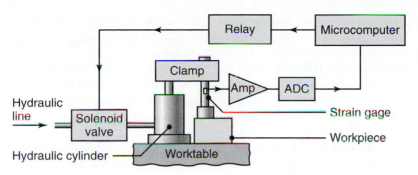

FIGURE 14.28 Schematic illustration of an adjustable-force clamping system. The clamping force is sensed by the strain gage, and the system automatically adjusts this force. ADC stands for analog-digital converter. *Source:* After P.K. Wright and D.A. Bourne.

clamping system, the strain gage attached to the clamp senses the magnitude of the clamping force; the system then adjusts this force to keep the workpiece securely clamped to the workpiece.

5. **Phase-change materials.** There are two basic methods to hold irregularly shaped or curved workpieces in a medium, other than hard tooling:

a. In the first, and older, method, a *low-melting-point metal* is used as the clamping medium. For example, an irregularly shaped workpiece is partially dipped into molten lead and allowed to set (like the wooden stick in an ice-cream bar, a process similar to *insert molding;* see Section 10.10.2 and Fig. 10.30). After setting, the assembly is clamped in a simple fixture. This method is particularly common in the aerospace industry. Note, however, the possible adverse effects of such shrouding materials as lead (see *liquid-metal embrittlement,* Section 3.4.2) on human health and the environment.

There is a similar application of this method in machining the honeycomb structures shown in Fig. 7.48. Note that because the walls of the hollow structure are very thin, the forces exerted by the cutting tool would easily distort or damage it. One method of stiffening this structure is to fill the cavities with water and freeze it. Thus, the hexagonal cavities are now supported by ice, whose strength is sufficient to resist the cutting forces. After machining, the ice is allowed to melt away.

b. In the second method, which is still in experimental stages, the supporting medium is either a *magnetorheological* (MR) or *electrorheological* (ER) fluid. In the MR application, magnetic particles (which are micrometer-sized or are nanoparticles; Sections 3.11.9, 11.2.1, and 11.8.1) are suspended in a nonmagnetic fluid. Surfactants are added to maintain dispersal of powders. After the workpiece is immersed in the fluid, an external magnetic field is applied, whereby the particles are polarized and the behavior of the fluid changes from a liquid to a solid. After the part is processed, it is retrieved after removing the external magnetic field. In the ER application, the fluid is a suspension of fine dielectric particles in a liquid with a low dielectric constant. After applying an electrical field, the liquid becomes a solid.

14.10 | Assembly, Disassembly, and Service

Some products are simple and have only two or three components to *assemble,* an operation that can be done with relative ease. Examples include the manufacture of an ordinary pencil with an eraser, a frying pan with a wooden handle, and an aluminum beverage can. Most products, however, consist of many parts, and their assembly requires considerable care and planning.

Traditionally, assembly has involved much *manual labor* and, thus, has contributed significantly to product cost. The total assembly operation is usually broken into individual assembly operations (**subassemblies**), with an operator assigned to carry out each step. Assembly costs are now typically 25 to 50% of the total cost of manufacturing, with the percentage of workers involved in assembly operations ranging from 20 to 60%. In the electronics industries, for example, some 40 to 60% of total wages are paid to assembly workers. As costs increase, the necessity for **automated assembly** becomes obvious.

Beginning with the hand assembly of muskets in the late 1700s and the introduction of *interchangeable parts* in the early 1800s (see Eli Whitney, Section 4.9.1), assembly methods have seen continued improvement. The first application of large-scale modern assembly was for the flywheel magnetos for the Model T Ford; this activity eventually led to *mass production* of the automobile itself.

The choice of an assembly method and system depends on the required production rate, the total quantity to be produced, the product's market life, labor availability, and cost. As indicated throughout this text, parts are manufactured within certain dimensional tolerance ranges. (See also Section 4.7.) Taking ball bearings as an example, we know that although they all have the same nominal dimensions, some balls in a particular lot will be a little smaller than others and some inner and outer races produced will be smaller than others in the lot.

There are two methods of assembly for such high-volume products:

1. In **random assembly,** parts are put together by selecting them randomly from the lots produced.

2. In **selective assembly,** the balls and the races are segregated by groups of sizes, from smallest to largest. The parts are then selected to mate properly. Thus, the smallest diameter balls are mated with inner races that have the largest outside diameters and with outer races that have the smallest inside diameters.

14.10.1 Assembly systems

There are three basic methods of assembly: *manual, high-speed automatic,* and *robotic*. These methods can be used individually or, as is the case for most applications in practice, in combination. Before assembly operations commence, an analysis of the product design (Fig. 14.29) should be made to determine the most appropriate and economical method or methods of assembly.

1. **Manual assembly** uses simple tools and is generally economical for relatively small lots. Because of the dexterity of the human hand and fingers and their capability for feedback through various senses, workers can assemble even

FIGURE 14.29 Stages in the design-for-assembly analysis. *Source:* After G. Boothroyd and P. Dewhurst.

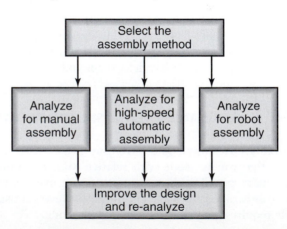

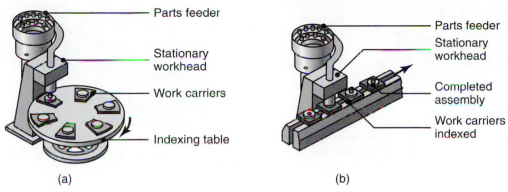

FIGURE 14.30 Transfer systems for automated assembly: (a) rotary indexing machine and (b) in-line indexing machine. *Source:* After G. Boothroyd.

complex parts without much difficulty. Note, for example, that aligning and placing a simple square peg into a square hole, involving small clearances, is a simple manual operation, but it can be difficult in automated assembly. There can, however, be potential problems of *cumulative trauma disorders* (*carpal tunnel syndrome*) associated with this method.

2. **High-speed automated assembly** uses *transfer mechanisms* designed specially for assembly operations. Two examples are shown in Fig. 14.30, in which individual assembly is carried out on products that are indexed for proper positioning during assembly. In **robotic assembly,** one or two general-purpose robots operate at a single workstation, or the robots operate at a multistation assembly system.

   There are three basic types of automated assembly systems: synchronous, nonsynchronous, and continuous.

   a. **Synchronous systems,** also called **indexing systems.** Individual parts and components are supplied and assembled at a constant rate at fixed individual stations. The rate of movement is based on the station that takes the longest time to complete its portion of the assembly. This system is used primarily for high-volume, high-speed assembly of small products.

      Transfer systems move the partial assemblies from workstation to workstation by various mechanical means, two typical transfer systems being **rotary indexing** and **in-line indexing systems** (Fig. 14.30). These systems can operate in either a fully automatic mode or a semiautomatic mode. However, a breakdown of one station will shut down the whole assembly operation. The part feeders supply the individual parts to be assembled and place them on other components, which are secured on work carriers or fixtures. The feeders move the individual parts (by vibratory or other means) through delivery chutes and ensure their proper orientation by various ingenious means (Fig. 14.31). Proper orientation of parts and avoiding jamming are essential in all automated assembly operations.

   b. **Nonsynchronous systems.** Each station operates independently, and any imbalance in product flow is accommodated in storage (**buffer**) between stations. The station continues operating until the next buffer is full or the previous buffer is empty. Also, if for some reason one station becomes inoperative, the assembly line continues to operate until all the parts in the buffer have been used up. Nonsynchronous systems are suitable for large assemblies with many parts to be assembled, such as motors. For types of

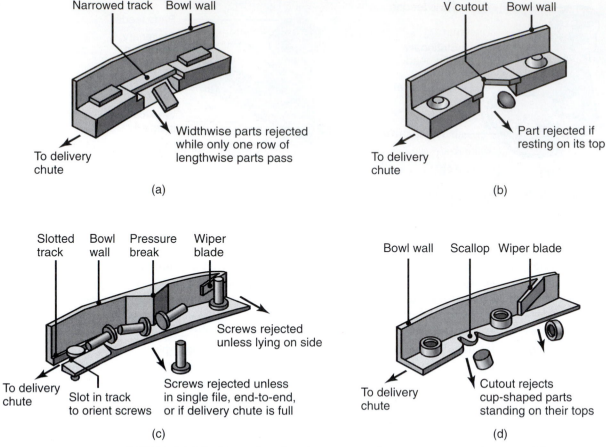

FIGURE 14.31 Examples of guides to ensure that parts are properly oriented for automated assembly. *Source:* After G. Boothroyd.

assembly in which the times required for individual assembly operations vary widely, the rate of output will be constrained by the slowest station.

c. **Continuous systems.** The product is assembled while moving at a constant speed on pallets or similar workpiece carriers. The parts to be assembled are brought to the product by various workheads, and their movements are synchronized with the continuous movement of the product. Typical applications of this system are in bottling and packaging plants, although this method has also been used on mass-production lines for automobiles and appliances.

Assembly systems are generally set up for a certain product line. They can, however, be modified for increased flexibility for product lines that have a variety of models, called flexible assembly systems (FAS). These systems use computer controls, interchangeable and programmable workheads and feeding devices, coded pallets, and automated guiding devices.

14.11 | Design Considerations

As in many aspects of manufacturing processes and systems, design is an integral part of the topics described in this chapter.

14.11.1 Design for fixturing

The proper design, construction, and operation of flexible workholding devices and fixtures are essential to the efficient operation of advanced manufacturing systems. The following is a list of the major design issues involved.

1. Workholding devices must position the workpiece automatically and accurately and must maintain its location precisely and with sufficient clamping force to withstand the forces involved in the particular manufacturing operation.

2. Fixtures must have sufficient stiffness to withstand, without excessive distortion, the normal and shear stresses developed at the fixture-workpiece interfaces.

3. The presence of loose chips and other debris between the locating surfaces of the workpiece and the fixture can be a serious problem. Chips are most likely to be present where cutting fluids are used, because chips tend to stick to the wet surfaces, by virtue of surface-tension forces.

4. A flexible fixture should accommodate parts to be made by different processes and for situations in which dimensions and surface features vary from part to part. This is even more important when the workpiece (a) is fragile or made of a brittle material, such as ceramics; (b) is made of a relatively soft and flexible material, such as plastic or rubber part; or (c) has a relatively soft coating on its contacting surfaces, such as a polymer coating.

5. Fixtures and clamps should have low profiles, so as to avoid collision with cutting tools. Collision avoidance is also an important factor in programming tool paths in machining operations; see Sections 14.3 and 14.4.

6. Flexible fixturing also must meet particular special requirements to function properly in manufacturing cells and flexible manufacturing systems. The time required to load and unload parts on machinery should be minimal in order to reduce cycle times.

7. Parts should be designed to allow for easy locating and clamping within a fixture. Flanges, flats, or other locating surfaces should be incorporated into the part design in order to simplify fixture design and to aid in part transfer among various machinery.

14.11.2 Design for assembly, disassembly, and service

Design for assembly.   Although the functions of a product and its design for manufacture have been matters of major interest, *design for assembly* (DFA) has attracted special attention (particularly design for automated assembly), because of the need to reduce assembly costs. In *manual assembly*, a major advantage is that humans can easily pick the correct part from a collection of parts or pick one from a collection of identical parts, such as from a nearby bin. Human vision, intelligence, and dexterity allow for proper orientation and assembly of very complex systems. In *high-speed automated assembly*, however, automatic handling generally requires that parts be separated from the bulk, conveyed by hoppers or vibratory feeders (see Fig. 14.30), and assembled in the proper locations and orientations.

Based on analyses of assembly operations as well as on experience, several guidelines for DFA have been developed through the years (see also Fig. 1.4):

1. The number and variety of parts in a product should be minimized, and multiple functions should be incorporated into a single part; also, consider subassemblies that could serve as modules.

2. Parts should have a high degree of symmetry (such as round or square) or a high degree of asymmetry (such as oval or rectangular) so that they cannot be installed

incorrectly and so that they do not require location, alignment, or adjustment. Parts should be designed for easy insertion into other components.

3. Designs should allow parts to be assembled without obstructions or without a direct line of sight.

4. Designs should, as much as possible, avoid the need for fasteners such as bolts, nuts, and screws; other methods such as snap fits should be considered (see Fig. 12.55). If fasteners must be used, their variety should be minimized, and they should be located and spaced in such a manner that tools can be used without obstruction.

5. Part designs should consider such factors as size, shape, weight, flexibility, abrasiveness, and tangling with other parts.

6. Assemblies should not be turned over for insertion of parts, so that they can be inserted from a single direction, preferably vertically and from above to take advantage of gravity.

7. Products should be designed, or existing products redesigned, so that there are no physical obstructions to the free movement of the parts during assembly (see Fig. 1.4); thus, for example, sharp external and internal corners should be replaced with chamfers, tapers, or radii.

8. Parts that may appear to be similar but are different should be color coded.

Design guidelines for *robotic assembly* have rules similar to those for manual and high-speed automated assembly. *Compliant end effectors* and *dexterous manipulators* have increased the flexibility of robots. Some additional guidelines are

1. Parts should be designed so that they can be gripped and manipulated by the same gripper (end effector) of the robot (see Fig. 14.18), thus avoiding the need for different grippers. Parts should be made available to the gripper in the correct orientation.

2. Assembly that involves threaded fasteners, such as bolts, nuts, and screws, may be difficult for robots to perform, but they can easily handle self-threading screws (for sheet metal, plastics, and wooden parts), snap fits, rivets, welds, and adhesives.

Evaluating assembly efficiency. Significant effort has been directed toward the development of analytical and computer-based tools to estimate the efficiency of assembly operations. These tools provide a basis for comparisons of different product designs and the selection of design attributes that make assembly easier.

To evaluate assembly efficiency, each component of an assembly is evaluated with respect to its features that can affect assembly, and a baseline estimated time required for assembling the part to the assembly is obtained. Note that such an evaluation can also be made for existing products. The assembly efficiency, ν, is given by

$$\nu = \frac{Nt}{t_{\text{tot}}} \tag{14.1}$$

where N is the number of parts and t_{tot} is the total assembly time. t is the ideal assembly time for a small part that presents no difficulties in handling, orientation or assembly and is commonly taken as 3 seconds. Using Eq. (14.1), competing designs can be evaluated with respect to design for assembly. It has been noted that products that are in need of redesign to facilitate assembly usually have assembly efficiencies of around 5–10%, whereas well-designed parts have assembly efficiencies of around 25%. It should be noted that assembly efficiencies near 100% are unlikely to be achieved in practice, because the 3-second baseline is usually not practical.

Design for disassembly. The manner and ease with which a product may be taken apart for maintenance or replacement of its parts is another important consideration in product design. Although there is no established set of guidelines, the general approach to design for disassembly requires the consideration of factors that are similar to those for design for assembly. Analysis of computer or physical models of products and their components can indicate potential problems in disassembly, such as obstructions, lack of line of sight, narrow and long passageways, and difficulty of firmly gripping and guiding objects.

An important aspect of design for disassembly is how, after its life cycle, a product is to be taken apart for *recycling,* especially with respect to its more valuable components. Note, for example, that depending on (a) their design and location, (b) the type of tools used for disassembly and (c) whether the tools are manual or power tools, rivets will take longer to remove than screws or snap fits, and that a bonded layer of valuable material on a component would be very difficult, if not impossible, and uneconomical to remove for recycling or reuse. Obviously, the longer it takes to disassemble components, the higher is the cost of doing so, which indeed may become prohibitive. Consequently, the time required for disassembly has to be studied and measured. Although it depends on the manner in which disassembly takes place, some examples of the time required are as follows: cutting wire, 0.25 s; disconnecting wire, 1.5 s; removing snap fits and clips, 1–3 s; and removing screws and bolts, 0.15–0.6 s per revolution.

Design for service. Design for assembly and disassembly must include the ease with which a product can be serviced and, if necessary, repaired. *Design for service* essentially is based on the concept that the elements that are most likely to require servicing should be placed at the outer layers of the product.

14.12 | Economic Considerations

As described in greater detail in Chapter 16, and as also described throughout the preceding chapters, there are numerous considerations involved in determining the overall economics of production operations. Because all production systems are essentially a combination of machines and people, important factors influencing the final decisions include the types and cost of machinery, cost of its operation, the skill level and amount of labor required, and production quantity. It was also noted that lot size and production rate greatly influence the economics of production. Small quantities per year can be manufactured in job shops; however, the type of machinery in typical job shops generally requires skilled labor to operate. Furthermore, production volume and production rate in job shops are low, and as a result, cost per part can be high. (See Fig. 14.3.)

At the other extreme, there is the production of very large quantities of parts, using conventional flow lines and transfer lines and involving special-purpose machinery and equipment, specialized tooling, and computer control systems. Although all these components constitute major investments, the level of labor skill required and the labor costs are both relatively low, because of the high level of automation implemented. Recall that these production systems are organized for a specific type of product and hence lack flexibility.

Because most manufacturing operations fall between these two extremes, an appropriate decision has to be made regarding the desirable level of automation to be implemented. In many situations, selective automation rather than total automation of a facility has been found to be cost effective. Generally, the higher the level of skill available in the workforce, the lower is the need for automation, provided that the higher labor costs involved are justified and assuming that there is a sufficient number of qualified workers available. Conversely, if a manufacturing facility has already been automated, the skill level required is lower.

In addition, the manufacture of some products may have a large labor component, and thus their production is *labor intensive;* this is especially the case with products that require extensive assembly. Examples of labor-intensive products include aircraft, locomotives, bicycles, pianos, furniture, toys, shoes, and garments. This high labor requirement is a major reason that so many household as well as high-tech products are now made or assembled in countries where labor costs are low, such as Mexico, China, and Pacific-rim countries. (See also Sections 1.10 and 16.9.)

In manual assembly, relatively simple tools are used, and the process is economical for relatively small lots. Because of the dexterity of the human hand and fingers and their capability for feedback through various senses, workers can manually assemble even complex parts without much difficulty. As described in Section 14.7, labor cost and benefit considerations are also significant aspects of robot selection and use.

CASE STUDY | Robotic Deburring of Plastic Toboggans

Robotor Technologie produces high-quality toboggans and car seats made of plastic (using injection molding described in Section 10.10.2) or blow molding (Section 10.10.3). After molding and while the part is cooling, holes have to be cut in the toboggans (and in back rests of child seats, which is another product line with a similar deburring need). The holes then have to be deburred (Section 9.8). The deburring operation is ideal for a robot but is very difficult to automate. If a rotating burr tool is used, fumes and particles are generated that pose a health risk, and a nonrotating cutter requires that the robot allow deviations in its programmed path to accommodate shrinkage variations in the molded parts.

Robotor Technologie found a solution: a float-mounted tool (see *compliant end effectors,* Section 14.7.1) that can accommodate various cutter blades. To remove burrs, the blade must maintain the correct cutting angle and a constant cutting force. This is achieved with a KUKA KR-15 robot, which carries out the cutting and deburring operations in a single step while also compensating for the plastic shrinkage (Fig. 14.32).

FIGURE 14.32 Robotic deburring of a blow-molded toboggan. *Source:* Courtesy of KUKA Robotics, Inc. and Roboter Technologie, GmbH.

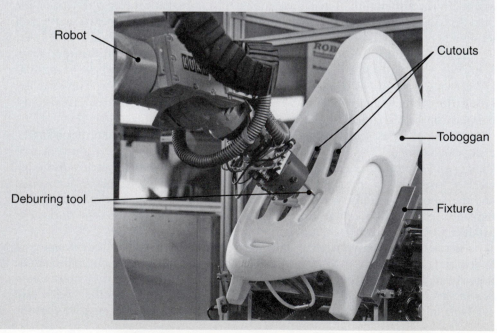

As soon as the blow-molded part leaves the molding machine, an operator removes the flash and places the part on a rotary indexing table (see Fig. 14.4). When a part is rotated into the robot's workspace, the robot cuts and deburrs the holes. Once the side parts have been cut and deburred, the fixture tilts the toboggan to a vertical position so that its top can be accessed.

An automatic tool changer (see Section 8.11) is used during the processing of each toboggan to switch from a convex cutter (used for the cut-outs) to a straight-edge cutter to produce a smooth exterior contour. The complex shape of a toboggan or child seat is a good example of the flexibility achievable by robots.

The robot completes the machining operation in 40–50 seconds, which can be compared to the cycle time of the blow molding process of 120 seconds; thus the implementation of the robot is consistent with the goals of a pull system (see Section 15.12).

The robot was successfully applied in a hazardous and dirty environment (see Section 14.7) and has eliminated tasks that were associated with work-related injuries associated with fume exposure and stresses on the wrist in manual deburring. In addition, because of the higher quality of deburring and elimination of rejected parts, the robot cell paid for itself within three months.

Source: Courtesy of KUKA Robotics, Inc. and Roboter Technologie, GmbH.

SUMMARY

- There are several levels of automation, from simple automation to complex systems. Automation has been successfully implemented in all manufacturing processes, material handling, inspection, assembly, and packaging. Production volume and production rate are major factors in selecting the most economic level of automation for a process or operation. (Sections 14.1, 14.2)

- True automation began with the numerical control of machines, achieving flexibility of operation, lower cost, and ease of making different parts with lower operator skill required. (Sections 14.3 and 14.4)

- Manufacturing operations are further optimized by adaptive-control techniques, which continuously monitor the operation and automatically make appropriate adjustments in the processing parameters. (Section 14.5)

- Advances in material handling include the implementation of industrial robots and automated guided vehicles. (Sections 14.6 and 14.7)

- Sensors are essential in the implementation of modern manufacturing technologies and computer-integrated manufacturing. A wide variety of sensors, based on various principles, has been developed and successfully installed. (Section 14.8)

- Flexible fixturing and automated assembly techniques greatly reduce the need for worker intervention and thus lower manufacturing costs. Effective and economic implementation of these techniques requires that design for assembly, disassembly, and servicing be recognized as an important factor in the total design as well as in manufacturing operations. (Sections 14.9 and 14.10)

- As in all other manufacturing processes, there are certain design considerations and guidelines regarding the implementation of the topics described in this chapter. (Section 14.11)

- Economic considerations in automation include decisions regarding the appropriate level of automation to be implemented; such decisions, in turn, involve parameters such as production volume and rate. (Section 14.12)

BIBLIOGRAPHY

Blum. R.S., and Liu, Z., *Multi-Sensor Image Fusion and Its Applications*, CRC, 2005.

Bolhouse, V., *Fundamentals of Machine Vision*, Robotic Industries Association, 1997.

Boothroyd, G., *Assembly Automation and Product Design*, 2nd ed., Dekker, 2005.

Boothroyd, G., Dewhurst, P., and Knight, W., *Product Design for Manufacture and Assembly*, 2nd ed., Dekker, 2001.

Brooks, R.R., and Iyengar, S., *Multi-Sensor Fusion: Fundamentals and Applications with Software*, Prentice Hall, 1997.

Busch-Vishniac, I., *Electromechanical Sensors and Actuators*, Springer, 1999.

Chow, W., *Assembly Line Design: Methodology and Applications*, Dekker, 1990.

Craig, J.J., *Introduction to Robotics*, 3rd ed., Prentice Hall, 2003.

Davies, E.R., *Machine Vision: Theory, Algorithms, Practicalities*, 3rd ed., Morgan Kaufmann, 2004.

Fraden, J., *Handbook of Modern Sensors: Physics, Designs, and Applications*, 3rd ed., Springer, 2003.

Galbiati, L.J., *Machine Vision and Digital Image Processing Fundamentals*, Prentice Hall, 1997.

Hornberg, A., *Handbook of Machine Vision*, Wiley, 2006.

Ioannu, P.A., *Robust Adaptive Control*, Prentice Hall, 1995.

Lynch, M., *Computer Numerical Control for Machining*, McGraw-Hill, 1992.

Molloy, O., Warman, E.A., and Tilley, S., *Design for Manufacturing and Assembly: Concepts, Architectures and Implementation*, Kluwer, 1998.

Myler, H.R., *Fundamentals of Machine Vision*, Society of Photo-optical Instrumentation Engineers, 1998.

Nof, S.Y., Wilhelm, W.E., and Warnecke, H.-J., *Industrial Assembly*, Chapman & Hall, 1998.

Rampersad, H.K., *Integral and Simultaneous Design for Robotic Assembly*, Wiley, 1995.

Rehg, J.A., *Introduction to Robotics in CIM Systems*, 5th ed., Prentice Hall, 2002.

Ripka, P., and Tipek, A., *Modern Sensors Handbook*, ISTE Publishing Co., 2007.

Smid, P., *CNC Programming Handbook*, 2nd ed., Industrial Press, 2002.

——, *CNC Programming Techniques*, Industrial Press, 2005.

Snyder, W.E., and Qi, H., *Machine Vision*, Cambridge, 2004.

Stenerson, J., and Curran, K.S., *Computer Numerical Control: Operation and Programming*, 3rd ed., Prentice Hall, 2005.

Umbaugh, S.E., *Computer Imaging*, CRC, 2005.

Valentino, J.V., and Goldenberg, J., *Introduction to Computer Numerical Control*, 3rd ed., Prentice Hall, 2002.

Van Doren, V., *Techniques for Adaptive Control*, Butterworth-Heinemann, 2002.

Wilson, J., *Sensor Technology Handbook*, Newnes, 2004.

Zuech, N., *Understanding and Applying Machine Vision*, 2nd ed., Dekker, 1999.

QUESTIONS

14.1 Describe the differences between mechanization and automation. Give several specific examples for each.

14.2 Why is automation generally regarded as evolutionary rather than revolutionary? Explain.

14.3 Are there activities in manufacturing operations that cannot be automated? Explain, and give specific examples.

14.4 Explain the difference between hard and soft automation. Why are they named as such?

14.5 Describe the principle of numerical control of machines. What factors led to the need for and development of numerical control? Name some typical applications of NC.

14.6 Explain the differences between direct numerical control and computer numerical control. What are their relative advantages?

14.7 Describe the principles of open-loop and closed-loop control circuits.

14.8 List and explain the advantages of computer-aided NC programming?

14.9 Describe the principle and the purposes of adaptive control. Give some examples of present applications in manufacturing and comment on other areas that you think can be implemented.

14.10 Explain the factors that have led to the development of automated guided vehicles. Do automated guided vehicles have any limitations? Explain.

14.11 List and discuss the factors that should be considered in choosing a suitable material-handling system for a particular manufacturing facility of your choice.

14.12 Make a list of the features of an industrial robot. Why are these features necessary?

14.13 Discuss the principles of various types of sensors, giving two applications for each type.

14.14 Describe the concept of design for assembly. Why it is now an important factor in manufacturing?

14.15 Is it possible to have partial automation in assembly operations? Explain.

14.16 Describe your thoughts on adaptive control in manufacturing operations.

14.17 What are the two types of robot joints? Give applications for each.

14.18 What are the advantages of flexible fixturing over other methods of fixturing? Are there any limitations to flexible fixturing? Explain.

14.19 Explain how robots are programmed to follow a certain path.

14.20 Giving specific examples, discuss your observations concerning Fig. 14.2.

14.21 What are the advantages and limitations of the two arrangements for power heads shown in Fig. 14.4?

14.22 Discuss methods of on-line gaging of workpiece diameters in turning operations other than that shown in Fig. 14.15. Explain the advantages and limitations of the various methods.

14.23 Is drilling and punching the only application for the point-to-point system shown in Fig. 14.10a? Are there others? Explain.

14.24 Describe possible applications for industrial robots not discussed in this chapter.

14.25 What determines the number of robots in an automated assembly line such as that shown in Fig. 14.23? Explain.

14.26 Describe situations in which the shape and size of the work envelope of a robot (see Fig. 14.20) can be critical.

14.27 Explain the functions of each of the components of the robot shown in Fig. 14.17a. Comment on their degrees of freedom.

14.28 Explain the differences between an automated guided vehicle and a self-guided vehicle.

14.29 It has been commonly acknowledged that, at the early stages of development and implementation of industrial robots, the usefulness and cost effectiveness of the robots were overestimated. What reasons can you think of to explain this situation?

14.30 Describe the type of manufacturing operations (see Fig. 14.2) that are likely to make the best use of a machining center described in Section 8.11. Comment on the influence of product quantity and part variety.

14.31 Give a specific example of a situation in which an open-loop control system would be desirable, and give a specific example of a situation in which a closed-loop system would be desirable. Explain.

14.32 Why should the level of automation in a manufacturing facility depend on production quantity and production rate? Explain.

14.33 Explain why sensors have become so essential in the development of automated manufacturing systems. Give some examples.

14.34 Give examples where there a clear need for flexible fixturing for holding workpieces. Are there any disadvantages to such flexible fixturing? Explain.

14.35 Describe situations in manufacturing for which you would not want to apply numerical control. Explain your reasons.

14.36 Table 14.2 shows a few examples of typical products for each category of production by volume. Add several other examples to this list.

14.37 Describe situations for which each of the three positioning methods shown in Fig. 14.6 would be desirable.

14.38 Describe applications of machine vision for specific parts, similar to the examples shown in Fig. 14.26.

14.39 Add examples of guides other than those shown in Fig. 14.31.

14.40 Sketch the work envelope of each of the robots shown in Fig. 14.19. Describe its implications in manufacturing operations.

14.41 Give several applications for the types of robots shown in Fig. 14.19.

14.42 Name some applications for which you would not use a vibratory feeder. Explain why vibratory feeding is not appropriate for these applications.

14.43 Give an example each of a metal-forming operation from Chapters 6 and 7 that would be suitable for adaptive control, similar to that shown in Fig. 14.15.

14.44 Give some applications for the systems shown in Fig. 14.26a and c.

14.45 Comment on your observations regarding the system shown in Fig. 14.1b.

14.46 Are there situations in which tactile sensors would not be suitable? Explain.

14.47 Give examples for which machine vision cannot be properly and reliably applied. Explain why machine vision may not be appropriate for these applications.

14.48 Comment on the effect of cutter wear on the profiles produced, such as those shown in Figs. 14.10b and 14.12.

14.49 Although future trends are difficult to predict with certainty, describe your thoughts as to what new developments in the topics covered in this chapter could possibly take place as we move through the late 2000s.

14.50 Describe the circumstances under which a (a) dedicated fixture, (b) modular fixture, and (c) flexible fixture would be preferable.

14.51 What is an anthropomorphic robot? What are its applications? Explain.

PROBLEMS

14.52 A spindle/bracket assembly uses the following parts: a steel spindle, two nylon bushings, a stamped steel bracket, and six screws and six nuts to mount the nylon bushings to the steel bracket and thereby support the spindle. Compare this assembly to the spindle/bracket assembly shown in Problem 7.98 and estimate the assembly efficiency for each design.

14.53 Disassemble an ordinary ball point pen. Carefully measure the time required to reassemble the pen and calculate the assembly efficiency. Repeat the exercise for a mechanical pencil.

14.54 Examine Fig. 14.11b and obtain an expression for the maximum error in approximating a circle with linear increments as a function of the circle radius and number of increments on the circle circumference.

14.55 A 35-kg worktable is controlled by a motor/gear combination that can develop a maximum force of 50 N. It is desired to move the worktable from its current location ($x = 0$) to a new location ($x = 100$ mm). Plot the resultant force and the position as a function of time for (a) an open-loop control system and (b) a closed loop control system. Let $k_p = 0.5$ N/mm and $k_v = 0.045$ Ns/mm.

14.56 Review Example 14.1 and develop open- and closed-loop control system equations for the force. Let the coefficient of friction be μ.

14.57 Develop closed-loop control system equations in order to have the worktable in Example 14.1 achieve a sinusoidal position given by $x = \sin \omega t$, where t is time. Assume the coefficient of friction is zero.

14.58 Assume that you are asked to give a quiz to students on the contents of this chapter. Prepare five quantitative problems and five qualitative questions, and supply the answers.

DESIGN

14.59 Design two different systems of mechanical grippers for widely different applications.

14.60 For a system similar to that shown in Fig. 14.28, design a flexible fixturing setup for a lathe chuck. (See Figs. 8.42 and 8.44.)

14.61 Add other examples to those shown in Fig. 1.4.

14.62 Give examples of products that are suitable for the type of production shown in Fig. 14.3.

14.63 Choose one machine each from Chapters 6 through 12 and design a system in which sensor fusion can be used effectively. How would you convince a prospective customer of the merits of such a system? Would the system be cost effective? Explain.

14.64 Does the type of material (metallic or nonmetallic) used for the parts shown in Fig. 14.31 have any influence on the effectiveness of the guides? Explain.

14.65 Think of a product and design a transfer line for it, similar to that shown in Fig. 14.5. Specify the types and numbers of machines required.

14.66 The basic principles of flexible fixturing were described in Section 14.9. Considering the wide variety of parts made, prepare a list of design guidelines for flexible fixturing. Make simple sketches illustrating each guideline for each type of fixturing, describing its ranges of applications and limitations.

14.67 Describe your thoughts on the usefulness and applications of modular fixturing, consisting of various individual clamps, pins, supports, and attachments mounted on a base plate.

14.68 Inspect several household products and sketch and describe the manner in which they have been assembled. Comment on any design changes you would make so that their assembly, disassembly, and servicing are simpler and faster.

14.69 Review the last design shown in Fig. 14.18a and design grippers that would be suitable for gripping the following products: (a) an egg; (b) an object made of soft rubber; (c) a metal ball with a very smooth and shiny surface; (d) a newspaper; and (e) tableware, such as forks, knives, and spoons.

14.70 Obtain an old toaster and disassemble it. Make recommendations on its redesign using the guidelines given in Section 14.11.2.

14.71 Comment on the design and materials used for the gripper shown in Fig. 14.25. Why do such grippers have low durability on the shop floor? Explain.

14.72 Comment on your observations regarding Fig. 14.28 and offer designs for similar applications in manufacturing. Also, comment on the usefulness of the designs in actual production on the shop floor.

14.73 Review the various toolholders used in the machining operations described in Chapter 8 and design sensor systems for them similar to those shown in Fig. 14.24. Comment on the features of the sensor systems and discuss any difficulties that may be associated with their use on the factory floor.

14.74 Design a guide that operates in the same manner as those shown in Fig. 14.31, but will align U-shaped parts so that they are inserted with the open end down.

14.75 Design a flexible fixture that uses powered workholding devices for a family of parts but that allows for a range of diameters and thicknesses.

14.76 Design a modular fixturing system for the part shown in the figure accompanying Problem 8.160.

14.77 Give some applications for the systems shown in Figs. 14.26a and c.

14.78 Review the specifications of various numerical-control machines and make a list of typical numbers for their (a) positioning accuracy, (b) repeat accuracy, and (c) resolution. Comment on your observations.

14.79 Consider the automated guided vehicles shown in Fig. 14.16. Is it possible for such systems to operate in multiple buildings? What difficulties would you expect that would have to be overcome? Explain.

CHAPTER 15

Computer-Integrated Manufacturing Systems

This chapter describes how computers are integrated into the manufacturing environment. Computer hardware, software, communications networks and protocols, and people are integrated into a modern and efficient manufacturing system. Topics covered include

- Computer-aided design through graphical description and analysis of parts.
- The use of computers to simulate manufacturing processes and systems.
- Group technology and database approaches to allow the rapid recovery of previous design and manufacturing experience and apply it to new situations in a straightforward manner.
- The principle of attended (manned) and unattended (unmanned) manufacturing cells and their features.
- The concept of holonic manufacturing and its applications.
- The principles of just-in-time manufacturing and lean manufacturing and their benefits.
- Importance and features of communication systems.
- Applications of artificial intelligence and expert systems in manufacturing.

15.1 | Introduction

The implementation, benefits, and limitations of mechanization, automation, and computer control of various stages of manufacturing operations have been described in the preceding chapter. This chapter concerns the *integration of manufacturing activities*, meaning that processes, machinery, equipment, operations, and their management are treated as a **manufacturing system.** Such a system allows the total control of the manufacturing facility, thereby increasing productivity, product quality, and product reliability, and reducing manufacturing costs.

In **computer-integrated manufacturing** (CIM), the traditionally separate functions of product design, research and development, production, assembly, inspection, and quality control are all interlinked. Integration requires that quantitative relationships among product design, materials, manufacturing processes, process and equipment capabilities, and related activities be well understood and established.

914

In this way, changes in, for example, materials, product types, production methods, or market demand can be properly and effectively accommodated.

Recall the statements that (a) quality must be built *into* the product, (b) higher quality does *not* necessarily mean higher cost, and (c) marketing poor-quality products can indeed be *very costly* to the manufacturer. High quality is far more attainable and less expensive through the proper *integration* of design and manufacturing than if the two were separate activities. As will be illustrated throughout this chapter, integration is successfully and effectively accomplished through **computer-aided design, engineering, manufacturing, process planning**, and **simulation of processes and systems.**

This chapter also describes and emphasizes how **flexibility** in machines, tooling, equipment, and production operations greatly enhances the ability to respond to market demands and product changes and ensures **on-time delivery** of high-quality products. As described in detail, important developments during the past 40 years or so (see Table 1.1) have had a major impact on modern manufacturing, especially in an increasingly competitive global marketplace. Among these important advances are **group technology, cellular manufacturing, flexible manufacturing systems**, and **just-in-time production** (also called **zero inventory, stockless production**, or **demand scheduling**). Furthermore, because of the extensive use of **computer controls** and hardware and software in integrated manufacturing, the planning and effective implementation of **communication networks** became an essential component of these activities.

The chapter concludes with a review of **artificial intelligence**, consisting of expert systems, natural-language processing, machine vision, artificial neural networks, and fuzzy logic, and how these developments impact manufacturing activities.

15.2 | Manufacturing Systems

As described throughout this text, manufacturing consists of a large number of interdependent activities with distinct entities, such as materials, tools, machines, controls, and people. Consequently, manufacturing should be regarded as a large and complex *system*, consisting of numerous diverse physical and human elements. Note that some of these factors are difficult to predict and to control, such as the supply and cost of raw materials, changes in market demand, economic conditions, global trends, as well as human behavior and performance.

Ideally, a system should be represented by **mathematical and physical models,** which can identify the nature and extent of the interdependence of the relevant variables involved. In a manufacturing system, a change or disturbance anywhere in the system requires that it adjust itself systemwide in order to continue functioning effectively and efficiently. For example, if the supply of a particular raw material is reduced (such as due to geopolitical maneuvers, wars, or strikes) and, consequently, its cost increases, alternative materials must be investigated and selected. However, this change must be made only after careful consideration of the effect it may have on processing conditions, production rate, product quality, and, especially, manufacturing costs, as also described throughout Chapter 16.

Similarly, the demand for a product may fluctuate randomly and rapidly due, for example, to its style, size, or capacity. Note, for example, the downsizing of automobiles during the 1980s in response to fuel shortages, the popularity of lower mileage sport-utility vehicles in the 1990s, and the current high demand for hybrid gas-electric vehicles due to high fuel costs and increased concerns for the environment. The manufacturing system must also be capable of producing the modified product on a short **lead time** and, preferably, with relatively small major capital investment in machinery and tooling. Lead time is defined as the length of time between the creation of the product as a concept (or receipt of an order for the product) and the time that the product first becomes available in the marketplace (or is delivered to the customer).

Computer simulation and modeling of such a complex system can be difficult, because of a lack of comprehensive and reliable data on some of the numerous variables involved. Moreover, it is not always easy to correctly predict and control some of these variables. The following are some examples of the problems that may be encountered:

1. The characteristics of a machine tool, its performance, and its response to random external disturbances cannot always be modeled precisely. (See also Section 15.7.)

2. Raw-material costs and properties may vary over a period of time and may be difficult to predict accurately.

3. Sensors may lack sufficient robustness to allow closed-loop control under all conditions.

4. Market demands and human behavior and performance are difficult to model reliably.

15.3 | Computer-Integrated Manufacturing

The various levels of automation in manufacturing operations, as described in Chapter 14, have been extended further through **information technology** (IT) and using an extensive network of interactive computers. *Computer-integrated manufacturing* is a broad term used to describe the computerized integration of product design, planning, production, distribution, and management. It should be appreciated that computer-integrated manufacturing is a *methodology* and a *goal,* rather than merely an assemblage of equipment and computers.

The effectiveness of CIM greatly depends on the use of a large-scale **integrated communications system,** involving computers, machines, equipment, and their controls. The difficulties that may arise in such systems are described in Section 15.14. Furthermore, because CIM ideally should involve the total operation of a company, it requires an extensive database containing technical as well as business information.

Implementation of CIM in existing manufacturing plants may begin with the use of modules in *selected phases* of a company's operation. For new plants, on the other hand, comprehensive and long-range strategic planning covering all phases of the operation is essential in order to fully benefit from CIM. Such plans must take into account considerations such as (a) the mission, goals, and culture of the organization; (b) the availability of resources; (c) emerging technologies relevant to the products made; and (d) the level of integration desired. It is apparent that if planned and implemented all at once, CIM can be prohibitively expensive, particularly for small and medium-size companies.

Subsystems. Computer-integrated manufacturing systems consist of *subsystems* that are integrated into a whole (Fig. 15.1). These subsystems consist of the following:

1. Business planning and support.
2. Product design.
3. Manufacturing process planning.
4. Process automation and control.
5. Factory-floor monitoring systems.

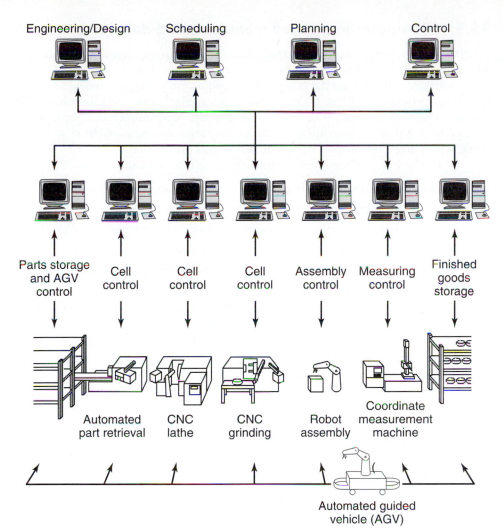

Engineering/Design   Scheduling   Planning   Control

Parts storage and AGV control   Cell control   Cell control   Cell control   Assembly control   Measuring control   Finished goods storage

Automated part retrieval   CNC lathe   CNC grinding   Robot assembly   Coordinate measurement machine

Automated guided vehicle (AGV)

FIGURE 15.1 A schematic illustration of a computer-integrated manufacturing system. *Source:* After U. Rembold.

The subsystems are designed, developed, and implemented in such a manner that the output of one subsystem serves as the input to another subsystem, as shown by the various arrows in Fig. 15.1. Organizationally, the subsystems are generally divided into two functions: (a) **business-planning functions,** which include activities such as forecasting, scheduling, material-requirements planning, invoicing, and accounting; and (b) **business-execution functions,** which include production and process control, material handling, testing, and inspection.

Benefits. The benefits of CIM include the following:

1. Responsiveness to shorter product life cycles (see Section 16.4), changing market demands, and global competition.

2. Emphasis on product quality and uniformity, implemented through better process control.

3. Better use of materials, machinery, and personnel, and the reduction of *work-in-progress* (WIP) inventory, all of which improve productivity and lower product cost.

4. Better control of the production, scheduling, and management of the total manufacturing operation, resulting in lower product cost.

15.3.1 Computer-integrated manufacturing databases

An effective computer-integrated manufacturing system requires a single, large database that is shared by members of the entire organization. *Databases* consist of real-time, detailed, and accurate data on product designs, products, machines, processes, materials, production, finances, purchasing, sales, marketing, and inventory. This vast array of data is stored in computer memory and randomly recalled or modified as necessary, either by individuals in the organization or by the CIM system itself while it is controlling various aspects of design and production.

A database typically consists of the following information:

1. **Product data** (part shape, dimensions, tolerances, and specifications).
2. **Data management attributes** (creator, revision level, and part number).
3. **Production data** (manufacturing processes used in making parts and products).
4. **Operational data** (scheduling, lot sizes, and assembly requirements).
5. **Resources data** (capital, machines, equipment, tooling, and personnel, and the capabilities of these resources).

Databases are built by individuals and through the use of various sensors in the machinery and equipment employed in production. Data from the latter are automatically collected by a **data acquisition system** (DAS) that can report, for example, the number of parts being produced per unit time, weight, dimensional accuracy, and surface finish, at specified rates of sampling. The components of DASs include microprocessors, transducers, and, often, analog-to-digital converters (ADCs). Data acquisition systems are also capable of analyzing and transferring the data to other computers for such purposes as data presentation, statistical analysis, and forecasting of product demand.

Several factors are important in the use and implementation of databases:

1. Databases should be up to date, accurate, user-friendly, and easily accessible and shared.
2. In the event that there is a malfunction in data retrieval, and the original correct data must be recovered and restored to the database.
3. Because they are used for various purposes and by numerous people, databases must be flexible and responsive to the needs of users with different backgrounds and needs.
4. Databases must easily be accessed by designers, manufacturing engineers, process planners, financial officers, and the management of the company, but they must be protected against tampering or unauthorized use.

15.4 | Computer-Aided Design and Engineering

Computer-aided design (CAD) involves the use of computers to create design drawings and geometric models of products and components (see also Fig. 1.8a) and is associated with **interactive computer graphics**, known as a **CAD system.** *Computer-aided engineering* (CAE) simplifies the creation of the database by allowing several applications to share the information in the database. These applications include, for example, (a) finite-element analysis of stresses, strains, deflections, and temperature distribution in structures and load-bearing members, (b) the generation, storage, and retrieval of NC data, and (c) the design of integrated circuits and various electronic devices.

In CAD, the user can generate drawings or sections of a drawing on the computer. The design is continuously displayed on the monitor and in different colors for its various components; the final drawing is printed or plotted if desired but is often stored as a digital file that can be accessed as needed on networked computers. When using a CAD system, the designer can conceptualize the object to be designed on the graphics screen and can consider alternative designs or quickly modify a particular design to meet specific design requirements.

Through the use of powerful software, such as CATIA (after *Computer-Aided Three-Dimensional Interactive Applications,* now in version 5), the design can be subjected to *engineering analysis,* which can identify potential problems, such as an excessive load or deflection, and interference at mating surfaces during assembly. In addition to the design's geometric and dimensional features, other information, such as a list of materials, specifications, and manufacturing instructions, is stored in the CAD database. Using such information, the designer can also analyze the economics of alternative designs.

15.4.1 Exchange specifications

Because of the availability of a wide variety of CAD systems with different characteristics, supplied by different vendors, proper communication and exchange of data between these systems is essential. (See also Section 15.14.) **Drawing exchange format** (DFX) was developed for use with Autodesk™ and has become a de facto standard, because of the long-term success of this software package. DFX is limited to transferring geometry information only. Similarly, **STL** (*STereo Lithography;* see Section 10.12.1) formats are used to export 3D geometries, used initially for rapid prototyping systems; recently, however, STL has become a format for data exchange between CAD systems.

The need for a single, neutral format for better compatibility and for the transfer of more information than geometry alone is currently filled mainly by the **Initial Graphics Exchange Specification** (IGES). Vendors need only provide translators for their own systems to preprocess outgoing data into the neutral format, and to postprocess incoming data from the neutral format into their system. IGES is used for translation in two directions (into and out of a system) and is also used widely for translation of 3D line and surface data. Because IGES is evolving, there are many variations of IGES in existence (now in version 5.3).

Another specification is a solid-model-based standard called **Product Data Exchange Specification** (PDES), which is based on the Standard for the Exchange of Product Model Data (STEP) developed by the International Standards Organization. PDES allows information on shape, design, manufacturing, quality assurance, life cycle, testing, maintenance, etc., to be transferred between CAD systems.

15.4.2 Elements of computer-aided design systems

The design process in a CAD system consists of four stages, described as follows:

1. **Geometric modeling.** In *geometric modeling,* a physical object or any of its parts is described mathematically or analytically. The designer first constructs a geometric model by giving commands that create or modify lines, surfaces, solids, dimensions, and text that, together, compose an accurate and complete two- or three-dimensional representation of the object. The results of these commands are then displayed; the images can be manipulated on the screen, and any section can be magnified in order to view its details. The models are stored in the database.

   The models can be presented differently, as described below.

   a.   In **line representation** (also called **wire-frame representation;** see Fig. 15.2), all edges of the model are visible as solid lines. However, this type of image can be ambiguous because line resolution may be a problem, particularly for complex shapes; hence, various colors are generally used for different

FIGURE 15.2   Various types of modeling for CAD.

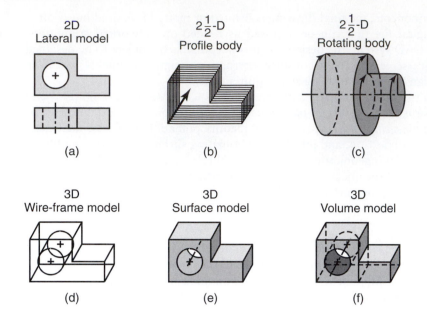

parts of the object. The three types of wire-frame representations are 2D, $2\frac{1}{2}$-D, and 3D. A 2D image shows the profile of the object, and a $2\frac{1}{2}$-D image can be obtained by *translational sweep,* that is, by moving the 2D object along the *z*-axis. For round objects, a $2\frac{1}{2}$-D model can be generated by simply *rotating* a 2D model around its axis.

b.   In the **surface model,** all visible surfaces are shown in the model and define surface features and edges of objects. CAD programs use *Bezier curves, B-splines,* or *nonuniform rational B-splines* (NURBS) for surface modeling. Each of these methods uses control points to define a polynomial curve or surface. A Bezier curve passes through the first and last vertex and uses the other control points to generate a blended curve. The drawback to Bezier curves is that modification of one control point will affect the entire curve. B-splines are a blended, piecewise polynomial curve, where modification of a control point affects the curve only in the area of the modification. Examples of two-dimensional Bezier curves and B-splines are given in Fig. 15.3. A NURBS is a special type of B-spline where each control point has a weight associated with it.

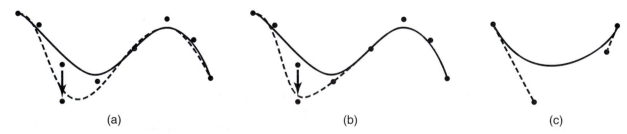

FIGURE 15.3   Types of splines: (a) a Bezier curve passes through the first and last control point and generates a curve from the other points; changing a control point modifies the entire curve, (b) a B-spline is constructed piecewise, so that changing a vertex affects the curve only in the vicinity of the changed control point, and (c) a third-order piecewise Bezier curve constructed through two adjacent control points, with two other control points defining the curve slope at the end points; a third-order piecewise Bezier curve is continuous, but its slope may be discontinuous.

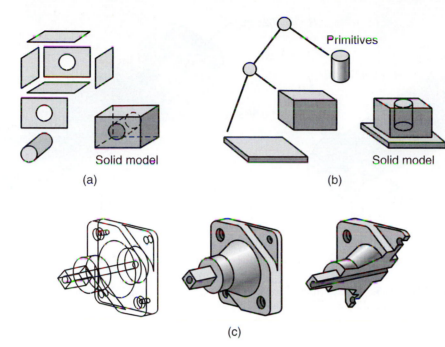

FIGURE 15.4

(a) Boundary representation of solids, showing the enclosing surfaces and the generated solid model, (b) a solid model represented as compositions of solid primitives, and (c) three different representations of the same part by CAD. *Source:* After P. Ranky.

c. In the **solid model,** all surfaces are shown, but the data describe the interior volume as well. Solid models can be constructed from *swept volumes* (Fig. 15.2b and c) or by the techniques shown in Fig. 15.4. In **boundary representation** (BREP), surfaces are combined to develop a solid model (Fig. 15.4a). In **constructive solid geometry** (CSG), simple shapes such as spheres, cubes, blocks, cylinders, and cones (called **primitives of solids**) are combined to develop a solid model (Fig. 15.4b). The user selects any combination of primitives and their sizes and combines them into the desired solid model. Although solid models have certain advantages, such as ease of design analysis and ease of preparation for manufacture of the part, they require more computer memory and processing time than the wire-frame and surface models.

   A special kind of solid model is a **parametric model,** where a part not only is stored in terms of a BREP or CSG definition, but also is derived from the dimensions and constraints that define the features (Fig. 15.5). Whenever a change is made, the part is re-created from these definitions, a feature that allows simple and straightforward updates and changes to be made to models.

d. The **octree representation** of a solid object (Fig. 15.6) is a type of model that is basically a three-dimensional analog to pixels on a television screen or monitor. Just as any area can be broken down into quadrants, any volume can be broken down into *octants,* which are then identified as solid, void, or partially filled. Partially filled *voxels* (from *volume pixels*) are broken into smaller octants and reclassified. With increasing resolution, exceptional part detail can be achieved. This process may appear to be somewhat cumbersome, but it allows for accurate description of complex surfaces and is used particularly in biomedical applications such as modeling bone or organ geometries.

e. A **skeleton** (Fig. 15.7) is commonly used for kinematic analysis of parts or assemblies. A skeleton is the family of lines, planes, and curves that describe

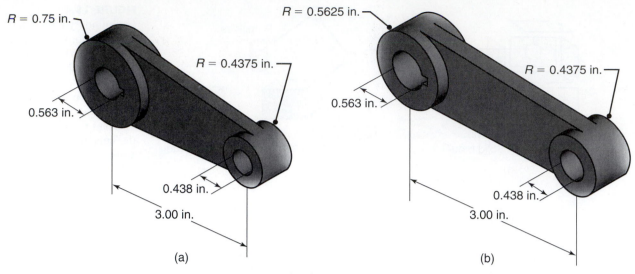

FIGURE 15.5 An example of parametric design: (a) original design and (b) modified design produced by modifying parameters in the data file of the part in (a). Dimensions of part features can easily be modified to quickly produce an updated solid model.

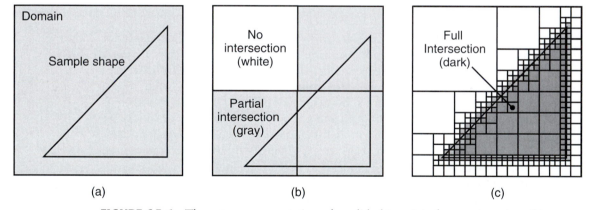

FIGURE 15.6 The octree representation of a solid object: (a) after one iteration, (b) two iterations, and (c) six iterations. Any volume can be broken down into octants, which are then identified as solid, void, or partially filled. Shown is two-dimensional, or quadtree, version, for representation of shapes in a plane.

a part without the detail of surface models. Conceptually, a skeleton can be constructed by fitting the largest circles (or spheres for three-dimensional objects) within the geometry; the skeleton is the set of points that connect the centers of the circles (or spheres). The circle radius data at each point also is stored. A current area of research involves using skeleton models instead of conventional surface or solid models.

2. **Design analysis and optimization.** After its geometric features have been determined, the design is subjected to an *engineering analysis,* a phase which may, for example, consist of analyzing stresses, strains, deflections, vibrations, heat transfer, temperature distribution, or dimensional tolerances. Several software packages are available that have the capabilities to compute these quantities accurately and rapidly. Because of the relative ease with which such analyses can be performed, designers can study a design more thoroughly before it moves on to production. Experiments and measurements in the field may

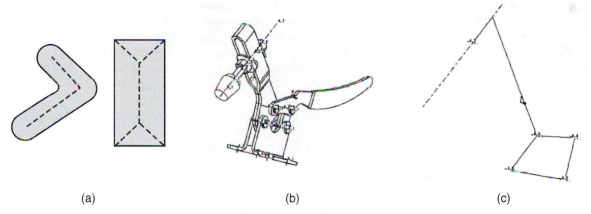

(a)   (b)   (c)

FIGURE 15.7   (a) Illustration of the skeleton data structure for two different solid objects; the skeleton is the dashed lines in the interior of the objects. (b) A general view of a clamp. (c) A skeleton model used for kinematic analysis of the clamp. *Source:* S.D. Lockhart and C.M. Johnson, *Engineering Design Communication*, Prentice Hall, 2000.

nonetheless still be necessary to determine and verify the actual effects of loads, temperature, and other variables on the designed components.

3. **Design review and evaluation.**   An important design stage is *review and evaluation* to check for any interference among various components. This step is necessary in order to avoid difficulties during assembly or use of the part and to determine whether moving members (such as linkages, gears, and cams) will operate and function as intended. Software with animation capabilities is available to identify potential problems with moving members and other dynamic situations. During the review and evaluation stage, the part is also precisely dimensioned and set within the full range of tolerance required for its production (Section 4.7).

4. **Documentation.**   After the preceding stages have been completed, a paper copy of the design may be produced, for documentation and reference. At this stage, detail and working drawings are also developed and printed. The CAD system is also capable of developing and drafting sectional views of the part, scaling the drawings, and performing transformations in order to present various views of the part. Since databases allow part drawing retrieval from any computer on the network, hard copies of drawings are not always produced or necessary.

5. **Database.**   Many components in products are either standard components that are mass produced according to a given design specification (such as bolts or gears) or are identical to parts used in previous designs. CAD systems now have a built-in database management system that allows designers to identify, view, and access parts from a library of stock parts. These parts can be parametrically modeled to allow cost-effective updating of the part geometry. Some databases with extensive parts libraries are commercially available; many vendors make their part libraries available on the Internet.

15.5 | Computer-Aided Manufacturing

Computer-aided manufacturing (CAM) involves the use of computers and computer technology to assist in all phases of manufacturing, including process and production planning, scheduling, manufacture, quality control, and management. Because of the obvious benefits, computer-aided design and computer-aided manufacturing are often combined into **CAD/CAM systems.** This combination allows information transfer from the design stage to the planning stage for the manufacture of a product, without the need to manually reenter the data on part geometry. The database developed during CAD is stored; it is then processed further by CAM into the necessary data and instructions for operating and controlling production machinery and material-handling equipment, as well as for performing automated testing and inspection for product quality (Section 4.8.3).

The emergence of CAD/CAM had a major impact on manufacturing operations by standardizing product development and by reducing design effort, evaluation, and prototype work. It has also made possible significant cost reductions and improved productivity. The two-engine Boeing 777 passenger airplane, for example, was designed completely by computer (**paperless design**), with 2000 workstations linked to eight computers. The plane was constructed directly from the CAD/CAM software developed (an enhanced CATIA system), and no prototypes or mock-ups were built, such as were required for previous models. The development cost for this aircraft was on the order of $6 billion.

An example of an important feature of CAD/CAM in machining is its capability to describe the *cutting-tool path* for various operations such as NC turning, milling, and drilling (see Sections 8.10 and 14.4). The instructions (*programs*) are computer generated, and they can be modified by the programmer to optimize the tool path. The engineer or technician can then display and visually check the tool path for possible tool collisions with clamps or fixtures or other interference. The tool path can be modified at any time to accommodate other part shapes or features. CAD/CAM systems are also capable of *coding and classifying parts* into groups that have similar shapes, using alphanumeric coding. (See the discussion of *group technology* in Section 15.8.)

15.6 | Computer-Aided Process Planning

For a manufacturing operation to be efficient, all its diverse activities must be planned and coordinated, a task that has traditionally been performed by process planners. *Process planning* involves selecting methods of production, tooling, fixtures, machinery, the sequence of operations, the standard processing time for each operation, and methods of assembly. These choices are all documented on a **routing sheet,** such as that shown in Fig. 15.8. Obviously, when prepared manually this task is highly labor intensive and time consuming, as well as greatly relying on the experience of the process planner.

Computer-aided process planning (CAPP) accomplishes this complex task by viewing the total operation as an integrated system, so that the individual operations and steps involved in making each part are coordinated with each other. CAPP is thus an essential adjunct to CAD and CAM. Although it requires extensive software and good coordination with CAD/CAM and various other aspects of integrated manufacturing systems (described throughout the rest of this chapter), CAPP is a powerful tool for efficiently planning and scheduling manufacturing operations. It is particularly effective in small-volume, high-variety parts production requiring machining, forming, and assembly operations.

15.6.1 Elements of computer-aided process-planning systems

There are two types of computer-aided process-planning systems: variant and generative process-planning systems, as described next.

1. In the **variant system,** also called the **derivative system,** the computer files contain *a standard process plan* for a particular part to be manufactured. The search

ROUTING SHEET		
CUSTOMER'S NAME: Midwest Valve Co.		PART NAME: Valve body
QUANTITY: 15		PART NO.: 302
Operation No.	Description of operation	Equipment
10	Inspect forging, check hardness	Rockwell tester
20	Rough machine flanges	Lathe No. 5
30	Finish machine flanges	Lathe No. 5
40	Bore and counterbore hole	Boring mill No. 1
50	Turn internal grooves	Boring mill No. 1
60	Drill and tap holes	Drill press No. 2
70	Grind flange end faces	Grinder No. 2
80	Grind bore	Internal grinder No. 1
90	Clean	Vapor degreaser
100	Inspect	Ultrasonic tester

FIGURE 15.8   A simple example of a routing sheet. These sheets may include additional information on materials, tooling, estimated time for each operation, processing parameters (such as cutting speeds and feeds), and various other details. Routing sheets travel with the part from operation to operation. Current practice is to store all relevant data in computers and to affix to the part a bar code or equivalent label that serves as a key into the information database.

for a standard plan is made in the database, using a specific code number for the part. The plan is based on the part's shape and its manufacturing characteristics. (See Section 15.8.) The standard plan can be retrieved, displayed for review, and printed or stored as that part's routing sheet. The process plan includes information such as the types of tools and machines to be used, the sequence of operations to be performed, the cutting speeds and feeds, and the time required for each sequence. Minor modifications to an existing process plan, which typically are necessary, can also be made. If the standard plan for a particular part is not in the computer files, a plan that is similar to it and that has a similar code number and an existing routing sheet is retrieved. If a routing sheet does not exist, one is prepared for the new part and stored in computer memory.

2. In the **generative system,** a process plan is automatically generated on the basis of the same logical procedures that would be followed by a traditional process planner in making that particular part. However, the generative system is complex, because it must contain comprehensive and detailed knowledge of the part's shape and dimensions, process capabilities, the selection of manufacturing methods, machinery, tools, and the sequence of operations to be performed. (Software with such capabilities, known as **expert systems,** are described in Section 15.15.) The generative system is capable of creating a new plan instead of having to use or modify an existing plan, as the variant system must do. Although it is currently used less commonly than is the variant system, the generative system has such advantages as (a) flexibility and consistency for process planning for new parts and (b) higher overall planning quality, because of the capability of the decision logic in the system to optimize the planning and to use up-to-date manufacturing technology.

The process-planning capabilities of computers can be integrated into the planning and control of production systems as a subsystem of computer-integrated manufacturing, as described in Section 15.3. Several functions can be performed using these activities, such as *capacity planning* for plants to meet production schedules and control of *inventory, purchasing,* and *production scheduling.*

Advantages of CAPP systems. The advantages of CAPP systems over traditional process-planning methods include

1. The standardization of process plans improves the productivity of process planners, reduces lead times and costs of planning, and improves the consistency of product quality and reliability.

2. Process plans can be prepared for parts that have similar shapes and features, and the plans can easily be retrieved to produce new parts.

3. Process plans can be modified to suit specific needs.

4. Routing sheets can be prepared more quickly, and are more legible and useful because of consistency among the sheets.

5. Other functions, such as cost estimation and work standards, can be incorporated into CAPP.

15.6.2 Material-requirements and manufacturing resource planning systems

Computer-based systems for managing inventories and delivery schedules of raw materials and tools are called *material-requirements planning* (MRP) systems. This activity, sometimes regarded as a method of inventory control, involves keeping complete records of inventories of materials, supplies, parts in various stages of production (*work in progress*), orders, purchasing, and scheduling. Several files of data are usually involved in a master production schedule. These files pertain to the raw materials needed (**bill of materials**), product structure levels (i.e., individual items that compose a product, such as components, subassemblies, and assemblies), and scheduling.

A further development is *manufacturing resource planning* (MRP-II) systems, which, through feedback, control all aspects of manufacturing planning. Although the system is complex, MRP-II is capable of final production scheduling, monitoring actual results in terms of performance and output, and comparing those results against the master production schedule.

15.6.3 Enterprise resource planning

Beginning in the 1990s, *enterprise resource planning* (ERP) became an important trend; it is basically an extension of MRP-II. Although there are variations, ERP has generally been defined as a method for effective planning and control of all the resources needed in a business enterprise to take orders for products, produce them, ship them to the customer, and service them. Accounting and billing are also incorporated in the ERP. Thus, ERP coordinates, optimizes, and dynamically integrates all information sources and the widely diverse technical and financial activities in a manufacturing organization. Its major goals are to improve productivity, reduce manufacturing cycle times, and optimize processes, benefiting not only the organization, but also the customer.

An effective implementation of ERP is, however, a difficult and challenging task because

1. Difficulties are encountered in attempting timely, effective, and reliable communication among all parties involved, especially in a global business enterprise (thus indicating the necessity of teamwork).

2. There is a need for modifying business practices in an age where information systems and e-commerce have become highly important to the success of a business organization.

3. Extensive and specific hardware and software requirements must be met to implement ERP.

15.7 | Computer Simulation of Manufacturing Processes and Systems

With increasing power and sophistication of computer hardware and software, an area that has grown rapidly is *computer simulation* of manufacturing processes and systems. Process simulation takes two basic forms:

1. A model of a specific manufacturing operation, intended to determine the viability of a process or to optimize or improve its performance.

2. A model of multiple processes and their interactions, to help process planners and plant designers during the layout of machinery and facilities.

Individual processes have been modeled using various mathematical schemes (see also Section 6.2.2). Finite-element analysis has been increasingly applied in software packages (**process simulation**) that are commercially available and are inexpensive. Typical problems addressed by such models include **process viability** (such as assessing the formability and behavior of a type of sheet metal in a certain pressworking operation) and **process optimization** [such as (a) analyzing the metal-flow pattern in forging a blank in a given die (Fig. 6.11) to identify potential defects, or (b) improving mold design in a casting operation to reduce or eliminate hot spots and minimize defects by promoting uniform cooling].

Simulation of an entire manufacturing system, involving multiple processes and equipment, helps plant engineers to organize machinery and to identify critical machinery (see *push and pull systems,* Section 15.12). Also, such models can assist manufacturing engineers with scheduling and routing by **discrete-event simulation.** Commercially available software packages are often used for these simulations, although the use of dedicated software programs written for a particular company and line of products is not unusual.

EXAMPLE 15.1 Simulation of plant-scale manufacturing

Several examples and case studies presented in this text have focused upon simulation of individual manufacturing processes. However, the availability of low-cost, high-performance computer systems and the development of advanced software have allowed the simulation of entire manufacturing systems and have led to the optimization of manufacturing and assembly operations.

(a)

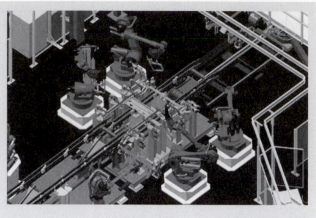

(b)

FIGURE 15.9 Simulation of plant-scale manufacturing operations: (a) the use of virtual manikins to evaluate the required motions and efficiency in manually assembling an automotive dashboard and (b) a robot welding line, where interactions of multiple robots and workpieces can be simulated to detect and avoid collisions and improve productivity. *Source:* Courtesy of Delmia Corporation.

As an example, *Digital Manufacturing Hub* software (Delmia Corporation) allows the simulation of manufacturing processes in three dimensions, including the use of human manikins, to identify safety hazards, manufacturing problems, or bottlenecks; improve machining accuracy; or optimize tooling organization. (See Fig. 15.9a.) Since simulation can be performed prior to building an assembly line, it can significantly reduce development times and cost. For example, Fig. 15.9b shows a simulation of a robotic welding line in an automotive plant, where the robot motions can be simulated and collisions between neighboring robots or other machinery detected in a virtual environment. The program for the robot can then be modified to prevent such collisions before the line is put into operation. While this example is a powerful demonstration of the utility of system simulation, the more common application is to optimize the sequence of operations and organization of machinery to reduce manufacturing costs.

The software also has the capability of conducting ergonomic analysis of various operations and machinery setups and, therefore, of identifying bottlenecks in the movement of parts, equipment, or personnel. The bottlenecks can then be relieved by the process planner by adjusting the automated or manual procedures at these locations. Using such techniques, a Daimler-Chrysler facility in Rastatt, Germany, was able to balance its production lines so that each worker is productive an average of 85 to 95% of the time.

Another application of systems simulation is planning of manufacturing operations to optimize production and to prepare for just-in-time production (Section 15.12). For example, if a car manufacturer needs to produce 1000 vehicles in a given period of time, production can be optimized by using certain strategies, such as by distributing the number of vehicles with sunroofs throughout the day or by grouping vehicles by color, so that the number of paint changes in paint booths is minimized. With respect to just-in-time production, software such as that produced by ILOG Corporation can plan and schedule plant operations far enough in advance to order material as it is needed, thus eliminating stockpiled inventory.

Source: Adapted from *Mechanical Engineering,* March 2001.

15.8 | Group Technology

As will be noted throughout this text, many parts have certain *similarities* in their shape and in their method of manufacture. Traditionally, parts were designed from scratch, and manufacturing planned accordingly, without a formal consideration of design or processing experience. *Group technology* (GT) is a concept that seeks to take advantage of the **design and processing similarities** among the parts to be produced. First developed in Europe in the early 1900s, this concept involves *classifying* and *coding* of parts into *families* (Fig. 15.10). The term *group technology* was first used in 1959, but it was not until the use of interactive computers became widespread in the 1970s that this technology developed significantly.

Surveys in manufacturing plants repeatedly have indicated the commonality of parts. A typical survey consists of (a) analyzing each product, (b) decomposing it into its basic components, and (c) identifying the similarities. One survey, for example, found that 90% of the 3000 parts produced by a particular company fell into only five major families of parts. A pump, for example, can be broken into such basic components as the motor, the housing, the shaft, the seals, and the flanges. In spite of the variety of pumps manufactured, each of these components is basically the same in terms of product design and manufacturing methods. Thus, all shafts can be categorized into one family of parts.

The group-technology approach becomes even more attractive in view of consumer demand for an ever-larger variety of products, each in smaller quantities, thus involving *batch production* (see Section 14.2.2). Maintaining high efficiency in batch operations can, however, be difficult and overall manufacturing efficiency can therefore be compromised by a reduction in production volume. A traditional *product flow* in a batch manufacturing operation is shown in Fig. 15.11a. Note that machines of the same type are arranged in groups, that is, groups of lathes, of milling machines, of drill presses, and of grinders. In such a layout, called **functional**

Part 1   Part 2

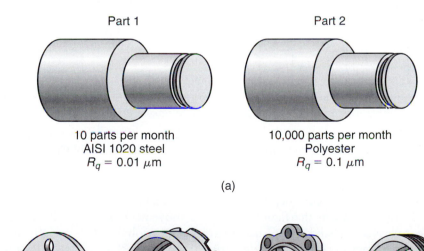

10 parts per month   10,000 parts per month
AISI 1020 steel   Polyester
$R_q = 0.01 \ \mu m$ $R_q = 0.1 \ \mu m$

(a)

FIGURE 15.10 (a) Two parts with identical geometries but with different manufacturing attributes. (b) Four parts with similar manufacturing attributes but with different geometries.

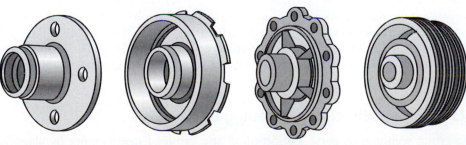

(b)

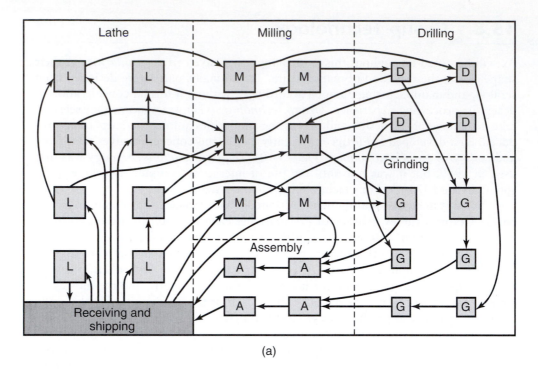

(a)

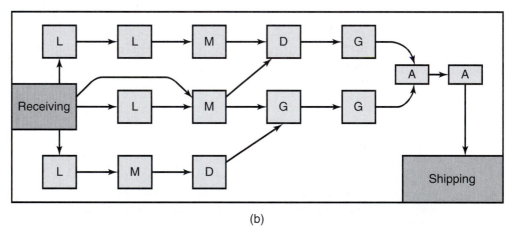

(b)

FIGURE 15.11 (a) Functional layout of machine tools in a traditional plant; arrows indicate the flow of materials and parts in various stages of completion. (b) Group-technology (cellular) layout. *Source:* After M.P. Groover.

layout, there is typically considerable random movement on the production floor, as shown by the arrows in the figure, indicating movement of materials and parts. Because it wastes time and effort, such an arrangement obviously is not efficient.

The machines in *cellular manufacturing* (Section 15.9) are arranged in a more efficient product flow line, called **group layout** (Fig. 15.11b). The manufacturing cell layout depends on the common features in parts; thus group technology is an essential feature of lean manufacturing (see Section 15.13).

15.8.1 Classification and coding of parts

In group technology, parts are identified and grouped into families by **classification and coding** (C/C) **systems.** This process is a critical and complex first step in

GT, and it is done according to the part's design and manufacturing attributes (see Fig. 15.10):

1. Design attributes pertain to *similarities* in geometric features and consist of the following:
    a. External and internal shapes and dimensions.
    b. Aspect ratio (length-to-width ratio or length-to-diameter ratio).
    c. Dimensional tolerance.
    d. Surface finish.
    e. Part function.

2. **Manufacturing attributes** pertain to *similarities* in the methods and the sequence of the operations performed on the part. The manufacturing attributes of a particular part or component consist of the following:
    a. Primary processes.
    b. Secondary processes and finishing operations.
    c. Process capabilities, such as dimensional tolerances and surface finish.
    d. Sequence of operations performed.
    e. Tools, dies, fixtures, and machinery.
    f. Production volume and production rate.

In reviewing these two lists, it can be appreciated that coding (see below) can be time consuming and that it requires considerable experience in the design and manufacture of products. In its simplest form, coding can be done by viewing the shapes of the parts in a generic way and then classifying the parts accordingly, such as in the categories of parts having rotational symmetry, parts having rectilinear shape, and parts having large surface-to-thickness ratios.

Parts may also be classified by studying their production flow throughout the total manufacturing cycle, called **production-flow analysis** (PFA). Recall from Section 15.6 that routing sheets clearly show process plans and the operations to be performed. However, a drawback of PFA is that a particular routing sheet does not necessarily indicate that the total operation is optimized. In fact, depending on the individual experience of, say, two process planners, the routing sheets for manufacturing the same part can be significantly different from each other. The beneficial role of computer-aided process planning to avoid or minimize such problems is thus obvious.

15.8.2 Coding

Coding of parts may be based on a particular company's own coding system or it may be based on one of several commercially available classification and coding systems (C/C system). Because product lines and organizational requirements vary widely, none of the C/C systems has been universally adopted. Whether it was developed in-house or is purchased, the system must be compatible with the company's other systems, such as NC machinery and CAPP systems. The **code structure** for part families typically consists of numbers, letters, or a combination of the two. Each specific component of a product is assigned a code, which may pertain to design attributes only (generally less than 12 digits) or to manufacturing attributes only. Most advanced systems include both attributes and use as many as 30 digits.

The three basic **levels of coding** described next have varying degrees of complexity.

1. **Hierarchical coding.** In this type of coding, also called **monocode**, the interpretation of each succeeding digit depends on the value of the preceding digit.

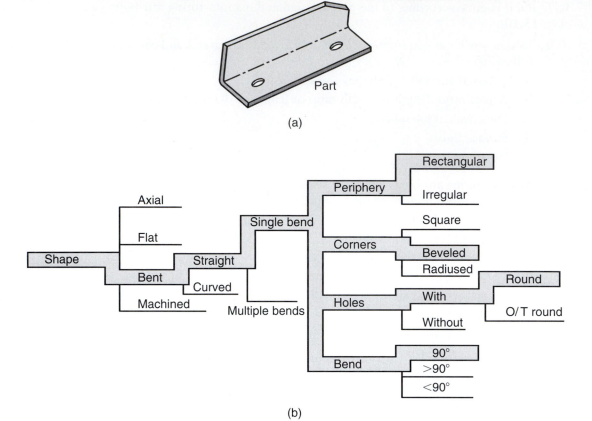

(a)

(b)

FIGURE 15.12 (a) A sheet-metal bracket and (b) its decision-tree classification. *Source:* After G.W. Millar.

Each symbol amplifies the information contained in the preceding digit, so that a single digit in the code cannot be interpreted alone. The advantage of hierarchical coding is that a short code can contain a large amount of information. This method is, however, difficult to apply in a computerized system.

2. **Polycodes.** Each digit in this code, also known as **chain-type code,** has its own interpretation and does not depend on the preceding digit. This structure tends to be relatively long, but it allows the identification of specific part **design and manufacturing** attributes and is well suited to computer implementation.

3. **Decision-tree coding.** This system, also called **hybrid codes,** is the most advanced, and it combines both design and manufacturing attributes (Fig. 15.12).

15.8.3 Coding systems

Three major industrial coding systems are described next.

1. The **Opitz system,** developed in the 1960s in Germany by H. Opitz (1905–1977), was the first comprehensive coding system presented. The basic code consists of nine digits (in the format 12345 6789) that represent design and manufacturing data (Fig. 15.13). Four additional codes (in the format ABCD) may also be used to identify the type and sequence of production operations. However, this system has two drawbacks: (a) It is possible to have different codes for parts that have similar manufacturing attributes, and (b) several parts with different shapes can have the same code.

Form Code					Supplementary Code			
1st digit Part class	2nd digit Main shape	3rd digit Rotational surface machining	4th digit Plane surface machining	5th digit Auxiliary holes, gear teeth, forming	Digit 1	2	3	4

The table body (rotational and nonrotational parts):

- 0 — $\frac{L}{D} \leqslant 0.5$
- 1 — $0.5 < \frac{L}{D} < 3$ — External shape, external shape elements — Internal shape, internal shape elements — Plane surface machining — Auxiliary holes
- 2 — $\frac{L}{D} \geqslant 3$ — Gear teeth
- 3 — $\frac{L}{D} \leqslant 2$ with deviation — Main shape — Rotational machining internal and external shape elements — Auxiliary holes, gear teeth, and forming
- 4 — $\frac{L}{D} > 2$ with deviation
- 5 — Special

Rotational parts (rows 0–5), Nonrotational parts (rows 6–9)

- 6 — $\frac{A}{B} \leqslant 3, \frac{A}{C} \geqslant 4$ Flat parts — Main shape — Principal bores — Plane surface machining — Auxiliary holes, gear teeth, and forming
- 7 — $\frac{A}{B} > 3$ Long parts — Main shape
- 8 — $\frac{A}{B} \leqslant 3, \frac{A}{C} \geqslant 4$ Cubic parts — Main shape
- 9 — Special

Supplementary code columns: Dimension, Material, Original shape of raw material, Accuracy

FIGURE 15.13 A classification and coding scheme using the Opitz system, consisting of five digits and a supplementary code of four digits, as shown along the top of the figure. *Source:* After H. Opitz.

2. The **MultiClass system,** illustrated in Fig. 15.14, was originally developed under the name MICLASS (Metal Institute Classification System of the Netherlands Organization for Applied Scientific Research) for the purpose of helping automate and standardize several design, production, and management functions in an organization. This system involves up to 30 digits and is used interactively with a computer that asks the user a number of questions. On the basis of the answers given, the computer automatically assigns a code number to the part.

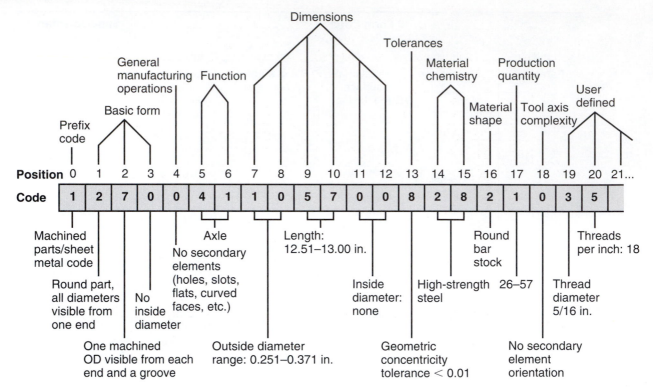

FIGURE 15.14 Typical MultiClass code for a machined part. *Source:* Courtesy of Organization for Industrial Research.

3. The **KK-3 system** is a general-purpose classification and coding system for parts that are to be machined or ground. It uses a 21-digit decimal system. This code is much longer than the two coding systems described above, but it also classifies dimensions and dimensional ratios (such as the length-to-diameter ratio of the part). The structure of a KK-3 system for rotational components is shown in Fig. 15.15.

15.8.4 Advantages of group technology

The major advantages of group technology may be summarized as follows.

1. Group technology makes possible the standardization of part designs and minimization of design duplication. New part designs can be developed using already existing and similar designs, thus saving a significant amount of time and effort. The product designer can quickly determine whether data on a similar part already exist in the computer files.

2. Data that reflect the experience of the designer and the process planner are stored in the database. A new and less experienced engineer can quickly benefit from that experience.

3. Manufacturing costs can more easily be estimated, and the relevant statistics on materials, processes, number of parts produced, and other factors can easily be obtained.

4. Process plans are standardized and scheduled more efficiently, orders are grouped for more efficient production, machine utilization is improved, setup times are reduced, and parts are produced more efficiently and with better and

Digit	Items	(Rotational component)
1	Part name	General classification
2		Detail classification
3	Materials	General classification
4		Detail classification
5	Major dimensions	Length
6		Diameter
7	Primary shapes and ratio of major dimensions	
8	External surface	External surface and outer primary shape
9		Concentric screw-threaded parts
10		Functional cut-off parts
11		Extraordinary shaped parts
12		Forming
13		Cylindrical surface
14	Internal surface	Internal primary shape
15		Internal curved surface
16		Internal flat and cylindrical surface
17	End surface	
18	Nonconcentric holes	Regularly located holes
19		Special holes
20	Noncutting process	
21	Accuracy	

(Digits 8–21 are grouped under the vertical label "Shape details and kinds of processes.")

FIGURE 15.15 The structure of a KK-3 system for rotational components. *Source:* Courtesy of Japan Society for the Promotion of Machine Industry.

more consistent product quality. Also, similar tools, fixtures, and machinery are shared in the production of a family of parts.

5. With the implementation of CAD/CAM, cellular manufacturing, and CIM, group technology is capable of greatly improving productivity and reducing costs in small-batch production (approaching those of mass production). Depending on the level of implementation, potential savings in each of the various design and manufacturing phases have been estimated to range from 5 to as much as 75%.

15.9 | Cellular Manufacturing

The concept of group technology can effectively be implemented in cellular manufacturing, which consists of one or more manufacturing cells. A **manufacturing cell** is a small unit consisting of one or more workstations. A workstation typically contains either one machine (**single-machine cell**) or several machines (**group-machine cell**), each of which performs a different operation on the part. The machines can be modified, retooled, and regrouped for different product lines within the same family of parts. Manufacturing cells are particularly effective in producing families of parts for which there is a relatively constant demand.

Cellular manufacturing has thus far been used primarily in material removal (Chapters 8 and 9) and in sheet-metal forming operations (Chapter 7), and to a lesser extent in polymer processing involving injection or compression molding. The machine tools commonly used in manufacturing cells are lathes, milling machines, drills, grinders, electrical-discharge machines, and machining centers; and for sheetmetal forming, they consist of shearing, punching, bending, and other forming machines.

Cellular manufacturing has some degree of automatic control for the following typical operations:

1. Loading and unloading of blanks (raw materials) and workpieces at workstations.
2. Changing of tools and dies at workstations.
3. Transfer of workpieces and tools and dies between workstations.
4. Scheduling and controlling of the total operation in the manufacturing cell.

Central to these activities is a *material-handling system* for transferring materials and parts among workstations. In attended (manned) machining cells, materials can be moved and transferred manually by the operator or by an industrial robot (Section 14.7) located centrally in the cell. Automated inspection and testing equipment also may be components of such a cell.

The significant benefits of cellular manufacturing are (a) the economics of reduced work in progress, (b) improved productivity, and (c) the ability to readily detect product quality problems right away. Furthermore, because of the variety of the machines and the processes involved, the operator becomes multifunctional and is not subjected to the tedium typically experienced in working on the same machine.

Manufacturing cell design. Because of their unique features, the design and implementation of manufacturing cells in traditional plants inevitably require the reorganization of the plant and the rearrangement of existing product flow lines. The machines may be arranged along a line, a U shape, an L shape, or in a loop (see also Fig. 14.23). For a group-machine cell, where the materials are handled by the operator, the U-shaped arrangement is convenient and efficient, because the operator can easily reach various machines. With mechanized or automated material handling, the linear arrangement and the loop layout are more efficient because operator access is not required and such arrangements are more easily achieved. The best arrangement of machines and material-handling equipment in a cell also involves consideration of such factors as the production rate, the type of product, and its shape, size, and weight.

Flexible manufacturing cells. Recall that in view of rapid changes in market demand and the need for more product variety in smaller quantities, flexibility of manufacturing operations is highly desirable. Manufacturing cells can be made flexible by using CNC machines and machining centers (see Section 8.11) and by means of industrial robots or other mechanized systems for handling materials (Sections 14.6 and 14.7). An example of attended *flexible manufacturing cell* (FMC) involving machining operations is illustrated in Fig. 15.16. In this illustration, an AGV moves parts between machines and inspection stations; machining centers fitted with automatic tool changers and tool magazines have an ability to perform a wide variety of operations (see Section 8.11).

A computer-controlled inspection station with a coordinate measurement machine can similarly inspect dimensions on a wide variety of parts. Therefore, the organization of these machines into a cell can allow the successful manufacture of very different parts. With computer integration, such a cell can manufacture parts in batch sizes as small as one part, with negligible delay between parts (really only the time required to download new machining instructions).

Flexible manufacturing cells are usually unattended (unmanned); their design and operation are thus more exacting than those for other cells. The selection of machines and robots, including the types and capacities of end effectors and their control systems, is critical to the proper functioning of the FMC. The likelihood of significant change in demand for part families should be considered during design of the cell, to ensure that the equipment involved has the necessary flexibility and

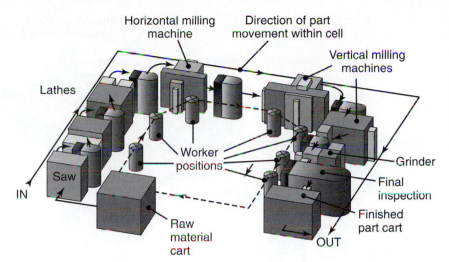

Horizontal milling machine

Direction of part movement within cell

Vertical milling machines

Lathes

Worker positions

Grinder

Final inspection

Finished part cart

Saw

IN

Raw material cart

OUT

FIGURE 15.16 Schematic illustration of an attended flexible manufacturing cell, showing various machine tools and an inspection station. Note the worker positions and the flow of parts in progress from machine to machine. *Source:* After J T. Black.

capacity. As is the case for other flexible manufacturing systems (described in Section 15.10), the cost of flexible manufacturing cells is very high. However, this disadvantage is outweighed by increased productivity (at least for batch production), flexibility, and controllability.

EXAMPLE 15.2 Manufacturing cells in small shops

Beginning in the mid-1990s, large companies began to increasingly outsource machining operations to their first-tier suppliers. This trend created pressure for small job shops to develop the capability to handle sophisticated projects, and it also has created a demand for larger firms to take on lower-volume, or turnkey, operations. These two basic market factors drove a number of trends in the design and implementation of flexible manufacturing cells:

- Manufacturing cells are being developed with a smaller number of machines, but greater versatility. Instead of four or five machines, many new manufacturing cells now consist of only one or two machines.

- Advanced CAD systems are now also used by smaller firms. While high-end CAD systems running on computer workstations are available at larger corporations, increasing computational power and software accessibility has allowed smaller firms to use less expensive, but still very powerful, PC-based packages.

- PC controllers have become more common and are driving a push to open architecture. Historically, machine manufacturers maintained full control over their control systems, and PC-based control was limited. Open architecture enables machine control to take place from a PC and, therefore, allows much greater flexibility in defining environments for machine operators.

- PC-based software is now available to translate solid-model part descriptions into surfaces for machining.

15.10 | Flexible Manufacturing Systems

In a *flexible manufacturing system* (FMS), all major elements of manufacturing are integrated into a highly automated system. First developed in the late 1960s, FMS consists of a number of manufacturing cells, each containing an industrial robot

FIGURE 15.17 A schematic illustration of a flexible manufacturing system, showing two machining centers, a coordinate measuring machine, and two automated guided vehicles.
Source: After J T. Black.

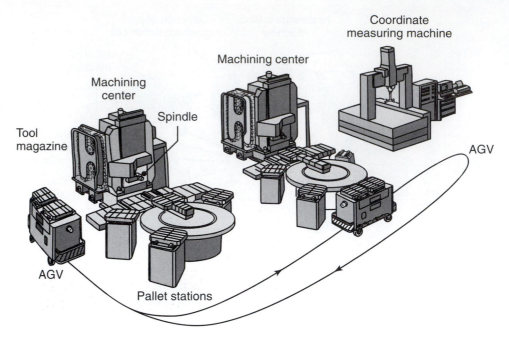

(serving several CNC machines) and an automated material-handling system, all interfaced with a central computer or fileserver. Different and specific computer instructions for the manufacturing process can be downloaded for each successive part that passes through a workstation. The flexibility of FMS is such that it can handle a variety of part configurations and produce them in any order. A general view of an FMS installation in a plant is shown in Fig. 15.17.

This highly automated system is capable of optimizing each step of the total manufacturing operation. These steps may specifically involve one or more processes and operations (such as machining, grinding, cutting, forming, powder metallurgy, heat treating, and finishing), as well as handling of incoming raw materials, part inspection, and assembly. The most common applications of FMS to date have been in machining and assembly operations. Flexible manufacturing systems represent the highest level of efficiency and productivity that has been achieved in manufacturing plants (Fig. 15.18).

FMS can be regarded as a system that combines the benefits of two other systems: (a) the highly productive, but inflexible, transfer lines (Section 14.2.4) and (b) job-shop production, which can fabricate a large variety of products on stand-alone machines but is inefficient. Table 15.1 compares some of the characteristics of transfer lines and FMS. Note that in FMS, the time required for changeover to a different part is very short, thus responsive to product and market-demand variations.

Elements of FMS. The computer control system of FMS is its brains, and the basic elements are (a) workstations, (b) automated handling and transport of materials and parts, and (c) control systems. The types of machines in workstations depend on the type of production. For machining operations (Chapter 8), they typically consist of a variety of three- to five-axis machining centers, CNC lathes, milling machines, drill presses, and grinders. Also included are various other equipment, such as those for automated inspection (including coordinate measuring machines), assembly, and cleaning. Other types of operations suitable for FMS include sheet-metal forming, punching and shearing, and forging systems for these operations incorporate furnaces, forging machines, trimming presses, heat-treating facilities, and cleaning equipment.

FIGURE 15.18   A general view of a flexible manufacturing system, showing several machining centers and an automated guided vehicle. *Source:* Courtesy of Cincinnati Machine, a UNOVA Company.

TABLE 15.1

Comparison of the Characteristics of Transfer Lines and Flexible Manufacturing Systems

Characteristic	Transfer line	FMS
Types of parts made	Generally few	Infinite
Lot size	>100	1–50
Part changing time	$\frac{1}{2}$ to 8 hr	1 min
Tool change	Manual	Automatic
Adaptive control	Difficult	Available
Inventory	High	Low
Production during breakdown	None	Partial
Efficiency	60–70%	85%
Justification for capital expenditure	Simple	Difficult

Material handling is controlled by a central computer and performed by automated guided vehicles, conveyors, and various transfer mechanisms. The system is capable of transporting raw materials, blanks, and parts in various stages of completion (work-in-progress) to any machine, in any order, and at any time. Prismatic parts are usually moved on specially designed **pallets,** and parts with rotational symmetry are moved by mechanical devices and robots.

Scheduling. Scheduling and process planning are crucial in FMS. Scheduling is *dynamic,* unlike in job shops, where a relatively rigid schedule is followed to perform a set of operations. Dynamic scheduling is capable of responding to quick changes in product type and thus is responsive to real-time decisions. The scheduling system for FMS specifies the types of operations to be performed on each part and identifies the machines or manufacturing cells to be used.

Because of the flexibility provided by FMS, no setup time is wasted in switching between manufacturing operations, and the system is capable of performing different operations in different orders and on different machines. However, the characteristics, performance, and reliability of each unit within the system must be continuously checked to ensure that parts moving from workstation to workstation are of acceptable quality and dimensional accuracy.

15.11 | Holonic Manufacturing

Holonic manufacturing is a more recent concept describing a unique organization of manufacturing units. The word *holonic* is from the Greek *holos*, meaning "whole," and the suffix *on*, meaning "a part of." Each component in a holonic manufacturing system is, at the same time, an independent entity (or *whole*) and a subservient *part* of a hierarchical organization. Because of its potential beneficial impact on manufacturing operations, we describe this system in some detail.

Holonic organizational systems have been studied since the 1960s, and there are a number of examples in biological systems. These systems are based on three fundamental observations:

1. Complex systems will evolve from simple systems much more rapidly if there are stable intermediate forms than if there are none. Also, stable and complex systems require a hierarchical system for evolution.

2. Holons simultaneously are self-contained wholes (to their subordinated parts) and the dependent parts of other systems. Holons are autonomous and self-reliant units, which have a degree of independence and can handle contingencies without asking higher levels in the hierarchical system for instructions. At the same time, holons are subject to control from multiple sources of higher system levels.

3. A holarchy consists of (a) autonomous wholes in charge of their parts, (b) dependent parts controlled by higher levels of a hierarchy, and (c) coordinate according to their local environment.

In biological systems, holarchies have the characteristics of stability regardless of disturbances, optimum use of available resources, and a high level of flexibility when their environment changes. A **manufacturing holon** is an autonomous and cooperative building block of a manufacturing system for production, storage, and transfer of objects or information. It consists of a control part and an optional physical-processing part. For example, a holon may be a combination of a CNC milling machine and an operator interacting via a suitable interface. A holon also may consist of other holons that provide the necessary processing, information, and human interfaces to the outside world, such as a group of manufacturing cells. Holarchies can be created and dissolved dynamically, depending on the current needs of the particular manufacturing operation.

A holonic-systems view is one of creating a working manufacturing environment from the bottom up. Maximum flexibility can be achieved by providing intelligence within holons to (a) support all production and control functions required to complete production tasks and (b) manage the underlying equipment and systems. The manufacturing system will dynamically reconfigure into operational hierarchies to optimally produce the desired products, with holons or elements being added or removed as needed.

Holarchical manufacturing systems rely on rapid and effective communication between holons, as opposed to traditional hierarchical control where individual processing power is essential (such as adaptive control or machine vision programming).

A large number of specific arrangements and software algorithms have been proposed for holarchical systems. Although a detailed description of these is beyond the scope of this book, the general sequence of events can be outlined as follows:

1. A factory consists of a number of resource holons, available as separate entities in a resource pool. For example, available holons may consist of a (a) CNC milling machine and operator, (b) CNC grinder and operator, and (c) CNC lathe and operator.

2. Upon receipt of an order or directive from higher levels in the factory hierarchical structure, an order holon is formed and it begins communicating and negotiating with the available resource holons.

3. The negotiations lead to a self-organized grouping of resource holons, assigned based on product requirements, resource holon availability, and customer requirements. For example, a given product may require the use of a CNC lathe, a CNC grinder, and an automated inspection station to organize it into a production holon.

4. In case of machine breakdown, lack of machine availability, or changing customer requirements, other holons from the resource pool can be added or removed as needed, allowing a reorganization of the production holon. Production bottlenecks can be identified and eliminated through communication and negotiation between the holons in the resource pool.

The last step is referred to as *plug and play,* a term borrowed from the computer industry where hardware components seamlessly integrate into a system.

15.12 | Just-in-Time Production

The *just-in-time production* (JIT) concept was originated in the United States decades ago but was first implemented on a large scale in 1953 at the Toyota Motor Company in Japan to eliminate waste of materials, machines, capital, manpower, and inventory throughout the manufacturing system. The JIT concept has the following goals:

1. Receive supplies just in time to be used.
2. Produce parts just in time to be made into subassemblies.
3. Produce subassemblies just in time to be assembled into finished products.
4. Produce and deliver finished products just in time to be sold.

In traditional manufacturing operations, the parts are made in batches, placed in inventory, and used when necessary. This approach, known as a **push system,** means that parts are made according to a schedule and are in inventory to be used if and when they are needed. In contrast, just-in-time manufacturing is a **pull system,** meaning that parts are produced to order, and the production is matched with demand. Consequently, there are no stockpiles, and the extra motions and expenses involved in stockpiling parts and then retrieving them from storage are eliminated.

Note that the ideal production quantity in JIT is one; hence this system is also called **zero inventory, stockless production,** and **demand scheduling.** Moreover, parts are inspected in real time either automatically or by the worker as they are being manufactured and are used within a short period of time. In this way, control is maintained continuously over production, immediately identifying defective parts or process variation. Adjustments can then be made rapidly to make a uniform, high-quality product (see also Section 16.3).

The ability to detect production problems has been likened to the level of water (representing inventory levels) in a lake covering a bed of boulders (representing production problems). When the water level is high (the high inventories associated with *push* production), the boulders are not exposed. By contrast, when the level is low (the low inventories associated with *pull* production), the boulders are exposed, and thus the problems can readily be identified and solved. This analogy indicates that high inventory levels can mask quality and production problems involving parts that already have been stockpiled.

Implementation of the JIT concept requires that all aspects of manufacturing operations be continuously monitored and reviewed, so that all operations and resources that *do not add* value are eliminated. This approach emphasizes (a) worker pride and dedication in producing high-quality products, (b) elimination of idle resources, and (c) teamwork among workers, engineers, and management to quickly solve any problems that arise during production or assembly.

An important aspect of JIT is the delivery of supplies and parts from outside sources or from other divisions of the company, so as to significantly reduce in-plant inventory. As a result, major reductions in storage facilities have taken place; in fact, the concept of building large warehouses for parts has now become obsolete. Suppliers are expected to deliver, often on a daily basis, preinspected goods as they are needed for production. Consequently, this approach requires reliable suppliers, close cooperation and trust between the company and its vendors, and a reliable system of transportation. Also important for smoother operation is reduction of the number of suppliers.

Kanban. JIT was first demonstrated in Japan under the name *kanban*, meaning "visible record." These records typically consist of two types of cards (kanbans): (a) the **production card,** which authorizes the *production* of one container or cart of identical, specified parts at a workstation, and (b) the **conveyance card,** or **move card,** which authorizes the transfer of one container or cart of parts from that particular workstation to the workstation where the parts will next be used. The cards, which are now bar-coded plastic tags or other devices, contain information on the place of issue, part type, part number, and the number of items in the container. The number of containers in circulation at any time is completely controlled and can be scheduled as desired for maximum production efficiency.

Advantages of JIT. The advantages of just-in-time production may be summarized as follows:

- Low inventory-carrying costs.
- Rapid detection of defects in the production or in the delivery of supplies (hence low scrap loss).
- Reduced need for inspection and reworking of parts.
- Production of high-quality parts at low cost.

Although there are significant variations, implementation of just-in-time production has resulted in reductions estimated at 20 to 40% in product cost; 60 to 80% in inventory; up to 90% in rejection rates; 90% in lead times; and 50% in scrap, rework, and warranty costs. Furthermore, increases of 30 to 50% in direct labor productivity and of 60% in indirect labor productivity have also been achieved.

15.13 | Lean Manufacturing

In a modern manufacturing environment, companies must be responsive to the needs of the customers and their specific requirements, and to fluctuating global market demands. At the same time, the manufacturing enterprise must be conducted with a minimum amount of wasted resources. This realization has lead to *lean production* or *lean manufacturing* strategies.

Lean manufacturing is a systematic approach to *identifying and eliminating waste* (that is, non-value-added activities) in every area of manufacturing, through continuous improvement, and emphasizing product flow in a pull system. When applied on a large scale, lean manufacturing generally is referred to as *agile manufacturing*. Lean production requires that a manufacturer review all its activities from the viewpoint of the customer, and optimizes processes to maximize added value. This viewpoint is critically important because it helps identify whether or not a particular activity (a) clearly adds value, (b) adds no value but cannot be avoided, or (c) adds no value but can be avoided.

The focus of lean production is on the entire process flow and not just the improvement of one or more individual operations. Typical wastes to be considered and reduced or preferably eliminated include the following:

- Eliminating inventory (using JIT methods) because inventory represents cost, leads to defects, and reduces responsiveness to changing market demands;
- Eliminating waiting time, which may be caused by unbalanced work loads, quality problems, and unplanned maintenance;
- Maximizing the efficiency of workers at all times;
- Eliminating unnecessary processes and steps, because they represent costs;
- Minimizing or eliminating product transportation, because it represents an activity that adds no value; this waste can be eliminated (such as by installing machining cells) or be minimized (such as by improving the plant layout);
- Performing time and motion studies to identify inefficient workers or unnecessary product movements; and
- Eliminating defects.

15.14 | Communications Networks in Manufacturing

In order to maintain a high level of coordination and efficiency of operation in integrated manufacturing, an extensive, high-speed, and interactive **communications network** is essential. The World Wide Web, or Internet, is well known and has profoundly impacted business, as well as the personal lives of hundreds of millions of users. Often, however, the Internet is not suitable for the manufacturing environment, because of security and performance issues involved. The **local area network** (LAN) has been a major advance in communications technology. This is a hardware and software system, in which logically related groups of machines and equipment communicate with each other. The network links these groups to each other, bringing different phases of manufacturing into a unified operation.

A local area network can be very large and complex, linking hundreds or even thousands of machines and devices in several buildings. Various network layouts (Fig. 15.19) of fiber-optic or copper cables are typically used, over distances ranging

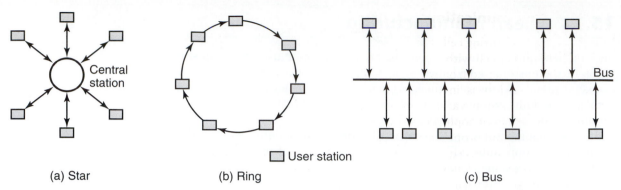

FIGURE 15.19   Three basic types of topology for a local area network (LAN). (a) The star topology is suitable for situations that are not subject to frequent configuration changes; all messages pass through a central station. (b) In the ring topology, all individual user stations are connected in a continuous ring; the message is forwarded from one station to the next until it reaches its assigned destination. Although the wiring is relatively simple, the failure of one station shuts down the entire network. (c) In the bus topology, all stations have independent access to the bus; this system is reliable and is easier to service than the other two. Because its arrangement is similar to the layout of the machines in a factory, its installation is relatively easy, and it can easily be rearranged when the machines are rearranged.

from a few meters to as much as 32 km (20 mi); for longer distances, **wide area networks** (WANs) are used. Different types of networks can be linked or integrated through "gateways" and "bridges," often using secure *file transfer protocols* (FTP) over Internet connections. A number of advanced network protocols are under development but have not yet been applied to manufacturing, including ipV6 and Internet2.

Access control to the network is important, as otherwise collisions can occur when several workstations are transmitting simultaneously; thus, continuous scanning of the transmitting medium is essential. In the 1970s, a *carrier sense multiple access with collision detection* (CSMA/CD) system was developed and implemented in **Ethernet.** Other access-control methods are **token ring** and **token bus,** in which a "token" (special message) is passed from device to device. The device that has the token is allowed to transmit, while all other devices only receive.

A trend is the use of **wireless local area networks** (WLANs). Conventional LANs require routing of wires, often through masonry walls or other permanent structures, and require computers or machinery to remain stationary. WLANs allow equipment such as mobile test stands or data collection devices (for example, barcode readers) to easily maintain a network connection. A communication standard (IEEE 802.11) currently defines frequencies and specifications of signals, and two radio frequency and one infrared methods. It should be noted, however, that wireless networks are slower than those that are hard wired, but their flexibility makes them desirable, especially for situations where slow tasks, such as machine monitoring, are the main application.

Personal area networks (PANs) also are available. They are based on communications standards such as Bluetooth, IrDA, and HomeRF and are designed to allow data and voice communication over short distances. For example, a short-range Bluetooth device will allow communication over a 10-m (32-ft) distance. This technology has been integrated into personal computers and cellular phones but is not yet as common for manufacturing applications. PANs are undergoing major changes, and communications standards are continually being refined.

15.14.1 Communications standards

Often, one manufacturing cell is built with machines and equipment purchased from one vendor, another cell with machines purchased from another vendor, a third cell with machines purchased from yet a third vendor, and so on. As a result, a variety of programmable devices is involved and driven by several computers and microprocessors, purchased at various times from different vendors and having various capacities and levels of sophistication. The computers in each cell may have their own specifications and proprietary standards, and they cannot communicate beyond the cell with others unless equipped with custom-built interfaces. This situation creates what is known as *islands of automation,* and in some cases, up to 50% of the cost of automation has been attributed to overcoming difficulties in communications among individual manufacturing cells and other parts of the organization.

Automated cells that could function only independently of each other, without a common base for information transfer, led to the need for a communications standard to improve communications and efficiency in computer-integrated manufacturing. After considerable effort and on the basis of existing national and international standards, a set of communications standards, known as **manufacturing automation protocol** (MAP), was developed. The capabilities and effectiveness of MAP were first demonstrated in 1984 with the successful interconnection of multiple devices.

The International Organization for Standardization/Open System Interconnect (ISO/OSI) reference model is now accepted worldwide. This model has a hierarchical structure in which communication between two users is divided into seven layers (Fig. 15.20). Each layer has a special task: (a) mechanical and electronic means of data transmission, (b) error detection and correction, (c) correct transmission of the message, (d) control of the dialog among users, (e) translation of the message into a common syntax, and (f) verification that the data transferred have been understood.

The operation of this system is complex. Basically, each standard-sized portion of the message or data to be transmitted from user *A* to user *B* moves sequentially through the successive layers, from 7 to 1, at user *A*'s end. More information is added to the original message as it travels through each layer. The complete packet

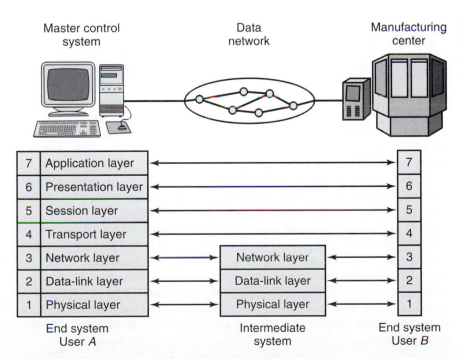

FIGURE 15.20 The ISO/OSI reference model for open communication. *Source:* After U. Rembold.

is transmitted through the physical communications medium to user *B* and then moves through the layers from layer 1 to layer 7 at user *B*'s end. Transmission takes place through fiber-optic cable, coaxial cable, microwaves, and similar devices.

Communication protocols were then extended to *office* automation with the development of **technical and office protocol** (TOP), which is based on the ISO/OSI reference model. In this way, total communication (MAP/TOP) can be established across the factory floor and among offices at all levels of an organization. **Internet tools** (hardware, software, and protocols) within a company now link all departments and functions into a self-contained and fully compatible **Intranet.** Several tools for implementing this linkage are available commercially; they are inexpensive and easy to install, integrate, and use.

15.15 | Artificial Intelligence

Artificial intelligence (AI) is an area of computer science concerned with systems that exhibit some characteristics usually associated with intelligence in human behavior, such as learning, reasoning, problem solving, and the understanding of language. The goal of AI is to simulate such human endeavors on the computer. The art of bringing relevant principles and tools of AI to bear on difficult applications is known as **knowledge engineering.**

Artificial intelligence has a major impact on the design, the automation, and the overall economics of manufacturing operations. This is, in large part, due to major advances in computer-memory expansion (VLSI chip design) and decreasing costs. Artificial-intelligence packages costing on the order of a few thousand dollars have been developed, which can run on personal computers, making AI accessible to office desks and shop floors.

Artificial-intelligence applications in manufacturing generally encompass the following:

1. Expert systems.
2. Natural language.
3. Machine (computer) vision.
4. Artificial neural networks.
5. Fuzzy logic.

15.15.1 Expert systems

Also called a **knowledge-based system,** an *expert system* (ES) is generally defined as an intelligent computer program that has the capability to solve difficult real-life problems, using **knowledge-base** and **inference** procedures (Fig. 15.21). The goal of an expert system is to develop the capability to conduct an intellectually demanding task in the way that a human expert would. The field of knowledge required to perform this task is called the **domain** of the expert system.

Expert systems use a knowledge base that contains facts, data, definitions, and assumptions. They also have the capacity to follow a **heuristic** approach, that is, to make good judgments on the basis of *discovery* and *revelation* and to make *high-probability* guesses, just as a human expert would. The knowledge base is expressed in computer codes, usually in the form of **if-then rules,** and can generate a series of questions. The mechanism for using these rules to solve problems is called an **inference engine.** Expert systems can also communicate with other computer software packages.

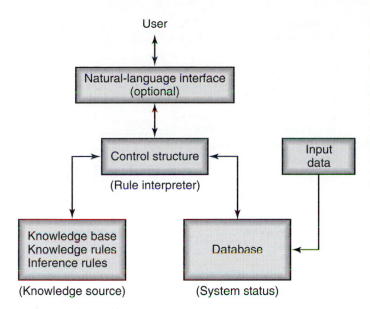

User

Natural-language interface
(optional)

Control structure
(Rule interpreter)

Input
data

Knowledge base
Knowledge rules
Inference rules

(Knowledge source)

Database

(System status)

FIGURE 15.21 The basic structure of an expert system. The knowledge base consists of knowledge rules (general information about the problem) and inference rules (the approach used to reach conclusions). The results are communicated to the user through the natural-language interface. *Source:* After K.W. Goff.

To construct expert systems for solving the complex design and manufacturing problems encountered, one needs (a) knowledge and (b) a mechanism for manipulating this knowledge to create solutions. The development of knowledge-based systems requires much time and effort because of (a) the difficulty involved in accurately modeling the many years of experience of an expert, or a team of experts, and (b) the complex inductive reasoning and decision-making capabilities of humans, including the capacity to learn from mistakes.

Expert systems operate on a *real-time* basis, and their short reaction times provide rapid responses to problems. The programming languages most commonly used for this application are C++, LISP, and PROLOG; other compiler languages such as FORTRAN (from the words formula translating) or BASIC can also be used. A significant development has been **expert-system software shells,** or **environments**. Also called framework systems, these software packages are essentially expert-system outlines that allow a person to write specific applications to suit special needs.

Several expert systems have been developed and used for such specialized applications as

1. Problem diagnosis in machines and equipment and determination of corrective actions.
2. Modeling and simulation of production facilities.
3. Computer-aided design, process planning, and production scheduling.
4. Management of a company's manufacturing strategy.

15.15.2 Natural-language processing

Obtaining information from a database in computer memory has traditionally required the use of computer programmers to translate questions expressed in natural language into "queries" in some machine language. Natural-language interfaces with database systems, which are in various stages of development, allow a user to obtain information by entering English-language commands in the form of simple, typed or spoken questions. Natural-language software shells are used in such applications as (a) scheduling of material flow in manufacturing and (b) analyzing information in databases.

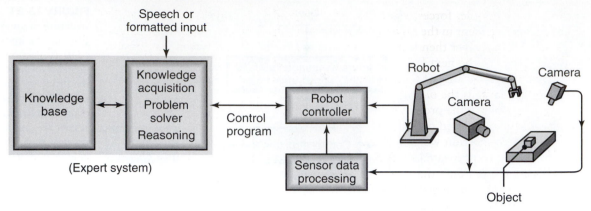

FIGURE 15.22   Expert system as applied to an industrial robot guided by machine vision using two cameras.

Major progress continues to be made in the development of computer software that will have speech-synthesis and recognition (**voice recognition**) capabilities, thus eliminating the need to type commands on keyboards. Although speech recognition software is now available for word processing and to allow the handicapped to use computers, higher-order understanding of the spoken word does not yet exist.

15.15.3 Machine vision

The basic features of *machine vision* are described and illustrated in Section 14.8 and Fig. 14.26. In systems that incorporate machine vision, computers and software implementing artificial intelligence are combined with cameras and other optical sensors. These machines then perform such operations as inspecting, identifying, and sorting parts, and guiding robots (*intelligent robots;* Fig. 15.22). Note that these are operations that otherwise would require human involvement and intervention.

15.15.4 Artificial neural networks

Although computers are much faster than the human brain at *sequential* tasks, humans are much better at pattern-based tasks that can be attacked with *parallel processing,* such as recognizing features (e.g., in faces and voices, even under noisy conditions), assessing situations quickly, and adjusting to new and dynamic conditions. These advantages are also due partly to the ability of humans to use, real time, their five **senses** (sight, hearing, smell, taste, and touch) simultaneously, a process called *data fusion. Artificial neural networks* (ANNs), a branch of artificial intelligence, attempts to gain some of these human capabilities through computer imitation of the way that data are processed by the human brain.

The human brain has about 100 billion linked **neurons** (cells that are the fundamental functional units of nervous tissue) and more than a thousand times that many *connections.* Each neuron performs only one simple task: It receives input signals from a fixed set of neurons, and when those input signals are related in a certain way (specific to that particular neuron), it generates an electrochemical output signal sent to a fixed set of neurons. It is now believed that human learning is accomplished by changes in the strengths of these signal connections between neurons.

A fully developed, feed-forward network is the most common type of ANN, and it is built according to this principle from several layers of processing elements (*simulating neurons*). The elements in the first (input) layer are fed with input data (such as, for example, forces, velocities, or voltages). Each element sums up all its inputs one per element in the input layer, and many per element in succeeding layers. Each element in a layer then transfers the data (according to a transfer function) to all the elements in the next layer. Each element in that next layer, however, receives a different signal, because of the different connection weights between the elements.

The last layer is the *output* layer, within which each element is compared with the desired output of the process being simulated. The difference between the desired output and the calculated output, referred to as *error,* is fed back to the network by changing the connection weights in such a way that it reduces this error. After this procedure has been repeated several times, the network has been "trained," and it can be used on input data that have not previously been presented to the system.

Other types of artificial neural networks are (a) *associative memories,* (b) *self-organizing ANN,* and (c) *adaptive-resonance ANN.* The feature common to these neural networks is that they must be trained with concrete and well-defined example problems. It is, therefore, very difficult to formulate input-output relations mathematically and to predict an ANN's behavior with untrained inputs.

Artificial neural networks have such applications as (a) noise reduction (in telephones), (b) speech recognition, and (c) process control in manufacturing operations. For example, they can be used to predict the surface finish of a workpiece obtained by end milling (see Fig. 8.1d) based on input parameters such as cutting force, torque, acoustic emission, and spindle speed.

15.15.5 Fuzzy logic

An element of artificial intelligence that has important applications in control systems and pattern recognition is *fuzzy logic.* Based on the observation that people can make good decisions on the basis of imprecise and nonnumerical information, **fuzzy models** were introduced in 1965 as mathematical means of representing information that is vague and imprecise, hence the term *fuzzy.* These models have the capability to recognize, represent, manipulate, interpret, and use data and information that are vague or lack precision. Fuzzy-logic methods deal with reasoning and decision making at a level higher than do neural networks. Typical linguistic examples of concepts used in fuzzy logic are words and terms such as *few, almost all, very, more or less, small, medium,* and *extremely.*

Fuzzy-logic technologies and devices have been developed and successfully applied in areas such as robotics and motion control, image processing and machine vision, machine learning, and the design of intelligent systems. Some applications of fuzzy logic include (a) automatic transmissions of automobiles; (b) washing machines that automatically adjust the washing cycle for load size, fabric type, and amount of dirt; and (c) helicopters that obey vocal commands to hover, land, or move go forward, up, left, or right.

SUMMARY

- Computer-integrated manufacturing systems are the most important means of improving productivity, responding to changing market demands, and better controlling manufacturing operations and management functions. (Sections 15.1–15.3)
- With rapid developments in advanced and sophisticated software, together with the capability for computer simulation and analysis, computer-aided product design and engineering are now much more efficient, detailed, and comprehensive. (Section 15.4)

- Computer-aided manufacturing is often combined with computer-aided design to transfer information from the design stage to the planning stage and to production. (Section 15.5)

- Continuing advances in manufacturing operations such as computer-aided process planning, computer simulation of manufacturing processes and systems, group technology, cellular manufacturing, flexible manufacturing systems, holonic manufacturing systems, and just-in-time manufacturing are all essential elements in improving productivity. (Sections 15.6–15.12)

- Lean manufacturing is intended to identify and eliminate waste, leading to improvements in quality and customer satisfaction and decreases in product cost. A key concept is to establish a pull system of inventory control using JIT principles. (Sections 15.12 and 15.13)

- Communications networks and their global standardization are critical to CIM strategies. (Section 15.14)

- Artificial intelligence has increasing applications in all aspects of manufacturing science, engineering, and technology. (Section 15.15)

BIBLIOGRAPHY

Amirouche, F.M.L., *Principles of Computer-Aided Design and Manufacturing,* 2nd ed., Prentice Hall, 2003.

Badiru, A.B., *Expert Systems Applications in Engineering and Manufacturing,* Prentice Hall, 1998.

Biekert, R., Berling, D., Evans, R.J., and Kelley, D.G., *CIM Technology: Fundamentals and Applications,* Goodheart Wilcox, 1998.

Black, J.T., and Hunter, S.L., *Lean Manufacturing: Systems and Cell Design,* Society of Manufacturing Engineers, 2003.

Burbidge, J.L., *Production Flow Analysis for Planning Group Technology,* Oxford, 1997.

Chang, T.-C., Wysk, R.A., and Wang, H.P., *Computer-Aided Manufacturing,* 3rd ed., Prentice Hall, 2005.

Cheng, T.C.E., and Podolsky, S., *Just-in-Time Manufacturing: An Introduction,* 2nd ed., Chapman & Hall, 1996.

Driankov, D., Hellendoorn, H., and Reinfrank, M., *Introduction to Fuzzy Control,* 2nd ed., Springer, 1996.

Fausett, L.V., *Fundamentals of Neural Networks,* Prentice Hall, 1994.

Gershwin, S.B., Chrissoleon, Y.D., Papadopoulos, T., and Smith, J.M., *Analysis and Modeling of Manufacturing Systems,* Springer, 2002.

Gu, P., and Norrie, D.H., *Intelligent Manufacturing Planning,* Chapman & Hall, 1995.

Hannam, R., *CIM: From Concept to Realisation,* Addison-Wesley, 1998.

Haykin, S.S., *Neural Networks: A Comprehensive Foundation,* 2nd ed., Prentice Hall, 1998.

Higgins, P., Roy, L.R., and Tierney, L., *Manufacturing Planning and Control: Beyond MRP II,* Chapman & Hall, 1996.

Hitomi, K., *Manufacturing Systems Engineering,* 2nd ed., Taylor & Francis, 1996.

Hyer, N., and Wemmerlov, U., *Reorganizing the Factory: Competing through Cellular Manufacturing,* Productivity Press, 2003.

Irani, S.A. (ed.), *Handbook of Cellular Manufacturing Systems,* Wiley-Interscience, 1999.

Jackson, P., *Introduction to Expert Systems,* 3rd ed., Addison-Wesley, 1998.

Kasabov, N.K., *Foundations of Neural Networks, Fuzzy Systems, and Knowledge Engineering,* MIT Press, 1996.

Krishnamoorty, C.S., and Rajeev, S., *Artificial Intelligence and Expert Systems for Engineers,* CRC, 1996.

Kusiak, A., *Computational Intelligence in Design and Manufacturing,* Wiley-Interscience, 2000.

———, *Engineering Design: Products, Processes, and Systems,* Academic Press, 1999.

Lee, K., *Principles of CAD/CAM Systems,* Addison-Wesley, 1999.

Leondes, C.T. (ed.), *Fuzzy Logic and Expert Systems Applications,* Academic Press, 1998.

Liebowitz, J. (ed.), *The Handbook of Applied Expert Systems,* CRC, 1997.

Louis, R.S., *Integrating Kanban with MRP II: Automating a Pull System for Enhanced JIT Inventory Management,* Productivity Press, 2005.

———, *Custom Kanban: Designing the System to Meet the Needs of Your Environment,* Productivity Press, 2006.

McMahon, C., and Browne, J., *CAD/CAM Principles, Practice and Manufacturing Management,* Addison-Wesley, 1999.

Monden, Y., *Toyota Production System: An Integrated Approach to Just-in-Time,* 3rd ed., Institute of Industrial Engineers, 1998.

Parsaei, H., Leep, H., and Jeon, G., *The Principles of Group Technology and Cellular Manufacturing Systems,* Wiley, 2006.

Popovic, D., and Bhatkar, V., *Methods and Tools for Applied Artificial Intelligence,* Dekker, 1994.

Rehg, J.A., *Introduction to Robotics in CIM Systems,* 5th ed., Prentice Hall, 2002.

——, and Kraebber, H.W., *Computer-Integrated Manufacturing,* 3rd ed., Prentice Hall, 2005.

Sandras, W.W., *Just-in-Time: Making It Happen,* Wiley, 1997.

Singh, N., *Systems Approach to Computer-Integrated Design and Manufacturing,* Wiley, 1995.

Singh, N., and Rajamani, D., *Cellular Manufacturing Systems: Design, Planning and Control,* Chapman & Hall, 1996.

Vajpayee, S.K., *Principles of Computer-Integrated Manufacturing,* Prentice Hall, 1995.

Vollmann, T.E., Berry, W.L., and Whybark, D.C., *Manufacturing Planning and Control Systems,* 4th ed., Irwin, 1997.

——, and Jacobs, F.R., *Manufacturing Planning and Control for Supply Chain Management,* McGraw-Hill, 2004.

Williams, D.J., *Manufacturing Systems,* 2nd ed., Chapman & Hall, 1994.

Wu, J.-K., *Neural Networks and Simulation Methods,* Dekker, 1994.

QUESTIONS

15.1 In what ways have computers had an impact on manufacturing? Describe in detail.

15.2 What advantages are there in viewing manufacturing as a system? What are the components of a manufacturing system? Explain.

15.3 Discuss the benefits of computer-integrated manufacturing operations.

15.4 What is a database? Why is it necessary in manufacturing? Why should the management of a manufacturing company have access to databases? Explain.

15.5 Explain how a CAD system operates.

15.6 What are the advantages of CAD systems over traditional methods of design? Do CAD systems have any limitations? Explain, with examples.

15.7 What is a NURB?

15.8 What are the advantages of a third-order piecewise Bezier curve over a B-spline or conventional Bezier curve?

15.9 Describe your understanding of the octree representation in Fig. 15.6.

15.10 Describe the purposes of process planning. How are computers used in such planning?

15.11 Explain the features of two types of CAPP systems described in this chapter.

15.12 Describe the features of a routing sheet. Why is a routing sheet necessary in manufacturing?

15.13 What is group technology? Why was it developed? Explain its advantages.

15.14 What does classification and coding mean with respect to group technology? Explain.

15.15 What is a manufacturing cell? Why was it developed?

15.16 What are the basic principles of flexible manufacturing systems. Why do they require major capital investment? Explain.

15.17 Why is a flexible manufacturing system capable of producing a wide range of lot sizes?

15.18 What are the benefits of just-in-time production? Why is it called a pull system?

15.19 Explain the function of a local area network.

15.20 Describe the advantages of a communications standard.

15.21 What is meant by the term *fuzzy logic*?

15.22 Explain the principle behind holonic manufacturing.

15.23 What are the differences between ring and star networks? What is the significance of these differences?

15.24 What is kanban? Why was it developed?

15.25 What is an FMC and what is an FMS? What are the differences between them? Explain.

15.26 Describe the elements of artificial intelligence. Why is machine vision a part of it?

15.27 Explain why humans will still be needed in the factory of the future regardless of how well it is automated.

15.28 How would you describe the principle of computer-aided manufacturing to an older worker in a manufacturing facility who has only elementary knowledge of computers?

15.29 Give examples of primitives of solids other than those illustrated in Fig. 15.4b.

15.30 Explain the logic behind the arrangements shown in Fig. 15.11b.

15.31 Describe your observations regarding Fig. 15.18.

15.32 What should be the characteristics of an effective guidance system for an automated guided vehicle?

15.33 Give specific examples in manufacturing for which artificial intelligence could be effective.

15.34 Describe your opinions concerning the voice-recognition capabilities of future machines and controls.

15.35 Would machining centers be suitable for just-in-time production? Explain.

15.36 Give an example of a push system and of a pull system to clarify the fundamental difference between the two methods.

15.37 Give a specific example in which the variant system of CAPP is desirable and an example in which the generative system is desirable.

15.38 Artificial neural networks are particularly useful where the problems are ill defined and the data are fuzzy. Give examples in manufacturing where this is the case.

15.39 Is there a minimum to the number of machines in a manufacturing cell? Explain.

15.40 List as many three-letter acronyms used in manufacturing (such as CNC, FMS, LAN, etc.) as you can, and give a brief definition of each.

15.41 Are there any disadvantages to zero inventory? Explain.

15.42 Why are robots a major component of a flexible manufacturing cell?

15.43 Is it possible to implement JIT in global manufacturing companies? Explain.

15.44 Describe the advantages and limitations of parametric design.

15.45 What are the advantages of hierarchical coding?

15.46 Assume that you are asked to rewrite Section 15.15 on artificial intelligence. Briefly outline your personal thoughts regarding this topic.

15.47 Assume that you own a manufacturing company and that you are aware you have not taken full advantage of the technological advances in manufacturing. Now, however, you would like to do so and you have the necessary capital. Describe how you would go about analyzing your company's needs and how you would plan to implement these technologies. Consider technical as well as human aspects.

15.48 With specific examples, describe your thoughts concerning the state of manufacturing in the United States as compared with its state in other industrialized countries.

15.49 It has been suggested by some that artificial intelligence systems will ultimately be able to replace the human brain. Do you agree? Explain.

PROBLEMS

15.50 Sketch skeletons for (a) a circle, (b) a square, and (c) an equilateral triangle. Indicate the radius information at every node and endpoint of the skeleton if the shapes have an area of 1 in^2.

15.51 List the primitives needed to describe the part shown in Fig. 15.12.

15.52 Assume that you are asked to give a quiz to students on the contents of this chapter. Prepare one quantitative problem and three qualitative questions and supply the answers.

DESIGN

15.53 Review various manufactured parts described in various chapters in this text and group them in a manner similar to that shown in Fig. 15.10. Comment on your observations.

15.54 Evaluate a process from a lean production perspective. For example, closely observe the following and identify, eliminate (when possible), or optimize the steps that produce waste when
(a)   Preparing breakfast for a group of eight.
(b)   Washing clothes or cars.
(c)   Using Internet browsing software.
(d)   Studying for an exam, or writing a report or term paper.

15.55 Consider a routing sheet for a product as discussed in Section 15.8. If a holonic manufacturing system is to automatically generate the manufacturing sequence, (a) what is the likelihood that it will be the same as that produced manually? (b) What is the likelihood that a holonic system will produce the same sequence every time? Explain.

15.56 Think of a specific product and make a decision-tree chart similar to the one shown in Fig. 15.12.

15.57 Describe the trends in product designs and features that have had a major impact on manufacturing. Give specific examples.

15.58   Think of a specific product and design a manufacturing cell for making it (see Fig. 15.16), describing the features of the machines and equipment involved. Explain how the cell arrangement would change, if at all, if design changes are introduced to the product.

15.59   Surveys have indicated that 95% of all the different parts made in the United States are produced in lots of 50 or less. Comment on this observation, and describe your thoughts regarding the implementation of the technologies outlined in Chapters 14 and 15.

15.60   Think of a simple product, and make a routing sheet for its production, similar to that shown in Fig. 15.8. If the same part is given to another person, what is the likelihood that the routing sheet developed will be basically the same? Explain.

15.61   Review Fig. 15.8 and basically suggest a routing sheet for the manufacture of each of the following: (a) an automotive connecting rod, (b) a compressor blade, (c) a glass bottle, (d) an injection-molding die, and (e) a bevel gear.

15.62   What types of production machines would not be suitable for a manufacturing cell? What design or production features make them unsuitable? Explain.

15.63   Demonstrate the benefits of holonic manufacturing by outlining a program that would uniformly space a group of 10 holons within a room for the case where (a) there is a master computer and the holons follow orders of the one computer and (b) there is no master computer, but each holon has intelligence.

CHAPTER 16

Product Design and Manufacturing in a Global Competitive Environment

Manufacturing high-quality, world-class products at the lowest possible cost requires an understanding of the often complex relationships among many factors. The factors that have a major impact on product design and manufacture covered in this chapter are

- Quality in design and the design of robust products.
- Consideration of a product's life cycle and design for sustainability.
- Importance of material and process selection on design.
- Economic significance of manufacturing operations, with a description of the important factors in costs associated with a product.

16.1 | Introduction

In an increasingly competitive global marketplace, manufacturing high-quality products at the lowest possible cost requires an understanding of the often complex relationships among many factors. We have seen that

1. Product design and selection of materials and manufacturing processes are all interrelated.

2. Designs are periodically modified to

   a. Improve product performance.

   b. Take advantage of available new materials or use less expensive materials.

   c. Make the products easier and faster to manufacture and assemble, hence cheaper.

   d. Strive for zero-based rejection and waste.

Because of the wide variety of materials and manufacturing processes available, the task of producing a high-quality product by selecting the best materials and the best processes, while minimizing costs, continues to be a major challenge, as well as an opportunity in the global marketplace. The term **world-class** is now widely used to indicate a certain high level of product quality, signifying the fact that products must meet international standards and be marketable and acceptable worldwide.

Recall also that the designation of world-class, like product quality, is not a fixed target for a company to reach, but rather a *moving target*, rising to higher and higher levels as time passes (*continuous improvement*).

This chapter begins with important considerations in **product design;** the concept of **robust design** and **Taguchi methods.** Product design involves numerous aspects of not only the basic design itself, but also the ease of manufacturing that product and the time required to produce it. Again, there are many opportunities to simplify designs, reduce the number of components, and reduce the size and/or certain dimensions of the components to save money, especially on costly materials.

Product quality and **life expectancy** are then discussed, outlining the relevant parameters involved, including the concept of *return on quality*. Increasingly important are **life-cycle assessment** and **life-cycle engineering** of products, services, and systems, particularly regarding their potential adverse impact on the environment. The major emphasis in **sustainable manufacturing** is to reduce or eliminate any adverse effects of manufacturing operations on the environment, and society in general, while allowing a company to be profitable.

Although the *selection of materials* for products has traditionally required much experience in order to meet specific performance requirements, there are now several databases and expert systems that facilitate the selection process. Also, in reviewing the materials used in existing products, there are numerous opportunities for material substitution to improve performance and, especially, to reduce cost.

In the production phase, it is imperative that the *capabilities of manufacturing processes* are properly assessed as a basis of, and an essential guide to, the final selection of an appropriate process or a sequence of processes. As described throughout this text and depending on the product design and the materials specified, there is usually more than one method of manufacturing a product, its components, and subassemblies. An improper selection will obviously have a major effect not only on product quality, but also its cost.

Although the *economics* of various types of manufacturing processes have been described at the end of individual chapters, this chapter takes a broader view and summarizes the important overall manufacturing cost factors. The **cost** of a product often, but not always, determines its marketability and its customer acceptance. Meeting this challenge requires not only a thorough and up-to-date knowledge of the characteristics of materials, state-of-the-art manufacturing processes, operations, and systems, and, with equal importance, economic factors, but also innovative and creative approaches to product design and manufacturing.

16.2 | Product Design and Robust Design

Although it is beyond the scope of this book to describe the principles and methodology of product design and various optimization techniques, we have highlighted throughout various chapters those aspects that are relevant to **design for manufacture and assembly** (DFMA), as well as to competitive manufacturing. The references listed in Table 16.1 give design guidelines for manufacturing processes that can be found in this text.

Major advances continue to be made in *design for manufacture and assembly,* for which software packages are widely available. Although their use requires considerable training and knowledge, the software helps designers quickly develop products that require fewer components, reduced assembly time, and reduced time for manufacture and thus reducing the total product cost.

TABLE 16.1

References to Various Topics in This Text	
Processes	**Design considerations**
Metal casting	Section 5.12
Bulk deformation	Various sections in Chapter 6
Sheet-metal forming	Section 7.9
Machining	Section 8.14
Abrasives	Section 9.16
Polymers	Section 10.13
Powder metallurgy and ceramics	Section 11.6 and Section 11.12
Joining	Section 12.17
Material properties	**Manufacturing characteristics of materials**
Tables 2.1, 2.3, and 2.5; Figs. 2.4, 2.6, 2.12, 2.13, and 2.27	Table 3.8
Tables 3.3, 3.4, 3.7, 3.9, and 3.10 through 3.14	Tables 5.2 and 5.3
Tables 5.4, 5.5, and 5.6; Fig. 5.13	Table 7.2
Table 7.3	Section 8.5
Table 8.6; Figs. 8.30, 8.31	Tables 11.5, 11.8, and 11.9
Table 9.1	Section 12.5
Tables 10.1, 10.4, 10.5, and 10.8; Fig. 10.16	Table 16.7
Tables 11.3, 11.4, 11.7, and 11.9; Figs. 11.22, 11.23, and 11.24	
Table 12.6	
Capabilities of manufacturing processes	**Dimensional tolerances and surface finish**
Fig. 4.20	Fig. 4.20
Tables 5.2, 5.7, and 5.8	Table 5.2
Table 6.1	Tables 8.7 and Fig. 8.26
Table 7.1	Fig. 9.27
Table 8.7; Fig. 8.26	Fig. 16.4
Table 9.4; Fig. 9.27	
Tables 10.6, 10.7, and 10.9	
Tables 11.5 and 11.8	
Tables 12.1, 12.2, and 12.5	
Table 13.5; Fig. 14.3; Table 15.1; Fig. 16.3	
General costs	**Material costs**
Table 5.2; Fig. 5.39	Table 10.4
Fig. 7.72	Table 11.9
Fig. 9.41	Table 16.4
Table 10.9	
Table 11.9	
Tables 12.1 and 12.2	
Table 16.8; Fig. 16.5	

16.2.1 Product design considerations

There are general product design considerations, in addition to the design guidelines outlined throughout this text. Some of the fundamental design considerations are:

1. Can the product design be simplified and the number of components reduced without adversely affecting intended functions and performance?

2. Have environmental considerations been considered and incorporated into material and process selection and product design?

3. Have all alternative designs been investigated?

4. Can unnecessary features of the product, or some of its components, be eliminated or combined with other features?

5. Have modular design and building-block concepts been considered for a family of similar products and for servicing and repair, upgrading, and installation options?

6. Can the design be made smaller and lighter?

7. Are the specified dimensional tolerances and surface finish excessively stringent? Can they be relaxed without any significant adverse effects?

8. Will the product be difficult or excessively time consuming to assemble and disassemble for maintenance, servicing, or recycling?

9. Have subassemblies been considered?

10. Has the use of fasteners, and their quantity and variety, been minimized?

11. Are some of the components commercially available?

12. Is the product safe for its intended application?

As some examples of the foregoing considerations, note how (1) the size of products such as radios, electronic equipment, cameras, computers, calculators, and cell phones has been greatly reduced, (2) repair of products is now often done by simply replacing subassemblies and modules, thus saving much labor, and (3) there are fewer traditional fasteners and more snap-on type of assemblies used in products.

16.2.2 Product design and quantity of materials

With the high production rates and reduced labor made possible by automation and computer integration, the cost of materials can often become a significant portion of a product's cost. Although the material cost cannot be reduced below the market level at the time of production, reductions can be made in the *quantity* of materials used in the components to be produced. The overall shape of the product is usually optimized during the design and prototype stages (see also *rapid prototyping*, Section 10.12), through techniques such as finite-element analysis, minimum-weight design, design optimization, and computer-aided design and manufacturing. Such methods have greatly facilitated design analysis, material selection, material usage, and overall optimization.

Typically, reductions in the amount of material used can be achieved by decreasing the component's volume. This approach may require the selection of materials that have high strength-to-weight or stiffness-to-weight ratios. (See Section 3.9.1.) Obviously, higher ratios can also be obtained by improving product design and by selecting different cross sections, such as ones that have a high moment of inertia (as in I-beams), or by using tubular or hollow components instead of solid bars.

Implementing such design changes and improvements and minimizing the amount of materials used can, however, lead to *thin* cross sections and consequently present significant problems in manufacturing. Consider, for example, the following examples.

1. Casting or molding of thin sections can present difficulties in mold filling and in maintaining the required dimensional accuracy and surface finish (Section 5.12).

2. Forging of thin sections requires high forces, due to effects such as friction and chilling of thin sections such as flash in impression die forging (Section 6.2.3).

3. Impact extrusion (Section 6.4.3) of thin-walled parts can be difficult, especially when high dimensional accuracy is required.

4. Formability of sheet metals (Section 7.7) is typically reduced as their thickness decreases; furthermore, reducing the sheet thickness can lead to wrinkling, due to compressive stresses developed in the plane of the sheet during forming (see Section 7.6 and Fig. 7.50).

5. Machining and grinding of thin workpieces can present difficulties such as part distortion, poor dimensional accuracy, and chatter; consequently, advanced machining processes may have to be considered (Sections 8.12 and 9.16).

6. Welding of thin sheets or slender structures can cause distortion, due to thermal gradients developed during the welding cycle (see Section 12.17).

Conversely, making parts that have *thick* cross sections can also have adverse effects. Some examples follow.

1. In processes such as die casting (Section 5.10.3) and injection molding (Section 10.10.2), the production rate can be lower with thick cross sections, because of the increased cycle time required to allow the part to cool in the mold prior to its removal.

2. Unless controlled, porosity can develop in thicker regions of castings (see Fig. 5.37).

3. The bendability of sheet metals decreases as their thickness increases (see Fig. 7.15b and Table 7.2).

4. In powder metallurgy, there can be significant variations in density and mechanical properties throughout thick parts (Fig. 11.6).

5. Welding of thick sections can present problems, such as residual stresses and incomplete penetration that compromise weld strength (Section 12.6).

6. In die-cast parts (Section 5.10.3), thinner sections will have higher strength per unit thickness than will thicker sections, because of the smaller grain size developed in the former.

7. Thick cross sections produced by hot rolling will have lower strength and dimensional accuracy and higher roughness than cold-rolled thinner sections (see Section 6.3).

8. Processing of polymer parts requires increased cycle times as their thickness or volume increases, because of the longer time required for the parts to cool sufficiently to be removed from the molds (see Section 10.13).

16.2.3 Robustness and robust design

An important consideration in product design is *robustness*. Originally introduced by G. Taguchi (see also Section 16.3.4), robustness may be defined as a design, a process, or a system that continues to function within acceptable parameters despite variations in its environment. These variations are called **noise,** defined as those factors that are difficult or impossible to control. Examples of noise in manufacturing

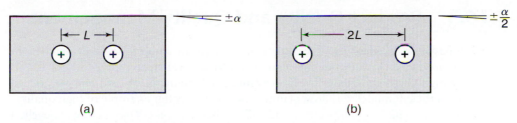

FIGURE 16.1   A simple example of robust design: (a) location of two mounting holes on a sheet-metal bracket where the deviation of the top and bottom surfaces of the bracket from being perfectly horizontal is $\pm\alpha$ and (b) new location of holes in a robust design, whereby the deviation is reduced to $\pm\alpha/2$.

operations include (a) variations in ambient temperature and humidity in a production facility, (b) random and unanticipated vibrations of the shop floor, (c) variations in the dimensions and surface and bulk properties of incoming batches of raw materials, and (d) the performance of different operators and machines at different times during the day or on different days. Note that a robust process does not change with the presence or changes in noise, and in a robust design, the part will function sufficiently well even if unanticipated events occur.

As a simple illustration, consider a sheet-metal mounting bracket to be attached to a wall with two bolts, as shown in Fig. 16.1a. The positioning of the two mounting holes will, as expected, include some error, such as caused by the type of manufacturing process employed and machine involved (such as an error due to punching, drilling, or piercing). This error may, in turn, prevent the top edge of the bracket from being horizontal. In a more robust design, shown in Fig. 16.1b, the mounting holes are now twice as far apart from each other. Even though the same manufacturing method is used and the production cost will remain the same, the robust design now has a variability that is one-half of that of the original design. In an even more robust design, the type of bolts used would be such that they will not loosen in case of vibration or during their use over a period of time (see *mechanical fastening*, Section 12.15).

EXAMPLE 16.1   An example of DFMA application

Numerous examples can be given regarding the benefits of applying design for manufacturing and assembly principles in the early stages of product concept and development. These principles can also be applied to the modification of existing designs and the selection of appropriate production methods. In this example, we consider the redesign of the pilot's instrument panel for a military helicopter designed and built by McDonnell Douglas.

The components of the panel consisted of sheet metal, extrusions, and rivets. Through the use of DFMA software and analysis of the panel, it was estimated that the redesign would lead to the following changes: (a) The number of parts would be reduced from 74 to 9; (b) the weight of the panel would be reduced from 3.00 to 2.74 kg; (c) fabrication time would be reduced from 305 to 20 hrs; (d) assembly time would be reduced from 149 to 8 hours; and (e) total time to produce the part would be reduced from 697 to 181 hrs. It was estimated that, as a result of redesign, the cost savings would be 74%. Based on these results, other components of the instrument panel were also subjected to similar analysis.

16.3 | Product Quality and Quality Management

In the past few decades, it has become a manufacturing imperative to produce high-quality products. However, quality can be a difficult and elusive concept. For example, is a high-performance sports car an example of high quality, or is a mass-produced and fuel-efficient model with low maintenance costs a better example of high quality? Each can fulfill its design objectives and delight their customers (see below). Clearly, a discussion of quality must allow for each of these examples.

16.3.1 Quality as a manufacturing goal

Product quality and the techniques employed in quality assurance and control are described in Section 4.9. Recall that the word *quality* is difficult to define precisely, partly because it includes not only well-defined technical considerations, but also human, and hence subjective, opinion. As defined by G. Taguchi (see Section 16.3.3), manufacturers are charged with the task of providing products that delight their customers. To do so, manufacturers should provide products with the following characteristics, which, as can be noted, also describe a product with high quality:

1. High reliability.
2. Perform the required functions well and safely.
3. Good appearance.
4. Inexpensive.
5. Upgradeable.
6. Available in the quantities desired when needed.
7. Robust over their intended life (see Section 16.2.3).

Product quality has always been a major consideration in manufacturing. In view of the global economy and competition, a major priority is the concept of *continuous improvement in quality,* as exemplified by the Japanese term **kaizen**, meaning *never-ending improvement.* It should, however, be noted that the level of quality that a manufacturer chooses for its products depends on the market for which the products are intended. Thus, for example, low-cost, low-quality products have their own market niche (as is easily proved by a visit to a large department store or a home-improvement center), just as there is a market, especially in prosperous times, for high-quality, expensive products, such as a Rolls-Royce automobile, audio equipment, and sporting goods such as tennis rackets, golf clubs, skis, and helmets.

We have seen that, through *concurrent engineering* (see Section 1.2), design and manufacturing engineers now have the capability to select and specify materials for the components of a particular product to be made. Quality considerations are always a part of such tasks. Let's take, for example, a simple case of selecting materials for a screwdriver stem. Based on the functions of a screwdriver, one can specify stem materials (from, for example, metals, plastics, elastomers, composite materials, or wood) that have desirable properties such as high yield and tensile strength, torsional stiffness, and resistance to wear and corrosion. A screwdriver made from such materials will perform better and last longer than one made of materials with lower mechanical properties. On the other hand, materials with better properties are generally more expensive and may be more difficult to process than others.

Consequently, the manufacturing operations and systems relevant to his particular item must be thoroughly reviewed to keep final product costs low; this can be done on the basis of numerous technical and economic considerations described and discussed throughout this text. Unfortunately, however, engineers are often faced with a dilemma, which succinctly has been stated as "good, fast, or cheap; pick any two."

Because of the significant costs that can be incurred in manufacturing operations, the concept of **return on quality** (ROQ) is important. The components of ROQ follow:

a. Because of its major influence on customer satisfaction, quality should be viewed as an *investment*.

b. There has to be a certain limit on how many resources should be expended on quality improvements.

c. The specific area for which the expenditure should be made toward quality improvement must properly be assessed.

d. The incremental improvement in quality must be carefully reviewed with regard to the additional costs involved.

Contrary to a common perception, high-quality products do not necessarily cost more. In most industries (such as automotive and aerospace as classic examples), the ROQ is minimized at a value of zero defects, while in others (such as the manufacture of integrated circuits), the cost of eliminating the final few defects is very high (leading to approaches such as fault-tolerant design). However, indirect factors may still encourage elimination of these final few defects. For example, although customer satisfaction is a qualitative factor and thus difficult to include in calculations, satisfaction is increased and customers are more likely to be retained when there are no defects in the product they purchase. Although there can be significant variations, it has been estimated that the *relative costs* involved in identifying and repairing defects in products grow by orders of magnitude, a concept called the *rule of ten* (see also Section 16.9.1), as follows:

Stage	Relative cost of repair
Fabrication of the part	1
Subassembly	10
Final assembly	100
At product distributor	1000
At the customer	10,000

16.3.2 Total quality management

Total quality management (TQM) is a system that emphasizes that quality must be designed and built into a product. It is a systems approach in that *both* management and employees must make a concerted effort to consistently manufacture high-quality products. Defect *prevention*, rather than defect *detection*, is the major goal.

Leadership and *teamwork* in the organization are essential. They ensure that the goal of **continuous improvement** in all aspects of manufacturing operations is imperative, because they reduce product variability and they improve customer satisfaction. The TQM concept also requires the control the *processes* and not the *parts produced*, so that process variability is reduced and no defective parts are allowed to continue through the production line.

Quality circle. This concept consists of regular meetings by groups of employees (workers, supervisors, managers) who discuss how to improve and maintain product quality at all stages of the manufacturing operation. Worker involvement, responsibility, creativity, and team effort are emphasized. Comprehensive training is provided so the worker can become aware of quality and be capable of analyzing statistical data, identifying causes of poor quality, and taking immediate action to correct the problem. Experience has indicated that quality circles are especially effective in lean manufacturing environments (see Section 15.13).

Quality engineering as a philosophy. Experts in quality control have put many of the quality-control concepts and methods into a larger perspective. Notable among these experts are W.E. Deming and G. Taguchi, whose philosophies of quality and product cost have had a major impact on modern manufacturing.

16.3.3 Deming methods

During World War II, W.E. Deming (1900–1993) and several others developed new methods of **statistical process control** (see Section 4.9.2) for war-industry manufacturing plants. The methods of statistical process control arose from the recognition that there were *variations* in (a) the performance of machines and people and (b) the quality and dimensions of incoming raw materials. These efforts involved not only statistical methods of analysis, but also a new way of looking at manufacturing operations, that is, from the perspective of improving quality while lowering costs.

Deming recognized that manufacturing organizations are systems consisting of management, workers, machines, and products. His basic ideas are summarized in the well-known Fourteen Points, given in Table 16.2. These points are not to be viewed as a checklist or menu of tasks to be performed; they are the *characteristics* that Deming recognized in companies that produce high-quality goods. He placed great emphasis on communication, direct worker involvement, and education in statistics and advanced manufacturing technology. His ideas have been widely accepted since the end of World War II.

16.3.4 Taguchi methods

In the G. Taguchi (1924–) methods, high quality and low costs are achieved by combining engineering and statistical methods to optimize product design and manufacturing processes. Taguchi methods refer to the approaches developed by Taguchi to develop high-quality products. Although it is challenging to actually satisfy all of these characteristics, *excellence in manufacturing* is a prerequisite.

Taguchi also contributed to the approaches that are used to document quality. It must be recognized that any deviation from the optimum state of a product represents

TABLE 16.2

Deming's Fourteen Points

1. Create constancy of purpose toward improvement of product and service.
2. Adopt the new philosophy.
3. Cease dependence on mass inspection to achieve quality.
4. End the practice of awarding business on the basis of price tag.
5. Improve constantly and forever the system of production and service, to improve quality and productivity, and thus constantly decrease cost.
6. Institute training for the requirements of a particular task, and document it for future training.
7. Institute leadership, as opposed to supervision.
8. Drive out fear so that everyone can work effectively.
9. Break down barriers between departments.
10. Eliminate slogans, exhortations, and targets for zero defects and new levels of productivity.
11. Eliminate quotas and management by numbers, numerical goals. Substitute leadership.
12. Remove barriers that rob the hourly worker of pride of workmanship.
13. Institute a vigorous program of education and self-improvement.
14. Put everybody in the company to work to accomplish the transformation.

a *financial loss*, because of such factors as reduced product life, performance, and economy. *Loss of quality* is defined as the financial loss to society after the product is shipped, with the following results:

1. Poor quality leads to customer dissatisfaction.
2. Costs are incurred in servicing and repairing defective products, some in the field.
3. The manufacturer's credibility in the marketplace is diminished.
4. The manufacturer eventually loses its share of the market.

The Taguchi methods of *quality engineering* emphasize the importance of

- **Enhancing cross-functional team interaction**, whereby design engineers and process or manufacturing engineers communicate with each other in a common language; they quantify the relationships between design requirements and manufacturing process selection.
- **Implementing experimental design**, in which the factors involved in a process or operation and their interactions are studied simultaneously.

In *experimental design,* the effects of controllable and uncontrollable variables on the product are identified. This approach minimizes variations in product dimensions and properties and ultimately brings the mean to the desired level. The methods used for experimental design are complex and involve the use of *factorial design* and *orthogonal arrays*, which reduce the number of experiments required. These methods are also capable of identifying the effects of variables that cannot be controlled (called *noise*), such as changes in environmental conditions and normal property variations of incoming materials.

The use of these methods results in (a) rapid identification of the controlling variables (observing *main effects*) and (b) the ability to determine the best method of process control. Thus, for example, variables affecting dimensional tolerances in machining a particular component can readily be identified and, whenever possible, the proper cutting speed, feed, cutting tool, and cutting fluids can be specified. The control of these variables may require new equipment or major modifications to existing equipment.

16.3.5 Taguchi loss function

An important concept introduced by Taguchi is that any deviation from a design objective constitutes a loss in quality. Consider, for example, the tolerancing standards given in Fig. 4.19, where there is a range of dimensions over which a part is acceptable. On the other hand, the Taguchi philosophy calls for a *minimization of deviation* from the design objective. Thus, using Fig. 4.19a as an example, a shaft with a diameter of 40.03 mm would normally be considered acceptable and would pass inspections. In the Taguchi approach, however, a part having this diameter represents a deviation from the design objective. Such deviations generally reduce the robustness and performance of products, especially in complex systems.

The *Taguchi loss function* was introduced because traditional accounting practices had no established methodologies of calculating losses on parts that met design specifications. In the traditional accounting approach, a part is defective and incurs a loss to the company when it exceeds its design tolerances; otherwise, there is no loss to the company.

The Taguchi loss function is thus a tool for comparing quality, based on minimizing variations. It calculates an increasing loss to the company when the component is further from the design objective. This function is defined as a parabola, where one point is the cost of replacement (including shipping, scrapping, and handling costs) at an extreme of the tolerances, while a second point corresponds to zero loss at the design objective.

The loss cost can be written mathematically as

$$\text{Loss cost} = k[(Y - T)^2 + \sigma^2] \tag{16.1}$$

where Y is the mean value from manufacturing, T is the target value from design, σ is the standard deviation of parts from manufacturing, and k is a constant, defined as

$$k = \frac{\text{Replacement cost}}{(\text{LSL} - T)^2} \tag{16.2}$$

where LSL is the lower specification limit. When the lower and upper specification limits, LSL and USL, respectively, are the same (that is, the tolerances are balanced), either of the specification limits can be used in this equation.

Note that the lower and upper specification limits are not necessarily related to the upper and lower control limits discussed in Section 4.9.2. LSL and USL are *design requirements,* which often do not take manufacturing capabilities into account. Often, designers choose specification limits or tolerances based on experience or as a maximum value that can be allowed and still have a functioning design; the desire is to allow the manufacturing engineers wide tolerances to ease manufacturing. On the other hand, upper and lower control limits do not necessarily reflect the designer's intention regarding tolerances. Ideally, the control limits discussed in Chapter 4 and the specification limits considered here should not be related. Instead, it should be the goal of manufacturers to meet design targets, as implied by Eq. (16.1).

EXAMPLE 16.2 Production of polymer tubing

High-quality polymer tubes are being produced for medical applications, where the target wall thickness is 2.6 mm, with an upper specification limit of 3.2 mm and a lower specification limit of 2.0 mm (2.6 ± 0.6 mm). If the units are defective, they are replaced at a shipping-included cost of $10.00. The current process produces parts with a mean of 2.6 mm and a standard deviation of 0.2 mm. The current volume is 10,000 pieces of tube per month. An improvement is being considered for the extruder heating system. This improvement will cut the variation in half, but it will cost $50,000. Determine the Taguchi loss function and the payback period for the original production process and incorporating the process improvement.

Solution. The quantities involved can be identified as follows: USL = 3.2 mm, LSL = 2.0 mm, T = 2.6 mm, σ = 0.2 mm, and Y = 2.6 mm. The quantity k is given by Eq. (16.2) as

$$k = \frac{(\$10.00)}{(3.2 - 2.6)^2} = \$27.28.$$

The loss cost is then

$$\text{Loss cost} = (27.78)[(2.6 - 2.6)^2 + 0.2^2] = \$1.11 \text{ per unit.}$$

After the improvement, the standard deviation is 0.1 mm; thus,

$$\text{Loss cost} = (27.78)[(2.6 - 2.6)^2 + 0.1^2] = \$0.28 \text{ per unit.}$$

The savings are then ($1.11 − $0.28)(10,000) = $8300 per month. Hence, the payback period for the investment is $50,000/($8300/month) = 6.02 months.

16.3.6 The ISO and QS Standards

With increasing international trade and global manufacturing, price-sensitive competition has become increasingly challenging to manufacturing organizations. Customers are increasingly demanding *high-quality products and services* at low prices and are looking for suppliers that can respond to this demand consistently and reliably. This trend has, in turn, created the need for international conformity and consensus regarding the establishment of methods for quality control, reliability, and safety of products. In addition to these considerations, equally important concerns regarding the environment and quality of life are addressed with new international standards. This section describes the standards that are relevant to product quality and environmental issues.

The ISO 9000 standard. First issued in 1987, and revised in 1994 and 2000, the ISO 9000 standard (*Quality Management and Quality Assurance Standards*) is a deliberately generic series of quality-system management standards. The ISO 9000 standard has permanently influenced the manner in which manufacturing companies conduct business in world trade and has become the world standard for quality.

The ISO 9000 series includes the following standards:

- ISO 9001—*Quality systems: Model for quality assurance in design/development, production, installation, and servicing.*
- ISO 9002—*Quality systems: Model for quality assurance in production and installation.*
- ISO 9003—*Quality systems: Model for quality assurance in final inspection and test.*
- ISO 9004—*Quality management and quality system elements: Guidelines.*

Companies voluntarily register for these standards and are issued certificates. Registration may be sought generally for ISO 9001 or 9002, and some companies have registration up to ISO 9003. The 9004 standard is simply a guideline and not a model or a basis for registration. For certification, a company's plants are visited and audited by accredited and independent third-party teams to certify that the standard's 20 key elements are in place and are functioning properly.

Depending on the extent to which a company fails to meet the requirements of the standard, registration may or may not be recommended at that time. The audit team does not advise or consult with the company on how to reconcile discrepancies but merely describes the nature of the noncompliance. Periodic audits are required to maintain certification. The certification process can take from six months to a year or more, and can cost tens of thousands of dollars, depending on the company's size, number of plants, and product line.

The ISO 9000 standard is not a product certification, but a **quality process certification**. Companies establish their own criteria and practices for quality. However, the documented quality system must be in compliance with the ISO 9000 standard. Thus, a company cannot write into the system any criterion that opposes the intent of the standard.

Registration symbolizes a company's commitment to conform to consistent practices, as specified by the company's own quality system (such as quality in design, development, production, installation, and servicing), including proper documentation of such practices. In this way, customers, including also government agencies, are assured that the supplier of the product or service (which may or may not be within the same country) is following specified practices. In fact, manufacturing companies are themselves assured of such practices regarding their own suppliers who have ISO 9000 registration; thus, suppliers must also be registered.

The QS 9000 standard. The QS 9000 standard, jointly developed by Chrysler, Ford, and General Motors, was first published in 1994. Prior to the development of QS 9000, each of the Big Three automotive companies had its own standard for quality system requirements. Tier I suppliers have been required to obtain third-party registration to QS 9000. Very often, QS 9000 has been described as an "ISO 9000 chassis with a lot of extras." This is a proper description, noting that all of the ISO 9000 clauses serve as the foundation of QS 9000.

The ISO 14000 standard. This is a family of standards, first published in 1996, pertaining to the international *Environmental Management Systems* (EMS), and it concerns the way an organization's activities affect the environment throughout the life of its products (see also Section 16.5). These activities (a) may be internal or external to the organization, (b) range from production to ultimate disposal of the product after its useful life, and (c) include effects on the environment, such as pollution, waste generation and disposal, noise, depletion of natural resources, and energy use.

A rapidly increasing number of companies in many countries have obtained certification for this standard. ISO 14000 has several sections: Guidelines for Environmental Auditing, Environmental Assessment, Environmental Labels and Declarations, and Environmental Management. ISO 14001: Environmental Management System Requirements consists of sections on General Requirements, Environmental Policy, Planning, Implementation and Operation, Checking and Corrective Action, and Management Review.

16.4 | Life-Cycle Engineering and Sustainable Manufacturing

The concept of a *product life cycle* was introduced in Section 1.4. *Life-cycle engineering* (LCE) is concerned with environmental factors, especially as they relate to design, optimization, and various technical considerations regarding each component of a product or process life cycle. A major aim of LCE is to consider the reuse and recycling of products from their earliest stage of design process (known as *green design,* or *green engineering*).

Cradle-to-grave. A traditional product life cycle, referred to as the *cradle-to-grave* model, can be defined as consecutive and interlinked stages of a product or service system, including the

1. Extraction of natural resources, including raw materials and energy.
2. Processing of raw materials.
3. Manufacture of products.
4. Transportation and distribution of the product to the customer.
5. Use, reuse, and maintenance of the product.
6. Disposal of the product.

Although life-cycle analysis and engineering is a comprehensive, powerful, and necessary tool, its total implementation can be expensive, challenging, and time consuming. This is largely because of uncertainties in the input data regarding materials, processes, long-term effects, costs, etc., and the time required to collect reliable data to properly assess the often-complex interrelationships among various components of the whole system. A number of software packages are available to expedite such analysis in certain specific industries, particularly the chemical and process industries, because of the higher potential for environmental damage in their operations.

Sustainable manufacturing. In recent years, we all have become increasingly aware that, ultimately, our natural resources are limited, thus clearly necessitating the need for conservation of materials and energy. The term *sustainable manufacturing* emphasizes the need for conserving these resources, particularly through maintenance and reuse. Recall that Section 1.4 described the importance of product life-cycle consideration, and the difference between cradle-to-grave and cradle-to-cradle approaches, as well as biological and industrial recycling.

It is important to remember that recycling often requires that individual components in a product must be taken apart, and that if much effort and time have to be expended in doing so, then recycling can become prohibitively expensive. The general guidelines to facilitate recycling are

1. Do not design products; instead design life cycles, where all the material and energy inputs are considered, as well as the product's destination when its design life is over.

2. Use materials that can easily be recycled whenever possible.

3. Reduce the number of parts and types of materials in products, and use as little material as possible.

4. Avoid mixing materials that are biologically recyclable with those that follow an industrial recycling life cycle.

5. Use modular design to facilitate disassembly.

6. For plastic parts use a single type of polymer as much as possible.

7. Mark plastic parts for ease of identification (as is done with plastic food containers and bottles).

8. Avoid using coatings, paints, and plating; use molded-in colors in plastic parts when necessary.

9. Avoid using adhesives, rivets, and other permanent joining methods in assembly; instead, use fasteners, especially snap-in fasteners,

---

EXAMPLE 16.3 Sustainable manufacturing in the production of Nike athletic shoes

Nike athletic shoes are assembled using adhesives (Section 12.14). Up to around 1990, the adhesives used contained petroleum-based solvents which pose health hazards to humans and contribute to petrochemical smog. In order to improve this situation, the company worked with chemical suppliers to successfully develop water-based adhesive technology, which is now used for the majority of the assembly operations. As a result, solvent use in all manufacturing operations in Nike's subcontracted facilities in Asia was reduced by 67% since 1995. In 1997, 834,000 gallons of hazardous solvents were replaced with 1290 tons of water-based adhesives.

As another example, the rubber outsoles of the shoe were made by a process that resulted in a significant amount of extra rubber around the periphery of the sole, called *flashing,* similar to the flash shown in Figs. 6.14c and 10.35c. With about 40 factories using thousands of molds and producing over a million outsoles a day, this flashing constituted the largest source of waste in the manufacturing process for the shoes. To reduce this waste, the company developed a technology that grinds the flashing into rubber powder, which was then added to the rubber mixture used to make subsequent outsoles. As a result, waste was reduced by 40%. Furthermore, it was found that the mixed rubber had better abrasion resistance, durability, and overall performance than the highest premium rubber.

16.5 | Selection of Materials for Products

Although the general criteria for selecting materials have been described in Section 1.5, this chapter will discuss them in greater detail.

16.5.1 General properties of materials

As described throughout Chapter 2, *mechanical properties* of materials include (a) strength (b) toughness, (c) ductility, (d) hardness, and (e) resistance to fatigue, creep, and impact. In addition, characteristics such as stiffness and dent resistance depend not only on the elastic modulus of the material, but also on the geometric features of the part. Friction and wear properties also depend on a combination of several factors, as described in Section 4.4.

Physical properties (Section 3.9) include density, melting point, specific heat, thermal and electrical conductivity, thermal expansion, and magnetic properties. The *chemical properties* (Section 3.9.7) that are of primary concern in manufacturing are oxidation and corrosion. The relevance of these properties to product design and manufacturing has been described in various chapters, and several tables relating to properties of metallic and nonmetallic materials are included therein. Note that Table 16.1 lists the sections in this text that are relevant to various material properties.

Selection of materials is now much easier and faster, because of the availability of computerized and extensive *databases,* which provide much greater accessibility to information than previously available. To facilitate the selection of materials and the determination of the parameters (described later in this section), expert-system software (*smart databases*) is also very helpful. With input such as product design and functional requirements, expert systems are capable of rapidly identifying appropriate materials for a particular specific application, just as an expert or a team of experts would.

Regardless of the method employed, the following considerations are important in materials selection for products:

1. Do the materials selected have properties that unnecessarily exceed minimum requirements and specifications?
2. Can some materials be replaced by others that are less expensive?
3. Do the materials selected have the appropriate manufacturing characteristics?
4. Are the raw materials (stock) to be ordered available in standard shapes, dimensions, tolerances, and surface finish?
5. Is the material supply reliable?

TABLE 16.3

Commercially Available Forms of Materials	
Material	Available as
Aluminum	B, F, I, P, S, T, W
Ceramics	B, p, s, T
Copper and brass	B, f, I, P, s, T, W
Elastomers	b, P, T
Glass	B, P, s, T, W
Graphite	B, P, s, T, W
Magnesium	B, I, P, S, T, w
Plastics	B, f, P, T, w
Precious metals	B, F, I, P, t, W
Steels and stainless steels	B, I, P, S, T, W
Zinc	F, I, P, W

Note: B = bar and rod; F = foil; I = ingots; P = plate and sheet; S = structural shapes;
T = tubing; W = wire.
Lowercase letters indicate limited availability. Most of the metals are also available in powder form,
including prealloyed powders.

6. Are there likely to be significant price increases or market fluctuations for the materials?

7. Can the materials be obtained in the required quantities in the desired time frame?

16.5.2 Shapes of commercially available materials

Materials are generally available in various forms (Table 16.3): casting, extrusion, forging, bar, plate, sheet, foil, rod, wire, and metal powders. Purchasing materials in shapes that require the least additional processing is an important consideration. Such characteristics as surface quality, dimensional tolerances, and straightness of the raw materials must also be taken into account. Obviously, the better and the more consistent these characteristics are, the less additional processing effort and time are required.

For example, if we want to produce simple shafts with good dimensional accuracy, surface finish, roundness, and straightness, we could purchase round bars that are already turned and centerless ground (Chapters 8 and 9) and meet these requirements. Unless the available facilities are capable of producing round bars economically, it is generally cheaper to purchase them from vendors. On the other hand, if we need a stepped shaft (with different diameters along its length), we could, for example, purchase a round bar with a diameter at least equal to the largest diameter of the stepped shaft and then turn it on a lathe. If the stock has broad dimensional tolerances, or is warped or out of round, we must purchase an even larger size to ensure proper dimensions of the final shaft.

As indicated throughout various chapters, each manufacturing operation produces parts that have a specific range of geometric features, shapes, dimensional tolerances, and surface finish. Consider the following examples.

1. Castings generally have less dimensional accuracy and poorer surface finish than parts made by processes such as cold extrusion or powder metallurgy (Chapters 5, 6, and 11).

2. Hot-rolled and hot-drawn products have a rougher surface finish and wider dimensional tolerances than cold-rolled and cold-drawn products (Chapter 6).

3. Extrusions have smaller cross-sectional dimensional tolerances than parts made by roll forming (Chapters 6 and 7).

4. The wall thickness of seamless tubing made by the tube-rolling process is less uniform than that of roll-formed and welded tubing (Chapters 6 and 12).

5. Round bars turned on a lathe have a rougher surface finish and wider dimensional tolerances than bars that are ground on cylindrical or centerless grinding machines (Chapters 8 and 9).

16.5.3 Manufacturing characteristics of materials

Manufacturing characteristics typically include castability, workability, formability, machinability, grindability, weldability, and hardenability by heat treatment. Because raw materials have to be formed, shaped, machined, ground, fabricated, or heat treated into individual components having specific shapes, dimensions, and surface finish, these properties are crucial to the proper selection of materials. Table 16.1 lists references to general manufacturing characteristics of materials.

Recall also that the quality of a raw material can greatly influence its manufacturing characteristics. The following are some examples.

1. A rod or bar with a longitudinal seam (lap) will develop cracks during simple upsetting or heading operations.

2. Bars with internal defects and inclusions may develop cracks during seamless-tube production.

3. Porous castings will produce poor surface finish when machined.

4. Blanks that are heat treated nonuniformly and bars that are not stress relieved will distort during subsequent operations, such as machining or drilling a hole.

5. Incoming stock that has variations in composition and microstructure cannot consistently be heat treated or machined.

6. Sheet-metal stock that has variations in its cold-worked condition or its thickness will exhibit uneven springback during bending and other forming operations.

7. If prelubricated sheet-metal blanks have nonuniform lubricant distribution and thickness over their surface, their formability, surface finish, and overall quality will be adversely affected.

16.5.4 Reliability of material supply

Geopolitical factors can significantly affect the supply of strategic materials, thus making supply unreliable and adversely affecting production. Other factors such as strikes, shortages, and the reluctance of suppliers to produce materials in a particular shape, quality, or quantity also affect reliability of supply.

16.5.5 Cost of materials and processing

Because of a raw material's processing history, the *unit cost* (cost per unit weight or volume) of the raw material depends not only on the material itself, but also on its shape, size, and condition. For example, because more operations are involved in the production of thin wire than in that of round rod (Section 6.5), the unit cost of wire is much higher. Similarly, powder metals are generally more expensive than bulk metals. Note also that the unit cost typically decreases as the quantity purchased increases (bulk discount).

TABLE 16.4

Approximate Cost Per Unit Volume for Wrought Metals and Plastics Relative to the Cost of Carbon Steel

Gold	60,000	Carbon steel	1
Silver	600	Magnesium alloys	2–4
Molybdenum alloys	200–250	Aluminum alloys	2–3
Nickel	35	Gray cast iron	1.2
Titanium alloys	20–40	Nylons, acetals, and silicon	1.1–2
Copper alloys	5–6	Rubber*	0.2–1
Stainless steels	2–9	Other plastics and elastomers*	0.2–2
High-strength low-alloy steels	1.4		

*As molding compounds.
Note: Costs vary significantly with the quantity of purchase, supply and demand, size and shape, and various other factors.

The cost per unit volume for wrought metals and plastics relative to that for carbon steel are given in Table 16.4. The benefit of using this unit can be seen by the following example: In the design of a cantilevered, rectangular steel beam that is to support a certain load at its end, a maximum deflection is specified. Using equations developed in mechanics of solids, and assuming that the weight of the beam can be neglected, appropriate cross-sectional dimensions of the beam can be determined. Since all dimensions are known, the volume of the beam can be calculated. The cost of the beam can then easily be calculated based on the cost of the material per unit volume. If the cost is given in terms of per unit weight, we then calculate the beam's weight and determine the cost.

The cost of a particular material is subject to fluctuations caused by factors as simple as supply and demand or as complex as geopolitics. If a product is no longer cost competitive, alternative and less costly materials can be selected. For example, the copper shortage in the 1940s led the U.S. government to mint pennies from zinc-plated steel. Similarly, when the price of copper increased substantially during the 1960s, the electrical wiring being installed in homes was, for a time, made of aluminum. It should be noted, however, that this material substitution led to the redesign of switches and outlets to avoid excessive heating at the junctions that was experienced.

When scrap is produced during manufacturing, as in sheet-metal fabrication, forging, and machining (see Table 16.5), the value of the scrap is deducted from the material's cost, in order to obtain net material cost. Recall that in machining, scrap (in the form of chips of various shapes; see Section 8.2.1) can be very high. On the other hand, operations such as rolling, shape rolling, ring rolling, and powder metallurgy (all of which are generally regarded as *net-shape* or *near-net-shape processes;* Section 1.6) produce the least scrap.

TABLE 16.5

Typical Scrap Produced in Various Manufacturing Processes

Process	Scrap (%)
Machining	10–60
Closed-die forging, hot	20–25
Sheet-metal forming	10–25
Extrusion, hot	15
Permanent-mold casting	10
Powder metallurgy	<5
Rolling and ring rolling	<1

As expected, the value of the scrap depends on the type of metal and on the demand for the scrap; typically, it is between 10 and 40% of the original cost of the material. Another factor in the value of scrap is whether or not the material is contaminated. Thus, metal chips collected from operations where cutting fluids have been used have less value than those from dry cutting.

16.6 | Substitution of Materials in Products

Although new products continually appear on the global market, the majority of the design and manufacturing effort is typically concerned with improving existing products. Major product improvements can result from (a) substitution of materials, (b) implementation of new designs, technologies, or improved processing techniques, (c) better control of the processing parameters, and (d) increased automation of plant operations. Access to reliable and extensive materials data is essential before a decision can be made to make substitutions.

Automotive and aircraft manufacturing are particular examples of major industries in which substitution of materials is an important and ongoing activity; other industries are sporting goods and medical products.

There are several reasons for substituting materials in products.

1. Reduce the costs of materials and processing.
2. Improve assembly and conversion to automated assembly.
3. Improve the performance of products, for example, by reducing weight and improving characteristics such as resistance to wear, fatigue, and corrosion.
4. Increase the stiffness-to-weight and strength-to-weight ratios of structures.
5. Reduce the need for maintenance and repair.
6. Reduce vulnerability to the unreliability of domestic and overseas supply of certain materials.
7. Improve compliance with legislation and regulations prohibiting the use of materials that have adverse environmental impact.
8. Improve the appearance of a product.
9. Reduce performance variations or environmental sensitivity in the product, such as by improving *robustness* (see Section 16.2.3).

From a review of this list, it is evident that there are numerous important technological and economic factors that must be considered. An important factor in substitution is, for example, the compatibility of the materials selected with respect to galvanic corrosion in metal pairs, as described in Section 3.9.7. As another example, note that in spite of some similarities, producing ceramic or plastic parts and producing metal parts involve basically different processes, machinery, assembly, production rates, as well as costs.

Substitution of materials in the automobile industry. Trends in the automobile industry provide several good examples of the effective substitution of materials over the years in order to achieve one or more of the foregoing objectives.

1. Several components in metal bodies have been replaced with plastic or reinforced-plastic parts.
2. Metal bumpers, fuel tanks, housings, clamps, and various other components have been replaced with plastic substitutes.

3. Some engine components have been replaced with ceramic and reinforced-plastic parts.

4. All-metal drive shafts have been replaced with composite-material drive shafts.

5. Cast-iron engine blocks have been replaced with cast-aluminum blocks; forged crankshafts with cast crankshafts; and forged connecting rods with cast, powder-metallurgy, or composite-material connecting rods. Some cast-aluminum pistons have been changed to forged steel pistons (see the Case Study in Chapter 12).

6. Steel structural elements have been replaced with extruded aluminum sections, and rolled sheet-steel body panels have been changed to rolled aluminum sheet (see Fig. 1.5). Because the automobile industry is a major consumer of both metallic and nonmetallic materials, there is constant and rigorous competition among suppliers, particularly for steel, aluminum, and plastics. The relative advantages and limitations of these principal materials, and their applications and costs are continually being investigated, including also recycling and other environmental considerations.

Substitution of materials in the aircraft industry. In the aircraft and aerospace industries, conventional aluminum alloys (2000 and 7000 series) for some components are being replaced with aluminum-lithium alloys, titanium alloys, and composite materials, particularly because of their higher strength-to-weight ratios. Forged parts are being replaced with powder-metallurgy parts that are manufactured with better control of impurities and microstructure. The P/M parts also require less machining and finishing operations, thus producing less scrap of expensive materials. Advanced composite materials and honeycomb structures are replacing traditional aluminum airframe components (Fig. 16.2), and metal-matrix composites are replacing some of the aluminum and titanium parts previously used in structural components.

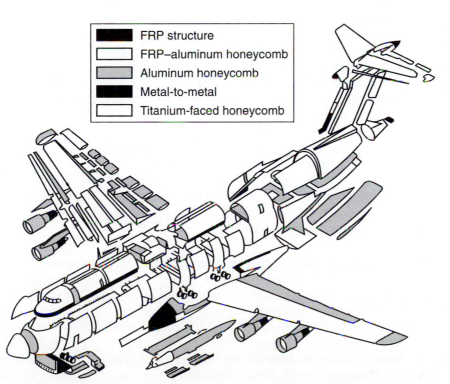

Legend:
- FRP structure
- FRP–aluminum honeycomb
- Aluminum honeycomb
- Metal-to-metal
- Titanium-faced honeycomb

FIGURE 16.2 Advanced materials used on the Lockheed C-5A transport aircraft. (FRP is fiber-reinforced plastic.)

EXAMPLE 16.4 Changes made in materials between C-5A and C-5B military cargo aircraft

Table 16.6 shows the changes made in materials for various components of the C-5A and C-5B military cargo aircraft and the reasons for the changes.

Source: After H.B. Allison.

TABLE 16.6

Changes in Materials from C-5A to C-5B Military Cargo Aircraft

Item	C-5A material	C-5B material	Reason for change
Wing panels	7075-T6511	7175-T73511	Durability
Main frame forgings	7075-F	7049-01	Stress-corrosion resistance
Machined frames	7075-T6	7049-T73	Stress-corrosion resistance
Frame straps	7075-T6 plate	7050-T7651 plate	Stress-corrosion resistance
Fuselage skin	7079-T6	7475-T61	Material availability
Fuselage under-floor end fittings	7075-T6 forging	7049-T73 forging	Stress-corrosion resistance
Wing/pylon attach fitting	4340 alloy steel	PH13-8Mo	Corrosion prevention
Aft ramps lock hooks	D6-AC	PH13-8Mo	Corrosion prevention
Hydraulic lines	AM350 stainless steel	21-6-9 stainless steel	Improved field repair
Fuselage fail-safe straps	Ti-6Al-4V	7475-T61 aluminum	Titanium strap debonding

16.7 | Capabilities of Manufacturing Processes

As indicated throughout this text, each manufacturing process has its particular advantages and limitations. References to the general characteristics and capabilities of manufacturing processes are listed in Table 16.1. Casting and injection molding, for example, can generally produce more complex shapes than can forging and powder metallurgy. On the other hand, forgings have toughness that is generally superior to that of castings and powder-metallurgy products, and they can be made into complex shapes by subsequent machining and finishing operations.

Recall that the shape of a product may be such that it can best be fabricated from several parts, by joining them with fasteners or with such techniques as brazing, welding, and adhesive bonding. However, the reverse may also be true for another product, in that manufacturing it in one piece may be more economical because of the significant assembly costs that may otherwise be involved. Other factors also must be considered in process selection, such as minimum section sizes and dimensions that can satisfactorily be produced (Fig. 16.3). For example, very thin sections can be produced by cold rolling, but processes such as sand casting and forging cannot produce thin sections.

Dimensional tolerances and surface finish. The *dimensional tolerances* and *surface finish* produced are important not only for the functioning of parts, machines, and instruments, but also in subsequent assembly operations. The capabilities of various manufacturing processes in this regard are qualitatively illustrated in Fig. 16.4, and references to the capabilities of manufacturing processes with respect to these features are given in Table 16.1.

In order to obtain closer dimensional tolerances and better surface finish, additional finishing operations, better control of processing parameters, and the use of

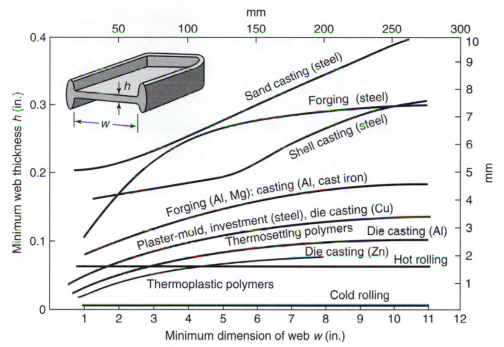

FIGURE 16.3 Minimum part dimensions obtainable by various manufacturing processes. *Source:* After J.A. Schey.

higher quality equipment and controls may be required. On the other hand, the closer the tolerance and the finer the surface finish specified, the higher is the cost of manufacturing (Fig. 16.5); because of the longer manufacturing time and greater number of processes involved (see Figs. 9.41 and 16.6). In the machining of aircraft structural members made of titanium alloys, for example, as much as 60% of the cost of machining the part may be expended in the final machining pass, in order to maintain specified tolerances and surface finish. Recall that parts should be made with as rough a surface finish and as wide a dimensional tolerance as will be functionally and aesthetically acceptable.

Production quantity or volume. Depending on the type of product, the *production quantity,* or *volume (lot size),* can vary widely. Note, for example, that paper clips, bolts, washers, spark plugs, bearings, and ballpoint pens are produced in very large quantities. On the other hand, jet engines for commercial aircraft, diesel engines for locomotives, machine tools, and propellers for cruise ships are manufactured in limited quantities. Production quantity plays a significant role in process and equipment selection. In fact, an entire manufacturing discipline is devoted to determining the optimum production quantity, called the *economic order quantity.*

Production rate. A significant factor in manufacturing process selection is the *production rate,* defined as the number of pieces to be produced per unit of time. Processes such as powder metallurgy, die casting, deep drawing, and roll forming are high-production-rate operations. By contrast, sand casting, conventional and electrochemical machining, spinning, superplastic forming, adhesive and diffusion bonding, and the processing of reinforced plastics are relatively slow operations. These production rates can, of course, be increased by automation and computer controls

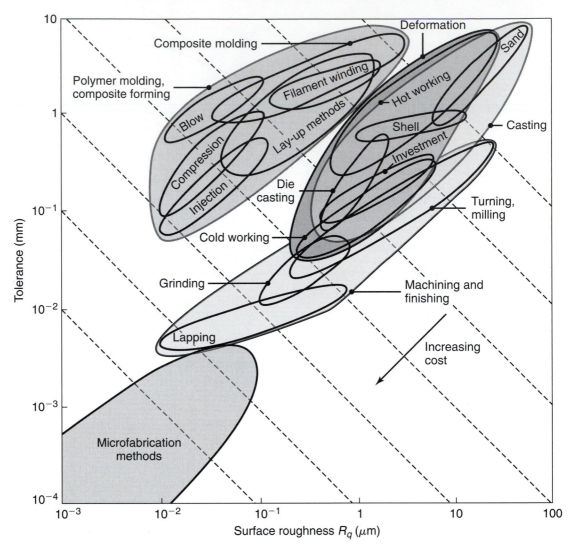

FIGURE 16.4   A plot of achievable dimensional tolerance versus surface roughness for various manufacturing operations; the dashed lines indicate cost factors. An increase in precision corresponding to the separation of two neighboring lines corresponds to a twofold increase in cost. *Source:* M.F. Ashby, *Materials Selection in Design,* 3rd ed., Butterworth-Heinemann, 2005.

or by using multiple machines. Note, however, that a lower production rate does not necessarily mean that a particular manufacturing process is inherently uneconomical.

Lead time.   *Lead time* is defined as the length of time between receipt of an order and the delivery of the product to the customer. The selection of a manufacturing process or processes is greatly influenced by the time required to start production. Such processes as forging, extrusion, die casting, roll forming, and sheet-metal forming typically require dies and tooling that can take a considerable amount of time to produce.

Lead times can range from weeks to months depending on the complexity of the shape of the die, its size, and the material from which the die is to be made (see Table 3.5). In contrast, operations such as machining and grinding have significant built-in flexibility. As described in Chapters 8 and 9, these processes use machinery, tooling, and abrasives that can be adapted to most requirements in a relatively short

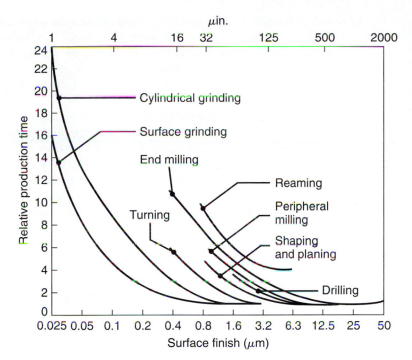

FIGURE 16.5 Relative production time as a function of surface finish obtained by various manufacturing processes. See also Fig. 9.41. *Source: American Machinist.*

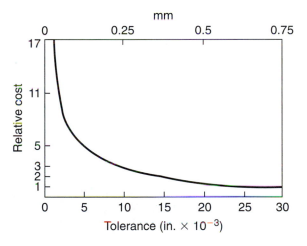

FIGURE 16.6 Relationship between relative manufacturing cost and dimensional tolerance. Note how rapidly cost increases as tolerance decreases.

time. Note also the capabilities of machining centers, flexible manufacturing cells, and flexible manufacturing systems (Chapters 8 and 15), which are able of responding quickly and effectively to product changes.

16.7.1 Robustness in manufacturing processes and machinery

Robustness has been described in Section 16.2.3 in terms of a design, or a process, or a system. In order to appreciate its importance in manufacturing processes, let's consider a simple injection-molded plastic gear, and let's assume that significant variations are being observed in quality as the products are being made. There are several well-understood variables in injection molding of plastics (Section 10.10.2), including the quality of the pellets (raw material), temperature, and time, all of which can be controlled.

As noted in Section 16.2.3, however, there are certain other variables (*noise*) that are difficult or impossible to control, such as variations in the ambient temperature and humidity in the plant, dust in the air entering the plant from an open door (and thus contaminating the pellets being fed into the hoppers of the injection-molding machine), and the variability in performance of different operators during different shifts.

In order to obtain sustained good quality, it is first necessary to understand the effects, if any, of each element of this noise on product quality. Examples of typical questions that would be asked would be (a) Why and how does the ambient temperature affect the quality of the injection-molded gears? (b) Why and how does a dust coating on the pellets affect its performance in the machine chamber and its components? (c) How different are the performances of different operators during different shifts? After supplying and analyzing answers to these questions, it will be possible to establish new operating parameters so that variations in, say, ambient temperature do not adversely affect gear quality.

16.8 | Selection of Manufacturing Processes

As described in various chapters, especially Chapters 14 and 15, most manufacturing processes are now largely automated and computer controlled so as to optimize all aspects of operations, increase product reliability and product quality, and reduce labor costs.

The selection of a manufacturing process or processes is dictated by various considerations (Table 16.7) such as,

1. Characteristics and relevant properties of the workpiece material.
2. Geometric features, shape, size, and part thickness and its variations.
3. Dimensional-tolerance and surface-finish requirements.
4. Functional requirements of the part.
5. Production quantity required.
6. Costs involved in various aspects of the total manufacturing operation.

As we have seen, some materials can be processed at room temperature, whereas others have to be processed at elevated temperatures and thus require furnaces and various related equipment. Some materials are less difficult to process, because they are soft and ductile; others may be hard, brittle, and abrasive and thus require special processing techniques and appropriate tool and die materials.

Materials have different characteristics of castability, forgeability, workability, machinability, and weldability; few (if any) materials have optimum characteristics in all the relevant categories. For example, a material that is castable or forgeable may later present difficulties in machining, grinding, or finishing operations that may be required in order to further improve the surface finish, dimensional accuracy, and other characteristics of the part.

A summary of the factors involved in process selection is given below, in the form of typical questions to be asked during this stage.

1. Have all alternative manufacturing processes been investigated?
2. What is the ecological impact of the processes, especially if incorporated on a plant (massive) scale?
3. Are the processing methods under consideration economical for the type of material, the shape to be produced, and the required production rate?

TABLE 16.7

General Applications of Manufacturing Processes for Various Metals and Alloys

	Carbon steels	Alloy steels	Stainless steels	Tool and die steels	Aluminum alloys	Magnesium alloys	Copper alloys	Nickel alloys	Titanium alloys	Refractory alloys
Casting										
Sand	A	A	A	B	A	A	A	A	B	A
Plaster	—	—	—	—	A	A	A	—	—	—
Ceramic	A	A	A	A	B	B	A	A	B	A
Investment	A	A	A	—	A	B	A	A	A	A
Permanent	B	B	—	—	A	A	A	—	—	—
Die	—	—	—	—	A	A	A	—	—	—
Forging, hot	A	A	A	B	A	A	A	A	A	A
Extrusion										
Hot	A	B	A	—	A	A	A	B	A	B
Cold	A	B	A	—	A	—	A	B	—	—
Impact	—	—	—	—	A	A	A	—	—	B
Rolling	A	A	A	A	A	A	A	A	A	B
Powder metals	A	A	A	A	A	A	A	A	A	A
Sheet-metal forming	A	A	A	—	A	A	A	A	A	B
Machining	A	B	A	B	A	A	A	B	B	A
Chemical	A	B	A	B	—	—	A	B	B	A
ECM	—	A	B	A	—	—	B	A	A	B
EDM	—	B	B	A	B	—	B	B	B	B
Grinding	A	A	A	A	A	A	A	A	A	A
Welding	A	A	A	—	A	A	A	A	A	A

Note: (A) Generally processed by this method; (B) can be processed by this method but may present some difficulties; (—) usually not processed by this method. Product quality and productivity depend greatly on the techniques and equipment used, operator skill, and proper control of processing variables.

4. Can the requirements for dimensional tolerances, surface finish, and product quality be met consistently?

5. Can the part be produced to final dimensions without the need for additional finishing operations?

6. Is the tooling required available in the plant? Can it be purchased as a standard item?

7. Is scrap produced? If so, can it be minimized? What is the value of the scrap?

8. Are all processing parameters optimized?

9. Have all the automation and computer-control possibilities been explored for all phases of the manufacturing cycle?

10. Can group technology be implemented for parts with similar geometric and manufacturing attributes?

11. Are inspection techniques and quality control being properly implemented?

12. Does each component of the product have to be manufactured in the plant? Are some of its parts commercially available as standard items from external sources?

16.9 | Manufacturing Costs and Cost Reduction

The cost of a product must be competitive with that of similar products, particularly in the global marketplace. The total cost of a product consists of several categories, such as material costs, tooling costs, fixed costs, variable costs, direct labor costs, and indirect labor costs. References to some cost data are listed in Table 16.1.

Manufacturing organizations use several methods of cost accounting; the accounting procedures can be complex and even controversial, and their selection depends on the particular company and its type of operations. Trends in costing systems (*cost justification*) include the following considerations: (a) the intangible benefits of quality improvements and inventory reduction; (b) life-cycle costs; (c) machine usage; (d) the cost of purchasing, compared with that of leasing, machinery; (e) the financial risks involved in implementing automation and in the new technologies available.

The costs to a manufacturer that are directly attributable to *product liability* (Section 1.9) have also been a matter of concern among all parties involved. Every modern product has a built-in added cost to cover possible liability claims. For example, it is estimated that liability suits against car manufacturers in the United States add about $500 to the indirect cost of an automobile, and that 20% of the price of a ladder is attributed to potential product liability costs.

The major cost factors are outlined below.

1. **Material costs.** *Material costs* are described throughout various sections in this text; see also Table 16.1.

2. **Tooling costs.** *Tooling costs* are the costs involved in making the tools, dies, molds, patterns, and special jigs and fixtures necessary for manufacturing a product. As expected, tooling costs greatly depend on the manufacturing process selected. Thus, for example, the tooling cost (a) for die casting is higher than that for sand casting, and (b) for machining or grinding is much lower than that for such processes as powder metallurgy, forging, or extrusion. Note, however, that some extrusion dies can easily be produced by electrical discharge machining, and that it can be more economical to extrude a complicated cross section than to

invest in expensive tooling required for roll forming, depending on the material and part size.

In machining operations, carbide tools are more expensive than are high-speed steel tools, but the life of carbide tools is much longer. If a part is to be manufactured by spinning, the tooling cost for conventional spinning is much lower than that of shear spinning. Tooling for rubber-pad forming processes is less expensive than that of the male and female die sets used for the deep drawing and stamping of sheet metals. It should be noted that high tooling costs may well be justified in the high-volume production of a single part (*die cost per piece*).

3. **Fixed costs.** *Fixed costs* include the costs of electric power, fuel, real estate taxes, rent, insurance, and capital (including depreciation and interest). These costs have to be met regardless of the number of parts made. Consequently, fixed costs are not sensitive to production quantity.

4. **Capital costs.** *Capital costs* are major expenses representing investments in buildings, land, machinery, tooling, and equipment. Note in Table 16.8 the wide range of processes in each category and the fact that some machines can cost millions of dollars. High production volumes are thus necessary to justify such large expenditures and to maintain product costs at or below competitive levels. Periodic equipment maintenance is essential to ensure high productivity (Section 14.2.7), as any breakdown of machinery leading to significant downtime can be very costly, typically from a few hundred to thousands of dollars per hour.

5. **Labor costs.** *Labor costs* are generally divided into direct and indirect costs. Direct labor costs are for the labor directly involved in manufacturing a part (*productive labor*). These costs include the cost of all labor, from the time raw materials are first handled to the time the product is finished (a period generally referred to as *floor-to-floor time*). Direct labor costs are calculated by multiplying the labor rate (hourly wage, including benefits) by the time spent

TABLE 16.8

Approximate Ranges of Machinery Base Prices

Type of machinery	Price range ($000)	Type of machinery	Price range ($000)
Broaching	10–300	Machining center	50–1000
Drilling	10–100	Mechanical press	20–250
Electrical discharge	30–150	Milling	10–250
Electromagnetic	50–150	Ring rolling	>500
Extruder	30–80	Robot	20–200
Fused deposition modeling	40–200	Roll forming	5–100
Gear shaping	100–200	Rubber forming	50–500
Grinding		Stereolithography	80–500
Cylindrical	40–150	Stretch forming	400–1000+
Surface	20–100	Transfer line	100–1000+
Headers	100–150	Welding	
Injection molding	30–200	Electron beam	75–1000
Jig boring	50–150	Gas tungsten arc	1–5
Horizontal boring mill	100–400	Laser beam	60–1000
Flexible manufacturing system	>1000	Resistance, spot	20–50
Lathe	10–100	Ultrasonic	50–200
Automatic	30–250		
Vertical turret	100–400		

Note: Prices vary significantly, depending on size, capacity, options, and level of automation and computer controls.

producing the part. Recall that the time required for producing a particular part depends not only on its size, shape complexity, dimensional accuracy, and surface finish, but also on the manufacturing characteristics of the workpiece material. For example, as can be seen in Table 8.10, the highest recommended cutting speeds for high-temperature alloys are lower than those for aluminum, cast iron, or copper alloys. Thus, the cost of machining aerospace materials is higher than that for machining the more common metals and alloys.

Labor costs vary greatly from country to country, as discussed in Section 1.10 and shown in Table 1.3. It is not surprising that many of the products one purchases today in Western countries are either made or assembled in countries where labor costs are low, particularly China, Taiwan, Mexico, and India. (See also Section 16.9.1.)

Manufacturing costs and production quantity. One of the significant factors in manufacturing costs is *production quantity*. Large production volumes obviously require high production rates, which in turn, requires the use of mass-production techniques that typically involve special machinery (*dedicated machinery*) and employ proportionally less direct labor. At the other extreme, small production volume usually means a larger direct labor involvement.

As described in Section 14.2.2, small batch production is usually done on general-purpose machines, such as lathes, milling machines, and hydraulic presses. The equipment is versatile, and parts with different shapes and sizes can be produced by appropriate changes in the tooling. Direct labor costs are, on the other hand, high because these machines are usually operated by skilled labor. For larger quantities (medium-batch production), these same general-purpose machines are computer controlled. To reduce labor costs further, machining centers and flexible manufacturing systems are important alternatives. For production quantities of 100,000 or more, the machines generally are designed for specific purposes and they perform a variety of specific operations with very little direct labor.

16.9.1 Cost reduction

Cost reduction first requires an assessment of how the costs described above are incurred and interrelated, with relative costs depending on numerous factors described in this chapter. Consequently, the unit cost of a product will vary widely, depending on design and manufacturing characteristics.

For example, some parts may be made from expensive materials but may require very little processing, such as minted gold coins. Note that the cost of materials relative to that of direct labor is high. By contrast, some products made of relatively inexpensive materials, such as carbon steels, may require several complex, expensive production steps to process. For instance, an electric motor is made of relatively inexpensive materials, but several different manufacturing processes are involved in the production of its components, such as housing, rotor, bearings, brushes, and wire windings.

An approximate, but typical, breakdown of costs in modern manufacturing is as follows:

Design	5%
Material	50%
Direct labor	15%
Overhead	30%

In the 1960s, labor accounted for as much as 40% of the total production cost. Today, and depending on the type of product and the level of automation, it can be as low as 5%, a good example of which is manufacturing of chips for computers and cell phones. Such a small labor cost is a result of the use of highly automated equipment, further indicating that moving production to countries with low wages may not be an economically viable choice. The situation, of course, changes as the labor cost of a product begins to increase.

Note in the preceding table the very small contribution of the design phase; yet it is this phase that generally has the greatest influence on the cost in other phases and thus success of a product in the marketplace. The engineering changes that are often made in the development of products can also have a significant influence on costs. In addition to the nature and extent of the changes made, the stage at which they are made is very significant. The cost of engineering changes made (from design stage to the final production) increases by orders of magnitude (rule of ten) when changes are made at later stages, as clearly indicated in the table in Section 16.3.

Cost reductions can be achieved by a thorough analysis of all the costs incurred in each step in the manufacture of a product. Recall that opportunities for cost reduction have been stated throughout this text, among which are the following:

1. Simplifying product design and reducing the number of subassemblies.
2. Reducing the amount of materials used.
3. Specifying broader dimensional tolerances and surface finish.
4. Using less expensive materials.
5. Investigating alternative methods of manufacturing.
6. Using more efficient machines, equipment, and automation and computer controls.

Although increased automation and implementation of up-to-date technology are obvious means of reducing some costs, this approach must be undertaken with due care. Only after a thorough *cost-benefit analysis,* with reliable data input and considerations of technical and human factors involved, can an appropriate decision be made. Also, because the implementation of advanced technology and modern computer-control machinery can be very expensive (depending on the type of product), the importance of the concept of **return on investment** (ROI) becomes self evident. (See also **return on quality,** Section 16.3.)

Note also that over a period of time, the prices of some products (such as calculators, computers, and digital watches) have decreased, while the prices of other products (such as automobiles, aircraft, houses, and books) have gone up. Such differences can generally be attributed to such changes as in the cost of labor, machinery, implementation of automation and computer controls, and global competition, as well as worldwide economic trends, such as demand, exchange rates, and tariffs.

SUMMARY

- Competitive aspects of production and costs are among the most significant considerations in manufacturing. Regardless of how well a product meets design specifications and quality standards, it must also meet economic criteria in order to be competitive in the global marketplace. (Section 16.1)
- Several guidelines for designing products for economical production have been established. (Section 16.2)

- Product quality and life expectancy are significant concerns because of their impact on customer satisfaction and the marketability of the product. (Section 16.3)

- The management philosophies of Deming and Taguchi provide a framework for improving quality and designing robust products. The quality loss function is a valuable tool to evaluate product quality. (Section 16.3)

- Life-cycle assessment and life-cycle engineering are among increasingly important considerations in manufacturing, particularly reducing any adverse impact the product might have on the environment. Sustainable manufacturing is another concept aimed at reducing waste of natural resources, including materials and energy. (Section 16.4)

- Selecting an appropriate material from among numerous candidates is a challenging aspect of manufacturing, and depends on factors such as properties, commercially available shapes, reliability of supply, and costs of materials and processing. (Section 16.5)

- Substitution of materials, modification of product design, and relaxing of dimensional tolerance and surface finish requirements are important methods of reducing costs. (Section 16.6)

- The capabilities of manufacturing processes vary widely; consequently, their proper selection for a particular product to meet specific design and functional requirements is critical. (Sections 16.7 and 16.8)

- The total cost of a product includes several elements. With available software, the least expensive material can be identified, without compromising design, service requirements, and quality. Although labor costs are continually becoming an increasingly smaller percentage of production costs, they can further be reduced through the implementation of highly automated and computer-controlled machinery. (Sections 16.9)

SUMMARY OF EQUATIONS

Taguchi loss function: Loss cost $= k[(Y - T)^2 + \sigma^2]$

where $k = \dfrac{\text{Replacement cost}}{(\text{LSL} - T)^2}$

BIBLIOGRAPHY

Anderson, D.M., *Design for Manufacturability & Concurrent Engineering*, CIM Press, 2003.

Ashby, M.F., *Materials Selection in Mechanical Design*, 3rd ed., Pergamon, 2005.

ASM Handbook, Vol. 20: *Materials Selection and Design*, ASM International, 1997.

Billatos, S., and Basaly, N., *Green Technology and Design for the Environment*, Taylor & Francis, 1997.

Boothroyd, G., Dewhurst, P., and Knight, W., *Product Design for Manufacture and Assembly*, 2nd ed., Dekker, 2001.

Bralla, J.G., *Design for Manufacturability Handbook*, 2nd ed., McGraw-Hill, 1999.

Cha, J., Jardim-Gonclaves, R., and Steiger-Garcao, A., *Concurrent Engineering*, Taylor & Francis, 2003.

Crowson, R., *Product Design and Factory Development*, 2nd ed., CRC, 2005.

Deming, W.E., *Out of the Crisis*, MIT Press, 1986.

Dettmer, W.H., *Breaking the Constraints to World-Class Performance*, ASQ Quality Press, 1998.

Giudice, F., La Rosa, G., and Risitano, A., *Product Design for the Environment*, CRC, 2006.

Harper, C.A. (ed.), *Handbook of Materials for Product Design*, McGraw-Hill, 2001.

Hartley, J.R., and Okamoto, S., *Concurrent Engineering: Shortening Lead Times, Raising Quality, and Lowering Costs*, Productivity Press, 1998.

Hundai, M. (ed.), *Mechanical Life Cycle Handbook*, CRC Press, 2001.

Imai, M., *Gemba Kaizen: A Commonsense, Low-Cost Approach to Management,* McGraw-Hill, 1997.

Madu, C. (ed.), *Handbook of Environmentally Conscious Manufacturing,* Springer, 2001.

Mahoney, R.M., *High-Mix Low-Volume Manufacturing,* Prentice Hall, 1997.

Mangonon, P.C., *The Principles of Materials Selection for Design,* Prentice Hall, 1999.

McDonough, W., and Braungart, M., *Cradle to Cradle: Rethinking the Way We Make Things,* North Point Press, 2002.

Poli, C., *Design for Manufacturing: A Structured Approach,* Butterworth-Heinemann, 2001.

Priest, J., and Sanchez, J., *Product Development and Design for Manufacturing,* 2nd ed., CRC, 2001.

Rhyder, R.F., *Manufacturing Process Design and Optimization,* Dekker, 1997.

Shina, S.G. (ed.), *Successful Implementation of Concurrent Engineering Products and Processes,* Wiley, 1997.

Stoll, H.W., *Product Design Methods and Practices,* Dekker, 1999.

Swift, K.G., and Booker, J.D., *Process Selection: From Design to Manufacture,* 2nd ed., Butterworth-Heinemann, 2003.

Taguchi, G., Chowdhury, S., and Wu, Y., *Taguchi's Quality Engineering Handbook,* Wiley, 2004.

Walker, J.M. (ed.), *Handbook of Manufacturing Engineering,* 2nd ed., Dekker, 2006.

Wang, J.X., *Engineering Robust Designs with Six Sigma,* Prentice Hall, 2005.

Wenzel, H., Hauschild, M., and Alting, L., *Environmental Assessment of Products,* Vol. 1, Springer, 2003.

Wenzel, H., and Hauschild, M., *Environmental Assessment of Products,* Vol. 2, Springer, 1997.

Wu, Y., and Wu, A., *Taguchi Methods for Robust Design,* American Society of Mechanical Engineers, 2000.

QUESTIONS

16.1 List and describe the major considerations involved in selecting materials for products.

16.2 Why is a knowledge of available shapes of materials important? Give five different examples.

16.3 Describe what is meant by the *manufacturing characteristics* of materials. Give three examples demonstrating the importance of this information.

16.4 Why is material substitution an important aspect of manufacturing engineering? Give five examples from your own experience or observations.

16.5 Why has material substitution been particularly critical in the automotive and aerospace industries?

16.6 What factors are involved in the selection of manufacturing processes? Explain why these factors are important.

16.7 What is meant by *process capabilities?* Select four different, specific manufacturing processes, and describe their capabilities.

16.8 Is production volume significant in process selection? Explain your answer.

16.9 Discuss the advantages of long lead times, if any, in production.

16.10 What is meant by an *economic order quantity?*

16.11 Describe the costs involved in manufacturing. Explain how you could reduce each of these costs.

16.12 What is a cradle-to-cradle approach? What are its benefits?

16.13 What is meant by *trade-off?* Why is it important in manufacturing?

16.14 Explain the difference between direct labor cost and indirect labor cost.

16.15 Explain why the larger the quantity per package of food products, the lower is the cost per unit weight.

16.16 Explain why the value of the scrap produced in a manufacturing process depends on the type of material.

16.17 Comment on the magnitude and range of scrap shown in Table 16.4.

16.18 Describe your observations concerning the information given in Table 16.6.

16.19 Other than the size of the machine, what factors are involved in the range of prices in each machine category shown in Table 16.7?

16.20 Explain how the high cost of some of the machinery listed in Table 16.7 can be justified.

16.21 Explain the reasons for the relative positions of the curves shown in Fig. 16.3.

16.22 What factors are involved in the shape of the curve shown in Fig. 16.6?

16.23 Make suggestions as to how to reduce the dependence of production time on surface finish (shown in Fig. 16.5).

16.24 Is it always desirable to purchase stock that is close to the final dimensions of a part to be manufactured? Explain your answer, and give some examples.

16.25   What course of action would you take if the supply of a raw material selected for a product line became unreliable?

16.26   Describe the potential problems involved in reducing the quantity of materials in products.

16.27   Present your thoughts concerning the replacement of aluminum beverage cans with steel ones.

16.28   There is a period, between the time that an employee is hired and the time that the employee finishes with training, during which the employee is paid and receives benefits, but produces nothing. Where should such costs be placed among the categories given in this chapter?

16.29   Why is there a strong desire in industry to practice near-net-shape manufacturing? Give several examples.

16.30   Estimate the position of the following processes in Fig. 16.4: (a) centerless grinding, (b) electrochemical machining, (c) chemical milling, and (d) extrusion.

16.31   From your own experience and observations, comment on the size, shape, and weight of specific products as they have changed over the years.

16.32   In Section 16.9.2, a breakdown of costs in today's manufacturing environment suggests that design costs contribute only 5% to total costs. Explain why this suggestion is reasonable.

16.33   Make a list of several (a) disposable and (b) reusable products. Discuss your observations, and explain how you would go about making more products that are reusable.

16.34   Describe your own concerns regarding life-cycle assessment of products.

PROBLEMS

16.35   A manufacturer is ring rolling ball-bearing races (see Fig. 6.43). The inner surface has a surface roughness specification of 0.10 ± 0.06 μm. Measurements taken from rolled rings indicate a mean roughness of 0.112 μm with a standard deviation of 0.02 μm. Fifty thousand rings per month are manufactured and the cost of discarding a defective ring is $5.00. It is known that by changing lubricants to a special emulsion, the mean roughness could be made essentially equal to the design specification. What additional cost per month can be justified for the lubricant?

16.36   For the data of Problem 16.35 assume that the lubricant change can cause the manufacturing process to achieve a roughness of 0.10 ± 0.01 μm. What additional cost per month for the lubricant can be justified? What if the lubricant did not add any new cost?

16.37   Assume that you are asked to give a quiz to students on the contents of this chapter. Prepare three quantitative problems and three qualitative questions, and supply the answers.

DESIGN

16.38   As you can see, Table 16.7 on manufacturing processes includes only metals and their alloys. On the basis of the information given in this book and other sources, prepare a similar table for nonmetallic materials, including ceramics, plastics, reinforced plastics, and metal-matrix and ceramic-matrix composite materials.

16.39   Review Fig. 1.3, and present your thoughts concerning the two flowcharts. Would you want to make any modifications to them and if so, what would the modifications be and why?

16.40   Over the years, numerous consumer products have become obsolete, or nearly so, such as rotary-dial telephones, analog radio tuners, turntables, and vacuum tubes. By contrast, many new products have entered the market. Make a comprehensive list of obsolete products and one of new products. Comment on the possible reasons for the changes you have observed. Discuss how

different manufacturing methods and systems have evolved in order to make the new products.

16.41   Select three different household products, and make a survey of the changes in their prices over the past 10 years. Discuss the reasons for the changes.

16.42   Figure 2.2a shows the shape of a typical tension-test specimen having a round cross section. Assuming that the starting material (stock) is a round rod and that only one specimen is needed, discuss the processes and the machinery by which the specimen can be made, including their relative advantages and limitations. Describe how the process you selected can be changed for economical production as the number of specimens required increases.

16.43   Table 16.3 lists several materials and their commercially available shapes. By contacting suppliers of materials, extend the list to include (a) titanium,

(b) superalloys, (c) lead, (d) tungsten, and (e) amorphous metals.

16.44 Select three different products commonly found in homes. State your opinions about (a) what materials were used in each product, and why, and (b) how the products were made, and why those particular manufacturing processes were used.

16.45 Inspect the components under the hood of your automobile. Identify several parts that have been produced to net-shape or near-net-shape condition. Comment on the design and production aspects of these parts and on how the manufacturer achieved the near-net-shape condition for the parts.

16.46 Comment on the differences, if any, between the designs, the materials, and the processing and assembly methods used to make such products as hand tools, ladders for professional use, and ladders for consumer use.

16.47 Other than powder metallurgy, which processes could be used (singly or in combination) in the making of the parts shown in Fig. 11.1? Would they be economical?

16.48 Discuss production and assembly methods that can be employed to build the presses used in sheet-metal forming operations described in Chapter 7.

16.49 The shape capabilities of some machining processes are shown in Fig. 8.40. Inspect the various shapes produced, and suggest alternative processes for producing them. Comment on the properties of workpiece materials that would influence your suggestions.

16.50 Consider the internal combustion engine in the tractor shown in Fig. 1.1. On the basis of the topics covered in this text, select any three individual components of such an engine, and describe the materials and processes that you would use in making those components. Remember that the parts must be manufactured at very large numbers and at minimum cost, yet maintain their quality, integrity, and reliability during service.

16.51 Discuss the trade-offs involved in selecting between the two materials for each of the applications listed:
(a) Sheet metal vs. reinforced plastic chairs
(b) Forged vs. cast crankshafts
(c) Forged vs. powder-metallurgy connecting rods
(d) Plastic vs. sheet-metal light-switch plates
(e) Glass vs. metal water pitchers
(f) Sheet-metal vs. cast hubcaps
(g) Steel vs. copper nails
(h) Wood vs. metal handles for hammers
Also, discuss the typical conditions to which these products are subjected in their normal use.

16.52 Discuss the manufacturing process or processes suitable for making the products listed in Question 16.51. Explain whether the products would require additional operations (such as coating, plating, heat treating, and finishing). If so, make recommendations and give the reasons for them.

16.53 Discuss the factors that influence the choice between the following pairs of processes to make the products indicated.

(a) Sand casting vs. die casting of a fractional electric-motor housing
(b) Machining vs. forming of a large-diameter bevel gear
(c) Forging vs. powder-metallurgy production of a cam
(d) Casting vs. stamping a sheet-metal frying pan
(e) Making outdoor summer furniture from aluminum tubing vs. cast iron
(f) Welding vs. casting of machine-tool structures
(g) Thread rolling vs. machining of a bolt for high-strength application
(h) Thermoforming a plastic vs. molding a thermoset to make the blade for an inexpensive household fan

16.54 The following figure shows a sheet-metal part made of steel:

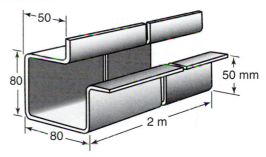

Discuss how this part could be made and how your selection of a manufacturing process may change (a) as the number of parts required increases from 10 to thousands and (b) as the length of the part increases from 2 to 20 m.

16.55 The part shown in the following figure is a carbon-steel segment (partial) gear.

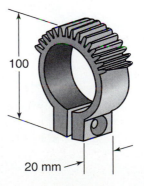

The smaller hole at the bottom is for clamping the part onto a round shaft, using a screw and a nut. Suggest a sequence of manufacturing processes to make this part. Consider such factors as the influence of the number of parts required, dimensional tolerances, and surface finish. Discuss such processes as machining from a bar stock, extrusion, forging, and powder metallurgy.

16.56 Many components in products have minimal effect on part robustness and quality. For example, the hinges in the glove compartment of an automobile do not really impact the owner's satisfaction, and the glove compartment is opened so few times that a robust design is easy to achieve. Would you advocate using Taguchi methods like the loss function on this type of component? Explain.

16.57–16.60 Review the product illustrated below and describe your thoughts on (a) the materials that could be used, your own selection, and your reason for it, (b) manufacturing processes, and why you would select them, and (c) based on your review, any design changes that you would like to recommend.

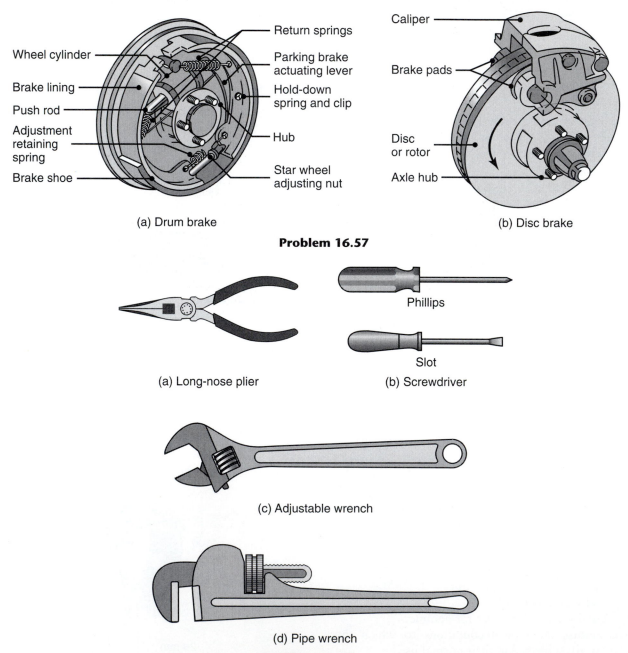

(a) Drum brake

(b) Disc brake

Problem 16.57

(a) Long-nose plier

(b) Screwdriver

Phillips

Slot

(c) Adjustable wrench

(d) Pipe wrench

Problem 16.58

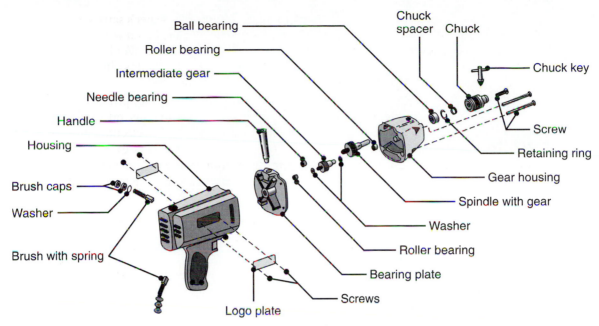

Ball bearing
Roller bearing
Intermediate gear
Needle bearing
Handle
Housing
Brush caps
Washer
Brush with spring
Logo plate
Chuck spacer
Chuck
Chuck key
Screw
Retaining ring
Gear housing
Spindle with gear
Washer
Roller bearing
Bearing plate
Screws

Problem 16.59

(a) Steel wheel   (b) Wire wheel   (c) Cast aluminum wheel

Problem 16.60

Answers to Selected Problems

2.47 $e_d = -0.92$, $e_l = 155$.

2.48 $K = 90,800$ psi.

2.49 UTS = 49,100 psi

2.50 UTS = 237 MPa.

2.51 $P = 3650$ N.

2.52 $P = 19,600$ lb.

2.55 $\sigma = -19,100$ psi, $\tau = 19,100$ psi.

2.57 For 1100-O, Work = 1562 Nm.

2.59 $\varepsilon = 0$.

2.63 $P = 157$ kN.

2.64 $l_f = 20.035$ in.

2.66 $l_f = 1.006$ m.

2.76 $D = 19.8$ mm.

2.80 For maximum shear stress, $P = 338$ kN.

2.82 $\sigma_1 = 60.0$ MPa.

2.83 $\Delta V = 70$ mm^3.

2.84 $e_l = 0.654$.

2.86 $h_f = 0.074$ in.

2.91 $\sigma_3 = -380$ MPa.

2.92 $\sigma = 284$ MPa, $\varepsilon = 0.478$.

2.94 von Mises: $\sigma_1 = 125$ MPa.

2.95 $t = 0.23$ mm

2.97 von Mises: $Y = 79.4$ MPa.

3.35 Al: $\tau = 9.5$ GPa, $\sigma = 7.9$ GPa.

3.36 $n = 7$.

3.37 9.34 million.

3.40 AISI 303 steel: $\delta = 6.5$ mm.

3.41 3.90×10^{15}.

4.57 Sine wave: $R_a/R_q = 0.90$.

4.60 $ID_f = 9.9$ mm.

4.62 $h = 0.073$ mm.

4.63 $d = 49.5$ mm.

4.65 $UCL_R = 13.468$, $LCL_R = 0.546$.

4.66 $UCL_{\bar{x}} = 41.556$, $LCL_{\bar{x}} = 39.444$.

4.67 $UCL_{\bar{x}} = 78.080$, $LCL_{\bar{x}} = 71.920$.

4.68 $UCL_{\bar{x}} = 0.7243$, $LCL_{\bar{x}} = 0.6223$.

4.69 $\bar{x} = 0.6733$, $\sigma = 0.0411$.

4.70 $UCL_{\bar{x}} = 132.46$, $LCL_{\bar{x}} = 117.53$.

5.57 $T_L = 1400°C$, $T_S = 1372°C$.

5.61 $P = 850$ lb.

5.68 $F_{hot} = 394$ kN, $F_{cold} = 506$ kN.

5.69 $L_{al} = 12.156$ in., $L_{iron} = 12.107$ in.

5.72 $P = 7.9$ lb.

5.76 $N = 405$ rpm.

5.77 $d = 0.73$ in.

5.79 $Q = 11.0$ in^3/s.

5.80 $d = 0.0196$ in., $t = 12$ min.

5.81 $t = 14.7$ s.

5.82 $99.0 \times 198.1 \times 396$ mm.

5.83 $t = 624$ s.

5.84 $N = 710$ rpm.

6.65 For $\mu = 0$, $W = 62,445$ in.-lb.

6.66 For $\mu = 0$, $\Delta T = 142°F$.

6.67 $\gamma = 78.6$

6.69 $\mu = 0.5$.

6.78 $F = 1.11$ MN.

6.82 $F = 24.75$ kip.

6.83 $F = 342$ lb.

6.85 $V_f = 668$ ft/min.

6.98 $F = 397$ kip.

6.99 $x_n = 8.64$ mm.

6.100 $F = 1.38$ MN, $P = 409$ kW.

6.101 Stand 3: 38%.

6.102 Stand 3, FS = 0, $V_r = 10.7$ m/s.

6.103 $F = 2.72 \times 10^6$ lb.

6.105 $\Delta T = 590°F$.

6.109 $F = 6.62$ MN.

6.110 $P = 6.4$ kW.

6.114 $D_f = 0.11$ in.

6.115 $Y = 50,000$ psi.

6.121 From Eq. (6.61), $F = 134$ lb.

7.65 (a) $\varepsilon_f = 0.51$.

7.66 $F_{max} = 4940$ lb.

7.67 $W = 34,600$ in.-lb.

7.68 $W = 35,700$ in.-lb.

7.69 $F = 32.2$ kip.

7.71 $F = 7980$ N.

7.72 OD = 21.1 mm.

7.74 $W = 0.134$ lb.

7.76 $F = 181,000$ lb.

7.77 $h/D_p = 1.55$.

7.79 $D = 61.7$ mm.

7.81 $t = 0.875$ mm.

7.83 $F = 7430$ lb, $n = 0.368$.

7.85 $P = 30.3$ hp.

7.87 $F = 79.8$ kN.

7.88 Single row: 32%

7.90 $d = 14$ in.

8.103 $t_o/t_c = 1.16$.

8.109 $\phi = 31.17°$.

8.110 $\phi = 28.2$, $\mu = 0.533$, $\gamma = 2.52$.

8.111 $T_{al} = 244°C$.

8.112 $\Delta \beta = 1.5°$.

8.113 $\phi = 27.3°$.

8.114 Reduced by 83%.

8.116 $f = 0.0022$ in./rev.

8.117 $T_2 = 3.3T_1$.

8.118 For $V = 400$ ft/min, $T = 54$ min.

8.120 $a = 0.34$.

8.121 (a) 44 s.

8.122 $t = 0.75$ min, MRR $= 3660$ mm^3/s.

8.123 $F_c = 158$ lb.

8.124 MRR $= 0.66$ in^3/min.

8.125 MRR $= 8840$ mm^3/min, $T = 1.40$ Nm.

8.127 $t_c = 0.05$ in.

8.129 MRR $= 8.1$ in^3/min, $t = 2.29$ min.

8.130 MRR $= 60$ mm^3/s, $t = 620$ s.

8.131 $t = 23$ s.

8.132 $t = 11.1$ s.

8.133 $N = 0.75$ rpm.

8.135 67.7%.

8.136 12% of the energy.

8.138 MRR $= 27,800$ mm^3/min.

8.140 $t_o = 1/8$ in.

8.141 $T = 1448°$F.

8.142 $P = 0.05$ hp.

8.143 For carbide, $t = 18.36$ min.

8.146 $f_1/f_2 = 1.96$.

8.148 $t = 11$ s.

8.149 MRR $= 6.75$ in^3/min.

8.151 for 0.0001, $\Delta T = 1.28°$F.

8.152 $V_o = 91$ m/min.

8.153 $V_o = 66.67$ m/min.

9.58 $t = 1.64 \times 10^{-4}$ in., $l = 0.0894$ in.

9.65 (a) $F_{ave} = 77.3$ N, (c) $F_{ave} = 312.7$ N.

9.66 $t = 4.17$ min.

9.67 $t = 81.8$ min.

9.68 $t = 6$ s.

9.71 $N = 4900$ rpm.

9.72 $P_{al} = 2.26$ hp.

9.73 For 5000 rpm, $\Delta T = 2010°$F.

9.74 $f = 0.44$ in./s.

9.79 $P = 13$ kW.

9.80 (a) 0.1265 in., (b) MRR $= 0.072$ in^3/min.

9.81 $l = 1.94$ mm, $F_c = 597$ N.

9.82 (a) $t_o = 13.9$ μs, $F_{ave} = 56.7$ N.

10.84 Fibers support 75% of load.

10.86 $t = 0.10$ in.

10.87 for PVC, $n = 1.707$.

10.88 $Q_d = 141,500$ mm^3/s.

10.89 $Q = 0.00833$ mm^3/min.

10.90 $N = 147$ rpm.

10.91 $\theta = 17.6°$.

10.92 $C_d = 0.116$ mm, $L_w = 0.19$ mm.

10.93 $t = 29.4$ s.

10.94 $t = 1.7$ hr.

10.100 $E_{carbon} = 91$ GPa, $E_{Kevlar} = 20.4$ GPa.

10.101 For carbon, $\sigma_f = 182$ MPa.

10.102 $E_m = 75$ MPa.

11.59 $N = 973$ million.

11.60 $V = 1770$ μm^3.

11.61 For the flake, SF $= 15.17$.

11.63 $V = 41.3$ cm^3.

11.67 $P = 27\%$, $L_o = 23.23$ mm.

11.68 $P = 14\%$, $L_o = 21.56$ mm.

11.72 $\rho = 8.99$ g/cm^3.

11.73 $F = 89$ tons.

11.74 $V = 185$ cm^3.

11.75 $F = 412$ kN, $C = 45$ cm^3.

11.78 For $n = 4$, UTS $= 80.9$ MPa, $E = 196.8$ GPa.

11.79 for $P = 1\%$, $k = 0.693$ W/mK.

11.80 $k_{ave} = 0.639$ W/mK.

12.87 $H = 2940$ J.

12.88 $\Delta T = 14,400$ K.

12.92 Gas tungsten arc: $\Delta T = 75°$C. Electron-beam: $\Delta T = 42°$C.

12.93 $I = 792$ A.

12.94 $I = 0.0228$ lb-ft^2.

12.96 For acrylic, $F_{max} = 1554$ N.

12.99 $v = 12.2$ mm/s

13.49 $Y_{500} = 90.0\%$

13.50 87.5 nm of resist, 7 nm of oxide.

13.51 266 nm.

13.54 $t = 10.91$ min.

13.56 $F = 4.37 \times 10^{-17}$ N.

13.57 For AR $= 200$, $\theta = 0.28°$. For AR $= 0.5$, $\theta = 63.4°$.

13.58 For KOH, $x = 0.1$ μm.

13.59 $t = 3$ days.

13.60 $t = 0.5$ μm.

13.62 $t = 150$ s.

13.65 For $d = 10$ mm, $v = 0.178$. For $d = 100$ μm, $v = 17.8$ m/s.

14.52 First part, $v = 11\%$.

16.35 Additional cost $= \$10,000$.

16.36 Additional cost $= \$31,000$.

Index